MW00397437

INTRODUCTION TO Criminal Justice

INTRODUCTION TO Criminal Justice

Tenth Edition

Robert M. Bohm
University of Central Florida

Keith N. Haley
Tiffin University

mheonline.com/advancedplacement

Send all inquiries to:
McGraw Hill
8787 Orion Place
Columbus, OH 43240

ISBN: 978-1-26-465082-8
MHID: 1-26-465082-5

Printed in the United States of America.

3 4 5 6 7 8 9 LWI 30 29 28 27 26 25 24

DEDICATION

To my wife, Linda Taconis, with love.

Robert M. Bohm

To my wife, Shelby, and daughter, Jill, with love.

Keith N. Haley

About the Authors

Robert M. Bohm is Professor Emeritus of Criminal Justice at the University of Central Florida. He also has been a faculty member in the Departments of Criminal Justice at the University of North Carolina at Charlotte (1989–1995) and at Jacksonville State University in Alabama (1979–1989). In 1973 and 1974, he worked for the Jackson County Department of Corrections in Kansas City, Missouri, first as a corrections officer and later as an instructor/counselor in the Model Inmate Employment Program, a Law Enforcement Assistance Administration sponsored–work-release project. He received his PhD in criminology from Florida State University in 1980. He has published numerous journal articles, book chapters, and books in the areas of criminal justice and criminology. In addition to being the co-author of *Introduction to Criminal Justice*, 10th ed. (McGraw-Hill), he is the author of *A Concise Introduction to Criminal Justice* (McGraw-Hill, 2008); *Deathquest: An Introduction to the Theory and Practice of Capital Punishment in the United States*, 5th ed.; *Ultimate Sanction: Understanding the Death Penalty Through Its Many Voices and Many Sides*; *The Past as Prologue: The Supreme Court's Pre-Modern Death Penalty Jurisprudence and Its Influence on the Supreme Court's Modern Death Penalty Decisions*; *Capital Punishment's Collateral Damage*; and *A Primer on Crime and Delinquency Theory*, 4th ed. (with Brenda L. Vogel). He is also editor of *The Death Penalty in America: Current Research* and *The Death Penalty Today* and co-editor (with James R. Acker and Charles S. Lanier) of *America's Experiment with Capital Punishment: Reflections on the Past, Present, and Future of the Ultimate Sanction*, 3rd ed., *Demystifying Crime and Criminal Justice*, 2nd ed. (with Jeffery T. Walker), and the *Routledge Handbook of Capital Punishment* (with Gavin M. Lee). Bohm served as president of the Academy of Criminal Justice Sciences in 1992–1993. In 1989, the Southern Criminal Justice Association selected him Outstanding Educator of the Year. In 1999, he was elected a Fellow of the Academy of Criminal Justice Sciences; in 2001, he was presented with the Academy's Founder's Award; and, in 2008, he received the Academy's Bruce Smith Sr. Award.

Keith N. Haley is Professor Emeritus of Criminal Justice in the School of Criminal Justice and Social Sciences at Tiffin University. He also has been Chair of Criminal Justice and Social Science Graduate Programs, Dean of the School of Criminal Justice, Chair of the Criminal Justice Department, Dean and Associate Vice President of the School of Off-Campus Learning, Associate Vice President for Special Projects, and has acted as the primary contact for the Tiffin University MBA program in Bucharest, Romania. He also has served as Coordinator of the Criminal Justice Programs at Collin County Community College in Texas; Director of the Criminal Justice Program at Redlands Community College in Oklahoma; Chair of the Criminal Justice Department at the University of Cincinnati; Executive Director of the Ohio Peace Officer Training Commission; and police officer in Dayton, Ohio. Haley received a bachelor of science degree in education from Wright State University and a master of science degree in criminal justice from Michigan State University. Haley is the author, co-author, and/or editor of 30 books (including revised editions), several book chapters, and many articles in criminal justice publications. He has served as a consultant to many public service, university, business, and industrial organizations on management, online learning, criminal justice research, and memory skills. The American Association of University Administrators presented the 2001 Nikolai N. Khaladjan International Award for Innovation in Higher Education to Haley for his leadership in the Tiffin University/University of Bucharest "Partnership for Justice" project, which established a graduate school of criminal justice administration at the University of Bucharest. Bohm and Haley's *Introduction to Criminal Justice* was translated into the Romanian language under the title of *Justitia Penala* and has been used in the University of Bucharest graduate program for criminal justice leaders.

Brief Contents

Contents

Part One The Foundations of Criminal Justice

Sam Lothridge/CBS/Getty Images

Ingram Publishing/SuperStock

stevecoleimages/E+/Getty Images

Image Source Plus/Alamy

Part Two Law Enforcement

The Print Collector/Print Collector/Getty Images

TIMOTHY A. CLARY/Staff/Getty Images

Alex Milan Tracy/Anadolu Agency/Getty Images

Part Three The Courts

Robert Daly/Getty Images

Justin Sullivan/Getty Images

Part **Four Corrections**

Scott Olson/Getty Images

Giles Clarke/Getty Images

Spencer Grant/PhotoEdit

Part **Five** Additional Issues in Criminal Justice

Mikael Karlsson/Alamy Stock Photo

Snap Stills/Shutterstock

Crime Stories

Careers in Criminal Justice

Preface

We are convinced that a good education in criminal justice—and in liberal arts generally—can improve significantly the performance of most criminal justice personnel by equipping them, as this text does, to think critically, act ethically, and solve problems effectively. *Introduction to Criminal Justice*, Tenth Edition, is not just for students interested in pursuing a career in criminal justice, however. It also is for students who simply want to learn more about this important social institution, which is vital to a free and democratic society.

All citizens need to know their legal rights and responsibilities. The better informed citizens are, the better able they are to protect themselves. A major theme of this book is that much of what the public "knows" about criminal justice in the United States is myth—that is, it is either wrong or significantly misunderstood. Consequently, in addition to presenting current, accurate information about criminal justice in the United States and generally accepted interpretations of historical and modern developments, this book "sets the record straight" in areas where, we believe, many people are being misled.

With the help of a quality education in criminal justice, people will feel more comfortable and better equipped to participate in criminal justice policy formulation. They also will be more effective in solving problems in their communities. They will have the critical thinking skills they need if they are to be constructive participants in a democratic nation and to have greater control over their own destinies.

The Tenth Edition

Since the events of September 11, 2001, many people who once had no fear of public transportation, especially air travel, now fear it, afraid that a bomb might explode on board or that an airplane, for example, will be commandeered by terrorists who will turn it into a weapon of mass destruction. People are even more suspicious of strangers, particularly of certain ethnic groups who before 9/11 caused little concern. In short, we now live in the Age of Terrorism.

The Tenth Edition of *Introduction to Criminal Justice* continues to examine the new, expanded role of criminal justice in the "war against terrorism," which we introduced in the Fourth Edition in 2004. As noted in the Fourth Edition, this war has caused the most dramatic transformation of the U.S. government in more than 50 years and has altered criminal justice at all levels of government in many ways. For example, the massive Department of Homeland Security was created, and, as in previous editions, its current organization is described. Also, following 9/11, the top priority of the Federal Bureau of Investigation (FBI) shifted from being a federal police agency to being an intelligence and counterterrorism agency. New laws, such as the USA PATRIOT Act, have been passed to provide governments with broad new powers to combat terrorism. These laws have changed many of the traditional rules of criminal justice. In short, in this Tenth Edition, the criminal justice response to terrorism continues to receive serious treatment in several chapters. A list of those additions and many others is provided later.

In addition to integrating detailed coverage of terrorism, this Tenth Edition examines other current issues and challenges that confront criminal justice professionals, such as gun control, technology, privacy, and budget shortfalls. The text has been updated throughout with the latest available statistics, research, and court cases, and like previous editions, presents supplementary features such as Myth versus Fact boxes, FYIs, and CJ Onlines in the text's margins. The Tenth Edition still has Critical Thinking questions at the end of each major section, as well as coverage of some of today's highest-profile criminal investigations and cases in 14 new or updated chapter-opening Crime Stories. We also have created "new knowledge," for example, by taking existing data and calculating new statistical insights about it. For instance, throughout the text, we calculate and report percentage changes over time to identify longer-term trends in the data. Following is a chapter-by-chapter list of new text material in this Tenth Edition:

- Chapter 1, Crime and Justice in the United States, presents (1) top broadcast television shows during a week in 2019, featuring criminal justice–oriented programs, (2) top crime news stories in the United States in 2018, (3) calls for police service in Portland, Oregon for 2018–2019, and (4) costs of criminal justice in 2016.
- Chapter 2, Crime and Its Consequences, introduces (1) a new example of concurrence—a legal element of crime, (2) new examples of the police manipulation of crime statistics, (3) recent developments in the NIBRS program, (4) an updated comparison of NCVS and UCR data, (5) updated information on the tangible and intangible costs of crime, (6) an updated list of what Americans fear about crime, and (7) strategies to avoid becoming a crime victim.
- Chapter 4, The Rule of Law, provides new information about (1) the history of Native American (Cherokee) criminal law and its administration, (2) the Fourth Amendment and cell phone records in *Carpenter v. United States* (2018), (3) the Supreme Court's incorporation of the excessive fines clause and making it applicable to the states in *Timbs v. Indiana* (2019), and (4) exonerations.
- Chapter 5, History and Structure of American Law Enforcement, features new material about (1) the history of Native American (Cherokee) law enforcement, (2) the Department of Homeland Security, and (3) private security.
- Chapter 6, Policing: Roles, Styles, and Functions, presents new material about (1) the FBI's biometric identification services' Next Generation Identification (NGI) system, (2) cybercrime, (3) limitations on civil asset forfeiture in *Timbs v. Indiana* (2019), and (4) "The First Step Act."
- Chapter 7, Policing America: Issues and Ethics, includes new information about (1) police attitudes toward the public, (2) Black Lives Matter, (3) Seattle police officer selection

standards, 2019, (4) 2020 police officer salaries in the United States, (5) police officer education levels in the United States, 2016, (6) affirmative action in policing, (7) a 2015 federal study of police and public contacts, (8) police use of excessive force, and (9) police corruption.

- Chapter 8, The Administration of Justice, presents updated information about (1) the race and gender of elected prosecutors, 2019, and (2) examples of a felony information, grand jury indictment, and subpoena.
- Chapter 10, Institutional Corrections, provides a new discussion of (1) the federal prison system (BOP), (2) private prison services, (3) privatization, (4) costs of incarceration, (5) prison inmate characteristics, (6) state prisons, (7) women's prisons, (8) elderly prisoners, (9) mental health in prisons, (10) HIV testing in prisons, (11) inmate work programs, (12) prison educational and vocational programs, and (13) cognitive behavioral therapy in prisons.
- Chapter 11, Prison Life, Inmate Rights, Release, Reentry and Recidivism, examines new information on (1) physical and sexual violence in prisons, (2) new forms of prison contraband, (3) pregnant inmates, (4) correctional officers, (5) clemency, (6) reentry, (7) recidivism, and (8) the First Step Act.
- Chapter 12, Community Corrections, introduces new material on (1) parole discharge and revocation and (2) electronic monitoring.
- Chapter 13, Juvenile Justice, includes new material on (1) teen or youth courts, (2) a recent evaluation of Florida's civil-citation program, (3) suicides of youth in custody, (4) juvenile probation, and (5) New York City's "Close to Home" program.

Organization

This book is divided into 14 chapters that are organized into five parts. Part One, The Foundations of Criminal Justice, introduces students to the concepts of crime and justice, as well as how crime affects society, how it can be explained, and the rule of law. Part Two, Law Enforcement, is dedicated to the law enforcement component of criminal justice–its history, structure, roles, ethical issues, and challenges. Part Three, The Courts, focuses on the administration of justice, sentencing, appeals, and the court process as a whole. Part Four, Corrections, introduces students to the corrections system, jails, prisons, alternative sanctions, community corrections, and release of prisoners back into the community. Part Five, Additional Issues in Criminal Justice, touches on some of the most pressing challenges in the system today: juvenile justice, the future of criminal and juvenile justice, new and proposed technology, terrorism, and more.

An Integrated Print and Digital Learning System

Working together, the authors and editors have developed a learning system designed to help students get the most out of their first criminal justice course. In this edition, Connect Criminal Justice, a highly interactive learning environment, coupled with proven pedagogical resources in the text, offer a learning system that is without peer in Introduction to Criminal Justice programs. In addition to the many changes already mentioned, we have included many new photographs to make the book even more inviting and relevant for students. As noted, we have added new and current chapter-opening *Crime Story* vignettes, giving the material a fresh flavor intended to motivate students to read on. New and carefully updated tables and figures highlight and amplify the text coverage; and chapter outlines, objectives, marginal definitions, and an end-of-book Glossary all help students master the material.

Other innovative learning tools include:

- *Myth vs. Fact.* These inserts debunk common misconceptions about the system and alert students to the need to question what they see in the media.
- *Thinking Critically.* These sections challenge students to think about and apply chapter concepts.
- *Careers in Criminal Justice.* These mini-biographies highlight some of the most exciting career options available to criminal justice majors and keep the book relevant for students.
- *CJ Online.* These inserts enable students to explore chapter topics on the Internet in a directed fashion.
- *FYI.* These sidebars present eye-opening additional information to retain students' interest and keep them thinking about what they are reading.

We are especially proud of our comprehensive end-of-chapter review sections in which we provide every kind of review and study tool students could need:

- *Summary*–an extremely effective study tool because it is organized into sections that mirror the chapter-opening objectives.
- *Key Terms*–a comprehensive list of the terms defined in the chapter, complete with page references to make it easy for students to review them in the chapter.
- *Review Questions*–study questions that allow students to test their knowledge and prepare for exams.
- *In the Field Activities*–unique experiential exercises that enable students to broaden their understanding of chapter material by taking it to the next level.
- *On the Web Exercises*–still more Internet-based exercises for today's students.
- *Critical Thinking Exercises*–unique scenario-based activities that challenge students to apply what they've learned in the chapter.

Criminal Justice Resources for Today's Instructors and Students

The tenth edition of *Introduction to Criminal Justice* is now available online with Connect, McGraw-Hill's integrated assignment and assessment platform. Connect also offers SmartBook® 2.0 for the new edition, which is the first adaptive reading experience proven to improve grades and help students study more effectively. All of the title's website and ancillary content also is available through Connect, including:

- A full Test Bank of multiple choice questions that test students on central concepts and ideas in each chapter.
- An Instructor's Manual for each chapter with full chapter outlines, sample test questions, and discussion topics.
- Lecture Slides for instructor use in class.

Test Builder in Connect

Available within Connect, Test Builder is a cloud-based tool that enables instructors to format tests that can be printed or administered within an LMS. Test Builder offers a modern, streamlined interface for easy content configuration that matches course needs, without requiring a download.

Test Builder allows you to:

- access all test bank content from a particular title.
- easily pinpoint the most relevant content through robust filtering options.
- manipulate the order of questions or scramble questions and/or answers.
- pin questions to a specific location within a test.
- determine your preferred treatment of algorithmic questions.
- choose the layout and spacing.
- add instructions and configure default settings.

Test Builder provides a secure interface for better protection of content and allows for just-in-time updates to flow directly into assessments.

Instructors: Student Success Starts with You

Tools to enhance your unique voice

Want to build your own course? No problem. Prefer to use our turnkey, prebuilt course? Easy. Want to make changes throughout the semester? Sure. And you'll save time with Connect's auto-grading too.

65%
Less Time Grading

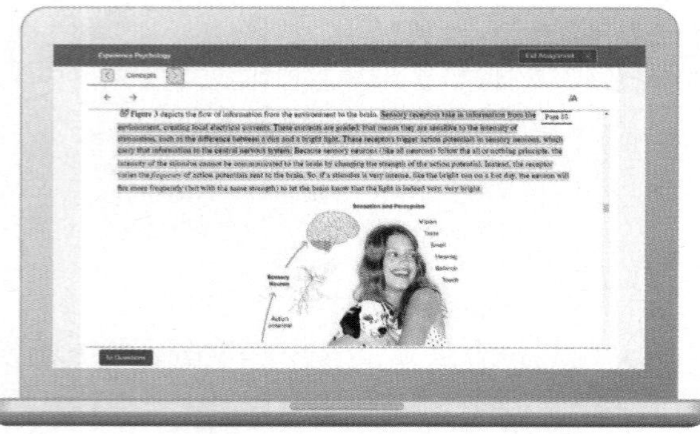

Laptop: McGraw Hill; Woman/dog: George Doyle/Getty Images

Study made personal

Incorporate adaptive study resources like SmartBook® 2.0 into your course and help your students be better prepared in less time. Learn more about the powerful personalized learning experience available in SmartBook 2.0 at **www.mheducation.com/highered/connect/smartbook**

Affordable solutions, added value

Make technology work for you with LMS integration for single sign-on access, mobile access to the digital textbook, and reports to quickly show you how each of your students is doing. And with our Inclusive Access program you can provide all these tools at a discount to your students. Ask your McGraw Hill representative for more information.

Padlock: Jobalou/Getty Images

Solutions for your challenges

A product isn't a solution. Real solutions are affordable, reliable, and come with training and ongoing support when you need it and how you want it. Visit **www.supportateverystep.com** for videos and resources both you and your students can use throughout the semester.

Checkmark: Jobalou/Getty Images

Students: Get Learning that Fits You

Effective tools for efficient studying

Connect is designed to make you more productive with simple, flexible, intuitive tools that maximize your study time and meet your individual learning needs. Get learning that works for you with Connect.

Study anytime, anywhere

Download the free ReadAnywhere app and access your online eBook or SmartBook 2.0 assignments when it's convenient, even if you're offline. And since the app automatically syncs with your eBook and SmartBook 2.0 assignments in Connect, all of your work is available every time you open it. Find out more at **www.mheducation.com/readanywhere**

"I really liked this app—it made it easy to study when you don't have your text-book in front of you."

- Jordan Cunningham, Eastern Washington University

Calendar: owattaphotos/Getty Images

Everything you need in one place

Your Connect course has everything you need—whether reading on your digital eBook or completing assignments for class, Connect makes it easy to get your work done.

Learning for everyone

McGraw Hill works directly with Accessibility Services Departments and faculty to meet the learning needs of all students. Please contact your Accessibility Services Office and ask them to email accessibility@mheducation.com, or visit **www.mheducation.com/about/accessibility** for more information.

Top: Jenner Images/Getty Images, Left: Hero Images/Getty Images, Right: Hero Images/Getty Images

Acknowledgments

This textbook, like any book, is the product of a collaborative effort. We would like to acknowledge and thank the many people who helped to make this and previous editions possible. First, our thanks go to **Kevin I. Minor** and **H. Preston Elrod**, both at Eastern Kentucky University, for their significant contributions to the chapters on corrections and the juvenile justice system, respectively. We also would like to thank the following colleagues for their substantial help with revisions: **James R. Acker**, State University of New York at Albany (Chapter 4, The Rule of Law); **John O. Smykla**, Florida Atlantic University (Chapter 12, Community Corrections); **Donna M. Bishop**, Northeastern University (Chapter 13, Juvenile Justice), and **Ellen Cohn**, Florida International University (general research and other assistance). In addition, for their insightful reviews, criticism, helpful suggestions, and information, we would like to thank:

Joel J. Allen
Oakland Community College

Brandon Applegate
University of South Carolina

Thomas Arnold
College of Lake County

Richard L. Ashbaugh
Clackamas Community College

Gregg Barak
Eastern Michigan University

David Barlow
Fayetteville State University

Michael Barrett
Ashland University

Denny Bebout
Central Ohio Technical College

Brenda Berretta
Middle Tennessee State University

Kristie Blevins
Eastern Kentucky University

Anita Blowers
University of North Carolina at Charlotte

John Boman
University of Florida

Robert J. Boyer
Luzerne County Community College

Jeffrey A. Brockhoff
El Paso Community College

William D. Burrell
Probation Services Division, State of New Jersey

Michael Cain
Coastal Bend Community College

Vincent J. Capozzella
Jefferson Community College

Kevin Cashen
Tiffin University

John Cavendish
Salem International University

Jonathan Cella
Central Texas Community College

Brenda Chappell
University of Central Oklahoma

Charles Chastain
University of Arkansas at Little Rock

David E. Choate
Arizona State University

Daryl Cullison
Columbus State Community College

Beth DeValve
Fayetteville State University

Hank DiMatteo
New Mexico State University

Vicky Dorworth
Montgomery College

Joyce K. Dozier
Wilmington College

Thomas E. Drerup
Clark State Community College

Matthew Drewry
West Hills Community College District

Mary Ann Eastep
University of Central Florida

Randy Eastep
Brevard Community College

Robert R. Eiggins
Cedarville College

Kelly Enos
Los Angeles Mission College

James Ferguson
Everest College

Linda L. Fleischer
Community College of Beaver County–Essex

John W. Flickinger
Tiffin University

Mike Flint
University of Central Florida

Kenneth A. Frayer
Schoolcraft College at Radcliff

Aric Steven Frazier
Vincennes University

David O. Friedrichs
University of Scranton

Rodney Friery
Jacksonville State University

Harold Frossard
Moraine Valley Community College

James N. Gilbert
University of Nebraska at Kearney

Brian J. Gorman
Towson University

Alex Greenberg
Niagara County Community College

Corajean Gregory
Mott Community College

Alejandro Gutierrez
West Hills Junior College

Bob Hale
Southeastern Louisiana University

David O. Harding
Ohio University at Chillicothe

Stuart Henry
San Diego State University

Sherry Herman
Catawba Valley Community College

Joseph Hogan
Central Texas College

Thomas E. Holdren
Muskingum Technical College

James Houston
Grand Valley State University

James L. Hudson
Clark State Community College

Miriam Huntley
South Piedmont Community College

Pearl Jacobs
Sacred Heart University

Stephanie James
Collin College

W. Richard Janikowski
University of Memphis

Paul Jefferson
Mount San Antonio College

Patricia Joffer
South Dakota State University

Robert Johnson
American University

Coy Hugh Johnston, Jr.
Arizona State University

Cleveland J. Jones, Jr.
Olive-Harvey City College

Mark A. Jones
Community College of Philadelphia

Michelle Jones
The Community College of Baltimore County–Catonsville

Lamar Jordan
Southern Utah University

Charles Klahm
Wayne State University

Douglas Klutz
University of Alabama

Don Knueve
Defence College

Andrew Kozal
Northwest State Community College

Peter C. Kratcoski
Kent State University

Jonathan M. Kremser
Kutztown University

Hamid R. Kusha
East Carolina University

Gregory C. Leavitt
Green River Community College

Vivian Lord
University of North Carolina at Charlotte

Karol Lucken
University of Central Florida

Richard Lumb
State University of New York at Brockport

Kathleen Maguire
Hindelang Criminal Justice Research Center

Stacy Mallicoat
California State University, Fullerton

Bradley Martin
University of Findlay

Richard M. Martin
Elgin Community College

Alida Merlo
Indiana University of Pennsylvania

John R. Michaud
Husson University

Holly Ventura Miller
University of North Florida

Dale Mooso
San Antonio College

James Newman
Rio Hondo College

Sarah Nordin
Solano Community College

Lisa S. Nored
University of Southern Mississippi–Hattiesburg

Les Obert
Casper College

Nancy Oesch
Everest University

Angela Ondrus
Owens Community College

Barry Lynn Parker
Palo Alto College

Jill Percy
ORANGELEGAL

Bill Pennington
Palo Alto College

Kelly E. Peterson
South Texas College, Mid-Valley Campus

Wayne Posner
East Los Angeles College

Mary Carolyn Purtill
San Joaquin Delta College

Jerome Randall
University of Central Florida

Selena M. Respass
Miami Dade College, School of Justice

Jim Reynolds
Florida Institute of Technology

Matthew Robinson
Appalachian State University

Joseph B. Sanborn Jr.
University of Central Florida

Martin D. Schwartz
George Washington University

Lance Selva
Middle Tennessee State University

Wayne Shelley
Sitting Bull College

Jo Ann M. Short
Northern Virginia Community College, Annandale Campus

David Slaughter
University of Central Florida

James E. Smith
West Valley College

Jeffrey B. Spelman
North Central Technical College

Domenick Stampone
Raritan Valley Community College

Clayton Steenberg
Arkansas State University at Mountain Home

Gene Stephens
University of South Carolina

James Stinchcomb
Miami-Dade Community College

David Streater
Catawba Valley Community College

David Striegel
Wor-Wic Community College

William L. Tafoya
University of New Haven

Mike Tatum
Brigham Young University–Idaho

Henry A. Townsend
Washtenaw Community College

Roger D. Turner
Shelby State Community College

Ronald E. Vogel
California State University Dominguez Hills

Arnold R. Waggoner
Rose State College

Katina Whorton
Delgado Community College

Harold Williamson
Northeastern Louisiana University

Jeremy R. Wilson
Catawba Valley Community College

Ross Wolf
University of Central Florida

Peter Wood
Eastern Michigan University

Tracy Woodard
University of North Florida

Jeffrey Zack
Fayetteville Technical Community College

We would also like to thank **Jack Bohm**, Bob's late father; **Richard Bohm**, of New York City; and **Lorie Klumb**, of Denver, Colorado, for their help with this book. Finally, we would like to express our appreciation to our families and friends for their understanding and patience over the many years we have worked on this project.

Constitutional Issues

You will find coverage of key constitutional issues throughout this book. Use this handy guide to coverage as you study *Introduction to Criminal Justice.*

Article I

Section 9—Chapter 8
Section 9.3—Chapter 4
Section 10.1—Chapter 4
Section 10.1—Chapter 2

Article III

Section 1—Chapter 8
Section 2—Chapter 8
Section 2.3—Chapter 4

The Bill of Rights

Amendment 1

Congress shall make no law respecting an establishment of religion, or prohibiting the free exercise thereof; or abridging the freedom of speech, or of the press; or the right of the people peaceably to assemble; and to petition the Government for a redress of grievances.

Amendment 2

A well-regulated militia, being necessary to the security of a free State, the right of the people to keep and bear arms shall not be infringed.

Amendment 3

No soldier shall, in time of peace be quartered in any house without the consent of the owner; nor in time of war but in a manner to be prescribed by law.

Amendment 4

The right of the people to be secure in their persons, houses, papers, and effects, against unreasonable searches and seizures, shall not be violated, and no warrants shall issue but upon probable cause, supported by oath or affirmation, and particularly describing the place to be searched, and the persons or things to be seized. Chapter 4.

Amendment 5

No person shall be held to answer for a capital or otherwise infamous crime, unless on a presentment or indictment of a Grand Jury, except in cases arising in the land or naval forces, or in the militia, when in actual service in time of war or public danger; nor shall any person be subject for the same offense to be twice put in jeopardy of life or limb; nor shall be compelled in any criminal case to be a witness against himself, nor be deprived of life, liberty, or property, without due process of law; nor shall private property be taken for public use without just compensation. Chapter 4.

Amendment 6

In all criminal prosecutions, the accused shall enjoy the right to a speedy and public trial, by an impartial jury of the State and district wherein the crime shall have been committed, which districts shall have been previously ascertained by law, and to be informed of the nature and cause of the accusation; to be confronted with the witnesses against him; to have compulsory process for obtaining witnesses in his favor, and to have the assistance of counsel for his defense. Chapter 4.

Amendment 7

In suits at common law, where the value in controversy shall exceed twenty dollars, the right of trial by jury shall be preserved, and no fact tried by a jury shall be otherwise re-examined in any court of the United States than according to the rules of the common laws.

Amendment 8

Excessive bail shall not be required, nor excessive fines imposed, nor cruel and unusual punishments inflicted. Chapter 4, Chapter 9, Chapter 11.

Amendment 9

The enumeration in the Constitution of certain rights shall not be construed to deny or disparage others retained by the people.

Amendment 10

The powers not delegated to the United States by the Constitution, nor prohibited by it to the States, are reserved to the States respectively, or to the people.

Post–Civil War Amendments

Amendment 14

No State shall make or enforce any law which shall abridge the privileges or immunities of citizens of the United States, nor shall any State deprive any person of life, liberty, or property, without due process of law; nor deny to any person within its jurisdiction the equal protection of the laws. Chapter 4, Chapter 9, Chapter 11.

Sam Lothridge/CBS/Getty Images

1 Crime and Justice in the United States

CHAPTER OUTLINE

Crime in the United States

Criminal Justice: An Institution of Social Control

Criminal Justice: The System
- Police
- Courts
- Corrections

Criminal Justice: The Nonsystem

Two Models of Criminal Justice
- The Crime Control Model
- The Due Process Model
- Crime Control versus Due Process

Costs of Criminal Justice

Myths About Crime and Criminal Justice

LEARNING OBJECTIVES

After completing this chapter, you should be able to:

1. Describe how the type of crime routinely presented by the media compares with crime routinely committed.
2. Identify institutions of social control, and explain what makes criminal justice an institution of social control.
3. Summarize how the criminal justice system responds to crime.
4. Explain why criminal justice in the United States is sometimes considered a nonsystem.
5. Point out major differences between Packer's crime control and due process models.
6. Describe the costs of criminal justice in the United States, and compare those costs among federal, state, and local governments.
7. Explain how myths about crime and criminal justice affect the criminal justice system.

Crime Story

On February 14, 2018, 19-year-old Nikolas Cruz (pictured) walked into Marjory Stoneman Douglas (MSD) High School in Parkland, Florida, and opened fire with an AR-15 assault rifle. After shooting more than 100 rounds in 6 minutes, 14 students and 3 staff members were dead, and 17 others were wounded. The 17 people killed at the high school represented 30% of the 56 school shooting victims killed in 2018—the deadliest year on record (as of 2018).

Cruz walked away from the carnage, blending in with other students, but was caught and arrested about an hour later. The state attorney in Broward County charged Cruz with 17 counts of premeditated murder and 17 counts of attempted murder and is seeking the death penalty. Cruz is being held without bail. His defense attorneys acknowledge that he was the killer and have focused their efforts on avoiding execution.

Several of the surviving students are determined that the deaths of their classmates and friends will not be forgotten following the requisite "thoughts and prayers," as has been the case after so many previous school shootings. Instead, these students organized the "Never Again MSD" movement and helped to coordinate the National School Walkout of March 14, and the March for Our Lives on March 24. The latter crusade attracted more than one million people across the country to rally for safe schools and an end to gun violence. They also have worked to get young people to register and vote, even though some of them were not old enough to vote themselves and have been lobbying for stricter gun laws and challenging the National Rifle Association and the lawmakers it supports.

The efforts of local, state, and national officials pale in comparison. The MSD High School Public Safety Commission, which was created to examine the tragedy, criticized Broward County Sheriff Scott Israel for a policy stipulating that deputies "may" confront active shooters rather than "shall" confront them. Deputy Scot Peterson, the school resource officer and first law enforcement on the scene, chose not to confront Cruz. Sheriff Israel called Peterson a "disgrace," telling reporters that Peterson should have rushed in, "addressed the killer, killed the killer." President Trump, responding to the massacre, said, "That's a case where somebody was outside, they're trained, they didn't act properly or under pressure or they were a coward." Peterson, in his defense, stated that at first he thought the shooting was firecrackers outside the school, and then he maintained that he could not tell from where the gunshots were coming. He asserted that he followed protocol by taking up a tactical position outside the building. However, the Commission determined that he lied, and that Peterson knew the shooter was inside the school but did not want to confront him. Peterson resigned from the Sheriff's department and is collecting a pension of more than $100,000 a year. As for Sheriff Israel, he justified his "non-confrontation" policy by explaining that he did not want his deputies charging into "suicide missions." New Florida Governor Ron DeSantis, in one of his first acts after taking office, suspended Israel, accusing him of "neglect of duty" and "incompetence." Israel has requested a hearing on the suspension before the state senate.

The report also was critical of school superintendent Robert Runcie for inadequate security at the school, not preparing school staff to respond to a mass shooting, and his PROMISE program, which was designed to prevent some student rule violators, such as Nikolas Cruz, from getting police records. Cruz had been expelled from the school the year before. Runcie has subsequently ordered safety upgrades. He also fired two school security monitors and reassigned three assistant principals and a security specialist. Runcie has retained his job for now. In his defense, the president of the NAACP Florida State Conference argued that removing Runcie "would be an extreme overreach, highly political and racist."

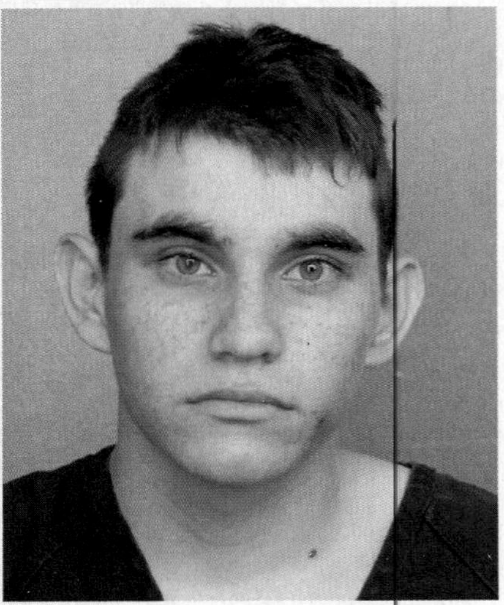

Broward's Sheriff's Office/Getty Images

Following the tragedy, some states approved gun control measures, such as keeping firearms from people convicted of domestic violence or considered suicidal, increasing background checks, and restricting the concealed carrying of firearms. The only federal legislation of note was the banning of bump stocks, which are devices that can make a weapon fire many rounds in a short period of time.

According to the Naval Postgraduate School Center for Homeland Defense and Security's K–12 School Shooting Database, the MSD High School shooting was one of 97 that occurred in 2018—the largest number of school shooting incidents in the United States in any year since data were first collected in 1970. Since December 14, 2012, the date of the Sandy Hook Elementary School shooting in Newtown, Connecticut, where 20-year-old Adam Lanza shot and killed 20 children between the ages of 6 and 7 and 6 adult staff members, and through 2018, 296 school shootings have occurred in the United States with 142 deaths and 300 injuries.

Among the topics discussed in this chapter is how the media shape people's understanding of crime and criminal justice. While the media coverage of the Sandy Hook Elementary School and MSD High School shootings received extensive media coverage, did you know that they were only two of the nearly 300 school shootings that occurred between 2013 and 2018? Why not? And while 34 students and 9 staff members were killed at Sandy Hook Elementary School and MSD High School, did you know that nearly 100 other students and staff members were killed between 2013 and 2018? Why not? The answers to those questions likely include the lack of extensive media coverage of the vast majority of school shootings. What determines which events the media focus on? Based on what you have learned here, do you think school shootings are a problem in the United States? Do you think they are a greater problem than you originally believed?

Crime in the United States

Every day we are confronted with reports of crime in newspapers, magazines, and radio and television news programs. We also see crime in TV docudramas; on such popular shows as the fictional *Criminal Minds, The Blacklist,* and *Bones*; on the long-running franchises *Law & Order* and *CSI*; and on reality-based shows such as *America's Most Wanted, Cops,* and *Unsolved Mysteries*. Crime shows are so popular on television they accounted for 7 of the top 25 broadcast shows for the week February 25 to March 3, 2019 (see Table 1.1). Furthermore, in some of the other top 25 series, such as *60 Minutes*, crime is frequently a

TABLE 1.1 Top 25 Broadcast Shows in Total Viewers for February 25–March 3, 2019

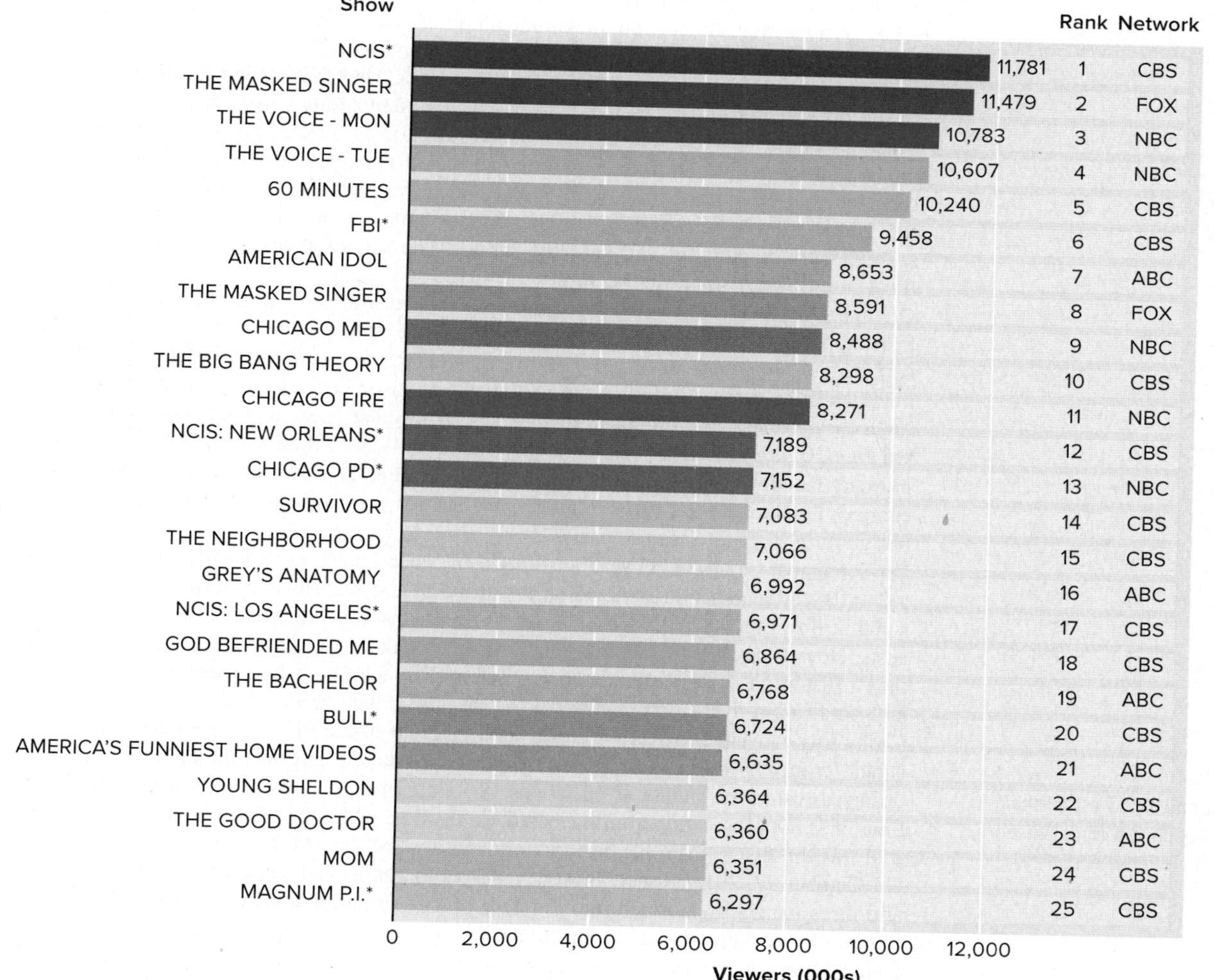

Show	Viewers (000s)	Rank	Network
NCIS*	11,781	1	CBS
THE MASKED SINGER	11,479	2	FOX
THE VOICE - MON	10,783	3	NBC
THE VOICE - TUE	10,607	4	NBC
60 MINUTES	10,240	5	CBS
FBI*	9,458	6	CBS
AMERICAN IDOL	8,653	7	ABC
THE MASKED SINGER	8,591	8	FOX
CHICAGO MED	8,488	9	NBC
THE BIG BANG THEORY	8,298	10	CBS
CHICAGO FIRE	8,271	11	NBC
NCIS: NEW ORLEANS*	7,189	12	CBS
CHICAGO PD*	7,152	13	NBC
SURVIVOR	7,083	14	CBS
THE NEIGHBORHOOD	7,066	15	CBS
GREY'S ANATOMY	6,992	16	ABC
NCIS: LOS ANGELES*	6,971	17	CBS
GOD BEFRIENDED ME	6,864	18	CBS
THE BACHELOR	6,768	19	ABC
BULL*	6,724	20	CBS
AMERICA'S FUNNIEST HOME VIDEOS	6,635	21	ABC
YOUNG SHELDON	6,364	22	CBS
THE GOOD DOCTOR	6,360	23	ABC
MOM	6,351	24	CBS
MAGNUM P.I.*	6,297	25	CBS

Note: * = Crime-related program
All numbers are live + same-day

Source: Adapted from the Nielsen Company, https://tvbythenumbers.zap2it.com/weekly-ratings/broadcast-top-25-networkrankings-feb-25-march-3-2019/ (accessed March 8, 2019).

Daily Television Time

According to a 2018 study, Americans aged 18 and older spend nearly 5 hours a day watching TV—about the same amount of time found in 2014 and 2009 studies. However, the 2018 study found a generation gap. Those aged 50 and older spent more than 6 hours a day watching TV, while young adults watched significantly less TV.

Source: Felix Richter, "The Generation Gap in TV Consumption," *Statista*, August 27, 2018, https://www.statista.com/chart/15224/daily-tv-consumption-by-us-adults/; David Hinckley, "Average American Watches 5 Hours of TV Per Day, Report Shows." *New York Daily News*, March 5, 2014, www.nydailynews.com/life-style/; Brian Stelter, "8 Hours a Day Spent on Screens, Study Finds," *The New York Times*, March 26, 2009, www.nytimes.com/2009/03/27/business/media/27adco.html?_r=1.

significant storyline element. There is also the Crime + Investigation channel, which prominently features crime and justice issues. And crime is a favorite subject of movies and novels. Unfortunately, some of us encounter crime more directly as victims. No wonder crime is a top concern of the American public.

We should keep in mind, however, that the crimes presented by the media are usually more sensational than the crimes routinely committed. Consider some of the top crime news stories in the United States in 2018:

- On February 5, 2018, Larry Nassar, the 54-year-old disgraced former USA Gymnastics and Michigan State University doctor, was sentenced in Eaton County, Michigan, to 40 to 125 years in prison after pleading guilty to 3 counts of first-degree criminal sexual conduct. He will serve that sentence concurrently (at the same time) with the 40-to-175-year sentence he received on January 24, 2018, in Ingham County, Michigan, after pleading guilty to 7 counts of first-degree criminal sexual conduct. He will serve those sentences after completing a separate 60-year sentence in federal prison that he received on December 7, 2017, for child pornography charges. At his state trials, more than 150 women and girls testified that he sexually abused them over the past 2 decades. He admitted to using his trusted medical position to assault and molest girls under the guise of medical treatment.[1]
- On February 14, 2018, 19-year-old Nikolas Cruz killed 14 students and 3 staff members and wounded 17 others when he opened fire with an AR-15 assault rifle at Marjory Stoneman Douglas High School in Parkland, Florida. Cruz, who had been expelled from the school the year before, walked away from the carnage, blending in with other students, but was arrested more than an hour later. The state attorney in Broward County charged Cruz with 17 counts of premeditated murder and 17 counts of attempted murder and will seek the death penalty. The 17 people killed at the high school represented 30% of the 56 school shooting victims in 2018–the deadliest year on record (as of 2018; see the Crime Story for more details).[2]
- On May 18, 2018, 17-year-old Dimitrios Pagourtzis used his father's legally owned sawed off shotgun and a .38 handgun to kill 9 students and 1 teacher and wound 13 others at Santa Fe High School, near Galveston, Texas. Pagourtzis admitted that he did not shoot people he liked and intended to kill the ones he targeted. One student reported that Pagourtzis was "really quiet and he wore like a trench coat almost every day." Law enforcement authorities later found explosive devices, including pipe bombs and pressure cookers, in and near the school. The school shooting was the 22nd in the United States since the beginning of 2018.[3]
- On June 28, 2018, 38-year-old Jarrod Ramos walked into the *Capital Gazette* newsroom in Annapolis, Maryland, opened fire, and killed five newspaper employees and wounded two others. The attack occurred a few years after Ramos had unsuccessfully sued the newspaper for defamation after the newspaper published an article about Ramos harassing a former classmate on social media. Ramos pleaded guilty to a harassment charge. On July 20, 2018, Ramos was indicted on 23 counts, including 5 counts of first-degree murder, 1 count of attempted first-degree murder, and 17 other assault and weapon charges. On January 22, 2019, a judge gave Ramos extra time to consider changing his not guilty plea to an insanity plea.[4]
- On September 25, 2018, 81-year-old Bill Cosby, once known as "America's Dad," received a 3-to-10-year sentence in state prison for drugging and sexually assaulting Andrea Constand at his home in 2004. Cosby also was ordered to pay a fine of $25,000 plus court costs. He will be classified as a "sexually violent predator," a designation that requires lifetime registration, lifetime mandatory sex offender counseling with a treatment provider, and notification to the community that a "sexually violent predator" lives in the area. Cosby was convicted in April of aggravated indecent assault. He had been accused of similar crimes by dozens of other women, but Constand's case was the only one that occurred within the **statute of limitations**. A statute of limitations is a law establishing a time limit for prosecuting a crime, based on the date when the offense occurred. Cosby's trial was the first high-profile celebrity criminal trial of the #MeToo era.[5]
- On October 27, 2018, Robert Bowers, a 46-year-old long-haul trucker armed with a rifle and three handguns, stormed into the Tree of Life Congregation Synagogue in the affluent Squirrel Hill neighborhood of Pittsburgh, Pennsylvania, shouting anti-Semitic slurs and killing 11 worshipers in a 20-minute attack. Two worshipers and four officers were injured. The murder victims ranged in age from 54 to 97. Bowers finally surrendered to police officers after suffering multiple gunshot wounds. Bowers was charged with 29 federal criminal counts, including 11 federal hate-crime charges. Another 11 counts of using a firearm to

statute of limitations A law establishing a time limit for prosecuting a crime, based on the date when the offense occurred.

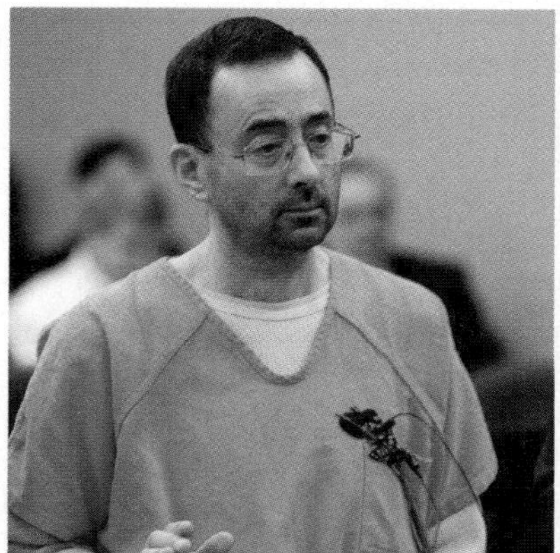
JEFF KOWALKSY/Contributor/Getty Images

Amy Beth Bennett/Sun Sentinel/TNS via Getty Images

Montgomery County District Attorney's Office/Getty Images

Criminal cases involving Larry Nassar, Nikolas Cruz, and Bill Cosby were among the top crime news stories of 2018. *What factors made these crimes so sensational?*

kill carry a maximum penalty of death. He had previously posted anti-Semitic rants on social media. In 2017, anti-Semitic incidents in the United States increased nearly 60%. The attack is believed to be the deadliest on the Jewish community in U.S. history.[6]

- On November 7, 2018, Ian David Long, a 28-year-old former Marine armed with a legally purchased .45-caliber Glock 21 handgun with an extended magazine, shot his way into the Borderline Bar and Grill in Thousand Oaks, California, about 40 miles northwest of Los Angeles. He killed 12 people, including a sheriff's sergeant, before shooting and killing himself. Many other people were wounded. Several hundred people were inside the bar for a "college country night." At the time, it was the 307th mass shooting in the United States in 2018.[7]

Crime in the News

You can read more about past and present crime news stories by visiting the ThoughtCo. Crime/Punishment website at https://www.thoughtco.com/crime-and-punishment-4132972.

Does this website provide a balanced picture of crime in the United States, or does it primarily provide sensational news coverage?

Some of these crime stories are likely to remain top news stories for 2019 and beyond. Furthermore, the fight against domestic terrorism occasioned by the tragedies of September 11, 2001, is likely to remain newsworthy for the foreseeable future. However, taken together, these sensational crime news stories do not provide a very accurate image of the types of crime by which the average citizen is victimized. Nor do such stories accurately depict the kinds of crime the police respond to on a daily basis.

To provide a more accurate idea of the kinds of crimes more typically committed, we reviewed a list of calls for police service in Portland, Oregon, from February 2018 to February 2019.[8] There were 832,123 calls for police service during the period selected. A close examination of the list of police calls (see Table 1.2) reveals that the most frequent type of call for

Portland Press Herald/Getty Images

As shown in Table 1.2, Portland, Oregon, police respond to unwanted person and welfare check calls more often than any other type of call for service. *Should police respond to these types of calls, or should some other agency be responsible for them? Defend your answer.*

TABLE 1.2 Distribution of Calls for Police Service

Police Calls	Percentage
Unwanted person[a]	11.6
Welfare check[b]	11.3
Disturbance[c]	10.6
Suspicious[d]	8.2
Theft[e]	7.4
Collision[f]	5.4
Alarm[g]	4.9
Stolen vehicle[h]	4.3
Area/premise check[i]	4.0
Assist[j]	3.9
Assault[k]	2.6
Threat[l]	2.1
Behavioral health[m]	2.0

[a]Calls related to subjects refusing to leave a location when asked; [b]Calls related to requests to check on the health or safety of a subject; [c]Calls related to nonspecific reports of a threat to public peace or safety; [d]Calls related to reports of suspicious persons, vehicles, or circumstances; [e]Calls related to the unlawful taking of property from the possession of another entity; [f]Calls related to reports of traffic collisions; [g]Calls initiated by the activation of an audible, silent, duress, and/or monitored alarm of a vehicle, residence, business, or other premise; [h]Calls related to the theft of a motor vehicle; [i]Calls related to requests to check a specific location for potential problems or to make sure the site is secure; [j]Calls to provide assistance to partner agencies, including Portland Fire and Rescue, American Medical Response (AMR), and other law-enforcement agencies; [k]Calls related to the unlawful attack by one person upon another. Includes stabbings, shootings, and other types of assaults; [l]Calls related to the placing of another person in reasonable fear of bodily harm through the use of words or other conduct; [m]Calls related to behavioral health, including crisis response, suicide attempts or threats, and transports to detox.

Notes: These data represent calls for police service to the Portland, Oregon Police Department from February 2018 to February 2019. There were a total of 832,123 calls.

Source: Police Foundation, Police Data Initiative. Portland, Oregon. Accessed April 4, 2019, policedatainitiative.org or https://www.portlandoregon.gov/police/76454.

service in Portland involves unwanted persons. This type of call accounted for 11.6% of the total. Other calls for service in Portland ranged from welfare checks (11.3%) to disturbances (10.6%); from suspicious persons, vehicles, or circumstances (8.2%) to thefts (7.4%); and from traffic collisions (5.4%) to alarms (4.9%). The calls for police service listed in Table 1.2 represent only those that accounted for at least 2% of the total calls. In all, there were 49 categories of calls for police service. There were also serious crimes or potential crimes not included in the list (because they accounted for less than 2% of the total calls): 3,284 robbery calls (0.4%), 1,491 domestic violence calls (0.2%), and 77 arson calls (0.009%). Police also are called to assist motorists and to provide escorts for funeral processions–to name just two additional services. Police services and responsibilities are discussed further in Chapter 6.

Here, it is important to observe that the calls police routinely respond to rarely involve the sensational crimes reported by the media. In many cases, they do not involve crimes at all. Critics argue that the news media have a dual obligation to (1) present news that reflects a more balanced picture of the overall crime problem and (2) reduce their presentation of sensational crimes, especially when such crimes are shown not so much to inform as to pander to the public's curiosity and its simultaneous attraction and repulsion toward heinous crimes. The more fundamental problem, however, is that the public's conception of crime is, to a large extent, shaped by the media, and what the media present, for the most part, misleads the public about the nature of crime.

THINKING CRITICALLY

1. Do you think the news media are obligated to present a balanced picture of the overall crime problem and reduce their presentation of sensational crimes? Why or why not?
2. How much do you think the media influence the public conception of crime? How do you think the media select crime stories on which to focus?
3. In what ways is crime entertainment?

Criminal Justice: An Institution of Social Control

In the United States, there are a variety of responses to crime. When a child commits a criminal act, even if that act does not come to the attention of the police, parents or school authorities nevertheless may punish the child for the offense (if they find out about it). Attempts to prevent crime by installing burglar alarms in automobiles and homes are other ways of responding to crime. Throughout this book, we focus on the criminal justice response to crime.

Like the family, schools, organized religion, the media, and the law, criminal justice is an **institution of social control** in the United States. A primary role of such institutions is to persuade people, through subtle and not-so-subtle means, to abide by the dominant values of society. Subtle means of persuasion include gossip and peer pressure, whereas expulsion and incarceration are examples of not-so-subtle means.

institution of social control An organization that persuades people, through subtle and not-so-subtle means, to abide by the dominant values of society.

As an institution of social control, criminal justice differs from the others in two important ways. First, the role of criminal justice is restricted officially to persuading people to abide by a limited range of social values whose violation constitutes crime. Thus, although courteous behavior is desired of all citizens, rude behavior is of no official concern to criminal justice unless it violates the criminal law. Dealing with noncriminal rude behavior is primarily the responsibility of the family. Second, criminal justice is generally society's "last line of defense" against people who refuse to abide by dominant social values and commit crimes. Usually, society turns to criminal justice only after other institutions of social control have failed.

FYI **Confidence in the Criminal Justice System**

In a 2018 public opinion poll, 22% of Americans responded they had a "great deal" (9%) or "quite a lot" (13%) of confidence in the criminal justice system, 41% had "some" confidence, 34% had "very little" confidence, 2% had no ("none") confidence, and 1% had no opinion.

Source: Gallup, "Confidence in Institutions." Accessed February 13, 2019, https://news.gallup.com/poll/1597/confidence-institutions.aspx.

THINKING CRITICALLY

1. Given what you know about crime in the United States, do you think that the criminal justice system is a strong institution of social control? Why?
2. Do you think that other institutions such as the family, schools, and organized religion are better institutions of social control than the criminal justice system? If so, which ones? Why?

Criminal Justice: The System

Criminal justice in the United States is administered by a loose confederation of more than 50,000 agencies of federal, state, and local governments. Those agencies consist of the police, the courts, and corrections. Together they are commonly referred to as the *criminal justice system*. Although there are differences in the ways the criminal justice system operates in different jurisdictions, there are also similarities. The term **jurisdiction**, as used here, means a politically defined geographical area (e.g., a city, a county, a state, or a nation).

jurisdiction A politically defined geographical area. The right or authority of a justice agency to act with regard to a particular subject matter, territory, or person. The authority of a court to hear and decide cases.

The following paragraphs provide a brief overview of a typical criminal justice response to criminal behavior. Figure 1.1 is a graphic representation of the process. It includes the variations for petty offenses, **misdemeanors** (less serious crimes), **felonies** (serious crimes), and juvenile offenses. Later chapters of this book provide a more detailed examination of the criminal justice response to crime and delinquency.

misdemeanor A less serious crime generally punishable by a fine or by incarceration in jail for not more than 1 year.

felony A serious offense punishable by confinement in prison for more than 1 year or by death.

Police

The criminal justice response to crime begins when a crime is reported to the police or, far less often, when the police themselves discover that a crime has been committed. Sometimes, solving the crime is easy—the victim or a witness knows the perpetrator or where to find him or her. Often, an arrest supported by witness statements and crime scene evidence is sufficient to close a case, especially with a less serious crime. More often, though, the police must conduct an in-depth investigation to determine what happened in a particular crime. Even when the police start with a known crime or a cooperative victim or witness, the investigation can be lengthy and difficult.

If police investigation of the crime is successful, a suspect is arrested. An **arrest** is the seizing and detaining of a person by lawful authority. After an arrest has been made, the

arrest The seizing or the taking of a person into custody by lawful authority, either actual physical custody, as when a suspect is handcuffed by a police officer, or constructive custody, as when a person peacefully submits to a police officer's control.

FIGURE 1.1 Overview of the Criminal Justice System

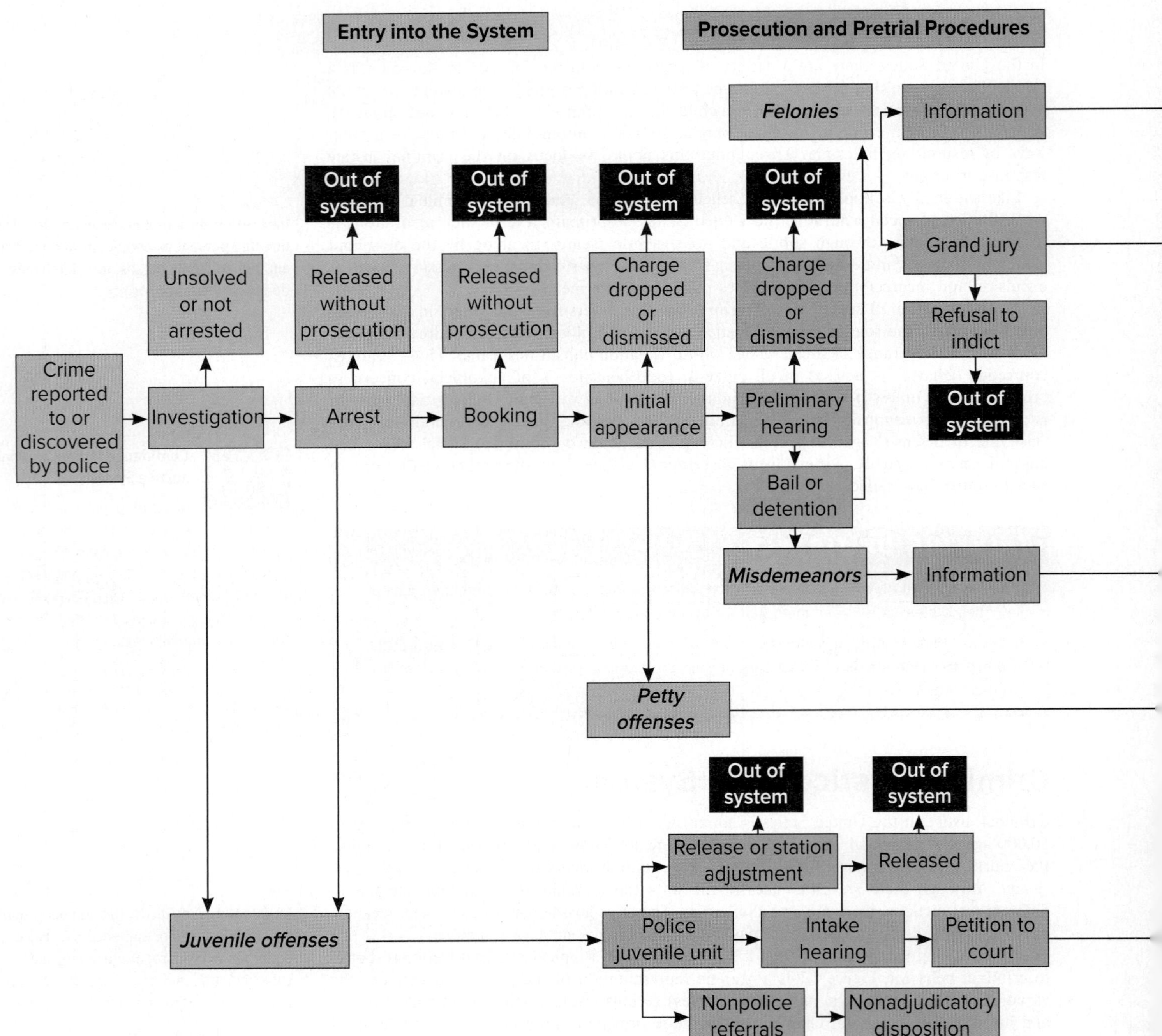

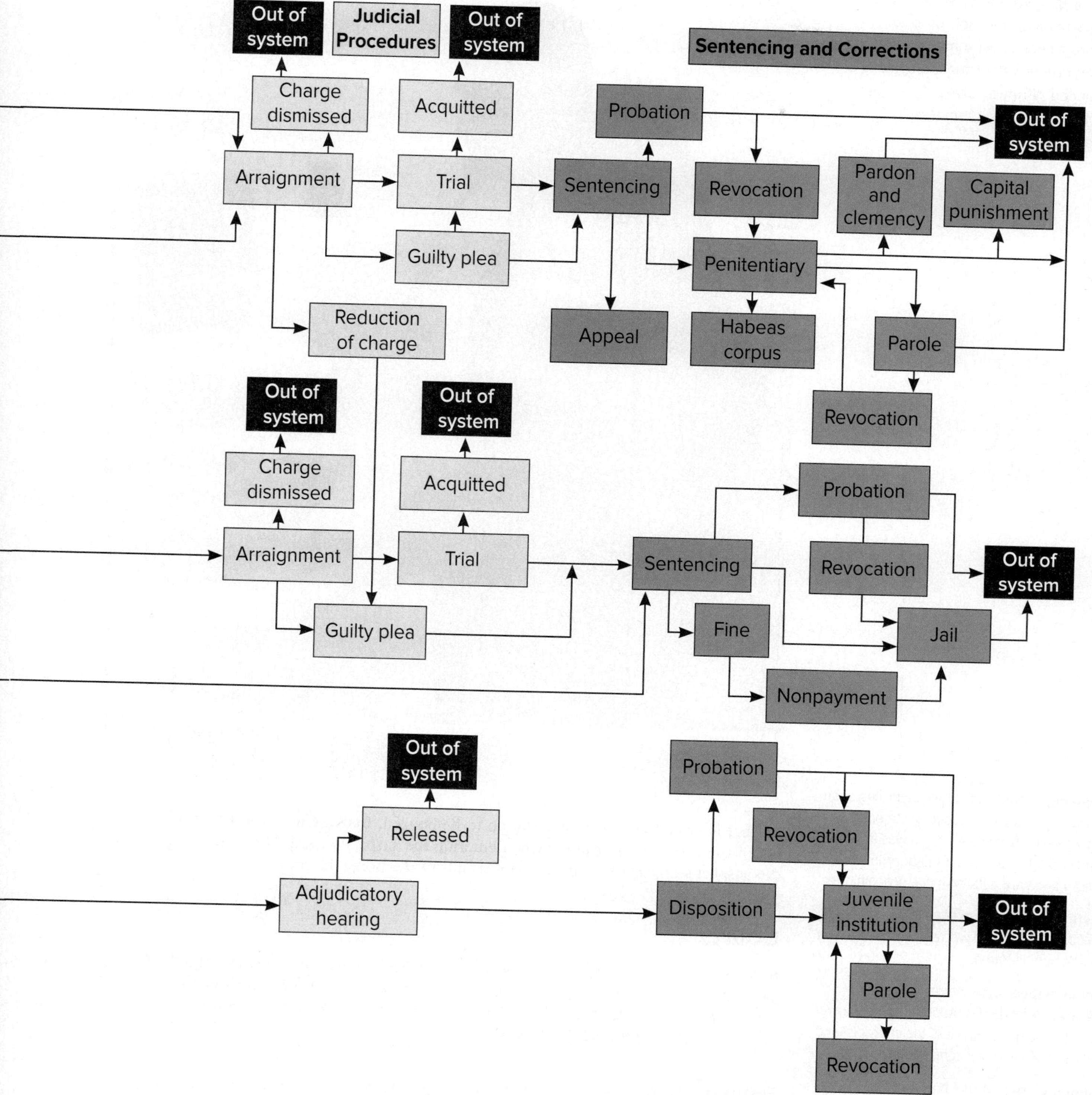

Judicial Procedures
Sentencing and Corrections
Out of system
Charge dismissed
Acquitted
Arraignment
Trial
Guilty plea
Reduction of charge
Sentencing
Probation
Revocation
Penitentiary
Pardon and clemency
Capital punishment
Appeal
Habeas corpus
Parole
Revocation
Charge dismissed
Acquitted
Arraignment
Trial
Guilty plea
Sentencing
Probation
Revocation
Jail
Fine
Nonpayment
Released
Adjudicatory hearing
Disposition
Probation
Revocation
Juvenile institution
Parole
Revocation

Getting fingerprints is generally a part of the booking into jail process. Pictured here is an inmate being fingerprinted with a technologically advanced "Printrat Livescan Finger Machine." *Why are inmates fingerprinted?*

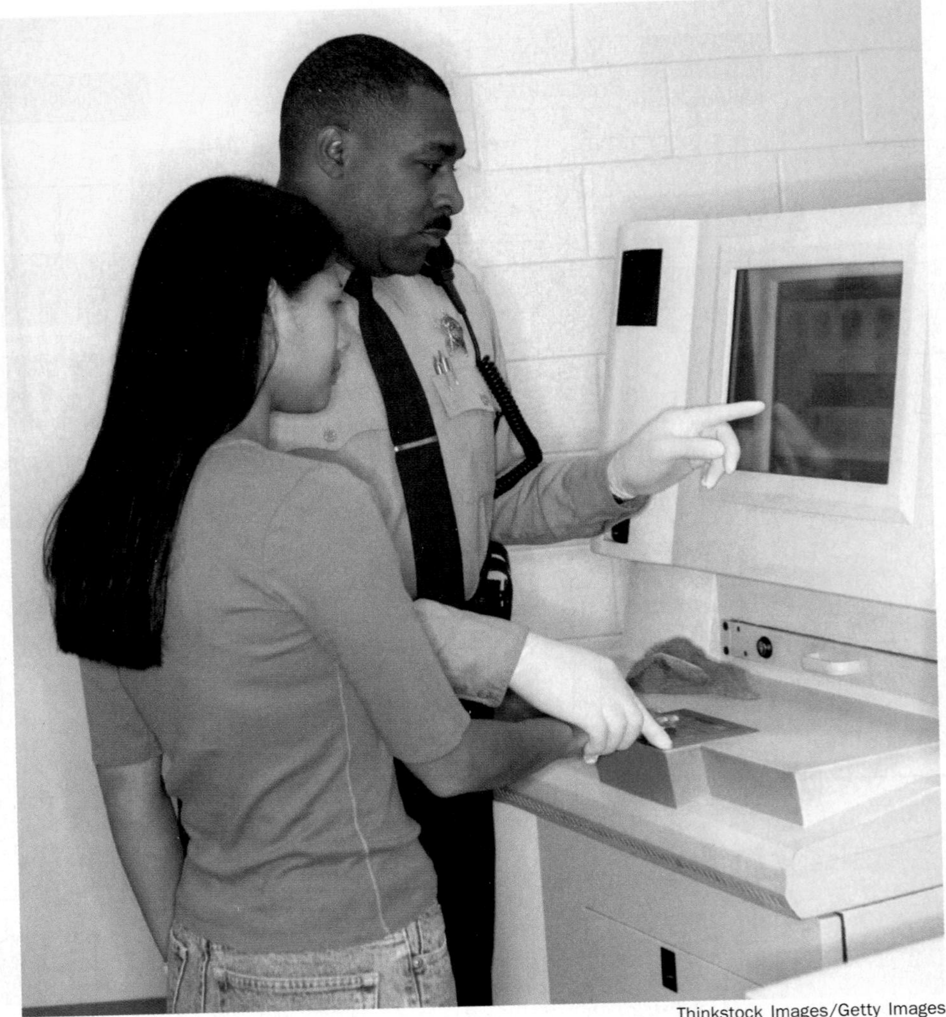

Thinkstock Images/Getty Images

booking The administrative recording of an arrest. Typically, the suspect's name, the charge(s) for which the person was arrested, and perhaps the suspect's fingerprints or photograph are entered in the police blotter.

defendant A person against whom a legal action is brought, a warrant is issued, or an indictment is found.

initial appearance A pretrial stage in which a defendant is brought before a lower court to be given notice of the charge(s) and advised of her or his constitutional rights.

summary trial A trial before a judge without a jury.

probable cause The amount of proof necessary for a reasonably intelligent person to believe that a crime has been committed or that items connected with criminal activity can be found in a particular place. It is the standard of proof needed to conduct a search or to make an arrest.

suspect is brought to the police station to be booked. **Booking** is the administrative recording of the arrest. It typically involves entering the suspect's name, the charge, and perhaps the suspect's fingerprints and/or photograph in the police blotter.

Courts

Soon after a suspect has been arrested and booked, a prosecutor reviews the facts of the case and the available evidence. Sometimes, a prosecutor reviews the case before arrest. The prosecutor decides whether to charge the suspect with a crime or crimes. If no charges are filed, the suspect must be released.

Pretrial Stages After the charge or charges have been filed, the suspect, who is now the **defendant**, is brought before a lower court judge for an initial appearance. At the **initial appearance**, the defendant is given formal notice of the charge or charges against him or her and advised of his or her constitutional rights (e.g., the right to counsel). In the case of a misdemeanor or an ordinance violation, a **summary trial** (an immediate trial without a jury) may be held. In the case of a felony, a hearing is held to determine whether the defendant should be released or whether there is probable cause to hold the defendant for a preliminary hearing. **Probable cause** is a standard of proof that requires trustworthy evidence sufficient to make a reasonable person believe that, more likely than not, the proposed action is justified. If the suspect is to be held for a preliminary hearing, bail is set if the judge believes release on bail is appropriate. **Bail**, usually a monetary guarantee deposited with the court, is meant to ensure that the defendant will appear at a later stage in the criminal justice process. In states that do not utilize preliminary hearings, an arraignment date is scheduled at the initial appearance.

In about half of all states, a preliminary hearing follows the initial appearance. Preliminary hearings are used only in felony cases. The purpose of the **preliminary hearing** is for a judge to determine whether there is probable cause to believe that the defendant committed the crime or crimes with which he or she is charged. If the judge finds probable cause, the defendant is bound over for possible indictment in a state with grand juries or for arraignment on a document called an *information* (see below) in a state without grand juries.

Grand juries are involved in felony prosecutions in about half the states. A **grand jury** is a group of citizens who meet in closed sessions for a specified period to investigate charges coming from preliminary hearings and to fulfill other responsibilities. Thus, a primary purpose of the grand jury is to determine whether there is probable cause to believe that the accused committed the crime or crimes with which the prosecutor has charged the person. The grand jury can either indict a suspect or issue a "true bill," or fail to indict a suspect or "issue no bill." If the grand jury fails to indict or issues no bill, the prosecution must be dropped. In states that do not use grand juries, prosecutors charge defendants with a document called an information. An **information** outlines the formal charge or charges, the law or laws that have been violated, and the evidence to support the charge or charges.

Once an indictment or information is filed with the trial court, the defendant is scheduled for arraignment. The primary purpose of **arraignment** is to hear the formal information or indictment and to allow the defendant to enter a plea. About 95% of criminal defendants plead guilty to the charges against them in an arrangement called *plea bargaining*. **Plea bargaining** is the practice whereby the prosecutor, the defense attorney, the defendant, and, in many jurisdictions, the judge agree on a specific sentence to be imposed if the accused pleads guilty to an agreed-on charge or charges instead of going to trial.

Trial If a defendant pleads not guilty or not guilty by reason of insanity, a trial date is set. Although all criminal defendants have a constitutional right to a trial (when imprisonment for 6 months or more is a possible outcome), only about 5% of all criminal cases are disposed of by trial. Approximately 2% of criminal cases involve jury trials. The remaining cases that are not resolved through plea bargaining are decided by a judge in a **bench trial** (without a jury). Thus, approximately 95% of all criminal cases are resolved through plea bargaining, about 2% through jury trials, and about 3% through bench trials. (See Figure 1.2.) In most jurisdictions, the choice between a jury trial and a bench trial is the defendant's to make.

If the judge or the jury finds the defendant guilty as charged, the judge begins to consider a sentence. In some jurisdictions, the jury participates to varying degrees in the sentencing process. The degree of jury participation depends on the jurisdiction and the crime. If the judge or jury finds the defendant not guilty, the defendant is released from the jurisdiction of the court and becomes a free person.

Judges cannot impose just any sentence. There are many factors that restrict sentencing decisions. The U.S. Constitution's Eighth Amendment prohibiting cruel and unusual punishments and various statutory provisions limit judges. Judges are guided by prevailing philosophical rationales, by organizational considerations, and by presentence investigation reports. They also are influenced by their own personal characteristics. Presentence investigation reports are used in the federal system and in the majority of states to help judges determine appropriate sentences.

Currently, five general types of punishment are in use in the United States: fines, probation, intermediate punishments, imprisonment, and death. Intermediate punishments are more restrictive than probation but less restrictive and less costly than imprisonment. **Probation** is a sentence in which the offender is retained in the community under the supervision of a probation agency rather than being incarcerated. Probation is the most frequently imposed criminal sentence in the United States. As long as a judge imposes one or a combination of the five punishments and the sentence length and type are within statutory limits, the judge is free to set any sentence.

bail Usually a monetary guarantee deposited with the court that is suposed to ensure that the suspect or defendant will appear at a later stage in the criminal justice process.

preliminary hearing A pretrial stage used in about one-half of all states and only in felony cases. Its purpose is for a judge to determine whether there is probable cause to support the charge or charges imposed by the prosecutor.

grand jury Generally a group of 12 to 23 citizens who meet in closed sessions to investigate charges coming from preliminary hearings or to engage in other responsibilities. A primary purpose of the grand jury is to determine whether there is probable cause to believe that the accused committed the crime or crimes.

information A document that outlines the formal charge(s) against a suspect, the law(s) that have been violated, and the evidence to support the charge(s).

arraignment A pretrial stage; its primary purpose is to hear the formal information or indictment and to allow the defendant to enter a plea.

plea bargaining The practice whereby the prosecutor, the defense attorney, the defendant, and—in many jurisdictions—the judge agree on a specific sentence to be imposed if the accused pleads guilty to an agreed-upon charge or charges instead of going to trial.

bench trial A trial before a judge without a jury.

FIGURE 1.2 Criminal Case Dispositions

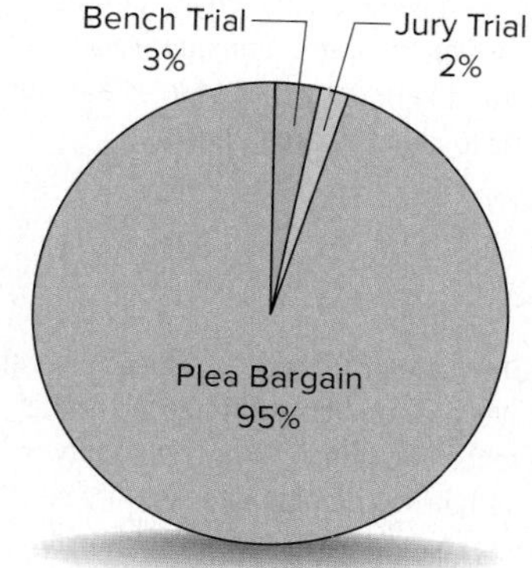

Careers in Criminal Justice

Law Enforcement/Security
BATF Agent
Border Patrol Agent
Campus Police Officer
Crime Prevention Specialist
Criminal Investigator
Criminal Profiler
Customs Officer
Deputy Sheriff
Deputy U.S. Marshal
Drug Enforcement Officer
Environmental Protection Agent
FBI Special Agent
Federal Agency Investigator
Fingerprint Technician
Forensic Scientist
Highway Patrol Officer
Immigration and Naturalization Service Officer
Insurance Fraud Investigator
Laboratory Technician
Loss Prevention Officer
Military Police Officer
Park Ranger
Police Administrator
Police Dispatcher
Police Officer
Polygraph Examiner
Postal Inspector
Private Investigator
Secret Service Agent
State Trooper

Courts/Legal
Arbitrator
Attorney General
Bailiff
Clerk of Court
Court Interpreter
Court Reporter
District Attorney
Judge
Jury Assignment Commissioner
Jury Coordinator
Juvenile Magistrate
Law Clerk
Law Librarian
Legal Researcher
Mediator
Paralegal
Public Defender
Public Information Officer
Trial Court Administrator
Victim Advocate

Teaching/Research
Agency Researcher
Community College, College, or University Lecturer or Professor

Corrections/Rehabilitation
Activity Therapy Administrator
Business Manager
Case Manager
Chaplain
Chemical Dependency Manager
Child Care Worker
Children's Services Counselor
Classification Officer
Client Service Coordinator
Clinical Social Worker
Community Liaison Officer
Correctional Officer
Dietary Officer
Drug Court Coordinator
Field Administrator
Fugitive Apprehension Officer
Home Detention Supervisor
Human Services Counselor
Job Placement Officer
Juvenile Detention Officer
Juvenile Probation Officer
Mental Health Clinician
Parole/Probation Officer
Presentence Investigator
Prison Industries Superintendent
Program Officer/Specialist
Programmer/Analyst
Psychologist
Recreation Coordinator
Rehabilitation Counselor
Researcher
Residence Supervisor
Sex Offender Therapist
Social Worker
Statistician
Substance Abuse Counselor
Teacher
Vocational Instructor
Warden or Superintendent
Youth Service Worker/Coordinator
Youth Supervisor

probation A sentence in which the offender, rather than being incarcerated, is retained in the community under the supervision of a probation agency and required to abide by certain rules and conditions to avoid incarceration.

parole A method of prison release whereby inmates are conditionally released at the discretion of a board or other authority before having completed their entire sentences; can also refer to the community supervision received upon release.

Appeals Defendants who are found guilty can appeal their convictions either on legal grounds or on constitutional grounds. Examples of legal grounds include defects in jury selection, improper admission of evidence at trial, and mistaken interpretations of law. Constitutional grounds include illegal search and seizure, improper questioning of the defendant by the police, identification of the defendant through a defective police lineup, and incompetent assistance of counsel.

The appellate court can *affirm* the verdict of the lower court and let it stand; modify the verdict of the lower court without totally reversing it; reverse the verdict of the lower court, which requires no further court action; or reverse the decision and *remand,* or return, the case to the court of original jurisdiction for either a retrial or resentencing.

Corrections

A defendant sentenced to prison may be eligible for parole (in those jurisdictions that grant parole) after serving a portion of his or her sentence. **Parole** is the conditional release of prisoners before they have served their full sentences. Generally, the decision to grant parole is made by a parole board. Once offenders have served their sentences, they are released from criminal justice authority.

THINKING CRITICALLY

1. Do you think the criminal justice system "works" in the United States? Why or why not?
2. What improvements do you think should be made to the criminal justice system?
3. Do you think judges should be limited in the sentences they are allowed to impose? Why or why not?

Criminal Justice: The Nonsystem

As noted earlier, the many police, court, and corrections agencies of the federal, state, and local governments, taken together, are commonly referred to as the criminal justice system. However, the depiction of criminal justice–or, more specifically, of the interrelationships and inner workings of its various components–as a "system" may be inappropriate and misleading for at least two reasons.

First, there is no single "criminal justice system" in the United States. Rather, as noted earlier, there is a loose confederation of many independent criminal justice agencies at all levels of government. This loose confederation is spread throughout the country with different, sometimes overlapping, jurisdictions. Although there are some similarities among many of those agencies, there are also significant differences. The only requirement they all share, a requirement that is the basis for their similarities, is that they follow procedures permitted by the U.S. Constitution.

Second, if a **system** is thought of as a smoothly operating set of arrangements and institutions directed toward the achievement of common goals, one is hard-pressed to call the operation of criminal justice in the United States a system. Instead, because there is considerable conflict and confusion among different agencies of criminal justice, a more accurate representation may be that of a criminal justice "nonsystem."

system A smoothly operating set of arrangements and institutions directed toward the achievement of common goals.

myth

The agencies that administer criminal justice in the United States form a unified system: the criminal justice system.

fact

There is no single "criminal justice system" in the United States. Instead, there is a loose confederation of many independent criminal justice agencies at all levels of government. Moreover, instead of operating together as a system, agencies of criminal justice in the United States interact but generally operate independently of one another, each agency often causing problems for the others.

For example, police commonly complain that criminal offenders who have been arrested after weeks or months of time-consuming and costly investigation are not prosecuted or are not prosecuted vigorously enough. Police often maintain that prosecutors are not working with them or are making their jobs more difficult than necessary. Prosecutors, however, often gripe about shoddy police work. Sometimes, they say that they are unable to prosecute a crime because of procedural errors committed by the police during the investigation or the arrest.

Even when a criminal offender is prosecuted, convicted, and sentenced to prison, police often argue that the sentence is not severe enough to fit the seriousness of the crime, or they complain when the offender is released from prison after serving only a portion of his or her sentence. In such situations, police frequently argue that the courts or the correctional agencies are undermining their efforts by putting criminals back on the streets too soon.

Conflicts between the courts and corrections sometimes occur when judges continue to impose prison sentences on criminal offenders, especially so-called petty offenders, even though the judges know that the prisons are under court orders to reduce overcrowding.

In addition, there is a mostly separate process for juvenile offenders. Criminal justice officials frequently complained that their jobs were made more difficult because of the practice, which used to be common in many states, of sealing juvenile court records. That practice withheld juvenile court records from the police, prosecutors, and judges even though the records may have been relevant and helpful in making arrests, prosecuting criminal cases, and determining appropriate sentences. Today, formerly confidential juvenile court records are made available to a wide variety of individuals, including prosecutors, law enforcement officers, social service personnel, school authorities, victims, and the public. However, access is not necessarily unlimited or automatic. Access still may be restricted to certain parts of the record and may require a court order.[9] A rationale for concealing juvenile court records is to prevent, as much as possible, the labeling of juvenile offenders as delinquents, which could make them delinquents. (Labeling theory is discussed in Chapter 3.)

In short, rather than operating together as a system, agencies of criminal justice in the United States generally operate independently of one another, each agency often causing problems for the others. Such conflicts may not be entirely undesirable, however, because

Criminal Justice System Needs Improvement

A 2018 national survey found that 76% of registered voters believe the U.S. criminal justice system needs significant improvements, while only 21% believe "it's working pretty well as it is." Sixty-eight percent of Republicans, 78% of Independents, 80% of Democrats, and 80% of women believe the system needs significant improvements.

Source: Robert Blizzard, "National Poll Results," Public Opinion Strategies, January 25, 2018, accessed June 3, 2019, https://www.politico.com/f/?id=00000161-2ccc-da2c-a963-efff82be0001.

they occur in a context of checks and balances by which the courts ensure that the law is enforced according to constitutional principles.

THINKING CRITICALLY

1. What do you think are some of the positive aspects of having a criminal justice nonsystem?
2. What do you think are some of the disadvantages of having a criminal justice nonsystem?

Two Models of Criminal Justice

In his influential 1968 book entitled *The Limits of the Criminal Sanction,* legal scholar Herbert Packer describes the criminal justice process in the United States as the outcome of competition between two value systems.[10] Those two value systems, which represent two ends of a value continuum, are the basis for two models of the operation of criminal justice—the crime control model and the due process model. Figure 1.3 depicts this continuum. From a political standpoint, the **crime control model** reflects traditional conservative political values, while the **due process model** embodies traditional liberal political values.[11] Consequently, when politically conservative values are dominant in society, as they have been for most of the past 50 years, the principles and policies of the crime control model seem to dominate the operation of criminal justice. During more politically liberal periods, such as the 1960s and 1970s, and the eight years of the Obama administration (2008–2016), the principles and policies of the due process model seem to direct criminal justice activity.

crime control model One of Packer's two models of the criminal justice process. Politically, it reflects traditional conservative values. In this model, the control of criminal behavior is the most important function of criminal justice.

due process model One of Packer's two models of the criminal justice process. Politically, it embodies traditional liberal values. In this model, the principal goal of criminal justice is at least as much to protect the innocent as it is to convict the guilty.

The models are ideal types, neither of which corresponds exactly to the actual day-to-day practice of criminal justice. Rather, they both provide a convenient way to understand and discuss the operation of criminal justice in the United States. In practice, the criminal justice process represents a series of conflicts and compromises between the value systems of the two models. In the following sections, we describe Packer's two models in detail.

FIGURE 1.3 Two Models of the Criminal Justice Process

Due Process Model	Crime Control Model
Traditional liberal values	Traditional conservative values

The Crime Control Model

In the crime control model, the control of criminal behavior is by far the most important function of criminal justice. Although the means by which crime is controlled are important in this view (illegal means are not advocated), they are less important than the ultimate goal of control. Consequently, the primary focus of this model is on the efficiency of the criminal justice process. Advocates of the crime control model want to make the process more efficient—to move cases through the process as quickly as possible and to bring them to a close. Packer characterizes the crime control model as "assembly-line justice." Bohm (one of your authors) has called it "McJustice."[12] To achieve "quick closure" in the processing of cases, a premium is placed on speed and finality. Speed requires that cases be handled informally and uniformly; finality depends on minimizing occasions for challenge, that is, appeals.

To appreciate the assembly-line or McJustice metaphors used by Packer and Bohm and to understand how treating cases uniformly speeds up the process and makes it more efficient, consider the way that McDonald's sells billions of hamburgers. When you order a Big Mac from McDonald's, you know exactly what you are going to get. All Big Macs are the same because they are made uniformly. Moreover, you can get a Big Mac in a matter of seconds most of the time. However, what happens when you order something different or something not already prepared, such as a hamburger with ketchup only? Your order is taken, and you are asked to stand to the side because your special order will take a few minutes. Your special order has slowed down the assembly line and reduced efficiency. This happens in criminal justice, too! If defendants ask for something special, such as a trial, the assembly line is slowed and efficiency is reduced.

As described in Chapter 8, even when criminal justice is operating at its best, it is a slow process. The time from arrest to final case disposition can typically be measured in weeks or months. If defendants opt for a jury trial, which is their right in most felony cases, the cases are handled formally and are treated as unique; no two cases are the same in their circumstances or in the way they are handled. If defendants are not satisfied with the outcome of their trials, then they have the right to appeal. Appeals may delay by years the final resolution of cases.

To increase efficiency–meaning speed and finality–crime control advocates prefer plea bargaining. As described previously and as you will see in Chapter 8, plea bargaining is an informal process that is used instead of trial. Plea bargains can be offered and accepted in a relatively short time. Also, cases are handled uniformly because the mechanics of a plea bargain are basically the same; only the substance of the deal differs. In addition, with successful plea bargains, there is no opportunity for challenge; there are no appeals. Thus, plea bargaining is the perfect mechanism for achieving the primary focus of the crime control model: efficiency.

Elaine Thompson/AP Photo

Gary Ridgway, the so-called Green River Killer, was allowed to plead guilty to 48 counts of murder in exchange for helping authorities find some of his victims' remains. The plea bargain allowed him to escape the death penalty. He was sentenced to 48 consecutive life sentences without parole instead. *Was the plea bargain in this case justified? Why or why not?*

The key to the operation of the crime control model is "a presumption of guilt." In other words, advocates of this model assume that if the police have expended the time and effort to arrest a suspect and the prosecutor has formally charged the suspect with a crime, then the suspect must be guilty. Why else would police arrest and prosecutors charge? Although the answers to that question are many (see the discussions in Chapters 7 and 8 of the extralegal factors that influence police and prosecutorial behavior), the fact remains that a presumption of guilt is accurate most of the time. That is, most people who are arrested and charged with a crime or crimes are, in fact, guilty. A problem–but not a significant one for crime control advocates–is that a presumption of guilt is not accurate all the time; miscarriages of justice do occur (see the discussion in Chapter 4). An equally important problem is that a presumption of guilt goes against one of the oldest and most cherished principles of American criminal justice–that a person is considered innocent until proven guilty.

Reduced to its barest essentials and operating at its highest level of efficiency, the crime control model consists of an administrative fact-finding process with two possible outcomes: a suspect's exoneration or the suspect's guilty plea.

The Due Process Model

Advocates of the due process model, by contrast, reject the informal fact-finding process as definitive of factual guilt. They insist instead on formal, adjudicative fact-finding processes in which cases against suspects are heard publicly by impartial trial courts. In the due process model, moreover, the factual guilt of suspects is not determined until the suspects have had a full opportunity to discredit the charges against them. For those reasons, Packer characterizes the due process model as "obstacle-course justice."

What motivates this careful and deliberate approach to the administration of justice is the realization that human beings sometimes make mistakes. The police sometimes arrest the wrong person, and prosecutors sometimes charge the wrong person. Thus, contrary to the crime control model, the demand for finality is low in the due process model, and the goal is at least as much to protect the innocent as it is to convict the guilty. Indeed, for due process model advocates, it is better to let a guilty person go free than it is to wrongly convict and punish an innocent person.

The due process model is based on the doctrine of legal guilt and the presumption of innocence. According to the **doctrine of legal guilt**, people are not to be held guilty of

doctrine of legal guilt The principle that people are not to be held guilty of crimes merely on a showing, based on reliable evidence, that in all probability they did in fact do what they are accused of doing. Legal guilt results only when factual guilt is determined in a procedurally regular fashion, as in a criminal trial, and when the procedural rules designed to protect suspects and defendants and to safeguard the integrity of the process are employed.

crimes merely on a showing, based on reliable evidence, that in all probability they did in fact do what they are accused of doing. In other words, it is not enough that people are factually guilty in the due process model; they must also be legally guilty. Legal guilt results only when factual guilt is determined in a procedurally regular fashion, as in a criminal trial, and when the procedural rules, or due process rights, designed to protect suspects and defendants and to safeguard the integrity of the process are employed. The conditions of legal guilt—that is, procedural, or due process, rights—are described in Chapter 4. They include:

- Freedom from unreasonable searches and seizures.
- Protection against double jeopardy.
- Protection against compelled self-incrimination.
- A speedy and public trial.
- An impartial jury of the state and district where the crime occurred.
- Notice of the nature and cause of the accusation.
- The right to confront opposing witnesses.
- Compulsory process for obtaining favorable witnesses.
- The right to counsel.
- The prohibition of cruel and unusual punishment.

In short, in the due process model, factual guilt is not enough. For people to be found guilty of crimes, they must be found *both* factually and legally guilty.

Due process advocates champion this obstacle-course model of justice because they are skeptical about the ideal of equality on which U.S. criminal justice is supposedly based. They recognize that there can be no equal justice where the kind of trial a person gets, or whether he or she gets a trial at all, depends substantially on how much money that person has. It is assumed that in an adversarial system of justice (as described in Chapter 8 and employed in the United States), an effective defense is largely a function of the resources that can be mustered on behalf of the accused. It also is assumed that there are gross inequalities in the financial means of criminal defendants. Most criminal defendants are indigent or poor, and because of their indigence, they are frequently denied an effective defense. Although procedural safeguards, or conditions of legal guilt, cannot by themselves correct the inequity in resources, they do provide indigent defendants, at least theoretically, with a better chance for justice than they would receive without them.

Fundamentally, the due process model defends the ideal of personal freedom and its protection. The model rests on the assumption that preventing tyranny by the government and its agents is the most important function of the criminal justice process.

Crime Control versus Due Process

As noted earlier, the model that dominates criminal justice policy in the United States at any particular time depends on the political climate. Until the election of Barack Obama—America's first African American president—in 2008, the United States was in the midst of a prolonged period—beginning in the mid-1970s—during which politically conservative values dominated the practice of criminal and juvenile justice. Thus, it should come as no surprise that the crime control model of criminal justice more closely resembled the actual practice of criminal and juvenile justice in the United States for the roughly three-and-one-half decades prior to Obama's first election. Before the mid-1970s, at the time Packer wrote and published his book, politically liberal values and, thus, the principles and policies of the due process model directed the operation of criminal and juvenile justice. However, because neither model ever completely represents the practices of criminal and juvenile justice, many elements of the crime control model were evident prior to the mid-1970s, as were many elements of the due process model during the next three-and-one-half decades.

During Obama's eight years in office, his administration's liberal political values seemed to tilt criminal justice policy toward principles and policies associated with the due process model of criminal justice. That said, any transformations made during Obama's eight years were relatively modest because, for one thing, criminal justice policies that have been entrenched for decades are hard to change, and, for another, the U.S. Supreme Court, arguably the key institution when it comes to criminal justice policy, had a conservative majority that was not particularly sympathetic to Obama's goals. So even though there was some shift to due process model policies and practices during the Obama administration, many elements of the crime control model remained.

With the 2016 election of Donald Trump, whatever Obama administration transformations in criminal justice policy toward the due process model are likely to be rolled back as conservative political values and the crime control model of criminal justice return to prominence, as Packer's model of criminal justice predicts. If President Trump has the opportunity to appoint two more conservative justices to replace the elderly, liberal-leaning Justices Ginsburg and Breyer following his appointment and the Senate's confirmation of Justices Neil Gorsuch and Brett Kavanaugh, politically conservative values and crime control model principles and policies likely will guide criminal justice policy in the United States for decades to come.

THINKING CRITICALLY

1. What do you think are some of the fundamental problems with the crime control model? What are the benefits of this model?
2. What do you think are some of the fundamental problems with the due process model? What are the benefits of this model?

Costs of Criminal Justice

Each year in the United States, an enormous amount of money is spent on criminal justice at the federal, state, and local levels. In 2016 (the latest year for which figures were available), a total of $284 billion was spent on civil and criminal justice. Nearly one-half of the $284 billion was spent on police protection, about 20% on judicial and legal services, and approximately 30% on corrections. Table 1.3 shows the breakdown of spending among

TABLE 1.3 Costs of Criminal Justice

In 2016, federal, state, and local governments spent $284 billion in direct expenditures for the criminal and civil justice systems.

POLICE PROTECTION	In Millions ($)
68% Local	94,207
11% States	15,003
21% Federal	28,886
100%	138,096
JUDICIAL/LEGAL SERVICES	
38% Local	23,431
37% States	22,826
25% Federal	14,892
100%	61,149
CORRECTIONS	
33% Local	28,977
59% States	49,040
8% Federal	6,952
100%	84,969

Note: Detail may not add to 100% because of rounding.

Source: Shelley Hyland, "Justice Expenditure and Employment Extracts, 2016 (preliminary)," Bureau of Justice Statistics, U.S. Department of Justice, https://www.bjs.gov/index.cfm?ty=pbdetail&iid=6728, filename jeee16t01.csv (date of version: 11/7/2019).

the three main segments of the criminal justice system and among the federal, state, and local levels. The $284 billion spent on civil and criminal justice in 2016 represents approximately $880 for every resident of the United States ($440 for police, $176 for judicial and legal services, and $264 for corrections). The $284 billion was an increase of about 7% from 2012 and approximately 10% from 2009.[13]

Criminal justice is primarily a state and local function; state and local governments spent about 80% of the 2016 total. Note that state and local governments share the costs of criminal justice by making police protection primarily a local function and corrections primarily a state function. In 2016, local governments spent 68% of the total spent on police protection, while state governments spent 58% of the total spent on corrections. The expense of judicial and legal services was evenly divided between state and local governments, with each level of government spending about 40% of the total on those services.

Although the bulk of government spending on criminal justice is at the state and local levels, the federal government uses its expenditures strategically to influence criminal justice policy at the other levels of government. For example, the federal government develops and tests new approaches to criminal justice and crime control. It then encourages state and local criminal justice agencies to duplicate effective programs and practices by awarding monetary grants to interested and willing agencies. Grants also are awarded to state and local criminal justice agencies to implement programs that address the federal government's crime control priorities, such as its emphasis on violent and drug-related crimes. In 2016, the federal government spent about 20% of the total expenditures on criminal and civil justice.

Also noteworthy is that despite the billions of dollars spent on criminal and civil justice at the state and local levels, as a percentage of all government expenditures, the amount spent on criminal justice represents only a tiny fraction–about 6% (2.7% for police protection, about 2.2% for corrections, and 1.3% for judicial and legal services). In other words, only about 6 cents of every state and local tax dollar are spent on criminal justice. Include federal tax dollars, and only about 4 cents of every tax dollar are spent on criminal justice–an amount that has remained relatively unchanged for approximately 25 years. Thus, compared with expenditures on other government services, such as social insurance, national defense, international relations, interest on debt at the federal level, public welfare, education, health and hospitals, and interest on debt at the state and local levels, spending on criminal justice remains a relatively low priority–a point apparently not missed by the American public.[14]

For at least four decades, public opinion polls have shown that more than half of all Americans believe that too little money is spent on crime control. Very few people think that too much is being spent.[15] What is not clear, however, because no data are available, is whether those people who believe more money should be spent to fight crime are willing to pay higher taxes to provide that money.

The data presented so far in this section provide a general overview of the aggregate costs of criminal justice in the United States. However, they do not reveal the expenses of individual-level justice, which vary greatly. On the one hand, administering justice to people who commit capital or death-eligible crimes costs, on average, between $1.4 and $7 million per case (the cost of the entire process in 2015 dollars); extraordinary cases can cost much more. For example, the state of Florida reportedly spent $10 million to administer justice to serial murderer Ted Bundy in 1989, and the federal government spent more than $100 million to execute mass murderer Timothy McVeigh in 2001.[16]

On the other hand, the routine crimes processed daily cost much less. A better idea of the costs of justice in more typical cases comes from an examination of the costs of each stage of the local criminal justice process. The results of such a study are presented in the following narrative.[17] A 49-year-old white male–let's call him Joe–was arrested in Orlando, Florida, in October 2011, for possession of cocaine. According to the arresting officer's report, Joe was observed running a stop sign. The officer pulled him over and asked for his driver's license. While doing so, the officer noticed 40 to 50 boxes of cigar packs on the passenger floorboard of the vehicle. The officer asked him if the cigar packs were his, and he said they were. The officer then asked Joe to step out of his car, and he complied. The officer asked Joe whether he had anything illegal on his person such as guns and drugs, and Joe stated, "No." The officer asked Joe if he could search

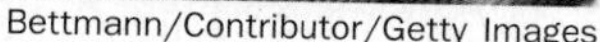

Bettmann/Contributor/Getty Images

Ralf-Finn Hestoft/Corbis/Getty Images

The state of Florida reportedly spent $10 million to administer justice to serial murderer Ted Bundy (left) in 1989, and the federal government spent more than $100 million to execute mass murderer Timothy McVeigh (right) in 2001. *Were the executions worth the expense? Why or why not?*

him, and Joe consented. The officer found no contraband on Joe's person. The officer then conducted a computer check of Joe's license, and it was valid. The officer returned Joe's license and told him that he was not going to issue a traffic citation. The officer then asked Joe whether there was anything illegal in his vehicle such as guns and drugs, and Joe stated, "No." The officer asked Joe if he could search his vehicle, and Joe stated, "Yeah, that's fine." The officer opened a box of cigars and found a square hard white substance that, based on the officer's training and work experience, appeared to be crack cocaine. The officer secured Joe with handcuffs and continued to search the vehicle. The officer lifted a stack of papers on the front passenger seat and discovered four pieces of a hard white substance that also appeared to be crack cocaine. The officer field-tested a sample of the suspected drug, and the test revealed that the substance was cocaine. The officer placed Joe under arrest for cocaine possession. Joe requested that his parents be contacted so they could pick up his car. Joe's father responded to the scene and took possession of the car. The officer transported Joe to the county jail for booking. Joe subsequently pled guilty and was sentenced to 4 months in jail. The jail sentence was rather modest considering the incident was Joe's 33rd offense on his lengthy rap sheet, which included arrests for burglary, grand theft, reckless driving, robbery, criminal mischief, possession of drug paraphernalia, dealing in stolen property, contempt of court, and violation of probation. The criminal justice-related costs to society for this crime were approximately $12,465. As shown in Figure 1.4, the bulk of the costs–98% of the total in this case–were incurred by corrections. The expenses to the police, defense, prosecution, and court were minor in comparison.

FIGURE 1.4 Total Costs of the Orlando, Florida, Cocaine Possession Case by Specific Criminal Justice Functions

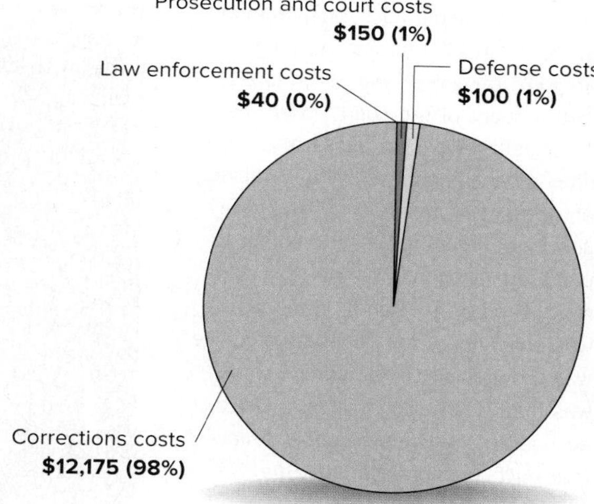

THINKING CRITICALLY

1. Do you think more money needs to be spent on criminal justice? Why or why not?
2. Were resources well spent in the case of the offender convicted of cocaine possession? Why or why not?

Myths About Crime and Criminal Justice

myths Beliefs based on emotion rather than analysis.

A major purpose of this book is to expose and correct misconceptions the American public has about crime and criminal justice. Much of the public's understanding of crime and criminal justice is wrong; it is based on myths. **Myths** are "simplistic and distorted beliefs based upon emotion rather than rigorous analysis" or "at worst . . . dangerous falsifications."[18] More specifically, myths are "credible, dramatic, socially constructed representation[s] of perceived realities that people accept as permanent, fixed knowledge of reality while forgetting (if they were ever aware of it) [their] tentative, imaginative, creative, and perhaps fictional qualities."[19] For example, during the Middle Ages in Europe, people commonly believed that guilt or innocence could be determined through *trial by ordeal*. The accused might be required to walk barefoot over hot coals, hold a piece of red-hot iron, or walk through fire. The absence of any injury was believed to be a sign from God that the person was innocent. Although we now may wonder how such a distorted and simplistic belief could have been taken as fact and used to determine a person's guilt or innocence, people did not consider the belief a myth during the time that it was official practice. The lesson to be learned from this example is that a belief that is taken as fact at one time may in retrospect be viewed as a myth. In this book, we attempt to place such myths about crime and criminal justice in perspective.[20]

Throughout this text, we present generally accepted beliefs about crime and the justice system that can be considered myths because they can be contradicted with facts. In some instances, it can be demonstrated that the perpetuation and acceptance of certain myths by the public, politicians, and criminal justice practitioners have contributed to the failure to significantly reduce predatory criminal behavior and to increase peace. It is also possible that acceptance of these myths as accurate representations of reality or as facts results in the waste of billions of dollars in the battle against crime.

During the Middle Ages, Holy Roman Emperor Otto III married the King of Aragon's daughter, who fell in love with a count of the court. When the count refused her advances, the emperor's wife accused him of making an attempt on her honor. The scene shows the count's wife attempting to prove his innocence by trial by fire (a red-hot rod in her hand). When she was not burned, the count's innocence was proved and the emperor's wife was burned alive for making the false accusation. *How could people believe that guilt or innocence could be proven in this way?*

The Justice of the Emperor Otto: Trial by Fire, 1471–73 (oil on panel) (see also 61050)/Bouts, Dirck (c. 1415–75)/Musees Royaux des Beaux-Arts de Belgique, Brussels, Belgium/Bridgeman Images

SUMMARY

1. **Describe how the type of crime routinely presented by the media compares with crime routinely committed.**

   Crime presented by the media is usually more sensational than crime routinely committed.

2. **Identify institutions of social control, and explain what makes criminal justice an institution of social control.**

   Institutions of social control include the family, schools, organized religion, the media, the law, and criminal justice. Such institutions attempt to persuade people to abide by the dominant values of society. Criminal justice is restricted to persuading people to abide by a limited range of social values, the violation of which constitutes crime.

3. **Summarize how the criminal justice system responds to crime.**

   The typical criminal justice response to the commission of a crime involves the following: investigation, arrest (if the investigation is successful), booking, the formal charging of the suspect, an initial appearance, a preliminary hearing (for a felony), either indictment by a grand jury followed by arraignment or arraignment on an information, either a plea bargain or a trial, sentencing, possible appeal, and punishment (if the defendant is found guilty).

4. **Explain why criminal justice in the United States is sometimes considered a nonsystem.**

   Criminal justice in the United States is sometimes considered a nonsystem for two major reasons. First, there is no single system but, instead, a loose confederation of more than 50,000 agencies on federal, state, and local levels. Second, rather than being a smoothly operating set of arrangements and institutions, the agencies of the criminal justice system interact with one another but generally operate independently, often causing problems for one another.

5. **Point out major differences between Packer's crime control and due process models.**

   From a political standpoint, the crime control model of criminal justice reflects traditional conservative values, while the due process model embodies traditional liberal values. In the crime control model, the control of criminal behavior is by far the most important function of criminal justice. Consequently, the primary focus of this model is on efficiency in the operation of the criminal justice process. The goal of the due process model, in contrast, is at least as much to protect the innocent as it is to convict the guilty. Fundamentally, the due process model defends the ideal of personal freedom and its protection and rests on the assumption that the prevention of tyranny on the part of government and its agents is the most important function of the criminal justice process.

6. **Describe the costs of criminal justice in the United States, and compare those costs among federal, state, and local governments.**

   An enormous amount of money is spent each year on criminal justice in the United States. In 2016, federal, state, and local governments spent a total of $284 billion on police protection ($138 billion), judicial/legal services ($61 billion), and corrections ($85 billion). The bulk of government spending on criminal justice is at the state and local levels, but the federal government spends money strategically to influence criminal justice policy at the other levels of government.

7. **Explain how myths about crime and criminal justice affect the criminal justice system.**

   The acceptance and perpetuation of myths, or simplistic beliefs based on emotion rather than rigorous analysis, can harm the criminal justice system by contributing to the failure to reduce crime and to the waste of money in the battle against crime.

KEY TERMS

statute of limitations 4
institution of social control 7
jurisdiction 7
misdemeanor 7
felony 7
arrest 7
booking 10
defendant 10
initial appearance 10
summary trial 10
probable cause 10
bail 11
preliminary hearing 11
grand jury 11
information 11
arraignment 11
plea bargaining 11
bench trial 11
probation 12
parole 12
system 13
crime control model 14
due process model 14
doctrine of legal guilt 15
myths 20

REVIEW QUESTIONS

1. What is the fundamental problem with the types of crime routinely presented by the media?
2. What was the most frequent type of call for police service in Portland, Oregon during the period examined in the text (see Table 1.2)?
3. What is an *institution of social control*?
4. Why is criminal justice sometimes considered society's "last line of defense"?
5. What three agencies make up the criminal justice system?
6. What is a *jurisdiction*?
7. What is the difference between a *misdemeanor* and a *felony*?

8. What is the difference between an *arrest* and a *booking*?
9. Who decides whether to charge a suspect with a crime?
10. What is a *defendant,* and when does a suspect become a defendant?
11. What is the difference between an *initial appearance* and a *preliminary hearing*?
12. Define *bench trial, summary trial, bail, grand jury, arraignment, plea bargaining,* and *parole.*
13. What is meant by *probable cause*?
14. Why are the conflicts between the different agencies of criminal justice not entirely undesirable?
15. Why does Packer use the metaphors of *assembly-line justice* and *obstacle-course justice* to characterize his crime control and due process models of criminal justice?
16. What are the bases (i.e., the presumptions or doctrines) of Packer's crime control and due process models of criminal justice?
17. Which levels of government—federal, state, local—bear most of the costs of criminal justice in the United States?
18. What is a lesson to be learned from myths about crime and criminal justice?

IN THE FIELD

1. **Crime and the Media** Watch a local television station's broadcast of the evening news for one or more days and record the crimes reported. Then obtain from your local police department a copy of the log of calls for police service for one of those days. Compare the crimes reported on the nightly news with the calls for police service. Describe similarities and differences between the two different sources of crime information. What have you learned?
2. **Costs of Crime** Follow a criminal case in your community and determine the costs of processing the case. You will have to contact the police, the prosecutor, the defense attorney, the judge, and other relevant participants. Remember to consider both monetary and psychological costs. After you have determined the costs, decide whether you think they were justified. Defend your answer.
3. **Costs of Justice** Only about 6% of state and local spending is for criminal justice. By contrast, states and localities spend more for education and public welfare and about the same for health care and hospitals. Divide into groups. Using the preceding information, debate within your group whether states and localities spend enough of their budgets on criminal justice. Share group results with the class.

ON THE WEB

1. **Criminal Justice in Other Countries** Learn about the criminal justice systems of other countries by visiting the U.S. Justice Department's website, *The World Factbook of Criminal Justice Systems,* at http://bjs.ojp.usdoj.gov/content/pub/html/wfcj.cfm.
2. **FBI's Most Wanted** Access the FBI's "Top Ten Most Wanted Fugitives" at www.fbi.gov/wanted/topten/. Read the descriptions of the fugitives. Write a report describing the characteristics they share. Also, try to determine what unique features qualify these fugitives, and not others, for the list.

CRITICAL THINKING EXERCISES

PLEA BARGAINING

1. Shirley Smith pleaded guilty to third-degree murder after admitting she had put rat poison in drinks her husband ingested at least 12 times during the course of their 13-month marriage. Sentenced to a maximum of 20 years in prison, she would have to serve at least 10 years before she could be considered for parole. The prosecutor defended the plea bargain against much public criticism. The prosecutor claimed that the costs of a murder trial and subsequent appeals were not paramount. However, the prosecutor did acknowledge that the case could have been the most expensive in county history, that it exhausted his entire $2 million budget for the fiscal year, and that it required a tax increase to cover the costs. As an elected official, the prosecutor attempted to seek justice while exercising a sense of fiscal responsibility.
   a. Do you think the prosecutor made the correct decision to plea bargain? Defend your answer.
   b. In potentially expensive cases, should the prosecutor seek a referendum on the matter (to determine whether residents are willing to pay additional taxes to try a defendant rather than accept a plea bargain)?

PRISON VERSUS REHABILITATION

2. The city council of a midsize East Coast city is locked in a debate concerning how to address the rising incidence of violent crime. John Fogarty, one of the most influential people in the city, is pushing for more police and stiffer penalties as the solution. He is the leader of a group that is proposing the construction of a new prison. Another group thinks putting more people in prison is not the answer. They believe that early intervention, education, and prevention programs will be most effective. There is not enough money to fund both sides' proposals.
   a. Which side would you support? Why?
   b. What do you think is the number-one crime problem in your community? List ways of dealing with that problem. What would be the most cost-effective way to lower the rate of that crime?

NOTES

1. Eric Levenson, "Larry Nassar Apologizes, Gets 40 to 125 Years for Decades of Sexual Abuse," *CNN*, February 5, 2018. Accessed March 10, 2019, https://www.cnn.com/2018/02/05/us/larry-nassar-sentence-eaton/index.html; Eric Levenson, "Larry Nassar Sentenced to up to 175 Years in Prison for Decades of Sexual Abuse," *CNN*, January 24, 2018. Accessed March 10, 2019, https://www.cnn.com/2018/01/24/us/larry-nassar-sentencing/index.html.
2. John Bacon, "'They have made change': 1 Year After Carnage in Parkland, Where Key Figures Are Now," *USA Today*, February 10, 2019. Accessed March 8, 2019, https://www.usatoday.com/story/news/nation/2019/02/10/parkland-one-year-after-shooting-where-key-figures-are-now/2721798002/; Laurel Wamsley, "Florida State Attorney Will Seek Death Penalty for Parkland Shooting Suspect," *NPR*, March 13, 2018. Accessed March 8, 2019, https://www.npr.org/sections/thetwo-way/2018/03/13/593206185/florida-ag-will-seek-death-penalty-for-parkland-shooting-suspect; "One Year Since Parkland, Some Stoneman Douglas Students Still Don't Feel Safe," *CBS NEWS*, February 14, 2019. Accessed March 8, 2019, https://www.cbsnews.com/news/parkland-shooting-anniversary-one-year-later-how-far-have-we-come/.
3. Jason Hanna, Dakin Andone, Keith Allen, and Steve Almasy, "Alleged Shooter at Texas High School Spared People He Liked, Court Document Says," *CNN*, May 19, 2018. Accessed March 10, 2019, https://www.cnn.com/2018/05/18/us/texas-school-shooting/index.html.
4. Dakin Andone, "Man Accused of Carrying Out Shooting at Maryland Newspaper Indicted," *CNN*, July 21, 2018. Accessed March 10, 2019, https://www.cnn.com/2018/07/21/us/jarrod-ramos-indicted-capital-gazette-shooting/index.html; "Man Accused of Killing 5 in Capital Gazette Shooting Gets Insanity Plea Deadline Extension," *WJLA*, January 23, 2019. Accessed March 10, 2019, https://wjla.com/news/local/capital-gazette-newspaper-shooting-suspect-gets-insanity-plea-deadline-extension.
5. Eric Levenson and Aaron Cooper, "Bill Cosby Sentenced to 3 to 10 Years in Prison for Sexual Assault," *CNN*, September 26, 2018. Accessed March 9, 2019, https://www.cnn.com/2018/09/25/us/bill-cosby-sentence-assault/index.html.
6. Ashley May and Josh Hafner, "Pittsburg Synagogue Shooting: What We Know, Questions That Remain," *USA Today*, October 29, 2018. Accessed March 8, 2019, https://www.usatoday.com/story/news/nation-now/2018/10/29/pittsburgh-synagogue-shooting-what-we-know/1804878002/; Dakin Adone, Jason Hanna, Joe Sterling, and Paul P. Murphy, "Hate Crime Charges Filed in Pittsburgh Shooting That Left 11 Dead," *CNN*, October 29, 2018. Accessed March 8, 2019, https://www.cnn.com/2018/10/27/us/pittsburgh-synagogue-active-shooter/index.html.
7. Alexander Smith, Pete Williams, Andrew Blankstein, Alastair Jamieson, and Corky Siemaszko, "Mass Shooting at Borderline Bar and Grill in Thousand Oaks, California," *NBC NEWS*, November 8, 2018. Accessed March 8, 2019, https://www.nbcnews.com/news/us-news/shooting-reported-borderline-bar-grill-thousand-oaks-california-n933831.
8. Our thanks to Garrett Johnson of the Police Foundation's Police Data Initiative for his help with these data.
9. Howard N. Snyder and Melissa Sickmund, *Juvenile Offenders and Victims: 2006 National Report* (Washington DC: U.S. Department of Justice, Office of Justice Programs, Office of Juvenile Justice and Delinquency Prevention, March 2006), 109, accessed January 5, 2009, www.ojjdp.ncjrs.gov/ojstatbb/nr2006/downloads/NR2006.pdf.
10. Herbert Packer, *The Limits of the Criminal Sanction* (Stanford, CA: Stanford University Press, 1968).
11. See, for example, Walter B. Miller, "Ideology and Criminal Justice Policy: Some Current Issues," *Journal of Criminal Law and Criminology* 64 (1973): 141–62.
12. Robert M. Bohm, "'McJustice': On the McDonaldization of Criminal Justice," *Justice Quarterly* 23 (2006): 127–46.
13. Unless otherwise indicated, the data in this section are from Shelley Hyland, "Justice Expenditure and Employment Extracts, 2016 (preliminary)," Bureau of Justice Statistics, U.S. Department of Justice, https://www.bjs.gov/index.cfm?ty=pbdetail&iid=6728, filename jeee16t01.csv (date of version: 11/7/2019); Tracey Kyckelhahn, "Justice Expenditure and Employment Extracts, 2012 (preliminary)," Bureau of Justice Statistics, U.S. Department of Justice, http://bjs.ojp.usdoj.gov/index.cfm?ty=pbdetail&iid=4335, filename jeeus1201.csv (date of version: 2/26/2015); Tracey Kyckelhahn, "Justice Expenditure and Employment in the United States, 2009 (preliminary)," Bureau of Justice Statistics, U.S. Department of Justice, http://www.bjs.ojp.usdojgov/index.cfm?ty=pbdetail&iid=4335, filename: cjee0901.csv (date of version: 5/30/2012).
14. Calculated from data in Kyckelhahn, "Justice Expenditure and Employment Extracts, 2012"; "Federal Budget Receipts and Outlays," in *The 2011 Statistical Abstract* (U.S. Census Bureau), Table 467, accessed January 26, 2011, www.census.gov/compendia/statab/2011/tables/11s0467.pdf; "State and Local Governments—Revenues and Expenditure by Function: 2006 and 2007," in *The 2011 Statistical Abstract* (U.S. Census Bureau), Table 434, accessed January 26, 2011, www.census.gov/compendia/statab/2011/tables/11s0434.pdf; Kristen A. Hughes, "Justice Expenditure and Employment in the United States, 2003," in *Bureau of Justice Statistics Bulletin* (Washington, DC: U.S. Government Printing Office, April 2006).
15. Andy Kiersz, "Trump Is Giving a Major Policy Speech—Here's What Americans Think About 13 Major Areas of Federal Spending," *Business Insider*, February 28, 2017. Accessed February 15, 2019, https://www.businessinsider.com/american-opinion-on-government-spending-and-budget-priorites-2017-2.
16. Robert M. Bohm, *Deathquest: An Introduction to the Theory and Practice of Capital Punishment in the United States,* 5th ed. (New York: Routledge, 2017), 277.
17. My thanks to Matt Landon for his help in collecting the data.
18. D. Nimmo and J. E. Combs, *Subliminal Politics: Myths and Mythmakers in America* (Englewood Cliffs, NJ: Prentice Hall, 1980), 6.
19. Ibid., 16.
20. Many of the myths presented in this book were taken from the following sources: Robert M. Bohm and Jeffery T. Walker, *Demystifying Crime and Criminal Justice*, 2nd ed. (New York, NY: Oxford University Press, 2013); Jeffrey H. Reiman, *The Rich Get Richer and the Poor Get Prison: Ideology, Class, and Criminal Justice*, 7th ed. (Boston: Allyn & Bacon, 2004); Harold E. Pepinsky and Paul Jesilow, *Myths That Cause Crime,* 2nd ed. (Cabin John, MD: Seven Locks, 1985); Kevin N. Wright, *The Great*

American Crime Myth (Westport, CT: Greenwood, 1987); also see Robert M. Bohm, "Myths About Criminology and Criminal Justice: A Review Essay," *Justice Quarterly* 4 (1987): 631–42; William Wilbanks, *The Myth of a Racist Criminal Justice System* (Belmont, CA: Wadsworth, 1987); Victor E. Kappeler and Gary W. Potter, *The Mythology of Crime and Criminal Justice*, 4th ed. (Long Grove, IL: Waveland Press, 2005); Jeffrey Reiman and Paul Leighton, *The Rich Get Richer and the Poor Get Prison: Ideology, Class, and Criminal Justice,* 10th ed. (Upper Saddle River, NJ: Pearson, 2012); Samuel Walker, *Sense and Nonsense About Crime, Drugs, and Communities: A Policy Guide,* 7th ed. (Belmont, CA: Wadsworth, 2010). For a discussion of why these myths exist, see Robert M. Bohm, "Crime, Criminal and Crime Control Policy Myths," *Justice Quarterly* 3 (1986): 193–214.

Design Elements: *Image of part of a court building with columns and statue*: Pixtal/AGE Fotostock RF; *Image of a door slightly open*: jadding/Shutterstock.com

Ingram Publishing/SuperStock

2

Crime and Its Consequences

CHAPTER OUTLINE

LEARNING OBJECTIVES

After completing this chapter, you should be able to:

1. Distinguish between a social definition and a legal definition of crime, and summarize the problems with each.
2. List the technical and ideal elements of a crime.
3. Identify some of the legal defenses or legal excuses for criminal responsibility.
4. Explain why crime and delinquency statistics are unreliable.
5. Identify the two major sources of crime statistics in the United States.
6. Describe the principal finding of the national crime victimization surveys.
7. Summarize the general finding of self-report crime surveys.
8. Identify the costs of crime.
9. Describe the characteristics of people most likely to fear crime.
10. List the characteristics of people who are the most likely and the least likely to be victims of crime.

Crime Story

On Saturday morning, October 27, 2018, 46-year-old Robert Bowers (pictured), armed with a Colt AR-15 rifle and three .357 Glock handguns, stormed into the Tree of Life Congregation synagogue in the affluent Squirrel Hill neighborhood of Pittsburgh, Pennsylvania, shouting anti-Semitic slurs and killing 11 worshippers in a 20-minute attack. Two worshippers and four or five officers were injured. The murder victims ranged in age from 54 to 97. Bowers finally surrendered to police officers after suffering multiple gunshot wounds. He told arresting officers that "all these Jews need to die" because they were "committing genocide to my people."

Bowers previously had posted anti-Semitic rants on social media. On his Gab profile, he claimed Jews were "the children of Satan." In one post, he accused the organization HIAS, the Hebrew Immigrant Aid Society, of transporting "invaders in that kill our people." He also posted about several conspiracy theories and his opposition to the migrant caravan, which he believed was being aided by Jews. Approximately four hours before the shooting, he wrote that he was not a Trump voter because Trump was too soft on Jews. He also suggested that President Trump surrounded himself with too many Jews (although he did not use the word Jews but instead used a derogatory term for Jews). In what was likely his last post just minutes before he entered the synagogue, Bowers declared, "I can't sit by and watch my people get slaughtered. Screw your optics, I'm going in."

Bowers had a troubled childhood. His mother divorced his father in 1972, when Bowers was less than a year old. She remarried in 1975, but apparently only lived with her new husband for a year. They did not divorce officially until 2004. Bowers' biological father committed suicide in 1979. He was facing charges for attempted rape in Squirrel Hill. In 1987, his mother's second husband was charged in Indiana for molesting two girls, ages 5 and 7. He pleaded guilty and served six years in prison. Bowers dropped out of high school in November of his senior year. He was a mediocre student. In the early 1990s, he got a job delivering baked goods, which he apparently liked. His coworkers, which included a couple of Jewish people, described him as "a guy who liked beer, Hooters, action films and guns, with a bit of an anti-government streak—not as a virulent anti-Semite primed to explode." Only later did he become a white nationalist and a follower of online right-wing provocateurs, especially Jim Quinn and Jack Corbin ("Pale Horse"). He quit the bakery in 2002, and worked sporadically, including a stint as a long-haul trucker. During this time, he lived with his grandfather and other relatives and shared material online about the Christian Identity movement, which is virulently anti-Semitic. He lived alone in an apartment at the time of the shootings.

Bowers originally was charged with 36 state crimes, including 11 homicides and 13 counts of ethnic intimidation. He also was charged with 29 federal crimes, including 11 federal hate-crimes and 11 counts of using a firearm to kill. The state charges will "remain in abeyance" while the federal case proceeds, but the state attorney did say that he would seek the death penalty. U.S. Attorney Scott Brady stated that he, too, would seek the death penalty. President Trump indicated his support for the penalty, when he opined, "When people do this, they should get the death penalty." Trump also told reporters, "If there was an armed guard inside the temple, they would have been able to stop him."

On January 29, 2019, a federal grand jury indicted Bowers on 44 charges, including more hate crimes. On February 11, 2019,

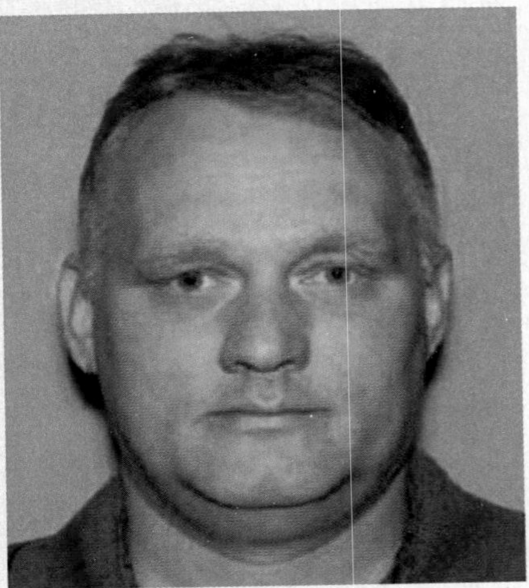

Pennsylvania Department of Transportation/AP Photos

Bowers pleaded not guilty in federal court to all charges. His lawyer, prominent death penalty attorney Judy Clarke, hopes that the case can be resolved without a trial. Clarke is known for negotiating plea deals that helped some of the country's most infamous killers avoid execution. Among her clients were Unabomber Ted Kaczynski, Atlanta Olympics bomber Eric Rudolph, and Arizona shooter Jared Lee Loughner, who killed 6 people and wounded 13 others, including U.S. Representative Gabrielle Giffords. She also represented Boston marathon bomber Dzhokhar Tsarnaev, who was sentenced to death. At the arraignment, U.S. Attorney Brady had not announced yet whether he would seek the death penalty.

In 2017, anti-Semitic incidents in the United States increased by nearly 60%. According to the Anti-Defamation League (ADL), in 2017, nearly 2,000 cases of harassment, vandalism, or physical assault were committed against Jews and Jewish institutions. In a statement, the ADL asserted, "It is simply unconscionable for Jews to be targeted during worship on a Sabbath morning and unthinkable that it would happen in the United States of America in this day and age." The attack on the Tree of Life Congregation synagogue in 2018 has been designated as the deadliest anti-Semitic hate crime in U.S. history.

Among the topics addressed in this chapter are hate-crime statistics, fear of crime, and victims of crime. All three topics are exemplified by the case of Robert Bowers and the Tree of Life synagogue attack. Hate crimes produce both victims and fear. Why do people commit hate crimes? Why are hate crimes against Jews increasing? What can criminal justice do about hate crimes? What more needs to be done about hate crimes?

Definitions of Crime

The object of criminal justice in the United States is to prevent and control crime. Thus, to understand criminal justice, we must understand crime. An appropriate definition of crime, however, remains one of the most critical unresolved issues in criminal justice today. One problem is that many dangerous and harmful behaviors are not defined as crimes, while many less dangerous and less harmful behaviors are. We begin, then, by examining how crime is defined and the problems with defining what a crime is.

Crime

The word "crime" is from the Latin *crimen*, meaning "accusation" or "fault."

Source: *Webster's New Twentieth Century Dictionary of the English Language Unabridged* (Parsippany, NJ: Williams Collins, 1980).

Social Definitions

The broadest definitions of crime are social definitions. A typical social definition of crime is behavior that violates the norms or social mores of society–or, more simply, antisocial behavior. A **norm** or **social more** is any standard or rule regarding what human beings should or should not think, say, or do under given circumstances. Because social definitions of crime are broad, they are less likely than narrower definitions to exclude behaviors that ought to be included. Nevertheless, there are several problems with social definitions of crime.

norm or social more Any standard or rule regarding what human beings should or should not think, say, or do under given circumstances.

First, norms or social mores vary from group to group within a single society. There is no uniform definition of antisocial behavior. Take, for example, the acts involved in gambling, prostitution, abortion, and homosexual behavior. As current public debates indicate, there is much controversy in the United States over whether those acts should be crimes. Even with acts about which there seems to be a consensus, such as murder and rape, there is no agreement on what constitutes such acts. For example, if a patient dies from a disease contracted from a doctor who did not wash his or her hands before examining the patient, has the doctor committed murder? Or, if a man has forcible sexual intercourse with a woman against her will but, before the act, at the woman's request, puts on a condom so that the woman will not get a sexually transmitted disease, has the man committed rape? Those examples illustrate the difficulty of determining what, in fact, constitutes antisocial behavior, let alone crime.

Second, norms or social mores are always subject to interpretation. Each norm's or social more's meaning has a history. Consider abortion, for example. For some people, abortion is the killing of a fetus or a human being. For other people, abortion is not killing because, for them, human life begins at birth and not at conception. For the latter group, the abortion issue concerns women's freedom to control their own bodies. For the former group, abortion constitutes an injustice to the helpless.

Third, norms or social mores change from time to time and from place to place. For example, the consumption of alcohol was prohibited in the United States during the 1920s and early 1930s but is only regulated today. Until the passage of the Harrison Act in 1914, it was legal in the United States to use opiates such as opium, heroin, and morphine without a doctor's prescription. Such use is prohibited today.

Casino gambling is allowed in some states but forbidden in other states. Prostitution is legal in a few counties in Nevada but illegal in the rest of the United States. Prior to the mid-1970s, a husband could rape his wife with impunity in all but a handful of states. Today, laws in every state prohibit a husband from raping or assaulting his wife.

A Legal Definition

legal definition of crime An intentional violation of the criminal law or penal code, committed without defense or excuse and penalized by the state.

In an attempt to avoid the problems with social definitions of crime, a legal definition of crime is used in criminal justice in the United States. A typical **legal definition of crime** is

ALPA PROD/Shutterstock

More and more states are legalizing casino gambling as a means of generating income. *Is this a desirable trend? Why or why not?*

this: an intentional violation of the criminal law or penal code, committed without defense or excuse and penalized by the state. The major advantage of a legal definition of crime, at least on the surface, is that it is narrower and less ambiguous than a social definition of crime. If a behavior violates the criminal law, then by definition it is a crime. However, although a legal definition eliminates some of the problems with social definitions of crime, a legal definition of crime has problems of its own.

First, some behaviors prohibited by the criminal law arguably should not be. This problem of **overcriminalization** arises primarily in the area of so-called victimless crimes. Lists of victimless crimes typically include gambling, prostitution involving consenting adults, homosexual acts between consenting adults, and the use of some illegal drugs, such as marijuana. Ultimately, whether those acts should or should not be prohibited by criminal law depends on whether they are truly victimless–an issue we will not debate here. Perhaps less controversial are some of the following illegal behaviors:

overcriminalization The prohibition by the criminal law of some behaviors that arguably should not be prohibited.

- It is illegal for a driver to be blindfolded while operating a vehicle in Alabama.
- In California, it is illegal to trip horses for entertainment, to possess bear gall bladders, or to peel an orange in your hotel room.
- It is illegal to throw shoes at weddings in Colorado.
- In Connecticut, it is illegal to walk across the street on your hands.
- Women in Florida may be fined for falling asleep under a hair dryer, as can the salon owner.
- Idaho state law makes it illegal for a man to give his sweetheart a box of candy weighing less than 50 pounds.
- It is illegal to take a bath in the wintertime in Indiana.
- Kisses may last for as much as, but no more than, 5 minutes in Iowa.
- In Michigan a woman isn't allowed to cut her own hair without her husband's permission.
- It is illegal to slurp soup in New Jersey.
- Beer and pretzels can't be served at the same time in any bar or restaurant in North Dakota.
- Violators in Oklahoma can be fined, arrested, or jailed for making ugly faces at a dog.
- The state law of Pennsylvania prohibits singing in the bathtub.
- In South Dakota, a woman over 50 is not allowed to go outside and strike up a conversation with a married man older than 20.
- In Tennessee, it is illegal to shoot any game other than whales from a moving automobile.
- In Texas, it is illegal to take more than three sips of beer at a time while standing.
- It is an offense in Washington State to pretend your parents are rich.[1]

nonenforcement The failure to routinely enforce prohibitions against certain behaviors.

A second problem with a legal definition of crime is that for some behaviors prohibited by criminal law, the law is not routinely enforced. **Nonenforcement** is common for many white-collar and government crimes. It is also common for blue laws, for example, those that require stores and other commercial establishments to be closed on Sundays. Many jurisdictions in the United States have blue laws, or they did until recently. The principal problem with the nonenforcement of prohibitions is that it causes disrespect for the law. People come to believe that because criminal laws are not routinely enforced, there is no need to routinely obey them.

undercriminalization The failure to prohibit some behaviors that arguably should be prohibited.

A third problem with a legal definition of crime is the problem of **undercriminalization**. That is, some behaviors that arguably should be prohibited by criminal law are not. Have you ever said to yourself that there ought to be a law against whatever it is you are upset about? Of course, most of the daily frustrations that people claim ought to be crimes probably should not be. Some people argue, however, that some very harmful and destructive actions or inactions that are not criminal should be. Examples include the government allowing employers (generally through the nonenforcement of laws) to maintain unsafe working conditions that cause employee deaths and injuries and corporations' intentional production of potentially hazardous products to maximize profits.[2]

Elements of Crime

A legal definition of crime is the basis of criminal justice in the United States. The legal definition of crime provided earlier in this chapter, however, is only a general definition. It does not specify all the elements necessary to make a behavior a crime. Technically and ideally, a crime has not been committed unless all seven of the following elements are present:[3]

1. Harm
2. Legality
3. *Actus reus*
4. *Mens rea*
5. Causation
6. Concurrence
7. Punishment

Only in a technical and ideal sense must all seven elements be present. In practice, a behavior is often considered a crime when one or more of the elements of crime are absent. We will examine each of the seven elements in turn, indicating exceptions to the technical and the ideal where relevant.

Sacramento Bee/Contributor/Getty Images

Ever since criminal sanctions were established for illegal drug use, some have argued for decriminalization by elimination or reduction of criminal penalties for possession or distribution of certain drugs. *Do you agree with this argument? Why or why not?*

Harm For crime to occur, there must be an external consequence, or **harm**. A mental or emotional state is not enough. Thus, thinking about committing a crime or being angry enough to commit a crime, without acting on the thought or the anger, is not a crime.

The harm may be physical or verbal. Physically striking another person without legal justification is an example of an act that does physical harm. An example of an act that does verbal harm is a threat to strike another person, whether or not the threat is carried out. Writing something false about another person that dishonors or injures that person is a physical harm called *libel*. The spoken equivalent of libel is called *slander*.

Whether the legal element of harm is present in all crimes is sometimes questioned. Some crimes, such as gambling, prostitution, marijuana consumption, and certain consensually committed sexual acts such as sodomy, have come to be called "victimless crimes" by those who argue that only those people involved in these behaviors are harmed, if at all. Other people maintain that the participants, their families, and the moral fabric of society are jeopardized by such behavior. In short, there is considerable debate as to whether so-called victimless crimes really are harmless.

harm The external consequence required to make an action a crime.

Legality The element of **legality** has two aspects. First, the harm must be legally forbidden for a behavior to be a crime. Thus, violations of union rules, school rules, religious rules, or any rules other than those of a political jurisdiction may be "wrong," but they are not crimes unless they are also prohibited by criminal law. Furthermore, rude behavior may be frowned upon, but it is not criminal.

Second, a criminal law must not be retroactive, or *ex post facto*. An ***ex post facto* law** (1) declares criminal an act that was not illegal when it was committed, (2) increases the punishment for a crime after it is committed, or (3) alters the rules of evidence in a particular case after the crime is committed. The first meaning is the most common. The U.S. Constitution (Article I, Section 10.1) forbids *ex post facto* laws.

legality The requirement (1) that a harm must be legally forbidden for the behavior to be a crime and (2) that the law must not be retroactive.

***ex post facto* law** A law that (1) declares criminal an act that was not illegal when it was committed, (2) increases the punishment for a crime after it is committed, or (3) alters the rules of evidence in a particular case after the crime is committed.

Actus reus The Latin term ***actus reus*** refers to criminal conduct—specifically, intentional or criminally negligent (reckless) action or inaction that causes harm. Crime involves not only things people do but also things they do not do. If people do not act in situations in which the law requires them to act, they are committing crimes. For example, parents are legally required to provide their children with adequate food, clothing, and shelter. If parents

actus reus Criminal conduct—specifically, intentional or criminally negligent (reckless) action or inaction that causes harm.

fail to provide those necessities–that is, if they fail to act when the law requires them to–they are committing a crime.

mens rea Criminal intent; a guilty state of mind.

negligence The failure to take reasonable precautions to prevent harm.

Mens rea The Latin term ***mens rea*** refers to criminal intent or a guilty state of mind. It is the mental aspect of a crime. Ideally, criminal conduct is limited to intentional or purposeful action or inaction and not to accidents. In practice, however, reckless actions or *negligence* may be criminal. **Negligence** is the failure to take reasonable precautions to prevent harm.

In some cases, offenders lack the capacity (sometimes called competence) to form *mens rea*. If they do not have that capacity, they are not to be held responsible for their criminal conduct. If they have a diminished capacity to form *mens rea*, they are to be held less than fully responsible. In other cases, offenders who have the capacity to form *mens rea* are not held responsible for their crimes or are held less responsible for them, either because they did not have *mens rea* when they acted or because there were extenuating circumstances when they did act with *mens rea*.

Legal Defenses for Criminal Responsibility In the United States, an offender is not considered responsible or is considered less responsible for an offense if he or she, for example, (1) acted under duress, (2) was underage, (3) was insane, (4) acted in self-defense or in defense of a third party, (5) was entrapped, or (6) acted out of necessity. Those conditions are legal defenses or legal excuses for criminal responsibility.

duress Force or coercion as an excuse for committing a crime.

If a person did not want to commit a crime but was forced or coerced to do so against his or her will, he or she committed the crime under **duress** and is generally excluded from criminal liability. Suppose that an intruder held a gun to the head of a loved one and threatened to kill that person if you did not rob a local convenience store and return immediately to give the intruder the money. If you committed the robbery to save the life of your loved one, you would probably not be held legally responsible for the crime because you committed it under duress. There were extenuating circumstances when you acted with *mens rea*. To prevent all offenders from claiming duress, the burden of proof is placed on the defendant.

Legal Infancy

On July 29, 2010, an Indianapolis, Indiana, 4-year-old boy picked up a loaded .45-caliber handgun that had been left on the kitchen table and shot and killed a 3-year-old girl in front of his siblings. Because of his age, he was not charged or held criminally responsible because, legally, he could not form *mens rea*, or the intent to kill.

Source: "Police: 4-Year-Old Boy Kills 3-Year-Old Playmate with Gun," accessed December 29, 2010, www.foxnews.com/us/2010/07/30/police-year-old-boy-kills-year-old-playmate-gun/.

Another legal excuse or legal defense against criminal responsibility is being underage. Although the age at which a person is considered legally responsible for his or her actions varies by jurisdiction, in most American jurisdictions, a child under the age of 7 is not held responsible for a crime. It is assumed that a child under 7 years of age does not have the capacity to form *mens rea*. A child under 7 years of age is considered a *legal infant* or of *legal nonage*. Such a child is protected by criminal law but not subject to it. Thus, if a 6-year-old child picks up a shotgun and shoots his or her parent, the child is unlikely to be charged with a crime. However, if a parent abuses a child, the criminal law protects the child by holding the abusive parent responsible for his or her actions.

juvenile delinquency A special category of offense created for youths who, in most U.S. jurisdictions, are persons between the ages of 7 and 18.

In most developed countries, children under 18 years of age are not considered entirely responsible for their criminal acts. It is assumed that their capacity to form *mens rea* is not fully developed. A special category of offense called **juvenile delinquency** has been created for those children. In most American jurisdictions, the upper age limit for juvenile delinquency is 18. The lower limit is usually 7. Criminal law generally treats anyone who is 18 or older as an adult. However, the upper age limit of juvenile delinquency is lower in some jurisdictions and sometimes varies with the sex of the offender. In some jurisdictions there is a legal borderland between the ages of 16 and 18. An offender in that age range may be treated as a juvenile or as an adult, depending on the severity of the offense. In some cases, an offense is considered heinous enough for a court to certify a juvenile, regardless of age, as an adult and to treat him or her accordingly. The subject of juvenile delinquency is discussed more fully in Chapter 13.

insanity Mental or psychological impairment or retardation as a defense against a criminal charge.

A third legal defense or legal excuse from criminal responsibility is insanity. **Insanity** is a legal term, not a medical one. It refers to mental or psychological impairment or retardation. Like many of the other legal defenses or excuses, an insanity defense rests on the assumption that someone who is insane at the time of a crime lacks the capacity, or has diminished capacity, to form *mens rea*. Thus, that person either should not be held responsible or should be held less responsible for a given crime.

In most western European nations, legal insanity is determined solely by the judgment and testimony of medical experts. British and American law, by contrast, provide guidelines for judges, juries, and medical experts to follow in determining whether a defendant is legally

FIGURE 2.1 Insanity Tests by State

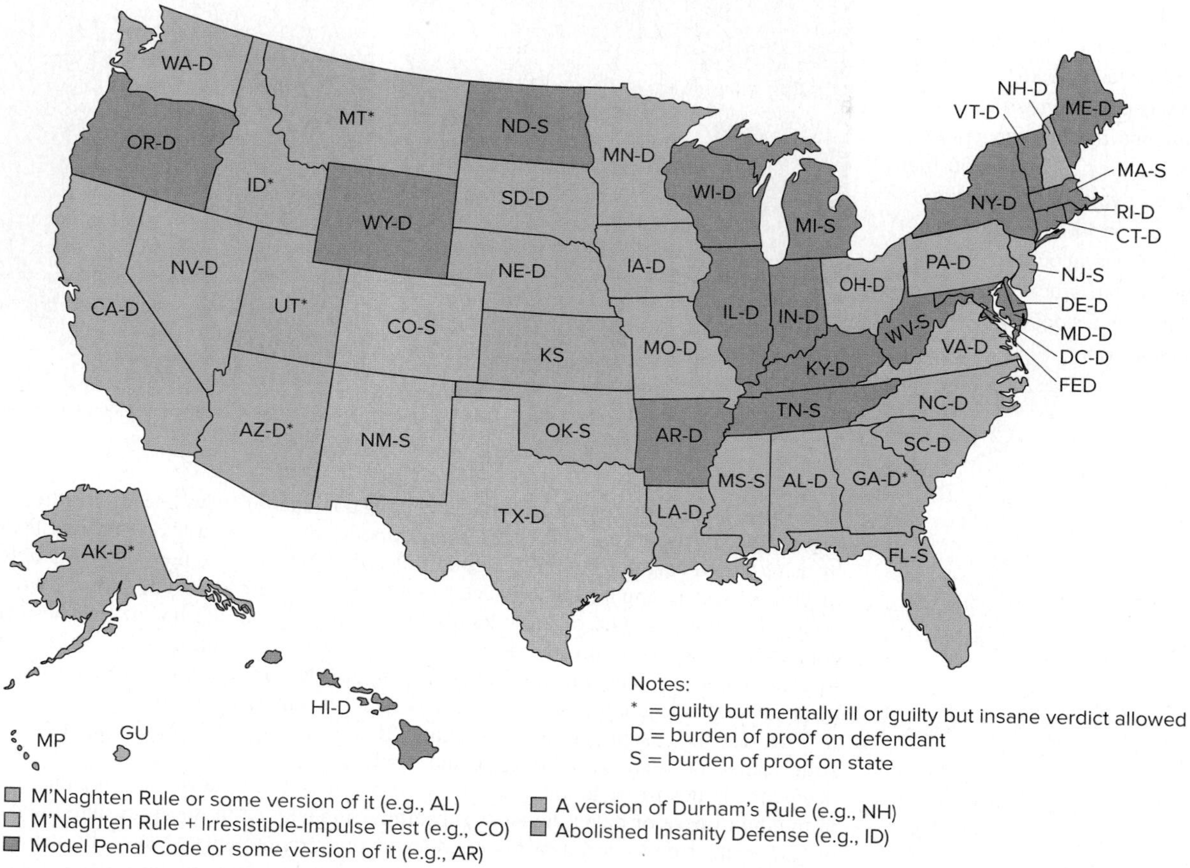

Notes:
* = guilty but mentally ill or guilty but insane verdict allowed
D = burden of proof on defendant
S = burden of proof on state

- M'Naghten Rule or some version of it (e.g., AL)
- M'Naghten Rule + Irresistible-Impulse Test (e.g., CO)
- Model Penal Code or some version of it (e.g., AR)
- A version of Durham's Rule (e.g., NH)
- Abolished Insanity Defense (e.g., ID)

Source: "The Insanity Defense Among the States," accessed February 16, 2019, http://criminal.findlaw.com/criminal-procedure/the-insanity-defense-among-the-states.html.

insane. The oldest of those guidelines is the M'Naghten rule, or some variation of it, which was first used in an English trial in 1843 and is now used in 21 states (see Figure 2.1).

Under the M'Naghten rule:

> Every man is to be presumed to be sane, and . . . to establish a defense on the ground of insanity, it must be clearly proved that, at the time of the committing of the act, the party accused was laboring under such a defect of reason, from disease of the mind, as not to know the nature and quality of the act he was doing; or if he did know it, that he did not know he was doing what was wrong.[4]

In short, according to the M'Naghten rule, which is also referred to as the "right-and-wrong test," a person is legally insane if, at the time of the commission of the act, he or she (1) did not know the nature and quality of the act or (2) did not know that the act was wrong. The burden of proof is on the defendant.

One problem with the M'Naghten rule is the difficulty of determining what a person's state of mind was at the time of the commission of the criminal act. The rule also has been criticized for its ambiguity. What is a "defect of reason," and by whose standards is the act a product of defective reason? Does "disease of the mind" refer to organic diseases, nonorganic diseases, or both? What does it mean to "know" the nature and quality of the act? Does it mean an intellectual awareness, an emotional appreciation, or both? Does "wrong" mean legally wrong, morally wrong, or both?

Perhaps the most serious problem with the M'Naghten rule is that it does not address the situation of a defendant who knew the difference between right and wrong but was unable to control his or her actions. To remedy that problem, four states have adopted the

Daniel M'Naghten

Daniel M'Naghten was acquitted of the murder of a person he had mistaken for his real target, Sir Robert Peel, then the Prime Minister of Great Britain. M'Naghten claimed that he was delusional at the time of the killing. Go to https://h2o.law.harvard.edu/collages/19272 and research the M'Naghten case.

Explain the insanity defense in that case and give your opinion about the court's ruling.

Jeffrey Markowitz/Contributor/Getty Images

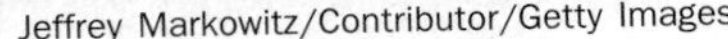

Jeffrey Markowitz/Contributor/Getty Images

In a 1994 trial in Virginia, attorneys for Lorena Bobbitt, who had sliced off her husband's penis with a kitchen knife while he was sleeping, successfully used the *irresistible-impulse* defense against charges of malicious wounding. She claimed that she had been subjected to physical and sexual abuse for years during her marriage. She was acquitted of the crime. *Was Bobbitt's act uncontrollable or uncontrolled? Defend your answer.*

irresistible-impulse or *control test* and use it in conjunction with the M'Naghten rule (see Figure 2.1). In those states, a defense against conviction on grounds of insanity is first made by using the M'Naghten rule. If the conditions of M'Naghten are met, the irresistible-impulse or control test is applied. If it is determined that the defendant knew that he or she was doing wrong at the time of the commission of the criminal act but nevertheless could not control his or her behavior, the defendant is entitled to an acquittal on the grounds of insanity. The major problem with the irresistible-impulse or control test is distinguishing between behavior that is uncontrollable and behavior that is simply uncontrolled.

The test for insanity used by another 21 states is the *substantial-capacity test* of the American Law Institute's Model Penal Code or some version of it. Under that test, a defendant is not to be found guilty of a crime "if at the time of such conduct as a result of mental disease or defect he lacks substantial capacity either to appreciate the criminality of his conduct or to conform his conduct to the requirements of law." By using the term *substantial capacity,* the test does not require that a defendant be completely unable to distinguish right from wrong. The test has been criticized for its use of the ambiguous terms *substantial capacity* and *appreciate.* It also does not resolve the problem of determining whether behavior is uncontrollable or uncontrolled.

myth

The availability of an insanity defense allows dangerous offenders to escape conviction and go free.

fact

Defendants found not guilty by reason of insanity rarely go free. Generally, they are confined to a mental institution until they are deemed by the committing court or some other judicial body to be sane or no longer dangerous. Research shows that the insanity defense is raised only in 0.3% of cases, and a finding of insanity is made in just 0.01% of cases.

Source: Erik Roskes, "Are Mass Killers 'Crazy?'" *The Crime Report*, https://thecrimereport.org/2012/09/18/2012-09-are-mass-killers-crazy/, September 18, 2012, accessed February 16, 2019.

A final insanity test used only in New Hampshire is a version of Durham's Rule and is referred to as "the product test." The *product test* is a two-prong test in which the defense must show that (1) the defendant suffered from a mental disease or defect and (2) the murder was a product of that disease or defect. A problem with the product test is that neither the New Hampshire legislature nor New Hampshire courts have defined the terms *mental disease* and *defect,* leaving interpretation entirely to juries.

Following the public uproar over the 1982 acquittal of John Hinckley, the would-be assassin of President Ronald Reagan, on the grounds that he was legally insane, five states, including three states that had otherwise abolished the insanity defense, enacted "guilty but insane" or "guilty but mentally ill" laws (see Figure 2.1). Defendants who are found guilty but insane generally receive sentences that include psychiatric treatment until they are cured. Then they are placed in the general prison population to serve the remainder of their sentences.

States are free to abolish insanity as a defense. The first state to do so was Montana in 1979. Idaho, Utah, and Kansas are the only other states that have eliminated any possibility of a criminal defendant being found not guilty by reason of insanity.[5] Figure 2.1 shows the insanity test used by each state (and Washington, DC), as well as which party, either the defendant or the state, has the burden of proof, and the states that allow guilty but mentally ill or guilty but insane verdicts. The federal government uses the M'Naghten rule, and the defendant has the burden of proof.[6]

A fourth legal defense or legal excuse from criminal responsibility is self-defense or the defense of a third party. Generally, people are relieved of criminal responsibility if they use only the amount of force reasonably necessary to defend themselves or others against an apparent threat of unlawful and immediate violence. When it comes to the protection of property, however, the use of force is much more limited. Deadly force is not allowed, but nondeadly force may be used to protect one's property. In 2005, Florida became the first state to pass the National Rifle Association–backed "castle doctrine" law or, as it is sometimes called, the "stand your ground" law. Since then, more than 30 states have adopted or strengthened such laws, though the wording varies among states. The law generally provides that someone attacked in his or her home can use reasonable force, including deadly force, to protect his or another's life without any duty to retreat from the attacker. In Florida, the "no duty to retreat" language also applies to street crimes. In some states, the law applies to other locations besides a home, such as a place where a person is a guest or a workplace. Twenty-two states specifically say "there is no duty to retreat (from) an attacker in any place in which one is lawfully present." Some versions provide criminal or civil immunity for someone who legally uses force in self-defense. Most of the laws presume that a person breaking into someone's house has the intent of a violent or forceful act. The law's name comes from the notion that "one's home is one's castle."[7] The reason people are not held legally responsible for acting in self-defense or in defense of a third party is that, because of extenuating circumstances, they do not act with *mens rea.*

Entrapment is a fifth legal defense or legal excuse from criminal responsibility. People are generally considered either not responsible or less responsible for their crimes if they were entrapped, or induced into committing them, by a law enforcement officer or by someone acting as an agent for a law enforcement officer, such as an informer or an undercover agent. A successful entrapment defense, however, requires proof that the law enforcement officer or his or her agent instigated the crime or created the intent to commit the crime in the mind of a person who was not already predisposed to committing it. Thus, it is not entrapment if a law enforcement officer merely affords someone an opportunity to commit a crime, as, for example, when an undercover agent poses as a drug addict and purchases drugs from a drug dealer.

entrapment A legal defense against criminal responsibility when a person, who was not already predisposed to it, is induced into committing a crime by a law enforcement officer or by his or her agent.

The final legal defense or legal excuse from criminal responsibility to be discussed here is necessity. A **necessity defense** can be used when a crime has been committed to prevent a greater or more serious crime. In such a situation, there are extenuating circumstances, even though the act was committed with *mens rea.* Although it is rarely used, the necessity defense has been invoked occasionally, especially in cases of "political" crimes. The necessity defense was used successfully by Amy Carter (daughter of former President Jimmy Carter), Jerry Rubin, and other activists who were charged with trespassing for protesting apartheid on the property of the South African embassy in Washington, DC. The court agreed with the protesters that apartheid was a greater crime than trespassing. Interestingly, the law does not recognize economic necessity as a defense against or an excuse from criminal responsibility. Therefore, the unemployed and hungry thief who steals groceries cannot successfully employ the necessity defense.

necessity defense A legal defense against criminal responsibility used when a crime has been committed to prevent a more serious crime.

Causation A fifth ideal legal element of crime is causation, or a causal relationship between the legally forbidden harm and the *actus reus.* In other words, the criminal act must lead directly to the harm without a long delay. In a Georgia case, for example, a father was accused of murdering his baby daughter. The murder charges were dropped, however, because too much time had passed between the night the 3½-month-old girl was shaken into a coma and her death 18 months later. Because of Georgia's year-and-a-day rule, the father was not charged with murder, but he still faced a charge of cruelty to children, which in Georgia carries a maximum sentence of 20 years. The purpose of the requirement of causation is to prevent people from facing the threat of criminal charges the rest of their lives.

The Year-and-a-Day Rule

The rule that a person cannot be prosecuted for murder if the victim dies more than a year and a day after the injury is based on thirteenth-century English common law.

Source: *Tennessee v. Rogers*, 992 S.W.2d 393 (1999); *United States v. Jackson*, 528 A.2d 1211, 1214 (D.C. 1987).

Concurrence Ideally, for any behavior to be considered a crime, there must be concurrence between the *actus reus* and the *mens rea.* In other words, the criminal conduct and the criminal intent must occur together. For example, suppose you are driving down the street, and you accidentally hit a pedestrian. You rush from your car to help the victim and discover that he is a hated enemy. You react by jumping up and down with joy and exclaiming how happy you are that you caused his injury. The conventional rule is that no crime has been committed because you had not formed *mens rea* either before or during the commission of the *actus reus.*

Punishment The last of the ideal legal elements of a crime is punishment. For a behavior to be considered a crime, there must be a statutory provision for punishment or at least the threat of punishment. Without the threat of punishment, a law is unenforceable and is therefore not a criminal law.

TABLE 2.1 Types and Definitions of Selected Crimes

VIOLENT CRIMES	Crimes that involve force or threat of force.
Murder	The unlawful killing of another human being with malice aforethought.
Manslaughter	The unlawful killing of another human being without malice aforethought.
Aggravated assault	An assault committed (1) with the intention of committing some additional crime, (2) with peculiar outrage or atrocity, or (3) with a dangerous or deadly weapon.
Forcible rape	The penetration, no matter how slight, of the vagina or anus with any body part or object, or oral penetration by a sex organ of another person, without the consent of the victim.
Robbery	Theft from a person, accompanied by violence, threat of violence, or putting the person in fear.
Kidnapping	The unlawful taking and carrying away of a human being by force and against his or her will.
PROPERTY CRIMES	Crimes that involve taking money or property, but usually without force or threat of force.
Larceny	The unlawful taking and carrying away of another person's property with the intent of depriving the owner of that property.
Burglary	Entering a building or occupied structure to commit a crime therein.
Embezzlement	The willful taking or converting to one's own use another person's money or property, which was lawfully acquired by the wrongdoer by reason of some office, employment, or position of trust.
Arson	Purposely setting fire to a house or other building.
Extortion/blackmail	The obtaining of property from another by wrongful use of actual or threatened force, violence, or fear, or under color of official right.
Receiving stolen property	Knowingly accepting, buying, or concealing goods that were illegally obtained by another person.
Fraud	The false representation of a matter of fact, whether by words or by conduct, by false or misleading allegations, or by concealment of that which should have been disclosed, which deceives and is intended to deceive, and causes legal harm.
Forgery	The fraudulent making of a false writing having apparent legal significance.
Counterfeiting	Under federal law, falsely making, forging, or altering any obligation or other security of the United States, with intent to defraud.
"MORALS" OFFENSES	Violations of virtue in sexual conduct (e.g., fornication, seduction, prostitution, adultery, illicit cohabitation, sodomy, bigamy, and incest).
PUBLIC ORDER OFFENSES	Violations that constitute a threat to public safety or peace (e.g., disorderly conduct, loitering, unlawful assembly, drug offenses, driving while intoxicated).
OFFENSES AGAINST THE GOVERNMENT	Crimes motivated by the desire to effect social change or to rebel against perceived unfair laws and governments (e.g., treason, sedition, hindering apprehension or prosecution of a felon, perjury, and bribery).
OFFENSES BY THE GOVERNMENT	Harms inflicted on people by their own governments or the governments of others (e.g., genocide and torture, police brutality, civil rights violations, and political bribe taking).
HATE CRIMES	Criminal offenses committed against a person, property, or society and motivated, in whole or in part, by the offender's bias against a race, a religion, an ethnic/national origin group, or a sexual-orientation group.
ORGANIZED CRIMES	Unlawful acts of members of highly organized and disciplined associations engaged in supplying illegal goods and services, such as gambling, prostitution, loan-sharking, narcotics, and labor racketeering.
WHITE-COLLAR AND CORPORATE CRIMES	Generally nonviolent offenses committed for financial gain by means of deception by entrepreneurs and other professionals who utilize their special occupational skills and opportunities (e.g., environmental pollution, manufacture and sale of unsafe products, price fixing, price gouging, and deceptive advertising).
OCCUPATIONAL CRIMES	Offenses committed through opportunities created in the course of a legal business or profession and crimes committed by professionals, such as lawyers and doctors, acting in their professional capacities.
"VICTIMLESS" CRIMES	Offenses involving a willing and private exchange of goods or services that are in strong demand but are illegal (e.g., gambling, prostitution, drug law violations, and homosexual acts between consenting adults).

Degrees or Categories of Crime

Crimes can be classified according to the degree or severity of the offense, according to the nature of the acts prohibited, or on some other basis, such as a statistical reporting scheme. One way crimes are distinguished by degree or severity of the offense is by dividing them into *felonies* and *misdemeanors.* A felony in one jurisdiction might be a misdemeanor in another jurisdiction, and vice versa. For example, in 2015, a theft in New Jersey of as little as $200 was a felony, while a theft in Wisconsin of 10 times that amount was a misdemeanor. In 2015, New Jersey's felony theft threshold of $200 was the lowest in the United States, while Wisconsin's felony theft threshold of $2,500 was the highest. In 2015, 30 states had felony theft thresholds of $1,000 or more, while 20 states had felony theft thresholds of less than $1,000.[8] Generally, a felony, as noted in Chapter 1, is considered a relatively serious offense punishable by death (in some states and the federal government), a fine, or confinement in a state or federal prison for more than 1 year. A misdemeanor, in contrast, is any lesser crime that is not a felony. Misdemeanors are usually punishable by no more than a $1,000 fine and 1 year of incarceration, generally in a county or city jail.

Another way of categorizing crimes is to distinguish between offenses that are *mala in se* and offenses that are *mala prohibita.* Crimes ***mala in se*** are "wrong in themselves." They are characterized by universality and timelessness. That is, they are crimes everywhere and have been crimes at all times. Examples are murder and rape. Crimes ***mala prohibita*** are offenses that are illegal because laws define them as such. They lack universality and timelessness. Examples are trespassing, gambling, and prostitution.

mala in se Wrong in themselves. A description applied to crimes that are characterized by universality and timelessness.

mala prohibita Offenses that are illegal because laws define them as such. They lack universality and timelessness.

For statistical reporting purposes, crimes are frequently classified as crimes against the person or violent crimes (e.g., murder, rape, assault); crimes against property or property crime (e.g., burglary, larceny, auto theft); and crimes against public decency, public order, and public justice or public order crimes (e.g., drunkenness, disorderly conduct, vagrancy).

Table 2.1 is a list of selected crimes and their definitions, grouped by type. The selection, placement, and definition of the crimes are somewhat arbitrary. There are many different types of crime, and some crimes can be placed in more than one category. Legal definitions of crime vary among jurisdictions and frequently list numerous degrees, conditions, and qualifications. A good source of legal crime definitions is *Black's Law Dictionary.*

THINKING CRITICALLY

1. Are there any acts that are currently legal that you think should be illegal? If so, what?
2. Do you think there should be other elements of crime besides the seven listed in this section? If so, name them.

Measurement of Crime

Near the end of 2018, 60% of Americans said there was more crime in the United States than there was in 2017, down from 68% in 2017 (25% stated there was less crime than a year ago; the remainder thought it was about the same or had no opinion). Forty-eight percent of Americans believed the crime problem in the United States was "extremely" or "very" serious, down from 59% in 2017. About 41% say that the crime problem is "moderately" serious, and 8% say it is "not too serious" or "not serious at all." At the same time, 39% of Americans believed there was more crime in their area than there was a year ago (42% believed there was less; 16% believed the level of criminal activity was the same). Nine percent of Americans believed the crime problem in their area was "extremely" or "very" serious, down from 12% in 2017.[9] Overall, a majority of Americans believe that crime in the United States is much worse than crime in their local area.

Many people who read the daily newspaper or watch the nightly news on television believe that crime is a pressing problem in the United States. But is it? How do you know how much crime is committed? How do you know whether crime is, in fact, increasing or decreasing? How do you know how serious the crime problem is? Besides what you learn from the media, perhaps you have been the victim of crime or know someone who has been. Although that information is important, it does not indicate whether your experience with crime or the experience of someone you know is typical. The fact is that what we and the media know about crime, by and large, is based on statistics supplied by government agencies.

The Distribution of Crime

Although social concern about crime is directed primarily at violent crimes, only 10% to 15% of all crimes committed annually, according to official statistics, are violent crimes. The remaining 85% to 90% are property and public order crimes.

Source: See, for example, Federal Bureau of Investigation, *Crime in the United States, 2017,* https://ucr.fbi.gov/crime-in-the-u.s/2017/crime-in-the-u.s.-2017/topic-pages/offenses-known-to-law-enforcement.

Mario Tama/Staff/Getty Images

Many people learn about crime from the media. *How do the media determine which crimes to report?*

Crime Statistics

The difficulty in relying on crime statistics to measure the prevalence of crime is that "statistics about crime and delinquency are probably the most unreliable and most difficult of all social statistics."[10] In other words, "it is impossible to determine with accuracy the amount of crime [or delinquency] in any given jurisdiction at any particular time."[11] Why? There are several reasons. First, "some behavior is labeled . . . 'crime' by one observer but not by another."[12] If a behavior is not labeled a crime, it is not counted. However, if a behavior is wrongly labeled a crime, then it may be wrongly counted as a crime. Both situations contribute to the inaccuracy of crime statistics. Second, a large proportion of crimes are undetected. Crimes that are not detected obviously cannot be counted. Third, some crimes may not be reported to the police. If they are not reported to the police, they are unlikely to be counted. Fourth, crimes that are reported to the police may not be officially recorded by them for various reasons (discussed later), or may be inaccurately recorded. Crimes that are not officially recorded by the police are called the **dark figure of crime** (see Figure 2.2).

dark figure of crime The number of crimes not officially recorded by the police.

For all of the foregoing reasons, any record of crimes—such as "offenses known to the police," arrests, convictions, or commitments to prison—can be considered at most a **crime index**, or an estimate of crimes committed. Unfortunately, no index or estimate of crimes is a reliable indicator of the actual amount of crime. The indexes or estimates vary independently of the true amount of crime, whatever that may be. Figure 2.2 portrays one possible relationship of a crime index to the dark figure of crime and the true amount of crime. It shows how great a discrepancy there can be between the index and the actual amount of crime.

crime index An estimate of crimes committed.

FIGURE 2.2 Dark Figure of Crime

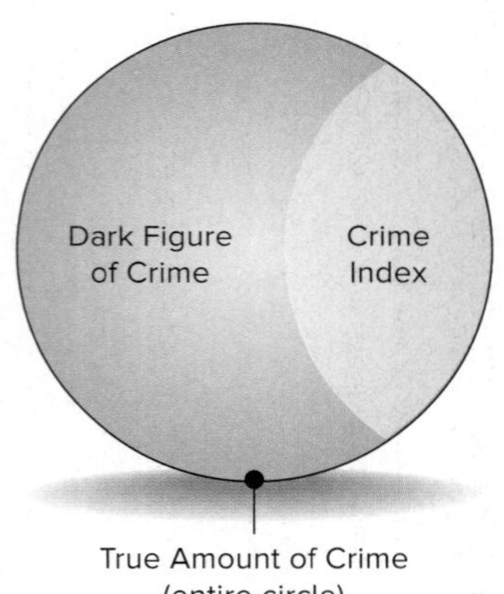

Adding to the confusion is the reality that any index of crime varies with changes in police practices, court policies, and public opinion—to name just three factors. Suppose, for example, that a large city is hosting a major convention and city leaders want to make a good impression on visitors. The mayor asks the police chief to order officers to "sweep" the streets of prostitutes. As a result of that police policy, there is a dramatic increase in arrests for prostitution and in the index measuring prostitution. Does the increase in the index mean that the true amount of prostitution increased in the city? The answer is that we do not know and, for that matter, can never know. All we do know is that the index measuring prostitution increased as a result of police practices.

Thus, despite what some government agencies or the media may suggest, we do not know, nor can we ever know, the true amount of crime. For the same reasons, we can never know for sure whether crime is increasing, decreasing, or remaining at the same level. The sophisticated student of crime knows only that indexes of crime are imperfect estimates that vary widely. Those variations, which are independent of variations in the true amount of crime, depend on such things as police practices, court policies, and public opinion. Therefore, comparisons of crime measures are an especially dubious exercise. Criminal justice officials and professors routinely compare crime measures in different jurisdictions and at different times. What they are doing, though they rarely acknowledge it, even when they are aware of it, is comparing indexes or estimates of crime. Although such comparisons tell us nothing about differences in true amounts of crime, they do provide insights into police practices, court policies, and public opinion.

offenses known to the police A crime index, reported in the FBI's uniform crime reports, composed of crimes that are both reported to and recorded by the police.

Probably the best index of crime—that is, the least inaccurate!—is **offenses known to the police**. That index, which is reported in the uniform crime reports (discussed later) from the Federal Bureau of Investigation (FBI), is composed of crimes that are both reported to and recorded by the police. The reason it is an inaccurate measure of the true amount of crime is that the number of offenses known to the police is always much smaller than the number of crimes actually committed. One reason is that victims do not report all crimes to the police. According to the national crime victimization survey in 2018, for example, victims did not report to the police 57% of all violent crime victimizations and 66% of all property crime victimizations.[13] There are many reasons for the nonreporting of crimes:[14]

1. Victims may consider the crime insignificant and not worth reporting.
2. They may hope to avoid embarrassing the offender, who may be a relative, school friend, or fellow employee.
3. They may wish to avoid the publicity that might result if the crime were reported.
4. They might have agreed to the crime, as in gambling offenses and some sexual offenses.
5. They may wish to avoid the inconvenience of calling the police (filling out a report, appearing in court, and so on).
6. They may be intimidated by (or afraid of) the offender.
7. They may dislike the police or be opposed to the punitive policies of the legal system.
8. They may feel that the police are so inefficient that they will be unable to catch the offender even if the offense is reported.

Another reason the number of offenses known to the police is necessarily much smaller than the number of crimes actually committed is that the police do not always officially record the crimes that are reported to them. In practice, police officers often use their discretion to handle informally an incident reported to them; that is, they do not make an official report of the incident. Or they may exercise discretion in enforcing the law (e.g., by not arresting the customer in a case of prostitution). The law is often vague, and officers may not know the law or how to enforce it. Still another reason is that some police officers feel they are too busy to fill out and file police reports. Also, some officers, feeling an obligation to protect the reputations of their cities or being pressured by politicians to "get the crime rate down," may manipulate statistics to show a decrease in crime.[15]

Reported Crime on Campus

Go to the U.S. Department of Education's Office of Postsecondary Education Campus Security Statistics website, http://ope.ed.gov/security/, for information and statistics on reported criminal offenses at colleges and universities across the United States.

What do the statistics say about crime at schools in your area?

Consider the following recent examples:

- An analysis by the *Los Angeles Times* found that the Los Angeles Police Department (LAPD) artificially lowered the city's crime levels from 2005 through 2012, by misclassifying an estimated 14,000 serious assaults as minor offenses. When the incidents were counted correctly, violent crime in Los Angeles was 7% higher than the LAPD reported during the period, and the number of serious assaults was 16% higher. Similar errors were found in separate analyses in 2013 and 2014. In the 2014 report, police auditors found that the number of aggravated assaults would have been 23% higher than reported if not for the errors.[16]
- CBS Denver television station reported that in 2018, about 1,000 Denver Police Department crime reports primarily from two districts over the last several years and about 1% of Denver crimes in 2017, appeared to have been downgraded from the way they were initially reported in a way that removed them from official reporting requirements. As a result, the crimes were not reported to the Colorado Bureau of Investigation and, thus, did not appear in any official state or national statistical totals. The types of cases reclassified included both property and violent crimes.[17]
- In 2018, Detroit's Police Chief James Craig claimed that "crime fell across the board in 2017 when compared with the year before, with almost all crime categories showing their second straight year of reductions." Most significant, according to Craig, was the 12% decrease in the city's homicide rate from the year before to 267 murders—"a low not seen in a half century." However, an investigative report discovered that the decrease in the homicide rate could be attributed in part to the incorrect categorization of homicides as "justified." Thus, if an assailant pulled a knife on a person, and the potential victim pulled a gun, shot, and killed the assailant, then the killing would be considered justified and would not be counted as a homicide for statistical reporting purposes. The Detroit Police Department lowered its homicide rate, in part, by misclassifying homicides as justified.[18]
- In 2017, the Nevada Policy Research Institute accused the Las Vegas Metropolitan Police Department of misleading the public over several years about violent and property crime rates. Prior to 2011, the LVMPD classified robberies as violent crimes, conforming to FBI standards. However, in 2011, the LVMPD began categorizing robberies as property crimes, even though robbery is a violent crime under Nevada law. The classification shift, which is not explained or mentioned in LVMPD annual reports, significantly lowered violent crime rates and increased property crime rates.[19]
- A review of Pittsburg Police Department data in 2016, by the *East Bay Times*, revealed the manipulation of crime statistics to mislead the public and burnish the

> city's image. The main problem is the Department's practice of reporting hundreds of crimes a year in a catchall category called "suspicious circumstances," which keeps them from being counted in FBI crime statistics. According to a former Pittsburg police lieutenant, the practice is "systematic and deliberate, and that officers are taught and pressured to classify certain cases–those with a lack of credible witnesses, workable leads or unlikely prosecutions–in a manner that treats them as if they were not crimes at all." He said that "supervisors would alter reports and pressure rank-and-file officers to follow the policy."[20]

Similar examples of police manipulating statistics to show reductions in crime have been reported for New York City, Chicago, Dallas, Atlanta, and Philadelphia.[21]

In addition, the number of offenses included in the index is much smaller than the actual number of crimes because of the way crimes are counted. For the FBI's uniform crime reports, when more than one crime is committed during a crime event, only the most serious is counted for statistical purposes. The seriousness of the crime is determined by the maximum legal penalty associated with it. Thus, for example, if a robber holds up 10 people in a tavern, takes their money, shoots and kills the bartender, and makes a getaway in a stolen car, only one crime–the murder–is counted for statistical purposes. However, the offender could legally be charged with several different crimes. (The practice of counting only the most serious offense in a multiple-crime event is being changed with the implementation of the National Incident-Based Reporting System, which is discussed later in this chapter.)

Despite the problems in recording crime events, *offenses known to the police* is a more accurate index of crime than arrest statistics, charging statistics, trial statistics, conviction statistics, sentencing statistics, and imprisonment statistics. As shown in Figure 2.3, the farther a crime index is from the initial commission of crime, the more inaccurate it is as a measure of the true amount of crime.

Crime Rates

crime rate A measure of the incidence of crime expressed as the number of crimes per unit of population or some other base.

When crime indexes are compared, rarely are total numbers of crimes used. Instead, crime is typically reported as rates. A **crime rate** is expressed as the number of crimes per unit of population or some other base. Crime rates are used instead of total numbers because they are more comparable. For example, suppose you wanted to compare the crime of murder in the United States for the census years 2000 and 2010. There were 15,517 murders

Los Angeles police officers respond to a bungled bank robbery. The heavily armed, masked robbers fired hundreds of shots in a gun battle getaway that left 2 dead and at least 11 hurt. *In this crime event, what crimes would be officially recorded for purposes of the FBI's uniform crime reports?*

Mike Meadows/AP Images

FIGURE 2.3 Indexes of Crime

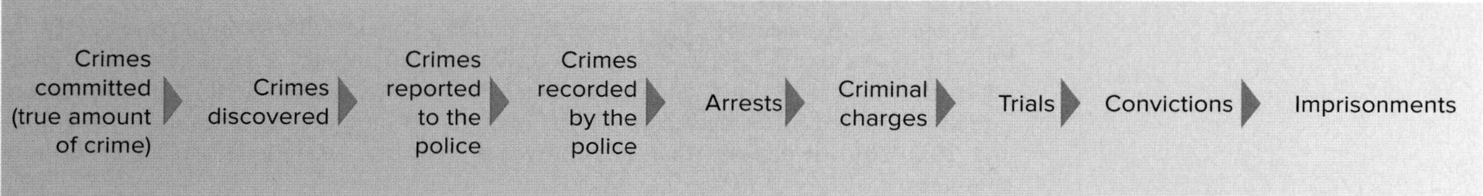

The farther away from the initial commission of a crime, the more inaccurate crime indexes are as measures of the true amount of crime.

and nonnegligent manslaughters reported to and recorded by the police in 2000; there were 14,748 murders and nonnegligent manslaughters reported to and recorded by the police in 2010. According to those data, the total number of murders and nonnegligent manslaughters reported to and recorded by the police in 2010 decreased about 5% from the number reported to and recorded by the police in 2000. Although this information may be helpful, it ignores the substantial increase in the population of the United States–and thus in the number of potential murderers and potential murder victims–between 2000 and 2010.

A better comparison–though, of course, not an accurate one–would be a comparison that takes into account the different population sizes. To enable such a comparison, a population base, such as "per 100,000 people," is arbitrarily chosen. Then the total number of murders and nonnegligent manslaughters for a particular year is divided by the total population of the United States for the same year. The result is multiplied by 100,000. When those calculations are made for the years 2000 and 2010, the rate of murders and nonnegligent manslaughters is 5.5 per 100,000 people for 2000 and 4.8 per 100,000 people for 2010 (see Figure 2.4). According to those figures, the rate of murder and nonnegligent manslaughter in 2010 was about 13% lower than it was in 2000. Thus, although both sets of data show that murders and nonnegligent manslaughters reported to and recorded by the police decreased between 2000 and 2010, the decrease is greater when the increase in the size of the population (which was approximately 10%) is taken into account. Crime rates provide a more accurate indication of increases or decreases in crime indexes than do total numbers of crimes. Remember, however, that what are being compared are indexes and not true amounts.

A variety of factors indirectly related to crime can affect crime rates. For example, burglary rates might increase, not because there are more burglaries but because more things are insured, and insurance companies require police reports before they will reimburse their policyholders. Changing demographic characteristics of the population also can have an effect. For example, because of the post–World War II baby boom (1945–1964), between 1963 and 1988 there were more people in the age group most prone to committing recorded crime (18- to 24-year-olds). All other things being equal, higher crime rates would be expected between 1963 and 1988 simply because there were more people in the age group that commits the most recorded crime. By the same token, a decrease in crime rates might be expected after 1988, all other things being equal, because the baby boom generation is no longer at those crime-prone ages. (That, in fact, occurred.) However, an increase in crime might be expected when their children reach the 18 to 24 age range (that has not occurred). Changes in the age structure of the population have been estimated to account for approximately 40% of the changes in the crime rate.[22]

Urbanization is another factor, especially with regard to violent crime. Violent crime is primarily a big-city phenomenon. Thus, violent crime rates might increase as more of the population moves from rural to urban areas or as what were once rural areas become more urban.

In 2015, data reported by the Federal Bureau of Investigation showed that the violent crime rate in the United States increased 3.0% from the year before.[23] The increase bucked a nearly quarter century downward trend in the violent crime rate that began

FIGURE 2.4 Calculating Crime Rates

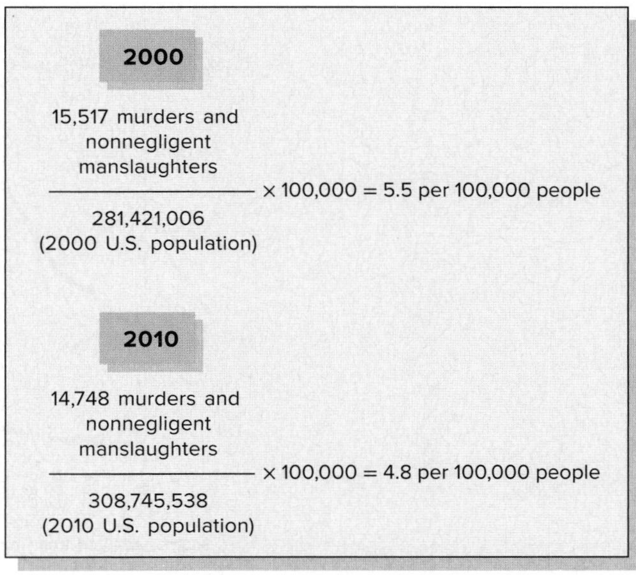

in 1991. In other words, except for the years 2004–2005, 2005–2006, 2011–2012, and 2014–2015, which had modest rate increases of 1.3%, 2.2%, 0.2%, and 3.0%, respectively, the violent crime rate in the United States declined every other year since 1991—a total decrease of about 50%.[24]

In the 3 years since the 2015 report, the nation's violent crime rate increased 3.45% in 2015–2016, decreased 2.8% in 2016–2017, and decreased another 3.9% in 2017–2018. Overall, from 2016 through 2018, and despite the 3.45% increase in 2015–2016, the violent crime rate in the United States decreased 1.3%. Consequently, despite the increases in 2014–2015 and 2015–2016, it appears that the downward trend in the nation's violent crime rate continues.[25]

Among the factors that police chiefs and academics have cited for the decline in the violent crime rate are these:

1. Aging-out of the crime-prone years by the postwar baby boom generation.
2. Fewer turf battles over crack cocaine distribution because of market maturation and consolidation.
3. Police efforts to disarm criminals and juveniles.
4. More police officers on the beat.
5. Smarter policing.
6. Tougher criminal justice legislation, such as the federal law that ties financial aid for prison building to a requirement that states keep violent offenders incarcerated for at least 85% of their sentences.
7. Increased interest in "grass-roots" crime prevention.
8. A better economy that provided jobs and gave cities more to invest in crime control.

Figure 2.5 displays the violent and property crime rates in the United States from 1960 through 2018.

Uniform Crime Reports

uniform crime reports (UCR) A collection of crime statistics and other law enforcement information gathered under a voluntary national program administered by the FBI.

A primary source of crime statistics in the United States is the **uniform crime reports (UCR)**, which are a collection of crime statistics and other law enforcement information published annually by the Federal Bureau of Investigation (FBI) under the title *Crime in the United States*. They are the result of a voluntary national program begun in the 1920s by the International Association of Chiefs of Police.[26] The program was turned over to the FBI

FIGURE 2.5 Violent and Property Crime Rates in the United States, 1960–2018

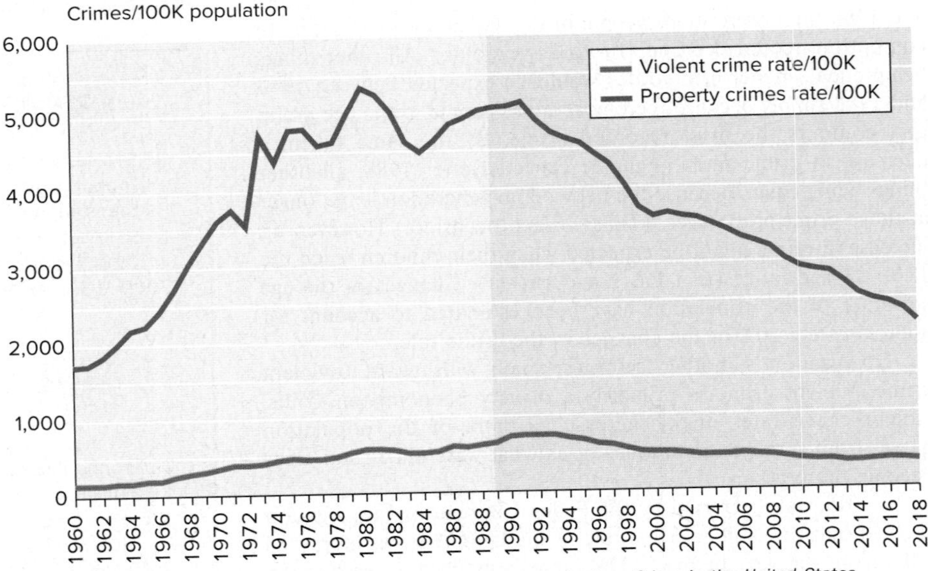

Source: Adapted from multiple editions of Federal Bureau of Investigation, *Crime in the United States*.

in 1930 by the attorney general, whose office Congress had authorized to serve as the national clearinghouse for crime-related statistics. In 2018, more than 18,000 city, university and college, county, state, tribal, and federal law enforcement agencies were active in the program; they represented nearly the entire U.S. population.[27]

Until June 2004, the UCR included two major indexes: (1) offenses known to the police, discussed earlier, and (2) statistics about persons arrested. The section on offenses known to the police, or offenses reported to the police, provided information about the **eight index crimes**, or Part I offenses:

eight index crimes The Part I offenses in the FBI's uniform crime reports. They were (1) murder and nonnegligent manslaughter, (2) forcible rape, (3) robbery, (4) aggravated assault, (5) burglary, (6) larceny-theft, (7) motor vehicle theft, and (8) arson, which was added in 1979.

1. Murder and nonnegligent manslaughter
2. Forcible rape
3. Robbery
4. Aggravated assault
5. Burglary
6. Larceny-theft
7. Motor vehicle theft
8. Arson (added in 1979; permanently in 1982)

The first four offenses were considered violent offenses; the last four were considered property offenses. Although the eight index crimes continue to be reported separately in the FBI's uniform crime reports, the crime index was discontinued and replaced in 2004 with a violent crime total and a property crime total until a better index could be developed. The problem was that the crime index was skewed upward by the offense with the highest number: larceny-theft, which currently accounts for nearly 62% of reported crime. This creates a bias against jurisdictions with high numbers of larceny-thefts but low numbers of other more serious but less frequently committed crimes such as murder and forcible rape.[28]

myth

When the media report that crime has increased or decreased from one year to the next, they are generally referring to increases or decreases in the true amount of crime.

fact

What the media are usually referring to when they report that crime has increased or decreased from one year to the next is an increase or decrease in the aggregate rate of the eight index crimes (i.e., the "crime index total") or now the violent and property crime totals, not the rates of other crimes or the true amount of crime.

In addition to several technical improvements to the UCR program, other recent developments include: (1) the change in the definition of rape, (2) the collection of human trafficking data, (3) the collection of federal crime data, (4) the collection of data on animal cruelty offenses, and (5) the collection of expanded data on law enforcement's use of force in shooting incidents. Each of these developments is discussed in the following paragraphs.

On January 6, 2012, the FBI changed its definition of forcible rape to "The penetration, no matter how slight, of the vagina or anus with any body part or object, or oral penetration by a sex organ of another person, without the consent of the victim." Before this change, and since 1927, the FBI had defined forcible rape as "the carnal knowledge of a female, forcibly and against her will." The old definition included only forcible male penile penetration of a female vagina. The new definition includes oral and anal penetration, rape of males, penetration of the vagina and anus with an object or body part other than the penis, rape of females by females, and nonforcible rape. An FBI press release noted that the new definition better reflects state criminal codes and victim experiences.[29]

In January 2013, the national UCR program began collecting offense and arrest data on human trafficking as authorized by the *William Wilberforce Trafficking Victims Protection Reauthorization Act of 2008*. To comply with the Act, the UCR program created two additional violent offenses: (1) Human Trafficking/Commercial Sex Acts and (2) Human Trafficking/Involuntary Servitude. Human Trafficking/Commercial Sex Acts refers to "inducing a person by force, fraud, or coercion to participate in commercial sex acts, or in which the person induced to perform such act(s) has not attained 18 years of age." Human Trafficking/Involuntary Servitude refers to "the obtaining of a person(s) through recruitment, harboring, transportation, or provision, and subjecting such persons by force, fraud, or coercion into involuntary servitude, peonage, debt bondage, or slavery (not to include commercial sex acts)." These are the first offenses added to the UCR's list of Part I offenses since 1982, when the collection of arson data was made permanent.[30]

Data on crimes motivated by bias against individuals because of race, religion, disability, sexual orientation, or ethnicity/national origin were included in the UCR for the first time in 1996. The UCR designated these as hate crimes or bias crimes. Beginning in 2005, the FBI began producing *Hate Crime Statistics* exclusively as a web publication separate from the UCR.[31] Agencies that participated in the 2017 data collection program procedure represented about 94% of the nation's population.[32] In 2017, 7,145 hate-crime incidents, involving 8,437 offenses, were reported to the FBI, about 17% more incidents than were reported in 2016.[33] All but 69 of the hate crimes reported in 2017 were single-bias incidents that

Hate Groups in the United States

At the end of 2018, 1,020 active hate groups operated in the United States, up more than 30% from the 784 groups in 2014, and up nearly 7% from the 954 groups in 2017. The 1,020 active hate groups in 2018 was the largest number recorded since 1999. Among the more active groups were the Ku Klux Klan, Neo-Nazis, White Nationalists, Racist Skinheads, Christian Identity, Black Separatists, Neo-Confederates, Anti-LGBT, Anti-Muslim, Anti-Immigrant, and General Hate.

Source: Southern Poverty Law Center, *Intelligence Report*, Spring 2019/Issue 166, accessed February 22, 2019, https://www.splcenter.org/sites/default/files/intelligence_report_166.pdf.

involved 8,126 offenses, 8,493 victims, and 6,307 known offenders.[34] Of all the single-bias hate crimes reported in 2017 (percentage change in the number of incidents from 2016 in parentheses) 58% (+18%) were motivated by race/ethnicity/ancestry bias, 22% (+23%) by religious bias, 16% (+5%) by sexual orientation bias, 2% (+66%, numbers were small) by disability bias, 2% (-4%, numbers were small) by gender-identity bias, and 0.6% (+48%, numbers were small) by gender bias. Three points about these data are worth emphasizing. First, the 8,126 hate-crime offenses reported to the police in 2017 represented only a miniscule percent (0.09%) of the nearly 9 million total offenses reported to the police in 2017. Second, as with all crime statistics, hate-crime statistics are only an index of the true number of hate crimes committed, whatever that number might be. Third, the general increase in hate-crime incidents, except for gender-identity incidents, from 2016 to 2017, may only or primarily be the result of an increase in the number of hate-crime incidents reported to or recorded by the police or of some other reason.[35]

In 2014, the UCR program for the first time began collecting federal crime data on three crimes: (1) human trafficking (defined earlier), (2) hate crime (defined earlier), and (3) criminal computer intrusion, which refers to "wrongfully gaining access to another person's or institution's computer software, hardware, or networks without authorized permissions or security clearances. (State, local, and tribal agencies were to report these offenses to the UCR program, beginning in 2016, as Hacking/Computer Invasion.)"[36]

In 2015, the FBI announced it would begin collecting expanded data on law enforcement's use of force in shooting incidents. The program became fully operational in January 2018. In a special publication, the UCR program will report data detailing what happened, who was involved, whether there were any injuries or deaths, and the circumstances surrounding each incident. Commenting on the lack of such data, then FBI Director Comey remarked, "It's ridiculous that I can't tell you how many people were shot by the police in the country." As is the case with all UCR program data, law enforcement agencies will be asked to voluntarily submit annual reports to the FBI, which likely will hamper the representativeness of the data. As of August 13, 2018, 3,839 law enforcement agencies were enrolled in the data collection effort.[37]

Beginning in January 2016, the UCR program began collecting data on animal cruelty offenses, as requested by the National Sheriff's Association and the Animal Welfare Institute. This new program was approved by the FBI director in September 2014. The FBI defines cruelty to animals as "Intentionally, knowingly, or recklessly taking an action that mistreats or kills any animal without just cause, such as torturing, tormenting, mutilation, maiming, poisoning, or abandonment." Included are four categories of crimes: simple or gross neglect; intentional abuse and torture; organized abuse, such as dog fighting and cock fighting; and animal sexual abuse.[38]

status offense An act that is not a crime when committed by adults but is illegal for minors (e.g., truancy or running away from home).

The other major crime index in the UCR is based on arrest statistics. Arrest data are provided for the previous eight index crimes, as well as 21 other crimes and status offenses. The 21 other crimes and status offenses were previously referred to as Part II offenses. A **status offense** is an act that is illegal for a juvenile but would not be a crime if committed by an adult (such as truancy or running away from home). Table 2.2 lists the now-combined former Part I and Part II offenses in the FBI's UCR.

According to the UCR, law enforcement agencies made about 10.3 million arrests nationwide for 28 separate index offenses (see Table 2.2) in 2018, approximately 3% fewer than the 10.5 million arrests made in 2017. Note that only about 5% of the arrests were for "index violent crimes" and approximately 11% were for "index property crimes." The number of arrests for violent crimes increased 0.2%, while the number of arrests for property crimes decreased about 7% in 2018 compared to 2017.[39] The index offense for which the most arrests were made in 2018 (approximately 1.65 million) was drug-abuse violations.[40] The second and third most arrests in 2018 were for driving under the influence (about 1 million) and larceny theft (about 890,000), respectively.[41] Arrestees generally were young (29% were under 25 years of age), male (73%), and white (69%).[42] Arrests of juveniles declined 11% in 2018 when compared to 2017; arrests of adults dropped about 2%.[43] The index crime for which women were most frequently arrested was larceny-theft. In 2018, the number of arrests of females was 1.7% lower than in 2017, and 15.1% lower than in 2009; the number of arrests of males was 3.2% lower in 2018 than it was in 2017, and 24.2% lower than it was in 2009.[44]

crime index offenses cleared The number of offenses for which at least one person has been arrested, charged with the commission of the offense, and turned over to the court for prosecution.

In addition to statistics on offenses known to the police and persons arrested, the UCR include statistics on crime index offenses cleared or "closed" by the police. **Crime index offenses cleared** (also called *clearance rates* or *percent cleared by arrest*) is a rough index of

TABLE 2.2 Former Part I and Part II Offenses of the FBI's Uniform Crime Reports

Part I Offenses—Index Crimes	Part II Offenses
VIOLENT CRIME	1. Other assaults (simple)
1. Murder and nonnegligent manslaughter	2. Forgery and counterfeiting
2. Forcible rape	3. Fraud
3. Robbery	4. Embezzlement
4. Aggravated assault	5. Stolen property: buying, receiving, possessing
PROPERTY CRIME	6. Vandalism
5. Burglary—breaking or entering	7. Weapons: carrying, possessing, etc.
6. Larceny-theft	8. Prostitution and commercialized vice
7. Motor vehicle theft	9. Sex offenses
8. Arson	10. Drug abuse violations
	11. Gambling
	12. Offenses against the family and children
	13. Driving under the influence
	14. Liquor laws
	15. Drunkenness
	16. Disorderly conduct
	17. Vagrancy
	18. All other offenses
	19. Suspicion
	20. Curfew and loitering laws

police performance in solving crimes. According to the UCR, offenses can be cleared in one of two ways: by arrest or by exceptional means. An offense that is cleared by arrest is one for which "at least one person is arrested, charged with the commission of the offense, and turned over to the court for prosecution."[45] The arrest of one person may clear several crimes, and the arrest of many persons may clear only one offense. An offense cleared by exceptional means is one for which a law enforcement agency, for reasons beyond its control, cannot arrest and formally charge an offender. Examples of exceptional clearances include the offender's death or the refusal of a victim to cooperate with the prosecution after the offender has been identified. To clear an offense by exceptional means, an offender must be identified; enough evidence must be available to support an arrest, a charge, and prosecution; the offender's exact location must be identified so that he or she can be taken into custody immediately; and a circumstance beyond the control of law enforcement must have occurred that prevented the offender's arrest, charging, and prosecution.[46] Clearances recorded in one year may be for offenses committed in previous years. Except for rape, clearance rates remain remarkably stable from year to year. In 2018, the police were able to clear about 62.3% of murders and nonnegligent manslaughters, 33.4% of rape offenses based on the revised definition; 30.4% of robberies, 52.5% of aggravated assaults, 13.9% of burglaries, 18.9% of larceny-thefts, and 13.8% of motor vehicle thefts.[47] In 2018, the police were able to clear 45.5% of violent crimes and 17.6% of property crimes.[48]

What about rape? The percentage of rapes cleared by arrest has been on a downward trajectory since the 1960s, attaining its lowest level since that time in 2018, at 33.4%. This has occurred despite the #MeToo movement, advances in DNA testing, and the new, more expansive definition of rape. One explanation is that fewer rapes are cleared by exceptional circumstances, as the police now keep those rape cases open indefinitely. Another explanation is that not enough resources are devoted to the investigation of sexual assaults at a time when an increasing number of victims are reporting them to the police.[49]

The UCR also provide statistics about law enforcement personnel, such as the number of full-time sworn officers in a particular jurisdiction and the number of law enforcement officers killed and assaulted (LEOKA) in the line of duty. In 2016, the FBI expanded its criteria for participating in the national LEOKA Program to include military police, civilian police, and Department of Defense law enforcement officers, who are killed or assaulted while performing law enforcement functions or duties. These data were first published in LEOKA, 2016, which was released in 2017.[50]

National Incident-Based Reporting System

In 1982, a joint task force of the Bureau of Justice Statistics (BJS) and the FBI was created to study and recommend ways to improve the quality of information contained in the UCR.[51] The result is the National Incident-Based Reporting System (NIBRS), which collected its first data in 1991. The FBI established the annual publication, *National Incident-Based Reporting System*, in 2013. Under the NIBRS, participating law enforcement authorities currently provide offense and arrest data on 24 broad categories of crime, covering 52 offenses (as compared to the former 8 UCR index offenses), and provide only arrest information on 11 other offenses (as compared to the former 20 Part II UCR offenses) (see Table 2.3).

Perhaps the greatest and most important difference between the NIBRS and the UCR is that the NIBRS contains more data on each crime, making it possible to examine crimes in much more detail. The NIBRS contains more than 50 different pieces of information about a crime, divided into six segments, or categories. It is hoped that the increased

TABLE 2.3 The National Incident-Based Reporting System

Group A Offenses	Group B Offenses
Animal cruelty	Bad checks
Arson	Curfew/loitering/vagrancy
Assault offenses	Disorderly conduct
Bribery	Driving under the influence
Burglary/breaking and entering	Drunkenness
Counterfeiting/forgery	Family offenses, nonviolent
Destruction/damage/vandalism	Liquor law violations
Drug/narcotic offenses	Peeping Tom
Embezzlement	Runaway*
Extortion/blackmail	Trespassing
Fraud offenses	All other offenses
Gambling offenses	
Homicide offenses	
Human trafficking offenses	
Kidnapping/abduction	
Larceny/theft offenses	
Motor vehicle theft	
Pornography/obscene material	
Prostitution offenses	
Robbery	
Sex offenses, forcible	
Sex offenses, nonforcible	
Stolen property offenses	
Weapons law violations	

*The UCR Program discontinued collecting this information in 2011, and it is not considered a crime for UCR purposes. However, some agencies still collect and report these data because the arrest of individuals for such occurrences sometimes clears associated offenses in incident reports.

TABLE 2.4 NIBRS Data Reporting Elements

ADMINISTRATIVE SEGMENT	VICTIM SEGMENT
1. ORI (originating agency identifier)	26. Victim sequence number
2. Incident number	27. Victim connected to UCR offense code
3. Cargo theft*	28. Type of victim
4. Incident date/report date indicator/hour cleared exceptionally	29. Type of officer activity/circumstance
5. Exceptional clearance date	30. Officer assignment type
6. Exceptional clearance offense code	31. Officer—ORI other jurisdiction
OFFENSE SEGMENT	32. Age, sex, race, and ethnicity of victim
7. UCR offense code	33. Resident status of victim
8. Offense attempted/completed	34. Aggravated assault/homicide circumstances
9. Offender suspected of using	35. Additional justifiable homicide circumstances
10. Bias motivation	36. Type of injury
11. Location type	37. Offender number to be related
12. Number of premises entered	38. Relationship of victim to offender
13. Method of entry	**OFFENDER SEGMENT**
14. Type of criminal activity/gang information	39. Offender sequence number
15. Type of weapon/force involved	40. Age, sex, race, and ethnicity of offender
16. Automatic weapon indicator	**ARRESTEE SEGMENT**
PROPERTY SEGMENT	41. Arrestee sequence number
17. Type of property loss/etc.	42. Arrestee transition number
18. Property description	43. Arrest date
19. Value of property	44. Type of arrest
20. Date recovered	45. Multiple arrestee segments indicator
21. Number of stolen motor vehicles	46. UCR arrest offense code
22. Number of recovered motor vehicles	47. Arrestee was armed with automatic weapon indicator
23. Suspected drug type	48. Age, sex, race, and ethnicity of arrestee
24. Estimated drug quantity	49. Resident status of arrestee
25. Type of drug measurement	50. Disposition of arrestee under 18
	51. Clearance indicator
	52. Clearance offense code

*A provision of the USA PATRIOT Improvement and Reauthorization Act of 2005 required a separate category in the UCR system for reports of cargo theft by federal, state, and local officials. Considered a "gateway" crime, cargo theft could be one part of a larger case about organized crime, drug trafficking, or funding for terrorism. Collecting cargo theft data could help measure the impact this type of crime has on both the economy and national security. Cargo theft data were first collected in 2013 and published as the special report *Cargo Theft Update*.

amount of information in the NIBRS will provide the basis for a much greater understanding of crime and its causes (or at least of crime reporting and recording behavior) than is possible with the data from the UCR. Table 2.4 lists the NIBRS data elements.

The NIBRS is continuously evolving, adding new offenses, offense subcategories, and offense elements. For example, although the FBI has collected victim-offender relationship data for crimes against persons and robbery since the NIBRS's inception, it intends to begin collecting data, specifically, on domestic and family violence, which it defines as:

> The use, attempted use, or threatened use of physical force or a weapon; or the use of coercion or intimidation; or committing a crime against property by a current or former spouse, parent, or guardian of the victim; a person with whom the victim shares a child in common; a person who is or has been in a social relationship of a romantic or intimate nature with the victim; a person who is cohabiting with or has cohabited with the victim as a spouse, parent, or guardian; or by a person who is or has been similarly situated to a spouse, parent, or guardian of the victim.

Also, on January 1, 2019, the FBI intended to replace the category of lover's quarrel in the NIBRS with domestic violence and add the category of ex-relationship. Only agencies that submit data to the NIBRS will be affected.[52]

On January 1, 2019, the FBI also intended to expand the definition of negligent manslaughter in the NIBRS to include impaired and/or distracted vehicle and vessel operators. The new definition is:

> The killing of another person through negligence. This offense includes killings from hunting accidents, gun cleaning, children playing with guns and arrests associated with driving under the influence, distracted driving (using cell/smartphone), and reckless driving traffic fatalities.[53]

Finally, in 2016, the FBI began collecting NIBRS data on two new fraud offenses: identity theft and hacking/computer invasion. Cyberspace also was added as a new location type. All three additions were first published in NIBRS, 2016, which was released in 2017.[54]

The BJS and the FBI hope that by January 1, 2021, the NIBRS will replace the UCR as the source of official FBI crime counts. In 2017, 6,998 law enforcement agencies in 39 states submitted NIBRS data. This is approximately 42% of all the agencies participating in the UCR Program and represents about 32% of the U.S. population. On the other hand, 11 states did not have a single agency to participate in the NIBRS program in 2017: Alaska, California, Florida, Georgia, Hawaii, Nevada, New Jersey, New Mexico, New York, North Carolina, and Wyoming. Five states (Alabama, Arizona, Illinois, Maryland, and Mississippi) had fewer than 6 agencies participating in 2017. So far, the biggest impediment to implementation of the NIBRS is that it is a "paperless" reporting system and thus requires the use of a computerized records management system. Many larger law enforcement agencies have older computer systems that require extensive and costly modifications. Many smaller agencies do not have computer systems. Although some agencies have received federal and state grants to upgrade or buy computer systems for the NIBRS, the amounts allocated have covered only a small part of the need. For example, the FBI reported in 2017 that it was working with the Bureau of Justice Statistics to fund NIBRS technology solutions for about 400 local, state, and tribal agencies.

Adding to the implementation problem are benefit and policy concerns.[55] Some law enforcement agencies question who, other than researchers, will benefit from their reporting NIBRS data. Others fear that because NIBRS reports multiple offenses within an incident, crime will appear to increase, thus causing a public relations nightmare for law enforcement officials. Some law enforcement administrators are concerned that the detailed incident reporting required for NIBRS will tie up patrol officers, keeping them from responding to the needs of the community.

National Crime Victimization Surveys

national crime victimization surveys (NCVS) A source of crime statistics based on interviews in which respondents are asked whether they have been victims of any of the FBI's index offenses (except murder, nonnegligent manslaughter, and arson) or other crimes during the past 6 months. If they have, they are asked to provide information about the experience.

The **national crime victimization surveys (NCVS)** are the other major source of crime statistics in the United States. The surveys provide a detailed picture of crime incidents, victims, and trends from the victim's perspective. Formerly called the national crime surveys (NCS), they have been conducted annually since 1972 by the Bureau of the Census for the U.S. Department of Justice's Bureau of Justice Statistics.[56] The NCVS, published under the title *Criminal Victimization in the United States,* were created not only as a basis for learning more about crime and its victims, but also as a means of complementing and assessing what is known about crime from the FBI's UCR. (From 1996 on, the NCVS are available only in electronic formats.)

From a nationally representative sample of 151,055 households in 2018, 242,928 respondents age 12 or older were asked in interviews whether they had been victims of any of the FBI's former index offenses (except murder, nonnegligent manslaughter, and arson) or any other crimes during the past 6 months.[57] If they had, they were asked to provide information about the experience. Because major changes were made in the format and methodology of the NCVS in 1992, adjustments have been made to the data before 1993 to make them comparable with data collected since the changes. Like the UCR, the NCVS are merely an index of crime and not an accurate measure of the true amount of crime that is committed.

Generally, the NCVS produce different results from the FBI's UCR. For nearly all offenses, the NCVS show more crimes being committed than the UCR do. This underestimation by the UCR may result from victims' failure to report crimes to the police or from failure by the police to report to the FBI all the crimes they know about. In addition, the NCVS count sexual assaults, while the UCR do not. The NCVS also use a broader definition

Careers in Criminal Justice

Social Science Statistician

My name is Tracy Snell. I am a statistician for the Bureau of Justice Statistics (BJS), U.S. Department of Justice. BJS collects, analyzes, publishes, and disseminates information on crime, criminal offenders, victims of crime, and the operation of justice systems at all levels of government.

BJS is divided into four units: Victimization Statistics; Law Enforcement, Adjudication, and Federal Statistics; Corrections Statistics; and Special Analysis and Methodology. I work in the corrections unit, and, over the years, I have worked on data collections pertaining to probationers, prison inmates, jail inmates, parolees, and correctional facilities. Currently, I am responsible for an annual collection of information on persons under sentence of death. I also recently finished work on a survey of inmates in state and federal correctional facilities. For this nationally representative survey, which BJS conducts every 5 to 7 years, we create the questionnaire, develop a sampling design, oversee the administration of the survey, edit and analyze the data, and prepare a file to be used by the general public.

I have a Bachelor of Science (BS) degree in psychology with a minor in political science from Denison University and a Masters of Public Policy (MPP) from the University of Michigan. While I was in graduate school, I did an internship with the Inspections Division of the New York City Police Department. I helped design and collect preliminary data for an evaluation of the Community Patrol Officer Program. Prior to coming to BJS, I worked as an information specialist at a clearinghouse for drugs and crime statistics, research, and information.

In addition to designing surveys and analyzing statistics, a social science statistician writes reports summarizing the information collected in our surveys, and we respond to information requests from policymakers, criminal justice practitioners, reporters, researchers, students, and inmates. We must be aware of current issues in the corrections and legal fields that will allow us to better understand the information that we collect and discuss the finer points in our findings.

Getting respondents to submit information to us in a timely fashion can be challenging. We are working with the staff of criminal justice agencies whose primary purpose is running a prison or jail, so our surveys are not their top priority. However, I find it very satisfying to provide accurate, reliable information that is often unavailable from any other source.

Tracy Snell

Does the job of social science statistician appeal to you? Why or why not?

of burglary than do the UCR. Still, the UCR count more of some kinds of offenses (such as assault) and count them differently. For example, the UCR count each report of a domestic assault at the same address separately; the NCVS count the repeated assaults as one victimization. The UCR count crimes reported by people and businesses that the NCVS do not reach. The NCVS does not count crimes against children aged 11 or younger or crimes against persons who are homeless or live in institutions, such as nursing homes and correctional institutions, or on military bases, as do the UCR.[58] Unlike the UCR, the NCVS rely on random samplings of victims and their memories of things that may have happened months ago, both of which are subject to some degree of error. Other problems with the NCVS are interviewers who may be biased or who may cheat, and respondents who may lie or exaggerate, or may respond without understanding the questions.

Differences in the data sources help explain the differences in the trends indicated by the NCVS and the UCR. For example, in 2018, the NCVS rate per 1,000 persons age 12 or older for serious violent crime was 4.3, and about 16% higher than the UCR rate of 3.7 per 1,000 residents. Rates for specific serious violent crimes in 2018 varied similarly. For example, the NCVS rate for rape was 0.7, while the UCR rate was 1.4 (based on the revised definition). In the case of robbery, the NCVS rate was 1.3, while the UCR rate was 2.8. Finally, the NCVS rate for aggravated assault was 2.3, while the UCR rate was 8.1. As for property crimes, the overall NCVS rate per 1,000 households for property crime in 2018 was 36.9, while the comparable UCR rate per 1,000 residents was 22.0. For the property crime of burglary, the NCVS rate was 6.6 compared to the UCR rate of 3.8. The NCVS rate for motor vehicle theft was 3.4, while the UCR rate was 2.3.[59]

Longer-term trends in the overall violent crime victimization rates of the NCVS and the UCR show that the two measures have nearly merged. So, for example, from 1993 through 2018, the rate of violent crime victimization reported in the NCVS decreased 71%, from

79.8 to 23.2 victimizations per 1,000 persons age 12 or older. From 1993 through 2018, the UCR violent victimization rate also decreased 71%, from 33.8 to 9.9 victimizations reported to police per 1,000 persons age 12 or older. The differences in the two measures probably can be attributed to the differences in what the two indexes measure.[60]

Self-Report Crime Surveys

self-report crime surveys Surveys in which subjects are asked whether they have committed crimes.

Whereas other tallies of crime rely on summary police reports, incident-based reports, or victim interviews, **self-report crime surveys** ask selected subjects whether they have committed crimes. Self-report crime surveys, like all crime measures, are indexes of crime; they are not accurate measures of the true amount of crime. To date, most self-report crime surveys conducted in the United States have been administered to schoolchildren, especially high school students. Some examples of such nationwide self-report crime survey efforts are the National Youth Survey, begun in 1975, and the effort to ascertain and gauge fluctuations in the levels of smoking, drinking, and illicit drug use among secondary school students, begun by the National Institute on Drug Abuse in 1975.

Earlier self-report crime surveys of adults interestingly enough found an enormous amount of hidden crime in the United States. Those self-report crime surveys indicated that more than 90% of all Americans had committed crimes for which they could have been found guilty and imprisoned.[61] This is not to say that all Americans are murderers, thieves, or rapists, for they are not, but only that crime serious enough to warrant an individual's imprisonment is more widespread among the U.S. population than many people might think or imagine. The most commonly reported offenses in self-report crime surveys are larceny, indecency, and tax evasion.[62] It is unlikely that the pervasiveness of crime in the population has lessened significantly since the earlier self-report crime surveys were conducted.

myth

Criminal activity is concentrated among certain groups of people.

fact

Early self-report crime surveys of adults found an enormous amount of hidden crime in the United States. They found that more than 90% of all Americans had committed crimes for which they could have been imprisoned.

One lesson that can be learned from the aforementioned survey findings is that people's criminality is better described as a continuum—that is, as having committed more crime or less crime, rather than simply being described as criminal or noncriminal. In society, there are probably few angels—that is, people who have never committed a crime. Likewise, there are probably few criminals whose whole lives are totally oriented toward the commission of crimes. Most people have committed a crime at some point in their lives, and some have committed crimes repeatedly. It probably makes more sense for us, then, to talk about relative degrees of criminality rather than to talk about all-encompassing criminality or its absence.

One of the criticisms of the National Youth Survey, and a problem with many self-report crime surveys, is that it asks about less serious offenses, such as cutting classes, disobeying parents, and stealing items worth less than $5, while omitting questions about serious crimes, such as robbery, burglary, and sexual assault. Self-report crime surveys also suffer from all the problems of other surveys, problems that were described in the last subsection—they produce results different from other surveys, and they are not an accurate measure of crime.

THINKING CRITICALLY

1. Of the various methods of measuring crime presented in this section, which one do you think is the most accurate? Why? Which one do you think is the least accurate? Why?
2. Do you think there are ways to get more victims of crime to report criminal incidents? If so, what would you suggest?

Costs of Crime

According to data from the Bureau of Justice Statistics and the NCVS, in 2014, the total economic loss to victims of crime in the United States was $12.3 billion, down about 18.5% from $15.1 billion in 2012.[63] Eighty-seven percent of the 2014 total economic loss is the result of property crime, and 13% is due to violent crime. The total includes costs from medical care, time lost from work, property loss and damage, costs to repair or replace property, and time spent dealing with the criminal justice system.[64] It does not include the cost of the criminal justice process (described in Chapter 1), increased insurance premiums, security devices bought for protection, losses to businesses (which are substantial), or corporate crime. Regarding corporate crimes, although no official measure of corporate crimes exists, estimates suggest that the cost of corporate crimes to victims far exceeds the cost of conventional crime, by hundreds of billions or, according to some estimates, a trillion dollars.[65] The crime of price-fixing, alone, in which competing companies explicitly agree to keep prices artificially high to maximize profits, is estimated to cost consumers about $60 billion a year.[66]

More recently, the Bureau of Justice Statistics estimated the total economic loss to crime victims in 2017. The reported loss to approximately 19 million victims was $8.5 billion, a decrease of nearly 31% from the $12.3 billion loss in 2014. However, the data from 2017 is not comparable to the data from 2014, or any previous years, because BJS statisticians decided to "top code" the data to meet the more rigorous Census Bureau's data quality standards. In this case, "top coding" involved the elimination (censoring) of the largest economic losses (those above an arbitrarily set upper limit) to preserve the anonymity of respondents.[67] All that said, the distribution of economic loss between violent and property crime should not be effected too adversely by the "top-coding" procedure. So, similar to the results from 2014, 91% of the 2017 total economic loss was the result of property crime, and 9% was due to violent crime.

The NCVS probably provides the best estimates of the annual costs of conventional crime. Those cost estimates, however, are deficient in two additional ways. First, they include only a limited number of personal and property crimes. And, as noted previously, they do not include the cost of the criminal justice process, increased insurance premiums, security devices bought for protection, losses to businesses, or corporate crime. Second, they report estimates only of relatively short-term and tangible costs. They do not include long-term and intangible costs associated with pain, suffering, and reduced quality of life.

To compensate for the deficiencies of the NCVS, a 1996 study was sponsored by the National Institute of Justice.[68] In addition to the more standard cost estimates in the NCVS, the study estimated long-term costs as well as the intangible costs of pain, suffering, and reduced quality of life. Intangible costs were calculated in a number of ways. For example, the costs of pain, suffering, and reduced quality of life for nonfatal injuries were estimated by analyzing jury awards to crime and burn victims. Only the portion of the jury award intended to compensate the victim for pain, suffering, and reduced quality of life was used; punitive damages were excluded from the estimates.

Furthermore, although the study included only "street crimes" and "domestic crimes," it expanded on the crime categories and information included in the NCVS by (1) including crimes committed against people under the age of 12, (2) using better information on domestic violence and sexual assault, (3) more fully accounting for repeat victimizations, and (4) including child abuse and drunk driving. Excluded from the study were crimes committed against business and government, personal fraud, white-collar crime, child neglect, and most "victimless" crimes, including drug offenses. The study provided a rough idea of the degree to which the NCVS underestimates the costs of crime.

The study estimated that the annual tangible costs of personal and property crime—including medical costs, lost earnings, and public program costs related to victim assistance—were $105 billion, or more than $400 per U.S. resident. When the intangible costs of pain, suffering, and reduced quality of life were added, the annual cost increased to an estimated $450 billion, or about $1,800 per U.S. resident.

Violent crime (including drunk driving and arson) accounted for $426 billion of the total, while property crime accounted for the remaining $24 billion. The study found that violence against children accounted for more than 20% of all tangible costs and more than 35% of all costs (including pain, suffering, and reduced quality of life).

Of the crimes included in the study, rape had the highest annual victim costs, at $127 billion a year (excluding child sexual abuse). Second was assault, with victim costs of $93 billion a year, followed by murder (excluding arson and drunk driving deaths), at $71 billion annually. Drunk driving (including fatalities) was next at $61 billion a year, and child abuse was estimated to cost $56 billion annually.

In 2017, the U.S. Government Accountability Office (GAO) was asked to revisit the annual costs of crime issue. Among its efforts, the GAO reviewed 27 studies published since and including the NIJ-sponsored study in 1996. It found annual cost estimates of crime that varied widely from a low of $450 billion in the 1996 study ($690 billion in 2016 dollars) to a high of $3.41 trillion (in 2016 dollars) in a 2012 study.

Researchers emphasized the challenges of estimating the annual costs of crime. The most significant obstacle is obtaining reliable data to monetize costs, especially intangible costs. Arguably, the most important contribution of the 2017 GAO report is the specification of the elements that ought to be included in any cost studies of crime (see Table 2.5). The elements in Table 2.5 were derived from the elements used in the 27 studies the GAO reviewed, as well as elements the researchers suggested ought to be used in future studies.

TABLE 2.5 Tangible and Intangible Costs of Crime

	Costs in Anticipation of Crime	Costs as a Direct Consequence of Crime	Costs in Response to Crime
Tangible	Expenditures to reduce likelihood of victimization (e.g., purchasing security systems and fencing) Crime prevention programs, government and nongovernment (e.g., education programs) Community expenditures (e.g., business investment) Insurance administration	Victim's lost wages for unpaid workdays Employer's lost productivity from victim missing workdays Victim's medical and mental health care treatment to recover from victimization Victim's property loss Cost to society for recovery and/or reimbursement of lost property Funeral and burial expenses of deceased victim	Police protection and investigative costs Costs to maintain courts to prosecute and try offenders Costs to defend accused offenders in court Victim service (i.e., compensation and support programs and volunteers' time) Non-criminal justice programs (e.g., hotlines, public service announcements, and community treatment programs) Incarceration costs Offender lost wages while incarcerated, and reduced future employment prospects Society's future lost tax revenue and productivity from offender as a result of incarceration
Intangible	Avoidance behavior (e.g., avoiding people and places) Fear of crime	Victim's pain, suffering, and lost quality of life Family of victim's loss of affection and enjoyment Second-generation costs (i.e., future social costs associated with crimes committed by earlier victims) Opportunity costs of offender's time spent in illegal activity, including incarceration, instead of working	Psychological cost to offender's family and loss of affection Overdeterrence cost (e.g., innocent individuals accused of offense, restriction of community's legitimate activities) Cost of developing, reviewing, and maintaining constitutional protections to avoid false accusations Cost of increasing detection rate to equal treatment of offenders

Source: United States Government Accountability Office, Report to Congressional Requester, "Costs of Crime: Experts Report Challenges Estimating Costs and Suggest Improvements to Better Inform Policy Decisions," GAO-17-732, September 2017, p. 7, accessed March 3, 2019, https://www.gao.gov/assets/690/687353.pdf.

As shown in Table 2.5, one of the intangible costs of crime is fear of crime, which is particularly challenging to measure. Yet, fear of crime makes people prisoners in their own homes, divides people, and destroys communities.

THINKING CRITICALLY

1. What would you estimate is the cost of crimes that go unreported?
2. Why are the highest costs of crimes the intangible costs?

Fear of Crime

A by-product of crime, beyond actual physical or material loss, is fear. For many crime victims, it is the most burdensome and lasting consequence of their victimization. However, fear of crime is also contagious. One does not have to be a victim of violent crime to be fearful of violent crime. In fact, research shows that people who have heard about other people's victimizations are nearly as fearful as the people who have been victimized themselves.[69]

What People Fear

In 2018, Gallup pollsters asked Americans about their fear of crime. They found that:

- 96% worried about crime and violence (51% a great deal, 27% a fair amount, and 18% only a little).
- 71% worried about having their personal, credit card, or financial information stolen by computer hackers.

- 69% worried about having the credit card information they have used at stores stolen by computer hackers.
- 67% worried about being a victim of identity theft.
- 64% worried about having their email, passwords, or electronic records hacked into.
- 40% worried about their home being burglarized when they were not there.
- 37% worried about having their car stolen or broken into.
- 33% were afraid to walk alone at night within a mile of where they lived.
- 32% worried about having a school-aged child of theirs physically harmed while attending school.
- 25% worried about getting mugged.
- 24% worried about being a victim of terrorism.
- 22% worried about their home being burglarized when they were there.
- 22% worried about being attacked while driving their car.
- 22% worried about being the victim of a hate crime.
- 20% worried about being sexually assaulted.
- 17% worried about getting murdered.
- 7% worried about being assaulted or killed by a coworker or other employee where they worked.[70]

Gallup Crime Polls

The Gallup Organization conducts public opinion polls on a variety of topics that affect Americans. You can learn more about the public's view of crime issues by visiting the Gallup website, https://news.gallup.com/poll/1603/crime.aspx.

What connection, if any, is there between public opinion toward crime and crime statistics?

Who Fears Crime

Fear of criminal victimization is neither evenly distributed across the population nor commensurate with the statistical probability of being the victim of crime. Following are the percentages of respondents in each demographic group who replied in 2016, that they personally worried "a great deal" about crime and violence.[71]

- *Gender*–Females (56%) were worried more than males (49%).
- *Race/Ethnicity*–Nonwhites (68%) were worried a great deal more than whites (46%).
- *Age*–People 55+ years old (58%) were worried more than people in other age categories (18 to 34 = 52% and 35 to 54 = 48%).
- *Education*–People who graduated from high school or who had less education (70%) were worried a great deal more than people who had some college (52%) or who graduated from college (32%).
- *Politics*–Republicans (53%) and Independents (53%) were slightly more worried than were Democrats (52%).
- *Income*–People whose annual income was less than $30,000 (66%) worried more than people whose income was $30,000 to $74,999 (57%) and a great deal more than people whose income was $75,000+ (36%).

myth

The people most fearful of crime are the people most vulnerable to crime.

fact

The people most fearful of crime are not necessarily members of groups with the highest rates of victimization. For example, the demographic group most afraid of crime—elderly women—is the least likely to be victimized.

How People Respond and Should Respond to a Fear of Crime

Fear of crime has many detrimental consequences. It makes people feel vulnerable and isolated, it reduces a person's general sense of well-being, it motivates people to buy safety devices with money that otherwise could be used to improve their quality of life, and it contributes to neighborhood decline and the crime problem. As criminologist Wesley Skogan explains:

> Fear . . . can work in conjunction with other factors to stimulate more rapid neighborhood decline. Together, the spread of fear and other local problems provide a form of positive feedback that can further increase levels of crime. These feedback processes include (1) physical and psychological withdrawal from community life; (2) a weakening of the informal social control processes that inhibit crime and disorder; (3) a decline in the organizational life and mobilization capacity of the neighborhood; (4) deteriorating business conditions; (5) the importation and domestic production of delinquency and deviance; and (6) further dramatic changes in the composition of the population. At the end lies a stage characterized by demographic collapse.[72]

In response to their fear of crime, Americans employ a number of different strategies. Among them are the following: (1) avoiding going to places or neighborhoods that they might otherwise have wanted to visit, (2) keeping a dog for protection, (3) having a burglar alarm

Safest and Most Dangerous Cities in America

In 2019, the safest city in the United States was McKinney, Texas, followed by Sunnyvale, California (tie); Olathe, Kansas (tie); Santa Clara, California; and Orange, California. The most dangerous city was Detroit, Michigan, followed by St. Louis, Missouri; Oakland, California; Memphis, Tennessee; and Birmingham, Alabama.

Sources: "Safest Cities in America - 2019 Edition," accessed February 1, 2020, finance.yahoo.com › news › safest-cities-america-2019-edition-1200005. . .; "The 10 Most Dangerous U.S. Cities," *Forbes*, accessed February 1, 2020, forbes.com.

Syracuse Newspapers/Gary Walts/The Image Works

Many people take extraordinary measures to protect themselves from crime. *Why are people so fearful of crime? Should they be? Why or why not?*

installed in their home, (4) buying a gun for self-protection or home protection, (5) carrying mace or pepper spray, (6) carrying a knife for defense, and (7) carrying a gun for defense.[73]

Crime experts list the following strategies that people should use to avoid becoming the victim of crime:

- **Trust yourself**. Many times, your senses give clues that something is threatening. Trust when something does not seem right.
- **Be aware of your surroundings**. No matter how safe you think an area might be, leaving the front door open, your valuables in your car, and your purse on top of your office desk are not good ideas. Neither is flaunting expensive jewelry and other belongings. Do not tempt would-be offenders. Do not walk through dark, isolated alleys, fields, or parking lots. Do not give would-be attackers opportunity and anonymity.
- **Pay attention to the people around you**. This advice is part of both listening to your instincts and being aware of your surroundings. You often can sense people's intentions just by the way that they look at you. Heed warning signs even when you are with people you know and trust.
- **Act confident and focused**. Just as you can sense people's feelings, others can sense yours as well. Would-be predators look for people who are meek, mild, weak, unfocused, and distracted. Would-be predators are looking for easy prey. Present yourself in an assertive manner. When walking down the street, make eye contact with people who look at you.
- **Understand that alcohol or drugs can cloud judgment**. Certain substances can dull your senses and slow down your reaction time to danger. They also can lower other people's inhibitions and make them more aggressive or belligerent.
- **Have an escape plan**. Wherever you are, and wherever you are going, know the layout of the place and visualize an escape route. This is not paranoid, it is being cautious. At home, knowing where the power switch is and your way in the dark can provide an advantage over intruders. Outside, knowing the layout of the town–where dangerous areas are, and where populated streets and venues are–can help you both prevent and escape an attacker. At work, knowing the building structure can give you an idea of where to flee.
- **If you cannot escape, resist**. You do not have to win the fight against an attacker. You just need to be able to survive it. By fighting back, you have more chance of injury, but you also have a better chance of survival.
- **React quickly to danger**. Response time is critical. Since the offender is counting on a surprise ambush to carry out his crime, you need to use the same element of surprise to escape or counterattack. This could mean running toward lights and people, or it could mean screaming or making noise with whatever you have to get other people's attention. Remember, escape first; resist second. If you must resist, use your energy effectively. Strike only vital targets, which are areas of the body where you can inflict the most pain and damage. This may disable the offender and let you get away. Some vital targets include the top center of the skull, base of skull, spine, eyes, temples, ears, windpipe, knees, and insteps.[74]

THINKING CRITICALLY

1. What steps could be taken, if any, to reduce people's overall fear of crime?
2. Why do you think there is no correlation between those who fear crime the most and the most likely victims of crime?
3. What factors contribute to people's fear of crime?
4. What are the costs, if any, of a fear of crime?

Victims of Crime

Findings from the 2018 NCVS reveal that in 2018, an estimated total of about 19 million crimes were threatened, attempted, or completed against U.S. residents age 12 or older–down from 21.2 million crimes in 2016, and 20.3 million crimes in 2010. (Note that for the first time, the 2011 NCVS employed a new measure of "series" victimizations to better capture repeated victimizations to the same person or household. All NCVS data from 1993 onward have been updated to reflect this change. The new measure is used in the reporting of raw numbers in this section; the old measure is used in the reporting of rates in the next section to make them comparable to rates from 1973 onward.) Approximately 6.4 million of the crimes in 2018 were violent crimes (rape, sexual assault, robbery, aggravated assault, and simple assault), up about 14% from 5.6 million in 2017, and up approximately 31% from the 4.9 million in 2010; about 13.5 million in 2018 were property crimes (burglary, motor vehicle theft, and other thefts), up 1.5% from the 13.3 million in 2017, but down about 12% from the 15.4 million in 2010.[75]

Victimization Trends

According to 2018 NCVS data, the violent and property crime victimization rates were 23.2 and 108.2 per 1,000 persons age 12 or older, respectively. Despite small, statistically insignificant increases in 2016, 2017, and 2018, the violent crime rate was 71% lower than the 1993 rate, a decline that continued a trend that began in 1994 (see Figure 2.6, top graph).[76] The 2018 property crime rate also decreased 71% from the 1993 rate, a decline that continued a trend that began in 1974 (see Figure 2.6, bottom graph).[77] Figure 2.6 shows that the decline in the violent crime rate is primarily a result of a decline in simple assaults, while the decrease in the property crime rate is almost entirely attributable to a decrease in thefts.

FIGURE 2.6 Trends in Violent and Property Crime Rates

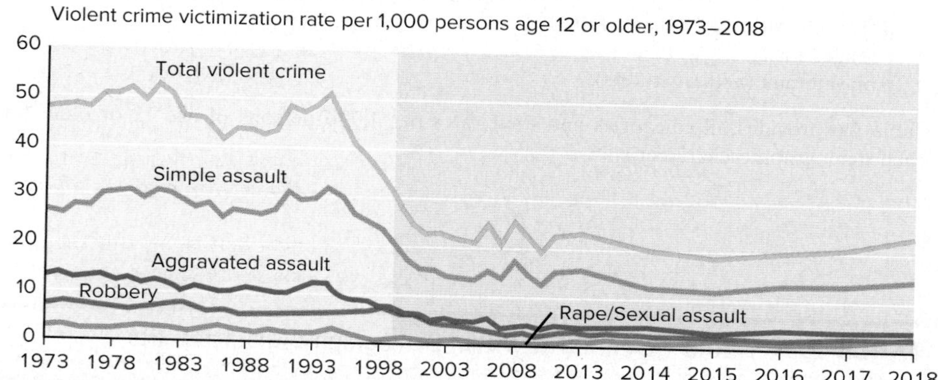

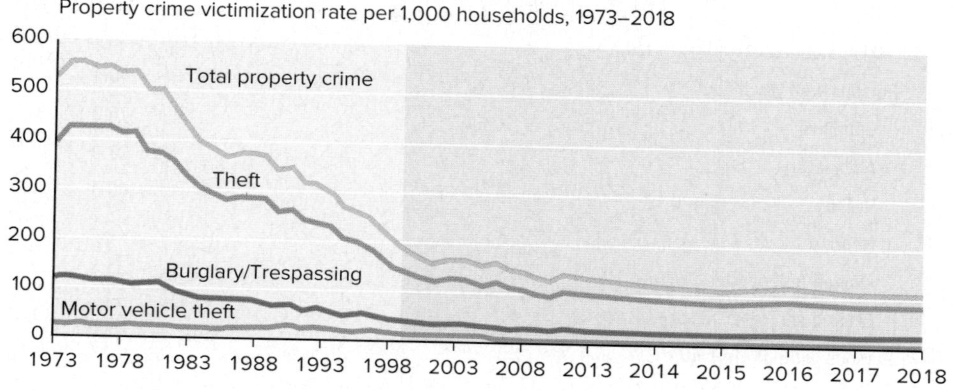

Note: From 1973 through June 1992, data were collected under the National Crime Survey (NCS) and made comparable to data collected under the redesigned methods of the NCVS that began in July 1992.

Sources: Michael R. Rand, "Criminal Victimization, 2007," in *National Crime Victimization Survey*, U.S. Department of Justice, Bureau of Justice Statistics (Washington, DC: U.S. Government Printing Office, December 2008), 2–3, www.ojp.usdoj.gov/bjs/pub/pdf/cv07.pdf; Jennifer L. Truman and Michael R. Rand, "Criminal Victimization, 2009," in *Bureau of Justice Statistics Bulletin*, National Crime Victimization Survey, U.S. Department of Justice (Washington, DC: U.S. Government Printing Office, October 2010), 2, http://bjs.ojp.usdoj.gov/content/pub/pdf /CV09.pdf.; Jennifer L. Truman and Rachel E. Morgan, "Criminal Victimization, 2015," accessed November 4, 2016, www.bjs.gov/content/pub/pdf/cv15.pdf; Rachel E. Morgan and Jennifer L. Truman, "Criminal Victimization, 2017," 4, Table 1, 6, Table 3, accessed February 27, 2019, https://www.bjs.gov/content/pub/pdf/cv17.pdf; Rachel E. Morgan and Barbara A. Oudekerk, "Criminal Victimization, 2018," 4, Table 1, 5, Table 3, accessed February 1, 2020, https://www.bjs.gov/content/pub/pdf/cv18.pdf.

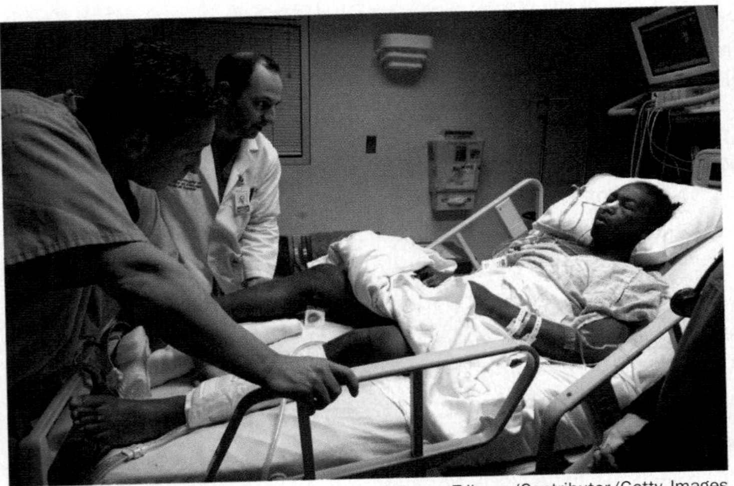

Chicago Tribune/Contributor/Getty Images

Young, economically disadvantaged black males are the most likely victims of personal violent crime. *Why, and what can be done about it?*

Who the Victims Are

Although each year millions of people are victimized by crime, victimization—like the fear of crime—is not spread evenly throughout the population. Certain types of people are much more likely to be crime victims. The types of people that were most vulnerable to victimization in the past continue to be the most vulnerable.

Based on 2018 NCVS data (and the new measure of "series" victimizations), the demographic groups with the highest rates of violent crime victimization per 1,000 persons age 12 or older (from highest to lowest overall rate) were:[78]

- Persons with cognitive disabilities (disability status)
- Separated persons (marital status)
- Native Hawaiians and Other Pacific Islanders, American Indians and Alaska Natives, and persons of two or more races (race/ethnicity)
- Persons with household incomes of less than $25,000 (household income)
- Persons age 18–24 (age)
- Born U.S. citizens (citizen status)
- Females (sex)
- Non-veterans (veteran status)

Table 2.6 provides all categories and their rates per 1,000 persons of age 12 or older for each demographic characteristic.

TABLE 2.6 2018 Violent Crime Victim Demographic Characteristics

Victim Demographic Characteristic	Rate per 1,000
Total	23.2
Disability status (2017)	
Persons with disabilities	40.4
Cognitive	76.0
Ambulatory	28.9
Vision	43.5
Hearing	23.2
Limited independent living	31.8
Limited self-care	34.4
Persons without disabilities	17.7
Marital status	
Never married	33.5
Married	12.1
Widowed	12.5
Divorced	39.1
Separated	58.2

TABLE 2.6 *(Concluded)*

Victim Demographic Characteristic	Rate per 1,000
Race/ethnicity	
White (non-Hispanic)	24.7
Black (non-Hispanic)	20.4
Hispanic	18.6
Asian	16.2
Native Hawaiians and Other Pacific Islanders, American Indians and Alaska Natives, and persons of two or more races	49.2
Household income	
Less than $25,000	40.8
$25,000–$49,999	23.5
$50,000–$99,999	16.5
$100,000–$199,999	19.2
$200,000 or more	16.3
Age	
12–17	34.2
18–24	35.9
25–34	31.8
35–49	25.2
50–64	18.3
65 or older	6.5
Citizenship status	
Born U.S. citizen	25.1
Naturalized U.S. citizen	11.6
Non-U.S. citizen	12.5
Sex	
Male	22.1
Female	24.3
Veteran status	
Veteran	20.7
Non-veteran	22.2

Sources: Adapted from Tables 9 and 10 in Rachel E. Morgan and Barbara A. Oudekerk, "Criminal Victimization, 2018," *Bureau of Justice Statistics Bulletin*, U.S. Department of Justice, September 2019, accessed October 17, 2019, https://www.bjs.gov/content/pub/pdf/cv18.pdf; Table 9 in Rachel E. Morgan and Jennifer L. Truman, "Criminal Victimization, 2017," *Bureau of Justice Statistics Bulletin*, U.S. Department of Justice, December 2018, accessed October 17, 2019, https://www.bjs.gov/content/pub/pdf/cv17.pdf.

THINKING CRITICALLY

1. Why do you think violent and property crime victimization rates have steadily declined since 1993 (see Figure 2.6)?
2. Why do you think that certain types of people are more likely than others to become crime victims?

SUMMARY

1. Distinguish between a social definition and a legal definition of crime, and summarize the problems with each.

A typical social definition of crime is behavior that violates the norms or social mores of society or, more simply, antisocial behavior. A typical legal definition of crime is an intentional violation of the criminal law or penal code, committed without defense or excuse and penalized by the state. There are problems with both definitions. Problems with the social definition are that (1) there are no uniform norms or social mores of behavior accepted by all of society, (2) norms or social mores of behavior are subject to interpretation, and (3) norms or social mores change from time to time and from place to place. Problems with the legal definition of crime are (1) overcriminalization, (2) nonenforcement, and (3) undercriminalization.

2. List the technical and ideal elements of a crime.

The technical and ideal elements of a crime are (1) harm, (2) legality, (3) *actus reus,* (4) *mens rea,* (5) causation, (6) concurrence, and (7) punishment.

3. Identify some of the legal defenses or legal excuses for criminal responsibility.

Among the legal defenses or legal excuses for criminal responsibility in the United States are these: the defendant (1) acted under duress, (2) was underage, (3) was insane, (4) acted in self-defense or in defense of a third party, (5) was entrapped, or (6) acted out of necessity.

4. Explain why crime and delinquency statistics are unreliable.

Among the reasons why crime and delinquency statistics are unreliable are the following: (1) some behaviors are labeled crime by one observer but not by another, (2) a large proportion of crimes are undetected, (3) not all crimes are reported to the police, and (4) not all crimes that are reported are officially recorded by the police.

5. Identify the two major sources of crime statistics in the United States.

The two major sources of crime statistics in the United States are the uniform crime reports (UCR) compiled by the FBI and the national crime victimization surveys (NCVS) compiled by the Bureau of Justice Statistics.

6. Describe the principal finding of the national crime victimization surveys.

The national crime victimization surveys (NCVS) produce different results from the FBI's uniform crime reports (UCR). The NCVS generally show more crimes being committed than the UCR do. The NCVS count more of some kinds of crimes and count them differently.

7. Summarize the general finding of self-report crime surveys.

Self-report crime surveys show that the amount of hidden crime in the United States is enormous; more than 90% of all Americans have committed crimes for which they could have been imprisoned.

8. Identify the costs of crime.

According to data from the NCVS, in 2014, the total economic loss to victims of crime in the United States was $12.3 billion. This figure does not include the tangible costs of the criminal justice process; security devices bought for protection; losses to businesses; losses from corporate crimes; or the intangible costs of pain, suffering, and reduced quality of life. When all costs are totaled, it is estimated that crime costs between $690 billion and $3.41 trillion (in 2016 dollars).

9. Describe the characteristics of people most worried about crime.

In general, those most worried about crime are female, nonwhite, people age 55+, people who graduated from high school or had less education, Republicans and Independents (but only slightly more than Democrats), and people whose annual income is less than $30,000.

10. List the characteristics of people who are the most likely to be victims of violent crime.

According to data from the 2018 NCVS, the most likely victims of violent crime are persons with cognitive disabilities (disability status); separated persons (marital status); Native Hawaiians and Other Pacific Islanders, American Indians and Alaska Natives, and persons of two or more races (race/ethnicity); persons with household incomes of less than $25,000 (household income); persons age 18–24 (age); born U.S. citizens (citizen status), females (sex); and non-veterans (veteran status).

KEY TERMS

norm or social more 27
legal definition of crime 27
overcriminalization 28
nonenforcement 28
undercriminalization 28
harm 29
legality 29
ex post facto law 29
actus reus 29
mens rea 30
negligence 30
duress 30
juvenile delinquency 30
insanity 30
entrapment 33
necessity defense 33
mala in se 35
mala prohibita 35
dark figure of crime 36
crime index 36
offenses known to the police 36
crime rate 38
uniform crime reports (UCR) 40
eight index crimes 41
status offense 42
crime index offenses cleared 42
national crime victimization surveys (NCVS) 46
self-report crime surveys 48

REVIEW QUESTIONS

1. What is a *norm* or *social more*?
2. What is the *harm* of crime?
3. What is the *mens rea* of crime?
4. What is *legal infancy* or *legal nonage*?
5. What is *juvenile delinquency*?
6. What is *insanity*?
7. Does the availability of the insanity defense allow a large number of dangerous criminals to go free?
8. What is the difference between a *felony* and a *misdemeanor*?
9. What percentages of all crime are violent crime, property crime, and public order crime?
10. Why aren't any records or indexes of crime reliable measures of the true amount of crime?
11. What are "offenses known to the police," and what are advantages and disadvantages of using it?
12. What is a *crime rate,* and why are crime rates used?
13. What is a *status offense*?
14. When the media report that crime has increased or decreased from one year to the next, to what are they referring?
15. What is the NIBRS?
16. What are some of the detrimental consequences of a fear of crime?
17. According to crime experts, what strategies should people use to avoid becoming a crime victim?
18. What long-term trends in violent and property crime victimization rates are revealed by the national crime victimization surveys?

IN THE FIELD

1. **Crime Victimization Survey** As an individual project or a group project, construct and conduct a crime victimization survey. (Use the national crime victimization survey as a model.) Either orally or in writing, present and discuss the results. Be sure to discuss problems encountered in constructing the survey and problems with the accuracy and trustworthiness of responses.
2. **Self-Report Crime Survey** As an individual project or a group project, construct and conduct a self-report crime survey. (Make sure you tell respondents not to put their names or any other identifying information on the survey. Also, be sure to include both serious and less serious crimes.) Either orally or in writing, present and discuss the results. Be sure to discuss problems encountered in constructing the survey and problems with the accuracy and trustworthiness of responses.

ON THE WEB

1. **Crime Statistics** Examine the latest edition of the FBI's uniform crime reports at www.fbi.gov/about-us/cjis/ucr. Then choose one of the former eight index crimes, and search the site for statistics about that crime. Write a brief report summarizing the results of your research.
2. **Victimization Data** Examine the latest edition of the NCVS at the Bureau of Justice Statistics site at https://www.bjs.gov/index.cfm?ty=dcdetail&iid=245. Scroll to Publications & Products and select the latest edition of the NCVS. Open the pdf. Compare the latest victimization data with the data presented in your textbook. Write a brief report summarizing any changes.

CRITICAL THINKING EXERCISES

THE TWINKIE DEFENSE

1. Dan White had been elected as a city supervisor of San Francisco. White resigned on November 10, 1978. Four days later, he changed his mind and asked Mayor George Moscone to reappoint him to his former position. The mayor refused. On the morning of November 27, White confronted the mayor and demanded to be reappointed. When the mayor refused, White shot the mayor five times. White reloaded the gun, walked across the hall to City Supervisor Harvey Milk's office, and shot him four times. Milk was a leader in San Francisco's gay community with whom White frequently clashed. After shooting Moscone and Milk, White fled but shortly thereafter went to the police and confessed. He was charged with first-degree murder. During his trial, several psychologists called by the defense testified that White's behavior was the result of long-term depression exacerbated by a craving for junk food. They testified that White's judgment and ability to control his behavior were

altered by the huge amount of sugar he had consumed the night before the killings. The so-called Twinkie defense worked. White was convicted of voluntary manslaughter instead of first-degree murder, and on May 21, 1979, he was sentenced to a prison term of 5 years to 7 years, 8 months. White was released from prison after serving 5 years, 1 month, and 9 days of his sentence. He committed suicide on October 21, 1985.

a. What is the legal rationale for accepted legal defenses against or excuses from criminal responsibility? Do you agree with the rationale?
b. Should all legal defenses or excuses be abolished? Why or why not?

THE SLEEPWALKING DEFENSE

2. Twenty-eight-year-old Justin Cox told a Florida jury that he was sleepwalking when he touched a 12-year-old girl's upper thigh at a friend's house one night in 2006. He stated that he would never do anything like that if he was awake, and he apologized to the girl when he awoke with her under him. Following a night of drinking, Cox ended up at a friend's house and went to sleep alone in a bedroom. The victim, who was spending the night with her girlfriend, had fallen asleep on the living room couch. At some point, Cox wandered into the living room, sat down on the couch, put the victim's legs over his lap, and fondled her. When she moved and woke him up, Cox said he was "freaked" about what was happening and apologized to the girl. She started to cry and asked him to leave the house, which he did. Until investigators informed him, Cox stated that he did not know he had done more than put his hand on her leg. Cox was arrested for lewd and lascivious acts, placed on house arrest, and made to wear an ankle monitor.

Cox's sleepwalking defense was unusual, but it had been used successfully in previous cases. However, juries are usually skeptical of it. Cox's attorney said, "It's almost like an insanity defense where you are not in your reality and therefore you are not conscious." Cox had a family history of sleepwalking and had sleepwalked since he was a child. After evaluating Cox, a doctor testified for the defense that at the time of the incident, Cox was sleepwalking. The state's sleepwalking expert said it was possible Cox was sleepwalking at the time of the crime, but not a "high medical probability." The jury believed the defense's doctor and, on May 8, 2008, acquitted Cox of the charges.

a. In this case, do you think the sleepwalking defense was reasonable? Why or why not?
b. Do you think the jury made the right decision? Why or why not?
c. Do you think it was fair to place Cox on house arrest and make him wear an ankle monitor for nearly 2 years? Why or why not?

FIGHTING ILLEGAL DRUGS

3. Activists are advocating a different strategy in the fight against illegal drugs. In agreement with the American Medical Association, they maintain that drug addiction is a disease. They believe it should be treated as a public health problem, not as a law enforcement problem. They propose a change in strategy from prohibition to decriminalization or legalization. According to these activists, decriminalization or legalization would do the following:

- Reduce profits from drug trafficking, crime caused by the need to support drug habits, and the prison population.
- Allow money used for incarceration to fund new models of drug control.
- Allow drugs such as marijuana to be regulated and taxed.
- Reduce the negative health consequences from using drugs of unknown potency and purity.
- Improve pain control options for certain medical conditions.
- Free physicians from fear of entrapment for prescribing certain drugs.
- Shift the focus of criminal justice to more serious crime problems.

a. One of the seven elements of a crime is harm. Do you think that illegal drug use is harmful? Explain.
b. If the use of addictive drugs is illegal, should addictive substances such as nicotine (in cigarettes) or caffeine (in coffee) also be made illegal? Explain.

NOTES

1. Loony Laws, accessed February 16, 2019, www.loonylaws.com/. The website notes: "These laws are culled from various print and online sources. There is a good chance that many of these have since been struck from the books, if they actually existed in the first place."
2. For additional examples and further discussion of this issue, see Robert M. Bohm, "Some Relationships That Arguably Should Be Criminal Although They Are Not: On the Political Economy of Crime," in *Political Crime in Contemporary America: A Critical Approach,* ed. K. D. Tunnell (New York: Garland, 1993), 3–29; Gregg Barak, ed., *Crimes by the Capitalist State: An Introduction to State Criminality* (New York: State University of New York Press, 1991); Gregg Barak and Robert M. Bohm, "The Crimes of the Homeless or the Crime of Homelessness? On the Dialectics of Criminalization, Decriminalization, and Victimization," *Contemporary Crises* 13 (1989): 275–88; James W. Coleman, *The Criminal Elite: Understanding White Collar Crime,* 6th ed. (New York: Worth, 2005); David O. Friedrichs, *Trusted Criminals: White Collar Crime in Contemporary Society*, 4th ed. (Belmont, CA: Wadsworth, 2010); Gary S. Green, *Occupational Crime,* 2nd ed. (Chicago: Nelson-Hall, 1996); David R. Simon, *Elite Deviance,* 9th ed. (Boston: Allyn & Bacon, 2007).
3. These elements and the discussion that follows are based on material from Edwin H. Sutherland and Donald R. Cressey, *Criminology,* 9th ed. (Philadelphia: J. B. Lippincott, 1974), 13–15.
4. *M'Naghten's Case,* 8 Eng. Rep. 718 (1843).
5. FindLaw, "The Insanity Defense Among the States," accessed December 29, 2010, criminal.findlaw.com/crimes/more-criminal-topics/insanity-defense/the-insanity-defense-among-the-states.html.

6. "18 U.S. Code sec. 7 - Insanity defense," www.law.cornell.edu/uscode/text/18/17 (accessed January 20, 2016).
7. Joe Lambe, "How Far Will 'Castle Doctrine' Defense Go?" *The Kansas City Star,* December 28, 2007, A1; Christopher Reinhart, "Castle Doctrine and Self-Defense," www.cga.ct.gov/2007/rpt/2007-R-0052.htm; Scott Sonner, "Nev. Tests 'Stand Your Ground' Laws," *The Orlando Sentinel*, May 27, 2015, A3.
8. National Conference of State Legislatures, "Making Sense of Sentencing: State Systems and Policies," June 2015. Accessed February 8, 2019, http://www.ncsl.org/documents/cj/sentencing.pdf.
9. Gallup, Crime, accessed February 16, 2019, https://news.gallup.com/poll/1603/crime.aspx (polls conducted October 1-10, 2018 and October 5-11, 2017).
10. Sutherland and Cressey, *Criminology*, 25.
11. Ibid. The remainder of the discussion in this section is based on material from the aforementioned source, 25-30.
12. Ibid.
13. Calculated from data in Rachel E. Morgan and Barbara A. Oudekerk, "Criminal Victimization, 2018," *Bureau of Justice Statistics Bulletin*, U.S. Department of Justice (Washington, DC: U.S. Government Printing Office, September 2019, NCJ 253043), 8, Table 5, accessed October 14, 2019, https://www.bjs.gov/content/pub/pdf/cv18.pdf.
14. Sutherland and Cressey, *Criminology*, 27-28.
15. D. Seidman and M. Couzens, "Getting the Crime Rate Down: Political Pressure and Crime Reporting," *Law and Human Behavior* 8 (1974), 327-42. See also L. DeFleur, "Biasing Influences on Drug Arrest Records: Implications for Deviance Research," *American Sociological Review* 40 (1975): 88-103; W. L. Selke and H. E. Pepinsky, "The Politics of Police Reporting in Indianapolis, 1948-1978," *Law and Human Behavior* 6 (1982): 327-42.
16. Ben Poston, Joel Rubin, and Anthony Pesce, "LAPD Underreported Serious Assaults, Skewing Crime Stats for 8 Years," *Los Angeles Times*, October 15, 2015. Accessed February 16, 2019, https://www.latimes.com/local/cityhall/la-me-crime-stats-20151015-story.html.
17. Brian Maass, "CBS4 Investigation: Hundreds of DPD Crime Reports Downgraded," *4 CBS Denver*, January 22, 2018. Accessed February 16, 2019, https://denver.cbslocal.com/2018/01/22/denver-police-crime-reports/.
18. Violet Ikonomova, "Putting Detroit's Improving Crime Picture into Context," *Detroit Metro Times*, January 5, 2018. Accessed February 17, 2019, https://www.metrotimes.com/news-hits/archives/2018/01/05/putting-detroits-improving-crime-picture-into-context.
19. Wesley Juhl, "Report Questions Violent Crime Statistics Released by Las Vegas Police," *Las Vegas Review-Journal*, January 24, 2017. Accessed February 17, 2019, https://www.reviewjournal.com/crime/report-questions-violent-crime-statistics-released-by-las-vegas-police/.
20. Matthias Gafni, "Pittsburg: Did Police Fabricate Crime Stats to Burnish City's Image," *East Bay Times*, April 26, 2016. Accessed February 17, 2019, https://www.eastbaytimes.com/2016/04/26/pittsburg-did-police-fabricate-crime-stats-to-burnish-citys-image/.
21. For New York City, see www.nytimes.com/2012/06/29/nyregion/new-york-police-department-manipulates-crime-reports-study-finds.html?_r=1. For Chicago, see www.economist.com/blogs/democracyinamerica/2014/05/crime-statistics-chicago; www.chicagomag.com/Chicago-Magazine/June-2014/Chicago-crime-statistics/. For Dallas, see www.dallasnews.com/news/crime/headlines/20091215-Dallas-undercount-of-violent-assaults-builds-744.ece. For Atlanta, see Wes Smith, "Report: Atlanta Hushed Crimes," *The Orlando Sentinel*, February 21, 2004, Al. For Philadelphia, see "Philly Underreports Crime," *The Orlando Sentinel*, September 15, 2000, A11.
22. Darrell Steffensmeir, "Is the Crime Rate Really Falling? An 'Aging' U.S. Population and Its Effect on the Nation's Crime Rate, 1980-1984," *Journal of Research in Crime and Delinquency* 24 (1987): 23-48.
23. Federal Bureau of Investigation, *2015 Crime in the United States*, accessed November 2, 2016, https://ucr.fbi.gov/crime-in-the-u.s/2015/crime-in-the-u.s.-2015/tables/table-1.
24. Calculated from data in Steffensmeir, "Is the Crime Rate Really Falling?"; Federal Bureau of Investigation, *Crime in the United States 2009*, accessed November 2, 2016, https://www2.fbi.gov/ucr/cius2009/offenses/violent_crime/index.html. Note that the rate for 2001 does include the murder and nonnegligent manslaughters that occurred as a result of the September 11 tragedy.
25. Calculated from data in *2018 Crime in the United States*, Table 1, accessed October 15, 2019, https://ucr.fbi.gov/crime-in-the-u.s/2018/crime-in-the-u.s.-2018/tables/table-1.
26. Federal Bureau of Investigation, *2018 Crime in the United States*, accessed October 15, 2019, https://ucr.fbi.gov/crime-in-the-u.s/2018/crime-in-the-u.s.-2018/home.
27. Ibid., https://www.fbi.gov/file-repository/ucr/about-the-ucr-program.pdf/view.
28. Ibid., https://ucr.fbi.gov/crime-in-the-u.s/2018/crime-in-the-u.s.-2018/tables/table-1.
29. Federal Bureau of Investigation, "Attorney General Eric Holder Announces Revisions to the Uniform Crime Report's Definition of Rape," U.S. Department of Justice (January 6, 2012), accessed July 3, 2012, www.fbi.gov/news/pressrel/press-releases/attorney-general-eric-holder-announces-revisions-to-the-uniform-crime-reports-definition-of-rape. Also see Federal Bureau of Investigation, Crime in the United States 2014, "Rape Addendum" accessed February 3, 2016, www.fbi.gov/about-us/cjis/ucr/crime-in-the-u.s/2014/crime-in-the-u.s.-2014/resource-pages/rape-addendum.
30. Federal Bureau of Investigation, "Human Trafficking in the Uniform Crime Reporting (UCR) Program," accessed February 3, 2016, www.fbi.gov/about-us/cjis/ucr/human-trafficking.
31. Federal Bureau of Investigation, *2014 Hate Crime Statistics*, "About Hate Crime Statistics," accessed January 20, 2016, www.fbi.gov/about-us/cjis/ucr/hate-crime/2014/resource-pages/about-hate-crime.
32. Calculated from data in Federal Bureau of Investigation, *2016 and 2017 Hate Crime Statistics*, accessed February 18, 2019, https://ucr.fbi.gov/hate-crime/2017/topic-pages/tables/participation.xls.
33. Calculated from data in Federal Bureau of Investigation, *Hate Crime Statistics 2016 and 2017*, accessed February 18, 2019.
34. Federal Bureau of Investigation, *Hate Crime Statistics, 2017*, accessed February 18, 2019, https://ucr.fbi.gov/hate-crime/2017/topic-pages/tables/table-1.xls.
35. Ibid.
36. Federal Bureau of Investigation, *2014 Crime in the United States*, "Federal Crime Data, 2014," accessed February 3, 2016, www.fbi.gov/about-us/cjis/ucr/crime-in-the-u.s/2014/crime-in-the-u.s.-2014/additional-reports/federal-crime-data/federal-crime-data.pdf.

37. "FBI to Report on Officer-Involved Shootings as Overall Crime Rate Drops," September 9, 2015 (accessed February 4, 2016), http://on.rt.com/6skk; Mark Berman. "FBI Director: We Will Collect and Publish More Info About Fatal Police Shooting," September 28, 2015 (accessed February 4, 2016), www.washingtonpost.com/news/post-nation/wp/2015/09/28/fbi-director-we-will-collect-and-publish-more-information-about-fatal-police-shootings/; Federal Bureau of Investigation, *2017 Crime in the United States*, accessed February 18, 2019, https://www.fbi.gov/file-repository/ucr/recent-developments-in-the-ucr-program.pdf/view.

38. Scott Heiser, "Tracking Animal Crimes Data in the FBI's Uniform Crime Reporting (UCR) Program - A Huge Step Forward," September 17, 2014 (accessed February 4, 2016), http://aldf.org/blog/tracking-animal-crimes-data-in-the-fbis-uniform-crime-reporting-ucr-program-a-huge-step-forward/; Colby Itkowitz, "The Feds Move In on Animal Cruelty," *The Orlando Sentinel* (January 8, 2016), p. A7.

39. Federal Bureau of Investigation, *2018 Crime in the United States*, "Persons Arrested," accessed October 15, 2019, https://ucr.fbi.gov/crime-in-the-u.s/2018/crime-in-the-u.s.-2018/topic-pages/persons-arrested.

40. Ibid., https://ucr.fbi.gov/crime-in-the-u.s/2017/crime-in-the-u.s.-2017/topic-pages/persons-arrested.

41. Ibid.

42. Ibid. Age calculated from https://ucr.fbi.gov/crime-in-the-u.s/2017/crime-in-the-u.s.-2017/tables/table-38, accessed February 21, 2019.

43. Ibid.

44. Ibid., Tables 37 and 33.

45. Federal Bureau of Investigation, *2017 Crime in the United States*, "Clearances," accessed February 21, 2019, https://ucr.fbi.gov/crime-in-the-u.s/2017/crime-in-the-u.s.-2017/topic-pages/clearances.

46. Federal Bureau of Investigation, *2018 Crime in the United States*, "Clearances," accessed October 15, 2019, https://ucr.fbi.gov/crime-in-the-u.s/2018/crime-in-the-u.s.-2018/topic-pages/clearances.

47. Ibid.

48. Ibid.

49. Jim Mustian and Michael R. Sisak, "'Clearance Rate' for Rape Cases Fell Last Year to Its Lowest Point since At Least the 1960s, According to FBI Data," *Chicago Tribune*, December 27, 2018. Accessed February 24, 2019, https://www.chicagotribune.com/ct-rape-investigations-20181227-story.html.

50. Federal Bureau of Investigation, "Recent Developments in the UCR Program," *2017 Crime in the United States.* Accessed February 20, 2019, https://www.fbi.gov/file-repository/ucr/recent-developments-in-the-ucr-program.pdf/view.

51. Unless indicated otherwise, all information in this section is from Brian A. Reaves, "Using NIBRS Data to Analyze Violent Crime," in *Bureau of Justice Statistics Technical Report*, U.S. Department of Justice (Washington, DC: U.S. Government Printing Office, October 1993); Federal Bureau of Investigation, *2017 National Incident-Based Reporting System*, accessed February 22, 2019, https://ucr.fbi.gov/nibrs/2017.

52. Federal Bureau of Investigation, "Recent Developments in the UCR Program," *2017 Crime in the United States.* Accessed February 20, 2019, https://www.fbi.gov/file-repository/ucr/recent-developments-in-the-ucr-program.pdf/view.

53. Ibid.

54. Ibid.

55. U.S. Department of Justice, "Implementing the National Incident-Based Reporting System: A Project Status Report" (Washington, DC: U.S. Government Printing Office, July 1997).

56. Jennifer L. Truman and Michael Planty, "Criminal Victimization, 2011," in *Bureau of Justice Statistics Bulletin*, U.S. Department of Justice, Office of Justice Programs (Annapolis, MD: Bureau of Justice Statistics Clearinghouse, October 2012), accessed October 17, 2012, http://bjs.ojp.usdoj.gov/index.cfm?ty=pbdetail&iid=4494.

57. Rachel E. Morgan and Barbara A. Oudekerk, "Criminal Victimization, 2018," *Bureau of Justice Statistics Bulletin*, U.S. Department of Justice, September 2019, accessed October 15, 2019, https://www.bjs.gov/content/pub/pdf/cv18.pdf.

58. Ibid.

59. Ibid., p. 5, Table 4.

60. The differences in the two measures probably can be attributed to the differences in what the two indexes measure. Ibid., p. 2.

61. See, for example, J. S. Wallerstein and C. J. Wylie, "Our Law-Abiding Lawbreakers," *Probation* 25 (1947): 107–12; I. Silver, "Introduction," in *The Challenge of Crime in a Free Society* (New York: Avon, 1968). See also C. Tittle, W. Villemez, and D. Smith, "The Myth of Social Class and Criminality," *American Sociological Review* 43 (1978): 643–56. Juvenile delinquency is also widespread. See, for example, Jerald Bachman, Lloyd Johnston, and Patrick O'Malley, *Monitoring the Future* (Ann Arbor: University of Michigan, Institute for Social Research, 1992); Martin Gold, "Undetected Delinquent Behavior," *Journal of Research in Crime and Delinquency* 3 (1966): 27–46; Martin Gold, *Delinquent Behavior in an American City* (Belmont, CA: Brooks/Cole, 1970); Maynard Erickson and LaMar Empey, "Court Records, Undetected Delinquency, and Decision Making," *Journal of Criminal Law, Criminology, and Police Science* 54 (1963): 446–69; James Short and F. Ivan Nye, "Extent of Unrecorded Delinquency," *Journal of Criminal Law, Criminology, and Police Science* 49 (1958): 296–302.

62. Thomas Gabor, *Everybody Does It! Crimes by the Public* (Toronto: University of Toronto Press, 1994).

63. Bureau of Justice Statistics, "National Crime Victimization Survey, 2014 and 2012," Special Tabulation. (Thanks to Jennifer Truman for providing the data.) Crimes included in the total include rape/sexual assault, robbery, assault, purse snatching, pocket picking, household burglary, motor vehicle theft, and theft.

64. Ibid.

65. Gregg Barak, *Theft of a Nation: Wall Street Looting and Federal Regulatory Colluding* (Lanham, MD: Rowman & Littlefield, 2012); Steve Tombs and David Whyte, *The Corporate Criminal* (New York: Routledge, 2015), pp. 14–16.

66. David R. Simon, *Elite Deviance,* 8th ed. (Boston: Allyn & Bacon, 2006), 97, 107.

67. Thanks to BJS Statistician Rachel Morgan for the data and the explanation. Personal communication, May 15, 2019.

68. Ted R. Miller, Mark A. Cohen, and Brian Wiersema, "Victim Costs and Consequences: A New Look," in *National Institute of Justice Report*, U.S. Department of Justice (Washington, DC: U.S. Government Printing Office, February 1996). Also see Kathryn E. McCollister, Michael T. French, and Hai Fang, "The Cost of Crime to Society: New Crime-Specific Estimates for Policy and Program Evaluation," *Drug and Alcohol Dependence* 108 (2010): 98–109, accessed February 12, 2016, www.ncbi.nlm.nih.gov/pmc/articles/PMC2835847/.

69. Wesley Skogan, "Fear of Crime and Neighborhood Change," in *Crime and Justice: A Review of Research,* ed. Albert J. Reiss Jr. and Michael Tonry, vol. 8, *Communities and Crime* (Chicago: University of Chicago Press, 1986).

70. "Crime," *Gallup Historical Trends–Gallup News*, accessed December 26, 2018, https://news.gallup.com/poll/1603/crime.aspx.

71. Alyssa Davis, "In U.S., Concern About Crime Climbs to 15-Year High," *Gallup*, April 6, 2016. Accessed February 25, 2019, https://news.gallup.com/poll/190475/americans-concern-crime-climbs-year-high.aspx?g_source=Politics&g_medium=lead&g_campaign=tiles.

72. Skogan, "Fear of Crime and Neighborhood Change," 215.

73. *Sourcebook of Criminal Justice Statistics Online*, accessed November 19, 2008, www.albany.edu/sourcebook/pdf /t2402007.pdf.

74. Dulce Zamora, "How to Protect Yourself Against Crime: Experts Give Advice on Ways to Fend Off Criminals–and Avoid Danger in the First Place," *WebMD*, accessed February 26, 2019, https://www.webmd.com/a-to-z-guides/features/how-protect-yourself-against-crime#1.

75. Rachel E. Morgan and Barbara A. Oudekerk, "Criminal Victimization, 2018," *Bureau of Justice Statistics Bulletin*, U.S. Department of Justice (Washington, DC: Government Printing Office, September 2019), 4, Table 1, 5, Table 3, accessed October 16, 2019, https://www.bjs.gov/content/pub/pdf/cv18.pdf; Rachel E. Morgan and Jennifer L. Truman, "Criminal Victimization, 2017," *Bureau of Justice Statistics Bulletin*, U.S. Department of Justice (Washington, DC: Government Printing Office, December 2018), accessed February 27, 2019, https://www.bjs.gov/content/pub/pdf/cv17.pdf; Jennifer L. Truman and Michael Planty, "Criminal Victimization, 2011."

76. Rachel E. Morgan and Barbara A. Ouderkerk, "Criminal Victimization, 2018."

77. Ibid.; Calculated from data in Ibid. and Craig A. Perkins, Patsy A. Klaus, Lisa D. Bastian, and Robyn L. Cohen, "Criminal Victimization in the United States, 1993," *Bureau of Justice Statistics, U.S. Department of Justice* (Washington, DC: Government Printing Office, May 1996), accessed February 27, 2019, https://www.bjs.gov/content/pub/pdf/cvus93.pdf.

78. Adapted from Tables 9 and 10 in Rachel E. Morgan and Barbara A. Oudekerk, "Criminal Victimization, 2018," and Table 9 in Rachel E. Morgan and Jennifer L. Truman, "Criminal Victimization, 2017," Bureau of Justice Statistics Bulletin, U.S. Department of Justice, December 2018, https://www.bjs.gov/content/pub/pdf/cv17.pdf.

Design Elements: *Image of part of a court building with columns and statue*: Pixtal/AGE Fotostock RF; *Image of a door slightly open*: jadding/Shutterstock.com

stevecoleimages/E+/Getty Images

3 Explaining Crime

CHAPTER OUTLINE

LEARNING OBJECTIVES

After completing this chapter, you should be able to:

1. Define criminological theory.
2. State the causes of crime according to classical and neoclassical criminologists and their policy implications.
3. Describe the biological theories of crime causation and their policy implications.
4. Describe the different psychological theories of crime causation and their policy implications.
5. Explain sociological theories of crime causation and their policy implications.
6. Distinguish major differences among classical, positivist, and critical theories of crime causation.
7. Describe how critical theorists would explain the causes of crime and their policy implications.

Crime Story

On May 25, 2018, 66-year-old movie mogul Harvey Weinstein pictured was arrested and charged in New York with rape and several counts of sexual abuse against women. He surrendered to police and was released the next day on $1 million bail, after agreeing to wear a GPS tracker and surrender his passport. On May 31, 2018, a New York grand jury indicted Weinstein on charges of rape and a criminal sexual act—forcibly performing oral sex on a woman—moving the case toward trial. If convicted of all charges, Weinstein could be sentenced to life in prison. On June 6, 2018, Weinstein pleaded not guilty to all charges in the New York Supreme Court (the name of New York's trial court of general jurisdiction). On July 2, Weinstein was charged with sexually assaulting a third woman. This was in addition to other investigations in Los Angeles, London, and by the U.S. government. On July 9, Weinstein pleaded not guilty in the third sexual assault case in the New York Supreme Court. His criminal trial was scheduled to begin on June 3, 2019, but was postponed until January 6, 2020 (and is ongoing at this writing). Weinstein also faces several civil cases.

Weinstein's road to perdition began nearly four decades ago when he started coercing young women to have sex or engage in other sexual behavior in exchange for promises to help promote their careers. Weinstein's pattern of sexual harassment first became public on October 5, 2017, when the *New York Times* published a story revealing his long history of predatory behavior. Actresses Rose McGowan, who allegedly turned down a $1 million offer from Weinstein to remain silent, and Ashley Judd, who allegedly was "blacklisted" by Weinstein for refusing his advances, were two of the first women to come forward with accusations against him. In doing so, they helped galvanize the #MeToo movement. About a week after the *Times* story appeared, on October 15, actress Alyssa Milano suggested on Twitter that to show the problem's magnitude, anyone who had been "sexually harassed or assaulted" should respond to her Tweet with "Me Too." Half a million people replied during the first 24 hours. Weinstein responded to the *Times* story by apologizing and admitting that he "has caused a lot of pain" but disputed the accusations that he had sexually harassed female employees over nearly three decades. He stated that he was taking a leave of absence from The Weinstein Company (formerly Miramax Films) and was seeing a therapist. Weinstein's lawyer told *The Hollywood Reporter* that his client was preparing to sue the *New York Times*.

Following publication of the *Times* story, more than 80 women came forward with tales of Weinstein's debauchery, including several rape accusations that occurred as long ago as the 1970s. Most of these claims cannot be prosecuted because of the statute of limitations. Weinstein also was accused of blacklisting actresses who refused his sexual advances. Despite these testimonies, in an article published in the *New Yorker* magazine on October 10, 2017, Weinstein's spokeswoman had the temerity to assert, "Any allegations of non-consensual sex are unequivocally denied by Mr. Weinstein." Among women to accuse Weinstein were Paz de la Huerta, Daryl Hannah, Salma Hayak, Lena Headey, Angelina Jolie, Lupita Nyong'o, Gwyneth Paltrow, Annabella Sciorra, Mira Sorvino, and Uma Thurman.

Without being convicted of a crime, Weinstein's downfall has been swift and harsh. On October 10, 2017, Weinstein's wife Georgina Chapman announced that she was leaving him and that her priority is her young children. On October 15, the Academy of Motion Picture Arts and Sciences' board voted to expel him. The organization awards the Oscars. On October 16, the Producers Guild of America's board terminated Weinstein's membership. On October 30, the same organization banned Weinstein for life. On October 18, Harvard University announced it was stripping Weinstein of the Du Bois medal he received in 2014, for his

Erik Pendzich/Shutterstock

contributions to African-American culture. On October 19, the British Film Institute announced that it was withdrawing the BFI Fellowship Weinstein was awarded in 2002. On November 7, the Television Academy expelled Weinstein from its organization. And on February 2, 2018, the British Academy of Film and Television Arts formally terminated Weinstein's membership.

On October 6, 2017, a spokesperson for the Weinstein Company stated that the company was taking the allegations "extremely seriously" and was starting an inquiry. On October 8, 2017, and "in light of new information about misconduct," his company's board fired him. On October 17, 2017, Weinstein resigned from his company's board but still retained 22% of his company's stock. On February 11, 2018, New York state prosecutors, following a four-month investigation, filed a lawsuit against the Weinstein Company, alleging that the studio failed to protect employees from Harvey Weinstein's harassment and abuse. The lawsuit claimed that Weinstein sexually harassed and abused female employees at the studio for years, including making verbal threats against their lives and using female staff as "wing women" to facilitate sexual conquests. His lawyer conceded that Weinstein's behavior is "not without fault" but declared that there was "no criminality." On March 20, 2018, Weinstein's former company filed for bankruptcy and ended all of its nondisclosure agreements that Weinstein forced on his victims to silence them.

This chapter addresses the causes of criminal behavior from a variety of theoretical perspectives. How might Harvey Weinstein's criminal behavior be explained? Was it caused by biological, psychological, or sociological factors or was it caused by a combination of all three factors? How should society respond to his crimes? Should Weinstein be sentenced to life in prison? Why or why not? How can society prevent such crimes? Answers to these questions demonstrate the difficulty of determining why people commit crimes and the appropriate responses to them.

Introduction to Criminological Theory

theory An assumption (or set of assumptions) that attempts to explain why or how things are related to each other.

A **theory** is an assumption (or set of assumptions) that attempts to explain why or how things are related to each other. A theory of crime attempts to explain why or how a certain thing (or things) is related to criminal behavior. For example, some theories assume that crime is part of human nature, that some human beings are born evil. In those theories, human nature is examined in relation to crime. Other theories assume that crime is caused by biological things (e.g., chromosome abnormalities, hormone imbalances), psychological things (e.g., below-normal intelligence, satisfaction of basic needs), sociological things (e.g., social disorganization, inadequate socialization), economic things (e.g., unemployment, economic inequality), or some combination. In this chapter, we examine a variety of crime theories and discuss the policy implications of each. (Unless indicated otherwise, the term *crime* includes delinquency.)

criminological theory The explanation of criminal behavior, as well as the behavior of police, attorneys, prosecutors, judges, correctional personnel, victims, and other actors in the criminal justice process.

Criminological theory is important because most of what is done in criminal justice is based on criminological theory, whether we or the people who propose and implement policies based on the theory know it or not. The failure to understand the theoretical basis of criminal justice policies leads to at least two undesirable consequences. First, if criminal justice policymakers do not know the theory or theories on which their proposed policies are based, then they will be unaware of the problems that are likely to undermine the success of the policies. Much time and money could be saved if criminal justice policies were based on a thorough theoretical understanding. Second, criminal justice policies invariably intrude on people's lives (e.g., people are arrested and imprisoned). If people's lives are going to be disrupted by criminal justice policies, it seems only fair that there be very good reasons for the disruption.

Technically, criminological theory refers not only to explanations of criminal behavior but also to explanations of police behavior and the behavior of attorneys, prosecutors, judges, correctional personnel, victims, and other actors in the criminal justice system. However, in this chapter, our focus is on theories of crime causation. Table 3.1 outlines the crime causation theories presented in this chapter.

THINKING CRITICALLY

1. What is a theory?
2. Why is it important to understand the various theories of criminal behavior?

Classical and Neoclassical Approaches to Explaining Crime

The causes of crime have long been the subject of much speculation, theorizing, research, and debate among scholars and the public. Each theory of crime has been influenced by the religious, philosophical, political, economic, social, and scientific trends of the time. One of the earliest secular approaches to explaining the causes of crime was the classical theory, developed in Europe at the time of profound social and intellectual change. Before classical theory, crime was generally equated with sin and was considered the work of demons or the devil.

classical theory A product of the Enlightenment, based on the assumption that people exercise free will and are thus completely responsible for their actions. In classical theory, human behavior, including criminal behavior, is motivated by a hedonistic rationality in which actors weigh the potential pleasure of an action against the possible pain associated with it.

Classical Theory

Classical theory is a product of the Enlightenment period, or the Age of Reason, a period of history that began in the late 1500s and lasted until the late 1700s. The Enlightenment thinkers, including members of the classical school of criminology, promoted a new,

TABLE 3.1 Theories of Crime Causation

Theories	Theorists	Causes	Policy Implications
CLASSICAL AND NEOCLASSICAL THEORIES			
Theories	**Theorists**	**Causes**	**Policy Implications**
Classical	Beccaria, Bentham	Free-willed individuals commit crime because they rationally calculate that crime will give them more pleasure than pain.	Deterrence: Establish social contract. Enact laws that are clear, simple, unbiased, and reflect the population's consensus. Impose punishments that are proportionate to the crime, prompt, certain, public, necessary, the least possible in the given circumstances, and dictated by law, not judges' discretion. Educate the public. Eliminate corruption from the administration of justice. Reward virtue.
Neoclassical	Rush	Moral insanity	Consider premeditation and mitigating circumstances in legal proceedings. Allow testimony about diminished responsibility. Sentence offenders to rehabilitative environments.
POSITIVIST THEORIES			
Theories	**Theorists**	**Causes**	**Policy Implications**
Biological	Lombroso, Sheldon	Biological inferiority or biochemical processes cause people to commit crimes.	Isolate, sterilize, or execute offenders. For specific problems, brain surgery, chemical treatment, improved diets, and better mother and child health care.
Psychological			
Intelligence	Goddard	Mental inferiority (low IQ) causes people to commit crimes.	Isolate, sterilize, or execute offenders.
Psychoanalytic	Freud	Crime is a symptom of more deep-seated problems.	Provide psychotherapy or psychoanalysis.
Humanistic	Maslow	Crime is a means by which individuals can satisfy their basic human needs.	Help people satisfy their basic needs legally.
	Halleck	Crime is an adaptation to helplessness caused by oppression.	Eliminate sources of oppression. Provide legal ways of coping with feelings of helplessness caused by oppression; psychotherapy.
Sociological			
Durkheim	Durkheim	Crime is a social fact. It is a "normal" aspect of society, although different types of societies should have greater or lesser degrees of it. Crime is also functional for society.	Contain crime within reasonable boundaries.
Chicago School	Park, Burgess, Shaw, McKay	Delinquency is caused by detachment from conventional groups, which is caused by social disorganization.	Organize and empower neighborhood residents.
Anomie	Merton, Cohen	For Merton, crime is caused by anomie, which is the contradiction between cultural goals and the social structure's capacity to provide the institutionalized means to achieve those goals. For Cohen, gang delinquency is caused by anomie, which is the inability to conform to middle-class values or to achieve status among peers legally.	Reduce aspirations. Increase legitimate opportunities. Do both.
Learning	Tarde, Sutherland, Glaser, Burgess, Akers, Jeffery	Crime is committed because it is positively reinforced, negatively reinforced, or imitated.	Provide law-abiding models. Regulate associations. Eliminate crime's rewards. Reward law-abiding behavior. Punish criminal behavior effectively.
Control	Reiss, Toby, Nye, Reckless, Hirschi	Crime is a result of improper socialization.	Properly socialize children so that they develop self-control and a strong moral bond to society.

(*continued*)

TABLE 3.1 Theories of Crime Causation (*continued*)

CRITICAL THEORIES			
Theories	**Theorists**	**Causes**	**Policy Implications**
Labeling	Becker, Lemert	Does not explain the initial cause of crime and delinquency (primary deviance); explains only secondary deviance, which is the commission of crime after the first criminal act with the acceptance of a criminal label.	Do not label. Employ radical nonintervention. Employ reintegrative shaming.
Conflict	Vold, Turk	Crime is caused by relative powerlessness.	Dominant groups give up power to subordinate groups. Dominant groups become more effective rulers and subordinate groups better subjects.
Radical	Quinney, Chambliss, Platt	Competition among wealthy people and among poor people, as well as between rich and poor (the class struggle), and the practice of taking advantage of other people.	Define crime as a violation of basic human rights. Replace the criminal justice system with "popular" or "socialist" justice. Create a socialist society appreciative of human diversity.
British or Left Realism	Young	Relative deprivation is a potent, although not exclusive, cause of crime.	Employ police power to protect people in working-class communities.
Peacemaking	Quinney, Pepinsky	Same as radical (different prescription for change).	Transform human beings so that they are able to experience empathy with those less fortunate and respond to other people's needs. Reduce hierarchical structures. Create communities of caring people. Champion universal social justice.
Feminist theory	Daly, Chesney-Lind, Simpson	Patriarchy (men's control over women's labor and sexuality) is the cause of crime.	Abolish patriarchal structures and relationships. Champion equality for women in all areas.
Postmodernism	Henry, Milovanovic	Crime is caused by processes that result in the denial of responsibility for other people and to other people.	Similar to peacemaking criminological theory.

scientific view of the world. In so doing, they rejected the then-dominant religious view of the world, which was based on revelation and the authority of the Church. The Enlightenment thinkers assumed that human beings could understand the world through science—the human capacity to observe and to reason. Moreover, they believed that if people could understand the world and its functioning, they could change it. The Enlightenment thinkers rejected the belief that either the nature of the world or the behavior of the people in it was divinely ordained or predetermined.

Instead, the Enlightenment thinkers believed that people exercise *free will*, or the ability to choose any course of action, for which they are completely responsible. Human behavior, according to the English philospher Jeremy Bentham, is motivated by a *hedonistic rationality* in which a person weighs the potential pleasure of an action against the possible pain associated with it. In that view, human beings commit crime because they rationally calculate that the crime will give them more pleasure than pain.

Classical criminologists, as Enlightenment thinkers, were concerned with protecting the rights of humankind from the corruption and excesses of the existing legal institutions. Horrible and severe punishments were common both before and during the Enlightenment. For example, in England during the eighteenth century, almost 150 offenses (some authorities claim more than 200 offenses) carried the death penalty, including stealing turnips, associating with gypsies, cutting down a tree, and picking pockets. Barbarous punishments were not the only problem. At the time, crime was rampant, yet types of crime were poorly defined. What we today call *due process of law* was either absent or ignored. Torture was employed routinely to extract confessions. Judgeships were typically sold to wealthy persons by the sovereign, and judges had almost total discretion. Consequently, there was little consistency in the application of the law or in the punishments imposed.

It was within that historical context that Cesare Beccaria, perhaps the best known of the classical criminologists, wrote and published anonymously in 1764 his truly revolutionary

Cesare Beccaria and the Death Penalty

Cesare Beccaria, the best known of the classical criminologists, opposed the death penalty. In *An Essay on Crimes and Punishments,* he wrote: "The death penalty cannot be useful, because of the example of barbarity it gives . . . . It seems to me absurd that the laws, which are an expression of the public will, which detest and punish homicide, should themselves commit it, and that to deter citizens from murder, they order a public one."

work, *An Essay on Crimes and Punishments* (*Dei Delitti e delle Pene*). His book generally is acknowledged to have had an enormous practical influence on the establishment of a more humane system of criminal law and procedure.[1] In the book, Beccaria sets forth most of what we now call classical criminological theory.

According to Beccaria, the only justified rationale for laws and punishments is the principle of **utility**; that is, "the greatest happiness shared by the greatest number."[2] The basis of society, as well as the origin of punishments and the right to punish, is the **social contract**. The social contract is an imaginary agreement entered into by persons who sacrifice the minimum amount of their liberty necessary to prevent anarchy and chaos.

utility The principle that a policy should provide "the greatest happiness shared by the greatest number."

social contract An imaginary agreement to sacrifice the minimum amount of liberty necessary to prevent anarchy and chaos.

Beccaria believed that the only legitimate purpose of punishment is deterrence, both special and general.[3] **Special** or **specific deterrence** is the prevention of the punished persons from committing crime again. **General deterrence** is the use of the punishment of specific individuals to prevent people in general or society at large from engaging in crime. To be both effective and just, Beccaria argued, punishments must be "public, prompt, certain, necessary, the least possible in the given circumstances, proportionate to the crime, dictated by the laws."[4] It is important to emphasize, however, that Beccaria promoted crime prevention over punishment.

special or specific deterrence The prevention of individuals from committing crimes again by punishing them.

general deterrence The attempt to prevent people in general or society at large from engaging in crime by punishing specific individuals and making examples of them.

In addition to the establishment of a social contract and the punishment of people who violate it, Beccaria recommended four other ways to prevent or to deter crime.[5] The first was to enact laws that are clear, simple, and unbiased and that reflect the consensus of the population. The second was to educate the public. Beccaria assumed that the more educated people are, the less likely they are to commit crimes. The third was to eliminate corruption from the administration of justice. Beccaria believed that if the people who dispense justice are themselves corrupt, people lose respect for the justice system and become more likely to commit crimes. The fourth was to reward virtue. Beccaria asserted that punishing crime is not enough; it is also important to reward law-abiding behavior. Such rewards might include public recognition of especially meritorious behavior or, perhaps, an annual tax deduction for people who have not been convicted of a crime.

Bettmann/Contributor/Getty Images

Cesare Beccaria believed that the only justified rationale for laws and punishments is the principle of utility; that is, "the greatest happiness shared by the greatest number." *Do you agree with him? Why or why not?*

The application of classical theory was supposed to make criminal law fairer and easier to administer. To those ends, judges would not select sentences. They could impose only the sentences dictated by legislatures for specific crimes. All offenders would be treated alike, and similar crimes would be treated similarly. Individual differences among offenders and unique or mitigating circumstances about the crime would be ignored. A problem is that all offenders are not alike, and similar crimes are not always as similar as they might appear on the surface. Should first offenders be treated the same as those who commit crime repeatedly? Should juveniles be treated the same as adults? Should the insane be treated the same as the sane? Should a crime of passion be treated the same as the intentional commission of a crime? The classical school's answer to all of those difficult questions would be a simple yes.

Despite those problems, Beccaria's ideas were very influential. France, for example, adopted many of Beccaria's principles in its Code of 1791—in particular, the principle of equal punishments for the same crimes. However, because classical theory ignored both individual differences among offenders and mitigating circumstances, it was difficult to apply the law in practice. Because of that difficulty, as well as new developments in the emerging behavioral sciences, modifications of classical theory and its application were introduced in the early 1800s.

Neoclassical Theory

Several modifications of classical theory are collectively referred to as **neoclassical theory**. The principal difference between the two theories has to do with classical theory's assumption about free will. In the neoclassical revision, it was conceded that certain factors, such as insanity, might inhibit the exercise of free will. Benjamin Rush, who is considered the father of American psychiatry, was the first to make the connection between "moral insanity" and crime. Thus, the idea of premeditation was introduced as a measure of the degree of free will exercised. Also, mitigating circumstances were considered legitimate grounds for an argument of diminished responsibility.

neoclassical theory A modification of classical theory in which it was conceded that certain factors, such as insanity, might inhibit the exercise of free will.

Those modifications of classical theory had two practical effects on criminal justice policy. First, they provided a reason for nonlegal experts such as medical doctors to testify in court as to the degree of diminished responsibility of an offender. Second, offenders began to be sentenced to punishments that were considered rehabilitative. The idea was that certain environments, for example, environments free of vice and crime, were more conducive than others to the exercise of rational choice.

The reason we have placed so much emphasis on the classical school of criminology and its neoclassical revisions is that, together, they are essentially the model on which criminal justice in the United States is based today. During the past 40 years or so, at least in part because the public frequently perceived as too lenient the sentences imposed by judges for certain crimes, such measures as legislatively imposed sentencing guidelines have limited the sentencing authority of judges in many jurisdictions. Public outrage over the decisions of other criminal justice officials has led to similar measures. For example, parole has been abolished in the federal jurisdiction and in some states because many people believe that parole boards release dangerous criminals from prison too soon.

The revival of classical and neoclassical theories during the past four decades, and the introduction of a more modern version called *rational choice theory*, are also a probable reaction to the allegation of some criminologists and public officials that criminologists have failed to discover the causes of crime. As a result of that belief, there had been a renewed effort, until recently, to deter crimes by sentencing more offenders to prison for longer periods of time and, in many jurisdictions, by imposing capital punishment for heinous crimes. Ironically, one reason the theory of the classical school lost favor in the nineteenth century was the belief that punishment was not a particularly effective method of preventing or controlling crime.

THINKING CRITICALLY

1. Name four of the ways that classical criminologist Cesare Beccaria thought were best to prevent or deter crime. Do you agree with Beccaria? Why or why not?
2. What are the main differences between classical and neoclassical theories?

Positivist Approaches to Explaining Crime

The theory of the positivist school of criminology grew out of positive philosophy and the logic and basic methodology of empirical or experimental science. Positive philosophy was an explicit rejection of the critical and "negative" philosophy of the Enlightenment thinkers. Among the founders of positivism was Auguste Comte, who also has been credited with founding sociology. Comte acknowledged that the Enlightenment thinkers had contributed to progress by helping to break up the old system and by paving the way for a new one.[6] However, Comte argued that the ideas of the Enlightenment period had outlived their usefulness and had become obstructive.

Hulton Archive/Stringer/Getty Images

In his *Politique Positive* (1854), Comte proclaimed himself Pope of the new positive religion. *What did he mean by that?*

At about the same time that positivist philosophy was developing, experimentation with animals was becoming an increasingly accepted way of learning about human beings in physiology, medicine, psychology, and psychiatry. Human beings were beginning to appear to science as one of many creatures with no special connection to God. Human beings were beginning to be understood, not as free-willed, self-determining creatures who could do anything they wanted, but rather as beings whose action was determined by biological and cultural factors.

Positivism was a major break with the classical and neoclassical theories that had preceded it. The following are key assumptions of the positivist school of thought:

1. Human behavior is determined and not a matter of free will. Consequently, positivists focus on cause-and-effect relationships.
2. Criminals are fundamentally different from noncriminals. Positivists search for such differences by scientific methods.
3. Social scientists (including criminologists) can be objective, or value-neutral, in their work.

Careers in Criminal Justice

Criminologist

My name is Alida Merlo, and I am a professor of criminology at Indiana University of Pennsylvania. I have taught criminal justice and criminology for more than 35 years. I have a PhD in sociology from Fordham University, a master's degree in criminal justice from Northeastern University, and a bachelor's degree from Youngstown State University. As an undergraduate, I majored in sociology and corrections.

I became interested in criminology while taking an undergraduate course in juvenile delinquency. That course inspired me to pursue a career in criminal justice as a probation officer and then as an intake supervisor for the Mahoning County Juvenile Court in Youngstown, Ohio. These experiences led me to enroll in graduate school to learn more about crime causation and the criminal justice process.

During graduate school, I realized I would like to teach criminology and criminal justice. I was attracted to the profession for a variety of reasons: It would enable me to do research, teach, develop programs, and pursue professional activities and community service. Teaching and studying criminal justice and criminology seemed a wonderful opportunity.

Being a criminologist involves developing and teaching courses; reading and contributing to the literature in the field; serving on departmental and university committees; advising students about courses, research, career options, and graduate education; meeting prospective students and their families; applying for research funding; conducting research; sharing research findings with colleagues at state, regional, national, and international meetings; being involved in professional organizations; and serving on community boards or assisting in community programs.

One great joy of teaching criminology and criminal justice is assisting students as they develop and blossom in the discipline. Many of my students work in the field. Former students have been awarded doctoral degrees and teach criminal justice. It is exciting to have played a part in their professional development.

One disadvantage of being a criminologist is the inability to communicate to the public and to policymakers the realities of criminal justice and the implications of one's research. Confronting the fact that policies are often developed and enforced with little enlightenment is frustrating, but I am confident that criminologists will play a greater role in the future.

Courtesy of Alida Merlo

What qualities does a criminologist need to succeed? Why?

4. Crime frequently is caused by multiple factors.
5. Society is based on consensus but not on a social contract.

As the social sciences developed and social scientists directed their attention to the problem of crime, they adopted, for the most part, positivist assumptions. For example, theories of crime were (and continue to be) based on biological positivism, psychological positivism, sociological positivism, and so on. However, as theories based on positivist assumptions were developed, it became apparent to close observers that there were problems, not only with the theories but with the positivist assumptions as well. We will briefly discuss five of those problems.[7] In subsequent subsections, we also will describe problems peculiar to specific positivist theories of crime causation.

The first problem with positivism is overprediction: Positivist theories generally account for too much crime. They also do not explain exceptions very well. For example, a positivist theory that suggests crime is caused by poverty overpredicts because not all poor people commit crime. The theory also cannot explain adequately why many poor people do not commit crimes.

Second, positivist theories generally ignore the criminalization process, the process by which certain behaviors are made illegal. They separate the study of crime from a theory of the law and the state and take the legal definition of crime for granted. Ignored is the question of why certain behaviors are defined as criminal while other, similar behaviors are not.

A third problem with positivism is its consensual worldview, the belief that most people agree most of the time about what is good and bad and right and wrong. Such a view ignores a multitude of fundamental conflicts of value and interest in society. It also tends to lead to a blind acceptance of the status quo.

A fourth problem is positivism's belief in determinism, the idea that choice of action is not free but is determined by causes independent of a person's will. Positivists generally assume that humans only adapt or react—but humans also create. How else could we explain new social arrangements or ways of thinking? A belief in determinism allows positivists to present an absolute situation uncomplicated by the ability to choose.

Finally, a fifth problem with positivist theories is the belief in the ability of social scientists (criminologists) to be objective, or value-neutral, in their work. Positivists fail to recognize that to describe and evaluate such human actions as criminal behavior is fundamentally a moral endeavor and, therefore, subject to bias.

FYI Charles Darwin

In *The Descent of Man* (1871), British naturalist Charles Darwin suggested that some people were "less highly evolved or developed than others," that some people "were nearer their apelike ancestors than others in traits, abilities, and dispositions."

Biological Theories

Biological theories of crime causation (biological positivism) are based on the belief that criminals are physiologically different from noncriminals. Early biological theories assumed that structure determined function. In other words, criminals behave differently because, structurally, they are different. Today's biocriminologists are more likely to assume that biochemistry determines function or, more precisely, that the difference between criminals and noncriminals is the result of a complex interaction between biochemical and environmental factors. To test biological theories, efforts are made to demonstrate, through measurement and statistical analysis, that there are or are not significant structural or biochemical differences between criminals and noncriminals.

biological inferiority According to biological theories, a criminal's innate physiological makeup produces certain physical or genetic characteristics that distinguish criminals from noncriminals.

Historically, the cause of crime, from this perspective, was **biological inferiority**. Biological inferiority in criminals was assumed to produce certain physical or genetic characteristics that distinguished criminals from noncriminals. It is important to emphasize that in these theories, the physical or genetic characteristics themselves did not cause crime; they were only the symptoms, or *stigmata*, of the more fundamental inferiority. The concept of biological inferiority has lost favor among today's biocriminologists who generally prefer to emphasize the biological differences between criminals and noncriminals without adding the value judgment. In any event, several different methodologies have been employed to detect physical differences between criminals and noncriminals. They are criminal anthropology; study of body types; heredity studies, including family trees, statistical comparisons, twin studies, and adoption studies; and, in the last 15 to 20 years or so, studies based on new scientific technologies that allow, for example, the examination of brain chemistry processes.

criminal anthropology The study of "criminal" human beings.

Criminal Anthropology

Criminal anthropology is the study of "criminal" human beings. It is associated with the work of an Italian army doctor and later, university professor, Cesare Lombroso. Lombroso first published his theory of a physical criminal type in 1876.

Lombroso's theory consists of the following propositions:[8]

1. Criminals are, by birth, a distinct type.
2. That type can be recognized by physical characteristics, or stigmata, such as enormous jaws, high cheekbones, and insensitivity to pain.
3. The criminal type is clearly distinguished in a person with more than five stigmata, perhaps exists in a person with three to five stigmata, and does not necessarily exist in a person with fewer than three stigmata.
4. Physical stigmata do not cause crime; they only indicate an individual who is predisposed to crime. Such a person is either an **atavist**—that is, a reversion to a savage type—or a result of degeneration.
5. Because of their personal natures, such persons cannot desist from crime unless they experience very favorable lives.

atavist A person who reverts to a savage type.

Lombroso's theory was popular in the United States until about 1915, although variations of his theory are still being taught today. The major problem with Lombroso's criminal anthropology is the assumption that certain physical characteristics are indicative of biological inferiority. Unless there is independent evidence to support that assumption, other than the association of the physical characteristics with criminality, then the result is circular reasoning. In other words, crime is caused by biological inferiority, which is itself indicated by the physical characteristics associated with criminality.

Body-Type Theory Body-type theory is an extension of Lombroso's criminal anthropology. William Sheldon, whose work in the 1940s was based on earlier work by Ernst Kretchmer in the 1920s, is perhaps the best known of the body-type theorists. According to Sheldon, human beings can be divided into three basic body types, or *somatotypes*, which correspond to three basic temperaments.[9] The three body types are the endomorphic (soft, fat), the mesomorphic (athletically built), and the ectomorphic (tall, skinny).

Sheldon argued that everyone has elements of all three types, but that one type usually predominates. In a study of 200 Boston delinquents between 1939 and 1949, Sheldon found that delinquents were more mesomorphic than nondelinquents and that serious delinquents were more mesomorphic than less serious delinquents. Subsequent studies by the Gluecks in the 1950s and by Cortes in the 1970s also found an association between mesomorphy and delinquency.

The major criticism of the body-type theory is that differences in behavior are indicative of the social selection process and not biological inferiority. In other words, delinquents are more likely to be mesomorphic than nondelinquents, because, for example, mesomorphs are more likely to be selected for gang membership. Also, the finding that delinquents are more likely than nondelinquents to be mesomorphic contradicts, at least with regard to physique, the theory's general assumption that criminals (or delinquents) are biologically inferior.

In any event, if one believes that crime is the product of biological inferiority, then the policy implications are limited. Either criminals are isolated from the rest of the population by imprisoning them, for example, or they are executed. If they are isolated, they also may need to be sterilized to ensure that they do not reproduce.

Heredity Studies A variety of methods has been employed to test the proposition that criminals are genetically different from noncriminals.

Family Tree Studies Perhaps the earliest method was the use of family trees in which a family known to have many "criminals" was compared with a family tree of "noncriminals." (See FYI "The Family Tree Method.") However, a finding that criminality appears in successive

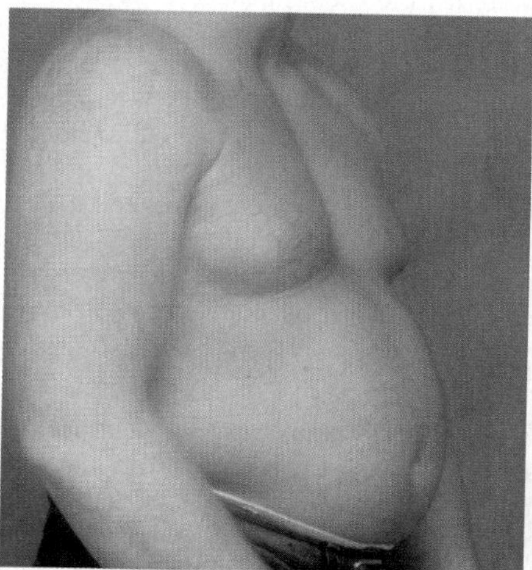
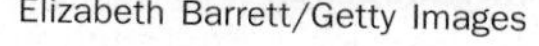
Elizabeth Barrett/Getty Images

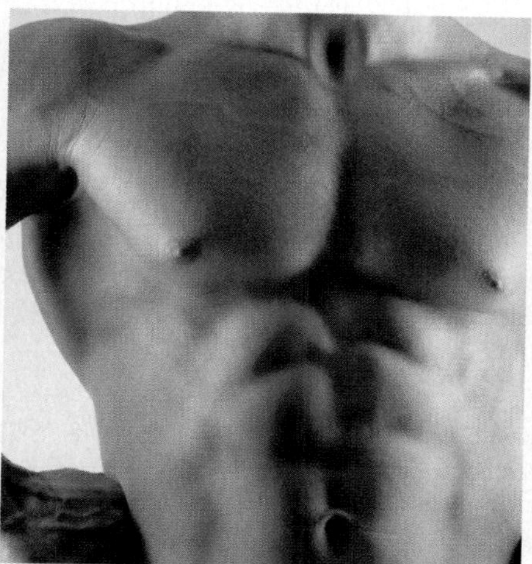
Anthony Saint James/Getty Images

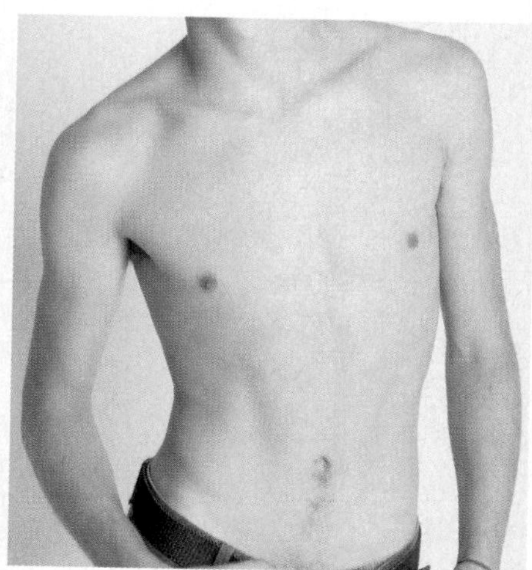
Image Source/Getty Images

William Sheldon, a well-known body-type theorist, divided humans into three basic body types. Left: endomorphic (soft, fat); center: mesomorphic (athletically built); right: ectomorphic (tall, skinny). *Why would delinquents more likely have a mesomorphic body type rather than an endomorphic or ectomorphic body type?*

The Family Tree Method

Dugdale (1877) and Estabrook (1916) both used the family tree method to study 709 (Dugdale) and 2,820 (Estabrook) members of the Juke family (a pseudonym), which presumably had a total of 171 criminals—mostly thieves and prostitutes. Even though only about 6% of the Jukes had criminal records and approximately 13% of them were paupers, proponents of the eugenics movement used the studies by Dugdale and Estabrook, and other studies like them, "to demonstrate scientifically that a large number of rural poor whites were 'genetic defectives.'"

Sources: Richard L. Dugdale, *The Jukes: A Study in Crime, Pauperism, Disease and Heredity* (New York: Putnam, 1877); Arthur H. Estabrook, *The Jukes in 1915* (Washington, DC: Carnegie Institute of Washington, 1916); Matt Wray and Annalee Newitz (eds.), *White Trash: Race and Class in America* (New York: Routledge, 1997), 2.

generations does not prove that criminality is inherited or is the product of a hereditary defect. For example, the use of a fork in eating has been a trait of many families for generations, but that does not prove that the use of a fork is inherited. In short, the family tree method cannot adequately separate hereditary influences from environmental influences.

Statistical Comparisons A second method used to test the proposition that crime is inherited or is the product of a hereditary defect is statistical comparison. The rationale is that if criminality exhibits the same degree of family resemblance as other physical traits, such as eye or hair color, then criminality, like those other traits, must be inherited. Although there is some evidence to support the notion, statistical comparisons also fail to separate adequately hereditary influences from environmental influences.

Twin Studies A third, more sophisticated method of testing the proposition that crime is inherited or is the result of a hereditary defect is the use of twin studies. Heredity is assumed to be the same in identical twins because they are the product of a single egg. Heredity is assumed to be different in fraternal twins because they are the product of two eggs fertilized by two sperm. The logic of the method is that if there is greater similarity in behavior between identical twins than between fraternal twins, the behavior must be due to heredity because environments are much the same. More than a half-century of this methodology has revealed that identical twins are more likely to demonstrate concordance (both twins having criminal records) than are fraternal twins, thus supporting the hereditary link. A problem with the twin studies, however, is the potential confounding of genetic and environmental influences. Identical twins tend to be treated more alike by others, spend much more time together, and have a greater sense of shared identity than do fraternal twins.

Adoption Studies A fourth method, the most recent and most sophisticated method of examining the inheritability of criminality, is the adoption study. The first such study was conducted in the 1970s. In this method, the criminal records of adopted children (almost always boys) who were adopted at a relatively early age are compared with the criminal records of both their biological parents and their adoptive parents (almost always fathers). The rationale is that if the criminal records of adopted boys are more like those of their biological fathers than like those of their adoptive fathers, the criminality of the adopted boys can be assumed to be the result of heredity.

The findings of the adoption studies reveal that the percentage of adoptees who are criminal is greater when the biological father has a criminal record than when the adoptive father has one. However, there also is an interactive effect. A greater percentage of adoptees have criminal records when both fathers have criminal records than when only one of them does. Like the twin studies, the adoption studies presumably demonstrate the influence of heredity but cannot adequately separate it from the influence of the environment. A problem with the adoption studies is the difficulty of interpreting the relative influences of heredity and environment, especially when the adoption does not take place shortly after birth or when, as is commonly the case, the adoption agency attempts to find an adoptive home that matches the biological home in family income and socioeconomic status.

If criminals are genetically different from noncriminals or the product of genetic defect, then the policy implications are the same as for other theories that propose that criminals are biologically inferior: isolate, sterilize, or execute. However, in the future, new technologies may make possible genetic engineering–that is, the removal or alteration of defective genes.

Modern Biocriminology Ongoing research has revealed numerous biological factors associated either directly or indirectly with criminal or delinquent behavior. Among such factors are certain chemical, mineral, and vitamin deficiencies in the diet, diets high in sugar and carbohydrates, hypoglycemia (low blood sugar level), certain allergies, ingestion of food dyes and lead, exposure to radiation from fluorescent tubes, and all sorts of brain dysfunctions such as attention deficit/hyperactivity disorder.[10] This section focuses on a few more of the biological factors linked to criminality and delinquency: disorders of the limbic system and other parts of the brain, brain chemical dysfunctions, minimal brain damage, and endocrine abnormalities.

FIGURE 3.1 The Limbic System

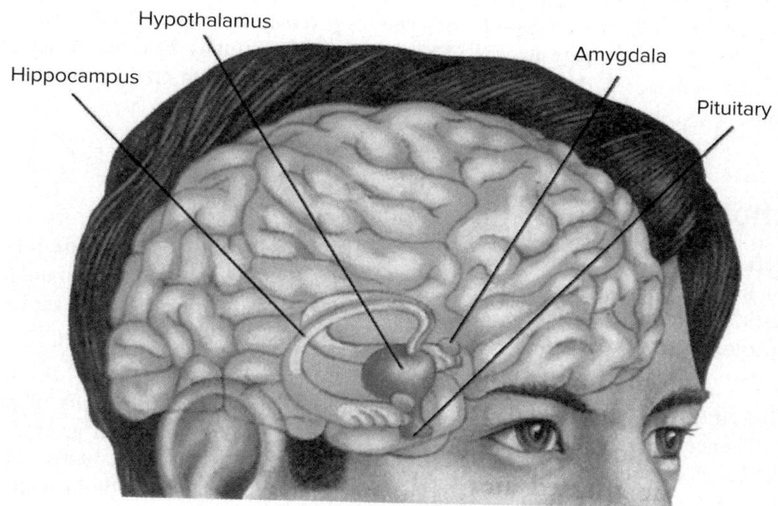

Limbic System Disorders At least some unprovoked violent criminal behavior is believed to be caused by tumors and other destructive or inflammatory processes of the limbic system.[11] The **limbic system** (see Figure 3.1) is a structure surrounding the brain stem that, in part, controls the life functions of heartbeat, breathing, and sleep. It also is believed to moderate expressions of violence; such emotions as anger, rage, and fear; and sexual response. Violent criminal behavior also has been linked to disorders in other parts of the brain. Recent evidence suggests that chronic violent offenders have much higher levels of brain disorder when compared to the general population. Surgical removal of the affected area sometimes eliminates expressions of violence. A problem with that type of intervention, however, is that it can cause unpredictable and undesirable behavior changes and, of course, is irreversible. Irreversibility is less a problem with newer chemical interventions, but the problem of unpredictable and undesirable behavior changes remains.

limbic system A structure surrounding the brain stem that, in part, controls the life functions of heartbeat, breathing, and sleep. It also is believed to moderate expressions of violence; such emotions as anger, rage, and fear; and sexual response.

Chemical Dysfunctions Some criminal behaviors are believed to be influenced by low levels of brain neurotransmitters (substances brain cells use to communicate).[12] For example, low levels of the brain neurotransmitter *serotonin* have been found in impulsive murderers and arsonists. Research also has found an association between low levels of another neurotransmitter, *norepinephrine*, and compulsive gambling. Another interesting discovery in this area may help explain cocaine use. Apparently, cocaine increases the level of the neurotransmitter *dopamine*, which activates the limbic system to produce pleasure. If such chemical deficiencies are linked to those behaviors, chemical treatment or improved diets might help. Neurotransmitters are products of the foods people eat.

Chemical Castration

In 1996, California became the first state to require chemical castration of repeat child molesters. Under the law, molesters who commit a second crime against a child under 13 must receive weekly injections of Depo-Provera. In 1997, Florida and Georgia passed similar legislation requiring repeat offenders to be chemically castrated, and Texas approved voluntary castration for repeat molesters. Iowa, Louisiana, Montana, Oregon, and Wisconsin also have chemical castration laws.

Sources: Cameron Cowan, "Is Chemical Castration a Viable Option for Child Sex Offenders?" accessed November 19, 2008, www.associatedcontent.com/article/308225/is_chemical_castration_a_viable_option.html?cat517; Madison Park, "Using Chemical Castration to Punish Child Sex Crimes," accessed October 3, 2012, www.cnn.com/2012/09/05/health/chemical-castration-science/index.html; Jamie Ducharme, "A Controversial Bill Would Allow Chemical Castration of Sex Offenders in Oklahoma," *Time*, February 3, 2018, accessed March 3, 2019, http://time.com/5132314/chemical-castration-sex-offenders/.

Endocrine Abnormalities Criminal behaviors have also been associated with endocrine, or hormone, abnormalities, especially those involving *testosterone* (a male sex hormone) and *progesterone* and *estrogen* (female sex hormones).[13] For example, administering estrogen to male sex offenders has been found to reduce their sexual drives. A similar effect has been achieved by administering the drug Depo-Provera, which reduces testosterone levels. However, a problem with Depo-Provera is that it is successful only for male sex offenders who cannot control their sexual urges. The drug does not seem to work on offenders whose sex crimes are premeditated. Studies also have found that a large number of crimes committed by females are committed during the menstrual or premenstrual periods of the female hormonal cycle. Those periods are characterized by a change in the estrogen-progesterone ratio.

In sum, there are probably no positivist criminologists today who would argue that biology or genetics makes people criminals. Nor, for that matter, are there many criminologists today who would deny that biology has some influence on criminal behavior. The position held by most criminologists today is that criminal behavior is the product of a complex interaction between biology and environmental or social conditions. What is inherited is not criminal behavior but rather the way in which the person responds to his or her environment. In short, biology or genetics gives an individual a *predisposition,* or a tendency, to behave in a certain way. Whether a person actually behaves in that way and whether that behavior is defined as a crime depend primarily on environmental or social conditions.

Psychological Theories

This section examines psychological theories of crime causation, namely, the relationship between intelligence and criminality and delinquency and discusses psychoanalytic and humanistic psychological theories. Learning or behavioral theories will be discussed along with sociological theories.

FYI **IQ and Crime**

One of the earliest promoters in the United States of the relationship between low IQ and crime was H. H. Goddard. In 1914, he published *Feeblemindedness: Its Causes and Consequences.* In the book, Goddard argues that criminals are feebleminded, an old-fashioned term that means below-normal intelligence.

Source: Henry H. Goddard, *Feeblemindedness: Its Causes and Consequences* (New York: Macmillan, 1914).

Intelligence and Crime

The idea that crime is the product primarily of people of low intelligence was popular in the United States from about 1914 until around 1930. It received some attention again during the mid-1970s and in the mid-1990s. The belief requires only a slight shift in thinking from the idea that criminals are biologically inferior to the idea that they are mentally inferior.

In 1931, Edwin Sutherland reviewed approximately 350 studies on the relationship between intelligence and delinquency and criminality.[14] The studies reported the results of intelligence tests of about 175,000 criminals and delinquents. Sutherland concluded from the review that although intelligence may play a role in individual cases, given the selection that takes place in arrest, conviction, and imprisonment, the distribution of the intelligence scores of criminals and delinquents is very similar to the distribution of the intelligence scores of the general population.

For the next 40 years or so, the issue of the relationship between intelligence and crime and delinquency appeared resolved. However, in the mid-1970s, two studies were published that resurrected the debate.[15] Those studies found that IQ was an important predictor of both official and self-reported juvenile delinquency, as important as social class or race. Both studies acknowledged the findings of Sutherland's earlier review. Both also noted that a decreasing number of delinquents had been reported as being of below-normal intelligence over the years. However, in both studies, it was maintained that the difference in intelligence between delinquents and nondelinquents had never disappeared and had stabilized at about eight IQ points. The studies failed to note, however, that the eight-point IQ difference found between delinquents and nondelinquents was generally within the normal range. The authors of the studies surmised that IQ influenced delinquency through its effect on school performance.

At this time, we cannot conclude with any degree of confidence that delinquents, as a group, are less intelligent than nondelinquents. We do know that most adult criminals are not of below-normal intelligence. Obviously, low-level intelligence cannot account for the dramatic increases or decreases in the crime rate over time unless one is prepared to conclude that the intelligence of offenders varies dramatically over time. Low-level intelligence certainly cannot account for complex white-collar and political crimes.

Nevertheless, to the degree, if any, that crime is caused by low-level intelligence, the policy implications are the same as for theories of biological inferiority: isolate, sterilize, or execute.

Psychoanalytic Theories

Psychoanalytic theories of crime causation are associated with the work of Sigmund Freud and his followers.[16] Freud did not theorize much about criminal behavior itself, but a theory of crime causation can be deduced from his more general theory of human behavior and its disorders. Had he contemplated the issue, Freud probably would have argued that crime, like other disorders, was a symptom of more deep-seated problems and that if the deep-seated problems could be resolved, the symptom of crime would disappear.

psychopaths, sociopaths, or antisocial personalities Persons characterized by no sense of guilt, no subjective conscience, and no sense of right and wrong. They have difficulty in forming relationships with other people; they cannot empathize with other people.

Freud believed that some people who had unresolved deep-seated problems were psychopaths (sociologists call them *sociopaths*). **Psychopaths, sociopaths, or antisocial personalities** are characterized by no sense of guilt, no subjective conscience, and no sense of right and wrong. They have difficulty in forming relationships with other people; they cannot empathize with other people. Table 3.2 provides an extended list of the characteristics of psychopaths.

The principal policy implication of considering crime symptomatic of deep-seated problems is to provide psychotherapy or psychoanalysis. *Psychoanalysis* is a procedure first developed by Freud that, among other things, attempts to make patients conscious or aware of unconscious and deep-seated problems in order to resolve the symptoms associated with them. Methods used include a variety of projective tests (such as the interpretation of Rorschach inkblots), dream interpretation, and free association. Another policy implication that derives logically from Freudian theory is to provide people with legal outlets to sublimate or redirect their sexual and aggressive drives. (Freud believed that all human beings are born with those two drives and that those drives are the primary sources of human motivation.)

Psychoanalysis, and the psychoanalytic theory on which it is based, are components of a medical model of crime causation that has, to varying degrees, informed criminal justice policy in the United States for more than a century. The general conception of this medical model is that criminals are biologically or, especially, psychologically "sick" and in need of treatment.

Despite the enduring popularity of this theory, a number of problems have been identified with it. First, the bulk of the research on the issue suggests that most criminals are not psychologically disturbed or, at least, are no more disturbed than the rest of the population.[17]

Second, if a person who commits a crime has a psychological disturbance, this does not mean that the psychological disturbance caused the crime. Many people with psychological disturbances do not commit crimes, and many people without psychological disturbances do commit crimes.

Third, psychoanalytic theory generally ignores the environmental circumstances in which the problematic behavior occurs. The problem is considered a personal problem and not a social one.

Fourth, there are problems with psychoanalysis and other forms of psychotherapy. Psychotherapy rests on faith. Much of its theoretical structure is scientifically untestable. The emphasis of psychotherapy as an approach to

Universal History Archive/UIG/Shutterstock

In "Criminality from a Sense of Guilt" (1915), Sigmund Freud, founder of psychoanalysis, suggested that some people commit crimes in order to be caught and punished—not for the crime for which they had been caught, but for something that they had done in the past about which they felt guilty and for which they were not caught or punished. *Do you think Freud was right? Why or why not?*

TABLE 3.2 Characteristics of the Psychopath

1. Superficial charm and good "intelligence."
2. Absence of delusions and other signs of irrational "thinking."
3. Absence of "nervousness" or psychoneurotic manifestations.
4. Unreliability.
5. Untruthfulness and insincerity.
6. Lack of remorse or shame.
7. Inadequately motivated antisocial behavior.
8. Poor judgment and failure to learn by experience.
9. Pathologic egocentricity and incapacity for love.
10. General poverty in major affective reactions.
11. Specific loss of insight.
12. Unresponsiveness in general interpersonal relations.
13. Fantastic and uninviting behavior, with drink and sometimes without.
14. Suicide rarely carried out.
15. Sex life impersonal, trivial, and poorly integrated.
16. Failure to follow any life plan.

Source: From Hervey Cleckley's "The Mask of Sanity," *Institutions, Etc.: A Journal of Progressive Human Services* 8, no. 9 (September 1985), 21.

CJ ONLINE

Serial Killers

Serial killers are characterized as psychopaths or sociopaths. Go to https://www.dirjournal.com/blogs/americas-famous-serial-killers/ to read more about some of the most well-known cases involving serial killers in the United States.

Were you able to note any of the characteristics of a psychopath listed in Table 3.2 in these articles?

EUGENE GARCIA/AFP/Getty Images

Jeffrey Dahmer was accused by the prosecutor in his case of being "an evil psychopath who lured his victims and murdered them in cold blood." He was convicted of 15 counts of murder and sentenced to serve 15 consecutive life sentences or a total of 957 years in prison. Dahmer was killed in prison in 1994. *Was Dahmer a psychopath? Why or why not?*

rehabilitation is on the individual offender and not on the offender in interaction with the environment in which the criminal behavior occurs. The behaviors that are treated in psychotherapy are not criminal; they are the deep-seated problems. It is assumed that criminal behavior is symptomatic of those problems. That assumption may not be true. Many people who do not engage in crime have deep-seated problems, and many people who do not have those problems do engage in crime.

Humanistic Psychological Theory Humanistic psychological theory, as described here, refers primarily to the work of Abraham Maslow and Seymour Halleck. The theories of Maslow and Halleck are fundamentally psychoanalytic, but they are called humanistic because they assume that human beings are basically good even though they are sometimes influenced by society to act badly. By contrast, Freudian theory assumes that human beings are inherently bad and motivated by sexual and aggressive drives.

Maslow did not apply his theory to crime itself, so we must infer from Maslow's work what we think he would have said about the causes of crime had he addressed the subject. Abraham Maslow believed that human beings are motivated by a hierarchy of basic needs (see Figure 3.2):

1. Physiological (food, water, and procreational sex)
2. Safety (security; stability; freedom from fear, anxiety, chaos, etc.)
3. Belongingness and love (friendship, love, affection, acceptance)
4. Esteem (self-esteem and the esteem of others)
5. Self-actualization (being true to one's nature, becoming everything that one is capable of becoming)[18]

According to Maslow, during a given period, a person's life is dominated by a particular need. It remains dominated by that need until the need has been relatively satisfied, at which time a new need emerges to dominate that person's life. From this view, crime may be understood as a means by which individuals satisfy their basic human needs. They choose crime because they cannot satisfy their needs legally or, for whatever reason, choose not to satisfy their needs legally. An obvious policy implication of the theory is to help people satisfy their basic human needs in legitimate ways. That may require governments to ensure adequate food, shelter, and medical care for those in need or to provide educational or vocational opportunities for those who are unable to obtain them. Those strategies are implied by several sociological theories as well.

FIGURE 3.2 Maslow's Hierarchy of Needs

Need for Self-Actualization

Esteem Needs

Belongingness and Love Needs

Safety Needs

Physiological Needs

Source: Adapted from Abraham Maslow, *Toward a Psychology of Being* (Van Nostrand, 1968).

Seymour L. Halleck views crime as one of several adaptations to the helplessness caused by oppression.[19] For Halleck, there are two general types of oppression, *objective* and *subjective*. Each has two subtypes. The subtypes of objective oppression are (1) social oppression (e.g., oppression resulting from racial discrimination) and (2) the oppression that occurs in two-person interactions (e.g., a parent's unfair restriction of the activities of a child). The subtypes of subjective oppression are (1) oppression from within (guilt) and (2) projected or misunderstood oppression (a person's feeling of being oppressed when, in fact, he or she is not).

For Halleck, the emotional experience of either type of oppression is helplessness, to which the person sometimes adapts by resorting to criminal behavior. Halleck suggests that criminal adaptation is more likely when alternative adaptations, such as conformity, activism, or mental illness, are not possible or are blocked by other people. He also maintains that criminal behavior is sometimes chosen as an adaptation over other possible alternatives because it offers gratifications or psychological advantages that could not be achieved otherwise. Halleck's psychological advantages of crime are listed in Table 3.3.

There are at least three crime policy implications of Halleck's theory. First, sources of social oppression should be eliminated wherever possible. Affirmative action programs, which attempt to rectify historic patterns of discrimination in such areas as employment, are an example of such efforts. Second, alternative, legal ways of coping with oppression must be provided. An example is the opportunity to file claims with the Equal Employment Opportunity Commission (EEOC) in individual cases of employment discrimination. Third, psychotherapy should be provided for subjective oppressions. Psychotherapy could make the individual aware of oppressive sources of guilt or sources of misunderstood oppression so that the individual could better cope with them.

TABLE 3.3 Halleck's 14 Psychological Advantages of Crime

1. Changing one's environment by engaging in crime has greater and more desirable adaptational advantages than changing one's environment through illness or conformity.
2. Crime involves activity, and people feel less helpless when they are active.
3. Crime promises change in a favorable direction regardless of how minor the crime may be.
4. An offender is a relatively free person during the planning and commission of a crime; frequently, he or she is beholden to no one but himself or herself.
5. Crime can be exciting.
6. Crime oftentimes requires an offender to maximize his or her talents and abilities, which he or she might not otherwise use.
7. Feelings of inner oppression and stress can sometimes be relieved through criminal activity.
8. Crime increases external pressures, which can divert an offender's attention from his or her own chronic personal problems.
9. A person is less likely to blame himself or herself for his or her failures, if he or she convinces himself or herself that his or her failures are the fault of others or are external to himself or herself.
10. The criminal role is an excellent rationalization for inadequacy.
11. Being "bad" (crime) has a more esteemed social status than being "mad" (mental illness).
12. Americans have mixed attitudes about crime. It is both condemned and glamorized.
13. Offenders sometimes can form close and relatively nonoppressive relationships with other offenders through their criminal activity.
14. Crime can be pleasurable and gratify needs.

Source: Adapted from Seymour L. Halleck, *Psychiatry and the Dilemmas of Crime* (New York: Harper and Row, 1967), 76–80.

A major problem with the theories of Maslow and Halleck is that they do not go far enough. Neither Maslow nor Halleck asks the basic questions: Why can't people satisfy their basic needs legally, or why do they choose not to? Why don't societies ensure that basic needs can be satisfied legally so that the choice to satisfy them illegally makes no sense? Similarly, why does society oppress so many people, and why aren't more effective measures taken to greatly reduce that oppression?

Sociological Theories

Sociologists emphasize that human beings live in social groups and that those groups and the social structure they create (e.g., political and economic systems) influence behavior. Most sociological theories of crime causation assume that a criminal's behavior is determined by his or her social environment, which includes families, friends, neighborhoods, and so on. Most sociological theories of crime explicitly reject the notion of the born criminal.

The Contributions of Durkheim Many of the sociological theories of crime (actually delinquency) causation have their roots in the work of the French sociologist Émile Durkheim. Durkheim rejected the idea that the world is simply the product of individual actions. His basic premise is that society is more than a simple aggregate of individuals; it is a reality *sui generis* (unique).[20] Rejecting the idea that social phenomena, such as crime, can be explained solely by the biology or psychology of individuals, Durkheim argued that society is not the direct reflection of the characteristics of its individual members because individuals cannot always choose. For Durkheim, social laws and institutions are "social facts" that dominate individuals, and all that people can do is submit to them. The coercion may be formal (e.g., by means of law) or informal (e.g., by means of peer pressure). Durkheim maintained that with the aid of positive science, all that people can expect is to discover the direction or course of social laws so that they can adapt to them with the least amount of pain.

For Durkheim, crime, too, is a social fact. It is a normal aspect of society because it is found in all societies. Nevertheless, different types of societies should have greater or lesser

Bettmann/Getty Images

Émile Durkheim wrote in his volume, *Moral Education*, that, "The more detached from any collectivity man is, the more vulnerable to self-destruction he becomes." *Do you agree? Why? Why not?*

degrees of it. The cause of crime for Durkheim is **anomie**, that is, the dissociation of the individual from the **collective conscience**, or the general sense of morality of the times. He also believed that crime is functional for society by marking the boundaries of morality. In other words, people would not know what acceptable behavior is if crime did not exist. Crime is also functional because it provides a means of achieving necessary social change through, for example, civil disobedience and, under certain circumstances, because it directly contributes to social change, as in the repeal of Prohibition. Because he believed that crime was functional for society, Durkheim did not believe it could or should be eliminated. Instead, he advocated that crime should be contained within reasonable boundaries. He warned, however, that too much crime could destroy society.

The Theory of the Chicago School In the 1920s, members of the Department of Sociology at the University of Chicago tried to identify environmental factors associated with crime. Specifically, they attempted to uncover the relationship between a neighborhood's crime rate and the characteristics of the neighborhood. It was the first large-scale study of crime in the United States and was to serve as the basis for many future investigations into the causes of crime and delinquency.

The research of the Chicago School was based on a model taken from ecology, and as a result, that school is sometimes called the Chicago School of Human Ecology.[21] Ecology is a branch of biology in which the interrelationship of plants and animals is studied in their natural environment. Robert Park was the first of the Chicago theorists to propose this organic or biological analogy—that is, the similarity between the organization of plant and animal life in nature and the organization of human beings in societies.

anomie For Durkheim, the dissociation of the individual from the collective conscience.

collective conscience The general sense of morality of the times.

Park and his colleagues described the growth of American cities like Chicago in ecological terms, saying growth occurs through a process of invasion, dominance, and succession.[22] That is, a cultural or ethnic group *invades* a territory occupied by another group and *dominates* that new territory until it is displaced, or *succeeded*, by another group, and the cycle repeats itself.

This model of human ecology was used by other Chicago theorists, most notably Clifford R. Shaw and Henry D. McKay in their studies of juvenile delinquency in Chicago.[23] From the life histories of delinquents, Shaw and McKay confirmed that most of the delinquents were not much different from nondelinquents in their personality traits, physical condition, and intelligence.[24] However, Shaw and McKay did find that the areas of high delinquency were "socially disorganized." For the Chicago theorists, **social disorganization** is the condition in which the usual controls over delinquents are largely absent; delinquent behavior is often approved of by parents and neighbors; there are many opportunities for delinquent behavior; and there is little encouragement, training, or opportunity for legitimate employment.

social disorganization The condition in which the usual controls over delinquents are largely absent; delinquent behavior is often approved of by parents and neighbors; there are many opportunities for delinquent behavior; and there is little encouragement, training, or opportunity for legitimate employment.

In 1934, Shaw and his colleagues established the Chicago Area Project (CAP), which was designed to prevent delinquency through the organization and empowerment of neighborhood residents. Neighborhood centers, staffed and controlled by local residents, were established in six areas of Chicago. The centers had two primary functions. One was to coordinate community resources, such as schools, churches, labor unions, and industries, to solve community problems. The other function was to sponsor activity programs, such as scouting, summer camps, and sports leagues; to develop a positive interest by individuals in their own welfare; and to unite citizens to solve their own problems. Some aspects of CAP have operated continuously for more than 75 years. Although subjective assessments by individuals involved with the project tout its success, outside evaluations of the project suggest that, for the most part, it has had a negligible effect on delinquency.

One of the problems with the theory of the Chicago School is the presumption that social disorganization is a cause of delinquency. Both social disorganization and delinquency

Stephen Wilkes/Getty Images

From his analysis of the life histories of individual delinquents, Chicago sociologist Clifford Shaw discovered that many delinquent activities began as play activities at an early age. *What kind of play activities do you think are delinquent activities?*

may be the product of other, more basic factors. For example, one factor that contributes to the decline of city neighborhoods is the decades-old practice of *redlining*, in which banks refuse to lend money in an area because of the race or ethnicity of the inhabitants. Though illegal today, the practice continues. What usually happens in redlined areas is that neighborhood property values decline dramatically. Then land speculators and developers, typically in conjunction with political leaders, buy the land for urban renewal or gentrification and make fortunes in the process. In other words, political and economic elites may cause both social disorganization and delinquency–perhaps not intentionally, but by the conscious decisions they make about how a city will grow–making social disorganization appear to be the basic cause of delinquency.

Anomie Theory In an article published in 1938, Robert K. Merton observed that a major contradiction existed in the United States between cultural goals and the social structure.[25] He called the contradiction **anomie**, a concept first introduced by Durkheim. Specifically, Merton argued that in the United States the cultural goal of achieving wealth is deemed possible for all citizens, even though the social structure limits the legitimate "institutionalized means" available for obtaining the goal. For Merton, legitimate institutionalized means are the Protestant work ethic (hard work, education, and deferred gratification); illegitimate means are force and fraud. Because the social structure effectively limits the availability of legitimate institutionalized means, a strain is placed on people. Merton believed that strain could affect people in all social classes, but he acknowledged that it would most likely affect members of the lower class.

anomie For Merton, the contradiction between the cultural goal of achieving wealth and the social structure's inability to provide legitimate institutional means for achieving the goal. For Cohen, it is caused by the inability of juveniles to achieve status among peers by socially acceptable means.

Merton proposed that individuals adapt to the problem of anomie in one of several different ways: (1) conformity, (2) innovation, (3) ritualism, (4) retreatism, and (5) rebellion. According to Merton, most people adapt by conforming; they "play the game." Conformers pursue the cultural goal of wealth only through legitimate institutional means. Innovation is the adaptation at the root of most crime. After rejecting legitimate institutional means, innovators pursue the cultural goal of wealth through illegitimate means. Ritualism is the adaptation of the individual who "takes no chances," usually a member of the lower middle class. Ritualists do not actively pursue the cultural goal of wealth (they are willing to settle for less) but follow the legitimate institutional means anyway. Retreatists include alcoholics, drug addicts, psychotics, and other outcasts of society. Retreatists "drop out"; they do not pursue the cultural goal of wealth, so they do not employ legitimate institutional means. Last is the adaptation of rebellion. Rebels reject both the cultural goal of wealth and the legitimate institutional means of achieving it. They substitute both different goals and different means. Rebellion also can be a source of crime.

A. Ramey/PhotoEdit

According to Albert K. Cohen's version of anomie theory, juveniles who are unable to achieve status among their peers by socially acceptable means sometimes turn to gangs for social recognition. *What are some other reasons why juveniles join gangs?*

In summary, Merton believed that a source of some, but not all, crime and delinquency was anomie, a disjunction or contradiction between the cultural goal of achieving wealth and the social structure's ability to provide legitimate institutional means of achieving the goal.

Beginning in the mid-1950s, concern developed over the problem of juvenile gangs. Albert K. Cohen adapted Merton's anomie theory to his attempt to explain gang delinquency.[26] In attempting to explain such behavior, Cohen surmised that it was to gain status among peers. Thus, Cohen substituted the goal of status among peers for Merton's goal of achieving wealth.

For Cohen, anomie is experienced by juveniles who are unable to achieve status among peers by socially acceptable means, such as family name and position in the community or academic or athletic achievement. In response to the strain, either they can conform to middle-class values (generated primarily through the public school) and resign themselves to their inferior status among their peers, or they can rebel and establish their own value structures by turning middle-class values on their head. Juveniles who rebel in this way tend to find one another and to form groups or gangs to validate and reinforce their new values. Like Merton, Cohen believed that anomie can affect juveniles of any social class but that it disproportionately affects juveniles from the lower class.

FYI

Juvenile Delinquency Prevention and Control

When Robert Kennedy was attorney general of the United States under his brother, President John F. Kennedy, he read Cloward and Ohlin's book *Delinquency and Opportunity: A Theory of Delinquent Gangs* (1960). Kennedy asked Lloyd Ohlin to help shape a new federal policy on juvenile delinquency. That effort, based on anomie theory, produced the Juvenile Delinquency Prevention and Control Act of 1961. The act included a comprehensive action program to provide employment opportunities and work training, in combination with community organization and improved social services, to disadvantaged youths and their families.

Source: George B. Vold and Thomas J. Bernard, *Theoretical Criminology,* 3rd ed. (New York: Oxford, 1986), 201.

Richard Cloward and Lloyd Ohlin extended Merton's and Cohen's formulations of anomie theory by suggesting that not all gang delinquents adapt to anomie in the same way. Cloward and Ohlin argue that the type of adaptation made by juvenile gang members depends on the *illegitimate opportunity structure* available to them.[27] They identified three delinquent subcultures: the criminal, the violent, and the retreatist. According to Cloward and Ohlin, if illegitimate opportunity is available to them, most delinquents will form *criminal* gangs to make money. However, if neither illegitimate nor legitimate opportunities to make money are available, delinquents often become frustrated and dissatisfied and form *violent* gangs to vent their anger. Finally, there are delinquents who, for whatever reason, are unable to adapt by joining either criminal or violent gangs. They *retreat* from society, as in Merton's retreatist adaptation, and become alcoholics and drug addicts.

The policy implications of anomie theory are straightforward: reduce aspirations or increase legitimate opportunities, or do both. Increasing legitimate opportunities, already a cornerstone of the black civil rights movement, struck a responsive chord as the 1960s began. Examples of this strategy are affirmative action employment programs, expansion of vocational education programs, and government grants that enable low-income students to attend college. Reducing aspirations (i.e., desires to be wealthy) received little attention, however, because to attempt it would be to reject the "American dream," a principal source of motivation in a capitalist society.

Among the problems with the anomie theories of Merton, Cohen, and Cloward and Ohlin is their reliance on official statistics (police and court records) as measures of crime. Because these theorists relied on official statistics, their theories focus on lower-class crime and delinquency and ignore white-collar and government crimes.

Learning Theories Gabriel Tarde was one of the first theorists to believe that crime was something learned by normal people as they adapted to other people and the conditions of their environment. His theory was a product of his experience as a French lawyer and magistrate and was described in his book *Penal Philosophy*, published in 1890. Reflecting the state of knowledge about the learning process in his day, Tarde viewed all social phenomena as the product of imitation. Through **imitation,** or **modeling**, a person can learn new responses, such as criminal behavior, by observing others without performing any overt act or receiving direct reinforcement or reward.[28]

imitation or modeling A means by which a person can learn new responses by observing others without performing any overt act or receiving direct reinforcement or reward.

The first twentieth-century criminologist to forcefully argue that criminal behavior was learned was Edwin H. Sutherland. His theory of **differential association**, developed between 1934 and 1947, was that persons who become criminal do so because of contacts with criminal definitions and isolation from noncriminal definitions. Together with its more recent modifications, his theory remains one of the most influential theories of crime causation.[29] The nine propositions of Sutherland's theory are presented in Table 3.4.

differential association Sutherland's theory that persons who become criminal do so because of contacts with criminal definitions and isolation from anticriminal patterns.

Borrowing a premise from the theory of the Chicago School, Sutherland maintained that differential associations would not produce criminality if it were not for *differential social organization.* In other words, the degree to which communities promote or inhibit criminal associations varies with the way or the degree to which they are organized (i.e., the extent of *culture conflict*).

Since its final formulation in 1947, modifications and additions have been made to Sutherland's theory as new developments in learning theory have emerged. For example, Daniel Glaser modified Sutherland's theory by introducing *role theory* and by arguing that criminal behavior could be learned by identifying with criminal roles and not just by associating with criminals.[30] Thus, a person could imitate the behavior of a drug dealer without actually having met one. Glaser obviously believed that the media had a greater influence on the learning of criminal behavior than Sutherland believed they had.

Robert L. Burgess and Ronald L. Akers, as well as C. Ray Jeffery, adapted the principles of *operant conditioning* and *behavior modification*, developed by B. F. Skinner, and the principles of *modeling*, as developed by Albert Bandura, to the explanation of criminal behavior. Burgess, Akers, and Jeffery integrated psychological concepts with sociological ones. Although they referred to their theories by different names, we will use the more general term *learning theory.*

TABLE 3.4 Sutherland's Nine Propositions of Differential Association

1. Criminal behavior is learned.
2. The learning of criminal behavior occurs as people interact and communicate with each other.
3. The learning of criminal behavior occurs primarily within familiar personal groups.
4. The learning of criminal behavior includes (a) techniques of committing the crime, which can be very simple or very complicated, and (b) the specific motives, drives, rationalizations, and attitudes.
5. Definitions of the legal codes as favorable and unfavorable provide the specific direction from which motives and drives are learned.
6. The principle of differential association is: A person becomes criminal when definitions favorable to law violation exceed definitions unfavorable to law violation.
7. Differential associations may vary in frequency, duration, priority, and intensity.
8. The mechanisms used in any type of learning also are used in the learning of criminal behavior by association with criminal and anticriminal patterns.
9. Criminal behavior is an expression of general needs and values but is not explained by those general needs and values because noncriminal behavior also is an expression of the same needs and values.

Source: Adapted from Edwin H. Sutherland and Donald R. Cressey, *Criminology*, 9th ed. (Philadelphia: Lippincott, 1974), 75–77.

learning theory A theory that explains criminal behavior and its prevention with the concepts of positive reinforcement, negative reinforcement, extinction, punishment, and modeling, or imitation.

positive reinforcement The presentation of a stimulus that increases or maintains a response.

negative reinforcement The removal or reduction of a stimulus whose removal or reduction increases or maintains a response.

extinction A process in which behavior that previously was positively reinforced is no longer reinforced.

punishment The presentation of an aversive stimulus to reduce a response.

Learning theory explains criminal behavior and its prevention with the concepts of *positive reinforcement, negative reinforcement, extinction, punishment,* and *modeling,* or *imitation.* In this view, crime is committed because it is positively reinforced, negatively reinforced, or imitated. We described the imitation, or modeling, of criminal behavior in our earlier discussion of Tarde. Here, we will focus on the other concepts.[31]

Positive reinforcement is the presentation of a stimulus that increases or maintains a response. The stimulus, or *reward,* can be either material, such as money, or psychological, such as pleasure. People steal (a response) because of the rewards–for example, the objects or money–that they receive. They use drugs (at least at first) because of the rewards, the pleasure, that the drugs give them.

Negative reinforcement is the removal or reduction of a stimulus whose removal or reduction increases or maintains a response. The stimulus in negative reinforcement is referred to as an *aversive stimulus.* Aversive stimuli, for most people, include pain and fear. Stealing may be negatively reinforced by removing or reducing the aversive stimuli of the fear and pain of poverty. For drug addicts, the use of addictive drugs is negatively reinforced because the drugs remove or reduce the aversive stimulus of the pain of drug withdrawal. In short, both positive and negative reinforcement explain why a behavior, such as crime, is maintained or increases. Both types of reinforcement can affect simultaneously the same behavior. In other words, people may commit crime, in this view, both because the crime is rewarded and because it removes an aversive stimulus.

According to learning theory, criminal behavior is reduced, but not necessarily eliminated, through *extinction* or *punishment.* **Extinction** is a process in which behavior that previously was positively reinforced is no longer reinforced. In other words, the rewards have been removed. Thus, if burglars were to continually come up empty-handed in their quests–not to receive rewards for their efforts–they would no longer continue to commit burglary. **Punishment** is the presentation of an aversive stimulus to reduce a response. It is the principal method used in the United States, and in other countries, to prevent crime or, at least, reduce it. For example, one of the reasons offenders are imprisoned is to punish them for their crimes.

Among the policy implications of learning theory is to punish criminal behavior effectively; that is, according to learning theory principles. For a variety of reasons, punishment is not used effectively in criminal justice in the United States. For example, to employ punishment effectively, one must prevent escape. Escape is a natural reaction to the presentation of an aversive stimulus like imprisonment. In the United States, the chances of an offender's escaping punishment are great. Probation probably does not function as an aversive stimulus, and most offenders, especially first-time offenders, are not incarcerated.

To be effective, punishment must be applied consistently and immediately. As for immediacy, the process of criminal justice in the United States generally precludes punishment immediately after a criminal act is committed. The process is a slow and methodical one. Consistent application of punishment is rare because most criminal offenders are not caught.

In addition, extended periods of punishment should be avoided, or the effectiveness of the punishment will be reduced. The United States currently imprisons more of its offenders for longer periods than any other country in the world, though in many cases, inmates actually serve only a fraction of their original sentences. A related issue is that punishment is far less effective when the intensity with which the aversive stimulus is presented is increased gradually than when the stimulus is introduced at full intensity. Prolonged imprisonment is a gradual process of punishment that lacks the full intensity and immediacy of corporal punishment, for example. Figure 3.3 shows how the corporal punishment of caning is done in Singapore.

Roslan Rahman/AFP/Getty Images

In 1994, then 18-year-old American student Michael Fay pled guilty to vandalism charges in Singapore and was sentenced to 3 months in jail, a $2,200 fine, and 6 strokes of the cane. *Should the flogging of convicted offenders be allowed in the United States? Why or why not?*

To be effective, punishment must also be combined with extinction. That is, the rewards that maintain the behavior must be removed. In the United States, after imprisonment, offenders are generally returned to the environments in which their crimes originally were committed and rewarded.

Finally, for punishment to be effective, it must be combined with the positive reinforcement of alternative, prosocial behaviors. Rarely does this occur.

We must emphasize that for learning theorists, positive reinforcement is a much more effective and preferred method of manipulating behavior than is punishment because positive reinforcement does not suffer the disadvantages associated with punishment. That point is often overlooked by criminal justice decision makers. Among the disadvantages of punishment are the effort to escape punishment by means other than law-abiding behavior; the development of negative self-concepts by offenders, who come to view themselves (instead of their criminal behaviors) as bad, making rehabilitation difficult; and the causing of aggression.

Among the problems with a learning theory of crime causation is that it ignores the criminalization process. It fails to consider why the normal learned behaviors of some groups are criminalized while the normal learned behaviors of other groups are not. For example, why is marijuana consumption illegal, while cigarette or alcohol consumption is not? Learning theory ignores the effect that political and economic power have on the definition of criminal behavior.

FIGURE 3.3 How Caning Is Done in Singapore

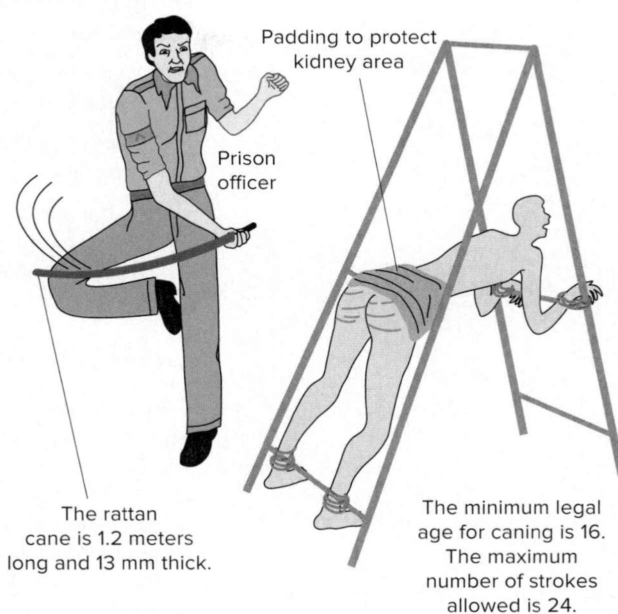

Social Control Theories The key question for social control theorists is not why people commit crime and delinquency, but rather why they do not. Why do people conform? From the perspective of **social control theory**, people are expected to commit crime and delinquency unless they are prevented from doing so. They will commit crime, that is, unless they are properly socialized.

social control theory A view in which people are expected to commit crime and delinquency unless they are prevented from doing so.

Like many of the other sociological theories of crime causation, social control theories have their origins in the work of Durkheim. It was not until the 1950s, however, that social control theories began to emerge to challenge other, more dominant theories, such as anomie and differential association. Among the early social control theorists were Albert J. Reiss, Jackson Toby, F. Ivan Nye, and Walter C. Reckless. Despite the important contributions of those early theorists, modern social control theory in its most detailed elaboration is attributed to the work of Travis Hirschi. Hirschi's 1969 book, *Causes of Delinquency*, has had a great influence on current criminological thinking.

As did proponents of earlier social control theories, Hirschi argued that delinquency (his was not a theory of adult criminality) should be expected if a juvenile is not properly socialized. For Hirschi, proper socialization involves the establishment of a strong moral bond between the juvenile and society. This *bond to society* consists of (1) *attachment* to others, (2) *commitment* to conventional lines of action, (3) *involvement* in conventional activities, and (4) *belief* in the moral order and law. Thus, delinquent behavior is likely to occur if there is (1) inadequate attachment, particularly to parents and school; (2) inadequate commitment, particularly to educational and occupational success; (3) inadequate involvement in such conventional activities as scouting and sports; and (4) inadequate belief, particularly in the legitimacy and morality of the law. For Hirschi, the units of social control most important in the establishment of the bond are the family, the school, and the law.

In a more recent book, Michael Gottfredson and Travis Hirschi argue that the principal cause of many deviant behaviors, including crime and delinquency, is ineffective child-rearing, which produces people with low self-control.[32] Low self-control impairs a person's ability to accurately calculate the consequences of his or her actions and is characterized by impulsivity, insensitivity, physical risk taking, shortsightedness, and lack of verbal skills. This theory posits that everyone has a predisposition toward criminality; therefore, low self-control makes it difficult to resist.

One of the appealing aspects of Hirschi's social control theory–and Gottfredson and Hirschi's self-control theory–is their seemingly commonsense policy implications. To prevent delinquency, juveniles must be properly socialized; they must develop a strong moral bond to society. As part of that strategy, children must be reared properly so that they develop a high level of self-control.

Although social control theory is currently very influential in the thinking of many criminologists, it has not escaped extensive criticism. Perhaps the major problem, at least for some criminologists, is the theory's assumption that delinquency will occur if not prevented. Some criminologists find it troublesome that the theory rejects altogether the idea of delinquent motivation. Another problem with social control theory is that it does not explain how juveniles are socialized. For example, how are attachments to others produced and changed? Finally, Hirschi's argument, as stated, does not allow for delinquency by juveniles who are properly socialized, nor does it allow for conformity by juveniles who are not properly socialized.

> **THINKING CRITICALLY**
>
> 1. What are the five key assumptions of the positivist school of thought?
> 2. How much crime do you think is related to biological factors? Which factors are most important?
> 3. Explain psychoanalytic and humanistic psychological theory. What are some of the problems associated with those theories?
> 4. Which one of the sociological theories do you think explains the most crime? Defend your choice.

Critical Approaches to Explaining Crime

Critical theories are, in part, a product of a different conception of American society that began to emerge toward the end of the 1950s. Concepts such as racism, sexism, capitalism, imperialism, monopoly, exploitation, and oppression were beginning to be employed with greater frequency to describe the social landscape. In the 1960s, the period of naive acceptance of the status quo and the belief in the purely benevolent actions of government and corporations ended for many social scientists, and critical theory emerged.

Not surprisingly, the basic assumptions of critical theories differ both from those of classical and neoclassical theories and from those of positivist theories. First, unlike classical and neoclassical theories, which assume that human beings have free will, and positivist theories, which assume that human beings are determined, critical theories assume that human beings are both determined *and* determining. In other words, critical theories assume that human beings are the creators of the institutions and structures that ultimately dominate and constrain them. Second, in contrast to both classical and neoclassical theories and positivist theories, critical theories assume that conflict is the norm, that society is characterized primarily by conflict over moral values rather than consensus. Finally, unlike positivist theorists, many critical theorists assume that everything they do is value laden by virtue of their being human; that is, they believe it is impossible to be objective or value neutral in anything a person does.

Labeling Theory

labeling theory A theory that emphasizes the criminalization process as the cause of some crime.

criminalization process The way people and actions are defined as criminal.

The focus of **labeling theory** is the **criminalization process**–the way people and actions are defined as criminal–rather than the positivist concern with the peculiarities of the criminal actor. From this perspective, the distinguishing feature of all "criminals" is that they have been the object of a negative social reaction. In other words, the state and its agents have designated them as different and "bad."

Note that labeling theorists attempt to explain only what Edwin Lemert called secondary deviance.[33] For our purposes, secondary deviance is the commission of crime subsequent to the first criminal act and the acceptance of a criminal label. Secondary deviance begins with an initial criminal act, or what Lemert called primary deviance. The causes of initial criminal acts are unspecified. Nevertheless, if society reacts negatively to an initial criminal act, especially through official agents of the state, the offender is likely to be stigmatized, or negatively labeled. It is possible, even likely, that there will be no reaction at all to an initial criminal act or that the offender will not accept or internalize the negative label. However, if the negative label is successfully applied to the offender, the label may become a self-fulfilling prophecy in which the offender's self-image is defined by the label. Secondary deviance is the prophecy fulfilled.

As is well known, once a person is labeled and stereotyped as a "criminal," he or she probably will be shunned by law-abiding society, have difficulty finding a good job, lose some civil rights (if convicted of a felony), and suffer a variety of other disabilities. The criminal (or delinquent) label is conferred by all of the agencies of criminal justice–the police, the courts, and the correctional apparatus–as well as by the media, the schools, churches, and other social institutions. The irony is that in its attempt to reduce crime and delinquency, society inadvertently may be increasing it by labeling people and producing secondary deviance.

A policy implication of labeling theory is simply not to label or to employ *radical nonintervention*.[34] This might be accomplished through decriminalization (the elimination of many behaviors from the scope of the criminal law), diversion (removing offenders from involvement in the criminal justice process), greater due-process protections (replacing discretion with the rule of law), and deinstitutionalization (a policy of reducing jail and prison populations and construction).[35]

An alternative to the nonintervention strategy is John Braithwaite's *reintegrative shaming*.[36] In this strategy, disappointment is expressed for the offender's actions, and the offender is shamed and punished. What is more important, however, is that following the expression of disappointment and shame is a concerted effort on the part of the community to forgive the offender and reintegrate him or her back into society. Braithwaite contends that the practice of reintegrative shaming is one of the principal reasons for Japan's relatively low crime rate.

A problem with labeling theory is that it tends to overemphasize the importance of the official labeling process.[37] On the one hand, the impression is given that innocent people are arbitrarily stigmatized by an oppressive society and that as a result, they begin a life of crime. That probably does not happen very much. On the other hand, the impression is given that offenders resist the criminal label and accept it only when they are no longer capable of fighting it. However, in some communities, the criminal label, or some variation of it, is actively sought. Perhaps the most telling problem with labeling theory is the question of whether stigmatizing someone as criminal or delinquent causes more crime and delinquency than it prevents. To date, the answer is unknown.

FYI

Shaming as a Crime Prevention Strategy

Current crime prevention programs based on the theoretical proposition that shame will prevent crime include programs that publicize the names and faces of people arrested for trying to buy or sell sex or require convicted DUI offenders to put bumper stickers on their cars identifying them as such. For the most part, these strategies ignore reintegration.

myth

Most offenders resist being labeled *criminal* and accept the label only when they are no longer capable of fighting it.

fact

In some communities, the label *criminal*, or some variation of it, is actively sought.

Conflict Theory

Unlike classical and neoclassical and positivist theories, which assume that society is characterized primarily by consensus, **conflict theory** assumes that society is based primarily on conflict between competing interest groups—for instance, the rich against the poor, management against labor, whites against minorities, men against women, and adults against children. In many cases, competing interest groups are not equal in power and resources. Consequently, one group is dominant, and the other is subordinate.

One of the earliest theorists in the United States to apply conflict theory to the study of crime was George B. Vold. For Vold and other conflict theorists, such as Austin Turk, many behaviors are defined as crimes because it is in the interest of dominant groups to do so.[38]

According to conflict theorists, dominant groups use the criminal law and the criminal justice system to control subordinate groups. However, the public image is quite different. The public image of the criminal law and the criminal justice system is that they are value-neutral institutions—that is, that neither institution has a vested interest in who "wins" a dispute. The public image is that the only concern of the criminal law and the criminal justice system is resolving disputes between competing interest groups justly and—more importantly—peacefully. According to conflict theorists, this public image legitimizes the authority and practices of dominant groups and allows them to achieve their own interests at the expense of less powerful groups. In this view, crime also serves the interests of dominant groups. It deflects the attention of subordinate group members from the many problems that dominant groups create for them and turns that attention to subordinate group members who are defined as criminal.

All behavior, including criminal behavior, in this view, occurs because people act in ways consistent with their social positions. Whether white-collar crime or ordinary street crime, crime is a response to a person's social situation. The reason members of subordinate groups appear in official criminal statistics more frequently than members of dominant groups is that the dominant groups have more control over the definition of criminality. Thus, they are better able to ensure that the responses of subordinate group members to their social situations will be defined and reacted to as criminal. For conflict theorists, the amount of crime in a society is a function of the extent of conflict generated by **power differentials**, or the ability of some groups to dominate other groups in that society. Crime, in short, is caused by **relative powerlessness**, the inability to dominate other groups.

There are two principal policy implications of conflict theory. One is for dominant groups to give up some of their power to subordinate groups, making the weaker more powerful and reducing conflict. Increasing equality in that way might be accomplished by redistributing wealth through a more progressive tax system, for example. Another way to

conflict theory A theory that assumes that society is based primarily on conflict between competing interest groups and that criminal law and the criminal justice system are used to control subordinate groups. Crime is caused by relative powerlessness.

power differentials The ability of some groups to dominate other groups in a society.

relative powerlessness In conflict theory, the inability to dominate other groups in society.

Ronen Tivony/SOPA Images/Shutterstock

A. Ramey/PhotoEdit

For conflict theorists, conflict between competing interest groups can sometimes lead to crime. *What kinds of group conflicts are more likely to lead to crime?*

increase equality, at least in the political arena, would be to strictly limit or eliminate altogether the contributions of wealthy people and corporations to political candidates. The other policy implication is for dominant group members to become more effective rulers and subordinate group members, better subjects. To do so, dominant group members would have to do a better job of convincing subordinate group members that the current unfair distribution of power in society is in their mutual interests.

A problem with conflict theory is that it generally fails to specify the sources of power in society. When those sources are identified, they are usually attributed to the personal characteristics of elites; that is, people with power are said to be smarter, better educated, luckier, and better able to defer gratification. Conflict theorists seem to ignore that power in society comes primarily from the ownership of private property.

Another criticism of conflict theory is that it is basically reformist in its policy implications. Conflict theorists generally assume that crime, as well as other social problems, can be corrected by existing social institutions. For example, if only the agencies of criminal justice were more effective, a conflict theorist might argue, crime would be reduced greatly. Historical evidence suggests that this assumption may not be true.

Radical Theory

The social and political turmoil in the United States during the 1960s and 1970s created a renewed interest in Marxist theory. Although Karl Marx wrote very little about crime and criminal justice, **radical theories** of crime causation are generally based on Marx's ideas. Among the first criminologists in the United States to employ Marxist theory to explain crime and justice were Richard Quinney, William J. Chambliss, and Anthony M. Platt.

radical theories Theories of crime causation that are generally based on a Marxist theory of class struggle.

Radical criminologists argue that *capitalism* is an economic system that requires people to compete against each other in the individualistic pursuit of material wealth. A defining characteristic of a capitalist society is that a very small percentage of people are the big winners in the competitive struggle for material wealth. Figure 3.4 shows the distribution of wealth in the United States in 2016.

Karl Marx

In an article entitled "Population, Crime and Pauperism," which appeared in the *New York Daily News* September 16, 1859, Karl Marx observed that "there must be something rotten in the very core of a social system which increases its wealth without decreasing its misery; and increases in crimes even more than in numbers."

The winners do everything in their considerable power to keep from becoming losers, including taking advantage of other people, preying on them. Their power is considerable, by virtue of their ownership of material wealth. (The really big winners are members of the *ruling class.*) Losers—relatively speaking, members of the *working class* and the *nonworking class*—in an effort to become winners, usually do what the winners do: prey on weaker people. Radical criminologists believe that the more unevenly wealth is distributed in a society, the more likely people are to be able to find persons weaker than themselves.

It is important to understand that for radical criminologists, the destructive effects of capitalism, such as crime, are not caused alone by income or property inequality or by poverty. Rather, the competition among wealthy people and among poor people and between rich and poor people—the **class struggle**—and the practice of taking advantage of other people cause crime. They also cause income or property inequality, poverty, and many of the other problems that are characteristic of a capitalist society. Crime in capitalist societies, according to radical criminologists, is a rational response to the circumstances in which people find themselves in the competitive class struggle to acquire material wealth. "Senseless" violent crime, which is most often committed by poor people against each other, is frequently a product of the demoralizing and brutalizing conditions under which many people are forced to live.

Radical criminologists argue that noncapitalist societies should have different types of crime and much lower rates of crime, as traditionally defined, "because the less intense class struggle should reduce the forces leading to and the functions of crime."[39]

Radical criminologists also define crime as a violation of human rights. As Tony Platt explains:

> A radical perspective defines crime as a violation of politically defined human rights: the truly egalitarian rights to decent food and shelter, to human dignity and self-determination, rather than the so-called right to compete for an unequal share of wealth and power.[40]

A radical definition of crime includes "imperialism, racism, capitalism, sexism and other systems of exploitation which contribute to human misery and deprive people of their human potentiality."[41] Although many behaviors currently proscribed by criminal law would be included in this radical definition, other behaviors now considered crimes would be excluded, such as prostitution, gambling, and drug use, and some behaviors not now considered crimes would be added, such as racism, sexism, and imperialism.

Consequently, the policy implications of radical theory include demonstrating that the current legal definition of crime supports the ruling class in a capitalist system and redefining crime as a violation of human rights. For nearly all radical criminologists—anarchists are exceptions—the solution to the crime problem is a benevolent socialist society governed by democratically elected representatives of the population. Arguably, the biggest difference

FIGURE 3.4 Financial Wealth Distribution in the United States, 2016

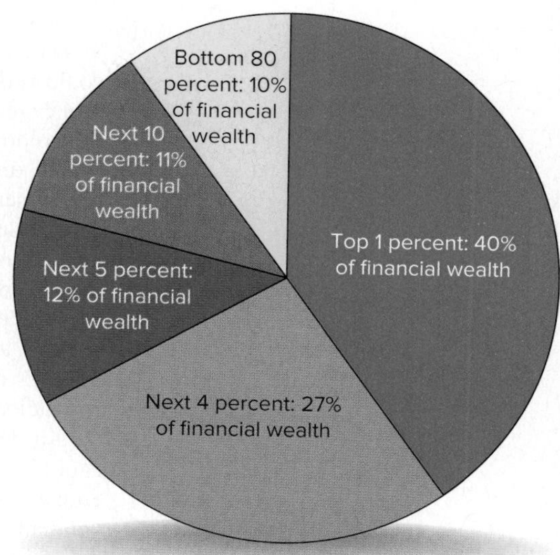

Source: Adapted from Table 2 in Edward N. Wolff, "Has Middle Class Wealth Recovered?" Presented at the Allied Social Sciences Association (ASSA) Meeting, January 6, 2018, accessed March 4, 2019, https://www.aeaweb.org/conference/2018/preliminary/paper/5ZFEEf69.

class struggle For radical criminologists, the competition among wealthy people and among poor people and between rich people and poor people, which causes crime.

Ulrich Baumgarten/Contributor/Getty Images

Radical criminologists define crime as a violation of politically defined human rights, such as the right to shelter. *Do you think that Americans should have a right to shelter? Why or why not?*

between the envisioned socialist society and the current capitalist society in the United States is that governments in the socialist society would regulate the economy to promote public welfare. Governments in capitalist America regulate the economy to promote the accumulation of material wealth by individuals. Radical criminologists stress that the difference is primarily a matter of priorities. Some accumulation of private material wealth would probably exist in the socialist society, and even capitalist America provides for the public welfare through such socialistic programs as Social Security, Medicare, and Medicaid. Another important difference between the two societies is that in the socialist society, human diversity in all areas of life would be not only tolerated but also appreciated by government agents (and, one would hope, by the rest of society). In the socialist society, the criminal law would be based on a slightly modified version of the positive sanction of the classical school: that every member of society has a right to do anything that is not prohibited by law without fearing anything but natural consequences. Acts prohibited by law would be those that violate basic human rights or the public welfare, interpreted as liberally as possible. The recognition of "hate crimes" by federal legislation in the United States is an example of an attempt to protect human diversity.

Radical criminologists maintain that creating such a socialist society would first require the development of political awareness among all people disadvantaged by the capitalist system. Such people must be made aware that they are in a class struggle. Once enough people are aware, according to radical criminologists, then only through *praxis* (human action based on theoretical understanding) will the new socialist society be achieved.

One objection to radical theory is that the radical definition of crime as the violation of human rights is too broad and vague. Although radical criminologists concede that it may be difficult to determine what a human right is, they generally assume that most people know when a human right has been violated.

Other criticisms of radical theory are that its adherents are pursuing a political agenda and thus are not objective in their work; that its causal model is wrong, that is, that social arrangements do not cause people to commit crime, as radical theorists argue, but rather that crime is committed by people who are born evil and remain evil; that it has not been tested satisfactorily; that it cannot be tested satisfactorily; and that it is utopian in its policy implications.

FYI **Utopians**

Early socialists criticized the rise of industry as the cause of great hardship among working people. During the early 1800s, socialists, who were sometimes called *utopians,* tried to foster communities with ideal social and economic conditions. Their name came from Sir Thomas More's *Utopia,* published in England in 1516. More's book describes an ideal society with justice and equality for all citizens.

Other Critical Theories

The law-and-order climate of the 1980s and 1990s was not a period in which critical theories of crime causation received much attention from government bureaucrats and criminal justice practitioners. Nevertheless, critical scholars continued to produce and to refine critical analyses of government efforts to understand and to address the crime problem. In this subsection, we briefly describe some of the new directions taken by critical theorists. These descriptions show the diversity of current critical thought.

British or Left Realism The focus of many critical criminologists has been on crimes committed by the powerful. While pursuing that area of study, however, they have tended either to ignore or to romanticize working-class crime and criminals. By the mid-1980s, a group of social scientists in Great Britain had begun to criticize that tendency and to argue that critical criminologists needed to redirect their attention to the fear and the very real victimization experienced by working-class people.[42] These **left realists** correctly observed that crimes against the working class were being perpetrated not only by the powerful but also by members of their own class. They admonished their critical colleagues to take crime seriously, especially street crime and domestic violence.

left realists A group of social scientists who argue that critical criminologists need to redirect their attention to the fear and the very real victimization experienced by working-class people.

relative deprivation Refers to inequalities (in resources, opportunities, material goods, etc.) that are defined by a person as unfair or unjust.

One of the leading exponents of left realism, Jock Young, has identified relative deprivation as a potent though not exclusive cause of crime.[43] **Relative deprivation** refers to inequalities (in resources, opportunities, material goods, etc.) that are defined by a person as unfair or unjust. Left realists have argued that police power must be employed to protect people living in working-class communities. Certain versions of community policing are examples of such policy.

Left realism has been criticized for holding a contradictory position regarding the police (and other agencies of criminal justice).[44] On the one hand, left realists want to give the police more power to combat crime, especially crime committed against the working class. On the other hand, they want to reduce the power of the police to intervene in people's lives and want to make the police more accountable for their actions. Another criticism of left realism is that its emphasis on the reform of criminal justice practice, rather than on radical change, makes it little different in that regard from conflict theory, classical and neoclassical theories, and many positivist theories.[45]

Peacemaking Criminology This perspective rejects the idea that predatory criminal violence can be reduced by repressive state violence. In this view, "wars" on crime make matters only worse. Consisting of a mixture of anarchism, humanism, socialism, and Native American and Eastern philosophies, **peacemaking criminology** suggests that the solutions to all social problems, including crime, are the transformation of human beings, mutual dependence, reduction of class structures, creation of communities of caring people, and universal social justice. For peacemaking criminologists, crime is suffering, and, therefore, to reduce crime, suffering must be reduced.[46] Policy emphasis is placed on the transformation of human beings through an inner rebirth or spiritual rejuvenation (inner peace) that enables individuals to experience empathy with those less fortunate and gives them a desire to respond to the needs of other people.

peacemaking criminology An approach that suggests that the solutions to all social problems, including crime, are the transformation of human beings, mutual dependence, reduction of class structures, creation of communities of caring people, and universal social justice.

Peacemaking criminology has been criticized for its extreme idealism and its emphasis on the transformation of individuals as a way of transforming society rather than on the transformation of society as a way of transforming individuals.

Feminist Theory Although feminism is not new, the application of feminist theory to the study of crime is.[47] Recognizing that the study of crime has always been male-centered, feminist criminologists seek a feminine perspective. Specifically, the focus of **feminist theory** is on women's experiences and ways of knowing, because, in the past, men's experiences have been taken as the norm and generalized to the population. As a result, women and girls have been omitted almost entirely from theories of crime and delinquency.

feminist theory A perspective on criminality that focuses on women's experiences and seeks to abolish patriarchy.

Three areas of crime and justice have commanded most of the attention of feminist theorists: (1) the victimization of women, (2) gender differences in crime, and (3) *gendered justice*, that is, differing treatment of female and male offenders and victims by the agencies and agents of criminal justice. Regarding gender differences in crime, two questions seem to dominate: (1) Do explanations of male criminality apply to women? (2) Why are women less likely than men to engage in crime?

Not all feminists share the same perspective on the issues noted. At least four different types of feminist thought have been identified: liberal, radical, Marxist, and socialist. (Conspicuously omitted from this typology are women of color, whose gender and race place them in a uniquely disadvantaged position.) Radical, Marxist, and socialist feminist thought share, to varying degrees, a belief that the problems of women, including female victimization, lie in the institution of **patriarchy**, men's control over women's labor and sexuality. The principal goal of most feminist theory is to abolish patriarchal relationships.

patriarchy Men's control over women's labor and sexuality.

Hero Images/Getty Images

Peacemaking criminologists also advocate restorative justice, compassion, and community action. *Should these principles replace the punitive justice that currently guides criminal justice in the United States? Why or why not?*

Liberal feminists are the exception. They do not generally view discrimination against women as the product of patriarchy. For them, the solution to women's subordinate position in society is the removal of obstacles to their full participation in social life. Liberal feminists seek equal opportunity, equal rights, and freedom of choice. As for crime, liberal feminists point to gender socialization (i.e., the creation of masculine and feminine identities) as the primary culprit.

One of the principal criticisms of feminist criminology is the focus on gender as a central organizing theme. Such a focus fails to appreciate differences among women—for example, differences between the experiences of black women and white women.[48] However, that criticism has been addressed by some feminist criminologists, who are more attentive to the intersection of race, class, sexual orientation, and gender because of the realization that "women's experiences of gender vary according to their position in racial and class hierarchies."[49] Another problem similar to the one associated with left realism is many feminists' contradictory position regarding the police (and the criminal justice system in general). On the one hand, several feminists call for greater use of the police to better protect women from abuse; on the other hand, some feminists concede that giving more power to the police, under present circumstances, will lead only to further discrimination and harassment of minority males and females.[50] A similar contradictory position is held by many feminists toward using the law to improve gender relations. The law, after all, is almost entirely the product of white males.

postmodernism An area of critical thought that, among other things, attempts to understand the creation of knowledge and how knowledge and language create hierarchy and domination.

Postmodernism **Postmodernism** originated in the late 1960s as a rejection of the "modern" or Enlightenment belief in scientific rationality as the route to knowledge and progress. Among the goals of this area of critical thought are to understand the creation of knowledge and how knowledge and language create hierarchy and domination. Postmodernist ideas began to be introduced in law and criminology during the late 1980s.[51] As applied to the area of crime and criminal justice, the major foci have been critical analyses of the privileged position of "the Law" and the construction of crime theories.

With regard to the Law, postmodernist criminologists reject the idea "that there is only one true interpretation of a law or for that matter the U.S. Constitution."[52] They argue, instead, that there is a plurality of interpretations that are dependent, in part, on the particular social context in which they arise. As in other critical criminologies, the law, from a postmodernist view, always has a human author and a political agenda.[53] As for crime theories, postmodernist criminologists typically abandon the usual notion of causation. From the postmodernist perspective:

> Crime is seen to be the culmination of certain processes that allow persons to believe that they are somehow not connected to other humans and society. These processes place others into categories or stereotypes and make them different or alien, denying them their humanity. These processes result in the denial of responsibility for other people and to other people.[54]

Postmodernist criminologists would, among other things, replace the prevailing description of the world with new conceptions, words, and phrases that convey alternative meanings, as Sutherland did when he introduced the concept of white-collar crime. These new descriptions would tell different stories about the world as experienced by historically subjugated people.[55] Postmodernist criminologists would also replace the formal criminal justice apparatus with informal social controls so that local groups and local communities handle the current functions of criminal justice.[56] This strategy is consistent with the one advocated by peacemaking criminologists. Postmodernist criminology has been criticized for its relativism and subjectivism or for being an "anarchy of knowledge": "If truth is not possible, how can we decide anything?"[57]

THINKING CRITICALLY

1. How would you explain labeling theory?
2. What is peacemaking criminology? Is this theory realistic?
3. Explain feminist theory and its key criticisms.
4. How might postmodernists explain acts of domestic terrorism?

SUMMARY

1. **Define criminological theory.**

Criminological theory is the explanation of the behavior of criminal offenders, as well as the behavior of police, attorneys, prosecutors, judges, correctional personnel, victims, and other actors in the criminal justice process. It helps us understand criminal behavior and the basis of policies proposed and implemented to prevent and control crime.

2. **State the causes of crime according to classical and neoclassical criminologists and their policy implications.**

Classical and neoclassical criminologists theorize that human beings are free-willed individuals who commit crime when they rationally calculate that the crime will give them more pleasure than pain. In an effort to deter crime, classical criminologists advocate the following policies: (1) establish a social contract; (2) enact laws that are clear, simple, unbiased, and reflect the consensus of the population; (3) impose punishments that are proportionate to the crime, prompt, certain, public, necessary, the least possible in the given circumstances, and dictated by law rather than by judges' discretion; (4) educate the public; (5) eliminate corruption from the administration of justice; and (6) reward virtue. Neoclassical criminologists introduced the concepts that mitigating circumstances might inhibit the exercises of free will and that punishment should be rehabilitative.

3. **Describe the biological theories of crime causation and their policy implications.**

The basic cause of crime for biological positivists has been biological inferiority, which is indicated by physical or genetic characteristics that distinguish criminals from noncriminals. The policy implications of biological theories of crime causation include a choice of isolation, sterilization, or execution. Biological theorists also advocate brain surgery, chemical treatment, improved diets, and better mother and child care.

4. **Describe the different psychological theories of crime causation and their policy implications.**

According to psychological theories, crime results from individuals' mental or emotional disturbances, inability to empathize with others, inability to legally satisfy their basic needs, or oppressive circumstances of life. To combat crime, psychological positivists would isolate, sterilize, or execute offenders not amenable to treatment. For treatable offenders, psychotherapy or psychoanalysis may prove effective. Other policy implications are to help people satisfy their basic needs legally, to eliminate sources of oppression, and to provide legal ways of coping with oppression.

5. **Explain sociological theories of crime causation and their policy implications.**

Sociological theories propose that crime is caused by *anomie*, or the dissociation of the individual from the *collective conscience* by *social disorganization*; by *anomie* resulting from a lack of opportunity to achieve aspirations; by the learning of criminal values and behaviors; and by the failure to properly socialize individuals. Among the policy implications of sociological theories of crime causation are containing crime within reasonable boundaries, organizing and empowering neighborhood residents, reducing aspirations, increasing legitimate opportunities, providing law-abiding models, regulating associations, eliminating crime's rewards, rewarding law-abiding behavior, punishing criminal behavior effectively, and properly socializing children so that they develop self-control and a strong moral bond to society.

6. **Distinguish major differences among classical, positivist, and critical theories of crime causation.**

Unlike classical theories, which assume that human beings have free will, and positivist theories, which assume that human beings are determined, critical theories assume that human beings are both determined and determining. In contrast to both classical and positivist theories, which assume that society is characterized primarily by consensus over moral values, critical theories assume that society is characterized primarily by conflict over moral values. Finally, unlike positivist theorists, who assume that social scientists can be objective or value neutral in their work, many critical theorists assume that everything they do is value laden by virtue of their being human and that it is impossible to be objective.

7. **Describe how critical theorists would explain the causes of crime and their policy implications.**

Depending on their perspective, critical theorists explain crime as the result of labeling and stigmatization; of relative powerlessness, the class struggle, and the practice of taking advantage of other people; or of patriarchy. Those who support labeling theory would address crime by avoiding labeling people as criminals or by employing radical nonintervention or reintegrative shaming. Conflict theorists would address crime by having dominant groups give up some of their power to subordinate groups or having dominant group members become more effective rulers and subordinate group members, better subjects. Radical theorists would define crime as a violation of basic human rights, replace the criminal justice system with popular or socialist justice, and (except for anarchists) create a socialist society appreciative of human diversity. Left realists would use police power to protect people living in working-class communities. Peacemaking criminologists would transform human beings so that they were able to empathize with those less fortunate and respond to other people's needs, would reduce hierarchical structures, would create communities of caring people, and would champion universal social justice. Feminist theorists would address crime by eliminating patriarchal structures and relationships and promoting greater equality for women. Postmodernist criminologists would replace the prevailing description of the world with new conceptions, words, and phrases that convey alternative meanings, as experienced by historically subjugated people. They also would replace the formal criminal justice system with informal social controls handled by local groups and local communities.

KEY TERMS

theory 64
criminological theory 64
classical theory 64
utility 67
social contract 67
special or specific deterrence 67
general deterrence 67
neoclassical theory 67
biological inferiority 70
criminal anthropology 70
atavist 70
limbic system 73
psychopaths, sociopaths, or antisocial personalities 74
anomie (Durkheim) 78
collective conscience 78
social disorganization 78
anomie (Merton and Cohen) 79
imitation or modeling 81
differential association 81
learning theory 82
positive reinforcement 82
negative reinforcement 82
extinction 82
punishment 82
social control theory 83
labeling theory 84
criminalization process 84
conflict theory 85
power differentials 85
relative powerlessness 85
radical theories 86
class struggle 87
left realists 88
relative deprivation 88
peacemaking criminology 89
feminist theory 89
patriarchy 89
postmodernism 90

REVIEW QUESTIONS

1. What are two undesirable consequences of the failure to understand the theoretical basis of criminal justice policies?
2. Before the Enlightenment and classical theory, what generally was believed to be the cause of crime?
3. Who was arguably the best known and most influential of the classical criminologists, and how did his ideas become known?
4. What is the difference between *special* or *specific deterrence* and *general deterrence*?
5. What are five problems with positivist theories?
6. What is the position held by most criminologists today regarding the relationship between biology and crime?
7. What does research indicate about the relationship between intelligence and both juvenile delinquency and adult criminality?
8. What are problems with psychological or psychoanalytic theories of crime and their policy implications?
9. In what ways did Durkheim believe that crime was functional for society?
10. What is a potential problem with the theory of the Chicago School?
11. What is a criticism of most anomie theories?
12. Who was the first twentieth-century criminologist to argue forcefully that crime was learned, and what is his theory called?
13. What is arguably the major problem with social control theory?
14. What is *secondary deviance*, and how does it occur, according to labeling theorists?
15. What are major criticisms of conflict theory? Radical theory?
16. What contributions to criminological theory have been made by British or left realism, peacemaking criminology, feminist theory, and postmodernism?

IN THE FIELD

1. **Professional Perspectives on Crime** Interview representatives of the criminal justice process–a police officer, a prosecutor, a defense attorney, a judge, a correctional officer, and a probation or parole officer. Ask each to explain why crimes are committed. Ask them how they would prevent or reduce crime. After completing the interviews, identify the theories and policy implications described in the textbook that best correspond to each response. Explain any differences in responses that emerged. Present your findings in a brief written report.
2. **Literary Perspectives on Crime** Read a nonfiction book about crime. Explain the criminal behavior, orally or in writing, using the theories in this chapter. Some good books are Truman Capote's *In Cold Blood*, Vincent Bugliosi's *Helter Skelter*, and Norman Mailer's *The Executioner's Song*.
3. **Explaining Crimes** Select several crimes from Table 2.1, and explain them by using the theories in Chapter 3.

ON THE WEB

1. **Comparing Crime Rates** Use the Internet to research crime rates of several capitalist and noncapitalist countries. (Remember that crime rates, as discussed in Chapter 2, are not accurate measures of the true amount of crime.) Then write a paragraph in which you support or oppose the proposition that capitalist countries have higher crime rates than noncapitalist countries.
2. **Criminologists** Go to www.asc41.com to learn more about professional, mostly academic criminologists. Read about the history of the American Society of Criminology. Examine the annual meeting information to see the topics on which criminologists are currently working. Explore specialty areas in criminology. Scan employment information to discover the types of jobs available.

CRITICAL THINKING EXERCISES

THE DAHMER CASE

1. Jeffrey Dahmer, a white male, was sentenced to 957 years in prison (Wisconsin had no death penalty) for the murder and dismemberment of 15 young males–mostly black and homosexual–in Milwaukee, Wisconsin.

Dahmer was arrested without a struggle at his apartment on Milwaukee's crime-infested west side in August 1991, after one of his victims escaped and notified the police. Dahmer was 31 years old and had recently been fired from his job at a chocolate factory. He immediately confessed to 11 murders. He told the police that he lured men from bars and shopping malls by promising them money to pose for pictures. He would then take them to his apartment, drug them, strangle them, and dismember their bodies. He boiled the heads of some of his victims to remove the flesh and had sex with the cadaver of at least one of them. Police found rotting body parts lying around his apartment, along with bottles of acid and chemical preservatives. Photographs of mutilated men were on a freezer that contained two severed heads. Another severed head was in the refrigerator. Dahmer told the police that he had saved a heart to eat later.

Dahmer's stepmother told the press that as a child he liked to use acid to scrape the meat off dead animals. When he was 18, his parents divorced. He lived with his mother until, one day, she took his little brother and disappeared, leaving him alone. He went to live with his grandmother, where he started to abuse alcohol. During the 6 years he lived with his grandmother, mysterious things were occurring in the basement and garage. Dahmer's father, a chemist, discovered bones and other body parts in containers. Dahmer told his father that he had been stripping the flesh from animals he found.

In 1988, Dahmer spent 10 months in prison for fondling a 13-year-old Laotian boy. On his release from prison, he was placed on probation, but his probation officer never visited him.

Dahmer's defense attorney claimed that his client was insane at the time of the murders. The jury rejected the insanity defense.

a. Which theory or theories of crime causation described in this chapter best explain Jeffrey Dahmer's criminal behavior?
b. What crime prevention and correctional policies described in this chapter should be employed with criminal offenders like Jeffrey Dahmer?

BREAKING THE LAW

2. Stacey Raines drives from Indianapolis to Fort Wayne, Indiana, twice a month as part of her job as a sales representative for a pharmaceutical company. Although the posted speed limit is 65 miles per hour, Stacey generally cruises at 80 to 85, intentionally exceeding the speed limit.

a. Which theory or theories of crime causation described in this chapter best explain the behavior of intentional speeders?
b. What crime prevention and correctional policies described in this chapter should be used with intentional speeders?

NOTES

1. Cesare Beccaria, *An Essay on Crimes and Punishments*, trans., with introduction, by Harry Paolucci (Indianapolis, IN: Bobbs-Merrill, 1975), ix.
2. Ibid., 8.
3. Ibid., 42.
4. Ibid., 99.
5. Ibid.
6. See Irving M. Zeitlin, *Ideology and the Development of Sociological Theory,* 7th ed. (Englewood Cliffs, NJ: Prentice Hall, 2000).
7. See Ian Taylor, Paul Walton, and Jock Young, *The New Criminology: For a Social Theory of Deviance* (New York: Harper & Row, 1974), 24–32.
8. George B. Vold and Thomas J. Bernard, *Theoretical Criminology*, 3rd ed. (New York: Oxford, 1986), 48.
9. See William H. Sheldon, *Varieties of Delinquent Youth* (New York: Harper, 1949).
10. Diana H. Fishbein, *The Science, Treatment, and Prevention of Antisocial Behaviors: Application to the Criminal Justice System* (Kingston, NJ: Civic Research Institute, 2000).
11. See Diana H. Fishbein, "Biological Perspectives in Criminology," *Criminology* 28 (1990): 27–72; Vold and Bernard, *Theoretical Criminology*, 87–92; Daniel J. Curran and Claire M. Renzetti, *Theories of Crime* (Boston: Allyn & Bacon, 1994), 54–63; James Q. Wilson and Richard J. Herrnstein, *Crime and Human Nature* (New York: Simon & Schuster, 1985), 75–81, 90–100.
12. See Curran and Renzetti, *Theories of Crime*, 78; Fishbein, "Biological Perspectives in Criminology," 38, 47.
13. See Fishbein, "Biological Perspectives in Criminology," 48, 53; *Preventing Delinquency*, vol. 1, National Institute for Juvenile Justice and Delinquency Prevention (Washington, DC: U.S. Government Printing Office, 1977); Curran and Renzetti, *Theories of Crime*, 73–77, 80–81.
14. Edwin H. Sutherland and Donald R. Cressey, *Criminology*, 9th ed. (Philadelphia: J. B. Lippincott, 1974), 152.
15. Robert Gordon, "Prevalence: The Rare Datum in Delinquency Measurement and Its Implications for the Theory of Delinquency," in *The Juvenile Justice System*, ed. Malcolm W. Klein (Beverly Hills, CA: Sage, 1976), 201–84; Travis Hirschi and Michael J. Hindelang, "Intelligence and Delinquency: A Revisionist Review," *American Sociological Review* 42 (1977): 572–87.
16. See Robert S. Woodworth and Mary R. Sheehan, *Contemporary Schools of Psychology*, 3rd ed. (New York: The Ronald Press, 1964).
17. Walter Bromberg and Charles B. Thompson, "The Relation of Psychosis, Mental Defect, and Personality Types to Crime," *Journal of Criminal Law and Criminology* 28 (1937): 70–89; Karl F. Schuessler and Donald R. Cressey, "Personality Characteristics of Criminals," *American Journal of Sociology* 55 (1950): 476–84; Gordon P. Waldo and Simon Dinitz, "Personality Attributes of the Criminal: An Analysis of Research Studies, 1950–1965," *Journal of Research in Crime and Delinquency* (1967): 185–202; John Monahan and Henry J. Steadman, "Crime and Mental Disorder: An Epidemiological Approach," in *Crime and Justice: A Review of Research*, vol. 4, ed. Michael Tonry and Norval Morris (Chicago: University of Chicago Press, 1983).

18. Abraham H. Maslow, *Motivation and Personality*, 2nd ed. (New York: Harper and Row, 1970).

19. Seymour L. Halleck, *Psychiatry and the Dilemmas of Crime* (New York: Harper & Row, 1967).

20. Émile Durkheim, *Rules of Sociological Method* (New York: Free Press, 1964).

21. Vold and Bernard, *Theoretical Criminology*, 160.

22. Robert E. Park, Ernest Burgess, and Roderick D. McKenzie, *The City* (Chicago: University of Chicago Press, 1928).

23. Clifford R. Shaw, *Delinquency Areas* (Chicago: University of Chicago Press, 1929); Clifford R. Shaw and Henry D. McKay, *Social Factors in Juvenile Delinquency* (Chicago: University of Chicago Press, 1931); Clifford R. Shaw and Henry D. McKay, *Juvenile Delinquency and Urban Areas* (Chicago: University of Chicago Press, 1942).

24. Clifford R. Shaw, *The Jackroller* (Chicago: University of Chicago Press, 1930); Clifford R. Shaw, *The National History of Delinquent Career* (Chicago: University of Chicago Press, 1931); Clifford R. Shaw, *Brothers in Crime* (Chicago: University of Chicago Press, 1938).

25. Robert K. Merton, "Social Structure and Anomie," *American Sociological Review* 3 (1938): 672–82.

26. Albert K. Cohen, *Delinquent Boys: The Culture of the Gang* (New York: Free Press, 1955).

27. Richard A. Cloward and Lloyd E. Ohlin, *Delinquency and Opportunity: A Theory of Delinquent Gangs* (New York: Free Press, 1960).

28. See, for example, Albert Bandura, *Social Learning Theory* (Englewood Cliffs, NJ: Prentice Hall, 1977).

29. See, for example, Edwin H. Sutherland and Donald R. Cressey, *Criminology*, 9th ed. (Philadelphia: J. B. Lippincott, 1974).

30. Daniel Glaser, "Criminality Theories and Behavioral Images," *American Journal of Sociology* 61 (1956): 433–44.

31. Definitions of learning theory concepts are from Howard Rachlin, *Introduction to Modern Behaviorism*, 2nd ed. (San Francisco: W. H. Freeman, 1976).

32. Michael R. Gottfredson and Travis Hirschi, *A General Theory of Crime* (Stanford, CA: Stanford University Press, 1990).

33. Edwin Lemert, *Social Pathology: A Systematic Approach to the Theory of Sociopathic Behavior* (New York: McGraw-Hill, 1951).

34. See Edwin M. Schur, *Radical Nonintervention* (Englewood Cliffs, NJ: Prentice Hall, 1973).

35. Robert J. Lilly, Francis T. Cullen, and Richard A. Ball, *Criminological Theory: Context and Consequences* (Thousand Oaks, CA: Sage, 2007), 136–38.

36. John Braithwaite, *Crime, Shame and Reintegration* (Cambridge: Cambridge University Press, 1989).

37. Vold and Bernard, *Theoretical Criminology*, 256.

38. George B. Vold, *Theoretical Criminology* (New York: Oxford, 1958).

39. William J. Chambliss, "Functional and Conflict Theories of Crime: The Heritage of Emile Durkheim and Karl Marx," in *Whose Law, What Order?*, ed. W. J. Chambliss and M. Mankoff (New York: Wiley, 1976), 9.

40. Tony Platt, "Prospects for a Radical Criminology in the USA," in *Critical Criminology*, ed. I. Taylor, et al. (Boston: Routledge & Kegan Paul, 1975), 103.

41. Ibid.; for a similar definition, see also Herman Schwendinger and Julia Schwendinger, "Defenders of Order or Guardians of Human Rights?" in *Critical Criminology*, ed. I. Taylor, et al. (Boston: Routledge & Kegan Paul, 1975), 113–46.

42. See, for example, Richard Kinsey, John Lea, and Jock Young, *Losing the Fight Against Crime* (London: Basil Blackwell, 1976); Roger Matthews and Jock Young, eds., *Confronting Crime* (London: Sage, 1986).

43. Jock Young, "Left Realism: The Basics," in *Thinking Critically About Crime*, ed. B. D. MacLean and D. Milovanovic (Vancouver, BC: Collective Press, 1997), 28–36.

44. Werner Einstadter and Stuart Henry, *Criminological Theory: An Analysis of Its Underlying Assumptions* (Fort Worth, TX: Harcourt Brace, 1995), 256.

45. Ibid., 257.

46. See Harold E. Pepinsky and Richard Quinney, eds., *Criminology as Peacemaking* (Bloomington, IN: Indiana University Press, 1991).

47. For two excellent reviews, see Kathleen Daly and Meda Chesney-Lind, "Feminism and Criminology," *Justice Quarterly* 5 (1988): 497–538; Sally S. Simpson, "Feminist Theory, Crime, and Justice," *Criminology* 27 (1989): 605–31.

48. See Einstadter and Henry, *Criminological Theory*, 275.

49. Jody Miller, "Feminist Criminology," in *Controversies in Critical Criminology*, ed. M. D. Schwartz and S. E. Hatty (Cincinnati, OH: Anderson, 2003), 15–28; Jeanne Flavin, "Feminism for the Mainstream Criminologist," *Journal of Criminal Justice* 29 (2001): 271–285.

50. See Einstadter and Henry, *Criminological Theory*, 275.

51. Ibid., 278.

52. Ibid., 287.

53. Bruce Arrigo and T. R. Young, "Chaos, Complexity, and Crime: Working Tools for a Postmodern Criminology," in *Thinking Critically About Crime*, ed. B. D. MacLean and D. Milovanovic (Vancouver, BC: Collective Press, 1997), 77. For an interesting postmodern analysis of the William Kennedy Smith rape trial, see Gregory M. Matoesian, *Law and the Language of Identity: Discourse in the William Kennedy Smith Rape Trial* (New York: Oxford University Press, 2001).

54. Einstadter and Henry, *Criminological Theory*, 291.

55. Arrigo and Young, "Chaos, Complexity, and Crime," 81–82; Stuart Henry and Dragan Milovanovic, *Constitutive Criminology: Beyond Postmodernism* (London: Sage, 1996).

56. Einstadter and Henry, *Criminological Theory*, 294.

57. Ibid., 280.

Design Elements: *Image of part of a court building with columns and statue*: Pixtal/AGE Fotostock RF; *Image of a door slightly open*: jadding/Shutterstock.com

Image Source Plus/Alamy

The Rule of Law

CHAPTER OUTLINE

Two Types of Law: Criminal Law and Civil Law

- Substantive versus Procedural Law
- Ideal Characteristics of the Criminal Law
- Criminal Law as a Political Phenomenon
- Creating Criminal Laws in the United States

Procedural Law: Rights of the Accused

- The Bill of Rights
- The Fourteenth Amendment and the Selective Incorporation of the Bill of Rights
- The Fourth Amendment
- The Fifth Amendment
- The Sixth Amendment
- The Eighth Amendment

Protecting the Accused from Miscarriages of Justice

LEARNING OBJECTIVES

After completing this chapter, you should be able to:

1. Distinguish between criminal law and civil law.
2. Distinguish between substantive law and procedural law.
3. List five features of "good" criminal laws.
4. Explain why criminal law is a political phenomenon.
5. Summarize the origins of American criminal law.
6. Describe the procedural rights in the Fourth Amendment.
7. Describe the procedural rights in the Fifth Amendment.
8. Describe the procedural rights in the Sixth Amendment.
9. Describe the procedural rights in the Eighth Amendment.
10. Explain why procedural rights are important to those accused of crimes.

Crime Story

For more than four decades, Clifford Williams Jr. and Hubert "Nathan" Myers (pictured) awaited the news they received on March 28, 2019. On that fateful day, Duval County, Florida Circuit Judge Angela Cox vacated their convictions and ruled that the uncle and nephew should be freed from prison for a murder that prosecutors now say they almost certainly did not commit. The rare finding overturned their 1976 convictions in a trial in Jacksonville. Williams was 33 when he was arrested; Myers was 18. They were 76 and 61, respectively, when they heard the good news. The order to vacate the convictions of Williams and his nephew Myers was the result of a recommendation made by Florida's first-ever conviction integrity review unit, created by State Attorney Melissa Nelson in 2018, just weeks after she took office. The purpose of such units is to investigate innocence claims after a defendant is convicted.

The conviction integrity review unit, with help from the Innocence Project, concluded that Williams and Myers did not receive effective assistance of counsel because their defense attorneys failed to present evidence to the jury that would have contradicted the prosecutors' single eyewitness that was instrumental in convicting the men of the murder of Jeanette Williams (no relation to Clifford) and the attempted murder of her girlfriend, Nina Marshall, in Williams' bedroom in 1976. The State Attorney's Office report stated, "There is no credible evidence of guilt, and likewise, there is credible evidence of innocence." The Report concluded: "In foregoing the forensics, the state relied on the testimony of one individual, and it is upon this testimony alone that these two men are serving life sentences, in the face of overwhelming contradictory forensic evidence and alibi testimony."

The only evidence that implicated the two men was Nina Marshall's testimony. Marshall and Williams were lying next to each other asleep in Williams' bed when both women were shot. Williams died instantly, and Marshall survived being shot in the neck. Marshall, who, along with Williams, knew the defendants socially, told investigators that Myers and Williams entered the bedroom and fired two guns at the women from the foot of the bed. However, police failed to discover any physical evidence, and prosecutors never presented any evidence in court that supported her account. From the moment investigators arrived at the scene, they questioned how the shooting could have occurred inside the room. The forensics indicated that the bullets came from one gun, fired from outside and through a window, rather than from inside by two guns, like Marshall had testified. In addition, just hours after the shooting, police failed to find gunshot residue on the hands of either Williams or Myers. Conviction Integrity Review Director and lead attorney Shelley Thibodeau, who reexamined the case, wrote in her report that every step along the way, she only found more evidence that supported Williams and Myers' innocence. For example, Myers passed a polygraph test administered as part of the review process.

So how were Williams and Myers the beneficiaries of such good fortune, albeit 43 years later? The short answer is that Myers was proactive. In January 2017, Myers read a newspaper story about Nelson's intention to create the conviction integrity review unit, so he wrote her a four-page letter explaining why he and Williams were innocent. In the letter, he noted that his defense attorney ignored about 40 alibi witnesses who would have testified that the two men were at a birthday party a block away at the time of the murder. He related that in prison, he discovered through records requests that forensic analysts had evidence that the shooting could not have occurred the way the eyewitness, Nina Marshall, claimed, and that investigators had missed a confession that tied the murder to another man, Nathaniel

Will Dickey/The Florida Times-Union via AP

Lawson. Years later, Lawson told at least five people that he killed Jeanette Williams and that the two men in prison were innocent. During police interviews, witnesses stated that Lawson was at the crime scene. On the night of the murder, police had stopped the truck in which Lawson was an occupant, but they only questioned two of the four people in the truck, and Lawson was not one of the two people they questioned. None of this evidence was presented to the jury. Why not? Again, the short answer is that police, prosecutors, and even Williams and Myers' defense attorneys chose to believe Marshall rather than all of the other contradictory evidence. When investigators reexamined the case, neither Lawson nor Marshall could be questioned because they had died in 1994 (Lawson) and 2001 (Marshall).

The prosecution hurried the case to trial just two months after the shooting. Defense attorneys Jim Harrison and Keith Vickers did not object and, as mentioned previously, ignored the alibi witnesses. The motive was presumably a $50 drug debt, but no evidence of that was presented at trial. After a mistrial, then-Assistant State Attorney Hank Coxe offered Myers a plea deal if he would testify against his uncle. Coxe claimed that he offered Myers five years, but Myers remembered it as a two-year deal. Either way, Myers refused to plead guilty. During a second, two-day trial, both men were convicted. Harrison and Vickers presented no witnesses but, instead, relied on making legal motions and their closing arguments. Coxe remembered that he was not worried about the ballistics tests or gunshot residue or anything else. He boasted, "When you have an eyewitness, you don't need all that." Harrison and Vickers, as noted previously, seemed to believe Marshall's version of the shooting. They only challenged her recollection of who did the shooting. Coxe framed the trial as a simple open-and-shut case for the jurors, who apparently agreed. It only took them two and a half hours to find both men guilty.

The jury recommended life sentences for both Williams and Myers, but Circuit Judge Cliff Shepard overrode the jury's recommendation for Williams and sentenced him to death, noting that he could not "conceive of any more heinous, atrocious or cruel act than to enter someone's home in the night while they are sleeping in their bed and shoot them to death." Four years after their initial sentencing, the Florida Supreme Court overturned Williams' death sentence, stating that Williams should have received a life sentence like Myers. Besides that decision, all of Williams and Myers' appeals had been unsuccessful. It took 43 years, but both men finally were exonerated.

This chapter examines the constitutional protections provided to people suspected or charged with crimes. Despite these protections, miscarriages of justice, such as the ones Clifford Williams and Nathan Myers experienced, occur with some regularity. How could Williams and Myers' nightmare have happened? Who is most to blame for this miscarriage of justice—the police, the prosecutors, his trial attorneys, the "expert" witness, the judge, or the jury? What can be done to prevent miscarriages of justice such as the one in this crime story from occurring again? The answer to these questions reveals the difficulties in administering justice fairly in the United States.

Two Types of Law: Criminal Law and Civil Law

As discussed in Chapter 2, the conventional, although not necessarily the best, definition of *crime* is "a violation of the criminal law." **Criminal law** is one of two general types of law practiced in the United States; the other is civil law. Criminal law is "a formal means of social control [that] involves the use of rules that are interpreted, and are enforceable, by the courts of a political community. . . . The function of the rules is to set limits to the conduct of the citizens, to guide the officials (police and other administrators), and to define conditions of deviance or unacceptable behavior."[1] The purpose of criminal justice is to enforce the criminal law.

A crime, as noted, is a violation of the criminal law, or of the **penal code** of a political jurisdiction. Although crime is committed against individuals, it is considered an offense against the state—that is, the political jurisdiction that enacted the law.[2] A **tort**, in contrast, is a violation of the **civil law** and is considered a private matter between individuals. Civil law includes the law of contracts and property as well as subjects such as administrative law (which deals with the rules and regulations created by government agencies) and the regulation of public utilities.

For legal purposes, a particular act may be considered an offense against an individual or the state or both. It is either a tort or a crime or both, depending on how it is handled. For example, a person who has committed an act of assault may be charged with a crime. If that person is convicted of the crime, the criminal court may order the offender to be imprisoned in the county jail for 6 months and to pay a fine of $2,000. Both the jail sentence and the fine are punishments, with the fine going to the state or local treasury (in federal court, to the national treasury). The criminal court also could order the offender to pay restitution to the victim. In that case, the offender would pay the victim a sum of money either directly or indirectly, through an intermediary. In addition, the victim may sue the offender in civil court for damages, such as medical expenses or wages lost because of injury. If the offender is found liable (responsible) for the damages because he or she has committed a tort (civil courts do not "convict"), the civil court also may order the offender to compensate the victim in the amount of $2,000 for damage to the victim's interests. The payment of compensation in the civil case is not punishment; it is for the purpose of "making the victim whole again."

criminal law One of two general types of law practiced in the United States (the other is civil law); "a formal means of social control [that uses] rules . . . interpreted [and enforced] by the courts . . . to set limits to the conduct of the citizens, to guide the officials, and to define . . . unacceptable behavior."

penal code The criminal law of a political jurisdiction.

tort A violation of the civil law.

civil law One of two general types of law practiced in the United States (the other is criminal law); a means of resolving conflicts between individuals. It includes personal injury claims (torts), the law of contracts and property, and subjects such as administrative law and the regulation of public utilities.

Substantive versus Procedural Law

substantive law The body of law that defines criminal offenses and their penalties.

procedural law The body of law that governs the ways substantive laws are administered; sometimes called *adjective* or *remedial* law.

due process of law The rights of people suspected of or charged with crimes. Also, the procedures followed by courts to ensure that a defendant's constitutional rights are not violated.

There are two types of criminal law: substantive and procedural. **Substantive law** is the body of law that defines criminal offenses and their penalties. Substantive laws, which are found in the various penal codes, govern what people legally may and may not do. Examples of substantive laws are those that prohibit and penalize murder, rape, robbery, and other crimes. **Procedural law**, sometimes called *adjective* or *remedial law*, governs the ways in which the substantive laws are to be administered. It covers such subjects as the way suspects legally can be arrested, searched, interrogated, tried, and punished. In other words, procedural law is concerned with **due process of law**, or the rights of people suspected of or charged with crimes. The last part of this chapter is devoted to a detailed description of procedural law.

Ideal Characteristics of the Criminal Law

Legal scholars identify five features that all "good" criminal laws ideally ought to possess. To the extent that those features are absent in criminal laws, the laws can be considered "bad" laws, and bad laws do exist. The five ideal features of good criminal laws are (1) politicality, (2) specificity, (3) regularity, (4) uniformity, and (5) penal sanction (see Figure 4.1).

politicality An ideal characteristic of criminal law, referring to its legitimate source. Only violations of rules made by the state, the political jurisdiction that enacted the laws, are crimes.

Politicality

Politicality refers to the legitimate source of criminal law. Only violations of rules made by the state (i.e., the political jurisdiction that enacted the laws) are crimes. Violations of rules made by other institutions–such as families, churches, schools, and employers–may be "bad," "sinful," or "socially unacceptable," but they are not crimes because they are not prohibited by the state.

specificity An ideal characteristic of criminal law, referring to its scope. Although civil law may be general in scope, criminal law should provide strict definitions of specific acts.

Specificity

Specificity refers to the scope of criminal law. Although civil law may be general in scope, criminal law should provide strict definitions of specific acts. The point is illustrated by an old case in which a person stole an airplane but was found not guilty of violating a criminal law that prohibited the taking of "self-propelled vehicles." The judge ruled that at the time the law was enacted, *vehicles* did not include airplanes. Ideally, as the Supreme Court ruled in *Papachristou v. City of Jacksonville* (1972), a statute or ordinance "is void for vagueness . . . [if] it fails to give a person of ordinary intelligence fair notice that his contemplated conduct is forbidden."

regularity An ideal characteristic of criminal law: the applicability of the law to all persons, regardless of social status.

Regularity

Regularity is the applicability of the criminal law to all persons. Ideally, anyone who commits a crime is answerable for it, regardless of the person's social status. Thus, ideally, when criminal laws are created, they should apply not only to the women who violate them, but also to the men; not only to the poor, but also to the rich. In practice, however, this ideal feature of law has been violated. Georgia's pre-Civil War criminal laws, for example, provided for a dual system of crime and punishment, with one set of laws for "slaves and free persons of color" and another for all other persons. Another example of the violation of this principle is illustrated by the case of *Michael M. v. Superior Court of Sonoma County* (1981). In this case, the U.S. Supreme Court upheld California's statutory rape law that made men alone criminally responsible for the act of illicit sexual intercourse with a minor female.

FIGURE 4.1 Ideal Characteristics of Criminal Law

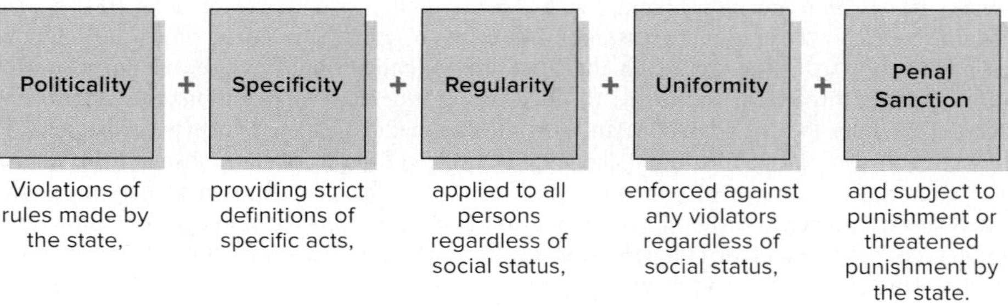

Uniformity **Uniformity** refers to the way in which the criminal law should be enforced. Ideally, the law should be administered without regard for the social status of the persons who have committed crimes or are accused of committing crimes. Thus, when violated, criminal laws should be enforced against both young and old, both rich and poor, and so on. However, as is the case with regularity, the principle of uniformity often is violated because some people consider the strict enforcement of the law unjust in some cases. For example, juveniles who are caught misbehaving in violation of the criminal law sometimes are ignored or treated leniently through the exercise of police or judicial discretion.

uniformity An ideal characteristic of criminal law: the enforcement of the laws against anyone who violates them, regardless of social status.

Penal Sanction The last ideal feature of criminal law is **penal sanction**, the principle that violators will be punished, or at least threatened with punishment, by the state. Conventional wisdom suggests that there would be no point in enacting criminal laws if their violation were not responded to with punishment or threat of punishment. Most people assume that sanctionless criminal laws would be ignored. Because all criminal laws carry sanctions, the power of sanctionless laws can be left to philosophers to debate. Table 4.1 shows the five general types of penal sanctions currently used in the United States, as well as the purpose and focus of each sanction. Combining different penal sanctions in the administration of justice is not uncommon.

penal sanction An ideal characteristic of criminal law: the principle that violators will be punished or at least threatened with punishment by the state.

Criminal Law as a Political Phenomenon

People sometimes forget that criminal law is a political phenomenon, that it is created by human beings to regulate the behavior of other human beings. Some people, for example, view the criminal law as divinely inspired, something that should not be questioned or challenged. That viewpoint probably comes from a belief in the biblical story of Moses receiving the Ten Commandments from God on Mount Sinai. However, as critical theorists are quick to point out, criminal law frequently promotes the interests of some groups over the interests of other groups. Thus, regardless of the law's source of inspiration, we must understand that what gets defined as criminal or delinquent behavior is the result of a political process in which rules are created to prohibit or to require certain behaviors. Nothing is criminal or delinquent in and of itself; only the response of the state makes it so.

myth

Law makes people behave.

fact

The existence of a law prohibiting a particular behavior does not necessarily prevent an individual from engaging in that behavior. Common sense suggests the implausibility of the notion. Ask yourself, if it were not for laws prohibiting murder, prostitution, or heroin use, for example, would you murder, engage in prostitution, or use heroin? How effective are speed limits in preventing you from exceeding them?

Origins of Laws Formal, written laws are a relatively recent phenomenon in human existence. The first were created about 5,000 years ago. They emerged with the institutions of property, marriage, and government. "Stateless" societies apparently managed without them for two primary reasons.[3] First, most stateless societies were governed by rigid customs to which citizens strictly adhered. Second, crimes of violence were considered private matters and usually were resolved through bloody personal revenge. Formal, written laws partially

TABLE 4.1 Five General Types of Penal Sanctions

Type	Purpose	Focus
Punishment	Prevent undesired conduct. Provide retribution ("an eye for an eye").	Offending conduct
Restitution	Make the victim "whole again" by having the offender directly or indirectly pay the victim.	Crime victim
Compensation	Make the victim "whole again" by having the state pay for damages to the victim.	Crime victim
Regulation	Control future conduct toward the best interests of the community (e.g., making it a crime or traffic violation to operate a motor vehicle with a blood alcohol content higher than a specified level).	The entire community
Treatment or rehabilitation	Change the offender's behavior and, perhaps, personality.	Criminal offender

Charles Walker/Topfoto/The Image Works

Bas relief depicting King Hammurabi with his code of laws. *Why do people believe that criminal laws come from God?*

replaced customs when nation-states appeared, although customs often remained the force behind the laws. Formal laws also replaced customs with the advent of writing, which allowed recorded legislation to replace the recollections of elders and priests.

The first known written laws (approximately 3000 B.C.) were found on clay tablets among the ruins of Ur, one of the city-states of Sumeria. Attributed to King Urukagina of Lagash, the laws were truly enlightened for their time and attempted to free poor people from abuse by the rich and everybody from abuse by the priests. For example, one law forbade the high priest from coming into the garden of a poor mother and taking wood or fruit from her to pay taxes. Laws also cut burial fees to one-fifth of what they had been and forbade the clergy and high officials from sharing among themselves the cattle that were sacrificed to the gods. By 2800 B.C., the growth of trade had forced the city-states of Sumeria to merge into an empire governed by a single, all-powerful king.

Around 2200 B.C., a war settlement between the Sumerians and the Akkadians produced the Babylonian civilization. Babylonia's best-known king was Hammurabi (1792–1750 B.C.), who ruled for 43 years. Hammurabi is famous for the first great code of laws. The Code of Hammurabi, like the laws of Moses later, presumably was a gift from God. Hammurabi was said to have received it from the sun god, Shamash, about 1780 B.C. There was a total of 285 laws in the code, arranged under the headings of personal property, real estate, trade and business, the family, injuries, and labor. The Code of Hammurabi combined very enlightened aims, such as "to prevent the strong from oppressing the weak, . . . to enlighten the land and to further the welfare of the people," with very barbaric punishments.

All the ancient nation-states or civilizations had formal legal codes. In addition to the laws of King Urukagina of Lagash (Sumeria) and the Code of Hammurabi (Babylonia), legal codes were established by the Egyptians, the Assyrians, the Hebrews, the Persians, the Indians, the Chinese, the Greeks (especially the codes of Lycurgus, Draco, Solon, and Plato), and the Romans (e.g., the Twelve Tables, the Justinian Code, and the Law of the Nations). The development and the content of those legal codes are of mostly historical interest. The criminal law of the United States, for the most part, is derived from the laws of England.

England's Contribution to American Criminal Law Before the Norman Conquest in 1066, England was populated by Anglo-Saxon tribes that regulated themselves through custom.[4] Wars between those tribes resulted in the taking of the tribal lands of the losers by the leader of the victorious tribe, who, by force, made the newly acquired land his own private property and himself the feudal lord. By the time of the Norman Conquest, there were about eight large and relatively independent feudal landholdings. In an effort to increase their power, the feudal lords took it on themselves to dispense justice among their subjects and began to require that disputes between subjects be settled in local courts rather than by relatives, as had previously been the custom.

When William I of Normandy conquered England in 1066 and proclaimed himself king, he declared that all land, and all land-based rights, including the administration of justice, were now vested in the king. King William also rewarded the Norman noblemen who had fought with him with large grants of formerly Anglo-Saxon land.

To make the dispensing of justice a profitable enterprise for the king and to make sure the local courts remained under his control, the institution of the *eyre* was created early in the twelfth century. The eyre was composed of traveling judges who represented the king and examined the activities of the local courts.

Of particular interest to the eyre was the resolution of cases of sufficient seriousness as to warrant the forfeiture of the offender's property as punishment. The notion of forfeiture was based on the feudal doctrine that the right to own private property rested on

FYI **Felony**

The term *felony* originally meant an offense serious enough "to break the relationship between [the landowner and his lord] and to cause the [land] holding to be forfeited to the lord."

Source: S. Francis Milson, *The Historical Foundations of the Common Law* (London: Butterworths, 1969), 355.

a relationship of good faith between the landowner and his lord. The Norman kings expanded the notion of forfeiture to include any violation of the "king's peace," which enabled the king to claim forfeited property for a variety of offenses, including such minor ones as trespassing. It was the responsibility of the judges in eyre to make sure the king received his portion of forfeited property.

A secondary responsibility of the eyre was to hear common pleas, which consisted primarily of disputes between ordinary citizens. Although common pleas could be handled in the local courts, which in many instances were still influenced by Anglo-Saxon customs, the Norman settlers frequently felt more comfortable having their cases heard in the king's courts, of which the eyre was one. It was the common-plea decisions made by judges in eyre that formed the body of legal precedent that became known as the *common law*; that is, the rules used to settle disputes throughout England. Thus, as the judges of eyre resolved common-plea disputes, they created precedents to be followed in similar cases. Because the common law was built case by case, it is sometimes also called *case law*. Many of the precedents that were created in medieval England became the basis of statutory law in modern England, as well as in the United States. In both countries, some of the early precedents still are used as the basis for settling disputes not covered by statutes.

Bettmann/Getty Images

The Magna Carta, signed by England's King John in 1215, placed limits on royal power and established the principle of the rule of law. *Which is preferable: unregulated royal power or the rule of law? Why?*

The efforts of the Norman kings to centralize their power over all of England were only partially successful. In 1215, powerful landholding nobles rebelled against the heavy taxation and autocratic rule of King John and forced him to sign the Magna Carta (the Great Charter). The primary purpose of the Magna Carta was to settle the dispute between the king and his nobles by placing checks on royal power. (It did little for the common person.) From that time forward, kings and queens of England were supposed to be governed by laws and customs rather than by their own wills, and the laws were supposed to be applied in a regular and fair way by the king or queen and his or her judges. Thus, the Magna Carta not only created the idea of the rule of law but also formed the basis of what would later be called *due process of law*.

Native American Criminal Law and Its Administration Before we discuss the creation of criminal laws in the United States, we briefly describe Native American criminal law and its administration. This subject is routinely neglected in most accounts of American criminal law and criminal justice, even though it was the first form of criminal law and criminal justice practiced in what we now call the United States. Although there were (and still are) hundreds of tribes with diverse organizational structures and cultures, the focus of this analysis is the Cherokees, who currently are the largest American tribe and, during the colonial period, were the largest tribe on the southern frontier of English America.[5]

Long before European colonists set foot on American soil, Cherokees lived in societies governed by a culture and a legal system grounded in the authority of the spirit world. Tribal authorities could apply the divinely ordained rules, but they were not empowered to formulate them. Once a year at an important religious ceremony a tribal priest would recite the ancient law to all tribe members, as part of the oral tradition. The law was simple, and all tribe members knew the law, so this part of the ceremony was mostly symbolic. The ceremony also included a rite of absolution, during which the priest would recite the crimes that had been committed during the previous year and appeal to the sacred fire to forgive them. All crimes except murder were forgiven. In this sacrifice made to friendship, vengeance and cruelty were to be forgotten. No one who had been guilty of an unpardonable offense could participate in the ceremony, and those who did participate had to be forgiven, no matter the offense. Cherokee criminal law was first and foremost a vehicle for achieving public consensus and harmony through what today would be called restorative justice (see the description in Chapter 14).

Four general types of offenses can be delineated in traditional Cherokee criminal law: (1) offenses against the supernatural or Spirit Beings, (2) offenses against the entire community, (3) offenses against a clan (a tribal subdivision akin to an extended family; there were/are seven clans), and (4) offenses against an individual. Each category of offense was "prosecuted" and "punished" by a separate authority. Offenses against the supernatural or Spirit Beings were, in the Cherokee view, automatically detected and were punished by divine retribution against either the individual or the entire community. A court-like tribal group comprised of a member from each of the seven clans addressed most offenses against the entire community. Public punishments, often whipping, ear or nose cropping, or hanging, followed admission of guilt for these types of offenses. Offenses against a clan were resolved according to a pre-established duty between clan members and the offense or offender and were avenged by individual members of the offended clan. Offenders generally were punished-in-kind, that is, "an eye for an eye and a tooth for a tooth." Finally, offenses against an individual required only a personal response, that is, revenge, and usually were also punished-in-kind by the victim or a member of the victim's clan. For all types of offenses, divine judgment and assistance with punishment might be sought.

As noted, a court-like tribal group handled offenses against the community. At the proceeding, the chief's "right-hand man" presented the charge or charges against the accused. There were no attorneys or juries. Court members were free to ask any questions of the accused, after which the accused was required to swear a sacred oath that he or she was innocent or guilty of the charges. According to Cherokee belief, violation of the oath would prevent the accused's ghost from passing to the Nightland and, instead, the accused's ghost would wander forever, haunting living relatives. Cherokees believed that the latter outcome was worst than any earthly punishment that could be imposed. Moreover, Cherokees took great pride in accepting a justified punishment and resented those who did not.

Murder was an offense against an individual and, therefore, did not require a trial. Blood revenge for the murder was a sacred duty and the responsibility of the oldest male relative of the victim's generation and clan, usually the victim's oldest brother. Failure to avenge the murder was subject to public scorn and disgrace. If the victim was not avenged, then his or her ghost could not pass from the earth to the Nightland. The clan member with the duty of revenge would determine guilt and could select any method to kill the murderer. A common punishment for murderers was to take them to the top of a steep and high cliff and throw them off headfirst onto the rocks below. If the avenging clan member needed assistance, he could seek it from other members of his clan. There were no degrees of murder; it did not matter whether the murder was deliberate or accidental. Every effort was made to kill the actual murderer, but if he were unavailable then another close relative of the murderer's clan would suffice. To avoid the killing of an innocent clan member, members of the murderer's clan might execute their own clan member. If a life was innocently or accidentally taken, the killer could flee to safety in one of four "free villages" or "sacred villages of refuge." As an alternative, a priest might offer the killer safety on sacred ground in any village.

This legal system remained unchanged until about 1710, when the Cherokees began significantly expanding trade with white colonists at Charles Town, the capital of the Carolina colony (now Charleston, South Carolina). By this time, more than 10,000 Cherokees lived in 60 or more scattered villages concentrated along rivers and streams in present-day Georgia and Tennessee. The next 100 years was a period of dramatic change for the Cherokees' traditional way of life, as they adapted to increasing domination by the white colonists, including adoption of their legal system. While a number of factors contributed to this change–economic, religious, and familial–arguably the most important factor was the attempt to preserve tribal lands. Both Presidents Washington and Jefferson urged the Cherokees to adopt the white man's legal system to save the Cherokee nation. In 1808, the Cherokees enacted their first written law. It was written in English because there was no written Cherokee language. Among its provisions was the establishment of the first Cherokee law enforcement agency–the "regulators" or "light horsemen"–and the abolition of clan revenge. In 1821, the first printed volume of Cherokee laws was published in English. The book is the earliest known publication of a Native American tribe's laws.

A written alphabet and syllabary, created by Sequoyah, an illiterate, mixed-blood Cherokee cripple, was introduced to the Cherokee people in 1822, and marked the beginning of a written Cherokee language. In 1828, the Cherokees acquired a printing press, type cast in the syllabary, and began publishing the *Cherokee Phoenix* newspaper. A goal of the newspaper was to promote the Cherokee legal system by publishing: (1) the texts of the laws, (2) editorials emphasizing law enforcement, (3) factual reports of crimes and punishments, (4) articles explaining the operation of the Cherokee legal system, and (5) law-oriented stories reprinted from other newspapers. The syllabary and newspaper brought literacy to the Cherokees seemingly overnight. Sequoyah's creation is considered the most significant single event in Cherokee history. Another singular event in Cherokee history occurred in 1838 and 1839, when the U.S. government, under President Andrew Jackson, forcibly removed most of the Cherokees (approximately 16,000) to Indian Territory (a part of present-day Oklahoma), where they joined about 6,000 western Cherokees, who had voluntarily migrated there. The Cherokees' autonomous legal system ended in 1898, when the Cherokees were forced to adopt the white man's legal system.

Morphart Creation/Shutterstock

Sequoyah and his alphabet. *How could Cherokee society have functioned for centuries without written law?*

Creating Criminal Laws in the United States

In the United States, criminal laws (or criminal statutes) are almost entirely a product of constitutional authority and the legislative bodies that enact them. They also are influenced by common law or case law interpretation and by administrative or regulatory agency decisions.

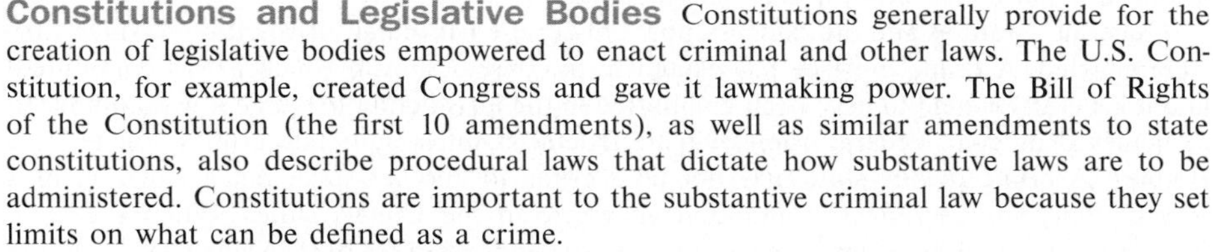

Constitutions and Legislative Bodies Constitutions generally provide for the creation of legislative bodies empowered to enact criminal and other laws. The U.S. Constitution, for example, created Congress and gave it lawmaking power. The Bill of Rights of the Constitution (the first 10 amendments), as well as similar amendments to state constitutions, also describe procedural laws that dictate how substantive laws are to be administered. Constitutions are important to the substantive criminal law because they set limits on what can be defined as a crime.

As noted, criminal laws are products of the lawmaking bodies created by constitutional authority. Thus, federal statutes are enacted by Congress, and state statutes are enacted by

Chip Somodevilla/Staff/Getty Images

Federal statutes are enacted by Congress. *How does Congress create statutes?*

CDC/Cade Martin

Ordinances are the laws of municipalities and are created by city councils, for example, often with more citizen input than laws created at the state or federal levels of government. *How much input should citizens have in the creation of criminal laws? Defend your answer.*

state legislatures. Laws created by municipalities, such as by city councils, generally are called *ordinances*. Both the federal criminal statutes and the criminal statutes of particular states, including the definitions of crimes and the penalties associated with them, can be found in penal codes, one for each jurisdiction.

Generally, statutes and ordinances apply only in the particular jurisdiction in which they were enacted. A crime must be prosecuted in the jurisdiction in which it was committed, and it generally is held to have been committed in the jurisdiction in which it was completed or achieved its goal. Federal crimes violate federal statutes, and state crimes violate state statutes. A crime in one state may not be a crime in another state, but a violation of a federal statute is a crime if committed anywhere in the United States. When a certain behavior violates both federal and state statutes, and possibly local ordinances, as is the case with many drug law violations, there is overlapping jurisdiction. In such cases, there is frequently confusion over which jurisdiction should assume responsibility for the enforcement of the law and the prosecution of the crime.

Common Law Common law, also called *case law*, is a by-product of decisions made by trial and appellate court judges, who produce case law whenever they render a decision in a particular case. The decision becomes a potential basis, or **precedent**, for deciding the outcomes of similar cases in the future. Although it is possible for the decision of any trial court judge to become a precedent, it is primarily the written decisions of appellate court judges that do. The reasons on which the decisions of appellate court judges are based are the only ones required to be in writing. This body of recorded decisions has become known as *common law*. Generally, whether a precedent is binding is determined by the court's location. (The different levels of courts in the United States will be described in detail in Chapter 8.)

precedent A decision that forms a potential basis for deciding the outcomes of similar cases in the future; a by-product of decisions made by trial and appellate court judges, who produce case law whenever they render a decision in a particular case.

stare decisis The principle of using precedents to guide future decisions in court cases; Latin for "to stand by decided cases."

The principle of using precedents to guide future decisions in court cases is called ***stare decisis***, short for *stare decisis et quieta non movere*–Latin for "to stand by and adhere to decisions and not disturb what is settled." Much of the time spent by criminal lawyers in preparing for a case is devoted to finding legal precedents that support their arguments. The successful outcome of a case depends largely on the success of lawyers in that endeavor.

Common Law

Many states have provisions in their statutes like Florida's: "The common law of England in relation to crimes . . . shall be of full force in this state where there is no existing provision by statute on the subject."

Source: § 775.01, Fla. Stat.

Although common law was an important source of criminal law in colonial America, it is less so today. Currently, what were originally common law crimes, as well as many new crimes, have been defined by statutes created by legislatures in nearly all states. There is no federal criminal common law. Nevertheless, as noted previously, common law or case law remains important for purposes of statutory interpretation.

Administrative or Regulatory Agency Decisions Administrative or regulatory agencies are the products of statutes enacted by the lawmaking bodies of different jurisdictions. Those agencies create rules, regulate and supervise activities in their areas of responsibility, and render decisions that have the force of law. Examples of federal administrative or regulatory agencies are the Federal Trade Commission (FTC), the Federal Communications Commission (FCC), the Nuclear Regulatory Commission (NRC), the Drug Enforcement Administration (DEA), and the Occupational Safety and Health Administration (OSHA). There are administrative or regulatory agencies at the state and local levels as well. Although violations of the rules and regulations of such agencies are generally handled through civil law proceedings, some violations—especially habitual violations—may be addressed through criminal proceedings if provided for by statute. In addition, legislatures often enact criminal statutes based on the recommendations of regulatory agencies.

CJ ONLINE

DEA

You can learn more about the programs, major operations, statistics, and so on, of the Drug Enforcement Administration (DEA) by visiting its website at https://www.dea.gov.

How much power should the recommendations of regulatory agencies like the DEA have in shaping criminal laws?

The Interdependency Among Sources of Legal Authority Although federal and state criminal statutes essentially are independent of one another, and although almost all of the action in the enforcement of criminal laws is at the state level, there is an important interdependency among sources of legal authority. For example, during the 1984 Republican National Convention in Dallas, Texas, Gregory Lee Johnson was part of a political protest of Reagan administration policies. As part of the protest, Johnson burned an American flag. Johnson was arrested and convicted of violating a Texas statute prohibiting the desecration of a venerated object. Several witnesses testified that the flag burning seriously offended them. A state court of appeals affirmed the conviction, but the Texas Court of Criminal Appeals reversed it, holding that to punish Johnson for burning the flag in this situation was inconsistent with the First Amendment. The U.S. Supreme Court agreed (see *Texas v. Johnson*, 1989). Provisions of the Constitution always take precedence over state statutes. However, if the state statute were not challenged, it would remain in effect in the particular state that enacted it.

THINKING CRITICALLY

1. Which of the five features of good criminal laws do you think are most important? Why?
2. Are there any other features that could or should be added to good criminal laws?

Procedural Law: Rights of the Accused

Most of the procedural or due process rights given to criminal suspects or defendants in the United States are found in the Bill of Rights. The Bill of Rights went into effect December 15, 1791. Other procedural rights are found in state constitutions and federal and state statutes. Probably the best systematic collection of due process rights is the *Federal Rules of Criminal Procedure*. Those rules apply only to federal crimes prosecuted in federal courts. Most states also have collections of rules regarding criminal procedures in state courts. Ohio, for example, has 60 such rules in its *Ohio Rules of Criminal Procedure*.

The Bill of Rights

The ink was barely dry on the new Constitution before critics attacked it for not protecting the rights of the people. The First Congress quickly proposed a set of 12 amendments and sent them to the states for ratification. By 1791, the states had ratified 10 of the amendments, which became known as the Bill of Rights (the first 10 amendments of the Constitution). Although the Bill of Rights originally applied only to the national government, almost all of its provisions also have been applied to the states through a series of U.S. Supreme Court decisions. Table 4.2 lists the 12 provisions in the Bill of Rights that are applicable to the criminal justice process. Note that only one of the provisions—the right to a grand jury indictment—is not yet applicable to the states.

FYI

Bill of Prohibitions or Bill of Rights?

Clancy and O'Brien argue, "The document, modeled after the French Declaration of the Rights of Man, is actually a bill of prohibitions, not rights; it's a list of practices that the government is prohibited from engaging in. And it reflects the outer limits of what the government *may* do; it was never intended to be a guide to what the government *should* do."

Source: Martin Clancy and Tim O'Brien, *Murder at the Supreme Court* (Amherst, NY: Prometheus, 2013), p. 157.

TABLE 4.2 The 12 Provisions in the Bill of Rights Applicable to the Criminal Justice Process

Procedural Right	Amendment
1. Freedom from unreasonable searches and seizures	Fourth
2. Grand jury indictment in felony cases*	Fifth
3. No double jeopardy	Fifth
4. No compelled self-incrimination	Fifth
5. Speedy and public trial	Sixth
6. Impartial jury of the state and district where crime occurred	Sixth
7. Notice of nature and cause of accusation	Sixth
8. Confront opposing witnesses	Sixth
9. Compulsory process for obtaining favorable witnesses	Sixth
10. Counsel	Sixth
11. No excessive bail and fines	Eighth
12. No cruel and unusual punishment	Eighth

Note: *This right has not been incorporated by and made applicable to the states.

The Fourteenth Amendment and the Selective Incorporation of the Bill of Rights

The Fourteenth Amendment was finally ratified by the required three-fourths of all states in 1868, shortly after the conclusion of the Civil War. In part, the amendment reads as follows:

> No State shall make or enforce any law which shall abridge the privileges or immunities of citizens of the United States, nor shall any State deprive any person of life, liberty, or property, without due process of law; nor deny to any person within its jurisdiction the equal protection of the laws.

ACLU

The American Civil Liberties Union is a nonprofit, nonpartisan, advocacy group devoted to the protection of civil liberties for all Americans. You can learn more about the organization and its defense of the Constitution—especially the Bill of Rights—by going to www.aclu.org. Review the "Issues" section, and look at issues such as capital punishment, criminal law reform, juvenile justice, prisoners' rights, racial justice, and smart justice.

Do you think that organizations like the ACLU are necessary? Why or why not?

One of the interesting and long-debated questions about the Fourteenth Amendment was whether its original purpose was to extend the procedural safeguards described in the Bill of Rights to people charged with crimes at the state level. Before the passage of the Fourteenth Amendment, the Bill of Rights applied only to people charged with federal crimes; individual states were not bound by its requirements. Some justices of the Supreme Court–for example, William Douglas (justice from 1939 to 1975), Hugo Black (justice from 1937 to 1971), and Frank Murphy (justice from 1940 to 1949)–believed that the Fourteenth Amendment was supposed to *incorporate* the Bill of Rights and make it applicable to the states. However, other justices, perhaps even a majority of them, did not. Thus, until the 1960s, the Supreme Court did not interpret the Fourteenth Amendment as incorporating the Bill of Rights.

There are at least three different explanations for the actions or, in this case, inactions of the Supreme Court.[6] First, there is little evidence that supporters of the Fourteenth Amendment intended it to incorporate the Bill of Rights. Second, by 1937, a series of court decisions had established the precedent that the due process clause of the Fourteenth Amendment did not require states to follow trial procedures mandated at the federal level by provisions in the Bill of Rights. The Supreme Court had held that due process was not violated if procedures followed in state courts were otherwise fair. Third, there was the states' rights issue. Because the administration of justice is primarily a state and local responsibility, many people resented what appeared to be unwarranted interference by the federal government in state and local matters. Indeed, the Constitution, for the most part, leaves questions about policing and administering justice to the states, unless a state's procedure violates a fundamental principle of justice.

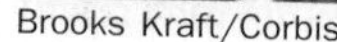

Brooks Kraft/Corbis

Wally McNamee/Contributor/Getty Images

Shepard Sherbell/Contributor/Getty Images

Henry Burroughs/AP Images

From left to right: Chief Justices John Roberts Jr., William Rehnquist, Warren Burger, and Earl Warren. Whereas the politically liberal Warren Court of the 1960s championed the rights of criminal suspects by extending procedural safeguards, the politically conservative Burger, Rehnquist, and Roberts Courts of the 1970s to the present have actively reversed or altered in other ways some of the work of the Warren Court. *How can the different direction in criminal procedure taken by the Burger, Rehnquist, and Roberts Courts be explained?*

Regardless of the reason, it was not until the early 1960s that the Supreme Court, then headed by Chief Justice Earl Warren, began to selectively incorporate most of the procedural safeguards contained in the Bill of Rights, making them applicable to the states.

Thus, it took nearly 100 years after the ratification of the Fourteenth Amendment for suspects charged with crimes at the state level to be afforded most of the same due process protections as people charged with crimes at the federal level. During the past 50 years, however, the composition of the Supreme Court has changed dramatically, and with the change in personnel, the Court's views of due process rights have changed as well. Whereas the politically liberal Warren Court of the 1960s championed the rights of criminal suspects by extending procedural safeguards, the politically conservative Burger, Rehnquist, and Roberts Courts of the 1970s to the present have actively reversed or altered in other ways some of the work of the Warren Court.[7]

FYI **"Evolving Standards of Decency"**

In *Trop v. Dulles* (1958), Chief Justice Warren wrote that the protections of the Bill of Rights "must draw [their] meaning from evolving standards of decency that mark the progress of a maturing society."

In the rest of this section, we consider the procedural rights in the Bill of Rights, which are found in the Fourth, Fifth, Sixth, and Eighth Amendments to the Constitution.[8] Before we do, however, note that the specific interpretation of each of the procedural or due process rights has evolved over time through dozens of Supreme Court and lower court decisions, or precedents. In this introductory examination, we limit our consideration of the legal development of those rights to what we believe are the most consequential cases.

The Fourth Amendment

The Fourth Amendment reads as follows:

> The right of the people to be secure in their persons, houses, papers, and effects, against unreasonable searches and seizures, shall not be violated, and no warrants shall issue, but upon probable cause, supported by Oath or affirmation, and particularly describing the place to be searched, and the person or things to be seized.

The Fourth Amendment (as well as other provisions of the Constitution) protects individual privacy against certain types of governmental interference. However, it does not provide a general constitutional "right to privacy," as many people wrongly believe. Nearly every governmental action interferes with personal privacy to some extent. Thus, the question in Fourth Amendment cases is limited to whether a governmental intrusion violates the Constitution.[9]

The procedural rights in the Fourth Amendment influence the operation of criminal justice in the United States nearly every day. They concern the legality of searches and seizures and the question of what to do with evidence that is illegally obtained. **Searches** are explorations or inspections, by law enforcement officers, of homes, premises, vehicles, or persons, for the purpose of discovering evidence of crimes or persons who are accused of crimes. A search occurs "when an expectation of privacy that society is prepared to consider reasonable is infringed [by the government]."[10] **Seizures** are the taking of persons or property into custody in response to violations of the criminal law. A seizure of property occurs "when there is some meaningful interference [by the government] with an individual's possessory interests in that property."[11]

searches Explorations or inspections, by law enforcement officers, of homes, premises, vehicles, or persons, for the purpose of discovering evidence of crimes or persons who are accused of crimes.

seizures The taking of persons or property into custody in response to violations of the criminal law.

In *United States v. Mendenhall* (1980), the Supreme Court created the following test for determining whether an encounter constitutes a Fourth Amendment seizure: "A person has been 'seized' within the meaning of the Fourth Amendment only if, in view of all the circumstances surrounding the incident, a reasonable person would have believed that he was not free to leave." The Court provided these examples of situations that might be construed as seizures, even if the person did not attempt to leave: (1) the threatening presence of several officers, (2) the display of a weapon by an officer, (3) some physical touching of the person, or (4) the use of language or a tone of voice that indicated that compliance with the officer's request might be compelled. In *California v. Hodari D.* (1991), the Court modified the plurality holding in *Mendenhall.* In *Hodari D.*, the suspect ran from the police, and an officer pursued, thereby creating a circumstance in which a "reasonable person would have believed that she or he was not free to leave" or to disobey the officer's command to halt. There was no physical touching of the suspect. The Court held that in cases involving a "show of authority," as distinguished from physical touching, no "seizure" occurs unless and until the suspect yields or submits to the assertion of authority.

According to the Supreme Court, the Fourth Amendment allows two kinds of searches and seizures: those made with a warrant and those made without a warrant. A **warrant** is a written order from a court directing law enforcement officers to conduct a search or to arrest a person. An **arrest** is the seizure of a person or the taking of a person into custody. An arrest can be either taking actual physical custody, as when a suspect is handcuffed by a police officer, or constructive custody, as when a person peacefully submits to a police officer's control. An arrest can occur without an officer's physically touching a suspect.

warrant A written order from a court directing law enforcement officers to conduct a search or to arrest a person.

arrest The seizing or the taking of a person into custody by lawful authority, either actual physical custody, as when a suspect is handcuffed by a police officer, or constructive custody, as when a person peacefully submits to a police officer's control.

The Fourth Amendment requires only that searches and seizures not be "unreasonable." Searches and seizures conducted with a legal warrant generally are considered reasonable. However, what is "reasonable" in warrantless searches remained vague for more than 100 years after the ratification of the amendment. It was not until a series of cases beginning in the 1960s that the Supreme Court began to provide a more precise definition of the term. Because the law concerning warrantless searches and seizures is complex, only a relatively brief and simplified overview will be provided in that section.

Searches and Seizures with a Warrant

First, law enforcement officers must have *probable cause* before a judicial officer can legally issue a search or arrest warrant. Probable cause for a search warrant requires substantial and trustworthy evidence to support two conclusions: (1) that the specific objects to be searched for are connected with criminal activity and (2) that the objects will be found in the place to be searched. In nearly all jurisdictions, law enforcement officers seeking a search warrant must specify in a signed *affidavit*, a written and sworn declaration, the facts that establish probable cause. The facts in the affidavit are the basis for determining later whether there was probable cause to issue the warrant in the first place. Some jurisdictions allow sworn oral testimony to establish probable cause. Figure 4.2 shows the FBI's application for a search warrant for Michael Cohen evidence in the Mueller Investigation.

The Fourth Amendment requires that a search warrant contain a particular description of the place to be searched and the person or things to be seized. Thus, the warrant must be specific enough that a law enforcement officer executing it would know where to search and what objects to seize, even if the officer was not originally involved in the case.

FIGURE 4.2 FBI's Application for a Search Warrant for Michael Cohen Evidence in the Mueller Investigation

AO 106 (Rev. 06/09) Application for a Search Warrant

UNITED STATES DISTRICT COURT

for the

Southern District of New York

In the Matter of the Search of
(Briefly describe the property to be searched or identify the person by name and address)
Four Premises and Two Electronic Devices, See Attached Affidavit and Riders

Case No. 18 MAG 2969

APPLICATION FOR A SEARCH WARRANT

I, a federal law enforcement officer or an attorney for the government, request a search warrant and state under penalty of perjury that I have reason to believe that on the following person or property *(identify the person or describe the property to be searched and give its location)*:
Four Premises and Two Electronic Devices, See Attached Affidavit and Riders

located in the Southern District of New York, there is now concealed *(identify the person or describe the property to be seized)*:

PLEASE SEE ATTACHED AFFIDAVIT AND RIDERS.

The basis for the search under Fed. R. Crim. P. 41(c) is *(check one or more)*:

- ☑ evidence of a crime;
- ☑ contraband, fruits of crime, or other items illegally possessed;
- ☑ property designed for use, intended for use, or used in committing a crime;
- ☐ a person to be arrested or a person who is unlawfully restrained.

The search is related to a violation of:

Code Section	*Offense Description*
18 U.S.C. s 371, 1005, 1014, 1343 and 1344, and 52 USC 30116 and 30109	Conspiracy, false bank entries, false statements to a financial institution, wire fraud, bank fraud, and illegal campaign contributions

The application is based on these facts:

PLEASE SEE ATTACHED AFFIDAVIT AND RIDER.

☑ Continued on the attached sheet.

☐ Delayed notice of ____ days (give exact ending date if more than 30 days: __________) is requested under 18 U.S.C. § 3103a, the basis of which is set forth on the attached sheet.

Printed name and title

Sworn to before me and signed in my presence.

Date: 04/08/2018

Judge's signature

City and state: NEW YORK, NEW YORK

Hon. Henry B. Pitman, U.S. Magistrate Judge
Printed name and title

Source: United States District Court for the Southern District of New York

However, absolute technical accuracy in the description of the place to be searched is not necessary. It is required only that an officer executing a warrant can find, perhaps by asking questions of neighborhood residents, the place to be searched.

A warrant also may be issued for the search of a person or an automobile, rather than a place. A warrant to search a person should provide the person's name or at least a detailed description. A warrant to search an automobile should include either the car's license number or its make and the name of its owner.

Search warrants are required to be executed in a reasonable amount of time. For example, federal law requires that a search be conducted within 10 days after the warrant is issued. The federal government and nearly half of the states also have laws limiting the time of day during which search warrants may be executed. In those jurisdictions, searches may be conducted only during daytime hours unless there are special circumstances.

Generally, before law enforcement officers may enter a place to conduct a search, they must first announce that they are law enforcement officers, that they possess a warrant, and that they are there to execute it. The major exceptions to this requirement are situations in which it is likely that the evidence would be destroyed immediately on notification or in which notification would pose a threat to officers. Judges in many jurisdictions also are authorized to issue "no-knock" warrants in some circumstances such as drug busts. However, if officers are refused entry after identifying themselves, they may then use force to gain entry, but only after they have given the occupant time to respond. In short, they cannot legally yell "police officers" and immediately kick down the door. The Supreme Court has held that under ordinary circumstances they must wait at least 15 to 20 seconds (see *United States v. Banks*, 2003). Finally, if in the course of conducting a legal search, law enforcement officers discover **contraband** (an illegal substance or object) or evidence of a crime not covered by the warrant, they may seize that contraband or evidence under the plain-view exception (discussed later) without getting a new warrant specifically covering it.

contraband An illegal substance or object.

Arrests with a Warrant

Most arrests are made without a warrant. Generally, an arrest warrant is legally required when law enforcement officers want to enter private premises to make an arrest unless there is consent or exigent circumstances (discussed later). An arrest warrant is issued only if substantial and trustworthy evidence supports these two conclusions: (1) a violation of the law has been committed and (2) the person to be arrested committed the violation.

CHARLES BENNETT/AP Images

Samar Kaukab sued the Illinois National Guard and O'Hare International Airport security officers for strip-searching her at the Chicago airport before a flight simply because she was wearing a hijab—a Muslim head scarf. *Were the security officers justified in conducting the strip-search? Why or why not?*

Searches and Seizures Without a Warrant

In guaranteeing freedom from illegal searches and seizures, the Fourth Amendment protects a person's privacy. Under most circumstances, the amendment requires a warrant signed by a judge to authorize a search for and seizure of evidence of criminal activity. However, Supreme Court interpretations of the Fourth Amendment have permitted warrantless searches and seizures in some circumstances. A person generally is protected from searches and seizures without a warrant in places, such as home or office, where he or she has a legitimate right to privacy. That same protection, however, does not extend to all places where a person has a legitimate right to be. For example, the Supreme Court has permitted the stopping and searching of automobiles under certain circumstances and with probable cause. Several doctrines concerning search and seizure without a warrant have developed over time.

Before 1969, when law enforcement officers arrested a suspect, they could legally search, without a warrant, the entire premises surrounding the arrest. That kind of search is called a *search incident to arrest*, and like a search with a warrant, it required probable cause. Evidence obtained through a search incident to arrest was admissible as long as the arrest was legal.

In 1969, in the case of *Chimel v. California*, the Supreme Court limited the scope of *searches incident to an arrest*. The Court restricted the physical area in which officers could conduct a search to the area within the suspect's immediate control. The

Court interpreted the area within the suspect's immediate control as an area near enough to the suspect to enable him or her to obtain a weapon or destroy evidence. The Court also ruled that it is permissible for officers, incident to an arrest, to protect themselves, to prevent a suspect's escape by searching the suspect for weapons, and to preserve evidence within the suspect's grabbing area.

The Supreme Court has continued to refine the scope of warrantless searches and seizures incident to an arrest. For example, in 1981, in *New York v. Belton*, the Court ruled that after police have made a lawful arrest of the occupant of an automobile, they may, incident to an arrest, search the automobile's entire passenger compartment and the contents of any opened or closed containers found in the compartment. The police had long been able to legally search the automobile's trunk, providing there was probable cause (see *Carroll v. United States*, 1925). In *United States v. Ross* (1982), the Court held that when police "have probable cause to search an entire vehicle, they may conduct a warrantless search of every part of the vehicle [including the trunk] and its contents, including all containers and packages, that may conceal the object of the search." Then, in 1991, in the case of *California v. Acevedo*, the Court ruled that the police, "in a search extending only to a container within an automobile, may search the container without a warrant where they have probable cause to believe that it holds contraband or evidence." The Court also will allow the police, with probable cause, to search an automobile passenger's belongings, if he or she is capable of concealing the object of the search (see *Wyoming v. Houghton*, 1999). However, in 2009, in *Arizona v. Gant*, the Court qualified its earlier decision in *New York v. Belton*. The Court held that police may search the passenger compartment of a vehicle incident to a recent occupant's arrest only if it is reasonable to believe that the arrestee might access the vehicle at the time of the search or that the vehicle contains evidence of the arrest offense. In this case, Gant was arrested for driving with a suspended license, handcuffed, and locked in a patrol car before officers searched his car and found cocaine in his jacket pocket. He was convicted of drug offenses. The Court noted that Gant could not have accessed his car at the time of the search and that he was arrested for driving with a suspended license—an offense for which police could not reasonably expect to find evidence in Gant's car.

In 2008, in *Brendlin v. California*, the Court ruled that when police make a traffic stop, a passenger in the car, like the driver, is seized for Fourth Amendment purposes and so may challenge the stop's constitutionality. In this case, Brendlin, a parole violator and passenger in a car that was stopped for a routine registration check, was recognized by one of the police officers, arrested, and searched. The driver and car were searched as well. The police found, among other things, methamphetamine paraphernalia. Brendlin was charged with possession and manufacture of the drug. He moved to suppress the evidence, arguing that the officers had neither probable cause nor reasonable suspicion to make the traffic stop, which was an unconstitutional seizure of his person. The Court agreed with Brendlin and remanded the case for further proceedings not inconsistent with its opinion.

In 2015, in *Rodriquez v. United States*, the Court held that police are not entitled to turn routine traffic stops into drug searches using trained dogs. The decision ends the practice whereby officers stop a car for a traffic violation and then call for a drug-sniffing dog to inspect the vehicle. Writing for the Court's majority, Justice Ginsberg wrote, "Police may not prolong detention of a car and driver beyond the time reasonably required to address the traffic violation."

Other Supreme Court decisions have established principles governing when private areas may be searched incident to an arrest. In 1968, for instance, in the case of *Harris v. United States*, the Court established the *plain-view doctrine*. Under this doctrine, the police may seize an item—evidence or contraband—without a warrant if they are lawfully in a position to view the item and if it is immediately apparent that the item is evidence or contraband. In 1990, the Court clarified its ruling in *Harris* by adding that the discovery of the item in plain view need not be "inadvertent." That is, the plain-view doctrine applies even when the police expect in advance to find the item in plain view (*Horton v. California*). Also, in 1990, in the case of *Maryland v. Buie*, the Court addressed the issue of *protective sweeps*. The Court held that when a warrantless arrest takes place in a suspect's home, officers, with reasonable suspicion, may make only a "cursory visual inspection" of areas that could harbor an accomplice or a person posing danger to them.

Even a warrantless search not incident to an arrest may be justified under the Supreme Court's *exigent circumstances* doctrine. It permits police to make warrantless searches in exigent, or emergency, situations. Such situations could include a need to prevent the imminent destruction of evidence, a need to prevent harm to individuals, or the hot pursuit of suspects.

E-mail and Privacy

E-mail does not have the same privacy protections as does "regular mail." Depending on where the e-mail is sent to and sent from, privacy rights vary greatly. However, under no circumstances can e-mail messages ever be considered confidential. Both government employees (federal, state, and local) and employees in the private sector have few, if any, privacy rights in the e-mail messages they send or receive. Employers are free to examine all electronic communications stored on their systems. The e-mail messages of government employees also may be available to the public through a freedom of information request. In addition, commercial service providers (such as gmail) are permitted to disclose messages to law enforcement officials when they "inadvertently" come across messages containing references to illegal activity. Finally, regardless of the setting (work or home), e-mail can be subpoenaed, discovered, and intercepted under existing wiretap laws.

Source: Erik C. Garcia, "E-mail and Privacy Rights," University of Buffalo, (on file with authors).

Frequently, law enforcement officers are not hampered by the warrant requirement because suspects often consent to a search. In other words, law enforcement officers who do not have enough evidence to obtain a search warrant, or who either cannot or do not want to take the time and trouble to obtain one, may simply ask a suspect whether they may conduct a search. If the suspect consents voluntarily, the search can be made legally. Law enforcement officers call this strategy "knock and talk." In 1973, in the case of *Schneckloth v. Bustamonte*, the Supreme Court upheld the legality of *consent searches*. The Court also ruled that officers do not have to tell suspects that they have a right to withhold consent unless they ask.

It is not surprising that consent searches have become the most common type of searches performed by law enforcement officers. They are used frequently in traffic stops and drug interdiction efforts at airports and bus terminals. In the 1980s, as a new tool in the war on drugs, several police departments adopted programs in which officers boarded buses and asked passengers to consent to searches. The practice was challenged in a 1985 Florida case in which a bus passenger had consented to having his luggage searched. When the police found cocaine, the passenger was arrested and subsequently convicted. The Florida Supreme Court ruled that the search was unconstitutional. In 1991, the U.S. Supreme Court reversed the Florida Supreme Court (in *Florida v. Bostick*) and held that the search was not unconstitutional and that law enforcement officers may make such a search without a warrant or suspicion of a crime–as long as the passenger feels free to refuse the search.

Critics argue that, in most cases, consent searches cannot be truly voluntary, even when permission is granted, because most people are intimidated by the police and would have a hard time refusing their request. Moreover, most people probably do not know that they may refuse a warrantless search except under the conditions described earlier.

FYI

Refusing a Stop or Search

According to a 2016 national survey, 73% of Americans support a law requiring police officers to tell citizens if they may refuse a stop or a requested search. All demographic groups support this proposal.

Source: Emily Ekins, "Policing in America: Understanding Public Attitudes Toward the Police. Results from a National Survey," Cato Institute, December 7, 2016, accessed June 15, 2019, https://www.cato.org/survey-reports/policing-america.

In the case of *Georgia v. Randolph* (2006), the U.S. Supreme Court restricted consent searches of a home. In this case, Randolph's estranged wife, concerned about his cocaine use, called the police and, when they arrived, told them they could find cocaine in their bedroom. The police asked her if they could conduct a search, and she consented. Randolph, a lawyer, who was present at the time, objected to the search, but the police ignored him, entered the house, and found cocaine in the bedroom. Randolph was arrested and subsequently convicted of drug possession. On appeal, the U.S. Supreme Court, citing the Fourth Amendment's central value of "respect for the privacy of the home," ruled that warrantless searches of a home, even when a co-habitant consented to the search, are prohibited if the other co-habitant is physically present and objects to the search. As a result of the decision, the incriminating evidence was suppressed (i.e., not admitted at the retrial). However, the consent of one co-occupant in the absence of another generally is considered sufficient to permit a home search by the police (see *United States v. Matlock*, 1974). A home search by the police also is permitted if a co-occupant objects to the search but is no longer present because he has been arrested and is in custody, and the home's other co-occupant consents (*Fernandez v. California*, 2014).

In a new twist on consent searches, Boston police instituted an anti-crime program called "Safe Homes" in which they ask parents or legal guardians in high-crime areas for permission to conduct a warrantless search of their children's bedrooms for guns. The police believe that parents who fear their children will become involved in gun violence will let police into their homes to search for guns. If parents refuse, the police will leave. Such searches will not be conducted in the homes of teens suspected in shootings or homicides so prosecutions are not jeopardized. If officers find drugs in the warrantless search, it will be up to them to decide to arrest; however, according to police brass, modest amounts of drugs such as marijuana simply will be confiscated. Civil libertarians are concerned.[12]

Arrests without a Warrant Officers may not enter a private home to make a warrantless arrest unless there is consent or the offense is a serious one and there are exigent circumstances, such as the likely destruction of evidence or the hot pursuit of a felony suspect. This is the same *exigent circumstances* doctrine that applies to warrantless searches and seizures.

A suspect who is arrested without a warrant and remains confined is entitled to have a judge determine whether there was probable cause for the arrest. Ordinarily, judges must make such a determination within 48 hours of arrest. The purpose of this proceeding is to ensure that the suspect's continuing custody is based on a judicial determination of probable cause and not merely on the police officer's judgment that probable cause supported an arrest.

FIGURE 4.3 Standards of Proof and Criminal Justice Activities

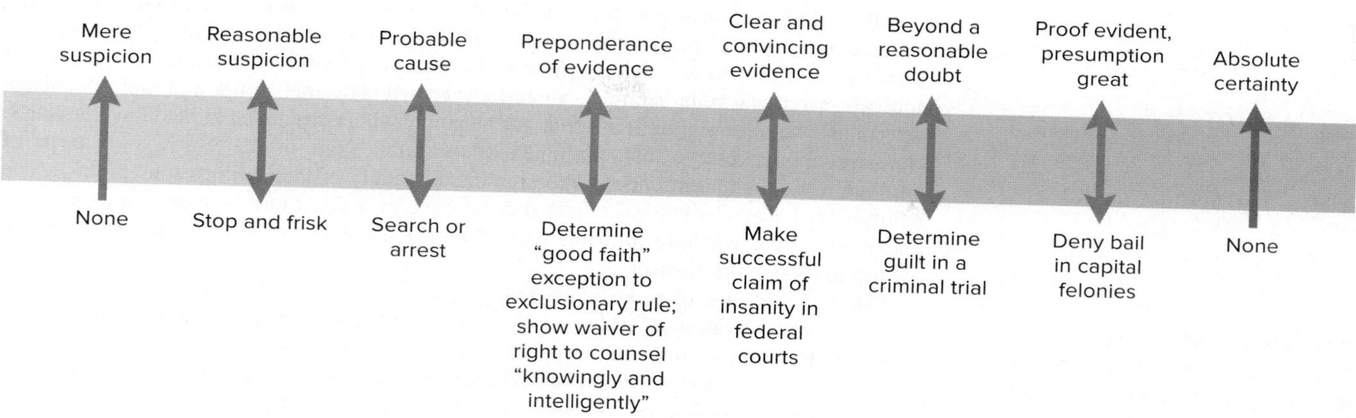

Standards of Proof As mentioned previously and as specified in the Fourth Amendment, neither search nor arrest warrants can be issued legally unless law enforcement officers convince a judge or a magistrate that there is probable cause to believe either that the specific items to be searched for are related to criminal activity and the items will be found in the place to be searched, or that a violation of the law has been committed and the person to be arrested committed the violation. Probable cause is one among a number of standards of proof for various criminal justice activities. The amount of proof necessary depends on the activity in question. Figure 4.3 shows various standards of proof, along a continuum of certainty, and the criminal justice activities that correspond to them.

Toward one end of the continuum is the standard of proof with the least certainty: *mere suspicion*. **Mere suspicion** is equivalent to a "gut feeling." In other words, a law enforcement officer may have a feeling that something is wrong—an uncanny knack that some experienced law enforcement officers possess—but be unable to state exactly what it is. With only mere suspicion, law enforcement officers cannot legally even stop a suspect.

mere suspicion The standard of proof with the least certainty; a "gut feeling." With mere suspicion, a law enforcement officer cannot legally even stop a suspect.

A standard of proof with greater certainty is *reasonable suspicion*. **Reasonable suspicion** is more than a gut feeling. It includes the ability to articulate reasons for the suspicion. For example, if a law enforcement officer observes a person in front of a bank wearing a heavy trench coat on a hot summer day, the officer might have a reasonable suspicion that something is wrong. The officer could state that idling in front of a bank while wearing a heavy trench coat on a hot summer day is suspect behavior. Until recently, an anonymous tip, as long as there were other reliable indicators, could be the basis for reasonable suspicion. However, in *Florida v. J.L.* (2000), the Supreme Court ruled that an uncorroborated anonymous tip was unconstitutional.[13] But the Court, in *Illinois v. Wardlow* (2000), confirmed that running from the police when they enter a high-crime area is reasonably suspicious behavior.[14] With reasonable suspicion, a law enforcement officer is legally permitted to stop and frisk a suspect (*Terry v. Ohio*, 1968). **Frisking** a suspect means conducting a search for weapons by patting the outside of a suspect's clothing, feeling for hard objects that might be weapons. Only if an officer feels something that may be a weapon may he or she search inside a pocket or an article of clothing. If evidence of a crime is discovered, the officer is permitted to make an arrest.

reasonable suspicion A standard of proof that is more than a gut feeling. It includes the ability to articulate reasons for the suspicion. With reasonable suspicion, a law enforcement officer is legally permitted to stop and frisk a suspect.

frisking Conducting a search for weapons by patting the outside of a suspect's clothing, feeling for hard objects that might be weapons.

The standard of proof needed to conduct a search or to make an arrest is *probable cause*. The conventional definition of **probable cause** is the amount of proof necessary for a reasonably intelligent person to believe that a crime has been committed or that items connected with criminal activity can be found in a particular place. Although its meaning is not entirely clear—what is "reasonably intelligent"?—probable cause has a greater degree of certainty than reasonable suspicion. For probable cause, law enforcement officers must have some tangible evidence that a crime has been committed, but that evidence does not have to be admissible at trial. Such evidence might include a tip from a reliable informant or the pungent aroma of marijuana in the air. As noted in Chapter 1, probable cause is also the standard of proof used in initial appearances and preliminary hearings.

probable cause The amount of proof necessary for a reasonably intelligent person to believe that a crime has been committed or that items connected with criminal activity can be found in a particular place. It is the standard of proof needed to conduct a search or to make an arrest.

The line between probable cause and reasonable suspicion, or even mere suspicion, is a fine one and a matter of interpretation. In practice, there are many gray areas. Consequently, criminal courts and the judicial officers who are authorized to approve search warrants have been given the responsibility of determining whether a standard of proof has been met in a particular situation. As noted, a judicial officer, for example, generally must approve search warrants before they can be executed. The way courts and judicial officers determine whether a standard of proof has been met will be discussed in detail in Chapter 8. Here we simply observe that, for much of the public, one of the frustrating aspects of criminal justice is that offenders who are factually guilty of their crimes sometimes escape punishment because a judicial officer did not have probable cause to issue a warrant, or a police officer did not have probable cause to make an arrest or have reasonable suspicion to stop and frisk the suspect.

preponderance of evidence Evidence that more likely than not outweighs the opposing evidence, or sufficient evidence to overcome doubt or speculation.

The next standard of proof along the continuum of legal certainty is *preponderance of evidence*. **Preponderance of evidence** is evidence that more likely than not outweighs the opposing evidence, or sufficient evidence to overcome doubt or speculation. It is the standard of proof necessary to find a defendant liable in a civil lawsuit. This standard also is used in determining whether the *inevitable-discovery rule* applies. That is, the prosecution must prove by a preponderance of the evidence that evidence actually uncovered as a result of a constitutional violation inevitably would have been discovered through lawful means, independent of the action constituting the violation. Finally, preponderance of evidence is the standard of proof in criminal proceedings by which the state must show that the right to counsel has been waived "knowingly and intelligently."

clear and convincing evidence The standard of proof required in some civil cases and, in federal courts, the standard of proof necessary for a defendant to make a successful claim of insanity.

Next along the continuum of certainty is **clear and convincing evidence**, which is evidence indicating that the thing to be proved is highly probable or reasonably certain. It is the standard of proof required in some civil cases and, in federal courts, the standard of proof necessary for a defendant to make a successful claim of insanity.

beyond a reasonable doubt The standard of proof necessary to find a defendant guilty in a criminal trial.

Of greater certainty still is proof **beyond a reasonable doubt**, the standard of proof necessary to find a defendant guilty in a criminal trial. "Reasonable doubt" as a standard of proof is a relatively recent concept. It appears to have been used for the first time in the Boston Massacre trials in 1770. Until then, no standards of proof existed in English colonies, and juries only had to return "true verdicts." Not until 1970 (in *In re Winship*) was reasonable doubt, as a standard of proof, made a constitutional requirement in all criminal cases, both federal and state.[15] Reasonable doubt is the amount of doubt about a defendant's guilt that a reasonable person might have after carefully examining all the evidence. In the case of *Sandoval v. California* (1994), the Court upheld the following definition of *reasonable doubt*:

> It is *not a mere possible doubt*; because everything relating to human affairs, and *depending on moral evidence*, is open to some possible or imaginary doubt. It is that state of the case which, after the entire comparison and consideration of all the evidence, leaves the minds of the jurors in that they cannot say they feel an abiding conviction, *to a moral certainty*, of the truth of the charge. [Emphasis in original.]

Thus, to convict a criminal defendant in a jury trial, a juror must be convinced of guilt by this standard. However, what is considered reasonable varies, and reasonableness is thus a matter of interpretation. Therefore, the procedural laws in most jurisdictions require that a panel of from 6 to 12 citizens all agree that a defendant is guilty beyond a reasonable doubt before that defendant can be convicted.

proof evident, presumption great The standard of proof required for a judicial officer to deny bail in cases involving capital felonies.

Of greater certainty than beyond a reasonable doubt—or at least as certain as beyond a reasonable doubt—is **proof evident, presumption great**, the standard of proof required for a judicial officer to deny bail in cases involving capital felonies. To meet the standard at a pretrial hearing, the burden is on the prosecution to show by testimony and/or real evidence that there is no question of the defendant's guilt. If there is some doubt about the testimony or evidence, or if there are contradictions or discrepancies in the testimony or evidence, then the standard has not been met and the defendant is entitled to reasonable bail as a matter of right. Although proof evident, presumption great has been considered the highest standard of proof known to American law, the Florida Supreme Court has declared that it is the same standard of proof as beyond a reasonable doubt.[16]

No criminal justice activity requires absolute certainty as a standard of proof, although standards of "beyond any doubt," "beyond all doubt," "no doubt," "no doubt about the guilt of the defendant," and "moral certainty" (a concept popular in the seventeenth century) have been proposed for use in capital cases.[17]

The Exclusionary Rule The **exclusionary rule** was created by the Supreme Court in 1914 in the case of *Weeks v. United States*. In *Weeks*, the Supreme Court held that illegally seized evidence must be excluded from trials in federal courts. In 1961, the Warren Court extended the exclusionary rule to state courts in the case of *Mapp v. Ohio*. The exclusionary rule originally had three primary purposes: (1) to protect individual rights from police misconduct, (2) to prevent police misconduct, and (3) to maintain judicial integrity (for citizens to have faith in the administration of justice, courts should not admit evidence that is tainted by the illegal activities of other criminal justice officials). Today, however, the principal purpose of the exclusionary rule is to deter the police from violating people's Fourth Amendment rights.

exclusionary rule The rule that illegally seized evidence must be excluded from trials in federal courts.

In practice, when suspects want to claim that incriminating evidence was obtained through an illegal search and seizure, that a confession was obtained without the required warnings or was involuntary, that an identification was made as a result of an invalid police lineup, or that evidence was in some other way illegally obtained, they attempt, through their attorneys, to show at a suppression hearing that the search and seizure, for example, violated the Fourth Amendment. If they are successful in their claims, the evidence that was obtained as a result of the illegal search and seizure will not be admitted at trial.

By the late 1970s, public opinion polls showed that Americans were becoming increasingly alarmed about the problem of crime and especially about what they perceived as the practice of allowing a substantial number of criminals to escape punishment because of legal technicalities. One so-called legal technicality that received much of the public's scorn was the exclusionary rule. In 1984, responding at least in part to public opinion, the Supreme Court, under Chief Justice Warren Burger, decided three cases that had the practical effect of weakening the exclusionary rule.

In two of the three cases, *United States v. Leon* and *Massachusetts v. Sheppard*, a *good-faith exception* to the *exclusionary rule* was recognized. The Court ruled that as long as the police act in good faith when they request a warrant, the evidence they collect may be used in court, even if the warrant is illegal or defective. In the *Leon* case, the judge's determination of probable cause turned out to be wrong. Prior to *Leon*, such an error by a judge would have been recognized as a violation of the Fourth Amendment, and the evidence seized with the warrant would have been excluded at trial. The Court reasoned that it was unfair to penalize law enforcement officers who conduct searches in which incriminating evidence is found, when those officers conduct the search in good faith that they have a legal warrant. In the *Sheppard* case, the judge had used the wrong form for the warrant. As in *Leon*, the Court reasoned that it was unfair to penalize law enforcement officers, and the public, just because there was a flaw in the warrant, when the officers had conducted a search in good faith and found incriminating evidence.

The third case, *Nix v. Williams*, established an *inevitable-discovery exception* to the *exclusionary rule*. The *Nix* case involved a murderer whom police had tricked into leading them to the hidden body of his victim. In *Nix*, the Court held that evidence obtained in violation of a defendant's rights can be used at trial if the prosecution can show, by a preponderance of the evidence, that the information ultimately or inevitably would have been discovered by lawful means.

The exclusionary rule was again weakened in 1995. In the case of *Arizona v. Evans*, the Supreme Court ruled that unlawful arrests based on computer errors do not always require the exclusion of evidence seized by police. In the *Arizona* case, the Court held that a good-faith exception to the exclusionary rule could be made as long as the errors of court employees and not the police caused the illegal seizure of evidence. In that case, a Phoenix man who had been stopped for a traffic violation was arrested because a computer record showed an outstanding arrest warrant for some traffic violations. In fact, the warrant had been dropped 17 days earlier, but the action had not been entered into the computer. After the arrest, marijuana was seized from the man's car, and he was arrested for illegal possession.

The exclusionary rule was further eroded in the 2006 case of *Hudson v. Michigan*. In *Hudson*, Detroit police executed a warrant for narcotics and weapons at Hudson's home, but, in doing so, violated the Fourth Amendment's "knock-and-announce rule." Recall that the "knock-and-announce rule" refers to the requirement that the police knock at the door and announce their identity, authority, and purpose before entering a residence to execute an arrest or search warrant. Hudson moved to suppress the evidence seized, and the trial court granted Hudson's motion. On appeal, the Michigan Court of Appeals reversed, and Hudson was convicted of drug possession. The Court affirmed the state court of appeals, holding that violation of the "knock-and-announce rule" does not require suppression of evidence found in a search, especially in this case, because the

myth

Many criminals escape punishment because of the exclusionary rule.

fact

Very few criminals escape punishment because of the exclusionary rule.

interests violated had nothing to do with the seizure of evidence. The relevant evidence was discovered not by a failure to knock and announce, according to the Court, but because of a subsequent search pursuant to a lawful warrant. Critics complain that the decision weakens and perhaps destroys much of the practical value of the "knock-and-announce requirement."

The USA PATRIOT Act[18] Less than 2 months after the terrorist attacks of 9/11, and with little debate or scrutiny, Congress passed the "Uniting and Strengthening America by Providing Appropriate Tools Required to Intercept and Obstruct Terrorism Act of 2001" or the USA PATRIOT Act, for short. Among a host of provisions, the law gives broad new powers to the Federal Bureau of Investigation (FBI), the Central Intelligence Agency (CIA), and other U.S. foreign intelligence agencies to spy on American citizens. It also eliminates checks and balances on those powers such as judicial oversight, public accountability, and the ability to challenge government searches in court. For example, the law allows the FBI to search private records (financial, medical, library, student–any recorded activity) *without a warrant and probable cause* and without having to reveal to anyone what it has done (referred to as "sneak-and-peek" searches). It is able to do this through the use of "national-security letters," which are administrative subpoenas. According to a recent audit by the Justice Department's Inspector General, the FBI used the "letters" to collect more data than allowed in dozens of cases from 2003 through 2006 and underreported to Congress how many "letters" were requested by more than 4,600. The Inspector General blamed agent error and shoddy record keeping for the problems, while then-FBI Director Mueller attributed the problem in part to banks, telecommunication companies, and other private businesses providing the FBI more personal client data than was requested. Also, under the act, the FBI no longer needs probable cause to conduct wiretaps of criminal suspects when "a significant purpose" is gathering intelligence.

To see the long list of federal terrorism crimes, go to Title 18, Part I, 113B, Section 2332b(g)(5) of the U.S. Code at https://www.law.cornell.edu/uscode/text/18/part-I/chapter-113B.

Do you agree that all the listed crimes should be considered terrorism crimes? Why or why not?

The act also creates the new crime of "domestic terrorism." Under the act, members of controversial activist groups, such as Operation Rescue or Greenpeace, or the World Trade Organization protesters could be charged with domestic terrorism if they committed specifically defined federal terrorism crimes. Even providing lodging or aid to such "terrorists" could initiate surveillance or prosecution. Another provision of the law allows the U.S. Attorney General to detain noncitizens in the United States after certifying that there are "reasonable grounds to believe" that the noncitizen endangers national security. If a foreign country will not accept such noncitizens who are to be deported, they can be detained indefinitely without trial.

Ironically, as critics point out, most of the PATRIOT Act's changes to surveillance laws were part of a long-standing law enforcement wish list that Congress had rejected repeatedly. The events of 9/11 changed that. Critics also argue that the law was hurriedly passed

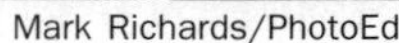
Mark Richards/PhotoEdit

JULIEN WARNAND/EPA-EFE/Shutterstock

Pascal Le Segretain/Sygma/Getty Images

Under the PATRIOT Act, activist groups such as Operation Rescue or Greenpeace or the World Trade Organization protesters could be charged with domestic terrorism if they committed specifically defined federal terrorism crimes. *Do you believe those groups engage in domestic terrorism? Why or why not?*

without determining whether problems with existing surveillance laws contributed to the terrorist attacks and whether the new law would help prevent further attacks. Many of the law's provisions do not even deal with terrorism. Particularly troublesome to many critics is the law's elimination of the checks and balances placed on the government's surveillance powers. Most of those checks and balances were created after it was learned in the 1970s that the government had misused those powers. Among other things, the FBI and the CIA had spied illegally on more than half a million U.S. citizens during the McCarthy era and later. Perhaps the most notorious example of this clandestine activity was the illegal surveillance of Martin Luther King Jr. during the 1960s. Critics took some comfort that several of the surveillance provisions of the PATRIOT Act were scheduled to expire on December 31, 2005.

However, on March 9, 2006, President George W. Bush signed into law the USA PATRIOT Act Improvement and Reauthorization Act of 2005 (after Congress had temporarily extended the original act). The legislation made permanent 14 of the 16 USA PATRIOT Act provisions set to expire and placed 4-year sunsets on the other two–the authority to conduct "roving" surveillance under the Foreign Intelligence Surveillance Act (FISA) and the authority to request production of business records under FISA. The new act also purportedly added dozens of additional safeguards to protect Americans' privacy and civil liberties. Critics contend the safeguards are inadequate. Among the provisions of the new legislation are the following:

- Authorized the Attorney General to reorganize the Department of Justice by placing the department's primary national security elements under the leadership of a new Assistant Attorney General for National Security. The new Assistant Attorney General will serve as the department's primary liaison to the new Director of National Intelligence, and the new division will gather expertise from across the department to create a focal point for providing advice on the numerous legal and policy issues raised by the department's national security missions.
- Provided tools to protect U.S. waterways and seaports from terrorists and thieves, including new or enhanced penalties for crimes such as smuggling goods into or out of the United States or bribing a public official to affect port security with the intent to commit international or domestic terrorism. Would-be terrorists will now face a U.S. Coast Guard empowered with new law enforcement tools for use at sea, including penalties for refusal to stop when ordered to do so and for transporting an explosive, biological agent, chemical weapon, or radioactive or nuclear materials knowing that the item is intended to be used to commit a terrorist act.
- Enhanced penalties for terrorism financing and closed a loophole in terrorist financing through hawalas, informal money transfer networks, rather than traditional financial institutions.
- The Combat Methamphetamine Act made certain drugs used in manufacturing "meth"–"scheduled listed chemical products"–harder to obtain in unlimited quantities and easier for law enforcement to track. It also enhanced penalties for the manufacture, smuggling, and selling of "meth."
- Eliminated confusion about the appropriate death penalty procedures for certain cases under the Controlled Substances Act and expanded on the authorities governing provision of counsel for death penalty–eligible defendants who are unable to afford counsel.
- Provided clear intent standards and tough penalties for terrorist attacks and other violence targeted at U.S. rail systems and other mass transportation systems regardless of whether they operate on land, on water, or through the air.
- Continued to allow investigators to use so-called Section 215 orders–court orders requiring production of business records or any "tangible things"–in all phases of national security investigations. Documents considered more sensitive, such as library, bookstore, medical, tax return, and gun sale records, require that applications to the FISA Court for Section 215 orders be signed by either the Director or the Deputy Director of the FBI. Recipients of Section 215 orders may seek judicial review and disclose receipt of the order to attorneys to obtain legal advice or assistance (something they were prohibited from doing under the original PATRIOT Act). Nevertheless, any employee who discloses a demand for such records, other than to the aforesaid attorney, can be imprisoned for 5 years under the new law.
- Amended the FISA Court's authority to issue an electronic surveillance order that attaches to a particular target rather than a particular phone or computer by

FISA Court

Since its creation in 1978, a secretive federal court known as the Foreign Intelligence Surveillance Act Court, or FISA Court, has approved thousands of Justice Department requests to conduct secret searches and surveillance of people in the United States who are suspected of having links to foreign agents or powers, often involving terrorism and espionage. The legislation that created the now 11-member FISA Court also established a 3-member Foreign Intelligence Surveillance Court of Review, which has never been used because the FISA Court almost always grants the government's requests, and suspects are never notified that they are the target of a search or surveillance. The U.S. Supreme Court has the final authority to review cases from the Court of Review. The Supreme Court's Chief Justice appoints all members of both courts. The 11 FISA Court judges are selected from among U.S. district court judges, and the three judges of the Court of Review are selected from the U.S. district courts or courts of appeals. Judges serve for a maximum of 7 years.

Source: Federal Judicial Center, "Foreign Intelligence Surveillance Court," https://www.fjc.gov/history/courts/foreign-intelligence-surveillance-court-and-court-review-1978-present.

increasing the level of detail needed to obtain an order and, in most cases, requiring the government to provide notice to the court within 10 days that surveillance had been directed at a new facility or place.
- To avoid adverse consequences such as endangering an individual's life or physical safety, allowed investigators to obtain court permission to delay giving notice that a search warrant had been executed for a presumptive limit of 30 days and extensions of 90 days.
- Limited the PATRIOT Act's broad definition of domestic terrorism to specific federal terrorism crimes, instead of any acts "dangerous to human life" to "influence the policy of a government by intimidation or coercion."[19]

In 2007 and 2008, Congress enacted the Amendment to the Foreign Intelligence Surveillance Act of 1978, which is referred to as the Protect America Act of 2007, and the Foreign Intelligence Surveillance Act of 1978 Amendments Act of 2008, which is also called the FISA Amendments Act of 2008.[20] The 2008 legislation, which President Bush signed into law July 10, 2008, reauthorizes many of the provisions in the 2007 legislation. The 2007 act, which was signed into law by President Bush on August 5, 2007, clarifies the use of electronic surveillance of non-U.S. persons outside the United States. Under the new laws, a court order is no longer required to collect foreign intelligence information against such a target located overseas. That authority now rests solely with the Attorney General and the Director of National Intelligence. The law thus removes an obstacle from U.S. intelligence agencies to obtain real-time information about the intent of enemies located outside the United States.

Specifically, the law allows U.S. intelligence agencies to collect foreign-to-foreign phone calls and e-mails as well as all international communications where one party is in the United States, so long as no one particular person in the United States is being targeted. It also requires, under penalty of law, third parties such as telecommunication companies and electronic communication service providers to provide information, facilities, and assistance necessary to conduct the surveillance. Finally, it protects those third parties from past or future lawsuits arising from the assistance they provide the government. Several companies, such as AT&T and Verizon, had been sued for allegedly violating their customers' privacy.

The law does empower the FISA Court to review the efforts of U.S. intelligence agencies to gather foreign intelligence, although, according to critics, it will have no information about how extensive the breach of American privacy is, or the authority to remedy it. Critics contend that the law allowed the Bush administration, especially the National Security Agency (NSA), to resume a once-secret warrantless wiretapping program it created after 9/11 but was brought under court oversight in January 2007. The ACLU calls the Protect America Act of 2007 the "Police America Act" and filed a lawsuit challenging the FISA Amendments Act of 2008, claiming the new law violates Americans' rights of free speech and privacy under the First and Fourth Amendments.

In 2009, legislation was introduced in both the Senate and the House aimed to revise and reform the PATRIOT Act and the FISA Court. Among the provisions of the legislation were the reform of the National Security Letter process, revision of the guidelines for business records orders, elimination of the catch-all provision for "sneak-and-peek" searches, the addition of new safeguards for FISA roving wiretaps, and the repeal of retroactive immunity for telecommunications companies. The legislation was not enacted.

In January 2010, the Department of Justice Office of Inspector General released a report critical of the FBI's use of "exigent letters" and other means to obtain telephone records from three unnamed phone companies. The 300-page report concluded that many of the FBI's practices "violated FBI guidelines, Department policy," and the Electronic Communications Privacy Act. The report also found that "the FBI sought and acquired reporters' telephone toll billing records and calling activity information" through improper means. The report concluded that "the FBI's initial attempts at corrective action were seriously deficient, ill-conceived, and poorly executed." Several recommendations for improvement were made.

In March 2010, after months of debate, Congress voted to extend the three expiring provisions of the USA PATRIOT Act for 1 year with no alteration. The provisions, concerning business records, roving wiretaps, and "lone wolf" investigations (intelligence investigations of lone terrorists not connected to a foreign nation or organization), give federal law enforcement agencies broad powers to gather information on Americans. Both the Senate

Barton Gellman/Getty Images

Is Edward Snowden a traitor or a hero?

and House Judiciary committees proposed bills to renew these provisions with reforms that would establish greater oversight, but neither bill went to a floor vote. In May 2011, Congress passed a 4-year extension of the three expiring provisions again with no alterations. The extended provisions were set to expire on June 1, 2015.

In the meantime, in June 2013, Edward Snowden, a private contractor and systems analyst for the NSA, leaked top-secret documents showing that the NSA was illegally spying on American citizens.[21] Snowden had taken tens of thousands of documents—many of them stamped Top Secret—from NSA internal servers. The leaked documents revealed that the NSA had secretly attached intercepts to the undersea fiber optic cables that crisscrossed the world. This allowed them to read much of the world's communications. The NSA could bug anyone, including world leaders such as Germany's Chancellor Angela Merkel, who WikiLeaks reported had been the target of tapped phone calls for years. The German newsweekly *Der Spiegel* reported that the NSA targeted at least 122 world leaders.

In theory, the spy agency was supposed to collect only "signals intelligence" (intelligence derived from electronic signals and systems) on foreign targets. In practice, the NSA was collecting, without acknowledgment or consent, metadata (e.g., phone records, e-mail headers, and subject lines) from millions of Americans. (The NSA also was intercepting 200 million text messages a day worldwide through a program called Dishfire.) Although the FISA Court could compel telecom and Internet providers to hand over data, that was unnecessary. Under a secret program called Prism, the NSA, and giant telecom and Internet corporations, such as Verizon, Google, Microsoft, Facebook, and Apple, among others, collaborated to provide the NSA with direct access to the contents of private telecom and Internet company servers, allowing it, for example, to get copies of Facebook messages, Skype conversations, or Gmail mailboxes. This arrangement relieved the government of the need to intercept private communications or to get FISA approval. The NSA even installed back doors into online encryption software, such as those used to make secure bank payments. People who originated the communications never knew that the NSA had access to them. Thus, the NSA had hijacked telephonic communications and the Internet, and Snowden became the source of the biggest intelligence leak in history.

Following the leak, U.S. government officials claimed that they had interpreted Section 215 of the PATRIOT Act in a way that allowed such bulk collection of data. They reasoned that all records were needed in the event that one day some of them might prove useful in foiling a terrorist plot. However, in May 2015, the Second U.S. Circuit Court of Appeals, in the case of *ACLU v. Clapper*, held that the bulk collection of phone records was illegal and that Section 215 did not support the bulk collection of metadata. A majority in Congress agreed with the appellate court's decision and, on June 2, 2015, President Obama signed into law the USA Freedom Act (officially called the "Uniting and Strengthening America by Fulfilling Rights and Ensuring Effective Discipline Over Monitoring Act of 2015"), which, among other provisions, bans the bulk collection of all records under Section 125 of the PATRIOT Act. Under the new law, the NSA must obtain a warrant before it can seize records about phone numbers suspected of belonging to terrorist suspects in the United States (as it was supposed to do in the first place). The USA Freedom Act is the first major legislative rebuff of domestic surveillance operations in the post-9/11 era. The NSA announced it would stop collecting the data and shut down the program on November 29, 2015. The other two expiring provisions–authorizing roving wiretaps and "lone wolf" investigations–were extended through 2019.

Clearly, the catastrophic events of 9/11 have had a dramatic effect on American law enforcement and administration of justice. However, whether the changes that have been implemented and proposed in the wake of those events–such as the USA PATRIOT Act, the USA PATRIOT Act Improvement and Reauthorization Act of 2005, the Protect America Act of 2007, the FISA Amendments Act of 2008, and the USA Freedom Act of 2015–are desirable is a fair subject of debate in a free and democratic country.

The Fourth Amendment: The Future Of all the due process guarantees in the Bill of Rights, those in the Fourth Amendment are the ones likely to require the most interpretation by the Supreme Court in the future. With advances in the technology of surveillance, the Court will have to determine the legality of increasingly more intrusive ways of gathering evidence. The star of a 1983 science fiction movie was a police helicopter named *Blue Thunder*. The helicopter was able to hover silently outside apartment buildings, record what was being said inside the apartments, and take pictures of what was being done. Although the movie was fictional, it will probably not be long before law enforcement has such equipment–if it does not have at least some of that equipment already. Will evidence obtained by means of the futuristic surveillance technology of *Blue Thunder* violate the Fourth Amendment prohibition against unreasonable searches and seizures?

The Supreme Court likely provided an answer to that question in *Kyllo v. United States* (2001). Danny Kyllo was convicted on a federal drug charge after federal agents, suspicious that Kyllo was growing marijuana in his home, used a thermal-imaging device to determine whether the heat coming from his house was consistent with the high-intensity lamps typically used in growing marijuana indoors. Based partly on the thermal imaging, a warrant was issued to search Kyllo's home, where agents found more than 100 marijuana plants growing. The Supreme Court, in a 5–4 ruling, held that the use of the thermal-imaging device before the warrant was issued was an impermissible search of Kyllo's home, violating the Fourth Amendment's prohibition of unreasonable searches and seizures. According to the Court, law enforcement agents must first obtain a search warrant before using high-tech devices to gather information from inside a home. At least for now, then, the use of futurist technology in the surveillance of a home, without a warrant, is legally prohibited. The use of a helicopter for visual aerial observation, however, is not (see *Florida v. Riley*, 1989).

What about the use of unmanned surveillance aircraft (i.e., Predator drones)? In June 2011, the sheriff of Nelson County, North Dakota, was looking for six missing cows on the Brossart family farm, but three men with rifles chased him off. Knowing that the gunmen could be anywhere on the 3,000-acre farm and fearing an armed standoff, the sheriff called in the state Highway Patrol, a regional SWAT team, a bomb squad, ambulances, deputy sheriffs from three other counties–and a Predator B drone from Grand Forks Air Force Base. The drone flew 2 miles above the farm and, using high-resolution cameras, heat sensors, and sophisticated radar, found the three suspects and determined they were unarmed. Law enforcement officers swooped in on the suspects and made the first known arrests of U.S. citizens with help from a Predator drone.

The drone belonged to the U.S. Customs and Border Protection Agency, which operates eight drones on the northern and southwest borders to search for illegal immigrants and smugglers. Congress first authorized the border protection agency to buy drones in 2005. Local North Dakota police used drones to fly at least two dozen surveillance flights between June and December 2011. In 2015, North Dakota became the first state to legalize the use of armed drones by law enforcement. The law limits armaments to "less than lethal" weapons, such as tear gas, rubber bullets, beanbags, pepper spray, and Tasers.[22] The FBI and the DEA also have used drones for domestic investigations. Even though for decades U.S.

Aerial drone conducting surveillance. Do you believe visual aerial observation with drones without a warrant should be legally allowed? Why or why not?

U.S. Customs and Border Protection

courts have allowed law enforcement to conduct aerial surveillance without a warrant, the use of drones for this task has occurred without any public acknowledgment or debate. Privacy advocates contend that the use of drones for domestic aerial surveillance allows law enforcement agencies to snoop on U.S. citizens in ways not anticipated in prior court decisions. That the drones may be armed presents a host of other issues. Recently, Florida, Maine, North Dakota, and Virginia have enacted laws requiring a warrant for police use of drones.[23]

The use of new surveillance technology to track an automobile was the subject of the Court's 2012 decision in *United States v. Jones*. In a unanimous decision, the Court held that the police could not attach a GPS (global-positioning system) device to the bumper of a suspect's automobile to track his movements for nearly a month without a valid warrant. To do so violated the Fourth Amendment's ban on unreasonable searches. By implication, the decision also may be applied to other technological tracking devices, such as surveillance cameras, facial recognition cameras, and cell phones. The Court rejected the government's view that Americans driving on the public streets have waived their right to privacy and can be monitored at will. The respondent in the case, Antoine Jones, was convicted of running a drug-dealing operation in the Washington, DC area based in part on data retrieved from tracking his Jeep.

FYI **Police Use of Drones Increasing**

As of May 2018, at least 910 state and local public safety agencies have purchased drones. Of those, 599 are law enforcement agencies.

Source: Center for the Study of the Drone at Bard College, "Public Safety Drones: An Update," May 28, 2018, accessed April 9, 2019, https://dronecenter.bard.edu/public-safety-drones-update/.

In 2018, the Court addressed cell phones. In *Carpenter v. United States*, the Court held that the government generally needs a warrant to access from cell phone companies historical cell phone records that provide a comprehensive chronicle of the user's past movements. In this case, the FBI used Carpenter's cell phone records to place him at the scene of several robberies. The Court's majority declined to grant the state unrestricted access to a wireless carrier's database of physical location information during criminal investigations. However, the Court did not prohibit the state from gaining access to cell phone records without a warrant in ongoing emergencies, such as bomb threats, active shootings, or child abductions.

The Fifth Amendment

The Fifth Amendment reads as follows:

> No person shall be held to answer for a capital, or otherwise infamous crime, unless on a presentment or indictment of a Grand Jury, except in cases arising in the land or naval forces, or in the Militia, when in actual service in time of War or public danger; nor shall any person be subject for the same offence to be twice put in jeopardy of life or limb, nor shall be compelled in any criminal case to be a witness against himself, nor be deprived of life, liberty, or property, without due process of law; nor shall private property be taken for public use without just compensation.

double jeopardy The trying of a defendant a second time for the same offense when jeopardy attaches in the first trial and a mistrial was not declared.

Right to Grand Jury Indictment and Protection Against Double Jeopardy The Fifth Amendment right to a grand jury indictment in felony cases, to be described in detail in Chapter 8, is the last of the Bill of Rights guarantees that has not yet been extended to the states (see *Hurtado v. California*, 1884). However, the Fifth Amendment protection against **double jeopardy** has been (see *Benton v. Maryland*, 1969). The protection provides that no person shall "be subject for the same offence to be twice put in jeopardy of life or limb."

When most people think of double jeopardy, they probably think of the classic case in which a defendant cannot be retried for the same crime or a related crime after he or she has been acquitted or convicted by a jury. However, the protection against double jeopardy can apply even without an acquittal or conviction. Technically, it does not apply until jeopardy has attached. If a trial ends before jeopardy has attached, the prosecution has the right to retry the defendant for the same charge in a new trial. But, when does jeopardy attach? In jury trials, jeopardy attaches when the entire jury has been selected and sworn in. In a bench trial (a trial before a judge without a jury), jeopardy attaches when the first witness has been sworn in. In cases that are resolved through a guilty plea, jeopardy attaches when the court unconditionally accepts the defendant's plea. Even after jeopardy has attached, however, the prosecution generally is not barred from retrying a defendant when a mistrial has been declared or when a defendant appeals and is granted a new trial.

The theoretical rationale behind the protection against double jeopardy is that the state should have one and only one chance to convict a defendant charged with a crime. Otherwise, the state could endlessly harass its citizens, as sometimes happens in countries without this protection.

self-incrimination Being a witness against oneself. If forced, it is a violation of the Fifth Amendment.

confession An admission by a person accused of a crime that he or she committed the offense charged.

doctrine of fundamental fairness The rule that makes confessions inadmissible in criminal trials if they were obtained by means of either psychological manipulation or "third-degree" methods.

Protection Against Compelled Self-Incrimination Arguably, the most important procedural safeguard in the Fifth Amendment is the protection against compelled **self-incrimination**. The protection guarantees that in criminal cases, suspects or defendants cannot be forced to be witnesses against themselves. The protection is based on the belief that confessions may not be truthful if they are not made voluntarily. It also expresses an intolerance for certain methods used to extract confessions, even if the confessions ultimately prove to be reliable. A **confession** is an admission by a person accused of a crime that he or she committed the offense as charged. According to the Supreme Court's **doctrine of fundamental fairness**, confessions are inadmissible in criminal trials if they were obtained by means of either psychological manipulation or "third-degree" methods–for example, beatings, subjection to unreasonably long periods of questioning, or other physical tactics.

Although the Fifth Amendment protection against compelled self-incrimination has long been observed in federal trials, it was not until the 1960s in the case of *Malloy v. Hogan* (1964) that the Fifth Amendment protection against compelled self-incrimination was extended to trials in state courts. In *Miranda v. Arizona* (1966), the Court broadened the protection against compelled self-incrimination to cover nearly all custodial police interrogations. (Custodial police interrogations are questionings that take place after an arrest or the functional equivalent of an arrest; they may or may not take place at the police station.) In *Miranda*, the Court added that confessions obtained without suspects being notified of their specific rights could not be admitted as evidence. Perhaps even more important, it established specific procedural safeguards that had to be followed to avoid violation of the protection against compelled self-incrimination. The Court said:

> Procedural safeguards must be employed to protect the privilege [against self-incrimination], and unless other fully effective means are adopted to notify the person of his right of silence and to assure that the exercise of the right will be scrupulously honored, the following measures are required. [The suspect] must be warned prior to any questioning (1) that he has the right to remain silent, (2) that anything he says can be used against him in a court of law, (3) that he has the right to the presence of an attorney, and (4) that if he cannot afford an attorney one will be appointed for him prior to any questioning if he so desires.

FYI

Coerced Confessions

The first time the Supreme Court held that a coerced confession, brutally beaten out of the suspect, was inadmissible in a state trial was in 1936, in the case of *Brown v. Mississippi*. However, in the *Brown* case, the Court did not find that the coerced confession violated the Fifth Amendment protection against self-incrimination. Rather, the Court found that it violated the Fourteenth Amendment right to due process.

If suspects indicate, before or during questioning, that they wish to remain silent, the interrogation must cease; if they state that they want an attorney, the questioning must cease until an attorney is present. Where an interrogation is conducted without the presence of an attorney and a statement is taken, a heavy burden rests on the government to demonstrate that a suspect knowingly and intelligently waived his or her right to counsel. However, if an individual being questioned is not yet in custody, the *Miranda* warnings do not have to be given. Also, volunteered confessions do not violate *Miranda* or the Fifth Amendment. In 1980, in the case of *Rhode Island v. Innis*, the Supreme Court expanded the meaning of interrogation under *Miranda* beyond express questioning "to any words or actions on the part of the police (other than those normally attendant to arrest and custody) that the police should know are reasonably likely to elicit an incriminating response from the suspect."

The Fifth Amendment protection against compelled self-incrimination also has been weakened by the Supreme Court. For example, in *New York v. Quarles* (1984), the Supreme Court created a public-safety exception to the Fifth Amendment protection. Also, in *Arizona v. Fulminante* (1991), the Court ruled that improper use of a coerced confession is a harmless trial error if other evidence is strong enough to convict the defendant. The burden of proof is on the state to show that a coerced confession is harmless error. The case involved a defendant who had been sentenced to death for killing his 11-year-old stepdaughter. While in prison for another crime, the defendant confessed to an FBI informant after the informant promised to protect the defendant from other inmates. Prior to *Fulminante*, such a conviction most likely would have been reversed on appeal because of the use of the coerced confession.

However, in *Dickerson v. United States* (2000), the Court reaffirmed the importance of *Miranda*, even when it inconveniences law enforcement officers. In *Dickerson*, a bank robbery suspect asked the Court to throw out incriminating statements he made to FBI agents because he was not given *Miranda* warnings prior to questioning. Prosecutors argued that the suspect made voluntary statements, which were admissible under a law approved by Congress in 1968. The law gives federal judges authority to admit statements from suspects if the judges believe that the statements are voluntary. The Court disagreed and held that Congress did not have the authority to supersede the Supreme Court's interpretation of the Constitution. Therefore, the incriminating statements made by the bank robbery suspect were inadmissible.[24]

In 2010, the Court backed off its strict enforcement of its *Miranda* decision by ruling that crime suspects' words can be used against them if they fail to clearly tell the police that they do not want to talk. In the past, the government had the burden of showing that crime suspects had "knowingly and intelligently" waived their rights. Many departments required suspects to sign a waiver of their *Miranda* rights before they were questioned. However, in *Berghuis v. Thompkins*, the Court shifted the burden to criminal suspects. The Court ruled that the police are not required to obtain a written waiver before questioning suspects and that suspects must unambiguously invoke their right to remain silent.

Pool/Pool/Getty Images

Lee Malvo asserted his Fifth Amendment right against self-incrimination at a pretrial hearing in the Washington, DC, sniper case, when he was asked whether he would testify. He also "took the Fifth" when he was asked whether he knew John Allen Muhammad, Malvo's alleged co-conspirator. *Should Malvo have been compelled to answer the questions? Why or why not?*

The Fifth Amendment protection against compelled self-incrimination also applies to trial procedures. If defendants do not voluntarily take the stand to testify, not only do they have a right to refuse to answer any questions put to them by the prosecution during a trial (by "pleading the fifth"), but they also have the right not to take the witness stand in the first place. Moreover, the prosecution is forbidden to comment on the defendant's silence or refusal to take the witness stand. This protection rests on a basic legal principle: The government bears the burden of proof. Defendants are not obligated to help the government prove they committed a crime. In 1964 and 1965, those Fifth Amendment rights were extended to defendants being tried in state courts in the cases of *Malloy v. Hogan* and *Griffin v. California*, respectively.

The Sixth Amendment

The Sixth Amendment reads as follows:

> In all criminal prosecutions, the accused shall enjoy the right to a speedy and public trial, by an impartial jury of the State and district wherein the crime shall have been committed; which district shall have been previously ascertained by law, and to be informed of the nature and cause of the accusation; to be confronted with the witnesses against him; to have compulsory process for obtaining witnesses in his favor, and to have the assistance of counsel for his defence.

Right to a Speedy and Public Trial The Sixth Amendment right to a speedy and public trial applies directly to trials in federal courts. It was extended to trials in state courts in 1967, in the case of *Klopfer v. North Carolina* (right to a speedy trial), and in 1948, in the case of *In re Oliver* (right to a public trial). Delays in a trial can severely hamper a defendant's case if favorable witnesses have died, have moved and cannot be found, or have forgotten what they saw. Delays also can adversely affect defendants by forcing them to remain in jail for long periods of time while awaiting trial. A long wait in jail can be a very stressful and sometimes dangerous experience.

In *United States v. Marion* (1971), the Supreme Court held that the right to a speedy trial "is activated only when a criminal prosecution has begun and extends only to those persons who have been 'accused' in the course of that prosecution." It added that "invocation of the right need not await indictment, information, or other formal charge but begins with the actual restraints imposed by arrest if those restraints precede the formal preferring of charges." In determining what constitutes a speedy trial, the Supreme Court has created a balancing test that weighs both the defendant's and the prosecution's behavior (see *Barker v. Wingo*, 1972). Thus, the reason for the delay in a trial is critical. For example, a search for a missing witness probably would be considered an acceptable reason for delay. Court congestion, on the other hand, typically would not.

The acceptable length of delay in a trial also depends partly on the nature of the charge. In *Barker*, the Court held that "the delay that can be tolerated for an ordinary street crime is considerably less than for a serious, complex conspiracy charge." There has been great variation in the length of delay tolerated by specific courts.

However, Florida rules of criminal procedure, for example, require prosecutors to bring a case to trial within 175 days of arrest. Federal courts, on the other hand, are regulated by the Speedy Trial act of 1974, which specifies two separate time limits: one for the period between arrest and charging and the other for the period between charging and trial. The act stipulates that, generally, a delay between arrest and charging (i.e., the filing of an indictment or information) may be no more than 30 days, and a delay between charging and trial may be no more than 70 days. The act also specifies periods of delay that do not count–for example, delay due to the unavailability of an essential witness or continuances (i.e., postponements) that serve "the ends of justice." If the delay, excluding periods of delay that do not count, is longer than the number of days allowed, the court must dismiss the charges. A dismissal with prejudice, which is given when there are no good reasons for the delay, prevents the re-prosecution of the case. A dismissal without prejudice gives the prosecutor the option of prosecuting the case again.

The Sixth Amendment right to a public trial means that a trial must be open to the public, but it need not be open to all who want to attend. Obviously, the number of people who can attend a trial depends on the size of the courtroom. The right would be violated only if the trial was held, for example, in a prison or in a closed judge's chambers against a defendant's wishes. Defendants have no right to a private trial.

A trial may be closed to the public if the defendant's right to a public trial is outweighed by "a compelling state interest." However, before a trial is closed, "the party seeking to close the hearing must advance an overriding interest that is likely to be prejudiced, the closure must be no broader than necessary to protect that interest, the trial court must consider reasonable alternatives to closing the proceeding, and it must make findings adequate to support the closure" (*Waller v. Georgia*, 1984). In some cases, parts of a trial may be closed–for example, to protect the identity of an undercover informant during his or her testimony.

Right to Impartial Jury of the State and District Wherein the Crime Shall Have Been Committed The right to an impartial jury promises not only that a jury will be unbiased but also that there will be a jury trial. As interpreted by the Supreme Court, this right means that defendants charged with felonies or with misdemeanors punishable by more than 6 months' imprisonment are entitled to be tried before a jury. The right was extended to the states in 1968, in the case of *Duncan v. Louisiana.* Most states also allow defendants to be tried by a jury for less serious misdemeanors, but states are not constitutionally required to do so.

Jury Trials

The fundamental right to a jury trial, itself, is provided in Article 3, Section 2.3, of the U.S. Constitution: "The Trial of all Crimes, except in Cases of impeachment, shall be by Jury, and such Trial shall be held in the State where the said crimes shall have been committed; but when not committed within any State, the Trial shall be at such Place or Places as the Congress may by Law have directed."

For practical purposes, the right to an impartial jury is achieved by providing a representative jury; that is, a jury randomly selected from a fair cross-section of the community. However, whether members of such a jury will be impartial in a particular case is a question that defies an easy answer. Juries will be discussed more extensively in Chapter 8.

Finally, the Sixth Amendment guarantees the specific **venue**, or the place of trial. (Venue also is mentioned in the Constitution in Article 3, Section 2: "Trial shall be held in the State where the said Crimes shall have been committed. . . .") The venue of a trial must be geographically appropriate. Generally, a crime must be tried in the jurisdiction–the politically defined geographical area–in which it was committed. However, if a defense attorney believes that a client cannot get a fair trial in the appropriate venue because of adverse publicity or for some other reason, the attorney can ask the court for a change of venue. If the change of venue is granted, the trial will be moved to another location (within the state in cases of state law violations), where, presumably, the adverse publicity or other factors are not as great. In 1994, the state of Florida passed a law allowing a jury selected in one county to hear a trial in another county. Before the law, trials–not juries–were moved, often causing hardship to victims' families.

venue The place of the trial. It must be geographically appropriate.

Right to Be Informed of the Nature and Cause of the Accusation The right to notice and a hearing is the very core of what is meant by due process. In *Twining v. New Jersey* (1908), for example, the Supreme Court held that "due process requires . . . that there shall be notice and opportunity for hearing given the parties. . . . [T]hese two fundamental conditions . . . seem to be universally prescribed in all systems of law established by civilized countries." Later, in *In re Oliver* (1948), the Court opined, "A person's right to reasonable notice of a charge against him, and an opportunity to be heard in his defense–a right to his day in court–are basic in our system of jurisprudence." The reason for the right is to prevent the practice, common in some countries, of holding suspects indefinitely without telling them why they are being held.

Right to Confront Opposing Witnesses The Sixth Amendment right to confront opposing witnesses was extended to trials in state courts in 1965, in the case of *Pointer v. Texas.* In essence, it means that defendants have a right to be present during their trials (otherwise, they could not confront opposing witnesses) and to cross-examine witnesses against them.

The Sixth Amendment's "confrontational clause" has been the subject of several recent Supreme Court cases. In the first and most important case, *Crawford v. Washington* (2004), the Court's majority decided that the only exceptions to a defendant's right to confront and cross-examine witnesses about testimonial evidence are those that were recognized at the nation's founding: (1) declarations made by a speaker who was both on the brink of death and aware that he or she was dying and (2) when the defendant engaged in conduct designed to prevent the witness from testifying. The Court's decision reversed an earlier precedent (*Ohio v. Roberts*, 1980), which did not bar admission of an unavailable witness's statement if the statement was deemed reliable. The issue in the second case, *Giles v. California* (2008), was whether a defendant forfeits his right to confront a witness against him when a judge determines that a wrongful act by the defendant (in this case, the defendant's killing of the witness) made the witness unavailable to testify at trial. The Court sided with the defendant because, according to the Court, the defendant did not intend to kill the witness to prevent her from testifying. In the third case, *Melendez-Diaz v. Massachusetts* (2009), the prosecution introduced notarized certificates of state laboratory analysts stating that material seized by the police and connected to Melendez-Diaz was cocaine of a certain quantity. Melendez-Diaz objected, arguing that *Crawford v. Washington* required the analysts to testify in person. The Court again sided with the defendant, ruling that the admission of the certificates violated Melendez-Diaz's right to confront the witnesses against him. In the fourth case, *Bullcoming v. New Mexico* (2011), Bullcoming was arrested on a DWI (driving while intoxicated) charge. The main evidence against him was a forensic laboratory report certifying that Bullcoming's blood-alcohol concentration exceeded the limit for aggravated DWI. At trial, the analyst who signed the certification was unavailable, so the prosecution called another analyst who was familiar with the laboratory's testing procedures but did not participate or observe the test on Bullcoming's blood sample. Bullcoming objected, and again the Court sided with the defendant and ruled that Bullcoming had the right to be confronted at trial with the analyst who made the certification, unless Bullcoming had an opportunity, before trial, to cross-examine that particular analyst. In the fifth case, *Williams v. Illinois* (2012), a forensic specialist testified that she matched a DNA profile produced by an outside lab to a profile the state lab produced using a sample of Williams's blood. Williams objected because he did not have an opportunity to confront the lab technician who conducted the original DNA test. This time the Court rejected the defendant's claim, holding that an expert witness is allowed to discuss others' testimonial statements, even without first-hand knowledge of the facts on which the statements are based, as long as those facts themselves are not admitted as evidence (the expert witness only testified about the match). Finally, in the sixth case, *Ohio v. Clark* (2015), the Court ruled that statements made by a 3-year-old physically abused boy to his teacher that identified his abuser as his mother's boyfriend could be presented at trial by the teacher, without requiring the child to testify. The Court opined that allowing the teacher to testify for the child does not violate the Sixth Amendment's confrontation clause because the statements were not made for the primary purpose of creating evidence for prosecution.

subpoena A written order issued by a court that requires a person to appear at a certain time and place to give testimony. It can also require that documents and objects be made available for examination by the court.

DAVE WEAVER/AP Images

Disorderly defendant removed from courtroom. *Are there other ways to control a disorderly defendant short of removing him or her from the courtroom? If yes, what are they?*

One last point about the confrontational clause is that the right to be present during the trial may be forfeited by a defendant's disruptive behavior. Thus, if a defendant continues to scream, use profanity, or refuse to sit quietly after being warned by the judge, the judge may have the defendant removed from the trial (see *Illinois v. Allen*, 1970).

Right to Compulsory Process for Obtaining Favorable Witnesses This right ensures a defendant the use of the subpoena power of the court to compel the testimony of any witnesses who may have information useful to the defense. A **subpoena** is a written order issued by a court that requires a person to appear at a certain time and place to give testimony. It also can require that documents and objects be made available for examination by the court. Even though the right to compulsory process for obtaining favorable witnesses already was applicable in many states because of its inclusion in state constitutions and laws, the Supreme Court officially extended it to state trials in 1967, in the case of *Washington v. Texas*.

Right to Counsel The Sixth Amendment right to privately retained and paid-for counsel has existed in federal courts since the ratification of the Bill of Rights. (The terms *counsel*, *attorney*, and *lawyer* are interchangeable.) Criminal defendants in state courts did not gain the right until 1954. In the case of *Chandler v. Fretag*, the Supreme Court held that the right to a privately retained lawyer is "unqualified"; that is, as long as a criminal defendant (or suspect) can afford to hire an attorney, he or she has the right to be represented by that attorney, not only at trial, but at any stage of the criminal justice process. But what if a criminal defendant is indigent, lacking the funds to hire an attorney? The Supreme Court first extended the right to court-appointed counsel to indigents in *Powell v. Alabama* (1932). The right, however, was extended only to indigents in death penalty cases who were "incapable adequately of making [their] own defense because of ignorance, feeblemindedness, illiteracy or the like." Moreover, the Court's decision in *Powell* was based on the Fourteenth Amendment right to due process and not on the Sixth Amendment right to counsel. It was not until 1938, in the case of *Johnson v. Zerbst*, that the Supreme Court first extended the Sixth Amendment right to court-appointed counsel to indigent defendants facing felony charges in federal trials. Another 25 years passed before the right to court-appointed counsel was extended to indigent defendants facing felony charges in state courts. That right was granted in the famous case of *Gideon v. Wainwright* (1963). In 1972, in the case of *Argersinger v. Hamlin*, the Court extended the Sixth Amendment right to court-appointed counsel to defendants in misdemeanor trials in which a sentence to jail might result. Thus, as a result of those decisions, no person may be imprisoned for any offense, whether classified as petty, misdemeanor, or felony, unless he or she is represented by counsel. If the person cannot afford to hire an attorney, then the court is required to appoint one. However, if there is no possibility of incarceration, then a defendant has no right to state-furnished counsel (see *Scott v. Illinois*, 1979). Later, in *Alabama v. Shelton* (2002), the Court added that indigents have a right to court-appointed counsel even when a defendant is given "a suspended sentence that may 'end up in the actual deprivation of a person's liberty.'"

In other Supreme Court decisions, the Sixth Amendment right to counsel has been extended to indigents at additional *critical stages* (described in Chapter 8 and elsewhere in this book) and other circumstances in the administration of justice, "where substantial rights of the accused may be affected." Those include (by date of Supreme Court decision):

1. Arraignment, under most circumstances (*Hamilton v. Alabama*, 1961).
2. The plea bargaining process (*Carnley v. Cochran*, 1962).
3. Initial appearances where defendants may be compelled to make decisions that may later be formally used against them (*White v. Maryland*, 1963).
4. Interrogations after formal charges (*Massiah v. United States*, 1964).
5. Postcharge police lineups (*United States v. Wade*, 1967; *Gilbert v. California*, 1967).
6. Sentencing (*Mempa v. Rhay*, 1967).
7. Preliminary hearings (*Coleman v. Alabama*, 1970).
8. A psychiatric examination used by the prosecution to show that a murder defendant remains dangerous and should receive the death penalty (*Estelle v. Smith*, 1981).

To date, the Court has not extended the Sixth Amendment right to counsel to preindictment lineups, booking, grand jury investigations, or appeals after the first one.

The Sixth Amendment not only guarantees the right to counsel in the areas to which it has been extended, it also guarantees the right to the "effective assistance of counsel" (see *McMann v. Richardson*, 1970). However, it was not until 1984, in the case of *Strickland v. Washington*, that the Supreme Court first established standards to define "ineffective assistance of counsel." The Court ruled that two facts must be proved to show that counsel was ineffective: (1) that counsel's performance was "deficient," meaning that counsel was not a "reasonably competent attorney" or that his or her performance was below the standard commonly expected, and (2) that the deficiencies in the attorney's performance were prejudicial to the defense, meaning that there is a "reasonable probability that, but for the counsel's unprofessional errors, the result of the proceeding would have been different." In other words, not only must it be shown that an attorney was incompetent, it also must be shown that there is a reasonable probability that the incompetence led to the final result. Thus, if the defendant was clearly guilty of the crime with which he or she was charged, it would most likely be impossible to win a claim of "ineffective assistance of counsel."

Two cases decided by the Court in 2012 extended the right to the effective assistance of counsel to plea bargains that lapse or are rejected. In *Missouri v. Frye*, the Court found that Frye's defense counsel had been ineffective because he failed to communicate written plea offers to Frye before they expired. The Court concluded that his lawyer's deficient performance caused Frye prejudice because Frye ultimately pleaded guilty to a felony instead of the previously offered misdemeanor. In the second case, *Lafler v. Cooper*, a favorable plea offer was reported to the defendant but, on advice of counsel, was rejected. After the plea offer was rejected, Cooper received a full and fair trial before a jury. Following a guilty verdict, Cooper received a harsher sentence than that offered in the rejected plea bargain. The Court ruled that counsel's plea offer advice fell below an objective standard of reasonableness because Cooper was advised to reject the plea offer on the grounds that he could not be convicted at trial. The Court's remedy was for the state to reoffer the plea agreement to Cooper.

Finally, the right to counsel may be waived, but only if the waiver is made knowingly, intelligently, and voluntarily. Thus, the Sixth Amendment also has been interpreted to mean that defendants have the right to represent themselves; that is, to conduct the defense *pro se* (see *Faretta v. California*, 1975). However, the wisdom of serving as one's own lawyer is captured by the adage attributed to Abraham Lincoln: "He who represents himself has a fool for a client." Note that if defendants choose to represent themselves, they cannot claim later, on appeal, that their defense suffered from ineffective assistance of counsel.

The Eighth Amendment

The Eighth Amendment reads as follows:

> Excessive bail shall not be required, nor excessive fines imposed, nor cruel and unusual punishments inflicted.

Protection Against Excessive Bail and Fines The Eighth Amendment to the Constitution does not require that bail be granted to all suspects or defendants, only that the amount of bail not be excessive. What constitutes excessive bail is determined by several factors, including the nature and circumstances of the offense, the weight of evidence against the suspect or defendant, the character of the suspect or defendant, the suspect or defendant's ties to the community, and the ability of the suspect or defendant to pay bail. The subject of bail will be discussed more fully in Chapter 8.

Until recently, the Eighth Amendment protection against excessive fines was one of two Bill of Rights guarantees that had not been extended to the states. The other was the right to a grand jury indictment in felony cases. However, on February 20, 2019, in the case of *Timbs v. Indiana*, the Supreme Court incorporated the excessive fines clause and made it applicable to the states under the Fourteenth Amendment's Due Process Clause. The decision also applies to fees and forfeitures. But what is an "excessive" fine? The key is that the fine should be proportional to the harm done. The Court has held that excessive fines are "so grossly excessive as to amount to a deprivation of property without due process of law." So, for example, in a conviction for illegal possession of a small amount of marijuana (in a state where it remains illegal), a defendant having to forfeit his or her home might be considered an excessive fine.

Bills of Attainder

The Constitution—in Article 1, Sections 9.3 and 10.1—prohibits bills of attainder. Under English common law, these legal documents stripped convicted offenders of their citizenship and allowed the Crown to confiscate all of their property. Bills of attainder also extended to include offenders' families, who were judged "attainted" and could not inherit the offender's property. Recent civil forfeitures of property in drug violations seem to circumvent the constitutional prohibition. In some cases, property has been confiscated even though the suspect was never arrested or convicted of a crime.

Protection Against Cruel and Unusual Punishments The final prohibition of the Eighth Amendment is against "cruel and unusual punishments." That prohibition was extended to trials in state courts in 1962, in *Robinson v. California*. Generally, discussions of this issue involve the practice of capital punishment, or the death penalty, which will be discussed in detail in Chapter 9. Here we will provide only a brief history of the definition of cruel and unusual punishments.

For approximately 120 years after the adoption of the Bill of Rights, the Supreme Court employed a fixed, historical meaning for "cruel and unusual punishments." In other words, the Court interpreted the concept's meaning in light of the practices that were authorized and were in use at the time the Eighth Amendment was adopted (1791). Thus, only the most barbarous punishments and tortures were prohibited. Capital punishment itself was not prohibited because there was explicit reference to it in the Fifth Amendment and it was in use when the Eighth Amendment was adopted.

Bettmann/Contributor/Getty Images

During the Inquisition, heretics were subjected to fire torture on the wheel. *Do you consider this cruel and unusual punishment in violation of the Eighth Amendment? Why or why not?*

The Court, in *Wilkerson v. Utah* (1878), provided examples of punishments that were prohibited by the Eighth Amendment because they involved "torture" or "unnecessary cruelty." They included punishments in which the criminal "was emboweled alive, beheaded, and quartered." In another case, *In re Kemmler* (1890), the Court expanded the meaning of cruel and unusual punishments to include punishments that "involve torture or lingering death . . . something more than the mere extinguishment of life." The Court also provided some examples of punishments that would be prohibited under that standard: "burning at the stake, crucifixion, breaking on the wheel, or the like."

In 1910, in the noncapital case of *Weems v. United States*, the Supreme Court abandoned its fixed, historical definition of cruel and unusual punishments and created a new one. Weems was a U.S. government official in the Philippines who was convicted of making two false accounting entries, amounting to 612 pesos ($11.81 in 2019 dollars).[25] He was sentenced to 15 years of hard labor and was forced to wear chains on his ankles and wrists. After completing his sentence, he was to be under surveillance for life, and he was to lose his voting rights as well. Weems argued that his punishment was disproportionate to his crime and, therefore, cruel and unusual.

The Court agreed with Weems and, breaking with tradition, held "(1) that the meaning of the Eighth Amendment is not restricted to the intent of the Framers, (2) that the Eighth Amendment bars punishments that are excessive, and (3) that what is excessive is not fixed in time but changes with evolving social conditions." Thus, the Court no longer interpreted the concept of cruel and unusual punishments in the context of punishments in use when the Eighth Amendment was adopted. Instead, it chose to interpret the concept in the context of "evolving social conditions."

The Court further clarified its position nearly 50 years later in another noncapital case, *Trop v. Dulles* (1958). As punishment for desertion during World War II, Trop was stripped of his U.S. citizenship. In reviewing the case on appeal, the Court ruled that the punishment was cruel and unusual because it was an affront to basic human dignity. Noting that the "dignity of man" was "the basic concept underlying the Eighth Amendment," the Court held that Trop's punishment exceeded "the limits of civilized standards." Referring to the earlier *Weems* case, the Court emphasized that "the limits of civilized standards . . . draws its meaning from the evolving standards of decency that mark the progress of a maturing society." Those "evolving standards of decency" are, in turn, determined by "objective indicators, such as the enactments of legislatures as expressions of 'the will of the people,' the decisions of juries, and the subjective moral judgments of members of the Supreme

Court itself." In short, it appears that a punishment enacted by a legislature and imposed by a judge or jury will *not* be considered cruel and unusual as long as the U.S. Supreme Court determines that (1) it is not grossly disproportionate to the magnitude of the crime, (2) it has been imposed for the same offense in other jurisdictions, and (3) it has been imposed for other offenses in the same jurisdiction (see *Solem v. Helm*, 1983; *Harmelin v. Michigan*, 1991; *Ewing v. California*, 2003; *Lockyer v. Andrade*, 2003).

THINKING CRITICALLY

1. Which of the amendments within the Bill of Rights do you think are the most protective of the rights of the accused? Why?
2. Why is the Bill of Rights subject to interpretation by the Supreme Court?

Protecting the Accused from Miscarriages of Justice

The legal system of the United States is unique in the world in the number of procedural rights that it provides people suspected or accused of crimes. The primary reason for procedural rights is to protect innocent people, as much as possible, from being arrested, charged, convicted, or punished for crimes they did not commit. One of the basic tenets of our legal system is that a person is considered innocent until proven guilty. However, even with arguably the most highly developed system of due process rights in the world, people continue to be victims of miscarriages of justice. For example, attorney Barry Scheck and his colleagues report, "Of the first eighteen thousand results [of DNA tests] at the FBI and other crime laboratories, at least five thousand prime suspects were excluded *before* their cases were tried." In other words, more than 25% of the prime suspects were wrongly accused. Scheck is a co-founder of the Innocence Project at the Cardoza School of Law in New York City.[26]

Unfortunately, there is no official record of miscarriages of justice, so it is impossible to determine precisely how many actually occur each year. Nevertheless, in an effort to provide some idea of the extent of the problem, a study was conducted of wrongful convictions–miscarriages of justice at just one of the stages in the administration of justice.[27] In the study, *wrongful convictions* were defined as cases in which a person "[is] convicted of a felony but later . . . found innocent beyond a reasonable doubt, generally due to a confession by the actual offender, evidence that had been available but was not sufficiently used at the time of conviction, new evidence that was not previously available, and other factors."

The conclusions of the study were based on the findings of a survey. All attorneys general in the United States and its territories were surveyed, and in Ohio, all presiding judges of common pleas courts, all county prosecutors, all county public defenders, all county

FYI

Convicting the Innocent

Benjamin Franklin, echoing Voltaire, wrote, "That it is better 100 guilty Persons should escape than that one innocent Person should suffer." According to a 2016 public opinion poll, 60% of Americans say it would be worse to have 20,000 people in prison who are actually innocent, while 40% say it would be worse to have 20,000 people not in prison who are actually guilty. Majorities across all demographic groups prioritized protecting the innocent.

Sources: Bartleby.com, Number 953, accessed April 21, 2019, www.bartleby.com/73/953.html; Emily Ekins, "Policing in America: Understanding Public Attitudes Toward the Police. Results from a National Survey," Cato Institute, December 7, 2016, accessed June 15, 2019, https://www.cato.org/survey-reports/policing-america.

Roberto Miranda is one of at least 165 inmates since 1973 who have been released from death row because of evidence of their innocence. Mr. Miranda spent 14 years on Nevada's death row. *What, if anything, does the state of Nevada owe Mr. Miranda?*

David Leeson/Dallas Morning News/The Image Works

sheriffs, and the chiefs of police of seven major cities also were surveyed. The authors of the study conservatively estimated that approximately 0.5% of all felony convictions are in error. In other words, of every 1,000 persons convicted of felonies in the United States, about 5 are probably innocent. The authors believe that the frequency of error is probably higher in less serious felony and misdemeanor cases.

Although an error rate of 0.5% may not seem high, consider that in 2018, approximately 10.3 million people were arrested in the United States.[28] Assuming conservatively that 50% of all people arrested are convicted[29]—about 5 million convictions in 2018—then approximately 25,000 people probably were wrongfully convicted!

Eyewitness misidentification is the most important factor contributing to wrongful convictions. For example, a study of DNA exonerations by the Innocence Project found that more than 80% of wrongful convictions could be attributed, at least in part, to eyewitness or victim misidentification.[30] The second and third most important contributing factors are police and prosecutorial errors, respectively. They accounted for nearly 65% of the DNA exonerations in the Innocence Project study.[31] Overzealous police officers and prosecutors, convinced that a suspect or defendant is guilty, may prompt witnesses, suggest to witnesses what may have occurred at the time of the crime, conceal or fabricate evidence, or even commit perjury. Another factor contributing to wrongful convictions is guilty pleas made "voluntarily" by innocent defendants. Innocent defendants are more likely to plead guilty to crimes they did not commit when they are faced with multiple charges and when the probability of severe punishment is great. They also are more likely to plead guilty to crimes they did not commit when they are mentally incompetent.

When the charge is a less serious one, innocent people who are unable to post bail sometimes admit guilt to be released from jail immediately. For many people, release from jail is more important than a minor criminal record. Besides, it often is difficult to prove one's innocence. (Remember that in the United States the prosecution is required to prove, beyond a reasonable doubt, that defendants are guilty. Defendants are not required to prove their innocence.) Problems faced by innocent people wrongly accused of crimes include inability to establish an alibi; misidentification by witnesses who swear they saw the defendant commit the crime; a lawyer who lacks the skill, time, or resources to mount a good defense; and a lawyer who is unconvinced of the defendant's innocence. Inadequate legal representation is one of the most important factors in wrongful convictions in death penalty cases.[32] It accounted for nearly a third of the wrongful convictions discovered in the Innocence Project study.[33] Other factors contributing to wrongful convictions are community pressures, especially in interracial and rape cases; false accusations; knowledge of a defendant's prior criminal record; judicial errors, bias, or neglect of duty; errors made by medical examiners and forensic experts; and errors in criminal record keeping and computerized information systems.[34] In short, numerous factors can cause wrongful convictions. And remember, the foregoing discussion addresses only wrongful convictions; it does not consider wrongful arrests or other miscarriages of justice.

While thousands of innocent people in the United States each year are wrongly convicted of crimes they did not commit, only a tiny fraction of them are ever exonerated. **Exonerations** refer to cases in which a person was wrongly convicted of a crime and later cleared of all the charges based on new evidence of innocence.

One of the best sources of information on exonerations in the United States is The National Registry of Exonerations, which was founded in 2012, as a joint project of the Newkirk Center for Science & Society at the University of California Irvine, the University of Michigan Law School, and the Michigan State University College of Law, in conjunction with the Center on Wrongful Convictions at Northwestern University School of Law.[35] The Registry provides detailed information about every known exoneration in the United States since 1989, and a more limited database of known exonerations prior to 1989.

As of April 14, 2019, the Registry listed a total of 2,418 exonerations since 1989. Those 2,418 exonerees collectively spent more than 21,000 years in prison, or an average of about 9 years per exoneree, for crimes they did not commit. Richard Phillips holds the distinction of serving the most time in prison for a crime he did not commit—45 years and 2 months. Phillips was wrongly convicted of murder in Detroit, Michigan in 1972, when he was 26 years old; he was exonerated in 2018, at nearly 72 years of age.

In 2018, the latest year for which complete data were available, the Registry listed 151 exonerations. The 2018 exonerees collectively spent 1,639 years in prison, or an average of nearly 11 years in prison per exoneree, for crimes they did not commit. Of the 151 exonerations, 146 occurred in 28 states, and there were 5 federal cases. Illinois

myth

Innocent people in the United States are almost never convicted of crimes they did not commit.

fact

Evidence suggests that thousands of people each year are likely convicted of crimes they did not commit. (See the discussion in this section.)

Death Row Reversals

From 1973 through March 28, 2019, 165 inmates in 28 states have been freed from death row because of problems or errors in the legal process. Common reasons for reversals include (1) key witnesses lied or recanted their testimony, (2) police overlooked or withheld important evidence, (3) DNA testing showed someone else committed the crime, (4) the defense lawyer was incompetent or negligent, and (5) prosecutors withheld exculpatory evidence from the defense.

Sources: Death Penalty Information Center, "The Innocence List." Accessed April 10, 2019, https://deathpenaltyinfo.org/innocence-and-death-penalty#inn-st; Jonathan Alter, "The Death Penalty on Trial," *Newsweek*, June 12, 2000.

exonerations Cases in which a person was wrongly convicted of a crime and later cleared of all the charges based on new evidence of innocence.

Exonerations

The National Registry of Exonerations provides detailed information about every known exoneration in the United States since 1989. Visit the Registry's website at http://www.law.umich.edu/special/exoneration/Pages/mission.aspx.

After examining the information on the website, do you think the U.S. legal system does an adequate job of preventing miscarriages of justice?

had the most exonerations by far with 49, while New York and Texas tied for second with 16 each. Thirty-one of the exonerations in Illinois occurred in Cook County (Chicago), where corrupt police officers, led by Sergeant Ronald Watts, for more than a decade planted drugs and weapons on people who refused to pay bribes to them. The extortion plot was discovered by federal investigators who corroborated the complaints of dozens of people framed by Watts and his officers. (Additional exonerations in this case were granted in 2016, 2017, and early 2019.) Conviction Integrity Units (CIUs) were involved in 58 exonerations in 2018, and Innocence Organizations (IOs) were involved in a record 86. Together, CIUs and IOs were responsible for 99 or two-thirds of the 2018 exonerations. (CIUs and IOs collaborated on 45 of the exonerations.) CIUs are divisions of prosecutorial offices that work to prevent, identify, and remedy false convictions. IOs are nongovernmental organizations committed to investigating and remedying wrongful convictions.

Of the 151 exonerations in 2018, two-thirds of them involved violent felonies, including 68 homicides (2 with death sentences), 17 sex crimes (7 for child sex abuse and 10 for sexual assault of an adult), and 16 cases divided among robbery, attempted murder, burglary or unlawful entry, assault, arson, kidnapping, menacing, and violent attempt. The other one-third of the exonerations involved nonviolent offenses, including 33 for drug crimes and 17 cases divided among gun possession, fraud, sex offender registration, traffic offenses, and stalking. In 46% of the exonerations no crime was actually committed. Nearly a third of the exonerations involved convictions based on guilty pleas, and 15% of the exonerations were based in whole or in part on DNA evidence.

Nearly 74% of the exonerations involved perjury or a false accusation. Misconduct by government officials was the reason for about 70% of the exonerations (79% of homicide exonerations); mistaken eyewitness identifications for approximately 20% of the exonerations; and false confessions for nearly 13% of the exonerations. The most common type of official misconduct involved police or prosecutors (or both) concealing exculpatory evidence or evidence that establishes a criminal defendant's innocence. Examples of possible exculpatory evidence are physical evidence, evidentiary documents (such as a defendant's recorded statements to police or reports of medical examinations or scientific tests), and lists of witnesses including potential law enforcement witnesses who previously have been disciplined for perjury or other offenses. A prosecutor's concealment or misrepresentation of evidence typically is referred to as a "Brady violation"; see *Brady v. Maryland* (1963). The National Registry of Exonerations indicates that cases overturned because of perjury and official prosecutor or police misconduct more than doubled from 2008 to 2018. Other types of official misconduct included police officers threatening witnesses, forensic analysts falsifying test results, and child welfare workers pressuring children to claim sexual abuse where none occurred.

What, if anything, can be done about miscarriages of justice? Scheck and his colleagues suggest the following reforms:

- *DNA testing.* Allow postconviction DNA testing nationwide. Test DNA on unsolved crimes where evidence exists.
- *Witness IDs.* An independent, trained examiner who does not know the suspect should conduct live lineups and videotape lineups, and handle photo IDs and photo spreads to ensure investigators don't influence witnesses and thereby to ensure neutrality.
- *Confessions.* Videotape all interrogations.
- *Informants.* A committee of prosecutors should screen all informant testimony before permission to use at trial. All deals between prosecutors and informants must be in writing.
- *Forensics.* Crime labs should function and be funded separately from police, prosecution, or defense. Strengthen accreditation programs for labs, and establish postgraduate forensic programs at universities.
- *Police, prosecutors.* Establish disciplinary committees to deal with legal misconduct by police and prosecution.
- *Defense attorneys.* Increase fees to attract competent lawyers. Public defenders' pay should equal prosecutors' pay.
- *Wrongful convictions.* Establish innocence commissions (and conviction integrity units) to investigate wrongful convictions. Create and fund innocence projects at law schools to represent clients. Provide compensation to those who clearly were wrongly convicted. Have a moratorium on the death penalty.[36]

Despite miscarriages of justice, many people still resent the provision of procedural safeguards to criminal suspects. The accusation is frequently made that procedural rights protect criminals and penalize victims—that many criminals escape conviction and punishment because of procedural technicalities. For example, a driving force behind the good-faith and inevitable-discovery exceptions was the belief that a substantial number of criminal offenders escaped punishment because of the exclusionary rule. The available evidence, however, does not support the belief. One of the most thorough studies of the effect of the exclusionary rule was conducted by the National Institute of Justice (NIJ).[37] The NIJ study examined felony cases in California between 1976 and 1979—a period during which the American public was becoming increasingly alarmed about the problem of crime and especially about what was perceived as the practice of allowing a substantial number of criminals to escape punishment because of legal technicalities. The study found that only a tiny fraction (fewer than 0.5%) of the felony cases reaching the courts were dismissed because of the exclusionary rule. It is important to emphasize that the study examined only the cases that reached the courts. It excluded cases that prosecutors elected not to pursue to trial because they assumed that the exclusionary rule would make the cases impossible to win. However, studies show that although there is some variation among jurisdictions, fewer than 1% of cases overall are dropped by prosecutors before trial because of search and seizure problems.[38] Interestingly, 71.5% of the California cases affected by the exclusionary rule involved drug charges. The problem in most of the drug cases was that in the absence of complaining witnesses, overaggressive law enforcement officers had to engage in illegal behavior to obtain evidence.

A study of the effect of the exclusionary rule at the federal level was conducted by the General Accounting Office (GAO).[39] The GAO examined 2,804 cases handled by 38 different U.S. attorneys in July and August of 1978. The GAO found results similar to those found by the NIJ in California. In only 1.3% of the nearly 3,000 cases was evidence excluded in the federal courts. As was the case with the NIJ study, the GAO study included only cases that went to trial. However, as noted earlier, evidence shows that, overall, fewer than 1% of cases are dropped by prosecutors before trial because of search and seizure problems. Moreover, having evidence excluded from trial does not necessarily mean that a case is impossible to win and that the defendant will escape punishment. A defendant may still be convicted on the basis of evidence that was legally obtained.

NIJ

Learn more about the National Institute of Justice and its various research findings by accessing its website at https://www.nij.gov/Pages/welcome.aspx.

Why is it important to have research organizations such as the NIJ?

The *Miranda* mandates, like the exclusionary rule, also are viewed by many people as legal technicalities that allow guilty criminals to escape punishment. That view is fortified by Supreme Court Justice Byron White's dissent in *Miranda*: "In some unknown number of cases the rule will return a killer, a rapist or other criminal to the streets." No doubt, Justice White's warning is true, but the evidence suggests that only a very small percentage of cases are lost as a result of illegal confessions. In one large survey, for example, fewer than 1% of all cases were thrown out because of confessions illegally obtained.[40]

In another study of decisions made by the Indiana Court of Appeals or the Indiana Supreme Court from November 6, 1980, through August 1, 1986, the researchers found that in only 12 of 2,354 cases (0.51%) was a conviction overturned because of the failure of the police to correctly implement the *Miranda* safeguards.[41] In only 213 of the 2,354 cases (9%) was a claim even made about improper interrogation procedures by the police, and in 201 of those 213 cases, the conviction was affirmed by the appellate court, resulting in a reversal rate of 5.6% for the cases raising a *Miranda* question.

myth

Many criminals escape punishment because of the Supreme Court's decision in *Miranda v. Arizona*.

fact

Very few criminals escape punishment because of that decision.

The authors of that study speculated on possible reasons for the low rate of successful appeals. One was that the police routinely comply with the *Miranda* decision. In fact, most police support *Miranda* and the other reforms because it makes them appear more professional. The second possible reason was that the police are able to solve most cases without having to question suspects. Studies show that the *Miranda* warnings rarely stop suspects from confessing anyway. Nearly 80% of suspects waive their *Miranda* rights. As many as 75% of suspects attempt to clear themselves in the eyes of the police and end up incriminating themselves instead; other suspects simply do not understand that they have a right to remain silent.[42] Third, the police are able to evade *Miranda* by using strategies that are more sophisticated, such as skillfully suggesting that suspects volunteer confessions or casually talking with suspects in the back of squad cars.[43] And, fourth, prosecutors, knowing that they cannot win cases involving illegal interrogations, screen them out before trial or settle them through alternative means, such as plea bargaining. However, as with the exclusionary rule, fewer than 1% of cases overall are dismissed or handled in other ways by prosecutors because of *Miranda*.[44] In short, the available evidence suggests that the effects of both *Miranda* and the exclusionary rule in Fourth and Fifth Amendment contexts have been minor.[45] Also worth noting is that nearly all of the exclusionary rule and *Miranda* studies were conducted decades ago; no recent studies could be found, primarily, we suppose, because the issues are considered settled.

THINKING CRITICALLY

1. Do you think miscarriages of justice are on the increase? Decrease? Why or why not?
2. Do you think that anything can be done to combat miscarriages of justice?

SUMMARY

1. Distinguish between criminal law and civil law.

There are two general types of law practiced in the United States–criminal and civil. Criminal law is a formal means of social control that involves the use of rules that are interpreted, and are enforceable, by the courts of a political community. The violation of a criminal law is a crime and is considered an offense against the state. Civil law is a means of resolving conflicts between individuals. The violation of a civil law is a tort–an injury, damage, or wrongful act–and is considered a private matter between individuals.

2. Distinguish between substantive law and procedural law.

There are two types of criminal law–substantive and procedural. Substantive law defines criminal offenses and their penalties. Procedural law specifies the ways in which substantive laws are administered. Procedural law is concerned with due process of law–the rights of people suspected of or charged with crimes.

3. List five features of "good" criminal laws.

Ideally, good criminal laws should possess five features: (1) politicality, (2) specificity, (3) regularity, (4) uniformity, and (5) penal sanction.

4. Explain why criminal law is a political phenomenon.

Criminal law is the result of a political process in which rules are created by human beings to prohibit or regulate the behavior of other human beings. Nothing is criminal in and of itself; only the response of the state makes it so.

5. Summarize the origins of American criminal law.

The criminal law of the United States is, for the most part, derived from the laws of England and is the product of constitutions and legislative bodies, common law, and, if provided for by statute, some administrative or regulatory agency rules and decisions.

6. Describe the procedural rights in the Fourth Amendment.

The Fourth Amendment protects persons from unreasonable searches and seizures (including arrests). Under most circumstances, it requires that a judge issue a search warrant authorizing law officers to search for and seize evidence of criminal activity, but the warrant can be issued only when there is probable cause. In 1914, the Supreme Court adopted the exclusionary rule, which barred evidence seized illegally from being used in a criminal trial; in 1961, the rule was made applicable to the states. Subsequent Supreme Court decisions have narrowed the application of the exclusionary rule. The Fourth Amendment also protects persons from warrantless searches and seizures in places where they have a legitimate right to expect privacy. The protection, however, does not extend to every place where a person has a legitimate right to be. The Court has permitted stopping and searching an automobile when there is probable cause to believe the car is carrying something illegal.

7. Describe the procedural rights in the Fifth Amendment.

The Fifth Amendment provides many procedural protections, the most important of which is the protection against compelled self-incrimination. This protection was extended to most police custodial interrogations in the 1966 case of *Miranda v. Arizona*. According to *Miranda*, police custody is threatening and confessions obtained during custody can be admitted into evidence only if suspects have been (1) advised of their constitutional right to remain silent, (2) warned that what they say can be used against them in a trial, (3) informed of the right to have an attorney, and (4) told that if they cannot afford an attorney, one will be appointed for them prior to questioning, if they so desire. Suspects may waive their *Miranda* rights, but only if the waiver is made knowingly, intelligently, and voluntarily. Other due process rights in the Fifth Amendment are the right to a grand jury indictment in felony cases (in federal court) and protection against double jeopardy.

8. Describe the procedural rights in the Sixth Amendment.

Many due process rights are provided by the Sixth Amendment: the right to a speedy and public trial, the right to an impartial jury of the state and district where the crime occurred, the right to be informed of the nature and cause of the accusation, the right to confront opposing witnesses, the right to compulsory process for obtaining favorable witnesses, and the right to counsel. In the 1963 case of *Gideon v. Wainwright*, the Supreme Court extended the right to court-appointed counsel to any poor state defendant charged with a felony.

9. **Describe the procedural rights in the Eighth Amendment.**
The Eighth Amendment protects against "cruel and unusual punishments." The Supreme Court has rarely ruled on this provision, generally approving a punishment as long as it has been enacted by a legislature; it has been imposed by a judge or jury; and the Court determines that (1) it is not grossly disproportionate to the magnitude of the crime, (2) it has been imposed for the same offense in other jurisdictions, and (3) it has been imposed for other offenses in the same jurisdiction. The Eighth Amendment also protects against excessive bail and fines.

10. **Explain why procedural rights are important to those accused of crimes.**
The primary reason for procedural rights is to protect innocent people, as much as possible, from being arrested, charged, convicted, or punished for crimes they did not commit. However, even with the most highly developed system of procedural, or due process, rights in the world, criminal defendants in the United States still face miscarriages of justice.

KEY TERMS

criminal law 97
penal code 97
tort 97
civil law 97
substantive law 98
procedural law 98
due process of law 98
politicality 98
specificity 98
regularity 98
uniformity 99
penal sanction 99
precedent 104
stare decisis 105
searches 108
seizures 108
warrant 108
arrest 108
contraband 110
mere suspicion 113
reasonable suspicion 113
frisking 113
probable cause 113
preponderance of evidence 114
clear and convincing evidence 114
beyond a reasonable doubt 114
proof evident, presumption great 114
exclusionary rule 115
double jeopardy 122
self-incrimination 122
confession 122
doctrine of fundamental fairness 122
venue 125
subpoena 126
exonerations 131

REVIEW QUESTIONS

1. How does one know whether a particular offense is a *crime* or a *tort*?
2. How did the institution of the *eyre* contribute to the development of American criminal law?
3. What is the importance of the Magna Carta for American criminal law?
4. How did the early Cherokees deal with murder?
5. To what jurisdiction do federal and state criminal statutes (and local ordinances) apply?
6. What is *stare decisis*?
7. Why did it take nearly 100 years after the ratification of the Fourteenth Amendment before suspects charged with crimes at the state level were afforded most of the same due process protections as people charged with crimes at the federal level?
8. What are *searches* and *seizures*?
9. What is an *arrest*?
10. What two conclusions must be supported by substantial and trustworthy evidence before either a search warrant or an arrest warrant is issued? (The two conclusions are different for each type of warrant.)
11. In *Chimel v. California* (1969), what limitations did the U.S. Supreme Court place on searches incident to an arrest?
12. What is *probable cause*?
13. Today, what is the principal purpose of the exclusionary rule?
14. What are the major provisions of the USA PATRIOT Act?
15. To what critical stages in the administration of justice has the Sixth Amendment right to counsel been extended, and to what critical stages has it not been extended?
16. What two conditions must be met to show that counsel was ineffective?
17. What are some of the factors that contribute to wrongful convictions?
18. Do many criminals escape conviction and punishment because of procedural technicalities, such as the exclusionary rule or the *Miranda* mandates?

IN THE FIELD

1. **Make a Law** By yourself or as part of a group, create a law. Choose a behavior that is currently not against the law in your community, and write a statute to prohibit it. Make sure that all five features of "good" criminal laws are included. (If this is a group exercise, decide by majority vote any issue for which there is not a consensus.) Critique the outcome.
2. **Exclusionary Rule** Make an oral or a written evaluation of the good-faith and inevitable-discovery exceptions to the exclusionary rule. Has the Supreme Court gone too far in modifying the exclusionary rule? Defend your answer.

ON THE WEB

1. **Historical Perspectives** "The Timetable of World Legal History," found at www.duhaime.org/LegalResources/LawMuseum/LawArticle-44/Duhaimes-Timetable-of-World-Legal-History.aspx, provides brief descriptions of important historical legal developments. (This website also provides links to other sources of legal information.) Choose among the available topics (e.g., the actual text of the Magna Carta), and write a summary of the information you find.
2. **Supreme Court Decisions** Access the Supreme Court website at www.supremecourtus.gov. Then select subjects of interest (especially Fourth, Fifth, Sixth, and Eighth Amendment cases), and read the Supreme Court's most recent decisions.

CRITICAL THINKING EXERCISES

MEGAN'S LAW

1. The sexual assault and murder of 7-year-old Megan Kanka in New Jersey October 31, 1994, struck a national nerve. Megan was assaulted and killed by a neighbor, Jesse Timmendequas, who had twice been convicted of similar sex offenses and was on parole. In response to the crime and public uproar, the state of New Jersey enacted "Megan's Law." The law, as originally passed, required sex offenders, upon their release from prison, to register with New Jersey law enforcement authorities, who are to notify the public about their release. The public was to be provided with the offender's name, a recent photograph, a physical description, a list of the offenses for which he or she was convicted, the offender's current address and place of employment or school, and the offender's automobile license plate number. The Supreme Court has upheld Megan's Law.

   Currently, all 50 states and the federal government have Megan's laws that require sex offenders released from prison to register with local law enforcement authorities. Many of those laws, like New Jersey's, require that law enforcement officials use the information to notify schools and day-care centers and, in some cases, the sex offender's neighbors. In 1997, California enacted a law allowing citizens access to a CD-ROM with detailed information on 64,000 sex offenders living in California who had committed a broad range of sex crimes since 1944. Megan's laws are not uniform across states. For example, not all states require active community notification; many of them (including the U.S. government) only make the information available to the public.

   In 1996, President Clinton signed into law the Pam Lyncher Sexual Offender Tracking and Identification Act, which called for a national registry of sex offenders, to be completed by the end of 1998. The national registry allows state officials to submit queries, such as the name of a job applicant at a day-care center, and to determine whether the applicant is a registered sex offender in any of the participating states.

   a. Is Megan's Law a good law? (Consider the ideal characteristics of the criminal law.)
   b. Is Megan's Law fair to sex offenders who have served their prison sentences (i.e., "paid their debt to society")?
   c. What rights does a sex offender have after being released from prison?
   d. What rights does a community have to protect itself from known sex offenders who have been released from prison?
   e. When the rights of an individual and the rights of a community conflict, whose rights should take precedence? Why?

SURVEILLANCE CAMERAS

2. In October 2012, a U.S. District Court judge ruled that police are allowed in some circumstances to install hidden surveillance cameras on private property without obtaining a search warrant. The case involved two defendants, Manuel Mendoza and Marco Magana, both of Green Bay, WI, who were charged with federal drug crimes after a DEA agent claimed to have discovered more than 1,000 marijuana plants grown on Magana's property. The defendants faced possible life imprisonment and fines of up to $10 million. The defendants asked the judge to throw out the video evidence on Fourth Amendment grounds, arguing that "No Trespassing" signs were posted throughout the heavily wooded, 22-acre property, and that it also had a locked gate.

   However, according to the judge, it was reasonable for Drug Enforcement Administration agents to enter the rural property without permission—and without a warrant—to install multiple "covert digital surveillance cameras" in hopes of uncovering evidence that marijuana plants were being grown. The judge noted that the U.S. Supreme Court has upheld the use of technology as a substitute for ordinary police surveillance and cited *Oliver v. United States* (1984), in which the Court held that "open fields" could be searched without a warrant because they are not covered by the Fourth Amendment.

   Magana's lawyer argued, "That one's actions could be recorded on their own property, even if the property is not within the curtilage [the land immediately surrounding a residence, which the Court has ruled has greater privacy protections], is contrary to society's concept of privacy." He added, "The owner and his guest . . . had reason to believe that their activities on the property were not subject to video surveillance as it would constitute a violation of privacy." Despite the argument, both Mendoza and Magana were convicted and, on April 22, 2013, a federal judge in Milwaukee sentenced both of them to prison: Mendoza to 4 years and 2 months and Magana to 5 years.

   a. Do you agree with the federal judge's ruling? Why or why not?
   b. If the U.S. Supreme Court elects to hear the case, how do you think it will rule in light of its other decisions on the use of technology and warrantless searches?

NOTES

1. Jay A. Sigler, *Understanding Criminal Law* (Boston: Little, Brown, 1981), 3.
2. The discussion in the remainder of this section is based on material from Edwin H. Sutherland and Donald R. Cressey, *Criminology,* 9th ed. (Philadelphia: J. B. Lippincott, 1974), 8.
3. Most of the material in this section comes from Will Durant, *Our Oriental Heritage,* part 1, *The Story of Civilization* (New York: Simon & Schuster, 1954).
4. Most of the material in this section comes from Raymond J. Michalowski, *Order, Law, and Crime: An Introduction to Criminology* (New York: Random House, 1985).
5. Information in this section is from Rennard Strickland, *Fire and the Spirits: Cherokee Law from Clan to Court* (Norman, OK: University of Oklahoma Press, 1975).
6. See Archibald Cox, *The Court and the Constitution* (Boston: Houghton Mifflin, 1987), 239–49.
7. For an examination of the influence of the Burger and Rehnquist Courts on criminal procedure, see Mary Margaret Weddington and W. Richard Janikowski, "The Rehnquist Court: The Counter-Revolution That Wasn't: Part II, The Counter-Revolution That Is," *Criminal Justice Review* 21 (1997): 231–50.
8. In addition to the Supreme Court cases themselves, much of the information in the remainder of this chapter is from the following sources: John N. Ferdico, Henry F. Fradella, and Christopher D. Totten, *Criminal Procedure for the Criminal Justice Professional,* 10th ed. (Belmont, CA: Wadsworth, 2009); Yale A. Kamisar, Wayne R. LaFave, Jerold H. Israel, and Nancy J. King, *Basic Criminal Procedure,* 11th ed. (St. Paul, MN: West, 2005); Stephen J. Schulhofer, Carol S. Steiker, and Sanford H. Kadish, *Criminal Law and Its Processes: Cases and Materials,* 8th ed. (New York: Aspen, 2007); Jerold H. Israel and Wayne R. LaFave, *Nutshell on Criminal Procedure in a Nutshell: Constitutional Limitations,* 7th ed. (St. Paul, MN: West, 2006); John M. Scheb and John M. Scheb II, *Criminal Law and Procedure,* 5th ed. (Belmont, CA: Wadsworth, 2004).
9. *Katz v. United States,* 389 U.S. 347 (1967).
10. *United States v. Jacobsen,* 466 U.S. 109 (1984).
11. Ibid.
12. "Boston Police Will Search Kids' Rooms—With Parents' Consent," *The Orlando Sentinel,* November 18, 2007, A2.
13. *Florida v. J. L.,* 529 U.S. 266 (2000).
14. *Illinois v. Wardlow,* 528 U.S. 119 (2000).
15. R. Erik Lillquist, "Absolute Certainty and the Death Penalty," August 23, 2004, Seton Hall Public Law Research Paper No. 10, available at SSRN: http:/ssrn.com/abstract=581286.
16. See Ira Still, "Criminal Defendants on Trial: What You Must Know as a Defendant Charged with a Crime," www.istilldefendliberty.com/id51.html (retrieved August 2, 2012); *State v. Arthur,* 390 So. 2d 717 (Fla. 1980).
17. Ibid.
18. Unless indicated otherwise, information in this section is from Electronic Privacy Information Center, www.epic.org/privacy/terrorism/hr3162.html, which provides the full text of the USA PATRIOT Act; American Civil Liberties Union, "Surveillance Under the USA PATRIOT Act," April 3, 2003, www.aclu.org/SafeandFree/SafeandFree.cfm?ID512263&c5206; Electronic Frontier Foundation, "EFF Analysis of the Provisions of the USA PATRIOT Act," October 31, 2001, www.eff.org/Privacy/Surveillance/Terrorism_militias/20011031_eff_usa_patriot_analysis.htm; Lara Jakes Jordan, "FBI Admits Further Privacy Violations," *Orlando Sentinel,* March 6, 2008, A3; Electronic Privacy Information Center, "USA PATRIOT Act," accessed January 2, 2010, http://epic.org/privacy/terrorism/usapatriot/.
19. Department of Justice, "Fact Sheet: USA PATRIOT Act Improvement and Reauthorization Act of 2005," www.usdoj.gov/opa/pr/2006/March/06_opa_113.html; The American Civil Liberties Union, "The Patriot Act: Where It Stands," http://action.aclu.org/reformthepatriotact/whereitstands.html.
20. Information about the Protect America Act of 2007 and the FISA Amendments Act of 1978 is from the following sources: www.lifeandliberty.gov/docs/text-of-paa.pdf; The White House, www.whitehouse.gov/news/releases/2007/08/20070806-5.html; American Civil Liberties Union, www.aclu.org/safefree/nsaspying/31203res20070807.html; www.govtrack.us/congress/bill.xpd?bill=h110-3773; http://74.125.113.132/search?q=cache:oTvSepr7yV8J:www.eff.org/files/filenode/att/FISAINTRO_001_xml.pdf+FISA+Amendments+Act+of+1978&hl=en&ct=clnk&cd=4&gl=us.
21. Jonathan Stempel, "NSA's Phone Spying Program Ruled Illegal by Appeals Court," *Reuters,* May 7, 2015. Accessed on January 24, 2016, www.reuters.com/article/us-usa-security-nsa-idUSKBN0NS1IN20150507; "ACLU v. Clapper – Challenge to NSA Mass Call-Tracking Program," October 29, 2015. Accessed January 24, 2016, www.aclu.org/cases/aclu-v-clapper-challenge-nsa-mass-call-tracking-program?redirect=national-security/aclu-v-clapper-challenge-nsa-mass-phone-call-tracking; Erin Kelly, "Senate Approves USA Freedom Act," *USA Today,* June 2, 2015. Accessed January 24, 2016, www.usatoday.com/story/news/politics/2015/06/02/patriot-act-usa-freedom-act-senate-vote/28345747/; Spencer Ackerman, "NSA Reform Bill Imperiled As It Competes with Alternative Effort in the Senate," *The Guardian,* April 28, 2015. Accessed January 24, 2016, www.theguardian.com/us-news/2015/apr/28/house-nsa-reform-bill-senate-usa-freedom-act; "USA Freedom Act," accessed February 19, 2016, http://judiciary.house.gov/index.cfm/usa-freedom-act; Ellen Nakashima, "NSA's Bulk Collection of Americans' Phone Records Ends Sunday," November 27, 2015. Accessed February 19, 2016, www.washingtonpost.com/world/national-security/nsas-bulk-collection-of-americans-phone-records-ends-sunday/2015/11/27/75dc62e2-9546-11e5-a2d6-f57908580b1f_story.html; "Edward Snowden Biography," accessed February 19, 2016, www.biography.com/people/edward-snowden-21262897.
22. Laura Wagner, "North Dakota Legalizes Armed Police Drones," NPR, August 27, 2015. Accessed January 24, 2016, www.npr.org/sections/thetwo-way/2015/08/27/435301160/north-dakota-legalizes-armed-police-drones.
23. Jake Laperruque and David Janovsky, "These Police Drones Are Watching You," *POGO,* September 25, 2018. Accessed April 8, 2019, http://www.pogo.org/analysis/2018/09/these-police-drones-are-watching-you/.
24. *Dickerson v. United States,* 530 U.S. 428 (2000).
25. See Raymond Paternoster, *Capital Punishment in America* (New York: Lexington, 1991), 51.
26. Barry Scheck, Peter Neufeld, and Jim Dwyer, *Actual Innocence: When Justice Goes Wrong and How to Make It Right* (New York: Signet, 2001), xx.
27. C. Ronald Huff, Arye Rattner, and Edward Sagarin, "Guilty Until Proven Innocent: Wrongful Conviction and Public Policy," *Crime and Delinquency* 32 (1986): 518–44. Also see Marvin Zalman,

"Qualitatively Estimating the Incidence of Wrongful Convictions," *Criminal Law Bulletin* 48 (2012): 221–79; Brian Forst, *Errors of Justice: Nature, Sources and Remedies* (Cambridge, UK: Cambridge University Press, 2004); James R. Acker and Allison D. Redlich, *Wrongful Conviction: Law, Science, and Policy* (Durham, NC: Carolina Academic Press, 2011); C. Ronald Huff and Martin Killias (Eds.) *Wrongful Convictions & Miscarriages of Justice: Causes and Remedies in North American and European Criminal Justice Systems* (New York and London: Routledge, 2013); Allison D. Redlich, James R. Acker, Robert J. Norris, and Catherine L. Bonventure (Eds.) *Examining Wrongful Convictions: Stepping Back, Moving Forward* (Durham, NC: Carolina Academic Press, 2014); Marvin Zalman & Julia Carrano, *Wrongful Conviction and Criminal Justice Reform: Making Justice* (New York & London: Routledge, 2014); Saundra D. Westervelt and John A. Humphrey (Eds.) *Wrongly Convicted: Perspectives on Failed Justice* (New Brunswick, NJ and London: Rutgers University Press, 2002); Jim Petro and Nancy Petro, *False Justice: Eight Myths that Convict the Innocent*, Revised Edition (New York and London: Routledge, 2015).

28. Federal Bureau of Investigation, *2018 Crime in the United States*, "Persons Arrested," accessed October 18, 2019, https://ucr.fbi.gov/crime-in-the-u.s/2018/crime-in-the-u.s.-2018/topic-pages/persons-arrested.

29. See Huff, Rattner, and Sagarin, "Guilty Until Proven Innocent," 523.

30. Scheck, Neufeld, and Dwyer, *Actual Innocence*, 95.

31. Ibid., 226.

32. Marcia Coyle, Fred Strasser, and Marianne Lavelle, "Fatal Defense," *The National Law Journal* 12 (1990): 30–44.

33. Scheck, Neufeld, and Dwyer, *Actual Innocence*, 242.

34. Huff, Rattner, and Sagarin, "Guilty Until Proven Innocent," 530–33; Scheck, Neufeld, and Dwyer, *Actual Innocence*.

35. Unless indicated otherwise, all information in this section is from The National Registry of Exonerations. Accessed April 13, 2019, http://www.law.umich.edu/special/exoneration/Pages/mission.aspx.

36. Scheck, Neufeld, and Dwyer, *Actual Innocence*, 351–57.

37. The National Institute of Justice, *The Effects of the Exclusionary Rule: A Study in California* (Washington, DC: U.S. Department of Justice, 1983).

38. Marvin Zalman, *Criminal Procedure: Constitution and Society*, 4th ed. (Upper Saddle River, NJ: Prentice Hall, 2005); American Bar Association Special Committee on Criminal Justice in a Free Society, "Criminal Justice in Crisis." 1988, www.druglibrary.org/special/king/cjic.htm; F. Feeney, F. Dill, and A. Weir, *Arrests Without Conviction: How Often They Occur and Why* (Washington, DC: U.S. Department of Justice, National Institute of Justice, 1983); P. Nardulli, "The Societal Cost of the Exclusionary Rule: An Empirical Assessment," *American Bar Foundation Research Journal* (1983): 585–609; *Report of the Comptroller General of the United States, Impact of the Exclusionary Rule on Federal Criminal Prosecutions* (Washington, DC: U.S. General Accounting Office, 1979); K. Brosi, *A Cross City Comparison of Felony Case Processing* (Washington, DC: U.S. Department of Justice, Law Enforcement Assistance Administration, 1979); B. Forst, J. Lucianovic, and S. Cox, *What Happens After Arrest: A Court Perspective of Police Operations in the District of Columbia* (Washington, DC: U.S. Department of Justice, Law Enforcement Assistance Administration, 1978).

39. *Report of the Comptroller General of the United States.*

40. Cited in Tamar Jacoby, "Fighting Crime by the Rules: Why Cops Like Miranda," *Newsweek*, July 18, 1988, 53.

41. Karen L. Guy and Robert G. Huckabee, "Going Free on a Technicality: Another Look at the Effect of the *Miranda* Decision on the Criminal Justice Process," *Criminal Justice Research Bulletin* 4 (1988): 1–3.

42. Jacoby, "Fighting Crime by the Rules"; Scheck, Neufeld, and Dwyer, *Actual Innocence*, 117.

43. Guy and Huckabee, "Going Free on a Technicality."

44. See, for example, George C. Thomas III and Richard Leo, "The Effects of *Miranda v. Arizona*: 'Embedded' in Our National Culture?" in *Crime and Justice: A Review of Research*, ed. Michael Tonry (Chicago: University of Chicago Press, 2002), 203–71.

45. For a different view, see Paul G. Cassell and Bret S. Hayman, "Police Interrogation in the 1990s: An Empirical Study of the Effects of *Miranda*," *UCLA Law Review* 43 (1996): 839–931; George C. Thomas III, "Is *Miranda* a Real-World Failure? A Plea for More and Better Empirical Evidence," *UCLA Law Review* 43 (1996): 821–37.

Design Elements: *Image of part of a court building with columns and statue*: Pixtal/AGE Fotostock RF; *Image of a door slightly open*: jadding/Shutterstock.com

5

The Print Collector/Print Collector/Getty Images

History and Structure of American Law Enforcement

CHAPTER OUTLINE

LEARNING OBJECTIVES

After completing this chapter, you should be able to:

1. Briefly describe the jurisdictional limitations of American law enforcement.
2. Trace the English origins of American law enforcement.
3. Discuss the early development of American law enforcement.
4. Describe the major developments that have occurred in American policing.
5. Describe the structure of American law enforcement.
6. Explain the relationship between the Federal Bureau of Investigation and the Department of Homeland Security.
7. Discuss the development and growth of private security in the United States.

Crime Story

On February 12, 2019, following a 12-week trial in U.S. district court in Brooklyn, New York, a federal jury convicted Sinaloa drug cartel kingpin Joaquin Archivaldo Guzman Loera (pictured), also known as "El Chapo" and "El Rapido," of being a principal operator of a continuing criminal enterprise—a charge that included 26 drug-related violations and one murder conspiracy. He also was convicted of all 10 counts of a superseding indictment, involving narcotics trafficking, using a firearm in furtherance of his drug crimes, and participating in a money laundering conspiracy. Because of an extradition agreement with Mexico, Guzman Loera did not face the death penalty. On February 17, 2019, Guzman Loera was sentenced to life in prison plus 30 years and was ordered to pay $12.6 billion in forfeiture of the cartel's illegal drug-trafficking proceeds.

Guzman Loera was arrested in February 2014. He had eluded Mexican and U.S. law enforcement agents for 13 years after escaping in a laundry cart from a maximum-security Mexican prison in 2001. Following his 2014 arrest and subsequent imprisonment, Guzman Loera escaped from prison a second time on July 11, 2015, by riding a modified motorcycle on tracks through a lighted and ventilated nearly mile-long tunnel. He was rearrested on January 8, 2016, and returned to the same federal maximum-security prison from which he escaped. On January 19, 2017, the day before President-elect Trump's inauguration and following his extradition from Mexico, he was brought to the Manhattan, New York, jail by a team of law enforcement officers from the Drug Enforcement Administration (DEA), the Immigration and Customs Enforcement (ICE), and the U.S. Marshals.

Guzman Loera's arrest and successful prosecution was the result of collaboration among local, state, federal, and international law enforcement partners led by the Organized Crime Drug Enforcement Task Force (OCDETF). The OCDETF is an umbrella agency headquartered in Washington, DC, that was established in 1982, to coordinate a comprehensive attack against the supply of illegal drugs in the United States and to reduce the violence and other criminal activity associated with the drug trade. The OCDETF combines the resources and expertise of the DEA and numerous federal agencies to target drug trafficking, weapons trafficking, and money laundering organizations. Among the law enforcement agencies that investigated the case were the Department of Justice's (DOJ's) Special Operations Division; the DOJ's Criminal Division's Narcotic and Dangerous Drug Section; DEA New York; DEA Miami; FBI Washington Field Office; FBI New York Field Office; FBI Miami Field Office; ICE's Homeland Security Investigations (HSI) New York; ICE's HIS Nogales; Bureau of Alcohol, Tobacco, Firearms and Explosives; U.S. Marshals Service; IRS Criminal Investigation; U.S. Bureau of Prisons; New York Police Department; and New York State Police. Also cooperating in the investigation were Mexican, Ecuadorian, Netherlands, Dominican, and Columbian law enforcement authorities. The DOJ's Office of International Affairs was instrumental in securing Guzman Loera's extradition to the United States.

Trial evidence included testimony from 14 cooperating witnesses; narcotics seizures of more than 140 tons of cocaine and heroin; weapons, including AK-47s and a rocket-propelled grenade launcher; ledgers; text messages; videos; photographs, and intercepted recordings that detailed the Sinaloa drug trafficking operation over a 25-year period from January 1989 until December 2014. The cartel smuggled the drugs, which also included marijuana and methamphetamine, from Central and South America to wholesale distributors in Arizona, Atlanta, Chicago, Los Angeles, Miami, New York, and elsewhere. The drugs

OMAR TORRES/AFP/Getty Images

were transported using submarines, carbon fiber airplanes, trains with secret compartments, and transnational underground tunnels. Guzman Loera and other cartel members communicated without fear of being intercepted by law enforcement or competitors using a sophisticated encrypted communications network of encrypted cell phones and encrypted apps. Billions of dollars generated from drug sales were transported secretly back to Mexico. Guzman Loera used "sicarios," or hit men, to enforce the cartel's control of territories and to eliminate competitors. He also participated in the violence himself.

A vast network of corrupt officials, including local law enforcement officers, prison guards, state officials, high-ranking members of the armed services, and politicians, facilitated the cartel's operation. Guzman Loera bribed these corrupt officials with millions of dollars to warn him and other cartel members of impending law enforcement operations and to look the other way to trafficking activities such as the shipments of drugs, weapons, and bulk cash.

Despite the considerable resources expended and the self-congratulatory boasting of the law enforcement officials involved in the operation, Guzman Loera's arrest and conviction are not likely to have much lasting effect on the illegal drug trade. According to a U.S. Customs and Border Protection report: "The removal of key personnel does not have a discernable impact on drug flows" into the United States. In an interview with actor Sean Penn in October 2015, Guzman Loera seemed to concur with the report: "The day I don't exist, it's not going to decrease [the flow of illegal drugs into the United States] in any way at all."

A focus of this chapter is the structure of American law enforcement. As this crime story illustrates, American law enforcement agencies at different levels of government, sometimes in cooperation with the law enforcement agencies of foreign governments, frequently join forces in large-scale operations such as illegal drug enforcement activities. Such operations raise many questions. In such operations, which law enforcement agency takes the lead? How are investigation strategies determined? Who determines what resources are required of each agency involved? Who determines how resources are utilized? How is intelligence shared? Who determines how forfeited assets are divided? Who receives credit for successes and blame for failures? These questions and their answers indicate the difficulty of coordinating multiple law enforcement agencies in any operation, especially large-scale illegal drug enforcement.

The Limited Authority of American Law Enforcement

The United States has more than 18,000 public law enforcement agencies at the federal, state, and local levels of government. The vast majority of those agencies, however, are local and serve municipalities, townships, villages, and counties. The authority of each agency—whether it is the FBI, a state highway patrol, or a county sheriff's department—is carefully limited by law. The territory within which an agency may operate also is restricted. The city police, for example, may not patrol or answer calls for service outside the city's boundaries unless cooperative pacts have been developed. **Jurisdiction**, which is defined as a specific geographical area, also means the right or authority of a justice agency to act with regard to a particular subject matter, territory, or person. It includes the laws a particular police agency is permitted to enforce and the duties it is allowed to perform. The Oklahoma Highway Patrol, for example, has investigative and enforcement responsibilities only in traffic matters, while the Kentucky State Police have a broader jurisdiction that includes the authority to conduct criminal investigations throughout the state. Each of the approximately 70 federal law enforcement agencies, large and small, has a specific jurisdiction, although one criminal event may involve crimes that give several federal agencies concurrent jurisdiction. For example, in a bank robbery, if mail of any sort is taken, both the Postal Inspection Service and the FBI are likely to investigate the case.

jurisdiction A politically defined geographical area. The right or authority of a justice agency to act with regard to a particular subject matter, territory, or person. The authority of a court to hear and decide cases.

Aaron Roeth

The only police contact most citizens have is in a traffic situation in a local or state jurisdiction. *Should citizens have more contact with the police in non-law enforcement situations? Why or why not?*

Beyond the statutes that create and direct law enforcement agencies, the procedural law derived from U.S. Supreme Court decisions also imposes limitations on the authority of those agencies. Giving arrested suspects the familiar *Miranda* warnings before questioning is a good example of the Court's role in limiting the authority of the police. In addition, police civilian review boards, departmental policies and procedures, and civil liability suits against officers who have abused their authority curtail the power of the police in the United States.

Thus, there is a great difference between law enforcement with limited authority, operating under the rule of law in a democratic nation, and law enforcement in countries where the law is by decree and the police are simply a tool of those in power. Even in comparison with other democratic nations of the world, however, the United States has remarkably more police agencies that operate under far more restrictions on their authority. To understand the origin of those unique qualities of law enforcement in the United States, it is necessary to look first at the history of law enforcement in England, the nation that provided the model for most of American criminal justice.

THINKING CRITICALLY

1. Why do you think the United States has so many law enforcement agencies?
2. Why do you think it is important that law enforcement agencies have limited authority?

English Roots

If you are the victim of a crime, you might expect that a uniformed patrol officer will respond quickly to your call and that a plainclothes detective will soon follow up on the investigation. Because there are thousands of police departments in local communities across the nation, you also might take for granted that the police handling your case are paid public servants employed by your city or county. Such was not always the case in the United States—or in England, where the basic concepts of American law enforcement and criminal justice originated. The criminal justice system in England took hundreds of years to develop, but eventually the idea arose of a locally controlled uniformed police force with follow-up plainclothes investigators.

The Tithing System

Before the twelfth century in England, justice was primarily a private matter based on revenge and retribution.[1] Victims of a crime had to pursue perpetrators without assistance from the king or his agents. Disputes often were settled by blood feuds in which families would wage war on each other. By the twelfth century, a system of group protection had begun to develop. Often referred to as the **tithing system** or the frankpledge system, it afforded some improvements over past practices. Ten families, or a *tithing*, were required to become a group and agree to follow the law, keep the peace in their areas, and bring law violators to justice. Over even larger areas, ten tithings were grouped together to form a *hundred*, and one or several hundred constituted a *shire*, which was similar to a modern American county. The shire was under the direction of the **shire reeve** (later called the *sheriff*), the forerunner of the American sheriff. The shire reeve received some assistance from elected constables at the town and village levels, who organized able-bodied citizens into **posses** to chase and apprehend offenders.[2] County law enforcement agencies in the United States still sometimes use posses to apprehend law violators. The Maricopa County (Arizona) Sheriff's Department, for example, has a 3,000-member volunteer posse, whose members are trained and are often former deputies.[3]

tithing system A private self-help protection system in early medieval England, in which a group of 10 families, or a *tithing*, agreed to follow the law, keep the peace in their areas, and bring law violators to justice.

shire reeve In medieval England, the chief law enforcement officer in a territorial area called a *shire*; later called the *sheriff*.

posses Groups of able-bodied citizens of a community, called into service by a sheriff or constable to chase and apprehend offenders.

The Constable-Watch System

The Statute of Winchester, passed in 1285, formalized the **constable-watch system** of protection. The statute provided for one man from each parish to be selected as **constable**, or chief peacekeeper. The statute further granted constables the power to draft citizens as watchmen and require them to guard the city at night. Watchmen were not paid for their

constable-watch system A system of protection in early England in which citizens, under the direction of a constable, or chief peacekeeper, were required to guard the city and to pursue criminals.

constable The peacekeeper in charge of protection in early English towns.

efforts and, as a result, often were found sleeping or sitting in a pub rather than performing their duties. In addition, the statute required all male citizens between the ages of 15 and 60 to maintain weapons and to join in the *hue and cry*, meaning to come to the aid of the constable or the watchman when either called for help. If they did not come when called, the male citizens were subject to criminal penalties for aiding the offender. This system of community law enforcement lasted well into the 1700s.

Two features of this system are worthy of note. First, the people were the police, and second, the organization of the protection system was local. These two ideas were transported to the American colonies centuries later.

The Bow Street Runners

In 1748, Henry Fielding, a London magistrate and author of the novel *Tom Jones*, founded a group of professional law enforcement agents to apprehend criminals and recover stolen property in the entertainment district of London, known as Bow Street Covent Garden. This publicly funded detective force, named the Bow Street Runners, was by far the most effective official law enforcement organization of its day. Efforts to duplicate it in other parts of London proved unsuccessful, but Fielding's work in organizing the first British detective force and his writing addressing the shortcomings of the criminal justice system had a great deal of influence. They helped pave the way for a more professional and better-organized response to the crime problems that were dramatically increasing in London by the end of the eighteenth century.[4]

The London Metropolitan Police

Because of the Industrial Revolution, urban populations in cities like London swelled with an influx of people from the countryside looking for work in factories. A major result of this social transformation was that England began experiencing increasing poverty, public disorder, and crime. There was no clear consensus about what to do. Several efforts to establish a central police force for London had been opposed by people who believed that police of any kind were a throwback to the absolute power formerly wielded by English kings. They also were fearful that a central police force for London would resemble the police of eighteenth-century Paris—the first modern police force—which was notorious for its extensive network of police spies and its intrusion into the lives of Parisian citizens and other inhabitants. Nevertheless, Parliament eventually responded, in 1829, with the London Metropolitan Police Act. It created a 1,000-officer police force with professional standards to replace the patchwork of community law enforcement systems then in use. Members of the London Police became known as *bobbies*, or *peelers*, after Robert Peel, the British Home Secretary who had prodded Parliament to create the police force.

The Police of Paris

The London Metropolitan Police, as well as other European police forces, were modeled after the police of eighteenth-century Paris, which is considered the first modern police force in Europe.

Source: Alan Williams, *The Police of Paris, 1718–1789* (Baton Rouge and London: Louisiana State University Press, 1979), xvi.

Topham/Fotomas/The Image Works

The London Metropolitan Police discover another victim of Jack the Ripper. *What do you suppose were some of the unique problems encountered by the first bobbies, or peelers?*

TABLE 5.1 Robert Peel's Principles of Policing

1. The police must be stable, efficient, and organized along military lines.
2. The police must be under governmental control.
3. The absence of crime will best prove the efficiency of police.
4. The distribution of crime news is essential.
5. The deployment of police strength both by time and area is essential.
6. No quality is more indispensable to a policeman than a perfect command of temper; a quiet, determined manner has more effect than violent action.
7. Good appearance commands respect.
8. The securing and training of proper persons is at the root of efficiency.
9. Public security demands that every police officer be given a number.
10. Police headquarters should be centrally located and easily accessible to the people.
11. Policemen should be hired on a probationary basis.
12. Police records are necessary to the correct distribution of police strength.

To ensure discipline, the London Police were organized according to military rank and structure and were under the command of two magistrates, who were later called commissioners. According to Peel, the main function of the police was to prevent crime not by force, but by preventive patrol of the community. Londoners, who resented such close scrutiny, as noted, did not at first welcome this police presence in the community. Eventually, though, the bobbies (the term was originally derogatory) showed that the police could have a positive effect on the quality of life in the community. The London Police are credited in large measure with the general increase in orderliness on the streets of Victorian society.[5] Peel's military approach to policing and some of his other principles remain in effect today throughout the world. **Peel's Principles of Policing** are outlined in Table 5.1.[6]

Peel's Principles of Policing A dozen standards proposed by Robert Peel, the author of the legislation resulting in the formation of the London Metropolitan Police Department. The standards are still applicable to today's law enforcement.

THINKING CRITICALLY

1. Do you think any of the early English systems of law enforcement (e.g., tithing) could work today? Why or why not?
2. Which one of Peel's Principles of Policing do you think is most important? Why?

The Development of American Law Enforcement

The United States has more police departments than any other nation in the world. The major reason for this is that local control is highly regarded in the United States. Thus, like many other services, even small communities that can barely afford police service provide it locally. This practice is primarily responsible for the disparity in the quality of American police personnel and service. The struggle to improve American law enforcement began even before formal police departments came into existence.

Early American Law Enforcement

The chance for a better life free of government intervention was key in the decision of many colonists to cross the Atlantic and settle in the New World. American colonists from England brought with them the constable-watch system with which they were familiar if not completely satisfied. Boston established a night watch as early as 1634. Except for the military's intervention in major disturbances, the watch system, at least in the cities, was the means of preventing crime and apprehending criminals for the next two centuries. As in England, the people were the police. Citizens could pay for watch replacements, and often the worst of the lot ended up protecting the community. In fact, Boston and other cities frequently deployed the most elderly citizens and occasionally sentenced minor offenders to serve on the watch.[7] Later, in rural and southern areas of the country, the office of sheriff was established and the power of the posse was used to maintain order and apprehend offenders. In essence, two forms of protection began to evolve—the watch in the villages, towns, and cities and the sheriff in the rural areas, unincorporated areas, and counties. Communities in the North often had both systems.

FYI

The "Leatherheads"

In Dutch-influenced New York in the seventeenth century, the first paid officers on the night watch were known as *leatherheads* because they wore leather helmets similar in appearance to the helmets worn by today's firefighters. The leatherheads were not known for their attention to duty and often spent entirely too much of their watch schedule inside.

Source: Carl Sifakis, "Leatherheads: First New York Police," in *The Encyclopedia of American Crimes* (New York: Smithmark Publishers Inc., 1992).

Law Enforcement in the Cities

As had happened in England, the growth of the Industrial Revolution lured people away from the farms to cities. Large groups of newcomers, sometimes immigrants from other countries, settled near factories. Factory workers put in long days, often in unsafe and unhealthy working conditions. Some workers organized strikes, seeking better working conditions, but the strikes were quickly suppressed. As the populations of cities swelled, living conditions in some areas became overcrowded and unhealthy. Major episodes of urban violence occurred in the first half of the nineteenth century because of the social and economic changes transforming American cities. Racial and ethnic tensions often reached a boiling point, resulting in mob disturbances that lasted for days. A particular source of trouble was the drinking establishments located throughout working-class districts of cities. Regular heavy drinking led to fights, brawls, and even full-scale riots.

Unlike London, which organized its police force in 1829, American citizens resisted the formation of police departments, relying instead on the constable-watch system, whose members lit streetlights, patrolled the streets to maintain order, and arrested some suspicious people. Constables often had daytime duties, which included investigating health hazards, carrying out orders of the court, clearing the streets of debris, and apprehending criminals against whom complaints had been filed. Neither the night watch nor the constables tried to prevent or discover crime, nor did they wear any kind of uniform. This weak protection system was unable to contain the increasing level of lawlessness.

Municipal Police Forces

In 1844, New York City combined its day and night watches to form the first paid, unified police force in the United States. Close ties developed between the police and local political leaders. As with the first police in London, citizens were suspicious of the constant presence of police officers in their neighborhoods. Also, citizens had little respect for the New York police because they thought they were political hacks appointed by local officials who wanted to control the police for their own gain. During the next several years, the struggle to control the police in New York built to a fever pitch.[8]

In 1853, the New York state legislature formed the Municipal Police Department, but within 4 years that force was so corrupt from taking bribes to overlook crime that the legislature decided to abolish it. It was replaced by the Metropolitan Police, which was administered by five commissioners appointed by the governor. The commissioners then selected one superintendent. Each commissioner was to oversee the others, as well as the superintendent, and keep them all honest. In the minds of the legislature, the new structure was an improvement that would prevent corruption in the top level of the department. But when the Metropolitan Police Board called on Mayor Fernando Wood to abolish the Municipal Police, he refused. Even after New York's highest court upheld a decision to disband the Municipals, the mayor refused. The Metropolitans even tried to arrest Mayor Wood, but that failed attempt resulted in a pitched battle between the two police forces.

Library of Congress/Contributor/Getty Images

New York had a watch system as early as 1658. *Why did the watch system of policing last so long?*

Historical/Contributor/Getty Images

A street arrest in 1878 Manhattan, New York. *What problems did the police of this era encounter?*

When the National Guard was called in, Mayor Wood submitted to arrest but was immediately released on bail.

During the summer of 1857, the two police forces often fought over whether to arrest certain criminals. A particularly troubling practice was one police force releasing from custody the criminals arrested by the other force. Lawbreakers operated freely during the dispute between the two police forces. Criminal gangs had a free hand to commit robberies and burglaries during most of that summer. The public became enraged over this neglect of duty and the increased danger on the streets of New York City. Only when another court order upheld the decision to disband the Municipal Police did Mayor Wood finally comply.

Following the course charted by New York City, other large cities in the United States soon established their own police departments. In 1855, Boston combined its day and night watches to form a city police department. By the end of the decade, police departments had been formed in many major cities east of the Mississippi. The officers' duties did not vary substantially from the duties of those who had served on the watch. After the Civil War, however, peace officers began to take on the trappings of today's police. They began to wear uniforms and carry nightsticks and even firearms, although many citizens resisted giving this much authority to the police.

Tangle of Politics and Policing Until the 1920s in most American cities, party politics prevented the development of professional police departments. Local political leaders understood that controlling the police was a means of maintaining their own political power and of allowing criminal friends and political allies to violate the law with impunity. In fact, in some cities, the police were clearly extensions of the local party machine, which attempted to dominate all activity in a community. If local politicians gave police applicants a job, it became the hired officers' job to get out the vote so the politicians could keep their positions. The system was so corrupt in some cities that police officers bought their

Bettmann/Getty Images

By the early 1900s, most American cities had organized, uniformed police forces similar to the police force of Newport, Rhode Island, pictured circa 1910. *How do current police officers differ from those depicted in the photo?*

jobs, their promotions, and their special assignments. In collaboration with local politicians, but often on their own, the police were more than willing to ignore violations of the law if the lawbreakers gave them money, valuables, or privileges.

A Brief History of Blacks in Policing For most of American history, blacks who have wanted to be police officers have faced blatant discrimination and generally have been denied the opportunity.[9] The first black police officers in the United States were "free men of color." They were hired around 1805 to serve as members of the New Orleans city watch system. They were hired primarily because other people did not want the job. In addition to serving on the watch, they were responsible for catching runaway slaves and generally policing black slaves in New Orleans.

By 1830, policing had become more important in New Orleans, and the "free men of color" lost their jobs on the city police force to others who wanted them. Not until after the Civil War were black Americans allowed to be police officers again. During Reconstruction, black Americans were elected to political office and hired as police officers throughout the South. This did not last long. By 1877, the backlash to Reconstruction drove black Americans and their white Republican allies from elective offices, and black police officers throughout the South lost their jobs. By 1890, most southern cities had all-white police departments. The few black police officers in the southern cities that retained them generally could not arrest white people and were limited to patrolling only areas and communities where other black Americans lived. By 1910, there were fewer than 600 black police officers in the entire United States, and most of them were employed in northern cities.

The hiring of black police officers did not begin again in most southern cities until the 1940s and 1950s. They were hired primarily to patrol black communities, to prevent crime, and to improve race relations. Still, few black Americans ever rose to command positions in their departments. Indeed, prior to the 1950s, only two black Americans had ever been promoted to the command position of captain: Octave Rey of New Orleans and John Scott of Chicago. Both served relatively short tenures in the position: Rey from 1868 to 1877 and Scott from 1940 to 1946.

Law Enforcement in the States and on the Frontier

The development of law enforcement on the state level and in the frontier territories often was peculiar to the individual location. Without large population centers that required the control of disorderly crowds, law enforcement was more likely to respond to specific situations–for example, by rounding up cattle rustlers or capturing escaped slaves. Still, out of this kind of limited law enforcement activity, the basic organizational structure of police units with broader responsibilities was born.

ANA/The Image Works

The plantation slave patrols have been called "the first distinctively American police system." *Have any elements of the slave patrols influenced contemporary American policing? If yes, what are they?*

Southern Slave Patrols In the South, the earliest form of policing was the plantation **slave patrols**.[10] They have been called "the first distinctively American police system."[11] The slave patrols were created to enforce the infamous slave codes, the first of which was enacted by the South Carolina legislature in 1712. Eventually all the Southern colonies enacted slave codes. The slave codes protected the slaveholders' property rights in human beings, while holding slaves responsible for their crimes and other acts that were not crimes if committed by free persons. Under some slave codes, enslaved people could not hold meetings, leave the plantation without permission from the master, travel without a pass, learn to read and write, carry a firearm, trade, or gamble. Both the slave codes and the slave patrols were created in part because of a fear of bloody slave revolts, such as had already occurred in Virginia and other parts of the South.

slave patrols The earliest form of policing in the South. They were a product of the slave codes. The plantation slave patrols have been called "the first distinctively American police system."

The most publicized slave revolt was the Nat Turner Rebellion of 1831 in Virginia. Turner and five other slaves killed Joseph Travis, Turner's owner, and his family. Approximately 70 more rebels joined Turner, whose immediate plan was to capture the county seat, where munitions were stored. Turner was unsuccessful in his plan, but during the siege, he and his rebels killed 57 whites. Turner was tried, convicted, and hanged, along with 16 other rebels. In response to the revolt, white mobs lynched nearly 200 blacks, most of whom were innocent.[12]

Slave patrols generally consisted of three men on horseback who covered a beat of 15 square miles. They were responsible for catching runaway slaves, preventing slave uprisings, and maintaining discipline among the slaves. To maintain discipline, the patrols often whipped and terrorized black slaves who were caught after dark without passes. The slave patrols also helped enforce the laws prohibiting literacy, trade, and gambling among slaves. Although the law required that all white males perform patrol services, the large plantation owners usually hired poor, landless whites to substitute for them. The slave patrols lasted until the end of the Civil War in 1865. After the Civil War, the Ku Klux Klan served the purpose of controlling blacks just as the slave patrols had before the Civil War.

Frontier Law Enforcement In the remote and unpopulated areas of the nation, and particularly on the expanding frontier, justice often was in the hands of the people in a more direct way. Vigilantism frequently was the only way that people could maintain order and defend themselves against renegades and thugs.[13] Even when formal law enforcement procedures were provided by the sheriff or a marshal, courts in many communities were held only once or twice a year, leaving many cases unresolved. This idea of self-protection remains very popular in the South and the West, where firearms laws in many states permit people to carry loaded weapons in a vehicle or even on their persons if they have completed a qualification and licensing procedure.

Many Native American tribes also lived on the frontier and, for centuries, followed a tradition of blood revenge for most offenses (see the description in Chapter 4). Thus, for the most part, Native American law enforcement also was in the hands of the people. The Cherokees, for example, had such a tradition, which did not change until 1808, when they enacted their first written laws. Among the laws was the establishment of the first Cherokee law enforcement agency: the "regulators" or "light horsemen." The law provided for companies consisting of six men, who had an all-encompassing criminal justice function. Their duties were to (1) supervise and protect children and widows as heirs, (2) investigate and locate criminals, (3) try criminals upon statements of witnesses, and (4) decide upon and administer punishment. The regulators provided justice for the Cherokees until 1825, when they were replaced with marshals, sheriffs, and constables with the power to arrest, similar to western peace officers.[14]

State Police Agencies Self-protection did not prove sufficient as populations and their accompanying problems increased. As early as 1823, mounted militia units in Texas protected American settlers throughout that territory. Called *rangers*, these mounted militia fought Native Americans and Mexican bandits. The Texas Rangers were officially formed in 1835, and the organization remains in existence today as an elite and effective unit of the Texas Department of Public Safety.[15]

Source: Winchester 1894.

The Texas Rangers, organized in the early 1800s to fight Native Americans, patrol the Mexican border, and track down rustlers, were the first form of state police. *Why and how do you think the Rangers have endured for so long?*

The inefficiency and unwillingness of some sheriffs and constables to control crime, along with an emerging crime problem that exceeded the local community's ability to deal with it, prompted other states to form state law enforcement agencies. In 1905, Pennsylvania established the first modern state law enforcement organization with the authority to enforce the law statewide, an authority that made it unpopular in some communities where enforcement of state laws had been decidedly lax.[16] The Pennsylvania state police officially had been created to deal with crime in rural areas, but in its early years it frequently responded to industrial discord. The event that led directly to the formation of the state agency was the 1902 anthracite coal strike, which caused a national crisis and the intervention of President Theodore Roosevelt. Industrialists believed municipal police departments and the state militia were too unreliable during strikes because officers were overly sympathetic to workers with whom they often shared community ties and social origins. Industrialists assumed that a centralized mobile force, recruited statewide with ties to no particular community, would eliminate any sympathy between officers and workers.[17] The authority of state police agencies was extended with the advent of the automobile and the addition of miles of state highways. Some form of state law enforcement agency existed in every state by the 1930s.

Texas Rangers

To learn more about the history of the Texas Rangers, visit the Texas Ranger Hall of Fame and Museum website at www.texasranger.org/index.htm.

Why do you think the Texas Rangers have elite status?

Professionalism and Reform

You will recall that the people themselves were once the police, as they served on the watch. Being an adult citizen was about the only qualification. No training was required, and it was common practice for citizens who did not want to serve to hire replacements, sometimes hiring sentenced offenders. Because of the few services and the little order the watch provided, not much else seems to have been required. Even when organized police forces were developed in the 1840s and 1850s in the United States, qualifications for the job mattered little beyond the right political connections or the ability to purchase one's position outright.

Not until the latter part of the nineteenth century did qualifications for the position of police officer begin to evolve. In the 1880s, Cincinnati posted two qualifications to be a police officer.[18] First, an applicant had to be a person of high moral character—an improvement over earlier times. Three citizens had to vouch for the applicant's character at a city council meeting. If deemed acceptable by the council, the applicant was immediately taken to a gymnasium and tested for the second qualification, foot speed.

Both Cincinnati and New York began police academies in the 1880s, but the curriculum was meager and recruits were not required to pass any examinations to prove their competence. The lack of adequate standards and training for police officers was recognized as a major stumbling block to improved policing. A group of reformers within policing allied themselves with the Progressives, a movement for political, social, and economic change. Among the reformers was August Vollmer, who became chief of police of Berkeley, California, in 1909. During his tenure as chief from 1909 to 1932, Vollmer attempted to create a professional model of policing. With Vollmer and a succession of internal reformers who followed, a new era of professional policing began.

Vollmer and his followers advocated training and education as two of the key ingredients of professionalism in policing. He also believed strongly that the police should stay out of politics and that politics should stay out of policing. Vollmer believed that the major function of the police was fighting crime, and he saw great promise in professionalizing law enforcement by emphasizing that role.[19] He began to hire college graduates for the Berkeley Police Department, and he held college classes on police administration.

Within a few decades, this professional model, sometimes called the reform model, had taken root in police departments across the country. To eliminate political influences, gain control of officers, and establish crime-fighting priorities, departments made major changes in organization and operation. Those changes included the following:

- Narrowing of the police function from social service and the maintenance of order to law enforcement only.
- Centralization of authority, with the power of precinct captains and commanders checked.
- Creation of specialized, centrally based crime-fighting units, as for burglary.
- A shift from neighborhood foot patrol to motorized patrol.
- Implementation of patrol allocation systems based on such variables as crime rates, calls for service, and response times.

Bettmann/Contributor/Getty Images

August Vollmer. *Do you agree with Vollmer's idea that police should focus on law enforcement and leave social services and the maintenance of order to others? Why or why not?*

- Reliance on technology, such as police radios, to both control and aid the policing function.
- Recruitment of police officers through psychological screening and civil service testing.
- Specific training in law enforcement techniques.

Policewomen It took a long time for policewomen to gain the opportunity to perform the same roles and duties as their male counterparts. From the early 1900s until 1972, when the Equal Employment Opportunity Commission began to assist women police officers in obtaining equal employment status with male officers, policewomen were responsible for protection and crime prevention work with women and juveniles, particularly with girls. The Los Angeles Police Department created the City Mother's Bureau in 1914 and hired policewomen to work with delinquent and predelinquent children whose mothers did not want formal intervention by a law enforcement agency. Policewomen also were used to monitor, investigate, and punish young girls whose behavior flouted social and sexual conventions of the times.

The first woman to have full police power (1905) was Lola Baldwin of Portland, Oregon. The first uniformed policewoman was Alice Stebbins Wells, who was hired by the Los Angeles Police Department in 1910. By 1916, 16 other police departments had hired policewomen as a result of the success in Los Angeles.[20]

Conflicting Roles

Throughout their history, Americans have never been sure precisely what role they want their police officers to play. Much of the ambivalence has to do with American heritage, which makes many Americans suspicious of government authority. At one time or another, local police have acted as peacekeepers, social workers, crime fighters, and public servants, completing any task that was requested. Often, the police have been asked to take on all those roles simultaneously.

For most of the nineteenth century, distrust of government was so strong and the need to maintain order in the cities so critical that the police operated almost exclusively as peacekeepers and social service agents, with little or no concern for enforcing the law beyond what was absolutely necessary to maintain tranquility.[21] In this role, the police in many American cities administered the laws that provided for public relief and support of the poor. They fed the hungry and housed the homeless at the request of the politicians who controlled them. Later, other social service agents, such as social workers, began to replace them, and a reform effort developed to remove policing from the direct control of corrupt politicians. As a result, the police began to focus on crime-fighting as early as the 1920s. Having the police enforce the law fairly and objectively was thought to be a major way of professionalizing law enforcement. This approach also fit the professional model of policing advocated by Vollmer and other reformers.

TopFoto/The Image Works

Brutal tactics by some police officers to suppress civil rights protests during the 1960s led to calls for improved standards of police conduct and training. *Have new standards of conduct and training ended brutal police tactics?*

By the end of the 1960s, strong doubts about the role of the police emerged again. The role they had been playing encouraged them to ferret out crime and criminals through such practices as aggressive patrol, undercover operations, and electronic surveillance. In some neighborhoods, the police came to be viewed as armies of occupation. Some confrontations between police and citizens resulted in violence. The civil rights movement produced a series of demonstrations and civil disorders in more than 100 cities across America, beginning in 1964. As in the labor struggles of the late nineteenth and early twentieth centuries, the police were called in to restore order. Some police officers suppressed the demonstrations with brutal tactics. The anti-Vietnam War movement during the 1960s sparked protests all over the country, especially on college campuses. Again, police officers were called on to maintain and sometimes to restore order. Thousands of students were sprayed with tear gas, and some were beaten and even killed by police.

By the end of the 1960s, it was clear that police standards and training had to be improved. To many observers, fast response and proactive patrols did not seem effective in reducing crime, and officers increasingly were seeing their work world through the windshield of a cruiser. The likelihood of establishing rapport with the people they served was remote as officers dashed from one crime scene to another.

Four blue-ribbon commissions studied the police in the United States. The four commissions and the years in which they released their reports are:

National Advisory Commission on Civil Disorders, 1967
President's Commission on Law Enforcement and the Administration of Justice, 1967
National Advisory Commission on Criminal Justice Standards and Goals, 1973
American Bar Association's Standards Relating to Urban Police Function, 1973

All four reports made the same major recommendations: They pointed out the critical role police officers play in American society, called for careful selection of law enforcement officers, and recommended extensive and continuous training. The reports also recommended better police management and supervision, as well as internal and external methods of maintaining integrity in police departments.

In an attempt to follow many of the specific recommendations of the reform commissions' reports, police selection became an expensive and elaborate process. It was designed to identify candidates who had the qualities to be effective law enforcement officers: integrity, intelligence, interpersonal skills, mental stability, adequate physical strength, and agility. Attempts also were made to eliminate discriminatory employment practices that had prevented minorities and women from entering and advancing in law enforcement. Finally, it became more common for police officers to attend college, and some police agencies began to set a minimum number of college credit hours as an employment qualification.

myth

Random patrol, as opposed to directed patrol, reduces crime. It is important to have police out in patrol cars scouting neighborhoods and business districts.

fact

There is not much value to such random patrols other than perhaps helping people feel safe. They probably would feel even safer if the police were walking a beat. However, little research supports the idea that officers who ride around for 3 to 5 hours of their shifts are repressing crime. Even being available to respond to calls from the public is not a strong argument for such patrols. Only a small percentage of reported crimes and other incidents require a rapid response.

Community Policing

By the 1970s, research began to show that a rapid response to crime does not necessarily lead to more arrests and that having more police officers using methods made popular under the professional or reform model does not significantly reduce crime.[22] What was emerging was the view that unattended disorderly behavior in neighborhoods—such as unruly groups of youths, prostitution, vandalism, drunk and disorderly vagrants, and aggressive street people—is a signal to more serious criminals that residents do not care what goes on in their community and that the criminals can move in and operate with impunity.

The 1970s and 1980s saw some experimentation with community- and neighborhood-based policing projects.[23] Those projects got mixed results, and many were abandoned because of high costs, administrative neglect, and citizen apathy. However, higher crime rates, continued community deterioration, and recognition of the failure to control crime caused law enforcement to again question the role it was playing. The enforcer role still was not working well enough. It appeared senseless simply to respond to calls for service and arrive at scenes of crime and disorder time and time again without resolving the problems or having any lasting effect on the lives of the residents of the community. Out of this failure and frustration came the contemporary concept of **community policing**.

community policing A contemporary approach to policing that actively involves the community in a working partnership to control and reduce crime.

Under a community policing philosophy, the people of a community and the police form a lasting partnership in which they jointly approach the problems of maintaining order, providing services, and fighting crime.[24] If the police show they care about the minor problems associated with community disorder, two positive changes are likely to occur: Citizens will develop better relations with the police as they turn to them for solutions to the disorder, and criminals will see that residents and the police have a commitment to keeping all crime out of the neighborhood. Once again, the emphasis has shifted from fighting crime to keeping peace and delivering social services. The goal is eradicating the causes of crime in a community, not simply responding to symptoms.

In the early 1990s, many communities across the nation began implementing community policing strategies. Community policing called for a shift from incident-based crime fighting to a problem-oriented approach in which police would be prepared to handle a broad range of troublesome situations in a city's neighborhoods. There was greater emphasis on foot patrol so that officers could come to know and be known by the residents of a neighborhood. Those citizens would then be more willing to help the police identify and solve problems in the neighborhood. Many other aspects of community policing are discussed more fully in Chapter 6.

CompStat

At about the same time that community policing was becoming popular in many American cities, a new policing strategy was being implemented in New York City.[25] By the beginning of the new millennium, a third of the nation's largest police departments had adopted it, and another 25% were planning to do so. The new strategy was called CompStat, an abbreviation of "compare stats" or "computer statistics meetings." **CompStat** is a technological and management system that aims to make the police better organized and more effective crime fighters. It combines innovative crime analysis and geographic information systems, that is, crime mapping (described in Chapter 6), with the latest management principles.

CompStat A technological and management system that aims to make the police better organized and more effective crime fighters. It combines innovative crime analysis and geographic information systems; that is, crime mapping, with the latest management principles.

CompStat is based on four interrelated crime-reduction principles: (1) provide accurate and timely crime data to all levels of the police organization, (2) choose the most effective strategies for specific problems, (3) implement those strategies by the rapid deployment of personnel and resources, and (4) diligently evaluate the results and make adjustments to the strategy as necessary. Crime analysts–who collect data, analyze the information, and then map it to show trends or trouble spots–identify problems. Armed with this information, precinct commanders are responsible for formulating a response and solving the problem. Failure to get the job done results in harsh reprimands from top administrators, and repeated failures can lead to removal from command.

Supporters of CompStat claim that it has reduced crime, and FBI statistics show that crime rates have declined in those cities that have implemented it. However, the simultaneous decrease in crime rates reported by the FBI and the implementation of CompStat may be nothing more than a coincidence. In fact, a few studies reveal that crime rates were already declining in cities before CompStat was implemented. Critics contend that CompStat is incompatible with community policing. Whereas community policing is based on the decentralization of decision-making authority and the empowerment of patrol officers to make decisions in their communities, CompStat concentrates decision-making power among command staff, who issue orders to the rank and file. Centralized command and control are key features of the traditional model of police organizations, and therein lies the appeal of CompStat to police administrators who are uncomfortable giving up too much control. CompStat allows the chief of police to judge the performance of precinct commanders and allows precinct commanders to hold their officers accountable. CompStat returns the control of everyday policing to police administrators and requires minimal disruption to the traditional police organization. At the same time, it allows police administrators to tout their use of innovative technologies and problem-solving techniques. It will be interesting to see whether the future of policing is community policing, CompStat, or some other system.

THINKING CRITICALLY

1. Which of the major changes in the organization and operation of police departments listed in the section Professionalism and Reform do you think brought about the most significant change? Why?
2. What do you think are the key benefits of community policing? Why?
3. Which system of policing do you believe will best serve the interests of the American public: community policing or CompStat? Why?

History of Four Federal Law Enforcement Agencies

Since the United States was formed, the American public has held a healthy skepticism about a centralized police system. This is why law enforcement in the United States, unlike in many other countries, is primarily a state and local matter. However, the creation of a federal system of government and laws necessitated a national law enforcement presence. The result has been dozens of federal law enforcement agencies. Although space limitations preclude an examination of all these agencies' histories, the histories of four of the more prominent ones–the U.S. Marshals Service, the Secret Service, the Federal Bureau of Investigation, and the Drug Enforcement Administration–are described in the following sections.

The U.S. Marshals Service

The first federal law enforcement agents in the United States were the U.S. Marshals, a product of the Judiciary Act of 1789.[26] The act fleshed out details of the new federal judicial system as provided for in the U.S. Constitution. Duties of the federal Marshals and their deputies included protecting the federal courts, supporting their operation, and enforcing federal court decisions and federal laws. In supporting the operation of the federal courts, U.S. Marshals served summonses, subpoenas, writs, warrants, and other process (i.e., proceedings in any action or prosecution) issued by the courts; arrested people suspected of committing federal crimes; were responsible for all federal prisoners; disbursed funds as ordered by the federal courts; paid the fees and expenses of court clerks, U.S. Attorneys, jurors, and witnesses; rented courtrooms and jail space and hired bailiffs, court criers, and janitors; and made sure that prisoners were present, jurors were available, and witnesses were punctual. U.S. Marshals also were charged with carrying out the lawful orders of Congress and the president. The position of U.S. Marshal was modeled after the position of county sheriff. In Virginia, between 1619 and about 1634, local sheriffs were called provost marshals or marshals. The same was true in Georgia between 1733 and 1773.

President George Washington personally selected the first 13 Marshals–one for each state. The president still nominates U.S. Marshals who must be confirmed by the Senate. Washington wanted men who would support the federal government without jeopardizing states' rights. Most of his appointees had a previous association with him, including service under his command during the Revolutionary War. The first Marshals helped to establish the federal judicial system and place the new federal government on sound footing because of their local ties, which made the exercise of federal power a little more palatable to the American public. Throughout their history, U.S. Marshals have been required to live within the districts they served. As civilian law enforcers, the availability of the U.S. Marshals frequently prevented military intervention in state and local affairs.

The U.S. Marshals represented the federal government's interests at the local level and performed a variety of non-law enforcement duties needed to keep the central government functioning effectively. For example, they conducted the first national census in 1790 and continued to do so until 1870. They also distributed presidential proclamations and collected statistical information on commerce and manufacturing. Until 1861, they reported directly to the secretary of state; in 1861, Congress assigned their supervision to the attorney general. Nevertheless, until the 1960s and the establishment of a centrally administered U.S. Marshals Service with control over district budgets and the hiring of deputies, the U.S. Marshals operated with little supervision. Working with federal judges and U.S. Attorneys, prior to 1960, U.S. Marshals were relatively free to determine how they would enforce the law.

One of the Marshals' first law enforcement duties–one they still perform today–was to conduct executions authorized by the federal courts. U.S. Marshal Henry Dearborn of Maine conducted the first federal execution in 1790. He executed Thomas Bird for a murder committed at sea. Another early duty of the U.S. Marshals was to enforce the Sedition Act of 1798. The act punished unlawful combinations against the government and publishing "false, scandalous, and malicious writing" about the government. Prior to the creation of the U.S. Secret Service in 1865, the Treasury Department used U.S. Marshals and their deputies to investigate and pursue counterfeiters nationwide. U.S. Marshals also were charged with enforcing the Fugitive Slave Act of 1850. The Marshals arrested fugitive slaves and returned them to their Southern masters. During the Civil War, U.S. Marshals confiscated property used to support the Confederacy, and they helped capture Confederate spies.

Following the Civil War, U.S. Marshals and their deputies were instrumental in keeping law and order in the "Wild West." One of the most infamous incidents involving the Marshals occurred in Tombstone, Arizona, in 1881. The gunfight at the O.K. Corral pitted U.S. Marshal Virgil Earp and his deputies, brothers Wyatt and Morgan Earp and John "Doc" Holliday, against the Clanton gang. The U.S. Marshals became a part of the newly created Justice Department in 1870. During the Pullman Railroad Strike of 1894, President Grover Cleveland and the federal courts ordered the U.S. Marshals to help U.S. Army troops break the strike and keep the trains rolling.

U.S. Marshals helping to break the Pullman railroad strike of 1894. *Was this an appropriate use of the U.S. Marshals? Why or why not?*

Sarin Images/Granger, NYC

In 1896, U.S. Marshals began to receive an annual salary for the first time. They previously had worked under a fee system in which they would collect set amounts for performing specific tasks. Getting paid under the fee system frequently was an ordeal. During World War I, U.S. Marshals helped protect the home front from enemy aliens, spies, and saboteurs. They also arrested draft dodgers and people who tried to disrupt Selective Service operations. With the ratification of the Eighteenth Amendment in 1919, which prohibited the manufacture, sale, and transportation of intoxicating beverages in the United States, the U.S. Marshals Service assumed the primary responsibility for enforcing the Prohibition laws. They continued in that role until 1927, when the Treasury Department gave the responsibility to the newly created Bureau of Prohibition. After that, the Marshals, along with other federal agencies, assisted in Prohibition efforts.

In the 1960s, U.S. Marshals helped enforce desegregation orders. For example, when James Meredith, a black man, enrolled in the University of Mississippi in 1962, deputies protected him 24 hours a day for an entire year. Following passage of the Organized Crime Control Act of 1970, the U.S. Marshals Service was given responsibility for the Witness Security Program. In 1979, the U.S. attorney general transferred primary jurisdiction for the apprehension of escaped federal prisoners from the FBI to the U.S. Marshals Service. In 1985, U.S. Marshals were given the task of managing and disposing of properties seized and forfeited by federal law enforcement agencies and U.S. Attorneys nationwide. In 1996, following a series of bombings, the U.S. Marshals Service was charged with protecting abortion clinics and doctors.

U.S. Marshals Service

To learn more about the U.S. Marshals Service, visit its website at www.usmarshals.gov/.

Based on what you have learned from the website, do you think you would be interested in a career with the U.S. Marshals Service? Why or why not?

Throughout their more than 220-year history, U.S. Marshals and their deputies have been "general practitioners within the law enforcement community," capable of responding quickly to new problems. Unlike other federal law enforcement agencies, the U.S. Marshals have not been restricted by legislation to specific, well-defined duties and jurisdictions. Today, their major responsibilities include:

- Judicial security.
- Fugitive investigations and apprehensions.
- Witness security.
- Prisoner services (e.g., detaining presentenced federal prisoners).
- Transporting federal prisoners and criminal aliens.
- Managing and disposing of seized and forfeited property.
- Serving federal court criminal and some civil process.
- Conducting special operations (e.g., providing security assistance when Minuteman and cruise missiles are moved between military facilities).

Careers in Criminal Justice

U.S. Marshal

My name is Frank Tallini. I am a deputy U.S. Marshal with the U.S. Marshals Service. As a deputy U.S. Marshal, I am charged with a variety of duties, such as ensuring the security of the federal judiciary, apprehending violent fugitives, and assisting with the operation of the Witness Security Program. Because the U.S. Marshals Service serves as the primary custodian of seized property for the DOJ Asset Forfeiture Program, my tasks also include managing and selling seized assets found to have been obtained through illegal activity. These items, which can include homes, vehicles, and boats, are subsequently sold to the public. I have personally stood on the steps of a federal courthouse and auctioned off an entire container ship and a mansion that sold for $3.5 million, among other assets. The proceeds from these U.S. Marshals' sales compensate victims and support future law enforcement efforts.

Other missions for which the U.S. Marshals are responsible are prisoner operations and prisoner transportation. The Marshals are responsible for the custody of all federal prisoners who have been remanded to federal custody for their trials. On average, more than 55,000 federal prisoners are in U.S. Marshals' custody on a daily basis. If found guilty, U.S. Marshals personnel transport the prisoners to their designated Federal Bureau of Prisons facilities. We move prisoners for other reasons, too, such as to testify at a trial, transfer to another institution, or to receive medical care or evaluations. U.S. Marshals move over 1,000 prisoners, on average, every day.

In addition, the U.S. Marshals Service is the lead federal agency to investigate, locate, and apprehend noncompliant sex offenders, as a result of the Adam Walsh Child Protection and Safety Act. Another law, the Justice for Victims of Trafficking Act of 2015, allows the Marshals to assist in the search for missing children, regardless of whether a fugitive or sex offender is involved.

Qualifications for the position of deputy U.S. Marshal are numerous and generally include possessing a bachelor's degree, being a U.S. citizen, and being in excellent physical condition. You also must be of outstanding character and able to successfully pass a thorough background investigation. Though not a prerequisite for employment consideration, many of my fellow deputies have worked for other law enforcement agencies prior to joining the U.S. Marshals. Some are military veterans. Prior to joining the U.S. Marshals, I was a member of the United States Army's 1st Infantry Division. I believe that my time in the military helped prepare me for a career in law enforcement because of the self-discipline and leadership abilities I gained from my experiences.

Courtesy of Frank Tallini, Deputy United States Marshall

My favorite part of the job is contributing to locating and apprehending some of the 347 violent fugitives the U.S. Marshals Service arrest, on average, each and every day. I also am honored to be part of the same storied agency whose ranks include such legendary lawmen as Wyatt Earp and Bass Reeves. For anyone seeking to follow in these lawmen's footsteps, I would tell them the work is demanding and the hours frequently are long, but I believe there's no other career quite as rewarding.

The Secret Service

In 1865, the United States Secret Service was created as a branch of the Treasury Department to combat the counterfeiting of U.S. currency.[27] During the mid-nineteenth century, approximately 1,600 state banks designed and printed their own bills, making it difficult to distinguish between counterfeit bills and the more than 7,000 uniquely designed legitimate bills. Counterfeiting was a serious problem. It was estimated that one-third to one-half of all currency in circulation was counterfeit. By comparison, the counterfeit rate today is a fraction of a percent. To resolve the counterfeiting problem, a national currency was adopted in 1862, but it was soon counterfeited extensively, too. The enforcement of anti-counterfeiting laws was clearly necessary, and the Secret Service was and continues to be effective in suppressing the problem.

In 1867, Secret Service responsibilities were expanded to include "detecting persons perpetrating frauds against the government." This resulted in investigations into the Ku Klux Klan, nonconforming distillers, smugglers, mail robbers, land fraudsters, and a number of other federal law violators. In 1984, Congress enacted legislation further

Counterfeiting Prevention

In 2017, the U.S. Secret Service prevented the circulation of more than $73 million in counterfeit U.S. currency. Agents arrested 1,548 criminals as a result of counterfeit investigations and suppressed 101 counterfeit manufacturing plants.

Source: United States Secret Service, 2017 Annual Report, p. 17, accessed April 19, 2019, https://www.secretservice.gov/data/press/reports/CMR-2017_Annual_Report_online.pdf.

AP Images

U.S. Secret Service agent Tim McCarthy "took a bullet" protecting President Ronald Reagan from would-be assassin John Hinckley Jr. in 1981. *Could you be a U.S. Secret Service agent?*

expanding the investigative responsibilities of the Secret Service to violations relating to credit and debit card fraud, federal-interest computer fraud, and fraudulent identification documents.

In 1894, the Secret Service began informal part-time protection of President Grover Cleveland, and, in 1902, a year after the assassination of President William McKinley, it was given full-time responsibility for the protection of the U.S. president. In 1951, Congress enacted legislation permanently authorizing Secret Service protection of the president, his or her immediate family, the president-elect, and the vice president, if he or she wishes. Ten years later, Congress authorized Secret Service protection of former presidents for a reasonable period of time. In 1962, Congress expanded Secret Service protection to include the vice president or the next officer to succeed the president and the vice president-elect. Congress passed legislation in 1963 to provide Secret Service protection of Mrs. John F. Kennedy and her minor children for 2 years. In 1965, Congress authorized Secret Service protection of former presidents and their spouses during their lifetime and minor children until age 16. Following the assassination of Robert F. Kennedy in 1968, Congress expanded Secret Service protection to major presidential and vice presidential candidates and nominees; it also authorized protection of widows of presidents until death or remarriage, and their children until age 16.

In 1922, President Warren Harding requested the creation of the White House Police, which was placed under the supervision of the Secret Service in 1930. The White House Police was renamed the Executive Protection Service in 1970, and the Secret Service Uniformed Division in 1977. The Treasury Police Force was merged into the Secret Service Uniformed Division in 1986. The Secret Service was transferred from the Treasury Department to the Department of Homeland Security in 2003. Today, "the mission of the United States Secret Service is to safeguard the nation's financial infrastructure and payment systems to preserve the integrity of the economy, and to protect national leaders, visiting heads of state and government, designated sites and National Special Security Events."

Secret Service

To learn more about the U.S. Secret Service, visit its website at www.secretservice.gov/.

Based on what you have learned from the website, do you think that you would be interested in a career with the Secret Service? Why or why not?

The Federal Bureau of Investigation

When he assumed the presidency following President McKinley's assassination in 1901, Theodore Roosevelt, who had served as New York City Police Commissioner from 1895 to 1897, began his crusades to break up big-business monopolies in the East and to stop land theft in the West.[28] He successfully employed Secret Service agents in that effort. Four years later, Roosevelt appointed Charles Bonaparte as U.S. Attorney General. As head of the Justice Department, Bonaparte had only a few special agents of his own and a group of Examiners, who were trained as accountants and charged with reviewing the federal courts' financial transactions. Since its establishment in 1870, the Justice Department had to hire private detectives and later investigators from other federal agencies to investigate federal crimes.

By 1907, the Justice Department primarily relied on Secret Service agents to conduct its investigations. These Secret Service agents reported to the Chief of the Secret Service and not to the Attorney General. Bonaparte did not like the arrangement and wanted complete control of investigations under his jurisdiction. In 1908, big-business and land interests were successful in getting Congress to pass a law prohibiting the Justice Department and all other executive agencies, except the Treasury Department, from hiring Secret Service agents to conduct investigations. The law was intended to thwart President Roosevelt's reform agenda. A month after the law was passed, Roosevelt ordered Bonaparte to appoint a force of what turned out to be 34 Special Agents within the Justice Department. Ten of his new appointees were former Secret Service agents. The primary purpose of the new force was to investigate violations of the Sherman Anti-Trust Act, which was passed in 1890, and was intended to prevent business monopolies from artificially raising prices by restriction of trade or supply. On July 26, 1908, Bonaparte ordered his new agents to report to

Chief Examiner Stanley W. Finch. This act is considered the beginning of the FBI. The force of 34 agents became a permanent part of the Justice Department in 1909, following the recommendations of both Attorney General Bonaparte and President Roosevelt. Later in 1909, George Wickersham, who succeeded Bonaparte as Attorney General, named the force the Bureau of Investigation and the Chief Examiner as the Chief of the Bureau of Investigation.

When the Bureau was created, there were few federal crimes. Investigations were limited mostly to crimes involving national banking, bankruptcy, naturalization, antitrust, land fraud, and peonage, the system by which debtors or legal prisoners were held in servitude to labor for their creditors or for persons who leased their services from the state. The Bureau began to expand in 1910, after Congress passed the Mann ("White Slave") Act. The Mann Act made it a crime to transport women across state lines for immoral purposes.

When the U.S. entered World War I in 1917, President Woodrow Wilson enlarged the Bureau's responsibility to include crimes of espionage and sabotage and violations of the Selective Service Act. The Bureau also assisted the Labor Department in the investigation of enemy aliens. In 1919, William J. Flynn, a former chief of the Secret Service, became the head of the Bureau of Investigation and was the first to use the title of Director. Also in 1919, Congress passed the National Motor Vehicle Theft Act, which further expanded the Bureau's investigative responsibilities.

In 1921, President Warren Harding's Attorney General Harry M. Daugherty appointed William J. Burns Director of the Bureau. Burns, like Flynn, had been chief of the Secret Service but gained notoriety by running the William J. Burns International Detective Agency. Because of his involvement in the infamous Teapot Dome Scandal, Attorney General Harlan Fiske Stone asked Burns to resign from the Bureau in 1924. Burns's short-lived career as director of the Bureau is perhaps best remembered for his appointment of a 26-year-old graduate of George Washington University Law School named John Edgar Hoover to the position of assistant director of the Bureau. Hoover had worked for the Justice Department since 1917, where he headed enemy alien operations during World War I and assisted in the investigation of suspected anarchists and communists in the General Intelligence Division under Attorney General A. Mitchell Palmer. Following Burns's resignation in 1924, Attorney General Stone appointed then 29-year-old J. Edgar Hoover as Director of the Bureau of Investigation, a position he would hold for the next 48 years.

Under Hoover's leadership, the Bureau of Investigation became a major factor in policing. A few spectacular and well-publicized crimes in the early 1930s, coupled with the problem of Prohibition and gangland killings in Chicago in the 1920s, fueled a public panic about a national crime emergency. The pivotal event was probably the Lindbergh baby kidnapping in 1932. Because of Charles Lindbergh's fame, the kidnapping received international attention. In response, Congress quickly passed a federal kidnapping statute dubbed the "Lindbergh Law." Prior to 1932, Hoover's only major accomplishment was getting the Bureau designated as the national clearinghouse and publisher of the new Uniform Crime Reports (UCR) in 1930. However, following passage of the Lindbergh Law, the Bureau, at Hoover's direction, mounted a massive publicity campaign that emphasized the threat of crime and the Bureau's role as the guardian of law and order. Through Bureau press releases about the killing of John Dillinger in 1934, and, in the next few years, the killing or apprehension of Pretty Boy Floyd, Baby Face Nelson, Ma Barker, and Alvin "Creepy" Karpis, a mythology was created about the Bureau's success in fighting crime. This mythology, which was promoted and exploited by Hoover over his long FBI career, appealed to many Americans, who found the sensationalized crime stories about "G-Men" and "Public Enemies" a titillating diversion from life's demoralizing daily drudge during the Great Depression.

In the next few years, the FBI expanded in size and prominence and gained increasing influence over local policing. This influence began in 1930, when the Bureau became responsible for the Uniform Crime Reports system. It received added momentum in 1932, when the Bureau established its own crime lab and, in 1935, when it founded the National Police Academy. In the midst of what appeared to be a mounting crime wave, few people objected to the establishment of a "national police force." As a result, in 1935, under Hoover's leadership, the Bureau of Investigation became the Federal Bureau of Investigation (FBI), and thanks to his aggressive public relations department, Hoover managed to win for himself the image of the nation's "top cop." The FBI's influence over local policing increased further in 1940, when it was given responsibility for coordinating domestic security during World War II.

The Federal Bureau of Investigation

To learn more about the Federal Bureau of Investigation, visit its website at www.fbi.gov/.

Based on what you have learned from the website, do you think that you would be interested in a career with the FBI? Why or why not?

FBI Director J. Edgar Hoover was considered by many as the nation's "top cop." *Did FBI Director J. Edgar Hoover have too much power? Why or why not?*

Hulton Archive/Getty Images

When the war ended and the Cold War began, the FBI continued its domestic security responsibilities. For example, it was given the job of investigating allegations of disloyalty among federal employees and was relentless in combating the communist threat, which Hoover always equated with U.S. labor union activity. Hoover began to consider himself as internal security czar, who was not subordinate to the attorney general but rather a coequal consultant and advisor. The Bureau also began devoting a larger portion of its resources to helping state and local law enforcement agencies.

In the 1960s, Congress passed new laws giving the FBI the authority to fight civil rights violations, racketeering, and gambling. However, under Hoover, the FBI "dragged its feet" in the field of civil rights, primarily because Hoover, a virulent racist, maintained, despite evidence to the contrary, that civil rights organizations such as the Southern Christian Leadership Conference (SCLC) had been infiltrated and were being led by Communists. The leader of the SCLC was Martin Luther King Jr. Hoover also was less than enthusiastic about enforcement of the civil rights laws because he did not want to jeopardize mutually beneficial relationships with powerful southern congressmen and local law enforcement agencies, whose officers often were sympathetic to Ku Klux Klan activities.

As for organized crime, new laws passed by Congress in 1968 enabled the FBI to engage in court-ordered electronic surveillance, and together with increased undercover work, to successfully develop cases against nearly all the heads of the U.S. organized crime families. Ironically, until 1957, Hoover insisted that organized crime in the United States (at least an Italian-dominated national syndicate) did not exist. According to extensive documentation, the reason for Hoover's denial of organized crime's existence was that the "Mafia" had evidence of Hoover's gambling debts and his homosexuality. However, in 1957, the major U.S. organized crime families held a conference in Apalachin, New York, and were detected by New York state police officers. This event proved that organized crime existed in the United States, and Mafia-deniers, including Hoover, were forced to admit as much. In 1961, Attorney General Robert Kennedy created an Organized Crime and Racketeering Section in the Department of Justice to coordinate activities against organized crime by the FBI and other department agencies.

As just shown with regard to civil rights and organized crime, the FBI and its long-time director had a darker, more sinister side. This was epitomized by the FBI's infamous covert domestic counterintelligence programs ("COINTELPROS"), which were used against dissidents and their organizations from 1956 through 1971. The purpose of COINTELPROS, according to Hoover, was to "expose, disrupt, misdirect, discredit, and otherwise neutralize" specific groups and individuals. To impede constitutionally protected political activity against groups and individuals who opposed government domestic and foreign policy, the FBI used surveillance, infiltration, harassment, intimidation, sabotage, provocation, media manipulation, and other often-illegal tactics, including complicity in the alleged assassination of Black Panther leader Fred Hampton. COINTELPROS' targets included the Communist Party; the Socialist Workers Party; the National Association for the Advancement of Colored People (NAACP); the American Civil Liberties Union (ACLU); the National Lawyers Guild; the American Friends Service Committee (a Quaker service organization that received the Nobel Peace Prize in 1947); the American Indian Movement; Black Nationalist groups, such as the Black Panther Party; White hate groups, such as the Ku Klux Klan; and many members of the New Left, including the Students for a Democratic Society (SDS) and numerous antiwar, antiracist, feminist, lesbian and gay, environmentalist, and other groups. It also targeted individuals such as civil rights leader Martin Luther King Jr., whom the Bureau set out to destroy in 1963, and civil rights leader and labor organizer Cesar Chavez. Some of the aforementioned groups, such as the ACLU, had been under FBI surveillance since the 1920s because of their criticism of the Bureau.

Hoover was able to freely pursue these clandestine and often-illegal activities because he was able to successfully insulate himself and the Bureau from executive and legislative control. He did this by amassing secret files on the conduct and associations of presidents

Hulton Archive/Getty Images, Katherine Young/Getty Images, Bettmann/Contributor/Getty Images, Michael Ochs Archives/Getty Images

Silver Screen Collection/Getty Images, George Rinhart/Contributor/Getty Images, Popperfoto/Getty Images, Michael Ochs Archives/Stringer/Getty Images

Top (L/R): Helen Keller, Felix Frankfurter, Joe Namath, and Marlon Brando. Bottom (L/R): Paul Newman, Rock Hudson. Joe Louis, and Muhammad Ali. *Why did the FBI keep secret files on these individuals?*

and legislators that might prove embarrassing to them if revealed. He also kept extensive investigative files on thousands of other individuals who had been involved in controversial causes and dissident organizations, including deaf and blind educator Helen Keller; U.S. Supreme Court Justice Felix Frankfurter; football player Joe Namath; actors Marlon Brando, Paul Newman, and Rock Hudson; and boxers Joe Louis and Muhammad Ali. Yet, as an enduring monument to his government service, adoration, and power, the mammoth FBI headquarters in Washington, DC, the preoccupation of his last years, was named the J. Edgar Hoover Building. The building, formally dedicated in 1975, dwarfs the Justice Department headquarters building and dominates the inaugural route between the Capitol and the White House.

The day after Hoover's death in 1972, President Richard Nixon appointed L. Patrick Gray III as the FBI's acting director. Gray, who most recently had been the Justice Department's assistant attorney general for the Civil Division, allowed the Bureau to become a part of the Watergate cover-up, authorized and approved illegal break-ins and burglaries, and even coached Deputy Attorney General Richard Kleindienst on his testimony before the Senate Judiciary Committee. The Justice Department had been charged with compromising its case against the International Telephone and Telegraph Company (ITT) in exchange for promised campaign contributions and other favors. When Gray's personal involvement in these nefarious activities became public, he resigned and withdrew his name from Senate consideration to be director. Hours after Gray resigned in 1973, William Ruckelshaus, a former congressman and the first head of the Environmental Protection Agency, was appointed acting director and served in that capacity for 3 months until Clarence Kelley was appointed director. Kelley, who had been an FBI agent from 1940 to 1961, was Kansas City, Missouri, police chief at the time of his appointment. Kelley labored to restore public trust in the FBI; he also established three national priorities for the FBI: foreign counterintelligence, organized crime, and white-collar crime. To accomplish his priorities, Kelley intensified the Bureau's recruitment of accountants, women, and minorities.

In 1978, Kelley resigned as FBI director and was replaced by former federal Judge William H. Webster. Webster made terrorism a fourth FBI national priority in 1982, following a series of worldwide terrorist incidents. Also in 1982, the attorney general gave the FBI concurrent jurisdiction with the Drug Enforcement Administration (DEA) over the War on Drugs. The FBI also served as lead security agency at the 1984 Los Angeles Olympics. In the mid-1980s, the FBI was successful in solving several espionage cases, the most serious of which involved John Walker and his spy ring. Under Webster's leadership, the FBI also attacked public corruption and white-collar crime nationwide. FBI operations led to convictions of members of Congress (ABSCAM), the judiciary (GREYLORD), defense procurement officials (ILLWIND), and state legislators in California and South Carolina. FBI investigations in the 1980s successfully uncovered massive fraud in the savings and loan debacle, too. Webster left the Bureau in 1987 to become director of the Central Intelligence Agency (CIA). He was replaced temporarily by FBI Executive Assistant Director John E. Otto, who during his 5-month tenure made drug investigations the FBI's fifth national priority.

Later in 1987, federal Judge William S. Sessions was appointed as the fourth FBI director. Following the fall of the Berlin Wall in 1989, and a steep rise in violent crime over the preceding 10 years, Sessions designated the investigation of domestic violent crimes as the FBI's sixth national priority. To address the new priority, he reassigned 300 special agents from foreign counterintelligence responsibilities to domestic violent crime investigations. By 1991, the FBI had instituted "Operation Safe Streets" in Washington, DC, which involved the coordination of federal, state, and local police task forces in the targeting of fugitives and gangs. With the FBI's assistance, the program would soon be expanded nationwide.

At about the same time, the FBI crime laboratory revolutionized violent criminal identification by successfully employing DNA technology. Under Sessions's leadership, the FBI refocused resources to combat a new wave of large-scale insider bank fraud and other financial crimes, complex health care frauds, and newly created environmental crimes. National security priorities also were refocused from the threats of communism and nuclear war to protecting U.S. information and technologies; the proliferation of biological, chemical, and nuclear weapons; and the theft of economic trade secrets and proprietary information. Also under Sessions's watch, the FBI's image was tarnished by the mishandling of two crisis situations: one in 1992, at Ruby Ridge, Idaho, where the wife of fugitive

Randall Weaver was accidentally shot and killed by an FBI sniper, and the other in 1993, at Waco, Texas, where 74 members of the Branch Davidian religious sect, including women and children, died as a result of the government's misguided attack of their compound. During the summer of 1993, President Bill Clinton removed Director Sessions from office when he refused to resign following allegations of ethics violations involving the misuse of government planes and limousines. President Clinton appointed Deputy Director Floyd I. Clarke as acting FBI director.

In the fall of 1993, Louis J. Freeh was sworn in as the fifth director of the FBI. Freeh had been a federal judge at the time of his appointment and a former FBI agent. Freeh's primary goal was to forge strong international police partnerships to fight evolving crime problems at home and abroad. He was instrumental in the establishment of the first International Law Enforcement Academy in Budapest, Hungary, in 1995. Between 1993 and 1996, the FBI conducted successful investigations into the 1993 World Trade Center bombing in New York City, the 1995 bombing of the Murrah Federal Building in Oklahoma City, the UNABOMBER Theodore Kaczynski in 1996, and the arrests of Russian crime boss Vyacheslav Ivankov in 1995 and Mexican drug-trafficker Juan Garciá-Ábrego in 1996. The Bureau under Freeh also created the Critical Incident Response Group (CIRG) in response to the tragedies at Ruby Ridge, Idaho, and Waco, Texas. To deal with crime in cyberspace, the Bureau under Freeh established the Computer Investigations and Infrastructure Threat Assessment Center (CITAC) and employed its Computer Analysis and Response Teams (CART) to successfully investigate and prevent computer crimes. In 1998, the FBI under Freeh instituted its National Infrastructure Protection Center (NIPC) to monitor the spread of computer viruses, worms, and other malicious programs and to warn government and businesses about these threats to their computers. Freeh resigned from the Bureau in the summer of 2001 amid criticism that the FBI needed stronger leadership—especially after allegations that 25-year FBI agent Robert Hanssen had been a spy for the Soviet Union and Russia since 1985, the FBI bungling of the investigation of Los Alamos National Laboratory scientist Wen Ho Lee, and allegations of incompetence at the FBI crime laboratory.

On September 4, 2001, President George W. Bush appointed U.S. Attorney Robert S. Mueller to succeed Director Freeh, as the sixth director of the FBI. Mueller's mandate as FBI director was to refine the Bureau's information technology infrastructure, to improve its records management system, and to upgrade FBI foreign counterintelligence analysis and security because of the damage done by former special agent and convicted spy Robert Hanssen. However, only days after Mueller took office, the terrorist attacks of September 11 occurred, and Mueller's mandate changed. The new mission of the FBI became "to protect and defend the United States against terrorist and foreign intelligence threats, to uphold and enforce the criminal laws of the United States, and to provide leadership and criminal justice services to federal, state, municipal, and international agencies and partners."

FBI Fights Terrorism

To learn more about the FBI's efforts to combat terrorism, visit its website at www.fbi.gov. Then, under the heading "What We Investigate," click on "Terrorism."

From what you have learned, do you think the FBI is effective in combating terrorism?

On September 4, 2013, James B. Comey was sworn in as the seventh director of the FBI, succeeding Director Mueller. At the time of his appointment, he was serving as general counsel for a Connecticut-based investment fund. Prior to that, he had served as deputy attorney general at the Department of Justice and U.S. Attorney for the Southern District of New York.

On May 9, 2017, President Trump fired Director Comey citing Comey's handling of the probe into Hillary Clinton's use of a private email server. At the time, Director Comey was leading the investigation into whether Trump campaign members colluded with Russians who hacked the 2016 presidential election.

To replace Comey, President Trump appointed Andrew McCabe as FBI acting director. McCabe served for about three months from May 9, 2017 to August 2, 2017, after which he resumed his previous role as deputy director, overseeing all FBI domestic and international investigative and intelligence activities. However, on March 16, 2018, just hours before he was to retire, then-Attorney General Jeff Sessions fired McCabe for "lacking candor under oath" in misleading investigators about his authorization of a conversation between FBI employees and *The Wall Street Journal* concerning investigations of Hillary Clinton. McCabe and others claimed that a vindictive President Trump fired McCabe because he defended Comey following Comey's firing.

SHAWN THEW/EPAEFE/Shutterstock

FBI Director Christopher A. Wray. *What attributes are needed to succeed in this position?*

In any event, on August 2, 2017, President Trump nominated and the Senate confirmed Christopher Wray to be the eighth director of the FBI. Director Wray previously had served as associate deputy attorney general and principal associate deputy attorney general in the Office of the Deputy Attorney General in Washington, DC. In 2003, President Bush nominated and the Senate confirmed Wray to be the assistant attorney general for DOJ's Criminal Division, where he supervised counterterrorism sections and major national and international criminal investigations and prosecutions. At the time of his nomination to be FBI Director, Wray was in private law practice, specializing in government investigations and white-collar crime.

As of this writing, the FBI's priorities are these:

1. Protect the United States from terrorist attack.
2. Protect the United States against foreign intelligence operations and espionage.
3. Protect the United States against cyber-based attacks and high-technology crimes.
4. Combat public corruption at all levels.
5. Protect civil rights.
6. Combat transnational/national criminal organizations and enterprises.
7. Combat major white-collar crime.
8. Combat significant violent crime.

Careers in Criminal Justice

FBI Special Agent

My name is Linda Dionne, and I am an FBI Special Agent in the Public Corruption Unit (PCU) of the Washington, DC, field office. I have a Bachelor of Arts degree in Criminal Justice from California State University at San Bernardino. While working on a Certificate in Crime and Intelligence Analysis through the California State University system and the California Department of Justice, I began to volunteer with the Riverside County Sheriff's Department. I eventually obtained a full-time position as an analyst there. Later I went to the Santa Clara Police Department as a Certified Crime and Intelligence Analyst. I designed and directed their crime analysis unit for 4 years. I left Santa Clara in 1999 to become an FBI agent.

Like all FBI Special Agents, I received training at the FBI Academy. The program was 16 weeks of rigorous and intellectually challenging work (now it is 21 weeks). I had to study 12 to 15 major subject areas, I often had 3 hours of homework for the next day's classes, and I also had an exam in another course the next day. Some agents realized that the FBI was not what they thought it was. Others found it difficult to be away from family and friends.

The Washington, DC, field office is the second-largest field office in the FBI. The PCU investigates allegations of bribery involving public officials, including extortion or using the mail to defraud the public. Examples include the issuance of licenses, permits, contracts, or zoning variances; judicial case fixing; and law enforcement corruption. The PCU's responsibilities are divided into two squads. My squad's responsibility is the District of Columbia government. The other squad's responsibility is the executive branch of the federal government in the Washington, DC area. Twelve agents are assigned to my squad, along with several non-agent support specialists.

Prior to transferring to the PCU, I worked on a special inquiry squad, conducting background investigations for White House staff and presidential appointees. It was in this capacity that I had the opportunity to interview Attorney General Janet Reno, as well as other prominent politicians.

To become an FBI Special Agent, applicants must fulfill a long list of requirements. For example, applicants must be between the ages of 23 and 36. They must have a bachelor's degree from a U.S.-accredited college or university and be able to meet the physical requirements. They need to have a minimum of 2 years of successful work experience in their chosen field before applying to be an agent. However, prior work experience need not be in law enforcement. When hired, new

AP Images

FBI Special Agents start at a GS-10, step 1 pay grade, which in 2019 was $48,973 per year before locality pay adjustment and overtime. My advice to someone who wants to become an FBI agent is to maintain the highest standards of conduct in your life. Your first job out of college does not have to be in criminal justice, but be sure to be successful in whatever you do. I thoroughly enjoy my career with the FBI.

After reading this account, what do you think is a key quality of a good FBI agent?

The Drug Enforcement Administration

President Richard Nixon created the Drug Enforcement Administration (DEA) by executive order in 1973.[29] His goal was to establish a single unified command to wage "an all-out global war on the drug menace." The DEA traces its history through several Treasury Department bureaus: the Bureau of Internal Revenue (1915–1927), the Bureau of Prohibition (1927–1930), the Bureau of Narcotics (1930–1968), and the Justice Department's Bureau of Narcotics and Dangerous Drugs (1968–1973).

The federal law that inaugurated America's War on Drugs was the Harrison Narcotics Tax Act of 1914. The act provided that all persons who produced, imported, manufactured, compounded, dealt in, dispensed, sold, distributed, or gave away opium or coca leaves, their salts, derivatives (such as morphine, heroin, and cocaine), or preparations had to register with the Bureau of Internal Revenue, pay a special tax, and keep records of all transactions. The act further authorized the commissioner of internal revenue, with the approval of the secretary of the treasury, to appoint such agents as necessary to enforce the provisions of the act. The act stipulated that any person who violated the law could be fined not more than $2,000 or be imprisoned for not more than 5 years, or both. On its face, the Harrison Act was a tax law and not a prohibition law, but the Treasury Department interpreted the law to mean that it was illegal for a doctor to prescribe any of the aforementioned drugs to an addict to maintain his or her use and comfort. The U.S. Supreme Court made that interpretation official in 1919 in *Webb v. U.S.* In 1922, the Court in *U.S. v. Behrman* added that a narcotic prescription for an addict was illegal, even if the drugs were prescribed as part of a cure program. (Cocaine was included, although it is not a narcotic.) These decisions made it nearly impossible for addicts to legally obtain their drugs. And despite the Court's reversing its *Behrman* decision in *Lindner v. U.S.* in 1925 (holding that addicts were entitled to medical care), the damage was done because physicians refused to treat addicts under any circumstances. As a result, a well-developed illegal drug marketplace arose to cater to addicts' needs.

How large the hardcore drug addict problem was before passage of the law is controversial. However, several indicators suggest that the use of the drugs among Americans was relatively widespread. First, in the eighteenth, nineteenth, and early twentieth centuries, a booming so-called patent medicine (even though they were not patented) and elixir industry flourished. The active ingredient in many of these medicines and elixirs was the prohibited drugs. In 1804, about 90 brands of elixirs were advertised; by 1905, the list had increased to more than 28,000. As for advertising, following the Civil War, the patent medicine industry was the leader in national advertising, with some individual proprietors spending more than $1 million a year. A second indicator of the widespread use of the drugs is an ad in the 1897 Sears Roebuck catalog that offered "hypodermic kits, which included a syringe, two needles, two vials, and a carrying case for as little as $1.50, with extra needles available at 25 cents each or $2.75 per dozen." A third indicator is the law itself. It is unlikely Congress would have passed a tax act unless it believed that revenue from the tax would be substantial. Finally, a fourth indicator comes from an editorial in *American Medicine*, published 6 months after the Harrison Act was signed into law. The editorial also sounded a warning about the legislation: "Narcotic drug addiction is one of the gravest and most important questions confronting the medical profession today. Instead of improving conditions the laws recently passed have made the problem more complex." The complex problems to which the editorial referred were made explicit in an editorial published in the *New York Medical Journal* just 6 weeks after the Harrison Act went into effect:

> As was expected . . . the immediate effects of the Harrison antinarcotic law were seen in the flocking of drug habitues to hospitals and sanatoriums. Sporadic crimes of violence were reported too, due usually to desperate efforts by addicts to obtain drugs, but occasionally to a delirious state induced by sudden withdrawal. . . . The really serious results of this legislation, however, will only appear gradually and will not always be recognized as such. These will be the failures of promising careers, the disrupting of happy families, the commission of crimes which will never be traced to their real cause, and the influx into hospitals to the mentally disordered of many who would otherwise live socially competent lives.

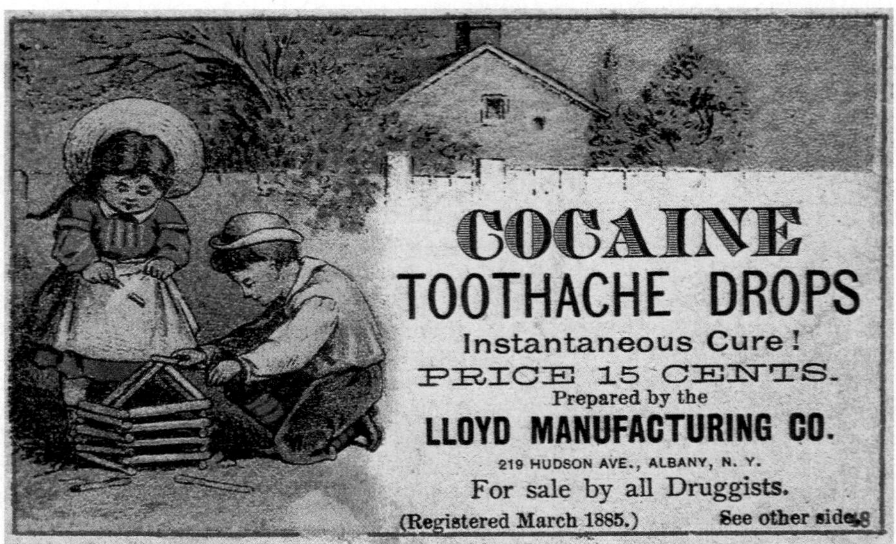

The Granger Collection, NYC

A patent-medicine advertisement. *Why were narcotics so popular with the American public?*

The Bureau of Prohibition originated in 1920 as the Prohibition Unit of the Bureau of Internal Revenue. The purpose of the Prohibition Unit was to enforce the National Prohibition Act of 1919 (also known as the Volstead Act). The act was passed to support the U.S. Constitution's newly ratified Eighteenth Amendment, which prohibited the manufacture, sale, and transportation of alcoholic beverages. In 1927, Congress passed the Bureau of Prohibition Act, which created the Bureau of Prohibition and the Bureau of Customs as independent agencies within the Treasury Department. The Bureau of Prohibition's most famous agent was Eliot Ness of *The Untouchables* fame. In 1930, the largely ineffective and corrupt Bureau was transferred from the Treasury Department to the Justice Department. With the ratification of the Twenty-First Amendment in 1933, the failed national experiment with alcohol prohibition was abandoned and with it, its primary enforcement agency. Commenting on the experiment's demise, early supporter of Prohibition John D. Rockefeller Jr. had this to say:

> When Prohibition was introduced, I hoped that it would be widely supported by public opinion and the day would soon come when the evil effects of alcohol would be recognized. I have slowly and reluctantly come to believe that this has not been the result. Instead, drinking has generally increased; the speakeasy has replaced the saloon; a vast army of lawbreakers has appeared; many of our best citizens have openly ignored Prohibition; respect for the law has been greatly lessened; and crime has increased to a level never seen before.

In 1930, Congress created the Federal Bureau of Narcotics (FBN) in the Treasury Department following the collapse of the Department's Narcotics Division the year before amid evidence of corruption. The first and only commissioner of the FBN was Harry J. Anslinger, who held the post for 32 years. Before his appointment to the FBN, Anslinger was the assistant commissioner in the Bureau of Prohibition. Anslinger is considered the United States' first "drug czar" and is best known for his sensational campaign to demonize marijuana, which he used to elevate himself to national prominence. To fuel his national anti-marijuana campaign, Anslinger maintained a "gore file" of reefer madness exploitation stories that linked the drug to heinous offenses featuring ax murderers and crazed black men sexually assaulting white women. Anslinger's campaign resulted in the Marijuana Tax Act passed by Congress in 1937.

Like the Harrison Narcotics Tax Act of 1914, the Marijuana Tax Act of 1937, on its face, was not intended to prohibit the popular and therapeutic use of marijuana. The ostensible purpose of the legislation was to levy a token tax on anyone who imported, manufactured, produced, compounded, sold, dealt, dispensed, prescribed, administered, or gave away marijuana or any of its derivatives. The act granted Commissioner Anslinger and his Bureau absolute administrative, regulatory, and enforcement authority. For most individuals, the tax was either $1 or $3 a year or a fraction thereof; for importers, manufacturers, and compounders, the tax was $24 a year or fraction thereof. Those who provided the drug, including physicians, also were required to maintain detailed records of their transactions (names, addresses, dates, amounts, and so on) that had to be made available on request to Bureau agents for inspection.

Also like the Harrison Act, the ostensible purpose of the Marijuana Tax Act was belied by the punishment provisions of the law: 5 years imprisonment, a $2,000 fine, or both. The penalties curiously are severe for failing to pay a tax that, even if collected, would produce only a tiny amount of revenue for the government. (A later version of the act made it possible to impose a life sentence for selling just one marijuana cigarette to a minor.) Another telltale sign of the act's "real" purpose was the onerous record-keeping requirement that had a chilling effect on anyone who wanted to legally provide the drug. Finally, the act erroneously classified marijuana as a narcotic, thus placing it in the same category and under the same controls as opium and coca products. (Marijuana is still included in the same category as heroin today.)

Buyenlarge/Contributor/Getty Images

Advertisement for the 1936 movie *Reefer Madness*. *Why did the FBN sensationalize the effects of marijuana consumption? Was it a good idea? Why or why not?*

Although Anslinger is best known for criminalizing marijuana, he also was instrumental in strengthening the Harrison Narcotics Tax Act of 1914 and lobbying for severe penalties for illegal drug usage generally. For example, in the 1950s, federal laws were passed that set mandatory sentences for drug-related offenses, including marijuana. A first-offense for possession of marijuana, for instance, carried a minimum sentence of 2 to 10 years with a fine of up to $20,000. Still, the main focus of the FBN during Anslinger's long tenure was combating opium and heroin smuggling. To that end, he opened offices in France, Italy, Turkey, Lebanon, Thailand, and other countries involved in the illegal drug trade. However, Anslinger's efforts in this area were handicapped by U.S. foreign policy considerations that shielded U.S. allies. For example, during the Vietnam War, investigations of large-scale smuggling operations in allied countries such as Thailand were never completed.

In 1968, the Justice Department's Bureau of Narcotics and Dangerous Drugs (BNDD) was formed by combining the Treasury Department's Bureau of Narcotics with the Food and Drug Administration's Bureau of Drug Abuse Control. The Food and Drug Administration was under the Department of Health, Education, and Welfare. The Bureau of Narcotics was responsible for the control of marijuana and narcotics such as heroin, while the Bureau of Drug Abuse Control was charged with the control of other dangerous drugs, including depressants, stimulants, and hallucinogens, such as LSD. The only director of the BNDD was John E. Ingersoll, who had been the police chief of Charlotte, North Carolina, immediately before his appointment. Under Ingersoll's leadership, the BNDD became the primary U.S. drug law enforcement agency. The Bureau's goals were fourfold: (1) to consolidate the authority and preserve the experience and manpower of the Bureau of Narcotics and the Bureau of Drug Abuse Control; (2) to work with state and local governments in their crackdown on illegal trade in drugs and narcotics, and to help train local agents and investigators; (3) to maintain worldwide operations, working closely with other nations, to suppress the trade in illicit narcotics and marijuana; and (4) to conduct an extensive campaign of research and a nationwide public education program on drug abuse and its tragic effects.

In 1970, under the BNDD, the first joint narcotics task force, comprising federal, state, and local law enforcement officers, was formed in New York to conduct complex drug investigations into the heroin trade. In 1971, the BNDD was given authority to enforce what became the Diversion Control Program, which investigated the large-scale diversion of such legitimate drugs as amphetamines and barbiturates to illicit markets. The BNDD also was responsible for the successful 1972 French Connection heroin investigation.

The United States' first "drug czar," Harry J. Anslinger. *In what ways were FBN Director Anslinger and FBI Director Hoover alike, and in what ways were they different?*

Bettmann/Contributor/Getty Images

In 1973, the short-lived BNDD became a part of the newly created Drug Enforcement Administration (DEA) within the Justice Department. In addition to the BNDD, the DEA combined the Justice Department's Office of National Narcotics Intelligence and the Office of Drug Abuse Law Enforcement, the Treasury Department's Drug Investigation Unit of the U.S. Customs Service, and the Narcotics Advance Management Research Team in the Executive Office of the President. The official rationale for combining the various drug enforcement agencies was (1) the growing availability of illegal drugs in most areas of the United States, (2) the lack of coordination and the perceived lack of cooperation between U.S. Customs and the BNDD, and (3) the need for better intelligence gathering on drug-trafficking organizations. The anticipated benefits of the new DEA included:

1. Putting an end to the interagency rivalries that have undermined federal drug law enforcement, especially the rivalry between the BNDD and the U.S. Customs Service.
2. Giving the FBI its first significant role in drug enforcement by requiring that the DEA draw on the FBI's expertise in combating organized crime's role in the trafficking of illicit drugs.
3. Providing a focal point for coordinating federal drug enforcement efforts with those of state and local authorities, as well as with foreign police forces.
4. Placing a single administrator in charge of federal drug law enforcement in order to make the new DEA more accountable than its component parts had ever been, thereby safeguarding against corruption and enforcement abuses.
5. Consolidating drug enforcement operations in the DEA and establishing the Narcotics Division in Justice to maximize coordination between federal investigation and prosecution efforts and eliminate rivalries within each sphere.
6. Establishing the DEA as a superagency to provide the momentum needed to coordinate all federal efforts related to drug enforcement outside the Justice Department, especially the gathering of intelligence on international narcotics smuggling.

The official version of the DEA's origins omits the DEA's link to the Watergate scandal that ultimately led to President Nixon's humiliating resignation from office. Many of the key participants in the DEA's creation were key conspirators in the Watergate affair. The Nixon White House wanted to establish its own domestic-intelligence system and private police force so it could control and eliminate Nixon administration enemies. The war on heroin provided the needed cover, and the Offices of National Narcotics Intelligence (ONNI) and Drug Abuse Law Enforcement (ODALE) became the vehicles. (No information could be found on the Narcotics Advance Management Research Team.) The new offices, which were created in 1972 by executive order, were placed in the Justice Department instead of the White House to satisfy concerns of BNDD Director Ingersoll and Deputy Attorney General Richard Kleindienst. ONNI and ODALE agents installed illegal "national security" wiretaps and committed burglaries, warrantless raids, and other crimes on the orders of John Ehrlichman, who was counsel and Assistant to the President for Domestic Affairs, and other high-ranking Nixon administration officials.

Besides the Watergate break-in, one of the more notorious operations of ONNI agents, called "plumbers," was the burglary of Daniel Ellsberg's psychiatrist's office. The "plumbers" were a special investigative unit charged with fixing "leaks" to the press, something with which the Nixon White House was obsessed. Ellsberg was a former military analyst, who "leaked" to *The New York Times* and other newspapers the *Pentagon Papers*, a top-secret Pentagon study that revealed faulty government decision making about the Vietnam War that was embarrassing to the Kennedy, Johnson, and Nixon administrations. The plumbers were seeking information to discredit Ellsberg, who had also been the subject of illegal wiretapping.

Today, the mission of the Drug Enforcement Administration (DEA) is:

> to enforce the controlled substances laws and regulations of the United States and bring to the criminal and civil justice system of the United States, or any other competent jurisdiction, those organizations and principal members of organizations, involved in the growing, manufacture, or distribution of controlled substances appearing in or destined for illicit traffic in the United States; and to recommend and support nonenforcement programs aimed at reducing the availability of illicit controlled substances on the domestic and international markets.

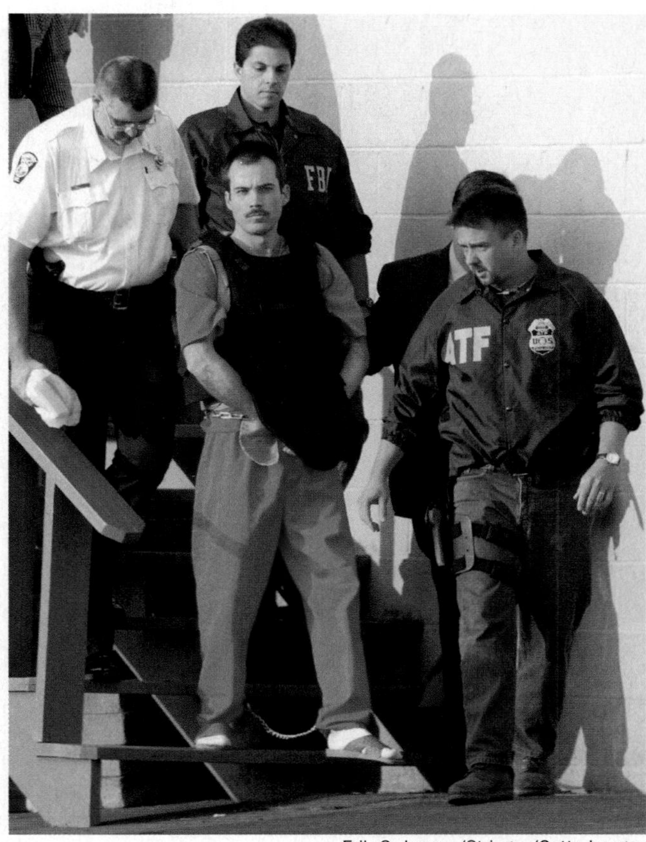

Erik S. Lesser/Stringer/Getty Images

Multiple bombing suspect Eric Robert Rudolph is escorted by law enforcement officials from the Cherokee County Courthouse and Jail in Murphy, North Carolina, June 2, 2003, to a federal court hearing in Asheville, North Carolina. *What causes interagency rivalries, and how can they be reduced?*

In carrying out its mission, the agency's primary responsibilities include:

- Investigation and preparation for the prosecution of major violators of controlled substance laws operating at interstate and international levels.
- Investigation and preparation for prosecution of criminals and drug gangs who perpetrate violence in our communities and terrorize citizens through fear and intimidation.
- Management of a national drug intelligence program in cooperation with federal, state, local, and foreign officials to collect, analyze, and disseminate strategic and operational drug intelligence information.
- Seizure and forfeiture of assets derived from, traceable to, or intended to be used for illicit drug trafficking.
- Enforcement of the provisions of the Controlled Substances Act as they pertain to the manufacture, distribution, and dispensing of legally produced controlled substances.
- Coordination and cooperation with federal, state, and local law enforcement officials on mutual drug enforcement efforts and enhancement of such efforts through exploitation of potential interstate and international investigations beyond local or limited federal jurisdictions and resources.
- Coordination and cooperation with federal, state, and local agencies, and with foreign governments, in programs designed to reduce the availability of illicit abuse-type drugs on the U.S. market through nonenforcement methods such as crop eradication, crop substitution, and training of foreign officials.

Drug Enforcement Administration

To learn more about the Drug Enforcement Administration, visit its website at https://www.dea.gov.

Based on what you have learned from the website, do you think you would be interested in a career with the DEA? Why or why not?

- Responsibility, under the policy guidance of the Secretary of State and U.S. Ambassadors, for all programs associated with drug law enforcement counterparts in foreign countries.
- Liaison with the United Nations, Interpol, and other organizations on matters relating to international drug control programs.

A detailed critique of America's War on Drugs is presented in Chapter 6.

THINKING CRITICALLY

1. Which of the four federal law enforcement agencies described in this section—the U.S. Marshals Service, the Secret Service, the FBI, and the DEA—is the most important, and why?
2. Do you think it is a good idea for any individual to head a federal law enforcement agency as long as J. Edgar Hoover or Harry J. Anslinger did? Why or why not?
3. Why do you suppose the DEA continues to include marijuana in the same category as heroin?

The Structure of American Law Enforcement

One Law Enforcement Officer for Every 419 U.S. Residents

In 2018, the United States had about one city, county, or state law enforcement officer for every 419 residents.

Source: Calculated from Table 74 in Federal Bureau of Investigation, *2018 Crime in the United States,* "Police Employee Data," accessed February 7, 2020, https://ucr.fbi.gov/crime-in-the-u.s/2018/crime-in-the-u.s.-2018/topic-pages/tables/table-74.

Describing American law enforcement and its structure is especially difficult today because of its ongoing restructuring and transformation, from community policing at the local level to the Department of Homeland Security at the federal level and privatization at all levels. It also is difficult to describe because law enforcement agencies are so diverse. To begin with, you must decide which law enforcement agency you are talking about.

For example, Oklahoma Highway Patrol officers cruise the highways and back roads, enforcing traffic laws, investigating accidents, and assisting motorists over seemingly endless miles of paved and unpaved routes. They do not ordinarily investigate criminal violations unless the violations are on state property. In contrast, a sheriff and two deputies in rural Decatur County, Kansas, conduct criminal investigations, serve subpoenas, and investigate accidents. In the towns of Homer, Oakwood, and Ridge Farm, Illinois, only one employee, the chief of police, works in each department, and that person is responsible for all law enforcement, public order, and service duties. About 50 sworn law enforcement officers at the Ohio State University in Columbus also are a part of American law enforcement.[30]

Altogether, tens of thousands of law enforcement officers at the federal, state, county, and municipal levels protect life and property and serve their respective publics. They are employed by government, private enterprise, and quasi-governmental entities. Their responsibilities are specific and sometimes unique to the kind of organization that employs them. Examples of these organizations are airports, transit authorities, hospitals, and parks.

At the state level, there are highway patrols, bureaus of investigation, park rangers, watercraft officers, and other law enforcement agencies and personnel with limited jurisdictions. Colleges and universities employ police officers, and some of those forces are comparable to many medium-sized police departments in the United States.

At the federal level, there are more than 70 law enforcement agencies if all of the small agencies with very specific jurisdictions are included. The FBI, the U.S. Secret Service, and the DEA are three of the better-known agencies. The U.S. Marshals Service, the Bureau of Alcohol, Tobacco, Firearms, and Explosives (ATF), and the U.S. Immigration and Customs Enforcement (ICE) are other federal law enforcement agencies, as are the Criminal Investigation Division of the Internal Revenue Service, the United States Postal Inspection Service, the U.S. Customs and Border Protection (CBP), and several dozen other agencies. As of December 9, 2018, federal law enforcement agencies employed nearly 163,000 law enforcement personnel.[31]

Don Tremain/Getty Images

Before entering some schools, students and their bags are checked every day by special law enforcement officers with limited jurisdiction. *Is this practice necessary? Why or why not?*

As the aforementioned list of law enforcement agencies suggests, explaining the law enforcement mandate and its execution in the United States is difficult. The structure of American police services is different from those of other countries. Japan and many other nations have only one police department. The United States has more than 18,000 public law enforcement agencies. Figure 5.1 shows the percentage of public law enforcement agencies in the United States.

You already have learned that law enforcement in America is fragmented, locally controlled, and limited in authority; to that, you also can add the terms *structurally* and *functionally different*. Virtually no two police agencies in America are structured alike or function in the same way. Police officers themselves are young and old; well trained and ill prepared; educated and uninformed; full-time and part-time; rural, urban, and suburban; generalists and specialists; paid and volunteer; and public and private. These differences lead to the following generalizations about law enforcement in the United States:

1. The quality of police services varies greatly among states and localities across the nation.
2. There is no consensus on professional standards for police personnel, equipment, and practices.
3. Expenditures for police services vary greatly among communities.
4. Obtaining police services from the appropriate agency often is confusing for crime victims and other clients.

FIGURE 5.1 Public Law Enforcement Agencies in the United States

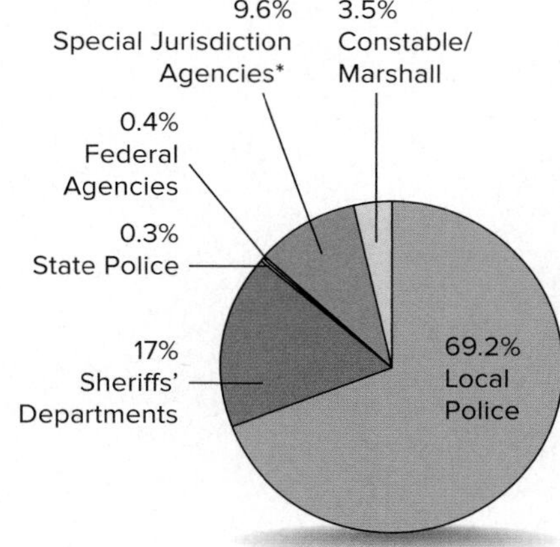

*Such as game protection agencies, water conservancies, and mental health institutions.

Source: Calculated from GOVTECH NAVIGATOR, "Total U.S. Law Enforcement Agencies," accessed April 20, 2019, https://www.govtech.com/navigator/numbers/us-law-enforcement-agencies_56.html. Seventy-three federal law enforcement agencies were added.

Local Policing and Its Duties

If a person knows a law enforcement agent at all, it is probably a local police officer. The officer may have given the person a traffic ticket or investigated an automobile accident. The officer may have conducted a crime prevention survey. Children meet local police officers through Drug Abuse Resistance Education (D.A.R.E.) in public or private schools. Almost everyone has seen the beat cop drive by in a patrol car. Some people have reported thefts or burglaries, but it is doubtful that even they understand what local police officers in America really do, besides what they see on television and in movies.

Municipal Police Departments More than 12,000 municipal police departments in the United States employed about 468,000 full-time sworn police officers in 2016 (the latest year for which data were available)—a decrease of about 2% from the number employed in 2013.[32] Police departments come in all sizes, but most of them are small in the number of officers employed. The overwhelming majority of police departments in America employ fewer than 50 sworn officers. Figure 5.2 shows the number of sworn officers in local police agencies in the United States in 2016. As shown, nearly one-half of all local police departments in the United States employ fewer than 10 officers, and fewer than 0.5% employ more than 1,000 sworn personnel.

What are some of the characteristics of the sworn personnel who occupy the ranks of municipal police agencies in the United States? Most police officers are white males. In 2016, 71.5% of full-time sworn officers were white and 87.7% were male. By comparison, in 1987, about 90% of full-time sworn officers were white and slightly more than 92% were male. Generally, the larger the police agency, the more likely it is to employ minority officers. However, diversity has increased in all population categories since 1987. Females represented 12.3% of all sworn officers in the nation's local police departments in 2016, which was up from 7.6% of officers in 1987.[33] Females comprised about 3% of the police chiefs in 2016, including about 8.5% of the chiefs in jurisdictions with 250,000 or more residents.[34] Figure 5.3 provides a breakdown of police employment in local agencies in 2016 by gender, race, and ethnicity.

Policing the City

According to the *Dictionnaire* of the French Academy (4th ed., 1762), the word "police" is from the Greek *polis,* or city, and refers to the regulation or control of a city for the purpose of providing security and comfort for the inhabitants.

Source: Alan Williams, *The Police of Paris, 1718–1789* (Baton Rouge and London: Louisiana State University Press, 1979), 9 and 11.

Local Police Duties The local police are the workhorses of the law enforcement system in America. They have many duties and tasks that will never be included in police detective novels or in movies about law enforcement. Their duties have been categorized in several different ways. One general grouping lists these four categories of local police duties:

1. *Law enforcement*—examples are investigating a burglary, arresting a car thief, serving a warrant, or testifying in court.
2. *Order maintenance* or *peacekeeping*—examples are breaking up a fight, holding back a crowd at a sporting event, or intervening in a domestic dispute before it gets violent.
3. *Service*—examples are taking people to the hospital, escorting funeral processions, delivering mail for city officials, or chasing bats out of a caller's house.
4. *Information gathering*—examples are determining neighborhood reactions to a proposed liquor license in the community, investigating a missing-child case, or investigating and reporting on a dangerous road condition.

FIGURE 5.2 Number of Full-Time Sworn Personnel in Local Departments, 2016

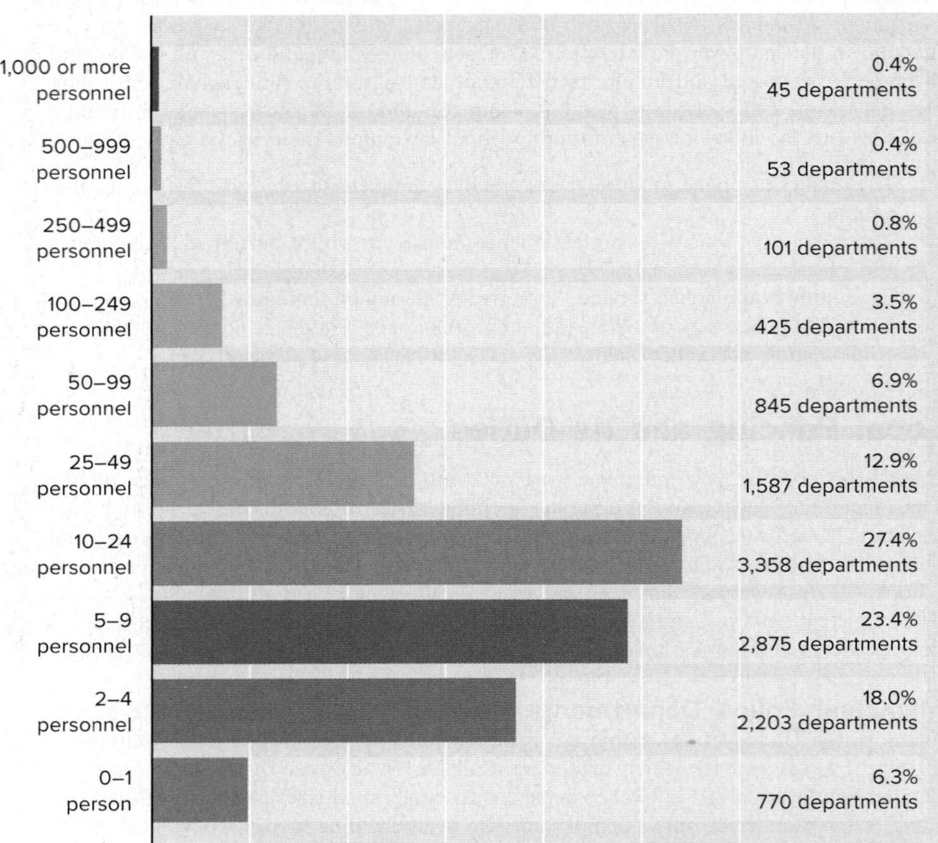

Source: Shelley S. Hyland and Elizabeth Davis, "Local Police Departments, 2016: Personnel," U.S. Department of Justice, *Bureau of Justice Statistics Bulletin*, October 2019, 3, Table 3, accessed October 26, 2019, https://www.bjs.gov/content/pub/pdf/lpd16p.pdf.

FIGURE 5.3 Characteristics of Local Full-Time Police Officers, 2016

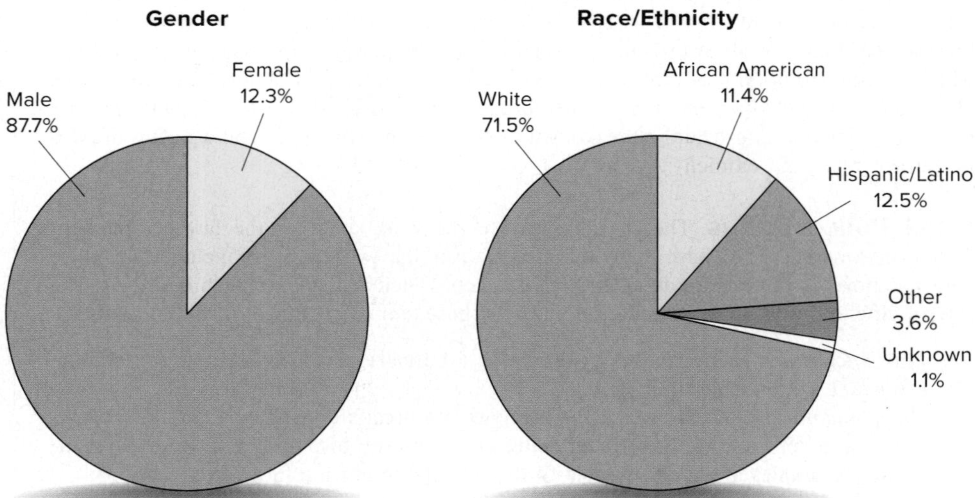

Source: Shelley S. Hyland and Elizabeth Davis, "Local Police Departments, 2016: Personnel," U.S. Department of Justice, *Bureau of Justice Statistics Bulletin*, October 2019, accessed October 26, 2019, https://www.bjs.gov/content/pub/pdf/lpd16p.pdf.

Arlene Gottfried/The Image Works

Local police departments, which make up the bulk of law enforcement agencies in America, are responsible for law enforcement, order maintenance, service, and information gathering. *How important is public relations for local police departments? Why?*

Some police academies teach recruits the duties of a police officer through the use of the acronym *PEPPAS*:

P–protect life and property (patrol a business district at night, keep citizens from a fire scene, recover and return lost property).
E–enforce the law (ensure traffic laws are obeyed, warn jaywalkers of the inherent danger, make out criminal complaints, seize illegal weapons).
P–prevent crime (give home security advice, patrol high-crime areas, work as a D.A.R.E. officer in schools).
P–preserve the peace (disband disorderly groups, have a visible presence at sporting events, intervene in neighbor conflicts).
A–arrest violators (apprehend fleeing suspects, give citations to alcohol permit-holders who sell to minors, conduct drug raids).
S–serve the public (give directions to travelers, deliver emergency messages, administer first aid).

There are literally dozens of other duties that the police of a city, town, or village carry out, and much of the work falls into the category of helping out when no one else seems to be available. Because the police are on duty 24 hours a day in nearly every community, they often are called on to perform services that have nothing to do with law enforcement. That round-the-clock availability also significantly affects the structure, work life, and activity of a police agency.

Organizational Structure How a police agency is structured depends on the size of the agency, the degree of specialization, the philosophy the leadership has chosen (such as community policing), the political context of the department (the form of municipal government), and the history and preferences of a particular community. Most medium- to large-staffed police agencies are subdivided into patrol, criminal investigation, traffic, juvenile, and technical and support services. Subspecialties include robbery, gangs, training, bombs, property, victims' services, jail, and mounted patrol. Table 5.2 lists the specialty units of the Seattle (Washington) Police Department (SPD). In 2017, the SPD had 1,448 sworn officers and 515 civilian employees.[35] The SPD usually requires officers to work at least their first 3 years in uniformed patrol assignments. Many officers choose to remain in patrol much longer or return to patrol after assignments in other units. Others choose to vary their assignments within the department. Some assignments outside of uniformed patrol are officer positions and some are detective positions. After 3 years in patrol, officers are eligible to attend a week-long detective school. Upon completion of the course, officers are placed on the Detective Eligibility List and are then available for assignment to a detective position. This is considered a lateral move, not a promotion.

Responsibilities of the First Modern Police Force: The Police of Paris

The responsibilities of the eighteenth-century Paris police—the first modern police force—were much more encompassing than their modern counterparts. The police of Paris provided many more municipal services than today's police, including deterrent patrol, investigation and intelligence (performed by police officials and a huge network of police spies), public services (such as street lighting, street cleaning, garbage collection, fire protection, operating the only pawn brokerage in the city, care of abandoned children, and procuring wet nurses for abandoned infants), inspection (e.g., of public works and buildings, horses, commercial and manufacturing establishments, and censorship of printed matter that challenged the Crown or Church), justice (e.g., holding court and resolving minor disputes), and police administration and communications.

Source: Alan Williams, *The Police of Paris, 1718–1789* (Baton Rouge and London: Louisiana State University Press, 1979), 66–136.

myth

The police spend most of their time and resources apprehending law violators and combating crime.

fact

Only about 10% of police time and resources are devoted to apprehending law violators and combating crime. Most of their time and resources are spent "keeping the peace," which means maintaining a police presence in the community, for example, by routine patrolling.

TABLE 5.2 Specialty Units of the Seattle Police Department

K-9	Traffic Enforcement	Special Activities (Seattle Center)
Auto Theft	Arson/Bomb Squad	Pawn Shop Detail
DUI Squad	Audit/Inspections	Traffic Collision Unit
Training	Anti-Crime Teams	Mounted (horses)
Bias Crimes	Community Police	Domestic Violence
Gang Unit	School Emphasis	Sexual Assault
Robbery	Vice	Burglary/Theft
Narcotics	Media Relations	Crime Analysis
Juvenile	Harbor (boats, divers)	Motorcycle
Homicide	Checks and Forgery	SWAT
C.S.I. Unit	Recruitment	Homeland Security
Background Unit	Violent Crimes	Fugitive Warrants
Internet Crimes	Child Exploitation	Cold Cases—Homicide
Missing Persons	Criminal Intelligence	Crisis Intervention Team

Source: No longer publicly posted. On file with authors.

To be promoted within the SPD, officers are required to take a civil service test, which is administered every other year. Tests are given for the rank of sergeant, lieutenant, and captain. The chief appoints assistant chiefs from the rank of captain. Officers may be promoted to sergeant after 5 years of experience with the SPD and passing the sergeant's test. Lieutenants must have at least 3 years' experience as sergeants, and captains must have at least 3 years' experience as lieutenants. A bachelor's degree may substitute for 1 year of experience but can be used only for one promotional exam.

In 2017, the Dallas (Texas) Police Department had 3,053 sworn officers and 605 civilian employees and, like the SPD, is large, sophisticated, and very specialized.[36] For example, it has a separate detective unit for each major category of crime. Evidence technicians collect and preserve evidence during the preliminary investigation of a crime. An entire contingent of officers is assigned to traffic regulation and enforcement duties. Bicycle patrol officers work the popular West End entertainment and restaurant section downtown. The Dallas police even have sworn officers who serve as crime analysts and collect, analyze, map, and report crime data to enable better prevention and repression of crime by means of scientific deployment of officers and other strategies (CompStat). Figure 5.4 presents the organizational structure of the Dallas Police Department.

Most police agencies in the United States do not have or need elaborate organizational structures. Police officers on the beat are generalists, and when special circumstances arise, such as a homicide or a complex financial investigation, they usually can rely on state bureaus of investigation to assist them. Moreover, local cooperation pacts among departments in a particular region often provide for sharing resources and specialized assistance when needed.

The infrequent need for homicide investigation skills in communities of fewer than 30,000 people, for example, makes it impractical to train one or more officers in the methods of conducting a thorough death investigation. An officer so trained might have to wait an entire career to put into practice the acquired skills, and it is most likely that by the time they were needed, the officer would have forgotten them. The lack of a trained specialist for the infrequent complex investigation, however, is one of the major reasons criminal investigation services in small communities are not equal to those in larger police departments.

The question has been raised whether larger, regional police departments would be more efficient providers of police services. However, as you already have discovered, policing in America is a local concern, and that is not likely to change.

The police are organized militarily with regard to accountability, discipline, rank, dress, and decorum. Many people believe that the military structure of a chain of command may

FIGURE 5.4 Dallas Police Department Organization Chart

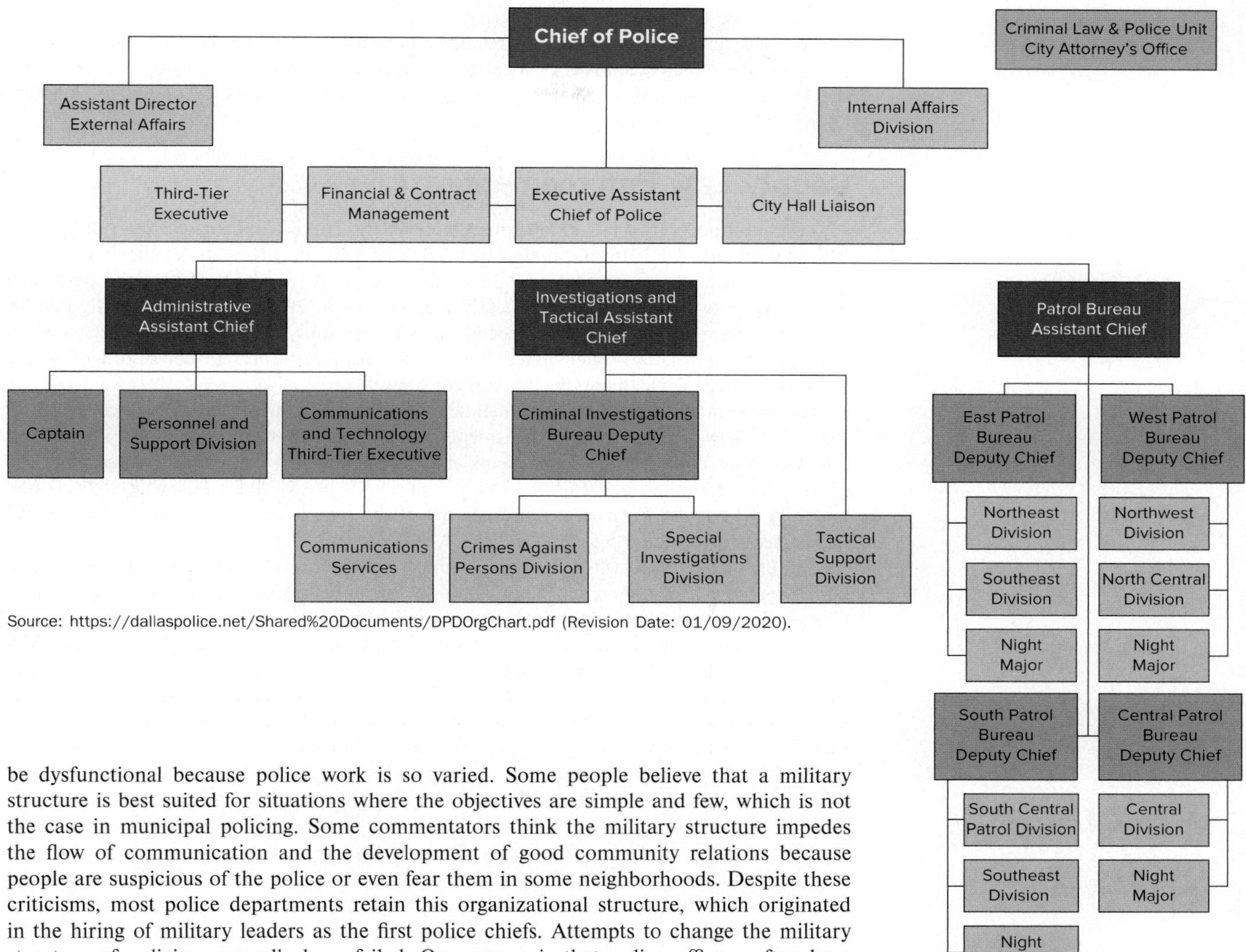

Source: https://dallaspolice.net/Shared%20Documents/DPDOrgChart.pdf (Revision Date: 01/09/2020).

be dysfunctional because police work is so varied. Some people believe that a military structure is best suited for situations where the objectives are simple and few, which is not the case in municipal policing. Some commentators think the military structure impedes the flow of communication and the development of good community relations because people are suspicious of the police or even fear them in some neighborhoods. Despite these criticisms, most police departments retain this organizational structure, which originated in the hiring of military leaders as the first police chiefs. Attempts to change the military structure of policing generally have failed. One reason is that police officers often have resisted any type of reorganization.

The Political Context of Policing A police department of any size is part of a larger government entity. Municipalities generally operate under one of four forms of municipal government[37]:

Strong mayor-council–Voters elect the mayor and the city council; the mayor appoints heads of departments.

Weak mayor-council–Voters elect the mayor and the city council; the city council appoints heads of departments.

City manager–Voters elect the city council and, in some cities, a mayor; the city council selects the city manager, who appoints heads of departments.

Commission–Voters elect a board of commissioners, who become the heads of departments; the commission or the voters may choose one commissioner to be mayor.

As you can see, the forms of municipal government vary in the amount of control citizens have over the municipality's leaders, the source of the executive authority of the chief of police, and the degree of insulation a chief of police has from interference by the executive head of the city (mayor or city manager) or the city council. Each form has advantages and disadvantages. At one time, it was thought that city manager government was the system under which the police were most likely to develop professionally, be free of political meddling from city lawmakers, and be insulated from local corruption. Although many progressive and effective police departments operate under a city manager form of government, other municipal forms of government have records of both success and failure in local police effectiveness and integrity.

You have probably noticed from reading newspapers, listening to radio, and watching television that chief executives of local police agencies have different titles, depending on the locale. Popular titles are chief of police (Kansas City), director of police (Dayton, Ohio), and commissioner (New York City).

County Law Enforcement

A substantial portion of law enforcement work in the United States is carried out by sheriffs' departments. In 2016, the nation had 3,063 sheriffs' departments, employing 359,843 full-time personnel–an increase of 2.26% compared to 2013. About 50% of the personnel in 2016 were sworn peace officers, nearly the same as in 2013. Sheriffs frequently employ part-time personnel who work as special deputies assisting with posses, disasters, county fairs, traffic control, and other duties. Sheriffs' departments represent about 20% of all the law enforcement departments in the United States.[38]

Like most municipal police departments, most sheriffs' departments in America are small. Figure 5.5 shows the number of departments and their respective sizes in 2016. Fifty-five and a half percent of all sheriffs' departments employed fewer than 25 sworn personnel.[39]

Sheriffs' personnel in 2016 were 75.8% white, 9.4% black, 10.5% Hispanic, 2.4% Other, and 1.8% Unknown. Females made up 13.6% of the sworn personnel working for sheriffs' departments (see Figure 5.6).

Sheriffs' departments often have employment qualifications similar to those of municipal police agencies. (Employment qualifications for police departments are described in Chapter 7.) A high school diploma or higher educational achievement was required by 98% of sheriffs' departments in 2013 (the latest year for which data were available; up from 89% in 2007). One percent of sheriffs' departments in 2013 required some college courses (down from 3% in 2007), and 5% of the departments required recruits to have a 2-year college degree (down from 7% in 2007). Fewer than one-half of 1% of departments required new recruits to have a 4-year college degree (the same as in 2007).[40]

In 2018, deputy sheriffs, on average, were paid between $30,143 and $73,072 a year. However, pay depended on the size of the population served, location, and experience level. For example, entry-level deputy sheriffs averaged $38,000, while the most experienced deputy sheriff averaged $55,000. In Los Angeles, deputy sheriffs earned much more than the $39,536 national average. Deputy sheriffs working in Los Angeles in 2018 earned an average of $83,015. Generally, salaries of police officers are higher than salaries of deputy sheriffs.[41]

FIGURE 5.5 Number of Sworn Personnel in Sheriffs' Departments, 2016

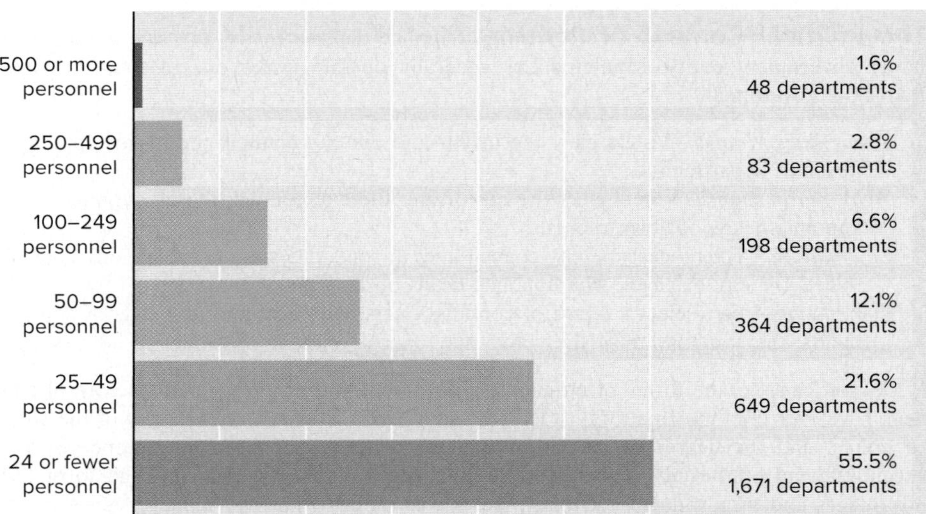

Source: Connor Brooks, "Sheriffs' Offices, 2016: Personnel," U.S. Department of Justice, *Bureau of Justice Statistics Bulletin*, October 2019, accessed October 26, 2019, https://www.bjs.gov/content/pub/pdf/so16p.pdf.

FIGURE 5.6 Characteristics of Sheriffs' Personnel, 2016

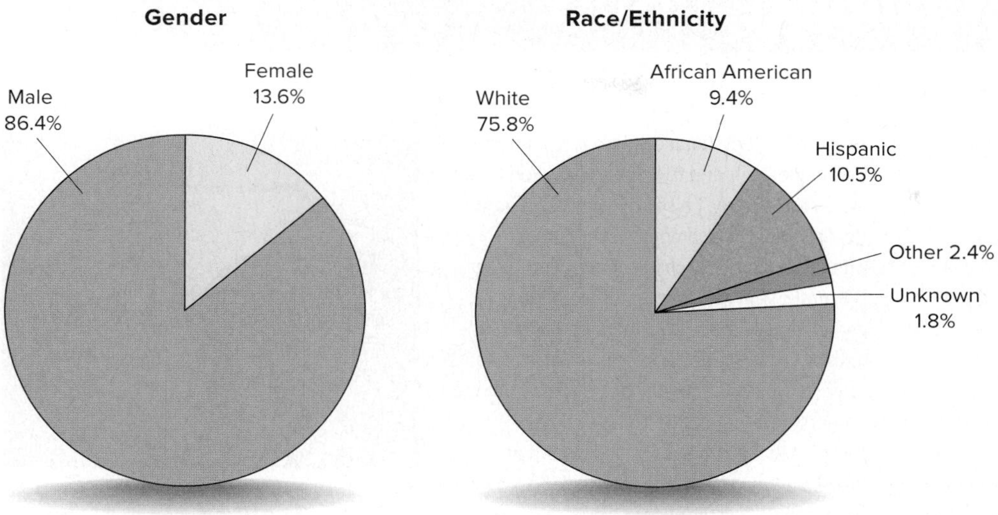

Gender

Male 86.4%
Female 13.6%

Race/Ethnicity

White 75.8%
African American 9.4%
Hispanic 10.5%
Other 2.4%
Unknown 1.8%

Source: Connor Brooks, "Sheriffs' Offices, 2016: Personnel," U.S. Department of Justice, *Bureau of Justice Statistics Bulletin*, October 2019, accessed October 26, 2019, https://www.bjs.gov/content/pub/pdf/so16p.pdf.

County Law Enforcement Functions The sheriff and department personnel perform functions that range from investigation to supervision of sentenced offenders. Even in the smallest departments, sheriffs are responsible for investigating crimes and enforcing the criminal and traffic laws of the state. They also perform many civil process services for the court, such as serving summonses, warrants, and various writs. In addition, they provide courtroom security and confine and transport prisoners. The larger the sheriff's department, the more confinement and corrections responsibilities it has. Sheriffs' departments frequently operate the county jail, which houses hundreds and even thousands of prisoners, depending on the particular county. In some counties, the sheriff's department shares law enforcement duties with a separate police department.

Female Sheriffs

Before 1992, no woman had ever been elected to the position of sheriff in the United States. The first two were elected in 1992: The first was Jackie Barrett in Fulton County, Georgia, and the second was Judy Pridgen in Saline County, Arkansas.

Source: Matthew J. Hickman and Brian A. Reaves, *Sheriffs' Offices, 2003,* Bureau of Justice Statistics (Washington, DC: U.S. Government Printing Office, 2006).

Politics and County Law Enforcement Most sheriffs are directly elected and depend on an elected board of county commissioners or supervisors for their funding and some oversight of their operations. Sheriffs generally have a freer hand in running their agencies than police chiefs do. In many counties, local politics govern the operation of the sheriff's department, and the sheriff must operate as a partisan politician to remain in office. The authority to appoint special deputies and to award patronage jobs contributes to the sheriff's power and influence in a county.

State Law Enforcement

Filling the complement of law enforcement agencies in a particular state are one or more state law enforcement agencies, which provide criminal and traffic law enforcement, as well as other services peculiar to the needs of that state government. In 2016, the 50 primary state law enforcement agencies had 91,097 full-time employees, of which about two-thirds were full-time sworn officers. The 91,097 full-time employees represented nearly a 3% increase from 2013. The California Highway Patrol is the largest state law enforcement agency with more than 11,000 personnel, of which nearly 70% are full-time sworn officers.[42]

For the most part, each state has chosen one of three models for providing law enforcement services at the state level. The first model is the **state police model** in which the agency and its officers have essentially the same enforcement powers as local police in the state and can work cases and enforce the law anywhere within the state's boundaries. One of the best-known state police agencies is the Texas Rangers, part of the Texas Department of Public Safety, which also employs state troopers to enforce criminal and traffic laws. The Rangers usually focus on special and complex investigations, such as the Branch Davidian case in Waco in 1993. A number of states have placed some restrictions on state police activities to avoid clashes with local politicians and local police agencies.

state police model A model of state law enforcement services in which the agency and its officers have the same law enforcement powers as local police but can exercise them anywhere within the state.

Careers in Criminal Justice

Deputy Sheriff

Courtesy of Terrence J. James

My name is Terrence J. James, and I am a deputy sheriff in Cumberland County, North Carolina. I have been with the Cumberland County Sheriff's Office (CCSO) for 12 years. Prior to joining the CCSO, I served 24 years in the U.S. Air Force. After leaving the Air Force, I earned a B.S. degree in criminal justice from Fayetteville State University in Fayetteville, North Carolina. I also have earned a Basic Law Enforcement Training and Crisis Intervention Training Certification from Fayetteville Technical Community College and Basic, Intermediate, and Advanced Law Enforcement Certification from the North Carolina Sheriff's Education and Training Standard Commission.

Among the many assignments I have held with the CCSO, the most rewarding and impactful was as a School Resource Officer. In this role, you are in a position to shape today's youth. The students and administrators rely on you to be more than just an officer upholding the laws. You are expected to be a leader by example, an advisor, a counselor, a supporter, and to most a friend to listen. In addition to being a road deputy, I also have served in units dealing with civil, community policing, sex offenders, fugitives, child support, and courts. Currently, I am assigned to the detention center, one of the most unheralded positions in the CCSO. The position calls for many talents to be successful. Empathy and integrity are the cornerstones of the position. I say that because the inmates, as well as their family members who visit the facility, test those two traits daily.

If you are thinking of becoming a deputy sheriff, you must realize that all of your actions and decisions from day one will be under scrutiny. You will have an enormous amount of responsibility, power, pressure, prestige, and privilege. Your actions or inactions will have repercussions. Failure to act may cause you or someone else to lose their life. This can make you fearful of the profession. During difficult times, you must rely on your training and experience. You must never stop learning your craft. You should seek out those who have more experience and further your education. Empathy is one of the most important skills to have as a deputy sheriff. You also must have a "thick skin" when dealing with bosses and the public. Perhaps the hardest part of the profession is not to take other people's actions personally. And, finally, always go home safely!!!

Chet Gordon/The Image Works

State police agencies in every state except Hawaii have statewide jurisdiction and may be set up according to state police, highway patrol, or department of public safety models. *Are state police agencies necessary? Why or why not?*

Careers in Criminal Justice

Kentucky Department of Fish and Wildlife Officer

My name is Myra Minton. I am a captain with the Kentucky Department of Fish and Wildlife Resources Law Enforcement Division. As a captain, I am responsible for the oversight of a Law Enforcement District and the officers that patrol there. A conservation officer is responsible for patrolling the fields, forests, lakes, and streams. During patrol, an officer looks for license and safety violations and enforces regulations that are in place to protect Kentucky's natural resources. An officer is also charged with enforcing boating regulations to maintain safety on the waterways of the Commonwealth in addition to other general peace officer duties. I graduated from Ohio Northern University with a bachelor of science degree in environmental studies before being hired by the Kentucky Department of Fish and Wildlife Resources Law Enforcement Division.

The requirements for becoming a Kentucky conservation officer include having either a 4-year degree, 4 years of law enforcement experience, or 4 years of experience in a fish and wildlife–related field. Candidates must be able to pass a written exam, a physical fitness test, a psychological exam, and a polygraph examination. Upon hiring, a recruit spends the next 30 weeks in two separate training programs. The Kentucky Department of Criminal Justice Training provides law enforcement agencies with a Basic Law Enforcement Training Program that spans 18 weeks. Upon successful completion of the Basic Law Enforcement Training Program, a conservation officer recruit attends a 12-week Fish and Wildlife Law Enforcement Academy. During this phase of training the recruits learn fish and wildlife laws, boating laws, boat operation, all-terrain vehicle (ATV) operation, wildlife identification, wildlife forensics, water survival, defensive tactics, and federal laws that pertain to their job.

As in most professions, there are high points and low points. A drawback with regard to the job as a conservation officer is that you work when most people are off. The busiest time is normally on a weekend or holiday, which can affect your personal life. But the realization that every day when you go to work there is something new makes and keeps the job interesting. You just never know what you are going to find.

Myra Minton

If you are interested in a job in fish and wildlife law enforcement, contact the agency you are interested in working for. Meet the people, ask to ride with an officer, establish a relationship, and find out what the requirements are.

Would you find the job as a fish and wildlife officer appealing? Why or why not?

The second model for state law enforcement services is the **highway patrol model** in which officers focus almost exclusively on highway traffic safety, enforcement of the state's traffic laws, and investigation of accidents on the state's roads and highways. Even highway patrols, however, may retain responsibility for investigating criminal violations on state property and in state institutions or for conducting drug interdictions.

highway patrol model A model of state law enforcement services in which officers focus on highway traffic safety, enforcement of the state's traffic laws, and the investigation of accidents on the state's roads, highways, and property.

States that employ the highway patrol model often have other state law enforcement agencies with narrow service mandates, such as these:

- Bureaus of criminal investigation (to investigate white-collar and organized crime, narcotics, and so on).
- State criminal identification services.
- Forest, game, and watercraft protection services.
- Alcoholic beverage control and enforcement.
- Crime laboratory and criminalistics services.
- Driver's license examinations.
- Drug interdiction activities.
- Peace officer training and certification.

Both state police and highway patrol agencies help regulate commercial traffic, conduct bomb investigations, protect the governor and the capitol grounds and buildings, and administer computer-based information networks for the state, which link up with the National Crime Information Center (NCIC) run by the FBI.

The third model for state law enforcement services is the *department of public safety model.* Departments of Public Safety (DPS) are often complex organizations composed of several agencies or divisions. For example, the Alabama Department of Public Safety is composed of six divisions: administrative, bureau of investigation, driver license, highway patrol, service, and special projects. The administrative division is responsible for financial services, inspections, legal, personnel, public information, recruiting, capitol police, and dignitary protection. The service division is charged with managing and maintaining the DPS facilities and infrastructure, supporting the mechanized fleet, communications, supply, aviation, and photographic services. The duties of the other divisions should be self-explanatory.[43]

Some tension always seems to exist between state police agencies and local law enforcement over legal jurisdiction and recognition for conducting investigations and making arrests. Recall that policing in America and the political system that governs it are local. Much of the resentment by locals over state interference is similar to the suspicions and doubts concerning federal involvement at the local level.

A significant function performed by a special category of state law enforcement officers is university or campus policing. Some of the large state and private universities and colleges have full-blown police agencies with many special subdivisions. They are very much like municipal police departments–and rightly so because a community's problems with crime and public order do not end at the university gate.

Federal Law Enforcement

Everyone has heard of a few of the better-known federal law enforcement agencies. The FBI, the U.S. Secret Service, and even the T-men and T-women of the Treasury Department have had their own television shows, creating wider public recognition of those agencies. The unrelenting War on Drugs has brought to the attention of the American public the activities of the DEA. There also are other, lesser-known federal police agencies. Specific statutes narrowly define their law enforcement jurisdictions, and their work is unlikely to come to the attention of most American citizens.

Three major differences exist between federal law enforcement and the local and state police agencies with which we are likely to be more familiar. First, federal agencies such as the FBI operate across the entire nation and even have agents serving abroad. Second, federal police agencies do not, as a rule, have the peacekeeping or order maintenance duties typical in local policing. Finally, some federal law enforcement agencies have extremely narrow jurisdictions. (The U.S. Supreme Court Police, for example, provide protective and investigative services for the Supreme Court only.)

As mentioned previously, at the end of 2018, the more than 70 federal law enforcement agencies employed nearly 163,000 personnel, a decrease of about 4% from the number of personnel in 2016.[44] Combined, those 70-plus agencies cost taxpayers about $29 billion in 2015, about the same amount as in 2012.[45] (Note that estimates of the number of federal law enforcement agencies vary. For example, the Bureau of Justice Statistics reports, "The 2008 Census of Federal Law Enforcement Officers collected data from 73 agencies, including 33 offices of inspectors general." Presumably, not all agencies responded to the survey. On the other hand, the Federal Law Enforcement Training Center (FLETC) website claims, "The FLETC serves as an interagency law enforcement training organization for 91 Federal agencies." The FLETC does not provide training for all federal law enforcement agencies (e.g., the FBI). Therefore, in this textbook, we utilize "more than 70" as the number of federal law enforcement agencies.)[46]

Table 5.3 shows some of the largest federal law enforcement agencies. U.S. Customs and Border Protection (CBP), a component of the U.S. Department of Homeland Security, is the largest of the agencies with 60,014 employees, including 19,555 border patrol agents (as of March 2019).[47] Compared to the number of employees and border patrol agents employed at year-end 2014, the 60,014 total employees in 2018 represent an increase of less than 1%, and the 19,555 border patrol agents in 2018 represent a decrease of slightly more than 6%.[48] CBP is responsible for protecting more than 5,000 miles of border with Canada, 1,900 miles of border with Mexico, and 95,000 miles of shoreline. In 2018, CBP apprehended 404,142 illegal immigrants, down 17% from 2014, and down almost 76% of the record number apprehended in 2000. Of the apprehensions in 2018, approximately 94% of them were of individuals from Mexico (38%), El Salvador (8%), Guatemala (29%), and Honduras (19%), and nearly all of the apprehensions were along the southwest border. About 2.5% of the individuals apprehended were wanted for serious crimes. In addition, CBP officers and agents seized 4,657 pounds of narcotics, a nearly 100% decrease from 2014, and $290,411 in undeclared currency, also a nearly 100% decrease from 2014.

TABLE 5.3 Largest Federal Law Enforcement Agencies, Department of Government, and Number of Personnel

Agency	Department	Number of Personnel	Year of Data
U.S. Customs and Border Protection	Homeland Security	60,014, including 19,555 Border Patrol Agents	2019
Federal Bureau of Prisons	Justice	35,469	2019
Federal Bureau of Investigation	Justice	34,694, including 12,927 Special Agents	2019
U.S. Immigration and Customs Enforcement	Homeland Security	More than 20,000	2019
National Park Service	Interior	More than 20,000	2019
Drug Enforcement Administration	Justice	10,169, including 4,924 Special Agents	2019
U.S. Secret Service	Homeland Security	About 6,500, including about 3,200 Special Agents and about 1,300 Uniformed Division Officers	2019
Bureau of Alcohol, Tobacco, Firearms, and Explosives	Justice	5,113, including 2,623 Special Agents and 828 Investigators	2017
U.S. Marshals Service	Justice	5,092, including 94 U.S. Marshals and 3,547 Deputy U.S. Marshals	2019
Internal Revenue Service, Criminal Investigation	Treasury	2,994, including 2,159 Special Agents	2017

Sources: U.S. Customs and Border Protection. "Snapshot: A Summary of CBP Facts and Figures," March 2019. Accessed April 22, 2019, https://www.cbp.gov/newsroom/stats/typical-day-fy2018; Federal Bureau of Prisons, "About Our Agency." Accessed April 22, 2019, https://www.bop.gov/about/agency/; Federal Bureau of Investigation "FBI Budget Request for Fiscal Year 2019." Accessed April 22, 2019, https://www.fbi.gov/news/testimony/fbi-budget-request-for-fiscal-year-2019.pdf; U.S. Immigration and Customs Enforcement. "Who We Are–Overview." Accessed April 22, 2019, https://www.ice.gov/about: www.ice.gov/about; U.S. Drug Enforcement Agency. "Staffing and Budget." Accessed April 22, 2019, https://www.dea.gov/staffing-and-budget; U.S. Secret Service. "Frequently Asked Questions." Accessed April 22, 2019, https://www.secretservice.gov/about/faqs/; U.S. Marshals. "Facts and Figures 2019." Accessed April 22, 2019, https://www.usmarshals.gov/duties/factsheets/facts.pdf; Bureau of Alcohol, Tobacco, Firearms and Explosives. "Fact Sheet: Staffing and Budget." Accessed April 22, 2019, https://www.atf.gov/resource-center/fact-sheet/fact-sheet-staffing-and-budget; National Park Service. "About Us: Organizational Structure of the National Park Service." Accessed April 22, 2019, https://www.nps.gov/aboutus/index.htm; Internal Revenue Service. IRS: Criminal Investigation: 2017 Annual Report, "IRS: CI 2017 Snapshot." Accessed April 22, 2019, https://www.irs.gov/pub/ci/2017_criminal_investigation_annual_report.pdf.

Some of the other federal law enforcement agencies already have been discussed in this chapter or will be discussed in later chapters. Table 5.4 provides a more extensive list of federal agencies and offices of inspectors general that employ full-time personnel with arrest and firearm authority.

Dan Callister/Shutterstock

Federal law enforcement agencies investigate violations of federal law, enforce laws that involve interstate crimes, and conduct activities to prevent and control domestic and international terrorism. *How do the jobs of federal law enforcement agents differ from the jobs of other law enforcement officers?*

TABLE 5.4 Federal Agencies and Offices of Inspectors General That Employ Full-Time Personnel with Arrest and Firearm Authority (from largest to smallest)

Federal Agencies	Offices of Inspectors General
U.S. Customs and Border Protection	U.S. Postal Service
Federal Bureau of Prisons	Department of Health and Human Services
Federal Bureau of Investigation	Department of Defense
U.S. Immigration and Customs Enforcement	Department of the Treasury, Tax Administration
U.S. Secret Service	Social Security Administration
Administrative Office of the U.S. Courts	Department of Housing and Urban Development
Drug Enforcement Administration	Department of Agriculture
U.S. Marshals Service	Department of Labor
Veterans Health Administration	Department of Homeland Security
Internal Revenue Service, Criminal Investigation	Department of Veterans Affairs
Bureau of Alcohol, Tobacco, Firearms, and Explosives	Department of Justice
U.S. Postal Inspection Service	Department of Transportation
U.S. Capitol Police	Department of Education
National Park Service—Rangers	General Services Administration
Bureau of Diplomatic Security	Department of the Interior
Pentagon Force Protection Agency	National Aeronautics and Space Administration
U.S. Forest Service	Department of Energy
U.S. Fish and Wildlife Service	Environmental Protection Agency
National Park Service—U.S. Park Police	Federal Deposit Insurance Corporation
National Nuclear Security Administration	Small Business Administration
U.S. Mint Police	Department of State
Amtrak Police	Office of Personnel Management
Bureau of Indian Affairs	Department of the Treasury
Bureau of Engraving and Printing	Tennessee Valley Authority
Environmental Protection Agency	Department of Commerce
Food and Drug Administration	U.S. Railroad Retirement Board
National Oceanic and Atmospheric Administration	Agency for International Development
Tennessee Valley Authority	Nuclear Regulatory Commission
Federal Reserve Board	Corporation for National and Community Service
U.S. Supreme Court	National Science Foundation
Bureau of Industry and Security	National Archives and Records Administration
National Institutes of Health	Government Printing Office
Library of Congress	Library of Congress
Federal Emergency Management Agency	
National Aeronautics and Space Administration	
Government Printing Office	
National Institute of Standards and Technology	
Smithsonian National Zoological Park	
Bureau of Reclamation	

Source: Brian A. Reaves, *Federal Law Enforcement Officers, 2008,* U.S. Department of Justice, Bureau of Justice Statistics Bulletin (Washington, DC: Government Printing Office, June 2012), 2, Table 1, 5, Table 2, 6, Table 3, http://bjs.ojp.usdoj.gov/content/pub/pdf/fleo08.pdf (accessed October 21, 2012). Agencies and offices are listed by the number of full-time personnel with arrest and firearm authority, from the largest to the smallest.

Training Federal Law Enforcement Officers The Federal Law Enforcement Training Center (FLETC) is the largest law enforcement-training establishment in the United States.[49] It provides some or all of the training for a majority of federal law enforcement agencies, as well as for many state, local, and international law enforcement agencies. Notable exceptions are the FBI and DEA, which train their special agents at their respective academies in Quantico, Virginia. Until 2003, when it became a part of the Department of Homeland Security, FLETC was a bureau of the Treasury Department.

The FLETC got its start in the late 1960s, when a federal government study disclosed that the training of most federal law enforcement personnel was inadequate at worst and substandard at best. With only a few exceptions, part-time instructors, on an irregular basis, conducted most training in inferior facilities. Much of the training duplicated the training of other federal agencies or was inconsistent with it. A government task force recommended that a federal law enforcement training center be established to provide the training for most federal law enforcement personnel. The center would have a professionally trained, full-time staff that offered consistent and high-quality programs in state-of-the-art facilities.

FLETC first opened in Washington, DC, in 1970. That year it graduated 848 students. In 1975, its first full year of operation at its current headquarters location on a 1,500-acre campus in Glenn County ("Glynco"), Georgia (near Brunswick, Georgia), it graduated more than 5,000 students. In 1989, the FLETC Office of Artesia Operations (OAO) in New Mexico opened to provide training for the Bureau of Indian Affairs and agencies with a large number of officers in the western United States. It also hosts the U.S. Border Patrol Academy. A temporary satellite training campus was opened in Charleston, South Carolina, in 1995, to train an increasing number of Immigration and Naturalization Service (INS) and border patrol agents. In 2003, the Charleston facility became the third FLETC residential campus and, in 2004, all of the border patrol training operations were moved to FLETC-Artesia. Besides some of the same training programs offered at FLETC-Glynco and FLETC-Artesia, the FLETC-Charleston facility specializes in maritime law enforcement training. A fourth training facility that was developed in 2002 in Cheltenham, Maryland, is used primarily for in-service and requalification training for officers and agents in the Washington, DC, area. It also serves as the new home for the U.S. Capitol Police Training Academy. FLETC also provides training at other temporary sites in the United States and in foreign countries. Currently, FLETC graduates approximately 60,000 students annually.

Federal Law Enforcement Training Center

To learn more about the FLETC, visit its website at www.fletc.gov. Access "Training Programs."

Which programs, if any, would be of interest to you? Why?

Because basic training requirements for federal officers vary by agency and by position within agencies, FLETC provides more than 150 different agency-specific training programs. About half the instructors are permanent employees, and the other half are federal officers on short-term assignment from their respective agencies. Depending on the agency, classroom instruction ranges from about 8 to 22 weeks for criminal investigators and from 4 to 26 weeks for patrol officers. Field training requirements range from 2 weeks to 6 months for patrol officers and up to 2 years for investigators.

THINKING CRITICALLY

1. Do you think one city, county, or state law enforcement officer for every 419 U.S. residents is enough? Why or why not?
2. What do you think are the pros and cons of working at the local, state, and federal levels of law enforcement?
3. Do you think that any one of the three major areas of law enforcement (local, state, federal) is most prestigious? Why?

The Department of Homeland Security

The U.S. Congress responded to the terrorist attacks of September 11, 2001 (described in Chapter 6), by enacting the Homeland Security Act of 2002.[50] Among other provisions, such as allowing commercial pilots to carry guns in cockpits, the act established the

Department of Homeland Security (DHS). According to the legislation, this new executive department was created to:

1. Prevent terrorist attacks within the United States.
2. Reduce the vulnerability of the United States to terrorism.
3. Minimize the damage, and assist in the recovery, from terrorist attacks that do occur within the United States.
4. Carry out all functions of entities transferred to the Department, including by acting as a focal point regarding natural and manmade crises and emergency planning.
5. Ensure that the functions of the agencies and subdivisions within the Department that are not related directly to securing the homeland are not diminished or neglected except by an explicit act of Congress.
6. Ensure that the overall economic security of the United States is not diminished by efforts, activities, and programs aimed at securing the homeland.
7. Monitor connections between illegal drug trafficking and terrorism, coordinate efforts to sever such connections, and otherwise contribute to efforts to interdict illegal drug trafficking.

The act also stipulates that "primary responsibility for investigating and prosecuting acts of terrorism shall be vested not in the Department, but rather in Federal, State, and local law enforcement agencies with jurisdiction over the acts in question."

The creation of the DHS represents the most dramatic transformation of the U.S. government since 1947, when President Harry S. Truman combined the various branches of the U.S. military into the Department of Defense (DOD). On an even grander scale, President George W. Bush combined 22 previously separate domestic agencies into the new department to protect the country from future threats. To head the new department, President Bush selected former Pennsylvania Governor Tom Ridge. On February 15, 2005, Michael Chertoff, former U.S. Court of Appeals judge, was sworn in as the second secretary of the DHS; on January 21, 2009, former Arizona Governor Janet Napolitano became the third secretary of the DHS; on December 23, 2013, Jeh Johnson, former general counsel for the Department of Defense, became the fourth secretary of the DHS; on January 20, 2017, John F. Kelly, former 4-star general and commander of the United States Southern Command, became the fifth secretary of the DHS; on December 6, 2017, Kirstjen Nielsen, former Deputy Chief of Staff in the Trump White House, was sworn in as the sixth secretary of DHS. Secretary Nielsen abruptly resigned her position on April 7, 2019. To fill her role, President Trump appointed Customs and Border Protection Commissioner Kevin McAleenan as acting DHS secretary.

In July 2005, then-secretary Michael Chertoff announced a six-point agenda for the DHS, which he developed to ensure that the Department's policies, operations, and

REUTERS/Mike Segar/Newscom

A UH-60 Black Hawk helicopter from the Department of Homeland Security's Bureau of Immigration and Customs Enforcement patrolling restricted airspace over the New York metropolitan area to detect unauthorized intrusions. *Is this an effective defensive strategy against terrorism? Why or why not?*

structures are coordinated in the best way to address both present and future threats to the United States. The agenda items listed below continue to guide the DHS:

1. Increase overall preparedness, particularly for catastrophic events.
2. Create better transportation security systems to move people and cargo more securely and efficiently.
3. Strengthen border security and interior enforcement and reform immigration processes.
4. Enhance information sharing with our partners.
5. Improve DHS financial management, human resource development, procurement, and information technology.
6. Realign the DHS organization to maximize mission performance.[51]

Currently, the DHS's vision "is to ensure a homeland that is safe, secure, and resilient against terrorism and other hazards." The DHS's mission is sixfold, five of which have related goals:

1. Prevent terrorism and enhance security.
   a. Prevent terrorist attacks.
   b. Prevent the unauthorized acquisition, importation, movement, or use of chemical, biological, radiological, and nuclear materials and capabilities within the United States.
   c. Reduce the vulnerability of critical infrastructure and key resources, essential leadership, and major events to terrorist attacks and other hazards.
2. Secure and manage our borders.
   a. Effectively secure U.S. air, land, and sea points of entry.
   b. Safeguard and streamline lawful trade and travel.
   c. Disrupt and dismantle transnational criminal and terrorist organizations.
3. Enforce and administer our immigration laws.
   a. Prioritize the identification and removal of criminal aliens who pose a threat to public safety and target employers who knowingly and repeatedly break the law.
4. Safeguard and secure cyberspace.
   a. Analyze and reduce cyber threats and vulnerabilities.
   b. Distribute threat warnings.
   c. Coordinate the response to cyber incidents to ensure that our computers, networks, and cyber systems remain safe.
5. Ensure resilience to disasters.
   a. Bolster information sharing and collaboration.
   b. Provide grants, plans, and training to our homeland security and law enforcement partners.
   c. Facilitate rebuilding and recovery along the Gulf Coast.
6. Focus on maturing and strengthening the homeland security enterprise itself.[52]

Department of Homeland Security

To learn more about the Department of Homeland Security, visit its website at www.dhs.gov/.

Based on what you have learned, do you think the DHS will be effective in preventing terrorism on American soil?

Department Components

The Department of Homeland Security, which has been reorganized several times since its inception, comprises the following major components.[53]

1. The *U.S. Citizenship and Immigration Services* (USCIS) administer the nation's lawful immigration system, safeguarding its integrity and promise by efficiently and fairly adjudicating requests for immigration benefits while protecting Americans, securing the homeland, and honoring our values.
2. The *United States Coast Guard* (USCG) protects the maritime economy and the environment, defends our maritime borders, and saves those in peril.
3. The *United States Customs and Border Protection* (CBP) has the priority mission of keeping terrorists and their weapons out of the United States. It also has a responsibility for securing and facilitating trade and travel while enforcing hundreds of U.S. regulations, including immigration and drug laws.
4. The *Cybersecurity and Infrastructure Security Agency* (CISA) leads the national effort to defend critical infrastructure against the threats of today, while working with partners across all levels of government and in the private sector to secure against the evolving risks of tomorrow.

5. The *Federal Emergency Management Agency* (FEMA) supports our citizens and first responders to ensure that as a nation we work together to build, sustain, and improve our capability to prepare for, protect against, respond to, recover from, and mitigate all hazards.
6. The *Federal Law Enforcement Training Center* (FLETC) provides career-long training to law enforcement professionals to help them fulfill their responsibilities safely and proficiently.
7. The *United States Immigration and Customs Enforcement* (ICE) promotes homeland security and public safety through the criminal and civil enforcement of federal laws governing border control, customs, trade, and immigration.
8. The *United States Secret Service* (USSS) safeguards the nation's financial infrastructure and payment systems to preserve the integrity of the economy, and protects national leaders, visiting heads of state and government, designated sites, and National Special Security Events.
9. The *Transportation Security Administration* (TSA) protects the nation's transportation systems to ensure freedom of movement for people and commerce.
10. The *Management Directorate* is responsible for budget, appropriations, expenditure of funds, accounting and finance; procurement; human resources and personnel; information technology systems; facilities, property, equipment, and other material resources; providing biometric identification services; and identification and tracking of performance measurements relating to the responsibilities of the Department.
11. The *Science and Technology Directorate* is the primary research and development arm of the Department. It provides federal, state, and local officials with the technology and capabilities to protect the homeland.
12. The *Countering Weapons of Mass Destruction Office* works to counter attempts by terrorists or other threat actors to carry out an attack against the United States or its interests using a weapon of mass destruction.
13. The *Office of Intelligence and Analysis* equips the Homeland Security Enterprise with the timely intelligence and information it needs to keep the homeland safe, secure, and resilient.
14. The *Office of Operations Coordination* provides information daily to the Secretary of Homeland Security, senior leaders, and the homeland security enterprise to enable decision-making; oversees the National Operations Center; and leads the Department's Continuity of Operations and Government Programs to enable continuation of primary mission essential functions in the event of a degraded or crisis operating environment.

FYI **TSA and Federal Air Marshals**

The Transportation Security Administration (TSA) was created in response to the terrorist attacks of 9/11. The TSA has more than 60,000 employees, including more than 43,000 transportation security officers, more than 1,200 transportation security inspectors, more than 2,300 behavior detection officers, more than 900 canine teams, and more than 360 explosives specialists. The TSA also employs federal air marshals and VIPR (Visible Intermodal Prevention and Response) teams, but does not provide the number of employees. It simply states that "thousands of federal air marshals are deployed every day."

Source: Transportation Security Administration. "Factsheet: TSA at a Glance." Accessed April 23, 2019, https://www.tsa.gov/sites/default/files/resources/tsaatglance_factsheet.pdf.

Office of the Secretary The *Office of the Secretary* oversees DHS efforts to counter terrorism and enhance security, secure and manage our borders while facilitating trade and travel, enforce and administer our immigration laws, safeguard and secure cyberspace, build resilience to disasters, and provide essential support for national and economic security–in coordination with federal, state, local, international, and private sector partners. The Office of the Secretary includes the following multiple offices that contribute to the overall Homeland Security mission.[54]

1. The *Office for Civil Rights and Civil Liberties* provides legal and policy advice to Department leadership on civil rights and civil liberties issues, investigates and resolves complaints, and provides leadership to Equal Employment Opportunity Programs.
2. The *Office of the Citizenship and Immigration Services Ombudsman* is dedicated to improving the quality of citizenship and immigration services delivered to the public by providing individual case assistance, as well as making recommendations to improve the administration of immigration benefits by U.S. Citizenship and Immigration Services.
3. The *Office of the Executive Secretary* provides all manner of direct support to the Secretary and Deputy Secretary, as well as related support to leadership and management across the Department.
4. The *Office of the General Counsel* integrates more than 2,500 attorneys from throughout the Department into an effective, client-oriented, full-service legal team. The Office of the General Counsel comprises a headquarters office with subsidiary divisions and the legal offices for nine Department components.

5. The *Joint Requirements Council* validates capability gaps, associated with operational requirements and proposed solution approaches to mitigate those gaps through the Joint Requirements Integration and Management System, leveraging opportunities for commonality to enhance operational effectiveness directly and better inform the DHS's main investment pillars.
6. The *Office of Legislative Affairs* serves as primary liaison to members of Congress and their staffs, the White House and Executive Branch, and to other federal agencies and governmental entities that have roles in assuring national security.
7. The *Office of the Military Advisor* provides counsel and support to the Secretary and Deputy Secretary in affairs relating to policy, procedures, preparedness activities, and operations between DHS and the Department of Defense.
8. *Partnership and Engagement* coordinates the DHS's outreach efforts with key stakeholders nationwide, ensuring a unified approach to external engagement.
9. The *Privacy Office* protects individuals by embedding and enforcing privacy protections and transparency in all DHS activities.
10. The *Office of Public Affairs* coordinates the public affairs activities of all of the Department's components and offices, and serves as the federal government's lead public information office during a national emergency or disaster.
11. The *Office of Strategy, Policy, and Plans* serves as a central resource to the Secretary and other Department leaders for strategic planning and analysis, and facilitation of decision-making on the full breadth of issues that may arise across the dynamic homeland security enterprise.

Advisory Panels and Committees Advisory panels and committees provide advice and recommendations on mission-related topics from academic engagement to privacy.[55]

1. *Academic Engagement* involves the academic community in homeland security efforts.
2. *Counternarcotics Coordinating Council* serves as an advisory body to the Director of the Office for Counternarcotics Enforcement, and to the Secretary on counternarcotics issues facing the Department.
3. *Counterterrorism Committees and Working Groups* provide advice and recommendations on counterterrorism issues.
4. *DHS Employee Committees and Working Groups* support cooperative and productive labor-management relations.
5. *Immigration Committees and Working Groups* provide advice and recommendations on immigration issues.
6. *Preparedness, Response, Recovery Committees and Working Groups* work across government and the private sector to prepare, respond to, and recover from large-scale emergencies.
7. *Privacy Advisory Committee* works to protect privacy rights, including the handling of personality identifiable information as well as data integrity and other privacy-related matters.

The organization chart of the Department of Homeland Security is shown in Figure 5.7.

Homeland Security and the FBI

Before the creation of the Department of Homeland Security, the FBI had primary responsibility for locating terrorist groups and preventing terrorist acts in the United States. It had many successes. According to FBI data, for example, the agency prevented 130 terrorist acts between 1980 and 1999.[56] However, following the al-Qaeda attacks on New York and Washington, the FBI was heavily criticized for missing clues and for intelligence failures.[57] For example, then-Senator John Edwards of North Carolina, who had oversight responsibility for the FBI as a member of both the Intelligence and Judiciary committees, remarked, "The FBI is clearly broken, and we can accept no further delay in the effort to fix it." He added, "The FBI should do what it does best: law enforcement [rather than] collecting information, fitting it into a bigger picture and sharing that information with people who can act on it." To root out terrorists within the United States, legislators on Capitol Hill called for the creation of a new domestic intelligence-gathering agency similar to the MI-5 in Great Britain.

FIGURE 5.7 Department of Homeland Security Organization Chart

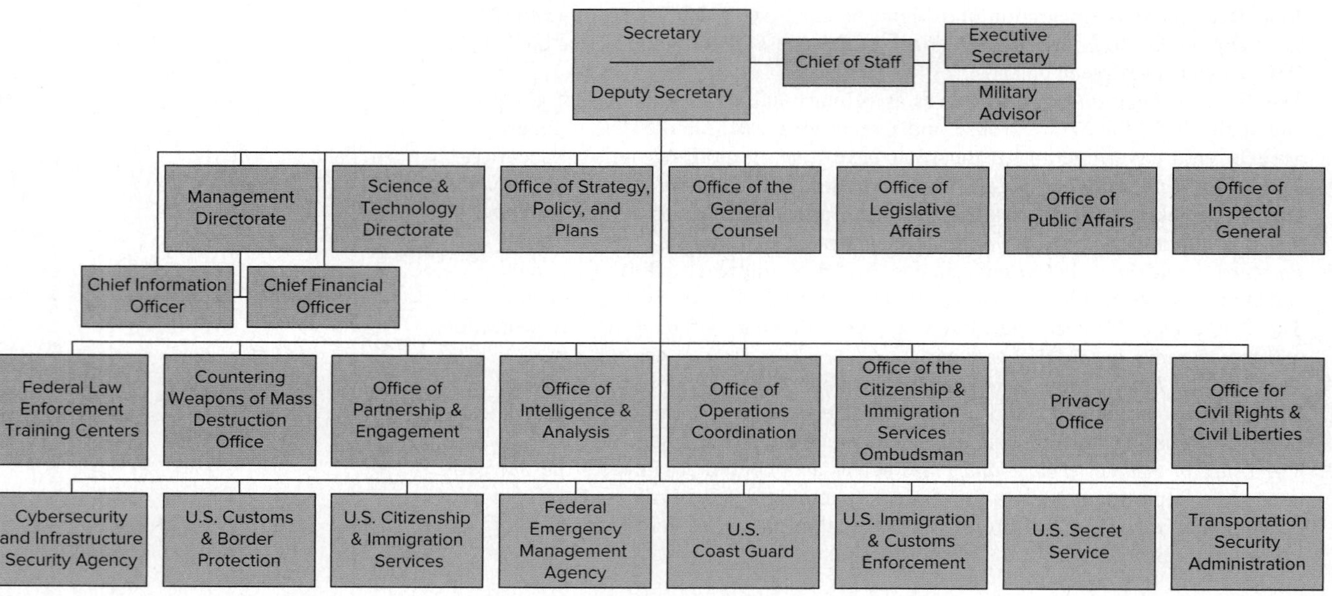

Source: https://www.dhs.gov/sites/default/files/publications/19_1205_dhs-organizational-chart.pdf (February 2020).

Then-FBI Director Robert Mueller defended his agency. He responded, "Establishing a new domestic intelligence agency would constitute a step backward in the war on terror, not a step forward." He maintained that rather than "create a new agency from whole cloth, the public would be better served by improving what the FBI is already doing."

Director Mueller won a reprieve for the FBI and quickly began implementing fundamental changes. First, he shifted the top priority of the FBI from being a federal police agency to being an intelligence and counterterrorism agency. In doing so, he no longer allowed local field offices to establish their own distinct agendas. As a result, by 2008, the FBI had referred 40% fewer criminal investigations to the Justice Department than it did during the previous two decades.

Second, he restructured the management hierarchy at FBI headquarters in Washington to support counterterrorism efforts. Figure 5.8 shows the FBI's post-9/11 organization chart with the new emphasis on counterterrorism.

Third, he reassigned about one-quarter of the FBI's then-11,000 agents to work on counterterrorism. That represents a doubling of the number of agents handling terrorism cases, a quadrupling of the number of strategic analysts at FBI headquarters, but a decrease of about one-third of all agents in criminal programs. Consequently, although the FBI official in charge of criminal investigations correctly predicted the mortgage crisis in 2004 and believed the FBI could have prevented it from spiraling out of control, by 2007, the FBI had only about 100 agents pursuing mortgage fraud. By comparison, during the savings and loan debacle of the 1980s and 1990s, the FBI had about 1,000 agents investigating banking fraud.[58]

Although the Bush administration refused to approve new agents to investigate financial crimes,[59] that changed under the Obama administration. Since 2008, the FBI has nearly tripled the number of special agents investigating mortgage fraud. By the end of the first decade of the twenty-first century, the FBI had more than 500 agents and analysts using intelligence to identify emerging health care fraud schemes. As the result of a new forensic accountant program, the FBI also had 250 forensic accountants trained to catch financial criminals. In 2009, the FBI established the Financial Intelligence Center to strengthen the Bureau's financial intelligence collection and analysis, and, in 2010, the FBI began embedding agents at the Securities and Exchange Commission to identify securities fraud more quickly and to push intelligence to the Bureau's field offices.[60]

Fourth, Director Mueller established a National Joint Terrorism Task Force at FBI headquarters that includes staffers from federal, state, and local agencies. They are responsible for coordinating the flow of information with task forces in each of the FBI's 56 field offices.

FIGURE 5.8 **Federal Bureau of Investigation Organization Chart**

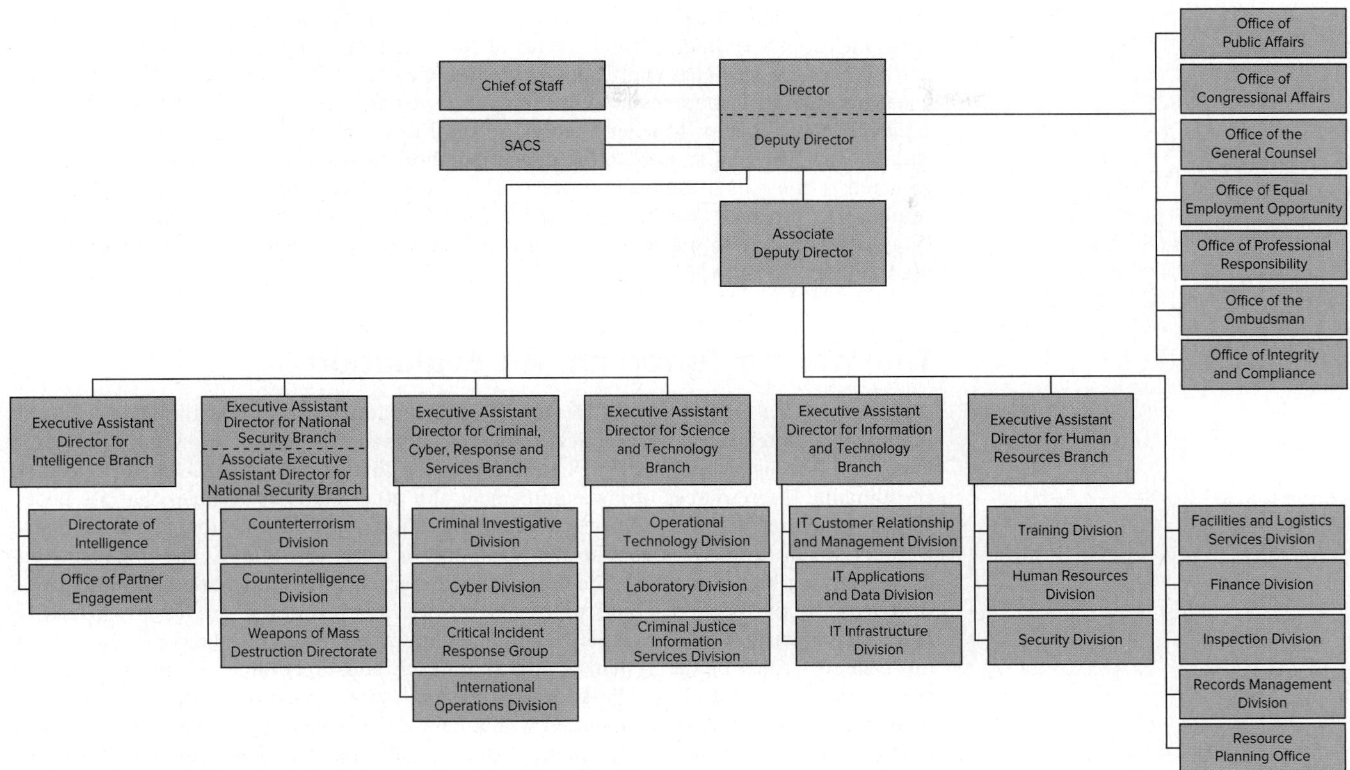

Source: www.fbi.gov/contact-us/fbi-headquarters/org_chart/organizational_chart.

Fifth, to more directly address the global threat of al-Qaeda and other terrorist groups, Director Mueller planned on opening FBI offices in Kabul, Afghanistan; Sarajevo, Bosnia; Jakarta, Indonesia; Tashkent, Uzbekistan; and Belgrade, Serbia. As of the end of 2018, the FBI had opened offices in Kabul, Sarajevo, and Jakarta but not in Tashkent or Belgrade. He also planned on expanding FBI offices in Ottawa, Canada; Seoul, South Korea; London, Berlin, and Moscow. Since the tragic events of 9/11, about 500 FBI agents and 200 support personnel have been working outside the United States on terrorism investigation. Already working outside the United States in 75 key cities worldwide were FBI agents known as permanent legal attachés, or "legats." The job of legats is to feed information gathered from interviews back to the United States for further investigation. Unlike CIA agents, legats, who will work more closely with the CIA, do not operate covertly and are involved more with investigations than with gathering intelligence.

In 2005, in response to a presidential directive to establish a "National Security Service" that combined the missions, capabilities, and resources of the FBI's counterterrorism, counterintelligence, and intelligence units, the FBI created a National Security Branch (NSB). In 2006, the Weapons of Mass Destruction (WMD) Directorate was established within the NSB to integrate WMD units that previously had been spread throughout the FBI. The NSB also includes the Terrorist Screening Center, whose role is to provide actionable intelligence to state and local law enforcement. In creating the NSB, the FBI has moved beyond case-focused intelligence to building a Bureau-wide intelligence collection, analysis, and dissemination program that combines intelligence from across the Bureau. The FBI now uses intelligence not just to pursue investigations, but to have greater awareness of national security threats and the total threat environment. The FBI now looks at information for its predictive value and shares that information–except that which it is legally proscribed from releasing–with its partners in law enforcement and the intelligence community.

An integral part of the FBI's information-sharing capabilities is the Guardian Terrorist Threat and Suspicious Incident Tracking System, which was introduced in 2002 and became available to all users on July 1, 2012. On that date, FBI personnel were granted access for use in investigative activities, and the FBI was no longer using its Automatic Case Support system and its paper-based records system for current case management.[61]

FYI **Rewards for Justice**

Rewards of greater than $25 million are available through the Rewards for Justice program for information that prevents or favorably resolves acts of international terrorism against U.S. citizens or property anywhere in the world. Since its inception in 1984, Rewards for Justice has paid more than $150 million to more than 100 people (as of April 24, 2019).

Source: Rewards for Justice, "Program Overview." Accessed April 24, 2019, https://rewardsforjustice.net/english/about-rfj/program-overview.html.

In sum, through the efforts of former-Director Mueller, the FBI remains an independent agency, albeit with a new top priority, and retains its traditional responsibility of intelligence gathering and analysis. However, it now closely coordinates its antiterrorism activities with personnel from the Office of the Director of National Intelligence (ODNI), the CIA, DOD, and DHS at the new National Counterterrorism Center (NCTC) and with state, local, and tribal partners in task forces around the country. Under the leadership of former-Director James Comey, Director Mueller's successor, the FBI's top priority remained counterterrorism, but the FBI now makes a more concerted effort to share with its partners intelligence gathered in the United States and overseas to provide a coordinated strategic and tactical response to threats.[62] We believe that Director Comey's successor, Director Christopher Wray will continue to make counterterrorism the FBI's top priority and to implement his predecessors' agendas.[63]

The War on Terrorism: An Evaluation

Advocates of the Department of Homeland Security are confident that the DHS will have the financial, intelligence, and tactical resources necessary to prevent and control domestic terrorism. At this writing, the DHS is assessing the threats against the United States and coordinating the resources of law enforcement and other kinds of agencies that are necessary to defeat terrorism at home. One of its first efforts was the creation of a color-coded warning system to alert citizens to the likelihood of a terrorist attack. The hope was that as the warning level was raised, the vigilance of Americans would increase and information would be discovered that would prevent a terrorist act. However, on January 27, 2011, Homeland Security Secretary Janet Napolitano announced the end to the color-coded warning system. Critics of the system argued that "each and every time the threat level was raised, very rarely did the public know the reason, how to proceed, or for how long to be on alert." The old system was replaced with a new, more targeted National Terror Advisory System that provides law enforcement and potential targets critical information without unnecessarily alarming or confusing the public. According to former-Secretary Napolitano, the new alerts "may recommend certain actions or suggest looking for specific suspicious behavior. And they will have a specified end date." One of the more obvious changes has occurred at airports, where public service recordings announcing the alert level are no longer made. The aviation threat had been on orange, or "high" alert, since 2006.[64]

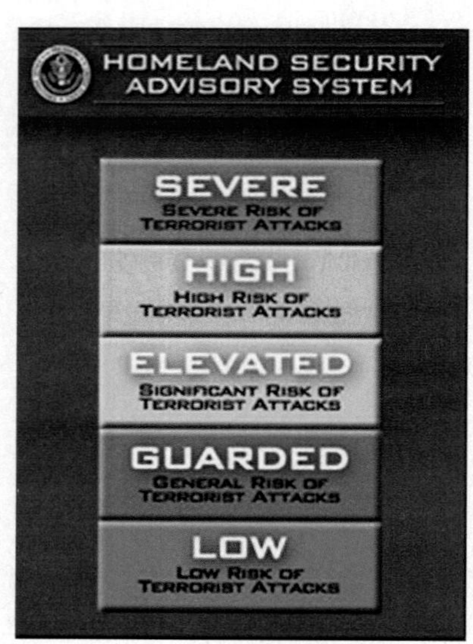

DOD/ZUMAPRESS/Newscom

The Homeland Security Department's color-coded national threat alert system was intended to help citizens better prepare for potential terrorist attacks. *Did the system also serve a public relations function? If so, what was it?*

The war on terrorism is an ongoing battle with no end in sight. Nevertheless, there already have been some successes. For example, since September 11, 2001, as a result of cooperation among law enforcement agencies, thousands of al-Qaeda members or their associates have been captured and detained in more than 100 countries.[65] In addition, most of al-Qaeda's top leadership, including Osama bin Laden on May 1, 2011, have been killed or captured.[66] On the other hand, a conflict between the al-Nasrah Front and al-Qaeda in Iraq, which changed its name to the Islamic State of Iraq and the Levant (ISIL) or later the Islamic State of Iraq and Syria (ISIS), resulted in the expulsion of ISIL from the al-Qaeda network in February 2014.[67] IISIL then greatly expanded its influence by seizing vast amounts of territory in Iraq and Syria and creating a self-declared Islamic caliphate, which encouraged terrorist fighters worldwide to flock to its ranks. The rise of ISIL or ISIS resulted in a powerful regional and international mobilization including the United States and 60 partners to counter it and its territorial gains.[68] By the second half of 2015, 40% of the territory ISIL controlled at the beginning of the year had been liberated. As ISIL or ISIS began losing territory, it, its affiliated groups, and self-radicalized individuals began staging mass-casualty attacks throughout the world.[69]

Although the United States, as of April 27, 2019, has not experienced a large-scale terrorist attack by a terrorist organization like the one committed on 9/11, the United States has not escaped terrorism. For example, on April 15, 2013, two apparently "lone wolf" terrorists exploded two bombs at the Boston marathon killing at least 3 people and wounding dozens more.[70] Also, on December 2, 2015, two more "lone wolf" terrorists committed a mass shooting and an attempted bombing at a San Bernardino County Department of Public Health training event and holiday party. Fourteen people were killed and 22 people were seriously injured. The terrorists were killed in a shootout with the FBI later that day.[71] On June 12, 2016, another "lone wolf" terrorist committed a mass shooting at a gay nightclub in Orlando, Florida. He killed 49 people and wounded 53 others before he was shot and killed by police officers after a three-hour standoff.[72]

In March 2019, ISIS lost its final stronghold in Syria, bringing an end to the so-called caliphate declared by the terrorist group in 2014. Despite the loss of territory, ISIS ideology lives on. In addition, ISIS still has leaders, fighters, and resources, and its black flag is still flown by affiliates in other parts of the world.[73] ISIS, al-Qaeda, and their affiliates have not been defeated. According to Ambassador Nathan Sales, Coordinator for Counterterrorism, They have proven to be "resilient, determined, and adaptable." "They have become more dispersed and clandestine," added Ambassador Sales, "turning to the Internet to inspire attacks by distant followers, and, as a result, have made themselves less susceptible to conventional military action." Moreover, continued Ambassador Sales, "the return or relocation of foreign terrorist fighters from the battlefield has contributed to a growing cadre of experienced, sophisticated, and connected terrorist networks, which can plan and execute terrorist attacks." In short, the nature of the terrorist threat confronting the United States and its allies globally has evolved and thus so must our counterterrorism efforts.[74]

ISIS Leader Killed

On October 26, 2019, ISIS leader Abu Bakr al-Baghdadi killed himself during a raid of his compound by the U.S. Army's elite Delta Force.

Source: Luis Martinez, "Pentagon Report Says al-Baghdadi Death Had Little Impact on ISIS Leadership and Operations," *ABC News*, February 4, 2020, accessed February 8, 2020, https://abcnews.go.com/Politics/pentagon-report-al-baghdadi-death-impact-isis-leadership/story?id=68755044.

THINKING CRITICALLY

1. Do you think the United States has too many law enforcement agencies? Why or why not?
2. With the creation of the DHS, is the FBI needed any longer? Defend your answer.
3. Do you consider mass killings such as those at the Boston marathon, the San Bernardino County Department of Public Health, and the gay nightclub in Orlando, Florida, terrorist acts? Why or why not?

American Private Security

Private security has been defined as "the nongovernmental, private-sector practice of protecting people, property, and information, conducting investigations, and otherwise safeguarding an organization's assets."[75] It has a role in "helping the private sector secure its business and critical infrastructure, whether from natural disaster, accidents or planned actions, such as terrorist attacks, vandalism, etc."[76]

A common way to categorize private security employment is to classify the agencies and personnel as either contract or proprietary. **Contract security** companies offer protective services for a fee to people, agencies, and companies that do not employ their own security personnel or that need extra protection. A state university, for example, may employ private security officers to work at a football game. Contract security employees are not peace officers. **Proprietary security** agents and personnel provide protective services for the entity that employs them. For example, the Ford Motor Company employs its own security forces at its large manufacturing plants. Most proprietary security officers are employed in retail/restaurants/food services (16.7%), casinos/hospitality/arenas/entertainment (15.7%), health care/medical centers/hospitals (12.9%), and government (10.9%).[77] About 40% of all private security officers are proprietary security officers.[78] Primarily for cost reasons, the number of contract security jobs is likely to increase faster than the number of proprietary security jobs.[79]

contract security Protective services that a private security firm provides to people, agencies, and companies that do not employ their own security personnel or that need extra protection.

proprietary security In-house protective services that a security staff provides for the entity that employs it.

Private security is a huge enterprise that complements public law enforcement in the United States. There were approximately 8,000 private security companies in the United States in 2018, compared to more than 18,000 public law enforcement agencies.[80] Although for many years more people worked in private security than in public law enforcement, that is no longer the case. In 2016, for example, about 1.1 million people worked as contract security guards, and about 1.1 million people worked in public law enforcement.[81] According to the Bureau of Labor Statistics, between 2016 and 2026, employment in private security is projected to increase by 6%, while employment in public law enforcement is projected to increase by 7%.[82] Until recently, substantially more money was being spent on private security than on public policing, but that gap has been narrowing somewhat because of the increases in spending on public law enforcement during the 1990s and the federalizing of airport security (through the Transportation Security Administration) in the wake of the 9/11 terrorist assault.

Table 5.5 lists the services provided by private security companies. Standing security officer and vehicle patrol services account for at least 75% of American private security's $25.5 billion in revenues in 2018.[83]

TABLE 5.5 Services Offered by Private Security Companies

1. Standing security officer and vehicle patrol services (at least 75% of total revenues)
2. Special event security
3. Risk analysis
4. Security consulting
5. Loss prevention
6. Investigators
7. Background screening
8. Facility design
9. Roving vehicle patrol services
10. Concierge services
11. Alarm services and security systems integration (many contract security companies do not actually perform these services in-house, but rather refer this type of work to a "partner" that specializes in providing the product or service)
12. Integrated guarding (video monitoring and vehicle patrol in combination with on-site manned guarding; or to take the place of on-site guarding)

More recent new offerings:

13. Systems integration
14. Drones
15. Security robots
16. Cyber security
17. Canine security
18. Cash management services

Source: Robert H. Perry Associates, U.S. Contract Security Industry, White Paper, 10th ed., July 2019, pp. 7–8, https://www.roberthperry.com/uploads/2018_White_Paper1.pdf (accessed April 28, 2019).

Private Security Officers

Private security officers, or guards, are hired to provide protection. Their duties vary and depend on the employers' particular needs. Private security officers generally specialize in one of the following areas:

- Protecting people, records, merchandise, money, and equipment in department stores; also working with undercover store detectives to prevent theft by customers or store employees and helping in the apprehension of shoplifting suspects before the police arrive.
- Patrolling the parking lots of shopping centers and theaters, sometimes on horseback or bicycles, to deter car theft and robberies.
- Maintaining order and protecting property, staff, and customers in office buildings, banks, and hospitals.
- Protecting people, freight, property, and equipment at air, sea, and rail terminals as well as other transportation facilities; also screening passengers and visitors for weapons and explosives using metal detectors and high-tech equipment, ensuring that nothing is stolen while being loaded or unloaded, and watching for fires and criminals.
- Protecting paintings and exhibits by inspecting people and packages entering and leaving public buildings such as museums or art galleries.
- Protecting information, products, computer codes, and defense secrets and checking the credentials of people and vehicles entering or leaving the premises of factories, laboratories, government buildings, data-processing centers, and military bases.
- Performing crowd control, supervising parking and seating, and directing traffic at universities, parks, and sports stadiums.
- Preventing access by minors, collecting cover charges at the door, maintaining order among customers, and protecting property and patrons while stationed at the entrance to bars and places of adult entertainment such as nightclubs.

- Protecting money and valuables during transit in armored cars; also protecting individuals responsible for making commercial bank deposits from theft or bodily injuries.
- Observing casino operations for irregular activities, such as cheating or theft, by employees or patrons.[84]

Private security officers typically are male and white. In 2018, for example, security guards and gaming surveillance officers were 78% male, 60% white, 31% black or African American, 17% Hispanic or Latino, and 4% Asian.[85]

Security guards generally receive relatively low wages. They typically work 8-hour shifts, 5 days a week. In 2018, security guards earned a median annual wage of $28,490 or approximately $14 per hour worked. The lowest 10% earned less than $20,290, and the highest 10% earned more than $49,650. Wages vary significantly depending on the area of the country, unionization, and whether or not security officers are working at a federal government facility where wage and benefits are mandated by the federal government.[86] Managers and corporate officers made more. Because of relatively low wages, private security officers frequently work part-time or have another "primary" job and use their security job wages to supplement their incomes.

Charles Rex Arbogast/AP Images

Private security is assuming an increasing role in maintaining order, investigating crime, and apprehending criminals. *Is this a positive trend? Why or why not?*

Reasons for Growth

A number of factors have stimulated the phenomenal growth of private security since the 1970s, until recently.

Companies and Municipalities Are Outsourcing Security Functions Many police departments are operating on limited budgets, have outdated equipment, lack special security expertise, and employ officers with low morale because of recent attacks on their fellow officers. Many insurance companies are demanding that companies and municipalities provide better protection. Therefore, with police departments unable to meet this demand, companies and municipalities increasingly are outsourcing their security needs, and often saving money in doing so. Because of competition in the industry, private security companies frequently are able to provide their clients with better-trained personnel and more state-of-the-art technology than are police departments.[87]

The largest growth in the contact security industry is likely to come from airport passenger screening and other airport security functions presently handled by the TSA. According to a recent *ABC News* report, many TSA agents are incredibly inept at their jobs. The report noted that Department of Homeland Security undercover agents successfully smuggled fake explosives and weapons through 96% of 70 checkpoints nationwide. Already 20 airports in the United States have outsourced security–San Francisco International Airport being the largest. Perhaps the most viable solution for the Department of Homeland Security is expansion of the Screening Partnership Program, which was created in 2001, and allows private airport screeners to work under supervision of the TSA.[88]

Increasing Crime and Terrorism Several major cities have experienced a higher crime rate, and people still fear "soft" terror attacks, as well as another major attack such as happened on 9/11. Continuous mass murders in schools, theaters, shopping malls, subways, hospitals, sporting events, and places of worship also have the public rattled. In response, people have demanded increased security. Private security companies are helping to meet this need.[89]

Cost Considerations Companies, and more recently, municipalities, trying to cut costs are eliminating their in-house security programs and using contract security companies. By hiring contract security companies, employers can avoid paying expensive retirement benefits.[90]

The Private Nature of Crimes in the Workplace A business depends on a positive reputation to remain competitive. Widespread employee theft, embezzlement scandals, and substance abuse harm an organization's public image and may cause potential customers to question the quality of a company's products and services. By employing private security personnel to prevent and repress crime in their facilities, businesses can either hide the crimes that occur or minimize the negative publicity.[91]

ASIS International

ASIS International is an organization for security professionals. Founded in 1955 as the American Society for Industrial Security, ASIS is dedicated to increasing the effectiveness and productivity of security professionals by developing educational programs and materials that address broad security interests. To learn more about ASIS, visit its website at www.asisonline.org/.

Why do you think a professional organization for private security companies and personnel is important?

Better Control and Attention to the Problem By employing in-house security personnel or by contracting with an outside firm, the management of a business can direct security personnel to do precisely what is needed to prevent crime, minimize substance abuse, and discipline wayward employees. Public police would have to combine the concerns of a business with the priorities of the citizens of the community.[92]

Fewer Constitutional Limitations Some of the constitutional restrictions that would limit the actions of public police officers working undercover to curtail drug trafficking in an industrial plant, for example, would not restrict private security personnel employed directly by that industry. U.S. Supreme Court prohibitions that restrict a public police officer's right to search and seize property, for instance, would not limit the actions of a private security agent.[93]

Issues Involving Private Security

A number of unresolved problems and issues impinge on the potential for development of the private security industry. Some of them put the industry at odds with public law enforcement.

Legal Status and Authority Private security officers' legal status and authority derive from the rights of the owner who employs them to protect property on the premises. These rights are essentially the same ones you have to protect your life and property at home.

If this view prevails, private security personnel face few constitutional limitations in investigating crime, obtaining evidence, employing reasonable force, searching personal property stored in corporate spaces, and interrogating suspects. Although this is not a unanimous view among courts, it is the most prevalent one. However, private security officers and their employers face the possibility of being held civilly or criminally liable for violating an individual's civil rights or for false arrest.[94]

Armed Employees Contrary to popular opinion, only about 10% of contract private security officers carry weapons. Security officers who do carry weapons today are more thoroughly vetted upon employment and receive extra training. Many contract security companies hire off-duty police officers to fill posts that require an armed security officer.[95]

However, as incidences continue to occur that could have been prevented by a weapon-carrying security officer, security customers are demanding that their contract security company meets this need. In an effort to limit their liability in issuing weapons and still meet the needs of their customers, many contract security companies provide personnel with nonlethal weapons. Popular are weapons that fire a pepper spray chemical that is accurate from a distance of 10 feet. These devices are equipped with cameras that record every encounter, to be used as evidence to support the officer's need to use this type of force.[96]

FYI

The Job of Private Security Officer Can Be Dangerous

In 2017, 50 private security guards were killed, compared to 46 public law enforcement officers, who were feloniously killed, and 47 public law enforcement, who were accidentally killed.

Sources: Bureau of Labor Statistics, U.S. Department of Labor, "Table A-1. Fatal occupational injuries by industry and event or exposure, all United States, 2017," https://www.bls.gov/iif/oshwc/cfoi/cftb0313.htm (accessed May 3, 2019); Federal Bureau of Investigation, U.S. Department of Justice, "2017 Law Enforcement Officers Killed & Assaulted," https://ucr.fbi.gov/leoka/2017 (accessed May 3, 2019).

Replacing Manned Guarding with Electronics Some companies are eliminating security officers altogether and replacing them with technology. However, most companies are simply reducing the number of security officers that they employ and supplementing them with "integrated guarding" (for example, video monitoring combined with standing security officers and/or roving security vehicle patrols) or other electronic security options. Oftentimes, electronic security is more cost effective.[97]

Public Policing in a Private Capacity Although some police departments prohibit moonlighting, thousands of police officers still work in a private capacity during their off-duty hours. Some police agencies even cooperate with private agencies in scheduling their officers for off-duty assignments. With regard to their legal status and authority, are these officers considered public police or private security personnel? The private organization that employs them believes that off-duty police officers are better qualified, have more authority to arrest, and will have a greater deterrent effect on the crimes and disturbances of the peace the employer is trying to prevent.

An equally important question is, Who is liable should moonlighting officers abuse their authority or make a mistake? At present, it seems police agencies that take an active role in scheduling off-duty assignments accept greater liability than the police departments that do not. Many agencies limit the assignments officers are allowed to accept and the number of hours they are permitted to work.[98]

Qualifications and Training Many superbly qualified people work in private security at all levels throughout the United States, but those people are not the norm. The minimum qualifications for private security personnel at all levels of employment are increasing, but they still lag far behind those of the public police.

Although many states have unique training and licensing requirements, nearly all feature the same minimum requirements, which are these: (1) being 18 years of age or older, (2) having no violent or felony convictions, (3) being of sound mind, and (4) having a GED or high school diploma.[99] The large private security companies that employ thousands of security guards have minimum training requirements that far exceed the minimum standards imposed by the states in which they operate. Also, the qualifications of proprietary security officers generally are higher than those of contract security officers, demonstrating corporate demand for high-quality security services even when they cost more.[100]

As of 2013 (the latest year for which data were available), 35 states required all security guards to be licensed, and 15 states did not. Of the 15 states that did not require individual security guards to be licensed, 6 of them required each security employer to register their guards with the state, state police, or Department of Public Safety.[101] Figure 5.9 shows the licensing status of all 50 states.

Diminished Public Responsibility The current mixture of public and private protection is a matter of concern to many. What does it say about a government's ability to govern and provide for the general welfare–let alone what it says about American society–that ever more frequently it is shifting responsibility for protecting life and property to private security enterprises? To some, it seems to mean that public police officers and the governments that employ them have defaulted on a major portion of the social contract.[102]

FIGURE 5.9 Licensing Status of Security Guards in the United States

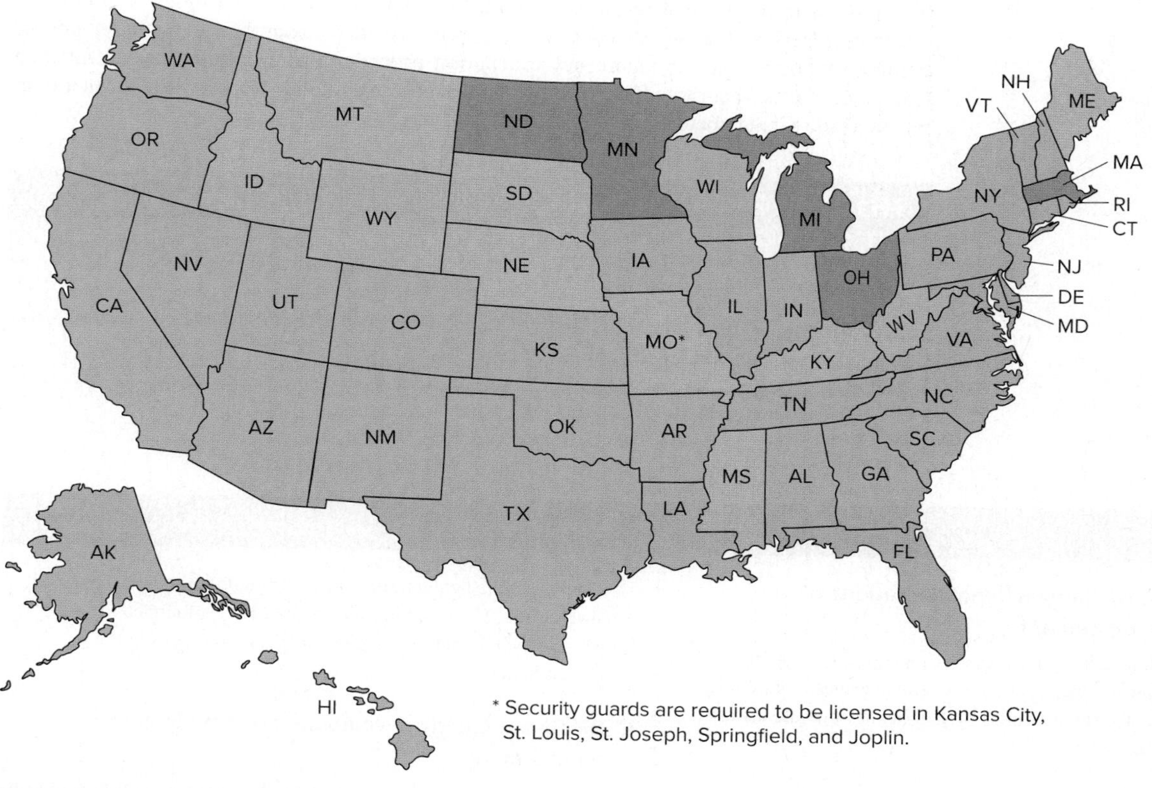

- States with licensing requirements
- States without licensing requirements
- States that require each security employer to register their guards with the state, state police, or Department of Public Safety

Source: Security Guard Training Tips, "Unarmed Security Guard State Requirements," 2013, http://securityguardtrainingtips.com/Unarmed-Security-Guard-State-Requirements.php (accessed May 1, 2019).

Private Security's Role in the Fight Against Terrorism Private security officers often are the first line of defense against terrorism in the United States and other nations. They guard government buildings, utilities, schools, courts, corporate headquarters, office complexes, laboratories, and transportation facilities, to name only a few. Security experts believe that 15% to 20% of the private security officers in the United States protect sites designated by the government as "critical infrastructure."[103]

The private security industry and its officers have been given more protection responsibilities as the threat of domestic terrorism in the United States has increased. Public law enforcement alone cannot begin to meet the protection responsibilities necessary to prevent terrorism. This means that the private security industry, comprising about 8,000 companies throughout America, will have to begin improving its selection standards and training if the nation's people, visitors, and assets are to be protected. As unbelievable as it is, hundreds of security guards employed to protect the Statue of Liberty in 2002 were found to have no licenses, and their ranks included ex-convicts. According to one report, tens of thousands of security guard applicants were found to have criminal backgrounds.[104]

Because it is so difficult to find enough capable private security officers, the proprietary security firms, motivated by large profits, have been employing hundreds if not thousands of unqualified and unmotivated applicants. For many of the guards, as noted previously, the position is a second or low-paying job that they can easily quit if they find a better-paying job or simply choose to do so. More troublesome is that foreign and domestic terrorists could easily obtain a private security position and inflict physical and emotional havoc on the nation by initiating some terrorist activity inside a vulnerable site.

Seldom is throwing money at a problem a solution in and of itself; yet, if the private security industry is to fulfill its obligation to protect the homeland, it is going to need more financial resources. It may be necessary, for example, to provide government subsidies for training and background checks and a significant increase in the amount of cooperation between the public police and the private security industry. That sort of investment and training can bring positive results. Following the 1993 World Trade Center bombing, the security officers at the World Trade Center were provided in-depth and regular follow-up training on such topics as emergency evacuation procedures and building layout. On September 11, 2001, security officers helped thousands of building workers safely out of the World Trade Center before the twin towers fell.

THINKING CRITICALLY

1. What do you think are some of the benefits and drawbacks of being a private security officer?
2. Do you think that stricter qualification standards should be established for private security personnel? Why or why not?

SUMMARY

1. **Briefly describe the jurisdictional limitations of American law enforcement.**

   The authority of public law enforcement agencies in the United States, whether they are local, state, or federal agencies, is carefully limited by law. The territory within which each may operate also is restricted.

2. **Trace the English origins of American law enforcement.**

   Many institutions of American law enforcement evolved from the English tradition. The medieval tithing system and the constable-watch system were early methods of community protection that led to the development of the positions of sheriff and constable. The Bow Street Runners in the city of London in the 1750s were an early group of crime fighters who patrolled neighborhoods and pursued lawbreakers. The London Metropolitan Police, founded in 1829, became the model for municipal police departments in the United States.

3. **Discuss the early development of American law enforcement.**

   Americans at first adopted the British system of community protection. When the constable-watch system proved inadequate in meeting the peacekeeping needs of the nation's major cities, municipal police forces were established in the mid-1800s. They soon became entangled with local politics. In the states and on the frontier, law enforcement reflected regional differences. In the South, the earliest policing was the plantation slave patrols. On the frontier, vigilantism and, later, local sheriffs or U.S. Marshals dealt with lawbreakers. Native Americans had their own law enforcement arrangements. In some states, state police agencies, such as the Texas Rangers, were established to enforce laws statewide.

4. Describe the major developments that have occurred in American policing.

During the period of professionalism and reform that lasted from about 1920 to 1970, the police became professional crime fighters, relying on the centralization of authority, motorized patrols, specialization, and technological aids. In the 1960s, the crime-fighting role of the police came into conflict with the social and political upheavals of the time, causing critics to call for improved standards and training. By the early 1990s, some police agencies began to turn to community policing, attempting to eliminate crime problems in neighborhoods and return to their role as peacekeepers. CompStat also was introduced as another policing strategy. At the start of the twenty-first century, the prevention and repression of domestic terrorism also has become a major priority of police in America.

5. Describe the structure of American law enforcement.

Law enforcement agencies are found at all levels of government in the United States. Most law enforcement officers work for local governments and are responsible for enforcing laws, maintaining order, providing service, and gathering information. In rural areas, the county sheriff's department is responsible for law enforcement. Every state except Hawaii has a state law enforcement agency. The law enforcement agencies of the federal government are concerned primarily with violations of federal laws, especially violations that cross state boundaries; maintaining homeland security; and preventing domestic terrorism.

6. Explain the relationship between the Federal Bureau of Investigation and the Department of Homeland Security.

Since 9/11, the FBI has undergone fundamental changes. The biggest change is that it has shifted its top priority from being a federal police agency to being an intelligence and counterterrorism agency. Although it remains an independent agency in the Justice Department, it will now closely coordinate its antiterrorism activities with the CIA and the DHS.

7. Discuss the development and growth of private security in the United States.

The private security industry has grown rapidly over the past 40+ years for a number of reasons: Companies and municipalities are cutting costs by outsourcing security functions; public fears about crime and terrorism are not being adequately addressed by public law enforcement; crimes in the workplace are often private, costly, and embarrassing; employers have better control of private security officers; and fewer constitutional limits restrict private security officers.

KEY TERMS

jurisdiction 141
tithing system 142
shire reeve 142
posses 142
constable-watch system 142
constable 142
Peel's Principles of Policing 144
slave patrols 147
community policing 151
CompStat 152
state police model 175
highway patrol model 177
contract security 189
proprietary security 189

REVIEW QUESTIONS

1. What is meant by *jurisdiction*?
2. What was the *tithing system*?
3. Who were the Bow Street Runners?
4. In what year was the London Metropolitan Police founded?
5. Who was Robert Peel?
6. What system of English policing did the colonists bring to America?
7. What were the *slave codes*?
8. What group is considered to be the first state police agency?
9. How did August Vollmer change policing?
10. How did police response to the demonstrations and civil disorders of the 1960s affect policing?
11. What is *community policing*?
12. What is *CompStat*?
13. Why were the U.S. Marshals Service, the Secret Service, the FBI, and the DEA created?
14. What are the four main functions of local police?
15. Why do county sheriffs have more political clout than police chiefs?
16. What are differences among a state police model, a highway patrol model, and a department of public safety model of state law enforcement?
17. Name some federal law enforcement agencies.
18. Distinguish between *contract* and *proprietary private security services.*
19. What are some of the unresolved problems and issues with private security?
20. What are some successes and problems with private security in the war on terrorism?

IN THE FIELD

1. **Your Local Law Enforcement** Identify all the local law enforcement agencies in your area. Divide up the list among your classmates, and arrange to visit your assigned agency. On the day you visit, find out how many calls the department received and/or how many crimes the agency personnel investigated in the 24-hour period prior to your visit, what types of calls were received or crimes were investigated, and how the agency handled the situation. Categorize the actions of the agency personnel into law enforcement, order maintenance, service, or information gathering. Identify the category with the most action. Share your findings with others in the class. What conclusions can you draw about the operation of the local law enforcement agencies in your area?
2. **Local and Private Police** Describe the possibilities you see for local police departments and private security agencies to work together more closely. To prepare for this activity, interview a local police official and a private security manager, either by telephone or in person, asking them what obstacles prevent closer cooperation between local policing and private security.

ON THE WEB

1. **Local Police Jobs** Go to the following links for sites dedicated to law enforcement careers. Select two municipal police departments, one large and one small, and one county sheriff's department that list each agency's employment qualifications on its web page:

   www.discoverpolicing.org

   www.911hotjobs.com

   http://jobs4police.myshopify.com

   www.lawenforcementjobs.com

   www.policecareer.com

   www.golawenforcement.com

   Then find the qualifications for a private security officer through other applicable career sites provided at www.securityjobs.net. Make a list of the similarities and differences between police and deputy sheriff qualifications and private security officer qualifications. Given your background and abilities, for which type of work would you be best suited? Why? Compile your findings in a two-page report and present it to the class.
2. **Federal Law Enforcement** To learn about the responsibilities of some of the lesser-known federal law enforcement agencies, access their websites: (a) U.S. Food and Drug Administration, Inspections, Compliance, Enforcement, and Criminal Investigations, https://www.fda.gov/inspections-compliance-enforcement-and-criminal-investigations/criminal-investigations; (b) National Park Service, United States Park Police, www.nps.gov/subjects/uspp/index.htm; (c) U.S. Fish and Wildlife Service, Office of Law Enforcement, www.fws.gov/le/; and (d) U.S. Capitol Police, https://www.uscp.gov. Which agency seems the most interesting? Why?

CRITICAL THINKING EXERCISES

NEIGHBORHOOD WATCH

1. You live in a middle-class community of single-family homes close to the center of a midsize city. Over the past 5 years, everyone in your neighborhood has noted the rise in burglaries, and many people feel that it is not safe to walk around the neighborhood after dark. You think that setting up a neighborhood watch would help lower the burglary rate and make people feel safer. Prepare an oral presentation of your ideas for a community meeting. Use the following questions as a guide.
   a. How would you go about organizing a night watch?
   b. How would you select volunteers?
   c. What training, if any, must volunteers have?
   d. How would you maintain interest and participation in the watch?

PUBLIC OFFICER OR PRIVATE CITIZEN?

2. An off-duty police officer was seated in a restaurant when two men entered, drew guns, and robbed the cashier. The officer made no attempt to prevent the robbery or apprehend the robbers. Later, the officer justified the conduct by stating that an officer, when off duty, is a private citizen with the same duties and rights as all private citizens. Do you agree? Explain.

NOTES

1. Material in this section was taken from T. A. Critchley, *A History of Police in England and Wales*, 2nd ed. rev. (Montclair, NJ: Patterson Smith, 1972).
2. Ibid.
3. Maricopa County Sheriff's office, accessed October 13, 2012, www.mcso.org/About/Sheriff.aspx.
4. Patrick Pringle, *Hue* and *Cry: The Story of Henry and John Fielding and Their Bow Street Runners* (New York: William Morrow, 1965).
5. Roger Hopkins Burke, *Criminal Justice Theory: An Introduction* (New York: Routledge, 2012), 22.
6. Material in this section was taken from A. C. Germann, F. Day, and R. Gallati, *Introduction to Law Enforcement and Criminal Justice* (Springfield, IL: Charles C. Thomas, 1962), 54–55.
7. Edward Savage, *Police Records and Recollections, or Boston by Daylight and Gaslights for Two Hundred and Forty Years* (Boston: John P. Dale, 1873).

8. This material came from Carl Sifakis, *The Encyclopedia of American Crime* (New York: Smithmark, 1992), 579–580.

9. Information in this section is from W. Marvin Dulaney, *Black Police in America* (Bloomington, IN: Indiana University Press, 1996).

10. Center for Research on Criminal Justice, *The Iron Fist and the Velvet Glove: An Analysis of the U.S. Police* (Berkeley, CA: Center for Research on Criminal Justice, 1975); Hubert Williams and Patrick V. Murphy, "The Evolving Strategy of Police: A Minority View," in *Perspectives on Policing* 13 (Washington, DC: U.S. Department of Justice, January 1990).

11. Dulaney, *Black Police in America,* 2.

12. John Duff and Peter Mitchell, *The Nat Turner Rebellion: The Historical Event and the Modern Controversy* (New York: Harper & Row, 1971).

13. Thad Sitton, *Texas High Sheriffs* (Austin, TX: Texas Monthly Press, 1988).

14. Rennard Strickland, *Fire and the Spirits: Cherokee Law from Clan to Court* (Norman, OK: University of Oklahoma Press, 1975), 58, 148, 217.

15. Adrian N. Anderson, Ralph A. Wooster, David G. Armstrong, and Jeanie R. Stanley, *Texas and Texans* (New York: Glencoe/McGraw-Hill, 1993).

16. Bruce Smith, *Police Systems in the United States* (New York: Harper & Row, 1960), 178–205.

17. Samuel Walker, *A Critical History of Police Reform: The Emergence of Professionalism* (Lexington, MA: Lexington, 1977), 76.

18. *Our Police* (Cincinnati, OH: Cincinnati Police Division, 1984).

19. Gene Carte and Elaine Carte, *Police Reform in the United States: The Era of August Vollmer* (Berkeley/Los Angeles, CA: University of California Press, 1975).

20. Roy Robert, John Crank, and Jack Kuykendall, *Police and Society* (Los Angeles: Roxbury Publishing, 1999), 432–433.

21. Malcolm K. Sparrow, Mark H. Moore, and David M. Kennedy, *Beyond 911* (New York: Basic Books, 1990).

22. U.S. Department of Justice, *Response Time Analysis: Executive Summary* (Washington, DC: U.S. Government Printing Office, 1978).

23. Samuel Walker, *The Police in America: An Introduction,* 2nd ed. (New York: McGraw-Hill, 1992).

24. Robert C. Trojanowicz and Bonnie Bucqueroux, *Community Policing: A Contemporary Perspective* (Cincinnati, OH: Anderson, 1989); Robert C. Trojanowicz and Bonnie Bucqueroux, *Community Policing: How to Get Started* (Cincinnati, OH: Anderson, 1994).

25. Material about CompStat is from James J. Willis, Stephen D. Mastrofski, and David Weisburd, "The Myth That COMPSTAT Reduces Crime and Transforms Police Organization," in *Demystifying Crime and Criminal Justice*, ed. R. M. Bohm and J. T. Walker (Los Angeles: Roxbury, 2006), 111–119.

26. Information on the U.S. Marshals Service is from www.usmarshals.gov/history/index.html and www.usmarshals.gov/history/broad_range.htm.

27. Information on the U.S. Secret Service, accessed October 29, 2008, is from www.ustreas.gov/usss/history.shtml, www.ustreas.gov/usss/faq.shtml#faq1, and www.ustreas.gov/usss/mission.shtml, and, accessed January 4, 2009, https://www.secretservice.gov/about/history/, accessed February 9, 2020).

28. Information on the FBI is from "The FBI," accessed October 29, 2008, www.fbi.gov/libref/historic/history/text.htm and https://www.fbi.gov/about, accessed February 9, 2020); Samuel Walker, *A Critical History of Police Reform* (Lexington, MA: Lexington, 2007); Frank J. Donner, *The Age of Surveillance* (New York: Alfred A. Knopf, 1980); Frank R. Hayde, *The Mafia and the Machine: The Story of the Kansas City Mob* (Fort Lee, NJ: Barricade Books, 2007).

29. Information on the DEA is from https://www.dea.gov/history and https://www.dea.gov/mission, accessed February 9, 2020). Other sources of information on America's War on Drugs are www.druglibrary.org/schaffer/history/e1910/harrisonact.htm; www.druglibrary.org/schaffer/Library/studies/cu/cu8.html; James A. Inciardi, *The War on Drugs: Heroin, Cocaine, Crime, and Public Policy* (Palo Alto, CA: Mayfield; 1986); www.druglibrary.org/SCHAFFER/library/studies/wick/wick1b.html; www.nationmaster.com/encyclopedia/Bureau-of-Narcotics; www.nationmaster.com/encyclopedia/Harry-J.-Anslinger; www.druglibrary.org/SCHAFFER/hemp/taxact/mjtaxact.htm; www.deamuseum.org/dea_history_book/pre1970.htm; https://www.dea.gov/sites/default/files/2018-07/1970-1975%20p%2030-39.pdf, accessed February 9, 2020); Edward Jay Epstein, *Agency of Fear: Opiates and Political Power in America* (New York: Putnam, 1977).

30. Federal Bureau of Investigation, *2017 Crime in the United States*, https://ucr.fbi.gov/crime-in-the-u.s/2017/crime-in-the-u.s.-2017/topic-pages/police-employee-data (accessed April 20, 2019).

31. Calculated from Table 11-1 in Federal Jobs Network, "Law Enforcement Hiring Agencies," accessed April 20, 2019, http://www.federaljobs.net/law_agencies.htm.

32. Calculated from data in Shelley Hyland, "Full-Time Employees in Law Enforcement Agencies, 1997–2016," *U.S. Department of Justice*, Bureau of Justice Statistics Statistical Brief, August 2018, 2, Table 2. Accessed April 21, 2019, https://www.bjs.gov/content/pub/pdf/ftelea9716.pdf.

33. Shelley S. Hyland and Elizabeth Davis, *Local Police Departments, 2016: Personnel, Policies, and Practices,* U.S. Department of Justice, *Bureau of Justice Statistics Bulletin*, October 2019, accessed October 26, 2019, https://www.bjs.gov/content/pub/pdf/lpd16p.pdf.

34. Ibid.

35. Federal Bureau of Investigation, *2017 Crime in the United States,* "Police Employment Data," Table 78, Washington, accessed April 21, 2019, https://ucr.fbi.gov/crime-in-the-u.s/2017/crime-in-the-u.s.-2017/tables/table-78/table-78-state-cuts/washington.xls.

36. Federal Bureau of Investigation, *2017 Crime in the United States,* "Police Employment Data," Table 78, Texas, accessed April 21, 2019, https://ucr.fbi.gov/crime-in-the-u.s/2017/crime-in-the-u.s.-2017/tables/table-78/table-78-state-cuts/texas.xls.

37. V. A. Leonard, *Police Organization and Management* (New York: Foundation Press, 1964).

38. Calculated from data in GOVTECH NAVIGATOR, "Total U.S. Law Enforcement Agencies," accessed April 21, 2019, https://www.govtech.com/navigator/numbers/us-law-enforcement-agencies_56.html and Shelley Hyland, "Full-Time Employees in Law Enforcement Agencies, 1997–2016," U.S. Department of Justice, *Bureau of Justice Statistics Statistical Brief*, August 2018, 2, Table 2. Accessed April 21, 2019, https://www.bjs.gov/content/pub/pdf/ftelea9716.pdf; Connor Brooks, "Sheriffs' Offices, 2016: Personnel," U.S. Department of Justice, *Bureau of Justice Statistics Bulletin*, October 2019, accessed October 26, 2019, https://www.bjs.gov/content/pub/pdf/so16p.pdf.

39. Connor Brooks, "Sheriffs' Offices, 2016: Personnel."

40. Ibid.; Bureau of Justice Statistics, 2007 Law Enforcement and Administrative Statistics Survey. (Data generously provided by Brian Reaves, January 28, 2011.)

41. Dana Severson, "Does a Deputy Sheriff Make More Money Than a Police Officer," *Chron*, June 29, 2018, accessed February 7, 2020, https://work.chron.com/deputy-sheriff-make-money-police-officer-21124.html.

42. Shelley Hyland, "Full-Time Employees in Law Enforcement Agencies, 1997–2016," U.S. Department of Justice, Bureau of Justice Statistics Statistical Brief, August 2018, 2, Table 2. Accessed April 21, 2019, https://www.bjs.gov/content/pub/pdf/ftelea9716.pdf; CA.GOV, CALIFORNIA HIGHWAY PATROL, accessed April 21, 2019, https://www.chp.ca.gov/home/about-us/organizational-chart.

43. Alabama Department of Public Safety, accessed October 13, 2012, https://www.alea.gov/dps, accessed February 9, 2020).

44. Calculated from Table 11-1 in Federal Jobs Network, "Law Enforcement Hiring Agencies," accessed April 21, 2019, http://www.federaljobs.net/law_agencies.htm.

45. Jennifer Bronson, "Justice Expenditure and Employment Extracts, 2015 (preliminary)," Bureau of Justice Statistics, U.S. Department of Justice, https://www.bjs.gov/index.cfm?ty=pbdetail&iid=6310, filename: jeee15t01.csv (date of version: 6/29/2018); Tracey Kyckelhahn, "Justice Expenditure and Employment Extracts, 2012 (preliminary)," Bureau of Justice Statistics, U.S. Department of Justice, http://bjs.ojp.usdoj.gov/index.cfm?ty=pbdetail&iid=4335 filename jeeus1201.csv (date of version: 2/26/2015).

46. Brian A. Reaves, *Federal Law Enforcement Officers, 2008,* U.S. Department of Justice, Bureau of Justice Statistics Bulletin (June 2012), http://bjs.ojp.usdoj.gov/content/pub/pdf/fleo08.pdf; Federal Law Enforcement Training Center, "About FLETC," www.fletc.gov/about-fletc (accessed October 16, 2012).

47. U.S. Customs and Border Protection, "Snapshot: A Summary of CBP Facts and Figures" March 2019. Accessed April 21, 2019, https://www.cbp.gov/newsroom/stats/typical-day-fy2018.

48. Calculated from data in Ibid. and U.S. Customs and Border Protection, "Snapshot: A Summary of CBP Facts and Figures," December 2015. Accessed January 25, 2016, www.cbp.gov/sites/default/files/documents/cbpsnapshot-121415.pdf.

49. Material on FLETC is from Brian A. Reaves and Lynn M. Bauer, *Federal Law Enforcement Officers, 2002,* U.S. Department of Justice, Bureau of Justice Statistics Bulletin (Washington, DC: U.S. Government Printing Office, August 2003), 11; Federal Law Enforcement Training Center, www.fletc.gov.

50. Unless indicated otherwise, material on the Department of Homeland Security is from the U.S. Department of Homeland Security website, https://www.dhs.gov/about-dhs.

51. Homeland Security, "Department Six-point Agenda." Accessed April 23, 2019, https://www.dhs.gov/department-six-point-agenda.

52. Homeland Security, "Our Mission," https://www.dhs.gov/our-mission (accessed April 23, 2019).

53. Homeland Security, "Operational and Support Components," November 20, 2018. Accessed April 23, 2019, https://www.dhs.gov/operational-and-support-components.

54. Homeland Security, "Office of the Secretary," March 20, 2019. Accessed April 23, 2019, https://www.dhs.gov/office-secretary.

55. Homeland Security, "Advisory Panels & Committees," November 14, 2018. Accessed April 24, 2019, https://www.dhs.gov/landing-page/advisory-panels-committees.

56. Federal Bureau of Investigation, "Terrorism in the United States, 1999," 10, www.fbi.gov/publications/terror/terrorism.htm.

57. Material on Homeland Security and the FBI is from the following sources: Rebecca Carr, "FBI Chief: My Agency Can Handle Spying Job," *The Orlando Sentinel*, December 20, 2002, A11; Eric Lichtblau, "Pressure on FBI Grows from Within," *The Orlando Sentinel*, December 2, 2002, A9; Curt Anderson, "FBI Seeks to Expand Presence Overseas," *The Orlando Sentinel*, March 29, 2003, A3; "Senate Confirms Ridge as Head of Homeland Dept," *USA Today*, January 23, 2003, 12A; "FBI Tackles Fewer Cases, Focuses on Fighting Terrorists," *The Orlando Sentinel*, March 7, 2008, A4.

58. Richard B. Schmitt, "FBI Forecast Mortgage Crisis, Failed to Block It," *The Orlando Sentinel*, August 26, 2008, A6.

59. Eric Lichtblau, David Johnston, and Ron Nixon, "Uptick in Financial Crime Catches FBI Flat-Footed," *The Orlando Sentinel*, October 19, 2008, A6.

60. The Federal Bureau of Investigation, "Major Financial Crime: Using Intelligence and Partnerships to Fight Fraud Smarter," April 4, 2012, www.fbi.gov/news/stories/2012/april/financial_040412/financial_040412 (accessed October 21, 2012).

61. U.S. Department of Justice, Office of the Inspector General, *Interim Report on the Federal Bureau of Investigation's Implementation of the Sentinel Project*, Report 12-38, September 7, 2012, www.justice.gov/oig/reports/2012/a1238.pdf (accessed October 21, 2012).

62. James B. Comey, "Threats to the Homeland," Statement Before the Senate Committee on Homeland Security and Governmental Affairs, October 8, 2015. Accessed April 24, 2019, https://www.fbi.gov/news/testimony/threats-to-the-homeland.

63. Christopher Wray, "Current Threats to the Homeland," Statement Before the Senate Homeland Security and Government Affairs Committee, September 27, 2017. Accessed April 25, 2019, https://www.fbi.gov/news/testimony/current-threats-to-the-homeland.

64. Alan Levin, "Napolitano Announces End to Color-Coded Terror Alerts," *USA Today*, accessed January 29, 2011, www.usatoday.com/news/washington/2011-01-27-terroralert27_ST_N.htm.

65. U.S. Department of State, "Patterns of Global Terrorism 2003," www.state.gov/s/ct/rls/pgtrpt/2003/c12153.htm (Introduction, p. v).

66. U.S. Department of State, "Country Reports on Terrorism 2011," accessed April 26, 2019, https://www.state.gov/j/ct/rls/crt/2011/195540.htm.

67. U.S. Department of State, "Country Reports on Terrorism 2013," accessed April 26, 2019, https://www.state.gov/j/ct/rls/crt/2013/224819.htm.

68. U.S. Department of State, "Country Reports on Terrorism 2014," accessed April 26, 2019, https://www.state.gov/j/ct/rls/crt/2014/239403.htm.

69. U.S. Department of State, "Country Reports on Terrorism 2015." Accessed April 27, 2019, https://www.state.gov/j/ct/rls/crt/2014/239403.htm.

70. Katharine Q. Seelye and Jess Bidgood, "Breaking Silence, Dzhokhar Tsarnaev Apologized for Boston Marathon Bombing," *The New York Times*, June 24, 2015. Accessed January 27, 2016, www.nytimes.com/2015/06/25/us/boston-marathon-bombing-dzhokhar-tsarnaev.html?_r=0.

71. "San Bernardino Shooting Updates," *Los Angeles Times*, December 9, 2015. Accessed January 27, 2016, www.latimes.com/local/lanow/la-me-ln-san-bernardino-shooting-live-updates-htmlstory.htm.

72. Ariel Zambelich and Alyson Hurt, "3 Hours in Orlando: Piercing Together an Attack and Its Aftermath," *NPR*, June 26, 2016. Accessed April 29, 2019, https://www.npr.org/2016/06/16/482322488/orlando-shooting-what-happened-update.

73. Ben Wedeman and Lauren Said-Moorhouse, "ISIS Has Lost Its Final Stronghold in Syria, the Syrian Democratic Forces Says," *CNN*, March 23, 2019. Accessed April 27, 2019, https://www.cnn.com/2019/03/23/middleeast/isis-caliphate-end-intl/index.html.

74. U.S. Department of State, "Country Reports on Terrorism 2017." Accessed April 27, 2019, https://www.state.gov/j/ct/rls/crt/2017/285830.htm.

75. Kevin Strom, Marcus Berzofsky, Bonnie Shook-Sa, Kelle Barrick, Crystal Daye, Nicole Horstmann, and Susan Kinsey, *The Private*

Security Industry: A Review of the Definitions, Available Data Sources, and Paths Moving Forward. U.S. Department of Justice, Bureau of Justice Statistics, Document No. 232781, December 2010, www.ncjrs.gov/pdffiles1/bjs/grants/232781.pdf (accessed October 21, 2012).

76. Ibid.

77. Bureau of Labor Statistics, U.S. Department of Labor, *Occupational Outlook Handbook*, 2016–17 edition, "Security Guards and Gaming Surveillance Officers." Accessed January 25, 2016, www.bls.gov/ooh/protective-service/security-guards.htm.

78. Ibid.

79. Ibid.

80. Robert H. Perry Associates, U.S. Contract Security Industry, White Paper, 10th ed., July 2019, p. 48. Accessed April 29, 2019, https://www.roberthperry.com/uploads/.

81. Calculated from data in previous sections of this chapter and in Bureau of Labor Statistics, U.S. Department of Labor, *Occupational Outlook Handbook*, Security Guards and Gaming Surveillance Officers, accessed April 28, 2019, https://www.bls.gov/ooh/protective-service/security-guards.htm#tab-6. Robert H. Perry and Associates estimated that in 2018, there were approximately 810,000 private security employees and about 900,000 public law enforcement employees.

82. Bureau of Labor Statistics, U.S. Department of Labor, *Occupational Outlook Handbook*, Police and Detectives. Accessed April 28, 2019, https://www.bls.gov/ooh/protective-service/home.htm.

83. Robert H. Perry Associates, U.S. Contract Security Industry, White Paper, 10th ed., July 2019, pp. 7–8, 48. Accessed April 28, 2019, https://www.roberthperry.com/uploads/.

84. Bureau of Labor Statistics, U.S. Department of Labor, *Occupational Outlook Handbook*, 2012–13 edition, "Security Guards and Gaming Surveillance Officers." Accessed October 21, 2012, http://www.bls.gov/ooh/protective-service/security-guards.htm.

85. Bureau of Labor Statistics, "Labor Force Statistics from the Current Population Survey." Accessed April 29, 2019, https://www.bls.gov/cps/cpsaat11.htm.

86. Bureau of Labor Statistics, U.S. Department of Labor, Occupational Outlook Handbook, "Pay." Accessed April 29, 2019, https://www.bls.gov/ooh/protective-service/security-guards.htm#tab-5; Robert H. Perry Associates, U.S. Contract Security Industry, White Paper, 10th ed., July 2019, p. 51. Accessed April 29, 2019, https://www.roberthperry.com/uploads/.

87. Robert H. Perry Associates, U.S. Contract Security Industry, White Paper, 10th ed., July 2019, pp. 54–55, 87–88. Accessed April 29, 2019, https://www.roberthperry.com/uploads/.

88. Ibid., pp. 54 and 87.

89. Ibid., pp. 54 and 88.

90. Ibid., p. 53.

91. Kevin Strom, Marcus Berzofsky, Bonnie Shook-Sa, Kelle Barrick, Crystal Daye, Nicole Horstmann, and Susan Kinsey, *The Private Security Industry: A Review of the Definitions, Available Data Sources, and Paths Moving Forward.* U.S. Department of Justice, Bureau of Justice Statistics, Document No. 232781, December 2010. Accessed October 21, 2012, www.ncjrs.gov/pdffiles1/bjs/grants/232781.pdf.

92. Ibid.

93. Ibid.

94. Ibid.

95. Robert H. Perry Associates, U.S. Contract Security Industry, White Paper, 10th ed., July 2019, p. 51. Accessed April 29, 2019, https://www.roberthperry.com/uploads/.

96. Ibid., p. 52.

97. Ibid., pp. 55–56.

98. Kevin Strom, Marcus Berzofsky, Bonnie Shook-Sa, Kelle Barrick, Crystal Daye, Nicole Horstmann, and Susan Kinsey, *The Private Security Industry: A Review of the Definitions, Available Data Sources, and Paths Moving Forward.* U.S. Department of Justice, Bureau of Justice Statistics, Document No. 232781, December 2010. Accessed October 21, 2012, www.ncjrs.gov/pdffiles1/bjs/grants/232781.pdf.

99. Security Guard Training Central, "The Complete Guide to USA Security Guard License Requirements," 2015–2019. Accessed May 1, 2019, https://www.securityguardtrainingcentral.com/complete-guide-usa-security-guard-license-requirements/.

100. Kevin Strom, Marcus Berzofsky, Bonnie Shook-Sa, Kelle Barrick, Crystal Daye, Nicole Horstmann, and Susan Kinsey, *The Private Security Industry: A Review of the Definitions, Available Data Sources, and Paths Moving Forward.* U.S. Department of Justice, Bureau of Justice Statistics, Document No. 232781, December 2010. Accessed October 21, 2012, www.ncjrs.gov/pdffiles1/bjs/grants/232781.pdf.

101. Security Guard Training Tips, "Unarmed Security Guard State Requirements," 2013. Accessed May 1, 2019, http://securityguardtrainingtips.com/Unarmed-Security-Guard-State-Requirements.php.

102. Kevin Strom, Marcus Berzofsky, Bonnie Shook-Sa, Kelle Barrick, Crystal Daye, Nicole Horstmann, and Susan Kinsey, *The Private Security Industry: A Review of the Definitions, Available Data Sources, and Paths Moving Forward.* U.S. Department of Justice, Bureau of Justice Statistics, Document No. 232781, December 2010. Accessed October 21, 2012, www.ncjrs.gov/pdffiles1/bjs/grants/232781.pdf.

103. Larry Margasak, "Low-Wage Security Has High Cost—Insecurity," *The Orlando Sentinel*, May 30, 2007, A3.

104. Ibid.

Design Elements: *Image of part of a court building with columns and statue*: Pixtal/AGE Fotostock RF; *Image of a door slightly open*: jadding/Shutterstock.com

6

TIMOTHY A. CLARY/Staff/Getty Images

Policing: Roles, Styles, and Functions

CHAPTER OUTLINE

Policing in America
- The Roles of the Police
- Characteristics of Police Work
- Operational Styles

Police Functions
- Patrol
- Investigation
- Traffic
- Drug Enforcement

Community Policing
- The Philosophy and Components of Community Policing

Terrorism and Homeland Security
- Definitions and Types of Terrorism
- The Law Enforcement Response to Terrorism
- How Prepared Is the United States to Defend Against Terrorism?

LEARNING OBJECTIVES

After completing this chapter, you should be able to:

1. Identify characteristics of police work.
2. Describe what studies of police operational styles show.
3. List the four major functions of police departments.
4. List the drug enforcement strategies of local police agencies.
5. Explain the main components of community policing.
6. Identify the four steps in a community policing approach to problem solving.
7. Define terrorism, and identify different types of terrorism.

Crime Story

On November 7, 2018, Ian David Long (pictured), 28, killed 11 people at the Borderline Bar & Grill in Thousand Oaks, California. He used his legally owned Glock 21 .45-caliber pistol with an extended magazine. Another 22 people were injured during the mass killing, which began at about 11:15 P.M. local time at the country music bar on Wednesday night's "College Country Night." The bar and weekly event are popular with students from nearby Pepperdine University, California Lutheran University, and California State University, Channel Islands. At the time of the shooting, the bar likely held hundreds of people.

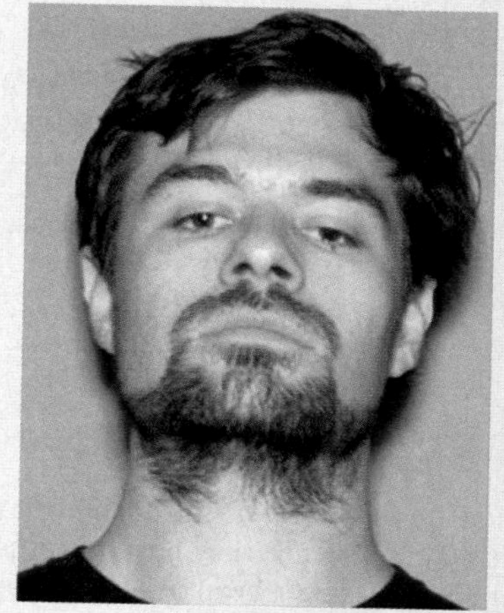

California Department of Motor Vehicles/AP Images

Ventura County Sheriff's Department/AP Images

Long lived near the bar in Newbury Park. He was a veteran of the U.S. Marine Corps, having served as a machine gunner in Afghanistan from November 2010 to June 2011. He attained the rank of corporal while on active duty and was honorably discharged in 2013. He might have been suffering from post-traumatic stress disorder. He had been the victim of a battery at a local bar in 2015, and in April 2018, deputies were called to his house for a "disturbance." The deputies talked with him and found him to be "somewhat irate" and "acting a little irrationally." A crisis intervention team and mental health specialist were called to the house, but after speaking with Long, they left the house without taking him into custody or otherwise intervening.

Dressed all in black and wearing a hood, Long reportedly approached the bar and shot a security guard outside, then entered the bar, threw several smoke bombs, and eventually shot and killed 11 people inside. The smoke bombs obstructed what employees and customers could see before he opened fire. To deal with the smoke, Long used a flashlight with a laser sight attached to his semi-automatic pistol as he fired. When the gunfire began, panicked students and other customers dived under tables, piled on top of each other, and scrambled to escape through windows, back doors, and fire escapes. Many of them hid in bathrooms or the attic. Several customers were cut from glass while jumping through broken windows trying to escape. A law enforcement spokesperson commented that young people, or people at nightclubs, have learned that these mass killings may happen, so they think about what they would do if they are caught in such a situation. That awareness probably saved a lot of lives, remarked the spokesperson. Also helping to save lives were about six off-duty law enforcement officers from various agencies that were inside the bar at the time of the shooting. Ironically, several of the customers at the Borderline that night were at last year's Route 91 Harvest Festival on the Las Vegas Strip, where, on October 1, 58-year-old Stephen Paddock opened fire from his 32nd-floor suite at the Mandalay Bay Resort and Casino into the crowd of 22,000 concertgoers across the street, killing 58 and wounding another 800. The mass shooting at the Route 91 Harvest Festival is considered the deadliest in modern American history.

Included among the dead at the Borderline Bar & Grill was 54-year-old Deputy Sheriff Sergeant Ron Helus (pictured), a 29-year law enforcement veteran, who, with a California Highway Patrol officer, were the first to arrive at the scene while shots still were being fired. Relying on active-shooter training that emphasizes a fast response, Sergeant Helus and the other officer entered the club with high-powered rifles and immediately exchanged gunfire with Long. Long was not struck by any of their rounds but took his own life after the firefight. Sergeant Helus, who was wearing a bulletproof vest, was shot five times and later died at a nearby hospital. However, according to the chief medical examiner, Sergeant Helus was not killed by Long's handgun that inflicted serious but potentially survivable wounds. Rather he was killed by "friendly fire." He was shot in the heart and killed by the California highway patrol officer, who had joined him in the chaotic gun battle. Sergeant Helus was scheduled to retire within a year or so. When he received the call about the shooting, he was on the phone with his wife. He quickly ended the call with "I love you, talk to you later." He had a son.

Among the topics addressed in this chapter are the characteristics of police work, and danger is one of those characteristics. Another topic examined in the chapter is terrorism, including domestic terrorism. Was the mass shooting at the Borderline Bar & Grill an act of domestic terrorism? Why or why not? Could the mass shooting at the Borderline Bar & Grill have been prevented? If so, how? Could the number of fatalities and those wounded at the Borderline Bar & Grill have been reduced? If so, how? Had Ian Long lived, should he have received the death penalty? Why or why not? (California currently has the death penalty.) Could you have charged into the Borderline Bar & Grill and confronted the shooter? How do you think the "friendly fire" death of Sergeant Helus will affect the California highway patrol officer who accidentally killed him? How would it affect you? Does the story about the mass shooting at the Borderline Bar & Grill and the death of Sergeant Helus affect your decision about becoming a law enforcement officer? The answers to these questions reveal some of the difficulties and hardships of being a law enforcement officer, especially a police officer or deputy sheriff, in America.

Policing in America

The police are at the forefront of the criminal justice process and, for most people, the only personal experience they have with that process is contact with a local police officer. Most people have never been in a courthouse for a criminal matter or in a jail or prison for any reason. This chapter examines what the police do and the qualities they need to do it.

The Roles of the Police

Our expectations of police behavior depend on where we live and when we consider the question. For example, we saw in the last chapter that Cincinnati wanted its police officers in the 1880s to be fleet-footed and honest. In Dallas, Miami, and New York City, citizens may expect police officers to have a working knowledge of Spanish. In Alaska, we would expect police officers to be self-reliant, enjoy the outdoors, and not mind working by themselves in lonely surroundings. In essence, what we expect from the police depends on how we view their role in society.

role The rights and responsibilities associated with a particular position in society.

role expectation The behavior and actions that people expect from a person in a particular role.

role conflict The psychological stress and frustration that results from trying to perform two or more incompatible responsibilities.

A **role** consists of the rights and responsibilities associated with a particular position in society. A related concept is **role expectation**, the behavior and actions that people expect from a person in a particular role. Suppose, for example, that teenagers living in a wealthy neighborhood have been caught drinking alcohol. Their parents probably expect police officers to warn the young people and bring them home. In a less affluent neighborhood, however, the expectation of community residents might be that the police will arrest the teenagers and bring them into juvenile court. This example illustrates a problem that often arises in our attempt to understand the police role in America. When the public's expectations differ from the official police role, the public may become disenchanted and sometimes hostile toward law enforcement officers. Such negative feelings cause officers personal frustration and role conflict. **Role conflict** is the psychological strain and stress that result from trying to perform two or more incompatible responsibilities. A common source of role conflict for the police is the expectation that they should be social or helping agents at the same time they are expected to be control agents by arresting law violators.

What we expect from police officers, then, depends on how we view the police role–a role that has been described as complex, ambiguous, changing, and repressive. Obviously, not everyone views the role of the police in the same way, but a definition that includes the majority of perspectives is possible. The police:

1. Are community leaders in public safety. (By nature, this makes the work potentially dangerous.)
2. Possess broad discretion.
3. Solve sociological and technological problems for people on a short-term basis.
4. Occasionally serve in a hostile or dangerous environment.[1]

Think about some of the common situations in which police officers find themselves when people call and want something "fixed." One example would be an officer's response

Spencer Ainsley/The Image Works

Police officers are expected to respond to traffic accidents. *How might such experiences affect them?*

to freeway accidents in which vehicles are overturned and burning and people are trapped inside. Such situations require leadership, informed and quick decisions, the solving of numerous immediate problems, and the use of extreme caution to prevent further injury to citizens or the police officer. Another example would be intervention in a long-running family dispute that has suddenly turned violent. Such a situation requires caution, quick thinking, and the solving of a number of problems in an effort to ensure the safety of all parties. Still another typical role of a police officer is to provide protection at protests and strikes. Those potentially volatile circumstances clearly illustrate the key elements of the police role. Of course, sometimes an officer's role simply may be to solve problems in the course of providing service; for example, when retrieving a citizen's dropped keys from below a sewer grate.

Characteristics of Police Work

Police work requires a combination of special characteristics. Personnel with the following qualities are best able to carry out the difficult service role mandated for law enforcement officers.

Quick Decision Making Sometimes, police officers must make on-the-spot decisions about whether to use force, how to maneuver a patrol car, or whether to stop a suspect. Making the wrong decision can be fatal for the officer or the other person. All of the work in a lengthy investigation can be ruined by a single procedural law violation if an officer unintentionally makes a wrong decision.

The Independent Nature of Police Work The position of peace officer in all states in the United States is a position of honor and trust. After patrol officers attend roll call, stand inspection, check out their equipment, and depart into the streets in their patrol cars, they work virtually unsupervised until the end of their tour of duty.

Figure 6.1 shows the Law Enforcement Officer Code of Ethics, which was written as a guide for working police officers. It offers some professional direction in a line of work with many opportunities to go astray. The independent nature of police work increases the chances of malfeasance and corruption–topics discussed in the next chapter.

"Dirty Work" Most people agree that police work needs to be done, but police work sometimes is distasteful–for example, dealing with people who have committed horrible acts and viewing mangled, broken, and decomposed bodies. Often, the police must deal

FIGURE 6.1 Law Enforcement Officer Code of Ethics

As a law enforcement officer, my fundamental duty is to serve the community; to safeguard lives and property; to protect the innocent against deception, the weak against oppression or intimidation and the peaceful against violence or disorder; and to respect the constitutional rights of all to liberty, equality, and justice.

I will keep my private life unsullied as an example to all and will behave in a manner that does not bring discredit to me or to my agency. I will maintain courageous calm in the face of danger, scorn or ridicule; develop self-restraint; and be constantly mindful of the welfare of others. Honest in thought and deed both in my personal and official life, I will be exemplary in obeying the law and the regulations of my department. Whatever I see or hear of a confidential nature or that is confided to me in my official capacity will be kept ever secret unless revelation is necessary in the performance of my duty.

I will never act officiously or permit personal feelings, prejudices, political beliefs, aspirations, animosities or friendships to influence my decisions. With no compromise for crime and with relentless prosecution of criminals, I will enforce the law courteously and appropriately without fear or favor, malice or ill will, never employing unnecessary force or violence and never accepting gratuities.

I recognize the badge of my office as a symbol of public faith, and I accept it as a public trust to be held so long as I am true to the ethics of police service. I will never engage in acts of corruption or bribery, nor will I condone such acts by other police officers. I will cooperate with all legally authorized agencies and their representatives in the pursuit of justice.

I know that I alone am responsible for my own standard of professional performance and will take every reasonable opportunity to enhance and improve my level of knowledge and competence.

I will constantly strive to achieve these objectives and ideals, dedicating myself before God to my chosen profession . . . law enforcement.

Source: The International Association of Chiefs of Police (IACP), "Law Enforcement Code of Ethics," accessed May 5, 2019, https://www.theiacp.org/resources/law-enforcement-code-of-ethics.

with people at their worst–angry, drunk, in trouble, victimized, violent, and so forth. The distasteful part of policing has been referred to as "dirty work."[2]

Danger Police officers in the United States spend a substantial amount of their time trying to resolve conflicts, frequently in hostile environments.[3] Table 6.1 identifies dangerous circumstances in which officers find themselves. Contrary to the media image, police officers often are afraid on the job, and far too many are injured or killed. The data reveal that disturbance calls (e.g., a family quarrel or a bar fight) and arrests of suspects are the most dangerous circumstances for police officers. Because of the danger they face, many departments require their field officers to wear body armor while on duty. In 2013 (the latest year for which data were available), 79% of local police departments required their field officers to wear body armor while on duty; 71% of departments required them to wear body armor at all times, while 8% of departments required them to wear body armor only in some circumstances.[4]

Despite the use of body armor and other precautions, each year police officers are killed while on duty. From 1972 through 2018, 6,215 law enforcement officers were killed in the line or duty (an average of 135 per year): 3,371 or 73 per year were feloniously killed, and 2,844 or 62 per year were accidentally killed.[5] In 2018, 55 officers were feloniously killed in the line of duty, many fewer than the 142 officers killed feloniously in 2001, 9 more than the 46 feloniously killed in 2017, and 17 fewer than the 72 officers killed feloniously in 2011–the most officers feloniously killed in any year since 2001. However, 2001 was an unusual year. Among the 142 officers feloniously killed were the 72 federal, state, and local officers killed during the tragedy of September 11–the most officers killed in the United States on a single day. In 2013, only 27 officers were feloniously killed in the line of duty,

TABLE 6.1 Law Enforcement Officers Assaulted in the United States by Circumstance, 2018

Circumstances at Scene of Incident	Total	Percentage of Total
Total	58,866	100%
Disturbance calls (family quarrel, bar fight, etc.)	18,232	30.9
Attempting other arrest	9,688	16.5
Handling, transporting, custody of prisoners	7,286	12.4
Investigating suspicious persons and circumstances	5,644	9.6
Traffic pursuits and stops	4,809	8.2
Handling person with mental illness	2,230	3.8
Burglary in progress or pursuing burglary suspects	762	1.3
Civil disorder (mass disobedience, riot, etc.)	590	1.0
Robbery in progress or pursuing robbery suspects	497	0.9
Ambush situation	300	0.5
All other	8,828	14.9

Source: Calculated from U.S. Department of Justice, Federal Bureau of Investigation, *2018 Law Enforcement Officers Assaulted*, accessed February 9, 2020, https://ucr.fbi.gov/leoka/2018/tables/table-83.xls.

which was the lowest recorded figure in more than 35 years.[6] Table 6.2 shows the circumstances in which police officers were feloniously killed in 2018. Accidents, such as automobile accidents, during the performance of official duties claimed the lives of an additional 51 officers in 2018, 4 more than the 47 in 2017, and 32 fewer than the 83 in 2007—the most officers accidentally killed since 2001.[7] The fewest officers accidentally killed in any year since 2001 were the 45 killed in 2014 and 2015.[8]

Killed in the Line of Duty

Today, more than one million sworn officers put their lives on the line for our protection each day. The first known line-of-duty death was that of U.S. Marshal Robert Forsyth, who was shot and killed January 11, 1794, while serving court papers in a civil suit. Wilmington, Delaware, police matron Mary T. Davis was the first female officer killed on duty. She was beaten to death in 1924 while guarding a prisoner in the city jail.

TABLE 6.2 Circumstances in Which Police Officers Were Feloniously Killed, 2018

55 police officers were feloniously killed in 2018; 51 were killed with firearms; 43 were wearing body armor.
23 officers were killed during investigative or enforcement activities.
- 8 were performing investigative activities.
- 6 were involved in tactical situations.
- 3 were investigating suspicious persons or circumstances.
- 3 were interacting with a wanted person.
- 2 were conducting traffic violation stops.
- 1 was handling a person with a mental illness.

11 officers were ambushed.
6 officers were involved in pursuits.
- 4 were involved in foot pursuits.
- 2 were involved in vehicular pursuits.

4 officers responded to crimes in progress.
- 2 were burglaries in progress.
- 1 was a report of a person with a firearm (no shots fired).
- 1 was a report of a crime against property.

3 officers were involved in arrest situations.
- All 3 were attempting to control/handcuff/restrain the offenders.

2 officers were on administrative assignments while performing a prisoner transport.
2 officers were assisting other law enforcement officers in foot pursuits.
2 officers responded to disorders/disturbances.
- 1 was responding to a disturbance call.
- 1 encountered a domestic violence situation.

1 officer was involved in unprovoked attack.
1 officer was performing traffic control (crash scene, directing traffic, etc.)

Source: U.S. Department of Justice, Federal Bureau of Investigation, "2018 Law Enforcement Officers Killed & Assaulted," accessed May 6, 2019, https://ucr.fbi.gov/leoka/2018/topic-pages/officers-feloniously-killed.

Operational Styles

operational styles The different overall approaches to the police job.

After police officers are trained and begin to gain experience and wisdom from their encounters with veteran police officers and citizens on the street, it is believed that they develop **operational styles** that characterize their overall approach to the police job. If these styles actually exist, it means that the effort of the police department to systematically train and deploy officers with the same philosophy and practical approach to policing in the community has not been entirely successful. The research on operational styles shows that they vary both between departments and among officers of the same department.

One of the earliest scholars to report on the existence of policing styles was political scientist James Q. Wilson, who found the following three organizational styles in a study of eight police departments:

1. *Legalistic style*–The emphasis is on violations of law and the use of threats or actual arrests to solve disputes in the community. In theory, the more arrests that are made, the safer a community will be. This style often is found in large metropolitan areas.
2. *Watchman style*–The emphasis is on informal means of resolving disputes and problems in a community. Keeping the peace is the paramount concern, and arrest is used only as a last resort to resolve any kind of disturbance of the peace. This style of policing most commonly is found in economically poorer communities.
3. *Service style*–The emphasis is on helping in the community, as opposed to enforcing the law. Referrals and diversion to community treatment agencies are more common than arrest and formal court action. The service style most likely is to be found in wealthy communities.[9]

Sociologist John Broderick, who also studied operational styles among the police, classified police officers by their degree of commitment to maintaining order and their respect for due process:

1. *Enforcers*–The emphasis is on order, with little respect for due process.
2. *Idealists*–The emphasis is on both social order and due process.
3. *Optimists*–The emphasis is on due process, with little priority given to social order.
4. *Realists*–Little emphasis is given to due process or social order.[10]

Another classification is based on the way officers use their authority and power in street police work. The two key ingredients of this scheme are passion and perspective. Passion is the ability to use force or the recognition that force is a legitimate means of resolving conflict; perspective is the ability to understand human suffering and to use force ethically and morally. According to political scientist William Muir's styles of policing, police officers include:

1. *Professionals*–Officers who have the necessary passion and perspective to be valuable police officers.
2. *Enforcers*–Officers who have passion for the job, for enforcing the law, and for taking decisive action; their inner drive or value system allows them to be comfortable using force to solve problems.
3. *Reciprocators*–Officers who lack the passion to do the job; they have a difficult time taking action, making arrests, and enforcing the law; their values make it difficult for them to use force to solve problems.
4. *Avoiders*–Officers who have neither passion nor perspective, resulting in no recognition of people's problems and no action to resolve them.[11]

Are there identifiable styles of policing? What value do these styles hold for us? In any area of human endeavor, classifications have been constructed. We have developed classifications for leaders, prisoners, quarterbacks, and teachers. These classifications give us a framework of analysis, a basis for discussion. But can they be substantiated when we go into a police agency to see if they actually exist?

Social scientist Ellen Hochstedler examined the issue of policing styles with 1,134 Dallas, Texas, police officers and was not able to confirm the officer styles identified in the literature by Broderick, Muir, and others. Her conclusion was that it is not possible to "pigeonhole" officers into one style because the way officers think and react to street situations varies, depending on the particular situation, the time, and the officers themselves.[12]

A problem with all of the previous studies, including Hochstedler's, is that they are at least 30 years old. Much has changed in the world of policing in the last 30-plus years, including the demographics of police personnel. Today's police departments, particularly in large cities, are likely to employ more females, racial minorities, and college-educated officers than police departments 30 years ago. Recently, criminologist Eugene Paoline examined police occupational styles in light of these changes. Paoline's classification scheme is based on police officers' attitudes toward citizens, supervisors, procedural guidelines, tactics, and police functions and includes seven different types of officers:

1. *Traditionalists*–Officers are distrustful of citizens and do not believe citizens will cooperate with them. They hold unfavorable views of supervisors, who are regarded as unsupportive and out of touch. They also hold unfavorable views of procedural guidelines. For example, they generally believe search-and-seizure laws sometimes should be overlooked. They favor aggressive patrol tactics and selective enforcement of the law. They endorse the crime-fighting aspects of the occupation and reject order maintenance and community policing functions.
2. *Law enforcers*–Officers are most distrustful of citizens and believe citizens are somewhat uncooperative. They relate more favorably to sergeants than senior personnel. They place a premium on due process safeguards and are positively oriented toward aggressive patrol tactics. They hold the strongest crime-fighting attitudes, moderately accept order maintenance functions, but show displeasure with community policing functions.
3. *Old-pros*–Officers hold favorable attitudes of citizens and supervisors. They believe search-and-seizure laws and other legal guidelines should be followed. They hold positive attitudes toward aggressive patrolling tactics and favor some degree of selective enforcement. They accept law enforcement, order maintenance, and community policing functions.
4. *Peacekeepers*–Officers are not distrustful of citizens, and their attitudes toward citizens are more favorable than unfavorable. They are very positive toward sergeants but view senior personnel less favorably. They view aggressive patrolling tactics negatively but have positive views toward selective enforcement. They embrace due process safeguards. They are heavily oriented toward order maintenance but place a much lower value on crime-fighting and slightly less value on community policing.
5. *Lay-lows*–Officers do not distrust citizens and view them as cooperative. They are not overly positive toward senior management but view sergeants much more favorably. They do not favor aggressive patrolling tactics but do favor selective enforcement. They have a narrow role orientation that focuses on law enforcement functions. They view order maintenance and community policing as ambiguous roles and time-consuming functions.
6. *Anti-organizational street cops*–Officers have very strong positive attitudes toward citizens and very negative attitudes toward supervisors. They do not strongly endorse aggressive patrolling tactics but favor selective enforcement. They strongly favor respecting citizens' due process rights. They have a positive attitude toward law enforcement but do not have overly favorable attitudes toward order maintenance and community policing.
7. *Dirty Harry enforcers*–Officers are somewhat distrustful of citizens but view supervisors favorably. They believe strongly in aggressive patrol tactics in crime and noncrime police functions and endorse, at times, the violation of citizens' rights. They also believe in selective enforcement. They are willing to perform law enforcement, order maintenance, and community policing functions.

Paoline's research shows, as did the previous studies before it, that the police do not have a monolithic culture and that police officers in the same department can possess significantly different attitudes toward their work.[13]

Enforcers or Protectors?

According to a recent national survey, 31% of police officers view themselves primarily as protectors, 8% view themselves primarily as enforcers, and 62% view themselves as both equally.

Source: Rich Morin, Kim Parker, Renee Stepler, and Andrew Mercer," Behind the Badge," Pew Research Center, January 11, 2017, accessed June 19, 2019, https://www.pewsocialtrends.org/2017/01/11/behind-the-badge/.

THINKING CRITICALLY

1. Which characteristics do you think are the most important for police officers to have? Why?
2. Could policing be made less dangerous? If so, how?
3. Which operational style of policing do you think is the most effective and the least effective? Defend your answer.
4. Which style of policing have you observed in your own community? Why do you suppose this style of policing is being employed?

Police Functions

The list of functions that police are expected to carry out is long and varies from place to place. In the following sections, we look at the major operations of police departments and the services they provide.

Patrol

Police administrators long have referred to patrol as the backbone of the department. It is unquestionably the most time-consuming and resource-intensive task of any police agency. More than half of the sworn personnel in any police department are assigned to patrol. In Houston, Chicago, and New York City, for example, patrol officers make up more than 65% of the sworn personnel in each department.

Patrol officers respond to burglar alarms, investigate traffic accidents, care for injured people, try to resolve domestic disputes, and engage in a host of other duties that keep them chasing radio calls across their own beats and the entire city and county when no other cars are available to respond. Precisely how to conduct patrol activities, however, is a matter of much debate in the nation today. Indeed, it seems that there are many ways to police a city.

Preventive Patrol For decades, police officers patrolled the streets with little direction. Between their responses to radio calls, they were told to be "systematically unsystematic" and observant in an attempt to both prevent and ferret out crime on their beats.

Steve Hamann/Shutterstock

Street patrol is the most resource-intensive task of any police agency. *Are there acceptable alternatives to street patrol? If yes, what are they?*

In many police departments, as much as 50% of an officer's time is uncommitted and available for patrolling the beats that make up a political jurisdiction. The simultaneous increases in the official crime rate and the size of police forces beginning in the 1960s caused police managers and academics to question the usefulness of what has come to be known as **preventive patrol** or *random patrol*. To test the usefulness of preventive patrol, the now-famous Kansas City (Missouri) Preventive Patrol Experiment was conducted in 1972.

preventive patrol Patrolling the streets with little direction; between responses to radio calls, officers are "systematically unsystematic" and observant in an attempt to both prevent and ferret out crime. Also known as *random patrol*.

The Kansas City, Missouri, Police Department and the Police Foundation set up an experiment in which 15 patrol districts were divided into three matched groups according to size, record of calls for service, and demographic characteristics. In the first group, the "control beats," the police department operated the same level of patrol used previously in those beats. In the second group of districts, the "proactive beats," the police department doubled or even tripled the number of patrol officers normally deployed in the area. In the third group of districts, the "reactive beats," the police department deployed no officers at all on preventive patrol. Officers only responded to calls for service and did no patrolling on their own. At the end of the 1-year study, the results showed no significant differences in crime rates among the three groups of patrol districts. In other words, a group of districts that had no officers on preventive patrol had the same crime rates as groups that had several times the normal level of staffing engaged in patrol activity. The number of officers made no difference in the number of burglaries, robberies, vehicle thefts, and other serious crimes experienced in the three groups of police districts. Perhaps even more important is that the citizens of Kansas City did not even notice that the levels of patrol in two of the three districts had been changed.[14]

myth

Adding more police officers will reduce crime.

fact

Short of having a police officer on every corner, evidence indicates no relationship between the number of police officers and the crime rate.

The law enforcement community was astounded by the results of the study, which showed that it made no difference whether patrol officers conducted preventive, or random, patrol. The research was immediately attacked on both philosophical and methodological grounds. How could anyone say that having patrol officers on the street made no difference?

One of the criticisms of the study was that no one in the community was told that there were no officers on patrol in reactive districts. What might have happened to the crime rates had the community known no officers were on patrol? Moreover, during the course of the study, marked police cars from other departments and districts crossed the reactive districts to answer calls but then left when the work was completed. Thus, there appeared to be a police presence even in the so-called reactive districts.

This study has forced police executives and academics to reconsider the whole issue of how patrol is conducted, once considered a closed issue. Police administrators have begun to entertain the possibility of reducing the number of officers on patrol. Innovations in patrol methods also have been proposed.

directed patrol Patrolling under guidance or orders on how to use patrol time.

GIS crime mapping A technique that involves the charting of crime patterns within a geographic area.

Directed Patrol In **directed patrol**, officers are given guidance or orders on how to use their patrol time. The guidance is often based on the results of crime analyses that identify problem areas. Evidence shows that directed patrol can reduce the incidence of targeted crimes such as thefts from autos and robberies.[15]

Police Views of Patrol

According to a recent national survey, 92% of police officers thought that requiring officers to patrol more frequently in high crime areas was "very useful" (58%) or "somewhat useful" (34%).

Source: Rich Morin, Kim Parker, Renee Stepler, and Andrew Mercer, "Behind the Badge," Pew Research Center, January 11, 2017, accessed June 19, 2019, https://www.pewsocialtrends.org/2017/01/11/behind-the-badge/.

Crime Mapping One technological innovation in crime analysis that has aided directed patrol is geographic information system (GIS) crime mapping. **GIS crime mapping** is a technique that involves the charting of crime patterns within a geographic area. Crime mapping makes it possible to keep a closer watch on crime and criminals through the generation of crime maps capable of displaying numerous fields of information. For example, if a series of armed robberies of dry cleaning stores had been committed over a period of several weeks in three adjacent police beats, police crime analysts would be able to record, analyze, and determine a definite pattern to these robberies and make a reasonable prediction as to when and where the next robbery in the series is likely to occur. The patrol and investigation forces could be deployed at a prescribed time to conduct surveillance of the prospective target dry cleaning store or stores with a good chance the robber can be arrested. This use of crime mapping is referred to as "resource reallocation" and is probably the most widely used crime-mapping application. Figure 6.2 is an example of two crime maps. Note that the two crime maps in Figure 6.2 show the overall decrease in the crime rate between 2013 and 2017 in Atlanta, Georgia, and its surrounding areas.

Crime mapping also is used as a tool to help evaluate the ability of police departments to resolve the problems in their communities. This is the primary purpose of the New York

FIGURE 6.2 Crime Map of Total Crime Index in Atlanta, Georgia, and Its Surrounding Areas, 2013 and 2017

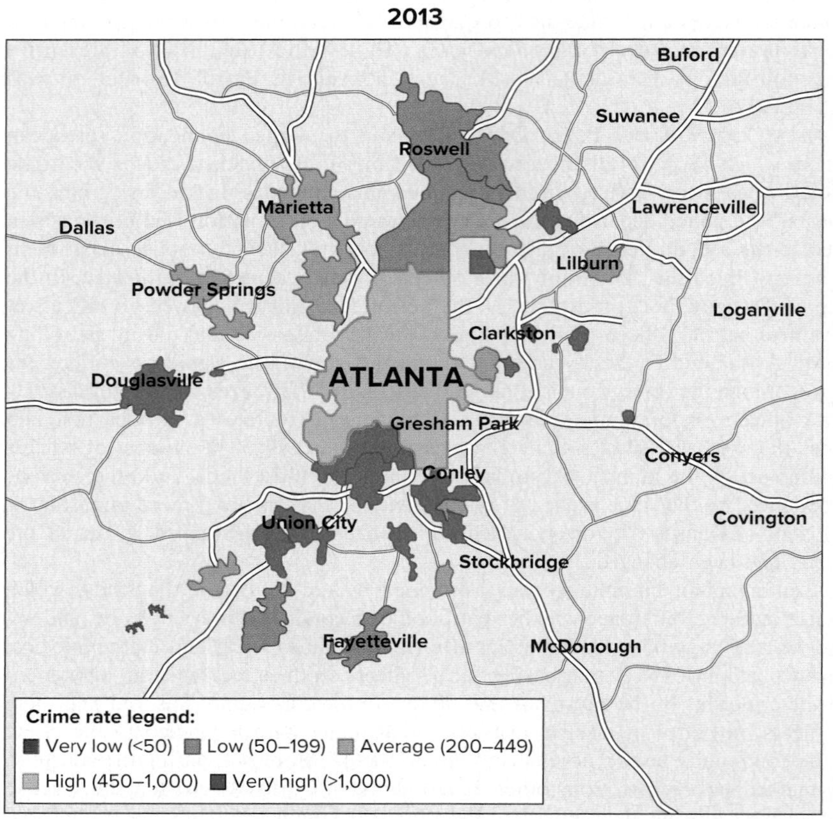

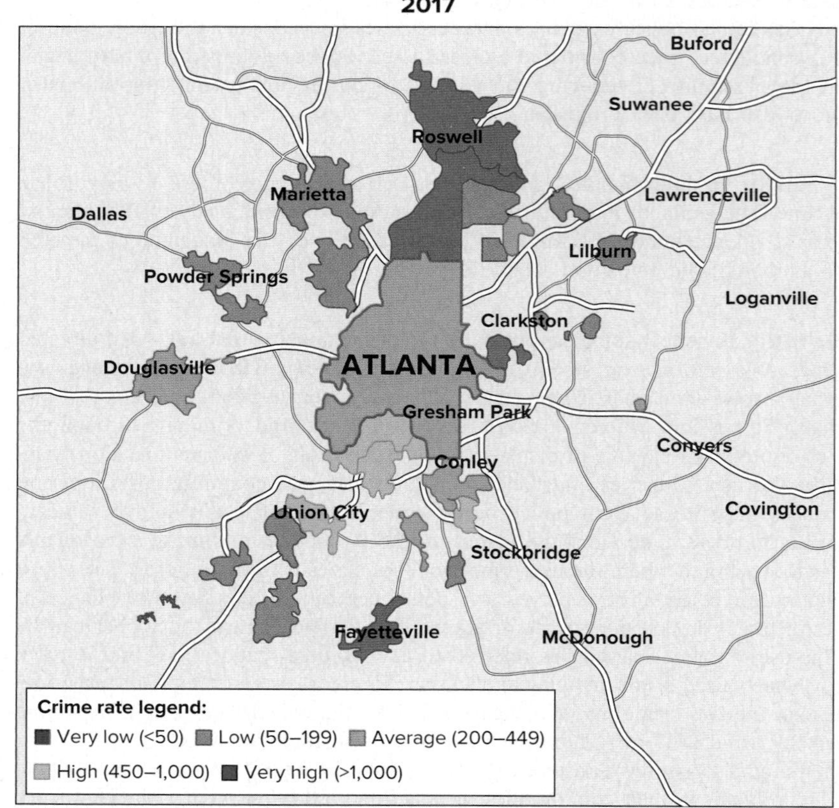

Source: http://www.city-data.com/crime/crime-Atlanta-Georgia.html (accessed May 21, 2019).

City Police Department's CompStat process, for example.[16] Begun in 1994, CompStat is a divisional unit responsible for statistical analysis of daily precinct crime reports frequently using crime mapping. The information produced by CompStat is used by the chief of police to judge the performance of precinct commanders and by precinct commanders to hold their officers accountable (see the discussion of CompStat in Chapter 5).

Crime mapping is likely to be used increasingly in crime scene investigations and the forensic sciences. For example, a GIS-based system has been created that can determine the origin of gunshots through sound triangulation. Crime mapping also will be combined with other technologies such as aerial photography so that geocoded data can be superimposed on aerial photographs rather than computer-generated maps. This should aid community policing efforts by making census data, liquor license locations, drug-market data, and probationer addresses, for example, readily available in a more useful form. Another technology that will be combined with crime mapping is global positioning system (GPS) technology. It would allow beat officers to track and monitor probationers and parolees in the area, for example. It currently is used in some departments to help manage the department's fleet of vehicles.[17]

GIS Crime Mapping

To learn more about GIS crime mapping and how it works, visit the GIS Lounge at www.gislounge.com.

Why is crime mapping important?

Aggressive Patrol In nearly all police departments, some patrol officers have used aggressive patrol tactics and have been rewarded as high performers because they made many arrests for both minor and serious offenses. When the entire patrol section is instructed to make numerous traffic stops and field interrogations, the practice is referred to as **aggressive patrol**. A **field interrogation** is a temporary detention in which officers stop and question pedestrians and motorists they find in suspicious circumstances. Such procedures have been found to reduce crime in targeted areas.[18]

aggressive patrol The practice of having an entire patrol section make numerous traffic stops and field interrogations.

field interrogation A temporary detention in which officers stop and question pedestrians and motorists they find in suspicious circumstances.

At least two problems can occur as a result of aggressive patrol. First, random traffic stops and field interrogations inconvenience innocent citizens. To avoid conflict, the police must be certain that those tactics are necessary, and they must explain the necessity to the public. Second, it often is difficult to get all officers on each work shift and in each patrol division motivated to use aggressive patrol tactics. Many officers are reluctant to carry out their duties in an aggressive way. Nevertheless, when crime rates are high and with research confirming that aggressive patrol can reduce crime, aggressive patrol tactics are likely to continue.

Foot Patrol For some time, there has been renewed interest in having police officers patrol their beats on foot. Is there value in this practice, or is it just nostalgia for a more romantic period in law enforcement? The use of motorized patrols has allowed the police to respond rapidly to citizen calls and to cover large geographical areas. Yet, officers working a busy shift, perhaps responding to more than two dozen calls, come to feel as if they are seeing the world through a windshield. Moreover, it is now generally accepted that rapid response time is useful in only a small portion of the incidents and crimes to which the police are asked to respond.

Challenging conventional wisdom about rapid response, two cities–Flint, Michigan, and Newark, New Jersey–launched substantial foot patrol programs. In Newark, the results of the foot patrol experiment showed that foot patrol had little or no effect on the level of crime. However, positive effects were identified:

1. Newark residents noticed whether foot patrol officers were present.
2. They were more satisfied with police service when foot patrol officers delivered it.
3. They were less afraid than citizens being served by motorized patrol.[19]

In Flint, Michigan, the extensive neighborhood foot patrol experiment also had positive results:

1. Flint residents had a decreased fear of crime.
2. Their satisfaction with police service increased.
3. There were moderate decreases in crime.
4. There were decreased numbers of calls for police service.

Justin Sullivan/Staff/Getty Images

Field interrogation has been found to reduce crime in targeted areas. *What are some of the problems with field interrogations?*

Citizens would wait to talk to their neighborhood foot patrol officer about a problem instead of calling the police department through 911 and speaking with an officer they were not likely to know. One astounding result of the Flint program was that the foot patrol officers became so popular that citizens saw them as real community leaders. They often became more influential than some elected officials. Evidence of the degree of satisfaction with the foot patrol program in Flint was that the community voted three times to continue and expand foot patrol at a time when the city was experiencing one of the nation's highest unemployment rates.[20] Perhaps even more important, the findings of foot patrol research provided the seeds of a much broader concept for law enforcement: community policing, which we will discuss later in this chapter.

Investigation

The role of the detective generally has been glorified by media sources in both fiction and nonfiction accounts. Homicide investigation, in particular, has captured the imagination of fiction readers worldwide. Most police officers aspire to be investigative specialists by attaining the position of detective. But it should be noted that detectives represent only one unit in a police department that conducts investigations. Investigators work in a variety of capacities in a police agency:

1. Traffic homicide and hit-and-run accident investigators in the traffic section.
2. Undercover investigators in narcotics, vice, and violent gang cases.
3. Internal affairs investigators conducting investigations of alleged crimes by police personnel.
4. Investigators conducting background checks of applicants to the police department.
5. Uniformed patrol officers investigating the crimes to which they have been dispatched or have encountered on their own while on patrol.
6. Detectives of criminal investigation divisions who conduct investigations into reports of criminal activity made by patrol officers.

What Is Criminal Investigation? Criminal investigation has been defined as a lawful search for people and things to reconstruct the circumstances of an illegal act, apprehend or determine the guilty party, and aid in the state's prosecution of the offender.[21] The criminal investigation process is generally divided into two parts: the preliminary, or initial, investigation and the continuing, or follow-up, investigation. Most of the time the preliminary investigation in both felony and misdemeanor cases is conducted by patrol officers, although for homicides and other complex, time-consuming investigations, trained investigators are dispatched to the crime scene immediately. The continuing investigation in serious crimes is ordinarily conducted by plainclothes detectives, although small and medium-sized agencies may require patrol officers or a patrol supervisor to follow up on serious criminal offenses.

For less serious crimes, many police departments use solvability-factor score sheets or software programs to assess information collected at crime scenes. The assessment, which is done by the responding officer, a case-screening officer, or a felony-review unit, determines which cases are likely to be solved, given the initial information obtained. Promising cases are turned over to detectives for follow-up investigation. The rest are often closed on the basis of the preliminary investigation and are reopened only if additional information is uncovered.[22]

Investigative Functions In any type of investigation in a police agency, all investigators share responsibility for a number of critical functions. They must:

1. Locate witnesses and suspects.
2. Arrest criminals.
3. Collect, preserve, and analyze evidence.
4. Interview witnesses.
5. Interrogate suspects.
6. Write reports.
7. Recover stolen property.
8. Seize contraband.
9. Prepare cases and testify in court.

ED BAILEY/AP Images

Criminal investigation is a time-consuming task that requires much attention to detail. *What aspects of criminal investigations are the most time-consuming and why?*

The specific application and context of those functions vary considerably, depending on whether the investigation is of the theft of expensive paintings, for example, or the rape of an elderly widow living alone.

The Role of the Detective At first glance, the role of the detective seems highly desirable. To a patrol officer who has been rotating work shifts for several years, seldom getting a weekend off, detectives in the police department seem to have a number of advantages:

1. They do not have to wear uniforms.
2. They have anonymity during work hours if they choose it.
3. They have steady work hours, often daytime hours with weekends off.
4. They have offices and desks.
5. They enjoy the prestige associated with the position.
6. In many agencies, detectives receive higher compensation and hold a higher rank.
7. Perhaps most important, they enjoy more freedom than patrol officers from the police radio, geographical boundaries, and close supervision.

All these advantages add up to a high-status position, both within the police department and in the eyes of the public.

Productivity Despite all the advantages of being a detective, investigators often are faced with insurmountable obstacles and stressful work conditions. Notifying the next of kin in a homicide is one of the worst tasks:

> Of all the dirty tasks that go with the dirty work of chasing a killer, notifying the next of kin is the job that homicide detectives hate most. It's worse than getting up at 3 A.M. on a February night to slog through a field of freezing mud toward a body that needed burying two days ago. Worse than staring into the flat cold eyes of a teenager who bragged about dragging a man through the streets to his death. Worse than visiting every sleazy dive in town until you finally find the one person who can put the murderer away and having that person say as cool as a debutante with a full dance card, "I don't want to get involved."[23]

Detectives have the cards stacked against them most of the time. Unless they discover, during the preliminary investigation, a named suspect or a description or other information that leads to a named suspect, the chances of solving the crime are low. Property crimes with no witnesses are particularly hard to solve. In 2018, for example, the clearance rates for crimes against persons were 62.3% for murder and nonnegligent manslaughter, 52.5% for aggravated assault, 33.4% for forcible rape (revised definition), and 30.4% for robbery. For crimes against property, the clearance rates were 13.9% for burglary, 18.9% for larceny-theft, 13.8% for motor vehicle theft, and 22.4% for arson. Clearances for crimes against persons generally are higher than for property crimes because crimes against persons receive more intensive investigative effort and because victims and witnesses frequently identify the perpetrators. In 2018, for example, the nationwide clearance rate for violent crimes was 45.5% and for property crimes, 17.6%.[24] As mentioned in Chapter 2, clearance rates remain remarkably stable from year to year.

myth

Improvements in detective work and criminal investigation will significantly raise clearance rates or lower the crime rate.

fact

"Cleared" crimes generally solve themselves. The offender either is discovered at the scene or can be identified by the victim or a witness. Investigation rarely solves "cold" or "stranger" crimes.

Studies have found that much of what a detective does is not needed and that an investigator's technical knowledge often does little to help solve cases.[25] In one study, for example, fewer than 10% of all arrests for robbery were the result of investigative work by detectives.[26] Nevertheless, police agencies retain detectives and plainclothes investigators for a number of reasons:

1. Detectives have interrogation and case-presentation skills that assist in prosecution.
2. Technical knowledge, such as knowing about burglary tools, does help in some investigations and prosecutions.
3. Law enforcement executives can assign detectives to a major, high-profile case to demonstrate to the public that they are committing resources to the matter.

The major studies of investigative effectiveness emphasize the value of improving the suspect-identification process. Once a suspect is identified by name or some other clearly distinguishing characteristic, the chances of making an arrest are increased substantially.

Identification Developments in Criminal Investigation Two of the most significant advances in criminal investigation have been the development of DNA profiling and Next Generation Identification (NGI). However, before examining these two investigative tools, it is instructive to consider the findings of a 2-year congressionally mandated study of forensic science and the crime lab system by the National Academy of Sciences.[27]

The study, released in 2009, discovered that the nation's forensic science system has serious deficiencies and that it needs major reforms and new research. Lacking are rigorous and mandatory certification programs for forensic scientists and strong standards and protocols for analyzing and reporting on evidence. Needed are more peer-reviewed, published studies establishing the scientific bases, accuracy, and reliability of many forensic methods. In addition, many forensic science labs require greater funding, staffing, and effective oversight.

The study revealed that with the exception of nuclear DNA analysis, no forensic method has been rigorously shown able to consistently, and with a high degree of certainty, demonstrate a connection between evidence and a specific individual or source. Yet, it is precisely this type of evidence that has been used to convict criminal defendants. Of the more than 230 people exonerated by DNA evidence, more than 50% of the cases involved faulty or invalidated forensic evidence. Highly suspect is evidence from ballistics, handwriting, bite marks, tool marks, shoe prints, and blood spatters—to name only a few techniques. Even fingerprint evidence is of concern. The researchers argue that zero-error-rate claims made about fingerprint analyses are not plausible; uniqueness does not guarantee that two individuals' prints are always so sufficiently different that they could not be confused. Recommended is the accumulation of data on how much a person's fingerprints vary from impression to impression, as well as the degree to which fingerprints vary across a population.

This is not to say that non-DNA forensic evidence is useless. It could, for example, provide valuable information to help narrow the range of possible suspects or sources. However, before this evidence is used to "prove" that a defendant is guilty, substantial research is needed to validate basic premises and techniques, assess limitations, and discern the sources and magnitude of error. The panel of researchers strongly urged Congress to establish a new, independent National Institute of Forensic Science to help solve these problems.

Although, to date, Congress has not created a National Institute of Forensic Science, in 2013, the Department of Justice established the National Commission on Forensic Science, in partnership with the National Institute of Standards and Technology, to improve the practice and reliability of forensic science. This unique partnership drew upon each agency's core strengths to promote scientific validity, reduce fragmentation, and improve federal coordination of forensic science. The Commission included federal, state, and local forensic service providers; research scientists and academics; law enforcement officials; prosecutors, defense attorneys, and judges; and other stakeholders from across the country.[28] In 2017, then-U.S. Attorney General Jeff Sessions inexplicably terminated the National Commission.[29]

DNA Profiling DNA (deoxyribonucleic acid) is a molecule present in all forms of life. A unique genetic profile can be derived from skin, blood, hair, semen, or other bodily substances found at the scene of a crime or on a victim. Not only can bodily substances found at a crime scene be matched with DNA samples from a suspect to give an extremely high probability of identifying the perpetrator, but it is believed that soon DNA from a sample as small as a flake of dandruff will yield a positive, unique identification with no need to consider mathematical probabilities.

DNA profiling has three distinct functions: linking or eliminating identified suspects to a crime; identifying "cold hits," whereby a sample from a crime scene is matched against numerous cases in a DNA database and a positive match is made; and clearing convicted rapists and murderers years after they began serving their sentences. DNA profiling would be very useful, for example, in cases where a murderer's blood was found at the scene of a crime after a deadly struggle or in a rape case where seminal fluid could be obtained from the victim. In approximately one-third of DNA examinations, the suspect's DNA cannot be matched with biological evidence from the crime scene. Thus, potential suspects can be eliminated from consideration early in the investigative process, allowing investigators to focus their efforts more effectively on other suspects or cases. Potential suspects also can be eliminated from an investigation years after the crime occurred, as happened in the JonBenet Ramsey murder case. The then 6-year-old beauty queen was killed in 1996, and her brother and parents remained suspects for more than a decade. In 2008, based on results obtained from a new technology called "touch DNA," the cloud of suspicion finally was removed from the Ramsey family. Touch DNA involves scraping genetic material from an object that otherwise could not be seen. In this case, newly discovered DNA from a few minute skin cells matched DNA found earlier and was not from the Ramsey family. Investigators unsuccessfully attempted to locate a match in the national DNA database, which at the time had more than five million offenders' profiles. In 2019, a pedophile Gary Olivia admitted to the murder, as did another pedophile John Mark Karr in 2008, but in neither case was evidence found. For now, no one has been charged in the case, and the murder remains unsolved.[30] Figure 6.3 shows how DNA profiling is performed.

DNA Evidence

The National Institute of Justice's "Forensic DNA" website provides a wealth of information about NIJ's role in funding DNA research and development projects. Visit the website at https://nij.gov/topics/forensics/evidence/dna/Pages/welcome.aspx.

How big of a role should DNA play in criminal investigations?

A serious issue at present is whether DNA databases ought to be assembled and from whom the samples should be taken. Many states permit the taking of DNA samples from arrested and convicted subjects. Some enthusiasts believe that DNA samples should be taken from all suspects in crimes, while a smaller number believe the samples should be collected from all people at birth. Another controversial issue is how long DNA samples should be kept. In December 2008, 17 judges on the European Court of Human Rights, Europe's highest human rights court, struck down a British law that allowed the government to store DNA and fingerprints of people with no criminal record. The law had allowed the government to keep samples until an individual died or reached the age of 100. Britain's DNA databases, with about six million samples, have been taken from arrestees, regardless of whether they have been charged, convicted, or acquitted and, occasionally, from crime victims. The court unanimously ruled that Britain's "blanket and indiscriminate" storage of DNA samples and fingerprints of people with no criminal record violated people's right to privacy–a protection under the Human Rights Convention to which the United Kingdom is a signatory. In response to the court ruling, the Protection of Freedoms Act was passed by both houses of Parliament in 2011, and received royal assent in 2012. Among other protections, the law provides destruction rules for an estimated 7 million DNA profiles of innocent people, so far, from Britain's National DNA Database.[31]

Currently, the most complete DNA database in the United States, with more than 18 million samples (as of March 2019), is the Combined DNA Index System (CODIS), which is managed by the Federal Bureau of Investigation (FBI).[32] CODIS comprises DNA profiles that have been entered into local, state, and other national databases. The profiles are from either biological evidence left at crime scenes or individuals convicted of violent crimes and other felonies. Undoubtedly, the more collected samples in a database, the more likely

DNA Investigations

As of March 2019, CODIS has produced approximately 460,000 hits assisting in about 449,000 investigations.

Source: Federal Bureau of Investigation, "CODIS - NDIS Statistics," accessed May 7, 2019, https://www.fbi.gov/services/laboratory/biometric-analysis/codis/ndis-statistics.

FIGURE 6.3 How DNA Profiling Is Performed

DNA, deoxyribonucleic acid, is the material that carries the genetic pattern that makes each person unique. Scientists in the laboratory can map DNA patterns in samples of skin, blood, hair, semen, or other body tissues or fluids. The DNA patterns can then be analyzed and compared.

There are two main DNA testing procedures used in criminal forensics.

1. Samples are taken of tissue or body fluids at crime scenes. Comparison samples are taken from victims and suspects.

RFLP (Restriction Fragment Length Polymorphism)

2. In the laboratory, DNA genetic material is extracted from the samples and mixed with enzymes to cut the DNA into fragments.
3. The DNA fragments are put in a special gel and exposed to an electrical charge to sort the fragments by size.
4. Genetic tracers are used to search out and lock onto specific fragments of the DNA.
5. The tracers reveal a pattern. Each evidence sample will have a pattern that can be compared with the sample from the victim and the sample from the suspect.

PCR (Polymerase Chain Reaction)

2. In the laboratory, DNA is extracted from the samples.
3. Part of the DNA molecule is amplified in a test tube to produce billions of copies of that part.
4. The amplified DNA is analyzed.
5. The analysis of the evidence sample can be compared with the analysis of the sample from the victim and the sample from the suspect.

Comparing the patterns in the samples results in a DNA profile representing distinctive features of the samples that may or may not match.

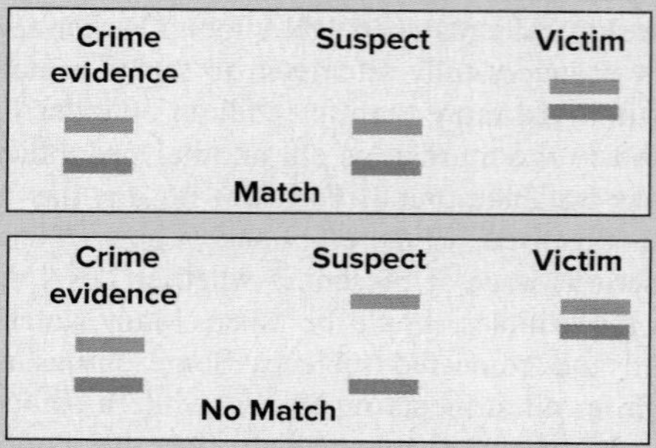

a match is going to be found. But privacy concerns and the potential for misuse of DNA samples are likely to hinder any more intrusive measures on the part of agents of the justice system.

New Generation Identification (NGI): Fingerprints, Palm Prints, Face Recognition, and Iris Image Recognition In 1999, the Integrated Automated Identification System (IAFIS) was launched by the FBI's Criminal Justice Information Services (CJIS). IAFIS allowed investigators to sort through thousands of sets of stored patent and latent fingerprints for a match with those of a crime suspect. A *patent fingerprint* is visible to the naked eye, while a *latent fingerprint* is not. A latent fingerprint is left on a surface by deposits of oils and/or perspiration from the finger and is detected with special techniques, such as dusting with fine powder and then lifting the pattern of powder with transparent tape. Prior to the IAFIS, many attempts to match prints would not have been made because the old process would have taken thousands of hours. The

IAFIS reduced the average response time for an electronic fingerprint submission to about 10 minutes. The IAFIS was able to process an average of approximately 162,000 ten-print submissions per day. Large metropolitan police agencies used it to identify 200 to 500 suspects a year who would have escaped apprehension before the implementation of the IAFIS.[33]

In February 2011, the FBI's CJIS improved its biometric identification services by introducing Advanced Fingerprint Identification Technology (AFIT) to replace the IAFIS. AFIT is the first increment of Next Generation Identification (NGI). AFIT enhances both patent and latent fingerprint processing services, increases the accuracy and daily fingerprint processing capacity, and boosts system availability. Compared to IAFIS, AFIT, with its new fingerprint-matching algorithm, improves matching accuracy from 92% to more than 99%, triples the speed of searches (when the searched repository contains the mate), produces fewer transaction rejects, and increases identification frequency. With AFIT, latent prints can be searched against the criminal, civil, and Unsolved Latent File (ULF) repositories, which can generate new investigative leads in unsolved and/or cold cases. With IAFIS, latent prints were searched against the criminal repository only.[34]

The second increment of NGI, also maintained and operated by the FBI's CJIS, is the Repository for Individuals of Special Concern (RISC), which was introduced in August 2011. RISC is a rapid search service available only to authorized law enforcement personnel through the use of a mobile fingerprint device. With a response time of less than 10 seconds, RISC provides on-scene access to a national repository of wants and warrants, including the Immigration Violator File (IVF) of the National Crime Information Center (NCIC), convicted sex offenders, and known or appropriately suspected terrorists. RISC can enhance the safety and situational awareness of law enforcement personnel.[35]

The National Palm Print System (NPPS) was introduced by the FBI's CJIS in May 2013, as part of the third increment of NGI. This powerful crime-solving tool contains searchable palm prints and supplemental fingerprints that can be directly added and deleted by law enforcement personnel nationwide. Law enforcement personnel can now search latent palm prints against the NPPS.[36]

"Rap Back," another part of NGI's third increment, uses fingerprints to identify individuals arrested and prosecuted for crimes so that the FBI can notify in a timely manner criminal justice and other authorized agencies about the ongoing criminal activity, if any, of these individuals. Prior to the introduction of "Rap Back," the national criminal history background check system provided only a one-time snapshot view of an individual's criminal history status. This service not only keeps tabs on individuals who hold positions of trust (for example, school teachers, daycare workers), or who are under criminal justice supervision or investigation, but it also eliminates the need for repeated background checks on a person from the same applicant agency.[37]

Another one of NGI's updated services is the Interstate Photo System (IPS), which is a face recognition service that allows law enforcement agencies to search criminal and noncriminal justice photographs to help with identifications. Prior to the NGI update, the FBI did not have the technical capacity to provide law enforcement with criminal photo search capability. Now, when an authorized law enforcement agency submits a photo image for face recognition, IPS automatically searches through its criminal identity group ("mugshot") repository and produces a list of potential candidates to serve as investigative leads; not positive identifications. Additional evaluation and investigation are needed for a positive identification. NGI enhanced IPS also allows law enforcement agencies to search for photos using text-based biographic or demographic data (for example, sex, race, age, hair color) rather than submitting a photo. This can be used to create photo lineups. It also allows photographs to be linked to other identity records in NGI, such as fingerprints.[38]

Finally, in September 2013, as another part of NGI, the FBI initiated an Iris Pilot project to evaluate the potential of building a criminal iris repository and of developing iris image recognition services. At the end of 2017, the FBI had collected more than 746,000 arrestee iris scans from 56 submitting facilities–an increase of nearly 72% from 2016–and had identified 372 wanted persons, who had committed such crimes as robbery, burglary, aggravated assault, larceny, and drug offenses.[39]

Number of Fingerprints: NGI vs. IAFIS

As of March 2019, NGI had approximately 144 million fingerprints in its repository: more than 76 million criminal fingerprints, about 64 million civil fingerprints, and nearly 3 million RISC fingerprints. By comparison, the IAFIS system had only a little more than 70 million criminal fingerprints in its repository.

Sources: Federal Bureau of Investigation, "NGI Monthly Fact Sheet," March 2019, accessed May 8, 2019, https://www.fbi.gov/file-repository/ngi-monthly-fact-sheet/view; Federal Bureau of Investigation, "Integrated Automated Fingerprint System," accessed October 29, 2012, www.fbi.gov/about-us/cjis/fingerprints_biometrics/iafis.

Cybercrime The use of computer technology to commit crime, or **cybercrime**, is of increasing concern to law enforcement officials. The U.S. President's Council of Economic Advisers estimated that malicious cyber activity cost the U.S. economy between $57 billion

cybercrime The use of computer technology to commit crime.

Combating Cybercrime

To learn more about what the U.S. government is doing to combat cybercrime, visit the U.S. Department of Justice's Cybersecurity Unit within the Computer Crime and Intellectual Property Section website at https://www.justice.gov/criminal-ccips/cybersecurity-unit.

Do you think that there is anything more the government can do to prevent cybercrime?

and $109 billion in 2016. The malicious cyber activity included "denial of service attacks, data and property destruction, business disruption (sometimes for the purpose of collecting ransoms) and theft of proprietary data, intellectual property, and sensitive financial and strategic information."[40] Many cybercrimes committed against businesses are not reported to law enforcement authorities because, for example, the business fears the loss of the public's confidence in the organization, the attention to vulnerability that a cybercrime report would attract, and the shame of not providing adequate security to protect trusted assets.

Cybercrime losses to U.S. consumers in 2017 were estimated to be nearly $17 billion.[41] It has been estimated that about two-thirds of victimized consumers do not report cybercrimes to law enforcement authorities. Reasons vary but include embarrassment, shame, the belief that law enforcement authorities cannot or will not do anything about it, and, in some cases, the victim's lack of awareness that she or he has been victimized.

Besides those already mentioned above, a variety of offenses can be committed using computer and Internet technology. Following are some of them:

- *Auction fraud.* Auction fraud involves fraud attributable to the misrepresentation of a product advertised for sale through an Internet auction site or the nondelivery of products purchased through an Internet auction site.
- *Child pornography/child sexual exploitation.* Computer telecommunications have become one of the most prevalent techniques used by pedophiles to share illegal photographic images of minors and to lure children into illicit sexual relationships. The Internet has dramatically increased the access of the preferential sex offenders to the population they seek to victimize and provides them greater access to a community of people who validate their sexual preferences.
- *Counterfeit cashier's check.* The counterfeit cashier's check scheme targets individuals who use Internet classified advertisements to sell merchandise. Typically, an interested party located outside the United States contacts a seller. The seller is told that the buyer has an associate in the United States who owes him money. As such, he will have the associate send the seller a cashier's check for the amount owed to the buyer. The amount of the cashier's check will be thousands of dollars more than the price of the merchandise, and the seller is told the excess amount will be used to pay the shipping costs associated with getting the merchandise to his location. The seller is instructed to deposit the check and, as soon as it clears, to wire the excess funds back to the buyer or to another associate identified as a shipping agent. In most instances, the money is sent to locations in West Africa (Nigeria). Because a cashier's check is used, a bank typically will release the funds immediately or after a 1- or 2-day hold. Falsely believing the check has cleared, the seller wires the money as instructed. In some cases, the buyer is able to convince the seller that some circumstance has arisen that necessitates the cancellation of the sale and is successful in conning the victim into sending the remainder of the money. Shortly thereafter, the victim's bank notifies him that the check was fraudulent, and the bank is holding the victim responsible for the full amount of the check.
- *Credit card fraud.* Credit/debit card numbers can be stolen from unsecured websites or can be obtained in an identity theft scheme.
- *Debt elimination.* Debt elimination schemes generally involve websites advertising a legal way to dispose of mortgage loans and credit card debts. Most often, all that is required of the participant is to send $1,500 to $2,000 to the subject, along with all the particulars of the participant's loan information and a special power of attorney authorizing the subject to enter into transactions regarding the title of the participant's home on his or her behalf. The subject then issues bonds and promissory notes to the lenders that purport to legally satisfy the participant's debts. In exchange, the participant then is required to pay a certain percentage of the value of the satisfied debts to the subject. The potential risk of identity theft crimes associated with the debt elimination scheme is extremely high because the participants provide all of their personal information to the subject.
- *Parcel courier e-mail scheme.* The parcel courier e-mail scheme involves the supposed use of various national and international parcel providers, such as DHL, UPS, FedEx, and the USPS. Often, the victim is e-mailed directly by the subject(s) following online bidding on auction sites. Most of the scams follow a general pattern that

includes the following elements. The subject instructs the buyer to provide shipping information, such as name and address. The subject informs the buyer that the item will be available at the selected parcel provider in the buyer's name and address, thereby, identifying the intended receiver. The selected parcel provider checks the item and purchase documents to guarantee everything is in order. The selected parcel provider sends the buyer delivery notification verifying his or her receipt of the item. The buyer is instructed by the subject to go to an electronic funds transfer medium, such as Western Union, and make a funds transfer in the subject's name and in the amount of the purchase price. After the funds transfer, the subject instructs the buyer to forward the selected parcel provider the funds transfer identification number, as well as his or her name and address associated with the transaction. The subject informs the buyer the parcel provider will verify payment information and complete the delivery process. Upon completion of delivery and inspection of the item(s) by the receiver, the buyer provides the parcel provider funds transfer information, thus allowing the seller to receive his funds.

- *Employment/business opportunities.* Employment/business opportunity schemes have surfaced in which bogus foreign-based companies are recruiting citizens in the United States on several employment-search websites for work-at-home employment opportunities. These positions often involve reselling or reshipping merchandise to destinations outside the United States. Prospective employees are required to provide personal information, as well as copies of their identification, such as a driver's license, birth certificate, or Social Security card. Those employees that are "hired" by these companies then are told that their salary will be paid by check from a U.S. company reported to be a creditor of the employer. This is done under the pretense that the employer does not have any banking set up in the United States. The amount of the check is significantly more than the employee is owed for salary and expenses, and the employee is instructed to deposit the check into his or her own account and then wire the overpayment back to the employer's bank, usually located in eastern Europe. The checks are later found to be fraudulent, often after the wire transfer has taken place.
- *Escrow services fraud.* In an effort to persuade a wary Internet auction participant, the perpetrator will propose the use of a third-party escrow service to facilitate the exchange of money and merchandise. The victim is unaware the perpetrator has actually compromised a true escrow site and, in actuality, created one that closely resembles a legitimate escrow service. The victim sends payment to the phony escrow and receives nothing in return. Or, the victim sends merchandise to the subject and waits for his or her payment through the escrow site, which is never received because it is not a legitimate service.
- *Identity theft.* Identity theft occurs when someone appropriates another's personal information without his or her knowledge to commit theft or fraud. Identity theft is a vehicle for perpetrating other types of fraud schemes. Typically, the victim is led to believe he or she is divulging sensitive personal information to a legitimate business, sometimes as a response to an e-mail solicitation to update billing or membership information, or as an application to a fraudulent Internet job posting.
- *Internet extortion.* Internet extortion involves hacking into and controlling various industry databases, promising to release control back to the company if funds are received, or the subjects are given web administrator jobs. Similarly, the subject will threaten to compromise information about consumers in the industry database unless funds are received.
- *Investment fraud.* Investment fraud is an offer using false or fraudulent claims to solicit investments or loans, or providing for the purchase, use, or trade of forged or counterfeit securities.
- *Lotteries.* The lottery scheme deals with persons randomly contacting e-mail addresses advising the e-mail recipients they have been selected as the winner of an international lottery. An agency name follows this body of text with a point of contact, phone number, fax number, and an e-mail address. An initial fee ranging from $1,000 to $5,000 often is requested to initiate the process, and additional fee requests follow after the process has begun.
- *Nigerian letter or "419."* Named for the violation of Section 419 of the Nigerian Criminal Code, the 419 scam combines the threat of impersonation fraud with a

Identity Theft

In 2017, an estimated 16.7 million Americans experienced some form of financial identity fraud, an increase of about 30% from 2014. Victims lost about $16.8 billion, an increase of 5% from 2014.

Sources: Al Pascual, Kyle Marcini, and Sarah Miller, "2018 Identity Fraud: Fraud Enters a New Era of Complexity," Javelin Strategy & Research, February 6, 2018, accessed May 10, 2019, https://www.javelinstrategy.com/coverage-area/2018-identity-fraud-fraud-enters-new-era-complexity; Herb Weisbuam, "Nearly 13 Million Americans Victimized by ID Thieves in 2014," *NBC News*, March 3, 2015. Accessed January 27, 2016, www.nbcnews.com/business/consumer/nearly-13-million-americans-victimized-id-thieves-2014-n316266.

variation of an advance fee scheme in which a letter, e-mail, or fax is received by the potential victim. The communication from individuals representing themselves as Nigerian or foreign government officials offers the recipient the "opportunity" to share in a percentage of millions of dollars, soliciting help in placing large sums of money in overseas bank accounts. Payment of taxes, bribes to government officials, and legal fees often are described in great detail with the promise that all expenses will be reimbursed as soon as the funds are out of the country. The recipient is encouraged to send information to the author, such as blank letterhead stationery, bank name and account numbers, and other identifying information using a facsimile number provided in the letter. The scheme relies on convincing a willing victim to send money to the author of the letter in several installments of increasing amounts for a variety of reasons.

- *Phishing/spoofing.* Phishing and spoofing are somewhat synonymous in that they refer to forged or faked electronic documents. Spoofing generally refers to the dissemination of e-mail that is forged to appear as though it was sent by someone other than the actual source. Phishing, often utilized in conjunction with a spoofed e-mail, is the act of sending an e-mail falsely claiming to be an established legitimate business in an attempt to dupe the unsuspecting recipient into divulging personal, sensitive information such as passwords, credit card numbers, and bank account information after directing the user to visit a specified website. The website, however, is not genuine and was set up only as an attempt to steal the user's information.
- *Ponzi/pyramid.* Ponzi or pyramid schemes are investment scams in which investors are promised abnormally high profits on their investments. No investment is actually made. Early investors are paid returns with the investment money received from the later investors. The system usually collapses. The later investors do not receive dividends and lose their initial investment.
- *Reshipping.* The "reshipping" scheme requires individuals in the United States, who sometimes are coconspirators and other times are unwitting accomplices, to receive packages at their residence and subsequently repackage the merchandise for shipment, usually abroad. "Reshippers" are being recruited in various ways, but the most prevalent are through employment offers and conversing with, and later befriending, unsuspecting victims through Internet Relay Chat Rooms. Unknown subjects post help-wanted advertisements at popular Internet job search sites, and respondents quickly reply to the online advertisement. As part of the application process, the prospective employee is required to complete an employment application in which he or she divulges sensitive personal information, such as date of birth and Social Security number that, unbeknownst to the victim employee, will be used to obtain credit in his or her name. The applicant is informed he or she has been hired and will be responsible for forwarding, or "reshipping," merchandise purchased in the United States to the company's overseas home office. The packages quickly begin to arrive and, as instructed, the employee dutifully forwards the packages to their overseas destination. Unbeknownst to the "reshipper," the recently received merchandise was purchased with fraudulent credit cards.
- *Spam.* With improved technology and worldwide Internet access, spam, or unsolicited bulk e-mail, is now a widely used medium for committing traditional white-collar crimes, including financial institution fraud, credit card fraud, and identity theft. Spam also can act as the vehicle for accessing computers and servers without authorization and transmitting viruses and botnets. The subjects masterminding this spam often provide hosting services and sell open proxy information, credit card information, and e-mail lists illegally.
- *Third-party receiver of funds.* In this scheme, the subject, usually foreign, posts work-at-home job offers on popular Internet employment sites, soliciting for assistance from U.S.

TIMOTHY A. CLARY/Staff/Getty Images

In 2009, 71-year-old investment manager Bernard Madoff, former chairman of the Nasdaq Stock Market, was sentenced to 150 years in prison for bilking mostly wealthy investors, corporations, nonprofit organizations, foundations, and charities of $50 billion in what is likely the largest "Ponzi" scheme in history. Had Madoff not faced requests for $7 billion in redemptions as a result of the worldwide economic crisis, his Ponzi scheme might not have been discovered. *Was the sentence imposed on Madoff appropriate? Why or why not?*

> citizens. The subject allegedly is posting Internet auctions but cannot receive the proceeds from these auctions directly because his or her location outside the United States makes receiving these funds difficult. The subjects ask a U.S. citizen to act as a third-party receiver of funds from victims who have purchased products from the subject via the Internet. The U.S. citizen, receiving the funds from the victims, then wires the money to the subject.[42]

Because so little hardware and expense is involved, any person with a computer and modem connection to the Internet has the potential to attack computer systems and people online. The knowledge to carry out these attacks often is available online at hacker websites, message boards, and chat rooms. The cybercriminal can be as unsophisticated as a teenage amateur hacker just out for some fun or a talented computer specialist possessing skills on par with technical experts employed by the nation's top security organizations. Specialization in cybercrime is becoming more common and includes the following roles:

1. Coders or programmers, who write the malware, exploits, and other tools necessary to commit the crime. Contrary to popular belief, coders are not protected by the First Amendment when they knowingly take part in a criminal enterprise–and they go to jail just like the rest of the enterprise.
2. Distributors or vendors, who trade and sell stolen data, and act as vouchers of the goods provided by the other specialties.
3. Techies, who maintain the criminal infrastructure, including servers, bulletproof Internet Service Providers (ISPs), and encryption, and who often have knowledge of common database languages and SQL servers.
4. Hackers, who search for and exploit application, system, and network vulnerabilities to gain administrator or payroll access.
5. Fraudsters, who create and deploy social engineering schemes, including phishing, spamming, and domain squatting.
6. Hosters, who provide "safe" hosting of illicit content servers and sites, often through elaborate botnet and proxy networks.
7. Cashers, who control drop accounts and provide those names and accounts to other criminals for a fee, and who also typically control full rings of money mules.
8. Money mules, who are divided into three types. First are "one and done mules"–people who get tricked by social engineering schemes to send money. Second are "career money mules," who make a living, or at least a substantial amount of fun money, by completing money transfers or wire transfers between bank accounts. Third are "premier mules," who are the top of the money mule world. These individuals actually are sent to the United States, often on work or student visas, with the purpose of moving money for criminals.
9. Tellers, who help with transferring and laundering illicit proceeds through digital currency services and between different world currencies.
10. Leaders, many of whom don't have any technical skills at all. They are the "people-people." They choose the targets; choose the people they want to work each role; decide who does what, when, and where; and take care of personnel and payment issues.[43]

To meet the challenge of cybercrime, the FBI has formed cyber squads in each of its 56 field offices with more than 1,000 advanced cyber-trained FBI special agents, intelligence analysts, and forensic examiners to combat the rising cybercrime threat. In addition, thousands of FBI agents have gone through and continue to go through basic cyber training, which is now required of every FBI special agent before he or she can graduate from the FBI Academy at Quantico, Virginia.[44] At present, unfortunately, most local and state law enforcement agencies in America are ill prepared to detect, investigate, and prosecute cybercriminals. However, cooperative efforts are now under way among law enforcement, business, high-tech, and national security organizations to better prepare the nation's police agencies to combat cybercrime.

Traffic

When loss of life, serious injury, suffering, and property damage are all considered, the regulation and control of vehicle and pedestrian traffic are important, if not the most important, police responsibilities. Each year, more than twice as many people are killed in automobile accidents on the streets and highways of the nation than are murdered.

Royalty-Free/CORBIS

Scientific traffic accident investigation requires technical expertise. *Should all patrol officers be trained to conduct traffic investigations? Why or why not?*

A large percentage of this highway death and suffering is attributable to alcohol. Enforcement of DUI (driving under the influence) laws is critical to the safety of a community. In addition, automobile insurance rates are based to some degree on a community's level of traffic enforcement. Thus, if the police neglect traffic regulation and enforcement, they are likely to hear about it from both insurance companies and premium payers.

Some of the debate about traffic enforcement concerns whether the major enforcers of traffic regulations should be specialized personnel or uniformed patrol officers. Some traffic responsibilities already are delegated to specialized personnel, such as enforcement of parking regulations and investigations of hit-and-run accidents and traffic fatalities. In some agencies, special **traffic accident investigation crews** are assigned to all traffic accident investigations. Otherwise, patrol officers investigate accidents and attend to other traffic-related duties as a normal part of their everyday workload.

Traffic units exist in nearly all medium-to-large police agencies. Some of their more important functions are:

- To educate motorists in a community about traffic safety and proper driving procedures.
- To enforce traffic laws, particularly when violations of those laws cause traffic accidents.
- To recommend traffic engineering changes that will enhance the flow of traffic and promote safety.

Enforcing traffic laws also may reduce criminal activity because stopping vehicles for traffic violations both day and night is likely to put police officers in contact with criminals.

Many veteran officers consider working in the traffic division "clean" police work because normally it does not involve responding to radio calls that take them to the scene of fights, domestic disturbances, or other distasteful incidents, such as those involving drunks. Traffic officers in large police agencies usually are well schooled in scientific accident investigation, a skill that makes them employable in the private sector, usually doing traffic reconstruction for insurance companies. The Traffic Institute at Northwestern University is one of the major schools that prepares officers for sophisticated accident investigation, although many state peace officer and highway patrol academies now have comparable training programs.

traffic accident investigation crews In some agencies, the special units assigned to all traffic accident investigations.

FYI **Public Opinion About Traffic Enforcement**

Only 19% of Americans think that police should make enforcing traffic laws a top priority.

Source: Emily Ekins, "Policing in America: Understanding Public Attitudes Toward the Police. Results from a National Survey," Cato Institute, December 7, 2016, accessed June 15, 2019, https://www.cato.org/survey-reports/policing-america.

Drug Enforcement

Illegal drug use in the United States is widespread and stable. According to the Substance Abuse and Mental Health Services Administration's *2017 National Survey on Drug Use and Health*, 30.5 million Americans age 12 and older (11.2% of the population age 12 or older) reported using an illegal drug in the month before the survey was conducted.[45] The percentage of Americans 12 and older using drugs in the month before the survey in 2017 (11.2%) was the highest percentage recorded since 2002; the lowest percentage recorded was 7.9% in 2004.[46] Illegal drugs included marijuana, cocaine (including crack), heroin, hallucinogens (e.g., LSD, PCP, and "Ecstasy" or MDMA), inhalants (e.g., amyl nitrite, cleaning fluids, gasoline, paint, and glue), and prescription psychotherapeutics (e.g., pain relievers, tranquilizers, stimulants, and sedatives) used nonmedically. A separate category was included for opioids (heroin or pain reliever misuse). In 2017, marijuana was the most commonly used illicit drug, with 26 million current (past month) users (85% of current illicit drug users). An estimated 6 million people reported the misuse of psychotherapeutic drugs in the past month in 2017, including 3.5 million people who misused opioids. Thus, the number of people who misused opiods in 2017 was second only to the number of marijuana users.[47] Table 6.3 provides the number of Americans age 12 or older who used a specific drug or category of drug at least once during their lifetime, during 2017, and during the month before the survey was conducted, and their percentage of the overall population.

Given the extent of illegal drug use in the United States, it is not surprising that drug enforcement has become an increasingly important responsibility of police departments

TABLE 6.3 Extent of Illicit Drug Use in U.S. Among Persons Aged 12 or Older, 2017

	During Lifetime		Past Year (2017)		Month Before Survey	
Drug	**Number**	**%**	**Number**	**%**	**Number**	**%**
Marijuana	122,943k	45.2	40,935k	15.0	25,997k	9.6
Cocaine	40,550k	14.9	5,943k	2.2	2,167k	0.8
Crack	9,599k	3.5	930k	0.3	473k	0.2
Hallucinogens	42,072k	15.5	5,125k	1.9	1,438k	0.5
Inhalants	25,187k	9.3	1,759k	0.6	556k	0.2
Heroin	5,295k	1.9	886k	0.3	494k	0.2
Methamphetamine	14,722k	5.4	1,633k	0.6	774k	0.3
Misuse of Psychotherapeutics	nr	nr	18,077k	6.6	5,956k	2.2
Opiods	nr	nr	11,401k	4.2	3,549k	1.3

k = Thousands (numbers in thousands); % = Percentage of overall population; nr = not reported due to measurement issues

Source: Center for Behavioral Health Statistics and Quality (2018), "2017 National Survey on Drug Use and Health: Detailed Tables," Substance Abuse and Mental Health Services Administration, Rockville, MD, Tables 1.1A and 1.1B, accessed May 11, 2019, https://www.samhsa.gov/data/sites/default/files/cbhsq-reports/NSDUHDetailedTabs2017/NSDUHDetailedTabs2017.pdf.

across the nation. Although the war on drugs is a priority of state and federal law enforcement agencies as well, many of the battles have been fought at the local level. Nearly all local police departments regularly engage in drug enforcement.

In 2013 (the latest year for which data were available), approximately one-half of all local police departments had one or more officers assigned full-time to a multiagency drug enforcement task force, about twice as many departments as in 2007.[48] Larger departments were more likely than smaller departments to participate in these task forces and assign full-time officers to them.

Although all levels of government are waging the war on drugs, the focus of the remainder of this section is the drug enforcement strategies of local police (and sheriffs') agencies. The particular strategies employed by individual agencies vary widely, but the most common strategies are street-level enforcement, mid-level investigations, major investigations, crop eradication, smuggling interdiction, problem-oriented and community policing strategies, drug demand reduction, and asset forfeiture. It is important to note at the outset that none of these strategies has had much of a long-term effect on the U.S. drug problem.

Street-Level Enforcement Patrol officers, officers assigned to special drug enforcement units, and plainclothes officers do most of the street-level enforcement. Their tactics include surveillance, interruption of suspected transactions, raids of "shooting galleries" and "crack houses," buy-and-bust operations, and "reverse stings" in which plainclothes officers offer to sell drugs to willing customers. Street-level enforcement is responsible for most drug arrests and seizures. However, typically, the people arrested are either drug users or small-time dealers, and the seizures are for small amounts of drugs. In 2017, a typical year, more arrests were made for drug abuse violations than for any other offense. About 1.6 million drug arrests were made in 2017 (about the same number of drug arrests made in each of the previous 4 years), which represented about 15% of all arrests reported to the FBI.[49]

Mid-Level Investigations Informants or undercover police officers are frequently employed in mid-level investigations. The primary purpose of these investigations is to identify and make cases against mid-level dealers. Although mid-level dealers generally occupy relatively low-level positions in a drug distribution network and are easily replaced, they often are the highest-ranking drug traffickers that local police agencies can catch. The basic tactic used in these investigations is for undercover officers to gain the confidence of street-level dealers and work their way up to mid-level dealers by requesting to buy larger quantities of drugs than the street-level dealer can provide. Another tactic is for an informant to simply introduce an undercover officer to a mid-level dealer. Street-level dealers often become informants when they are arrested and agree to inform on their suppliers in return for some consideration in prosecution.

Drug busts can be dangerous. *As a law enforcement officer, would you be willing to participate in drug busts? Why or why not?*

Lori Wolfe/The Herald-Dispatch/AP Images

Major Investigations The goal of major investigations is to arrest drug kingpins and shut down the organizations responsible for producing, importing, and distributing large quantities of illegal drugs. Major investigations are conducted primarily by federal and state law enforcement agencies, but sometimes the largest local police agencies are involved as well. Most local police agencies do not have the resources to engage in major investigations, which often require long-term commitments, extensive travel, and specialized expertise. In some cases, however, local police agencies participate in major investigations as part of a multiagency task force.

Crop Eradication Crop eradication is a tactic employed by federal, state, and local law enforcement agencies. However, the only crop targeted by local police agencies is cannabis (marijuana) because it is the only major illegal drug grown in the United States. Because cannabis eradication intervenes at the beginning of the trafficking process, in theory it should have the greatest potential for eliminating or at least significantly reducing marijuana availability and use. In practice, however, a number of problems reduce the effectiveness of the eradication strategy. First, large quantities of marijuana generally are grown in remote, largely inaccessible areas. This makes locating and destroying crops by hand difficult, time-consuming, labor-intensive, and dangerous. Second, spraying crops with chemicals can be hazardous to people, water supplies, animals, and vegetation. Third, even when crop eradication is successful in one area, a new crop easily can be grown in another area. The huge profits that can be made from a marijuana crop create a powerful incentive not to be deterred by eradication efforts. In 2017, the Drug Enforcement Administration (DEA) reported that about 3 million marijuana plants cultivated outdoors were eradicated, down from about 4 million plants in 2014, and down significantly from almost 10 million plants in 2010 and 6 million plants in 2011. In addition, about 1,400 indoor plants were eradicated in 2017, down from about 396,000 plants in 2014, and down significantly from about 462,000 plants in 2010 and 509,000 plants in 2011.[50]

myth

Drug interdiction and eradication are effective strategies.

fact

Despite the regular seizure of huge quantities of illegal drugs, federal, state, and local law enforcement efforts net only about 10% to 15% of the total supply. As for drug interdiction, if correctional officials cannot keep illegal drugs out of maximum security prisons, how effective is the government going to be in keeping illegal drugs from entering the country?

Smuggling Interdiction Federal law enforcement agencies have the primary responsibility for smuggling interdiction, but local police agencies can play a role. Local police officers in jurisdictions near the U.S. borders or those in jurisdictions with international airports or even small airports and airfields can be on the lookout for drug smugglers.

Problem-Oriented and Community Policing Strategies Problem-oriented policing is a strategy that focuses on the underlying problems that cause crime rather than focusing on each specific criminal event. The strategy involves identifying the underlying problems, analyzing them in detail, applying solutions to them, and then evaluating the effectiveness of the solutions. The problem-oriented approach to drug enforcement has been adopted by many local police agencies to address drug problems in public housing projects, drug abuse among teenagers, and drug abuse in abandoned buildings. The strategy makes sense because it promotes careful analysis of a community's unique drug problem before action is taken, advocates customizing responses and targeting resources, and recommends working with other public and private agencies in the problem-solving process.

Community policing, which is addressed in more detail later in the chapter, is both a philosophy and a set of methods. The philosophy is that citizens and the police must form a partnership and work collectively to identify problems, propose solutions, implement action, and evaluate results in the community. Methods include foot patrol, storefronts and other mini-stations in the community, door-to-door contact with citizens, community organizing, ombudsperson-like activities, provision of social services, and problem-oriented policing. Community policing can contribute to drug enforcement in several ways. For example, foot patrol and problem-oriented policing can reduce street-level dealing. Successful community policing may increase public support for drug enforcement efforts by encouraging citizens to report drug crimes and identify drug dealers. Community organizing may empower citizens to resist drug dealers and drug abusers who invade their communities. The provision of enhanced social services, whether provided by police officers or through police ombudspersons, may help individuals resist the temptation of illegal drugs.

Drug Demand Reduction Local police agencies can play a role in drug demand reduction strategies in at least four ways. First, through visible drug enforcement efforts, the police may discourage some people from using drugs in the first place because of a fear of arrest. Second, the police may stop some people from continuing to use drugs by arresting them so that they can get court-ordered treatment or by diverting them to drug treatment programs. Third, through public education programs, especially in the schools, the police may get young people to resist the temptations of drugs or the peer pressure to use them. Fourth, the police can lend their stature and credibility to efforts to increase funding for drug treatment, prevention, and education programs.

D.A.R.E. (Drug Abuse Resistance Education) is the nation's largest and best-known substance abuse prevention program. It was developed in 1983 by Los Angeles Police Chief Daryl Gates and is now taught in 75% of the school districts nationwide.[51] In its 2009 annual report, the claim was made that "during its first twenty years, D.A.R.E. contributed to the accomplishment of a 50% reduction of drug use in the United States."[52]

D.A.R.E. is a school-based collaborative effort among police departments, schools, parents, and community leaders to teach children how to recognize and resist the direct and subtle pressures that influence them to experiment with alcohol, tobacco, marijuana, and other drugs. In recent years, its mission has expanded to include Internet safety, bullying prevention, cyber-bullying prevention, prescription and over-the-counter drug abuse prevention, and gang involvement prevention. Usually addressing children in the fifth or sixth grade, a specially trained uniformed police officer comes to the school 1 day a week for 17 weeks and teaches the children for about an hour. The D.A.R.E. curriculum is integrated with other regular subjects. Although D.A.R.E. programs are hugely popular, no scientific study, among the many evaluations that have been conducted, has discovered any statistically significant difference in drug-usage rates between students who had taken D.A.R.E. and those who had not.

However, the 2013 annual report states that a rigorous scientific evaluation of the new "D.A.R.E. keepin' it REAL" (kiR) program for grades six through nine showed "a 32% to 44% reduction in marijuana, tobacco, and alcohol use; a 29% to 34% decrease in intent to accept substances; and a reduction and cessation of substance abuse." The report also noted that "kiR's cost-benefit ratio is 28:1, making it one of the most cost-beneficial school-based curriculum-driven drug prevention programs. This is a $28 return for every $1 invested in program delivery, yielding a net benefit to the concerned community of $3,600 per pupil."[53]

Asset Forfeiture A huge incentive for local police departments to participate in drug enforcement activities is *asset forfeiture*. Asset forfeiture is an ancient practice that is referred to in the Bible. It was part of English common law and helped instigate the American

D.A.R.E. programs are taught in 75% of the nation's school districts. *Why haven't these programs been more effective in reducing illegal drug use among the nation's youth?*

Mikael Karlsson/Alamy Stock Photo

Civil Forfeiture

Civil forfeiture is based on the legal fiction that the property that facilitates or is connected to a crime has itself committed a wrong and can be seized and tried in civil court. Such judicial hearings are referred to as *in rem* proceedings, meaning "against the thing."

Source: Scott Ehlers, *Policy Briefing: Asset Forfeiture*, The Drug Policy Foundation (Washington, DC: U.S. Government Printing Office, 1999), 4.

Revolution. Partly because of asset forfeiture abuses, the due process clause of the Fifth Amendment was included in the U.S. Constitution to guarantee that property could not be taken from citizens without a judicial hearing.

Congress enacted the first drug-related civil asset forfeiture law in 1970 as part of the Comprehensive Drug Abuse and Prevention Act. The law authorized the government to seize and forfeit illegal drugs, manufacturing and storage equipment, and vehicles used to transport drugs. A major rationale for the law was the belief that drug traffickers should not benefit financially from their illegal activities or be able to use property or money obtained illegally in future drug crimes. In 1978 and throughout the 1980s, Congress passed several more antidrug laws that enhanced the government's power to seize and forfeit property.

Asset forfeiture received a boost in 1984, when Congress created the Assets Forfeiture Fund (AFF) as part of the Comprehensive Crime Control Act. The law required the Attorney General to deposit all net forfeiture proceeds into the AFF for use by the Department of Justice and other federal law enforcement agencies. An important provision of the legislation was the directive to equitably share forfeiture proceeds with state, local, and tribal agencies and foreign governments that directly assisted law enforcement efforts in the seizure and forfeiture of assets.[54] According to several commentators, these two developments—the creation of the AFF and equitable sharing—initiated the modern era of policing and prosecuting for profit.

Note that the government can seize a person's property under both criminal and civil law. A reason civil asset forfeiture has been the preferred tool in drug enforcement is that property can be forfeited under criminal law only if the property owner has been convicted of a crime (beyond a reasonable doubt). Under civil forfeiture law, however, the government is required only to have probable cause to seize a person's property. Furthermore, until the passage of the Civil Asset Forfeiture Reform Act of 2000 (CAFRA), an owner under civil forfeiture law had to be proactive to get his or her property back: An owner had to prove in a civil proceeding by a preponderance of the evidence that the property was not used in a crime. CAFRA shifted the burden of proof so that, today, the government has to prove by a preponderance of evidence that the property should be forfeited. Despite the change, civil asset forfeiture remains the preferred method. From 1997 to 2013, only 13% of Justice Department forfeitures were criminal forfeitures, while 87% were civil forfeitures. In addition, 88% of civil asset forfeitures during this period took place "administratively," that is, they occurred automatically when a property owner failed to challenge a seizure in court. The seized property simply was presumed "guilty" without a neutral arbiter such as a judge determining whether it should be permanently taken from its owner.[55]

Assets can be forfeited under either federal or state law. When federal agencies are involved in a successful drug enforcement operation, they can return up to 80% of forfeited assets to other participating state or local law enforcement agencies (this is the "equitable sharing" part of the law). In 2018, state and local law enforcement agencies received $400 million through equitable sharing.[56] The amount of money or property returned depends on the particular agency's level of involvement in the case. Federal law requires that all funds returned to state and local law enforcement agencies should be used for law enforcement activities. Priority is given to supporting community policing activities, training, and law enforcement operations that are likely to contribute to further seizures and forfeitures.[57] However, each state has its own formula for how forfeited assets are to be distributed and used. For example, some states require all forfeitures to be used for drug enforcement activities; some states require forfeited assets to be given to the state educational system; and other states require the proceeds from forfeitures be deposited in the state treasury to be used at the legislature's discretion. The few data available for the federal government and a handful of states provide only broad categories of spending, making it difficult to determine individual expenditures. When expenditures have been provided by category, most known spending by state and local agencies has been listed under equipment, "other," and salaries and overtime. Only small amounts have been spent on substance abuse or crime prevention programs.[58]

Prior to CAFRA, critics pointed to many problems with civil asset forfeiture laws. First, as noted previously, civil asset forfeiture laws placed the burden of proof on the owner of the property seized to show that the property was not used in a drug-related crime. Not only was it difficult to prove a negative—that the property was *not* used in a drug-related crime—but the government also had to prove almost nothing. CAFRA shifted the burden of proof to the government.

Second, the costs involved in a civil asset forfeiture proceeding could be prohibitively high. For example, to contest a federal forfeiture, a property owner had to post a bond equal to 10% of the seized property's value. This requirement placed a burden on the poor. CAFRA abolished the bond requirement. The poor were further handicapped in civil asset forfeiture proceedings because the government was not obligated to provide counsel, as the Sixth Amendment required in criminal trials. In many cases, property owners simply forfeited the seized property because of the high costs of getting it back. In some cases, the costs of getting property back were higher than the value of the seized property itself. CAFRA provides for court-appointed counsel for indigent owners if (1) the indigent owner makes a "good-faith" claim and is already represented by court-appointed counsel in the related criminal case and (2) the subject of the forfeiture is the indigent's primary residence.

In addition, nonindigent claimants may be entitled to court and litigation costs provided they "substantially prevail" in the civil forfeiture action. Finally, before CAFRA, the government generally was exempt from claims stemming from seizure and possession of an owner's property in civil forfeiture actions. CAFRA now provides for compensation to victims of forfeiture in cases where their property was seized and they were not convicted of a crime giving rise to that forfeiture. One study found that approximately 80% of persons whose property was seized by the federal government for forfeiture were never even charged with a crime.[59]

Third, innocent owners could lose their property when someone else used it without their permission or knowledge to commit a drug crime. In the case of *Calero-Toledo v. Pearson Yacht Leasing Company* (1974), for example, law enforcement officers found a marijuana cigarette onboard a yacht rented by the Pearson Yacht Leasing Company. The yacht was forfeited to the government because it was used to transport a controlled substance. The U.S. Supreme Court upheld the forfeiture, establishing the principle that the government can seize an innocent owner's property in a civil proceeding. This seeming injustice was corrected somewhat in 1988, when Congress passed the Asset Forfeiture Amendments Act. The act created forfeiture exceptions for some, but not all, innocent owners and for violations involving the possession of personal-use quantities of drugs. However, innocent owners were not fully protected by the exception. For example, in 1999, a Wichita, Kansas, couple had their motel forfeited and sold because drugs had been sold on the property. The couple had tried to keep drug dealers off the property by installing floodlights and fences and calling the police, but their property was forfeited anyway. CAFRA created an innocent-owner defense applicable to all civil asset forfeiture statutes.

Still, critics argue that the standard of proof in civil forfeiture cases is too low. Because probable cause is the standard of proof necessary to seize property that was used to facilitate a drug crime, little more than hearsay evidence is required. Another problem with probable cause as the standard of proof in civil asset forfeiture proceedings is exemplified by the sheriff of Volusia County, Florida (near Daytona Beach), who defined probable cause for purposes of asset forfeiture as having more than $100 in cash.

Critics also maintain that civil asset forfeiture policy undermines the integrity of the police and the criminal justice system. The proceeds that can be had from civil asset forfeitures are a corrupting influence for many law enforcement officers. Sometimes officers are more concerned with seizing assets than they are with getting drugs off the streets. In some cases, law enforcement officers simply extort money and property from innocent people. Civil asset forfeiture laws sometimes promote other illegal practices such as racial profiling (described in Chapter 7). Sometimes resources are diverted from more serious crimes to drug cases that promise asset forfeitures.

Another major problem with civil asset forfeiture is that many law enforcement agencies have become dependent on it. A survey found that nearly 3,000 participating state and local law enforcement agencies nationwide considered civil asset forfeiture proceeds as a necessary budget supplement. This was true even at the federal level, where the Department of Justice "has urged its lawyers to increase their civil forfeiture efforts so as to meet the Department's annual budget targets."[60] As a result, growth in the AFF has been phenomenal. For example, in 1986, the AFF received $93.7 million in revenue from federal forfeitures; by 2014, the Fund had taken in $4.5 billion–a 4,667% increase. Between 2000 and 2013, annual DOJ equitable sharing payments to state and local law enforcement more than tripled, increasing from $198 million to $643 million. In all, the DOJ paid state and local agencies $4.7 billion in forfeiture proceeds from 2000 to 2013.[61]

FYI Americans Oppose Civil Asset Forfeiture

A 2016 poll found that 84% of Americans oppose civil asset forfeiture. Strong opposition was recorded for all demographic groups. Those groups prefer property only be seized after a person is convicted of a crime. 76% of Americans would not allow local departments to keep the assets.

Source: Emily Ekins, "Policing in America: Understanding Public Attitudes Toward Police. Results from a National Survey," Cato Institute, December 7, 2016, https://www.cato.org/survey-reports/policing-america (accessed February 26, 2019).

One final concern with asset forfeiture programs (to be discussed here) is their role in the militarization of state and local law enforcement. During the last 3 decades, state and local law enforcement agencies have used some of the hundreds of millions of dollars made available to them through asset forfeiture programs to purchase surplus military equipment from the Department of Defense. The money, or the "free-floating slush fund," as a former director of asset forfeiture at the Justice Department called it, enables law enforcement agencies to bypass the traditional budget process in which elected officials create law enforcement spending priorities and set their own priorities. (Since 1990, when Congress created the program, the Pentagon has transferred more than $6.5 billion worth of military gear to local law enforcement agencies.[62] In many cases, the Pentagon simply gives surplus military equipment to law enforcement agencies that want it.) Among weapons made available by the Pentagon are armored personnel carriers, such as the one used by the police in Ferguson, Missouri, and Mine-Resistant Ambush Protected (MRAP) vehicles, such as those used in Iraq to protect soldiers from roadside bombs. Those are in addition to weapons such as .50 caliber rifles and grenade launchers that are modified to shoot tear gas canisters. Proponents of police militarization argue that law enforcement officers face new and increasingly lethal situations that require military equipment for their safety—which may be true. However, an interesting question is: Would law enforcement agencies want surplus military equipment if it were not free or if it had to be bought within the regular budgetary process and not through asset forfeiture funds? If the latter, law enforcement agencies may not need surplus military equipment.

In short, although civil asset forfeiture helps fuel the war on drugs and can reduce the profits from the illegal drug trade, there are many reasons to question its fairness and utility, despite the significant reforms of CAFRA. Its harshest critics contend that civil asset forfeiture should be eliminated altogether because criminal asset forfeiture is available and has procedures that better protect the innocent. Recently, several states have enacted or are considering legislation that restricts or prohibits state, local, or tribal law enforcement agencies' participation in federal equitable sharing of asset forfeiture proceeds.[63] Many other states, perhaps most of them, however, may have difficulty rejecting the financial windfall that the equitable sharing of either civil or criminal asset forfeiture proceeds provides them.

U.S. Supreme Court Limits Civil Asset Forfeiture

On February 20, 2019, in the case of *Timbs v. Indiana*, the U.S. Supreme Court ruled in a 9-to-0 decision that the 8th Amendment, which bars "excessive fines," limits the ability of the federal government, as well as states and localities, to seize property through civil asset forfeiture. While not prohibiting the practice, the Court held that the value of the property seized and forfeited no longer may be disproportionate to the crime involved.

Source: Adam Liptak and Shaila Dewan, "Supreme Court Limits Police Powers to Seize Private Property," *The New York Times*, February 20, 2019. Accessed February 21, 2019, https://www.nytimes.com/2019/02/20/us/politics/civil-asset-forfeiture-supreme-court.html.

Criticisms of the War on Drugs In addition to the already mentioned problems with specific aspects of drug enforcement, the war on drugs itself has been severely criticized. First, critics contend that the government has exaggerated the dangers of illegal drug use to gain public support for the war on drugs and has generally ignored the harms caused by the drug laws themselves. For example, it has been estimated that about 20,000 people die each year from the consumption of all illegal drugs combined. While any death from illegal drug consumption is regrettable, the 20,000 deaths attributed to illegal drugs (with more than half caused by heroin overdoses) pale in comparison to the approximately 85,000 alcohol-related and 440,000 tobacco-related deaths each year.[64] Moreover, rarely are drug-related deaths the result of abuse or misuse of the drugs. Instead, most of the drug-related deaths are directly attributable to the drug laws that prohibit their use. For example, most heroin overdoses occur because the heroin is adulterated (i.e., the drug is "cut" or mixed with dangerous substances to increase the quantity of the drug available to sell or to produce a more readily ingestible form of the drug). Heroin is adulterated because it is illegal and unregulated; there is no quality control in its production and distribution. Also, because it is illegal, heroin addicts share needles, a practice that spreads disease and illness. It has been estimated that 25% of AIDS cases in the United States are a direct result of the unsafe and unsanitary conditions in which illegal drugs are consumed.[65] The same is true of cocaine. Because of the drug laws, the price of using powder cocaine is too high for many people, so they substitute the more affordable and more dangerous crack cocaine. A person is more likely to die or suffer injury from smoking the drug than he or she is from snorting it. As for marijuana, no one has ever died from using marijuana. The greatest danger to marijuana smokers is in smoking the drug after it has been adulterated by government control programs, such as the spraying of herbicides on marijuana crops.

Another way the government has tried to rally support for the drug war has been to try to persuade the public of a direct connection between illegal drug use and crime. However, as described in the Myth/Fact box on Illegal drug use, it is not drugs that cause crime, but rather the drug laws that cause crime.

myth

Illegal drug use causes crime.

fact

Illegal drug use is not a significant cause of crime. Although the effects of some illegal drugs may cause a very small number of violent crimes, the vast majority of drug-related crimes are caused by the drug laws that make the possession, distribution, cultivation, or manufacture of certain drugs a crime. The drug laws also cause crime by creating an illegal market in which drug prices are artificially and dramatically increased. This makes it necessary for some drug users, especially addicts, to steal or deal to get money to buy drugs. It also contributes to the violence of rival drug dealers seeking to monopolize a market. The illegality of desired drugs also puts otherwise law-abiding people in contact with members of the criminal underworld from whom they can learn criminal skills. Ironically, alcohol is the only drug that has been found to cause a significant amount of crime, and alcohol is legal for adults.

Second, critics have argued that the drug war is racist. Of the approximately 1.6 million people arrested in 2017 for drug abuse violations, approximately 70% were white and about 27% were black.[66] Yet, blacks are only about 11% of all lifetime illegal drug users, and 13% of all illegal drug users in 2017.[67] Furthermore, in many jurisdictions, the penalties are much more severe for possession of crack cocaine than they are for the possession of powder cocaine, although pharmacologically they are identical drugs. For example, until August 3, 2010, when the Fair Sentencing Act of 2010 was enacted, federal sentencing guidelines provided for a 100-to-1 sentencing disparity, which meant that conviction for possessing 5 grams of crack resulted in the same mandatory minimum 5-year sentence as conviction for possessing 500 grams of powder cocaine. On average, federal crack-cocaine defendants received sentences that were 50% longer than those received by federal powder-cocaine defendants. When Congress passed the Anti-Drug Abuse Act in 1986, crack was a relatively new drug believed to be more dangerous than powder cocaine. Among suspects arrested by the Drug Enforcement Administration in 2012, blacks accounted for 77% of the arrests for offenses involving crack cocaine but only 28% of the arrests for offenses involving

AFP/Stringer/Getty Images

About 70% of people arrested for possession of powder cocaine are white, while about 80% of people arrested for possession of crack cocaine are black. *Should law enforcement authorities distinguish between powder cocaine and crack cocaine when making arrests for possession? Why or why not?*

powder cocaine.[68] Ironically, relatively low-level crack-cocaine dealers have received longer prison sentences than wholesale-level powder-cocaine dealers from whom the crack dealer originally bought the powder to make the crack.

In two cases, the U.S. Supreme Court ameliorated somewhat the harsh federal sentencing guidelines for cocaine. In *United States v. Booker* (2005), the Court held that the formerly mandatory sentencing guidelines are now only advisory. Under the new rule, the guidelines are to serve as only one factor among others that must be considered in determining an appropriate sentence. If judges believe that deviating from the guidelines is reasonable under the circumstances, they may do so. In *Kimbrough v. United States* (2007), the Court reversed the Fourth Circuit of Appeals, which ruled that "a sentence outside the guidelines range is *per se* unreasonable when it is based on a disagreement with the sentencing disparity for crack and powder offenses." As a result, in December 2007, the U.S. Sentencing Commission voted unanimously to allow about 20,000 federal inmates to seek reductions in their crack sentences. The sentencing commission estimated the average reduction would likely be a little more than 2 years.

On August 3, 2010, as noted previously, President Obama signed into law the Fair Sentencing Act (FSA) of 2010. The law did not eliminate the sentencing disparity between crack- and powder-cocaine offenses, but it did dramatically reduce the disparity from the previous 100-to-1 ratio to a new 18-to-1 ratio. Another significant change the law made is to eliminate the mandatory minimum 5-year prison sentence for first-time offenders possessing crack cocaine. However, the law as passed did not apply retroactively as many advocates had hoped. In 2012, in *Dorsey v. United States*, the U.S. Supreme Court ruled 5–4 that "the new, lower sentences in the FSA may be applied to offenders sentenced after the FSA became law, even if their crimes were committed before its enactment." The decision, however, does not grant full retroactivity. Nevertheless, the sentences for more than 7,700 prisoners have been reduced as a result of the ruling.[69] Note that these developments only applied to federal cocaine cases; they did not apply to those that originate at the state level.

> **FYI**
>
> **Drug Offenders in Prison**
>
> In September 2017, about 47% of federal prisoners and, at year-end 2016, 15% of state prisoners were incarcerated for drug offenses. That compares with 58% of federal prisoners and 21% of state prisoners who were incarcerated for drug offenses in 1998.
>
> Sources: Jennifer Bronson and E. Ann Carson, *Prisoners in 2017*, April 2019, NCJ 252156, 22, Table 13 and 23, Table 14. Accessed May 13, 2019, https://www.bjs.gov/content/pub/pdf/p17.pdf; Allen J. Beck, *Prisoners in 1999*, in Bureau of Justice Statistics Bulletin, U.S. Department of Justice (Washington, DC: U.S. Government Printing Office, August 2000), 10, Table 15 and 12, Table 21.

Then, in April 2014, the U.S. Sentencing Commission voted unanimously to reduce sentencing guidelines for most federal drug trafficking offenders. In July 2014, the Commission voted, again unanimously, to make the sentencing reduction retroactive. Congress did not act to modify or disapprove the change, so the policy became effective on November 1, 2014.[70] In April 2016, the U.S. Sentencing Commission reported that 26,000 federal drug offenders (or about 70%) who sought a sentence reduction had received shorter prison sentences because of the 2014 sentencing guideline changes.[71]

Federal drug sentencing reform efforts received an added boost in December 2018, when President Trump signed into law "The First Step Act." The law includes provisions that will reduce the sentence length of most federal drug law violators. The law gives eligible federal prisoners time credits for successfully completing evidence-based recidivism reduction programming or productive activities. The law stipulates that a prisoner "shall earn 10 days of time credits for every 30 days of successful participation" in the programming or activities. In addition, a prisoner can earn another 5 days of time credits for his or participation in the programming or activities if he or she is "determined by the Bureau of Prisons to be at a minimum or low risk for recidivating, [and] who, over two consecutive assessments, has not increased his or her risk of recidivism."[72] What The First Step Act does not do is provide full retroactivity to the Fair Sentencing Act.

Third, as noted previously, critics point out that the war on drugs and the huge amount of money involved in the drug trade have corrupted many law enforcement personnel, who have been found guilty of drug dealing, providing protection for drug dealers, conspiracy, extortion, bribery, robbery, theft, and murder. Unfortunately, examples abound. A *New York Times* report disclosed that each year more than 100 law enforcement officers are prosecuted in state and federal courts on drug corruption charges.

Fourth, critics maintain that the war on drugs is hugely expensive, diverts resources from arguably more important projects, and has had little lasting effect on illegal drug use. Since 1980, the United States has spent hundreds of billions of dollars on federal, state, and local antidrug efforts—more than is spent on medical research into cancer, heart disease, or

AIDS, to name just three worthwhile projects. Furthermore, despite the huge amount of money and effort spent on the drug war, the prices of many illegal drugs have not increased and in many cases have fallen, and the purity of illegal drugs and their availability have not decreased but in many cases have increased. If the drug war is successfully reducing supply or demand, one would expect prices to increase, and the purity and availability of illegal drugs to decline.

Although the measurement of illegal drug use is fraught with all sorts of reliability problems, according to the Substance Abuse and Mental Health Services Administration's annual *National Survey on Drug Use and Health*, there was a significant decline in illegal drug use among both youths and adults between 1979 and 1992 (from 25 million past-month illegal drug users to 12 million past-month illegal drug users). However, the number of past-month illegal drug users has increased unevenly between 1992 and 2017. The estimated number of past-month illegal drug users aged 12 or older in the United States in 2017 was about 30.5 million–likely the largest number of past-month users since 1979. In 2017, the illegal drug most commonly used in the past month was marijuana, which was used by 26 million people or 85% of all illegal drug users. By contrast, the second most commonly used in the past month illegal drug was misused prescription pain relievers, which were used by 3.2 million people or 10.5% of past-month illegal drug users. In 2017, about 2 million adolescents age 12–17 were current illegal drug users or about 8% of that age group, approximately 8 million young adults age 18–25 were current illegal drug users or approximately 24% of that age group, and about 20 million adults age 26 or older were current illegal drug users or about 9.5% of that age group.[73] Critics are calling for an end to the drug war. Some critics want illegal drugs to be decriminalized or legalized. Other critics want the drug problem treated as a public health problem and not as a criminal justice problem. Perhaps it is time to seriously rethink America's strategy for dealing with illegal drug use.

FYI **Marijuana Decriminalization**

As of April 2019, 10 states (Alaska, California, Colorado, Maine, Massachusetts, Michigan, Nevada, Oregon, Vermont, and Washington) and the District of Columbia have decriminalized marijuana for recreational use. Thirty-four states and DC have decriminalized marijuana for medical use.

Source: Calculated from data in DISA, "Map of Marijuana Legality by State," https://disa.com/map-of-marijuana-legality-by-state (accessed May 13, 2019).

THINKING CRITICALLY

1. What do you think are the pros and cons of being an investigator/detective? Does this type of work sound attractive to you?
2. Do you think state and federal law enforcement agencies should be able to collect and store DNA and fingerprints of people who have not been convicted of crimes? Why or why not? If they should be allowed, for how long should the DNA and fingerprints be kept?
3. Do you think there are any ways to protect individuals and businesses from becoming victims of cybercrime? If so, what do you propose?
4. Do you think the United States should change its drug enforcement strategy? If so, in what ways? If not, why not?

Community Policing

For decades, police followed the professional model, which rested on three foundations: preventive patrol, quick response time, and follow-up investigation. Sensing that the professional model did not always operate as efficiently and effectively as it could, criminal justice researchers set out to review current procedures and evaluate alternative programs. One of the first and best known of these studies was the Kansas City, Missouri, Preventive Patrol Experiment, discussed earlier in this chapter. That study's conclusion was that preventive patrol did not necessarily prevent crime or reassure citizens. Following the study, some police departments assigned police units to proactive patrol, giving them specific assignments rather than having them randomly cruise the streets.

Another study, again with the Kansas City Police Department, examined the effects of police response time. The study found that police response time was unrelated to the probability of making an arrest. Researchers discovered that the time it takes a citizen to report a crime–not the speed with which police respond–was the major determinant of whether an on-scene arrest took place or witnesses could be located. In 90% of crimes, citizens wait 5 to 10 minutes to call the police, precluding catching the criminal at the scene.

myth

Shorter police response time contributes to more arrests.

fact

For most crimes, police response time is irrelevant. Approximately two-thirds of crimes are "cold"; the offender is gone long before the crime is discovered. In cases in which time counts, the critical delay often occurs in the time it takes the victim to call the police.

As preventive patrol and fast response time were being questioned, so, too, was follow-up investigation. A study by the RAND Corporation reviewed the criminal investigation process for effectiveness. The researchers concluded that the work of a criminal investigator alone rarely leads to an arrest and that the probability of arrest is determined largely by information that patrol officers obtain at the crime scene in their preliminary investigation.[74]

Criminal justice researchers continued their review of accepted police functions with the aim of making policing more effective by initiating new techniques and procedures. One of the interesting findings of the foot patrol research was that foot patrol officers were better able to deal with minor annoyances—such as rowdy youths, panhandlers, and abandoned cars—that irritate citizens.

In a theory called "broken windows," noted police scholars James Q. Wilson and George Kelling proposed that those minor annoyances are signs of crime and cause a fear of crime and that if they are not dealt with early, more serious and more costly problems are likely to occur.[75] Wilson and Kelling concluded that to help solve both minor and major problems in a neighborhood and to reduce crime and fear of crime, police officers must be in close, regular contact with citizens. That is, police and citizens should work cooperatively to build a strong sense of community and should share responsibility in the neighborhood to improve the overall quality of life within the community.

The Philosophy and Components of Community Policing

With community policing, citizens share responsibility for their community's safety. Citizens and the police work collectively to identify problems, propose solutions, implement action, and evaluate the results in the community. A community policing perspective differs in a number of ways from a traditional policing perspective. For example, in community policing, the police must share power with residents of a community, and critical decisions need to be made at the neighborhood level, not at a downtown police headquarters. Such decentralization of authority means that credit for bringing about a safer and more secure community must be shared with the people of the community, a tall order for any group of professionals to accept. Achieving the goals of community policing requires successful implementation of three essential and complementary components or operational strategies: community partnership, problem solving, and change management.[76]

Community Partnership Establishing and maintaining mutual trust between citizens of a community and the police is the main goal of the first component of community policing. Police have always recognized the need for cooperation with the community and have

Community police officers visit with a citizen. *How does such activity contribute to crime fighting?*

Tom Carter/PhotoEdit

encouraged members of the community to come forward with crime-fighting information. In addition, police have spoken to neighborhood groups, worked with local organizations, and provided special-unit services. How are those cooperative efforts different from the community partnership of community policing?

In community policing, the police become an integral part of the community culture, and the community, in turn, helps the police define future crime prevention strategies and allocate community protection services. Establishing a community partnership means adopting a policing perspective that exceeds the standard law enforcement emphasis. The police no longer view the community as a passive presence connected to the police by an isolated incident or series of incidents. The community's concerns with crime and disorder become the target of efforts by the police and the community working together.

For patrol officers, building police–community partnerships entails such activities as talking to local business owners to identify their concerns, visiting residents in their homes to offer advice on security, and helping to organize and support neighborhood watch groups and regular community meetings. It also involves ongoing communication with residents. For example, a patrol officer might canvas a neighborhood for information about a string of burglaries and then revisit those residents to inform them when the burglar is caught.

Problem Solving Problem solving requires a lot more thought, energy, and action than traditional incident-based police responses to crime and disorder. In full partnership, the police and a community's residents and business owners identify core problems, propose solutions, and implement a solution. Thus, community members identify the concerns that they feel are most threatening to their safety and well-being. Those areas of concern then become priorities for joint police–community interventions.

For this problem-solving process to operate effectively, the police need to devote time and attention to discovering a community's concerns, and they need to recognize the validity of those concerns. Police and neighborhood groups may not always agree on the specific problems that deserve attention first. For example, the police may regard robberies as the biggest problem in a particular neighborhood, while residents find derelicts who sleep in doorways, break bottles on sidewalks, and pick through garbage cans the number-one problem. In community policing, both problems should receive early attention from the police, other government agencies, and the community.

Some community policing advocates recommend a four-step problem-solving process referred to as SARA: *S*canning–identifying problems; *A*nalysis–understanding underlying conditions; *R*esponse–developing and implementing solutions; *A*ssessment–determining the solutions' effect. One useful tool in working toward a solution is known as the *crime triangle*. The crime triangle is a view of crime and disorder as an interaction among three variables: a victim, an offender, and a location. Solutions can be developed that affect one

The Crime Act and COPS

The Violent Crime Control and Law Enforcement Act of 1994—popularly known as the Crime Act—authorized $8.8 billion over 6 years for grants to local policing agencies to add 100,000 officers and promote community policing in innovative ways. To implement the law, the Office of Community Oriented Policing Services (COPS) was created in the U.S. Department of Justice. As of January 5, 2016, COPS had invested more than $14 billion to help advance community policing. This includes grants to more than 13,000 state, local, and tribal law enforcement agencies to fund the hiring and redeployment of more than 127,000 officers. For 2017, the COPS Hiring Program had approximately $137 million to provide funding to hire and re-hire entry level law enforcement officers in state, local, and tribal law enforcement agencies.

Source: U.S. Department of Justice, Community Oriented Policing Services, "U.S. Department of Justice's COPS Office Announces Winners of Its Second Annual 'Community Policing in Action' Photo Contest," January 5, 2016. Accessed January 27, 2016, www.cops.usdoj.gov/Default.asp?Item=2840; U.S. Department of Justice, Community Oriented Policing Services, "COPS Hiring Program (CHP)," accessed May 13, 2019, https://cops.usdoj.gov/chp.

Bernd Obermann/Getty Images

A community police officer talks to a boy at the Puerto Rico Day Parade. *What, if anything, is the police officer trying to accomplish?*

School Resource Officers

School resource officers (SROs) use a community policing approach to provide a safe environment for students and staff. They respond to calls for service within the school and work with school administrators and staff to prevent crime and disorder by monitoring crime trends, problem areas, cultural conflicts, and other concerns. The total number of SROs in the United States was last estimated in a 2007 survey. However, the National Association of School Resource Officers estimates that currently there are between 14,000 and 20,000 SROs nationwide. At year-end 2007, by contrast, there were about 13,000 full-time SROs.

Sources: National Association of School Resource Officers, "Frequently Asked Questions," accessed May 13, 2019, https://nasro.org/frequently-asked-questions/; Brian A. Reaves, *Local Police Departments, 2007,* U.S. Department of Justice, Bureau of Justice Statistics (Washington, DC: U.S. Government Printing Office, December 2010), Table 32.

or more of the three elements of the crime triangle. For example, suppose elderly residents are being threatened by speeding teenagers in automobiles as they walk across the streets of a suburban residential neighborhood. Using a crime triangle analysis might result in the following police-community solutions: using the juvenile court to alter the probation period of offending drivers (focus on the offender), installing speed bumps in the pavement or changing the cycles of traffic signals on opposite ends of the street so that motorists cannot build up speed (focus on the location), holding safety education classes at the senior center to educate elderly residents to use marked crosswalks (focus on the victim). More than likely, in community policing, a combination of those solutions would be used. Such a response to a community problem is much more thorough than merely having a squad car drive by the location when a citizen calls in a complaint.

Change Management Forging community policing partnerships and implementing problem-solving strategies necessitate assigning new responsibilities and adopting a flexible style of management. Traditionally, patrol officers have been accorded lower status in police organizations and have been dominated by the agency's command structure. Community policing, in contrast, emphasizes the value of the patrol function and the patrol officer as an individual. It requires the shifting of initiative, decision making, and responsibility downward within the police organization. The neighborhood police officer or deputy sheriff becomes responsible for managing the delivery of police services to the community or area to which he or she is permanently assigned. Patrol officers are the most familiar with the needs and concerns of their communities and are in the best position to forge the close ties with the community that lead to effective solutions to local problems.

Under community policing, police management must guide, rather than dominate, the actions of the patrol officer and must ensure that patrol officers have the necessary resources to solve the problems in their communities. Management must determine the guiding principles to convert the philosophy of the agency to community policing and then to evaluate the effectiveness of the strategies implemented.

THINKING CRITICALLY

1. Can you think of ways to make community policing even more effective?
2. Do you think there is a certain operational style that is most appropriate for community policing? If so, which one?

Terrorism and Homeland Security

Terrorism is one of the oldest forms of human conflict. Before societies began waging war against each other, individuals and small groups used acts of terror to achieve their goals, such as deposing existing leaders and frightening and repelling adversaries from territory they claimed for themselves.[77] Terrorism has been used by both right-wing and left-wing political organizations, by ethnic and nationalistic groups, by revolutionaries, and by the armies and secret police of established governments. However, according to the FBI, the modern era of terrorism did not begin until the late 1960s, and U.S. soil remained largely free from serious acts of international terrorism until the bombing of the World Trade Center in 1993.[78] Before then, the vast majority of deadly terrorist acts carried out in the United States were committed by domestic terrorists, such as Timothy McVeigh and Terry Nichols, who were responsible for the 1995 bombing of the Alfred P. Murrah Federal Building in Oklahoma City.[79] All that changed on September 11, 2001, when, according to Ambassador Francis X. Taylor, the U.S. Department of State's Coordinator for Counterterrorism, "the United States suffered its bloodiest day on American soil since the Civil War, and the world experienced the most devastating international terrorist attack in recorded history."[80] On that infamous day in September, 19 hijackers belonging to the al-Qaeda terrorist network commandeered four aircraft to commit the most audacious terrorist act in American history.

- Five terrorists hijacked American Airlines flight 11, which departed Boston for Los Angeles at 7:45 A.M. An hour later it was deliberately piloted into the North Tower of the World Trade Center in New York City.
- Five terrorists hijacked United Airlines flight 175, which departed Boston for Los Angeles at 7:58 A.M. At 9:05 A.M. the plane crashed into the South Tower of the World Trade

Center. Both towers collapsed shortly thereafter, killing approximately 3,000 people, including hundreds of firefighters and rescue personnel who were helping to evacuate the buildings.

- Four terrorists hijacked United Airlines flight 93, which departed Newark for San Francisco at 8:01 A.M. At 10:10 A.M. the plane crashed in Stony Creek Township, Pennsylvania, killing all 45 people onboard. The intended target of this hijacked plane is not known, but it is believed that passengers overpowered the terrorists, thus preventing the aircraft from being used as a missile.
- Five terrorists hijacked American Airlines flight 77, which departed Washington Dulles Airport for Los Angeles at 8:10 A.M. At 9:39 A.M. the plane was flown directly into the Pentagon in Arlington, Virginia, near Washington, DC. A total of 189 people were killed, including all who were onboard the plane.[81]

BILL WAUGH/AP Images

Before 9/11, the bombing of the Alfred P. Murrah Federal Building in Oklahoma City on April 19, 1995, by Timothy McVeigh was the worst incident of domestic terrorism in American history. *What are the similarities and differences between the bombing of the Murrah Federal Building by McVeigh and the 9/11 terrorist attacks on the World Trade Center and the Pentagon?*

Definitions and Types of Terrorism

There is no single, universally accepted definition of terrorism, and types of terrorism can be distinguished by perpetrators, motives, methods, and targets. Whatever the definition, the key elements of terrorism are fear, panic, violence, and disruption.[82] The FBI defines **terrorism** as "the systematic use of terror or unpredictable violence against governments, publics, or individuals to attain a political objective."[83] In the U.S. Code of Federal Regulations (28 CFR Section 0.85), terrorism is defined as "the unlawful use of force and violence against persons or property to intimidate or coerce a government, the civilian population, or any segment thereof, in furtherance of political or social objectives."[84] Title 22 of the U.S. Code, Section 2656f(d), defines terrorism as "premeditated, politically motivated violence perpetrated against noncombatant targets by subnational groups or clandestine agents."[85]

terrorism The systematic use of terror or unpredictable violence against governments, publics, or individuals to attain a political objective; the unlawful use of force and violence against persons or property to intimidate or coerce a government, the civilian population, or any segment thereof, in furtherance of political or social objectives; or premeditated, politically motivated violence perpetrated against noncombatant targets by subnational groups or clandestine agents, usually intended to influence an audience.

The FBI divides terrorism into two broad categories: *domestic terrorism* and *international terrorism*:

- **Domestic terrorism** is perpetrated by individuals and/or groups inspired by or associated primarily with U.S.-based movements that espouse extremist ideologies of a political, religious, social, racial, or environmental nature; for example, the June 8, 2014, Las Vegas shooting, during which two police officers inside a restaurant were killed in an ambush-style attack, which was committed by a married couple who held anti-government views and who intended to use the shooting to start a revolution.[86]
- **International terrorism** is perpetrated by individuals and/or groups inspired by or associated with designated foreign terrorist organizations or nations (state-sponsored); for example, the December 2, 2015, shooting in San Bernardino, California, that killed 14 people and wounded 22, which involved a married couple who radicalized for some time prior to the attack and were inspired by multiple extremist ideologies and foreign terrorist organizations.[87]

domestic terrorism Perpetrated by individuals and/or groups inspired by or associated with primarily U.S.-based movements that espouse extremist ideologies of a political, religious, social, racial, or environmental nature.

international terrorism Perpetrated by individuals and/or groups inspired by or associated with designated foreign terrorist organizations or nations (state-sponsored).

The FBI further divides domestic terrorism into "right-wing terrorism," "left-wing terrorism," "special interest terrorism," and "individual terrorism." *Right-wing terrorist groups*, such as the Aryan Nations and the World Church of the Creator, generally oppose government in general and government regulation in particular. Members typically are virulent racists who believe in racial supremacy and conspiracy theories.[88] According to the Southern Poverty Law Center (SPLC), the number of radical-right hate groups increased in 2017, to 954, up 4% from 917 in 2016. Until 2012, the number of radical-right hate groups had increased every year since 1999, when there were 457 such hate groups. From 1999 through 2017, the number of radical-right hate groups peaked at 1,018 in 2011. In addition, the SPLC counted 689 antigovernment "Patriot" groups in 2017, up 10.6% from the 623 such groups in 2016. The number of antigovernment "Patriot" groups peaked at an all-time high of 1,360 in 2012.[89]

Left-wing terrorist groups generally believe in revolutionary socialist doctrine and seek radical change outside the established political process. They want to liberate people from the dehumanizing effects of capitalism and imperialism. With the collapse of the Soviet

Union and the fall of communism in eastern Europe in the 1980s, left-wing terrorist groups no longer pose much of a threat to the United States.[90] That said, in 2018, the FBI still was concerned with black nationalists, who killed eight people in 2017, and who the FBI categorize as a far-left extremist group.[91]

The goal of *special interest terrorist groups* is to change the attitudes of the public about specific issues that are important to the group rather than to bring about fundamental political change. These groups generally are fringe elements of antinuclear, environmental, pro-life, animal rights, and other movements.[92] In the past, the FBI considered the Animal Liberation Front (ALF) and the Earth Liberation Front (ELF) two of the more active special interest terrorist groups.

Individual terrorist acts are committed by "lone wolf" extremists who operate alone or in small groups and defy detection as to both their identities and their plans for destruction. Timothy McVeigh, Eric Robert Rudolph, and Stephen Paddock are examples of this category of terrorist, as are homegrown violent extremists radicalized by ISIS and other radical Islamist groups.[93]

The FBI divides terrorist threats to U.S. interests in the United States and abroad into three categories: state sponsors of international terrorism, formalized terrorist organizations, and loosely affiliated extremists and rogue international terrorists.[94] As of this writing, the principal *state sponsors of terrorism*, according to the U.S. Department of State, are Iran (as of January 19, 1984), North Korea (as of November 20, 2017), Sudan (as of August 12, 1993), and Syria (as of December 29, 1979). For these countries, terrorism is a tool of foreign policy.[95] Three countries have been removed from the state sponsors of terrorism list: Iraq in 2004, Libya in 2006, and Cuba in 2015.

Formal terrorist organizations are autonomous; generally transnational; and have their own personnel, infrastructures, financial arrangements, and training facilities.[96] Table 6.4 lists the 68 foreign terrorist organizations recognized by the U.S. State Department as of April 15, 2019, which is an increase of 12 foreign terrorist organizations since January 27, 2016.

TABLE 6.4 68 Foreign Terrorist Organizations Recognized by the U.S. Department of State

- Abdallah Azzam Brigades (AAB)
- Abu Nidal Organization (ANO)
- Abu Sayyaf Group (ASG)
- Al-Aqsa Martyrs Brigade (AAMB)
- al-Ashtar Brigades (AAB)
- al-Nusrah Front
- Ansar al-Dine (AAD)
- al-Mulathamun Battalion
- Ansar al-Islam (AAI)
- Ansar al-Shari'a in Benghazi
- Ansar al-Shari'a in Darnah
- Ansar al-Shari'a in Tunisia
- Ansaru
- Army of Islam (AOI)
- Asbat al-Ansar (AAA)
- Aum Shinrikyo (AUM)
- Basque Fatherland and Liberty (ETA)
- Boko Haram
- Communist Party of Philippines/New People's Army (CPP/NPA)
- Continuity Irish Republican Army (CIRA)
- Gama'a al-Islamiyya (IG)
- Hamas
- Haqqani Network (HQN)
- Harakat ul-Jihad-i-Islami (HUJI)
- Harakat ul-Jihad-i-Islami/Bangladesh (HUJI-B)
- Harakat ul-Mujahideen (HUM)
- Hizballah
- Hizbul Mujahideen (HM)
- Indian Mujahideen (IM)
- ISIL-Khorasan (ISIL-K)
- ISIL Sinai Province (formerly Ansar Bayt al-Maqdis)
- ISIS-Bangladesh
- ISIS-Philippines
- ISIS-West Africa
- ISIS-Greater Sahara
- Islamic Jihad Union (IJU)
- Islamic Movement of Uzbekistan (IMU)
- Islamic Revolutionary Guard Corps (IRGC)
- Islamic State of Iraq and the Levant
- Islamic State of Iraq and the Levant's Branch in Libya (ISIL-Libya)
- Jaish-e-Mohammed (JEM)
- Jama'at Nusrat al-Islam wal-Muslimin (JNIM)
- Jaysh Rijal al-Tariq al Naqshabandi (JRTN)
- Jemaah Anshorut Tauhid (JAT)
- Jemaah Islamiya (JI)
- Jundallah
- Kahane Chai (Kach)
- Kata'ib Hizballah (KH)
- Kurdistan Workers' Party (PKK)
- Lashkar e-Tayyiba (LeT)
- Lashkar i Jhangvi (LJ)
- Liberation Tigers of Tamil Eelam (LTTE)
- Mujahidin Shura Council in the Environs of Jerusalem (MSC)
- National Liberation Army (ELN)
- Palestine Liberation Front (PLF)
- Palestine Islamic Jihad (PIJ)
- Popular Front for the Liberation of Palestine (PFLP)
- Popular Front for the Liberation of Palestine-General Command (PFLP-GC)
- Al-Qa'ida (AQ)
- Al-Qa'ida in the Arabian Peninsula (AQAP)
- Al-Qa'ida in the Indian Subcontinent
- Al-Qa'ida in the Islamic Maghreb (AQIM)
- Real IRA (RIRA)
- Revolutionary Armed Forces of Colombia (FARC)
- Revolutionary People's Liberation Party/Front (DHKP/C)
- Revolutionary Struggle (RS)
- Al-Shabaab (AS)
- Shining Path (SL)
- Tehrik-e Taliban Pakistan (TTP)

Source: U.S. Department of State "Foreign Terrorist Organizations," accessed May 14, 2019, https://www.state.gov/j/ct/rls/other/des/123085.htm.

JOSEPHINE SUNSHINE OVERAKER

Conspiracy to Commit Arson of United States Government Property and of Property Used in Interstate Commerce; Conspiracy to Commit Arson and Destruction of an Energy Facility; Attempted Arson of a Building; Arson of a Vehicle; Arson of a Building; Destruction of an Energy Facility

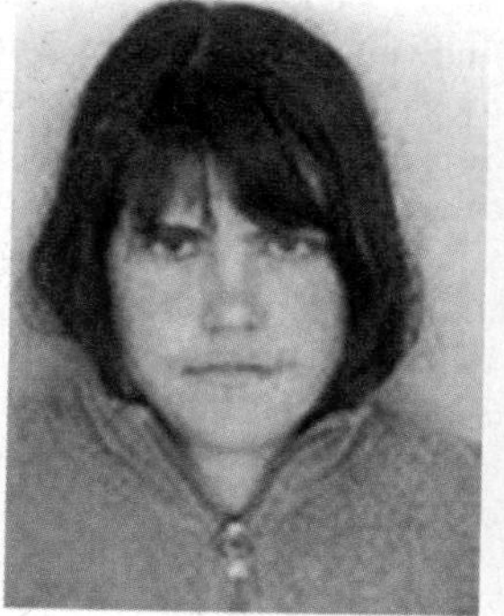

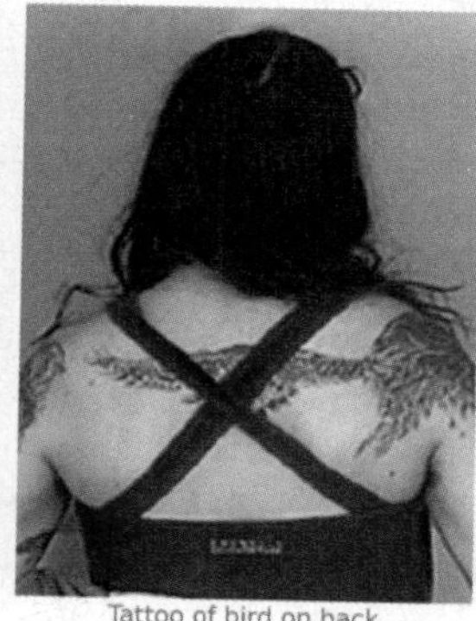

Tattoo of bird on back

Photograph Age Progressed to 41 Years Old

DESCRIPTION

Aliases: Lisa Rachelle Quintana, Lisa R. Quintana, Maria Rachelle Quintana, Maria Quintana, "Osha", "Jo", "China", "Josie", "Mo"

Date(s) of Birth Used: November 19, 1974, October 4, 1971, November 4, 1971	**Place of Birth:** Canada
Hair: Brown	**Eyes:** Brown
Height: 5'3"	**Weight:** 130 pounds
Sex: Female	**Race:** White
Occupation: Overaker may seek employment as a firefighter, a midwife, a sheep tender, or a masseuse.	**Nationality:** Canadian
Scars and Marks: Overaker has a large unknown tattoo on her upper left arm and a very large bird tattoo stretching from her right upper arm across her upper back. She has scars on her left ankle, right ankle, right calf, and right thigh.	**NCIC:** W105842105 / W258514859

REWARD

The FBI is offering a reward of up to $50,000 for information leading to the arrest of Josephine Sunshine Overaker.

REMARKS

Overaker is fluent in Spanish. She is known to use illegal narcotics. Overaker may have a light facial moustache. She was a vegan and may still be. Overaker is an American citizen.

CAUTION

On January 19, 2006, a federal grand jury in Eugene, Oregon, indicted Josephine Sunshine Overaker on multiple charges related to her alleged role in a domestic terrorism cell. Overaker was charged with two conspiracy violations related to seventeen incidents, five counts of arson, one count of attempted arson, and one count of destruction of an energy facility. These crimes occurred in Oregon, Washington, California, Colorado, and Wyoming, and date back to 1996. Many of the crimes she is accused of participating in were claimed to be committed by the Earth Liberation Front (ELF) or the Animal Liberation Front (ALF).

SHOULD BE CONSIDERED AN ESCAPE RISK

If you have any information concerning this person, please contact your local FBI office or the nearest American Embassy or Consulate.

Field Office: Portland

Federal Bureau of Investigation

Josephine Sunshine Overaker probably was an active member of the Earth Liberation Front (ELF) and/or the Animal Liberation Front (ALF). *Could you be a radical activist? Why or why not?*

Loosely affiliated extremists and *rogue international terrorists* include the World Trade Center bombers and rogue terrorists such as Osama bin Laden and members of al-Qaeda.[97]

Still another way of characterizing terrorism is by the methods used by terrorists. The traditional method has involved a conventional bomb, and it has continued to be the preferred method. However, today, the most feared method involves weapons of mass destruction (WMDs), such as nuclear, radiological, and biological devices. Agroterrorism is another method and is an attack on a food source or the distribution of food supplies. Cyberterrorism, dubbed "information warfare," may take various forms, including the hacking into or destroying of the nation's electronic infrastructure, thus rendering mass communications via the Internet inoperable. High-energy radio frequency (HERF) and electromagnetic pulse (EMP) weapons can shut down mass communication systems almost instantly.[98]

For purposes of this textbook, note that terrorists are criminals, according to the FBI's counterterrorism policy.[99] However, also remember that one person's terrorist is another person's freedom fighter. The British, for example, accused the American revolutionaries of being terrorists.[100]

The Law Enforcement Response to Terrorism

Terrorists' acts often are perpetrated by disaffected and angry individuals and groups who are far removed from the everyday American scene. Yet, through the media, American citizens are learning more about terrorism than ever before. Still, terrorism is a complex phenomenon. American law enforcement officers have no choice anymore. They must learn all they can to protect the nation and its people.

Since the tragedy of September 11, 2001, terrorism has been at the top of the national agenda. People are frightened without knowing exactly what to do about it. Local, state, and federal law enforcement agencies have new responsibilities for the prevention of both domestic and international terrorism and for reducing America's fear. To guide them, especially at the national level, President George W. Bush laid out four general policy principles:

1. Make no concessions to terrorists and strike no deals (even if U.S. citizens are held hostage).
2. Bring terrorists to justice for their crimes (no matter how long it takes).
3. Isolate and apply pressure on states that sponsor terrorism to force them to change their behavior.
4. Bolster the counterterrorist capabilities of those countries that work with the United States and require assistance.[101]

To prevent terrorism, billions of dollars are being committed and spent, and some civil liberties have been curtailed or severely restricted. The typical business or pleasure trip by airplane, for example, has become an even more taxing and intrusive experience. Even a simple trip to the library can make a person suspect should he or she select literary works that trigger the suspicion of a federal law enforcement agency. In a nation that was founded on the principles of freedom, openness, and anonymity for the most part, the new security, surveillance, and intelligence initiatives implemented since 9/11 have created concerns for many Americans.

Attempting to prevent terrorism is a daunting task. How can law enforcement officers protect every soft and hard target in America: all schools, all local water supplies, all food supply lines, all religious institutions, every public building in large and small communities, and all sources of energy? Although herculean does not adequately describe the charge, the protection of the American people and the nation's physical assets, as well as the reduction of fear from the threat of terrorism, is precisely what American law enforcement aims to do.

American law enforcement now is engaged in both defensive and offensive strategies to combat terrorism. However, in a free, open, and largely anonymous society, terrorists and potential terrorists know that the United States is fertile ground for terrorism. The police are faced with a dilemma. If the police are too aggressive trying to find and capture terrorists, American citizens are likely to become more fearful, anxious, and either resistant to or overly enthusiastic about repressive measures. If that occurs, terrorists win because fear, intimidation, disruption, and media attention to their cause are principal goals. However, if American law enforcement is not firm enough in its resolve and tactics to repress terrorism, terrorists will be able to again perpetrate a disaster on the public such as the events of 9/11, with the results of loss of life, serious injury, fear, and anger. Such an event would likely encourage even tighter domestic security procedures, which would be resented by many members of the public, and the entire terror cycle would start over again. Law enforcement officers walk a fine line between preserving liberty and tranquility and providing security that may be perceived as excessive and repressive.

To illustrate, on June 17, 2003, President Bush issued guidelines barring federal agents from using race or ethnicity in routine investigations, but the policy permits exceptions in investigations involving terrorism and national security issues. Consequently, if law enforcement and intelligence officials receive information that terrorists of a certain ethnic group plan to hijack a plane in a particular state in the near future, officials are allowed to heighten security on people of that ethnicity that board planes in that area (also see the description of the USA PATRIOT Act in Chapter 4).[102]

How Prepared Is the United States to Defend Against Terrorism?

The answer depends on whom you ask, and there are many people and groups willing to give an answer. As noted in the last chapter, there already have been some successes. Terrorist acts have been prevented; terrorists have been captured and brought to justice; Osama bin Laden and Abu Bakr al-Baghdadi have been killed; and the nation, as of this writing, has avoided another tragedy of 9/11 proportions. Nevertheless, U.S. borders remain porous, and U.S. assets are still vulnerable.

In 2003, two reports critical of the nation's preparedness to respond to terrorist attacks were released. The independent and bipartisan Council on Foreign Relations concluded that the United States was "drastically underfunding local emergency responders and remains dangerously unprepared to handle another September 11 event."[103] The nonprofit Partnership for Public Service determined that the United States is likely to be overwhelmed in the event of a bioterrorism attack because of serious shortages in skilled medical and scientific personnel.[104]

At the end of 2005, the "Final Report on 9/11 Commission Recommendations" was issued, and the news was not good. Congress created the bipartisan commission in 2002 to investigate aspects of the 9/11 terrorist attacks. Thomas Kean, the Republican chairperson of the 9/11 Commission, told NBC's *Meet the Press* that enacting the changes is "not a priority for the government right now." He added, "A lot of the things we need to do really to prevent another 9/11 just simply aren't being done by the president or by the Congress." Lee Hamilton, the commission's Democratic vice chairperson, predicted another attack would occur. He maintained, "It's not a question of if."[105] The conclusion, it seems, is that the United States has much more to do in its preparation to fight terrorism.

None of this was lost on the American public—at least for nearly 2 decades. In a 2015 opinion poll conducted in the aftermath of the Paris terrorist shootings, and months after the start of U.S. airstrikes against ISIS, 64% of Americans were worried (either "very worried" or "somewhat worried") about a domestic terrorist attack happening "soon," compared to 75% in 2003, prior to the Iraq war, and 58% in 2013, after the Boston Marathon bombings.[106] In a 2007 Harris Poll, substantial majorities of Americans responded that the United States was "likely" to experience the following types of terrorist attacks (percentages in parentheses): (1) "a suicide bomber in a shopping mall" (82%), "a chemical attack using a poison gas" (70%), "a biochemical attack using diseases such as anthrax or small pox" (69%), and "an attack on a nuclear power station" (62%). More than 40% of respondents thought the United States was "likely" to experience "another attack using airplanes like 9/11" (48%) and "a nuclear bomb exploding in a city" (42%). Americans' concern about terrorism apparently has declined significantly since the 2015 poll. In 2019 (the latest year for which data were available), 68% of Americans were satisfied (either "very" or "somewhat" satisfied) with "the nation's security from terrorism."[107]

As mentioned previously, the anticipated 9/11-caliber terrorist attack has not occurred (as of May 24, 2020), and since 2001, only about 6 deaths per year—104 deaths from 2001 through 2018—in the United States can be attributed to Islamist-linked terrorism.[108] In fact, from 2001 through 2018, the chance of being killed in the United States by a terrorist of any kind was about 1 in 40 million.[109] Even official and media concern about terrorism has

FYI

Who Is Winning the War on Terrorism?

In a 2018 opinion poll of 1,000 likely American voters, 56% thought the United States and its allies were winning the war on terrorism, while only 12% thought the terrorists were winning. This is a remarkable turnaround from 2015, when the same poll found that 46% thought terrorists were winning the war on terrorism, while only 26% thought the United States and its allies were winning. This was yet another turnaround from 2011, when 42% thought the United States and its allies were winning, and 9% thought the terrorists were winning. The 2011 views are almost identical to those found in an October 2001 poll.

Sources: "War on Terror Update," Rasmussen Reports, September 10, 2018, accessed May 15, 2019, http://www.rasmussenreports.com/public_content/politics/top_stories/war_on_terror_sep10; John Hayward. "Poll: Americans Think Obama Is Losing the War on Terror," October 13, 2015. Accessed January 27, 2016, www.breitbart.com/national-security/2015/10/13/poll-americans-think-obama-losing-war-terror/; *The Gallup Poll*, September 9, 2011, accessed November 3, 2012, www.gallup.com/poll/149381/Ten-Years-Later-Doubts-War-Terrorism.aspx.

waned over the years. For example, regular government warnings about an imminent attack are no longer issued. There has been no reason for them.[110] One should not conclude from the aforementioned that the U.S. government has ignored or discounted the terrorism threat. It has not. Since 9/11, more than $1 trillion dollars has been spent on counterterrorism.[111] In 2017 alone, the U.S. government spent approximately $175 billion on counterterrorism, including both military and law enforcement spending.[112] Thus, since 9/11, the United States has been either very lucky, more prepared than doomsayers have suggested or predicted, or both–probably both.

However, no one should be lulled into a false sense of security. International terrorism remains a threat, and likely will always be a threat, so it is incumbent on the United States to remain vigilant and prepared. Probably more concerning is the threat of domestic terrorism, for which the United States is most unprepared, as recent horrific incidents illustrate. Much more needs to be done to protect Americans from domestic terrorism, and that is the challenge that must be met by U.S. law enforcement personnel, especially the police.

THINKING CRITICALLY

1. What type of terrorism do you think poses the greatest threat to the United States and why?
2. Which of America's assets do you think are most vulnerable to terrorist attacks, and what do you think can be done to better defend them?
3. How concerned are you about another terrorist attack in the United States?

SUMMARY

1. **Identify characteristics of police work.**

   The role of the police officer is complex and requires a combination of special characteristics, which involve quick decision making, invisible work, "dirty work," and danger.

2. **What do studies of police operational styles show?**

   Studies of police operational styles show that the police do not have a monolithic culture and that police officers in the same department can possess significantly different attitudes toward their work.

3. **List the four major functions of police departments.**

   The four major functions of police departments in the United States are patrol, investigation, traffic, and drug enforcement.

4. **List the drug enforcement strategies of local police agencies.**

   Although the particular drug enforcement strategies employed by individual agencies vary widely, the most common strategies are street-level enforcement, mid-level investigations, major investigations, crop eradication, smuggling interdiction, problem-oriented and community policing strategies, drug demand reduction, and asset forfeiture.

5. **Explain the main components of community policing.**

   The three main components of community policing are community partnership, problem solving, and change management.

6. **Identify four steps in a community policing approach to problem solving.**

   Community policing relies heavily on problem solving. The four steps in a community policing approach to problem solving often are referred to as SARA: scanning–identifying problems, analysis–understanding underlying conditions, response–developing and implementing solutions, and assessment–determining the solutions' effect.

7. **Define terrorism, and identify different types of terrorism.**

   Terrorism has been defined as the systematic use of terror or unpredictable violence against governments, publics, or individuals to attain a political objective; the unlawful use of force and violence against persons or property to intimidate or coerce a government, the civilian population, or any segment thereof, in furtherance of political or social objectives; or premeditated, politically motivated violence perpetrated against noncombatant targets by subnational groups or clandestine agents, usually intended to influence an audience. Two broad categories of terrorism are domestic terrorism and international terrorism. Specific types of domestic terrorism are right-wing terrorism, left-wing terrorism, special interest terrorism, and individual terrorism. International terrorism is divided into state-sponsored terrorism, formalized terrorist organizations, and loosely affiliated extremists and rogue international terrorists.

KEY TERMS

role 202
role expectation 202
role conflict 202
operational styles 206
preventive patrol 209
directed patrol 209
GIS crime mapping 209
aggressive patrol 211
field interrogation 211
cybercrime 217
traffic accident investigation crews 222
terrorism 235
domestic terrorism 235
international terrorism 235

REVIEW QUESTIONS

1. What is a common source of role conflict for the police?
2. Distinguish among the four sets of operational styles identified by criminal justice scholars.
3. How are preventive patrol, directed patrol, and aggressive patrol different?
4. What are the major uses of GIS crime mapping in law enforcement?
5. What are some of the functions of a criminal investigator?
6. What are six services of Next Generation Identification (NGI)?
7. What are three distinct functions of DNA profiling?
8. What are some of the types of cybercrime?
9. What are some of the more important functions of traffic units?
10. What are some of the problems with the use of civil asset forfeiture in drug enforcement?
11. What is the philosophy of community policing?
12. What are the three complementary operational strategies of community policing?
13. How frequently has the United States been victimized by domestic and international terrorism?
14. What were President Bush's four general terrorism policy principles?

IN THE FIELD

1. **SARA Approach** In groups of three or four, select a local crime or disorder problem. Using the first three components of the SARA approach to problem solving (scanning, analysis, and response) and considering the elements of the crime triangle, formulate some options to deal with the problem. Present your options to the class, and let the whole class vote to determine which option is the best to solve the problem. Do you agree with the class vote? Why or why not? What do the results of this vote say about attitudes toward crime-related problem solving?
2. **Law Enforcement Resources** Look through several police periodicals, such as *Law and Order, Police Chief*, or other professional police magazines. What do the advertised products tell you about law enforcement? What law enforcement issues are addressed in the articles? Write a brief summary of your findings.

ON THE WEB

1. **Terrorists.** To learn more about terrorists, visit the FBI's website at https://www.fbi.gov. Select "Most Wanted" from the menu bar, and then select "Terrorism" from the "Most Wanted" menu bar. Choose 6 "Most Wanted [International] Terrorists" and 6 "Domestic Terrorism" subjects and create a profile for each type. Identify similarities and differences between the two types of terrorists. Present your findings in an essay or an oral presentation.
2. **Drug Enforcement** To learn more about illegal drugs and drug control policy, visit the website of the Office of National Drug Control Policy at www.whitehousedrugpolicy.gov. Based on what you have learned, write an essay or prepare an oral presentation on whether national drug control policy has been or will be effective in reducing drug use and abuse among the American public.

CRITICAL THINKING EXERCISES

FAMILIAL DNA SEARCHES

1. Police in Colorado are increasingly using a DNA crime-solving technique called "familial DNA search." The conventional way of using DNA to identify a crime's perpetrator is to gather blood, semen, or other genetic material at the crime scene and run it through a DNA database to see if it provides an exact match. A problem is that the perpetrator may not be in the database. In those cases, a familial DNA search may be used, wherein the database is searched for a near-match; that is, for a close relative of the perpetrator. Police can then use that information to help identify the person who committed the crime. Denver police used the technique to help catch a burglar who left a drop of blood on a seat when he broke a car window and stole $1.40 in change. California identified and apprehended the infamous "Golden State Killer" in 2018, more than 30 years after his last known crime spree, using the technique. The legality of the technique has not been tested in federal court. Only 12 states currently use familial DNA in criminal cases. Critics contend that a familial DNA search amounts to guilt by association and that it

could subject innocent people to arrest or hours of interrogation. Proponents argue that it would be unconscionable not to use the technique if it could prevent a rape, for example.

a. Do you think familial DNA searches should be a legal law enforcement investigative tool? Why or why not?

b. Do you think the federal courts will approve familial DNA searches? Why or why not?

COMMUNITY POLICING

2. You are a new community police officer assigned with seven other officers to a low-income, heterogeneous, high-turnover, high-crime neighborhood. How should you and your fellow officers address the following problems?

    a. Neighborhood residents, neighborhood business owners, and community leaders disagree about the most important problems of the neighborhood and what to do about them. What should you and your fellow officers do?

    b. You organize and heavily publicize a meeting to discuss neighborhood problems and their solutions. However, only about 10% to 20% of neighborhood residents attend, and most of them are white and more affluent homeowners who live in the better parts of the neighborhood. Few minorities or renters attend. What should you and your fellow officers do?

    c. Your precinct captain disagrees with neighborhood residents about the most important problems in the neighborhood. Your precinct captain's top priority is abandoned cars used for drug dealing. The top priority of neighborhood residents is the overall appearance of the neighborhood. What should you and your fellow officers do?

NOTES

1. Keith Haley, "Training," in *What Works in Policing? Operations and Administration Examined*, ed. Gary Cordner and Donna Hale (Cincinnati, OH: Anderson, 1992).
2. Lee Rainwater, "The Revolt of the Dirty Workers," *Transaction* (November 1967), 2.
3. Jerome Skolnick, *Justice Without Trial* (New York: Wiley, 1966).
4. Brian Reaves, *Local Police Departments, 2013: Equipment and Technology*, U.S. Department of Justice, Bureau of Justice Statistics (Washington, DC: U.S. Government Printing Office, July 2015), 2, Table 2, www.bjs.gov/content/pub/pdf/lpd13et.pdf.
5. Calculated from Sourcebook of Criminal Justice Statistics Online, "Law Enforcement Officers Killed, United States, 1961–2012," accessed May 7, 2019, https://www.albany.edu/sourcebook/pdf/t31542012.pdf; U.S. Department of Justice, Federal Bureau of Investigation, "2018 Law Enforcement Officers Killed & Assaulted," accessed May 7, 2019, https://ucr.fbi.gov/leoka/2018/topic-pages/officers-accidentally-killed.
6. U.S. Department of Justice, Federal Bureau of Investigation, "2018 Law Enforcement Officers Killed & Assaulted," accessed May 6, 2019, https://ucr.fbi.gov/leoka/2018/; U.S. Department of Justice, Federal Bureau of Investigation, "2017 Law Enforcement Officers Killed & Assaulted," accessed May 5, 2019, https://ucr.fbi.gov/leoka/2017/tables/table-1.xls; Federal Bureau of Investigation, "FBI Releases 2011 Preliminary Statistics for Law Enforcement Officers Killed in the Line of Duty" (May 14, 2012), www.fbi.gov/news/pressrel/press-releases/fbi-releases-2011-preliminary-statistics-for-law-enforcement-officers-killed-in-the-line-of-duty (accessed October 25, 2012); Department of Justice, Federal Bureau of Investigation, *Uniform Crime Reports*, 2001, www.fbi.gov/ucr/ucr.htm.
7. U.S. Department of Justice, Federal Bureau of Investigation, "2018 Law Enforcement Officers Killed & Assaulted," accessed May 6, 2019, https://ucr.fbi.gov/leoka/2018/topic-pages/officers-accidentally-killed; U.S. Department of Justice, Federal Bureau of Investigation, "2017 Law Enforcement Officers Killed & Assaulted," accessed May 5, 2019, https://ucr.fbi.gov/leoka/2017/topic-pages/accidental_topic_page_-2017.
8. Ibid.
9. James Q. Wilson, *Varieties of Police Behavior* (Cambridge, MA: Harvard University Press, 1968), 140–227.
10. John Broderick, *Police in a Time of Change* (Morristown, NJ: General Learning Press, 1977), 9–88.
11. William Muir Jr., *Police: Streetcorner Politicians* (Chicago: University of Chicago Press, 1977).
12. Ellen Hochstedler, "Testing Types: A Review and Test of Police Types," *Journal of Criminal Justice* 9 (1981): 451–66; also see Eugene A. Paoline, "The Myth of a Monolithic Police Culture," in *Demystifying Crime and Criminal Justice*, ed. Robert M. Bohm and Jeffery T. Walker (Los Angeles: Roxbury, 2006).
13. Eugene A. Paoline III, "Shedding Light on Police Culture: An Examination of Officers' Occupational Attitudes," *Police Quarterly* 7 (2004): 205–236.
14. G. Kelling, T. Pate, D. Dieckman, and C. Brown, *The Kansas City Preventive Patrol Experiment: A Summary Report* (Washington, DC: Police Foundation, 1974).
15. Gary Cordner and Donna Hale, "Patrol," in *What Works in Policing? Operations and Administration Examined*, ed. Gary Cordner and Donna Hale (Cincinnati, OH: Anderson, 1992): 143–58.
16. Daniel Mabrey, "Crime Mapping: Tracking the Hotspots," *Crime & Justice International* 18, no. 67 (2002): 31–32.
17. Ibid.
18. James Q. Wilson and Barbara Boland, "The Effect of Police on Crime," *Law and Society Review* 12 (1978): 367–84.
19. George Kelling, *The Newark Foot Patrol Experiment* (Washington, DC: Police Foundation, 1981).
20. Robert C. Trojanowicz, *The Neighborhood Foot Patrol Program in Flint, Michigan* (East Lansing, MI: National Neighborhood Foot Patrol Center, n.d.).
21. Bruce L. Berg and John J. Horgan, *Criminal Investigation*, 3rd ed. (Westerville, OH: Glencoe/McGraw-Hill, 1998).
22. Alfred Blumstein and Joan Petersilia, "NIJ and Its Research Program," in *25 Years of Criminal Justice Research* (Washington, DC: The National Institute of Justice, 1994), 13.
23. Christine Wicker, "Death Beat," in *Texas Crime, Texas Justice*, ed. Keith N. Haley and Mark A. Stallo (New York: McGraw-Hill, 1996).
24. U.S. Department of Justice, Federal Bureau of Investigation, "2018 Crime in the United States." Accessed February 10, 2020, https://ucr.fbi.gov/crime-in-the-u.s/2018/crime-in-the-u.s.-2018/topic-pages/clearances.

25. Mark Willman and John Snortum, "Detective Work: The Criminal Investigation Process in a Medium-Size Police Department," *Criminal Justice Review* 9 (1984): 33–39; V. Williams and R. Sumrall, "Productivity Measures in the Criminal Investigation Function," *Journal of Criminal Justice* 10 (1982): 111–22; I. Greenberg and R. Wasserman, *Managing Criminal Investigations* (Washington, DC: U.S. Department of Justice, 1979); P. Greenwood and J. Petersilia, *The Criminal Investigation Process*, vol. I, Summary and Policy Implications (Washington, DC: U.S. Department of Justice, 1975); B. Greenberg, C. Elliot, L. Kraft, and H. Procter, *Felony Investigation Decision Model: An Analysis of Investigative Elements of Information* (Washington, DC: U.S. Government Printing Office, 1975).

26. John Conklin, *Robbery and the Criminal Justice System* (New York: J. B. Lippincott, 1972), 149.

27. The National Academies, "'Badly Fragmented' Forensic Science System Needs Overhaul; Evidence to Support Reliability of Many Techniques Is Lacking," accessed January 8, 2011, www8.nationalacademies.org/onpinews/newsitem.aspx?RecordID=12589; Jason Felch, "Panel Urges Overhaul of Crime Labs," *Orlando Sentinel*, February 19, 2009, A3.

28. U.S. Department of Justice, National Commission on Forensic Science. Accessed March 10, 2016, www.justice.gov/ncfs.

29. Jessica S. Henry, "How Corruption in Forensic Science Is Harming the Criminal Justice System," *The Conversation*, January 25, 2019, https://theconversation.com/how-corruption-in-forensic-science-is-harming-the-criminal-justice-system-108975 (accessed January 25, 2019).

30. "Cops Clear JonBenet's Family," *Orlando Sentinel*, July 10, 2008, A2; Catherine Tsai, "DNA Offers Hope JonBenet's Killer May Be Found," *Orlando Sentinel*, July 11, 2008, A12; Ishani Ghose, "23 years later, 6-year-old beauty pageant star JonBenet Ramsey's brutal murder remains unsolved," meaww.com, April 16, 2019, accessed May 7, 2019, https://meaww.com/revisiting-the-cold-case-of-child-beauty-queen-jon-benet-ramsey-murder-unsolved-crime.

31. Paisley Dodds, "After Ruling, Much of Britain's DNA Database May Be DOA," *Orlando Sentinel*, December 5, 2008, A14; Home Office, "Protection of Freedoms Bill," GOV.UK, February 11, 2011, accessed May 7, 2019, https://www.gov.uk/government/publications/protection-of-freedoms-bill; legislation.gov.uk, "Protection of Freedoms Act 2012," accessed May 7, 2019, http://www.legislation.gov.uk/ukpga/2012/9/contents/enacted; GeneWatch, "The UK Police National DNA Database," accessed May 7, 2019, http://www.genewatch.org/sub-539478.

32. Federal Bureau of Investigation, *CODIS-NDIS Statistics*. Accessed May 7, 2019, https://www.fbi.gov/services/laboratory/biometric-analysis/codis/ndis-statistics.

33. Federal Bureau of Investigation, "Integrated Automated Fingerprint System," accessed October 29, 2012, www.fbi.gov/about-us/cjis/fingerprints_biometrics/iafis.

34. Federal Bureau of Investigation, Criminal Justice Information Services, "Next Generation Identification (NGI)," accessed May 7, 2019, https://www.fbi.gov/services/cjis/fingerprints-and-other-biometrics/ngi.

35. Ibid.

36. Ibid.

37. Ibid.

38. Ernest J. Babcock, "Privacy Impact Assessment for the Next Generation Identification (NGI) Interstate Photo System," FBI Criminal Justice Information Services, September 2015. Accessed May 10, 2019, https://www.fbi.gov/services/information-management/foipa/privacy-impact-assessments/interstate-photo-system.

39. Federal Bureau of Investigation, Criminal Justice Information Services, "Next Generation Identification (NGI)." Accessed May 10, 2019, https://www.fbi.gov/services/cjis/fingerprints-and-other-biometrics/ngi; U.S. Department of Justice, Federal Bureau of Investigation, Criminal Justice Information Services Division, "2017: The Year in Review," p. 11. Accessed May 10, 2019, https://www.eff.org/files/2018/02/11/2017_cjis_year_in_review.pdf; Colin Lecher and Russell Brandom, "The FBI Has Collected 430,000 Iris Scans in a So-Called 'Pilot Program'," *The Verge*, July 12, 2016. Accessed May 11, 2019, https://www.theverge.com/2016/7/12/12148044/fbi-iris-pilot-program-ngi-biometric-database-aclu-privacy-act.

40. The Council of Economic Advisers, Executive Office of the President of the United States, "The Cost of Malicious Cyber Activity to the U.S. Economy," February 2018, p. 1. Accessed May 10, 2019, https://www.whitehouse.gov/wp-content/uploads/2018/03/The-Cost-of-Malicious-Cyber-Activity-to-the-U.S.-Economy.pdf.

41. Al Pascual, Kyle Marcini, and Sarah Miller, "2018 Identity Fraud: Fraud Enters a New Era of Complexity," *Javelin Strategy & Research*, February 6, 2018. Accessed May 10, 2019, https://www.javelinstrategy.com/coverage-area/2018-identity-fraud-fraud-enters-new-era-complexity.

42. Internet Crime Complaint Center, "Internet Crime Schemes."

43. Steven R. Chabinsky, Deputy Assistant Director, Cyber Division, Federal Bureau of Investigation, speech given at GovSec/FOSE Conference, Washington, DC, March 23, 2010, accessed January 9, 2011, www.fbi.gov/news/speeches/the-cyber-threat-whos-doing-what-to-whom.

44. Ibid.

45. Center for Behavioral Health Statistics and Quality (2018), "2017 National Survey on Drug Use and Health: Detailed Tables," Substance Abuse and Mental Health Services Administration, Rockville, MD, Tables 1.1A and 1.1B, accessed May 11, 2019, https://www.samhsa.gov/data/sites/default/files/cbhsq-reports/NSDUHDetailedTabs2017/NSDUHDetailedTabs2017.pdf.

46. *Results from* Ibid. and Center for Behavioral Health Statistics and Quality (2015), "2014 National Survey on Drug Use and Health: Detailed Tables," Substance Abuse and Mental Health Services Administration, Rockville, MD, Table 7.3B, accessed May 11, 2019, https://www.samhsa.gov/data/sites/default/files/NSDUH-DetTabs2014/NSDUH-DetTabs2014.pdf.

47. Center for Behavioral Health Statistics and Quality (2018), "2017 National Survey on Drug Use and Health: Detailed Tables," Substance Abuse and Mental Health Services Administration, Rockville, MD, Tables 1.1A, accessed May 11, 2019, https://www.samhsa.gov/data/sites/default/files/cbhsq-reports/.

48. Unless indicated otherwise, information in this section on drug enforcement is from the following sources: Brian Reaves, *Local Police Departments, 2013: Personnel, Policies, and Practices,* Bureau of Justice Statistics, U.S. Department of Justice (Washington, DC: U.S. Government Printing Office, May 2015), 10, Table 11, www.bjs.gov/content/pub/pdf/lpd13ppp.pdf; Brian A. Reaves, *Local Police Departments 2007*, Bureau of Justice Statistics, U.S. Department of Justice (Washington, DC: U.S. Government Printing Office, December 2010), 29, Table 33, http://bjs.ojp.usdoj.gov/content/pub/pdf/lpd07.pdf; Peter Joseph Loughlin, "Does the Civil Asset Forfeiture Reform Act of 2000 Bring a Modicum of Sanity to the Federal Civil Asset Forfeiture System?" www.malet.com/does_the_civil_asset_forfeiture_.htm; United Nations, Office on Drugs and Crime, "Drug Control Strategies of United States Law Enforcement," H. Williams, www.unodc.org/unodc/en/data-and-analysis/bulletin/bulletin_1990-01-01_1_page004.html; Scott Ehlers, *Policy*

Briefing: Asset Forfeiture, The Drug Policy Foundation (Washington, DC: U.S. Government Printing Office, 1999); Samuel Walker, *Sense and Nonsense About Crime and Drugs: A Policy Guide*, 5th ed. (Belmont, CA: Wadsworth, 2001), chap. 13; Victor E. Kappeler, Mark Blumberg, and Gary W. Potter, *The Mythology of Crime and Criminal Justice*, 3rd ed. (Prospect Heights, IL: Waveland Press, 2000), chap. 8; U.S. Department of Health and Human Services, Office of Applied Studies, "Long-Term Trends in Illicit Drug Use," www.samhsa.gov/oas/nhsda/_2k1nhsda/vol1/_chapter9.htm#9.2.

49. U.S. Department of Justice, Federal Bureau of Investigation, *2017 Crime in the United States,* "Persons Arrested." Accessed May 11, 2019, https://ucr.fbi.gov/crime-in-the-u.s/2017/crime-in-the-u.s.-2017/tables/table-29.

50. U.S. Drug Enforcement Administration, "2017 Final Domestic Cannabis Eradication/Suppression Program Statistical Report," accessed May 11, 2019, https://www.dea.gov/sites/default/files/2018-07/cannabis_2017.pdf; U.S. Drug Enforcement Administration, "Domestic Cannabis Eradication/Suppression Program," accessed January 27, 2016, www.dea.gov/ops/cannabis.shtml; U.S. Drug Enforcement Administration, "Domestic Cannnabis Eradication/Suppression Program," accessed November 2, 2012, www.justice.gov/dea/ops/cannabis.shtml.

51. D.A.R.E. America, "About," https://dare.org/about/ (accessed May 11, 2019).

52. Drug Abuse Resistance Education, www.dare.com/home/default.asp; The Drug Reform Coordination Network, "A Different Look at D.A.R.E.," www.drcnet.org/DARE/section1.html and www.drcnet.org/DARE/section6.html.

53. D.A.R.E. 2013 Annual Report. Accessed January 27, 2016, www.dare.org/wp-content/uploads/2015/07/DARE_AnnualReport_2013web.pdf.

54. Marian R. Williams, Jefferson E. Holcomb, Tomislav V. Kovandzic, and Scott Bullock, "Policing for Profit: The Abuse of Civil Asset Forfeiture," Institute for Justice, March 2010. Accessed February 3, 2019, https://ij.org/wp-content/uploads/2015/03/assetforfeituretoemail.pdf.

55. Dick M. Carpenter II, Lisa Knepper, Angela C. Erickson, and Jennifer McDonald with contributions from Wesley Hottot and Keith Diggs, *Policing for Profit: The Abuse of Civil Asset Forfeiture*, 2nd ed., Institute for Justice, November 2015. Accessed February 3, 2019, https://ij.org/wp-content/uploads/2015/11/policing-for-profit-2nd-edition.pdf.

56. Adam Liptak and Shaila Dewan, "Supreme Court Limits Police Powers to Seize Private Property," *New York Times*, February 20, 2019, https://www.nytimes.com/2019/02/20/us/politics/civil-asset-forfeiture-supreme-court.html (accessed February 21, 2019).

57. U.S. Department of Justice, Office of the Inspector General, Audit of the Assets Forfeiture Fund and Seized Asset Deposit Fund Annual Financial Statements Fiscal Year 2018, December 2018, https://oig.justice.gov/reports/2018/a1905.pdf#page=1 (accessed February 4, 2019).

58. Dick M. Carpenter II, Lisa Knepper, Angela C. Erickson, and Jennifer McDonald with contributions from Wesley Hottot and Keith Diggs, *Policing for Profit: The Abuse of Civil Asset Forfeiture*, 2nd ed., Institute for Justice, November 2015. Accessed February 3, 2019, https://ij.org/wp-content/uploads/2015/11/policing-for-profit-2nd-edition.pdf.

59. Marian R. Williams, Jefferson E. Holcomb, Tomislav V. Kovandzic, and Scott Bullock, "Policing for Profit: The Abuse of Civil Asset Forfeiture," Institute for Justice, March 2010. Accessed February 3, 2019, https://ij.org/wp-content/uploads/2015/03/assetforfeituretoemail.pdf.

60. Dick M. Carpenter II, Lisa Knepper, Angela C. Erickson, and Jennifer McDonald with contributions from Wesley Hottot and Keith Diggs, *Policing for Profit: The Abuse of Civil Asset Forfeiture*, 2nd ed., Institute for Justice, November 2015. Accessed February 3, 2019, https://ij.org/wp-content/uploads/2015/11/policing-for-profit-2nd-edition.pdf; U.S. Department of Justice, Office of the Inspector General, Audit of the Assets Forfeiture Fund and Seized Asset Deposit Fund Annual Financial Statements Fiscal Year 2018, December 2018. Accessed February 4, 2019, https://oig.justice.gov/reports/2018/a1905.pdf#page=1.

61. Marian R. Williams, Jefferson E. Holcomb, Tomislav V. Kovandzic, and Scott Bullock, "Policing for Profit: The Abuse of Civil Asset Forfeiture," Institute for Justice, March 2010. Accessed February 3, 2019, https://ij.org/wp-content/uploads/2015/03/assetforfeituretoemail.pdf.

62. Joseph Tanfani and Kate Mather, "Sessions: 'Lifesaving Gear' Will Help Officers Do Jobs," *Orlando Sentinel*, August 29, 2017, A6.

63. U.S. Department of Justice, Office of the Inspector General, Audit of the Assets Forfeiture Fund and Seized Asset Deposit Fund Annual Financial Statements Fiscal Year 2018, December 2018. Accessed February 4, 2019, https://oig.justice.gov/reports/2018/a1905.pdf#page=1.

64. National Institute on Drug Abuse. Accessed March 11, 2016, https://teens.drugabuse.gov/national-drug-alcohol-facts-week/drug-facts-chat-day-drug-abuse.

65. National Institute on Drug Abuse, National Institutes of Health, *How Does Drug Abuse Affect the HIV Epidemic?* Accessed January 27, 2016, www.drugabuse.gov/publications/research-reports/hivaids/how-does-drug-abuse-affect-hiv-epidemic.

66. Calculated from data in Federal Bureau of Investigation, 2017 Crime in the United States, accessed May 11, 2019, https://ucr.fbi.gov/crime-in-the-u.s/2017/crime-in-the-u.s.-2017/tables/table-43.

67. Calculated from data in Center for Behavioral Health Statistics and Quality (2018), "2017 National Survey on Drug Use and Health: Detailed Tables," Substance Abuse and Mental Health Services Administration, Rockville, MD, Tables 1.29A and 1.30A, accessed May 11, 2019, https://www.samhsa.gov/data/sites/default/files/cbhsq-reports/.

68. Mark Motivans, Federal Justice Statistics, 2011–2012, January 2015, NCJ 248493. Accessed January 27, 2016, www.bjs.gov/content/pub/pdf/fjs1112.pdf.

69. Sarah Anderson, "Fair Sentencing Act Retroactivity: Addressing the Sentencing Disparity of Crack Cocaine vs. Powder Cocaine," *FreedomWorks*, August 6, 2018, https://www.freedomworks.org/content/fair-sentencing-act-retroactivity-addressing-sentencing-disparity-crack-cocaine-vs-powder (accessed May 12, 2019).

70. United States Sentencing Commission, "Policy Profile: Sensible Sentencing Reform: The 2014 Reduction of Drug Sentences." Accessed July 4, 2016, www.ussc.gov/sites/default/files/pdf/research-and-publications/backgrounders/profile_2014_drug_amendment.pdf.

71. "Sentence Reduction 26,000," *Orlando Sentinel,* April 15, 2016, p. A4.

72. CONGRESS.GOV., "H.R. 5682–FIRST STEP Act," May 23, 2018. Accessed May 12, 2019, https://www.congress.gov/bill/115th-congress/house-bill/5682/text#toc-H66997AD545454BF1B345D41C0309.]809B.

73. Jonaki Bose, Sarra L. Hedden, Rachel N. Lipari, and Eunice Park-Lee, "Key Substance Use and Mental Health Indicators in the United States: Results from the 2017 National Survey on Drug Use and Health." Accessed May 13, 2019, https://www.samhsa.gov/data/sites/default/files/cbhsq-reports/NSDUHFFR2017/NSDUHFFR2017.htm#illicit1.

74. Material on research studies based on information in Alfred Blumstein and Joan Petersilia, "NIJ and Its Research Program," in *25 Years of Criminal Justice Research* (Washington, DC: The National Institute of Justice, 1994), 10–14.

75. James Q. Wilson and George L. Kelling, "Broken Windows: The Police and Neighborhood Safety," *Atlantic Monthly*, 256 (1982), 29–38.

76. Material in this subsection is based on information in the following sources: Community Policing Consortium, *Understanding Community Policing: A Framework for Action*, monograph, January 1988, www.communitypolicing.org/conpubs.html; U.S. Department of Justice, National Institute of Corrections, *Community Justice: Striving for Safe, Secure, and Just Communities* (Louisville, KY: LIS, 1996).

77. Federal Bureau of Investigation, "Terrorism in the United States, 1999," 15, www.fbi.gov/publications/terror/terrorism.htm.

78. Ibid.

79. Ibid., 16.

80. U.S. State Department, "Patterns of Global Terrorism, 2001," May 2002, v, www.state.gov/s/ct/rls/pgtrpt/2000/pdf.

81. Ibid., 1.

82. Federal Bureau of Investigation, "Terrorism in the United States, 1999," 15.

83. Ibid., i.

84. U.S. Department of Justice, Federal Bureau of Investigation, "Terrorism 2002–2005," www.fbi.gov/publications/terror/terrorism2002_2005.htm.

85. U.S. State Department, "Patterns of Global Terrorism, 2002," April 2003, xiii, www.state.gov/s/ct/rls/pgtrpt/2002.

86. U.S. Department of Justice, Federal Bureau of Investigation, "What We Investigate, Terrorism," accessed May 13, 2019, https://www.fbi.gov/investigate/terrorism.

87. Ibid.

88. U.S. Department of Justice, Federal Bureau of Investigation, "Terrorism 2002–2005", 18.

89. Heidi Beirich and Susy Buchanan, "The Year in Hate and Extremism" (Montgomery, AL: Southern Poverty Law Center, Spring 2018).

90. U.S. Department of Justice, Federal Bureau of Investigation, "Terrorism 2002–2005," 19.

91. Kate Irby, "These Are the FBI's Top Domestic Terrorism Concerns for 2018," McClatchy DC Bureau, December 26, 2017, https://www.mcclatchydc.com/news/nation-world/national/national-security/article191127109.html (accessed May 13, 2019).

92. Ibid., 20.

93. Ibid., 25–26; Kate Irby, "These Are the FBI's Top Domestic Terrorism Concerns for 2018," McClatchy DC Bureau, December 26, 2017, accessed May 13, 2019, https://www.mcclatchydc.com/news/nation-world/national/national-security/article191127109.html.

94. Ibid., 23.

95. Ibid., xi; U.S. Department of State, "State Sponsors of Terrorism," accessed May 13, 2019, https://www.state.gov/j/ct/list/c14151.htm.

96. Federal Bureau of Investigation, "Terrorism in the United States, 1999," 23.

97. Ibid.

98. Ibid., 38–40.

99. Ibid., i.

100. See Howard Zinn, *A People's History of the United States* (New York: Harper Colophon, 1980), chap. 4.

101. U.S. State Department, *Patterns of Global Terrorism, 2002*, xi.

102. Eric Lichtblau, "Bush Issues Federal Ban on Racial Profiling," *The New York Times*, June 18, 2003.

103. Deborah Orin and David Kadison, "Government Still Napping as New 9/11 Looms," *New York Post Online Edition*, June 30, 2003.

104. David Johnston, "Report Calls U.S. Agencies Understaffed for Bioterror," *The New York Times*, July 6, 2003.

105. "U.S. Not 'Well-Prepared' for Terrorism," accessed November 28, 2009, www.cnn.com/2005/US/12/04/911.commission/index.html.

106. Pew Research Center, "Terrorism Worries Little Changed: Most Give Government Good Marks for Reducing Threat," January 12, 2015. Accessed January 27, 2016, www.people-press.org/2015/01/12/terrorism-worries-little-changed-most-give-government-good-marks-for-reducing-threat/.

107. Megan Brenan, "Americans Most Satisfied with Nation's Military, Security," *Gallup*, January 28, 2019. Accessed May 15, 2019, https://news.gallup.com/poll/246254/americans-satisfied-nation-military-security.aspx?g_source=link_NEWSV9&g_medium=TOPIC&g_campaign=item_&g_content=Americans%2520Most%2520Satisfied%2520With%2520Nation%27s%2520Military%2c%2520Security.

108. John Mueller and Mark G. Stewart, "Why Are Americans Still So Afraid of Islamist Terrorism," *The Washington Post*, March 23, 2018. Accessed May 15, 2019, https://www.washingtonpost.com/news/monkey-cage/wp/2018/03/23/why-are-americans-still-so-afraid-of-islamic-terrorism/?utm_term=.fb43ee4cb9ee.

109. Ibid.

110. Ibid.

111. Ibid.

112. Doyle McManus, "Here's a Secret: We May Be Winning the War on Terrorism," *Los Angeles Times*, January 16, 2019. Accessed May 15, 2019, https://www.latimes.com/nation/la-na-pol-mcmanus-column-20190116-story.html.

Design Elements: *Image of part of a court building with columns and statue*: Pixtal/AGE Fotostock RF; *Image of a door slightly open*: jadding/Shutterstock.com

Alex Milan Tracy/Anadolu Agency/Getty Images

7

Policing America: Issues and Ethics

CHAPTER OUTLINE

The Police and the Public
- Public Attitudes Toward the Police
- Police Attitudes Toward the Public

Police Recruitment and Selection
- Qualities of a Successful Police Officer
- The Police Recruitment Process
- Successful Recruiting Practices
- The Police Selection Process
- The Selection of a Law Enforcement Executive

Issues in Policing
- Discretion
- Job Stress
- Use of Force
- Police Corruption

Professionalizing Law Enforcement

LEARNING OBJECTIVES

After completing this chapter, you should be able to:

1. Describe the general attitude of the public toward the police and the police toward the public.
2. Summarize the steps in an effective police officer selection process.
3. Identify factors that affect the exercise of police discretion and methods of limiting discretion.
4. Describe two general ways that law enforcement agencies can reduce stress on the job.
5. Explain the circumstances under which police officers may be justified in using deadly force.
6. List some of the ways to control and reduce police corruption.

Crime Story

On January 18, 2019, former Chicago Police Officer Jason Van Dyke (pictured) was sentenced to 6 years and 9 months in prison for the killing of 17-year-old Laquan McDonald (pictured) about 4 years earlier. Van Dyke is white and McDonald was black. The killing sparked public outrage in Chicago and beyond and provided another iconic example for the Black Lives Matter movement. Van Dyke originally had been charged with first-degree murder, but on October 5, 2018, a jury found him guilty of a single count of second-degree murder and 16 counts of aggravated battery—one for each shot he fired at McDonald. The jury found him not guilty of official misconduct. Prosecutors asked the judge to sentence Van Dyke to a prison term of 18 to 20 years, while his defense team pleaded with the judge to release their client on probation. Second-degree murder carried a sentence of from 4 to 20 years, while aggravated battery carried a sentence of from 6 to 30 years. He could have been sentenced to up to 96 years in prison. Illinois state law requires that Van Dyke serve at least 50% of his sentence, and since he already had served about 3 months awaiting sentencing, he could be released from prison in a little more than 3 years.

Paul Beaty/AP Photo

ANTONIO PEREZ/POOL/EPA-EFE/Shutterstock

The tragedy began the evening of October 20, 2014, when police were called to a parking lot on the southwest side of Chicago. They had received reports of a person breaking into trucks and stealing radios. When they arrived at the scene, they found McDonald walking erratically in the street while holding a knife with a 3-inch blade. McDonald was a troubled teen, who had been in and out of the juvenile justice system and was a ward of the state at the time of the encounter. When Van Dyke pulled up to the scene, he got out of his squad car and almost immediately began shooting at McDonald. He continued to shoot at McDonald for 12.5 seconds after McDonald fell to the ground. Van Dyke was the only officer to shoot at McDonald. Officers at the scene reported that McDonald, who allegedly had PCP in his system, ignored repeated commands to drop his knife, lunged at Van Dyke with the knife, aggressively swung the knife toward other officers, and assaulted other officers. Remarkably, none of the 10 other officers was hurt. Van Dyke told investigators that McDonald raised the knife in a menacing manner before he opened fire, and that he backpedaled as McDonald approached. However, police dash-cam video did not support Van Dyke's account. The video appears to show McDonald walking away from police officers as Van Dyke opened fire. Police and city leaders initially resisted releasing the dashcam video, but a city activist and a journalist sued for its release. The video finally was released to the public in response to a court order 400 days after the shooting and the day Van Dyke was charged with first-degree murder. The video sparked citywide protests and sullied the reputations of several high-ranking political leaders, including the mayor, the police superintendent, and the Cook County state's attorney. Before the video's release, the city agreed to pay a $5 million settlement to McDonald's family.

Van Dyke's defense team gave the court dozens of letters from family, friends, and the public requesting the judge to show mercy. Van Dyke testified in his own defense. He related that McDonald's "face had no expression, his eyes were just bugging out of his head." He stated that McDonald "had these huge white eyes just staring right through me." Van Dyke said that he believed his fellow officers "were under attack." "The whole thing was shocking to me," he concluded.

Prosecutors called several witnesses to testify that Van Dyke had caused enormous anguish for the McDonald family, and that he had a long history of using excessive force. Prosecutors were able to convince the jury that Van Dyke intended to draw his gun and shoot McDonald before he got out of his squad car.

According to activists, also on trial was an unofficial "code of silence," which they contend persists in the Chicago Police Department. The "code of silence" refers to unwillingness among police officers to speak up against fellow officers when they act unlawfully. In this case, three Chicago police officers went beyond the "code of silence" when they filed false reports to protect Van Dyke. The three officers were charged with conspiracy, obstruction of justice, and official misconduct. In a bench trial on January 17, 2019, a judge found all three officers not guilty on all charges in the alleged cover up.

Among the topics examined in this chapter is police use of force, including excessive and deadly force. Do you believe that Van Dyke used excessive force with McDonald? Do you believe that Van Dyke murdered McDonald? Do you agree with the jury that Van Dyke did not engage in official misconduct? Why or why not? Do you agree with the sentence Van Dyke received? If not, what sentence should Van Dyke have received? Do you agree with the amount of time Van Dyke will spend in prison? Why or why not? Does McDonald share any of the blame for his death? If yes, why? What would have happened had the video never surfaced? What would have happened had there been no video? Do you think the Chicago Police Department has/had an unofficial "code of silence?" Why or why not? Do you agree that the three officers who falsified their reports were not guilty of all charges? Why or why not? Do you believe there is a blue-on-black violence problem in the United States? If yes, what can be done about it?

The Police and the Public

Police-Public Contact

In 2015, approximately 21% of U.S. residents age 16 or older—about 53.5 million persons—had some type of contact with the police during the previous 12 months. This was down from 26% of residents or 62.9 million people in 2011.

Source: Elizabeth Davis, Anthony Whyde, and Lynn Langton, "Contacts Between Police and the Public, 2015," in *Bureau of Justice Statistics Special Report*, U.S. Department of Justice (Washington, DC: USGPO, October 2018), accessed June 30, 2019, https://www.bjs.gov/content/pub/pdf/cpp15.pdf.

To carry out the duties of law enforcement, order maintenance, service, and information gathering successfully, the police must have the confidence, respect, trust, and cooperation of the public. The manner in which they carry out those functions, especially law enforcement and order maintenance, determines the community's level of confidence, respect, trust, and cooperation with the police. Citizens who possess those qualities are much more likely to help the police carry out their duties; citizens who lack those qualities may rebel against the police in particular and government in general. For example, groups that are more favorable toward the police are more likely to report a crime to the police. Reporting crimes to the police helps the police solve them and, hopefully, reduce crime more effectively and safely. Favorability toward the police also increases the perceived legitimacy of the police, and when the police are perceived as legitimate, people are more likely to obey the law. What often makes police-community relations difficult is that, on a variety of issues and attitudes, the police and the public view the world very differently.

Public Attitudes Toward the Police

What do people think of the police? The answer depends on what and whom you ask. It also depends on people's previous experience with the police. Research shows that citizens who have experienced positive contact with the police generally have positive attitudes toward the police.[1] Research also shows that attitudes toward the police are influenced by the experiences of family members and friends.[2]

The level of respect and confidence the public has for the police is not particularly high, especially among minorities. *Why is this the case?*

David McNew/Staff/Getty Images

Confidence in the Police Table 7.1 reveals that, in 2018, 54% of the public had "a great deal" (29%) or "quite a lot" (25%) of confidence in the police, 31% had "some" confidence, 14% had "very little," and 1% had "no confidence at all."[3] Over the past 25 years, the percentage of the public with "a great deal" or "quite a lot" of confidence in the police has ranged narrowly between 52% and 64% (see Table 7.1). The 64% figure was recorded in 2004, and the 52% figure was recorded in both 1993 and 2015 (see Table 7.1). Nothing remarkable seems to explain the 25-year high figure recorded in 2004, except, perhaps, residual admiration for the police following the 9/11 attacks. The 1993 poll, on the other hand, was conducted during the federal civil rights trial of four white police officers in the 1991 beating of Rodney King, and the 2015 poll was conducted during a period of lethal police-citizen interactions involving unarmed black men in places such as Ferguson, Missouri; Staten Island, New York; and North Charleston, South Carolina. Perhaps not surprisingly, the 2015 poll also recorded the highest percentage of the public that had "very little" or "no confidence" in the police (18%).

Overall, and when compared to other social institutions, confidence in the police is relatively high. In 2018, for example, the police ranked third among 15 other institutions with regard to confidence in the institution.[4] Only the military and small business ranked

TABLE 7.1 Confidence in the Police, 1993–2018

Year	Great Deal (%)	Quite a Lot (%)	Great Deal/ Quite a Lot (%)	Some (%)	Very Little (%)	None (Vol.) (%)	No Opinion (%)
2018	29	25	54	31	14	1	1
2017	31	26	57	28	14	1	1
2016	25	31	56	29	13	1	1
2015	25	27	52	30	16	2	1
2014	25	28	53	31	14	2	*
2013	26	31	57	30	12	1	1
2012	26	30	56	28	15	1	*
2011	25	31	56	30	11	2	*
2010	26	33	59	27	12	1	1
2009	28	31	59	29	10	1	*
2008	28	30	58	30	10	1	1
2007	23	31	54	33	12	1	1
2006	25	33	58	29	10	2	1
2005	28	35	63	29	7	1	—
2004	24	40	64	26	10	*	*
2003	29	32	61	29	9	1	*
2002	28	31	59	31	9	1	*
2001	26	31	57	31	11	1	*
2000	18	36	54	33	10	2	1
1999	24	33	57	33	10	*	1
1998	26	32	58	30	10	1	1
1997	27	32	59	30	10	1	*
1996	22	38	60	28	11	1	*
1995	26	32	58	30	10	1	1
1994	22	32	54	33	11	1	1
1993	22	30	52	35	11	1	1

*Less than 0.5%.

Source: Adapted from "Confidence in Institutions," *Gallup*, accessed June 11, 2019, https://news.gallup.com/poll/1597/confidence-institutions.aspx.

higher. Eighty-five percent of Americans had at least "some" confidence in the police (that is, they responded that they had either "a great deal," "quite a lot," or "some" confidence in the police), while 94% and 93% of Americans had at least "some" confidence in the military and small business, respectively. The other institutions for which Americans had at least "some" confidence in 2018 (the percentage with at least "some" confidence in parentheses) were: the U.S. Supreme Court (79%), banks (76%), the medical system (73%), the public schools (73%), the church or organized religion (71%), organized labor (71%), big business (68%), the criminal justice system (63%), newspapers (58%), the presidency (55%), television news (54%), and Congress (50%).[5]

To have confidence in the police to do their job with all that entails is one thing, but to have enough confidence in the police to report a crime to them is something else, especially since, as noted in Chapter 5, only about 10% of police time is devoted to apprehending law violators and combatting crime. According to a recent national survey, 72% of Americans would definitely report a crime to the police, as would a majority of every demographic group (see Table 7.2).[6] However, within groups, willingness to report a crime varied substantially. As shown in Table 7.2, although a majority of whites, blacks, and Hispanics would definitely report a crime to the police, whites were significantly more likely to do so than were blacks and Hispanics. Gender made little difference, as about 70% of both males and females would definitely report a crime. Willingness to report a crime increased with age, annual income, and education. Area of residence made little difference, as approximately 70% of each area's residents were willing to report a crime. Republicans were much more likely to report a crime than were Democrats or Independents, and Conservatives and Liberals were more likely to report a crime than were Libertarians and Communitarians. In the survey, ideology was measured using two indicators: size of government (smaller or larger) and traditional values (promotes or does not promote). Thus, conservatives were defined as people who favor a smaller government that promotes traditional values, libertarians favor a smaller government that does not promote traditional values, communitarians favor a larger government that promotes traditional values, and liberals favor a larger government that does not promote traditional values.

Respect for the Police in Their Area The public's willingness to report a crime to the police suggests that the public respects the police in their area, more so than they are confident that the police can get the job done. In a 2016 Gallup poll, 93% of Americans had "a great deal" (76%) or "some" (17%) respect for the police in their area; only 7% had "hardly any" respect for the police in their area.[7] Recall that, in 2018, only 54% of the public had "a great deal" or "quite a lot" of confidence in the police. The Gallup pollsters have asked Americans about their respect for the police in their area nine times since 1965. The 76% of Americans that had "a great deal" of respect for the police in 2016 was the second highest percentage recorded, only one percentage point less than the 77% high recorded in 1967.[8] The lowest percentage of Americans with "a great deal" of respect for the police (56%) was recorded in 2005. In each of the nine polls, a solid majority of Americans expressed "a great deal" of respect for the police in their area.

Americans with "a great deal" of respect for the police in their area surged 12 percentage points from the 64% recorded in 2015 to the 76% recorded in 2016, despite the racial tensions caused by the high-profile police shootings of unarmed black men. In the 2014 poll, only 61% of Americans had "a great deal" of respect for the police. (The last poll before the 2014 poll was conducted in 2005.) Not only did respect for the police increase by 11 percentage points among whites between 2015 and 2016, from 69% to 80%, but it also increased 14 percentage points among nonwhites, from 53% to 67%.[9] In all nine polls, whites expressed more respect for the police than had nonwhites. The remarkable 1-year change between 2015 and 2016 may reflect in part sympathy by both groups for the increasing number of on-duty police officers who had been shot and killed in retaliation for the police shootings of unarmed black men. It also could expose in part the dilution of the number of blacks in the nonwhite category. It is not clear why the pollsters decided to combine blacks and Hispanics in the nonwhite category, especially when they separated blacks and Hispanics in other polls. In any event, and whatever the reason, substantial increases in respect for the police between 2015 and 2016 also were recorded

TABLE 7.2 Percentage of Americans Who Would Definitely Report a Crime to the Police, by Demographic Characteristic

Characteristic	Percentage
All Americans	72
Race/Ethnicity	
White	78
Blacks	54
Hispanic	57
Gender	
Male	70
Female	74
Age	
18–29	54
30–44	63
45–64	82
65+	87
Annual Income	
Less than $30,000	62
$30,000–$49,999	73
$50,000–$99,999	76
$100,000+	84
Education	
High school or less	68
Some college	69
College graduate+	81
Residence	
City	69
Suburb	74
Rural area	72
Politics	
Democrat	69
Independent	67
Republican	82
Ideology	
Libertarian	70
Conservative	82
Liberal	77
Communitarian	71

Source: Adapted from Emily Ekins, "Policing in America: Understanding Public Attitudes Toward the Police. Results from a National Survey," Cato Institute, December 7, 2016, accessed June 11, 2019, https://www.cato.org/survey-reports/policing-america.

for age, residence, politics, and political ideology (see Table 7.3). Young adults, town or rural residents, and liberals accounted for the largest increases of 19 to 21 percentage points (see Table 7.3). Older adults and Republicans recorded the smallest increases of 4 percentage points, but their percentages already were among the highest (see Table 7.3).

TABLE 7.3 Americans' Respect for Police in Their Area, by Group, 2015 and 2016

How much respect do you have for the police in your area? Percentage that responded "a great deal."

Group	October, 2015	October, 2016	Change (% pts.)
All	64	76	+12
Race			
Whites	69	80	+11
Nonwhites	53	67	+14
Age			
18–34	50	69	+19
35–54	61	77	+16
55+	77	81	+4
Residence			
Large/small city	61	68	+7
Suburb	71	82	+11
Town/rural area	61	80	+19
Politics			
Republicans	82	86	+4
Independents	60	75	+15
Democrats	54	68	+14
Political ideology			
Conservatives	69	85	+16
Moderates	67	72	+5
Liberals	50	71	+21

Source: Adapted from Justin McCarthy, "Americans' Respect for Police Surges," *Gallup*, October 24, 2016, accessed June 18, 2019, https://news.gallup.com/poll/196610/americans-respect-police-surges.aspx.

In sum, although there was variation among subgroups, majorities of all groups had "a great deal" of respect for the police in their area.

Honesty and Ethical Standards of the Police Americans' attitudes about the honesty and ethical standards of the police track more closely to their confidence in the police than their respect for the police in their area. In 2018, 54% of Americans rated the honesty and ethical standards of the police as "very high" (15%) or "high" (39%), 32% rated their honesty and ethical standards as "average," and 13% rated their honesty and ethical standards as "low" (9%) or "very low" (4%).[10] Gallup has asked Americans about the honesty and ethical standards of the police in 35 polls between 1977 and 2018. The average "very high" or "high" rating over the 35 polls is 52%—only two percentage points lower than the 2018 rating. The range is between a low of 37% in 1977, and a high of 68% in 2001 (see Table 7.4). The 2001 poll was conducted only 2 months after the 9/11 tragedies. An examination of Table 7.4 shows that in only 1 year, 1977, is the percentage rating in the 30s, and in only five polls is the percentage rating in the 60s. Most of the percentage ratings are in the 40s (13) and 50s (16), and most of the 40s are recorded prior to 1999. Together with the average percentage rating of 52, these data indicate that only about half of Americans ever have rated the honesty and ethical standards of the police as "very high" or "high." Unfortunately, this measure does not provide a breakdown by demographic characteristics.

Still, as was the case with confidence in the police, when compared to other professions, the police rank relatively high. In 2018, for example, the police ranked fifth among 20 professions.[11] As noted previously, 54% of Americans ranked the police "very high" or "high" with regard to their honesty and ethical standards. Ranking higher were nurses

TABLE 7.4 Percentage of Americans Who Rated the Honesty and Ethical Standards of the Police as "Very High" or "High," by Year

Year	Percentage
2018	54
2017	56
2016	58
2015	56
2014	48
2013	54
2012	58
2011	54
2010	57
2009	63
2008	56
2007	53
2006	54
2005	61
2004	60
2003	59
2002 (November)	59
2002 (February)	61
2001	68
2000	55
1999	52
1998	49
1997	49
1996	49
1995	41
1994	46
1993	50
1992	42
1991	43
1990	49
1988	47
1985	47
1983	41
1981	44
1977	37

Source: Adapted from "Honesty/Ethics in Professions," *Gallup*, accessed June 18, 2019, https://news.gallup.com/poll/1654/honesty-ethics-professions.aspx.

(84%), medical doctors (67%), pharmacists (66%), and high school teachers (60%).[12] Ranking lower were accountants (42%), funeral directors (39%), clergy (37%), journalists (33%), building contractors (29%), bankers (27%), real estate agents (25%), labor union leaders (21%), lawyers (19%), business executives (17%), stockbrokers (14%), advertising practitioners (13%), telemarketers (9%), car salespeople (8%), and members of Congress (8%).[13]

Favorability Toward the Police In the aforementioned recent national survey, 64% of Americans had a favorable view of the police: 33% had a "very favorable" view, 31% had a "somewhat favorable" view, 14% had an "unfavorable" view, and 22% had neither positive nor negative views of the police.[14] These data, however, hide significant demographic differences among Americans (see Table 7.5). Some of the largest and arguably most troubling differences involve race and ethnicity. For example, while 68% of whites and 59% of

TABLE 7.5 Percentage of Americans Who Had a Favorable View of the Police, by Demographic Characteristic

Characteristic	Percentage
All Americans	64
Race/Ethnicity	
White	68
Black	40
Hispanic	59
Gender	
Male	65
Female	64
Age	
18–29	54
30–44	54
45–64	70
65+	82
Annual Income	
Less than $30,000	54
$30,000–$49,999	64
$50,000–$99,999	76
$100,000+	76
Education	
High school or less	59
Some college	64
College graduate+	73
Residence	
City	60
Suburb	69
Rural area	61
Politics	
Democrat	59
Independent	59
Republican	81
Ideology	
Libertarian	64
Conservative	80
Liberal	54
Communitarian	58

Source: Adapted from Emily Ekins, "Policing in America: Understanding Public Attitudes Toward the Police. Results from a National Survey," Cato Institute, December 7, 2016, accessed June 11, 2019, https://www.cato.org/survey-reports/policing-america.

Hispanics had favorable views of the police, only 40% of blacks did so. This distribution of views, moreover, has not changed much since the 1970s, which indicates that efforts to reform how police interact with the black community over the past half century have not been very successful. As for other demographic characteristics, 65% of males and 64% of females had favorable views of the police. Favorability toward the police increased with age, annual income, and education. Americans living in the suburbs (69%) had more favorable views of the police than Americans living in rural areas (61%) or in cities (60%). Republicans (81%) had more favorable views of the police than Independents (59%) or Democrats (59%). Finally, with regard to ideology, conservatives (80%) had more favorable views of the police than libertarians (64%), communitarians (58%), and liberals (54%). In sum, these data show that more than half of every demographic group, except blacks, had a favorable view of the police, and the type of person most likely to favor the police was an older (65+) white Republican conservative with a college degree or higher and an annual income of more than $50,000 who lived in the suburbs. A person's gender made no difference.

Are Blacks, as a Group, "Anti-Police?" Those who say, "no," acknowledge that only 40% of blacks had a favorable view of the police, but they also note that few blacks had an unfavorable view of the police. In the aforementioned survey, only 19% of blacks had an "unfavorable" view of the police, while 40% of blacks reported having "neutral" feelings toward the police.[15] Moreover, how can blacks be considered "anti-police," when majorities of blacks do not want to decrease the number of police officers in their neighborhoods and are sympathetic to the difficulty of police work, acknowledging that policing is a "very dangerous" job?[16]

Still, on nearly all measures (see Table 7.6), blacks were less approving of the police than were white and Hispanic Americans. As noted previously, this has been the case for

TABLE 7.6 Attitudes About the Police, by Race and Ethnicity, 2016

Attitude About Police	Percent Who Agree			
	White	Black	Hispanic	All
1. Tactics are too harsh	26	56	33	30
2. Too quick to use deadly force	35	73	54	42
3. Only use deadly force when necessary	65	27	46	58
4. Are courteous	62	43	49	57
5. Treat all racial groups equally	64	31	42	56
6. Competently enforce the law	60	40	50	59
7. Protect people like you from violent crime	60	38	49	56
8. Respond quickly to call for help	59	44	50	56
9. Used abusive language/profanity with you	15	24	25	17
10. Knows someone physically abused by police	20	41	29	21
11. Oppose racial profiling	62	77	62	63
12. Police are "above the law"	46	61	61	49
13. Held accountable for misconduct	57	36	49	54
14. Are honest and trustworthy	62	36	51	57
15. Gaining trust of local residents	57	33	46	53
16. Oppose use of military weapons and armored vehicles	53	58	51	54
17. Know someone stopped and searched by police	47	60	40	47
18. Had a satisfactory encounter with police in the past 5 years	70	50	66	67
19. Been stopped by police 3 or more times in the past 5 years	8	17	4	9
20. Have a very dangerous job	66	65	58	65
21. There is a "war on police"	64	46	52	61
22. People show too little respect for police these days	64	34	45	58

Source: Adapted from Emily Ekins, "Policing in America: Understanding Public Attitudes Toward the Police. Results from a National Survey," Cato Institute, December 7, 2016, accessed June 11, 2019, https://www.cato.org/survey-reports/policing-america.

at least the past 50 years. This holds true even when controlling for income and political affiliation. In other words, blacks with higher incomes (more than $60,000 a year) were only a little more favorable toward the police (48% vs. 41%) than blacks with lower incomes (less than $30,000 a year).[17] Likewise, black Republicans (44%) were no more likely than black Democrats (44%) to have a favorable view of the police.[18] These data suggest that if blacks become wealthier and Republican, they are not likely to become much more favorable toward the police. On a more positive note, research shows that black favorability toward the police may improve significantly by simultaneously improving blacks' perceptions of police use of force, impartiality, and professionalism during encounters.[19] Thus, it is unlikely that blacks as a group are "anti-police" *per se*, but rather that they object to the police treating blacks badly and illegally.

Police Attitudes Toward the Public

Data in this section are from a 2016 Pew Research Center national survey conducted by the National Police Research Platform.[20] The survey is one of the largest ever conducted with a nationally representative sample of nearly 8,000 police officers from departments with at least 100 officers.[21] The survey was conducted, as were some of the surveys in the previous section, during a volatile period in police-community relations exemplified by the high-profile lethal police shootings of unarmed black men and the retaliatory lethal shootings of on-duty police officers. Unless indicated otherwise, the results presented are for rank-and-file officers—the group that has the most direct daily contact with citizens—and not for sergeants or administrators.

Police Views of the Public The survey shows that the police have a complicated relationship with the public. On one hand, 79% of police officers acknowledged that, in the month prior to the survey and while on duty, someone had thanked them for their service (83% of black officers, 81% of white officers, and 73% of Hispanic officers); 67% of officers either "strongly agreed" (6%) or "agreed" (61%) that most people respect the police; 72% of officers either "strongly agreed" (16%) or "agreed" (56%) that they had reason to be trustful of most citizens; and 58% of officers admitted that their work "nearly always" (23%) or "often" (35%) made them feel proud (58% of white officers, 60% of black officers, and 63% of Hispanic officers). As for respect, recall that 93% of Americans had "a great deal" (76%) or "some" (17%) respect for the police in their area.

On the other hand, only 11% of officers thought that "all or most" of the people in the neighborhoods where they routinely worked shared their values and beliefs; 59% of officers thought that at least "some" people in those neighborhoods shared their values and beliefs. In addition, 84% of officers were either "nearly always or often" (42%) or "sometimes" (42%) worried about their personal safety on the job, 67% of officers maintained that, in the month prior to the survey, a member of their community had verbally abused them, 51% lamented that the job often frustrated them, 56% either "strongly agreed" (13%) or "agreed" (43%) that their job had made them more callous, and only 42% "nearly always" (9%) or "often" (33%) felt fulfilled by their job as a police officer (42% of white officers, 45% of black officers, and 47% of Hispanic officers). As for worrying about their personal safety on the job, only 33% of employed adults "nearly always or often" (14%) or "sometimes" (19%) worried about their personal safety on the job—51 percentage points lower than police officers. Regarding job fulfillment, 52% of employed adults acknowledged that their work often made them feel fulfilled—10 percentage points higher than police officers. Underscoring the complexity of their relationship with the public, more than half of all police officers (55%) disclosed that they experienced both praise and hostility from the public during the month prior to the survey.

Police Officers' Facebook Posts and Comments

To view public Facebook posts and comments made by police officers about race, religion, ethnicity, the acceptability of violent policing, etc., visit The Plain View Project website at https://www.plainviewproject.org.

What do these Facebook entries reveal about the posting and commenting police officers' attitudes toward the people and communities they serve? How prevalent do you suppose these attitudes are?

The amount of verbal abuse, frustration, and callousness experienced by police officers varies widely by demographic characteristic of the officer. For example, male officers (69%) were more likely than were female officers (60%) to have been verbally abused by a community member in the month before the survey; white (70%) and Hispanic officers (69%) were more likely than were black officers (53%) to have been verbally abused; and younger officers (ages 18 to 44) were more likely than were their older colleagues to have been verbally abused (75% vs. 58%). As for frustration on the job, 54% of white officers "nearly always" or "often" felt frustrated with their job, while only 47% of Hispanic officers and 41% of black officers felt the same way. Of those officers with five or more years on the force, 53% acknowledged that they "nearly always" or "often" felt frustrated by their job, while only 39% of new officers felt that way. By comparison, only 29% of all employed adults disclosed that their job "nearly always" or "often" frustrated them. Regarding callousness, younger officers and white officers were more likely than were older or black officers to admit that they had become more callous. Moreover, officers who reported becoming more callous also were more likely than were other officers to support aggressive or physically harsh tactics with some people or in some areas of the community, confess that they were often angered or frustrated with their jobs, had been involved in a physical or verbal dispute with a citizen in the past 12 months, or had fired their service weapon while on duty during their careers. Whether callousness is the cause or the consequence of the aforementioned situations is difficult to determine; it is likely both.

Police officers do not think the public understands the risks and challenges they face on the job. Only 13% of officers thought the public understood their risks and challenges "very well" (1%) or "somewhat well" (12%), while 86% of officers thought the public understands their risks and challenges "not too well" (46%) or "not well at all" (40%). White and Hispanic officers were more skeptical than were black officers on this issue: 42% of both white and Hispanic officers but only 29% of black officers did not think the public understands their risks and challenges very well. Younger officers were more skeptical than were their older colleagues: 44% of officers age 18 to 44 said the public did not understand their risks and challenges very well, while only 34% of officers aged 45 and older agreed with this view. Size of department matters, too: 45% of officers in departments with 1,000 or more officers did not believe the public understands, while only 35% of officers in departments with fewer than 1,000 officers shared this assessment. The public disagrees with the police on this issue: 83% of American adults countered that they did, indeed, understand the risks and challenges police officers face on the job, including 38% who believed they understand the risks and challenges very well. The difference between the police and the public on this issue is the single largest difference measured in the surveys.

To be effective at their jobs, 97% of police officers maintained that it is "very important" (72%) or "somewhat important" (25%) to have detailed knowledge of the people, places, and culture in the areas where they work. Moreover, the percentage of officers who believed that local knowledge is either "very important" or "somewhat important" varied little by race or gender. However, the percentage of officers who put a premium on local knowledge and considered it "very important" did vary significantly by race and gender. While 84% of black officers and 78% of Hispanic officers agreed that local knowledge is "very important," only 69% of white officers shared their evaluation. As for gender, 80% of female officers but only 71% of male officers believed that local knowledge is "very important."

When responding to the public, 41% of officers worried more that an officer "will not spend enough time diagnosing the situation before acting decisively," while 56% of officers worried more that an officer "will spend too much time diagnosing the situation before acting decisively." Black officers (61%) and department administrators (59%) were more likely than were white officers (37%) and Hispanic officers (44%) to worry more that officers will not spend enough time diagnosing a situation before acting. When an officer is faced with a law enforcement situation in which doing the morally right thing will require breaking a department rule, 40% of officers would advise another officer to "follow the department rule," while 57% of officers would advise another officer to "do the morally right thing, even it requires breaking a department rule." There was a significant racial divide on this issue: 63% of white officers would advise another officer to do the morally right thing, even if it means violating a department rule, while only 43% of black officers and 47% of Hispanic officers would give the same advice. The survey also discovered that 56% of officers either "strongly agreed" (17%) or "agreed" (39%) that in certain areas of the city it is more useful for an officer to be aggressive than to be courteous, and 44% of officers either "strongly agreed" (5%) or "agreed" (39%) that "some people can only be brought to reason the hard, physical way."

Discharging a Firearm

According to a recent national survey, 72% of police officers confessed that other than on a gun range or while training, they had never discharged their service firearm while on duty. Seventy percent of male officers, 89% of female officers, 69% of white officers, 79% of black officers, and 79% of Hispanic officers reported their lack of firearm use during their careers. The disconnect with the public on this issue is extreme, as 83% of Americans believed that typical police officers fire their service weapon while on duty at least once in their careers, and 31% believed that police officers discharge their weapon at least a few times a year. In fact, only 27% of all police officers had ever fired their service weapon while on duty.

Source: Rich Morin, Kim Parker, Renee Stepler, and Andrew Mercer, "Behind the Badge," Pew Research Center, January 11, 2017, accessed June 19, 2019, https://www.pewsocialtrends.org/2017/01/11/behind-the-badge/.

Black Lives Matter The survey found that police officers and the public differ significantly on whether the police killings of unarmed black men were isolated incidents or signs of broader problems between the police and the black community. Two-thirds of police officers (67%) thought the police killings of unarmed black men were isolated incidents, while 31% of officers viewed them as signs of broader problems between the police and the black community. The public saw the issue differently: 60% of the public considered the police killings of unarmed black men as signs of broader problems, while only 40% of the public assumed they are isolated incidents. The officers' and publics' race mattered on this issue. Seventy-two percent of white officers believed the fatal encounters were isolated incidents, while only 43% of black officers (37% of black female officers) shared their belief. Among the public, 44% of whites considered the fatal encounters as isolated events, while only 18% of blacks thought similarly.

The survey also discovered significant differences between and among police officers and the public on what each believed motivates the Black Lives Matter protests. Ninety-two percent of police officers viewed the protests as motivated "a great deal" (68%) or "some" (24%) by long-standing anti-police bias. While the percentage of white and black officers who held this view was similar (95% of white officers and 91% of black officers), they differed in how strongly they held the view: 72% of white officers believed "a great deal" and 23% believed "some" that anti-police bias explained the motivation for the protests, while 59% of black officers believed "a great deal" and 32% believed "some" about the anti-police motivation for the protests. As noted, the views of the public on this issue differed dramatically from the views of police officers. On this issue, 79% of the public responded that anti-police bias was the motivation for the protests, including 41% who saw this as a major motivation. Again, race mattered. While 85% of whites believed "a great deal" (47%) or "some" (38%) that the protests were motivated by anti-police bias, only 56% of blacks shared their belief (25% "a great deal" and 32% "some").

On a separate question, police officers and the public were queried about a second motivation for the Black Lives Matter protests: whether the protests were motivated by a "genuine desire to hold police accountable for their actions." Only 35% of officers believed the protests were motivated "a great deal" (10%) or "some" (25%) by such a desire. Here again, the race of the police officer mattered. While only 27% of white officers responded "a great deal" (5%) or "some" (22%) that the protests were motivated by a desire to hold police accountable, 69% of black officers answered "a great deal" (34%) or "some" (35%) that the protests were so motivated. On this question, the public sided more with black officers than with white officers: 65% of the public said accountability was a factor. As for race, 63% of whites thought "a great deal" (27%) or "some" (36%) that the protests were motivated by a desire to hold police accountable, and 79% of blacks agreed to "a great extent" (55%) or "some" (24%) that police accountability motivated the protests.

Additional survey results revealed that, because of the fatal incidents, 93% of police officers were more concerned about their safety, 86% believed their jobs are harder, 76% had been more reluctant to use force when it was appropriate, 75% contended interactions between police and blacks had become more tense, 72% had become less willing to stop and question people who seemed suspicious, and 70% believed policing had become more dangerous in recent years.

Previously, the question was asked: Are blacks, as a group, "anti-police?" Here, the question is reversed: Are the police, as a group, "anti-blacks?" In answering the question, consider the following: (1) the long history of police harassment, intimidation, and repression of blacks briefly described in Chapter 5, (2) the contemporary situation exemplified by the Black Lives Matter movement addressed above, and (3) the views of the police themselves about their relationships with blacks in the communities that they serve. Telling is how police officers rate the relationships between other officers in their department and various racial or ethnic groups in the communities that they serve. As shown in Table 7.7, 87% of police officers rated the relationship between departmental colleagues and whites in the communities they serve as "excellent" (21%) or "good" (66%), 73% of officers rated the relationship with Asians as "excellent" (15%) or "good" (58%), 67% of officers rated the relationship with Hispanics/Latinos as "excellent" (10%) or "good" (57%), while only 54% of officers rated the relationship with blacks as "excellent" (8%) or "good" (46%). Conversely, 43% of officers rated the relationship with blacks as "only fair" (25%) or "poor" (18%), 28% of officers rated the relationship with Hispanics/Latinos as "only fair" (21%) or "poor" (7%), 10% of officers rated the relationship with Asians as "only fair" (9%) or "poor" (1%), and 9% of officers rated the relationship with whites as "only fair" (8%) or "poor" (1%). Clearly, police officers viewed the relationship between blacks and their department colleagues in the communities they serve as significantly poorer than the relationships between their department colleagues and the other racial or ethnic groups, which underscores the long-standing antipathy between the police and blacks. Here again, the race of the police officer responding to the question mattered. While 60% of white and Hispanic officers responding to the question characterized police relations with blacks as "excellent" or "good," only 32% of responding black officers shared their view.

One of the more troublesome findings of the survey is that 80% of police officers believed that the United States "has made the changes needed to give blacks equal rights with whites" (that is, that racial justice had been achieved), while only 16% of police officers

TABLE 7.7 Relationships Between Police Officers and Racial or Ethnic Groups in the Communities That They Serve, 2016

	Percent of Police Officers Selecting				
Race/Ethnicity	Excellent	Good	Only Fair	Poor	Too Few/No Answer
White	21	66	8	1	4
Black	8	46	25	18	3
Hispanic/Latino	10	57	21	7	4
Asian	15	58	9	1	17

Source: Rich Morin, Kim Parker, Renee Stepler, and Andrew Mercer, "Behind the Badge," Pew Research Center, January 11, 2017, accessed June 19, 2019, https://www.pewsocialtrends.org/2017/01/11/behind-the-badge/.

believed that the United States "needs to continue making changes to give blacks equal rights with whites." Again, there were significant differences between officers based on race. While 92% of white officers believed the country had made the changes needed to secure black equal rights, only 29% of black officers shared that belief. Moreover, not only did the views of white and black officers differ on this issue, but white officers also differed significantly from whites overall, as only 57% of all white adults believed that racial justice had been achieved. Only 12% of black adults agreed that no more changes were needed.

At the departmental level, only 59% of officers remarked that their department "has taken steps to improve relations between police and blacks," and only 46% of officers acknowledged that their department "has modified its policies or procedures about the use of force." On these issues, there was significant variation by size of department. On the first issue, 66% of police officers in large departments responded that their department had "taken steps to improve relations with black residents," while only 35% of officers in small departments maintained that their department had made similar outreach efforts. On the second issue, 68% of officers in large departments replied that their department had modified their policies about the use of force, while only 19% of officers in small departments reported that their department had made this change.

As for the use of unnecessary force, 73% of officers agreed that the department rules governing the use of force were "about right," while 26% replied that they were "too restrictive," and only 1% believed that they were "not restrictive enough." Similarly, 85% of officers thought their department's use of force guidelines were either "very useful" (34%) or "somewhat useful" (51%) when officers were confronted with actual situations where force might be needed. Also, 84% of officers thought they should be required to intervene when they believed another officer was about to use unnecessary force.

Following the fatal encounters between the police and unarmed black citizens, a number of reforms in police tactics and procedures were recommended in an effort to prevent those types of situations from happening again. One of the reforms was the increased use of body cameras. The survey found that 66% of police officers favored the use of body cameras, but only 33% of officers thought that their use would make members of the public "more likely to cooperate with officers," 10% of officers thought that their use would make the public "less likely to cooperate with officers," and 56% of officers thought that their use would "make no difference." Some 44% of black officers thought wearing body cameras would make the public more likely to cooperate with officers, while only 31% of white officers agreed. However, 50% of officers thought wearing a body camera would make officers "more likely to act appropriately," 5% of officers thought it would make officers "less likely to act appropriately," and 44% of officers thought it would "make no difference." While 71% of black officers thought wearing body cameras would make officers act more appropriately, only 46% of white officers thought so. As for the public, 93% supported the police use of body cameras and saw more benefits from their use than did the police. In addition, 59% of the public thought that body cameras would make the public more likely to cooperate with officers, 5% said they would make the public less likely to obey officers, and 35% assumed that they would make no difference. Two-thirds of the public (66%) thought that wearing body cameras would make officers more likely to act appropriately, 6% replied they would make officers less likely to act appropriately, and 27% answered that they would not impact police behavior.

Training in preventing the use of unnecessary force was another recommended reform. Table 7.8 shows the percentages of police officers who received training in the prevention

TABLE 7.8 Percentages of Police Officers Receiving Training in Preventing the Use of Unnecessary Force in the Year Before the Survey, 2016

	Amount of Training			
Training Area	4 Hours+	Less Than 4 Hours	None	NA
De-escalating a situation	44%	32%	23%	1%
Firearms training involving shoot-don't shoot scenarios	53	31	15	1
Dealing with individuals having a mental health crisis	45	35	19	1
Non-lethal methods to control a combative or threatening individual	49	32	18	1
How to deal with people so they feel they've been treated fairly and respectfully	37	34	27	2
Bias and fairness	40	37	22	1

Source: Rich Morin, Kim Parker, Renee Stepler, and Andrew Mercer, "Behind the Badge," Pew Research Center, January 11, 2017, accessed June 19, 2019, https://www.pewsocialtrends.org/2017/01/11/behind-the-badge/.

Police Views of the Media

A recent national survey found that 81% of police officers believed that the media treats them unfairly.

Source: Rich Morin, Kim Parker, Renee Stepler, and Andrew Mercer, "Behind the Badge," Pew Research Center, January 11, 2017, accessed June 19, 2019, https://www.pewsocialtrends.org/2017/01/11/behind-the-badge/.

of using unnecessary force in the year before the survey. As Table 7.8 reveals, and despite the national attention on training and reforms to prevent the use of unnecessary force, fewer than half of all police officers reported having received at least four hours of training in five of six relevant training areas during the prior 12 months. In the one area in which more than half of all officers received four hours or more of training—firearms training involving shoot-don't shoot scenarios—only 53% of all police officers received training. Remarkably, between 15% and 27% of police officers received no training in the six areas.

In sum, these data clearly show that the police and the public view the world and each other very differently. Moreover, there are significant demographic, especially racial differences both among the public and among the police and between the public and the police. The public, for its part, seems to have more respect for the police than it has confidence in the police, high regard for police honesty and ethical standards, and a favorable view of the police. The police, for their part, have mixed views of the public, and, as mentioned in the narrative, a complicated relationship with the public. Two intriguing and provocative questions were posed in the narrative: (1) Are blacks, as a group, anti-police and (2) Are the police, as a group, anti-blacks? Regarding the first question, it appears that blacks, as a group, are not anti-police *per se*, but they do object to the long history of being treated badly and illegally by the police. As for the second question, the data indicate a real and concerning racial divide between the police and blacks. Particularly telling is that 80% of police officers (92% of white officers but only 29% of black officers) believe that racial justice in society has been achieved, and that there is no need to continue making changes to give blacks equal rights with whites. As for the public on this issue, 57% of white Americans agree with white police officers, but only 12% of black Americans agree with white officers. Regarding the public's opinion of the police overall, most of the public thinks the police do a fairly good job, but it also believes there is much room for improvement. One way to improve the police is to employ better police officers.

THINKING CRITICALLY

1. Why do you think racial groups, especially blacks, differ so greatly in their attitudes toward the police?
2. Why do you think the police differ so greatly in their attitudes toward the public, especially blacks?
3. What do you think could be done to improve public attitudes, especially the attitudes of blacks, toward the police?
4. What do your think could be done to improve police attitudes toward the public, especially blacks?
5. How do the perceptions presented in this section compare with your own?

Police Recruitment and Selection

Deciding whom to employ should be simple: Hire the type of police officer that the citizens of the community want. Of course, that approach assumes that the citizens of a community have some idea of what it takes to be a police officer. Then, there is the matter of which people to consult. Who should decide? The wealthy? The middle class? The poor? The politically conservative? The politically liberal? The young? The old? The business community? Community leaders? Crime victims? Those people most likely to be policed? Some consensus is needed on the type of police officer desired. Seeking that consensus in metropolitan communities is filled with conflict. Police administrators need to be very careful in choosing police officers who may well be with the agency for 20 years or more. A police department will never reach its full potential without selecting the best available personnel. Selection decisions have momentous long-term implications for a police department.[22]

Qualities of a Successful Police Officer

Given the complexity of the role of the police officer, it comes as no surprise that deciding what qualities the successful police officer needs is not easy. Indeed, police officers require a combination of qualities and abilities that is rare in any pool of applicants. Robert B. Mills, a pioneer in the psychological testing of police officers, believes that police applicants should possess the following psychological qualities:

- Motivation for a police career.
- Normal self-assertiveness.
- Emotional stability under stress.
- Sensitivity toward minority groups and social deviates.
- Collaborative leadership skills.
- A mature relationship with social authority.
- Flexibility.
- Integrity and honesty.
- An active and outgoing nature.[23]

The Berkeley, California, Police Department lists these qualities:

- Initiative.
- Ability to carry heavy responsibilities and handle emergencies alone.
- Social skills and ability to communicate effectively with persons of various cultural, economic, and ethnic backgrounds.
- Mental capacity to learn a wide variety of subjects quickly and correctly.
- Ability to adapt thinking to technological and social changes.
- Understanding of other human beings and the desire to help those in need.
- Emotional maturity to remain calm and objective and provide leadership in emotionally charged situations.
- Physical strength and endurance to perform these exacting duties.[24]

Three qualities seem to be of paramount importance. One commentator refers to them as the **three I's of police selection**: intelligence, integrity, and interaction skills. In short, police officers need to be bright enough to complete rigorous training. They should be honest enough to resist—and have a lifestyle that allows them to resist—the temptation of corrupting influences in law enforcement. They also should be able to communicate clearly and get along with people of diverse backgrounds.

three I's of police selection Three qualities of the American police officer that seem to be of paramount importance: intelligence, integrity, and interaction skills.

Nearly as important as the three I's, however, are common sense and compassion. In resolving conflicts and solving problems they encounter, police officers often must choose a course of action without much time to think about it. Common sense is a key quality, for example, in locating a suspect who has just fled a crime scene on foot or in deciding when to call off a high-speed vehicle pursuit that suddenly endangers innocent citizens and other police officers.

Police agencies also seek to employ officers with the core value of compassion. Without a genuine concern for serving one's fellow human beings, a police officer is not likely to sustain a high level of motivation over a long period of time. Many people the police meet on a daily basis simply need help, sometimes required by law; but more often, these officers are spurred by a compassion for helping people no matter what their need.

Other qualities, such as physical strength, endurance, and appearance, seem less important. If you were the one who needed to be dragged from a burning automobile, however, the physical strength of the police officer might be important to you.

Gregory Smith/Contributor/Getty Images

Police officers must have a variety of qualities to be successful on the job. *What qualities would you bring to the position?*

The Police Recruitment Process

Few occupations have selection processes as elaborate as the ones used in choosing police officers in most departments of the nation. Before choices are made, a wide net must be cast in the recruiting effort to come up with enough potential applicants to fill the vacancies for an academy class. Police departments, often working with city personnel agencies, generally are guided in their selection decisions by civil service regulations. Those regulations are developed either locally or at the state level. They guarantee a merit employment system with equal opportunity for all.

Because employment qualifications are supposed to be based on perceived needs in policing, law enforcement agencies must be careful not to set unnecessary restrictions that have no bearing on an officer's ability to complete training and perform successfully on the job. The addition of just one seemingly minor qualification, such as requiring four pull-ups instead of three during physical ability testing, or making the eyesight requirement slightly more stringent, may eliminate thousands of men and women from the selection process in a large metropolitan area. It is difficult enough to find capable police candidates without needlessly eliminating them from the selection process.

Recruitment and Pay Most police agencies finally have realized that the kind of officers they desire will not gravitate naturally to the doors of the department. The search for top-notch applicants is very competitive, and many chiefs and sheriffs believe that they have to look at larger pools of applicants than in the past to find the same number of qualified officers. The reasons for the increased difficulty in finding good police candidates involve social maturity and lifestyle issues. Problems with drugs and alcohol, sexually transmitted diseases, personal debt, and dependability have reduced the number of qualified police applicants. Table 7.9 provides the selection standards for Seattle police officers.

The major goal of the recruiting effort is to cast police work as an attractive and sustaining career, even to those who might initially be turned off by it. Research supports the allure of policing for many people who view a career in law enforcement as financially rewarding and status enhancing. In addition, the work itself is intrinsically satisfying because it is nonroutine, exciting, generally outdoors, and people oriented.[25]

Figure 7.1 shows 2020 police officer salaries in the United States by percentile. The average annual police officer salary in 2020 was $48,801 or $23.46 an hour. Salaries in 2020 ranged from a low of $25,500 to a high of $75,000. Four percent of police officers in 2020 earned between $25,500 and $29,999 a year, while 2% of police officers in 2020 earned between $70,500 and $75,000 a year. Seventy percent of police officers in 2020 earned less than $52,500 a year, while 30% of police officers earned $52,500 or more a year.

Table 7.10 shows 2020 police officer salaries in the United States by state. New York paid its police officers the most—an average annual salary of $53,405 or $25.68 an hour,

TABLE 7.9 Selection Standards for Seattle Police Officers, 2019

QUALIFICATIONS

Minimum Hiring Standards

The following are requirements of the State of Washington for law enforcement officers:

- Applicant must be at least 20.5 years of age at the time of taking the written exam. This requirement is necessary to ensure all police applicants will be able to legally enter all premises that SPD responds to. There is no maximum age limit.
- Proof of high school graduation or a certified GED.
- Applicant must pass the Criminal Justice Training Commission Fitness Ability Test.
- Applicant must be a U.S. citizen, or lawful permanent resident to be hired.
- Applicant must have a valid Washington State Driver's License prior to being hired. It is understood that out of state candidates won't have this at time of application, but they must get one prior to accepting a job. Driving is an essential function of this position with SPD.
- Military discharge under honorable conditions, if applicable (fair employment laws apply).

Additional Hiring Standards

The following Seattle Police Department hiring standards and information are some of the areas that are considered during an SPD background investigation. The Department is most interested in an applicant's life history as a complete picture.

SPD understands that, at times, people encounter challenges in their lives. During the background investigation process SPD is interested in learning greater detail about those challenges, as well as the lessons the applicant has learned and the changes the applicant has made as a result.

Please be forthcoming in your responses, as SPD expects honesty in this process. All disqualifications are reviewed by the HR Director or his designee for final approval.

Criminal Record

An applicant's criminal record, including all arrests, prosecutions, deferred prosecutions, "Alford" pleas, and non-conviction information will be thoroughly assessed and *may* be grounds for disqualification. The following *will* be disqualifying:

- Any adult felony conviction.
- Any misdemeanor or felony conviction while employed in a criminal justice and/or law enforcement capacity.
- Any domestic violence conviction.

Traffic Record

An applicant's driving record will be considered on a case-by-case basis with the past five (5) years being the most critical. The following will be disqualifying until the time parameters have been met:

- Driving under the influence (DUI), Negligent and Reckless Driving, or Hit & Run Driving within the past five (5) years of taking the exam.
- Suspension of your driver's license as a result of a DUI within the past five (5) years of taking the exam.

Employment History

An applicant's employment history, including any terminations or leaving an employer in lieu of termination, will be thoroughly assessed and *may* be grounds for disqualification.

Financial History

An applicant's credit history will be thoroughly assessed and related decision-making issues *may* be grounds for disqualification. The following are areas of concern:

- Failure to pay income tax
- Failure to pay child support

Professional Appearance

All applicants are expected to maintain a professional appearance at all times. SPD has sole discretion in determining what is considered professional, as it relates to the position the applicant is applying for. Any and all tattoos, branding (intentional burning of skin to create a design), voluntary disfigurement (marring or spoiling of the appearance or shape of a body part), or scarification (intentional cutting of the skin to create a design) shall be carefully reviewed by SPD on a case-by-case basis.

Residence

Applicants are not required to live in the City of Seattle.

Drug Use

An applicant's drug use will be looked at on a case-by-case basis. In order to be considered the most competitive candidate and to increase the likelihood of continuing on in the process, the closer the applicant is to the timeline listed for the drug(s) in question the better.

- **MARIJUANA:** An applicant has **not** used Marijuana within twelve (12) months prior to the date they took/take the related Police Officer Civil Service Exam, **and**
- **COCAINE/CRACK:** An applicant has **not** used cocaine or crack within the ten (10) years prior to the date they took/take the related Police Officer Civil Service Exam, **and**

(*continued*)

TABLE 7.9 Selection Standards for Seattle Police Officers, 2019 (*Continued*)

QUALIFICATIONS
• **CLUB DRUGS:** An applicant has **not** used club drugs, such as, but not limited to: Ketamine, GHB, Rohypnol, or MDMA (ecstasy) within the five (5) years prior to the date they took/take the related Police Officer Civil Service Exam, **and**
• **HALLUCINOGENS:** An applicant has **not** used any Hallucinogens; PCP, Angel Dust, Wet, Phencyclidine, LSD, Mushrooms, or Psylocybin, within the ten (10) years prior to the date they took/take the related Police Officer Civil Service Exam, **and**
• **OPIATES:** An applicant has **not** used Opium, Morphine, or Heroin within the ten (10) years prior to the date they took/take the related Police Officer Civil Service Exam, **and**
• **STIMULANTS:** An applicant has **not** used Methamphetamine, Crank, Crystal, Ice, Speed, Glass, or Amphetamine within the ten (10) years prior to the date they took/take the related Police Officer Civil Service Exam, **and**
• **AEROSOLS:** An applicant has **not** inhaled aerosols, sometimes referred to as Huffing (paint) or Whippits (Nitrous Oxide) or used Khat within the five (5) years prior to the date they took/take the related Police Officer Civil Service Exam, **and**
• An applicant has **not** used four (4) or more controlled substances within the ten (10) years prior to the date you took/take the related Police Officer Civil Service Exam, **and**
• An applicant has **not** used any illegal drug(s) while employed in a criminal justice and/or law enforcement capacity, **and**
• An applicant has **not** manufactured or cultivated illegal drug(s) for the purpose of the sales/marketing of the drug(s), **and**
• An applicant has **not** sold or facilitated the sale of illegal drugs.
The Seattle Police Department is an equal opportunity employer that values diversity in its workforce. At SPD we acknowledge and honor the fundamental value and dignity of all individuals and pledge ourselves to creating and maintaining an environment that respects diverse traditions, heritages, and experiences.

Source: "Qualifications," Seattle Police Department Jobs, accessed June 27, 2019, https://www.seattle.gov/police/police-jobs/qualifications.

FIGURE 7.1 Police Officer Salary, by Percentile, 2020

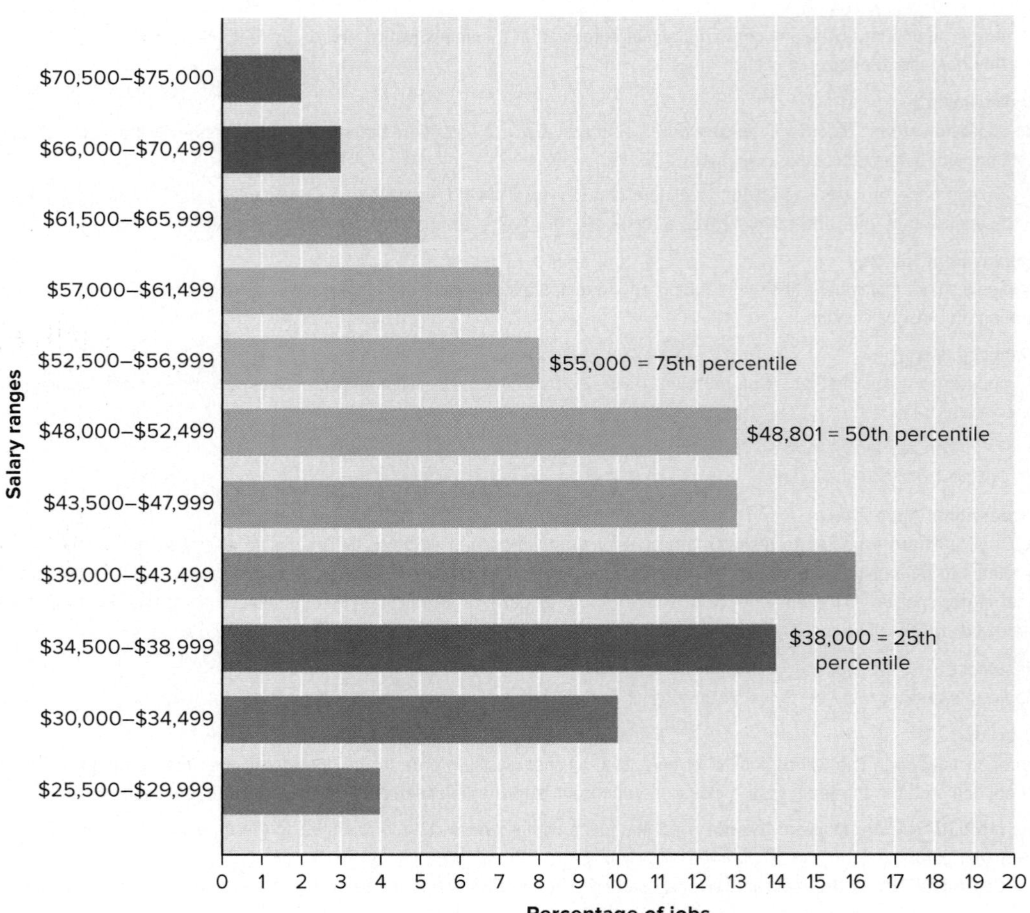

Source: Adapted from ZipRecruiter, "Police Officer Salary," accessed February 15, 2020, https://www.ziprecruiter.com/Salaries/Police-Officer-Salary.

TABLE 7.10 Police Officer Salaries, by State, 2020

State	Average Hourly Wage	Average Annual Salary
1. New York	$25.68	$53,405
2. Massachusetts	$25.46	$52,964
3. New Hampshire	$24.99	$51,975
4. Maryland	$23.78	$49,467
5. Alaska	$23.46	$48,801
6. Nevada	$23.46	$48,801
7. Montana	$23.46	$48,801
8. North Dakota	$23.46	$48,801
9. Wyoming	$23.46	$48,801
10. Idaho	$23.46	$48,801
11. Hawaii	$23.43	$48,472
12. Connecticut	$23.39	$48,657
13. Vermont	$23.38	$48,631
14. Washington	$23.34	$48,550
15. Nebraska	$23.29	$48,450
16. California	$23.07	$47,984
17. Virginia	$22.99	$47,809
18. Rhode Island	$22.81	$47,441
19. Arizona	$22.42	$46,638
20. New Jersey	$22.40	$46,600
21. West Virginia	$22.32	$46,435
22. Colorado	$22.30	$46,385
23. Pennsylvania	$22.20	$46,177
24. Minnesota	$22.19	$46,160
25. South Dakota	$22.17	$46,119
26. South Carolina	$22.14	$46,046
27. Oregon	$22.10	$45,973
28. Tennessee	$22.10	$45,964
29. Delaware	$22.05	$45,855
30. Utah	$21.94	$45,641
31. Kentucky	$21.85	$45,451
32. Ohio	$21.79	$45,327
33. Indiana	$21.68	$45,097
34. Oklahoma	$21.62	$44,977
35. Kansas	$21.56	$44,842
36. Wisconsin	$21.55	$44,834
37. Iowa	$21.49	$44,701
38. Louisiana	$21.48	$44,671
39. Maine	$21.41	$44,524
40. Texas	$21.15	$43,995
41. Georgia	$20.95	$43,572
42. Alabama	$20.94	$43,555

(*continued*)

TABLE 7.10 Police Officer Salaries, by State, 2020 (*Continued*)

State	Average Hourly Wage	Average Annual Salary
43. Arkansas	$20.88	$43,425
44. New Mexico	$20.79	$43,245
45. Illinois	$20.57	$42,790
46. Michigan	$20.56	$42,773
47. Mississippi	$20.45	$42,539
48. Missouri	$20.13	$41,873
49. Florida	$19.80	$41,190
50. North Carolina	$18.07	$37,578
National Average	$23.46	$48,801

Source: Adapted from ZipRecruiter, "What Is the Average Police Officer Salary by State," accessed February 15, 2020, https://www.ziprecruiter.com/Salaries/What-Is-the-Average-Police-Officer-Salary-by-State.

while North Carolina paid its police officers the least–an average annual salary of $37,578 or $18.07 an hour. Finally, Table 7.11 shows the 25 highest paying cities for police officer jobs in 2020. New York City paid its police officers the highest average annual salary in 2020 of $53,405 or $25.68 an hour. Note that three of the top five paying cities in 2020 were in New Jersey: Bridgewater, Newark, and Jersey City.

Affirmative Action Since the passage of the Civil Rights Act of 1964 and the threat of court challenges to the fairness of the police selection process, police agencies have struggled to find the best-qualified applicants and yet achieve satisfactory race, ethnicity, and gender representation within the ranks of the department. Failure to seriously pursue equitable representation has led to expensive lawsuits, consent decrees, and court-ordered quotas to achieve the desired diversity. Consequently, affirmative action has become a major concern in police selection and employment.

One of the major obstacles to the recruitment, hiring, retention, and promotion of police officers from underrepresented groups is legal challenges to affirmative action programs. Ironically, most of these challenges are not filed by members of underrepresented groups against police departments, but rather are "reverse discrimination" suits filed by white men against police departments (for example, *Sullivan v. City of Springfield, MA*, 2009; *Jones v.*

In recruiting new police officers, efforts are made to cast police work as an attractive and sustaining career. *What should people interested in becoming police officers be told about the job?*

Karsten Bidstrup/Getty Images

TABLE 7.11 25 Highest Paying Cities for Police Officer Jobs, 2020

City	Average Hourly Wage	Average Annual Salary
1. New York City, NY	$25.68	$53,405
2. Seattle, WA	$25.48	$53,001
3. Bridgewater, NJ	$24.71	$51,387
4. Newark, NJ	$24.33	$50,612
5. Jersey City, NJ	$24.32	$50,581
6. Yonkers, NY	$24.25	$50,443
7. Anchorage, AK	$24.16	$50,249
8. Minneapolis, MN	$24.03	$49,980
9. Chicago, IL	$24.02	$49,952
10. San Diego, CA	$23.98	$49,883
11. Long Beach, NY	$23.95	$49,824
12. Mount Vernon, NY	$23.68	$49,259
13. Gary, IN	$23.56	$49,012
14. Atlanta, GA	$23.55	$48,977
15. Dallas, TX	$23.52	$48,918
16. New Windsor, NY	$23.50	$48,877
17. Providence, RI	$23.44	$48,752
18. Philadelphia, PA	$23.41	$48,687
19. Chester, PA	$23.31	$48,486
20. Detroit, MI	$23.29	$48,443
21. Houston, TX	$23.28	$48,418
22. Milwaukee, WI	$23.23	$48,316
23. Gardena, CA	$23.04	$47,932
24. Bakersfield, CA	$22.95	$47,734
25. Ontario, CA	$22.90	$47,626

Source: ZipRecruiter, "Highest Paying Cities for Police Officer Jobs," accessed February 15, 2020, https://www.ziprecruiter.com/Salaries/Highest-Paying-Police-Officer-Salary-by-City.

City of Springfield, MA, 2009; *Klawitter v. City of Trenton, NJ*, 2007; *Brackett v. Civil Service Commission*–Suffolk County, MA, 2006).[26] Other problems that hamper recruitment from underrepresented groups are: (1) strained relations, lack of trust, and resentment of the police on the part of potential applicants, (2) the reputation, operational practices, or organizational culture of police agencies, and (3) lack of awareness of career opportunities with the police.[27] Additional problems with the hiring process include: (1) reliance on inadequately constructed examinations as part of the screening process that have the unintended consequence of excluding qualified individuals from underrepresented groups, (2) reliance on certain selection criteria and screening processes (for example, criminal background checks) that disproportionately impact individuals from underrepresented groups, (3) residency and citizenship requirements, (4) the length, complexity, and the cost of the application processes (which may include required civil service examinations and completing police academy training), and (5) the limited ability of police agencies to modify or adjust hiring and selection criteria to attract members of underrepresented groups because of fear of litigation, for example.[28] Finally, problems with the retention of officers from underrepresented groups are: (1) difficulties adjusting to a police agency's organizational culture and (2) difficulties with the promotion process due to a lack of transparency about the process, as well as a lack of mentoring relationships and professional development opportunities.[29]

How successful has affirmative action been in accomplishing the desired goal of race, ethnicity, and gender balance in police departments? Affirmative action *has* been relatively successful in increasing the percentage of minority members in policing. It has been less

Increasing the number of female officers is a major concern in police selection and employment. *What are some of the advantages of having a greater number of female police officers?*

The Star-Ledger/Aristide Economopoulos/The Image Works

successful in increasing the percentage of women. As noted previously, about 71.5% of the sworn officers in the nation's police departments in 2016 were white and 87.7% were males. However, as shown in Table 7.12, the percentage of black and Hispanic officers in local departments in 2016 was 11.4% and 12.5%, respectively–down from 12.2% and up from 11.6% in 2013, and down from 11.9% and up from 10.3% in 2007. The 2016 figures slightly underrepresent both the percentage of blacks and the percentage of Hispanics in the general population of the United States (13.0% for blacks and 17.8% for Hispanics). In addition, Table 7.12 shows the smaller the department, the larger the percentage of white officers it is likely to have. Women, who compose more than 50% of the United States population, represented only 12.3% of the police officers in local police departments in 2016; still, that was a slight increase from 12.2% in 2013, and 11.9% in 2007.[30]

TABLE 7.12 Race and Hispanic Origin of Full-Time Sworn Personnel in Local Police Departments, by Size of Population Served, 2016

Population Served	Total	White*	Black*	Hispanic	Other**	Unknown
All sizes	100%	71.5%	11.4%	12.5%	3.6%	1.1%
1,000,000 or more	100%	50.4	16.6	27.0	5.6	0.4
500,000–999,999	100%	59.7	21.5	10.9	6.8	1.1
250,000–499,999	100%	68.0	14.8	12.6	4.0	0.6
100,000–249,999	100%	72.9	11.7	10.0	3.4	2.0
50,000–99,999	100%	74.7	7.6	12.9	2.6	2.2
25,000–49,999	100%	86.9	6.0	5.2	1.3	0.6
10,000–24,999	100%	85.2	6.0	5.9	1.8	1.1
9,999 or less	100%	87.1	4.7	5.5	1.8	0.9

Note: Detail may not sum to total because of rounding.

*Excludes persons of Hispanic or Latino origin.

**Includes Asians, Native Hawaiians, Other Pacific Islanders, American Indians, Alaska Natives, or persons of two or more races.

Source: Shelley S. Hyland and Elizabeth Davis, "Local Police Departments, 2016: Personnel," U.S. Department of Justice, *Bureau of Justice Statistics Bulletin*, October, 2019, accessed October 27, 2019, https://www.bjs.gov/content/pub/pdf/lpd16p.pdf.

Affirmative action has been less successful in local police departments at the managerial and supervisory ranks. In 2016, 89.6% of police chiefs were white; only 4.0% were black and 3.1% were Hispanic.[31] The size of the department, however, made a difference. In the largest departments serving 250,000 or more residents, black chiefs were overrepresented, comprising 19.1% of the total. In those departments, 64.9% of the chiefs were white and 12.8% were Hispanic. In departments serving 100,000 or more residents, which employed 52% of all full-time sworn officers, 71% of chiefs were white, 15% were black, and 10% were Hispanic. As departments got smaller in residents served, the percentage of white chiefs increased. In departments serving less than 100,000 residents, which employed 48% of full-time sworn officers, 90% of chiefs were white, 4% were black, and 3% were Hispanic.

A similar pattern emerges in the case of intermediate supervisors (those below chief and above sergeant). Overall, 81.5% of intermediate supervisors in 2016 were white, 9.2% were black, and 6.4% were Hispanic.[32] However, in departments that served 250,000 or more residents, only 69.3% of intermediate supervisors were white, 14.3% were black, and 11.6% were Hispanic. In departments that served 9,999 or fewer residents, by comparison, 87.4% were white, 6.0% were black, and 3.9% were Hispanic.

In the case of first-line supervisors (sergeants), 77.0% were white, 9.4% were black, and 9.6% were Hispanic in 2016.[33] In departments that served 250,000 or more residents, 63.0% of sergeants were white, 15.3% were black, and 15.4% were Hispanic. In departments that served 9,999 or fewer residents, 85.9% of sergeants were white, 5.8% were black, and 5.3% were Hispanic.

Affirmative action has been even less successful for females at the managerial and supervisory ranks in local police departments. Overall, in 2016, only 2.9% of chiefs, 7.5% of intermediate supervisors, and 9.7% of sergeants were female.[34] Again the size of the department mattered. In departments serving 250,000 or more residents, 8.5% of chiefs, 12.7% of intermediate supervisors, and 14.3% of sergeants were female. In the smallest departments serving 9,999 or fewer residents, 2.6% of chiefs, 5.0% of intermediate supervisors, and 6.1% of sergeants were female. In sum, in the more than half century since the passage of the Civil Rights Act of 1964, some progress has been made in the effort to make local police officers look like the residents of the communities that they serve. However, as the aforementioned data indicate, much work remains to be done, especially at the higher ranks.

Women and Policing

The National Center for Women & Policing is a nationwide resource that seeks to increase the number of women in policing. You can visit its website at http://womenandpolicing.com.

What are some of the best ways to increase the number of women in policing?

Education Given the amount of discretion that law enforcement officers have and the kinds of sociological problems they deal with, selecting reasonably intelligent, educated officers seems a wise practice. According to a recent survey, a high school diploma or higher educational achievement was required of entry-level officers by 99.9% of local police departments in the United States in 2016.[35] However, 81.5% of departments required only a high school diploma. Some college courses were required of entry-level officers in 6.6% of local police departments in 2016, and 10.5% of the agencies required entry-level officers to have a minimum of 2 years of college. Just 1.3% of agencies required entry-level officers to have a 4-year college degree. Agency minimum education requirements primarily are dictated by state standards. Indeed, only 13% of agencies in 2016 chose to deviate from those standards and require, per department policy, more education than was required by state law. Agencies with collective bargaining agreements were more likely to require higher education requirements than state law: 18.5% of "union" agencies required more than the state's minimum education level compared to 7.2% of "non-union" agencies. Agencies in Colorado, Florida, and Illinois were more likely to require more than the state's minimum education requirements than agencies in other states. Some agencies provided higher pay for officers with more education. About one-quarter (22.8%) of agencies had a written policy that permitted the agency to hire exceptional candidates who lacked the minimum education required. Table 7.13 shows the minimum educational requirements of local police agencies in the United States in 2016.

According to the aforementioned survey, in 2016, 48.2% of all sworn police officers in the United States had only a high school diploma, 21.6% had a 2-year degree, 24.8% had a 4-year degree, 5.1% had a Master's degree, and 0.3% had a Doctorate. For chiefs and sheriffs, in 2016, 17.1% had only a high school diploma, 19.0% had a 2-year degree, 28.7% had a 4-year degree, 32.1% had a Master's degree, and 3.0% had a Doctorate.[36]

TABLE 7.13 Minimum Education Requirement of Local Police Agencies in the United States, by Officer Designation, 2016

Officer Designation	High School Diploma	Some College	2-Year Degree	4-Year Degree	Master's Degree
Entry-level officer	81.5%	6.6%	10.5%	1.3%	–
Lateral officer	81.7	6.7	10.1	1.5	–
Detective	81.4	7.0	10.1	1.4	–
Sergeant	68.1	11.4	17.2	3.2	–
Lieutenant	62.0	9.1	15.0	13.5	0.4
Command staff	55.4	7.3	13.6	22.9	0.7
Chief/Sheriff	44.9	5.8	8.2	35.9	5.2

Source: Adapted from Christie Gardiner, "Policing Around the Nation: Education, Philosophy, and Practice," California State University Fullerton Center for Public Policy and the Police Foundation, 16, Table 2, September 2017, accessed February 16, 2020, https://www.policefoundation.org/wp-content/uploads/2017/10/PF-Report-Policing-Around-the-Nation_10-2017_Final.pdf.

Among the perceived advantages of hiring college-educated officers are the following:

1. Are better report writers
2. Are better able to use modern technology efficiently
3. Are less resistant to organizational change and more open to new policing methods
4. Are better problem solvers
5. Are better able to solve complex crimes
6. Are better able to deal effectively with diverse community groups
7. Are more sensitive to cultural differences and community needs
8. Are better able to identify crime problems/trends[37]

This list of perceived advantages should impress police administrators and the public. These perceived advantages are good reasons for law enforcement executives to search for police applicants with college backgrounds.

Recognition that college-educated police officers generally are better performers than officers without that level of education is long overdue. And the idea is catching on. Minnesota's Board of Peace Officer Standards and Training (POST) now requires a 2- or 4-year college degree for licensing. The Ohio Peace Officer Training Commission (OPOTC) now has more than five dozen **college academies**, where students pursue a program that integrates an associate's degree curriculum in law enforcement or criminal justice with the state's required peace officer training. On receipt of the associate's degree, students sit for the peace officer certification exam. If they receive a passing score on the exam, they are eligible to be hired by any police agency and to go to work without any additional academy training.

college academies Schools where students pursue a program that integrates an associate's degree curriculum in law enforcement or criminal justice with the state's required peace officer training.

Successful Recruiting Practices

Where do you find the best-qualified police applicants? Some of the more successful recruiting practices have included going to colleges, neighborhood centers, and schools in minority communities; using television, radio, newspaper advertisements, and social media; and working with local employment offices. Demystifying the nature of police work and the selection process and shortening the time from application to final selection also have helped to attract and retain qualified candidates.

public safety officers Police department employees who perform many police services but do not have arrest powers.

Public Safety Officers Another promising recruitment strategy has been the employment of 18-year-olds as **public safety officers** (sometimes called community service officers or public service aides), who perform many police service functions but do not have arrest powers. By the time they are 21, the department has had an excellent opportunity to assess their qualifications and potential to be sworn officers.

police cadet program A program that provides persons aged 18 to 21 a chance to experience the challenges and rewards of a police career. Often cadets are paid and work part or full time. Cadet programs are designed to assist cadets in transitioning into the position of full-time police officer.

Police Cadets **Police cadet programs** provide persons aged 18 to 21 a chance to experience the challenges and rewards of a police career. Often cadets are paid and work part or full time. Cadet programs are designed to assist cadets in transitioning into the position of full-time police officer.

ZUMA Press Inc/Alamy Stock Photo

Pictured here are police cadets experiencing tear gas. *What are some of the advantages of police cadet programs?*

High School Tech Prep Programs Another program that is proving useful in attracting potentially capable police officers at an even earlier age is known as **tech prep (technical preparation)** for a criminal justice career. Area community colleges and high schools team up to offer 6 to 9 hours of college law enforcement courses in the 11th and 12th grades, as well as one or two training certifications, such as police dispatcher or local corrections officer. Students who graduate are eligible for employment at age 18. They become interested in law enforcement work early and are ideal police applicants when they become old enough to apply. Accurate law enforcement career information can be passed on to high school students through a tech prep program because the teachers are required to either currently work in law enforcement or have police experience in their backgrounds.

tech prep (technical preparation) A program in which area community colleges and high schools team up to offer 6 to 9 hours of college law enforcement courses in the 11th and 12th grades, as well as one or two training certifications, such as police dispatcher or local corrections officer. Students who graduate are eligible for police employment at age 18.

The Police Selection Process

In many communities, selection of police officers takes place through a merit system. A **merit system** of employment is established when an independent civil service commission, in cooperation with the city personnel section and the police department, sets employment qualifications, performance standards, and discipline procedures. Officers employed under such a system are hired and tenured, in theory, only if they meet and maintain the employment qualifications and performance standards set by the civil service commission. Officers in such a system cannot be fired without cause.

merit system A system of employment whereby an independent civil service commission, in cooperation with the city personnel section and the police department, sets employment qualifications, performance standards, and discipline procedures.

To find the best possible recruits to fill department vacancies, police agencies use a selection process that includes some or all of the following steps.

Short Application This brief form registers the interest of the applicant and allows the agency to screen for such things as minimum age, level of education, residency, and other easily discernible qualifications.

Detailed Application This document is a major source of information for the department and background investigators. The applicant is asked for complete education and work histories, military status, medical profile, references, a record of residence over many years, and other detailed information. Applicants also are asked to submit copies of credentials, military papers, and other certificates.

Medical Examination This exam determines if applicants are free of disease, abnormalities, and any other medical problems that would disqualify them for police work. This information is critical because retiring a young officer on a medical disability shortly after employment could cost the public hundreds of thousands of dollars.

Physical ability tests are common in police selection. *Why are they important?*

Enigma/Alamy Stock Photo

Physical Ability Test Physical ability tests are common in police selection despite having been challenged in the courts as having an adverse effect on the hiring of female applicants. Physical ability tests were initially a direct response to the elimination of height and weight standards, which also were discriminatory against female applicants. The first tests required exceptional speed and strength, such as going over walls that were taller than any of the walls in the cities that had such tests. Those tests were struck down by the courts as not being job related. Today, any physical ability tests must be based on a thorough analysis of the actual work of police officers.

Written Examination Police agencies once used intelligence tests in their selection process. Most agencies now use some type of aptitude, personality, general knowledge, reading comprehension, writing, or police skill exam. The courts have held that those tests must be true measures of the knowledge and abilities needed to perform police work successfully. Pre-employment tests have been the subject of much controversy in the courts.

Background Investigation Investigators in this process look for any factors in the applicants' backgrounds that would prevent them from performing successfully as police officers. Past drug use or excessive alcohol use, a poor driving record, employer problems, a bad credit history, criminal activity, and social immaturity are areas of concern in the background investigation. The investigator relies heavily on the detailed application, verifies its contents, explores any discrepancies, and develops additional leads to follow.

Recently, some police agencies have added the examination of social networking sites, such as Facebook, Twitter, and YouTube, as part of their background checks. According to the 2016 Law Enforcement Use of Social Media Survey, 58% of law enforcement agencies use social media for recruitment and applicant vetting.[38] Investigators look for information that might disqualify the applicant. Agencies are concerned mainly that testimony in a criminal or civil case could be impeached using information from an officer's personal social media page. Some law enforcement agencies require social-media passwords from job applicants; other agencies ask prospective employees to log on to their social-media accounts in front of the investigator to review the material; and still other agencies ask job applicants to provide their passwords for the background check.[39]

Psychological Testing Emotional stability and good mental health are critical to the ability to perform police work, which can be very stressful. Departments have been held liable for not screening their applicants for those psychological traits.[40]

Careers in Criminal Justice

Police Officer

My name is Robert Bour. I really enjoy being a police officer in Tiffin, Ohio, a small city of about 20,000 people. My assignment is in the patrol section, but I also work as a bike patrol officer, a field training officer, and a member of the SWAT team. As a SWAT officer, I go out mainly on search warrant and drug raids and in cases where someone has barricaded him- or herself in, and a life is in danger. In a small police department, you have the opportunity to be involved in a number of specialized assignments.

As a patrol officer, I have the opportunity to interact with the public a great deal. We have always done community policing here in one form or another. Small communities have to. You stop people on the street and say hello, chat with business owners, and meet as many residents as you can. People always have questions for you; often, they are about traffic laws and enforcement.

I also have to respond to calls for services, and these calls take precedence over everything else. I get one to three calls per hour on the evening and early morning shift, but there are rare times when I don't get any calls during the entire 10 hours of work.

My third area of activity involves self-initiated patrol activity. That includes traffic enforcement, of course, but also patrolling the alleys and streets around bars when they are closing in order to protect inebriates who may have passed out outside the establishment. The patrol also includes guarding against people who may be driving under the influence of alcohol. Tiffin has two universities, so sometimes college students may get a little more disorderly than they should and we have to respond.

I graduated in 1992 from a police academy that was hosted by a local community college. Before becoming a police officer in Tiffin, I worked as a sheriff's deputy and a village police officer, both full and part time. The Tiffin Police Department now requires a 2-year college degree as the minimum to become employed. At the time I came on, it was not necessary, but I intend to go back to college and finish my degree anyway.

I can offer a few suggestions for anyone who wants to be a police officer. First, you should have a desire to work with people from diverse backgrounds because that is what police work is. Second, you must genuinely want to help people, keeping in mind that you are not always able to help everybody; sometimes that is frustrating. Finally, you must think about your family. They must support you in being a cop because there is some danger involved in the job. Still, police work is one of the best jobs a person can have.

Courtesy of Robert Bour

What are the pros and cons of being a police officer?

Systematic psychological testing of police officers began in the 1950s. At first, the typical approach was to have the psychological evaluators look for disqualifying factors. The process included a pencil-and-paper test and a one-on-one interview with a psychologist. Today, the testing focus generally has shifted to a search for the positive psychological qualities required in police work. Current tests include multiple versions of both written and clinical evaluations.

The validity of psychological tests has been an issue for decades. Psychologists often are reluctant to rate with any specificity the police candidates they evaluate. Candidates considered "unacceptable" sometimes are classified as "uncertain" to avoid lawsuits. It is important to remember that understanding and predicting human behavior is an inexact art. So, it is easy to appreciate the reluctance of psychologists to be more specific.

Oral Interview/Oral Board This step is frequently the final one in the selection process. Members of the interview team have the results of the previous selection procedures, and they now have an opportunity to clear up inconsistencies and uncertainties that have been identified. The board normally restricts itself to evaluating the following qualities:

1. Appearance, poise, and bearing.
2. Ability to communicate orally and organize thoughts.
3. Attitude toward law enforcement and the job required of police officers.
4. Speech and the ability to articulate.
5. Attitude toward drug, narcotic, and alcohol use.
6. Sensitivity to racial and ethnic issues.[41]

Syracuse Newspapers/D. Blume/The Image Works

At the end of many police academy training programs, there is a state licensing or certification examination. *Do you believe those examinations are necessary? Why or why not?*

Academy Training The police academy is part of the selection process. Virtually every academy class in any sophisticated police department loses up to 10% of its students. Thus, to survive academy training, students must be committed to the process. Students undergo from nearly 700 to nearly 1,800 hours of academic, skill, and physical training and are tested virtually every week of the process. At the end of many academy training programs, students must take a state licensing or certification examination.

Probation Under local or state civil service requirements, employers may keep a new police officer on probation for 6 months to a year. The probation period gives the new police officer a chance to learn policing under the guidance of a well-qualified field training officer. Formal field training is a wise investment, and it ensures that new officers get as much knowledge and experience as possible before an agency commits to them for their careers.

A 6-month probation period no longer seems logical, however, because police academy training now often extends 5 or 6 months. The agency is, in effect, offering the police officer tenure in a matter of weeks after graduation from the academy. This practice defeats the purpose of probation, which was designed to allow the employer to see whether the newly trained officer can successfully perform the job.

The Selection of a Law Enforcement Executive

No less important than the selection of operations-level officers is the choice of the chief executive of a police agency. This executive might be a chief of police, a sheriff, or the head of a state law enforcement organization. A crucial decision in the selection process is whether to allow people from outside the agency to apply. In some police agencies in the United States, civil service regulations prohibit the selection of outside candidates. The rationale for this rule is that there must be qualified internal candidates. In addition, it is discouraging to hard-working and talented police administrators to be denied a chance to lead the agency they have spent many years serving.

Actual hiring decisions usually are shared by members of a selection committee. Frequently, an executive search firm also is employed. The selection committee typically consists of representatives of the local government, the police department, the search firm, and the community. Applicants are put through a rigorous process that includes several visits to the city, written exams, interviews, and assessment center testing in which candidates try to resolve real-world management problems. Once the interviews and testing have been completed, applicants are normally ranked, and the list is presented to the city manager, the mayor, or others so that a final selection can be made.

The pursuit of a police chief's job is very competitive. Often, several hundred candidates contend for a position even in a small suburban community. A typical police chief rarely serves longer than 10 years, and life in the chief's seat may not be very comfortable, particularly if a new chief intends to change things. Much of the political controversy and many of the social problems in major cities and counties end up at the door of the police department, so police chiefs must be politically savvy to survive. Many chiefs discover that they cannot please everybody, particularly if they are trying to change the department. Should police chiefs have protection under civil service? Most commentators say no, arguing that mayors and city managers ought to have the authority to pick the management teams that work immediately under their direction. A small number of cities give their police chiefs civil service protection to insulate them from unnecessary political interference.

The selection of a sheriff of one of the nation's counties is just as important as choosing a police chief. The difference in the two processes has to do with who does the selecting. In all but a few of the nation's counties, sheriffs are elected by the county's eligible voters. Not all sheriffs have a law enforcement background. In most states, they are not required to be licensed or certified peace officers because their deputies or municipal officers are. As a result, people from various occupations often succeed in being

FYI **Hiring Bad Police Chiefs and Sheriffs**

A recent *USA TODAY* investigative report identified 32 people who became police chiefs or sheriffs from 2008 through 2018, despite a finding of serious misconduct, usually at another department. At least eight of them were found guilty of a crime. Others amassed records of domestic violence, improperly withholding evidence, falsifying records or other conduct that could impact the public.

Source: James Pilcher, Aaron Hegarty, Eric Litke, and Mark Nichols, "Fired for a felony, again for perjury. Meet the new police chief." *USA TODAY*, October 14, 2019, accessed October 21, 2019, https://www.usatoday.com/in-depth/news/investigations/2019/04/24/police-officers-police-chiefs-sheriffs-misconduct-criminal-records-database/2214279002/.

elected sheriff. To be elected, sheriffs must be good politicians. They often have a much better idea of a community's priorities and wield more influence with prosecutors and in the legislature than chiefs of police. Sheriffs who do not exhibit this political acumen are not likely to be reelected.

THINKING CRITICALLY

1. What do you think are the most important qualities for police officers to have? Why?
2. How much formal education do you think police officers should have? Why?
3. Do you think law enforcement agencies should have access to a job applicant's social-media accounts? Why or why not?

Issues in Policing

The discussion of law enforcement thus far has made it clear that not all matters of policing in America are settled. This final section of the chapter highlights some of the issues that continue to be major topics of debate in law enforcement and have significant impact on the quality of life in neighborhoods and communities across the nation.

Discretion

Discretion is the exercise of individual judgment, instead of formal rules, in making decisions. No list of policies and procedures could possibly guide police officers in all of the situations in which they find themselves. Even the police officer writing a ticket for a parking meter violation exercises a considerable amount of discretion in deciding precisely what to do.[42] Police even have the discretion to ignore violations of the law when they deem it appropriate in the context of other priorities.

discretion The exercise of individual judgment, instead of formal rules, in making decisions.

The issue of police discretion is very controversial. Some believe that the discretion of police officers should be reduced. The movement to limit the discretion of police officers is the result of abuses of that discretion, such as physical abuse of citizens or unequal application of the law in making arrests. Other people argue that we should acknowledge that officers operate with great discretion and not attempt to limit it. Advocates of this view believe that better education and training would help officers exercise their judgment more wisely.

Patrol Officer Discretion Patrol officers frequently find it necessary to exercise their discretion. Within the geographical limits of their beats, they have the discretion to decide precisely where they will patrol when they are not answering radio calls. They decide whom to stop and question. For example, they may tell some children playing ball in the street to move, while they ignore others. Patrol officers decide for themselves which traffic violators are worth chasing through busy traffic and which ones are not. They even have the right not to arrest for a minor violation when, for example, they are on the way to investigate a more serious matter.

Some of the more critical situations involve decisions about stopping, searching, and arresting criminal suspects. Many citizens have been inconvenienced and some have been abused because of a police officer's poor use of discretion in those areas.

Police officers cannot make an arrest for every violation of law that comes to their attention—that is, they cannot provide **full enforcement**. The police do not have the resources to enforce the law fully, nor can they be everywhere at once. And even if full enforcement was possible, it may not be desirable. For example, persons intoxicated in front of their own homes may not need to be arrested but only to be told to go inside. Motorists slightly exceeding the speed limit need not be arrested if they are moving with the flow of traffic. Prostitution may be widely practiced in large metropolitan areas, but police officers have little to gain by searching hotels and motels to stamp it out, particularly when judges will turn the prostitutes right back out on the street. Generally, only when such an activity becomes a clear nuisance, is the subject of a public outcry, or threatens health and safety do the police department and its officers choose to take formal action.

full enforcement A practice in which the police make an arrest for every violation of law that comes to their attention.

Police officers can exercise discretion in arrest situations. *Should police officers be allowed to use discretion in arrest situations? Why or why not?*

Spencer Grant/PhotoEdit

selective enforcement The practice of relying on the judgment of the police leadership and rank-and-file officers to decide which laws to enforce.

The practice of relying on the judgment of the police leadership and rank-and-file officers to decide which laws to enforce is referred to as **selective enforcement**. The practice allows street police officers to decide important matters about peacekeeping and enforcement of the law. For most violations of the law, but not all felonies, a police officer usually can exercise a number of options:

1. Taking no action at all if the officer deems that appropriate for the situation.
2. Giving a verbal warning to stop the illegal action.
3. Issuing a written warning for the violation.
4. Issuing a citation to the perpetrator to appear in court.
5. Making a physical arrest in serious matters or in situations with repeat offenders.

Factors Affecting Discretion Dozens of studies have been conducted on the exercise of discretion by police patrol officers. A number of significant factors affect discretion:

- *The nature of the crime*–The more serious the crime, the more likely it is that police officers will formally report it. In cases involving lesser felonies, misdemeanors, and petty offenses, police officers are more likely to handle the offenses informally. A minor squabble between over-the-fence neighbors is an example of a matter that would probably be handled informally.
- *Departmental policies*–If the leadership of a police department gives an order or issues a policy demanding that particular incidents be handled in a prescribed way, then an officer is not supposed to exercise discretion but is to do as the order or policy directs. Thus, if a city has had many complaints about dangerous jaywalking in a certain downtown area, the chief of police may insist that citations be issued to those found jaywalking, even though citations were not issued in the past.
- *The relationship between the victim and the offender*–Particularly for minor offenses, the closer the relationship between the victim of an alleged offense and the suspected perpetrator, the more discretion the officer is able to exercise. For example, police officers are not likely to deal formally with a petty theft between two lovers if they believe that the victim will not prosecute his or her partner.
- *The amount of evidence*–If officers do not have enough evidence to substantiate an arrest or to gain a conviction in court, they are likely to handle the case in some way other than making an arrest.
- *The preference of the victim*–Sometimes the victim of a crime may simply want to talk the matter over with someone, and the police are available on a 24-hour basis. Also, if the officer senses that the victim of a minor assault does not wish to prosecute the perpetrator of the offense, the patrol officer will not make a formal complaint, and the complainant will most likely never know that a report was not made.

- *The demeanor of the suspect*–Suspects who are disrespectful and uncooperative may very well feel the full brunt of the law. Patrol officers often choose the most severe option possible in dealing with such suspects.
- *The legitimacy of the victim*–Patrol officers are bound to pass some kind of judgment on the legitimacy of the victim. An assault victim who is belligerent and intoxicated, for instance, will not be viewed favorably by the investigating officer. Criminals victimized by other criminals also are seen as less than fully authentic victims, no matter what the offense.
- *Socioeconomic status*–The more affluent the complainant, the more likely a patrol officer is to use formal procedures to report and investigate a crime.

Contrary to popular belief, the personal characteristics of an officer (such as race, gender, and education) do not seem to influence the exercise of discretion.

Discretion and Domestic Violence Police officers have intervened in domestic violence cases and other kinds of family disputes, sometimes off duty, since the inception of public policing. For the longest period of time, these interventions were viewed as peacekeeping activities when, in fact, they should have been treated as criminal matters. Many women and some men were hurt, and some killed, as a result of the restrictions on the police in making arrests for assault misdemeanors that were not made in their presence, as well as a view among the police that these calls were the private business of the family instead of real police matters. Traditionally, law enforcement has been less interested in arresting perpetrators of crimes when the victim and the perpetrator have a close relationship.

Approximately one million women are victims of domestic violence each year. Even today, with every state requiring the police to have domestic violence intervention training, some commentators believe that the police are not the best qualified of available community helpers to intervene. However, if crimes are committed in the form of physical abuse, the police are not only the best qualified to intervene but also are required by law to do so. Certainly, the availability of 24-hour service always has made the police the major responder to domestic violence calls.

The police in general do not relish the task of responding to domestic violence calls for several reasons. First, the calls can be dangerous, although generally no more dangerous than other disturbance calls. Nevertheless, officers are hurt each year by responding to domestic violence complaints. Second, police officers know from experience that many of the tense and hostile dynamics that exist between quarreling spouses, couples, and other family members have a way of dissipating over time or at least subsiding for a while. Third, the police know that they often have conducted investigations, even arrested the suspected batterer, and the victim has later chosen to drop charges. Finally, the police know that responding to the minor assault cases in domestic violence calls is not always the best thing for the family because the arrest creates its own complications that may, in fact, exacerbate the family crisis to a state of irreparable harm.

Police have responded to domestic violence in three distinct ways: mediate the dispute, separate domestic partners in minor disputes, and arrest the perpetrator of the assault. Which of these ways is the most effective? This question was put to the test in a 3-year study in the city of Minneapolis. In minor domestic dispute cases, Minneapolis police officers gave up their discretion in handling domestic violence calls. Instead of deciding for themselves the appropriate disposition for each call, they randomly chose arrest, separation, or mediation. The results of the study showed that the arrested perpetrators were about half as likely to repeat their violence against the original victim.[43] This study may have been the impetus for many states to implement a mandatory arrest domestic violence law. Subsequent studies, however, have not been able to clearly support the mandatory arrest disposition as the most effective way to handle the problem of domestic violence. Yet today nearly half of the states have a mandatory arrest law requiring the arrest of any suspect that has battered a spouse or domestic partner.[44] While victim safety and welfare are indeed the major reasons for police intervention, more research clearly is needed to determine the best approaches to handling domestic violence calls.

SuperStock/Alamy Stock Photo

Many states have mandatory arrest laws that require police officers to arrest any suspect that has battered a spouse or domestic partner. *Do you support mandatory arrest laws in domestic violence cases? Why or why not?*

racial profiling The stopping and/or detaining of individuals by law enforcement officers based solely on race.

Discretion and Racial Profiling Racial profiling is of growing concern to law enforcement officials and the public. Just how frequently this illegal practice occurs is difficult to discern, particularly because the term *racial profiling* seldom is defined in the discussions found in the national media. **Racial profiling** is a law enforcement infringement on a citizen's liberty based solely on race. It widely is believed that on freeways, highways, and streets throughout the nation, blacks and other minorities are stopped for traffic violations and field interrogations in numbers disproportionate to their representation in the population. It further is assumed that many of these stops are pretext stops in which the stop is justified by a minor equipment or moving traffic violation that might otherwise be ignored. Where the practice is considered widely experienced, it has been called "driving while black or brown" (DWBB). It is presumed that racial stereotyping and prejudice are at the root of such a practice.

Results of a recent federal study of police and public contacts show that in 2015, 8.6% of 223.3 million U.S. drivers age 16 or older were stopped by the police (see Table 7.14).[45] The police stopped a slightly higher percentage of black drivers (9.8% of all black drivers) than white drivers (8.6% of all white drivers) or Hispanic drivers (7.6% of all Hispanic drivers). The percentage of all drivers stopped by the police decreased from 12.3% in 2011 to 8.6% in 2015, as did the percentage of black, white, and Hispanic drivers stopped by the police. Still, the percentage of blacks stopped by the police in both 2015 and 2011 was higher than the percentage of whites or Hispanics stopped by the police. In 2011, 12.8% of all black drivers, 10.4% of all Hispanic drivers, and 9.8% of all white drivers were stopped by the police.[46] The 2015 survey also found that black drivers (18%) were more likely to be stopped by the police multiple times than were Hispanic drivers (13.6%) or white drivers (11.8%) (see Table 7.14).[47]

Whether drivers perceived a traffic stop as legitimate in the 2015 survey depended on whether police officers gave a reason for the stop. In more than 95% of all traffic stops, police officers gave a reason for the stop, and nearly 84% of drivers judged the stop to be legitimate. In the approximately 2% of stops during which police officers did not give a

TABLE 7.14 Police Traffic Stops and Race/Ethnicity of Driver, 2015

	Percentage			
Situation	**All**	**White**	**Black**	**Hispanic**
Traffic stop	8.6	8.6	9.8	7.6
Stopped multiple times	12.8	11.8	18.0	13.6
Police gave reason for stop	95.4	96.1	94.5	92.3
Police did not give reason for stop	2.1	1.8	3.1	3.4
Driver judged stop was legitimate				
Police gave reason for stop	83.7	86.2	72.7	80.4
Police did not give reason for stop	36.7	46.3	25.7	19.0
Police took no enforcement action	12.7	13.5	14.6	7.7
Police took enforcement action				
Warning (written or verbal)	36.1	38.0	33.6	32.9
Ticket	48.8	46.4	49.9	56.4
Search or arrest (person or vehicle)	3.7	3.3	4.4	5.0
Driver thought police behaved properly				
No enforcement action	92.2	92.7	88.3	90.4
Warning (written or verbal)	93.9	94.8	92.2	88.8
Ticket	89.8	90.8	85.1	88.2
Search or arrest (person or vehicle)	67.0	70.5	68.5	62.8

Source: Adapted from Elizabeth Davis, Anthony Whyde, and Lynn Langton, "Contacts Between Police and the Public," *Bureau of Justice Statistics Special Report*, U.S. Department of Justice (Washington, DC: USGPO, October 2018), accessed July 1, 2019, https://www.bjs.gov/content/pub/pdf/cpp15.pdf.

reason, only about 37% of drivers thought the stop was legitimate (see Table 7.14). The primary reason police officers gave for stopping a driver was speeding (40.9%), followed by vehicle defect (12.2%) and record check (9.8%). However, whether police officers gave a reason for a traffic stop and the perception of a stop's legitimacy varied by the driver's race and ethnicity. Police officers were more likely to give a reason to white drivers (96.1%), than black (94.5%), and Hispanic (92.3%) drivers. When officers gave a reason for the stop, 86.2% of whites, 80.4% of Hispanics, but only 72.7% of blacks judged the stop as legitimate. When officers did not give a reason for the stop, 46.3% of whites, but only 25.7% of blacks and 19% of Hispanics deemed the stop legitimate (see Table 7.14).

As shown in Table 7.14, the 2015 survey found that nearly 87% of traffic stops involved some sort of enforcement action. Approximately 49% of drivers received a ticket, about 36% received a warning, and nearly 4% were searched or arrested (some drivers received more than one enforcement action). Again, enforcement actions by police officers varied by the race and ethnicity of the driver. A larger percentage of white drivers (38%) than black drivers (33.6%) or Hispanic drivers (32.9%) received warnings. A larger percentage of Hispanic drivers (56.4%) and black drivers (49.9%) than white drivers (46.4%) received tickets, and a larger percentage of Hispanic drivers (5%) and black drivers (4.4%) than white drivers (3.3%) were searched or arrested.

The 2015 survey also revealed that among all drivers who were involved in a traffic stop, 94% of those who received a warning, 90% who received a ticket, and 67% who were searched or arrested felt police officers behaved properly. Ninety-two percent of drivers who received no enforcement action felt the same way. Again, there were race and ethnic differences. In the case of warnings, 94.8% of whites, 92.2% of blacks, and 88.8% of Hispanics felt that police officers behaved properly. As for drivers who were ticketed, 90.8% of whites, 88.2% of Hispanics, and 85.1% of blacks felt that police officers behaved properly. For drivers who were searched or arrested, 70.5% of whites, 68.5% of blacks, and 62.8% of Hispanics felt the police acted properly (see Table 7.14).

In sum, although these data do not support the contention that minorities, especially blacks, experience traffic stops at rates substantially higher than rates for whites, they do show that blacks are more likely than whites and Hispanics to be stopped multiple times. They also show that police officers are more likely to give a reason for a stop to whites than to blacks and Hispanics and, when police officers give a reason, blacks are less likely than whites and Hispanics to judge the stop as legitimate. When police officers engage in an enforcement action during a traffic stop, they are more likely to give a warning to whites than to blacks and Hispanics, a ticket to Hispanics and blacks than to whites, and to search or arrest Hispanics and blacks than whites. Finally, police officers are more likely to behave properly during enforcement actions with whites than with blacks and Hispanics. Less clear is whether these actions are a function of racial stereotyping and prejudice or legitimate legal factors.

FYI

Most Americans Oppose Racial Profiling

According to a 2016 national survey, 63% of Americans oppose racial profiling: 34% strongly oppose it and 29% somewhat oppose it. The remaining 37% support racial profiling: 10% strongly support it and 26 somewhat support it. Blacks (77%) are most opposed to racial profiling, although 62% of both Latinos and whites also oppose it. A slim majority of Republicans (51%) support racial profiling.

Source: Emily Ekins, "Policing in America: Understanding Public Attitudes Toward the Police. Results from a National Survey," Cato Institute, December 7, 2016, accessed June 15, 2019, https://www.cato.org/survey-reports/policing-america.

Racial profiling remains a hot topic in the U.S. Congress, state legislatures, county commissions, and city councils, as well as in the meeting rooms of civil rights and professional police organizations. The American Civil Liberties Union (ACLU), for example, has a national project to eliminate racial profiling and even provides citizens with a "Bust Card" that tells them how to respectfully interact with the police (acknowledging the difficulty of their job) even when falling victim to racial profiling. The National Association for the Advancement of Colored People (NAACP) also has a national project to eliminate racial profiling and, in 2014, released a groundbreaking report on the subject entitled *Born Suspect: Stop-and-Frisk Abuses and the Continued Fight to End Racial Profiling in America*. In 2014 (the latest year for which these data were available), 30 states had laws prohibiting racial profiling by law enforcement officers, but only 18 of those states required police departments to collect information on the race of motorists they stop.[48]

ACLU

To learn more about what is being done to combat racial profiling, visit the American Civil Liberties Union (ACLU) website on racial profiling at https://www.aclu.org/issues/racial-justice/race-and-criminal-justice/racial-profiling.

What can police departments do to prevent racial profiling?

Racial profiling to any degree is a blight on the record of professional law enforcement and on democracy. Some of the methods that have been prescribed to stop racial profiling include racial and cultural diversity training for police personnel, strong discipline for errant officers, videotaping all traffic stops, collecting data on the race of stopped motorists and pedestrians and the disposition of the encounter, and having police officers distribute business cards to all motorists and pedestrians they stop. The business card may reduce race-based stops because it would allow an officer to be easily identified at a later time. But city leaders throughout the United States are in a quandary as to precisely what to do to stop racial profiling.

Factors Limiting Discretion Several methods are employed to control the amount of discretion exercised by police officers. One method is close supervision by a police agency's management. For example, a department may require that officers consult a sergeant before engaging in a particular kind of action. Department directives or policies also limit the options police officers have in particular situations. Decisions of the U.S. Supreme Court, such as one restricting the use of deadly force to stop a fleeing felon, limit the options available to officers on the street. Finally, the threat of civil liability suits has reduced the discretion an officer has, for example, in the use of deadly force or in the pursuit of fleeing suspects in an automobile.

The debate over how much control should be placed on the exercise of police discretion is ongoing. Few other professionals have experienced a comparable attack on their authority to make decisions for the good of the clients they serve. The continuing attempt to limit discretion also seems out of place at a time when community policing is being widely advocated. Remember that community policing decentralizes authority and places it in the hands of the local beat officers and their supervisors. Community policing is bound to fail if citizens see that the police they work with every day do not have the authority and discretion to make the decisions that will ultimately improve the quality of life in a community.

Job Stress

Stress in the workplace is common today. According to a 2015 survey by the American Institute of Stress, job pressure is the number-one cause of stress in the United States. Money and work were cited by 76% of those surveyed as the leading cause of their stress, and 30% said they were "always" or "often" under stress at work. The annual costs to employers in stress-related health care and missed work was approximately $300 billion.[49] Given the nature of police work, no one is surprised to discover that a law enforcement officer's job is stressful. Police officers intervene in life's personal emergencies and great tragedies. Working extended shifts, for example, at the scene of the bombing of the federal building in Oklahoma City or the World Trade Center disaster would tax the resources of even the most resourceful police officer. Who would deny the stress involved in working deep undercover on a narcotics investigation over a period of several months? Some officers are able to manage stress on the job better than others.

job stress The harmful physical and emotional outcomes that occur when the requirements of a job do not match the capabilities, resources, or needs of the worker.

Job stress is defined as the harmful physical and emotional outcomes that occur when the requirements of a job do not match the worker's capabilities, resources, or needs. Poor health and injury are possible results of prolonged job stress. Police work has long been identified as one of the most stressful of all occupations, and many police

Working extended shifts at the World Trade Center disaster was very stressful for police officers and other first responders. *In what ways is a police officer's job more stressful than other jobs?*

ROBERTO SCHMIDT/AFP/Getty Images

officers suffer each year from the deleterious effects of a job that tests their physical and emotional limits.

Sources and Effects of Stress A number of conditions can lead to stress: (1) design of tasks–heavy lifting, long hours without breaks, and monotonous repetition of dangerous maneuvers; (2) management style–lack of participation by workers in decision making, poor communication, lack of family-friendly policies; (3) interpersonal relationships–poor social environment and lack of support from co-workers and supervisors; and (4) work roles–conflicting or uncertain job expectations, wearing too many hats, too much responsibility.[50] The signs that stress is becoming a problem with an officer are frequent headaches, difficulty in concentrating, short temper, upset stomach, job dissatisfaction, abuse of alcohol and drugs, and low morale. Individually and collectively these symptoms can have other origins, but job stress often is the source.

Copicide As if the work of confronting dangerous suspects and preserving the peace were not stressful enough, "copicide," or "death by cop," has entered the work life of some police officers. **Copicide** is a form of suicide in which a person gets fatally shot after intentionally provoking police officers.[51] A study of police shootings resulting in the death of a citizen in Los Angeles found that 10% could be attributed to copicide.[52] A more recent study of North American officer-involved shootings pegs the incidence of suicide by cop at 36%.[53] One commentator believes that "dozens of times each year during jittery hostage dramas and routine traffic stops, desperate people lure police officers into shooting them in a phenomenon known in law enforcement circles as 'suicide by cop.'"[54] The truth is that no one knows exactly how many times copicide incidents occur each year. Most police officers require a lot of time to emotionally recover from a fatal shooting in circumstances where they were fully authorized to use deadly force. To later discover that they were provoked into killing people who simply used police as a tool in a suicide scheme creates an extra emotional burden to bear.

copicide A form of suicide in which a person gets fatally shot after intentionally provoking police officers.

Stress Management and Reduction Fortunately, there are ways to manage and reduce stress without leaving police work. The "fixes" for stress come in two general categories: stress management and organizational change. Stress management now encompasses a variety of programs and procedures that include discussing stressful events with colleagues and mental health professionals, regular exercise, relaxation techniques such as structured visualization, a healthy diet that also eliminates caffeine and nicotine, enriched family support, religious support, prayer, meditation, and stress management classes that often involve spouses.

Organizational change also can reduce the potential for stress in the police work environment. Officers, for example, may be given more discretion in determining their work hours and shifts as long as the police agency is able to respond effectively to the workload requirements of the community. Flattening the organizational structure can help reduce stress by giving officers more discretion in carrying out the responsibilities of their job. Community policing is an effective paradigm for increasing officers' ability to control their work and perhaps minimize stress. Job redesign can assist assigning the right number and type of tasks to a police position when one job requires too much of an officer to maintain emotional stability and good health. Finally, excellent public safety equipment can minimize stress. Proper police weaponry, dependable vehicles, and the best of protective equipment such as high-grade body armor not only can protect officers but also can put their minds a little more at ease on those important concerns.[55]

Ralf-Finn Hestoft/Contributor/Getty Images

Police officers sometimes manage stress by seeking counseling. *Are there ways that you handle stress that you think would work particularly well in policing?*

Use of Force

FYI **Use of Force and Race/ Ethnicity**

According to a 2015 national survey, when police initiate contact, blacks (5.2%) and Hispanics (5.1%) are more likely to experience the threat or use of physical force than are whites (2.4%).

Source: Elizabeth Davis, Anthony Whyde, and Lynn Langton, "Contacts Between Police and the Public, 2015," in *Bureau of Justice Statistics Special Report*, U.S. Department of Justice (Washington, DC: USGPO, October 2018), accessed June 30, 2019, https://www.bjs.gov/content/pub/pdf/cpp15.pdf.

No issue in policing has caused as much controversy in recent decades as the use of force. In New York, Los Angeles, Chicago, Detroit, Miami, and many other cities, excessive-force charges against police officers have been made and documented and have resulted in the loss of public confidence in the police. Although the vast majority of police officers of this country go to work every day with no intention of using excessive force, far too many instances of brutality still occur.

A precise definition of brutality is not possible. However, the use of excessive physical force is undoubtedly a factor in everyone's definition. For many people, particularly members of racial and ethnic minorities, brutality also includes verbal abuse; profanity; harassment; threats of force; and unnecessary stopping, questioning, and searching of pedestrians or those in vehicles.

Excessive Force Why do the police have to use force as frequently as they do? A major responsibility of police officers is to arrest suspects so they can answer criminal charges. No criminal suspect wants his or her liberty taken away, so some of them resist arrest. Invariably, a few suspects are armed with some kind of weapon, and some are prepared to use that weapon against the police to foil the arrest. The police need to establish their authority to control such conflicts. The disrespect and physical resistance that frequently are the result of encounters with suspects have caused the police on occasion to use **excessive force**, which is a measure of coercion beyond that necessary to control participants in a conflict.

excessive force A measure of coercion beyond that necessary to control participants in a conflict.

Excessive force by a law enforcement officer violates the Fourth Amendment.[56] Whether police use of force under the Fourth Amendment is excessive is determined by answering the question of "whether the officers' actions [were] objectively reasonable in light of the totality of the circumstances."[57] Courts determine whether force was unreasonable by considering the following conditions: (1) the severity of the crime at issue, (2) whether the suspect posed an immediate threat to the safety of the officers or others, and (3) whether the suspect was actively resisting arrest or attempting to evade arrest by flight.[58] Courts are mindful that law enforcement officers frequently are forced to make split-second judgments about the amount of force that is necessary in a particular situation in tense, uncertain, and rapidly evolving circumstances.[59] Therefore, a law enforcement officer's use of force is unreasonable if, judging from the totality of circumstances at the time of the arrest, the officer uses greater force than was reasonably necessary to effectuate the arrest.[60]

Not only is the persistent use of excessive force by the police against citizens unethical, civilly wrong, and criminally illegal, but it also creates a situation in which nobody wins. Police may face criminal and civil prosecution in such cases, citizens build up layers of resentment against the police, and law enforcement agencies pay out millions of dollars in damages while losing respect in the eyes of the community.

One of the most egregious modern cases of police officers employing excessive force involves former Chicago Police Commander Jon Burge.[61] In June 2010, Burge was found guilty in a civil trial in federal district court of obstruction of justice and perjury for lying about alleged torture during interrogations. He could not be criminally prosecuted because the statute of limitations had expired on the cases. Burge was fired from the police force in 1993, because of mounting torture allegations. Burge, who died on September 19, 2018, and his "Midnight Crew" allegedly tortured more than 200 black men in police interrogation rooms during the 1970s, 1980s, and early 1990s. Dozens of these men were sent to prison on the basis of false confessions, including several who were sent to death row.

For example, after being tortured by Burge's detectives, Ronald Kitchen falsely confessed to the 1988 killing of two women and three children inside their South Side home. Kitchen was exonerated in 2009 after spending 21 years in prison—13 of them on death row. Kevin Fox, who was suspected of sexually assaulting and murdering his 3-year-old daughter, had been interrogated in a small, windowless room for 14 hours. His request for a lawyer was denied, and he was told that if he did not confess, arrangements would be made for inmates to rape him in jail. Detectives screamed at him, showed him a picture of his daughter bound and gagged with duct tape, and told him his wife was planning to divorce him.

Michael Tillman, serving a life sentence and seeking a new trial, was convicted of the rape and murder of a South Side woman in 1986. Tillman said that Burge's detectives coerced him to confess in a police interview room by putting a gun to his head and threatening him with death, repeatedly punching him in the stomach, holding a plastic bag over his head and suffocating him, and pouring 7-Up into his nose after forcing his head back in what his lawyers called a crude form of waterboarding.

In another instance, in 2009, Cortez Brown sought a new trial for his conviction in the 1990 killing of two men. He was sentenced to death but then-Governor George Ryan commuted his death sentence to life in prison in 2003. Brown maintains that Burge's detectives coerced him to falsely confess by beating him with their fists and a flashlight. In other cases, Burge and his detectives were accused of extracting false confessions with radiator burns, electric shocks, and putting a shotgun in a suspect's mouth.

Over the past 2 decades, the city of Chicago paid about $10 million to defend Burge, itself, and other defendants in the torture scandal in an attempt to avoid worse potential court damages in dozens of related civil lawsuits. (The Fraternal Order of Police also contributed to Burge's defense bills.) Nevertheless, in 2007, the city of Chicago agreed to pay nearly $20 million to settle lawsuits filed by four former death row inmates who claimed they were tortured by Burge and his detectives and falsely confessed. The four inmates—Aaron Patterson, Leroy Orange, Stanley Howard, and Madison Hobley—were pardoned by then-Governor Ryan in 2003. In January 2013, the city of Chicago agreed to pay Alton Logan $10.25 million to settle a federal wrongful conviction lawsuit. Logan was wrongly imprisoned for 26 years because of Burge's misconduct.[62] Burge's abuse and suppression of evidence cases collectively have cost Chicago approximately $60 million.[63] In addition, Chicago taxpayers paid more than $7 million for a report on Burge's alleged torturing of suspects. Special prosecutors took more than four years to investigate and complete the report released in 2006. Finally, during this entire time, the city had been paying Burge's pension of $3,768 a month for his years on the force.

Charles Rex Arbogast/ASSOCIATED PRESS

Jon Burge. Why does the use of excessive force in police interrogations continue in light of the Fifth Amendment's prohibition of coerced confessions?

Unfortunately, problems with the Chicago Police Department (CPD) did not end with revelations about Jon Burge. Partly in response to the killing of black teenager Laquan McDonald by a white police officer in October 2014 (see the Crime Story at the beginning of this chapter) and the significant and historic public outcry that followed, Chicago Mayor Rahm Emanuel created the Task Force on Police Accountability on December 1, 2015. The Task Force's report was released to the public in April 2016.[64] The Task Force addressed decades-old systemic institutional failures involving (1) death and injury at the hands of the police, (2) random but pervasive physical and verbal abuse by the police, (3) deprivation of basic human and constitutional rights by the police, and (4) lack of individual and systemic police accountability. The causes of these failures were identified as: (1) racism, (2) a mentality in the CPD that the ends justify the means, (3) the failure to make accountability a core value and imperative within CPD, and (4) a significant underinvestment in human capital.

One of the major findings of the Task Force was "the widely held belief [that] the police have no regard for the sanctity of life when it comes to people of color." This belief, moreover, is supported by CPD's own data. The data show that Chicago police officers shoot blacks at "alarming rates." Between 2008 and 2015, CPD officers were involved in 404 shootings. Among those shot by the police, 74% were black, 14% were Hispanic, 8% were white, and 0.25% were Asian. Citywide, Chicago is almost evenly split by race among whites (32%), blacks (33%), and Hispanics (29%). CPD data also show that CPD officers disproportionately used Tasers against blacks. Of the 1,886 Taser discharges between 2012 and 2015, 76% of them targeted blacks, 13% targeted Hispanics, 8% targeted whites, and 0.21% targeted Asians.

Beyond the use of force with guns and Tasers, police–community relations have been worsened further by the CPD's reliance on investigatory stops as a key part of its policing strategy. In 2013, for example, 46% of 100,676 traffic stops involved blacks, 22% involved Hispanics, and 27% involved whites. In addition, black and Hispanic drivers were searched about four times as often as white drivers, even though CPD's own data showed that contraband was discovered on white drivers twice as often as on black and Hispanic drivers. In the case of street stops, during the summer of 2014, CPD officers stopped more than 250,000 people in encounters that did not lead to an arrest. Of the 250,000 people stopped, 72% were black, 17% were Hispanic, 9% were white, and 1% was Asian. Many of those stops involved actual or threatened physical abuse by CPD officers.

The Task Force found that the public has lost faith in the oversight system. According to the report, real accountability is nearly impossible because serious structural and procedural flaws plague every stage of investigations and discipline. A major flaw with the oversight system is that the public has no meaningful input or power in the process. Another flaw is the unfair advantage provided to officers by the collective bargaining agreements with the police union. For example, the collective bargaining agreements (1) discourage the reporting of officer misconduct by requiring affidavits, (2) prohibit anonymous complaints, (3) require that accused officers be given the complainant's name early in the process, (4) allow officers to wait 24 hours before providing a statement following a shooting, which gives them the opportunity to confer with other officers, (5) permit officers to amend their statements after viewing video or hearing audio evidence, and (6) in many cases, require the city to ignore or destroy evidence of misconduct after a specified number of years. Other flaws with the oversight system involve the investigating agencies–the Independent Police Review Authority (IPRA) and the Bureau of Internal Affairs (BIA). Both agencies lack sufficient resources to do their jobs, neither agency is truly independent, and neither agency is held accountable for its work. Even when misconduct is discovered, officers rarely are subjected to meaningful consequences because the disciplinary process lacks transparency, is protracted, and is not subject to meaningful review (that is, no entity exists to police the oversight system). Between 2011 and 2015, neither the IPRA nor the BIA investigated 40% of filed complaints. Only 7% of the filed complaints were sustained, meaning that the allegations were supported by sufficient evidence to justify disciplinary action. Furthermore, in 2015, arbitrators reduced disciplinary recommendations in about 56% of cases and eliminated any discipline in approximately 16% of cases. In total, arbitrators reduced or eliminated discipline in 73% of cases.

Police Misconduct Costly

Chicago borrowed approximately $709 million to pay settlements for police misconduct cases from 2010 to 2017. Los Angeles payouts for police misconduct totaled more than $190 million from July 2005 to 2018.

Sources: Leonard Quart, "Letter from New York: Chicago's intractable violence," *The Berkshire Eagle*, April 25, 2019, accessed June 1, 2019, https://www.berkshireeagle.com/stories/leonard-quart-letter-from-new-york-chicagos-intractable-violence,571466; Emily Alpert Reyes and Ben Welsh, "L.A. is slammed with record costs for legal payouts," *Los Angeles Times*, June 27, 2018, accessed July 3, 2019, https://www.latimes.com/local/lanow/la-me-ln-city-payouts-20180627-story.html.

The Task Force noted that because of the large number of citizen complaints made about officers, an unacceptably high number of lawsuits are filed against the city and individual police officers every year. The number of officers who have been the subject of multiple citizen complaints is staggering. According to 2007–2015 data, more than 1,500 CPD officers have received 10 or more complaints; 65 officers have received 30 or more complaints. These data support the widespread perception that a deeply entrenched code of silence is supported not only by individual officers but also by the CPD administration. This perception is unfortunate because it paints all CPD officers with the same brush. The reality also is unfortunate because the failure of the CPD administration to meaningfully address the problems that cause the lawsuits ends up costing taxpayers tens of millions of dollars each year.

The Task Force identified a number of longstanding training issues that contribute to the aforementioned problems. First, Chicago is one of the most segregated cities in the United States. Although efforts have been made to recruit more officers from segregated (black and Hispanic) neighborhoods, fully integrating these new officers into the CPD force has been difficult. One problem is that for many of these minority recruits the training academy may be their first meaningful experience with someone from a different race or ethnicity. Second, as part of training, little commitment has been made to imparting constitutional policing strategies and tactics that provide appropriate balance to keeping communities safe and respecting basic constitutional and human rights. Third, firearms certification is the only mandatory training required of officers on an annual basis. Training on other subjects is voluntary, infrequent, and irregular. Much of it is delivered through "roll-call videos." That means that once officers have graduated from the academy, they can serve their entire careers without receiving any annual, mandatory training of any kind, besides firearms certification. Fourth, the community has no role in training, either in the academy or afterward, even though community involvement is considered an important element of training. Fifth, being an academy instructor is not sufficiently valued within the CPD, and some instructors teach while under investigation for alleged offenses.

The Task Force concluded that the community's lack of trust in the CPD, especially minority communities, is justified, and that the CPD is not doing enough to combat racial prejudice in its ranks. Diversity in the force, especially at supervisory levels, needs to be increased. Cultural sensitivity training should be mandatory. CPD officers need more training in working with the city's youth. The existing relationship between CPD officers and youth, especially minority youth, is best characterized as hostile.

Another problem is that the CPD often ignores human and civil rights. For example, in 2014, only 3 of every 1,000 arrestees were provided an attorney at any point while in police custody. In 2015, that number "doubled" to six. The city's youth are especially vulnerable and frequently are unaware of their rights.

These are not the only problems identified by the Task Force report, but they do provide an indication of the work that needs to be done. The Task Force provided numerous recommendations suggesting how to improve policing in Chicago. Whether the recommendations will be adopted and implemented, and whether they ultimately will prove successful, only time can tell. However, if past efforts to reform the police are any indication, prospects are not promising.

One more thing–the CPD is not unique in the problems identified by the Task Force. Police departments large and small labor under similar conditions and share many of the same problems. Consider, for example, findings from the 2015 investigation of the Ferguson, Missouri Police Department by the U.S. Department of Justice's Civil Rights Division:

> Many officers are quick to escalate encounters with subjects they perceive to be disobeying their orders or resisting arrest. They have come to rely on electric control weapons (ECWs), specifically Tasers, where less force–or no force at all–would do. They also release canines on unarmed subjects unreasonably and before attempting to use force less likely to cause injury. [In every canine bite incident for which racial information is available, the person bitten was African American.] Some incidents of excessive force result from stops or arrests that have no basis in law. Others are punitive and retaliatory. In addition, FPD records suggest a tendency to use unnecessary force against vulnerable groups such as people with mental health conditions or cognitive disabilities, and juvenile students. Furthermore . . . Ferguson's pattern of using excessive force disproportionately harms African-American members of the community. The overwhelming majority of force–almost 90%–is used against African Americans. [Author's note: African Americans comprise 67% of Ferguson's population of 21,000 residents. Ninety-three percent of Ferguson's 54 sworn police officers are white. Twenty-five percent of Ferguson's population lives below the federal poverty level.][65]

Use of Excessive Force

On November 9, 2015, the U.S. Supreme Court announced its ruling in *Mullenix v. Luna*. In an 8–1 decision, the Court held that law enforcement officers are immune from lawsuits unless it is "beyond debate" that their use of excessive force was clearly unreasonable.

Although past and recent incidents of excessive force have been disturbing, research reveals that police use of force does not occur as often as some people might think. For example, according to a government-sponsored survey, an estimated 985,300 persons in the United States experienced force or the threat of force by police at least once in 2015. The total represented an estimated 1.8% of the approximately 53.5 million people who experienced face-to-face police contact during 2015.[66] Force included threatening use of force, pushing or grabbing, handcuffing, hitting or kicking, using chemical or pepper spray, using an electroshock weapon, or pointing a gun. Males (2.7%), blacks (3.3%), and Hispanics (3.0%) were more likely than females (0.9%) and whites (1.3%) to have force used or threatened against them, as were individuals between the ages of 18 and 24 (3.2%), compared to those 16 or 17 (2.0%) and age 25 or older (25–44 = 2.4%, 45–64 = 2.4%, and 65+ = 0.1%). Some people reported that more than one type of force was used by police.

As shown in Table 7.15, in 2015, about 3.3% of persons age 16 or older, or approximately 30.2 million people, experienced force during their most recent police-initiated contact. Males (4.4%), blacks (5.2%), and Hispanics (5.1%) were more likely than females (1.8%) and whites (2.4%) to experience force during their most recent police-initiated contact, as were individuals between the ages of 18–24 (4.3%) and 25–44 (4.0%), compared to those 16 or 17 (2.5%), 45–64 (2.6%), and 65+ (0.1%). Of those persons age 16 or older that experienced force during their most recent police-initiated contact, 30.2% perceived the force to be necessary, while 48.4% perceived the force to be excessive. Males (50.1%), blacks (59.9%), and Hispanics (52.5%) were more likely than females (43.4%) and whites (42.7) to perceive the force to be excessive, as were all age groups with the exception of those age 16 or 17 (22.1%). Of those aged 18–24, 49.2% perceived the force to be excessive, as did those aged 25–44 (44.0%), 45–64 (58.5%), and 65+ (more than 99.9%).

Ben Margot/ASSOCIATED PRESS

There is much debate on how much control should be placed on the decisions of police officers in arrest situations. *How much control should be placed on them and why?*

TABLE 7.15 The Use of Force During the Most Recent Police-Initiated Contact, by Demographics, Type of Force, and Whether Action Was Perceived to Be Necessary or Excessive, 2015

Demographic Characteristic	Experienced Force	Force Perceived to Be	
		Necessary	Excessive
Total	3.3%	30.2%	48.4%
Sex			
Male	4.4	28.2	50.1
Female	1.8	36.2	43.4
Race/Ethnicity			
White	2.4	32.4	42.7
Black	5.2	32.0	59.9
Hispanic	5.1	20.8	52.5
Age			
16–17	2.5	35.3	22.1
18–24	4.3	28.2	49.2
25–44	4.0	34.1	44.0
45–64	2.6	24.4	58.5
65+	0.1	<0.1	>99.9
Type of Force			
Threat of force	0.3	<0.1	83.5
Handcuff	1.8	45.3	27.9
Push/grab/hit/kick	0.7	15.1	78.3
Pepper spray	<0.1	<0.1	<0.1
Shock	<0.1	<0.1	>99.9
Point gun	0.3	15.0	65.2

Source: Adapted from Elizabeth Davis, Anthony Whyde, and Lynn Langton, "Contacts Between Police and the Public," *Bureau of Justice Statistics Special Report*, U.S. Department of Justice (Washington, DC: USGPO, October 2018), accessed July 1, 2019, https://www.bjs.gov/content/pub/pdf/cpp15.pdf.

FYI

Black Millennials and Police Violence

More than half (54.43%) of black Millennials (blacks 18 to 34 years old) report that they or someone they know was harassed by, or experienced violence from, the police, compared with 32.8% of white Millennials and 24.8% of Latino Millennials. Black Millennials comprise 13% of all Millennials.

Source: Jon C. Rogowski and Cathy J. Cohen (2015), "Black Millennials in America: Documenting the Experiences, Voices and Political Future of Young Black Americans," Black Youth Project, accessed May 4, 2016 from http://blackyouthproject.com/wp-content/uploads/2015/11/BYP-millenials-report-10-27-15-FINAL.pdf.

As also shown in Table 7.15, of the approximately 30.2 million individuals who experienced the threat or use of force during their most recent police initiated contact in 2015, only 0.3% experienced the threat of force (less than 0.1% perceived the threat of force as necessary, while 83.5% perceived it as excessive); only 1.8% were handcuffed (45.3% perceived it as necessary and 27.9% as excessive); 0.7% were pushed, grabbed, hit, or kicked (15% perceived it as necessary and 78.3% as excessive); less than 0.1% were pepper sprayed (based on 10 or fewer cases); less than 0.1% were shocked (less than 0.1% perceived it as necessary and more than 99.9% perceived it as excessive); and 0.3% had a gun pointed at them (15% perceived it as necessary and 65.2% as excessive). In sum, these data show that the police rarely threaten force or use force in their contacts with persons aged 16 or older. However, on the rare instances when they threaten or use force, they are more likely to threaten or use force against males than females, blacks and Hispanics than whites, and those aged 18–44 than those younger or older than that. Finally, when the police threaten or use force, the people that experience that threat or use, with the exception of being handcuffed, are more likely to perceive it to be excessive rather than necessary.

Deadly Force The greatest concern over the use of force by the police has to do with the infliction of death or serious injury on citizens and criminal suspects. Since the U.S. Supreme Court's 1985 decision in *Tennessee v. Garner*, the use of deadly force has been severely restricted, and police shootings of suspects and citizens have been reduced. In the *Garner* case, an unarmed teenage boy was shot as he fled a house burglary, failing to heed the warning to stop given by a Memphis police officer. The boy later died of a gunshot wound to the head. He was found with $10 in his pocket that he had stolen from the home. The Memphis officer was acting in compliance with his department's policy on the use of deadly force and with the law in Tennessee and in most other states in the nation.

In the *Garner* decision, the Court ruled that "[A] police officer may not seize an unarmed, nondangerous suspect by shooting him dead." However, "[W]here the officer has probable cause to believe that the suspect poses a threat of serious physical harm, either to the officer or to others, it is not constitutionally unreasonable to prevent escape by using deadly force. . . . Thus," the Court continued, "if the suspect threatens the officer with a weapon or there is probable cause to believe that he has committed a crime involving the infliction or threatened infliction of serious physical harm, deadly force may be used if necessary to prevent escape, and if, where . . . feasible, some warning has been given."

Giving law enforcement officers the authority to use deadly force to stop a fleeing felon, even when they know the suspect is unarmed and not likely to be a danger to another person, derived from the common law in England and the United States that permitted such a practice. At the time the rule developed, however, unlike today, dozens of crimes were capital offenses, and the fleeing suspect, if apprehended and convicted, would have been executed. The *Garner* decision, no doubt, was long overdue, and it included a rule that many police agencies in the nation had adopted years earlier. The perspective that professional law enforcement agencies already had begun to adopt on deadly force was from the Model Penal Code, Section 307(2)(B). It reads:

The use of deadly force is not justifiable under this section unless:

1. The arrest is for a felony.
2. The person effecting the arrest is authorized to act as a peace officer or is assisting a person whom he believes to be authorized to act as a peace officer.
3. The actor believes that the force employed creates no substantial risk of injury to innocent persons.
4. The actor believes that: (a) The crime for which the arrest is made involved conduct including the use or threatened use of deadly force. (b) There is substantial risk that the person to be arrested will cause death or serious bodily harm if his or her apprehension is delayed.

Nearly all law enforcement agencies have adopted policies on police use of deadly force based on the *Garner* decision.[67] Most states also have enacted statutes governing police use of deadly force. A problem with these policies and statutes in all 50 states and the District of Columbia, according to a 2015 report by Amnesty International, is that they fail to comply with international law and standards.[68] Following are some of the other findings in the Amnesty report. The federal government, nine states, and the District of Columbia have no laws on the use of deadly force by law enforcement officers. Instead, they rely entirely on the *Garner* decision and departmental policies. Thirteen states have laws that do not comply with the standards set by U.S. constitutional law on police use of deadly force. No state laws require that deadly force be used only as a last resort, with nonviolent and less harmful means to be tried first, and no states limit the use of deadly force to only situations where there is an imminent threat to the life of, or possible serious injury to, the officer or others. Nine states allow deadly force to be used to suppress a riot, 22 states allow law enforcement officers to kill someone trying to escape from a prison or jail, and 20 states allow private citizens to use deadly force if they are carrying out law enforcement activities (e.g., assisting an officer in making an arrest). Only eight states require police to give a warning (when feasible) before using deadly force, and only three states provide that officers not create substantial risk to bystanders when using deadly force. Two states provide for training on the use of deadly force, and no state laws include accountability mechanisms. Table 7.16 lists these issues and the jurisdictions involved.

As noted previously, several black suspects recently have been shot and killed by white police officers, suggesting that perhaps something sinister is occurring. Charges of racism have been made, and a "Black Lives Matter" movement has been galvanized. Data on violent deaths recorded by the Centers for Disease Control show that 30.5% of the people killed by the police from 2004 through 2016 were black, even though blacks at the time comprised only about 13% of the population.[69] Evidence that many of the killings were not

TABLE 7.16 Issues with Laws Governing Police Use of Deadly Force

Issues	States
No laws on use of deadly force by law enforcement officers	Maryland, Massachusetts, Michigan, Ohio, South Carolina, Virginia, West Virginia, Wisconsin, Wyoming, District of Columbia, U.S. Government
Laws do not comply even with the lower standards set by U.S. constitutional law on the use of deadly force by law enforcement officers	Alabama, California, Delaware, Florida, Mississippi, Missouri, Montana, New Jersey, New York, Oregon, Rhode Island, South Dakota, Vermont
Laws require that use of deadly force may only be used as a last resort with nonviolent and less harmful means to be tried first	None
Laws limit the use of deadly force to only those situations where there is an imminent threat to life or serious injury to the officer or to others	None
Laws allow the use of deadly force to suppress a riot	Arizona, Delaware, Idaho, Mississippi, Nebraska, Pennsylvania, South Dakota, Vermont, Washington
Laws allow law enforcement officers to kill someone trying to escape from a prison or jail	Alabama, Colorado, Delaware, Georgia, Hawaii, Idaho, Indiana, Kentucky, Maine, Mississippi, Montana, Nebraska, New Hampshire, New Jersey, New Mexico, New York, North Carolina, North Dakota, Oklahoma, Pennsylvania, South Dakota, Washington
Laws allow private citizens (non-state actors) to use deadly force if they carry out law enforcement activities (e.g., assisting an officer in making an arrest)	Alabama, Arizona, California, Colorado, Connecticut, Indiana, Kansas, Kentucky, Louisiana, Maine, Mississippi, Nebraska, New Hampshire, New Jersey, New York, North Dakota, Pennsylvania, South Dakota, Texas, Washington
Laws require a warning to be given (when feasible) before deadly force is used	Connecticut, Florida, Indiana, Nevada, New Mexico, Tennessee, Utah, Washington
Laws provide that officers should create no "substantial risk" to bystanders when using deadly force	Delaware, Hawaii, New Jersey
Laws provide for training on the use of deadly force	Georgia, Tennessee
Laws include accountability mechanisms	None

Source: Adapted from Amnesty International, "Deadly Force: Police Use of Lethal Force in the United States," 2015, pp. 4–5, accessed January 31, 2016, www.amnestyusa.org/sites/default/files/aiusa_deadlyforcereportjune2015.pdf.

justified has led to wrongful death lawsuits and millions of dollars in financial settlements paid to the families of the victims. Following are eight recent examples:

- In September 2018, the Chicago City Council authorized a $16 million settlement to the family of innocent bystander Bettie Jones, 55, who was shot and killed by Chicago Police Officer Robert Rialmo on December 26, 2015, the same night Rialmo shot and killed bat-wielding teenager and Jones' neighbor Quintonio LeGrier. Jones was simply in the wrong place at the wrong time, said Finance Committee Chairman Edward Burke.
- In June 2017, the insurance company for the city of Ferguson, Missouri, paid $1.5 million to settle a wrongful death lawsuit by Michael Brown's parents. Brown, 18, was black and unarmed when he was fatally shot by white officer Darren Wilson on August 9, 2014. Wilson was cleared of wrongdoing.
- In April 2016, the city of Cleveland, Ohio, agreed to pay $6 million to settle the federal lawsuit filed by the family of Tamir Rice, a black 12-year-old killed by police gunfire in November 2014. Rice was shot and killed by a white police trainee, who along with his training officer, responded to a 911 call about a child in a city park brandishing what appeared to be a toy gun. The dispatcher apparently did not tell the officers that the individual was a child and the gun was a toy.
- In January 2016, the family of Samuel DuBose settled with the University of Cincinnati for $4.85 million plus free tuition for DuBose's 12 children. DuBose was shot and killed In July 2015 by a University of Cincinnati police officer during a traffic stop.

- In October 2015, the family of Walter Scott reached a $6.5 million settlement with the city of North Charleston, South Carolina, for the shooting death of Scott, a black man, by a white police officer in April 2015. A bystander's video appears to show the officer shooting Scott eight times while Scott was running away. The officer had pulled Scott over for driving with a broken taillight.
- In September 2015, the city of Baltimore, Maryland, agreed to pay $6.4 million in the Freddie Gray case. Gray died in April 2015, after he suffered a spinal injury while riding in a Baltimore police van. Gray was arrested for allegedly possessing an illegal switchblade.
- In July 2015, New York City agreed to pay $5.9 million to the family of Eric Garner, who, in July 2014, died at the hands of police after being put in a chokehold. The incident was captured on bystander video. Officers, who said they believed he was illegally selling individual cigarettes, stopped Garner outside a convenience store. Garner, a black asthmatic father of six children, could be seen yelling, "I can't breathe!" 11 times before losing consciousness.
- In April 2015, the Chicago City Council approved a $5 million settlement for the killing of black 17-year-old Laquan McDonald by a white police officer in October 2014. Police had been called to investigate reports that McDonald had a knife. A police dashboard camera video showed McDonald walking away from the officer when the officer shot McDonald 16 times in 13 seconds from within 10 feet.[70]

The shooting of unarmed blacks, as well as other similar incidents, has fueled the suspicion and perception that these types of events have become epidemic. Available evidence, however, is mixed, contradictory, and unreliable.

Consider what heretofore has been thought the best available evidence on the subject. Since 2003, the U.S. Justice Department's Bureau of Justice Statistics has had the responsibility for administering The Arrest-Related Deaths (ARD) program.[71] ARD program data come from "an annual national census of persons who die either during the process of arrest or while in the custody of state or local law enforcement personnel." Data collection is the responsibility of state reporting coordinators and comes from several sources: law enforcement, medical examiners or coroners, prosecutor's offices, Uniform Crime Reports (UCRs), the National Violent Deaths Reporting System, and open-source media searches. Participation in the program is voluntary. A recent evaluation concluded that, nationally, the ARD program captures only about 50% of all law enforcement homicides. So unreliable are the data on arrest-related deaths that former FBI Director James Comey, in a speech at Georgetown University in February 2015, bemoaned, "It's ridiculous that I can't tell you how many people were shot by the police in this country–last week, last year, the last decade–it's ridiculous."[72] Former President Obama also remarked on the problem in 2015: "Right now, we do not have a good sense, and local communities do not have a good sense, of how frequently there may be interactions with police and community members that result in a death, result in a shooting."[73]

To help fill this void in information, the *Washington Post* conducted a yearlong study of the police's use of deadly force in 2015.[74] The *Post* discovered that the police killed nearly 1,000 people in 2015, but fewer than 4% of the killings involved white police officers killing unarmed black men. The majority of people killed by the police in 2015 were white, had attacked someone with a weapon, brandished a weapon, were suicidal or mentally troubled, or ran when officers ordered them to halt. About 25% of those people killed by the police were mentally ill or experiencing an emotional crisis, and 90% of them were armed, usually with guns. Most of them were killed by officers who had not been trained to deal with the mentally ill. Black men accounted for 40% of the unarmed men shot by police (of all races and ethnicities) in 2015, even though black men make up only 6% of the U.S. population. The *Post* also found that of those people killed by police for less threatening behavior, approximately 60% of them were black or Hispanic. Body cameras, according to the *Post*, captured about 6% of fatal shootings by police in 2015. Furthermore, although the FBI is supposed to keep statistics on such shootings, the *Post* found that fewer than half of the nation's 18,000 police departments report their incidents to the agency.

The *Post* research found that over the previous decade an average of five officers per year had been indicted on felony charges; in 2015, 18 officers had been charged with felonies

Arrest-Related Deaths

According to preliminary data from the Bureau of Justice Statistics, an estimated 1,900 arrest-related deaths occurred in the United States from June 2015 through May 2016, or about 158 per month.

Source: Duren Banks, Paul Ruddle, Erin Kennedy, and Michael G. Planty, "Arrest-Related Deaths Program Redesign Study, 2015–20: Preliminary Findings," *Bureau of Justice Statistics Technical Report*, U.S. Department of Justice (Washington, DC: USGPO, Revised December 22, 2016).

including murder, manslaughter, and reckless discharge of a firearm. Rarely, however, are officers convicted. Only 11 of the 65 officers charged in fatal shootings during the past decade were convicted.

The vast majority of police officers do not want to kill the citizens they are sworn to protect, and police departments are attempting to reduce the police-shooting problem. For example, in 2009, Las Vegas police adopted a new use-of-force policy that required officers to put the highest premium on "the sanctity of human life." Four years after implementing the policy, officer-involved shootings had fallen by nearly half, suggesting that new language can change police culture. The *Post* study also discovered that nearly a third of police shootings are the result of a car chase that began with a traffic stop for a minor infraction. Some cities, such as New York and Boston, have tightened rules on pursuits, causing a significant decrease in the number of officers who shoot at vehicles. However, other research shows that while tightening pursuit policies reduces injuries caused by police, it also is associated with more crime.

FYI **Stricter Use-of-Force Policies**

Research shows that police officers in departments with stricter use-of-force policies kill fewer people and are less likely to be killed or seriously injured themselves.

Source: Samuel Sinyangwe, "Examining the Role of Use of Force Policies in Ending Police Violence" (September 20, 2016). Available at SSRN: https://ssrn.com/abstract=2841872 or http://dx.doi.org/10.2139/ssrn.2841872.

Regarding police pursuit policies, according to Highway Traffic Safety Administration records, about 360 people are killed each year in high-speed police chases, and about one-third of them are innocent bystanders. Criminologist Geoffrey Alpert, who has studied police pursuits since the 1980s, maintains that the actual number of fatalities is three or four times higher. However, no one knows for sure because there is no mandatory reporting for pursuit deaths. Moreover, bystanders killed after police stop chasing suspects–even seconds afterward–are not counted. Alpert estimates that 35% to 40% of all police chases end in crashes.[75]

We should not forget that citizens and criminal suspects also attack the police. Mentally ill persons, parties to a family dispute, and suspects trying to avoid arrest feloniously kill between 50 and 100 officers each year. More than 90% of officers are killed by assailants using firearms. In response, police officers exercise caution by wearing protective vests, proactively using what they learn in courses on self-defense (unarmed and armed), and attempting to defuse hostile situations through peaceful techniques they learn in training. When all else fails, they can use their sidearms to protect themselves.

Police Corruption

Almost from the beginning of formal policing in the United States, corruption of law enforcement officers has been a fact of life. Almost nothing is more distasteful to the public than a police officer or a whole department gone bad. Throughout history, police officers have bought their positions and promotions, sold protection, and ignored violations of the law for money.

The amount of law enforcement corruption and misconduct is staggering. A recent *USA TODAY* investigation found that from 2008 through 2018, at least 85,000 law enforcement officers in the United States had been investigated or disciplined for misconduct.[76] Officers had beaten members of the public, planted evidence and used their badges to harass women. They had lied, stolen, dealt drugs, driven drunk, and abused their spouses. The investigation documented at least 200,000 incidents of alleged misconduct, much of it previously unreported. More than 110,000 internal affairs investigations were conducted by hundreds of individual departments, and more than 30,000 officers were decertified by 44 state oversight agencies.

Why is policing so susceptible to bribery and other forms of corruption? Perhaps it has to do with the combination of two critical features of the police role in society. On the one hand, the police have authority to enforce laws and to use power to make sure that those laws are obeyed. On the other hand, they also have the discretion *not* to enforce the law. The combination of those two features makes the police vulnerable to bribes and other forms of corruption. Other features of police work add to the potential for corruption: low pay in relation to important responsibilities; cynicism about the courts' soft handling of criminals that the police spend so much time trying to apprehend; society's ambivalence about vice (most citizens want the laws on the books, but many of them are willing participants in vice); and the practice of recruiting officers from working-class and lower-class backgrounds, where skepticism about obeying the law might be more prevalent.

Some of those factors undoubtedly help explain the following five recent examples of corruption.

MICHAEL NELSON/European Pressphoto Agency /LOS ANGELES/CALIFORNIA/UNITED STATES/Newscom

Policing is susceptible to bribery and other forms of corruption. *Who is policing the police? Are controls adequate?*

- On July 2, 2019, Ludwig Paz, a retired New York Police detective who had been assigned to the vice squad, was sentenced to 4 to 12 years in prison for running a multimillion-dollar prostitution and gambling ring with seven active-duty police officers. Paz plead guilty to 2 counts of attempted enterprise corruption and 1 count of promoting prostitution in the third degree in order to "save his family" and reduce the sentences of his wife and two daughters, who were co-defendants. The former detective and his crew ran eight brothels in Brooklyn, Queens, and Nassau County and illegal lotteries at delis and beauty salons in Brooklyn and Queens.[77]
- On June 28, 2019, two Miami police officers–Kelvin Harris, 53, a veteran and James Archibald, 33, a rookie–were convicted by a federal jury of providing protection for drug traffickers who were really FBI undercover investigators posing as dealers moving loads of cocaine through the city. Harris had accepted $10,000 in bribery payments, while Archibald had accepted $6,500 in payoffs.[78]
- On June 27, 2019, in one of the biggest corruption cases in Hawaii history, former Honolulu Police Chief Louis Kealoha and his prosecutor wife Katherine Kealoha were convicted by federal jurors of framing her uncle in an attempt to settle a family feud over money. The defendants were convicted of federal conspiracy charges and three counts of obstruction of justice. They also face a second trial for a series of financial crimes, including bank fraud and identity theft. Katherine Kealoha faces a third trial for allegedly running a prescription drug trafficking ring with her younger brother, who is an anesthesiologist.[79]
- On June 7, 2018, the leader of Baltimore's Gun Trace Task Force (GTTF), Wayne Jenkins, 37, was sentenced to 25 years in prison in a corruption scandal prosecutors called "breathtaking." The ex-police sergeant pleaded guilty in January and admitted participating in at least 10 robberies of Baltimore citizens, planting drugs on innocent people, and reselling more than $1 million worth of drugs, including heroin, cocaine, and prescription painkillers, that he and his crew stole from suspects on a nearly daily basis. He sent innocent people to prison and caused at least one murder. The GTTF's corruption affected about 1,700 criminal cases, many of which have been overturned. Besides Jenkins, seven of the eight members of the GTTF were arrested and indicted. Five of the officers pleaded guilty and two of them were convicted at trial of robbery, extortion, and overtime fraud.[80]
- On June 29, 2018, a Chicago jury awarded Jacques Rivera $17 million for the more than 20 years in prison he spent for a murder he did not commit. In what his attorneys called one of the biggest scandals in Chicago history, former Chicago Police Department gang crimes detective Reynaldo Guevara and other officers framed Rivera. Guevara has a more than 2-decade history, from the 1980s through the early 2000s, of beating suspects into confessions, manipulating witnesses into selecting innocent people from lineups, fabricating and withholding evidence, and just plain lying to close unsolved cases. Guevara has been accused of framing more than 50 people for murders they did not commit. So far, 10 people have been exonerated. Including Rivera's payout, Guevara's misconduct has cost the city of Chicago $38 million.[81]

Unfortunately, police corruption is a recurring problem endemic to the police profession as the cases of Chicago Police Department officers Reynaldo Guevara and Jon Burge illustrate. What especially is disturbing about the cases of Guevara and Burge, and so many other police officers throughout the United States, is the police

CJ ONLINE

The Marshall Project and the Citizens Police Data Project

To read about more examples of police corruption, visit The Marshall Project website at https://www.themarshallproject.org/records/2528-police-corruption and the Citizens Police Data Project at https://cpdp.co.

Why do you think some police officers engage in illegal behavior?

department's brass, the city's prosecutors and judges, police oversight commissions, and even federal officials had ample warnings about their misconduct, numerous chances to make amends for the injustices they stand accused of committing, and opportunities to stop them from perpetrating more, but they did not. One other problem is that the true killers, in some cases, have escaped justice and remain in communities poised to kill again.

Types of Corruption In 1972, the Knapp Commission issued a report on corruption in the NYPD. Two types of corrupt officers were identified: "grass eaters" and "meat eaters." **Grass eaters** were officers who occasionally engaged in illegal or unethical activities, such as accepting small favors, gifts, or money, for ignoring violations of the law during the course of their duties. **Meat eaters**, in contrast, actively sought ways to make money illegally while on duty. For example, they would solicit bribes, commit burglaries, or manufacture false evidence for a prosecution.[82]

grass eaters Officers who occasionally engage in illegal and unethical activities, such as accepting small favors, gifts, or money for ignoring violations of the law during the course of their duties.

meat eaters Officers who actively seek ways to make money illegally while on duty.

More than 40 years ago, sociologist Ellwyn Stoddard identified a more complete list of types of police misconduct, with examples, in what he described as the "blue-coat code":

1. ***Bribery***–Accepting cash or gifts in exchange for nonenforcement of the law.
2. ***Chiseling***–Demanding discounts, free admission, and free food.
3. ***Extortion***–The threat of enforcement and arrest if a bribe is not given.
4. ***Favoritism***–Giving breaks on law enforcement, such as for traffic violations committed by families and friends of the police.
5. ***Mooching***–Accepting free food, drinks, and admission to entertainment.
6. ***Perjury***–Lying for other officers apprehended in illegal activity.
7. ***Prejudice***–Unequal enforcement of the law with respect to racial and ethnic minorities.
8. ***Premeditated theft***–Planned burglaries and thefts.
9. ***Shakedown***–Taking items from the scene of a theft or a burglary the officer is investigating.
10. ***Shopping***–Taking small, inexpensive items from a crime scene or an unsecured business or home.[83]

Controlling Corruption Corruption in law enforcement strikes at the core of the profession and takes a heavy toll. All peace officer positions are positions of honor and trust, and agencies invest money and time in selecting officers with integrity. To see this investment lost is disheartening. But more than anything else, public confidence and trust plummet after a widely publicized corruption case, such as the scandal in Chicago involving Jacques Rivera and other police officers. The following list describes some ways to control and reduce corruption in policing.

- ***High moral standards***–Selecting and maintaining officers with high moral standards is a step in the right direction. Some police agencies in the United States still hire convicted felons to do police work. In-depth academy and in-service training on ethical issues that officers are likely to face would prepare officers for the compromises they may be asked to make later in their careers.
- ***Police policies and discipline***–A police department should develop rigid policies that cover the wide range of activities that corruption comprises. Drug testing of officers, particularly those in narcotics-sensitive positions, may be necessary although unpopular. Policies mean nothing unless they are enforced. Discipline should be imposed, and prosecutions should go forward when officers are found guilty of violating established policies and laws.
- ***Proactive internal affairs unit***–The **internal affairs investigations unit** of a police department should ferret out illegal and unethical activity. Any internal affairs unit that waits for complaints probably is not going to receive many of them. First-line supervisors should know whether their subordinates are engaging in unethical and illegal violations of department rules and state laws. They also should be held responsible for the actions of their subordinates.
- ***Uniform enforcement of the law***–If a police agency makes it clear that no group of citizens, no matter what their affiliation with the police department, is going to receive special treatment from the police department, the incentive for offering bribes and other forms of corruption will be minimized. This process starts with

internal affairs investigations unit The police unit that ferrets out illegal and unethical activity engaged in by the police.

clear policies and procedures and must be backed up with discipline, when necessary.

- ***Outside review and special prosecutor***–Police leadership and police labor associations heavily resist any kind of outside review of their actions. However, both the Christopher Commission and the Knapp Commission are examples of outside reviews that brought about improvements in the agencies they investigated. Special prosecutors are recommended in serious cases to relieve the police and the government of any accusations of a whitewash.
- ***Court review and oversight***–Criminal prosecutions or civil liability suits deriving from police corruption cases can be very costly to a police agency. Such visible forms of oversight often result in adverse media coverage, civil liability awards, and higher insurance rates–all of which should encourage police agencies to control corruption.[84]
- ***Stop rehiring bad police officers***–Much of the information about how a police officer has performed is stored in the personnel records of individual departments and unavailable to hiring agencies. Among the reasons for the secrecy are concerns about unfairly jeopardizing police officers' jobs by placing them on a list based on possibly minor or unfounded accusations. "The Thin Blue Line," which refers to the camaraderie among police officers, who carry out the difficult law enforcement mission, and the powerful commitment to protect fellow officers. A third reason is the struggle by many agencies, particularly in small communities, to attract police recruits and executives. Many agencies are in no position to be selective when hiring officers. Regardless of the reason, bad police officers who have been terminated from one department often are rehired by other law enforcement agencies without the new agency knowing of the officers' past record.

Outside Review of Wrongdoing

A 2016 public opinion poll found that 79% of Americans would prefer that an "outside law enforcement agency take over the investigation" when an officer is suspected of criminal wrongdoing. Only 21% favor police departments conducting internal investigations of their own officers.

Source: Emily Ekins, "Policing in America: Understanding Public Attitudes Toward the Police. Results from a National Survey," Cato Institute, December 7, 2016, accessed June 15, 2019, https://www.cato.org/survey-reports/policing-america.

Problem police officers that move around among agencies are not limited to the rank-and-file but also include police chiefs and sheriffs. As noted previously, a recent *USA TODAY* investigative report identified 32 people who became police chiefs or sheriffs despite a finding of serious misconduct, usually at another department. At least eight of them were found guilty of a crime. Others amassed records of domestic violence, improperly withholding evidence, falsifying records or other conduct that could impact the public.[85]

What is needed is a central clearinghouse that documents police misconduct and disciplinary cases, including cases that result in termination or decertification. Attempts to create such a clearinghouse have been resisted by powerful forces across the nation, including police departments and their officers' unions, state agencies created to oversee police, legislators, prosecutors and judges, that actively work to keep police officer conduct–and misconduct–records secret.[86]

THINKING CRITICALLY

1. What do you think are the best ways for police officers to handle stress on the job?
2. What do you think are the best ways to reduce police corruption?

Professionalizing Law Enforcement

Many people would argue that policing in America already has reached professional status. Law enforcement is a valued service. Its agents make important decisions daily that substantially affect the lives of people and the quality of life in a community. The police officer's position is one of honor and trust. There are academy programs consisting of hundreds of hours of instruction, as well as law enforcement degree programs. Now there are even signs that law enforcement is attempting to police its own profession. Professional accreditation for police agencies is a rite of passage that is needed if law enforcement is to join the list of the most respected professions. Nevertheless, resistance to it and the other developments still is widespread.

Not everyone has the qualities to be a police officer. To allow into law enforcement those people with no desire to serve, low intelligence, a shady past, poor work habits, and no ability to communicate effectively is to court disaster for every department that does so–and for the entire profession.

Accreditation

On July 8, 2019, 730 police agencies in the United States were accredited by the Commission on Accreditation for Law Enforcement Agencies (CALEA). Another 96 police agencies in the United States were in the process of becoming accredited by conducting self-assessments.

Source: "CALEA, Commission on Accreditation for Law Enforcement Agencies," CALEA Client Database, accessed July 8, 2019, https://www.calea.org/calea-client-database.

Some police officers and their leaders resist 600 hours of initial training and do all they can to avoid continuing education and training. Real professionals seek advanced training.

Professionals in any field make mistakes, and a caring public should forgive most of them. In police work, there are incomplete interviews, evidence left at crime scenes, and bad reports written. In the long run, the consequences of such mistakes generally are insignificant as long as corrections are made. Mistakes also can be technological, such as the failure of a radar gun. No one should blame the police for technological mishaps that are not the result of negligence.

One kind of mistake, however, stands out more than any other: the condoning of racist and brutal tactics like the Los Angeles police officers' beating of Rodney King in 1991. The findings of the Christopher Commission confirmed that such tactics generally were condoned and even encouraged. The videotaped replay of that performance will be an embarrassment to professional policing for years to come. Police departments need to remove officers from the profession who would participate in or overlook such violence.

Many police officers go to work each day with a negative attitude, and some may take out their frustrations on the citizens they meet. Police officers need to treat their on-duty time as a professional performance and render the best service possible on any given day. If they treat the citizens they serve with respect and concern, officers will make great progress in improving the public's perception of law enforcement as a profession worthy of trust and admiration.

SUMMARY

1. Describe the general attitude of the public toward the police and the police toward the public.

The data show that the police and the public view the world and each other very differently. Moreover, there are significant demographic, especially racial, differences both among the public and among the police and between the public and the police. The data indicate a real and concerning racial divide between the police and blacks. The public, for its part, seems to have more respect for the police than it has confidence in the police, high regard for police honesty and ethical standards, and a favorable view of the police. The police, for their part, have mixed views of the public and a complicated relationship with the public. Regarding the public's opinion of the police overall, most of the public thinks the police do a fairly good job, but it also believes there is much room for improvement.

2. Summarize the steps in an effective police officer selection process.

Police applicants go through several different kinds of testing to become law enforcement officers. Steps in an effective police officer selection process include recruitment, short application, detailed application, medical examination, physical ability test, written examination, background investigation, psychological testing, oral interview/oral board, academy training, and a probationary employment period.

3. Identify factors that affect the exercise of police discretion and methods of limiting discretion.

Factors that affect the exercise of police discretion include the nature of the crime, departmental policies, the relationship between the victim and the offender, the amount of evidence, the preference of the victim, the demeanor of the suspect, the legitimacy of the victim, and the suspect's socioeconomic status. Methods of limiting police discretion are close supervision by a police agency's management, department directives and policies, U.S. Supreme Court decisions, and the threat of civil liability suits.

4. Describe two general ways that law enforcement agencies can reduce stress on the job.

Two general ways that law enforcement agencies can reduce job stress for police officers are (1) to employ *stress management* strategies, such as discussing stressful events with colleagues and mental health professionals, regular exercise, relaxation techniques, healthy diet, religious support, enriched family support, prayer, meditation, and stress management classes that often involve spouses, and (2) to implement *organizational change*, such as allowing officers more discretion in determining work hours and shifts, flattening the organizational structure and giving officers more discretion in carrying out their responsibilities, redesigning the job to assign unwanted or unnecessary tasks for a police officer to another position, and having excellent public safety equipment such as weaponry, vehicles, and body armor.

5. Explain the circumstances under which police officers may be justified in using deadly force.

The use of deadly force by a police officer may be justifiable if (1) the arrest is for a felony, (2) the person effecting the arrest is authorized to act as a peace officer or is assisting a person whom he or she believes to be authorized to act as a peace officer, (3) the officer believes that the force employed creates no substantial risk of injury to innocent persons, (4) the officer believes that the crime for which the arrest is made involved conduct including the use or threatened use of deadly force, and (5) the officer believes there is substantial risk that the person to be arrested will cause death or serious bodily harm if his or her apprehension is delayed.

6. List some of the ways to control and reduce police corruption.

Ways to control and reduce police corruption include selecting and maintaining officers with high moral standards, developing rigid departmental policies that cover the wide range of activities that corruption comprises, disciplining and prosecuting officers who are guilty of violating established policies and laws, utilizing a proactive internal affairs investigations unit, holding first-line supervisors responsible for the actions of their subordinates, employing outside review and special prosecutors, emphasizing to officers the costs to police agencies of criminal prosecutions and civil liability suits, and stop rehiring bad police officers.

KEY TERMS

three I's of police selection 261
college academies 270
public safety officers 270
police cadet program 270
tech prep (technical preparation) 271
merit system 271
discretion 275
full enforcement 275
selective enforcement 276
racial profiling 278
job stress 280
copicide 281
excessive force 282
grass eaters 292
meat eaters 292
internal affairs investigations unit 292

REVIEW QUESTIONS

1. Explain how the three I's of police selection (intelligence, integrity, and interaction skills) relate to the success of a police officer.
2. What are some advantages of hiring college-educated police officers?
3. What are some arguments in favor of and opposing the reduction of police discretion?
4. Why do police generally not like to respond to domestic violence calls?
5. What is *racial profiling* in law enforcement, and what are some of the methods that have been prescribed to stop it?
6. What are some of the conditions that can lead to police job stress?
7. What is meant by *copicide*?
8. What is meant by *excessive force*?
9. What are some types of police misconduct?

IN THE FIELD

1. **Neighborhood Survey** Conduct a survey about the police in your neighborhood. Use the same survey items as were used in Table 7.6. Ask your neighbors whether they agree or disagree with each item. Compare the results of your survey with the results in Table 7.6.
2. **Police Recruiting** Contact your local police department, and find out what it does to recruit police candidates. Does it run a police academy? Does it generally recruit officers from other jurisdictions? Compare your findings with those of others in the class and what you have learned from your textbook.

ON THE WEB

1. **Racial Profiling** Go to the website of the American Civil Liberties Union at https://www.aclu.org/report/driving-while-black-racial-profiling-our-nations-highways and look at its report on "Driving While Black: Racial Profiling on Our Nation's Highways" by Professor David A. Harris from the University of Toledo College of Law. Read Professor Harris's report and his five recommendations for ending racial profiling. Which of the five recommendations do you agree with and why? Which do you disagree with and why? Discuss these findings in class and see if other students share your views.
2. **Job Qualifications** Go to the websites of four large police departments (New York City at www.nypdrecruit.com; Los Angeles at www.lapdonline.org; Philadelphia at www.phillypolice.com/; Houston at www.houstonpolice.org/), and identify the major qualifications to be a police officer that are listed there. What qualifications are the same or similar among the agencies you examined? What qualifications are unique among those listed by the four agencies? Which employment qualifications are the most difficult to meet? Do you agree that all of the qualifications most difficult to meet are necessary? Would you add any qualifications? If so, what would they be? What would be the impact of your new qualifications on the recruitment and selection of police officers?

CRITICAL THINKING EXERCISES

STRESS

1. You are the commander of the operations division of a medium-sized police department in charge of the patrol, criminal investigation, and traffic sections.

a. What would you do if several officers from each of the sections came to you and said that they believed that job stress was hindering officer performance and endangering the health of several officers in each of the units?

b. How would you go about validating their claims of job stress in the work environment?

c. If it was determined that stressors such as shift changes that were too frequent, poor communication among the various ranks, and a lack of sufficient safety equipment were present, what would be your plan to improve working conditions and reduce job stress?

d. Provide a step-by-step summary of what you would do.

THE CASE OF PHILANDO CASTILLE

2. On July 6, 2016, Philando Castile, a 32-year-old school cafeteria worker, was shot and killed by St. Anthony, Minnesota, Police Officer Jeronimo Yanez. The entire episode was captured on audio from Yanez's squad car and on video. Officer Yanez pulled Castile over for a broken brake light and approached Castile's car on the driver's side. Also in the car were Castile's fiancée, Diamond Reynolds, and their 4-year-old daughter. After exchanging greetings, Yanez told Castile about the broken brake light. By that time, St. Anthony Police Officer Joseph Kauser, who had arrived as backup, approached Castille's car on the passenger side. Yanez asked Castile for his driver's license and proof of insurance. Castile provided Yanez with his proof of insurance card; Yanez looked at it, and put it into his pocket. Castile then told Yanez: "Sir, I have to tell you that I do have a firearm on me" (which he was licensed to have). Before Castile completed the sentence, Yanez interrupted and replied, "Okay" and placed his right hand on the holster of his gun. Yanez said, "Okay, don't reach for it, then." Castile responded: "I'm . . . I'm . . . [inaudible] reaching . . . ," before being again interrupted by Yanez, who said, "Don't pull it out." Castile responded, "I'm not pulling it out," and Reynolds said, "He's not pulling it out." Yanez then screamed: "Don't pull it out," and pulled his gun with his right hand. Yanez fired seven shots in the direction of Castile in rapid succession. The seventh shot was fired 62 seconds after Castile and Yanez exchanged greetings. Officer Kauser did not touch or remove his gun. Reynolds yelled, "You just killed my boyfriend!" Castile moaned and said, "I wasn't reaching for it." Those were his last words. Reynolds said, "He wasn't reaching for it." Before she completed her sentence, Yanez screamed, "Don't pull it out!" Reynolds responded, "He wasn't." Yanez yelled, "Don't move! F***." Reynolds started live-streaming onto Facebook about 40 seconds after the last shot. Reynolds said, "Stay with me, we got pulled over for a busted tail light in the back. And the police just he's, he's, he's covered. He, they just killed my boyfriend." Yanez said, "F***!" Reynolds responded, "He's licensed. He's carried, he is licensed to carry." Yanez said, "Ma'am, just keep your hands where they are." Reynolds, responded, "I will sir. No worries. I will." Yanez said, "F***" and added, "I told him not to reach for it. I told him to get his hand off of it." Reynolds replied, "He had, you told him to get his ID, sir, and his driver's license. Oh my God. Please don't tell me he's dead."

The next day, Officer Yanez told investigators his side of the story. Yanez said Castile told him he had a gun at the same time he reached down between his right leg and the center console of the vehicle. "And he put his hand around something," Yanez said. He said Castile's hand took a C-shape, "like putting my hand up to the butt of the gun." Yanez said he then lost view of Castile's hand. "I know he had an object and it was dark," he said. "And he was pulling it out with his right hand. And as he was pulling it out, a million things started going through my head. And I thought I was gonna die." Yanez said he thought Castile had the gun in his right hand and he had "no option" but to shoot.

On June 16, 2017, after five days of deliberations, a jury found Officer Jeronimo Yanez not guilty of all counts in the shooting of Philando Castile. He had been charged with 1 count of felony second-degree manslaughter in Castile's death and 2 felony counts of intentional discharge of a dangerous weapon over allegedly endangering Diamond Reynolds and her daughter. In announcing the charges, the county attorney said, "No reasonable officer knowing, seeing and hearing what Officer Yanez did at the time would have used deadly force under these circumstances. Officer Yanez pleaded not guilty. After the verdict, a spokesperson for the City of St. Anthony said that Yanez "will not return to active duty" and that "the public will be best served if Officer Yanez is no longer a police officer in our city." Yanez would receive a "voluntary separation agreement."[87]

a. Do you believe that Officer Yanez used excessive force? Why or why not?

b. Should Officer Yanez have handled the situation (a traffic stop for a broken tail light) differently? How?

c. Should Philando Castile have behaved differently? What should he have done differently?

d. Could Officer Kauser have done anything to prevent the tragedy? If yes, what?

e. Do you agree with the jury that Officer Yanez was not guilty on all counts? Why or why not?

POLICE USE OF DEADLY FORCE

3. Police officers in the United States kill more than 1,000 people a year, while British and German police together kill about 10. Within the United States, a California resident is four times more likely to be killed by the police than a New York state resident. Arizona and New Mexico residents are eight times more likely to be killed by the police than residents of Connecticut.

In the largest cities of the United States, race appears to be a dominant influence on the likelihood of being killed by the police. For example, in Houston, black residents are four times more likely than white residents to be killed by the police. Black residents of New York and Los Angeles are six to seven times more likely than white residents to die in police shootings. In Chicago, black residents are 18 times more likely than white residents to be victims of police use of deadly force. However, in some cases, regional differences in the overall incidence of police killings are greater than racial disparities. For example, whites in Houston have a higher likelihood of being killed by the police than do blacks in New York City.

a. What accounts for the geographic variations in the police use of deadly force?

b. What accounts for the racial disparities in the police use of deadly force?

c. What can be done to reduce the police use of deadly force, especially the geographical and racial disparities?

NOTES

1. W. S. Wilson Huang and Michael S. Vaughn, "Support and Confidence: Public Attitudes Toward the Police," in *Americans View Crime and Justice: A National Public Opinion Survey*, ed. Timothy J. Flanagan and Dennis R. Longmire (Thousand Oaks, CA: Sage, 1996), 31–45.
2. Ronald Weitzer and Steven A. Tuch, "Racially Biased Policing: Determinants of Citizen Perceptions," *Social Forces* 83 (2005): 1009–1030.
3. "Confidence in Institutions," *Gallup*, accessed June 11, 2019, https://news.gallup.com/poll/1597/confidence-institutions.aspx.
4. Lydia Saad, "Military, Small Business, Police Still Stir Most Confidence," *Gallup*, June 28, 2018. Accessed June 12, 2019, https://news.gallup.com/poll/236243/military-small-business-police-stir-confidence.aspx.
5. Ibid.
6. Emily Ekins, "Policing in America: Understanding Public Attitudes Toward the Police. Results from a National Survey," Cato Institute, December 7, 2016, accessed June 11, 2019, https://www.cato.org/survey-reports/policing-america.
7. Justin McCarthy, "Americans' Respect for Police Surges," *Gallup*, October 24, 2016, accessed June 17, 2019, https://news.gallup.com/poll/196610/americans-respect-police-surges.aspx.
8. Ibid.
9. Ibid.
10. "Honesty/Ethics in Professions," *Gallup*, accessed June 18, 2019, https://news.gallup.com/poll/1654/honesty-ethics-professions.aspx.
11. Megan Brenan, "Nurses Again Outpace Other Professions for Honesty, Ethics," *Gallup*, December 20, 2018, accessed June 19, 2019, https://news.gallup.com/poll/245597/nurses-again-outpace-professions-honesty-ethics.aspx.
12. Ibid.
13. Ibid.
14. Ekins, "Policing in America."
15. Ibid.
16. Ibid.
17. Ibid.
18. Ibid.
19. Ibid.
20. Rich Morin, Kim Parker, Renee Stepler, and Andrew Mercer, "Behind the Badge," Pew Research Center, January 11, 2017, accessed June 19, 2019, https://www.pewsocialtrends.org/2017/01/11/behind-the-badge/.
21. The views of the public reported in this section are from a Pew Research Center American Trends Panel survey conducted in 2016.
22. Larry Gaines and Victor Kappeler, "Police Selection," in *What Works in Policing: Operations and Administration Examined,* ed. Gary Cornder and Donna Hale (Cincinnati, OH: Anderson, 1992).
23. Robert B. Mills, "Psychological, Psychiatric, Polygraph, and Stress Evaluation," in *The Police Personnel System*, ed. Calvin Swank and James Conser (New York: Wiley, 1981).
24. O. W. Wilson and Roy McLaren, *Police Administration* (New York: McGraw-Hill, 1972), 261.
25. Albert Reiss, *The Police and the Public* (New Haven, CT: Yale University Press, 1971).
26. Norma M. Riccucci and Karina Saldiver, "The Status of Employment Discrimination Suits in Police and Fire Departments Across the United States," *Review of Public Personnel Administration* 34 (2014):263–288, accessed June 29, 2019, http://citeseerx.ist.psu.edu/viewdoc/download?doi=10.1.1.866.7252&rep=rep1&type=pdf.
27. "Advancing Diversity in Law Enforcement," U.S. Department of Justice, Equal Employment Opportunity Commission, October 2016, accessed June 27, 2019, https://www.eeoc.gov/eeoc/interagency/police-diversity-report.cfm#_ftn59.
28. Ibid.
29. Ibid.
30. Shelley S. Hyland and Elizabeth Davis, "Local Police Departments, 2016: Personnel," U.S. Department of Justice, Bureau of Justice Statistics Bulletin, October, 2019, accessed October 27, 2019, https://www.bjs.gov/content/pub/pdf/lpd16p.pdf; Brian A. Reaves, *Local Police Departments, 2013: Personnel, Policies, and Practices*, U.S. Department of Justice, Bureau of Justice Statistics, May 2015, NCJ 248677, accessed January 31, 2016, www.bjs.gov/content/pub/pdf/lpd13ppp.pdf; Brian A. Reaves, *Local Police Departments 2007*, U.S. Department of Justice, Bureau of Justice Statistics (Washington, DC: U.S. Government Printing Office, December 2010), http://bjs.ojp.usdoj.gov/content/pub/pdf/lpd07.pdf; U.S. Census Bureau.
31. Shelley S. Hyland and Elizabeth Davis, "Local Police Departments, 2016: Personnel," 8, Table 10.
32. Ibid.
33. Ibid.
34. Ibid., 7, Table 9.
35. Christie Gardiner, "Policing Around the Nation: Education, Philosophy, and Practice," California State University Fullerton Center for Public Policy and the Police Foundation, September 2017, accessed February 16, 2020, https://www.policefoundation.org/wp-content/uploads/2017/10/PF-Report-Policing-Around-the-Nation_10-2017_Final.pdf.
36. Ibid., p. 32, Figure 19.
37. Ibid., 23, Figure 9.
38. KiDeuk Kim, Ashlin Oglesby-Neal, and Edward Mohr, 2016 Law Enforcement Use of Social Media Survey, A Joint Publication by the International Association of Chiefs of Police and the Urban Institute, February 2017, accessed June 29, 2019, https://www.urban.org/sites/default/files/publication/88661/2016-law-enforcement-use-of-social-media-survey_5.pdf.
39. Melody Gutierrez, "Police Agencies Want Access to Applicants' Social-Media Passwords," *SFGATE*, September 2, 2014, accessed June 29, 2019, https://www.sfgate.com/news/article/Law-enforcement-employers-still-view-private-5722229.php.
40. *Hild v. Bruner*, 1980; *Bonsignore v. City of New York*, 1981.
41. Jack Gregory, "The Background Investigation and Oral Interview," in *The Police Personnel System*, ed. Calvin Swank and James Conser (New York: Wiley, 1981).
42. Jerome Skolnick, *Justice Without Trial* (New York: Wiley, 1996).
43. Lawrence W. Sherman and Richard A. Berk, *The Minneapolis Domestic Violence Experiment* (Washington, DC: The Police Foundation, 1984).
44. Meghan A. Novisky and Robert L. Peralta, "When Women Tell: Intimate Partner Violence and the Factors Related to Police Notification," *Violence Against Women* 21 (2014):65–86.

45. Elizabeth Davis, Anthony Whyde, and Lynn Langton, "Contacts Between Police and the Public, 2015," in *Bureau of Justice Statistics Special Report*, U.S. Department of Justice (Washington, DC: USGPO, October 2018), accessed June 30, 2019, https://www.bjs.gov/content/pub/pdf/cpp15.pdf.
46. Lynn Langton and Matthew Durose, "Police Behavior During Traffic and Street Stops, 2011," Bureau of Justice Statistics, September 2013, NCJ 242937, accessed January 31, 2016, www.bjs.gov/content/pub/pdf/pbtss11.pdf.
47. Elizabeth Davis, Anthony Whyde, and Lynn Langton, "Contacts Between Police and the Public, 2015," in *Bureau of Justice Statistics Special Report*, U.S. Department of Justice (Washington, DC: USGPO, October 2018), accessed June 30, 2019, https://www.bjs.gov/content/pub/pdf/cpp15.pdf.
48. NAACP, *Born Suspect: Stop-and-Frisk Abuses & the Continued Fight to End Racial Profiling in America*, p. 19, September 2014, accessed April 20, 2016, http://action.naacp.org/page/-/Criminal%20Justice/Born_Suspect_Report_final_web.pdf.
49. "Stress Statistics," Statistic Brain Research Institute, American Institute of Stress, N.Y., October 19, 2015, accessed January 31, 2016, www.statisticbrain.com/stress-statistics/.
50. National Institute of Occupational Safety and Health (NIOSH), "Stress at Work," www.cdc.gov/niosh/stresswk.html.
51. The Word Spy, www.LOGOPHILIA.com/WordSpy/c.html.
52. "10% of Police Shootings Found to Be 'Suicide by Cop,'" *Criminal Justice Newsletter* 29, no. 17 (September 1, 1998): 1–2.
53. Kris Mohandie, J. Reid Meloy, and Peter I. Collins, "Suicide by Cop Among Officer-Involved Shooting Cases," *Journal of Forensic Sciences* 54 (2009): 456–462.
54. Alan Feuer, "Drawing a Bead on a Baffling Endgame: Suicide by Cop," *The New York Times,* June 25, 1998.
55. NIOSH, "Stress at Work."
56. "Investigation of the Chicago Police Department," United States Department of Justice Civil Rights Division and United States Attorney's Office Northern District of Illinois, January 13, 2017, accessed June 1, 2019, https://www.justice.gov/opa/file/925846/download.
57. Ibid., citations omitted.
58. Ibid.
59. Ibid.
60. Ibid.
61. Information about Burge is from a series of *Chicago Tribune* articles retrieved on November 12, 2010, from http://articles.chicagotribune.com/keyword/jon-burge.
62. "Largest Legal Settlements Against Police," *HUFFPOST*, September 11, 2016, accessed July 3, 2019, https://www.huffpost.com/entry/largest-legal-settlements_b_8122202.
63. Ibid.
64. Information from The Task Force Report is from "Police Accountability Task Force, Recommendations for Reform: Restoring Trust Between the Chicago Police and the Communities They Serve, Executive Summary," April 2016, accessed April 27, 2016 from https://assets.documentcloud.org/documents/2801130/Chicago-Police-Accountability-Task-Force-Report.pdf.
65. United States Department of Justice, Civil Rights Division, "Investigation of the Ferguson Police Department," March 4, 2015, accessed May 4, 2016 from www.justice.gov/sites/default/files/opa/press-releases/attachments/2015/03/04/ferguson_police_department_report.pdf.
66. Elizabeth Davis, Anthony Whyde, and Lynn Langton, "Contacts Between Police and the Public, 2015."
67. William Terrill and Eugene A. Paoline, III, "Examining Less Lethal Force Policy and the Force Continuum: Results From a National Use-of-Force Study," *Police Quarterly* 16 (2013): 38–65.
68. Amnesty International, "Deadly Force: Police Use of Lethal Force in the United States," 2015, accessed January 31, 2016, www.amnestyusa.org/sites/default/files/aiusa_deadlyforcereportjune2015.pdf.
69. Calculated from National Violent Death Reporting System (NVDRS), Centers for Disease Control and Prevention, accessed July 3, 2019, https://wisqars.cdc.gov:8443/nvdrs/nvdrsDisplay.jsp. The number of states for which data were available varied by year.
70. Mark Berman and Wesley Lowery, "Cleveland to Pay $6 Million in Police Shooting Case," *Orlando Sentinel*, April 26, 2016, A3; "How the Tamar Rice Settlement with Cleveland Compares to Similar Cases Across the Nation," accessed April 26, 2016 from www.cleveland.com/metro/index.ssf/2016/04/how_the_tamir_rice_settlement.html#0.
71. Information on the ARD program is from Michael Planty, Andrea Burch, Duren Banks, Lance Couzens, Caroline Blanton, and Devon Cribb, "Arrest-Related Deaths Program: Data Quality Profile."
72. Cited in Tom McCarthy, "Police Killed More than Twice as Many People as Reported by US Government," March 4, 2015, accessed April 29, 2016 from www.theguardian.com/us-news/2015/mar/04/police-killed-people-fbi-data-justifiable-homicides.
73. Ibid.
74. See Kimberly Kindy, Marc Fisher, Julie Tate, and Jennifer Jenkins, "A Year of Reckoning: Police Fatally Shoot Nearly 1,000," *The Washington Post*, December 26, 2015, accessed April 29, 2016 from www.washingtonpost.com/sf/investigative/2015/12/26/a-year-of-reckoning-police-fatally-shoot-nearly-1000/.
75. Larry Copeland, "Deaths Lead Police to Question High-Speed Chase Policies," *USA TODAY*, April 23, 2010, accessed January 17, 2010, www.usatoday.com/news/nation/2010-04-22-police-chase-deaths_N.htm.
76. John Kelly and Mark Nichols, "We Found 85,000 Cops Who've Been Investigated for Misconduct. Now You Can Read Their Records," *USA TODAY*, October 14, 2019, accessed October 21, 2019, https://www.usatoday.com/in-depth/news/investigations/2019/04/24/usa-today-revealing-misconduct-records-police-cops/3223984002/.
77. Georgett Roberts and Lia Eustachewich, "Retired NYPD Detective Get 4 to 12 Years for Running Brothel," *New York Post*, July 2, 2019, accessed July 5, 2019, https://nypost.com/2019/07/02/retired-nypd-detective-gets-4-to-12-years-for-running-brothel/.
78. Jay Weaver, "Two Miami Cops Found Guilty of Protecting Drug Dealers, Now Face Long Prison Terms, *Miami Herald*, July 1, 2019, accessed July 4, 2019, https://www.miamiherald.com/news/local/crime/article232082547.html.
79. Nick Grube, "Jury Convicts Kealohas and 2 HPD Officers of Conspiracy, *Honolulu Civil Beat*, June 27, 2019, accessed July 4, 2019, https://www.civilbeat.org/2019/06/jury-convicts-kealohas-and-2-hpd-officers-of-conspiracy/.
80. Jessica Lussenhop, "Rogue Baltimore Police Unit Ringleader Wayne Jenkins Sentenced," *BBC News*, June 7, 2018, accessed July 4, 2019, https://www.bbc.com/news/world-us-canada-44402948.

81. Roseanne Tellez, "Jury Awards Jacques Rivera $17M After Being Framed by CPD," *CBS Chicago*, June 29, 2018, accessed July 5, 2019, https://chicago.cbslocal.com/2018/06/29/jaques-rivera-gets-17m-from-cpd/; Melissa Segura, "Another Man Who Accused a Retired Chicago Cop of Framing Him Has Been Exonerated," *BuzzFeed News*, January 16, 2019, accessed July 5, 2019, https://www.buzzfeednews.com/article/melissasegura/another-guevara-defendant-is-exonerated.

82. Knapp Commission, *Report on Police Corruption* (New York: George Braziller, 1972).

83. Ellwyn R. Stoddard, "The Informal 'Code' of Police Deviancy: A Group Approach to Blue-Coat Crime," *Journal of Criminal Law, Criminology, and Police Science* 59 (1968): 204.

84. Candace McCloy, "Lawsuits Against Police: What Impact Do They Have?" *Criminal Law Bulletin* 20 (1984): 49–56.

85. James Pilcher, Aaron Hegarty, Eric Litke, and Mark Nichols, "Fired for a Felony, Again for Perjury. Meet the New Police Chief," *USA TODAY*, October 14, 2019, accessed October 21, 2019, https://www.usatoday.com/in-depth/news/investigations/2019/04/24/police-officers-police-chiefs-sheriffs-misconduct-criminal-records-database/2214279002/.

86. "When Police Misconduct Occurs, Records Often Stay Secret. One Mom's Fight to Change That," *USA TODAY*, October 14, 2019, accessed October 21, 2019, https://www.usatoday.com/in-depth/news/2019/05/23/police-brutality-misconduct-california-cop-records-sb-1421/3555785002/.

87. Quoted nearly verbatim from Madison Park, "The 62-second Encounter Between Philando Castile and the Officer Who Killed Him," *CNN*, May 30, 2017, accessed July 6, 2019, https://www.cnn.com/2017/05/30/us/philando-castile-shooting-officer-trial-timeline/index.html.

Design Elements: *Image of part of a court building with columns and statue*: Pixtal/AGE Fotostock RF; *Image of a door slightly open*: jadding/Shutterstock.com

8

Robert Daly/Getty Images

The Administration of Justice

CHAPTER OUTLINE

LEARNING OBJECTIVES

After completing this chapter, you should be able to:

1. Identify the type of court structure in the United States, and describe its various components.
2. Summarize the purposes of courts.
3. Identify the most powerful actors in the administration of justice, and explain what makes them so powerful.
4. Summarize the types of attorneys available to a person charged with a crime.
5. Describe the responsibilities of a judge.
6. Describe the purposes of an initial appearance.
7. Explain what bail is, and describe the different methods of pretrial release.
8. Describe what a grand jury is, and explain its purposes.
9. Describe the purposes of the arraignment and the plea options of defendants.
10. Describe the interests served and not served by plea bargaining.
11. List and define the stages in a criminal trial.
12. Explain the different roles of judges in adversarial and inquisitorial trial systems.

Crime Story

On September 25, 2018, 81-year-old actor Bill Cosby (pictured) was sentenced by Montgomery County, Pennsylvania Judge Steven O'Neill to 3 to 10 years in a state prison for drugging and sexually assaulting Andrea Constand at his home in 2004. He also was fined $25,000 and ordered to pay $43,611 for the cost of the prosecution. In addition, Cosby was designated a "sexually violent predator," a status that requires lifetime registration, lifetime mandatory sex offender counseling with a treatment provider, and notification to the community that a "sexually violent predator" lives in the area. Prosecutors had sought a sentence of 5 to 10 years in state prison, claiming that he had shown no remorse for his actions. His defense attorney Joseph P. Green argued for a sentence of house arrest because of Cosby's advanced age and blindness. His attorneys said they planned to file an appeal.

Cosby's first trial, in June 2017, ended in a hung jury when the jury deadlocked and could not unanimously agree on a verdict. A different jury, at a second trial on April 26, 2018, deliberated for more than 14 hours over 2 days before finding Cosby guilty of 3 counts of aggravated indecent assault. Cosby could have been sentenced to up to 10 years on each count. The judge merged the 3 counts into one because they all stemmed from the same event. Responding to the verdict, Montgomery County District Attorney Kevin Steele told reporters, "What was revealed through this investigation was a man who had spent decades preying on women that he drugged and sexually assaulted, and a man who evaded this moment right here far too long." Steele added, "He used his celebrity, he used his wealth, he used his network of supporters to help him conceal his crimes." Cosby's attorney, Tom Mesereau, said he planned to appeal "very strongly." Attorney Gloria Allred, who represents many of the women who have accused Cosby of sexual improprieties, said, "This was the happiest she had been with a verdict in 42 years." She praised the sentence as a long journey to justice and declared, "After all is said and done, women were finally believed." Cosby's trial was the first celebrity sexual assault trial since the #MeToo movement began, and many observers saw it as a test of whether the cultural shift the movement initiated would translate in court.

Although dozens of women have accused Cosby of sexual misconduct, only Constand's allegations resulted in criminal charges because it was the only case that occurred within the statute of limitations. Thus, the case against Cosby revolved around Constand's testimony. Constand was a former employee with the Temple University basketball team. She testified that Cosby, a powerful university trustee, drugged her and sexually assaulted her when she visited his home to ask for career advice in 2004. Constand reported the assault to police in 2005, but then-Montgomery County District Attorney Bruce Castor decided not to file charges in the case. Constand and Cosby settled a civil lawsuit for $3.38 million in 2006. Not until 2015, when several women accused Cosby of sexual assault, did new Montgomery County prosecutor Steele file charges and have Cosby arrested.

The trial was a classic "he said, she said" case common to sexual assault trials. Prosecutors had little forensic evidence. At the retrial, five other Cosby accusers testified as "prior bad acts" witnesses and stated that Cosby had drugged and assaulted them decades ago. Prosecutors attempted to prove that Cosby's behavior toward Constand was part of a pattern. The other witnesses and an incriminating civil deposition supported Constand's account. Cosby's defense team aggressively attacked Constand's credibility, argued their sexual relations were consensual, and that Constand was a "con artist," who wanted a piece of Cosby's fortune. In closing arguments, defense attorney Kathleen Bliss argued that Cosby's legal team was taking a stand against "witch hunts, lynchings (and) McCarthyism." Cosby's spokesperson called the prosecution "the most racist and sexist trial in the history of the United States."

Montgomery County Correctional Facility/Getty Images

Before sentencing, Constand wrote a five-page victim impact statement to the court in which she explained, "Bill Cosby took my beautiful, healthy young spirit and crushed it. He robbed me of my health and vitality, my open nature and my trust in myself and others." The judge said Constand's statement influenced the sentence: "I have given great weight to the victim impact testimony in this case, and it was powerful." Cosby declined to speak to the court before the sentence was imposed.

Cosby is serving his sentence at SCI Phoenix, a maximum-security prison in Collegeville, near Philadelphia. Cosby's spokesperson has told reporters that Cosby says, "It's not so bad behind bars." "Despite the circumstances," Cosby relates, "This is an amazing experience." Cosby reiterated that he has "no remorse" and never will, comparing himself to Martin Luther King, Jr., Mahatma Gandhi, Nelson Mandela, and other "political prisoners." In a statement from prison, Cosby wrote:

> My political beliefs, my actions of trying to humanize all races, genders,

and religions landed me in this place surrounded by barb-wire fencing, a room made of steel and iron. So, I now have a temporary residence that resembles the quarters of some of the Greatest Political Prisoners . . . I stand upright as a Political Prisoner and I Smile. The Truth is Strong!

Among the topics discussed in this chapter are the American court structure, key actors in the court process, and the criminal trial. Do you agree with the sentence imposed on Bill Cosby? (For each of these questions, defend your answer.) Should Cosby's advanced age and blindness have made a difference in his sentence? Should Cosby have been retried after his first trial ended with a hung jury? Did the #MeToo movement have an unfair influence on Cosby's case? Should there be a statute of limitations on sexual assault? Did Cosby receive effective representation? Should Cosby have testified on his own behalf? Should victim impact statements be admissible at trial? Is Cosby a political prisoner? These questions and their answers illustrate why criminal trials and sentencing are so difficult.

The American Court Structure

The United States has a **dual court system**–a separate judicial system for each of the states and a separate federal system. Figure 8.1 displays this dual court system and routes of appeal from the various courts. The only place where the two systems connect is in the U.S. Supreme Court.

The authority of a court to hear and decide cases is called the court's **jurisdiction**. It is set by law and is limited by territory and type of case. A court of **original jurisdiction** has the authority to hear a case when it is first brought to court. Courts having the power to review a case for errors of law are courts of **appellate jurisdiction**. Courts having the power to hear any type of case are said to exercise **general jurisdiction**. Those with the power to hear only certain types of cases have **special jurisdiction**. **Subject matter jurisdiction** is the court's power to hear a particular type of case. **Personal jurisdiction** is the court's authority over the parties to a lawsuit.

dual court system The court system in the United States, consisting of one system of state and local courts and another system of federal courts.

jurisdiction A politically defined geographical area. The right or authority of a justice agency to act with regard to a particular subject matter, territory, or person. The authority of a court to hear and decide cases.

original jurisdiction The authority of a court to hear a case when it is first brought to court.

appellate jurisdiction The power of a court to review a case for errors of law.

general jurisdiction The power of a court to hear any type of case.

special jurisdiction The power of a court to hear only certain kinds of cases.

subject matter jurisdiction The power of a court to hear a particular type of case.

personal jurisdiction A court's authority over the parties to a lawsuit.

The Federal Courts

The authority for the federal court system is the U.S. Constitution, Article III, Section 1, which states, "The judicial Power of the United States, shall be vested in one Supreme Court, and in such inferior courts as the Congress may from time to time ordain and

FIGURE 8.1 Dual Court System of the United States

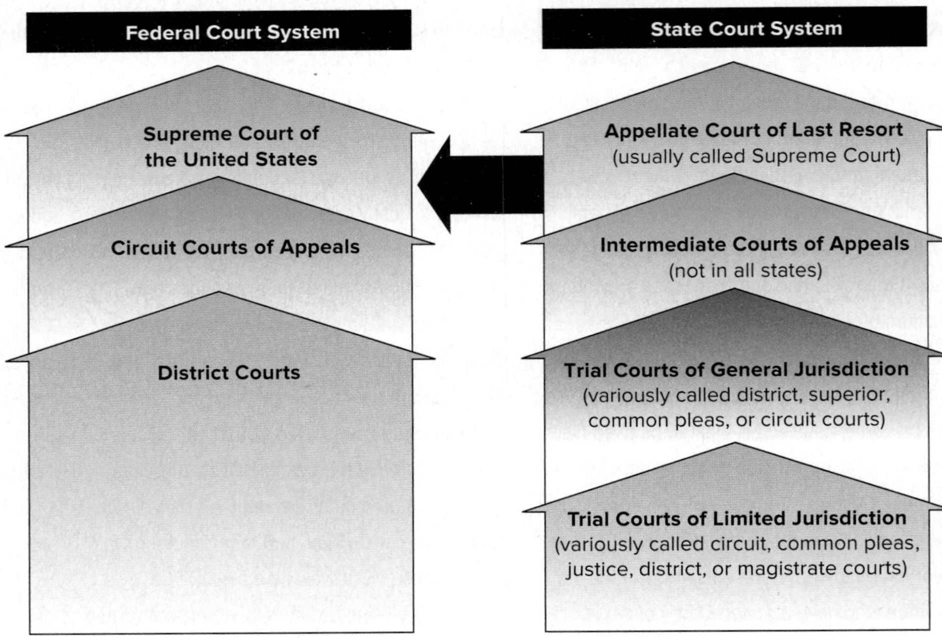

FYI The Judicial Conference of the United States

The Judicial Conference of the United States governs the federal court system. The conference is composed of 27 federal judges and is presided over by the Chief Justice of the U.S. Supreme Court. The conference meets twice a year to consider policies affecting the federal courts, to make recommendations to Congress on legislation affecting the judicial system, to propose amendments to the federal rules of practice and procedure, and to address administrative problems of the courts.

Source: Administrative Office of the U.S. Courts, Federal Judiciary website, www.uscourts.gov/judconf.html.

FIGURE 8.2 The Federal Court Structure

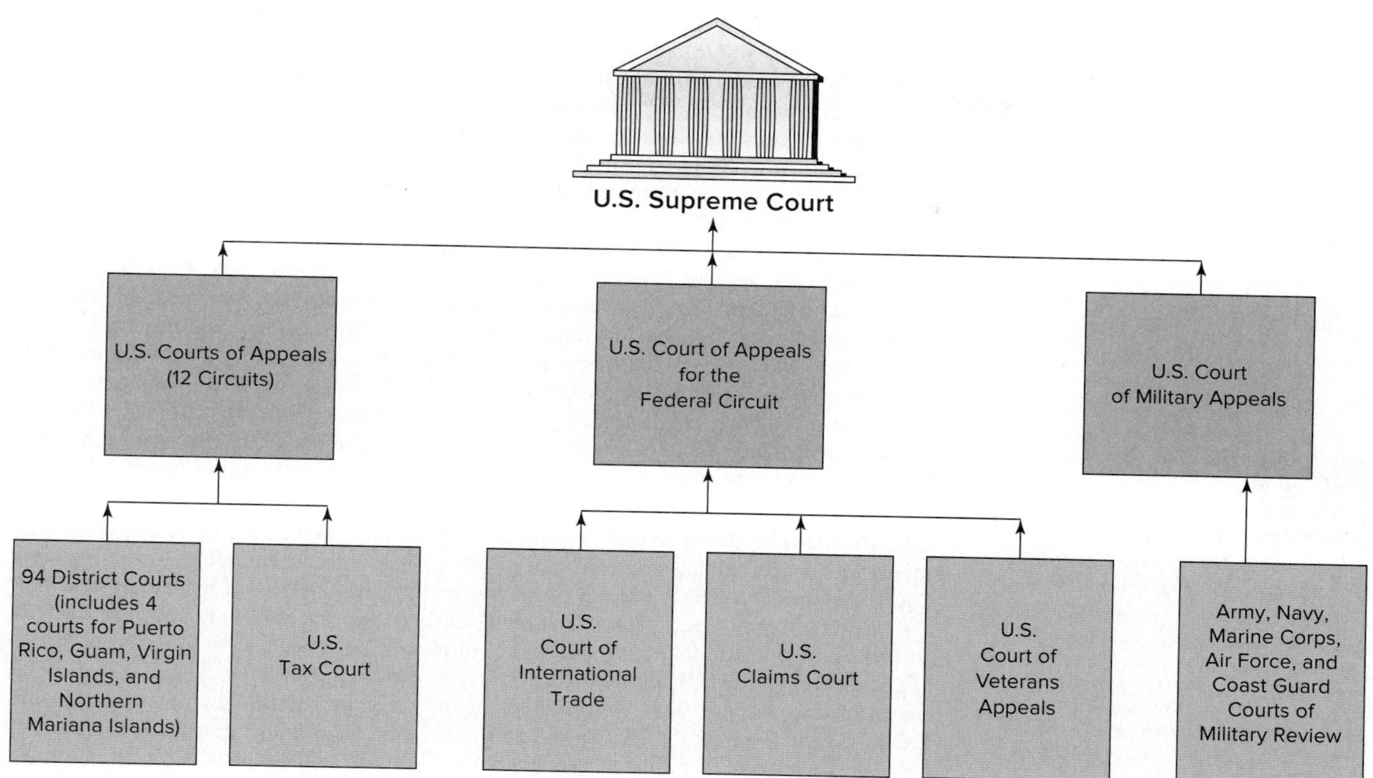

establish." The federal court system includes the Supreme Court, the federal courts of appeals, and the federal district courts. The federal court structure is shown in Figure 8.2.

U.S. District Courts Forming the base of the federal court structure are the U.S. district courts. These are courts of original jurisdiction, or courts where most violations of federal criminal and civil law are first adjudicated. Today, there are 94 district courts divided into 13 circuits, with at least one federal district court in each state, one each in the District of Columbia and the commonwealths of Puerto Rico and the Northern Mariana Islands, and one each in the U.S. territories of the Virgin Islands and Guam. In some states, the courts are divided into districts geographically. New York, for example, has northern, eastern, southern, and western district courts.

Two factors determine the jurisdiction of federal district courts: the subject matter of a case and the parties to a case. Federal district courts have subject matter jurisdiction over cases that involve federal laws, treaties with foreign nations, or interpretations of the Constitution. Cases involving admiralty or maritime law–the law of the sea, including ships, their crews, and disputes over actions and rights at sea–also come under federal district court jurisdiction.

Federal district courts have personal jurisdiction in cases if certain parties or persons are involved. These include (1) ambassadors and other representatives of foreign governments, (2) two or more state governments, (3) the U.S. government or one of its offices or agencies, (4) citizens of different states, (5) a state and a citizen of a different state, (6) citizens of the same state claiming lands under grants of different states, and (7) a state or its citizens and a foreign country or its citizens.

U.S. district courts are presided over by district court judges who are appointed by the president; are confirmed by the Senate; and except for the territorial judges who serve 10-year terms, serve for life (if they choose, do not resign, or are not impeached and convicted by Congress). At the end of 2018, there were 667 authorized federal district judgeships, of which 10 were temporary.[1] As of May 21, 2019, 122 of these judgeships were vacant.[2] Several district courts, such as the district court of eastern Oklahoma, had only one judge; the southern-district court of New York in New York City had the most with 28.[3] In 2019, the annual salary of a U.S. district court judge was $210,900.[4]

FYI **Federal Judge Impeached**

For only the eighth time in U.S. history, a federal judge was removed through the impeachment process. In December 2010, the Senate convicted U.S. District Judge G. Thomas Porteous of corruption and perjury for lying during the appointment process. As a Louisiana state judge, Porteous had accepted gifts from lawyers and friends to pay gambling debts—a fact that he never denied. Porteous is the first federal judge removed from office for pre-federal conduct.

Source: Michael E. Memoli, "Senate Votes to Remove Louisiana Federal Judge," *The Orlando Sentinel*, December 9, 2010, A7.

Usually a single federal judge presides over a trial, and trial by jury is allowed if requested by a defendant. A special panel of three judges hears some complex civil cases. The bulk of the workload of the U.S. district courts is devoted to civil cases, the number of which has risen dramatically in recent years, to 278,721 in 2018, an increase of 1.5% from 2017, but a slight decrease of 0.1% from 2009.[5] Federal criminal cases involve such crimes as bank robbery, counterfeiting, mail fraud, kidnapping, and civil rights abuses. Until 1980, the number of criminal cases in federal district courts remained relatively stable at about 30,000 a year. However, as a result of the federal government's war on drugs and an increase in illegal immigration cases, the number of criminal cases had nearly tripled to 89,098 in 2018, an increase of 12.7% from 2017, but a decrease of 9.7% from 2009.[6]

To ease the caseload of U.S. district court judges, Congress created the judicial office of federal magistrate in 1968. The title changed to magistrate judge in 1990. As their caseloads require and as funding from Congress permits, district judges may appoint magistrate judges to part-time or full-time positions. Magistrate judges have authority to issue warrants, conduct preliminary proceedings in criminal cases, such as initial appearances and arraignments, and hear cases involving petty offenses committed on federal lands. In most districts, magistrate judges handle pretrial motions and hearings in civil and criminal cases. While most civil cases are tried by district judges, magistrate judges also may preside over civil trials if all parties consent.[7] The behavior of most U.S. judges is governed by a code of conduct, presented in Figure 8.3.

FYI

Circuit Courts

The circuit courts are so named because early in the nation's history, federal judges traveled by horseback to each of the courts in a specified region, or "circuit," in a particular sequence. In short, they rode the circuit.

Source: Howard Ball, "The Federal Court System," in *Encyclopedia of the American Judicial System: Studies of the Principal Institutions and Processes of Law*, ed. R. J. Janosik (New York: Charles Scribner's Sons, 1987), 556.

Circuit Courts of Appeals A person or group that loses a case in district court may appeal to a federal circuit court of appeals or, in some instances, directly to the Supreme Court. Congress created the U.S. circuit courts of appeals in 1891 to reduce the case burden of the Supreme Court. The U.S. circuit courts of appeals have only appellate jurisdiction and review a case for errors of law, not of fact. Most appeals arise from the decisions of district courts, the U.S. Tax Court, and various territorial courts (see Figure 8.2). Federal courts of appeals also hear appeals of regulatory agency rulings, such as those by the Federal Trade Commission. An appeal to the U.S. circuit court of appeals is a matter of right—the court cannot refuse to hear the case. However, unless appealed to the Supreme

FIGURE 8.3 Code of Conduct for United States Judges

Canon 1	A Judge Should Uphold the Integrity and Independence of the Judiciary.
Canon 2	A Judge Should Avoid Impropriety and the Appearance of Impropriety in All Activities.
Canon 3	A Judge Should Perform the Duties of the Office Impartially and Diligently.
Canon 4	A Judge May Engage in Extra-Judicial Activities to Improve the Law, the Legal System, and the Administration of Justice.
Canon 5	A Judge Should Regulate Extra-Judicial Activities to Minimize the Risk of Conflict with Judicial Duties.
Canon 6	A Judge Should Regularly File Reports of Compensation Received for Law-Related and Extra-Judicial Activities.
Canon 7	A Judge Should Refrain from Political Activity.

The Code of Conduct for United States Judges was initially adopted by the Judicial Conference on April 5, 1973, and was known as the "Code of Judicial Conduct for United States Judges." At its March 1987 session, the Judicial Conference deleted the word "Judicial" from the name of the Code. Substantial revisions to the Code were adopted by the Judicial Conference at its September 1992 session. This Code applies to United States Circuit Judges, District Judges, Court of International Trade Judges, Court of Federal Claims Judges, Bankruptcy Judges, and Magistrate Judges. Certain provisions of this Code apply to special masters and commissioners. In addition, the Tax Court, Court of Appeals for Veterans Claims, and Court of Appeals for the Armed Forces have adopted this Code. Persons to whom the Code applies must arrange their affairs as soon as reasonably possible to comply with the Code and should do so in any event within one year of appointment.

Source: Administrative Office of the U.S. Courts, Federal Judiciary website, "Judges and Judgeships," www.uscourts.gov/guide/vol2/ch1.html.

FIGURE 8.4 The 13 U.S. Circuits

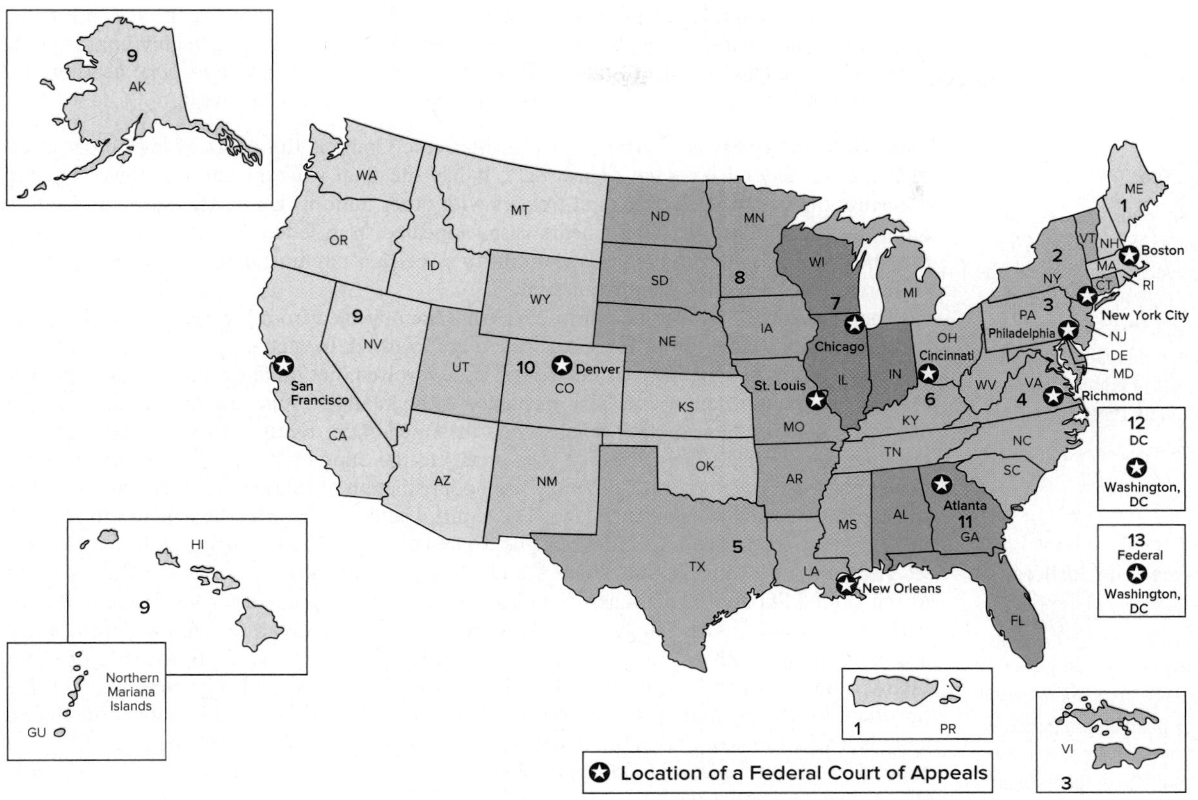

Court, decisions of the courts of appeals are final. U.S. courts of appeals heard 48,876 cases in 2018, a decrease of 1.9% from 2017, and 14.5% from 2009.[8]

There are currently 13 U.S. circuit courts of appeals (see Figure 8.4). Twelve of them have jurisdiction over cases from particular geographic areas. The Court of Appeals for the Federal Circuit, created in 1982, has national jurisdiction over specific types of cases, such as appeals from the U.S. Court of International Trade.[9] Like U.S. district judges, federal appellate judges are nominated by the president, confirmed by the Senate, and may serve for life. Currently, 179 judges serve on the 13 U.S. circuit courts (168 on the 12 regional courts and 11 on the

PAUL SAKUMA/AP Images

The 9th U.S. Circuit Court of Appeals in San Francisco, California (pictured here), has the most judges (29) of all the circuit courts of appeals. *What determines how many judges are selected to serve on a U.S. circuit court of appeals?*

federal circuit).[10] As of May 21, 2019, 5 of these judgeships were vacant.[11] The yearly salary of a U.S. circuit court of appeals judge was $223,700 in 2019.[12] The number of judges assigned to each court of appeals ranges from 6 (the 1st Circuit) to 29 (the 9th Circuit) and, normally, 3 judges sit as a panel.[13] Jury trials are not allowed in these courts. In highly controversial cases, all the judges in a circuit may sit together and hear a case. Those *en banc* hearings are rare; there probably are no more than 100 of them in all circuits in 1 year.

Court Fees

The fee for filing an appeal with the U.S. Supreme Court is $300. Fees are waived for indigent defendants. Approximately 90% of the criminal proceedings appealed to the Supreme Court involve indigent defendants. The fees do not include the costs of printing various documents required under other court rules.

Source: Clerk of the Supreme Court; 28 U.S.C. §1911.

The U.S. Supreme Court The U.S. Supreme Court is the court of last resort in all questions of federal law (see Figure 8.2). It has the final word in any case involving the Constitution, acts of Congress, and treaties with other nations. Under the Supreme Court's appellate jurisdiction, the Court hears cases appealed from federal courts of appeals, or it may hear appeals from federal district courts in certain circumstances in which an act of Congress has been held unconstitutional.

The Supreme Court also may hear cases that are appealed from the high court of a state, if claims under federal law or the Constitution are involved. In such cases, however, the Court has the authority to rule only on the federal issue involved, not on any issues of state law. For example, suppose a state tries a person charged with violating a state law. During the trial, the accused claims that the police violated Fourth Amendment rights with an illegal search at the time of the arrest. The defendant may appeal to the Supreme Court on the constitutional issue only. The Supreme Court generally has no jurisdiction to rule on the state issue (whether the accused actually violated state law). The Court would decide only whether Fourth Amendment rights were violated. Decisions of the Supreme Court are binding on all lower courts.

Supreme Court Justices

The Supreme Court Justices of the United States have varied biographical, legal, and educational backgrounds. Learn more about the current justices by going to the following websites: www.oyez.org/justices and www.supremecourt.gov/about/biographies.aspx.

What do the justices have in common? How do their backgrounds differ?

The Supreme Court is composed of a chief justice, officially known as the Chief Justice of the United States, and eight associate justices. They are appointed for life by the president with the consent of the Senate and, like other federal judges, can be removed from office against their will only by "impeachment for, and Conviction of, Treason, Bribery, or other high Crimes and Misdemeanors." The Chief Justice, who is specifically nominated by the president for the position, presides over the Court's public sessions and private conferences, assigns justices to write opinions (when the Chief Justice has voted with the majority), and supervises the entire federal judiciary. (When the Chief Justice has voted with the minority, the Associate Justice who has the greatest seniority, that is, has been on the Court the longest, assigns who writes the opinion.) In 2019, the salary of the Chief Justice was $270,700; the salaries of the Associate Justices were $258,900.[14]

For a case to be heard by the Supreme Court, at least four of the nine justices must vote to hear the case (the "rule of four"). When the required number of votes has been achieved, the Court issues a **writ of *certiorari*** to the lower court whose decision is being appealed, ordering it to send the records of the case forward for review. The Court is limited by law and custom in the types of cases for which it issues writs of *certiorari*. The Court will issue a "writ" only if the defendant in the case has exhausted all other avenues of appeal and the case involves a substantial federal question as defined by the appellate court. A substantial federal question, as noted, is one in which there is an alleged violation of either the U.S. Constitution or federal law.

writ of *certiorari* A written order, from an appellate court to a lower court whose decision is being appealed, to send the records of the case forward for review.

When the Supreme Court decides a case it has accepted on appeal, it can take one of these actions:

1. Affirm the verdict or decision of the lower court and "let it stand."
2. Modify the verdict or decision of the lower court, without totally reversing it.
3. Reverse the verdict or decision of the lower court, requiring no further court action.
4. Reverse the verdict or decision of the lower court and remand the case to the court of original jurisdiction, for either retrial or resentencing.

In some cases, the Supreme Court has ordered trial courts to resentence defendants whose original sentences violated the Eighth Amendment prohibition against cruel and unusual punishment. In other cases, prison authorities have been ordered to remedy unconstitutional conditions of imprisonment.

Appeals to the Supreme Court are heard at the discretion of the Court, in contrast to appeals to the U.S. circuit courts, which review cases as a matter of right. In 2017 for example, the Supreme Court heard only 69 of the 6,315 cases filed for review and disposed of 63 of the cases, 59 with signed opinions.[15] In other words, in 2017, the Court opted not to review 99% of the appealed cases—a figure typical of Court practice. Generally, the Supreme Court's refusal to hear a case ends the process of direct appeal. In certain circumstances, an imprisoned defendant whose appeal has been denied may still try to have

Shutterstock

U.S. Supreme Court Justices (left to right): Associate Justices Stephen G. Breyer, Neil M. Gorsuch, Clarence Thomas, and Sonia Sotomayor; Chief Justice John Roberts Jr.; and Associate Justices Elena Kagan, Ruth Bader Ginsburg, Brett M. Kavanaugh, and Samuel A. Alito.

the Supreme Court review his or her case on constitutional grounds by filing a writ of *habeas corpus*. A **writ of *habeas corpus***–which is guaranteed by Article I, Section 9, of the Constitution, the Federal *Habeas Corpus* Act, and state *habeas corpus* laws–is a court order directing a law officer to produce a prisoner in court to determine if the prisoner is being legally detained or imprisoned. The *habeas corpus* proceeding does not test whether the prisoner is guilty or innocent.

writ of *habeas corpus* An order from a court to an officer of the law to produce a prisoner in court to determine if the prisoner is being legally detained or imprisoned.

The Supreme Court also has original jurisdiction. However, Article III, Section 2.2, of the Constitution limits the Court's original jurisdiction to two types of cases: (1) cases involving representatives of foreign governments and (2) certain cases in which a state is a party. Many cases have involved two states and the federal government. When Maryland and Virginia argued over oyster fishing rights and when a dispute broke out between California and Arizona over the control of water from the Colorado River, the Supreme Court had original jurisdiction in the matters. Original jurisdiction cases are a very small part of the Court's yearly workload. Most of the cases the Court decides fall under its appellate jurisdiction.

The State Courts

The state courts have general power to decide nearly every type of case, subject only to the limitations of the U.S. Constitution, their own state constitutions, and state law. State and local courts are the courts with which citizens most often have contact. These courts handle most criminal matters and the majority of day-to-day legal matters. The laws of each state determine the organization, function, and even the names of its courts. Thus, no two state court systems are exactly alike. For discussion purposes, it is useful to distinguish four levels of state courts: trial courts of limited jurisdiction, trial courts of general jurisdiction, intermediate appellate courts, and state courts of last resort.[16]

State Courts

To learn more about state courts, go to the National Center for State Courts website at https://www.ncsc.org and read more about these courts.

Do you think variation in the names, functions, and organization of state court systems should be eliminated, and that all states courts should be identical in those areas? Why or why not?

Trial Courts of Limited Jurisdiction At the base of the state court structure (see Figure 8.1) are the approximately 14,000 to 16,000 trial courts of limited jurisdiction, sometimes referred to generally as "inferior trial courts" or simply as "lower courts."[17] Depending on the jurisdiction, those courts are called city courts, municipal courts, county courts, circuit courts, courts of common pleas, justice-of-the-peace courts, district courts, or magistrate courts. (Technically, most of the lower courts are not really part of the state judicial structure because they are the creation of, and funded by, either city or county governments.) In several states, judges of the lower courts are not required to have any formal legal training.

The lower courts typically deal with minor cases, such as ordinance and traffic violations, some misdemeanors, and–in many jurisdictions–civil cases involving less than $1,000. For those types of offenses, the lower courts in many states are allowed to conduct **summary** or **bench trials**, or trials without a jury. Typically, the greatest penalty that can be imposed is a fine of $1,000 and a maximum of 12 months in jail. Unlike trial courts of general jurisdiction, the lower courts are not courts of record in which detailed transcripts of the proceedings are made. Because they are not courts of record, an appeal from such a lower court requires a **trial *de novo*** in which a trial court of general jurisdiction must rehear the entire case.

summary or bench trials A trial before a judge without a jury.

trial *de novo* A trial in which an entire case is reheard by a trial court of general jurisdiction because there is an appeal and there is no written transcript of the earlier proceeding.

In addition to handling minor cases, the lower courts in most states hear the formal charges against persons accused of felonies, set bail, appoint counsel for indigent defendants, and conduct preliminary hearings for crimes that must be adjudicated at a higher level. The legal proceedings in these courts typically are less formal, and many cases are resolved without defense attorneys. Lower courts process and quickly dispose of large numbers of cases, approximately 56 million in 2015.[18]

FYI Confidence in State Court System

A 2018 public opinion poll found that 76% of registered voters had a "great deal" or "some" confidence in the state court system; 22% had "not much" or "no confidence at all" in the state court system. This level of confidence was relatively consistent regardless of political party affiliation.

Source: National Center for State Courts, "2018 State of the State Courts—Survey Analysis," December 3, 2018, accessed May 28, 2019, https://www.ncsc.org/~/media/Files/PDF/Topics/Public%20Trust%20and%20Confidence/SoSC_2018_Survey_Analysis.ashx.

Trial Courts of General Jurisdiction Variously called district courts, superior courts, and circuit courts, depending on the jurisdiction, the more than 3,000 trial courts of general jurisdiction have the authority to try all civil and criminal cases and to hear appeals from lower courts. They are courts of record (formal transcripts of the proceedings are made), and judges and lawyers in those courts have formal legal training. Trial courts of general jurisdiction are funded by the state. In 2016, trial courts of general jurisdiction processed about 84 million cases, a decrease of 19% from the nearly 104 million cases processed in 2007.[19]

Drug Courts and Other Specialty Courts Augmenting trial courts of general jurisdiction are specialty courts that have been created during the past three decades to deal with increases in certain types of crimes or chronic social problems.[20] "Drug courts" were the first of these new "problem-solving" courts, the first of which was established in Dade County (Miami), Florida, in 1989. Within a decade, all but 10 states had at least one such court. Drug courts were created to (1) help handle the dramatic increase in drug cases resulting from the war on drugs that has been overwhelming the trial courts of general jurisdiction and (2) use the court's authority to reduce crime by changing defendants' drug-using behavior. In exchange for the possibility of dismissed charges or reduced sentences, defendants accept diversion to drug treatment programs during the judicial process. Drug court judges preside over drug court proceedings; monitor the progress of defendants by means of frequent status hearings; and prescribe sanctions and rewards as appropriate in collaboration with prosecutors, defense attorneys, treatment providers, and others.

Research indicates that the key to a successful drug court program is a judge who interacts with program participants in a respectful, fair, attentive, enthusiastic, consistent, predictable, caring, and knowledgeable way. Not all judges make good drug court judges. In addition to choosing judges with the right personality characteristics, the most effective drug court programs require participants to have (1) at least two judicial status hearings per month, (2) at least two urine drug tests per month, (3) at least one clinical case management session per week, and (4) a minimum of 35 days of formal drug-abuse treatment sessions. They also make participants aware of the alternative sanctions they face if they fail the program or do not have regular contact with program personnel or the judge; they provide participants and staff with a written schedule of sanctions for rule infractions (although staff retain the discretion not to impose sanctions if there are good reasons); and they maintain only one point of entry into the program either at preadjudication or postadjudication, but not both.

Joe Burbank/REUTERS/POOL/Newscom

Noelle Bush, daughter of former Florida Governor Jeb Bush, is shown with her brother George P. Bush in Orange County, Florida, drug court during her status hearing. Noelle was charged with contempt of court and spent 43 hours in jail for violating her court-ordered drug treatment plan. She was ordered to continue in the drug program. *What are some problems with drug courts?*

Research shows that compared to participants in other programs for drug-involved offenders, active drug court participants, especially those who graduate from the program, (1) are less likely to relapse to drug use and, when they do relapse, they use drugs significantly less often than they did before participating in the program; (2) are less likely to commit self-acknowledged criminal acts; (3) are less likely to be rearrested for substantial amounts of time after completing the program; and (4) are more likely to reap psychosocial benefits not related to drug use or criminal behavior, such as less family conflict, being enrolled in school, and needing less assistance with employment, educational services, and financial issues. Drug courts also are cost effective. It has been estimated that local communities realize about $2 in net economic benefits for every $1 invested in drug courts. However, to garner even greater financial benefits for communities, drug courts likely must start targeting more serious offenders.

Not surprisingly, drug courts are popular. In 1999, the National District Attorney's Association and the National Sheriff's Association officially endorsed drug courts. In 2019, there were 4,168 drug courts in the United States[21]—an increase of nearly 33% since 2015, and about 2,489% since 1997. Drug courts are located in all 50 states, the District of Columbia, Guam, Northern Mariana Islands, Puerto Rico, and a number of Native American Tribal Courts. California had the most drug courts (413) in 2019, followed by New York (316) and Ohio (234).[22] In recent years, drug courts have served annually about 150,000 people throughout the United States and its territories.[23]

Drug courts are so popular and the need for them is perceived to be so great that more specialized drug courts have been established. The first of these "sub-specialty" drug courts was a women's drug court, which opened in Kalamazoo, Michigan, in 1992. That was followed in 1995 by the first juvenile drug court in Visalia, California, and the first family drug court in Reno, Nevada. Juvenile drug courts operate within juvenile courts and handle delinquency cases or status offenders who have alcohol and/or drug problems. Family drug courts deal with selected abuse, neglect, and dependency cases in which parental substance abuse is a primary factor. Other specialized drug courts include:

- Adult drug courts (for nonviolent substance-abusing adult offenders)
- Campus drug courts (for students with substance abuse-related disciplinary cases that would otherwise result in suspension or expulsion from college)
- DUI/DWI (driving under the influence of drugs or alcohol/driving while intoxicated) drug courts (for alcohol-/drug-dependent offenders arrested for DUI or DWI)
- Integrated-treatment courts (for juveniles and their families/guardians who need services to help juveniles make better choices about their lives and drug and alcohol use)
- Reentry drug courts (for drug-involved offenders being released from local or state correctional facilities)
- "Tribal healing to wellness" drug courts (for Native American alcohol- and/or drug-related offenders living in Native American communities)
- Veterans treatment courts (for military veterans or active duty military personnel charged with crimes caused or influenced by a moderate-to-severe substance use disorder and/or serious and persistent mental health disorder)

Drug Courts

To learn more about drug courts, go to the National Drug Court Resource Center website at www.ndcrc.org and read more about this specialty court.

Do you think drug courts are a good idea? Why or why not?

As of June 30, 2015, drug courts also have provided the general model for 1,310 other, nondrug specialty courts.[24] These other problem-solving courts include:

- Child-support courts (for defendants with child-support issues)
- Collections courts (for offenders, particularly probationers, who are having problems paying or are unwilling to pay court costs and restitution)
- Community courts (for low-level, nonviolent offenders who are required, as punishment for their crimes, to help in the restoring of distressed neighborhoods by cleaning streets and removing graffiti)
- Domestic-violence courts (for defendants facing domestic-violence charges, and victims of domestic violence who need help)
- Environmental courts (for defendants charged with housing, community health, solid waste, fire, building, and zoning violations)
- Gambling courts (for defendants suffering from a pathological or compulsive gambling disorder)
- Gun courts (for defendants charged with illegal firearm possession)
- Homeless courts (for homeless individuals with records of misdemeanor offenses, who, because of their records, are ineligible for government aid and, in some jurisdictions, such things as driver's licenses)
- Mental health courts (for mentally ill or developmentally disabled individuals who have been charged with nonviolent misdemeanors and are in need of treatment)
- Prostitution courts (for defendants charged with sex-work offenses)
- Reentry courts (for parolees or other persons released conditionally from jail or prison who have service needs that must be addressed to achieve successful reintegration into the community)
- Truancy courts (for children trying to overcome the underlying causes of truancy)
- Veterans courts (for military veterans, primarily those returning from Iraq and Afghanistan but also Vietnam, who, because of posttraumatic stress disorder, brain trauma, and chemical dependency, commit crimes)[25]

Mental Health Courts

More than one in eight prisoners in the United States have a serious mental illness. Judge Charlotte Cooksey, who founded a mental health court in Baltimore in 2002, argues, "Our jails have become de facto psychiatric hospitals." The few studies performed suggest that mental health courts work. In Allegheny County, Pennsylvania, for example, the recidivism rate for participants was just 14% after 6 years, compared with 67% in the general prison population.

Source: "Mental Health Courts Offer Unique Approach to Changing Lives," *Justice Matters* 13, no. 1 (2010), www.courts.state.md.us/publications/ejusticematters/2010/winter/mentalhealthcourts.html.

Teen courts (also called peer courts or youth courts in which youths who commit minor offenses receive consequences for their behavior from a jury of their peers) are not included in the 1,310 total of other, nondrug specialty courts because a majority of teen courts do not operate under the judicial branch. All of these specialty courts share three common characteristics: (1) they focus on one type of case, (2) court personnel receive specialized training for the particular type of case, and (3) judges closely monitor compliance with court dispositions.

Intermediate Appellate Courts In some of the geographically smaller and less-populous states, there is only one appellate court, the state court of last resort, usually called the state supreme court. Many states, however, have created intermediate appellate courts to reduce the overwhelming case burden of the state supreme court. As of 2017, 41 states had at least one intermediate appellate court; 9 states (Delaware, Maine, Montana, New Hampshire, Rhode Island, South Dakota, Vermont, West Virginia, and Wyoming) and the District of Columbia did not.[26] North Dakota has an intermediate appellate court that serves on a temporary basis at the request of the North Dakota Supreme Court.

The intermediate appellate courts have no trial jurisdiction. They hear only appeals in both civil and criminal cases from the trial courts of general jurisdiction. An appeals court is charged with reviewing a case for errors of law and ensuring that legal procedures were followed. The decision rendered by the appeals court is based on a review of the trial court's official transcript and any other legally relevant information that may be submitted. Brief oral arguments by the attorneys for both sides also are allowed.

Like their federal counterparts, the intermediate appellate courts cannot refuse to hear any legally appealed case. Intermediate appellate courts range in size from 1 to 99 judges, although the states with the largest number of intermediate appellate judges–California (99), Texas (80), Ohio (69), Florida (64), Illinois (54), Louisiana (53), and New York (53)–are divided into regional district courts. The largest intermediate appellate court with statewide jurisdiction is New Jersey, which has 33 judges.[27] Normally, three judges sit as a panel to decide cases.

State Courts of Last Resort In most states, the state court of last resort is referred to as the state supreme court, although, as noted earlier, some states use different names. In Massachusetts and Maine, for example, the state court of last resort is called the supreme judicial court; in Maryland and New York, it is called the court of appeals. Oklahoma and Texas have two courts of last resort: the court of criminal appeals (for criminal appeals) and the supreme court (for civil cases). The court of last resort in 15 states or the chief justice of the court of last resort in 34 states, the District of Columbia, and Puerto Rico is the designated head of a state's judicial branch. In one state, Utah, the Judicial Council is the designated head of the judicial branch.[28]

As previously noted, the primary responsibility of state courts of last resort is to hear appeals from either trial courts of general jurisdiction (in those states without intermediate appellate courts) or intermediate appellate courts. In states with intermediate courts of appeal, the state court of last resort, like the U.S. Supreme Court, has discretion in which cases it will hear. And, like the U.S. Supreme Court, most state courts of last resort have original jurisdiction over a few types of cases.

Depending on the state, the number of judges that serve on the state court of last resort ranges from five to nine, though more than half the states have seven.[29] However, unlike judges of intermediate courts of appeal, judges of the state court of last resort are not divided into panels to hear cases. Instead, all of the judges hear all of the cases; that is, they sit *en banc*. State courts of last resort have the final word on matters involving interpretation of state law. Although defendants dissatisfied with a verdict rendered in a state court of last resort may appeal the decision to the Supreme Court, the Supreme Court will hear the appeal only if it involves an alleged violation of the Constitution or federal law. In deciding cases, most state courts of last resort follow procedures similar to those employed by the Supreme Court.

FYI

Native American Courts

In 1820, the Cherokees created a judiciary modeled after the Anglo-American judiciary. It consisted of eight district courts, each with a judge, and four circuit courts with one circuit judge that had jurisdiction over two districts. A Supreme Court was added in 1823, to hear cases on appeal. Originally, the four circuit judges presided over the Supreme Court with judges disqualified in cases appealed from their own district. However, in 1828, elected Supreme Court judges replaced the circuit judges.

Source: Rennard Strickland, *Fire and the Spirits: Cherokee Law from Clan to Court* (Norman, OK: University of Oklahoma Press, 1975), 64, 216, 221.

THINKING CRITICALLY

1. What are some of the benefits of a dual court system?
2. Why is it difficult for a case to make it all the way to the Supreme Court? Should it be that difficult? Why or why not?

Purposes of Courts

Ted Rubin, former juvenile court judge and noted expert on juvenile justice, court, and rehabilitation issues, outlines 10 purposes of courts.[30] The first is to "do justice." However, whether justice is done usually depends on the interests and viewpoints of the parties involved in a dispute. Typically, the "winning" party believes that justice has been done, while the "loser" thinks otherwise.

A second purpose of courts is "to appear to do justice." Even when a decision rendered by a court seems unjust to some people, it still is important that the court appear to "do justice." The appearance of justice is accomplished primarily by providing due process of law. **Due process of law**, as noted in Chapter 4, refers to the procedures followed by courts to ensure that a defendant's constitutional rights are not violated.

due process of law The rights of people suspected of or charged with crimes. Also, the procedures followed by courts to ensure that a defendant's constitutional rights are not violated.

A third purpose of courts is "to provide a forum where disputes between people can be resolved justly and peacefully." Until the creation of courts of law, disputes often were settled violently through blood feuds and other acts of revenge. Aggrieved parties would gather their extended families and friends and make war against the families and friends of the person or persons who had presumably violated their rights. Sometimes those feuds would span generations, as did the legendary battles between the Hatfields and the McCoys. Courts were instituted, at least in part, to prevent those calamities. Thus, regardless of which side in a dispute "wins," what is important, from the standpoint of the court, is that the dispute be resolved justly and, perhaps more important, peacefully.

A fourth purpose of courts is "to censure wrongdoing." To "censure" means to condemn or to blame. In this context, it refers to condemning or blaming people who have violated the law.

Purposes five through eight involve specific outcomes that courts hope to achieve by their actions. They are **incapacitation**, or the removal or restriction of the freedom of those found to have violated criminal laws; **punishment**, or the imposition of a penalty for criminal wrongdoing; **rehabilitation**, or the attempt to "correct" the personality and behavior of convicted offenders through educational, vocational, or therapeutic treatment and to return them to society as law-abiding citizens; and **general deterrence**, or the attempt to prevent people in general from engaging in crime by punishing specific individuals and making examples of them. (Omitted from Rubin's list is special or specific deterrence.) These purposes are discussed at greater length in Chapter 9 under the goals of sentencing.

incapacitation The removal or restriction of the freedom of those found to have violated criminal laws.

punishment The presentation of an aversive stimulus to reduce a response.

rehabilitation The attempt to "correct" the personality and behavior of convicted offenders through educational, vocational, or therapeutic treatment and to return them to society as law-abiding citizens.

general deterrence The attempt to prevent people in general or society at large from engaging in crime by punishing specific individuals and making examples of them.

A ninth purpose of courts is to determine legal status. For example, courts determine marital status by dissolving marriages and granting divorces. Similarly, parental status is determined by approving adoptions.

A tenth and critically important purpose of courts is to protect individual citizens from arbitrary government action. Recourse through the courts is available to citizens who have been abused by government agencies and agents. Examples of such abuses include illegal invasion of a person's privacy; interference with a person's First Amendment rights of religious choice, speech, press, and assembly; and denial of public employment because of race, gender, or age. In summary, by custom and law, courts have become an integral and seemingly indispensable part of modern life.

THINKING CRITICALLY

1. The first purpose of courts is to "do justice." Given what you know about the American court system, do you think that justice is done? Why or why not?
2. Of the remaining nine purposes of courts, which ones do you think can be most easily achieved? Why?

Key Actors in the Court Process

The three key actors in the court process are the prosecutor, the defense attorney, and the judge. In this section, we examine the roles those three officers of the court play in the administration of justice.[31]

Tracie Van Auken/EPA/Shutterstock

District Attorney Kevin Steele responds to reporters' questions about the Bill Cosby sexual assault case. *Was it appropriate for Steele to respond to reporters' questions? Why or why not?*

The Prosecutor

Because most crimes violate state laws, they fall under the authority, or jurisdiction, of the state court system and its prosecutors. The prosecutor is a community's chief law enforcement official and is responsible primarily for the protection of society.[32] Depending on the state, the prosecutor may be referred to as the district attorney, the prosecuting attorney, the county attorney, the state's attorney, the commonwealth's attorney, or the solicitor. In large cities, assistant district attorneys perform the day-to-day work of the prosecutor's office.

Whatever the name, the prosecutor is the most powerful actor in the administration of justice. As former Supreme Court Justice and U.S. Attorney General Robert Jackson wrote in 1940, "The prosecutor has more control over life, liberty, and reputation than any other person in America."[33] Not only do prosecutors conduct the final screening of each person arrested for a criminal offense, deciding whether there is enough evidence to support a conviction, but in most jurisdictions they also have unreviewable discretion in deciding whether to charge a person with a crime and whether to prosecute the case. In other words, regardless of the amount (or lack) of incriminating evidence, and without having to provide any reasons to anyone, prosecutors have the authority to charge or not charge a person with a crime and to prosecute or not prosecute the case. If they decide to prosecute, they also determine what the charge or charges will be. (The charge or charges may or may not be the same as the one or ones for which the person was arrested.) In most cities, prosecutors refuse to formally charge a person in about one-half of all felony arrests. Prosecutors are not required to prosecute a person for all the charges that the evidence will support. However, the more charges they bring against a suspect, the more leverage prosecutors have in plea bargaining, which is discussed later in this section. Regardless of the reason, when prosecutors elect not to prosecute, they enter a notation of ***nolle prosequi (nol. pros.)*** on the official record of the case and formally announce in court the decision to dismiss the charge or charges. Insufficient evidence is the most frequent reason given by prosecutors for not prosecuting cases.

nolle prosequi (nol. pros.) The notation placed on the official record of a case when prosecutors elect not to prosecute.

The Decision to Charge and Prosecute The exercise of prosecutorial discretion in charging contrasts sharply with the way prosecutors are supposed to behave in their professional capacities. Ideally, prosecutors are supposed to charge an offender with a crime and to prosecute the case if after full investigation three, and only three, conditions are met:

1. They find that a crime has been committed.
2. A perpetrator can be identified.
3. There is sufficient evidence to support a guilty verdict.

Prosecutors are not supposed to charge suspects with more criminal charges or for more serious crimes than can be supported reasonably by the evidence. They are not supposed to be deterred from prosecution because juries in their jurisdiction frequently have refused to convict persons of particular kinds of crimes. Conversely, they are not supposed to prosecute simply because an aroused public demands it. Prosecutors are not supposed to be influenced by the personal or political advantages or disadvantages that might be involved in prosecuting or not prosecuting a case. Nor, for that matter, are they supposed to be swayed by their desire to enhance their records of successful convictions. It would be naive, however, to believe that those factors do not have at least some influence on prosecutors' decisions to pursue or to drop criminal cases.

Prosecutors sometimes choose not to charge or prosecute criminal cases for any of the following nine additional reasons. The first is their belief that an offense did not cause sufficient harm. This decision is usually a practical one. Given limited time and resources, overworked and understaffed prosecutors often have to choose which cases go forward and which do not.

A second reason involves the relationship between the statutory punishment and the offender or the offense. In today's legal climate of increasingly harsh sentencing laws, a prosecutor may feel that the statutory punishment for a crime is too severe for a particular offender (e.g., a first-time offender) or for a particular offense. In an effort to be fair, at least in their own minds, prosecutors in such cases impose their own sense of justice.

FYI

U.S. Attorneys

Violations of federal laws are prosecuted by the U.S. Justice Department. The Justice Department is headed by the U.S. attorney general and staffed by 93 U.S. attorneys, each of whom is nominated to office by the president and confirmed by the Senate. One U.S. attorney is assigned to each of the 94 federal district court jurisdictions (Guam and the Northern Mariana Islands share 1 attorney). There were approximately 5,800 assistant U.S. attorneys in 2019.

Source: National Association of Assistant U.S. Attorneys (personal correspondence, May 23, 2019).

A third reason for not prosecuting, even when the three ideal conditions are met, is an improper motive on the part of a complainant. The prosecutor may feel that a criminal charge has been made for the wrong reasons. For example, if a prosecutor was convinced that a woman who had caught her husband cheating lied when charging her husband with beating her, the prosecutor might elect not to charge the husband with assault or some other crime.

A fourth reason prosecutors sometimes choose not to prosecute a case is that the public has violated the particular law with impunity for a long time with few complaints. In the case of "blue laws," for example, which may require stores to be closed on Sundays, prosecutors would occasionally get complaints from outraged churchgoers. In some cases, prosecutors simply ignored the complaints because the law, although on the books for decades, had not been enforced for years and few complaints had been received. In most states, incidentally, blue laws have been declared unconstitutional.

A fifth reason prosecutors often choose not to prosecute a case, even though, ideally, prosecution is required, is that a victim may refuse to testify. In rape cases, for example, prosecutors realize that it is nearly impossible to secure a conviction without the testimony of the victim. Thus, if the victim refuses to testify, the prosecutor may decide to drop the case, knowing that the chances of obtaining a conviction are reduced dramatically without the cooperation of the victim.

A sixth reason for not prosecuting has to do with humanitarian concerns for the welfare of the victim or the offender. In child sexual molestation cases, for example, conviction depends on the victim's testimony. Prosecutors may decline to prosecute because of the possible psychological injury to a child who is forced to testify. When offenders are suffering from mental illness, prosecutors may decide that diverting the offender to a mental health facility rather than prosecuting for a crime is in the best interests of all concerned.

Seventh, prosecutors sometimes do not prosecute a case otherwise worthy of prosecution because the accused person cooperates in the apprehension or conviction of other criminal offenders. In drug cases, for example, prosecutors often "cut deals" (make promises not to prosecute) with users or low-level dealers in order to identify "higher-ups" in the drug distribution network. Prosecutors make similar deals with low-level operatives in organized crime.

An eighth reason prosecutors sometimes choose not to prosecute is that the accused is wanted for prosecution of a more serious crime in another jurisdiction. Thus, there is little reason to expend resources prosecuting a case if an offender likely is to receive greater punishment for a crime committed elsewhere. It is easier and cheaper simply to extradite, or to deliver, the offender to the other jurisdiction.

Finally, if an offender is on parole when he or she commits a new crime, prosecutors may not prosecute the new crime because they consider it more cost effective to simply have the parole revoked and send the offender back to prison.

The Decision to Plea-Bargain Unreviewable discretion in deciding whether to charge and prosecute citizens for their crimes is the principal reason the prosecutor is considered the most powerful figure in the administration of justice, but it is by no means the prosecutor's only source of power. Probably the most strategic source of power available to prosecutors is their authority to decide which cases to "plea bargain." **Plea bargaining** or **plea negotiating** refers to the practice whereby the prosecutor, the defense attorney, the defendant, and–in many jurisdictions–the judge agree to a negotiated plea. For example, they may agree on a specific sentence to be imposed if the accused pleads guilty to an agreed-upon charge or charges instead of going to trial. It is the prosecutor alone, however, who chooses what lesser plea, if any, will be accepted instead of going to trial. Contrary to popular belief, justice in the United States is dispensed mostly through plea bargaining. Criminal trials are relatively rare events; plea bargaining is routine. About 95% of all convictions in felony cases are the result of guilty pleas. Guilty pleas account for an even higher percentage of convictions in misdemeanor cases.[34] Plea bargaining and the different types of plea bargains are discussed in more detail later in this chapter.

plea bargaining or plea negotiating The practice whereby the prosecutor, the defense attorney, the defendant, and—in many jurisdictions—the judge agree on a specific sentence to be imposed if the accused pleads guilty to an agreed-upon charge or charges instead of going to trial.

Recommending the Amount of Bail In addition to having control over charging, prosecuting, and plea bargaining, another source of power for the prosecutor in many jurisdictions is the responsibility of recommending the amount of bail. Although the final

Actor Robert Downey Jr. departs Riverside County Superior Court in 2001, after pleading no contest to a felony count for cocaine possession and a misdemeanor count for being under the influence of the drug in a plea bargain that allowed him to continue live-in drug treatment rather than face jail time. *What factors might influence a prosecutor's plea-bargaining decision? Are all those factors legitimate?*

Jim Ruymen/REUTERS/Newscom

decision on the amount or even the opportunity for bail rests with the judge, the prosecutor makes the initial recommendation. By recommending a very high bail amount, an amount that a suspect is unlikely to be able to raise, a prosecutor can pressure a suspect to accept a plea bargain. Bail is discussed later in this chapter.

rules of discovery Rules that mandate that a prosecutor provide defense counsel with any exculpatory evidence (evidence favorable to the accused that has an effect on guilt or punishment) in the prosecutor's possession.

Rules of Discovery Perhaps the only weakness in a prosecutor's arsenal of weapons is the legal rules of discovery. The **rules of discovery** mandate that a prosecutor provide defense counsel with any exculpatory evidence in the prosecutor's possession. Exculpatory evidence is evidence favorable to the accused that has an effect on guilt or punishment. Examples of possible exculpatory evidence are physical evidence, evidentiary documents (such as a defendant's recorded statements to police and reports of medical examinations or scientific tests), and lists of witnesses, including potential law enforcement witnesses who previously have been disciplined for perjury and other offenses). A prosecutor's concealment or misrepresentation of evidence, commonly referred to as a "Brady violation" (see *Brady v. Maryland*, 1963), as noted in Chapter 4, is grounds for an appellate court's reversal of a conviction. Defense attorneys, however, are under no constitutional obligation to provide prosecutors with incriminating evidence. Nevertheless, most states and the federal system have by statute or by court rules given the prosecution some discovery rights (e.g., the right to notice of the defense's intent to use an alibi or an insanity defense). The rationale for the rules of discovery is that, ideally, the prosecutor's job is to see that justice is done and not necessarily to win cases.

Selection and Career Prospects of Prosecutors Given the power of prosecutors in the administration of justice, the public can only hope that prosecutors wield their power wisely and justly. We believe that many of them do. Unfortunately, political considerations and aspirations may be too enticing to some prosecutors, causing them to violate the canons of their position. The partisan political process through which the typical prosecutor is elected increases the chances that political influence will affect the prosecutor's decisions. In most jurisdictions, prosecutors run for office on either the Democratic or Republican ticket, although in some jurisdictions, they are elected in nonpartisan races–that is, without party affiliation. At the end of 2018, 43 chief prosecutors or Attorneys General were elected by a state's voters, 5 were appointed by the governor (Alaska, Hawaii, New Hampshire, New Jersey, and Wyoming), 1 was elected by the state legislature (Maine), and 1 was elected by the state supreme court (Tennessee). Ninety percent of chief prosecutors or Attorneys General were elected or appointed to 4-year terms.[35] Although prosecutors can be removed from office for criminal acts or for incompetence, removal is rare.

The potential for political influence is increased even further by prosecutors' frequent use of the position as a stepping-stone to higher political office (several mayors, governors, and presidents have been former prosecutors). There are very few career prosecutors. Those who are not elected to higher political office and those who choose not to seek it typically go into private law practice.

Gender, Race, and Ethnicity of Prosecutors Throughout U.S. history, nearly all elected prosecutors have been white, especially white males. In 2019, for example, 95% of elected prosecutors were white, the same percentage as in 2015.[36] Seventy-three percent of the elected prosecutors in 2019 were white males, down from 79% of elected prosecutors in 2015. In 2019, 24% of elected prosecutors were female, an increase from the 17% of female prosecutors in 2015. In 2019, white females were 22% of elected prosecutors (an increase from 16% in 2015), females of color were 2% (an increase from 1% in 2015), and males of color were 3% (a decrease from 4% in 2015). Note that the 1% increase of female prosecutors of color was accompanied by a 1% decrease of male prosecutors of color.

A reason why the race/ethnicity and gender of elected prosecutors do not change very much is that prosecutors run for office unopposed 80% of the time. However, when elections are contested, change is possible. In competitive elections in 2018, for example, white males were 69% of the candidates, but only 59% of the winners, while females and people of color were 31% of candidates and 41% of the winners. From 2015 through the summer of 2019, the overrepresentation of white male prosecutors decreased in 34 states. Yet, only one state–New Mexico–even approaches equitable representation. People of color comprise 62% of New Mexico's population and 47% of its elected prosecutors. By comparison, people of color are 63% of California's population, but California's elected prosecutors are 90% white. Similarly, in Arizona, New York, and Florida, people of color comprise more than 40% of the population and, in all three states, more than 90% of elected prosecutors are white. In New York, more than 95% of elected prosecutors are white. In Washington, Nevada, Louisiana, and South Carolina, more than 90% of prosecutors are male. In 12 other states, including Oklahoma, whose female incarceration rate is twice the national average, more than 80% of prosecutors are male.

First African American State Attorney in Florida

In 2016, Aramis Ayala was elected as Florida's first African American state attorney. However, on May 28, 2019, she announced that she would not seek a second term.

Source: Monivette Cordeiro and Jeff Weiner, "Aramis Ayala won't seek re-election as Orange-Oceola state attorney; Belvin Perry may enter race," Orlandosentinel.com, May 28, 2019, accessed November 10, 2019, https://www.orlandosentinel.com/news/breaking-news/os-ne-aramis-ayala-no-re-election-run-orange-osceola-state-attorney-20190528-z65rv7rmqjdqfoyxsd6rp6junu-story.html.

Assistant District Attorneys The workhorses of the big-city prosecutor's office, as already noted, are the assistant district attorneys (or deputy district attorneys), who are hired by the prosecutor. Generally, assistant district attorneys are hired right out of law school or after a brief and usually unsuccessful stint in private practice. Most of them remain in the prosecutor's office for only 2 to 4 years and then go into, or back into, private law practice. Reasons for leaving the prosecutor's office include low pay, little chance for advancement, physical and psychological pressures, boredom, and disillusionment with the criminal adjudication process. Assistant district attorneys provide an important social service, and the job is a good way of gaining legal experience. Lawyers who occupy the position, however, usually do not consider it anything more than temporary employment before they go into private practice.

The Defense Attorney

The Sixth Amendment to the U.S. Constitution and several modern Supreme Court decisions, discussed in Chapter 4, guarantee the right to the "effective assistance" of counsel to people charged with crimes.[37] (The terms *counsel*, *attorney*, and *lawyer* are interchangeable.)

The right to counsel extends not only to representation at trial but also to other critical stages in the criminal justice process, "where substantial rights of the accused may be affected." Thus, defendants have a right to counsel during custodial interrogations, preliminary hearings, and police lineups. They also have a right to counsel at certain posttrial proceedings, such as their first (and only the first) appeal that is a matter of right, and probation and parole revocation hearings. The Supreme Court also has extended the right to counsel to juveniles in juvenile court proceedings.

A defendant may waive the right to counsel and appear on his or her own behalf. However, given the technical nature of criminal cases and the stakes involved (an individual's freedom!), anyone arrested for a crime is well advised to secure the assistance of counsel at the earliest opportunity. The phone call routinely given to arrested suspects at the police station is for the specific purpose of obtaining counsel. If a suspect cannot afford an attorney and is accused of either a felony or a misdemeanor for which

Careers in Criminal Justice

Assistant State Attorney

My name is Wilson Green, and I am an assistant state attorney (prosecutor) in Orlando, Florida. I received a bachelor of arts degree from Vanderbilt, a master of divinity degree from Reformed Theological Seminary, and a juris doctor degree from the University of Mississippi, with some U.S. Navy time along the way. I interviewed for and began my current job about 20 years ago here at the State Attorney's Office.

I am in the homicide unit. Most of us deal with either misdemeanor or felony cases that are set for trial. Most cases are resolved by defendants pleading guilty (or no contest) and being sentenced by a judge, while others proceed to a jury trial. The prosecutor is assigned cases as they come to our office from law enforcement agencies and after the defendants have been formally charged by other attorneys in our office.

The first step is to "work up" each case—to review the entire file (police reports, witness statements, photos, tape recordings, lab reports, etc.) and to send out appropriate "discovery" to the defendant's attorney. Witnesses must be listed and subpoenaed for trial. Requests for further investigation may be needed as the prosecutor anticipates what the defense will probably be. Often, the defense attorney will take depositions of state witnesses before trial, which we attend. There may also be pretrial hearings on legal issues, as well as plea negotiations.

Jury trials generally last a day or so or up to a week or more. A prosecutor may have as many as a hundred cases pending at one time (spread over a period of a few months), with new ones coming in as fast as others are resolved. The stress level can be high at times, due to the caseload, problem cases, witness problems, cases being postponed at the last minute, not knowing which case will actually go to trial until the last minute, and having to make appropriate plea offers. On the plus side, prosecutors (who are to be "ministers of justice," not just win cases) have a beneficial role in society; have much discretion in handling their cases; are in the courtroom on a regular basis; and are in touch with the "real world" through their contact with police agencies, witnesses, victims, families of victims, judges, and juries. They can use both common sense and legal training. Each case seems to have something new.

Courtesy of Wilson Green

Those considering becoming prosecutors might think about sampling criminal justice and prelaw undergraduate courses, taking advantage of internships with law enforcement agencies or prosecutors' offices, or simply observing court proceedings as a citizen. Use any employment opportunities as a way to gain commonsense experience that will help in a later career. Take advantage of any group or public speaking opportunities as well.

What do you think is the most challenging part of a prosecutor's job? Explain.

imprisonment could be the result of conviction, the state is required to provide an attorney at the state's expense.

The media and the public sometimes vilify lawyers for defending people unquestionably guilty of crimes. That lawyers sometimes succeed in getting guilty persons "off" by the skillful use of legal technicalities only makes matters worse. However, what some people fail to understand is that in the American system of justice, it is not the role of defense lawyers to decide their clients' guilt or innocence. Rather, the role of defense lawyers is to provide the best possible legal counsel and advocacy within the legal and ethical limits of the profession. Legal and ethical codes forbid lawyers, for example, to mislead the court by providing false information or by using perjured testimony (false testimony under oath). The American system of justice is based on the premise that a person is innocent until proven guilty. In the attempt to ensure, as far as possible, that innocent people are not found guilty of crimes, all persons charged with crimes are entitled to a rigorous defense. The constitutional right to counsel and our adversarial system of justice would be meaningless if lawyers refused to defend clients that they "knew" were guilty.

FYI **Public Opinion and Lawyers**

In 2018, only 19% of Americans rated the honesty and ethical standards of lawyers "high" or "very high," 51% rated lawyers' honesty and ethical standards "average," and 28% rated them "low" or "very low."

Source: Megan Brenan, "Nurses Again Outpace Other Professions for Honesty, Ethics," *Gallup*, December 20, 2018. Accessed May 25, 2019, https://news.gallup.com/poll/245597/nurses-again-outpace-professions-honesty-ethics.aspx.

Not all lawyers are adequately trained to practice in the specialized field of criminal law. Law schools generally require only a one-semester course in criminal law and a one-semester course in criminal procedure, though additional courses may be taken as electives in a typical six-semester or 3-year program. Many lawyers prefer to practice other, often more lucrative, areas of law, such as corporate, tax, or tort law. Compared with those other areas of legal practice, the practice of criminal law generally provides its practitioner less income, prestige, and status in the community.

For those reasons, criminal defendants in search of counsel are limited by practical considerations. They can choose a privately retained lawyer who specializes in criminal law, if they can afford one. Or, if they are indigent and cannot afford one, they will receive a court-appointed attorney (who may or may not be skilled in the practice of criminal law), a public defender, or a "contract" lawyer. In large cities, there is little problem in locating a criminal lawyer. They are listed online. In rural areas, however, finding a lawyer who specializes in criminal law may be more difficult. Because of the lower volume of criminal cases in rural areas, attorneys who practice in those areas generally must practice all kinds of law to make a living; they cannot afford to specialize. Outside large cities, then, a criminal defendant may have no better option than to rely on the services of a court-appointed attorney, a public defender, or a contract lawyer, whichever is provided in that particular jurisdiction. Court-appointed attorneys, public defenders, and contract lawyers are discussed further later in this section.

Criminal Lawyers In discussing privately retained criminal lawyers, it is useful to think of a continuum. At one end are a very few nationally known, highly paid, and successful criminal lawyers, such as Mark Geragos, Gloria Allred, and Benjamin Brafman. Those criminal lawyers generally take only three kinds of cases: (1) those that are sensational or highly publicized, (2) those that promise to make new law, or (3) those that involve large fees. A little further along the continuum is another small group of criminal lawyers, those who make a very comfortable living in the large cities of this country by defending professional criminals, such as organized crime members, gamblers, pornographers, and drug dealers. Toward the other end of the continuum are the vast majority of criminal lawyers who practice in the large cities of this country.

Most criminal lawyers must struggle to earn a decent living. They have to handle a large volume of cases to make even a reasonable salary. To attract clients, some criminal lawyers engage in the unethical practice of paying kickbacks to ambulance drivers, police officers, jailers, and bail bond agents for client referrals. Some criminal lawyers also engage in the unethical practice of soliciting new clients in courthouse hallways.

Although a few criminal lawyers are known for their criminal trial expertise, most relatively successful criminal lawyers gain their reputations from their ability to "fix" cases; that is, their ability to produce the best possible outcome for a client, given the circumstances of the case. Fixing cases usually involves plea bargaining but also may include the strategic use of motions for continuances and for changes in the judge assigned to the case. Who a lawyer knows often is more important than what the lawyer knows. In other words, a close personal relationship with the prosecutor or a hearing before the "right" judge often is far

Female Lawyers

In 1971, just 3% of all lawyers in the United States were women. By 1980, the percentage of female lawyers had increased to 8%, and by 2019, 38% of all lawyers were women. In 1961, only 4% of the nearly 16,500 first-year law students were women. In 2017, about 51% of the approximately 108,000 first-year law students were women. Still, women make up only about 22.7% of the non-equity partners and about 19% of the equity partners in the nation's approximately 50,000 major law firms. The percentage of associates is much higher (about 46%). In 2018, female lawyers' average weekly salary ($1,762) was 80% of male lawyers' average weekly salary ($2,202), down from 89.7% in 2015, and 86.6% in 2011. In 1981, Sandra Day O'Connor became the first female justice of the U.S. Supreme Court, followed by Ruth Bader Ginsburg in 1993, Sonia Sotomayor in 2009, and Elena Kagan in 2010. Also, in 1993, Janet Reno became the first female Attorney General of the United States, followed by Loretta Lynch in 2015, who also was the first black female Attorney General of the United States.

Sources: American Bar Association, Commission on Women in the Profession, "A Current Glance at Women in the Law April 2019," accessed May 25, 2019, https://www.americanbar.org/content/dam/aba/administrative/women/current_glance_2019.pdf; "Despite Gains, Women Lawyers Still Face Bias," *The Orlando Sentinel*, August 18, 1996, A20; "Where My Girls At? Women Are Finally Catching Up to Men in Law School Enrollment," *The Princeton Review*, www.princetonreview.com/law/research/articles/decide/gender.asp.

Jeff Chiu/AP Images

D Dipasupil/Contributor/Getty Images

Broadimage/Shutterstock

Mark Geragos, Gloria Allred, and Benjamin Brafman (from left to right) are among the very few nationally known, highly paid, and successful criminal lawyers. *What do you think distinguishes these lawyers from all other lawyers?*

more important to the case's outcome (e.g., a favorable plea bargain) than is an attorney's legal ability.

Besides legal expertise and good relationships with other actors in the adjudication process, the only other commodity criminal lawyers have to sell is their time. Thus, for the typical criminal lawyer, time is valuable. As a result, criminal lawyers try to avoid time-consuming court battles and are motivated to get their clients to plead guilty to reduced charges.

Another reason most criminal lawyers prefer plea bargaining to trials is that they are more likely to receive a fee, however small, from their typically poor clients for arranging the plea. When cases go to trial and the legal fee has not been paid in advance, there is the possibility that the lawyer will receive no compensation or inadequate compensation from his or her client, especially if the case is lost.

To prevent nonpayment for services, criminal lawyers sometimes engage in unethical, if not illegal, behavior. Sometimes, criminal lawyers plea-bargain cases that they know they can win at trial. They do this because they know that their clients do not have the money to compensate them for the time they would put into a trial. If clients have only a few hundred dollars, some unscrupulous criminal lawyers reason that those clients can afford only "bargain" justice. Another unethical tactic employed by some criminal lawyers to secure a fee from their clients is to seek delays in a case and allow their clients to remain in jail until they have been paid for services yet to be performed. Getting a client to pay a legal fee is not a legitimate basis for seeking a court delay.

Rules of Conduct

As noted previously, the American public generally does not rate the honesty and ethical standards of lawyers very high. Go to the American Bar Association's website at https://www.americanbar.org/groups/professional_responsibility/publications/model_rules_of_professional_conduct/ and review the Model Rules of Professional Conduct. Although individual states have their own rules, most states modeled their rules after these.

From what you have read and heard, which of the rules do you think are most commonly violated?

Criminal lawyers often spend more time at the county jail and the courthouse, where their clients are, than in their offices, which generally are spartan compared with the plush offices of their corporate counterparts. Their clients sometimes tell them about grisly crimes that have been committed. Because of attorney–client privilege, the lawyer in possession of this information cannot reveal it, under penalty of disbarment (the revocation of his or her license to practice law). In short, many criminal lawyers are considered somewhat less than respectable by their professional colleagues and much of the general public because (1) their clients are "criminals," (2) they deal with some of the more unsavory aspects of human existence, and (3) a few may engage in unethical behavior to earn a living.

The Court-Appointed Lawyer In some jurisdictions, criminal suspects or defendants who cannot afford to hire an attorney are provided with court-appointed lawyers. The court-appointed lawyer is usually selected in one of two ways. In some jurisdictions, lawyers volunteer to represent indigent offenders and are appointed by judges on a rotating basis from a list or from lawyers present in the courtroom. In other jurisdictions, lawyers are appointed by a judge from a list of attorneys who are members of the county bar association.

Defense lawyers spend much of their time at the local jail. *Do you think that the jail setting influences a lawyer's opinion of his or her client?*

Andrew Lichtenstein/Contributor/Getty Images

In the past, appointed lawyers frequently represented indigent clients *pro bono* (without pay). Today, however, the vast majority of appointed lawyers are paid by the jurisdiction. Hourly fees usually are much less than would be charged by privately retained counsel. Seldom is money available to hire investigators or to secure the services of expert witnesses, the lack of which weakens the defense that can be provided a client.

In addition, although appointed attorneys may be experts in some areas of law, they may not be familiar with the intricacies of a criminal defense. They may have never before represented a client charged with a crime. Nevertheless, many lawyers view their appointments as a public service and do their best to represent their clients professionally. Some, however, take cases grudgingly (they can refuse an appointment only for very good reasons) and regard an appointment as a financially unrewarding and unpleasant experience. They perform well enough only to escape charges of malpractice. In many cases, an appointed attorney first meets his or her client in the courtroom, where they have a brief conference before a guilty plea is arranged. When they go to trial, they often fail their clients. An estimated 5,000 to 50,000 people may be convicted wrongfully in American courts every year.[38]

The Public Defender In many jurisdictions today, people who are charged with crimes and cannot afford an attorney are provided with public defenders. Public defenders are paid a fixed salary by a jurisdiction (city, county, state, or federal) to defend indigents charged with crimes. Frequently, public defenders are assigned to courtrooms instead of to specific clients. They defend all of the indigent clients who appear in their courtrooms. As a result, a defendant may have a different public defender at different stages in the process (e.g., preliminary hearing, arraignment, and trial). Public defenders commonly spend only 5 to 10 minutes with their clients. In addition, this bureaucratic arrangement makes the public defender a part of a courtroom work group that includes the judge, the prosecutor, and the other courtroom actors with whom the public defender interacts daily. This work group, in turn, may be more interested in processing cases efficiently and maintaining cooperative relationships with each other than in pursuing adversarial justice. Nevertheless, despite the method of assignment, the impersonal nature of the defense, and the potential conflict of interest, most indigent clients in criminal cases prefer public defenders to court-appointed attorneys because public defenders practice criminal law full time.

Like prosecutors, many public defenders regard their position not as a career but as a valuable learning experience that will eventually lead to private law practice.

First Public Defender's Office

In 1914, the County of Los Angeles opened the country's first public defender's office.

Source: Law Office of the Los Angeles County Public Defender. Accessed May 9, 2016, http://pd.co.la.ca.us/About_history.html.

The Contract Lawyer A relatively new and increasingly popular way of providing for indigent defense is the contract system. In this system, private attorneys, law firms, and bar associations bid for the right to represent a jurisdiction's indigent defendants. Terms of contracts differ, but in the typical contract, counsel agrees to represent either all or a specified number of indigent defendants in a jurisdiction during a certain period of time, in exchange for a fixed dollar amount. Contracts are awarded on the basis of costs to the jurisdiction, qualifications of bidders, and other factors.

In certain situations, some jurisdictions employ a combination of indigent defense systems. For example, jurisdictions that regularly use public defenders sometimes contract out the defense of cases where there is a conflict of interest (e.g., multiple-defendant cases) or cases that require special expertise (such as some death penalty cases).

Indigent Defense Systems In 2013, all 50 states, the U.S. government, and the District of Columbia had indigent defense systems. At the federal level, two types of defender organizations provided indigent defense: public defender organizations (FPDOs) and community defender organizations (CDOs).[39] FPDOs are headed by a federal public defender, who is a federal officer appointed by the court of appeals for the circuit in which the district is located to a four-year term, with no limit on the number of terms. FPDOs are funded by an annual budget and all employees are federal employees. CDOs, on the other hand, are private, nonprofit organizations selected by district courts, subject to their circuit council's approval. Once selected, CDOs keep their designation indefinitely and lose it only if the district and circuit court agree to end the designation. CDOs are funded by annual grants, and none of their employees are government employees.

The need for indigent defense at the federal level has grown dramatically during the past several decades, as has the cost of funding it. For example, the percentage of federal defendants who qualified for appointed counsel increased from 30% in 1963, to nearly 90% in 2014. Likewise, the number of federal cases assigned to appointed counsel in 1963 was about 10,000, compared to more than 200,000 cases in 2014. Finally, the budget for federal indigent defense has grown from about $1 million in 1964 to more than $1 billion in 2014.

As for the states and the District of Columbia, one or more of five different types of indigent defense systems are provided.[40] The most common types are governmental public defender offices, contract systems, and assigned or appointed counsel systems. All three of these indigent defense systems have been described in previous subsections of this chapter under the general heading of "The Defense Attorney." Two other, less commonly used types of systems are governmental conflict public defender offices and nongovernmental public defender offices. Governmental conflict public defender offices provide representation in cases involving legal conflicts. They are publicly operated governmental offices, and staff are government employees. Nongovernmental public defender offices provide representation through written contracts between some governmental organization and a nonprofit organization. Staff are not government employees. Table 8.1 displays the type or types of indigent defense systems of the various states and the District of Columbia in 2013 (the latest year for which these data were available). As the table shows, governmental public defender offices are the most common type of indigent defense systems with 25 states and the District of Columbia employing them, followed by contract systems with 16 states, assigned or appointed counsel systems with 14 states and the District of Columbia, governmental conflict public defender offices with 6 states, and nongovernmental public defender offices with 5 states. Table 8.1 also reveals that Iowa was the only state to utilize all five of the indigent defense systems, while Hawaii, Kentucky, Minnesota, Missouri, New Jersey, and Virginia employed only one of the systems–a governmental public defender office.

In 2013, 28 states and the District of Columbia had state-administered indigent defense systems that were either completely funded and administered by the state, or were funded by the state and county but administered by the state. In the remaining 22 states, indigent defense systems were administered by counties and funded by the county or by a combination of county and state.

At the state and county level, approximately 80% of criminal defendants cannot afford an attorney. However, according to Fordham Law School Professor John Pfaff, state and county spending on indigent defense has been decreasing. Pfaff estimates that about $2.3 billion is spent annually on indigent defense at the state and county level, which is approximately 1% of the annual criminal justice budget for states and counties in the United States.[41] Thus, although millions of dollars are spent on indigent defense systems nationwide, many state and county indigent defense systems remain woefully underfunded. Orleans Parish, Louisiana, is an extreme example. In 2014, the public defender's office employed 51 lawyers who were expected to handle 22,000 cases, or an average of 431 cases per lawyer. Each lawyer reportedly spent an average of about 7 minutes on each case. Tina Peng, an Orleans Parish public defender, wrote in a *Washington Post* op-ed that her caseload was "unconstitutionally high." She lamented, "I miss filing important motions. . . I am unable to properly prepare for every trial. I plead some of my clients to felony convictions on the day I meet them. If I don't follow up to make sure clients are released when they should be, they can sit in jail for unnecessary weeks and months." The office could not hire investigators and did not have adequate support staff. So desperate was the Orleans Parish Public Defender's Office that in 2015, it resorted to a crowdfunding campaign to augment its budget. As a last resort, the office refused to accept certain cases and put those defendants on a waiting list, where they languished in jail without lawyers.[42] Louisiana is by no means the only state with indigent defense problems. For example, Mississippi has an indigent defense system funded entirely by counties. Some of those counties spend less than $2 per defendant! In many Mississippi counties, criminal defendants remain incarcerated in jails for months while awaiting the appointment of an attorney.[43] Public defender offices in 43 states and the District of Columbia have provisions for recouping costs from their indigent clients. For example, South Dakota charges each indigent defendant $92 an hour for his or her public defender. Such fees motivate indigent defendants to waive their right to counsel.[44] Remember, indigent defendants have a Sixth Amendment right to appointed counsel. That right is

FYI **Determining Indigency**

The typical criteria for determining indigency include some combination of financial ability based on at least one of the following: income, assets value, employment status, amount of debt, personal expenses, receipt of public assistance, or number of dependents; federal poverty guidelines; nature of charge; cost of private counsel; age; education; residence in a public institution such as a mental health or correctional institution; financial ability of family members; and ability to post bond. Judges and public defenders most commonly are responsible for determining indigency.

Source: Suzanne M. Strong, "State-Administered Indigent Defense Systems, 2013." U.S. Department of Justice, *Bureau of Justice Statistics Special Report*, November 2016, 9, Table 6. Accessed January 5, 2017, www.bjs.gov/content/pub/pdf/saids13.pdf.

TABLE 8.1 State-Administered Indigent Defense Systems, by State, 2013

State	Number of Methods	Govt Public Defender Office	Govt Conflict Public Defender Office	Nongovt Public Defender Office	Contract System	Assigned or Appointed Counsel System
Total		**26**	**6**	**5**	**16**	**15**
Alaska	3	Yes	Yes	No	Yes	No
Arkansas	4	Yes	Yes	No	Yes	Yes
Colorado	2	Yes	Yes	No	No	No
Connecticut	2	Yes	No	No	Yes	No
Delaware	2	Yes	Yes	No	No	No
District of Columbia	2	Yes	No	No	No	Yes
Hawaii	1	Yes	No	No	No	No
Iowa	5	Yes	Yes	Yes	Yes	Yes
Kentucky	1	Yes	No	No	No	No
Louisiana	2	Yes	No	No	Yes	No
Maine	2	No	No	No	Yes	Yes
Maryland	2	Yes	No	No	No	Yes
Massachusetts	2	Yes	No	No	No	Yes
Minnesota	1	Yes	No	No	No	No
Missouri	1	Yes	No	No	No	No
Montana	3	Yes	Yes	No	Yes	No
New Hampshire	3	No	No	Yes	Yes	Yes
New Jersey	1	Yes	No	No	No	No
New Mexico	2	Yes	No	No	Yes	No
North Carolina	4	Yes	No	Yes	Yes	Yes
North Dakota	3	Yes	No	No	Yes	Yes
Oregon	4	Yes	No	Yes	Yes	Yes
Rhode Island	2	Yes	No	No	/	Yes
South Carolina	3	Yes	No	No	Yes	Yes
Vermont	3	Yes	No	No	Yes	Yes
Virginia	1	Yes	No	No	No	No
West Virginia	2	No	No	Yes	No	Yes
Wisconsin	3	Yes	No	No	Yes	Yes
Wyoming	2	Yes	No	No	Yes	No

Note: / = Not reported.

Source: Adapted from Suzanne M. Strong, "State-Administered Indigent Defense Systems, 2013." U.S. Department of Justice, *Bureau of Justice Statistics Special Report*, November 2016, 23, Appendix Table 1. Accessed January 4, 2017, www.bjs.gov/content/pub/pdf/saids13.pdf.

violated by the requirement to pay defense fees. If defendants could pay defense fees, then they would not be indigent.

The trend in indigent defense systems is toward greater centralized control through increased state funding and the establishment of state oversight bodies or commissions, usually with a director and/or a chief public defender. In 2013, 21 states and the District of Columbia had advisory boards or commissions. The primary purpose of the state oversight bodies or commissions is to ensure that indigent defense remains independent of political and judicial influence. However, in 2013, the governor in 18 states was involved in appointing advisory board members and, in 7 of the 18 states, the governor had sole responsibility for appointing advisory board members. Other functions of advisory boards or commissions include monitoring costs and caseloads, creating standards, and enforcing compliance with standards.

The Judge

According to the Bible, the idea of judges was first suggested to Moses by his father-in-law, Jethro. Following their exodus from Egypt, the Hebrews, having been slaves, had no formal government. Because Moses was the only one among them with evident authority, he served as arbiter of domestic controversies–a time-consuming and exhausting job. Jethro recognized the problem with this arrangement and suggested to Moses that he create a kind of theocracy. Moses would be the supreme judge and would represent the people before God. He would also teach the people laws governing everyday behavior. The actual administration of the laws, which came from God through Moses, was to be handled "by able men, . . . such as fear God, men who are trustworthy and who hate a bribe." These "judges" were to be established in a hierarchical order, so that the highest would be rulers over thousands of people; beneath them would be, successively, rulers of hundreds, fifties, and tens.[45]

For most people, black-robed judges are the embodiment of justice, and although they generally are associated with trials, judges actually have a variety of responsibilities in the criminal justice process.[46] Among their nontrial duties are determining probable cause, signing warrants, informing suspects of their rights, setting and revoking bail, arraigning defendants, and accepting guilty pleas. Judges spend much of the workday in their chambers (offices), negotiating procedures and dispositions with prosecutors and defense attorneys. The principal responsibility of judges in all their duties is to ensure that suspects and defendants are treated fairly and in accordance with due process of law.

Furthermore, in jurisdictions without professional court administrators, judges are responsible for the management of their own courtrooms and staff. Judges in some jurisdictions also are responsible for the entire courthouse and its personnel, with the added duties of supervising building maintenance, budgeting, and labor relations.

In jury trials, judges are responsible for allowing the jury a fair chance to reach a verdict on the evidence presented. Judges must ensure that their behavior does not improperly affect the outcome of the case. Before juries retire to deliberate and reach a verdict, judges instruct them on the relevant law. This involves interpreting legal precedents and applying them to the unique circumstances of the case. Those jury instructions are reviewable by an appellate court.

Characteristics of Judges In the United States, judges are overwhelmingly white and male. They generally come from upper-middle-class backgrounds, are Protestant, are better educated than most citizens, and are 50 years of age or older. A majority of them were in private legal practice before becoming judges. Most were born in the communities in which they preside and attended college and law school in that state.

Selection of Judges States vary in the ways they select judges. The two most common selection methods are election and merit selection. Other, less common, ways of selecting judges are gubernatorial appointment and legislative election.

Like prosecutors, many state judges are elected to their positions in either partisan (with political party designation) or nonpartisan elections. This selection process exposes judges, like prosecutors, to potential charges of political influence. For many years, individuals campaigning for elected judgeships were required by judicial codes of conduct to refrain from revealing their positions on controversial political and legal issues. This prohibition set judicial campaigns apart from those of all other elected officials. The prohibition was intended to preserve fair and impartial justice and public trust in the judiciary. In 2002, in *Republican Party of Minnesota v. White*, the U.S. Supreme Court ruled that the prohibition was unconstitutional in violation of the First Amendment's protection of speech, especially speech about the qualifications of candidates for public office. The Court's decision has raised many concerns. For example, former U.S. Supreme Court Justice John Paul Stevens contended, "electing judges is like allowing football fans to select their referees–an unwise practice." Stevens noted, "a campaign promise to 'be tough on crime' or to 'enforce the death penalty' is evidence of bias that should disqualify a candidate from sitting in criminal cases." He concluded, "making the retention of judicial office dependent on the popularity of the judge inevitably affects the decisional process in high-visibility cases, no matter how conscientious the judge may be."[47]

Careers in Criminal Justice

Superior Court Judge

My name is Rainey R. Brandt. I am a judge at the Superior Court of the District of Columbia. The Superior Court is a quasi federal courthouse that serves as the state court for Washington, DC. It was created by an act of Congress in the 1970s because Washington, DC, is not a state. The Superior Court serves some 10,000 people daily handling all matters ranging from criminal, civil actions, probate, family matters, and tax.

I have been a judge since 2012. My educational path to becoming a judge started at American University, where I earned a BS, MS, and Ph.D. While in graduate school, I studied criminal justice. I have taught in the Department of Justice, Law and Criminology at American University as an adjunct associate professor for several years. Additionally, I have a law degree from the Columbus School of Law at the Catholic University of America. After law school, I clerked for two wonderful judges who taught me the importance of a positive work ethic and strong writing skills. Thereafter, I worked as a senior court manager (special counsel) at the Superior Court. That job allowed me to use both my academic education and the law degree. My responsibilities spanned many different topics such as prisoner health care and sentencing guidelines, and allowed me interactions with all of our criminal justice partners like the prosecutors, defense attorneys, law enforcement, and bureau of prisons.

After doing that job for a long time, I ascended to the bench. Currently, I am on a criminal assignment determining whether newly arrested defendants remain detained pending trial or whether release is appropriate. I see defendants charged with major felonies such as assault while armed, armed robbery, etc. and see ones charged with misdemeanors like shoplifting and simple assault. The volume is large and steady, so I have to think quickly, carefully, and legally. That is the biggest challenge to my work because a judge always has to be mindful of public safety while balancing the rights of the defendant. The best part of being a judge is that daily challenge. Being a judge carries an incredible weight and duty to enforce the law and protect the community.

Judge Rainey R. Brandt

If being a judge is a career you aspire to, the best advice I can give is to start early on cultivating connections by interning in offices that develop your communication and writing skills.

Joshua Sudock/ZUMA Press/Newscom

Judges spend much of their workday in their chambers (offices). *How does this differ from what the public generally believes about judges?*

TABLE 8.2 Salaries of State Court Judges, 2019

	Mean	Median	Range
Chief, Highest Court	$182,841	$181,290	$136,000 to $265,508
Associate Justice, Court of Last Resort	$176,714	$172,716	$136,000 to $253,189
Judge, Intermediate Appellate Courts	$170,237	$168,059	$132,838 to $237,365
Judge, General Jurisdiction Trial Courts	$158,748	$155,677	$125,499 to $208,000

Source: National Center for State Courts, "Survey of Judicial Salaries" 2019, accessed May 25, 2019, https://www.ncsc.org/~/media/Microsites/Files/Judicial%20Salaries/Judicial-Salary-Tracker-Jan-2019.ashx. Used by permission.

Electing judges also may discourage some of the best lawyers from becoming judges. Successful lawyers with lucrative practices may not want to interrupt their careers to take the chance of becoming a judge, only to lose the position in the next election. Besides, successful lawyers generally earn much more money than judges. Table 8.2 summarizes the 2019 salaries of state court judges. The table provides mean and median salaries of the chief and associate justices of a state's highest court, intermediate appellate court judges, and general jurisdiction trial court judges. Table 8.3 shows 2019 judicial salaries for each state and the District of Columbia.

TABLE 8.3 2019 Salaries for Appellate and General Jurisdiction Judges, and Rank

	Highest Court		Intermediate Appellate Court		General Jurisdiction Court	
	Salary	Rank	Salary	Rank	Salary	Rank
Alabama	$172,716	28	$184,244	9	$138,991	41
Alaska	$205,176	8	$193,836	6	$189,720	6
Arizona	$159,685	36	$154,534	30	$149,383	30
Arkansas	$174,925	26	$169,672	18	$168,096	16
California	$253,189	1	$237,365	1	$207,424	3
Colorado	$182,671	18	$175,434	13	$168,202	15
Connecticut	$185,610	15	$174,323	16	$167,634	17
Delaware	$196,245	12			$184,444	8
District of Columbia	$220,600	5			$208,000	1
Florida	$220,600	5	$169,554	19	$160,688	22
Georgia	$175,600	25	$174,500	15	$173,714	12
Hawaii	$227,664	4	$210,780	4	$205,080	4
Idaho	$151,400	42	$141,400	37	$135,400	43
Illinois	$234,391	2	$220,605	2	$202,433	5
Indiana	$177,244	21	$172,296	17	$147,164	34
Iowa	$174,808	27	$158,420	29	$147,494	33
Kansas	$142,089	46	$137,502	38	$125,499	51
Kentucky	$138,890	48	$133,299	39	$127,733	48
Louisiana	$170,325	30	$159,347	27	$153,143	27
Maine	$138,070	49			$129,397	46
Maryland	$181,433	19	$168,633	20	$159,433	24
Massachusetts	$200,984	10	$190,087	8	$184,694	7
Michigan	$164,610	34	$160,695	25	$146,721	35
Minnesota	$177,697	20	$167,438	22	$157,179	25

TABLE 8.3 2019 Salaries for Appellate and General Jurisdiction Judges, and Rank (continued)

	Highest Court		Intermediate Appellate Court		General Jurisdiction Court	
	Salary	Rank	Salary	Rank	Salary	Rank
Mississippi	$152,250	41	$144,827	35	$136,000	42
Missouri	$176,157	23	$161,038	24	$151,840	28
Montana	$144,061	45			$132,558	45
Nebraska	$176,299	22	$167,484	21	$163,077	21
Nevada	$170,000	31	$165,000	23	$160,000	23
New Hampshire	$175,837	24			$164,911	20
New Jersey	$201,842	9	$191,534	7	$181,000	10
New Mexico	$139,819	47	$132,838	40	$126,187	49
New York	$230,200	3	$219,200	3	$208,000	1
North Carolina	$149,115	43	$142,947	36	$135,236	44
North Dakota	$157,009	38			$143,869	36
Ohio	$172,200	29	$160,500	26	$147,600	32
Oklahoma	$154,174	39	$146,059	33	$139,298	40
Oregon	$154,040	40	$150,980	31	$142,136	37
Pennsylvania	$211,027	7	$199,114	5	$183,184	9
Rhode Island	$183,872	16			$165,545	19
South Carolina	$148,794	44	$145,074	34	$141,354	39
South Dakota	$136,893	50			$127,862	47
Tennessee	$188,952	14	$182,664	10	$176,364	11
Texas	$168,000	32	$158,500	28	$149,000	31
Utah	$182,950	17	$174,600	14	$166,300	18
Vermont	$163,757	35			$155,677	26
Virginia	$197,827	11	$181,610	11	$171,120	14
Washington	$190,415	13	$181,263	12	$172,571	13
West Virginia	$136,000	51			$126,000	50
Wisconsin	$159,297	37	$150,280	32	$141,773	38
Wyoming	$165,000	33			$150,000	29
Mean	**$176,714**		**$170,237**		**$158,748**	
Median	**$172,716**		**$168,059**		**$155,677**	
Range	**$136,000 to $253,189**		**$132,838 to $237,365**		**$125,499 to $208,000**	

Source: National Center for State Courts, "Survey of Judicial Salaries" 2019, accessed May 25, 2019, https://www.ncsc.org/~/media/Microsites/Files/Judicial%20Salaries/Judicial-Salary-Tracker-Jan-2019.ashx.

In an attempt to reduce the appearance of possible political influence, several states have adopted merit selection, sometimes referred to as the "Missouri Plan," as their method of selecting judges. First used in Missouri in 1940, merit selection is a process in which the governor appoints judges from a list of qualified lawyers compiled by a nonpartisan nominating commission composed of both lawyers and other citizens. After serving a short term on the bench, usually 1 year, the appointed judges face the voters in an uncontested election. Voters are instructed to vote yes or no on whether the judge should be retained in office. If a majority vote yes, the judge remains in office for a full term (usually 6 years; the range is 4 years to life). In Illinois, a judge must receive at least 60% of the votes. Toward the end of the term, the judge must face the voters again

FIGURE 8.5 Methods of Selecting State Supreme Court Judges, 2019

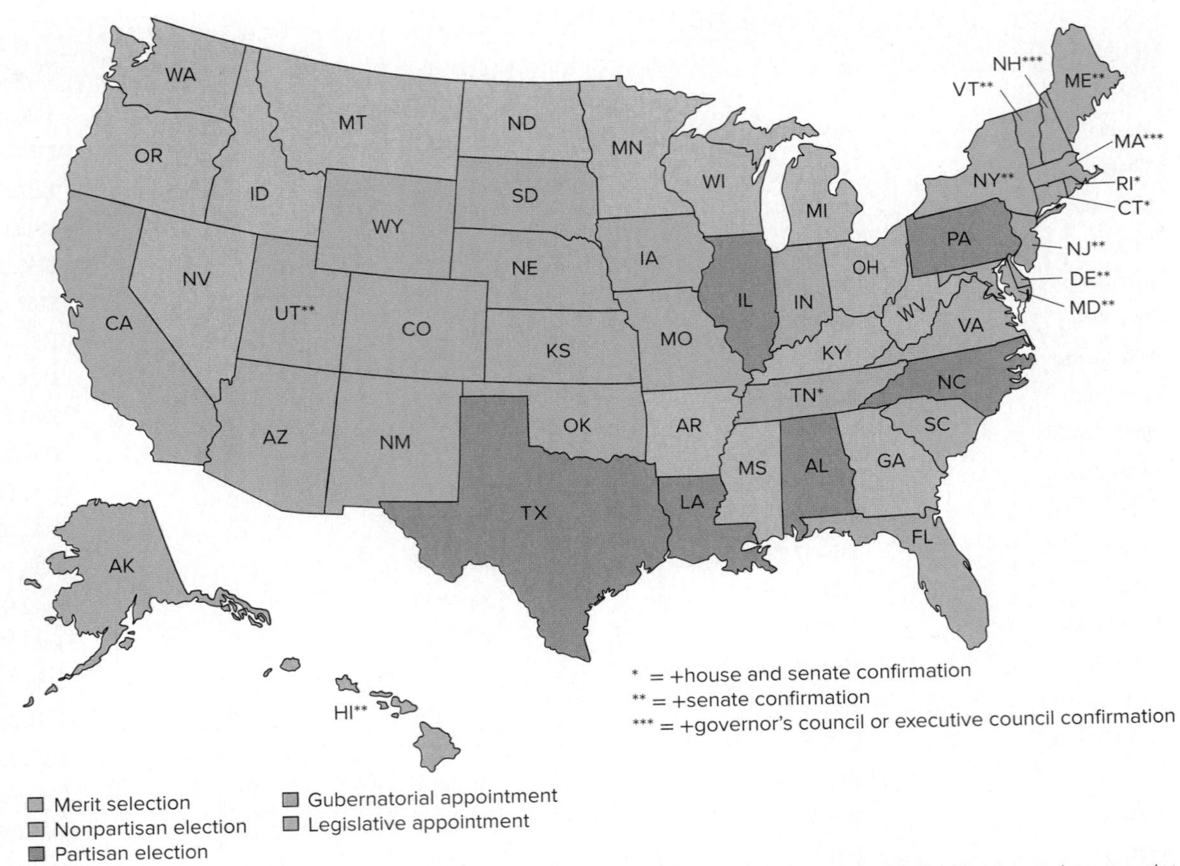

Source: Brennan Center for Justice, Judicial Selection: An Interactive Map, accessed May 26, 2019, http://judicialselectionmap.brennancenter.org/?court=Supreme&phase.

in the same kind of election. The merit plan does not entirely eliminate political influence from the selection process, but it seems to have less potential for influence than the direct popular election of judges. Figures 8.5 and 8.6, respectively, display the methods states use to choose their supreme court justices and trial court judges. As shown in Figure 8.5, merit selection (52%)–some states with confirmation requirements–and nonpartisan election (30%) are used by more than 80% of states for the selection of state supreme court justices, while nonpartisan election (38%) and merit selection (32%)–some states with confirmation requirements–are employed by 70% of states for the selection of state trial court judges. Other methods, such as partisan election, gubernatorial appointment without nominating commission, and legislative appointment, are used by only a few states.

Judicial Knowledge

To learn more about what judges should know to do their jobs well, visit The National Judicial College website at https://www.judges.org/2019courses/ and review the courses in the curriculum.

From your review of courses, what is important for judges to know?

Qualifications and Training Although in most jurisdictions lower-court judges are not required to be lawyers or to possess any special educational or professional training, nearly all states require judges who sit on the benches of appellate courts and trial courts of general jurisdiction to be licensed attorneys and members of the state bar association. However, being a lawyer is not the same as being a judge. Many judges come to the bench without any practical experience with criminal law or procedure. Consequently, many states now require new appellate court and trial court judges to attend state-sponsored judicial training seminars. In addition, more than 8,000 judges a year take 1-week to 4-week summer courses offered by the National Judicial College, founded in 1963 and located at the University of Nevada, Reno. Another 2,000-plus judges annually take courses from the National

FIGURE 8.6 Methods of Selecting State Trial Court Judges, 2019

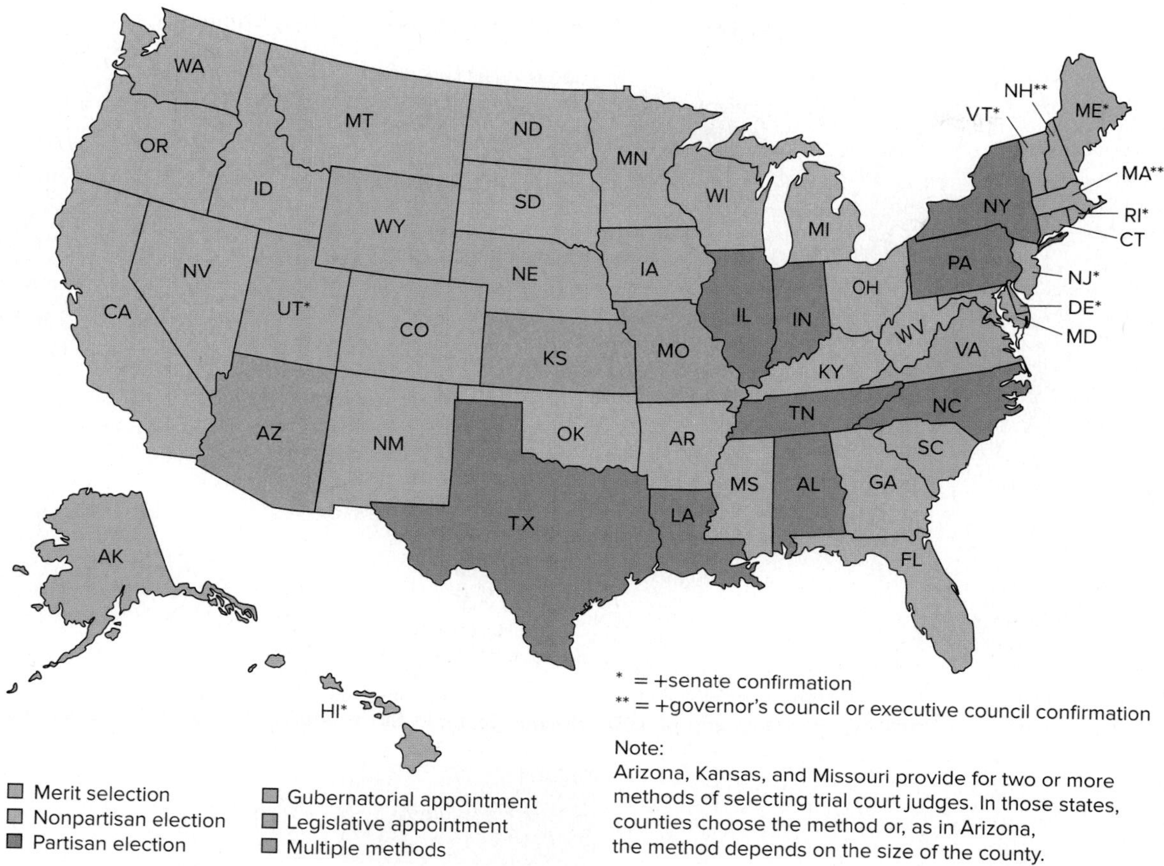

Source: Brennan Center for Justice, Judicial Selection: An Interactive Map, accessed May 26, 2019, http://judicialselectionmap.brennancenter.org/?court=Trial&phase.

Judicial College online.[48] Despite those efforts, most judges still learn the intricacies of their profession on the job.

THINKING CRITICALLY

1. Do you think prosecutors have too much power? Why or why not?
2. Of the different types of attorneys a defendant might have, which one do you think will do the best job of defending the accused? Why?
3. Which method of selecting judges do you think is best? Why?

Pretrial Stages

As described in Chapter 2, probably less than one-half of the crimes committed each year are reported to the police and, of those, only a fraction are officially recorded. Of the crimes that are recorded by the police, only about 20% are "cleared by arrest." Still, an arrest by no means guarantees prosecution and conviction. As a result of initial prosecutorial screening, for example, about 25% of all arrests are rejected, diverted, or referred to other jurisdictions. Another 20% or so of all people arrested are released later for various reasons during one of the pretrial stages. Thus a powerful "funneling" or screening process in the administration of justice eliminates about one-half of all persons arrested. Figures 8.7 and 8.8 illustrate this funneling or screening process.

FIGURE 8.7 Funneling Effect

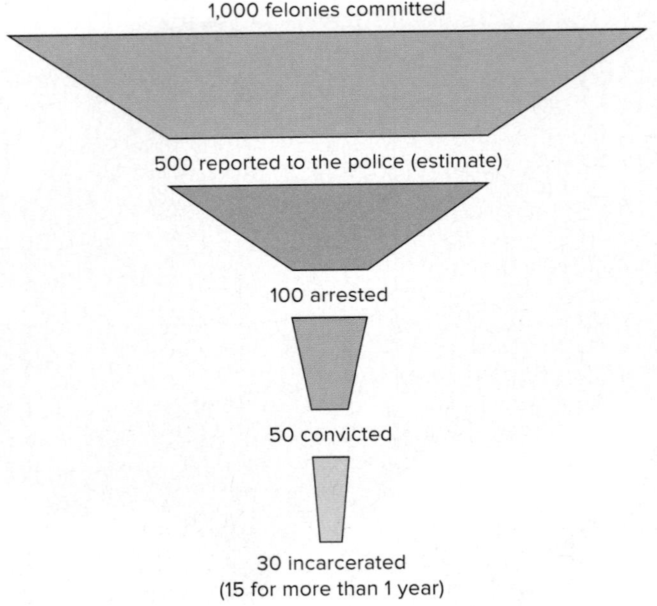

FIGURE 8.8 Outcome of 100 Criminal Cases in State Courts of Large Urban Counties

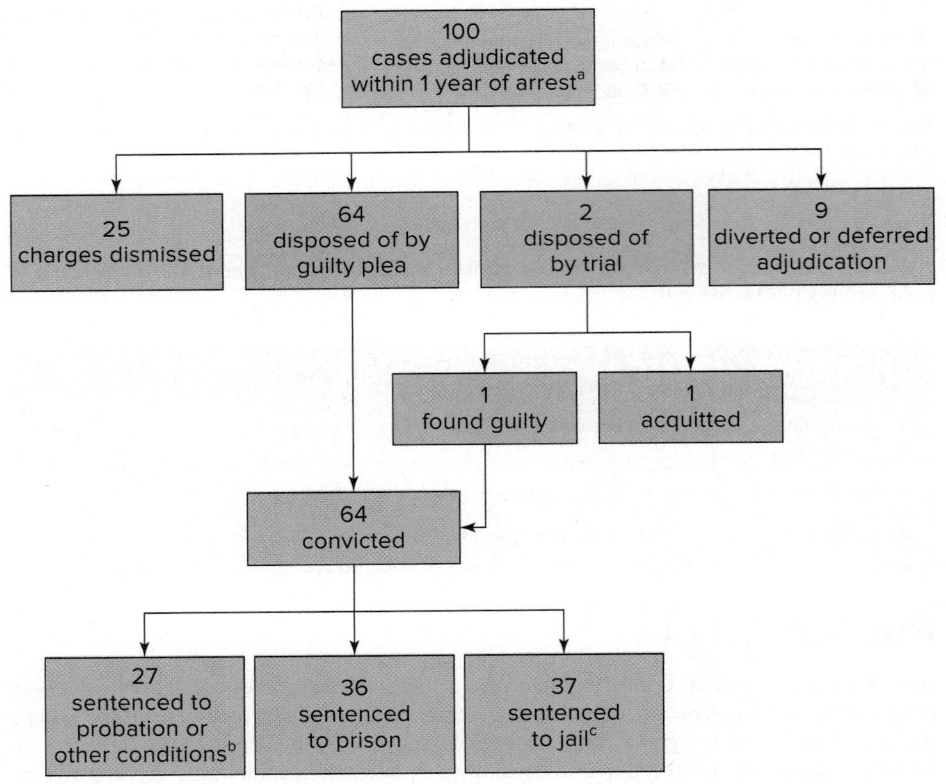

[a]From May 2009.
[b]Other conditions include fines, community service, and treatment.
[c]Includes split sentences (jail and probation).
Source: Adapted from Brian A. Reaves, "Felony Defendants in Large Urban Counties, 2009—Statistical Tables," in *Bureau of Justice Statistics*, U.S. Department of Justice (Washington, DC: U.S. Government Printing Office,December 2013), accessed May 9, 2016, www.bjs.gov/content/pub/pdf/fdluc09.pdf.

The pretrial stages do not have the same names or order in every jurisdiction. So what follows should be considered only a general overview. States are required to provide only a prompt, neutral review of the evidence to determine whether there is probable cause that the suspects/defendants committed the crime or crimes with which they are charged. *Probable cause*, which was described in more detail in Chapter 4, is an abstract term that basically means that a law enforcement officer or a judge has trustworthy evidence that would make a reasonable person believe that, more likely than not, the proposed action, such as an arrest, is justified. Figure 1.1 in Chapter 1 provided a simplified view of the caseflow through the criminal justice process, highlighting pretrial stages.[49] Following a brief consideration of caseflow management, we describe the pretrial stages typical of many jurisdictions.

Caseflow Management

Increasingly large caseloads in the state and federal courts are requiring new management skills that enable the courts to process a huge volume of civil and criminal cases as efficiently as possible. Caseflow management has been called "the conceptual heart of judicial administration in the new millennium."[50] Today's court administrators (see the Careers in Criminal Justice box) are being influenced by basic business principles. Among the subjects being learned about managing jury trials, for example, are "How to Relieve Juror Boredom," "Routine Use of Anonymous Juries," "Juror Questions to Witnesses," and "Post-Verdict Debriefings with the Trial Judge."[51] There is even a manual for Managing Notorious Cases.[52]

FYI **Trial Court Administrator Salaries**

As of 2019, state trial court administrators earned a median annual salary of $150,822. The mean annual salary was $159,399. Annual salaries ranged from a low of $107,000 to a high of $288,888.

Source: National Center for State Courts, "Survey of Judicial Salaries" (2019), accessed May 26, 2019, https://www.ncsc.org/~/media/Microsites/Files/Judicial%20Salaries/Judicial-Salary-Tracker-Jan-2019.ashx.

From Arrest Through Initial Appearance

Soon after most suspects are arrested, they are taken to the police station to be "booked." **Booking** is the process in which suspects' names, the charges for which they were arrested, and perhaps their fingerprints or photographs are entered on the police blotter. Following booking, a prosecutor is asked to review the facts of the case and, considering the available evidence, to decide whether a suspect should be charged with a crime or crimes (sometimes prosecutors review a case prior to the arrest). As a result of the review, the prosecutor may tell the police that they do not have a case or that the case is weak and requires further investigation and additional evidence.

booking The administrative recording of an arrest. Typically, the suspect's name, the charge(s) for which the person was arrested, and perhaps the suspect's fingerprints or photograph are entered in the police blotter.

However, if the prosecutor decides that a suspect is "chargeable," the prosecutor prepares a charging document. The crime or crimes with which the suspect is charged may or may not be the same crime or crimes for which the suspect was originally arrested.

There are three primary kinds of charging documents: (1) a complaint, (2) an information, and (3) a grand jury indictment. If the offense is either a misdemeanor or an ordinance violation, then the prosecutor in many jurisdictions prepares a complaint. A **complaint** is a charge that an offense has been committed by a person or persons named or described. Complaints must be supported by the oath or affirmation of either the arresting officer or the victim. If the offense is a felony, an information is used in those states that do not rely on a grand jury indictment. An **information** outlines the formal charge or charges, the law or laws that have been violated, and the evidence to support the charge or charges. A **grand jury indictment**, on the other hand, is a written accusation by a grand jury charging that one or more persons have committed a crime. Informations and grand juries are described later in this section.

complaint A charging document specifying that an offense has been committed by a person or persons named or described; usually used for misdemeanors and ordinance violations.

information A document that outlines the formal charge(s) against a suspect, the law(s) that have been violated, and the evidence to support the charge(s).

grand jury indictment A written accusation by a grand jury charging that one or more persons have committed a crime.

On rare occasions, the police obtain an arrest warrant from a lower-court judge before making an arrest. An **arrest warrant** is a written order directing law enforcement officers to arrest a person. The charge or charges against a suspect are specified on the warrant. Police officers more frequently make an arrest and then apply for an arrest warrant. Some jurisdictions have created joint police-prosecution teams so that decisions about arrest and chargeability can be coordinated and can be made early on. If no charges are filed, the suspect must be released.

arrest warrant A written order directing law enforcement officers to arrest a person. The charge or charges against a suspect are specified on the warrant.

After the charge or charges have been filed, suspects, who are now *defendants*, are brought before a lower-court judge for an initial appearance, where they are given formal notice of the charges against them and advised of their constitutional rights (e.g., the right to counsel). For misdemeanors or ordinance violations, a summary trial may be held at the initial appearance. About 75% of misdemeanants and ordinance violators plead guilty at the initial appearance and are sentenced on the spot. For felonies, a hearing is

Careers in Criminal Justice

Trial Court Administrator

My name is Todd Nuccio. I am the trial court administrator for the 26th Judicial District in North Carolina. The district encompasses Charlotte and greater Mecklenburg County. It is the largest jurisdiction in the state, serving approximately 700,000 people.

I have a bachelor of arts degree in political science from Northern Illinois University and a master of science degree in judicial administration from the University of Denver, College of Law. I recently received another bachelor of arts degree in psychology from the University of North Carolina at Charlotte because I have come to learn that 90% of successfully managing an organization depends on one's ability to understand and motivate people.

During my final undergraduate semester, a constitutional law professor told me about a program that combined aspects of public administration and law.

Since graduation, I have held the position of assistant court administrator in the 7th Judicial District of Iowa, the trial court administrator position in the 12th Judicial District in North Carolina, and my current position here in Charlotte.

The responsibilities of a trial court administrator are similar to the more commonly known position of hospital administrator. The only difference is that the hospital administrator manages operations for doctors, whereas the trial court administrator manages operations for lawyers serving in the capacity of judges.

Court administration involves everything that goes into making a court system good or bad. The court manager's mission is to improve the administration of justice by (1) working with other court officials to facilitate change; (2) initiating and coordinating discussions that form the basis of consensus; and (3) identifying problems and recommending solutions that lead to a more accessible, accountable, effective, and efficient court system.

Todd Nuccio

My primary responsibilities include developing local rules and administrative policies to facilitate calendaring and administrative activities, research, strategic planning, caseflow and jury management, budgeting, personnel oversight, facilities management, grant procurement, program development, project management, coordination of alternative dispute-resolution programs, and oversight of the drug treatment and family court operations and collections department.

What I like best about my job is the variety. There are always new programs to develop and projects to manage.

What do you think are the pros and cons of being a trial court administrator?

held to determine whether the suspect should be released or whether there is probable cause to hold the suspect for a preliminary hearing. If the suspect is to be held for a preliminary hearing, bail may be set if the judge believes release on bail is appropriate. In states that do not utilize preliminary hearings, an arraignment date is scheduled at the initial appearance.

One of the critical questions about the initial appearance is how long suspects may be held in jail before being brought before a judge. In some countries the answer to the question is "indefinitely," but in the United States there is a limit so that innocent persons are not left in jail too long. If suspects are freed after having posted station-house bail (based on a bail fee schedule for minor offenses posted at the police station), then the initial appearance may be several days after the arrest. However, if suspects remain in custody (the usual scenario), they must be brought for an initial appearance "without unnecessary delay." In 1975, in *Gerstein v. Pugh*, the Supreme Court held that a "prompt" judicial hearing is required in a warrantless arrest to determine whether the officer had probable cause to make the arrest. The vast majority of arrests are warrantless. If suspects are not brought before a judge promptly, then they are to be released. The one exception to this requirement is suspects arrested on a Friday night and not brought before a judge until Monday morning.

A problem with the Supreme Court's promptness requirement is that the Court did not define *prompt*. The norm in most jurisdictions was between 24 and 72 hours. Even after the ruling, however, some jurisdictions held suspects much longer than that. For example, in 1975, a district court judge in Birmingham, Alabama, ruled that because of abusive holding practices in that city, suspects arrested on a felony charge could not be held longer than 24 hours. In 1984, the same judge set a new limit of 8 hours for holding suspects, because his previous order had been ignored. The judge added, however, that in certain cases a limited time extension could be obtained. In 1991, in *County of Riverside v. McLaughlin*, the Supreme Court finally clarified the situation by ruling that anyone arrested without a warrant may be held no longer than 48 hours before a judge decides whether the arrest was justified.

Bail and Other Methods of Pretrial Release

A **bail bond** or **bail** is usually a monetary guarantee deposited with the court that is supposed to ensure that the suspect or defendant will appear at a later stage in the criminal justice process. In other words, it allows suspects or defendants to remain free while awaiting the next stage in the adjudication process. It is *not* a fine or a penalty; it is only an incentive to appear. Opportunities for bail follow arrest, initial appearance, preliminary hearing, arraignment, and conviction.

bail bond or bail Usually a monetary guarantee deposited with the court that is supposed to ensure that the suspect or defendant will appear at a later stage in the criminal justice process.

Bail

The granting of bail and pretrial release to criminal suspects was common practice in England by the twelfth century. It was recognized officially and legally regulated by the Statute of Westminster in 1275.

Source: "Bail: An Ancient Practice Reexamined," *Yale Law Journal* 70 (1961), 966–977.

Although bail-setting practices vary widely by jurisdiction, the amount of bail, assuming that bail is granted, generally depends on the likelihood that the suspect or defendant will appear in court as required. If the suspect or defendant has strong ties to the community–for example, a house, a family, or a job–then the amount of bail will be relatively low (depending on the seriousness of the offense and other factors). However, if the suspect or defendant has few or no ties to the community, the amount of bail will be relatively high, or the judge may refuse to grant bail.

Next to likelihood of appearance, the most important factor in a judge's determination of the amount of bail–and, perhaps, whether bail is granted at all–is the seriousness of the crime. Generally, the more serious the crime, the higher the bail amount. A third influence on the amount of bail is prior criminal record. A prior criminal record usually increases the amount of bail. Jail conditions are a fourth influencing factor–at least on whether bail is set. If the jail is overcrowded, as are many of the jails in the United States, judges are more likely to grant bail, particularly in borderline cases.

In rare cases, when a judge believes that a suspect or defendant would pose a threat to the community if released, the judge can refuse to set bail. Holding suspects or defendants in jail without giving them an opportunity to post bail because of the threat they pose to society is called **preventive detention**. Preventive detention is used most often in cases involving violent crimes, drug crimes, and immigration offenses. However, in *U.S. v. Salerno* (1987), the Supreme Court made clear that before a judge can legally deny bail for reasons of preventive detention, a suspect or defendant must have the opportunity for a hearing on the decision. At the hearing, the individual circumstances of the suspect or defendant (such as community ties, convictions, and past dangerous tendencies) must be considered. At least 22 states, the District of Columbia, and the federal jurisdiction have enacted laws or have constitutional provisions that provide for some form of preventive detention,[53] although the laws and provisions, at least at the state level, are rarely implemented. The reason is that they are unnecessary. Judges traditionally have imposed high bail amounts on dangerous suspects or defendants. Preventive detention has been employed more frequently at the federal level.

preventive detention Holding suspects or defendants in jail without giving them an opportunity to post bail because of the threat they pose to society.

Preventive detention will reduce violent crime.

Preventive detention is not new and always has been used by judges in the United States, even though it was not referred to by that name. Judges simply set bail at a level beyond the financial means of a person suspected of violent crime.

For suspects or defendants who cannot afford to post bail, professional bond agents, who are private entrepreneurs, are available to post it for them for a nonrefundable fee. The fee is typically 10% of the required amount. Thus, a bond agent would collect $500 for posting a $5,000 bond. If a suspect or a defendant is considered a greater than average risk, then the bond agent can require collateral–something of value like money or property–in addition to the fee.

Bail bond agents are under no obligation to post a surety bond if they believe that a suspect or a defendant is a bad risk. If bond agents believe that a client might flee and not appear as required, they have the right to revoke the bond without refunding the fee. Bail bond agents and their associates (such as modern-day bounty hunters known as "skip tracers") are allowed, without a warrant, to track down clients who fail to appear and to return them forcibly, if necessary, to the jurisdiction from which they fled. If the client crosses state lines, the bond agent does not have to seek extradition, as a law enforcement official would, to bring the client back. Clients sign extradition waivers as a condition of receiving bail.

Bail Agents

You can learn more about how bail agents work by going to the website of Professional Bail Agents of the United States at www.pbus.com, where you can review the code of ethics for bail agents.

What does the code of ethics tell you about the different responsibilities of a bail agent?

Nagel Photography/Shutterstock

Without the cooperation of the bail bond industry, jail populations would be unmanageable. *Is that reason enough to allow the bail bond industry to exist?*

In practice, most bail bond agents assume little risk. Many of them use part of their fee (generally 30%) to secure a surety bond from a major insurance company, which then assumes financial liability if the bail is forfeited. However, even this practice is unnecessary in many jurisdictions because the courts do not collect forfeited bonds. Judges are able to use their discretion to simply vacate outstanding bonds and, by doing so, relieve the bail bond agent or the insurance company of any financial obligation. Judges vacate bonds because they realize that without the cooperation of bail bond agents, the courts would be faced with an unmanageably large jail population and prohibitively high pretrial detention costs. If the bond is not vacated, the bond agent will generally stop looking for the fugitive after 2 years because, by then, there is little chance that the bond agent will get his or her money back.[54]

Suspects or defendants who post their own bail ("full cash bond," 100% of the amount required), or have family members or friends do it for them, get it all back after they appear. Sometimes the court will accept property instead of cash. If the suspect or defendant does not appear, the bail is forfeited and the judge issues a **bench warrant**, or ***capias***, authorizing the suspect's or defendant's arrest. It is estimated that about 10,000 criminal defendants skip bail each year, costing local governments tens of millions of dollars. Part of the problem is that overburdened courts are not pursuing many criminal defendants on bond who fail to appear. Another problem is that new companies are writing bonds for more high-risk defendants. However, not all bail skippers flee. Some are just irresponsible and do not come to court until someone drags them in. Others change their address so often they do not receive the notices to appear. Still others do not appear because they are in jail or prison in other counties or states.[55] Once arrested, the absconder must be brought before the judge who issued the warrant and cannot be released on bail again. Failure to appear also constitutes a new offense, "bond jumping," that carries criminal penalties.

bench warrant or *capias* A document that authorizes a suspect's or defendant's arrest for not appearing in court as required.

Courts in Illinois, Kentucky, and Pennsylvania, as well as the federal courts, allow defendants to post 10% of the required bond (deposit bond) directly with the court, thereby circumventing the need for bail bond agents. When a defendant makes all required court appearances, 90% of the amount posted is refunded (the remaining 10% covers administrative costs). Defendants who fail to appear still are liable for the entire bail amount.

As noted in Chapter 4, the Eighth Amendment to the Constitution does not require that bail be granted to all suspects or defendants. It requires only that the amount of bail, when granted, not be excessive. What constitutes excessive bail is determined by several factors, including the nature and circumstances of the offense, the weight of evidence against the suspect/defendant, the character of the suspect/defendant, and the ability of the suspect/defendant to provide bail. Table 8.4 shows selected presumptive bail amounts for the Superior Court of Santa Barbara County, California, in 2019.

The bail system has been criticized for unfairly discriminating against the poor, who are the least likely to have the assets for their own bail and are least able to pay the fee to a bail bond agent. About 85% of all suspects or defendants are released before the final disposition of their cases, but in some jurisdictions, as many as 90% of them are held in jail because they are unable to afford bail. Nationwide, one-third of defendants remain in jail prior to trial solely because they are unable to pay a cash bond, and most of these people are among the poorest one-third of Americans.[56] In most jurisdictions, defendants held in jail because they are unable to post bail (nearly 65% of the jail population nationwide in 2017)[57] are mixed with the rest of the jail population, that is, convicted offenders. They also are treated in the same way as the rest of the jail population. One of the tragedies of this practice is the occasional brutalization (e.g., rape) of the jailed indigent defendant whose case is later dismissed for lack of evidence or who is found not guilty at trial. Although brutalization of any prisoner is horrible and reprehensible, it is particularly so when the prisoner is an innocent person.

myth

The most dangerous defendants released on bail are the ones least likely to appear as required.

fact

Defendants accused of minor offenses and released on bail have the highest rates of failure to appear.

Jailed indigent defendants also are likely to lose their jobs and have their personal lives disrupted in other ways as a result of their detention. Studies consistently show that jailed defendants, who are frequently brought into court in jail clothes and handcuffs, are more

TABLE 8.4 Selected Bail Amounts for the Superior Court of Santa Barbara County, California, 2019

Offense[a]	Presumptive Bail Amount[b]
Violation of Protective Order: Domestic Violence	$50,000
Murder with special circumstance	NOT BAILABLE
All other murders	$2,000,000
Manslaughter (Involuntary)	$25,000
Vehicular Manslaughter (DUI w/gross negligence)	$100,000
Kidnapping	$100,000
Robbery (First Degree)	$100,000
Carjacking	$100,000
Rape	$100,000
Corporal Punishment or Injury of Child	$50,000
Child Stealing	$40,000
Failure of Convicted Sex Offender to Register	$20,000
Elder Abuse	$50,000
Aggravated Arson	$500,000
Burglary (Residential)	$50,000
Burglary (All others)	$20,000
Theft of Motor Vehicle	$35,000
Organized Retail Theft	$20,000
Solicit Murder	$1,000,000
Attempted Murder	$1,000,000
Driving Under the Influence of Alcohol or Drugs (if a felony)	$100,000
Sale of Controlled Substances (Up to 1 kilogram)	$30,000
Manufacture of Any Controlled Substance	$75,000
Resisting Arrest/Threatening Officer	$25,000
Special Allegation Bail Enhancements	
Serious Felony for Benefit of Street Gang	$100,000
Felony that is a Hate Crime	$30,000
One Strike	Schedule Times Two or $50,000, Whichever Greater
Two Strikes (current offense neither serious nor violent)	$100,000
Two Strikes (current offense is serious or violent)	$1,000,000

[a]Bail for unlisted felony offenses: $20,000.

[b]Is applicable to warrantless arrestees until the matter is reviewed by a judge.

Source: Superior Court of California, County of Santa Barbara, "2019 Felony Bail Schedule," March 2019, accessed May 26, 2019, https://www.sbcourts.org/dv/bail/FelonyBailSchedule.pdf.

likely to be indicted and convicted and are sentenced more severely than defendants who have been released pending the next stage in the process. However, it is not clear whether those disparities are the result of pretrial detention or of the selection process in which suspects or defendants charged with more serious crimes and with prior criminal records are more likely to be denied bail.

When the crime is minor and suspects or defendants have ties to the community, they generally are released on their own recognizance. **Release on own recognizance (ROR)** is simply a release secured by a suspect's written promise to appear in court.

release on own recognizance (ROR) A release secured by a suspect's written promise to appear in court.

conditional release A form of release that requires that a suspect/defendant maintain contact with a pretrial release program or undergo regular drug monitoring or treatment.

unsecured bond An arrangement in which bail is set but no money is paid to the court.

Another nonfinancial means of release, more restrictive than ROR, is **conditional release**. This form of release (sometimes called *supervised release*) usually requires that a suspect/defendant maintain contact with a pretrial release program or undergo regular drug monitoring or treatment. Some conditional release programs also require a third-party custody agreement (i.e., a promise by a reputable person to monitor the person released). Another way that suspects or defendants are released without a financial requirement is by **unsecured bond**. Under this arrangement, bail is set, but no money is paid to the court. Suspects or defendants are liable for the full amount of bail if they do not appear as required.

Information

If the decision is made to prosecute a defendant, in states that do not use grand juries, the prosecutor drafts a document called an *information*. The information outlines the formal charge or charges, the law or laws that have been violated, and the evidence to support the charge or charges. The information generally is filed with the court at the preliminary hearing or, if the preliminary hearing is waived, at the arraignment. Figure 8.9 displays a felony information.

Preliminary Hearing

preliminary hearing A pretrial stage used in about one-half of all states and only in felony cases. Its purpose is for a judge to determine whether there is probable cause to support the charge or charges imposed by the prosecutor.

The purpose of the **preliminary hearing**, used in about one-half of all states, is for a judge to determine whether there is probable cause to support the charge or charges imposed by the prosecutor. Preliminary hearings are used only in felony cases, and defendants may waive the right to the hearing. A preliminary hearing is similar to a criminal trial in two ways but also differs from a criminal trial in two ways. It is similar in that defendants can be represented by legal counsel and can call witnesses on their behalf. It differs in that the judge must determine only that there is probable cause that the defendant committed the crime or crimes with which he or she is charged. At criminal trials, guilt must be determined "beyond a reasonable doubt." Also, at preliminary hearings, unlike criminal trials, defendants have no right to be heard by a jury.

If the judge determines that there is probable cause that a defendant committed the crime or crimes with which he or she is charged, then the defendant is bound over for possible indictment in states with grand juries or for arraignment on an information in states without grand juries. In grand jury states, even if the judge at the preliminary hearing rules that there is insufficient evidence to proceed, the case is not necessarily dropped. The prosecutor could take the case directly to the grand jury. If the case is not dropped for lack of evidence, then the judge may set bail again or may continue the previous bail to ensure that the defendant appears at the next stage in the process.

Although judges at preliminary hearings are supposed to examine the facts of the case before making a probable cause determination, in practice they seldom do. In big cities, judges generally do not have the time to inquire into the facts of a case. Consequently, at most preliminary hearings, judges simply assume that if a police officer made an arrest and a prosecutor charged the defendant with a crime or crimes, then there must be probable cause that the defendant, in fact, committed the crime or crimes. Few cases are dismissed for lack of probable cause at preliminary hearings.

Grand Jury

grand jury Generally a group of 12 to 23 citizens who meet in closed sessions to investigate charges coming from preliminary hearings or to engage in other responsibilities. A primary purpose of the grand jury is to determine whether there is probable cause to believe that the accused committed the crime or crimes.

A **grand jury** is generally a group of 12 to 23 citizens who, for a specific period of time (usually 3 months), meet in closed sessions to investigate charges coming from preliminary hearings or to engage in other responsibilities. Thus, a primary purpose of the grand jury is to determine whether there is probable cause to believe that the accused is guilty of the charge or charges brought by the prosecutor. Grand juries do not convene every day. Many federal grand juries meet only once a week, and some may meet only twice a month. State practice varies, but a state grand jury may meet only twice a month or even only once a month. Sometimes a grand jury does not convene unless it is asked to do so by a prosecutor with cases to hear.

FIGURE 8.9 Felony Information

SECOND JUDICIAL DISTRICT COURT
COUNTY OF ALBANY, STATE OF WYOMING

THE STATE OF WYOMING,) Criminal Action No.2015-8294
)
Plaintiff,)
)
-vs-)
)
BRAD HEROD,)
)
Defendant.)

FELONY INFORMATION

The State of Wyoming, by and through E. Kurt Britzius, Deputy County and Prosecuting Attorney in and for Albany County, Wyoming, and in the name and by the authority of the State of Wyoming, informs the Court and gives the Court to understand that, BRAD HEROD, late of the county aforesaid, on or between September 10, 2015 to December 9, 2015, in the County of Albany and in the State of Wyoming, did commit the offense of *Influencing/Intimidating/Impeding Witnesses,* in that the Defendant, by force or threats, attempted to influence, intimidate or impede witnesses in Docket 15-8276, an active criminal case in this Judicial District, in violation of and punishable pursuant to Wyoming Statute § 6-5-305(a), as more particularly set forth in the *Affidavit of Probable Cause* by Officer Josh Anderson of the Laramie Police Department field herein and incorporated herein by reference, and against the peace and the dignity of the State of Wyoming.

RESPECTFULLY SUBMITTED this 31st day of December, 2015.

E. Kurt Britzius,
Deputy Albany County & Prosceuting. Attorney
Supreme Court Registration #6-4274
Albany County Courthouse
525 Grand Avenue, Suite 100
Laramie, WY 82070
(307) 721-2552 voice
(307)721-2554fax

Source: Second Judicial District Court, County of Albany, State of Wyoming.

Before appearing before a grand jury, the prosecutor drafts an **indictment**, a document that outlines the charge or charges against a defendant. Figure 8.10 displays a grand jury indictment. Because the grand jury has to determine only whether there is probable cause that a defendant committed the crime or crimes with which he or she is charged, only the prosecution's evidence and witnesses are heard. In most jurisdictions, neither the defendant nor the defendant's counsel has a right to be present during the proceedings. Furthermore, in grand jury proceedings, unlike criminal trials, prosecutors are allowed to present *hearsay* evidence (information learned from someone other than the witness who is testifying). They also may use illegally obtained evidence because the exclusionary rule does not apply to grand jury proceedings. Prosecutors can subpoena witnesses to testify. A

indictment A document that outlines the charge or charges against a defendant.

FIGURE 8.10 Grand Jury Indictment

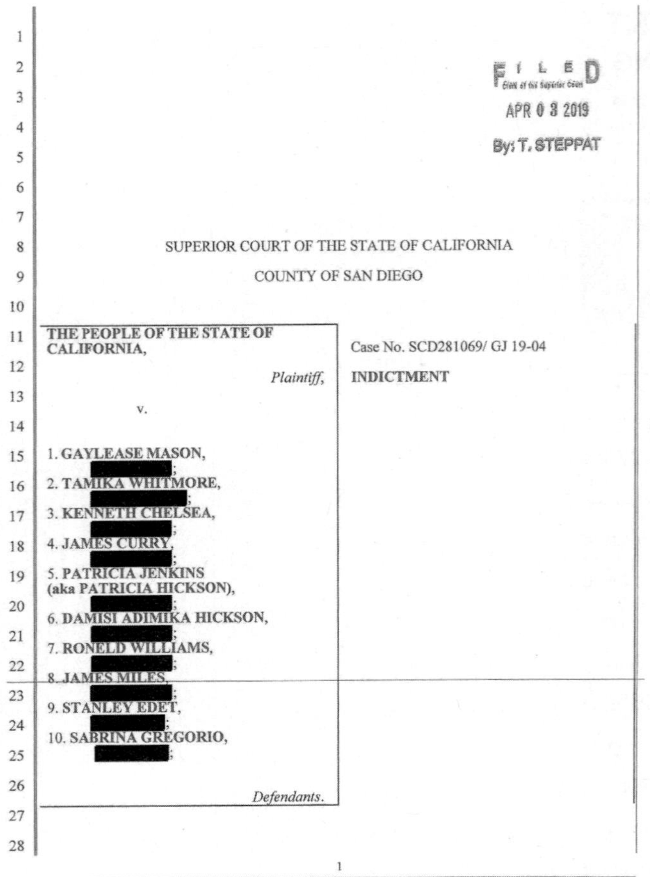

FILED
Clerk of the Superior Court
APR 03 2019
By: T. STEPPAT

SUPERIOR COURT OF THE STATE OF CALIFORNIA

COUNTY OF SAN DIEGO

THE PEOPLE OF THE STATE OF CALIFORNIA, Plaintiff,	Case No. SCD281069/ GJ 19-04
v.	INDICTMENT
1. GAYLEASE MASON, [redacted]; 2. TAMIKA WHITMORE, [redacted]; 3. KENNETH CHELSEA, [redacted]; 4. JAMES CURRY, [redacted]; 5. PATRICIA JENKINS (aka PATRICIA HICKSON), [redacted]; 6. DAMISI ADIMIKA HICKSON, [redacted]; 7. RONELD WILLIAMS, [redacted]; 8. JAMES MILES, [redacted]; 9. STANLEY EDET, [redacted]; 10. SABRINA GREGORIO, [redacted]; Defendants.	

1

Indictment (GJ 19-04/Case No. SCD281069)

A Special Statewide Grand Jury of the County of San Diego, State of California, hereby accuses by this Indictment, the following defendants of committing, in the Counties of Orange, San Diego, San Bernardino, Riverside, and Los Angeles, the following crimes:

COUNT 1 - VEHICLE 1

GRAND THEFT OF AN AUTOMOBILE

On or between March 17, 2015, and June 23, 2015, in the County of Orange, defendants GAYLEASE MASON, KENNETH CHELSEA, and JAMES MILES committed the crime of GRAND THEFT OF AN AUTOMOBILE, in violation of Penal Code section 487(d)(1), a felony, in that they did unlawfully take an automobile, the property of another, to wit: 2014 Ford Taurus from Exeter Finance.

COUNT 2 - VEHICLE 1

GRAND THEFT

On or about July 17, 2015, in the County of Los Angeles, defendant TAMIKA WHITMORE and KENNETH CHELSEA committed the crime of GRAND THEFT, in violation of Penal Code section 487(a), a felony, in that she did unlawfully take money and/or personal property of another of a value exceeding Nine Hundred Fifty Dollars ($950), to wit: $17,000 in sale auto sale proceeds from Santa Monica SUV.

COUNT 3 - VEHICLE 8

GRAND THEFT OF AN AUTOMOBILE

On or about August 20, 2015, in the County of San Bernardino, defendants KENNETH CHELSEA and JAMES CURRY committed the crime of GRAND THEFT OF AN AUTOMOBILE, in violation of Penal Code section 487(d)(1), a felony, in that they did unlawfully take an automobile, the property of another, to wit: 2014 Mercedes CLA 250 from Exeter Finance.

COUNT 4 - VEHICLE 8

GRAND THEFT

On or about September 5, 2015, in the County of Los Angeles, defendant JAMES CURRY committed the crime of GRAND THEFT, in violation of Penal Code section 487(a), a felony, in

2

Indictment (GJ 19-04/Case No. SCD281069)

subpoena is a written order to testify issued by a court officer. Figure 8.11 displays a subpoena for a grand jury. A witness who refuses to testify can be held in contempt and can be jailed until he or she provides the requested information. In practice, however, a witness jailed for contempt generally is held only as long as the grand jury is in session and no longer than 18 months in the federal jurisdiction.

After hearing the prosecutor's evidence and witnesses, the grand jury makes its probable cause determination and, usually on a majority vote, either indicts (issues a *true bill*) or fails to indict (issues *no bill*). If the grand jury fails to indict, then in most jurisdictions, the prosecution must be dropped. However, in some jurisdictions, the case can be brought before another grand jury.

COUNT 35 - VEHICLE W8

GRAND THEFT OF AN AUTOMOBILE

On or about June 11, 2016, in the County of Los Angeles, defendant STANLEY EDET committed the crime of GRAND THEFT OF AN AUTOMOBILE, in violation of Penal Code section 487(d)(1), a felony, in that he did unlawfully take an automobile, the property of another, to wit: 2010 Cadillac Escalade from WELLS FARGO.

* *

A TRUE BILL, and with a finding that the requirements of Penal Code section 923(c) have been satisfied.

Joanne H. Leslie — 15
Foreperson of the Grand Jury — **Juror #**
Dated: Apr. 3, 2019

10

Indictment (GJ 19-04/Case No. SCD281069)

Source: Superior Court of the State of California, County of San Diego.

All but two states and the District of Columbia use grand juries to indict. Twenty-three states plus the District of Columbia require that indictments be used to charge certain crimes, generally serious crimes. Twenty-five states make the use of indictments optional. In those states, charges may be brought by an information. Connecticut and Pennsylvania have abolished the use of grand juries to return indictments, but they have retained them to investigate crimes and the conduct of public affairs.[58] The most common civil matter they investigate is the operation and condition of local jails and other confinement facilities. In the federal system, a grand jury indictment is required in all felony prosecutions unless the defendant waives that right.[59]

subpoena A written order issued by a court that requires a person to appear at a certain time and place to give testimony. It can also require that documents and objects be made available for examination by the court.

FIGURE 8.11 Grand Jury Subpoena

AO 88 (Rev. 02/14) Subpoena to Appear and Testify at a Hearing or Trial in a Civil Action

UNITED STATES DISTRICT COURT

for the

Plaintiff	)	
v.	)	Civil Action No.
Defendant	)	

SUBPOENA TO APPEAR AND TESTIFY AT A HEARING OR TRIAL IN A CIVIL ACTION

To:

(Name of person to whom this subpoena is directed)

YOU ARE COMMANDED to appear in the United States district court at the time, date, and place set forth below to testify at a hearing or trial in this civil action. When you arrive, you must remain at the court until the judge or a court officer allows you to leave.

Place:	Courtroom No.:
	Date and Time:

You must also bring with you the following documents, electronically stored information, or objects *(leave blank if not applicable)*:

The following provisions of Fed. R. Civ. P. 45 are attached – Rule 45(c), relating to the place of compliance; Rule 45(d), relating to your protection as a person subject to a subpoena; and Rule 45(e) and (g), relating to your duty to respond to this subpoena and the potential consequences of not doing so.

Date:

CLERK OF COURT

OR

Signature of Clerk or Deputy Clerk *Attorney's signature*

The name, address, e-mail address, and telephone number of the attorney representing *(name of party)*
, who issues or requests this subpoena, are:

Notice to the person who issues or requests this subpoena

If this subpoena commands the production of documents, electronically stored information, or tangible things before trial, a notice and a copy of the subpoena must be served on each party in this case before it is served on the person to whom it is directed. Fed. R. Civ. P. 45(a)(4).

Source: Administrative Office of the U.S. Courts on behalf of the Federal Judiciary.

FYI Grand Jury Indictments Required

The 23 states where grand jury indictments are required for certain serious crimes are Alabama, Alaska, Delaware, Florida, Kentucky, Louisiana, Maine, Massachusetts, Minnesota, Mississippi, Missouri, New Hampshire, New Jersey, New York, North Carolina, North Dakota, Ohio, Rhode Island, South Carolina, Tennessee, Texas, Virginia, and West Virginia.

Source: Daniel Taylor, "Which States Use Criminal Grand Juries," FindLaw, November 26, 2014, accessed May 27, 2019, https://blogs.findlaw.com/blotter/2014/11/which-states-use-criminal-grand-juries.html.

In practice, the grand jury system is criticized for merely providing a rubber stamp for whatever the prosecutor wants to do. In other words, in cases where the prosecutor wants an indictment, the grand jury likely is to indict. Likewise, in cases where the prosecutor does not want to indict, the grand jury tends to fail to indict. The reason prosecutors are so successful with grand juries is that they manage the entire proceedings (remember, only the prosecution's evidence and witnesses are heard). So it should not be surprising that suspects waive the right to a grand jury hearing in about 80% of cases. Defendants also may waive the right to a grand jury hearing to speed up their trial date.

You might be wondering why prosecutors would want a grand jury to fail to indict after they have gone to the trouble of bringing a case to the grand jury in the first place. The reason is political. To avoid losing marginal cases or looking cowardly by dropping charges, prosecutors can bring a case to the grand jury, have the grand jury fail to indict, and then blame the grand jury for its failure. The strategy deflects criticism from the prosecutor to the anonymous members of the grand jury. Sometimes prosecutors delay preliminary hearings to await a grand jury action to avoid disclosing evidence in open court that might help defense attorneys at trial. Because of the way the grand jury is used today, some critics of the system suggest that it ought to be abolished.

Arraignment

arraignment A pretrial stage; its primary purpose is to hear the formal information or indictment and to allow the defendant to enter a plea.

nolo contendere Latin for "no contest." When defendants plead *nolo*, they do not admit guilt but are willing to accept punishment.

The primary purpose of an **arraignment** is to hear the formal information or indictment and to allow the defendant to enter a plea. The two most common pleas are "guilty" and "not guilty." "Not guilty" is the most common plea at arraignments. However, some states and the federal courts allow defendants to plead *nolo contendere* to the charges against them. ***Nolo contendere*** is Latin for "no contest." When defendants plead "*nolo*," they do not admit guilt but are willing to accept punishment anyway. The *nolo* plea is used for strategic purposes. If a defendant does not admit guilt in a criminal trial, admission of guilt cannot be the basis for a subsequent civil lawsuit. If there is a subsequent civil lawsuit, the lack of an admission of guilt in a criminal trial may allow the defendant to avoid a penalty of treble, or triple, damages. Furthermore, in some states, defendants can stand mute or can plead "not guilty by reason of insanity." Standing mute at arraignment is interpreted as pleading "not guilty." In states that do not accept a plea of "not guilty by reason of insanity," defendants plead "not guilty" and assume the burden of proving insanity at trial.

If a defendant pleads guilty, the judge must determine whether the plea was made voluntarily and whether the defendant is fully aware of the consequences of his or her action. If the judge doubts either of those conditions, then the judge can refuse to accept the guilty plea and can enter in the record a plea of "not guilty" for the defendant.

At arraignment, the judge also determines whether a defendant is competent to stand trial. Defendants can seek a delay before trial to consult further with their attorneys. A defendant who does not have an attorney already can ask for one to be appointed. Finally, defendants sometimes attempt to have their cases dismissed at arraignment. For example, they may assert that the state lacks sufficient evidence or that improper arrest procedures were used.

myth

All guilty pleas are the result of plea bargaining.

fact

Although the vast majority of guilty pleas probably are a product of plea bargaining—we do not know for sure—some defendants simply plead guilty to the original charges and do not bargain.

THINKING CRITICALLY

1. Do you agree with the criticism that the poor are unfairly discriminated against because they are the least able to pay their own bail and the least able to pay the fee to a bail bonds agent? Why or why not?
2. Do you think a preliminary hearing is necessary? Should it be abolished? Why or why not?
3. Are grand jury proceedings fair to the suspect? Why or why not?

Plea Bargaining

As noted earlier in this chapter, justice in the United States is dispensed mostly through plea bargaining.[60] About 95% of all convictions in felony cases are the result of guilty pleas; criminal trials are relatively rare. We already saw in Figure 8.8 the typical outcome of 100 felony arrests and the prevalence of plea bargaining.

There are three basic types of plea bargains. First, the defendant may be allowed to plead guilty to a lesser offense. For example, a defendant may be allowed to plead guilty to manslaughter rather than to first-degree murder. Second, at the request of the prosecutor, a defendant who pleads guilty may receive a lighter sentence than typically would be given for the crime. Note, however, that the prosecutor can only recommend the sentence; the judge does not have to grant it. Third, a defendant may plead guilty to one charge in return for the prosecutor's promise to drop other charges that could be brought.

The bargain a prosecutor will strike generally depends on three factors. The most important factor is the seriousness of the offense. Generally, the more serious the crime, the more difficult it is to win concessions from the prosecutor. A second factor is the defendant's criminal record. Defendants with criminal records usually receive fewer concessions from prosecutors. The final factor is the strength of the prosecutor's case. The stronger the case, the stronger is the position of the prosecutor in plea negotiations.

Surprisingly, there is neither a constitutional basis nor a statutory basis for plea bargaining. It is a custom that developed because of the mutual interests it serves. The custom received formal recognition from the Supreme Court in 1970 in the case of *Brady v. United States*. In that case, the Court upheld the use of plea bargaining because of the "mutuality of advantage" it provided the defendant and the state. In two later cases, the Court provided safeguards for the bargaining process. In the 1971 case of *Santobello v. New York*, the Court held that the "deal" offered by a prosecutor in a plea negotiation must be kept. In 1976, in the case of *Henderson v. Morgan*, the Court held that to be valid, a guilty plea must be based on full knowledge of its implications and must be made voluntarily. However, prosecutors are under no obligation to plea-bargain. Once defendants plead guilty and are sentenced, they are almost always stuck with the bargain, even if they have a change of heart. Nevertheless, in most jurisdictions, defendants are allowed to withdraw their guilty pleas before sentencing.

As just noted, the reason justice in the United States is administered primarily through plea bargaining is that the process seemingly serves the interests of all the court participants by, among other things, reducing uncertainty. Uncertainty is a characteristic of all criminal trials because neither the duration of the trial, which may be a matter of minutes or of months, nor the outcome of the trial ever can be predicted with any degree of accuracy. Plea bargaining eliminates those two areas of uncertainty by eliminating the need for a trial.

Plea bargaining also serves the interests of the individual participants in the administration of justice. Prosecutors, for example, are guaranteed high conviction rates. For prosecutors and, apparently, the general public, a conviction is a conviction, whether it is obtained through plea bargaining or as the result of a trial. A prosecutor's conviction rate is one of the principal indicators of job performance, and job performance certainly helps determine whether the prosecutor will fulfill his or her aspirations for higher political office.

Plea bargaining also serves the interests of judges by reducing their court caseloads, thus allowing more time to be spent on difficult cases. In addition, if a large proportion of the approximately 95% of felony cases that are handled each year by plea bargaining were to go to trial instead, the administration of justice in the United States would be even slower than it already is. Nevertheless, federal judges and the judges in seven states legally are prohibited from participating in plea negotiations. In other states, law limits their role in the process.

Plea bargaining serves the interests of criminal defense attorneys by allowing them to spend less time on each case. It also allows them to avoid trials. Trials are relatively expensive events. Because most criminal defendants are poor, they usually are unable to pay a large legal fee. Thus, when criminal defense attorneys go to trial, they frequently are unable to recoup all of their expenses. Plea bargaining provides many criminal defense attorneys with the more profitable option of charging smaller fees for lesser services and handling a larger volume of cases.

Even most criminal defendants are served by plea bargaining. A guilty plea generally results in either no prison sentence or a lesser prison sentence than the defendant might receive if found guilty at trial. Plea bargaining also often allows defendants to escape conviction of socially stigmatizing crimes, such as child abuse. By "copping" a plea to assault rather than to statutory rape, for example, a defendant can avoid the embarrassing publicity of a trial and the wrath of fellow inmates or of society in general.

Two types of criminal defendants are not served by the practice of plea bargaining. The first are innocent, indigent, highly visible defendants who fear being found guilty of crimes they did not commit and receiving harsh sentences. Unscrupulous defense attorneys

Plea Bargaining

Plea bargaining became a common practice in state courts shortly after the Civil War. The practice was instituted at the federal level during Prohibition in the 1930s as a result of the tremendous number of liquor law violations.

Sources: Albert W. Alschuler, "Plea Bargaining and Its History," *Law and Society Review* 13 (1979), 211–245; John F. Padgett, "Plea Bargaining and Prohibition in the Federal Courts, 1908–1934," *Law and Society Review* 24 (1990), 413–450.

myth

Abolishing plea bargaining would reduce the level of serious crime.

fact

Despite the administrative nightmare it would cause, the abolition of plea bargaining probably would only shift discretion to another area. It unlikely would have any effect on crime.

Three Strikes Law

Although habitual-offender statutes have existed in many jurisdictions for decades, the first state to enact a "three strikes and you're out" law was Washington in 1993. The Washington law, called the Persistent Offender Accountability Act, allows three-time felons to be imprisoned for life without parole. In August 1996, the Washington Supreme Court upheld the law as constitutional. As of December 2018, 31 states and the federal government had three-strike or habitual-offender laws on their books. The states are: Arizona, Arkansas, California, Colorado, Connecticut, Delaware, Florida, Georgia, Indiana, Kansas, Louisiana, Maryland, Massachusetts, Missouri, Montana, Nevada, New Hampshire, New Jersey, New Mexico, New York, North Carolina, North Dakota, Pennsylvania, South Carolina, Tennessee, Texas, Utah, Vermont, Virginia, Washington, and Wisconsin.

Sources: "Three Strikes Laws in Different States," LegalMatch, accessed May 27, 2019, https://www.legalmatch.com/law-library/article/three-strikes-laws-in-different-states.html; Mary Randolph, "Three Strikes Law," Criminal Defense Lawyer, accessed May 27, 2019, https://www.criminaldefenselawyer.com/resources/three-strikes-law.htm; New Hampshire, "Sentencing," accessed May 27, 2019, http://www.ncrp.info/StateFactSheets.aspx?state=NH.

sometimes pressure such defendants into waiving their constitutional right to trial. The second type is the habitual offender. In this context, a habitual offender is a person who has been convicted under a state's habitual-offender statute (sometimes called a "three strikes and you're out" law). Most such statutes provide that upon conviction of a third felony, a defendant must receive life imprisonment. Although habitual-offender statutes would seem to imprison offenders for life, they actually are used mostly as bargaining chips by prosecutors in plea negotiations and not as they were intended.

A problem with those statutes is illustrated by the Supreme Court case of *Bordenkircher v. Hayes* (1978). The defendant, who previously had been convicted of two minor felonies, was arrested and charged with forging an $88 check. The prosecutor in the case told the defendant that if he did not plead guilty to the charge and accept a 5-year prison sentence, which on its face seemed very harsh, then the prosecutor would invoke the state's habitual-offender statute. The statute required the judge to impose a sentence of life imprisonment if the defendant was found guilty at trial. The defendant elected to play "you bet your life" and turned down the prosecutor's plea offer. At trial, the defendant was found guilty of forging the check and was sentenced to life imprisonment. Clearly, the defendant in this case was not served by plea bargaining or, perhaps, was not served by refusing the prosecutor's offer. In either case, with the possible exception of habitual offenders and innocent people, plea bargaining serves the interests of all the actors in the administration of justice. It does so by allowing cases to be disposed of predictably, quickly, and with little of the adversarial conflict associated with criminal trials.

THINKING CRITICALLY

1. Do you think that the plea-bargaining process is beneficial? Why or why not?
2. Why is murder the crime least likely by far to be resolved through a guilty plea?
3. How could the plea-bargaining process be improved?

The Criminal Trial

One of the distinctive features of criminal justice in the United States is trial by a jury of one's peers.[61] The principal purpose of jury trials–and of criminal trials without juries–is to discover the truth of whether defendants are guilty or innocent of the crimes with which they are charged. The process by which truth is sought is an adversarial one regulated by very specific procedures and rules. The adversaries in a criminal trial are the state (represented by the prosecutor) and the defendant (usually represented by defense counsel). The burden of proof is on the prosecution to show, beyond a reasonable doubt, that the defendant is guilty. The goal of defense counsel is to discredit the prosecution's case and to create reasonable doubt about the defendant's guilt. It is the responsibility of the jury (in jury trials) or the judge (in trials without juries) to determine and assign guilt.

bench or summary trial A trial before a judge without a jury.

Although all criminal defendants have a constitutional right to a jury trial (when imprisonment for 6 months or more is a possible outcome), only about 2% of all criminal cases are disposed of in this way. Approximately 95% of cases are resolved through a guilty plea, and a judge decides the remaining cases in a **bench trial** (without a jury). In most jurisdictions, defendants may choose whether they want to exercise their right to a jury trial or whether they prefer a bench trial. The principal reason that so few criminal cases are decided by criminal trials is undoubtedly the advantages associated with plea bargaining.

Criminal justice in the United States is dispensed primarily through criminal trials.

Criminal trials are relatively rare. Approximately 95% of all criminal cases are resolved through guilty pleas. Only about 5% are decided by bench or jury trials.

The Jury

Trial by an impartial jury of one's peers is an exalted American tradition and a Sixth Amendment right.[62] Its principal purposes are:

1. To protect citizens against arbitrary law enforcement.
2. To prevent government oppression.
3. To protect citizens from overzealous or corrupt prosecutors and from eccentric or biased judges.

But jury trials in the United States, as noted above, are relatively rare. So on those rare occasions when a jury tries defendants, the jury seldom is composed of their peers. Until the mid-twentieth century, many states excluded women and people of color from jury service. Even today, class, gender, and racial biases enter into the jury selection process.

In many jurisdictions, jury pools are selected from voter registration lists. About 30% of eligible voters do not register; in some jurisdictions, the rate is as high as 60%. People not registered to vote are excluded from jury service. Studies show that the poor, the poorly educated, the young, and people of color are least likely to register to vote and, as a result, are least likely to be called for jury service. To remedy this problem, some jurisdictions now use multiple-source lists for obtaining jurors. In addition to voter registration lists, their sources include lists of licensed drivers, lists of utility users, and names listed in the telephone directory. Appellate courts have ruled that master jury lists must reflect an impartial and representative cross-section of the population. People of color and women cannot be excluded systematically from juries solely because of race, ethnicity, or gender. But this does not mean that people of color and women must be included on all juries, only that they cannot be denied the opportunity of being chosen for jury service.

FYI

Trial by Jury

Trial by a jury of one's peers originated in England as a way of limiting the power of the king. When the Magna Carta was signed in 1215, it contained the following provision: "No freeman shall be taken, or imprisoned, or disseized, or outlawed, or exiled, or in any way harmed—nor will we go upon or send upon him—save by the lawful judgment of his peers or by the law of the land."

Source: "Magna Carta," www.britannia.com/history/magna2.html.

From the master list of all eligible jurors (sometimes called the master wheel or the jury wheel), a sufficient number of people are randomly chosen to make up the jury pool, or **venire**. Those chosen are summoned for service by the sheriff. However, not all those summoned actually will serve on the venire. Potential jurors generally must be U.S. citizens, residents of the locality of the trial, of a certain minimum age, and able to understand English. Convicted felons and insane persons almost always are excluded. Most jurisdictions also require that jurors be of "good character" and be "well-informed," which eliminates other potential jurors. In addition, members of other groups often escape jury service. Professionals such as doctors, lawyers, and teachers; some elected officials; military personnel on active duty; and law enforcement personnel frequently are not called for jury duty because their professional services are considered indispensable or because they are connected to the criminal justice process. Many jurisdictions allow citizens to be excused from jury service if it would cause them physical or economic difficulties.

venire The pool from which jurors are selected.

Compensation for jury service varies widely among states.[63] In 2016 or 2017, depending on the state, about 20 states, the District of Columbia, and the federal courts paid jurors a flat daily rate, ranging from $10 in Alabama, Maine, and Tennessee to $50 in Arkansas. Technically, New Mexico paid the highest flat daily rate of $60, which was based on the state minimum wage of $7.50 for an 8-hour day. Some of these flat-rate jurisdictions pay for half a day of service. Approximately 20 states use a graduated rate system in which jurors receive either no fee or a significantly reduced fee usually for 1 to 3 days of service; after that, they receive an increased fee. In some states, the rate varies depending on the county. A few states pay for mileage and reimburse for expenses.

In 1986, Massachusetts became one of the first states to employ a graduated rate system. Under the Massachusetts system, jurors receive no fee for the first 3 days of service, and $50 per day thereafter. Massachusetts law also requires employers to pay their employees on jury service their regular pay for the first 3 days of service. Unemployed jurors or jurors with child care needs in Massachusetts do not receive compensation for the first 3 days of service, but they are entitled to be reimbursed for actual out-of-pocket expenses they would not ordinarily incur such as mileage, public transportation fees, and child care expenses when necessary, at a rate not to exceed $50 per day of service. Eight states—Colorado, Connecticut, Iowa, Maryland, Massachusetts, Missouri, North Dakota, and Oklahoma–pay jurors $50 a day under a graduated rate system. Ohio's base pay rate varies by county.

Some states are moving from local funding of jurors to state funding to equalize what some people consider unfair pay differentials among counties. Some states still have state-mandated minimum funding for jury service and encourage counties to supplement the mandated minimum. Illinois, for example, has the lowest state-mandated minimum of $4 per day with county supplements as high as $40 per day. Some counties in Wisconsin augment the state base rate of $16 per day. Not all employers pay for time lost from work. Consequently, only about 25% of adult Americans have ever served on juries; still, that is up from only 6% in 1977.[64]

voir dire The process in which potential jurors who might be biased or unable to render a fair verdict are screened out.

Trial Before Judge or Jury?

A 2014 public opinion poll discovered that 58% of American adults trusted a jury more than a judge to give a fair verdict, while 22% of American adults trusted a judge more than a jury to render a fair verdict. Twenty percent of Americans were unsure.

Source: Rasmussen Reports, "58% Still Trust a Jury's Verdict More Than a Judge's," February 19, 2014. Accessed May 10, 2016, www.rasmussenreports.com/public_content/lifestyle/general_lifestyle/february_2014/58_still_trust_a_jury_s_verdict_more_than_a_judge_s.

From the venire, as many as 30 people (in felony prosecutions that mandate 12 jurors) are selected randomly by the court clerk for the jury panel from which the actual trial jury is selected. To ensure a fair trial, potential trial jurors go through ***voir dire***, a process in which persons who might be biased or unable to render a fair verdict are screened out. During *voir dire*, which means "to speak the truth," the defense, the prosecution, and the judge question jurors about their backgrounds and knowledge of the case and the defendant. If it appears that a juror might be biased or unable to render a fair verdict, the juror can be challenged "for cause" by either the defense or the prosecution. If the judge agrees, the juror is dismissed from jury service. In death penalty trials, for example, death penalty opponents can be excluded from juries for cause if they are opposed to the death penalty under any circumstances (*Lockhart v. McCree*, 1986). Generally, there is no limit to the number of jurors who can be eliminated for cause. In practice, however, few potential jurors are eliminated for cause, except in high-profile trials, such as the O. J. Simpson murder trial in 1995.

Another way that either the defense or the prosecution can eliminate potential jurors from jury service is by the use of *peremptory challenges*, which allow either prosecutors or defense attorneys to excuse jurors without having to provide a reason. Peremptory challenges frequently are used to eliminate jurors whose characteristics place them in a group likely to be unfavorable to the case of either the prosecution or the defense. For example, in death penalty cases, prosecutors often use their peremptory challenges to eliminate people of color and women from the juries, because, statistically, people of color and women are less likely to favor capital punishment. However, prosecutors must be careful in their use of peremptory challenges for such purposes because the Supreme Court has forbidden the use of peremptory challenges to exclude potential jurors solely on account of their race (*Batson v. Kentucky*, 1989) or gender (*J. E. B. v. Alabama*, 1994). The number of peremptory challenges is limited by statute. In most jurisdictions, the prosecution is allowed from 6 to 8 peremptory challenges and the defense from 8 to 10. *Voir dire* continues until the required number of jurors has been selected. The *voir dire* process may take an hour or two or, in rare cases, months.

Traditionally, a jury in a criminal trial, sometimes called a petit jury to distinguish it from a grand jury, consists of 12 citizens plus one or two alternates, who will replace any jurors unable to continue because of illness, accident, or personal emergency. Recently, however, primarily to reduce expenses, some states have gone to 6-, 7-, and 8-member juries in noncapital criminal cases. The Supreme Court will not allow criminal trial juries with 5 or fewer members, and 12-member juries still are required in all states in capital (death penalty) cases.

Recently, attempts have been made to reduce the burden of jury service. In many jurisdictions, jurors can now call a number to find out whether they will be needed on a particular day during their term of service. Some jurisdictions have instituted "1 day/1 trial" jury systems, which require jurors to serve either for 1 day or for the duration of one trial. Once they have served, they are exempt from jury service for 1 or 2 years. Each year in the United States, approximately 2 million jurors serve in about 200,000 criminal and civil trials.

The Trial Process

Before a criminal trial formally begins, attorneys in about 10% of felony cases file pretrial motions. A motion is an application to a court, requesting a judge to order a particular action. Motions also can be made during and after a trial. Common pretrial motions are to obtain discovery of the prosecution's evidence and to have some of the prosecution's evidence suppressed (e.g., to have a confession ruled inadmissible because of *Miranda* violations).

The following is a general description of the stages in a criminal trial.[65] Figure 8.12 shows that process and includes some pretrial stages as well (not all stages occur in every trial). After the jury has been sworn in (if the case is tried before a jury) and the court clerk has read the criminal complaint, the prosecution begins the trial with an opening statement outlining its case. Next is the opening statement by the defense. However, the defense is not required to make an opening statement. In some jurisdictions, the defense is allowed to defer its opening statement until after the prosecution has presented its case. Opening statements rarely are made in bench trials.

The prosecution then submits its evidence and questions its witnesses. The prosecution must establish beyond a reasonable doubt each element of the crime. The elements of a

FIGURE 8.12 Stages in a Criminal Trial

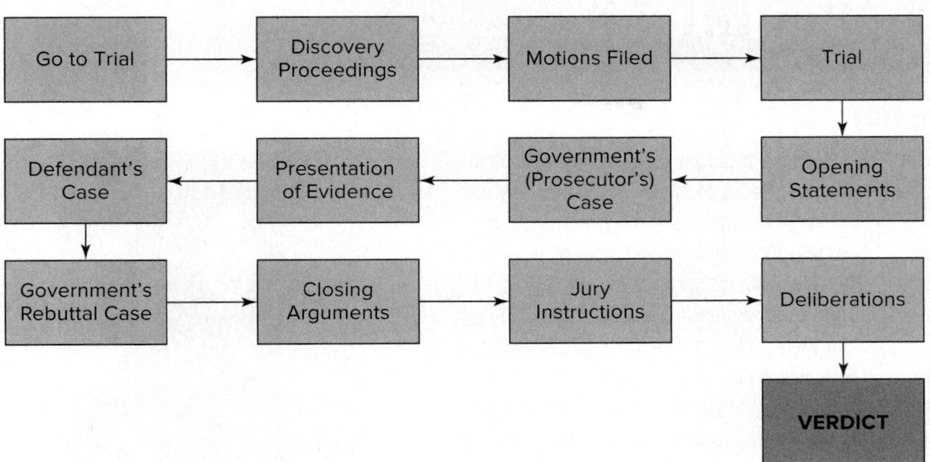

crime vary by offense and jurisdiction. In other words, two jurisdictions (e.g., two states) may have different elements for the same crime. The specific elements of each crime are provided in state and federal statutes. Some of the elements common to most crimes were described in Chapter 2 and include *actus reus*, *mens rea*, and concurrence.

If the defense believes that the prosecution has failed to make its case, then the defense may choose to *rest*, that is, not to defend against the charge or charges. At this point, the defense in most states is allowed to request a *directed verdict* or to make a motion for dismissal. If the judge agrees with the defense that the prosecution's evidence is insufficient for conviction, then the judge can either direct the jury to acquit the defendant or "take the case from the jury" and grant the motion for dismissal.

If the defense does not seek a dismissal or a dismissal is not granted, the defense follows the prosecution with its witnesses and any contrary evidence. Then the prosecution and the defense take turns offering rebuttals to the other side's evidence, cross-examining witnesses, and reexamining their own witnesses. Following the rebuttal period, the prosecution summarizes its case. The defense then summarizes its case and makes its closing statement. The closing statement by the prosecution ends the adversarial portion of the trial.

Normally, after the closing statement by the prosecution, the judge instructs, or *charges*, the jury concerning the principles of law the jurors are to utilize in determining guilt or innocence. The judge also explains to the jury the charges, the rules of evidence, and the possible verdicts. In some jurisdictions, the judge summarizes the evidence presented from notes taken during the trial. The jury then retires to deliberate until it reaches a verdict. In a room where it has complete privacy, the jury elects from its members a foreperson to preside over the subsequent deliberations. Jurors are not allowed to discuss the case with anyone other than another juror. In some cases, a jury is sequestered at night in a hotel or motel to prevent any chance of outside influence.

To find a defendant guilty as charged, the jury must be convinced "beyond a reasonable doubt" that the defendant committed the crime. Some juries reach a verdict in a matter of minutes; some juries have taken weeks or more. If the jury finds the defendant guilty as charged, as it does in two-thirds of criminal cases, the judge begins to consider a sentence. In some jurisdictions, the jury participates to varying degrees in the sentencing process. If the jury finds the defendant not guilty, the defendant is released from the jurisdiction of the court and is a free person. After the verdict has been read in the courtroom, either the defense or the prosecution may ask that the jury be polled individually, with each juror stating publicly how he or she voted.

In the federal courts and in every state except Oregon, a unanimous verdict is required.[66] If, after serious deliberation, even one juror cannot agree with the others on a verdict, the result is a **hung jury**. Even in jurisdictions that do not require a unanimous verdict, a hung or deadlocked jury results when a jury cannot reach a verdict by the required voting margin. Judges hate hung juries. Not only do they fail to produce a decisive trial outcome, either a conviction or an acquittal, but they also result in a huge waste of time and money for all parties involved. In the fourteenth century, judges routinely would confine deadlocked

hung jury The result when jurors cannot agree on a verdict. The judge declares a mistrial. The prosecutor must decide whether to retry the case.

Careers in Criminal Justice

Court Reporter/Stenographer

Khalid Musrati

My name is Jazzmin Dashiell, and I am a freelance court reporter (stenographer) in Central Florida. I began court reporting school a few months after my high school graduation. Most stenography schools require a high school diploma or equivalent to enroll in the program.

With stenography, there are a few routes you can take once you complete school and state certification (if applicable). There are official stenographers who are employed within the state or federal government to make a written record of proceedings taken in their particular jurisdiction. There are also freelance stenographers who are generally hired by an attorney or law firm to create an official verbatim written record of out-of-court proceedings (that is, depositions, examinations under oath, and arbitrations). Another option after completing court reporting school is captioning for the hard-of-hearing in either a classroom setting, event, or for broadcast television.

The amount of time it takes to get through steno school can vary greatly. Most programs suggest completion in 2 years. The first 6 to 12 months are spent learning the theory. "Theory" is how we describe the phonetic language that we use on our steno machines. If you've been in a courtroom setting or watched a crime drama, you may have gotten a glimpse of the machine or "writer" that we use. Once you've successfully learned the theory you then must practice to increase your writing speed up to 225 words per minute (WPM). This is where the program seems to get difficult for many people, as you now must dedicate yourself and your patience to diligent practice to pass your last question-and-answer exam at 225 WPM with accuracy of 95% or higher.

As a stenographer we are responsible for accurately taking down all words spoken in the room, which can be somewhat of a challenge in a room full of people. Not only are we taking the words being spoken, we are also capturing which words were spoken by whom. If people in the room are speaking out of turn or speaking over one another, this can pose a bit of a conundrum for the stenographer, as we only have one set of hands to take down people speaking at once. But even with these less than stellar situations, stenographers are still able to tune in and capture what is being said.

The duty of a stenographer doesn't end with only taking down what is being said, we also have the responsibility of editing our "notes" and creating the official record, or transcript, of the proceedings. Our notes are taken in their rough form and uploaded into our CAT (computeraided translation) software to edit. The steno notes are then translated into English to allow us to edit the transcript in English as it was taken down in steno. Through CAT software we are also able to provide realtime translation, which is helpful for attorneys by allowing them to view the words on an iPad or computer screen in front of them immediately as they are spoken. This gives them the opportunity to see how the testimony appears in black and white and the ability to further clarify if needed.

Once the proceedings have concluded, we go through our notes and edit them to reconcile punctuation, speaker designations, and grammatical errors that may have occurred in the rough form. Since we are taking down exactly what has been spoken, our duty is not to insert proper grammar or interpret what has been said. We provide a readable form of verbatim speech. Becoming a stenographer has been a great career choice for me. Stenography provides me day-to-day variety, a flexible schedule, and exceptional earning potential. As long as there's a need for people to read what has been said in a particular setting, there will always be a need for stenographers.

juries "without meat, drink, fire, or candle, or conversation with others, until they were agreed."[67] Judges continued to force deadlocked juries to reach a verdict well into the twentieth century, for example, by requiring them to deliberate all night, threatening to deprive them of food from Saturday to Monday, or warning them on a frigid winter day that water and heat would be turned off in the jury room until they reached a unanimous verdict.[68] Judges have special instructions for deadlocked juries. They are called "Allen charges" because the U.S. Supreme Court created them in 1896 in the case of *Allen v. United States*.[69] These instructions also have been referred to as "shotgun charges," "hammer charges," "the third-degree charges," "dynamite charges," and "nitroglycerine charges." As the names imply, the purpose of the instructions is to blast a verdict out of a deadlocked jury.[70] Following is a typical Allen charge:

> You have informed the Court of your inability to reach a verdict in this case. At the outset, the Court wishes you to know that although you have a duty to reach a verdict, if that is not possible, the Court has neither the power nor the desire to compel

agreement upon a verdict. The purpose of these remarks is to point out to you the importance and the desirability of reaching a verdict in this case, provided, however, that you as individual jurors can do so without surrendering or sacrificing your conscientious scruples or personal convictions. You will recall that upon assuming your duties in this case each of you took an oath. The oath places upon each of you as individuals the responsibility of arriving at a true verdict upon the basis of your opinion and not merely upon acquiescence in the conclusions of your fellow jurors. However, it by no means follows that opinions may not be changed by conference in the jury room. The very object of the jury system is to reach a verdict by a comparison of views and by consideration of the proofs with your fellow jurors. During your deliberations you should be open-minded and consider the issues with proper deference to and respect for the opinions of each other and you should not hesitate to re-examine your own views in the light of such discussions. You should consider also that this case must at some time be terminated; that you are selected in the same manner and from the same source from which any future jury must be selected; that there is no reason to suppose that the case will ever be submitted to twelve persons more intelligent, more impartial or more competent to decide it, or that more or clearer evidence will ever be produced on one side or the other. You may retire now, taking as much time as is necessary for further deliberations upon the issues submitted to you for determination.[71]

If the jury remains deadlocked, the judge declares a mistrial, and the prosecutor must decide whether to retry the case. Recent data show that in the small number of jury trials (fewer than 3% of all criminal case dispositions), hung juries occur infrequently: only in about 6% of state court criminal jury trials and in fewer than 3% of federal court criminal jury trials. The percentage of hung juries in the state courts ranged from 0% to nearly 19%. In state court criminal trials that end in hung juries, possible outcomes include case dismissal, guilty plea, another jury trial, or a bench trial. The average percentage of hung juries in criminal trials has not varied much in the past 40 years.[72]

Adversarial Versus Inquisitorial Trial Systems

As noted previously, the United States employs an adversarial trial system, which is one of two basic types of trial systems employed throughout the world. The other and more widely used type is the inquisitorial trial system.[73] An **inquisitorial trial system** is a legal system where the court or a part of the court is actively involved in investigating the facts of the case, as opposed to an adversarial system where the role of the court is primarily that of an impartial referee between the prosecution and the defense. Figure 8.13 shows the countries with either type of system. In general, adversarial trial systems are used in England and the countries it once ruled (e.g., the United States, except Louisiana; Canada, except Quebec; India; Australia; and New Zealand), while inquisitorial trial systems generally are found in the countries of continental Europe and their former colonies (e.g., France, Germany, Italy, Spain, and Mexico). A third system based on Islamic legal tradition combines elements of both adversarial and inquisitorial systems, although inquisitorial elements seem to dominate.

inquisitorial trial system A legal system where the court or a part of the court is actively involved in investigating the facts of the case, as opposed to an adversarial system where the role of the court is primarily that of an impartial referee between the prosecution and the defense.

It is important to understand that both trial system models are ideal types, and no country's trial system should be considered a pure model of either type. All countries have at least some elements of both models. Nevertheless, some countries' trial systems are predominantly adversarial, and other countries' trial systems are principally inquisitorial. Significant differences also exist among adversarial trial systems and among inquisitorial trial systems.

As described in the previous section, the adversarial system aims to determine the truth about a crime through an open competition between the prosecution and the defense. Each side in the competition attempts to make the most compelling argument. Critics of the adversarial system contend that the goal of winning often supersedes the search for truth. In the inquisitorial system, on the other hand, the goal also is the search for the truth about a crime but not by a competition, as in the adversarial system, but by an extensive investigation and examination of all evidence. In both systems, defendants are guaranteed the right to a fair trial and are protected from self-incrimination.

FIGURE 8.13 Countries with Different Types of Trial Systems

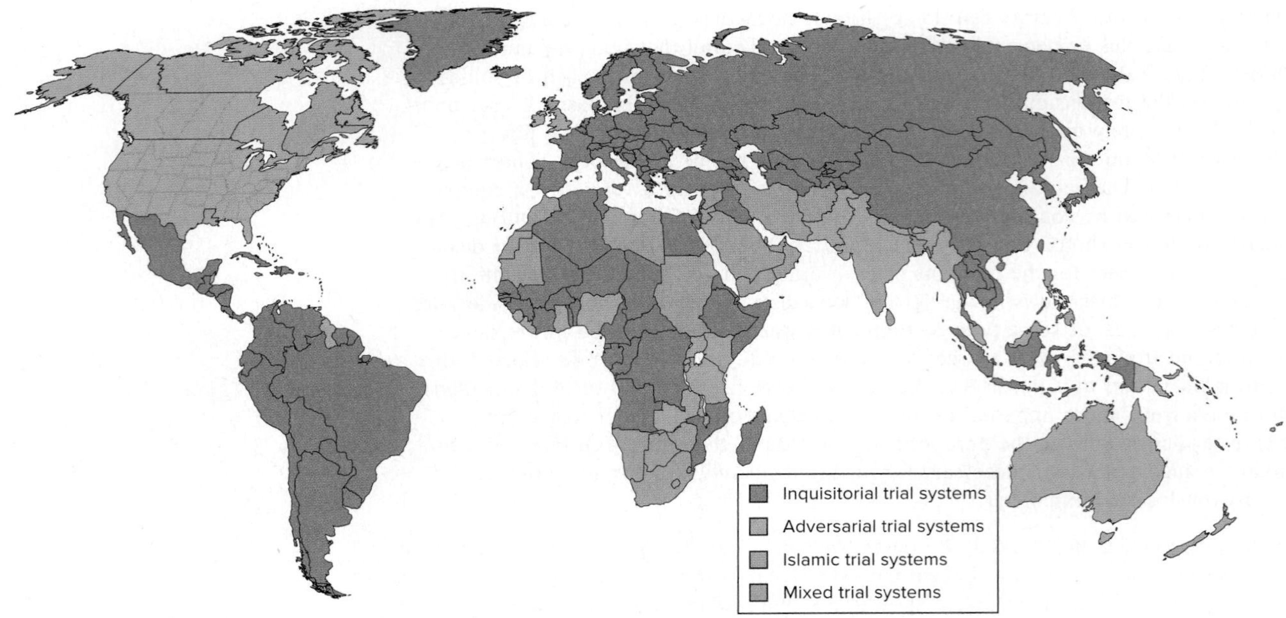

Source: http://upload.wikimedia.org/wikipedia/commons/2/21/LegalSystemsOfTheWorldMap.png.

Historical Origins of Inquisitorial and Adversarial Systems

The inquisitorial system originated in continental Europe around the twelfth and thirteenth centuries when the Catholic Church and individual secular jurisdictions established professional bureaucracies to prosecute and adjudicate crime, authorized the initiation of legal process by public officials, and adopted a system of legal proof and legal torture. The adversarial system originated in the eighteenth century when English law and judges gradually authorized the participation of professional defense attorneys in felony trials, which gradually led to the adoption of common law rules of evidence and to the redefinition of the role of judges as passive umpires, among other changes.

Source: Maximo Langer, "In the Beginning Was Fortescue: On the Intellectual Origins of the Adversarial and Inquisitorial Systems and Common and Civil Law in Comparative Criminal Procedure," UCLA School of Law Research Paper No. 16-03, December 13, 2015, accessed May 30, 2019, https://ssrn.com/abstract=2703126.

Roles of the Judge, Prosecutor, and Defense Attorney

A significant difference between the two systems has to do with roles of the three main actors in the trial process: the judge, the prosecutor, and the defense attorney. In the adversarial system, judges, prosecutors, and defense attorneys have separate, well-defined roles and responsibilities. Trial judges, who in the American system are selected or elected from among practicing attorneys, have a relatively passive role and act as referees, making sure that the defense and prosecution follow the rules and that due process is respected. Defense attorneys and prosecutors have a pivotal role and are responsible for preparing and presenting their own cases by collecting and presenting their own evidence, examining their own witnesses, and cross-examining opposing witnesses.

The inquisitorial system, by contrast, relies on professional judges, who, in some countries are called prosecutors or inquisitors and are a part of a career judiciary that is trained specifically for the bench. They have an active and overarching role, typically directing and supervising the police's pretrial collection and preparation of evidence and sometimes conducting investigations themselves, interrogating witnesses, presenting evidence at trial, and ensuring an outcome based on the merits of the case. The extensive judge-directed pretrial investigation is intended to prevent innocent people from being brought to trial. In some inquisitorial systems, judges have little charging discretion and must prosecute a case if sufficient evidence is available. Traditionally, defendants in inquisitorial systems do not have the option of pleading guilty, whereas in some adversarial systems, most criminal defendants–about 95% of them in the United States–plead guilty. Because of the investigation's thoroughness in inquisitorial systems, a written record of evidence is available to the defense and prosecution well in advance of the trial. In adversarial systems, there is little disclosure of the defense's case prior to trial; in inquisitorial systems, there is full disclosure of both prosecution and defense cases prior to trial.

During the trial, judges in inquisitorial systems have the responsibility of examining defendants and witnesses, while prosecutors and defense attorneys are limited to asking supplementary questions (e.g., regarding interpretation of alleged facts) and nominating additional witnesses. In the inquisitorial system, judges determine which witnesses, including expert witnesses, will testify, which eliminates "witness coaching" and notorious "battles of the experts" that occur in adversarial systems. Also, little emphasis is placed on cross-examination in inquisitorial systems because during cross-examination in adversarial systems clever attorneys oftentimes make honest witnesses contradict themselves and, therefore, unfairly discredit their testimony. In Germany, which uses an inquisitorial system, witnesses are allowed to present narrative testimony without having to respond to questions from the defense and prosecution. Another major difference is that in adversarial systems, no inference of guilt can legitimately be made from a defendant's refusal to testify, unlike in inquisitorial systems where guilt can legitimately be inferred by a defendant's silence. Proceedings in inquisitorial systems are about fact-finding and are less formal and confrontational than in adversarial systems. One reason that attorneys play such a minor role in inquisitorial trials is the recognition that there frequently is great disparity in the skills and resources of opposing attorneys, and such disparity unfairly affects outcomes. As for outcomes, in adversarial systems, judges make use of precedents and *stare decisis* in deciding cases. In inquisitorial systems, judges exercise more discretion and seldom rely on previous cases in pronouncing judgment. They attempt to apply the law to the current situation. Finally, while both systems aim to determine guilt, adversarial systems seem more interested in legal guilt, while inquisitorial systems appear more interested in factual guilt.

Role of Juries

At least theoretically, the jury plays an important role in adversarial trials (in actuality, as noted previously, only about 2% of criminal cases in the United States are disposed of through jury trials). Inquisitional trials, by contrast, do not employ juries as understood in the adversary process. Instead, in inquisitorial systems, a combination of lay and professional judges decides serious cases; professional judges, alone, decide less serious cases. The use of lay judges in the decision-making process is intended to broaden the perspectives of the professional judges. Also, the decision-making process in the adversarial system generally is shrouded in secrecy, while the decision-making process in the inquisitorial system is more transparent–written records explaining findings of fact and law accompany verdicts.

Rules of Evidence

The adversarial system is governed by elaborate rules of evidence, while the rules of evidence in the inquisitorial system are less technical and restrictive. For example, judges in inquisitorial systems are not limited to the evidence presented at trial, whereas judges and juries in adversarial systems generally are. The rules governing evidence admissibility and exclusion in adversarial systems were created primarily because of lay juries whose members may not have the wherewithal to evaluate prejudicial evidence. In the absence of lay juries in the inquisitorial system, concern is less with the admissibility or exclusion of evidence and more on the value of the evidence. For instance, hearsay evidence generally is inadmissible in an adversarial trial but allowed in an inquisitorial trial. In the latter, the major consideration is not that the evidence is hearsay but rather the value of the hearsay evidence. Because judges have no direct control over investigations in adversarial systems, rules of evidence allow judges to control the conduct of investigations by not admitting tainted evidence–"fruit of the poisonous tree." In inquisitorial systems, no evidence is automatically excluded because judges control investigations, but admitted evidence is subject to appeals for factual error.

These differences in the handling of evidence give rise to the accusation that judges in adversarial systems only search for the formal truth or legal guilt (based on the evidence presented at trial), while judges in inquisitorial systems search for the material truth or factual guilt (based on all available evidence–both exculpatory evidence that favors the defendant and incriminating evidence that disfavors the defendant). In an inquisitorial system, active judges have the advantage of knowing what evidence they need and what questions to ask. Passive judges in an adversarial system must wait for the evidence to be presented to them and, sometimes, the jury. They are not likely to receive incriminating evidence from the defense or exculpatory evidence from the prosecution.

Role of Victims

The role of victims also is different in the two systems. In the adversarial system, victims generally are not parties to the proceedings (the crime is committed against the state). In the inquisitorial system, victims are parties to the proceedings. Some countries with inquisitorial systems allow victims to play a role in the pretrial investigative process, giving them the right to suggest lines of questioning or to participate in interviews. At trials in some inquisitorial systems, victims have independent standing and are allowed attorneys to represent them.

TABLE 8.5 Pros and Cons of an Inquisitorial System

Pros

1. It reduces the advantage of wealth within the justice system.
2. It reduces emotional judgments against people.
3. It reduces bias within the justice system.
4. It is a system that exempts no one (no special treatment).
5. It is a system that features independent review.
6. It must resolve all factual uncertainties before arriving at a resolution.
7. It allows the judicial system to play a substantial role in the proceedings.
8. It makes distortion of evidence easier to detect.
9. It balances out the availability of resources.
10. It allows defendants to tell their story.

Cons

1. It does not remove completely bias from the justice system.
2. It lengthens the time required to reach an outcome in the justice system.
3. It reduces the opportunities for defendants to defend themselves.
4. It is a system where truth is decided before trial.
5. It requires investigators to ask the right questions to achieve a fair outcome.
6. It does not provide a right to silence like an adversarial system.
7. It can place the final judgment in the hands of a single individual.
8. It requires people to trust the government to achieve a just result.
9. It is a system that can be influenced heavily by corruption.

Source: Crystal Ayres, "19 Big Pros and Cons of Inquisitorial System," *ConnectUS*, March 16, 2019, accessed May 28, 2019, https://connectusfund.org/19-big-pros-and-cons-of-inquisitorial-system.

Conclusion

The inquisitorial system has been criticized for the multidimensional role of the judge: (1) as the judge who conducts the trial and determines guilt or innocence; (2) as the prosecutor who charges the defendant, presents evidence, and examines witnesses; and (3) as the police who investigate the crime and collect evidence. Critics contend that this multidimensional role could create prejudice against the defendant. For example, even though judges in inquisitorial systems are supposed to search for evidence in favor of the defendant, they have less of an incentive to do so than the defendant and his or her attorneys in adversarial systems. Still, in an adversarial system, police and prosecutors generally do not seek exculpatory evidence, as the judge does in inquisitorial systems. Table 8.5 lists pros and cons of an inquisitorial system. Clearly, both systems have their advantages and disadvantages. As a result, the contemporary trend seems to be toward mixed trial systems that combine elements of both adversarial and inquisitorial systems.

THINKING CRITICALLY

1. What are some of the benefits of having a trial by jury?
2. How could the jury system be improved so there is a more diverse mix of people?
3. Do you think the media's sensationalizing of crime affects the ultimate verdict in a criminal case? Defend your answer.
4. What are the advantages and disadvantages of the inquisitorial system compared to the adversarial system?
5. Which features of the inquisitorial system should be incorporated into the United States' adversarial system, and why?

SUMMARY

1. **Identify the type of court structure in the United States, and describe its various components.**

   The United States has a dual court system—a separate judicial system for each of the states and a separate federal system. The only place where the two systems "connect" is in the U.S. Supreme Court. The federal court system is composed of U.S. district courts, U.S. circuit courts of appeals, and the U.S. Supreme Court. The state court system consists of trial courts of limited jurisdiction, trial courts of general jurisdiction, intermediate appellate courts (in most states), and state courts of last resort.

2. **Summarize the purposes of courts.**

   The purposes of courts are (1) to do justice, (2) to appear to do justice, (3) to provide a forum where disputes between people can be resolved justly and peacefully, (4) to censure wrongdoing, (5) to incapacitate criminal offenders, (6) to punish offenders, (7) to rehabilitate offenders, (8) to deter people from committing crimes, (9) to determine legal status, and (10) to protect individual citizens from arbitrary government action.

3. **Identify the most powerful actors in the administration of justice, and explain what makes them so powerful.**

   Prosecutors are the most powerful actors in the administration of justice because they conduct the final screening of all persons arrested for criminal offenses, deciding whether there is enough evidence to support a conviction, and because, in most jurisdictions, they have unreviewable discretion in deciding whether to charge a person with a crime and prosecute the case.

4. **Summarize the types of attorneys available to a person charged with a crime.**

   People charged with crimes may have privately retained counsel, or if indigent, they may have court-appointed attorneys, public defenders, or "contract" lawyers, depending on which is provided by the jurisdiction.

5. **Describe the responsibilities of a judge.**

   Judges have a variety of responsibilities in the criminal justice process. Among their nontrial duties are determining probable cause, signing warrants, informing suspects of their rights, setting and revoking bail, arraigning defendants, and accepting guilty pleas. Judges spend much of the workday in their chambers, negotiating procedures and dispositions with prosecutors and defense attorneys. The principal responsibility of judges in all of those duties is to ensure that suspects and defendants are treated fairly and in accordance with due process of law. In jury trials, judges are responsible for allowing the jury a fair chance to reach a verdict on the evidence presented. A judge must ensure that his or her behavior does not improperly affect the outcome of the case. Before juries retire to deliberate and reach a verdict, judges instruct them on the relevant law. In addition, in jurisdictions without professional court administrators, each judge is responsible for the management of his or her own courthouse and its personnel, with the added duties of supervising building maintenance, budgets, and labor relations.

6. **Describe the purposes of an initial appearance.**

   At the initial appearance—the first pretrial stage—defendants are given formal notice of the charges against them and are advised of their constitutional rights. For a misdemeanor or an ordinance violation, a summary trial may be held. For a felony, a hearing is held to determine whether the suspect should be released or whether the suspect should be held for a preliminary hearing.

7. **Explain what bail is, and describe the different methods of pretrial release.**

   Bail is usually a monetary guarantee deposited with the court that is supposed to ensure that the suspect or defendant will appear at a subsequent stage in the criminal justice process. Different pretrial release options include station-house bail, surety bonds, full cash bonds, deposit bonds, release on own recognizance (ROR), conditional release, and unsecured bonds.

8. **Describe what a grand jury is, and explain its purposes.**

   A grand jury is a group of generally 12 to 23 citizens who, for a specific period of time, meet in closed sessions to investigate charges coming from preliminary hearings or to engage in other responsibilities. A primary purpose of the grand jury is to determine whether there is probable cause to believe that the accused committed the crime or crimes with which he or she is charged by the prosecutor. Other purposes of a grand jury are to protect citizens from unfounded charges and to consider the misconduct of government officials.

9. **Describe the purposes of the arraignment and the plea options of defendants.**

   The primary purpose of arraignment is to hear the formal information or grand jury indictment and to allow defendants to enter a plea. Plea options include "guilty," "not guilty," and in some states and the federal courts, *nolo contendere*. In some states, defendants can also stand mute or can plead "not guilty by reason of insanity."

10. **Describe the interests served and not served by plea bargaining.**

    Plea bargaining seemingly serves the interests of all the court participants by, among other things, reducing uncertainty about the length or outcome of trials. Plea bargains serve prosecutors by guaranteeing them high conviction rates; judges by reducing court caseloads; defense attorneys by allowing them to avoid trials and spend less time on each case; and even some criminal offenders by enabling them to escape a prison sentence altogether, to receive a lesser sentence than they might have received if convicted at trial, or to escape conviction of socially stigmatizing crimes. Two types of criminal offenders are not served by plea bargaining: (1) innocent, indigent defendants who fear being found guilty of crimes they did not commit and receiving harsh sentences and (2) habitual offenders.

11. **List and define the stages in a criminal trial.**

    The stages in a criminal trial are as follows: (1) in jury trials, selection and swearing in of the jury; (2) opening statements by the prosecution and the defense; (3) presentation of the prosecution's case; (4) presentation of the defense's case; (5) rebuttals, cross-examination, and reexamination of witnesses; (6) closing arguments by the defense and the prosecution; (7) the judge's instructing, or charging, the jury; and (8) deliberation and verdict.

12. Explain the different roles of judges in adversarial and inquisitorial trial systems.

In adversarial systems, trial judges, who in the American system are selected or elected from among practicing attorneys, have a relatively passive role and act as referees, making sure that the defense and prosecution follow the rules and that due process is respected. The inquisitorial system, by contrast, relies on professional judges, who, in some countries are called prosecutors or inquisitors and are a part of a career judiciary that is trained specifically for the bench. They have an active and overarching role, typically directing and supervising the police's pretrial collection and preparation of evidence and sometimes conducting investigations themselves, interrogating witnesses, presenting evidence at trial, and ensuring an outcome based on the merits of the case.

KEY TERMS

dual court system 302
jurisdiction 302
original jurisdiction 302
appellate jurisdiction 302
general jurisdiction 302
special jurisdiction 302
subject matter jurisdiction 302
personal jurisdiction 302
writ of *certiorari* 306
writ of *habeas corpus* 307
summary or bench trial 307
trial *de novo* 307
due process of law 311
incapacitation 311
punishment 311
rehabilitation 311
general deterrence 311
nolle prosequi (nol. pros.) 312
plea bargaining or plea negotiating 313
rules of discovery 314
booking 329
complaint 329
information 329
grand jury indictment 329
arrest warrant 329
bail bond or bail 331
preventive detention 331
bench warrant or *capias* 332
release on own recognizance (ROR) 333
conditional release 334
unsecured bond 334
preliminary hearing 334
grand jury 334
indictment 335
subpoena 336
arraignment 338
nolo contendere 338
bench trial 340
venire 341
voir dire 342
hung jury 343
inquisitorial trial system 345

REVIEW QUESTIONS

1. What is the difference between *original* and *appellate jurisdiction*? Between *general* and *special jurisdiction*? Between *subject matter* and *personal jurisdiction*?
2. Under what circumstances will the U.S. Supreme Court issue a writ of *certiorari*?
3. Ideally, what are the three conditions that must be met before a prosecutor charges a person with a crime and prosecutes the case?
4. Why do prosecutors sometimes choose not to prosecute criminal cases?
5. In general, when does an individual accused of a crime have the right to counsel?
6. By what methods are judges selected?
7. Describe the "funneling" or screening process in the administration of justice.
8. When do suspects officially become defendants?
9. How long may suspects who are arrested without a warrant be held in jail before being brought before a judge for an initial appearance?
10. In what two ways are preliminary hearings similar to criminal trials, and in what two ways do preliminary hearings differ from criminal trials?
11. What is the primary purpose of a grand jury?
12. What are three basic types of plea bargains?
13. What are three principal purposes of jury trials?
14. What is *voir dire*, and what is its purpose?
15. What are the stages of a criminal trial?
16. What type of trial system is used most widely in the world, and why is it used in some countries but not in others?

IN THE FIELD

1. **Bail Bonds** Visit the office of a local bail bonds agent. It is generally near the courthouse and well marked. Ask the bonds agent to describe the job. Ask about major problems with the business and satisfactions of the job. Specific questions could be: (1) For what type of offender is it most risky to provide bail? (2) How does a bail transaction work? (3) Do bonds agents actually have to give money to the court when they put up bail?
2. **Court Proceedings** Visit several different types of courts, such as a lower court, a trial court, an appellate court, state courts, and federal courts. Observe the proceedings, and describe how they differ from or are similar to those described in this chapter and to each other.
3. **Report on a Criminal Trial** Scan a local newspaper for a story of a criminal trial. Write a report that includes information about (1) the type of court in which the trial is being held and why it has jurisdiction, (2) the type of case being tried (misdemeanor or felony), and (3) the outcome of the case.

ON THE WEB

1. **District Courts** Go to the United States Courts website at www.uscourts.gov and find the directory of district courts. Look up the number of districts into which your state is divided, where the courts for the districts are located, and how many judgeships are authorized for each district. Report the results to your class or instructor.
2. **State Courts** Go to the National Center for State Courts website at www.ncsc.org/Information-and-Resources/Browse-by-State/StateCourtWebsites.aspx and examine the material for your state. Write a report about your state court system.

CRITICAL THINKING EXERCISES

WHAT WOULD YOU DO?

1. As a defense attorney, what would you do under the following circumstances?
   a. Your client tells you that he committed the crime or crimes for which he is being prosecuted.
   b. Your client tells you about a serious crime that will be committed sometime next week.
   c. Your client tells you that if you lose the case, his friends will harm your family.
   d. You learn that your client, who has paid you nothing so far for your services, will not be able to pay your fee.
   e. Your client insists on testifying, even though you believe that it is not in his best interests to do so.

JURY NULLIFICATION

2. In John Grisham's 1992 novel, *A Time to Kill*, the defendant, a black man whose young daughter was viciously raped by two white men, is on trial for gunning down the two men on the courthouse steps in full view of many bystanders. Even though it was obvious to all that the defendant had killed the two men, the jury in the case returned a not-guilty verdict, and the defendant was allowed to walk free. This is an example of *jury nullification*, the power of a jury in a criminal case to acquit a defendant despite overwhelming evidence. The jury can acquit for any reason or for no reason at all, and the decision of the jury cannot be appealed. Jury nullification is one of the problems cited by critics who call for the abolition of the present American jury system. Another complaint about the jury system is the inability of some jurors in some trials to understand legal arguments, the evidence presented, or the instructions of the judge. Critics of the jury system suggest replacing jury trials with bench trials or with trials before a panel of judges or substituting professionally trained jurors for the current "amateur" jurors.
   a. Should the American jury system be abolished? Why or why not?
   b. Do you believe that any of the alternatives suggested by jury critics would produce a better or more just system? Defend your answer.

NOTES

1. United States Courts, Authorized Judgeships, accessed May 21, 2019, https://www.uscourts.gov/sites/default/files/districtauth.pdf.
2. United States Courts, Judicial Vacancies, accessed May 21, 2019, https://www.uscourts.gov/judges-judgeships/judicial-vacancies.
3. United States Courts, Authorized Judgeships, accessed May 21, 2019, https://www.uscourts.gov/sites/default/files/districtauth.pdf.
4. United States Courts, Judicial Compensation, accessed May 21, 2019, https://www.uscourts.gov/judges-judgeships/judicial-compensation.
5. United States Courts, Caseload Statistics Data Tables, accessed May 21, 2019, https://www.uscourts.gov/statistics-reports/caseload-statistics-data-tables (stfj_jci_1231.2018.xlsx).
6. Ibid.
7. United States Courts, About Federal Judges, accessed May 21, 2019, https://www.uscourts.gov/judges-judgeships/about-federal-judges.
8. United States Courts, Caseload Statistics Data Tables, accessed May 21, 2019, https://www.uscourts.gov/statistics-reports/caseload-statistics-data-tables (stfj_jci_1231.2018.xlsx).
9. United States Courts, accessed February 6, 2013, www.uscourts.gov/FederalCourts/UnderstandingtheFederalCourts/FederalCourtsStructure.aspx.
10. United States Courts, Authorized Judgeships, U.S. Courts of Appeals: Additional Authorized Judgeships, accessed May 21, 2019, https://www.uscourts.gov/sites/default/files/appealsauth.pdf.
11. United States Courts, Judicial Vacancies, accessed May 21, 2019, https://www.uscourts.gov/judges-judgeships/judicial-vacancies.
12. United States Courts, Judicial Compensation, accessed May 21, 2019, https://www.uscourts.gov/judges-judgeships/judicial-compensation.
13. United States Courts, Authorized Judgeships, U.S. Courts of Appeals: Additional Authorized Judgeships, accessed May 21, 2019, https://www.uscourts.gov/sites/default/files/appealsauth.pdf.
14. United States Courts, Judicial Compensation, accessed May 21, 2019, https://www.uscourts.gov/judges-judgeships/judicial-compensation.
15. Supreme Court of the United States, "2018 Year-End Report on the Federal Judiciary," p. 12, accessed May 21, 2019, https://www.supremecourt.gov/publicinfo/year-end/2018year-endreport.pdf.
16. Material on the state courts is based on information from the following sources: Richard Y. Schauffler, Robert C. LaFountain, Neal B. Kauder, and Shauna M. Strickland, *Examining the Work of State Courts, 2004. A National Perspective from the Court Statistics Project* (Williamsburg, VA: National Center for State Courts, 2005); David B. Rottman, Carol R. Flango, Melissa T. Cantrell, Randall Hansen, and Neil LaFountain, *State Court Organization 1998*, U.S. Department of Justice, Bureau of Justice Statistics (Washington, DC: U.S. Government Printing Office, June 2000); David W. Neubauer, *America's Courts and the Criminal Justice System*, 4th ed. (Pacific Grove, CA: Brooks/Cole, 1992); Christopher Smith, *Courts, Politics, and the Judicial Process* (Chicago: Nelson-Hall, 1993); N. Gary Holten and Lawson L. Lamar, *The Criminal Courts: Structures, Personnel, and Processes* (New York: McGraw-Hill, 1991); Lawrence Baum, *American Courts,* 3rd ed. (Boston: Houghton Mifflin, 1994). U.S. Department of Justice, Bureau of Justice Statistics, *Report to the Nation on Crime and Justice*, 2nd ed. (Washington, DC: U.S. Government Printing Office, 1988); H. Ted Rubin, *The Courts: Fulcrum of the Justice System* (Santa Monica, CA: Goodyear, 1976); James Eisenstein, Roy Flemming, and Peter Nardulli, *The Contours of Justice: Communities and Their Courts* (Boston: Little, Brown, 1988); Malcolm Feeley, *The Process Is the Punishment: Handling Cases in Lower Criminal Court* (New York: Russell Sage, 1979); Harry P. Stumpf and John H. Culver, *The Politics of State Courts* (New York: Longman, 1992); Paul Wice, *Chaos in the Courthouse: The Inner Workings of the Urban Criminal Courts* (New York: Praeger, 1985).
17. National Center for State Courts, Trends in State Courts, "Limited Jurisdiction Courts—Challenges, Opportunities, and Strategies for

Action," accessed May 21, 2019, https://www.ncsc.org/sitecore/content/microsites/future-trends-2012/home/Courts-and-the-Community/3-6-Limited-Jurisdiction-Courts.aspx.

18. National Center for State Courts, Court Statistics Project, "Examining the Work of State Courts: An Overview of 2015 State Court Caseloads," p. 2, accessed May 21, 2019, http://www.courtstatistics.org/~/media/Microsites/Files/CSP/EWSC%202015.ashx.

19. National Center for State Courts, Court Statistics Project, "Total Incoming Cases in State Courts, 2007–2016," accessed May 21, 2019, http://www.courtstatistics.org/NCSC-Analysis/~/media//7F3DA5FEF1BF4BE1BE2BDE6BA0E86C60.ashx.

20. Unless otherwise indicated, material about specialty courts is from Shelli B. Rossman and Janine M. Zweig, "The Multisite Adult Drug Court Evaluation," National Association of Drug Court Professionals (May 2012), accessed September 15, 2012, www.nadcp.org/sites/default/files/nadcp/Multisite%20Adult%20Drug%20Court%20Evaluation%20-%20NADCP.pdf; National Drug Court Institute, "Research Findings," accessed January 30, 2011, www.ndci.org/research; *BJA Drug Court Clearinghouse Project at American University: Summary of Drug Court Activity by State and County*, July 14, 2009, accessed January 30, 2011, www1.spa.american.edu/justice/documents/2150.pdf; National Association of Drug Court Professionals, "About NADCP," accessed January 30, 2011, www.nadcp.org/learn/about-nadcp; C. West Huddleston III, Douglas B. Marlowe, and Rachel Casebolt, *Painting the Current Picture: A National Report Card on Drug Courts and Other Problem Solving Court Programs in the United States,* vol. 2, no. 1 (Rockville, MD: National Drug Court Institute, May 2008); C. West Huddleston III, Karen Freeman-Wilson, Douglas B. Marlowe, and Aaron Roussell, *Painting the Current Picture: A National Report Card on Drug Courts and Other Problem Solving Court Programs in the United States,* vol. 1, no. 2 (Rockville, MD: National Drug Court Institute, May 2005); "Drug Courts: Overview of Growth, Characteristics, and Results," July 31, 1997, GAO/GGD-97-106, www.ncjrs.org/txtfiles/D.C.ourts.txt; Rottman et al., *State Court Organization 1998*, 207.

21. National Drug Court Resource Center, "Drug Treatment Court Programs in the United States," accessed May 21, 2019, https://ndcrc.org/database/.

22. Ibid.

23. DrugRehab.com, "Drug Courts in the United States: Avoiding Jail, Improving Lives," May 30, 2018, accessed May 21, 2019, https://www.drugrehab.com/featured/drug-courts/.

24. National Drug Court Resource Center, "Drug Court Counts," accessed January 27, 2016, www.ndcrc.org/content/drug-court-counts.

25. Ibid.

26. The Council of State Governments, Book of the States 2018, Chapter 5: State Judicial Branch, Table 5.2, accessed May 22, 2019, http://knowledgecenter.csg.org/kc/system/files/5.2.2018.pdf.

27. Ibid.

28. S. Strickland, R. Schauffler, R. LaFountain, and K. Holt, eds., State Court Organization. Last updated January 9, 2015. National Center for State Courts, www.ncsc.org/sco (accessed January 28, 2016).

29. The Council of State Governments, Book of the States 2018, Chapter 5: State Judicial Branch, Table 5.1, accessed May 22, 2019, http://knowledgecenter.csg.org/kc/system/files/5.1.2018%20Infographic.pdf.

30. Rubin, *The Courts*.

31. In addition to the other sources cited, material on the key actors in the court process is from Neubauer, *America's Courts*; David W. Neubauer, *Judicial Process: Law, Courts and Politics in the United States* (Pacific Grove, CA: Brooks/Cole, 1991); Smith, *Courts, Politics, and Judicial Process*; Holten and Lamar, *The Criminal Courts*; Baum, *American Courts*.

32. Material about prosecutors is also taken from David Heilbroner, *Rough Justice: Days and Nights of a Young D.A.* (New York: Pantheon, 1990).

33. Robert H. Jackson, "The Federal Prosecutor," *Journal of the American Judicature Society* 24 (1940): 18–20.

34. See, for example, Sean Rosenmerkel, Matthew Durose, and Donald Farole Jr. "Felony Sentences in State Courts, 2006—Statistical Tables," U.S. Department of Justice, Bureau of Justice Statistics Statistical Tables, December 2009, 25, Table 4.1, accessed November 7, 2012, http://bjs.ojp.usdoj.gov/content/pub/pdf/fssc06st.pdf.

35. Calculated from data in The Council of State Governments, The Book of States 2018, "The Attorneys General 2018," Table 4.19, accessed May 24, 2019, http://knowledgecenter.csg.org/kc/system/files/4.19.2018.pdf.

36. "Tipping the Scales: Challengers Take on the Old Boys' Club of Elected Prosecutors," Reflective Democracy Campaign, October 2019, accessed November 10, 2019, https://wholeads.us/tipping-the-scales-read-the-report/.

37. In addition to the other sources cited, material on defense attorneys is from Elizabeth Loftus and E. Ketcham, *For the Defense* (New York: St. Martin's, 1991); Paul Wice, *Judges and Lawyers: The Human Side of Justice* (New York: HarperCollins, 1991); Paul Wice, *Criminal Lawyers: An Endangered Species* (Newbury Park, CA: Sage, 1978); Seymour Wishman, *Confessions of a Criminal Lawyer* (New York: Penguin, 1982); Lisa J. McIntyre, *The Public Defender: The Practice of Law in the Shadows of Repute* (Chicago: University of Chicago Press, 1987).

38. Based on figures provided in Marvin Zalman, "Qualitatively Estimating the Incidence of Wrongful Convictions," *Criminal Law Bulletin* 48 (2012): 221–79; Marvin Zalman, Brad W. Smith, and Angie Kiger, "Officials' Estimates of the Frequency of 'Actual Innocence' Convictions," *Justice Quarterly* 25 (2008): 72–100; Robert J. Ramsey and James Frank, "Wrongful Conviction: Perceptions of Criminal Justice Professionals Regarding the Frequency of Wrongful Conviction and the Extent of System Errors," *Crime & Delinquency* 53 (2007): 436–70; C. Ronald Huff, Arye Rattner, and Edward Sagarin, "Guilty Until Proven Innocent: Wrongful Conviction and Public Policy," *Crime and Delinquency* 32 (1986): 518–44.

39. Information about federal defender offices is from *Federal Indigent Defense 2015: The Independence Imperative*. A Report by the National Association of Criminal Defense Lawyers, accessed January 4, 2017, www.nacdl.org/federalindigentdefense2015.

40. Information about indigent defense systems for the states and the District of Columbia is from Suzanne M. Strong, "State-Administered Indigent Defense Systems, 2013." U.S. Department of Justice, *Bureau of Justice Statistics Special Report*, November 2016, accessed January 4, 2007, www.bjs.gov/content/pub/pdf/saids13.pdf.

41. Zoukis, "Indigent Defense in America: An Affront to Justice," Human Rights Defense Center, Criminal Legal News, March 16, 2018, accessed May 25, 2019, https://www.criminallegalnews.org/news/2018/mar/16/indigent-defense-america-affront-justice/.

42. Ibid.

43. Ibid.

44. Ibid.

45. Exodus 18:1–27.

46. In addition to the other sources cited, material on judges is from Paul Ryan, Allan Ashman, Bruce D. Sales, and Sandra Shane-DuBow, *American Trial Judges* (New York: Free Press, 1980);

Robert Satter, *Doing Justice: A Trial Judge at Work* (New York: Simon & Schuster, 1990); Wice, *Judges and Lawyers*; Wice, *Chaos in the Courthouse.*

47. Debbie Salamone and Gerald Shields, "Justice to ABA: Races Not Popularity Contests," *The Orlando Sentinel*, August 4, 1996, A-23.

48. The National Judicial College, "A Legacy of Learning," accessed May 26, 2019, https://www.judges.org/about/the-njc-experience/history/.

49. In addition to the other sources cited, material on pretrial stages is from Neubauer, *America's Courts*; Holten and Lamar, *The Criminal Courts.*

50. David C. Steelman, James E. McMillan, and John A. Goerdt, *Caseflow Management: The Heart of Court Management in the New Millennium* (Williamsburg, VA: National Center for State Courts, 2000).

51. G. Thomas Munsterman, Paula L. Hannaford, and G. Marc Whitehead, eds., *Jury Trial Innovations* (Williamsburg, VA: National Center for State Courts, 1997).

52. Timothy R. Murphy, Paula L. Hannaford, Geneva Kay Loveland, and G. Thomas Munsterman, *Managing Notorious Cases* (Williamsburg, VA: National Center for State Courts, 1998).

53. "Moving Beyond Money: A Primer on Bail Reform," Criminal Justice Policy Program at Harvard Law School, October 2016, p. 25, accessed May 26, 2019, http://cjpp.law.harvard.edu/assets/FINAL-Primer-on-Bail-Reform.pdf.

54. Debbie Salamone Wickhan, "On the Run," *The Orlando Sentinel*, January 7, 2001, A-1.

55. Ibid.; Richard Willing, "Skipped Bail Is Taking Huge Bite," October 13, 2004, *USA Today,* www.usatoday.com/_news/nation/2004-10-13-bail-bonds_x.htm.

56. "Moving Beyond Money: A Primer on Bail Reform," Criminal Justice Policy Program at Harvard Law School, October 2016, p. 6, accessed May 26, 2019, http://cjpp.law.harvard.edu/assets/FINAL-Primer-on-Bail-Reform.pdf.

57. Zhen Zeng, "Jail Inmates in 2017," U.S. Department of Justice, Bureau of Justice Statistics, April 2019, p. 6, Table 4, accessed June 1, 2019, https://www.bjs.gov/content/pub/pdf/ji17.pdf.

58. Daniel Taylor, "Which States Use Criminal Grand Juries," FindLaw, November 26, 2014, accessed May 27, 2019, https://blogs.findlaw.com/blotter/2014/11/which-states-use-criminal-grand-juries.html.

59. Federal Grand Jury, "Grand Jury Functions," accessed November 11, 2012, http://campus.udayton.edu/~grandjur/stategj/funcsgj.htm.

60. In addition to the other sources cited, material on plea bargaining is from Neubauer, *America's Courts*; Neubauer, *Judicial Process*; Holten and Lamar, *The Criminal Courts*; Baum, *American Courts*.

61. In addition to the other sources cited, material on criminal trials is from Neubauer, *America's Courts;* Neubauer, *Judicial Process*; Smith, "Officials' Estimates"; Holten and Lamar, *The Criminal Courts*.

62. In addition to the other sources cited, material on juries is from Rottman et al., *State Court Organization*; Holten and Lamar, *The Criminal Courts*; Valerie P. Hans and Neil Vidmar, *Judging the Jury* (New York: Plenum, 1986); Harry Kalvan Jr. and Hans Zeisel, *The American Jury* (Boston: Little, Brown, 1996); James P. Levine, *Juries and Politics* (Pacific Grove, CA: Brooks/Cole, 1992).

63. National Center for State Courts, "Trial Juries: Exemptions, Term of Service, and Fees," accessed May 27, 2019, http://data.ncsc.org/QvAJAXZfc/opendoc.htm?document=Public%20App/SCO.qvw&host=QVS@qlikviewisa&anonymous=true&bookmark=Document%5CBM180; Matrix of Mnemosyne, "Juror Compensation in America," accessed January 28, 2016, http://matrixbookstore.biz/trial_jury.htm.

64. "Just Under Three in Five Americans Believe Juries Can Be Fair and Impartial All or Most of the Time," *The Harris Poll* #9, January 21, 2008, www.harrisinteractive.com/harris_poll/printerfriend/Index.asp?PID=861; C. Flango, A. McDowell, C. Campbell, and N. Kauder, *Future Trends in State Courts 2008* (Williamsburg, VA: National Center for State Courts, 2008), 3.

65. In addition to the other sources cited, material on the trial process is from Satter, *Doing Justice.*

66. Andrew Selsky, "Split-jury verdicts divide Oregon," *The Orlando Sentinel*, May 12, 2018, p. A18.

67. Cited in Scott E. Sundby, *A Life and Death Decision: A Jury Weighs the Death Penalty* (New York: Palgrave Macmillan, 2005), 154.

68. Ibid.

69. Joan M. Cheever, *Back from the Dead: One Woman's Search for the Men Who Walked Off America's Death Row* (West Sussex, UK: Wiley & Sons, 2006), 287, n. 3.

70. Sundby, *A Life and Death Decision.*

71. Ibid., 205–6, note 17.

72. *Examining the Work of State Courts, 2004*; Jeffrey Rosen, "One Angry Woman: Why Are Hung Juries on the Rise?" *The New Yorker*, February 24 & March 3, 1997, 55.

73. Material in this section is from the following sources: Franklin Strier, "What Can the American Adversary System Learn from an Inquisitorial System of Justice," *Judicature* 76 (1992): 109–11, 161; Lirieka Meintjes van der Walt, "Comparative Method: Comparing Legal Systems and/or Legal Cultures," *Speculum Juris* 1 (2006): 51–64, http://ufh.academia.edu/LiriekaMeintjesvanDerWalt/Papers/969527/Comparative_method_comparing_legal_systems_or_legal_cultures (retrieved August 7, 2012); The Ministry of Justice, "Appendix B: A Comparison of the Inquisitorial and Adversarial Systems," www.justice.govt.nz/publications/global-publications/a/alternative-pre-trial-and-trial-processes-for-child-witnesses-in-new-zealands-criminal-justice-system/appendix-b-a-comparison-of-the-inquisitorial-and-adversarial-systems (retrieved August 15, 2012); Abyssinia Law, "Systems of Criminal Procedure," www.abyssinialaw.com/index.php/study-online/440-systems-of-criminal-procedure?catid=944%3Acriminal-procedure (retrieved August 8, 2012); Roger Hopkins Burke, *Criminal Justice Theory: An Introduction* (London and New York: Routledge, 2012), 119–21.

Design Elements: *Image of part of a court building with columns and statue*: Pixtal/AGE Fotostock RF; *Image of a door slightly open*: jadding/Shutterstock.com

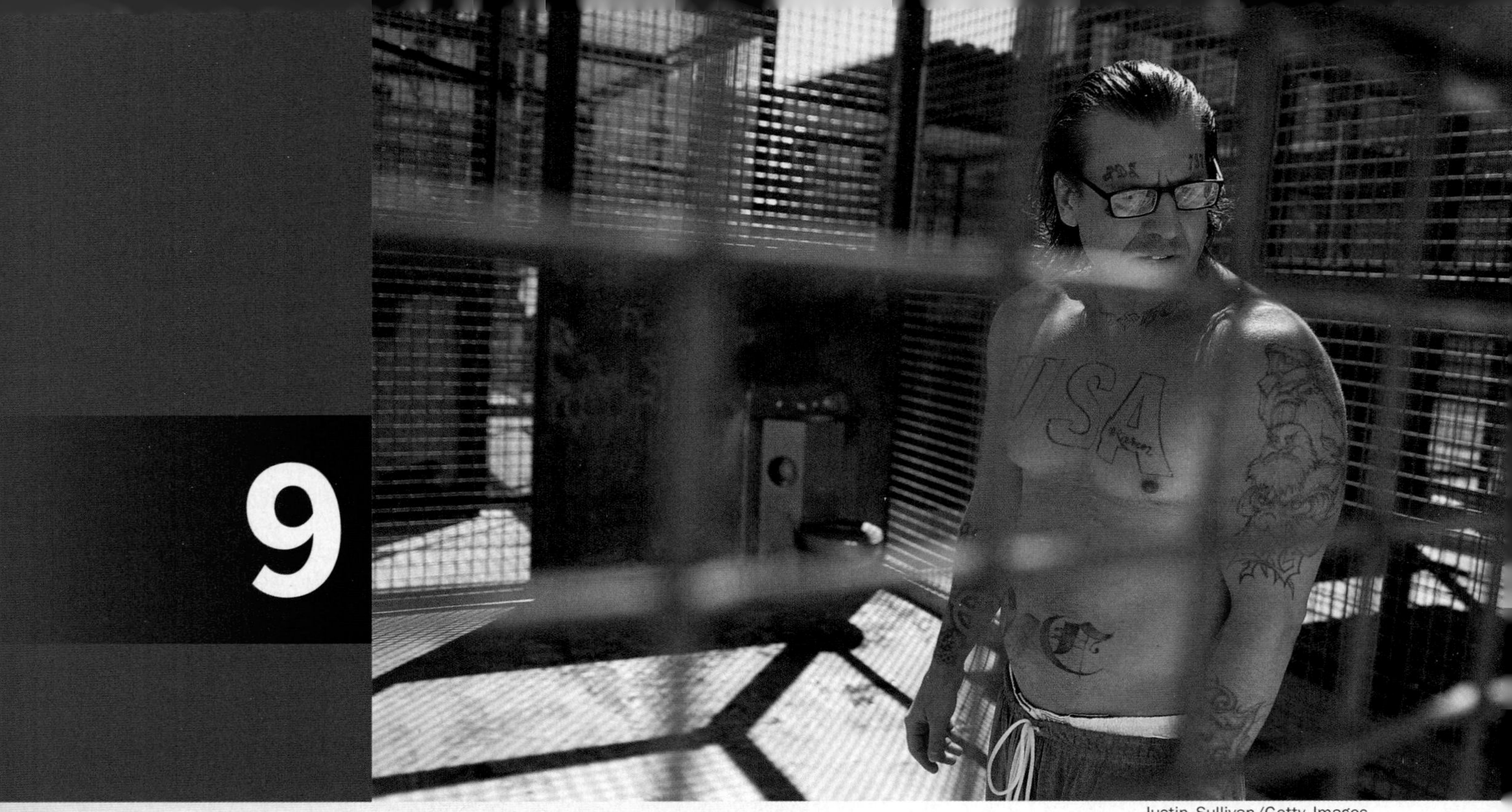

Justin Sullivan/Getty Images

9 Sentencing, Appeals, and the Death Penalty

CHAPTER OUTLINE

Sentencing
- Statutory Provisions
- Philosophical Rationales
- Organizational Considerations
- Presentence Investigation Reports
- Personal Characteristics of Judges

Appeals

The Death Penalty
- A Brief History of the Death Penalty in the United States
- Enter the Supreme Court
- The Procedural Reforms Approved in *Gregg*
- Prospects for the Future

LEARNING OBJECTIVES

After completing this chapter, you should be able to:

1. Identify the general factors that influence a judge's sentencing decisions.
2. Describe how judges tailor sentences to fit the crime and the offender.
3. Distinguish between indeterminate and determinate sentences.
4. Explain the three basic types of determinate sentences.
5. List five rationales or justifications for criminal punishment.
6. Explain the purposes of presentence investigation reports.
7. List the legal bases for appeal.
8. Identify the type of crime for which death may be a punishment.
9. Summarize the three major procedural reforms the U.S. Supreme Court approved for death penalty cases in the *Gregg* decision.

Crime Story

On August 14, 2018, Nebraska executed 60-year-old Carey Dean Moore (pictured) at Nebraska State Penitentiary in Lincoln. He was the first person executed in Nebraska in 21 years, and the first person executed in Nebraska by lethal injection. Prison authorities used a previously untried four-drug combination of diazepam, fentanyl, cisatracurium, and potassium chloride. He was the first inmate executed using the drug fentanyl—a powerful narcotic painkiller that has contributed to the drug overdose epidemic in the United States. Moore's execution proceeded despite two federal lawsuits filed the previous week by drug companies attempting to keep their products from being used. The first of the four execution drugs was administered at 10:24 a.m., and Moore was declared dead 23 minutes later. The state supreme court set his eighth and final execution date in July. His previously scheduled executions all were stayed for a variety of legal reasons, including a challenge to the electric chair's constitutionality. He was not interested in an eighth stay, telling family, friends, and reporters that he was tired of living in the hell of death row. Moore had spent 38 years on Nebraska's death row and was its longest serving inmate. He arrived on death row in 1980, when he was 22 years old, just 4 years after the state reinstated the death penalty in 1976.

Moore had been sentenced to death on June 20, 1980, for the murders of Reuel Eugene Van Ness on August 22, 1979, and Maynard D. Helgeland 5 days later, on August 27, 1979. Both victims were Omaha cabdrivers, Korean War veterans, fathers, and 47 years old. Van Ness was married with 10 children and stepchildren and drove a cab to supplement his income as an auto mechanic and from construction. Moore shot him three times in the back and stole $140. Moore's 14-year-old brother Donald was with him, and they used the money to buy marijuana and porn magazines. Van Ness bled to death. For his part, Donald was sentenced to life in prison but was later paroled. He twice violated parole and was sent back to prison, where he was the day his brother was executed. Helgeland, the second victim, drove a cab because it was something he could do with two prosthetic legs. He lost his legs to frostbite as a result of his alcohol addiction, which also cost him his marriage and damaged his relationships with his three children. However, at the time of his death, he had been sober for a year and was trying to reconnect with his children. Moore shot Helgeland three times in the head but got no money because after he was shot, Helgeland slumped over on his wallet, and Moore was too squeamish to move his body. Moore never denied his guilt in the murders, admitting that he killed the cabdrivers for drug money, and because he did not want to leave any witnesses. Moore confessed that he killed cabdrivers because his mother was a cabdriver, and she always had cash. The media labeled him the "cabdriver killer."

Moore had a troubled background. He had 11 siblings and an alcoholic father who, his siblings testified in court, beat them frequently and severely. His siblings testified that his father often sent him and his twin brother, Harry David, out to beg for money from neighbors, to steal cigarettes and, once, when they were 5 years old, to set a car on fire. The twin brothers were inseparable delinquents. They burned cars and homes, stole from neighbors, and burglarized businesses. At age 10, they were separated, when a judge sent them to different foster and group homes, from which they promptly got kicked out under the mistaken belief that they would be reunited. They were reunited briefly as adults in the 1980s, when Harry David was sentenced to the Nebraska State Penitentiary for burglary, and his brother already was on death row. Moore told

Nebraska Department of Correctional Services/ AP Images

reporters that he did not remember much about his childhood and refused to blame either his father or his upbringing for his crimes. He said that other people have terrible childhoods, but they do not commit murder. In the month before his execution, he apologized for the murders. He wrote in part:

> I am sorry for what I had done to these families, even more than anyone can imagine. I am thankful for God's forgiveness for my actions and my sins, and I pray these families will forgive me somehow; it is easy to cause hurt but it takes great strength to forgive.

Moore became a born-again Christian in prison, believed that he had received God's forgiveness for his crimes, and that heaven awaited him. Harry David and his daughter Taylor shared a last meal with Moore of pizza from Pizza Hut, strawberry cheesecake, and Pepsi. Taylor told reporters that she did not excuse her uncle for his horrible crimes but that he had changed long ago. She added that the uncle she knew was kind, funny, and loving.

The topics addressed in this chapter are sentencing, appeals, and the death penalty. Do you agree with the death sentence imposed on Moore? (For each of these questions, defend your answer.) Should Moore have been executed using an untried four-drug combination? Should the drug fentanyl have been used in Moore's execution? Should Moore's execution have proceeded while two federal lawsuits filed by drug companies were pending? Should Moore have been executed after serving 38 years on death row? Should his death sentence have been commuted to LWOP? Should Moore's troubled background have mitigated his death sentence? Should it matter that Moore had changed while on death row and became a born-again Christian? Moore's story, these questions, and their answers provide some idea of why the death penalty is such a controversial issue.

Sentencing

If a criminal defendant pleads guilty or is found guilty by a judge or jury, then the judge must impose a sentence.[1] In a few jurisdictions, sentencing is the responsibility of the jury for certain types of offenses (e.g., capital crimes). In most misdemeanor cases and many felony cases, judges frequently sentence defendants immediately after they plead guilty or no contest, or are found guilty after trial. When sentences include incarceration, judges may not impose sentences until some days or weeks later, in separately scheduled sentencing hearings. Sentencing hearings often follow investigations by probation officers, who prepare presentence reports for judges to review.[2]

Sentencing is arguably a judge's most difficult responsibility. Judges cannot impose just any sentence. They are limited by the U.S. Constitution's Eighth Amendment prohibition of cruel and unusual punishments (discussed in Chapter 4) and statutory provisions; guided by prevailing philosophical rationales, organizational considerations, and presentence investigation reports; and influenced by their own personal characteristics.

FYI **Time from Conviction to Sentencing**

In the federal court system, sentencing typically takes place about 75 days after a guilty plea or guilty verdict if the defendant is in custody, or after about 90 days if the defendant is out of custody.

Source: "A Federal Criminal Case Timeline," accessed June 2, 2019, https://vae.fd.org/sites/vae.fd.org/files/FedCrimTimeline.pdf.

Statutory Provisions

As described in Chapter 4, state and federal legislative bodies enact penal codes that specify appropriate punishments for each statutory offense or class of offense, such as a class B felony or class C felony. Currently, five general types of punishment are in use in the United States: fines, probation, intermediate punishments (various punishments that are more restrictive than probation but less restrictive and costly than imprisonment), imprisonment, and death. As long as judges impose one or a combination of those five punishments, and the sentence type and length are within statutory limits, judges are free to set any sentence they want.

Thus, within limits, judges are free to tailor the punishment to fit the crime and the offender. As noted, judges can impose a combination sentence of, for example, imprisonment, probation, and a fine. They can suspend the imprisonment portion of a combination sentence. Or they can suspend the entire sentence if the offender stays out of trouble; makes **restitution**, which is paying money or providing services to victims, their survivors, or the community to make up for the injury inflicted; or seeks medical treatment. If the offender already has spent weeks, months, or sometimes even years in jail awaiting trial, judges can give the offender credit for jail time and deduct that time from any prison sentence. When jail time is not deducted from the sentence, it is called "dead time." In some cases, the sentence that a judge intends to impose closely matches the time an offender already has spent in jail awaiting trial. In such cases, the judge may impose a sentence of "time served" and release the offender. When an offender is convicted of two or more crimes, judges can order the prison sentences to run concurrently (together) or consecutively (one after the other). Judges also can delay sentencing and retain the right to impose a sentence at a later date if conditions warrant.

The sentence of death generally is limited to offenders convicted of "aggravated" murder, and because most criminal offenders are poor, fines are seldom imposed for serious crimes. (When they are, it is generally for symbolic reasons.) Thus, in practice, judges have three sentencing options–probation, intermediate punishments, and imprisonment. Later chapters in this text cover those options. The death penalty is discussed at the end of this chapter.

restitution Money paid or services provided by a convicted offender to victims, their survivors, or the community to make up for the injury inflicted.

The type of sentence imposed on an offender can be a highly volatile issue. Also controversial is the length of the sentence imposed. Table 9.1 shows the average prison sentence length imposed by state court judges in 2016, by offense. The average prison sentence length for all offenses in 2016 was 6.4 years. The longest average prison sentence length in 2016 was 40.6 years for murder, while the shortest average prison sentence length in 2016 was 3.7 years for larceny-theft.

TABLE 9.1 Average Prison Sentence Length, by Offense, 2016

Offense	Average Prison Sentence Length (years)
All offenses	6.4
Violent	10.2
Murder[a]	40.6
Negligent manslaughter	10.1
Rape/sexual assault	12.2
Robbery	9.0
Assault	5.6
Other violent[b]	7.2
Property	4.8
Burglary	5.8
Larceny-theft	3.7
Motor vehicle theft	4.0
Fraud[c]	4.4
Other property[d]	4.5
Drug	5.3
Possession	4.0
Trafficking	6.7
Other drug[e]	4.9
Public Order	4.2
Weapons	4.6
Other public order[f]	4.0

Note: Average prison sentence length excludes time in jail and reflects the total maximum sentence that prisoners received. It is based on data from 44 states and excludes state prisoners with sentences of 1 year or less; those with missing values for most serious offense or calculated time served; those released by transfer, appeal, or detainer, and those who escaped. Data include 2,755 deaths in 2016.

[a]Includes non-negligent manslaughter.

[b]Includes kidnapping, blackmail, extortion, hit and run with injury, and other unknown violent offenses.

[c]Includes forgery and embezzlement.

[d]Includes arson, receiving and trafficking of stolen property, destruction of property, trespassing, and other unknown property offenses.

[e]Includes forging prescriptions, possession of drug paraphernalia, and other unspecified offenses.

[f]Includes DUIs/DWIs; court offenses; commercialized vice, morals, and decency offenses; liquor law violations; and other public order offenses.

Source: Adapted from Danielle Kaeble, "Time Served in State Prison, 2016," in *Bureau of Justice Statistics Bulletin*, U.S. Department of Justice (Washington, DC: U.S. Government Printing Office, November 2018), 4, Table 3, accessed June 2, 2019, https://www.bjs.gov/content/pub/pdf/tssp16.pdf.

The Longest Sentences?

The longest prison sentence in the United States is believed to be a 30,000-year sentence imposed on 8-time felon Charles Scott Robinson, 30, on December 14, 1994, in Oklahoma County, Oklahoma, for sexually assaulting a 3-year-old girl. Also contending for longest sentence is the 161 consecutive life terms without possibility of parole imposed on Terry Nichols, 49, on August 9, 2004, in Pittsburg County, Oklahoma, for 161 counts of murder in the Oklahoma City bombing in 1995.

Sources: Ed Godfrey, "Jury Sentences 8-Time Felon to 30,000 Years in Prison," *The Oklahoman*, December 15, 1994, accessed June 2, 2019, https://oklahoman.com/article/2486866/jury-sentences-8-time-felon-to-30000-years-in-prison; CNN Library, "Terry Nichols Fast Facts," *CNN*, April 2, 2019, accessed June 2, 2019, https://www.cnn.com/2013/03/25/us/terry-nichols-fast-facts/index.html.

Judges in states that have indeterminate sentencing statutes generally have more discretion in sentencing than judges in states with determinate sentencing laws. An **indeterminate sentence** has a fixed minimum and maximum term of incarceration rather than a set period. Sentences of 10 to 20 years in prison or of not less than 5 years and not more than 25 years in prison are examples of indeterminate sentences. The amount of the term that actually is served is determined by a parole board. In 2016, 44% of adults entered parole as a result of a parole board decision, an increase from a third of adults entering parole in 2006.[3]

indeterminate sentence A sentence with a fixed minimum and maximum term of incarceration, rather than a set period.

Indeterminate sentences were a principal tool in the effort to rehabilitate offenders in the United States from about 1875 to about 1975. They are based on the idea that correctional personnel must be given the flexibility necessary to successfully treat offenders and return them to society as law-abiding members. The rationale underlying indeterminate sentencing is that the time needed for "correcting" different offenders varies so greatly that a range in sentence length provides a better opportunity to achieve successful rehabilitation.

Beginning in the early 1970s, social scientists and politicians began to question whether the rehabilitation of most criminal offenders was even possible. Skepticism about offender

"Three strikes and you're out" laws have been used against offenders whose third felony was a relatively minor offense. *Are "three strikes" laws good laws? Why or why not?*

DAMIAN DOVARGANES/AP Images

determinate sentence A sentence with a fixed period of incarceration, which eliminates the decision-making responsibility of parole boards.

flat-time sentencing Sentencing in which judges may choose between probation and imprisonment but have little discretion in setting the length of a prison sentence. Once an offender is imprisoned, there is no possibility of reduction in the length of the sentence.

good time Time deducted from an inmate's sentence by prison authorities for good behavior and other meritorious activities in prison.

mandatory sentencing Sentencing in which a specified number of years of imprisonment (usually within a range) is provided for particular crimes.

Americans Reject Mandatory Minimum Sentences

A 2018 national survey found that 87% of registered voters strongly support replacing mandatory minimum sentences for nonviolent offenders with a system that allows judges more discretion; 11% do not. Support is strong among Republicans (83%–16%), Independents (88%–10%), and Democrats (89%–10%).

Sources: Robert Blizzard, "National Poll Results," Public Opinion Strategies, January 25, 2018, accessed June 3, 2019, https://www.politico.com/f/?id=00000161-2ccc-da2c-a963-efff82be0001.

rehabilitation, a public outcry to do something about crime, and a general distrust of decisions made by parole boards continued to grow. By the mid-1970s, several state legislatures had abandoned or at least deemphasized the goal of rehabilitation and had begun to replace indeterminate sentencing with determinate sentencing. Maine was the first state to replace indeterminate sentencing with determinate sentencing in 1975. It abolished parole at the same time.[4]

A **determinate sentence** has a fixed period of incarceration, which eliminates the decision-making responsibility of parole boards. The hope of determinate sentencing is that it will at least get criminals off the street for longer periods of time. Some people also consider a determinate sentence more humane because prisoners know exactly when they will be released, something that they do not know with an indeterminate sentence. Several states and the federal government have developed guidelines for determinate sentencing; other states have established sentencing commissions to do so.

There are three basic types of determinate sentences: flat-time, mandatory, and presumptive. With **flat-time sentencing**, judges may choose between probation and imprisonment but have little discretion in setting the length of a prison sentence. Once an offender is imprisoned, there is no possibility of a reduction in the length of the sentence. Thus, parole and **good time**–the number of days deducted from a sentence by prison authorities for good behavior or for other reasons–are not options under flat-time sentencing. Before New York imposed the first indeterminate sentence in the United States in 1924, nearly all sentences to prison in the United States were flat-time sentences. Flat-time sentences rarely are imposed today.

With **mandatory sentencing**, the second type of determinate sentencing, a specified number of years of imprisonment, usually within a range, is provided for particular crimes. Mandatory sentencing usually allows credit for good time but does not allow release on parole. Beginning in the 1980s, two principal variations of mandatory sentencing emerged. The first was mandatory minimum sentences, which require that offenders serve a specified amount of prison time. Mandatory minimum sentences most frequently are imposed on offenders who commit certain types of offenses such as drug offenses, offenses committed with weapons, and offenses committed by repeat or habitual ("three strikes and you're out") offenders. All states and the federal government have one or more mandatory minimum sentencing laws.

Similar to mandatory minimum sentences are sentences based on truth-in-sentencing laws. First enacted in the state of Washington in 1984, truth-in-sentencing laws require offenders to serve a substantial portion of their prison sentence, usually 85% of it. Most truth-in-sentencing laws target violent offenders and restrict or eliminate parole eligibility and good-time credits. Probably because of incentive grants authorized by Congress in 1994 to build or expand correctional facilities, nearly all states and the District of Columbia enacted truth-in-sentencing laws modeled after the federal government's, which requires that 85% of a prison sentence must be served for certain offenses.

In practice, however, offenders sentenced under truth-in-sentencing laws rarely serve 85% of their sentences. Table 9.2 shows the percent of time served by persons released from

TABLE 9.2 Percent of Prison Sentence Served, by Offense, 2016

Offense	Percent of Prison Sentence Served
All offenses	45.5
Violent	53.7
Murder[a]	57.2
Negligent manslaughter	58.4
Rape/sexual assault	61.9
Robbery	57.7
Assault	47.9
Other violent[b]	47.0
Property	42.4
Burglary	43.1
Larceny-theft	43.7
Motor vehicle theft	41.5
Fraud[c]	39.9
Other property[d]	40.5
Drug	40.6
Possession	37.5
Trafficking	40.9
Other drug[e]	43.2
Public Order	44.5
Weapons	46.6
Other public order[f]	43.6

Number of releases = 375,739

[a]Includes non-negligent manslaughter.

[b]Includes kidnapping, blackmail, extortion, hit and run with injury, and other unknown violent offenses.

[c]Includes forgery and embezzlement.

[d]Includes arson, receiving and trafficking of stolen property, destruction of property, trespassing, and other unknown property offenses.

[e]Includes forging prescriptions, possession of drug paraphernalia, and other unspecified offenses.

[f]Includes DUIs/DWIs; court offenses; commercialized vice, morals, and decency offenses; liquor law violations; and other public order offenses.

Source: Adapted from Danielle Kaeble, "Time Served in State Prison, 2016," in *Bureau of Justice Statistics Bulletin*, U.S. Department of Justice (Washington, DC: U.S. Government Printing Office, November 2018), 4, Table 3, accessed June 2, 2019, https://www.bjs.gov/content/pub/pdf/tssp16.pdf.

state prisons in 2016, by offense. As the table indicates, persons released from state prisons in 2016 served an average of 45.5% of their maximum sentence length. However, the amount of sentence served varied by offense. On the high end, persons who had been convicted of rape or sexual assault served nearly 62% of their sentences before being released, while, on the low end, persons who had been convicted of drug possession had served only 37.5% of their sentences. For both offenses (and for all offenses displayed in Table 9.2), persons released from state prisons in 2016 served well short of 85% of their sentences.

The third type of determinate sentencing is presumptive sentencing. **Presumptive sentencing** allows a judge to retain some sentencing discretion, subject to appellate review. In presumptive sentencing, the legislature determines a sentence range for each crime, usually based on the seriousness of the crime and the criminal history of the offender. The judge is expected to impose the typical sentence, specified by statute, unless mitigating or aggravating circumstances justify a sentence below or above the range set by the legislature. Any sentence that deviates from the norm, however, must be explained in writing and is subject to appellate review. Generally, with presumptive sentencing, credit is given for good time, but there is no opportunity of parole. Presumptive sentencing is a compromise between legislatively mandated determinate sentences and their indeterminate counterparts. Figure 9.1 displays Minnesota's presumptive sentencing guidelines grid.

presumptive sentencing Sentencing that allows a judge to retain some sentencing discretion, subject to appellate review. The legislature determines a sentence range for each crime.

FIGURE 9.1 Sample Sentencing Guidelines

Sentencing Guidelines Grid

Presumptive sentence lengths are in months. Italicized numbers within the grid denote the discretionary range within which a court may sentence without the sentence being deemed a departure. Offenders with stayed felony sentences may be subject to local confinement.

	Severity Level of Conviction Offense (Example offenses listed in italics)	Criminal History Score 0	1	2	3	4	5	6 or more
11	*Murder, 2nd Degree (intentional murder; drive-by-shootings)*	306 *261–367*	326 *278–391*	346 *295–415*	366 *312–439*	386 *329–463*	406 *346–480*[2]	426 *363–480*[2]
10	*Murder, 3rd Degree Murder, 2nd Degree (unintentional murder)*	150 *128–180*	165 *141–198*	180 *153–216*	195 *166–234*	210 *179–252*	225 *192–270*	240 *204–288*
9	*Assault, 1st Degree Controlled Substance Crime, 1st Degree*	86 *74–103*	98 *84–117*	110 *94–132*	122 *104–146*	134 *114–160*	146 *125–175*	158 *135–189*
8	*Aggravated Robbery, 1st Degree Controlled Substance Crime, 2nd Degree*	48 *41–57*	58 *50–69*	68 *58–81*	78 *67–93*	88 *75–105*	98 *84–117*	108 *92–129*
7	*Felony DWI; Financial Exploitation of a Vulnerable Adult*	36	42	48	54 *46–64*	60 *51–72*	66 *57–79*	72 *62–84*[2, 3]
6	*Controlled Substance Crime, 3rd Degree*	21	27	33	39 *34–46*	45 *39–54*	51 *44–61*	57 *49–68*
5	*Residential Burglary Simple Robbery*	18	23	28	33 *29–39*	38 *33–45*	43 *37–51*	48 *41–57*
4	*Nonresidential Burglary*	12[1]	15	18	21	24 *21–28*	27 *23–32*	30 *26–36*
3	*Theft Crimes (Over $5,000)*	12[1]	13	15	17	19 *17–22*	21 *18–25*	23 *20–27*
2	*Theft Crimes ($5,000 or less) Check Forgery ($251–$2,500)*	12[1]	12[1]	13	15	17	19	21 *18–25*
1	*Sale of Simulated Controlled Substance*	12[1]	12[1]	12[1]	13	15	17	19 *17–22*

(Light box) Presumptive commitment to state imprisonment. First-degree murder has a mandatory life sentence and is excluded from the Guidelines under Minn. Stat. § 609.185. See section 2.E, for policies regarding those sentences controlled by law.

(Shaded box) Presumptive stayed sentence; at the discretion of the court, up to one year of confinement and other non-jail sanctions can be imposed as conditions of probation. However, certain offenses in the shaded area of the Grid always carry a presumptive commitment to state prison. See sections 2.C and 2.E.

[1] 12[1] = One year and one day.

[2] Minn. Stat. § 244.09 requires that the Guidelines provide a range for sentences that are presumptive commitment to state imprisonment of 15% lower and 20% higher than the fixed duration displayed, provided that the minimum sentence is not less than one year and one day and the maximum sentence is not more than the statutory maximum. See section 2.C.1–2.

[3] The stat. max. for Financial Exploitation of Vulnerable Adult is 240 months; the standard range of 20% higher than the fixed duration applies at CHS 6 or more. (The range is 62–86.)

Source: Minnesota Sentencing Guidelines Commission, accessed June 3, 2019, http://mn.gov/msgcstat/documents/NewGuidelines/2018/StandardGrid.pdf.

Presumptive sentences also may be based on *sentencing guidelines* developed, not by legislatures, but by *sentencing commissions* composed of both criminal justice professionals and private citizens. Sentencing guidelines are a different way of restricting the sentencing discretion of judges. In 1984, Congress created the nine-member U.S. Sentencing Commission, which is charged with creating and amending federal sentencing guidelines.

Until recently, and the challenge "smart justice" policies are bringing to older "law and order" inspired policies, state legislatures, as noted, increasingly replaced indeterminate sentences with determinate ones. That trend, however, has not escaped criticism. For example, it has been argued that the consequences of determinate sentencing include longer prison sentences and overcrowded prisons. Whether it is the result of a shift in sentencing philosophy or some other factor or factors, there is no question that the United States, until about 2009, had been experiencing a dramatic increase in the number of people sentenced to prison and in the length of prison sentences. A result has been a crisis of prison overcrowding. For some time now, the United States has had one of the highest imprisonment rates in the world.

U.S. Sentencing Commission

To learn more about the U.S. Sentencing Commission, visit its website, www.ussc.gov.

Why do you think sentencing commissions are necessary?

A related criticism of determinate sentencing is that it produces an unusually harsh prison system. For example, because of prison overcrowding, many states all but abandoned even the pretense of rehabilitating offenders. Prisons increasingly have become places where offenders simply are "warehoused." In addition, because of the abolition of good time and parole under some determinate sentencing schemes, prison authorities are having a more difficult time maintaining discipline and control of their institutions. Eliminating good time and parole removed two of the most important incentives that prison authorities would use to get inmates to behave and to follow prison rules. Also, because of the perceived harshness of some of the determinate sentencing schemes, some judges simply have ignored the guidelines. Other judges have ignored sentencing guidelines because they believe they are too lenient. In short, many judges resent sentencing guidelines and refer to their use as "justice by computer."

Too-Lenient Sentence Overturned on Appeal

On August 29, 2018, 10th U.S. Circuit Court of Appeals Judge Carlos Lucero overturned Matthew Sample's sentence of 5 years of probation, stating that U.S. District Court Judge Judith Herrera incorrectly sentenced Sample because he had a lot of money. Sample had pleaded guilty in 2017, to one count of "fraud and swindles" and one count of wire fraud for stealing more than $1 million from clients. Judge Lucero wrote, "Our system of justice has no sentencing discount for wealth." Federal sentencing guidelines indicate that Sample should have been sentenced to about 7 years in federal prison.

Source: "Federal appeals court: Swindler's sentence too lenient," AP, August 29, 2018, accessed June 3, 2019, https://www.apnews.com/1e9179e2cac24e50b05077cdba1f22c4.

A third criticism of determinate sentencing is that it merely shifts sentencing discretion from judges to legislatures and prosecutors (through plea bargaining). Whether this shift in sentencing responsibility is desirable is a matter of debate. On the one hand, prosecutors generally exercise their discretion in secret, whereas judges exercise discretion in the open. Also, prosecutors and legislators generally are subject to more political influence than are judges.

Yet, on the other hand, one of the major criticisms of indeterminate sentencing and a principal reason for the adoption of determinate sentencing schemes by some states is judicial disparity in sentencing. Judges vary widely in the sentences they impose for similar crimes and offenders. For example, in one study, 41 New York state judges were asked to review files of actual cases and to indicate the sentences they would impose. Sentences for the same crime were quite different. In one case, a heroin addict robbed an elderly man at gunpoint. The assailant was unemployed, lived with his pregnant wife, and had a minor

JEAN-LOUP SENSE/Staff/Getty Images

In response to overcrowding, some states erected tents on prison grounds, as in this facility. *Do you think that the use of tents to deal with prison overcrowding is acceptable? Why or why not?*

FIGURE 9.2 Sentencing Choices of 41 New York State Judges

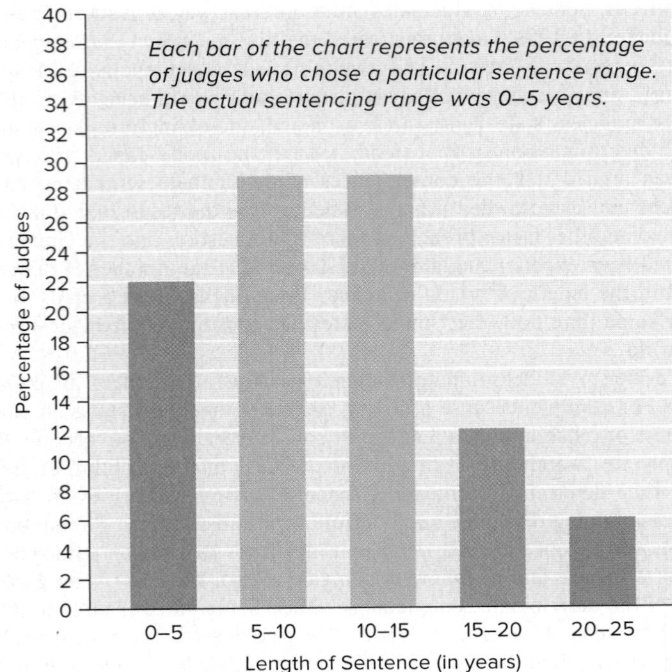

Source: Adapted from "Sentence Disparity in New York: The Response of Forty-One Judges," *The New York Times*, March 30, 1979, B3.

criminal record. He was convicted of first-degree robbery, and under New York's indeterminate sentencing statute, the actual sentence was between 0 and 5 years. When the 41 judges were asked what sentence they would impose in the case, 22% of them chose the actual sentence (between 0 and 5 years), 29% chose a sentence of 5 to 10 years, another 29% selected a sentence of 10 to 15 years, 12% opted for a sentence of 15 to 20 years, and 7% of them chose a sentence of 20 to 25 years.[5] Figure 9.2 displays the judges' choices. Obviously, different judges view the same circumstances very differently. Critics have charged that disparity in sentencing has resulted in discrimination against people of color and the poor.

A fourth, related criticism of determinate sentencing in those jurisdictions that retain good time is that sentencing discretion, at least to some degree, actually shifts from legislators and prosecutors to correctional personnel. By charging inmates with violations of prison rules, correctional personnel can reduce (if the charges are upheld) the amount of good time earned by inmates and, by doing so, increase an inmate's time served.

A fifth criticism of determinate sentencing is that it is virtually impossible for legislatures or sentencing commissions to define in advance all of the factors that ought to be considered in determining a criminal sentence. You may recall from the discussion in Chapter 3 that this was a problem with one of the crime prevention implications of the classical school (equal punishment for equal crime) and the major reason for neoclassical reforms.

Philosophical Rationales

criminal sanctions or **criminal punishment** Penalties that are imposed for violating the criminal law.

At the beginning of Chapter 2, the goals of criminal justice in the United States were identified as the prevention and the control of crime. Those also are the goals of **criminal sanctions** or **criminal punishment**–the penalties that are imposed for violating the criminal law. What always has been at issue, however, is how best to achieve those goals. This decision is the main problem faced by legislators who determine what the criminal sanctions will be in general and by judges who make sentencing decisions in individual cases. Historically, four major rationales or justifications have been given for the punishment imposed by the criminal courts: retribution, incapacitation, deterrence, and rehabilitation. A fifth rationale, restoration, also has been receiving greater attention.

Frequently, judges impose sentences for all five reasons, but at certain times in history, one or more of the reasons have been seen as less important than the others. Today, for example, punishment is imposed less for rehabilitative purposes than it once was because of the prevalent view that we do not know how to change the behavior of criminal offenders. We will now examine each of the rationales for criminal punishment.

Retribution From biblical times through the eighteenth century, **retribution** was the dominant justification for punishment. It implies repayment for an offense committed. Although it probably always has played some role in sentencing decisions, it now is increasingly popular with the public as a rationale for punishment. However, *retribution* is an imprecise term that has been defined in many ways.[6] Nevertheless, when people say that criminal punishment should be imposed for retribution, what most of them want probably is either *revenge* or *just deserts*. **Revenge** is the justification for punishment expressed by the biblical phrase, "An eye for an eye, and a tooth for a tooth." People who seek revenge want to pay back offenders by making them suffer for what they have done. The concept of **just deserts** is another justification in which punishment is seen as a payback, one that is based on something more than vindictive revenge. It supposedly does not contain the emotional element of vengeance. Just deserts draws part of its meaning from the idea, attributed to the German philosopher Immanuel Kant (1724–1804), that offenders should be punished automatically, simply because they have committed a crime—that is, they "deserve" it. Another aspect of just deserts is proportionality of punishment. That is, a punishment should fit the crime and should not be more nor less than the offender deserves.

retribution A justification for punishment that implies repayment for an offense committed.

revenge The punishment rationale expressed by the biblical phrase, "An eye for an eye, and a tooth for a tooth." People who seek revenge want to pay back offenders by making them suffer for what they have done.

just deserts The punishment rationale based on the idea that offenders should be punished automatically, simply because they have committed a crime—they "deserve" it—and the idea that the punishment should fit the crime.

Based on the assumption that the desire for revenge is a basic human emotion, retributivists generally believe that state-authorized punishment greatly reduces the likelihood that individual citizens will take it upon themselves to pay back offenders for what they have done. Vigilante justice is thereby avoided. Retributivists also believe that if offenders are not punished for their crimes, then other people will lose respect for the criminal law and will not obey it.

Finally, retribution is the only rationale for criminal punishment that specifically addresses what has happened in the past; that is, to pay back offenders for their crimes. All the other rationales focus on the future and seek to influence it, for example, to restrain or prevent an offender from committing future crimes.

Incapacitation **Incapacitation** is the removal or restriction of the freedom of those found to have violated criminal laws. Incapacitation makes it virtually impossible for offenders to commit crimes during the period of restraint. Banishment or exile was once used to achieve incapacitation. Even today, foreign nationals are deported after conviction of certain crimes. Capital punishment is the ultimate means of incapacitation. An executed offender can never commit a crime again.

incapacitation The removal or restriction of the freedom of those found to have violated criminal laws.

Currently, incapacitation is achieved primarily through imprisonment, which keeps inmates from committing further crimes (at least outside the prison). Some states, as noted previously, have habitual-offender statutes, "three strikes and you're out" laws, or "life without opportunity of parole" (LWOP) laws that are intended to incapacitate for life repeat felons or, in the case of LWOP laws, offenders who commit capital crimes. Every state except Alaska has an LWOP law. In 2016, 53,290 prisoners in the United States were serving LWOP sentences, an 8.6% increase from the 49,081 prisoners serving LWOP sentences in 2012.[7] The number of prisoners in the United States serving LWOP sentences contrasts sharply with many European countries, such as Germany, where the longest sentence imposed for a single offense, including murder, is 15 years.[8]

Deterrence As described in Chapter 3, in keeping with their goal of achieving the "greatest happiness for the greatest number," Beccaria and other classical theorists believed that the only legitimate purpose for punishment is the prevention or deterrence of crime. They generally viewed punishment for purely retributive reasons as a pointless exercise.

There are two forms of deterrence. **Special** or **specific deterrence** is the prevention of individuals from committing crime again by punishing them. **General deterrence** is the prevention of people in general from engaging in crime by punishing specific individuals and making examples of them.

special or **specific deterrence** The prevention of individuals from committing crimes again by punishing them.

general deterrence The attempt to prevent people in general or society at large from engaging in crime by punishing specific individuals and making examples of them.

Historically, many people have believed that public executions have a general deterrent effect. *Do you agree? Why or why not?*

Bettmann/Contributor/Getty Images

One of the problems with general deterrence as a rationale for punishment is that even though it makes intuitive sense, social science is unable to measure its effects. Only those people who have not been deterred come to the attention of social scientists and criminal justice personnel.

rehabilitation The attempt to "correct" the personality and behavior of convicted offenders through educational, vocational, or therapeutic treatment and to return them to society as law-abiding citizens.

Rehabilitation

A 2018 national survey found that 85% of registered voters agreed that the main goal of our criminal justice system should be rehabilitating people to become productive, law-abiding citizens; 13% disagreed. Significant majorities of Republicans (79%), Independents (83%), and Democrats (92%) agreed with this approach.

Source: Robert Blizzard, "National Poll Results," Public Opinion Strategies, January 25, 2018, accessed June 3, 2019, https://www.politico.com/f/?id=00000161-2ccc-da2c-a963-efff82be0001.

Rehabilitation For nearly 100 years, between the 1870s and 1970s, the primary rationale for punishing criminal offenders was **rehabilitation**, which is the attempt to correct the personality and behavior of convicted offenders through educational, vocational, or therapeutic treatment. The goal was to return them to society as law-abiding citizens. However, the goal of rehabilitating offenders was challenged on the grounds that we simply do not know how to correct or cure criminal offenders because the causes of crime are not fully understood. Beginning in the mid-1970s, the goal of rehabilitation was abandoned altogether in some states, or at least deemphasized, in favor of the goals of retribution and incapacitation. In other states, attempts at rehabilitation continue in an institutional context that seems to favor retribution and incapacitation (as it probably always has). Some critics have suggested that punishment and rehabilitation are incompatible ways of preventing and controlling crime and that rehabilitation generally cannot be achieved in a prison setting. This criticism, however, apparently has not influenced judges, who continue to send criminal offenders to prison to rehabilitate them.

Restoration and Victims' Rights Until recently, victims of crime and their survivors generally have been forgotten or neglected in criminal justice. They have not been important or respected participants in the adjudication process except, perhaps, as witnesses to their own or their loved ones' victimization. Beginning in the 1980s, however, because of increased scholarly attention to their plight and a fledgling victims' rights movement, attempts have been made to change the situation. Today, in many jurisdictions, a greater effort is being made to do something for victims and their survivors—to restore them, as much as possible, to their previous state and to make them "whole" again.

In the early 1980s, only four states had laws that protected the basic rights of crime victims in the criminal justice system. Now, every state has such laws. In fact, there has been an explosion of activity in this area over the past 35 years. States have enacted more than 30,000 crime victim-related statutes; 32 state victims' rights constitutional amendments have been passed; and the federal government has passed legislation providing basic rights and services to federal crime victims, such as victim assistance and victim compensation programs.[9]

The federal government, through its Crime Victims Fund, also has made millions of dollars available for state crime victim compensation, local victim assistance programs, and national training and technical assistance. The Crime Victims Fund is derived from fines, forfeited bonds, and penalties paid by federal criminal offenders, as well as gifts, bequests, and donations. Since the fund was established in 1984, and through 2017, about $95 billion have been awarded, and approximately 15 to 17.5 million victims have been helped each year.[10]

The core rights for victims of crime include:

- The right to be treated with fairness, dignity, sensitivity, and respect.
- The right to attend and be present at criminal justice proceedings.
- The right to be heard in the criminal justice process, including the right to confer with the prosecutor and submit a victim impact statement at sentencing, parole, and other similar proceedings.
- The right to be informed of proceedings and events in the criminal justice process, including the release or escape of the offender, legal rights and remedies, and available benefits and services, and access to records, referrals, and other information.
- The right to protection from intimidation and harassment.
- The right to restitution from the offender.
- The right to privacy.
- The right to apply for crime victim compensation.
- The right to the expeditious return of personal property seized as evidence whenever possible.
- The right to a speedy trial and other proceedings free from unreasonable delay.
- The right to enforcement of these rights and access to other available remedies.[11]

Victim Assistance

The U.S. Department of Justice established the Office for Victims of Crime (OVC) in 1983. OVC provides federal funding for victim assistance and compensation programs throughout the country. It also develops policies and works with criminal justice professionals in order to support crime victims. Visit the OVC website at http://ojp.gov/ovc/.

Do you think that enough is being done to support crime victims?

In the sentencing process, the U.S. Supreme Court ruled in 1991 in *Payne v. Tennessee* that judges and juries may consider *victim-impact statements* in their sentencing decisions. **Victim-impact statements** are descriptions of the harm and suffering that a crime has caused victims and their survivors. Before the Court's 1991 decision, victim-impact statements were considered irrelevant and potentially inflammatory and were not allowed.

victim-impact statements Descriptions of the harm and suffering that a crime has caused victims and their survivors.

Despite the many rights that have been granted crime victims in the United States, there are many problems with the victims' rights laws and their implementation. Victims' rights vary greatly among states and at the federal level. Even in states that have constitutional rights for victims, those rights often are ignored, suffer from arbitrary implementation, or depend on the whims of criminal justice officials. Victims from other cultures and those with disabilities frequently are not informed of their rights or given the opportunity to participate in criminal and juvenile justice proceedings. There also are gaps in the laws. For example, many states do not provide comprehensive rights for the victims of crimes committed by juveniles. Victims' rights in tribal, military, and administrative proceedings generally are nonexistent.[12]

David Maialetti-Pool/Getty Images

Andrea Constand makes a victim impact statement to the court at Bill Cosby's sentencing hearing. *Should victim-impact statements be allowed at sentencing hearings? Why or why not?*

Careers in Criminal Justice

Victim Advocate

My name is Dori DeJong, and I am a victim advocate with the Denver District Attorney's Office in Denver, Colorado. I have a bachelor's degree in education from the University of Northern Colorado, a master's degree in social work, and a juris doctor degree from the University of Denver. Prior to working in the district attorney's office, I worked as a counselor at a children's psychiatric hospital, as a children's counselor at a battered women's shelter, and as a legal advocate for a community agency that provided assistance to victims of domestic violence. I decided to become a victim advocate because it was a great way to combine my educational background in both social work and law.

A typical day as a victim advocate includes contacting the victims of crimes (domestic violence, stranger assault, sex assault, child abuse, etc.) and explaining the criminal justice system to them; providing details about their specific case; giving notification of their rights as a victim; referring victims to various outside assistance agencies; and obtaining their input as to the best possible outcome of the case in which they are involved. In addition, I coordinate and participate in the meetings between the victims and the assistant district attorneys to assist in the preparation of the victims for trial and to answer any questions they have about the process. I accompany the victims to the courthouse on the day of the trial to provide encouragement, support, and understanding of the process in order for them to get through a very difficult experience. I also send update letters and victim-impact statements to victims, notify other agencies and jurisdictions of active cases, run criminal background checks on defendants to identify habitual offenders, and maintain statistics on domestic violence cases. I have regular contact and communication with police officers, detectives, district attorneys, investigators, and other victim advocates. I also communicate with staff from outside assistance agencies to help sustain continuing support for the victim.

The most positive aspect of the job is the ability to help people and feel that you are really making a difference in someone's life. It is so encouraging when you have a client call you, months or even years after you've assisted them, to thank you and tell you what a difference you've made in their lives and their children's lives. It is also rewarding to be a part of the prosecution team that holds criminals responsible and accountable for their behavior.

Being a victim advocate can sometimes be difficult and frustrating because you work with people who are in crisis and often are not appreciative of your involvement in their lives. Sometimes their anger about the incident is misdirected at the victim advocate and the system, which is trying to provide justice.

Dori DeJong

The job of a victim advocate is interesting, challenging, and often fast paced—and can be very rewarding. However, the job can also be overwhelming, because it frequently involves dealing with offensive criminal defendants and their actions, and the sometimes inequitable system of justice. If you are considering a job as a victim advocate, I would encourage you to volunteer or do an internship in an agency that works with victims. This invaluable experience will provide you with the knowledge to determine if this career suits you.

Do you have any characteristics and/or abilities that you think would make you particularly suited to being a victim's advocate? If so, what?

Another effort at restoration places equal emphasis on victims' rights and needs and the successful reintegration (and, in some cases, initial integration) of offenders into the community. Unlike retribution, which focuses almost entirely on offenders and their punishments, restorative justice seeks to restore the health of the community, repair the harm done, meet victims' needs, and require the offender to contribute to those repairs. Following Braithwaite's theory of reintegrative shaming, described in Chapter 3, in restorative justice the criminal act is condemned, offenders are held accountable, participants are involved, and repentant offenders are encouraged to earn their way back into society.[13]

Restitution and community service are two examples of restorative practices (they are used for other purposes as well). Restitution is a court requirement that a victim's convicted offender pay money or provide services to victims, survivors, or the community that has been victimized. Typically, laws specify that restitution may be ordered to cover medical expenses, lost wages, counseling expenses, lost or damaged property, funeral expenses, and other direct out-of-pocket expenses. The 1984 Federal Comprehensive Crime Control Act requires that when offenders convicted of federal violations are sentenced to probation, they must pay a fine, make restitution, perform community service, or do all three. In 2019, approximately two-thirds of the states provided for the availability of some sort of court-ordered restitution to be paid to crime victims.[14] Unfortunately, a problem with restitution is that most offenders have neither the financial means nor the abilities to provide adequate restitution. Also, studies show that nearly half of crime victims are not awarded any sort of restitution–usually because the victim fails to request it.[15]

Organizational Considerations

A judge's sentence also is guided by organizational considerations. We already have discussed at some length the practice of plea bargaining and have shown that without it, the judicial process could not function. For practical reasons, judges almost always impose the sentence agreed on during plea negotiations. If they did not, plea bargaining would not work. That is, if defendants could not be sure that judges would impose the agreed-on sentence, there would be no reason for them to plead guilty. They might as well take their chances at trial.

Another organizational consideration is the capacity of the system. As we already have noted, many of the prisons in this country and some entire state prison systems are overcrowded and are under federal court order to reduce the problem. Judges in jurisdictions with overcrowded prisons generally are less inclined to sentence offenders to prison.

A third organizational consideration is the cost–benefit question. Every sentence involves some monetary and social cost. Judges must be sensitive to this issue and must balance the costs of the sentence they impose with the benefits that might be derived from it.

Presentence Investigation Reports

A purpose of **presentence investigation reports** (PSIs or PSIRs), used in the federal system and by the majority of states, is to help judges determine the appropriate sentence for particular defendants. They also are used in classifying probationers, parolees, and prisoners according to their treatment needs and security risk. Generally, a PSI is prepared by a probation officer, who conducts as thorough of a background check as possible on a defendant. In some jurisdictions, probation officers recommend a sentence based on the information in the PSI. In other jurisdictions, they simply write the report and do not make a sentencing recommendation. Studies show that judges follow the sentencing recommendations in PSIs most of the time, although they are not required to do so.

presentence investigation reports Reports, often called PSIs or PSIRs, that are used in the federal system and the majority of states to help judges determine the appropriate sentence. They also are used in classifying probationers, parolees, and prisoners according to their treatment needs and security risk.

In most jurisdictions, after the PSI has been submitted to the judge, a sentencing hearing is held at which the convicted defendant has the right to address the court before the sentence is imposed. This procedure is called **allocution**. During allocution, a defendant is identified as the person found guilty and has a right to deny or explain information contained in the PSI if his or her sentence is based on it. The defendant also has the opportunity to plead for a **pardon**, which is a "forgiveness" for the crime committed that stops further criminal processing. He or she also may attempt to have the sentencing process stopped or may explain why a sentence should not be pronounced. However, during allocution, defendants are not entitled to argue about whether they are guilty. Among the claims that a convicted offender can make at allocution are the following:

allocution The procedure at a sentencing hearing in which the convicted defendant has the right to address the court before the sentence is imposed. During allocution, a defendant is identified as the person found guilty and has a right to deny or explain information contained in the PSI if his or her sentence is based on it.

pardon A "forgiveness" for the crime committed that stops further criminal processing.

1. That he or she is not the person who was found guilty at trial.
2. That a pardon has been granted for the crime in question.
3. That he or she has gone insane since the verdict was rendered. Rules of due process prohibit the sentencing of convicted offenders if they do not understand why they are being punished. Punishment must be deferred until they are no longer insane.
4. That she is pregnant. The sentence of a pregnant offender must be deferred or adjusted, especially in a capital case.

Personal Characteristics of Judges

Although extralegal factors are not supposed to influence a judge's sentencing decision, studies show that they invariably do. Judges, after all, are human beings with all of the human frailties and prejudices of other human beings. Among the personal characteristics of judges that have been found to affect their sentencing decisions are these:

1. Their socioeconomic backgrounds.
2. The law schools they attended.

3. Their prior experiences both in and out of the courtroom.
4. The number of offenders they defended earlier in their careers.
5. Their biases concerning various crimes.
6. Their emotional reactions and prejudices toward the defendants.
7. Their own personalities.
8. Their marital and sexual relations.

In summary, a judge's sentencing decision is the result of the complex interplay of several different factors. Those factors include statutory provisions, philosophical rationales, organizational considerations, presentence investigation reports, results of the allocution, and personal characteristics of the judge.

THINKING CRITICALLY

1. What do you think are some of the most important issues to consider when sentencing a convicted criminal?
2. Do you think that victims should play more or less of a role in sentencing?

Appeals

As described previously, defendants can appeal their convictions either on legal grounds (e.g., defects in jury selection, improper admission of evidence at trial, mistaken interpretations of law) or on constitutional grounds (e.g., illegal search and seizure, improper questioning of the defendant by the police, identification of the defendant through a defective police lineup, incompetent assistance of counsel). However, they are not entitled to present new evidence or testimony on appeal if that evidence or testimony could have been presented at trial. If new evidence is discovered that was unknown or unknowable to the defense at trial, then an appeal sometimes can be made on the basis of that new evidence. Still, in *Herrera v. Collins*, a 1993 death penalty case, the Supreme Court ruled that, absent constitutional grounds, new evidence of innocence is no reason for a federal court to order a new state trial. In any event, because the defendant already has been found guilty, the presumption of innocence no longer applies during the appellate process, and the burden of showing why the conviction should be overturned shifts to the defendant.

Generally, notice of intent to appeal must be filed within 30 to 90 days after conviction. Also, within a specified period of time, an *affidavit of errors* specifying the alleged defects in the trial or pretrial proceedings must be submitted. If those two steps are followed, the appellate court must review the case. Nevertheless, very few appeals are successful. Nearly 80% of state trial court decisions are affirmed on appeal. (See Chapter 8 for more on the appellate courts.)

THINKING CRITICALLY

1. Why do you think so few appeals are successful?
2. Do you think the appeals process works effectively? Why or why not?

The Death Penalty

Before concluding this chapter, we will examine in some detail the death penalty in the United States, because, as the Supreme Court has acknowledged, "death is different."[16] Other sentencing options are discussed in later chapters of this book. As a punishment for the most heinous of crimes, the death penalty, or capital punishment, differs from all other criminal sanctions, not only in the nature of the penalty itself (the termination of life) but also in the legal procedures that lead to it. However, before describing the unique way in which capital punishment is administered in the United States, we provide some background about the penalty.

Careers in Criminal Justice

Court Librarian

My name is Evelyn Ortiz Smykla, and I am a court librarian at the satellite library located in the Southern District of Alabama for the 11th Circuit U.S. Court of Appeals, Atlanta, Georgia. Along with my responsibilities for the Southern District of Alabama, I am also responsible for library needs at the Northern District of Florida courthouse located in Pensacola, Florida. I am one of eight satellite librarians in the three states that compose the jurisdiction of the 11th Circuit U.S. Court of Appeals (Alabama, Georgia, and Florida). I am the only part-time satellite professional librarian in the Circuit.

I have been a professional librarian for more than 25 years and have spent the last 10 years with the 11th Circuit serving the needs of appellate, district, and bankruptcy judges and assisting probation and pretrial officers and other court personnel. I work as a solo librarian in what in the profession is referred to as a "special library." In this position, my tasks are varied, from processing materials on a daily basis to researching cases, ordering materials, putting together and disseminating legislative histories, and assisting court personnel with technology and online resources and needs. In addition to meeting library needs, I also am responsible for preparing judges' chambers for incoming new judges and those judges moving to new locations within the courthouses. I work closely with judges and their law clerks to ensure that they have all the materials and sources required to meet their needs in a timely manner because oftentimes their needs on the bench are immediate.

I also mentor, orient, and train incoming new law clerks and help career law clerks with their new online research needs. My day is a very busy one, and my responsibility to the professionals I serve is not taken lightly.

Requirements for this position are a master's degree from an accredited library school and, in some cases, an additional JD degree, but the JD degree is not mandatory. Also required is prior experience in a university law library, law firm, or law department in a large public research library. Unlike getting a job with most libraries, employment with the federal government requires a background check.

Why did I choose to work with the courts? There is not a day that passes when I am not helping someone who needs information about a specific area of law, assistance with definitions for certain law terms, online research help, aid with case law to assist a family member in prison, help with quotes or medical terms, or just a comforting word. Court librarians teach skills, concepts, and research. There is a certain gratification one gets when, as a last resort, you are the one who finds the answer to a very difficult question that can make a difference in another individual's life.

Evelyn Ortiz Smykla

Would you find a job as a court librarian appealing? Why or why not?

A Brief History of the Death Penalty in the United States

When the first European settlers arrived in America, they brought with them the legal systems from their native countries, which included the penalty of death for a variety of offenses. For example, the English Penal Code at the time, which was adopted by the British colonies, listed more than 50 capital offenses, but actual practice varied from colony to colony. In the Massachusetts Bay Colony, 12 crimes carried the death penalty:

- Idolatry
- Witchcraft
- Blasphemy
- Rape
- Statutory rape
- Perjury in a trial involving a possible death sentence
- Rebellion
- Murder
- Assault in sudden anger
- Adultery
- Buggery (sodomy)
- Kidnapping

In the statute, each crime was accompanied by an appropriate biblical quotation justifying the capital punishment. Later, the colony added arson, treason, and grand larceny to the list of capital offenses. In contrast, the Quakers adopted much milder laws. The Royal Charter for South Jersey (1646), for example, did not permit capital punishment for any crime, and in Pennsylvania, William Penn's Great Act of 1682 limited the death penalty to treason and murder. Most colonies, however, followed the much harsher British Code.

The earliest recorded lawful execution in America was in 1608, in the Jamestown colony of Virginia. Captain George Kendall, a councillor for the colony, was executed for being a spy for Spain. Since Kendall, about 20,000 legal executions have been performed in the United States under civil authority. However, only about 3% of those people executed since 1608 have been women. Ninety percent of the women executed were executed under local, as opposed to state, authority, and the majority (87%) were executed before 1866. The first woman executed was Jane Champion in the Virginia colony in 1632. She was hanged for murdering and concealing the death of her child, who had not been fathered by her husband. Since 1962, only 16 women have been executed in the United States (as of February 23, 2020).[17]

In addition, about 2% of those executed in the United States since 1608 have been juveniles—those whose offenses were committed before their 18th birthdays. The first juvenile executed in America was Thomas Graunger in the Plymouth Colony in 1642, for the crime of bestiality. Between 1990 and 2019, the United States was one of only ten countries that had executed anyone under 18 years of age at the time of the crime; the others were China, the Democratic Republic of Congo, Iran, Nigeria, Pakistan, Saudi Arabia, South Sudan, Sudan, and Yemen. The United States (as of March 1, 2005) no longer executes juveniles. The United States had executed 22 juveniles since 1976.[18]

Enter the Supreme Court

For more than 150 years, the U.S. Supreme Court ("the Court") has exercised its responsibility to regulate capital punishment in the United States and its territories. Among the principal issues the Supreme Court considered in relation to capital punishment before 1968 concerned the means of administering the death penalty. The Court upheld the constitutionality of shooting (*Wilkerson v. Utah*, 1878), electrocution (*In re Kemmler*, 1890), and a second electrocution after the first attempt had failed to kill the offender (*Louisiana ex rel. Francis v. Resweber*, 1947). Currently, there are five methods of execution authorized: lethal injection, electrocution, lethal gas, hanging, and firing squad. However, lethal injection is now the primary method of execution used in all executing states, as well as the U.S. government and the U.S. military. In 2008, in *Baze v. Rees*, the Court upheld the constitutionality of lethal injection.

For a century after the ratification of the Eighth Amendment, hanging was the only legally authorized method of execution in the United States. The one exception was that spies, traitors, and deserters convicted under federal statutes could be shot. Hanging is no longer authorized as the sole or principal method of execution in any executing jurisdiction. *Why was hanging as a method of execution abandoned?*

AFP/Getty Images

Between 1968 and 1972, a series of lawsuits challenged various aspects of capital punishment, as well as the constitutionality of the punishment itself. During this period, an informal moratorium on executions was observed, pending the outcome of the litigation, and no death row inmates were executed. Some of the suits were successful, and some of them were not. Finally, on June 29, 1972, the Supreme Court set aside death sentences for the first (and only) time in its history. In its decisions in *Furman v. Georgia*, *Jackson v. Georgia*, and *Branch v. Texas* (hereafter referred to together as the *Furman* decision), the Court held that the capital punishment statutes in those three cases were unconstitutional because they gave the jury complete discretion to decide whether to impose the death penalty or a lesser punishment in capital cases. Although nine separate opinions were written—a very rare occurrence—the majority of five justices (Douglas, Brennan, Stewart, White, and Marshall) pointed out that the death penalty had been imposed arbitrarily, infrequently, and often selectively against people of color. According to the majority, those statutes constituted "cruel and unusual punishment" under the Eighth and Fourteenth Amendments. (The four dissenters were Chief Justice Burger and Justices Blackmun, Powell, and Rehnquist.) Note that the Supreme Court did not rule that the death penalty itself was unconstitutional, only the way in which it was being administered.

The practical effect of the *Furman* decision was that the Supreme Court voided the death penalty laws of some 35 states and the federal government, and more than 600 men and women had their death sentences vacated and commuted to a term of imprisonment. Although opponents of capital punishment were elated that the United States finally had joined all the other Western industrialized nations in abolishing capital punishment either in fact or in practice, the joy was short-lived. By the fall of 1974, 30 states had enacted new death penalty statutes that were designed to meet the Court's objections.

The new death penalty laws took two forms. Some states removed all discretion from the process by mandating capital punishment upon conviction for certain crimes (mandatory statutes). Other states provided specific guidelines that judges and juries were to use in deciding if death was the appropriate sentence in a particular case (guided-discretion statutes).

The constitutionality of the new death penalty statutes was quickly challenged, and on July 2, 1976, the Supreme Court announced its rulings in five test cases. In *Woodson v. North Carolina* and *Roberts v. Louisiana*, the Court rejected mandatory statutes that automatically imposed death sentences for defined capital offenses. However, in *Gregg v. Georgia*, *Jurek v. Texas*, and *Proffitt v. Florida* (hereafter referred to together as the *Gregg* decision), the Court approved several different forms of guided-discretion statutes. Those statutes, the Court wrote, struck a reasonable balance between giving the jury some guidance and allowing it to consider the background and character of the defendant and the circumstances of the crime.

The most dramatic effect of the *Gregg* decision was the resumption of executions on January 17, 1977, when the state of Utah executed Gary Gilmore (at his own request) by firing squad. Since then, 1,516 people have been executed in 34 states and by the federal government, which has executed three (as of February 20, 2020).[19] More than half of the 1,516 executions have taken place in just three states—Texas (569), Virginia (113), and Oklahoma (112). Figure 9.3 shows the states in which executions have taken place. Note that Texas alone accounts for more than one-third of all post-*Furman* executions. Texas has executed five times as many offenders as any other state.

PAUL BUCK/Stringer/Getty Images

During the twentieth century, more people were executed by electrocution than by any other method. *Why have states abandoned electrocution as an execution method?*

FIGURE 9.3 State-by-State Count of Inmates Executed in 34 States Since Executions Resumed in 1977

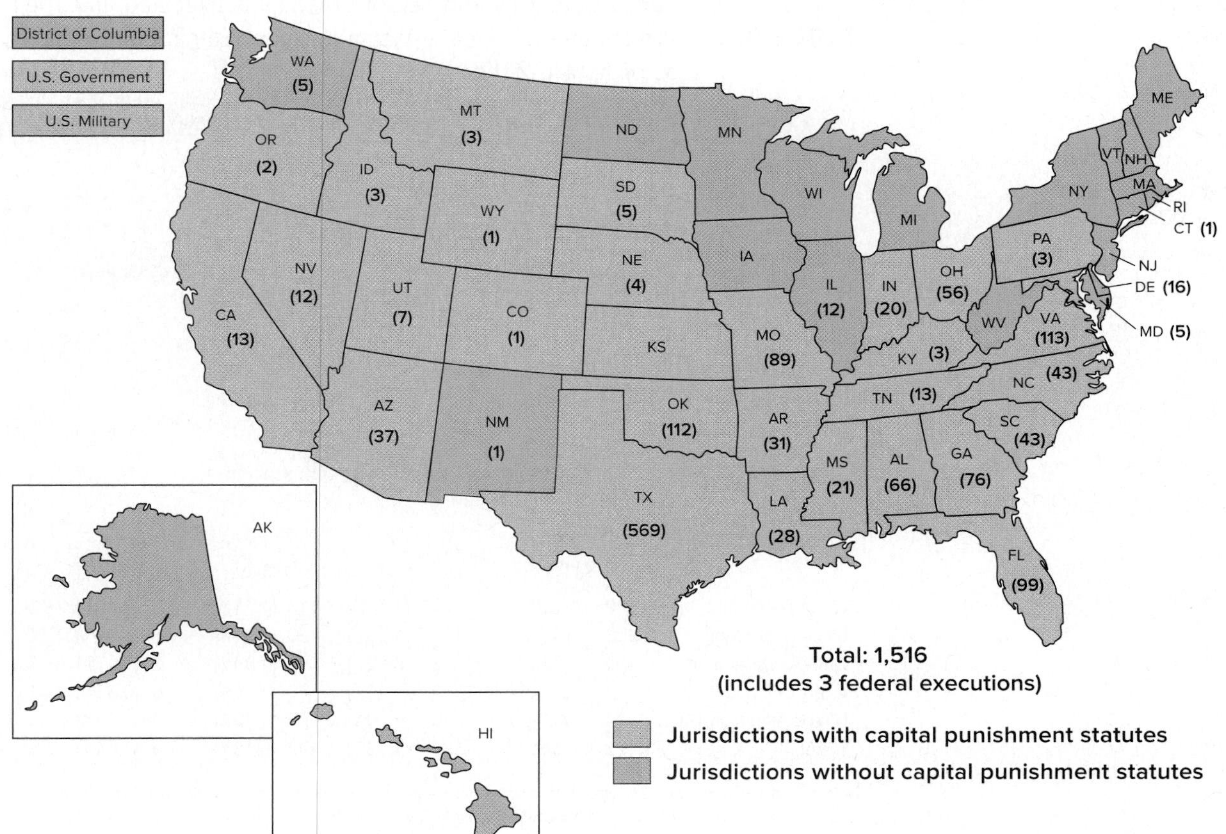

Source: Adapted from Death Penalty Information Center, February 23, 2020, https://files.deathpenaltyinfo.org/documents/pdf/FactSheet.f1582296575.pdf.

DOUGLAS C. PIZAC/AP Images

The most celebrated recent execution by firing squad was the January 17, 1977, execution of Gary Gilmore in Utah. It no longer is authorized as the sole or principal method of execution in any executing jurisdiction. *Does execution by firing squad have any advantages over other methods of execution? If so, what are they?*

Table 9.3 shows the race or ethnicity and gender of the defendants executed and their victims, as well as defendant-victim racial or ethnic combinations. As shown, nearly all the people executed since Gilmore's execution in 1977 have been male, whereas the gender of the victims is divided nearly evenly between males and females. As for race, about 56% of all people executed under post-*Furman* statutes have been white; about 34% have been black. Thus, the percentage of blacks who have been executed far exceeds their proportion of the general population (about 13%). Particularly interesting is that more than 75% of the victims of those executed have been white. What makes this finding interesting is that murders, including capital murders (all post-*Furman* executions have been for capital murders; see the following discussion), tend to be intraracial crimes. However, the death penalty is imposed primarily on the killers of white people, regardless of the race or ethnicity of the offender. The figures on defendant–victim racial or ethnic combinations (see Table 9.3) further support this conclusion. Note that about 52% of executions have involved white killers of white victims, and about 19% of executions have involved black killers of white victims. However, only about 12% of executions have been of black killers of black victims, and there have been only 20 executions of white killers of black persons (less than 1.5%).

Currently (as of February 23, 2020), 31 jurisdictions have capital punishment statutes (29 states, the U.S. government, and the U.S. military); 22 jurisdictions do not have capital punishment statutes (21 states and the District of Columbia). (In 2004, a part of New York's death penalty statute was declared unconstitutional; in 2007, the court ruled that its prior holding applied to the last remaining person on the state's death row. The legislature has voted down attempts to restore the statute. Nine other states have abolished their death penalty since 2007: New Jersey in 2007, New Mexico in 2009, Illinois in 2011, Connecticut in

TABLE 9.3 Race or Ethnicity and Sex of Defendants Executed and Their Victims and Defendant–Victim Racial or Ethnic Combinations, as of April 1, 2019

Gender of Defendants Executed			Gender of Victims		
Total number 1,493			Total number 2,182		
Female	16	(1.07%)	Female	1,069	(48.99%)
Male	1,477	(98.93%)	Male	1,113	(51.01%)
Race of Defendants Executed			**Race of Victims**		
White	832	(55.73%)	White	1,648	(75.53%)
Black	512	(34.29%)	Black	335	(15.35%)
Latino/a	126	(8.44%)	Latino/a	153	(7.01%)
Native American	16	(1.07%)	Native American	5	(0.23%)
Asian	7	(0.47%)	Asian	41	(1.88%)

Defendant–Victim Racial Combinations

	White Victim	Black Victim	Latino/a Victim	Asian Victim	Native American Victim
White Defendant	769 (51.51%)	20 (1.34%)	18 (1.21%)	6 (0.40%)	0 (0%)
Black Defendant	290 (19.42%)	173 (11.59%)	20 (1.34%)	16 (1.07%)	0 (0%)
Latino/a Defendant	54 (3.62%)	3 (0.20%)	61 (4.09%)	2 (0.13%)	0 (0%)
Asian Defendant	2 (0.13%)	0 (0%)	0 (0%)	5 (0.33%)	0 (0%)
Native American Def.	14 (0.94%)	0 (0%)	0 (0%)	0 (0%)	2 (0.13%)
TOTAL	1,129 (75.62%)	196 (13.13%)	99 (6.63%)	29 (1.94%)	2 (0.13%)

Note: In addition, there were 38 defendants executed for the murders of multiple victims of different races. Of those, 21 defendants (55.26%) were white, 11 black (28.95), and 6 Latino (15.79%).

Source: NAACP Legal Defense and Educational Fund, "Death Row, U.S.A.," April 1, 2019, accessed February 23, 2020, https://www.naacpldf.org/wp-content/uploads/DRUSAFall2019.pdf.

2012, Maryland in 2013, Nebraska in 2015, Delaware in 2016, Washington in 2018, and New Hampshire in 2019. However, in 2016, Nebraska reinstated its death penalty.) Figure 9.3 shows the jurisdictions with and without capital punishment statutes.

In decisions since *Gregg*, the Supreme Court has limited the crimes for which death is considered appropriate and has further refined death penalty jurisprudence. In 1977, in the cases of *Coker v. Georgia*, *Eberheart v. Georgia*, and *Hooks v. Georgia*, the Court held that rape of an adult female when the victim was not killed (in *Coker*), kidnapping when the victim was not killed (in *Eberheart*), and armed robbery (in *Hooks*) do not warrant death. In 2008, in *Kennedy v. Louisiana*, the Court ruled that the rape of a child when the victim was not killed also did not warrant death. Those four decisions effectively limit the death penalty to those offenders convicted of capital, or aggravated, murder.

In 1986, in *Ford v. Wainwright*, the Court barred states from executing inmates who have developed mental illness while on death row; in 2002, in *Atkins v. Virginia*, the Court held that it is cruel and unusual punishment to execute the intellectually challenged. In the 2005 case of *Roper v. Simmons*, the Court, again referring to "evolving standards of decency," effectively limited capital punishment to offenders who are 18 years of age or older at the time of their offenses. Another death penalty decision of the Supreme Court is the 1987 case of *McCleskey v. Kemp*, in which the Court held that state death penalty statutes are constitutional even when statistics indicate that they have been applied in racially biased ways. The Court ruled that racial discrimination must be shown in individual cases. The *McCleskey* decision was particularly disheartening to opponents of capital punishment because, for them, the case seemed like the last best chance of having the Supreme Court declare the death penalty unconstitutional once and for all.

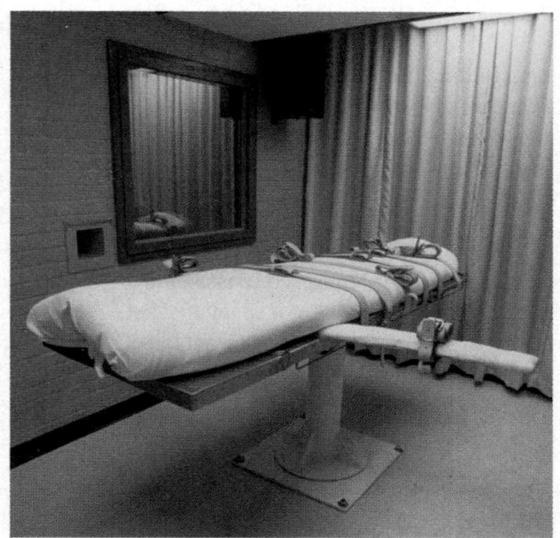

Gregory Smith/Contributor/Getty Images

Lethal injection has become the sole or principal method of execution in all executing jurisdictions in the United States. *Why?*

myth

Capital punishment is no longer administered in a way that is racially discriminatory and legally impermissible.

fact

Dozens of scientific studies clearly show that the death penalty continues to be administered in a legally impermissible and discriminatory fashion against blacks and the killers of whites.

Source: For a summary of those studies, see Robert M. Bohm, *DeathQuest: An Introduction to the Theory and Practice of Capital Punishment in the United States*, 5th ed. (New York and London: Routledge, 2017), chapter 10.

The 1994 federal crime bill (the Violent Crime Control and Law Enforcement Act) expanded the number of federal crimes punishable by death to about 50. (Estimates vary depending on whether statutes or offenses are counted, and how offenses are counted.) The bill reinstated the death penalty for federal crimes for which previous death penalty provisions could not pass constitutional muster. The new law brought the earlier statutes into compliance with guidelines established by the Supreme Court. As of year-end 2016, the number of federal crimes punishable by death had been reduced to 41.[20] The U.S. government executed Timothy McVeigh and Juan Raul Garza in 2001 and Louis Jones Jr. in 2003. They were the first (and last) federal executions in nearly 40 years. Prior to those three, the last execution by the U.S. government was March 15, 1963, when Victor H. Feguer was hanged at Iowa State Penitentiary.

The Procedural Reforms Approved in *Gregg*

It is important to emphasize that the Supreme Court approved the new death penalty statutes in *Gregg* "on their face." That is, the Court assumed, without any evidence, that the new guided-discretion statutes would eliminate the arbitrariness and discrimination that the Court found objectionable in its *Furman* decision. The Court was particularly optimistic about the following procedural reforms: bifurcated trials, guidelines for judges and juries, and automatic appellate review.

bifurcated trial A two-stage trial (unlike the one-stage trial in other felony cases) consisting of a guilt phase and a separate penalty phase.

Bifurcated Trials A **bifurcated trial** is a two-stage trial—unlike the one-stage trial in other felony cases—consisting of a guilt phase and a separate penalty phase. If, in the guilt phase, the defendant is found guilty as charged, then at the penalty phase, the jury must determine whether the sentence will be death or life in prison (there are no other choices except, in most death penalty states, life imprisonment without opportunity of parole).

Some states require the selection of two separate juries in capital trials, one for the guilt phase and one for the penalty phase. During *voir dire* in capital cases, each side is generally allowed more peremptory challenges, necessitating a larger panel from which to select the jury. California, for example, allows 20 peremptory challenges in capital cases and only 10 in noncapital cases. During both phases of a bifurcated capital trial, evidence may be introduced and witnesses may be called to testify. In short, all of the procedures of due process apply to both phases of the bifurcated trial.

aggravating factors, aggravating circumstances, or **special circumstances** In death sentencing, facts or situations that increase the blameworthiness for a criminal act.

mitigating factors, mitigating circumstances, or **extenuating circumstances** In death sentencing, facts or situations that do not justify or excuse a criminal act but reduce the degree of blameworthiness and thus may reduce the punishment.

Guidelines for Judges and Juries What the Court found especially appealing about the guided-discretion statutes approved in *Gregg* is that judges and juries are provided with standards that presumably restrict, but do not eliminate, their sentencing discretion. Specifically, judges and juries, in most states, are provided with lists of aggravating and, at least in some states, mitigating factors. **Aggravating factors, aggravating circumstances,** or **special circumstances** are facts or situations that increase the blameworthiness for a criminal act. **Mitigating factors, mitigating circumstances,** or **extenuating circumstances** are facts or situations that do not justify or excuse a criminal act but reduce the degree of blameworthiness and thus may reduce the punishment. The Court has since ruled (in *Lockett v. Ohio*, 1978; *Bell v. Ohio*, 1978; and *Hitchcock v. Dugger*, 1987) that judges and juries must consider any mitigating circumstance offered by the defense, whether it is listed in the statute or not. Table 9.4 lists the aggravating and mitigating factors in Florida's current death penalty statute. The factors are typical of most states that provide them.

TABLE 9.4 Aggravating and Mitigating Factors in Florida's Death Penalty Statute

Aggravating Factors
1. The capital felony was committed by a person previously convicted of a felony and under sentence of imprisonment or placed on community control or on felony probation.
2. The defendant was previously convicted of another capital felony or of a felony involving the use or threat of violence to the person.
3. The defendant knowingly created a great risk of death to many persons.
4. The capital felony was committed while the defendant was engaged, or was an accomplice, in the commission of, or an attempt to commit, or flight after committing or attempting to commit, any: robbery; sexual battery; aggravated child abuse; abuse of an elderly person or disabled adult resulting in great bodily harm, permanent disability, or permanent disfigurement; arson; burglary; kidnapping; aircraft piracy; or unlawful throwing, placing, or discharging of a destructive device or bomb.
5. The capital felony was committed for the purpose of avoiding or preventing a lawful arrest or effecting an escape from custody.
6. The capital felony was committed for pecuniary gain.
7. The capital felony was committed to disrupt or hinder the lawful exercise of any governmental function or the enforcement of laws.
8. The capital felony was especially heinous, atrocious, or cruel.
9. The capital felony was a homicide and was committed in a cold, calculated, and premeditated manner without any pretense of moral or legal justification.
10. The victim of the capital felony was a law enforcement officer engaged in the performance of his or her official duties.
11. The victim of the capital felony was an elected or appointed public official engaged in the performance of his or her official duties if the motive for the capital felony was related, in whole or in part, to the victim's official capacity.
12. The victim of the capital felony was a person less than 12 years of age.
13. The victim of the capital felony was particularly vulnerable due to advanced age or disability, or because the defendant stood in a position of familial or custodial authority over the victim.
14. The capital felony was committed by a criminal street gang member, as defined in § 874.03.
15. The capital felony was committed by a person designated as a sexual predator pursuant to § 775.21 or a person previously designated as a sexual predator who had the sexual predator designation removed.
16. The capital felony was committed by a person subject to an injunction issued pursuant to § 741.30 [domestic violence] or § 784.046 [repeat violence, sexual violence, or dating violence], or a foreign protection order accorded full faith and credit pursuant to § 741.315 [domestic violence], and was committed against the petitioners who obtained the injunction or protection order or any spouse, child, sibling, or parent of the petitioner.
Mitigating Factors
1. The defendant has no significant history of prior criminal activity.
2. The capital felony was committed while the defendant was under the influence of extreme mental or emotional disturbance.
3. The victim was a participant in the defendant's conduct or consented to the act.
4. The defendant was an accomplice in the capital felony committed by another person, and his or her participation was relatively minor.
5. The defendant acted under extreme duress or under the substantial domination of another person.
6. The capacity of the defendant to appreciate the criminality of his or her conduct or to conform his or her conduct to the requirements of law was substantially impaired.
7. The age of the defendant at the time of the crime.
8. The existence of any other factors in the defendant's background that would mitigate against imposition of the death penalty.

Source: Online Sunshine, The 2018 Florida Statutes, Section 921.141, accessed June 6, 2019, http://www.leg.state.fl.us/STATUTES/index.cfm?App_mode=Display_Statute&Search_String=&URL=0900-0999/0921/Sections/0921.141.html.

Under Florida's death penalty statute, which is an "aggravating-versus-mitigating" type, at least one aggravating factor must be found before death may be considered as a penalty. If one or more aggravating factors are found, it is weighed against any mitigating factors. If the aggravating factors outweigh the mitigating factors, the sentence is death. However, if the mitigating factors outweigh the aggravating factors, the sentence is life imprisonment without opportunity of parole.

Another type of guided-discretion statute is Georgia's "aggravating-only" statute. In Georgia, if a jury finds at least one statutory aggravating factor, then it may, but need not, recommend death. The jury also may consider any mitigating factors, although mitigating factors are not listed in the statute, as they are in some states.

A third type of guided-discretion statute is Texas's "structured-discretion" statute. In Texas, for a defendant to be convicted of capital murder, the state must prove beyond a reasonable doubt that the defendant committed murder with at least one of nine statutorily enumerated aggravating circumstances. If the state fails in that effort, then the sentencing authority may still convict the defendant of murder or any other lesser-included offense. If the defendant is convicted of capital murder, then during the sentencing phase of the trial, the state and the defendant or the defendant's counsel may present evidence as to any matter that the court deems relevant to sentence—that is, any aggravating or mitigating factors—as long as they have not been secured in violation of the U.S. or Texas constitutions. The court then submits the following issues to the jury:

1. Whether there is a probability that the defendant would commit criminal acts of violence that would constitute a continuing threat to society.
2. If raised by the evidence, whether the defendant actually caused the death of the deceased or did not actually cause the death of the deceased but intended to kill the deceased or another or anticipated that a human life would be taken.

During penalty deliberations, juries in Texas must consider all evidence admitted at the guilt and penalty phases. Then, they must consider the two aforementioned issues. To answer "yes" to the issues, all jurors must answer "yes"; to answer "no" to the issues, 10 or more jurors must answer "no." If the two issues are answered in the affirmative, jurors are then asked if there is a sufficient mitigating factor or factors to warrant that a sentence of life imprisonment without parole rather than a death sentence be imposed. To answer "no" to this issue, all jurors must answer "no"; to answer "yes," 10 or more jurors must agree. If the jury returns an affirmative finding on the first two issues and a negative finding on the third issue, then the court must sentence the defendant to death. However, if the jury returns a negative finding on either of the first two issues or an affirmative finding on the third issue, then the court must sentence the defendant to life imprisonment without parole.

Automatic Appellate Review The third procedural feature of most of the new death penalty statutes is automatic appellate review. Currently, 30 of the 31 states with death penalty statutes provide for automatic appellate review of all death sentences, regardless of the defendant's wishes. South Carolina allows the defendant to waive sentence review if the defendant is deemed competent by the court; also, the federal jurisdiction does not provide for automatic appellate review. Most of the 31 states automatically review both the conviction and the sentence.[21] Generally, the automatic review is conducted by the state's highest appellate court. If either the conviction or the sentence is overturned, then the case is sent back to the trial court for additional proceedings or for retrial. It is possible that the death sentence may be reimposed as a result of this process.

Some states are very specific in defining the review function of the appellate courts, while other states are not. Although the Supreme Court does not require it (*Pulley v. Harris*, 1984), some states have provided a proportionality review. In a **proportionality review**, the appellate court compares the sentence in the case it is reviewing with penalties imposed in similar cases in the state. The object of proportionality review is to reduce, as much as possible, disparity in death penalty sentencing.

proportionality review A review in which the appellate court compares the sentence in the case it is reviewing with penalties imposed in similar cases in the state. The object is to reduce, as much as possible, disparity in death penalty sentencing.

In addition to the automatic appellate review, there is a dual system of collateral review for capital defendants. In other words, capital defendants may appeal their convictions and sentences through both the state and the federal appellate systems. Table 9.5 shows the general appeals process in death penalty cases.

TABLE 9.5 The Appellate Process in Capital Cases

Stage 1: Direct appeal
Step 1: Trial and sentence in state court
Step 2: Direct appeal to state appeals court (state supreme court)
Step 3: U.S. Supreme Court for writ of *certiorari* (discretionary)
Stage 2: State postconviction review
Step 1: Trial court
Step 2: State court of appeals (state supreme court)
Step 3: U.S. Supreme Court for writ of *certiorari* (discretionary)
Stage 3: Federal postconviction review
Step 1: Petition for writ of *habeas corpus* in U.S. District Court
Step 2: U.S. Circuit Court of Appeals
Step 3: U.S. Supreme Court for writ of *certiorari* (discretionary)

Source: Robert M. Bohm, *DeathQuest: An Introduction to the Theory and Practice of Capital Punishment in the United States*, 5th ed. (New York and London: Routledge, 2017), 101.

The size of the death row population in the United States does not fluctuate very much from year to year. *What factors keep it from dramatically increasing in size?*

Shepard Sherbell/Contributor/Getty Images

Some death row inmates whose appeals have been denied by the U.S. Supreme Court may still try to have the Supreme Court review their cases on constitutional grounds by filing a writ of *habeas corpus*. Recall that a writ of *habeas corpus* is a court order directing a law officer to produce a prisoner in court to determine whether the prisoner is being legally detained or imprisoned. Critics maintain that abuse of the writ has contributed to the long delays in executions (currently averaging more than 15 years after sentencing[22]) and to the high costs associated with capital punishment.

Congress passed the Antiterrorism and Effective Death Penalty Act of 1996, in part to speed up the process and reduce costs. President Clinton signed it into law on April 24, 1996. The law requires that second or subsequent *habeas* petitions be dismissed when the claim already had been made in a previous petition. It also requires that new claims be dismissed, unless the Supreme Court hands down a new rule of constitutional law and makes it retroactive to cases on collateral review. Under the act, the only other way the Supreme Court will hear a claim made for the first time is when the claim is based on new evidence not previously available. Even then, the new evidence must be of sufficient weight, by a clear and convincing standard of proof, to convince a judge or jury that the capital defendant was not guilty of the crime or crimes for which he or she was convicted.

The act also made the federal appellate courts "gatekeepers" for second or subsequent *habeas corpus* petitions. Thus, to file a second or subsequent claim under the new law, a capital defendant must first file a motion in the appropriate appellate court announcing his or her intention. A panel of three judges must then hear the motion within 30 days. The judges must decide whether the petitioner has a legitimate claim under the act. If the claim is denied, the law prohibits any review of the panel's decision, either by a rehearing or writ of *certiorari* to the Supreme Court. So far, the Supreme Court has upheld the constitutionality of the law.

Some people argue that the appellate reviews are unnecessary delaying tactics (at least those beyond the automatic review). However, the outcomes of the reviews suggest otherwise. Nationally, between 1973 and 2013, nearly 38% of the initial convictions or sentences in capital cases were overturned on appeal,[23] and, contrary to popular belief, those reversals generally were not the result of so-called legal technicalities. They were the product of "such fundamental constitutional errors" as denial of the right to an impartial jury, problems

FIGURE 9.4 The Total Number of Death Row Inmates, by Race and Gender, 2019

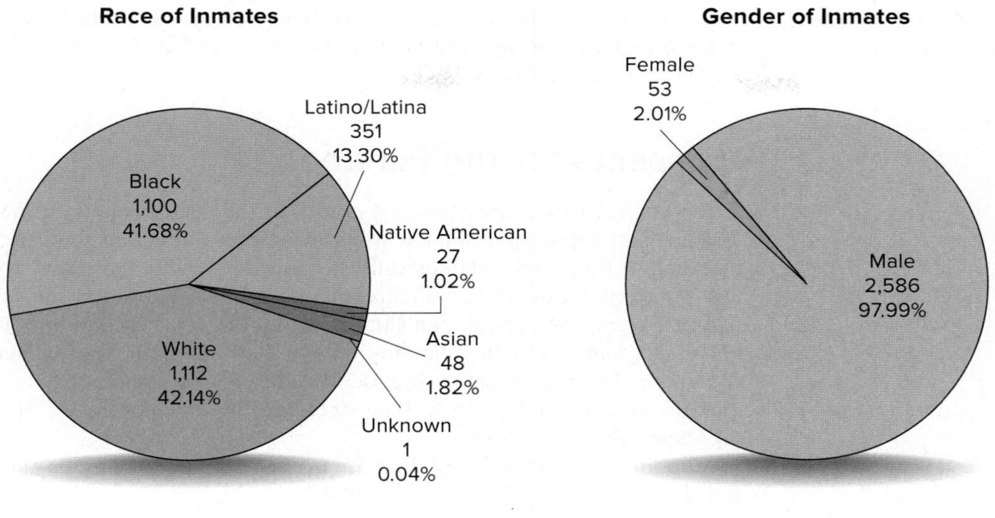

Race of Inmates

Black 1,100 41.68%
Latino/Latina 351 13.30%
Native American 27 1.02%
Asian 48 1.82%
Unknown 1 0.04%
White 1,112 42.14%

Total 2,639

Gender of Inmates

Female 53 2.01%
Male 2,586 97.99%

Total 2,639

Source: Adapted from "Death Row, U.S.A.," NAACP Legal Defense and Educational Fund Fall 2019. Data depicted are as of October 1, 2019, accessed February 23, 2020, https://www.naacpldf.org/wp-content/uploads/DRUSAFall2019.pdf.

of tainted evidence and coerced confessions, ineffective assistance of counsel, and prosecutors' references to defendants who refuse to testify.[24]

The number of persons currently on death rows in the United States is 2,639 (as of October 1, 2019).[25] Figure 9.4 shows the race, ethnic, and gender distributions of the death row population. The size of the death row population in the United States does not fluctuate very much from year to year, despite the relatively few executions each year (the largest number since 1977 was 98 in 1999).[26] One reason is that the number of new death sentences has been declining in recent years (see Figure 9.5).[27] Another reason, as noted earlier, is that since January 1, 1973, nearly

FIGURE 9.5 Death Sentences and Executions in the United States, 1976–2019

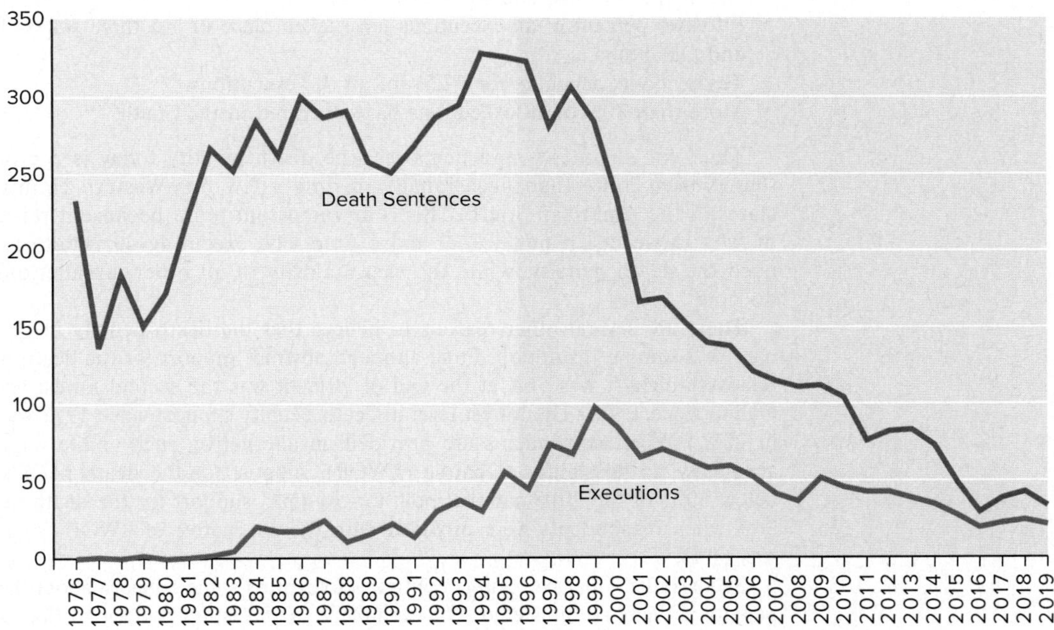

Sources: Adapted from Death Penalty Information Center, "Death Sentencing," accessed February 23, 2020, https://files.deathpenaltyinfo.org/documents/pdf/FactSheet.f1582296575.pdf; "Number of Executions Since 1976," accessed February 23, 2020, https://files.deathpenaltyinfo.org/documents/pdf/FactSheet.f1582296575.pdf.

3,000 of the more than 8,000 defendants sentenced to death (nearly 38%) have been removed from death row by having their convictions or sentences reversed. In addition, since January 1, 1973, 392 death row inmates have received **commutations**–reductions in sentences, granted by a state's governor or the president of the United States–and approximately 509 have died of natural causes or have been killed.[28]

commutations Reduction of the original sentence given by executive authority, usually a state's governor.

Prospects for the Future

These are interesting times when it comes to the death penalty in the United States. It appears that capital punishment is receiving more attention than usual. Although less than half of the world's nations still have a death penalty, those that do have one seldom use it. Among Western, industrialized nations, the United States stands alone as the only nation to employ capital punishment. However, even within the United States, 22 (21 states and D.C.) jurisdictions do not have a death penalty, and among the 31 (29 states, U.S. government, and U.S. military) jurisdictions that do have one, only a handful of them use it more than occasionally, and almost all of them are located geographically in the South.

Consider the distribution of the 1,516 executions conducted in the United States between January 17, 1977, and February 21, 2020[29]:

- The 1,516 executions have occurred in 35 of what were at one time 38 death penalty jurisdictions (New Jersey had no executions and abolished the death penalty in 2007; New Mexico had 1 execution but abolished the death penalty in 2009; Illinois had 12 executions but abolished the death penalty in 2011; Connecticut had 1 execution but abolished the death penalty in 2012; Maryland had 5 executions but abolished the death penalty in 2013; Nebraska had 3 executions but abolished the death penalty in 2015, before reinstating it in 2016; Delaware had 16 executions but abolished the death penalty in 2016; Washington had 5 executions but abolished the death penalty in 2018; and New Hampshire had no executions and abolished the death penalty in 2019).
- Two jurisdictions with death penalty statutes (Kansas and the U.S. military) have not had a single execution.
- Fifteen of the 35 "executing" jurisdictions (43%) have held fewer than 10 executions.
- Only 20 executing jurisdictions (57%) have conducted 10 or more executions.
- Sixty-five percent of all executions have taken place in just five states–Texas, Virginia, Oklahoma, Florida, and Missouri.
- Fifty-two percent of all executions have taken place in just three states–Texas, Virginia, and Oklahoma.
- Texas, alone, accounts for 37.5% of all the executions.
- More than 80% of all executions have occurred in the South.[30]

Thus, for all intents and purposes, the death penalty today is a criminal sanction that is used more than occasionally in only a few non-Western countries and a few states in the American South. This is an important point because it raises the question of why those death penalty–or more precisely, executing–jurisdictions in the world need the death penalty, while the vast majority of all other jurisdictions in the world do not?

There are several other reasons to believe that the death penalty in the United States may be a waning institution. First, although abstract support for the death penalty remains relatively high–it was 56% at the end of 2019–it was the second lowest level recorded by Gallup since 1972. (The lowest level in death penalty support since 1972 was 55% recorded in 2017.) When respondents are provided an alternative, such as life imprisonment with absolutely no possibility of parole (LWOP), support for the death penalty typically falls below 50%. In the 2019 national poll, for example, support for the death penalty was only 36% when respondents were provided with the alternative of LWOP (support for LWOP was 60%).[31]

Second, the American public continues to express some concern about the way the death penalty is being administered. For example, a 2018 Gallup poll found that only 49% of the American public believed that the death penalty is applied fairly–a new low, while 45% believed that the death penalty is not applied fairly.[32] Furthermore, 71% of Americans responded in a 2015 Pew Research Center poll that there is some risk of executing innocent people, even though most people believe that the execution of innocent people is a rare occurrence.[33] The same poll revealed that only 26% of Americans believe that there are adequate safeguards in place to make sure that the execution of an innocent person does not occur.[34]

A third factor is the positions taken by respected organizations within the United States, such as the American Bar Association (ABA) and organized religions. In 1997, the ABA adopted a resolution that requested death penalty jurisdictions to refrain from using the sanction until greater fairness and due process could be ensured. The leaders of most organized religions in the United States–whether Catholic, Protestant, or Jewish–openly oppose capital punishment. On August 2, 2018, the Vatican announced that "Pope Francis had changed the Catechism of the Catholic Church about the death penalty, saying it can *never* be sanctioned because it 'attacks' the inherent dignity of all humans" (emphasis added).[35]

A fourth factor is world opinion. In Europe, the death penalty is viewed as a violation of human rights.[36] A condition for admittance into the European Union (EU) is the abolition of the death penalty. That is why Turkey abolished its death penalty in 2004. Admittance into the 40-nation Council of Europe also requires the renouncing of the death penalty. Georgia, a former republic of the Soviet Union, abolished its death penalty in 1997 so that it could join the Council. Figure 9.6 lists the principal executing countries of the world in 2018. The United Nations Commission on Human Rights has condemned repeatedly the death penalty in the United States, urging the U.S. government to stop all executions until it brings states into compliance with international standards and laws.

However, capital punishment in some states has proven stubbornly resilient. There are reasons to believe that in those U.S. states the death penalty will remain a legal sanction for the foreseeable future. One reason is that death penalty support among the American public, at least according to the major opinion polls, remains relatively strong. It is unlikely that the practice of capital punishment could be sustained if a majority of American citizens were to oppose it. In no year for which polls are available has a majority of Americans opposed the death penalty (the first national death penalty opinion poll was conducted in December 1936).[37]

Innocent people are never executed.

fact

As many as 25 people (and likely more) may have been executed in error in the United States since 1976.

Sources: Robert M. Bohm, *DeathQuest: An Introduction to the Theory and Practice of Capital Punishment in the United States,* 5th ed. (New York and London: Routledge, 2017), 313; Talia Roitberg Harmon and William S. Lofquist, "Too Late for Luck: A Comparison of Post-*Furman* Exonerations and Executions of the Innocent." *Crime and Delinquency* 51 (2005), 498–520; Patrick Lehner, "Abolition Now!!!" www.abolition-now.com; Death Penalty Information Center, www.deathpenaltyinfo.org /innocothers.html#executed.

FIGURE 9.6 Countries with the Most Executions in 2018

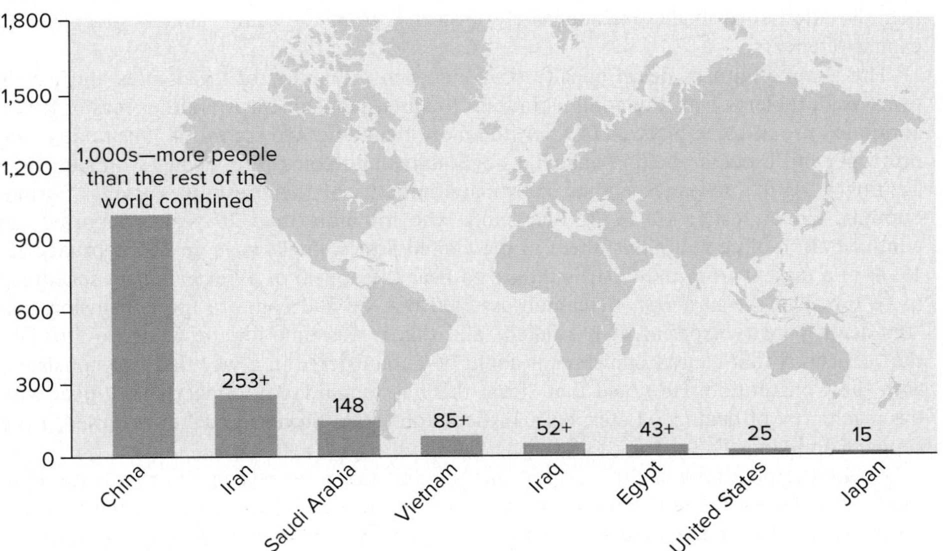

Note: These figures include only cases known to Amnesty International; the true figures certainly were higher. In 2018, at least 690 people were executed in 20 countries—a 58% decrease from the 1,634 reported executions in 2015 (does not include executions in China).

Source: Death Penalty Information Center, "The Death Penalty: An International Perspective," accessed June 8, 2019, https://deathpenaltyinfo.org/death-penalty-international-perspective#interexec.

According to public opinion polls, a majority of Americans support the death penalty. *Why?*

Jim Bourg/REUTERS/Newscom

Although life imprisonment without opportunity of parole seems to be a popular alternative to the death penalty in polls, a problem with the LWOP alternative is that many people are very skeptical about the ability of correctional authorities to keep capital murderers imprisoned for life. Thus, although 60% of the public may say it prefers LWOP to capital punishment, in practice, it may be reluctant to make the substitution because it fears that the alternative might not adequately protect it from the future actions of convicted capital offenders.

The abiding faith of death penalty proponents in the ability of legislatures and courts to fix any problems with the administration of capital punishment is another reason for its continued use in some places. However, the more than 4-decade record of "fine-tuning" the death penalty process remains ongoing. Legislatures and courts are having a difficult time "getting it right," despite spending inordinate amounts of their resources trying. Former Supreme Court Justice Harry A. Blackmun, who for more than 20 years supported the administration of capital punishment in the United States, finally gave up. On February 22, 1994, in a dissent from the Court's refusal to hear the appeal of a Texas inmate scheduled to be executed the next day, Blackmun asserted that he had come to the conclusion that "the death penalty experiment has failed" and that it was time for the Court to abandon the "delusion" that capital punishment could be administered in a way that was consistent with the Constitution. He noted that "from this day forward, I no longer shall tinker with the machinery of death."[38] Later, both Justices John Paul Stevens and Lewis Powell made similar admissions.[39]

As for the positions against capital punishment taken by respected organizations in the United States, "true believers" in the death penalty could not care less what others think, especially in the case of organizations such as the American Bar Association. This holds true for world opinion as well. In the case of organized religions, the situation is probably more complex. Although most people who consider themselves religious and are affiliated with religions whose leadership opposes capital punishment probably respect the views of their leaders, they obviously live their daily lives and hold beliefs about capital punishment (and other issues such as abortion) based on other values.

Some death penalty opponents believe that a principal reason for the continuing support of capital punishment is that most people know very little about the subject, and what they think they know is based almost entirely on myth. It is assumed that if people were educated about capital punishment, most of them would oppose it. Unfortunately, research suggests that educating the public about the death penalty may not have the effect that abolitionists desire. Although information about the death penalty can reduce support for the sanction—sometimes significantly—rarely is the support reduced to less than a majority, and any reduction in support may be only temporary.[40]

A final factor (to be discussed here) that likely sustains death penalty support is the symbolic value it has for politicians and criminal justice officials. Politicians use support for the death penalty as a symbol of their toughness on crime. Opposition to capital punishment invariably is interpreted as symbolic of softness on crime. Criminal justice officials and much of the public often equate support for capital punishment with support for law enforcement in general. It is ironic that although capital punishment has virtually no effect on crime, the death penalty continues to be a favored political "silver bullet"—a simplistic solution to the crime problem used by aspiring politicians and law enforcement officials.

In short, the reasons provided for supporting capital punishment do not stand up well to critical scrutiny, but the American public has not been deterred from supporting it anyway. Together with the movement to replace indeterminate sentencing with determinate sentencing and to abolish parole, the death penalty is part of the "law and order" agenda popular in the United States since the mid-1970s. Whether this direction in criminal justice has run its course is anyone's guess.

THINKING CRITICALLY

1. Do you think the death penalty helps to prevent crime in the United States? Why or why not?
2. Do you think the death penalty should continue to be legal in this country? Why or why not?

SUMMARY

1. Identify the general factors that influence a judge's sentencing decisions.

In sentencing, judges are limited by statutory provisions; guided by prevailing philosophical rationales, organizational considerations, and presentence investigation reports; and influenced by their own personal characteristics.

2. Describe how judges tailor sentences to fit the crime and the offender.

Judges have several ways to tailor sentences to fit the crime and the offender. They can impose a combination sentence of, for example, imprisonment, probation, and a fine. They can suspend the imprisonment portion of a combination sentence, or they can suspend the entire sentence if the offender stays out of trouble, makes restitution to the victim, or seeks medical treatment. Judges can give offenders credit for time spent in jail while awaiting trial, deducting that time from any prison sentence. A judge may even impose a sentence of "time served" and release the offender. When an offender is convicted of two or more crimes, a judge can order the prison sentences to run concurrently or consecutively. Judges also can delay sentencing and retain the right to impose a sentence at a later date if conditions warrant it.

3. Distinguish between indeterminate and determinate sentences.

An indeterminate sentence has a fixed minimum and maximum term of incarceration, rather than a set period. A determinate sentence, in contrast, has a fixed period of incarceration and eliminates the decision-making responsibility of parole boards.

4. Explain the three basic types of determinate sentences.

There are three basic types of determinate sentences: flat-time, mandatory, and presumptive. With flat-time sentencing, judges may choose between probation and imprisonment but have little discretion in setting the length of a prison sentence. With mandatory sentencing, a specified number of years of imprisonment (usually within a range) is provided for particular crimes. Mandatory sentencing generally allows credit for good time but does not allow release on parole. Presumptive sentencing allows a judge to retain some sentencing discretion (subject to appellate review). It requires a judge to impose the normal sentence, specified by statute, on a "normal" offender who has committed a "normal" crime. However, if the crime or the offender is not normal—if there are mitigating or aggravating circumstances—then the judge is allowed to deviate from the presumptive sentence.

5. **List five rationales or justifications for criminal punishment.**

Five rationales or justifications for criminal punishment are retribution, incapacitation, deterrence, rehabilitation, and restoration.

6. **Explain the purposes of presentence investigation reports.**

Presentence investigation reports (PSIs) help judges determine the appropriate sentences for particular defendants. PSIs also are used in the classification of probationers, parolees, and prisoners according to their treatment needs and their security risks.

7. **List the legal bases for appeal.**

Defendants can appeal their convictions either on legal grounds (such as defects in jury selection, improper admission of evidence at trial, mistaken interpretations of law) or on constitutional grounds (such as illegal search and seizure, improper questioning of the defendant by the police, identification of the defendant through a defective police lineup, and incompetent assistance of counsel).

8. **Identify the type of crime for which death may be a punishment.**

In the United States, death is the ultimate punishment. At the state level, death can be imposed only for the crime of aggravated murder and a few other seldom-committed offenses.

9. **Summarize the three major procedural reforms the U.S. Supreme Court approved for death penalty cases in the *Gregg* decision.**

The three major procedural reforms the Court approved in *Gregg* were bifurcated trials, guidelines for judges and juries to follow, and automatic appellate review.

KEY TERMS

restitution 356
indeterminate sentence 357
determinate sentence 358
flat-time sentencing 358
good time 358
mandatory sentencing 358
presumptive sentencing 359
criminal sanctions or criminal punishment 362
retribution 363
revenge 363
just deserts 363
incapacitation 363
special or specific deterrence 363
general deterrence 363
rehabilitation 364
victim-impact statements 365
presentence investigation reports 367
allocution 367
pardon 367
bifurcated trial 373
aggravating factors, aggravating circumstances, or special circumstances 374
mitigating factors, mitigating circumstances, or extenuating circumstances 374
proportionality review 375
commutations 378

REVIEW QUESTIONS

1. What are five general types of punishment currently being used in the United States?
2. What are some criticisms of determinate sentencing?
3. Which rationale for criminal punishment is the only one that specifically addresses what has happened in the past, and what are its two major forms?
4. What are three organizational considerations that may influence a judge's sentencing decision?
5. What is *allocution*?
6. What two steps must be taken before an appellate court will hear an appeal?
7. What was the landmark 1972 decision in which the Supreme Court set aside death sentences for the first (and only) time in its history?
8. What are the five methods of execution currently available in the United States?

IN THE FIELD

1. **Sentencing** Select a criminal case that currently is receiving publicity in your community. Conduct an informal survey, asking respondents what sentence they believe would be appropriate in the case. Ask them why they chose the sentence. Determine whether respondents tended to agree with each other. If they did not tend to agree, speculate on the reasons for the disagreement.
2. **Death Penalty Opinion** With family members or friends, discuss the death penalty. Ask them why they hold their particular positions (in favor, opposed, undecided). Also, ask them under what circumstances they would change their positions. Would any of the mitigating circumstances listed in the second Critical Thinking exercise that follows cause any of them to change their positions?

ON THE WEB

1. **Death Penalty** Access the Death Penalty Information Center website at www.deathpenaltyinfo.org and the PROCON Death Penalty website at https://deathpenalty.procon.org. Choose the same topic included in both sources, and review the information provided. Write a brief summary of the information you discovered and how it affected your view of capital punishment.

2. **Mandatory Minimums** Access the website of Families Against Mandatory Minimums at www.famm.org. After studying information provided at the site, write an essay or prepare an oral presentation critiquing mandatory minimum sentences.

CRITICAL THINKING EXERCISES

CREATIVE SENTENCES

1. For causing the fatal wreck that killed Army Sgt. Thomas E. Towers Jr., 22-year-old Andrew Gaudioso was ordered to send the soldier's family a postcard every week for 15 years. According to Towers's father, "At first I thought I wanted prison for Gaudioso. Then I thought it would be better to force him some way to remember–at least once a week–what he did. I think this does that." The unusual sentence does not specify what Gaudioso should write on the postcard, which must be presented to his probation officer with the 35 cents postage paid. Gaudioso, who had traces of street drugs in his blood when he veered left of center at more than 80 mph in the rain and smashed almost head-on into Towers's car, must also pass drug tests. If he fails to send the weekly postcards, or if he fails the drug tests, Gaudioso could be sent to prison for 15 years.
   a. What do you think of this creative sentence?
   b. Do you think that creative sentences like this one should be allowed? Why or why not?
   c. Is the punishment for Gaudioso proportional to the crime? Should it be?
   d. Do you think Gaudioso's sentence will deter others? Why or why not?

DEATH SENTENCING

2. You are a juror in a death penalty case. The defendant in the case already has been found guilty of capital murder during the guilt phase of the trial. During the penalty phase, you have to determine whether the defendant is to be sentenced to death or to life imprisonment without opportunity of parole (LWOP). The judge has instructed you (and the rest of the jury) to consider the aggravating and mitigating circumstances of the case. If the aggravating circumstances outweigh the mitigating circumstances, you are expected to vote for death. However, if the mitigating circumstances outweigh the aggravating circumstances, you are expected to vote for LWOP. The lone aggravating circumstance in the case is that the defendant committed the capital murder during the commission of a robbery. Though they are not aggravating circumstances, the defendant also has two prior convictions for robbery and one prior conviction for the sale of illegal drugs. Under which of the following mitigating circumstances would you vote for LWOP or death in this case? Explain the reasons for your decision.
   a. The defendant is a female with children.
   b. The defendant acted under extreme duress or under the substantial domination of another person.
   c. The defendant was seriously abused, both mentally and physically, as a child.
   d. The capacity of the defendant to appreciate the criminality of his or her conduct to the requirements of law was substantially impaired.

NOTES

1. In addition to the other sources cited, material on sentencing and appeals is from David W. Neubauer and Henry F. Fradella, *America's Courts and the Criminal Justice System*, 10th ed. (Belmont, CA: Wadsworth, 2011); Cassia Spohn, *How Do Judges Decide? The Search for Fairness and Justice in Punishment*, 2nd ed. (Thousand Oaks, CA: Sage, 2009); Christopher Smith, *Courts, Politics, and the Judicial Process* (Chicago: Nelson-Hall, 1993); N. Gary Holten and Lawson L. Lamar, *The Criminal Courts: Structures, Personnel, and Processes* (New York: McGraw-Hill, 1991); Lawrence Baum, *American Courts*, 3rd ed. (Boston: Houghton Mifflin, 1994); David Garland, *Punishment and Modern Society: A Study in Social Theory* (Chicago: University of Chicago Press, 1990); Paul Wice, *Chaos in the Courthouse: The Inner Workings of the Urban Criminal Courts* (New York: Praeger, 1985); John Paul Ryan, Allan Ashman, Bruce D. Sales, and Sandra Shane-DuBow, *American Trial Judges* (New York: Free Press, 1980); Robert Satter, *Doing Justice: A Trial Judge at Work* (New York: Simon & Schuster, 1990); Herbert Packer, *The Limits of the Criminal Sanction* (Stanford, CA: Stanford University Press, 1968).
2. Sara J. Berman, "What Happens at Sentencing," NOLO, accessed June 2, 2019, https://www.nolo.com/legal-encyclopedia/what-happens-sentencing.html.
3. Danielle Kaeble, "Probation and Parole in the United States, 2016," in *Bureau of Justice Statistics Bulletin*, U.S. Department of Justice (Washington DC: U.S. Government Printing Office, April 2018), 20, Appendix Table 6, accessed June 2, 2019, https://www.bjs.gov/content/pub/pdf/ppus16.pdf; Lauren E. Glaze and Thomas

P. Bonczar, "Probation and Parole in the United States, 2006," in *Bureau of Justice Statistics Bulletin*, U.S. Department of Justice (Washington, DC: U.S. Government Printing Office, December 2007, revised July 2008), 6, accessed December 3, 2008, www.ojp.usdoj.gov/bjs/pub/pdf/ppus06.pdf.

4. Robert Carter, "Determinate Sentences," in *Encyclopedia of American Prisons,* ed. M. D. McShane and F. P. Williams III (New York: Garland, 1996), 147–49.

5. "Sentence Disparity in New York: The Response of Forty-One Judges," *The New York Times*, March 30, 1979, B3.

6. See Robert M. Bohm, "Retribution and Capital Punishment: Toward a Better Understanding of Death Penalty Opinion," *Journal of Criminal Justice* 20 (1992): 227–35.

7. The Sentencing Project, "Trends in U.S. Corrections," accessed June 3, 2019, https://www.sentencingproject.org/wp-content/uploads/2016/01/Trends-in-US-Corrections.pdf.

8. Diana D'Abruzzo, "Shining a Light Inside Prisons," Arnold Ventures, accessed June 5, 2019, https://www.arnoldventures.org/stories/shining-a-light-inside-prisons?cm_ven=ExactTarget&cm_cat=JPC+-+6.5.2019&cm_pla=All+Subscribers&cm_ite=https%3a%2f%2fwww.arnoldventures.org%2fstories%2fshining-a-light-inside-prisons&cm_lm=rbohm@mail.ucf.edu&cm_ainfo=&&utm_source=%20urban_newsletters&&utm_medium=news-JPC&&utm_term=JPC&&.

9. "Issues: Constitutional Amendments, State Amendments," The National Center for Victims of Crime, accessed February 1, 2016, www.victimsofcrime.org/our-programs/public-policy/amendments; "New Directions from the Field: Victims' Rights and Services for the 21st Century," in *OVC Bulletin*, U.S. Department of Justice, Office for Victims of Crime (Washington, DC: U.S. Government Printing Office, December 2004).

10. Calculated from data in "The Crime Victims Fund," 2017 OVC Report to the Nation: Fiscal Years 2015-2016, accessed June 5, 2019, https://www.ovc.gov/pubs/reporttonation2017/; "The Crime Victims Fund," 2013 OVC Report to the Nation: Fiscal Years 2011–2012, accessed February 7, 2016, www.ncjrs.gov/ovc_archives/reporttonation2013/crim_vict_fnd.html; "The Crime Victims Fund," 2015 OVC Report to the Nation: Fiscal Years 2014–2015, accessed February 7, 2016, www.ovc.gov/pubs/reporttonation2015/crime-victims-fund.html; U.S. Department of Justice, Office for Victims of Crime, Grants & Funding, Types of Funding, VOCA Nationwide Performance Reports, accessed February 1, 2013, www.ojp.usdoj.gov/ovc/grants/vocareps.html; *2009 OVC Report to the Nation: Fiscal Years 2007–2008*, U.S. Department of Justice, Office for Victims of Crime, accessed February 7, 2011, www.ojp.usdoj.gov/ovc/welcovc/reporttonation2009/ReporttoNation09full.pdf; *Report to the Nation 2003: Fiscal Years 2001 and 2002*, U.S. Department of Justice, Office for Victims of Crime, www.ojp.usdoj.gov/ovc/welcovc/reporttonation2005/welcome.html; *Report to the Nation 2005: Fiscal Years 2003 and 2004*, U.S. Department of Justice, Office for Victims of Crime, www.ojp.usdoj.gov/ovc/welcovc/reporttonation2003/_welcome.html.

11. Office of Justice Programs, Victim Law, "About Victims' Rights," accessed June 4, 2019, https://victimlaw.org/victimlaw/pages/victimsRight.jsp.

12. "New Directions from the Field: Victims' Rights and Services for the 21st Century."

13. "Restorative Justice: An Interview with Visiting Fellow Thomas Quinn," *National Institute of Justice Journal* 235 (March 1998): 10.

14. Matthiesen, Wickert & Lehrer, S.C., "Subrogation of Criminal Restitution in All 50 States," Last updated 4/25/19, accessed June 5, 2019, https://www.mwl-law.com/wp-content/uploads/2018/02/CRIMINAL-RESTITUTION-CHART.pdf.

15. Ibid.

16. Unless indicated otherwise, material about the death penalty is from Robert M. Bohm, *DeathQuest: An Introduction to the Theory and Practice of Capital Punishment in the United States,* 5th ed. (New York and London: Routledge, 2017); James R. Acker, Robert M. Bohm, and Charles S. Lanier, eds., *America's Experiment with Capital Punishment: Reflections on the Past, Present, and Future of the Ultimate Penal Sanction,* 3rd ed. (Durham, NC: Carolina Academic Press, 2014); Hugo Alan Bedau, *The Death Penalty in America; Current Controversies* (New York: Oxford University Press, 1997); Hugo Adam Bedau, ed., *The Death Penalty in America,* 3rd ed. (London: Oxford University Press, 1982); William J. Bowers with Glenn L. Pierce and John McDevitt, *Legal Homicide: Death as Punishment in America, 1864–1982* (Boston: Northeastern University Press, 1984); Raymond Paternoster, *Capital Punishment in America* (New York: Lexington, 1991); Robert M. Bohm, "Humanism and the Death Penalty, with Special Emphasis on the Post-Furman Experience," *Justice Quarterly* 6 (1989): 173–95; Victoria Schneider and John Ortiz Smykla, "A Summary Analysis of Executions in the United States, 1608–1987: The Espy File," in *The Death Penalty in America: Current Research,* ed. R. M. Bohm (Cincinnati: Anderson, 1991), 1–19.

17. Death Penalty Information Center, "Women and the Death Penalty," accessed February 23, 2020, https://deathpenaltyinfo.org/women-and-death-penalty.

18. Death Penalty Information Center; "Execution of Juveniles in the U.S. and other Countries," accessed June 5, 2019, https://deathpenaltyinfo.org/execution-juveniles-us-and-other-countries.

19. Death Penalty Information Center, "Number of Executions by State and Region Since 1976," accessed February 23, 2020, https://files.deathpenaltyinfo.org/documents/pdf/FactSheet.f1582296575.pdf.

20. Elizabeth Davis and Tracy L. Snell, "Capital Punishment, 2016," *Bureau of Justice Statistics*, U.S. Department of Justice (April 2018), Appendix Table 2, accessed June 5, 2019, https://www.bjs.gov/content/pub/pdf/cp16sb.pdf.

21. Tracy L. Snell, "Capital Punishment, 2005," in *Bureau of Justice Statistics Bulletin*, U.S. Department of Justice (Washington, DC: U.S. Government Printing Office, December 2006), 3, www.ojp.usdoj.gov/bjs/_abstract/cp05.htm. Review of South Carolina death penalty statute conducted on January 13, 2013.

22. See Justice Breyer's dissent in *Glossip v. Gross*, 576 U.S.___ (2015).

23. Calculated from data in Tracy L. Snell, "Capital Punishment, 2013—Statistical Tables," p. 20, Table 17. Accessed May 12, 2016, www.bjs.gov/content/pub/pdf/cp13st.pdf.

24. Paternoster, *Capital Punishment in America*, 208–9; Barry Scheck, Peter Neufeld, and Jim Dwyer, *Actual Innocence: When Justice Goes Wrong and How to Make It Right* (New York: Penguin Putnam, 2001); James S. Liebman, Jeffrey Fagan, and Valerie West, "A Broken System: Error Rates in Capital Cases, 1973–1995," The Justice Project, www.justice.policy.net/jpreport.html.

25. "Death Row U.S.A. (Fall 2019)," Criminal Justice Project, NAACP Legal Defense and Educational Fund, Inc., October 1, 2019, accessed February 23, 2020, https://www.naacpldf.org/wp-content/uploads/DRUSAFall2019.pdf.

26. Death Penalty Information Center, "Death Row Inmates by State and Size of Death Row by Year," accessed June 7, 2019, https://deathpenaltyinfo.org/death-row-inmates-state-and-size-death-row-year?scid=9&did=188.

27. Death Penalty Information Center, "Death Sentences in the United States from 1977 by State and by Year," accessed June 7, 2019, https://deathpenaltyinfo.org/death-sentences-united-states-1977-present.

28. Snell, "Capital Punishment, 2013—Statistical Tables," 19, Table 16.

29. Death Penalty Information Center, "Number of Executions by State and Region Since 1976."

30. Calculated from data in ibid; accessed February 24, 2020.

31. Jeffrey M. Jones, "Americans Now Support Life in Prison over Death Penalty," *Gallup*, November 25, 2019, accessed February 24, 2020, https://news.gallup.com/poll/268514/americans-support-life-prison-death-penalty.aspx.

32. Justin McCarthy, "New Low of 49% in U.S. Say Death Penalty Applied Fairly," *Gallup*, October 22, 2018, accessed June 8, 2019, https://news.gallup.com/poll/243794/new-low-say-death-penalty-applied-fairly.aspx.

33. "Less Support for Death Penalty, Especially Among Democrats," Pew Research Center, April 16, 2015, accessed June 8, 2019, https://www.people-press.org/2015/04/16/less-support-for-death-penalty-especially-among-democrats/.

34. Ibid.

35. Ines San Martin, "Pope Francis changes teaching on death penalty, it's 'inadmissible'," *CRUX*, August 2, 2018, accessed June 8, 2019, https://cruxnow.com/vatican/2018/08/02/pope-francis-changes-teaching-on-death-penalty-its-inadmissible/.

36. "The Shadow over America," *Newsweek*, May 29, 2000, 27.

37. Bohm, *DeathQuest*, p. 323.

38. *Callins v. Collins*, 510 U.S. 1141 (1994).

39. Bohm, *DeathQuest*, p. 92.

40. Robert M. Bohm, Louise J. Clark, and Adrian F. Aveni, "Knowledge and Death Penalty Opinion: A Test of the Marshall Hypotheses," *Journal of Research in Crime and Delinquency* 28 (1991): 360–87; Robert M. Bohm and Ronald E. Vogel, "A Comparison of Factors Associated with Uninformed and Informed Death Penalty Opinions," *Journal of Criminal Justice* 23 (1994): 125–43; Robert M. Bohm, Ronald E. Vogel, and Albert A. Maisto, "Knowledge and Death Penalty Opinion: A Panel Study," *Journal of Criminal Justice* 21 (1993): 29–45; Robert M. Bohm and Brenda L. Vogel, "More than Ten Years After: The Long-Term Stability of Informed Death Penalty Opinions," *Journal of Criminal Justice* 32 (2004): 307–27.

Design Elements: *Image of part of a court building with columns and statue*: Pixtal/AGE Fotostock RF; *Image of a door slightly open*: jadding/Shutterstock.com

Scott Olson/Getty Images

Institutional Corrections

CHAPTER OUTLINE

LEARNING OBJECTIVES

After completing this chapter, you should be able to:

1. Summarize the purposes of confinement in Europe before it became a major way of punishing criminals.
2. Describe how offenders were punished before the large-scale use of confinement.
3. Explain why confinement began to be used as a major way of punishing offenders in Europe.
4. Describe the recent trends in the use of incarceration in the United States.
5. List some of the characteristics of the incarcerated population in the United States.
6. Describe how incarceration facilities are structured, organized, and administered by the government in the United States.
7. Name some of the common types of correctional facilities in the United States.
8. Identify some of the procedures that institutions employ to maintain security and order.
9. List the services and programs that commonly are available to inmates.

Crime Story

The United States operates arguably the most expensive prison on Earth: The Pentagon detention center at Guantánamo Bay Naval Base, Cuba, which is officially named "Camp Delta" and frequently referred to as "Gitmo" (pictured). The prison was built in response to the terrorist attacks on September 11, 2001, to hold foreign terrorists or what Pentagon officials called "the worst of the worst enemies in the war on terror." The prison originally was built to hold 600 inmates. The first prisoners arrived from Afghanistan on January 11, 2002, to a tent city of open-air cells guarded by Marines. Approximately 780 prisoners have been incarcerated at Gitmo, and most of them have been released without charge.

In 2010, the prison held 171 inmates at an annual cost of $139 million a year—about $800,000 per inmate—more than 10 times the $78,000 average annual cost of incarcerating a prisoner in a maximum-security federal prison in the United States. By May 2016, the prison held 80 inmates at an annual cost of approximately $445 million, or more than $5 million per inmate—more than 64 times the $78,000 average annual cost of incarceration in a maximum-security federal prison in the United States. In July 2019, the prison held 40 inmates. Fifteen of them are considered "highest value" inmates and are housed in the ultrasecretive maximum-security facility "Camp 7" at an annual cost of approximately $120 million, or about $8 million per inmate—more than 102 times the $78,000 average annual cost of incarceration in a maximum-security federal prison in the United States. The remaining 25 non-"highest value" prisoners are held in medium-security confinement facilities that have been upgraded or rebuilt three times. Generally, the costs of incarcerating inmates at Gitmo increase as the number of inmates decline because fixed costs do not change that much.

The facility continues to be funded as an open-ended military necessity, even though the last prisoner arrived March 14, 2008. Driving up the costs are the prison's location and supposedly temporary nature. At Gitmo, everything arrives by barge or aircraft—from paper clips to bulldozers, as well as the revolving staff of 1,850, including guards (who get combat pay), lawyers, interrogators, and contractors. Food, alone, costs about $40 a day for each Gitmo prisoner. As noted, many of the buildings at the prison, for example, the original "temporary" dwellings where the military police currently reside, were built to last 5 years and are falling apart after nearly 2 decades in the salt air and tropical sun. Top Gitmo officials are seeking hundreds of millions of dollars for construction projects needed to improve troops' quality of life and their safety. In recent years, Congress has appropriated $12.4 million for a new dining facility for troops, $8 million for updated medical facilities for prisoners, $14 million to expand war court facilities, and $115 million to build permanent barracks with dormitory-style living to house about one-half of the enlisted personnel. Another new barracks will be needed soon. Top Gitmo officials' number-one funding priority is to replace "Camp 7" with "Camp 8" because "Camp 7" is deteriorating rapidly and hemorrhaging money. "Camp 8" will cost about $69 million.

One of President Obama's campaign promises before his first election in 2008 was to shut down Gitmo—a promise he was unable to keep. Obama was thwarted in his efforts by members of Congress, who would not allocate funding to buy, build, or convert a prison in the United States for terrorists; transfer prisoners to the United States for trials in federal courts instead of trying them at Gitmo by military commissions; or transfer

Joe Raedle/Getty Images

prisoners to other countries willing to accept them. As for fears about incarcerating Gitmo prisoners in U.S. prisons, in 2016, 443 individuals convicted of terrorism-related charges were being held in U.S. prisons, and none of them had escaped.

The Trump administration has told the Pentagon to plan on keeping Gitmo open for 25 years. It is considering sending to Gitmo nearly 1,000 suspected Islamic State fighters currently in the custody of U.S.-backed Syrian Democratic Forces in eastern Syria after the withdrawal of U.S. troops from Syria. They would be the first new prisoners sent to Gitmo since 2008.

Gitmo has been compared to the former Spandau Prison in Germany, which held Nazi war criminals from 1947 until its last prisoner died in 1987. Then the prison was demolished so that it would not become a neo-Nazi shrine. At one time, Spandau Prison, just like Gitmo, was considered the world's most expensive prison to operate.

This chapter addresses institutional corrections, and Gitmo is a federal correctional institution, although it is not the responsibility of the Federal Bureau of Prisons. Are the exorbitant costs of incarceration at Gitmo justified? Should Gitmo be closed, as former President Obama wanted? If yes, what should be done with the remaining inmates? If no, should Gitmo remain in operation until the last inmate dies? Should the suspected Islamic State fighters currently in Syria be sent to Gitmo after U.S. troops are withdrawn from Syria, as the Trump administration proposes? If no, what should be done with them? If yes, should Gitmo remain in operation until the last inmate dies?

Historical Overview of Institutional Corrections

Students often wonder why they must learn about the history of institutional corrections. One reason is that it is impossible to fully understand (and improve) the present state of affairs without knowledge of the past; the present developed out of the past. To paraphrase the philosopher George Santayana, people who fail to remember the past are destined to repeat its mistakes. Another reason is that nothing helps us see how institutional corrections is linked to our larger society and culture better than the study of history.

myth

Throughout history, imprisonment has been the primary sentence for lawbreakers, and it still is today.

fact

Viewed historically, imprisonment is a relatively recent sentence for lawbreaking. Even today in the United States, the number of people in prison is small compared with the number on probation or under other types of supervision in the community.

European Background

In Europe, institutional confinement did not become a major punishment for criminals until the 1600s and 1700s. (In the United States, institutional confinement was not used extensively as a punishment until the 1800s.) As a practice, though, institutional confinement has existed since ancient times. Before the 1600s, however, it usually served functions other than punishment for criminal behavior. For example, confinement was used to:

1. Detain people before trial.
2. Hold prisoners awaiting other sanctions, such as death and corporal punishment.
3. Coerce payment of debts and fines.
4. Hold and punish slaves.
5. Achieve religious indoctrination and spiritual reformation (as during the Inquisition).
6. Quarantine disease (as during the bubonic plague).[1]

Mamertine Prison

Although it surely did not resemble today's prisons, one of the earliest known prisons was the Mamertine Prison, built around 64 B.C. under the sewers of Rome.

Source: Robert Johnson, *Hard Time: Understanding and Reforming the Prison*, 3rd ed. (Belmont, CA: Wadsworth, 2002), 19.

Forerunners of Modern Incarceration Unlike modern incarceration, which strives to change the offender's character and is carried out away from public view, popular early punishments for crime, which predated the large-scale use of imprisonment, were directed more at the offender's body and property; one basic goal was to inflict pain.[2] Furthermore, those punishments commonly were carried out in public to humiliate the offender and to deter onlookers from crime. Examples of such early punishments are fines, confiscation of property, and diverse methods of corporal and capital punishment. Some popular methods of corporal and capital punishment were beheading, stoning, hanging, crucifixion, boiling and burning, flogging, branding, and placement in the stocks or pillory.[3] As this brief list illustrates, the eventual shift to incarceration reduced the severity and violence of punishment.

Two additional forerunners of modern incarceration were banishment and transportation. In essence, they were alternatives to the more severe corporal punishments or capital punishment. Originating in ancient times, **banishment** required offenders to leave the community and live elsewhere, commonly in the wilderness. The modern version of banishment is long-term incarceration (e.g., life imprisonment without opportunity for parole). As population and urban growth displaced frontiers across Europe and as demands for cheap labor increased with the rise of Western capitalism, **transportation** of offenders from their home nation to one of that nation's colonies gradually replaced banishment. England, for instance, was transporting hundreds of convicts a year to North America by the early 1600s.[4] Transportation fell into disuse as European colonies gained independence.

banishment A punishment, originating in ancient times, that required offenders to leave the community and live elsewhere, commonly in the wilderness.

transportation A punishment in which offenders were transported from their home nation to one of that nation's colonies to work.

workhouses European forerunners of the modern U.S. prison, where offenders were sent to learn discipline and regular work habits.

The closest European forerunners of the modern U.S. prison were known as **workhouses** or *houses of correction*. Offenders were sent to them to learn discipline and regular work habits. The fruits of inmate labor also were expected to pay for facility upkeep and even to yield a profit. One of the first and most famous workhouses, the London Bridewell, opened in the 1550s, and workhouses spread through other parts of Europe thereafter. Such facilities were used extensively throughout the next three centuries, coexisting with such responses

Bettmann/Contributor/Getty Images

Besides being painful, placement in the stocks or pillory was intended to humiliate and shame offenders. *Is a greater emphasis on the shame of punishment needed today? If so, how should it be accomplished?*

to crime as transportation, corporal punishment, and capital punishment. In fact, crowding in workhouses was a major impetus for the development of transportation as a punishment.

Reform Initiatives As described in Chapter 3, the Enlightenment was a time of faith in science and reason, as well as a period of humanistic reform. The Enlightenment thinkers and reformers of the 1700s and 1800s described the penal system of their day with such terms as *excessive, disorderly, inefficient, arbitrary, capricious, discriminatory* (against the poor), and *unjust*. Three reformers who were important to initiatives in corrections were Cesare Beccaria (1738–1794), John Howard (1726–1790), and Jeremy Bentham (1748–1832).

Criminologists Graeme Newman and Pietro Marongiu contend that Milanese philosopher Cesare Beccaria's famous book, *On Crimes and Punishments* (1764), although often acclaimed for its originality, actually brought together the reformist principles espoused by other thinkers of the era, such as Montesquieu and Voltaire.[5] One of those principles concerned replacing the discretionary and arbitrary administration of justice with a system of detailed written laws describing the behaviors that constitute crime and the associated punishments. Beccaria believed that people need to know exactly what punishments are prescribed for various offenses if the law is to deter criminal behavior. As part of his quest to deter crime, Beccaria declared that the punishment should fit the crime in two senses: (1) The severity of punishment should parallel the severity of harm resulting from the crime, and (2) the punishment should be severe enough to outweigh the pleasure obtainable from the crime. Furthermore, he believed that punishment needed to be certain and swift to deter crime. Certainty implies that the likelihood of getting caught and punished is perceived as high. Swiftness implies that punishment will not be delayed after commission of the crime.

Beccaria did not ground his thinking firmly in empirical observations and did little to actively campaign for the reforms he advocated.[6] The work of John Howard, an English sheriff and social activist, presents an interesting contrast in that regard. Howard's 1777 book, *The State of the Prisons in England and Wales*, was based on his visits to penal institutions in various parts of Europe. The crowding, overall poor living conditions, and disorderly and abusive practices he observed in those facilities appalled him. He advocated that penal environments be made safe, humane, and orderly. Howard's opinion was that incarceration should do more than punish–that it also should instill discipline and reform inmates. Toward that end, he proposed an orderly institutional routine of religious teaching, hard work, and solitary confinement to promote introspection and penance.[7] Howard's work inspired the growing popularity of the term *penitentiary* to refer to penal confinement facilities.

In **penology**, the study of prison management and the treatment of offenders, English philosopher Jeremy Bentham perhaps is best remembered for his idea that order and reform

myth

The reason punishment fails to adequately deter crime in the United States is that it is not severe enough.

fact

The United States has a higher rate of imprisonment and longer sentences than virtually any other nation. It also is one of the few advanced, industrialized nations to have retained the death penalty. It is hard indeed to support the argument that our punishment is not severe enough. Certainty and swiftness, however, are lacking, and that is the failure to which Beccaria probably would point.

penology The study of prison management and the treatment of offenders.

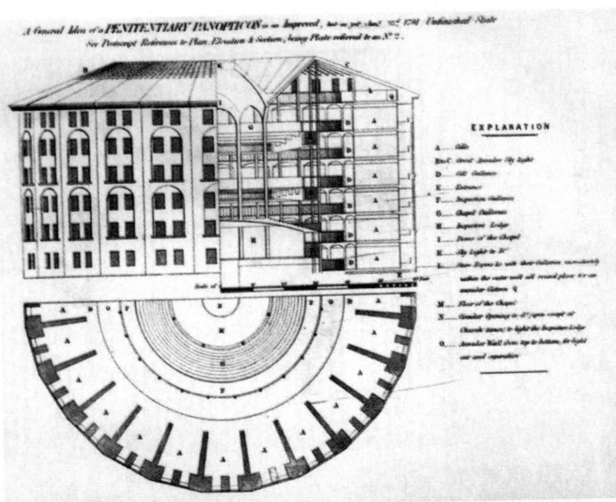

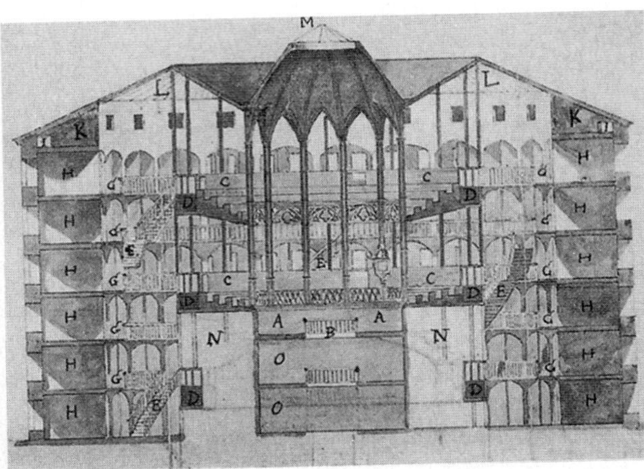

British Library/The Image Works; Historia/Shutterstock; Mary Evans/Higginbotham Collection/The Image Works

Stateville Correctional Center in Illinois is similar in design to Bentham's panopticon plan. *Why hasn't this prison design been used more widely in the United States?*

could be achieved in a prison through architectural design. His **panopticon** ("all-seeing" or "inspection-house") prison design consisted of a round building with tiers of cells lining the circumference and facing a central inspection tower so that staff from the tower could watch prisoners. Although no facilities completely true to Bentham's panopticon plan were ever constructed, structures similar in design were erected at Illinois's Stateville Penitentiary (now Stateville Correctional Center), which opened in 1925.

In sum, the historical roots of the modern prison lie in Europe. It was in America, however, that the penitentiary concept was first put into wide practice.

Developments in the United States

In colonial America, penal practice was loose, decentralized, and unsystematic, combining private retaliation against wrongdoing with fines, banishment, harsh corporal punishments, and capital punishment. Local jails were scattered about the colonies, but they were used primarily for temporary holding rather than for punishment.[8] Some people, such as William Penn, the famous Quaker and founder of Pennsylvania, promoted incarceration as a humane alternative to the physically brutal punishments that were common. However, that idea was largely ignored because there was no stable, central governmental authority to coordinate and finance (through tax revenue) the large-scale confinement of offenders.

panopticon A prison design consisting of a round building with tiers of cells lining the inner circumference and facing a central inspection tower.

The Penitentiary Movement In the aftermath of the American Revolution, it rapidly became apparent that the colonial system of justice would not suffice. Economic chaos and civil disorder followed the war. Combined with population growth and the transition from an agricultural society to an industrial one, they created the need for a strong, centralized government to achieve political and economic stability. The rise of the penitentiary occurred in that context.[9] Philosophically, it was guided by Enlightenment principles. In 1790, the Walnut Street Jail in Philadelphia was converted from a simple holding facility to a prison to which offenders could be sentenced for their crimes. It commonly is regarded as the nation's first state prison. In a system consistent with Howard's plan, its inmates labored in solitary cells and received large doses of religious teaching. Later in the 1790s, New York opened Newgate Prison. Other states quickly followed suit, and the penitentiary movement was born. By 1830, Pennsylvania and New York had constructed additional prisons to supplement their original ones.

Pennsylvania and New York pioneered the penitentiary movement by developing two competing systems of confinement.[10] The **Pennsylvania system**, sometimes called the separate system, required that inmates be kept in solitary cells so that they could study religious writings, reflect on their misdeeds, and perform handicraft work. In the New

Pennsylvania system An early system of U.S. penology in which inmates were kept in solitary cells so that they could study religious writings, reflect on their misdeeds, and perform handicraft work.

York system, or the **Auburn system** (named after Auburn Penitentiary and also referred to as the congregate or silent system), inmates worked and ate together in silence during the day and were returned to solitary cells for the evening. Ultimately, the Auburn system prevailed over the Pennsylvania system as the model followed by other states. It avoided the harmful psychological effects of total solitary confinement and allowed more inmates to be housed in less space because cells could be smaller. In addition, the Auburn system's congregate work principle was more congruent with the system of factory production emerging in wider society than was the outdated craft principle of the separate system. If prison labor was to be profitable, it seemed that the Auburn plan was the one to use.

It is interesting that although penitentiary construction flourished and the United States became the model nation in penology during the first half of the nineteenth century, there was serious discontent with the penitentiary by the end of the Civil War. There were few signs penitentiaries were deterring crime, reforming offenders, or turning great profits from inmate labor. In fact, prisons were becoming increasingly expensive to run, and opposition was growing to selling prisoner-made goods on the open market. With faith in the penitentiary declining, the stage was set for a new movement—a movement that, rather than challenging the fundamental value of incarceration as a punishment, sought to improve the method of incarceration.

Bettmann/Contributor/Getty Images

The Elmira Reformatory, which opened in 1876 in Elmira, New York, was the first institution for men that was based on reformatory principles. *What caused the change in penal philosophy?*

The Reformatory Movement The reformatory movement got its start at the 1870 meeting of the National Prison Association in Cincinnati. The principles adopted there were championed by such leaders in the field as Enoch Wines (1806–1879) and Zebulon Brockway (1827–1920).[11] A new type of institution, the reformatory, was designed for younger, less hardened offenders between 16 and 30 years of age. Based on a military model of regimentation, it emphasized academic and vocational training in addition to work. A classification system was introduced in which inmates' progress toward reformation was rated. The sentences for determinate periods of time (e.g., 5 years) were replaced with indeterminate terms in which inmates served sentences within given ranges (e.g., between 2 and 8 years). Parole or early release could be granted for favorable progress in reformation.

It has been observed that indeterminate sentences and the possibility of parole facilitate greater control over inmates than determinate sentences do. Many inmates are interested, above all else, in gaining their freedom. The message conveyed by indeterminate sentences and the possibility of parole is this: "Conform to institutional expectations or do more time."

Auburn system An early system of penology, originating at Auburn Penitentiary in New York, in which inmates worked and ate together in silence during the day and were placed in solitary cells for the evening.

Institutions for Women Until the reformatory era, there was little effort to establish separate facilities for women. Women prisoners usually were confined in segregated areas of male prisons and generally received inferior treatment. The reformatory movement, reflecting its assumptions about differences between categories of inmates and its emphasis on classification, helped feminize punishment.[12] The first women's prison organized according to the reformatory model opened in Indiana in 1873. By the 1930s, several other women's reformatories were in operation, mainly in the Northeast and the Midwest. Most employed cottages or a campus and a family-style living plan, not the cell-block plan of men's prisons. Most concentrated on molding inmates to fulfill stereotypical domestic roles upon release, such as cleaning and cooking.

Twentieth-Century Prisons Criminologist John Irwin has provided a useful typology for summarizing imprisonment in the last century.[13] According to Irwin, three types of institutions have been dominant. Each has dominated a different part of the century. The dominant type for about the first 3 decades was the "big house." In Irwin's words:

> The Big House was a walled prison with large cell blocks that contained stacks of three or more tiers of one- or two-man cells. On the average, it held 2,500 men. Sometimes a single cell block housed over 1,000 prisoners in six tiers of cells. Most of these prisons were built over many decades and had a mixture of old and new cell blocks. Some of the older cell blocks were quite primitive.[14]

FYI **Native American Prisons**

In the 1870s, the Cherokees constructed a national prison at Tahlequah, Oklahoma. (In 1838, 16,000 Cherokees were forcibly removed from their ancestral lands in the Southeastern United States over the "Trail of Tears" to what became the state of Oklahoma. Four thousand Cherokees died on the journey.) The purpose of the national prison was not only to incarcerate the dangerous criminal but also to teach the criminal a trade to enable him to make an economic contribution to the tribe. The success of the national prison encouraged a movement to build district jails to hold prisoners awaiting trial, but none was ever built. Instead, marshals and sheriffs guarded prisoners who accompanied them on horseback throughout the district in a form of police custody.

Source: Rennard Strickland, *Fire and the Spirits: Cherokee Law from Clan to Court* (Norman, OK: University of Oklahoma Press, 1975), 67, 168–169, 172–173.

Big-house prisons, which consisted of large cell blocks containing stacks of cells, were the dominant prison design of the early twentieth century. *What are some problems with the big-house prison design?*

Photographs in the Carol M. Highsmith Archive, Library of Congress, Prints and Photographs Div

Note that big houses were not new prisons distinct from earlier penitentiaries and reformatories. They were the old penitentiaries and reformatories expanded in size to accommodate larger inmate populations. Originally, big-house prisons exploited inmate labor through various links to the free market. Industrial prisons predominated in the North, while plantation prisons dominated much of the South. With the rise of organized labor and the coming of the Great Depression, free-market inmate labor systems fell into demise during the 1920s and 1930s. Big houses became warehouses oriented toward custody and repression of inmates.

What Irwin calls the "correctional institution" arose during the 1940s and became the dominant type of prison in the 1950s. Correctional institutions generally were smaller and more modern in appearance than big houses. However, correctional institutions did not replace big houses; they simply supplemented them, although correctional-institution principles spread to many big houses. Correctional institutions emerged as penologists turned to the field of medicine as a model for their work. During that phase of corrections, a so-called **medical model** came to be used because crime was viewed as symptomatic of personal illness in need of treatment. Under the medical model, shortly after being sentenced to prison, inmates were subjected to psychological assessment and diagnosis during classification processes. Assessment and diagnosis were followed by treatment designed to address the offender's supposed illness. The main kinds of treatment, according to Irwin, were academic and vocational education and therapeutic counseling. After institutional treatment came parole, which amounted to follow-up treatment in the community. Importantly, the ways of achieving control over inmate behavior shifted from the custodial repression typical of the big house to more subtle methods of indirect coercion: Inmates knew that failure to participate in treatment and exhibit "progress" in prison meant that parole would be delayed.

medical model A theory of institutional corrections, popular during the 1940s and 1950s, in which crime was seen as symptomatic of personal illness in need of treatment.

During the 1960s and 1970s, both the effectiveness and the fairness of coerced prison rehabilitation programming began to be challenged,[15] and the correctional institution's dominance began to wane. In Irwin's view, the third type of prison, the "contemporary violent prison," arose by default as the correctional institution faded. Many of the treatment-program control mechanisms of the correctional institution were eliminated. Further, many of the repressive measures used to control inmates in the big house became illegal after the rise of the inmates' rights movement during the 1960s (to be discussed later). In essence, what emerged in many prisons was a power vacuum that was filled with inmate gang violence and interracial hatred.

The Federal Prison System The Federal Prison System was founded in 1891, when Congress passed the "Three Prisons Act." The legislation established the first three prisons

(U.S. Penitentiaries) at Leavenworth, Kansas; Atlanta, Georgia; and McNeil Island, Washington. The three prisons received minimal oversight by the Department of Justice.[16] The Justice Department created the Federal Bureau of Prisons (BOP) in 1930, during the Hoover Administration. The person most responsible for the BOP's formation was Assistant Attorney General Mabel Walker Willebrandt. In the 1920s, Ms. Willebrandt was tasked with the creation of separate institutions for juvenile offenders and women. The juvenile institution, the Federal Reformatory at Chillicothe, Ohio, was authorized in 1923, and opened in 1926. The women's institution, the Federal Reformatory at Alderson, West Virginia, was authorized in 1924, and opened in 1928.[17] With these new institutions, Ms. Willebrandt realized that a centralized administration and standardized regulations were needed, so she successfully lobbied for a new Justice Department agency–the BOP–to oversee the Federal Prison System.[18] At the time of the BOP's creation, there were 11 federal prisons (this source previously reported 7 federal prisons), each separately funded and each operated under policies and procedures established by its warden.[19]

By the end of 1930, the BOP operated 14 facilities for approximately 13,000 inmates, and, by the end of 1940, the BOP had grown to 24 facilities with 24,360 inmates.[20] In 1932, the BOP opened the first penitentiary entirely built by the Bureau at Lewisburg, Pennsylvania. USP Lewisburg featured an original design and introduced many new correctional ideas, such as housing inmates with different custody levels in the same institution.[21] The following year, the BOP opened its first medical facility at Springfield, Missouri, and began its enduring partnership with the U.S. Public Health Service.[22] Then, in 1934, the BOP opened its first maximum-security prison, USP Alcatraz, on Alcatraz Island in San Francisco Bay. Alcatraz was designed to confine "the most violent, disruptive, and escape-prone inmates in the Federal System."[23]

With only a few exceptions, the number of federal inmates did not fluctuate much between 1940 and 1980, when the inmate population stood at 24,252.[24] However, between 1940 and 1980, the number of facilities nearly doubled from 24 to 44, primarily because the BOP changed its philosophy from operating large facilities that confined inmates of many custody levels to operating smaller facilities that confined inmates with similar custody needs.[25] During the 1980s, the number of federal inmates increased dramatically because of federal law enforcement efforts and new criminal justice legislation, such as The Sentencing Reform Act of 1984, which established determinate sentencing, abolished parole, and reduced good time.[26] Also, in 1986, 1988, and 1990, several mandatory minimum sentencing provisions were enacted.[27] As a result, from 1980 to 1989, the number of federal inmates more than doubled from about 24,000 to nearly 58,000.[28] During the 1990s, the inmate population more than doubled again to about 136,000 at the end of 1999, primarily because of increased efforts to combat illegal drugs and illegal immigration.[29] For the same reasons, the federal inmate population continued to increase from 2000 until 2013, when

Photographs in the Carol M. Highsmith Archive, Library of Congress, Prints and Photographs Div

Alcatraz. On March 21, 1963, the BOP closed Alcatraz after 29 years of operation. *Was Alcatraz a good idea? Why or why not?*

it peaked at 219,298.[30] On July 18, 2019, the BOP inmate population was 180,248–a nearly 18% decrease from the peak.[31]

Privatization As will be readily apparent in the next section of this chapter, the three and a half decades between 1973 and 2008, are likely to be remembered for the largest incarceration boom in American history and for desperate attempts to deal with prison crowding by developing alternatives to traditional incarceration.

The principal alternative to traditional confinement has undoubtedly been **privatization**, the involvement of the private sector in the construction and operation of confinement facilities. The private sector has a long tradition in institutional corrections. For instance, such diverse services as food, legal aid, medical and psychiatric care, and education have long been provided through private vendors.

Currently, a handful of American corporations annually make billions of dollars providing products and services to American prisons, generally with little oversight. Consider the following examples:

- Two corporations–Securus Technologies and Global Tel Link (GTL)–manage approximately 80% of inmate phone calls. Both corporations are known for charging inmates outrageous fees. According to court documents, in 2016, Securus charged inmates as much as $24.95 for a 15-minute intrastate call ($5.35 for the first minute). Inmates also were charged extra to open and close phone accounts and to fund them in the first place.[32]

privatization The involvement of the private sector in the construction and the operation of confinement facilities.

- Corizon Health is believed to be the largest private provider of medical services to prisons. In Arizona, for example, Corizon is paid $15.16 per inmate per day for medical staffing. However, according to a whistleblower complaint in 2018, Corizon violated state regulations, purposefully misled state auditors, falsified records, and put patient lives at risk. Although Corizon's contract with the state guaranteed 90% staffing fulfillment, medical units typically were only 50% staffed. Moreover, many of the nurses who were hired had little or no experience. They were promised two weeks of training but often received only two days of training. Many of the nurses could not start an IV or operate an oxygen tank. Staffing shortages contributed to inmates being given the wrong medication or no medication at all. Diabetics frequently did not get their insulin when it was needed. Medical units often lacked necessary resources, such as working oxygen tanks or an EKG machine. Administrators routinely refused to send inmates to hospitals or another facility that could provide proper care. In 2019, the Arizona Department of Corrections selected a new health care vendor, Centurion Managed Care.[33]
- Two corporations–Aramark Correctional Services and Trinity Services Group–are two of the largest private prison food services providers. In 2013, Aramark was awarded a 3-year $145 million contract to provide food services to Michigan prisons, replacing 370 state employees. Persistent problems arose almost from the beginning, including inadequate staffing levels, meal shortages, unauthorized menu substitutions, maggots in the kitchen, the smuggling of drugs and other contraband by Aramark employees, and Aramark workers engaging in sex acts with inmates. In 2015, Aramark, after only 2 years of their 3-year contract, was replaced by Trinity, which received a 3-year $158 million contact. However, the problems with Aramark continued under Trinity, so, in 2018, Michigan Governor Rick Snyder decided to end the state's 4-year experiment with privatizing its prison food service and return to using state employees.[34]

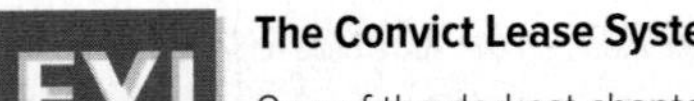

The Convict Lease System

One of the darkest chapters in the history of American corrections involves the convict lease system that was adopted by many Southern states following the Civil War. It perpetuated slavery in the states that embraced it. Counties and states leased thousands of prisoners (mostly black) to private individuals and companies to work (and die) in cotton fields, mines, and forests of the Deep South. Governments made millions of dollars in lease payments, and those who leased prisoners and brutally exploited them made millions more. The practice did not end until the 1930s.

Source: Fletcher M. Green, "Some Aspects of the Convict Lease System in the Southern States," in *Essays in Southern History*, vol. 31 (Durham: University of North Carolina Press, 1949).

- The largest private provider of transportation for jails and prisons is Prisoner Transportation Services (PTS). In 2016, PTS priced its services to Nevada at $1.05 per adult per mile. Rates for minors or those with mental disabilities were higher. The minimum trip fee was $350. During the past few years, 14 women, who are transported with men, claimed to have been sexually assaulted by transportation guards, and at least four people have died while being transported in PTS vehicles.[35]

As noted in the FYI above, there is a rich history of private labor contracting in the operation of prisons, and it is now witnessing something of a revival in certain jurisdictions. Also, the private sector has operated juvenile institutions for many years. But mounting prison populations (until 2008), combined with space and budget limitations, have helped give privatization new twists. One of those twists entails having the private sector finance construction of institutions under what amounts to a lease-purchase agreement. The Potosi Correctional Center in Missouri was constructed under such a strategy. In another twist, the state contracts with private companies like Corrections Corporation of America (now CoreCivic) to have them operate prisons. One of the earliest privately operated state prisons for adult felons, Kentucky's minimum-security Marion Adjustment Center, was opened in January 1986.

At the beginning of 2018, 27 states and the federal prison system reported that 121,420 prison inmates (8.2% of all state and federal prison inmates) were held in privately operated facilities. Private facilities confined 7.2% of all state prisoners and 15.1% of all federal prisoners. Even though the number of state and federal prisoners held in private facilities between the beginning of 2017 and the beginning of 2018 decreased 0.3% and 19.3%, respectively, since 2000, the number of state prison inmates held in private facilities has increased about 24%, from 5.8% to 7.2% of state prison inmates, while the number of federal prison inmates in private facilities has increased about 41%, from 10.7% to 15.1% of federal prison inmates. The federal prison system with 27,569 inmates, Texas with 13,692 inmates, and Florida with 12,176 inmates had the largest number of prison inmates held in private facilities at the beginning of 2018. States with the largest percentage of their prison populations in private facilities at the beginning of 2018 were Montana (38.1%), Hawaii (28.5%), Tennessee (26.3%), Oklahoma (26.1%), and Arizona (19.7%).[36]

In 2012 (the latest year for which data were available), private contractors operated about 11% of all adult state prisons (136/1,223), 42% of all juvenile facilities (272/651), and 11% of all federal prisons (15/138).[37] The vast majority of privately operated prisons were either minimum- or medium-security facilities. Few of them were maximum-security institutions, which gives rise to the criticism that private prisons are relatively successful because they do not house the most difficult and dangerous inmates.[38] The number of detention facilities operated by private contractors has decreased since 2012.

State and federal prisons represent only a small part of private contactors' correctional-institution inventory. Private contractors operate many more community-based and juvenile institutions. They also operate county and local jails; detention facilities for the U.S. Marshals Service and the Bureau of Immigration and Customs Enforcement; and correctional institutions in other countries, such as Canada, Australia, South Africa, and the United Kingdom.

Two corporations dominate the industry: CoreCivic (formerly Corrections Corporation of America) and The GEO Group, Inc. (formerly Wackenhut Corrections Corporation). CoreCivic founded the private corrections industry in 1983, and has its principal executive offices in Nashville, Tennessee.[39] The GEO Group's U.S. Corrections & Detention component was established in 1984, and its headquarters are in Boca Raton, Florida.[40] In 1997, The GEO Group became the first private company hired by the BOP to manage one of its major facilities: the 2,048-bed Taft Correctional Institution in Taft, California. In 2018, The GEO Group was the largest private prison owner in the United States based on total revenue, number of correctional and detention facilities, and total design capacity (beds). The GEO Group and CoreCivic, respectively, are the fifth and the sixth largest correctional systems in the United States, behind only the federal system and the systems of Texas, California, and Florida.

The GEO Group, as noted, has become the nation's largest private prison owner in the United States, recently surpassing CoreCivic, primarily in two ways: (1) the acquisition of other smaller private prison and related-service companies and (2) by lobbying and making campaign contributions at all levels of government. Since 2005, The GEO Group has purchased 10 private prison or related-service companies for approximately $2.4 billion.[41] As for its political activities, in 2017, The GEO Group spent $1.7 million on lobbying, "the highest amount on record for a private prison contractor." In the first quarter of 2018, The GEO Group spent an additional $380,000 on lobbying.[42] During the 2018 election cycle, The GEO Group contributed about $1.3 million dollars: $80,645 or 6% went to Democrats, $733,942 or 58% went to Republicans, and $480,000 or about 38% went to Political Action Committees (PACs) that mostly supported Republican candidates.[43]

In 2018, The GEO Group and CoreCivic reported, respectively, about $2.3 billion and $1.8 billion in total revenue. Among their U.S. customers were various state, county, and local governments and, at the federal level, the BOP, the U.S. Marshals Service (USMS), and U.S. Immigration and Customs Enforcement (ICE). CoreCivic reported that payments by state, county, and local governments constituted 39% of its total revenue or $706.8 million in 2018, and that no state, county, or local partner accounted for 10% or more of total

FYI **Federal Policy Reversal on Private Prisons**

On August 18, 2016, then-Deputy Attorney General Sally Yates sent a memo to the Acting Director of the Bureau of Prisons ordering the gradual phase out of all privately operated prisons in the BOP. She wrote, "Private prisons served an important role during a difficult period, but time has shown that they compare poorly to our own Bureau facilities. They simply do not provide the same level of correctional services, programs, and resources; they do not save substantially on costs; and as noted in a recent report by the Department's Office of Inspector General, they do not maintain the same level of safety and security. The rehabilitative services that the Bureau provides, such as educational programs and job training, have proved difficult to replicate and outsource—and these services are essential to reducing recidivism and improving public safety." On January 30, 2017, President Trump fired then-acting Attorney General Yates for another matter, and, on February 21, 2017, in a one-paragraph memo, then-Attorney General Jeff Sessions rescinded Yates' order on private prisons, directing the BOP "to return to its previous approach."

Sources: "Justice Department memo announcing the end of its use of private prisons," *The Washington Post*. Accessed February 3, 2019, https://apps.washingtonpost.com/g/documents/national/justice-department-memo-announcing-announcing-the-end-of-its-use-of-private-prisons/2127/; Evan Perez and Jeremy Diamond, "Trump fires acting AG after she declines to defend travel ban," *CNN Politics*. Accessed February 3, 2019, https://www.cnn.com/2017/01/30/politics/donald-trump-immigration-order-department-of-justice/index.html; "Justice Department will again use private prisons," *The Washington Post*. Accessed February 3, 2019, https://www.washingtonpost.com/world/national-security/justice-department-will-again-use-private-prisons/2017/02/23/da395d02-fa0e-11e6-be05-1a3817ac21a5_story.html?noredirect=on&utm_term=.37b5a449e047.

revenue. The aforementioned federal agencies accounted for 48% or $890.5 million of CoreCivic's total revenue in 2018: 25% was from ICE, 17% from USMS, and 6% from BOP. The GEO Group did not report comparable data.

In 2018, The GEO Group owned and/or operated 69 correctional and detention facilities in the United States with a total design capacity of 74,746 beds. These numbers represent an increase of 19% in facilities and 12.4% in beds from 2015, when The GEO Group reported owning and/or operating 58 correctional and detention facilities in the United States with about 66,500 beds. CoreCivic, in 2018, operated 51 correctional and detention facilities in the United States, 44 of which it owned, with a total design capacity of 72,833 beds. In 2015, CoreCivic operated and/or managed 77 facilities in the United States with approximately 88,500 beds, which were decreases from 2015, of 33.8% in facilities and nearly 25% in beds.[44] In its annual report, CoreCivic also noted that, as of December 31, 2018, it had approximately 9,800 beds at eight prison facilities that were vacant and immediately available for use. The GEO Group did not report comparable data.

To expand their businesses and to hedge against a decrease in prison and jail populations, private prison companies have branched out into other services. For example, in 2018, The GEO Group received 64% of its total revenue (or approximately $1.5 billion) from its U.S. Corrections & Detention business segment, but it also earned 25% of its total revenue (about $580 million) from its GEO Care business unit. In 2018, GEO Care served nearly 700,000 individuals and an average daily census of 200,000 participants in community reentry, youth services treatment, and electronic monitoring programs.

GEO Reentry supplies resources necessary to help individuals productively transition back into society. GEO has approximately 50 reentry centers (or "halfway houses," which are described in Chapter 12) that provide federal and state parolees and probationers with temporary housing, employment assistance, rehabilitation, substance abuse counseling, and vocational and education programs. Since 1996, GEO also has operated non-residential day reporting centers (described in Chapter 12) for state and county correctional agencies. Among other services, the day reporting centers offer evidence-based cognitive behavioral treatment programs that provide behavioral assessments, treatment, supervision, and education. In 2018, GEO operated 66 of these non-residential reentry centers with the capacity to serve more than 5,200 parolees and probationers throughout the nation.

The GEO Care's Abraxas division began in Pennsylvania in 1973 to provide residential, shelter care, and alternative education programs for youths in the juvenile justice system (which is discussed more fully in Chapter 13). Programs are tailored to youths in need of mental health, behavioral health, and drug and alcohol treatment. In 2018, GEO had about 1,200 youth service beds in 10 residential facilities, as well as non-residential programs.

Through its wholly owned subsidiary, BI Incorporated, which was founded in 1978, and purchased by The GEO Group in 2011 for $415 million, GEO Care has become the leading provider of community supervision and electronic monitoring services (which are discussed further in Chapter 12) to federal, state, and local governments with a presence in all 50 states. In 2018, BI supervised about 144,000 parolees, probationers, and pretrial defendants using electronic monitoring technologies. Also, through the Intensive Supervision and Appearance Program (intensive supervision probation and parole also is discussed more fully in Chapter 12), a core component of the Department of Homeland Security's Alternatives to Detention program, in 2018, BI provided 49,000 immigrants with case management, community supervision, and monitoring services for U.S. Immigration and Customs Enforcement.

In 2007, The GEO Group established its in-house transportation division, GEO Transport, Inc. (GTI), which, in 2018, securely transported more than 783,000 inmates for federal, state, and local customers in the United States and internationally. GTI operates a fleet of about 350 customized and U.S. Department of Transportation-compliant land and air vehicles.

In 2017, CoreCivic started a reentry business of evidence-based and customizable reentry programs, including academic education, vocational training, substance abuse treatment, life skills training, and faith-based programming. CoreCivic's proprietary reentry process and cognitive/behavioral curriculum, "Go Further," promises a comprehensive approach to addressing the barriers to a successful return to society. In 2018, through its CoreCivic Community division, CoreCivic owned and operated 26 residential reentry centers with a total design capacity of 5,214 beds, making CoreCivic "the second largest community corrections owner and operator in the United States."

In addition to providing fundamental residential services, CoreCivic's correctional, detention, and residential reentry facilities provide a range of rehabilitation and educational programs, including basic education, faith-based services, life skills and employment training, and substance abuse treatment. These services are intended to help reduce recidivism and to prepare offenders for their successful reentry into society upon their release. CoreCivic also offers or makes available to offenders certain health care (including medical, dental, and mental health services), food services, and work and recreational programs.

CoreCivic, like The GEO Group, also has inmate transportation (TransCor), electronic monitoring, and immigrant detention businesses.

The GEO Group and CoreCivic have seen businesses expand significantly as a result of increased ICE detention of immigrants. These two companies constitute almost a duopoly within the private prison system. They received the lion's share of the estimated $807 million paid by ICE to private prison corporations in fiscal year 2018. How have they been so fortunate? In 2016, The GEO Group and CoreCivic through subsidiaries contributed $250,000 each to President Trump's inauguration, and The GEO Group contributed another $281,360 to President Trump's presidential campaign and $225,000 to Rebuilding America Now, a super PAC that supported President Trump's campaign. In addition, The GEO Group spent $1.2 million on federal lobbying in 2016, compared to $120,000 in 2004. Between 2008 and 2014, CoreCivic spent nearly $10 million just to lobby the House appropriations subcommittee that controls immigration-detention funding. Perhaps not coincidentally, CoreCivic has seen a 935% increase in federal contract awards from 2016 to 2017, to $388 million, and The GEO Group has received half a billion dollars in federal payments.[45]

Because of government secrecy, the exact size and scope of the immigration prison industry is unknown. The $807 million figure is based on pricing data for 19 privately owned or operated detention centers. The true number of ICE detention centers probably is between 200 and more than 600, and the percentage of them that are privately owned or operated is unknown (estimates suggest 10%). Other corporations that profit from ICE detention are LaSalle Corrections, Management & Training Corporation, and Immigration Centers of America. More than a record high 50,000 undocumented immigrants currently are being held in detention by ICE, and the majority are in private facilities.[46]

The GEO Group detained approximately 11,000 immigrants at 17 prisons in 2018, at an average cost of about $101 per prisoner per day. By comparison, ICE's overall projected daily rate for adult beds in fiscal year 2018 was $121.90. The Government Accountability Office, however, warned that ICE "consistently lowballs its detention costs through dubious accounting."[47]

ICE's internal detention standards set pay for "voluntary" immigrant labor at only "at least $1.00 (USD) per day." ICE contends that "voluntary" immigrant labor decreases idleness, improves morale, and decreases disciplinary incidents and, thus, mitigates the negative impact of confinement, which, ironically, is supposed to be administrative and not punitive. Lawsuits over the past few years suggest that immigrant labor within private prisons is not as "voluntary" as the corporations claim but, instead, is "slave labor." A lawsuit against The GEO Group in Colorado, for example, argues that immigrant detainees were forced to mop floors, clean windows, wipe down mattresses, and clean dining areas under the threat of solitary confinement. Plaintiffs' attorneys asserted that the work detainees were forced to perform allowed The GEO Group to maintain its entire facility with just one janitor on the payroll. A lawsuit against The GEO Group in California charges "systematic and unlawful wage theft, unjust enrichment, and forced labor," including an arrangement in which the corporation requires work to "buy the basic necessities–including food, water, and hygiene products–that GEO refuses to provide for them." A lawsuit in Washington State against The GEO Group charges that GEO pays immigrant laborers in "snack food" instead of $1 a day. Lawsuits against The GEO Group also claim that GEO uses immigration detainees to run "virtually all non-security functions" in its detention centers. The GEO Group has vigorously contested the suits, winning some and losing some.[48]

Proponents of states' contracting to have the private sector finance construction and operate prisons often point to efficiency, flexibility, and cost effectiveness. Opponents frequently worry about liability issues, about creating a profit motive for incarcerating people, and about the incentive to trim inmate services and programs to maximize profits.

Among the advantages of private prisons cited by proponents are the following[49]:

1. Private prisons can be constructed and opened more quickly. Following contract awards, private prisons can be constructed within 12 to 18 months; similar facilities constructed by governments usually take 36 to 48 months to be brought "online."
2. Construction cost savings of 25% or more and operating cost savings of 10–15% are common for private facilities.

3. The quality of services and programs in private facilities frequently is superior to that in publicly operated facilities.
4. Private contractors frequently can deliver correctional services and products more cost effectively than public employees can. Private companies often can negotiate lower rates for necessary items and save money in many other ways.
5. Privatization can reduce substantially legal liability costs of operating prisons and jails.
6. Necessary changes in the nature and scope of prisoner programs can be made quicker and easier in private facilities.
7. The existence of private facilities in a jurisdiction sometimes encourages improvements in the public facilities in that jurisdiction.
8. Prisoners in private facilities have a broader array of legal remedies when challenging conditions of confinement than prisoners in public facilities.
9. For-profit prisons create economic opportunities. The privatization of prisons creates job opportunities on numerous levels for a community. Jobs are available in the prison, and service industry jobs are required to support that population. In total, the private prison economy in the United States has an estimated annual economic impact of approximately $80 billion.
10. Private prisons can reduce overcrowding in publicly operated federal and state prisons. By opening for-profit prisons, per-facility population levels can be lowered, so prisoners can experience a better quality of life.
11. Prisoners from for-profit prisons may reoffend at lower rates. Although prisoners in for-profit prisons may serve longer sentences, they may spend less overall time behind bars because they have access to more resources inside and outside of prison.
12. Private prisons can be used for more than housing prisoners. One of the most common uses for for-profit detention facilities is to accommodate immigration detention needs. They also can be used or altered to accommodate many different community needs. Some have been turned into museums or converted into administrative offices. In Portland, Oregon, a vacant 525-bed prison facility has been used for filming television shows and movies while serving as a headquarters for the local anti-prison movement.
13. Privatization in prisons has a history of proven results. For decades, everything from food preparation to health care services has been contracted successfully to private companies in the United States.
14. Private prisons can provide an entry-level law enforcement opportunity. Even with a criminal justice degree, finding work in law enforcement can be difficult. Many agencies want experienced personnel that can begin working immediately. For-profit facilities offer an entry-level position for correctional officers where that experience can be obtained. It is an opportunity to learn new skills, apply the knowledge from a criminal justice degree for the first time, and get comfortable. That can create some challenges because of inexperience, but the benefits often outweigh the risks.

myth

Private prisons are the corrupt heart of mass incarceration.

fact

Less than 8% of all incarcerated people are held in private prisons. Private prisons are essentially a parasite on the massive publicly owned system—not the root of it.

Source: Wendy Sawyer and Peter Wagner, "Mass Incarceration: The Whole Pie 2019," Prison Policy Initiative, March 19, 2019, accessed July 21, 2019, https://www.prisonpolicy.org/reports/pie2019.html.

Critics of the private prison industry paint a very different picture. They argue that the supposed cost savings of private prisons are illusory and that cost-savings studies invariably exclude corporate subsidies and many operating costs. They also contend that favorable contracts allow private prison companies to bilk governments for services not performed. Another point to consider is that, recently, both The GEO Group and CoreCivic changed their corporate structures and have become Real Estate Investment Trusts (REITs) for tax avoidance purposes. Among the disadvantages of private prisons cited by opponents are the following[50]:

1. Desperate for economic development, many communities offer to help fund private prison companies through tax breaks, infrastructure subsidies, and other benefits. Research has shown that taxpayers subsidized 78% of CoreCivic and 69% of The GEO Group's prisons. That money, of course, could have been used for education, health care, or a wide variety of other beneficial projects.
2. Private prison companies could create a system of dependency. When governments are reliant on private companies to provide needed services, the potential for a destructive dependency becomes possible. For-profit companies could use that dependency as leverage to negotiate higher compensation rates. A common method of negotiation is to offer services at lower costs, create a monopoly for those services, and then increase prices to maximize profits. This is a very real possibility if prison privatization continues to be utilized.

3. Prison privatization reduces the level of transparency. Government and public-sector agencies are expected to operate with a certain level of transparency. That makes the government minimally accountable for its actions to the people it serves. For-profit companies are not held to the same standard. The privatization of prisons could create a system where inmates are not treated ethically, but no one would ever know because the company running the facility would not be required to report anything.
4. Private prison companies protect their interests by lobbying and contributing to political campaigns. Since for-profit companies need prisoners to make money, they may lobby legislative bodies for longer standard sentencing guidelines or mandatory-minimum sentences, for example. Some for-profit companies may even lobby local law enforcement and prosecutors to charge people with higher-level crimes on the chance that they will receive a longer sentence that can be served within their prison.
5. Prison privatization creates the potential for bribery and corruption. In February 2009, two Pennsylvania judges, Mark Ciavarella and Michael Conahan, pleaded guilty in federal court to a "Kids for Cash" corruption scandal in which they took $2.6 million as payments from the operator of privately run juvenile detention centers to send juvenile offenders to his centers.[51]
6. Private prison company contracts typically specify a per diem rate of pay based on the average daily census for the month. They also guarantee payment for a minimum number of inmates. This results in paying for nonexistent inmates, referred to as "ghost inmates." Moreover, many contracts have a clause that significantly increases the per diem if the facility exceeds 90% of inmate capacity. That creates an incentive for governments to maintain inmate capacity below 90%, which produces more "ghost inmates" and makes taxpayers pick up the tab for empty beds. Private prison companies profit either way.
7. Private prison company contracts customarily include staffing-level requirements; some of them specify the type of staff (e.g., supervisors, shift commanders, line officers) as well. Companies have been found to bill for "ghost employees," similar to the "ghost inmates" just described. They also have been found to bill for supervisors when, in fact, the employees were line officers.
8. Private prison companies tend to limit training opportunities. According to *Time*, for-profit prisons generally achieve their cost-savings by cutting down on staff costs. That often means limiting the training opportunities available to correctional officers and administrative staff. Fewer training hours combined with higher staffing ratios can lead to higher levels of stress, which create a much higher risk for everyone while providing very little fiscal benefit.
9. Many private prison company contracts require the company to conduct background checks on their prison job applicants, but the contracts rarely specify what conditions would disqualify an applicant for a position. Sometimes, the background checks simply are not conducted.
10. Research shows that private prison executives are paid 10 to 20 times more per year than the heads of state departments of correction. This trend holds true for many lower positions as well. Private prison companies also have boards of directors, who receive generous annual retainers and extra compensation for attending meetings and serving on committees. Executives of the private prison companies also receive generous severance packages, "change-in-control" payments (e.g., after being acquired by another company or after significant changes to the board of directors), deferred compensation plans, and stock options–to name a few added sources of compensation.
11. Correctional officers, on average, earn less working at private prisons. In the United States, correctional officers in public (state) prisons were paid an average of $16.67 per hour in 2019. In many instances, they are eligible for overtime and are treated as public service workers, which generally make them eligible for pensions, good leave benefits, decent health care, and other public-sector benefits. In 2019, private prison officers averaged $15.30 per hour (The GEO Group) and $15.69 per hour (CoreCivic) and may not receive any other benefits. Private prison officer jobs also may be classified as salaried positions to avoid the need to pay overtime.
12. Other hidden overhead costs of for-profit prisons include the legal costs of preparing contracts with various governments, fees for securities lawyers to prepare Security Exchange Commission (SEC) filings, fees for Wall Street investment banks (private prison companies have received initial financing from venture capitalists),

merger and acquisition costs, business-development costs, "customer-acquisition" (i.e., inmate-acquisition) costs, the costs of constructing and operating headquarters in several countries, and so on.

13. For-profit prisons may not offer cost-savings advantages. Although the primary advantage of prison privatization supposedly is lower per-prisoner costs, this is not necessarily always the case. According to *The New York Times*, the inmates in private prisons may cost up to $1,600 more per person, per year, compared to prisons that are operated by the state.
14. People become commodities for private prison companies. Prisons do not make money unless they have a prison population to maintain. That means prisoners' for-profit prisons tend to serve longer sentences and have less access to options such as probation or early release. A recent study of ICE detainees found that ICE holds immigrants in for-profit detention facilities an average of 87 days versus an average of 33.3 days in public detention facilities.[52]
15. Private prisons can choose which prisoners they take. Public prisons often are more expensive to operate because they are forced to take all prisoners, including those that are high security risks. For-profit prison companies have the luxury of choosing prisoners that maximize their profits. If a low-risk prisoner becomes a high-risk prisoner under the supervision of a private prison, most companies' contracts stipulate that they can "replace" the high-risk prisoner. That is another reason why private prisons generally are less expensive to operate than are public prisons.
16. Private prison companies have no obligation to the community where the prison is located. For-profit prisons operate on contracts. Most communities are responsible for the actual facilities that are being used, not the private prison company. That means the for-profit company may not be responsible for repairs or upgrades that are needed. It also means that the prison company can decide to leave if it determines that the prison is not profitable enough. Should that happen, a community is stuck with a useless facility, no jobs, and plenty of unpaid bills.
17. Private prison employees face higher risks of inmate violence. Studies show that, compared to public prison employees, private prisons employees experience 50% more inmate violence. Rates of inmate-on-inmate violence in private prisons are even higher. Many private prisons have staffing waivers, which allow them to under-staff their facility compared to public prisons. Some private facilities operate with an officer-to-prisoner ratio ten or more times higher than comparable public facilities. As another way to save money, private prison companies often rely on inmates to self-govern and reduce violence levels on their own.

Despite the supposed advantages and the many concerns about private prisons, the future of correctional privatization ultimately may depend on the answer to a philosophical question: To what degree should governments cede their correctional responsibilities to the private sector?

Cycles in History The history of institutional corrections has evolved in cycles of accumulation. Developments viewed as innovative replacements for old practices almost always contain vestiges of the old practices, and the old practices seldom disappear when "new" ones are introduced. The new is implemented alongside the old and contains elements of the old. Penitentiaries were not torn down when reformatories were introduced, nor were big houses abolished when correctional institutions arose. Similarly, many jurisdictions moved away from indeterminate sentences toward determinate ones. Those determinate sentences, although often seen as innovative, are hard to distinguish from the fixed sentences that preceded the move to indeterminate sentencing. An interesting question is whether penological history represents progress or the coming back around of what went around before.

THINKING CRITICALLY

1. Do you think any of the early forerunners to modern corrections (such as banishment, etc.) could be used today? Why or why not?
2. Do you think that private prisons have any merit? Why or why not?

The Incarceration Boom

From 1925 until 1972, the incarceration rate in the United States was fairly steady; however, from 1972 through 2008, the incarceration rate increased each and every year. From 1972 through 2008, the incarceration rate in the United States increased from 93 prisoners per 100,000 residents to 506 prisoners per 100,000 residents–an increase of 444%![53] Then, in 2009, the rate decreased for the first time since 1972; it has decreased each year from 2009 through 2017. Figure 10.1 shows the incarceration rate from the mid-1920s through 2017. The incarceration rate of 440 prisoners per 100,000 U.S. residents in 2017 was the lowest rate since 1997; yet, the U.S. still has one of the highest incarceration rates in the world (more about this later).

Recent Trends

There were 329,821 inmates in state and federal prisons at the end of 1980 (305,458 state prison inmates and 24,363 federal prison inmates). By mid-decade, the total number of prison inmates had increased 52% to 502,752 inmates, and by the end of the 1980s, there were a total of 712,967 inmates, which was an increase of 116% over 1980.[54] That represents an average increase of about 9% per year.

Nearly 3 decades later, at year-end 2017, the adult prison population stood at 1,489,363 (1,306,305 state prison inmates and 183,058 federal prison inmates), an increase of about 109% over the beginning of 1990, and an increase of nearly 352% over 1980. About 12% of all inmates at year-end 2017 were in federal prisons. The remaining approximately 88% were in state prisons.[55] The total number of prison inmates at year-end 2017 was about 8% less than the record high of 1,615,487 inmates at year-end 2009. (Between 1980 and year-end 2009, the state prison population increased 360% and the federal prison population increased 754%.) In other words, between 1980 and year-end 2009, the adult prison population (both state and federal) had more than quadrupled.[56]

However, the growth of the prison population slowed somewhat in the 1990s to an average of about 6.5% per year (from approximately 9% per year during the 1980s). Between year-end 2007 and year-end 2017, the state and federal prison population decreased 6.7% (the state prison population decreased 6.5%, while the federal prison population

FIGURE 10.1 **Sentenced Prisoners in State and Federal Institutions**

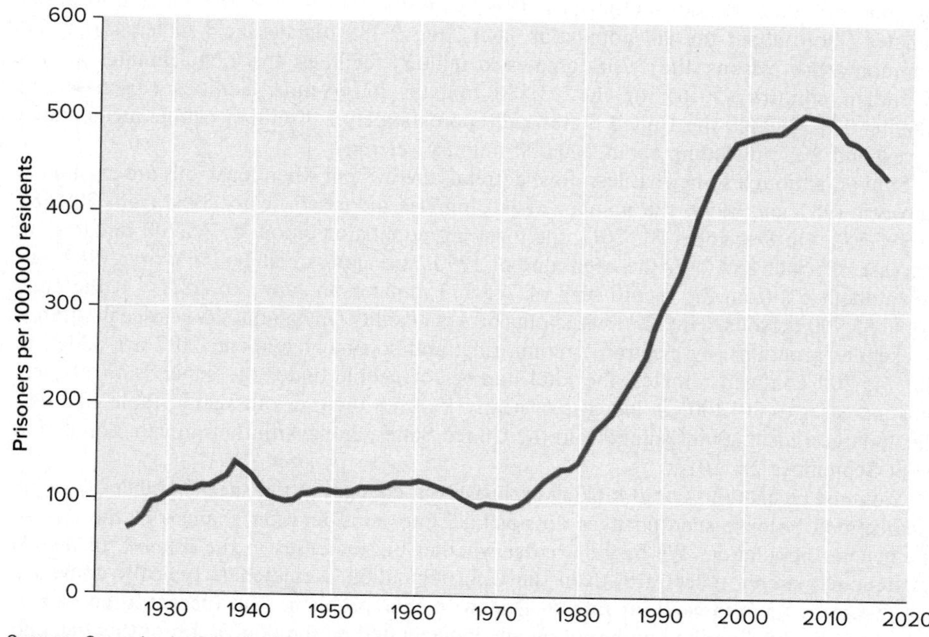

Sources: *Sourcebook of Criminal Justice Statistics Online* Table 6.28.2012, accessed July 16, 2019, https://www.albany.edu/sourcebook/pdf/t6282012.pdf; Jennifer Bronson and E. Ann Carson, "Prisoners in 2017, in *Bureau of Justice Statistics Bulletin*, U.S. Department of Justice (Washington, DC: U.S. Government Printing Office, April 2019), 9, Table 5, accessed July 16, 2019, https://www.bjs.gov/content/pub/pdf/p17.pdf.

TABLE 10.1 Jurisdictions with the Largest and Smallest Numbers of Prison Inmates and Incarceration Rates per 100,000 Residents at the End of 2017

Prison Population	Number of Inmates	Prison Population	Incarceration Rate
5 LARGEST:			
Federal	183,058	Louisiana	719
Texas	162,523	Oklahoma	704
California	131,039	Mississippi	619
Florida	98,504	Arkansas	598
Georgia	53,667	Arizona	569
5 SMALLEST:			
Vermont*	1,546	Federal	51
North Dakota	1,723	Massachusetts	120
Maine	2,404	Maine	134
New Hampshire	2,750	Rhode Island*	170
Rhode Island*	2,861	Vermont*	180

*Includes jail and prison population.

Source: Calculated from Jennifer Bronson and E. Ann Carson. "Prisoner in 2017," in *Bureau of Justice Statistics Bulletin*, U.S. Department of Justice (Washington, DC: U.S. Government Printing Office, April 2019), 4, Table 2 and 11, Table 6. Accessed July 18, 2019, https://www.bjs.gov/content/pub/pdf/p17.pdf.

decreased 8.3%).[57] Table 10.1 shows the jurisdictions with the largest and smallest numbers of prison inmates and the highest and lowest incarceration rates per 100,000 residents at year-end 2017. (We will discuss incarceration rates later in this section.)

The approximately 1.5 million state and federal prison inmates incarcerated in the United States and its territories at year-end 2017 include 121,420 prisoners held in privately operated facilities and 80,917 state and federal prisoners held in local jails or other facilities operated by county or local authorities. This figure does not include the 745,200 local jail inmates (more about the jail population later); the 9,488 inmates held in territorial and commonwealth prisons; the 1,258 inmates in military facilities; the 2,540 inmates in jails in Indian country (2016); or the 43,580 inmates in juvenile facilities (discussed in Chapter 13). Overall, the United States had approximately 2.3 million people incarcerated at year-end 2017, including about 200,000 military veterans.[58]

Similar, although somewhat less drastic, trends are evident when local jails are examined. Between 1982 and 1990, the number of jail inmates increased nearly 89%, from 209,582 to 395,553. On December 31, 2017, the local jail population stood at 745,200 inmates, an increase of about 88% over the beginning of 1990, and approximately 256% over 1982, but down about 5% from the record high of 785,533 inmates on June 30, 2008.[59] If the additional 55,900 persons being supervised outside a jail facility (in community service programs, weekender programs, by electronic monitoring, and so on) at year-end 2017 are added to the 745,200 confined inmates, the total number of people under the supervision of local jails increases to 801,100, an increase of nearly 1% from the previous year.[60] Table 10.2 lists the 10 largest local jail jurisdictions in the United States, along with their inmate population as of September 28, 2017.

incarceration rate A figure derived by dividing the number of people incarcerated by the population of the area and multiplying the result by 100,000; used to compare incarceration levels of units with different population sizes.

We must be cautious about looking exclusively at changes in the sheer number of people incarcerated because such changes do not take into consideration changes in the size of the general population. We might wonder whether big increases in the number of people incarcerated simply reflect growth in the U.S. population. Researchers typically convert a raw figure to an **incarceration rate** to deal with that problem. The incarceration rate is calculated by dividing the number of people incarcerated by the population of the area and multiplying the result by 100,000:

$$\text{Incarceration rate} = \text{Number incarcerated/Population} \times 100{,}000$$

TABLE 10.2 The 10 Largest Local Jail Jurisdictions with Their Inmate Population as of September 28, 2017

Jurisdiction	Location	Inmate Population
Los Angeles County	Los Angeles, CA	19,836
Rikers Island	New York, NY	13,849
Harris County Jail	Houston, TX	10,000
Cook County Jail	Chicago, IL	9,900
Maricopa County Jail	Phoenix, AZ	9,265
Curran-Fromhold Correctional Facility	Philadelphia, PA	8,811
Metro West Detention Center	Miami, FL	7,050
Dallas County Jail	Dallas, TX	6,385
Orange County Jail	Orange County, CA	6,000
Shelby County Jail	Memphis, TN	5,765

Source: Adapted from John Misachi, "The Largest Jails in The United States." *worldatlas*, September 28, 2017, accessed July 19, 2019, https://www.worldatlas.com/articles/the-largest-jails-in-the-united-states.html.

As shown in Figure 10.1, the U.S. prison incarceration rate was comparatively stable from the period before 1930 until the mid-1970s, when the dramatic upward climb began. Between 1980 and year-end 2007, the prison incarceration rate rose about 267%, from 138 to an all-time high of 506 prisoners per 100,000 residents. However, in 2008, the incarceration rate did not increase but remained at 506 prisoners per 100,000 residents. Then, in 2009, the incarceration rate declined for only the second time since 1968, from the high of 506 prisoners per 100,000 residents in 2007 and 2008 to 504 prisoners per 100,000 residents in 2009. The rate declined again to 500 prisoners per 100,000 residents in 2010, 492 prisoners per 100,000 residents in 2011, 479 prisoners per 100,000 residents in 2012, 479 prisoners per 100,000 residents in 2013 (stayed the same), 471 prisoners per 100,000 residents in 2014, 459 prisoners per 100,000 residents in 2015, 450 prisoners per 100,000 residents in 2016, and 440 prisoners per 100,000 residents in 2017 (the rate also declined in 2000).[61] Likewise, between 1982 and mid-year 2007, the jail incarceration rate rose from 90 to an all-time high of 259 prisoners per 100,000 residents, or 188%. However, in 2008, the jail incarceration rate declined for only the third time since the early 1980s, from the high of 259 prisoners per 100,000 residents in 2007 to 258 prisoners per 100,000 residents in 2008. The jail incarceration rate declined again to 250 prisoners per 100,000 residents in 2009, 242 prisoners per 100,000 residents in 2010, 236 prisoners per 100,000 residents in 2011 (the rate also declined in 1999 and 2001).[62] In 2012, the jail incarceration rate increased slightly to 237 prisoners per 100,000 residents, but declined again in 2013 to 231 prisoners per 100,000 residents. A slight increase in the rate to 233 prisoners per 100,000 residents was recorded in 2014, another decline to 225 prisoners per 100,000 residents in 2015, followed by a small increase to 229 prisoners per 100,000 residents in both 2016 and 2017.[63] The steady decline in the prison incarceration rate appears to be a trend. That does not seem to be the case with the recent small increases and decreases in the jail incarceration rate.

Incarceration rates also differ significantly among nations. The data in Figure 10.2 show the prison population rates for 35 selected nations in 2018. For most of the 21st century, the United States has had the rate of incarceration in the world.[64] An article on the front page of the April 23, 2008, *The New York Times* titled "Inmate Count in U.S. Dwarfs Other Nations" noted that the United States has less than 5% of the world's population but almost a quarter of the world's prisoners.[65]

Annual incarceration rates such as those depicted in Figure 10.2 and discussed earlier reflect the numbers of people admitted to institutions, the lengths of time those people serve, and (for purposes of international comparisons) the nations' levels of crime. Nations with higher levels of crime, with more people admitted to prison, and with prisoners serving longer terms can be expected to have higher annual rates of incarceration. Thus, sociologist James Lynch argues that the type of international comparisons depicted in Figure 10.2 may be misleading because the United States has a more serious crime problem than most other nations.[66] Lynch contends that if the analysis is controlled for nations' levels of crime, the

myth

Releasing "nonviolent drug offenders" would end mass incarceration in the United States.

fact

Far more people are incarcerated for violent and property offenses than for drug offenses alone. To end mass incarceration, reforms will have to go further than the "low hanging fruit" of nonviolent drug offenses.

Source: Wendy Sawyer and Peter Wagner, "Mass Incarceration: The Whole Pie 2019," Prison Policy Initiative, March 19, 2019, accessed July 21, 2019, https://www.prisonpolicy.org/reports/pie2019.html.

Incarceration Rates

For the incarceration rates of 222 nations, access World Prison Brief, Institute for Criminal Policy Research at www.prisonstudies.org/highest-to-lowest/prison_population_rate?field_region_taxonomy_tid=All.

Why do you think incarceration rates vary so greatly among different nations?

FIGURE 10.2 **Incarceration Rates for Selected Nations in 2018**

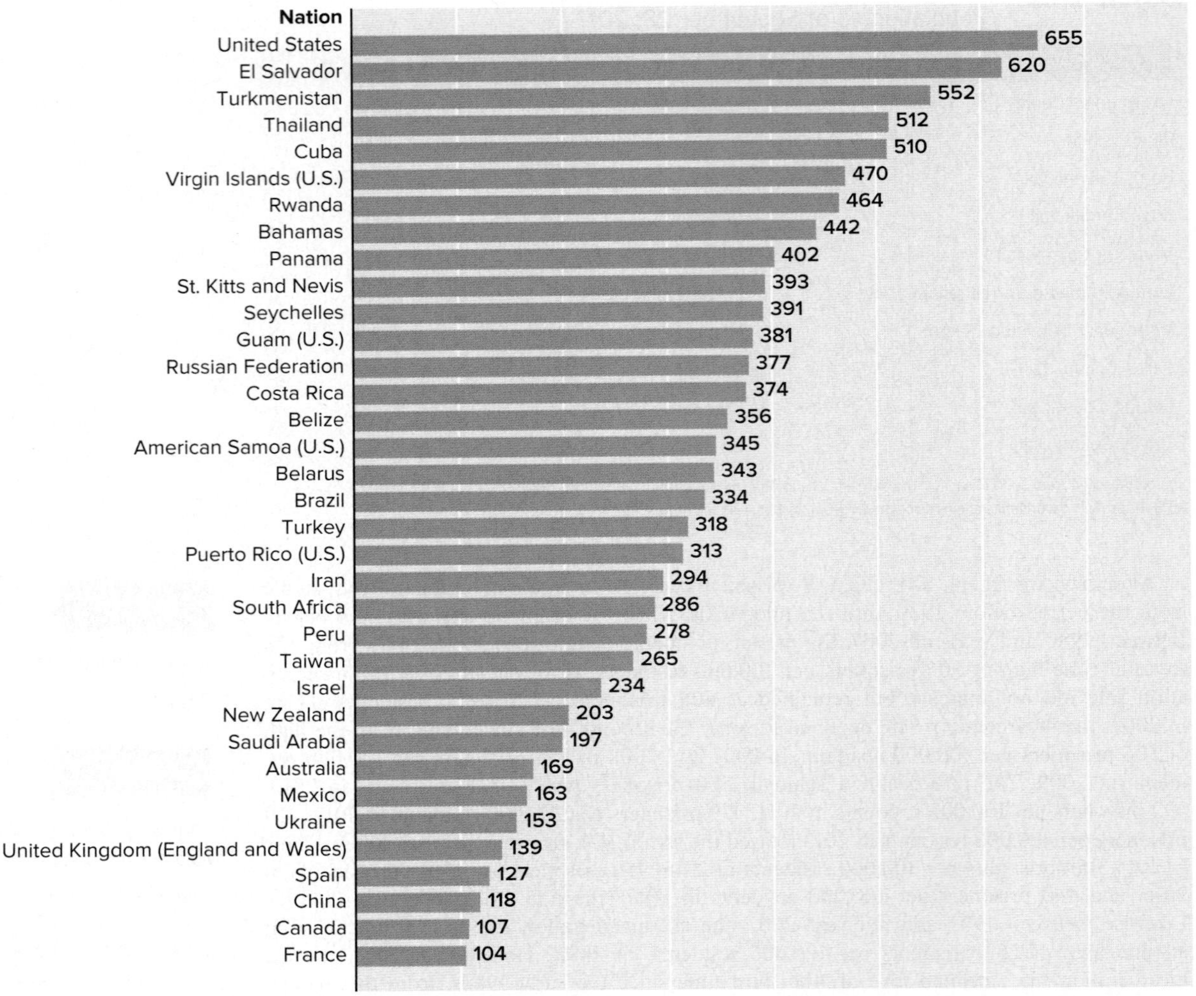

Nation	
United States	655
El Salvador	620
Turkmenistan	552
Thailand	512
Cuba	510
Virgin Islands (U.S.)	470
Rwanda	464
Bahamas	442
Panama	402
St. Kitts and Nevis	393
Seychelles	391
Guam (U.S.)	381
Russian Federation	377
Costa Rica	374
Belize	356
American Samoa (U.S.)	345
Belarus	343
Brazil	334
Turkey	318
Puerto Rico (U.S.)	313
Iran	294
South Africa	286
Peru	278
Taiwan	265
Israel	234
New Zealand	203
Saudi Arabia	197
Australia	169
Mexico	163
Ukraine	153
United Kingdom (England and Wales)	139
Spain	127
China	118
Canada	107
France	104

Source: World Prison Brief, Institute for Criminal Policy Research, accessed July 19, 2019, http://www.prisonstudies.org/highest-to-lowest/prison_population_rate?field_region_taxonomy_tid=All.

differences between the United States and other nations in the use of imprisonment are smaller. Still, Americans are incarcerated for such crimes as writing bad checks or using marijuana, which would rarely result in prison sentences in other countries.[67]

Corrections Spending

In 2018, spending for corrections was the fourth largest general fund expenditure category for states at 3.1%, behind only education at 29.7%, Medicaid at 29.7%, and transportation at 8.0%.

Source: "Summary: NASBO State Expenditure Report," National Association of State Budget Officers, November 15, 2018, accessed July 19, 2019, https://higherlogicdownload.s3.amazonaws.com/NASBO/9d2d2db1-c943-4f1b-b750-0fca152d64c2/UploadedImages/Issue%20Briefs%20/2018_State_Expenditure_Report_Summary.pdf.

Cost Estimates Total spending on state and federal corrections in 2017 was approximately $66.7 billion (an increase of about $17.7 billion or approximately 36% from 2012).[68] The Federal Bureau of Prisons spent approximately $6.9 billion and, as shown in Table 10.3, the 50 states spent a total of about $60 billion. The District of Columbia spent about $248 million.[69] Among $24.2 billion in specific expenditures of state and federal corrections in 2017 were (from highest to lowest): (1) health care, $12.3 billion; (2) construction, $3.3 billion; (3) food, $2.1 billion; (4) interest payments, $1.9 billion; (5) utilities, $1.7 billion; (6) commissary, $1.6 billion; and (7) telephone calls, $1.3 billion.[70]

California spent by far the largest amount on corrections in 2017—about $13.5 billion—which was 3.5 times more than the second highest spending state, Texas, which spent approximately $3.8 billion in 2017 (see Table 10.3). The $13.5 billion spent on corrections by California was 5.2% of its entire 2017 budget, which was the largest percentage of its budget spent on corrections by any state in 2017 (see Table 10.3). South Dakota spent the least amount on corrections in 2017, just $106 million, which was 2.5% of its 2017 budget (see Table 10.3). West Virginia spent the smallest percentage of its entire 2017 budget on corrections, only 1.2% (see Table 10.3). Table 10.3 also shows the percentage change in state

TABLE 10.3 2017 State Corrections Budgets, Budgets as a Percent of Total State Expenditures, and Percentage Change Between 2016 and 2017

State	2017 Budget ($ in millions)	As % of Total State Expenditures	% Change 2016–2017
1. California	$13,500	5.2	6.3
2. Texas	3,808	3.4	1.1
3. New York	2,973	1.9	-0.4
4. Pennsylvania	2,813	3.5	6.2
5. Florida	2,796	3.7	2.8
6. North Carolina	2,209	4.6	8.8
7. Michigan	2,202	4.0	1.6
8. Ohio	2,025	3.0	2.9
9. Georgia	1,843	3.7	10.9
10. New Jersey	1,594	2.7	2.0
11. Maryland	1,581	3.6	1.0
12. Massachusetts	1,437	2.6	3.0
13. Virginia	1,428	2.8	4.5
14. Illinois	1,398	2.1	27.9
15. Wisconsin	1,196	2.5	−2.0
16. Washington	1,133	2.5	9.5
17. Arizona	1,131	2.7	−6.8
18. Oregon	1,070	2.7	5.0
19. Tennessee	941	2.9	3.2
20. Colorado	861	2.4	4.0
21. Louisiana	827	2.9	2.5
22. Indiana	806	2.5	1.3
23. Missouri	688	2.7	1.6
24. Kentucky	682	2.1	4.9
25. Connecticut	656	2.1	-6.4
26. South Carolina	632	2.6	6.9
27. Alabama	614	2.3	2.8
28. Minnesota	586	1.6	4.6
29. Oklahoma	576	2.5	2.3
30. Arkansas	509	2.0	−2.1
31. Iowa	445	2.0	−0.7
32. Utah	396	2.8	−13.5
33. Kansas	377	2.4	−1.6
34. Alaska	367	3.8	0.5
35. Nebraska	348	2.9	2.1
36. Mississippi	337	1.7	−6.4
37. New Mexico	334	1.7	0.3
38. Nevada	317	2.3	7.8
39. Delaware	314	2.9	5.0
40. Idaho	297	4.0	6.1

(continued)

TABLE 10.3 2017 State Corrections Budgets, Budgets as a Percent of Total State Expenditures, and Percentage Change Between 2016 and 2017 (*continued*)

State	2017 Budget ($ in millions)	As % of Total State Expenditures	% Change 2016–2017
41. Hawaii	271	1.8	6.3
42. Rhode Island	218	2.5	2.8
43. Montana	214	3.1	0.0
44. West Virginia	202	1.2	0.5
45. Maine	187	2.3	2.7
46. Vermont	156	2.8	4.0
47. New Hampshire	142	2.4	21.4
48. Wyoming	131	3.0	−6.4
49. North Dakota	118	1.7	2.6
50. South Dakota	106	2.5	−3.6
Total	$59,792		
Average		3.1%	3.9%

Source: National Association of State Budget Officers, 2018 State Expenditure Report, 62–63, Table 32, 63, Table 33, 65, Table 35, accessed July 20, 2019, https://higherlogicdownload.s3.amazonaws.com/NASBO/9d2d2db1-c943-4f1b-b750-0fca152d64c2/UploadedImages/SER%20Archive/2018_State_Expenditure_Report_S.pdf.

expenditures on corrections between 2016 and 2017. As Table 10.3 reveals, states varied widely in their year-over-year expenditures. For example, Illinois reported the largest increase in corrections spending of 27.9%, while Utah had the largest decrease in corrections spending of 13.5%. Overall, the states increased corrections spending between 2016 and 2017, an average of 3.9%. Forty-one states increased spending, and 9 states decreased spending on corrections between 2016 and 2017.

The average daily cost of incarceration per state inmate in 2017 was $109.26 or $39,882 per inmate per year, an increase of nearly 30% from $84.10 a day or $30,696.50 a year in 2012. As shown in Table 10.4, Vermont had the highest daily cost at $322.65 or $117,768 per inmate a year, and Mississippi had the lowest daily cost at $47.06 or $17,177 per inmate a year.

Understand that the costs per inmate presented in Table 10.4 are averages. Inmate costs vary by a prison's security level (which is discussed later in this chapter). For example, North Carolina's prisons are divided into three security levels. The average daily cost per inmate in North Carolina's prisons in 2018 was $99.23, and the annual cost per inmate was $36,219 (the same as in 2017, and as indicated in Table 10.4). However, the cost per inmate in North Carolina's lowest security prisons in 2018 was $86.92 a day or $31,726 a year. The cost per inmate in North Carolina's middle security prisons in 2018 was $102.46 a day or $37,398 a year, and the cost per inmate in North Carolina's highest security prisons in 2018 was $116.75 a day or $42,614 a year.[71] Thus, the difference in inmate costs between North Carolina's lowest security and highest security prisons in 2018 was $29.83 a day or $10,888 a year.

As professors for many years, we have heard students' remark that many people's lives are better off in prison than they are free in society. Students argue that inmates are sheltered, get a bed, three meals a day, free medical care, and so on. What they often ignore is that the shelter is cramped and often stifling in summer, the bed is a thin mattress on a steel frame, the meals are poor, and the medical care is substandard. They also tend to ignore nonmonetary aspects of incarceration such as the constant threat of violence, lack of privacy, absence of heterosexual relationships, and other deprivations and degradations, all of which are described in the next chapter. Perhaps what students mean when they confidently state that people are better off in prison than free in society is that inmates receive goods and services in prison that they otherwise could not afford if they were free in society. If we limit the statement's meaning to this narrow financial issue, what do the data show? In 2017, the federal poverty level for an individual was $12,060, and the poverty rate was 12.3%, meaning that the income of approximately 40 million people in the United

TABLE 10.4 Annual Cost and Average Daily Cost per State Prison Inmate, by State, 2017

State	Annual Cost ($)	Cost per Day ($)
1. Vermont	117,768	322.65
2. North Dakota	88,390	242.16
3. Rhode Island	81,289	222.71
4. New Mexico	79,792	218.61
5. California	70,812	194.01
6. Massachusetts	66,593	182.45
7. Washington	63,953	175.21
8. Nevada	59,055	161.79
9. New York	54,000	147.95
10. Wyoming	53,688	147.09
11. Connecticut	50,262	137.70
12. Alaska	49,800	136.44
13. Pennsylvania	48,341	132.44
14. New Jersey	48,070	131.70
15. Utah	47,407	129.88
16. Maine	43,773	119.93
17. Montana	42,763	117.16
18. Delaware	40,650	111.37
19. Minnesota	40,569	111.15
20. Wisconsin	39,551	108.36
21. Oregon	39,420	108.00
22. Nebraska	38,822	106.36
23. Maryland	38,383	105.16
24. Illinois	38,268	104.84
25. New Hampshire	37,740	103.40
26. North Carolina	36,219	99.23
27. Iowa	34,985	95.85
28. Tennessee	33,595	92.04
29. Colorado	30,817	84.43
30. Michigan	30,401	83.29
31. Virginia	29,967	82.10
32. Ohio	26,382	72.28
33. West Virginia	26,081	71.45
34. Kansas	25,841	70.80
35. Kentucky	25,594	70.12
36. Arizona	25,021	68.55
37. South Carolina	23,711	64.96
38. Arkansas	22,104	60.56
39. Georgia	22,087	60.51
40. Louisiana	21,848	59.86
41. Idaho	21,716	59.50
42. Missouri	21,480	58.85

(continued)

TABLE 10.4 Annual Cost and Average Daily Cost per State Prison Inmate, by State, 2017 (*continued*)

State	Annual Cost ($)	Cost per Day ($)
43. Texas	21,168	57.99
44. Oklahoma	21,101	57.81
45. Florida	20,440	56.00
46. Indiana	19,706	53.99
47. South Dakota	19,465	53.33
48. Alabama	19,005	52.07
49. Hawaii	19,005	52.07
50. Mississippi	17,177	47.06
Average	$39,882	$109.26

Sources: Calculated from data in National Institute of Corrections, "State Statistics Information," accessed July 20, 2019, https://nicic.gov/state-statistics-information?location; National Association of State Budget Officers, 2018 State Expenditure Report, 62–63, Table 32, accessed July 20, 2019, https://higherlogicdownload.s3.amazonaws.com/NASBO/9d2d2db1-c943-4f1b-b750-0fca152d64c2/UploadedImages/SER%20Archive/2018_State_Expenditure_Report_S.pdf.

States fell below the federal poverty level. As shown in Table 10.4, in 2017, Mississippi spent $17,177 per inmate per year–the lowest amount spent of all 50 states. Yet, Mississippi spent about $5,000 per inmate more than the federal poverty level for an individual. Of course, not all of the money allocated per inmate is spent to the benefit of the inmate, but still inmates do receive the amenities noted above. Moreover, in 2017, the real median earnings of all male workers in the United States were $44,408; they were $31,610 for all female workers. As shown in Table 10.4, 15 states spent more per inmate per year than was earned by a male worker, and 28 states spent more per inmate per year than was earned by a female worker. With the caveat above that not all funds spent on inmates benefit the inmates, these data seem to suggest that, at least financially, perhaps millions of people in the United States would be better off in prison than free in society. That statement, of course, ignores the deprivations and degradations of incarceration mentioned above and, perhaps more importantly, begs the question of how much freedom is worth? For anyone who assumes that the aforementioned argument implies that state legislatures should significantly reduce corrections' budgets, remember that state legislatures do not spend lavishly on prisons and prison inmates; rather they spend as little as possible so as to not violate laws governing minimum standards.[72]

The Crowding Issue Crowding always has been a problem in American prisons, but it has become especially troublesome over the past 3-plus decades. The increase in prison construction across the nation, while staggering, has failed to keep pace with the increase in prison populations that has been produced partly by the war on drugs. At year-end 2017, for example, 13 states and the federal prison system reported operating at more than 100% of their highest capacity, a decrease from the 18 states at year-end 2015, and the 24 states that were operating at more than 100% of capacity at year-end 2012. At year-end 2017, the federal prison system was operating at 114.1% over its highest capacity (compared to 120% over its highest capacity at year-end 2015, and 138% over its highest capacity in 2012). Among the states that were operating over their designated maximum capacity at year-end 2017 were Nebraska (127% over capacity), Iowa (115.1%), Delaware (110.3%), Colorado (108.1%), and Washington (105.4%). Moreover, these capacity figures do not include prison inmates held in local jails, in other states, or in private facilities because of insufficient prison space.[73]

Several states have at least one prison under court order to rectify crowded conditions, and some states have their entire prison systems under court order to relieve prison overcrowding. California, until recently, was an example of the latter. On May 23, 2011, in the case of *Brown v. Plata*, the U.S. Supreme Court declared that California's degraded prison conditions led to "needless suffering and death." The ruling upheld an order by three federal judges forcing state officials to significantly reduce the prison population by more than 30,000 inmates within 2 years. The order required state officials to shrink the prison system to about 112,000 inmates, or 137.5% of capacity by the end of 2013. To comply with the order, the state adopted a two-prong plan. First, the prison population was to be reduced through normal attrition of the existing population. Second, new nonviolent, nonserious, nonsexual offenders were to be placed under county jurisdiction for incarceration in local jail facilities. The state intended to give additional funding to the counties to deal with the increased correctional population.[74]

By 2014, California had failed to meet the Court's order, so the federal judges appointed a "compliance officer" empowered to release state prison inmates if California failed to meet new Court-ordered deadlines. The Court ordered California to reduce the prison population to 143% of capacity or 116,651 inmates by June 30, 2014; to 141.5% of capacity or 115,427 inmates by February 28, 2015; and to 137.5% of capacity or 112,164 inmates after that.[75] At year-end 2014, California's prison population was 136,088, an increase of 0.1% over its prison population at the end of 2013 (which was 1.1% higher than it was at the end of 2012), and clearly more than the 116,651 inmates mandated by June 30, 2014, or the final mandate of 112,164 inmates after February 28, 2015.[76] However, by March 2015, California's prison population was 112,300, or 135.8% of capacity. A combination of reforms, inmate transfers, and new construction brought the prison population below the mandated threshold.[77] At year-end 2017, California's prison population was 118,058, or 131.5% of design capacity and 97.2% of operational capacity—well within compliance.[78] Remaining in compliance will be difficult.

Crowded prisons often are volatile prisons, and efforts to address problems related to crowding frequently end up diverting resources from inmate services and programs. How did we arrive at this state of affairs? The most obvious explanation is that a massive outbreak of crime in the United States has fueled the growth of the prison population. But as criminologist Nils Christie and a number of other observers have pointed out, that explanation is simply not supported by the data, which show relatively stable, and in some cases even declining, rates of crime for much of the period of the incarceration boom.[79] Likewise, there has not been an increase in the proportion of young adults, who constitute the most prison-prone age group, in the general population. Criminologist Michael Tonry, in a cross-national analysis, attributes high incarceration rates to high levels of income inequality, low levels of trust and legitimacy, weak welfare states, politicized as opposed to professionalized criminal justice systems (e.g., the popular election of judges and prosecutors), and conflictual rather than consensual political cultures.[80]

Law professor Franklin Zimring offers an interesting public opinion explanation.[81] Zimring claims that members of the public will think punishment for crime is too soft and will demand more imprisonment as long as they think crime is too high. The problem is that prisons generally do an unsatisfactory job of controlling crime, so the public continues to perceive crime as high, despite increases in the prison population. The inability of prisons to control crime fuels the public demand for still more punishment (such as more imprisonment). Zimring uses the analogy of a person who finds that the medicine he or she has been taking for a headache is not helping and decides to increase the dose of the same medication.

We can expand a bit on Zimring's perspective. Over the past 200 years, Americans have developed a tradition of strong reliance on the prison to control crime. It never has done very well. As Zimring observes, the typical response to high crime and high recidivism is to conclude that criminals are not being punished enough and to gradually increase the use of imprisonment. In pursuing that strategy, we get caught in a loop. We continually are forced to direct the greatest portions of our overall correctional budgets toward imprisonment. Relatively few resources are left to develop effective programs in community corrections and crime prevention programs that might reduce reliance on imprisonment. Data show, for example, that about two-thirds of all money spent on corrections is used to finance institutions. Roughly a quarter of all persons under correctional supervision are incarcerated. Therefore, about three-quarters of the correctional population

myth

Six percent of America's criminals commit 70% of all violent crimes. Thus, crime control could be improved dramatically if only those 6% were imprisoned.

fact

This well-publicized statistic comes from a misinterpretation of two studies focusing on criminal activity by boys born in Philadelphia. Criminologist Marvin Wolfgang of the University of Pennsylvania discovered that 6% of the boys born in Philadelphia in 1945 were responsible for more than half of the serious crimes committed by the entire group. Of those born in 1958, 7.5% committed 69% of the serious crimes. Those who cite these outdated studies focus erroneously on 6% of "criminals"—as if future high-rate offenders could be predicted—instead of 6% of all male children born in a given year, whose future criminal behavior is also not predictable.

Source: *Seeking Justice: Crime and Punishment in America* (New York: Edna McConnell Clark Foundation, 1997), 12.

Dale Wetzel/AP Images

The incarceration boom of the past 3-plus decades, until recently, has required the construction of dozens of new prisons. *Was this construction boom necessary? Why or why not?*

must be accommodated with one-third of the resources. The lack of resources devoted to community corrections and crime prevention helps ensure that programs in those areas will fail to control crime. So there always will be an abundant supply of offenders to feed the prison population and escalate the cost of maintaining that population. The irony is clear and substantial. While crowding in correctional institutions suggests the need for effective alternatives in community corrections and crime prevention, that same crowding and the resources it consumes preclude such alternatives. Ineffectiveness in community corrections and crime prevention simply makes the crowding worse, thereby consuming even more resources.

Prison Inmate Characteristics

Who are the people in prison, and why are they there? Table 10.5 presents the characteristics of state and federal inmates from the National Inmate Survey 2016 (which is the most recent data on state inmate characteristics available at this writing). Before examining the prison inmate characteristics, however, two points are important to note. First, the vast majority of prison inmates are state inmates. For example, of the approximately 1.5 million state and federal prison inmates in 2016, 87.5% were state inmates, and only 12.5% were federal inmates.[82] Second, the characteristics of both state and federal inmates are remarkably stable over time; that is, they do not change much from one year to the next.

Current Federal Inmate Characteristics

For current federal inmate characteristics, visit the Federal Bureau of Prisons website at https://www.bop.gov/about/statistics/ and select the Inmate Statistics tab.

How do current federal inmate characteristics differ from those presented in Table 10.5?

As Table 10.5 shows, in 2016, more than 90% of both state and federal prison inmates were male. Because the general U.S. population is nearly evenly split between males and females (see Table 10.5), males are disproportionately overrepresented in prison—by a wide margin. Put differently, in 2016, the imprisonment rate for men age 18 or older was 1,109 per 100,000 U.S. residents, compared with 82 per 100,000 U.S. residents for women.[83]

Also disproportionately overrepresented in both state and federal prisons are non-Hispanic blacks and inmates of two or more races and, to a lesser extent, Hispanics. Underrepresented in both state and federal prisons are non-Hispanic whites; American Indians and Alaska Natives, and Asians, Native Hawaiians, and Other Pacific Islanders. As shown in Table 10.5, there were approximately 2.7 times more non-Hispanic blacks in state prisons and about 2.6 times more non-Hispanic blacks in federal prisons than in the general U.S. population in 2016 (see Table 10.5). At the same time, there were approximately 5.3 times more inmates of two or more races in state prisons and about 3.1 times more inmates of two or more races in federal prisons than in the general population in 2016 (see Table 10.5). In the case of Hispanics, in 2016, their percentage in state prisons was only slightly higher than their percentage in the general U.S. population (1.2 times higher), while their percentage in federal prisons was approximately 2.1 times higher than their percentage in the general U.S. population (see Table 10.5).

As noted, non-Hispanic whites are underrepresented in state and federal prisons. If non-Hispanic whites were imprisoned at a rate commensurate with their proportion of the general U.S. population in 2016, there would be 1.9 times more non-Hispanic whites in state prisons and 2.9 times more non-Hispanic whites in federal prisons (see Table 10.5). Likewise, if American Indians and Alaska natives were imprisoned at a rate equal to their percentages in the general U.S. population in 2016, there would be about 2.0 times more of these groups in state prisons and approximately 2.4 times more of them in federal prisons (see Table 10.5). Finally, if Asians, Native Hawaiians, and other Pacific Islanders were imprisoned at the same rate as their percentage in the general U.S. population in 2016, about 6.3 times more of them would be expected in state prisons and approximately 3.8 times more of them would be expected in federal prisons.

The overrepresentation of blacks in prison is a very heated issue in criminal justice, and research has not established a consensus on the reasons for that overrepresentation. The weight of the evidence suggests that offense seriousness and prior criminal record generally exert a stronger impact on decisions to imprison than extralegal factors such as race.[84] However, in her award-winning best seller, *The New Jim Crow: Mass Incarceration in the Age of Colorblindness*, Michelle Alexander offers a more nefarious reason for the overrepresentation of blacks in U.S. prisons, especially state prisons. She claims that "mass incarceration [in the United States] is, metaphorically, the New Jim Crow . . . a stunningly comprehensive and well-disguised system of racialized social control [and] the most damaging

TABLE 10.5 Selected Characteristics of the State and Federal Prison Population, 2016

Characteristics	State Prison Inmates	Federal Prison Inmates	U.S. Population
SEX			
Male	92.8%	93.8%	49.2%
Female	7.2	6.2	50.8
RACE/ETHNICITY			
White, non-Hispanic	32.1%	21.1%	61.2%
Black, non-Hispanic	33.7	32.0	12.5
Hispanic	20.7	37.2	17.9
American Indian/Alaska Native	1.4	1.7	0.7
Asian/Native Hawaiian/Other Pacific Islander	0.9	1.5	5.7
Two or more races	11.2	6.5	2.1
AGE AT TIME OF SURVEY			
18–24	10.2%	4.8%	9.9%
25–34	32.1	28.0	13.3
35–44	26.3	34.5	13.3
45–54	18.6	21.5	14.6
55 or older	12.8	11.2	24.9
MARITAL STATUS			
Married	13.9%	21.6%	49.8%
Widowed	2.8	1.8	2.6
Separated	4.8	5.6	1.8
Divorced	19.3	18.1	9.6
Never married	59.1	52.8	36.2
EDUCATION			
Less than high school	62.0%	56.8%	13.0%
High school graduate	22.6	21.8	27.5
Some college	11.1	13.8	29.2
College degree or more	3.6	7.6	30.3
CITIZENSHIP			
U.S. citizen	95.6%	75.0%	92.9%
Non-U.S. citizen	4.4	25.0	7.1
MILITARY SERVICE			
Yes	7.9%	5.4%	7.6%
No	92.1	94.6	92.4

Sources: Calculated from data in Mariel Alper and Lauren Glaxe, "Source and Use of Firearms Involved in Crimes: Survey of Prison Inmates, 2016," U.S. Department of Justice, *Bureau of Justice Statistics Special Report* (Washington, DC: USGPO, January 2019), 3, Table 1, accessed February 26, 2020, https://www.bjs.gov/content/pub/pdf/suficspi16.pdf; "Age and Sex Composition: 2010," 2010 Census Briefs, United States Census Bureau (sex, 2010); United States Census Bureau, American FactFinder (race and ethnicity, marital status, education, 2016), accessed February 29, 2020, https://factfinder.census.gov/faces/nav/jsf/pages/searchresults.xhtml?refresh=t; "State Demographics by Citizenship Status," Ballotpedia, accessed February 28, 2020, https://ballotpedia.org/State_demographics_by_citizenship_status; Jennifer Schultz, "Veterans By the Numbers," National Conference of State Legislatures, accessed February 28, 2020, https://www.ncsl.org/blog/2017/11/10/veterans-by-the-numbers.aspx.

The largest proportion of state prisoners are male and black. *Why, and what can be done about it?*

Mark Peterson/Contributor/Getty Images

manifestation of the backlash against the Civil Rights Movement." She adds that "no other country in the world imprisons so many of its racial or ethnic minorities." She notes, "The United States imprisons a larger percentage of its black population than South Africa did at the height of apartheid."[85]

Another important distinguishing feature of the prison population, especially the state prison population, is economic status. Even though the characteristic is not included in Table 10.5, economic status is as significant as sex, race/ethnicity, or any other characteristic for that matter. Prisoners, especially state prisoners, are overwhelmingly poor. Economic status likely is confounded with race and ethnicity. That is, prisons are occupied predominately by poor blacks, poor whites, and poor Hispanics. Very few prisoners are wealthy or even middle class, regardless of whether they are black, white, or Hispanic.

In 2016, as in most years, nearly 70% of both state and federal prison inmates were between the ages of 18 and 45 (see Table 10.5). With the exception of federal inmates aged 18–24, whose percentage in the general U.S. population was about twice as high as it was in federal prisons (and thus they were underrepresented in federal prisons), inmates in the 18–45 age group were overrepresented in both the state and the federal prison populations. Inmates 55 or older also were underrepresented in the prison population, compared to the general U.S. population–by a margin of approximately 1.9 in state prisons and about 2.2 in federal prisons (see Table 10.5).

Prison inmates also can be differentiated from the general U.S. population by marital status. The majority of both state and federal prison inmates are single (never married). In 2016, state prison inmates were about 1.6 times more likely to be single and federal prison inmates were approximately 1.5 times more likely to be single than the general U.S. population (see Table 10.5). State and federal prison inmates also were more likely to be separated and divorced than the general U.S. population in 2016 (see Table 10.5). Thus, both state and federal prison inmates are less likely to be married than the general U.S. population. In 2016, state and federal prison inmates, respectively, were nearly 3.6 times and 2.3 times less likely to be married than the general U.S. population (see Table 10.5).

Education is another defining characteristic of the prison population, compared to the general U.S. population. Both state and federal prison inmates, as a group, are less educated than the general U.S. population. In 2016, state and federal prison inmates, respectively, were nearly 5 times and about 4.4 times less likely to be high school graduates than the general U.S. population (see Table 10.5). In addition, in 2016, the general U.S. population was more than 8 times more likely to have a college degree or more than state prison inmates and nearly 4 times more likely to have a college degree or more than federal prison inmates (see Table 10.5). Note that federal prison inmates generally are more educated than state prison inmates. In 2016, for example, federal prison inmates were about twice as likely to have a college degree or more than federal prison inmates (see Table 10.5). Also, in 2016, both state and federal prison inmates were only slightly less likely to be high school graduates than the general U.S. population, while the general U.S. population was approximately 2.6 times more likely to have some college than state prison inmates and about 2.1 times more likely to have some college than federal prison inmates (see Table 10.5).

Both state and federal prison inmates are likely to be U.S. citizens. In 2016, state prison inmates were slightly more likely to be U.S. citizens than the general U.S. population, while only three-fourths of federal prison inmates were U.S. citizens (see Table 10.5). The 25% of federal prison inmates that were not U.S. citizens in 2016 (see Table 10.5) likely can be attributed in large measure to U.S. policy on illegal immigration.

Military service is another characteristic shown in Table 10.5. More than 90% of both state and federal prison inmates have not served in the military, about the same percentage as the general U.S. population (see Table 10.5).

Table 10.6 provides one last set of comparisons; this time, between state and federal prison inmates on the most serious offense for which prison inmates are incarcerated. As

TABLE 10.6 Most Serious Offenses for Which State and Federal Inmates Were Serving Sentences, 2016

Most Serious Offense	State Prison Inmates	Federal Prison Inmates
Violent	55.1%	12.3%
Homicide[a]	15.8	2.2
Rape/sexual assault	12.0	1.4
Robbery	12.4	6.3
Assault	12.3	1.7
Other[b]	2.6	0.7
Property	15.4%	7.0%
Burglary	7.3	0.1
Other[c]	8.1	6.9
Drug	14.9%	47.2%
Trafficking[d]	10.8	42.4
Possession	3.8	2.0
Other	0.3	2.8
Public Order	13.0%	31.0%
Weapons[e]	3.6	13.0
Other[f]	9.4	18.0
Other	0.3%	1.0%
Unknown	1.2%	1.3%

[a]Includes murder and both negligent and non-negligent manslaughter.

[b]Includes kidnapping, blackmail, extortion, hit-and-run driving with bodily injury, child abuse, and criminal endangerment.

[c]Includes larceny, theft, motor vehicle theft, arson, fraud, stolen property, destruction of property, vandalism, hit-and-run driving with no bodily injury, criminal tampering, trespassing, entering without breaking, and possession of burglary tools.

[d]Includes possession with intent to distribute.

[e]Includes being armed while commiting a crime; possession of ammunition, concealed weapons, rearms and explosive devices; selling or tracking weapons; and other weapons offenses. Among federal prisoners, weapons offense include violations of federal rearms and explosives.

[f]Includes commercialized vice, immigration crimes, DUI, violations of probation/parole, and other public-order offenses.

Sources: Calculated from data in Mariel Alper and Lauren Glaxe, "Source and Use of Firearms Involved in Crimes: Survey of Prison Inmates, 2016," U.S. Department of Justice, *Bureau of Justice Statistics Special Report* (Washington, DC: USGPO, January 2019), 6, Table 4, accessed February 26, 2020, https://www.bjs.gov/content/pub/pdf/suficspi16.pdf; "Age and Sex Composition: 2010," 2010 Census Briefs, United States Census Bureau (sex, 2010); United States Census Bureau, American FactFinder (race and ethnicity, marital status, education, 2016), accessed February 29, 2020, https://factfinder.census.gov/faces/nav/jsf/pages/searchresults.xhtml?refresh=t; "State Demographics by Citizenship Status," Ballotpedia, accessed February 28, 2020, https://ballotpedia.org/State_demographics_by_citizenship_status; Jennifer Schultz, "Veterans By the Numbers," National Conference of State Legislatures, accessed February 28, 2020, https://www.ncsl.org/blog/2017/11/10/veterans-by-the-numbers.aspx.

FYI **Survey of Prison Inmates, 2016**

According to the Survey of Prison Inmates, 2016, conducted by the U.S. Department of Justice's Bureau of Justice Statistics, state prison inmates are most likely to be male (93%); non-Hispanic black or Hispanic (54%); 25–44 years old (58%); never married (59%); have less than a high school education (62%); a U.S. citizen (96%); no military service (92%); and a violent offender (55%). Federal prison inmates are likely to be male (94%); non-Hispanic black or Hispanic (69%); 25–44 years old (63%); never married (53%); have less than a high school education (57%); a U.S. citizen (75%); no military service (95%); and a drug offender or a public order offender (78%).

Source: See Tables 10.5 and 10.6.

shown in Table 10.6, in 2016, 55.1% of state prison inmates were incarcerated for violent offenses, compared to only 12.3% of federal prison inmates. On the other hand, 47.2% of federal prison inmates were incarcerated for drug offenses (mostly trafficking), compared to only 14.9% of state prison inmates. State prison inmates also were more than twice as likely as federal prison inmates to be incarcerated for property offenses (15.4% versus 7.0%), while federal prison inmates were about 2.4 times more likely than state prison inmates to be incarcerated for public order offenses (31.0% versus 13.0%) (see Table 10.6). A more detailed breakdown of the offense categories is shown in Table 10.6. Besides the most serious offense for which inmates are incarcerated and the other differences highlighted, the overall profile of federal prison inmates is very similar to the overall profile of state prison inmates.

THINKING CRITICALLY

1. Why do you think prison and jail incarceration rates have decreased in recent years?
2. Should more money be spent on institutional corrections? Why or why not?
3. What do the characteristics of prison inmates say about American society as a whole?

Incarceration Facilities

In the United States, the organizational and administrative structure of institutional corrections is diffuse and decentralized. Primary administrative responsibility for facilities lies with the executive branch of government, but the legislative and judicial branches also are involved. For example, the legislative branch appropriates resources and passes statutes that affect sentence length. The judicial branch sentences offenders to facilities and oversees the legality of institutional practices.

Incarceration facilities exist at all three levels of government (federal, state, and local), and power and decision-making responsibility are widely distributed both among and within levels. Within broad guidelines, the federal level and the various state and local (county and city) jurisdictions have much autonomy to organize and carry out incarceration practices. As a general rule, the federal government operates its own prison system, each state operates its own prison system, and local jurisdictions operate their own jail systems. Decentralization and autonomy notwithstanding, there are interrelationships between levels. For example, federal requirements affect the operation of state prisons, and local jails are affected by both federal and state regulations.

Federal Bureau of Prisons (BOP)

Federal institutions are administered by the Federal Bureau of Prisons (BOP), which was established within the U.S. Justice Department in 1930. The BOP has come to serve as a source of innovation and professionalization in the field of institutional corrections. The BOP's mission is "to protect society by confining offenders in the controlled environments of prisons and community-based facilities that are safe, humane, cost-efficient, and appropriately secure, and that provide work and other self-improvement opportunities to assist offenders in becoming law-abiding citizens."[86] The administrative organization of the bureau is shown in Figure 10.3.

The BOP is comprised of a headquarters, 6 regional offices, 2 staff training centers, 110 institutions (prisons), 22 residential reentry management offices, and 11 additional correctional institutions that are under contract with private corporations (see Figure 10.4).[87] The BOP's headquarters is in Washington, DC, and the 6 regional offices are in Philadelphia, Pennsylvania; Annapolis Junction, Maryland (near Baltimore); Atlanta, Georgia; Grand Prairie, Texas (near Dallas); Kansas City, Kansas; and Stockton, California. The two staff training centers are in Aurora (Colorado) and Glynco (Georgia) at the Federal Law Enforcement Training Center. The other facilities are scattered throughout the United States (see Figure 10.4).

FIGURE 10.3 Organization of Federal Bureau of Prisons

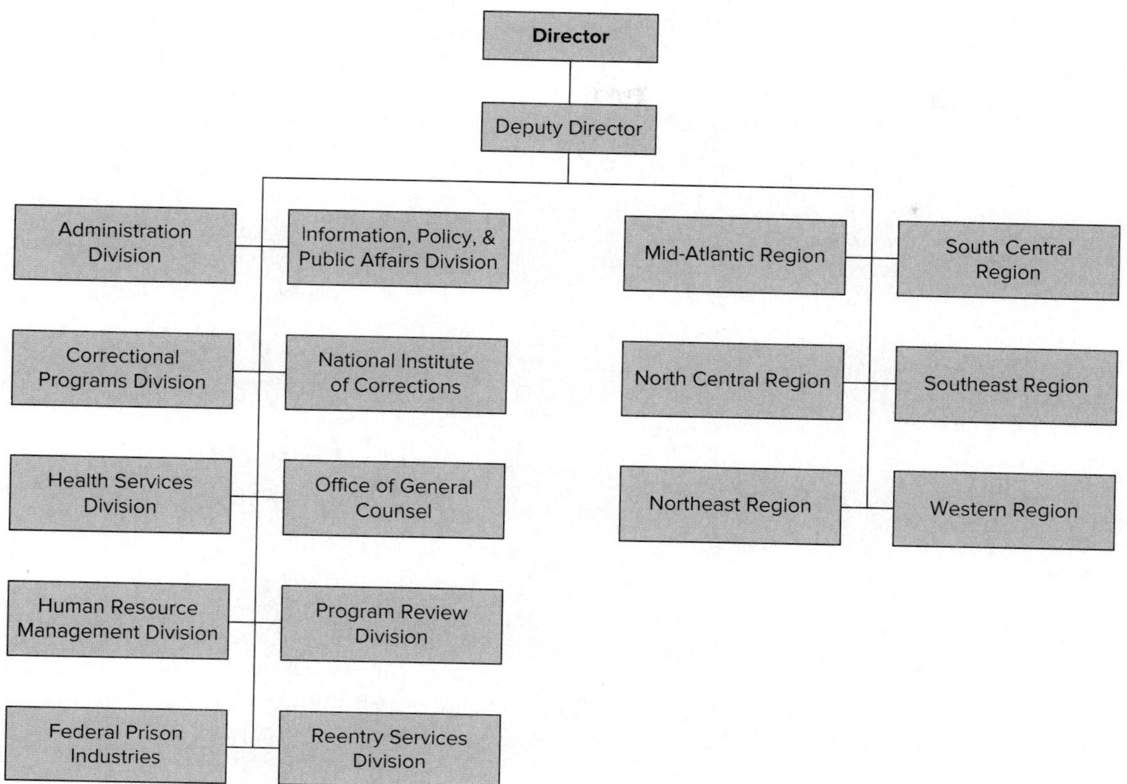

Source: Federal Bureau of Prisons, "About Our Agency–Mission," accessed July 22, 2019, https://www.bop.gov/about/agency/agency_pillars.jsp.

The BOP's facilities hold inmates convicted of violating the U.S. Penal Code. In 2001, the BOP assumed the added responsibility of incarcerating the District of Columbia's sentenced felons because of federal legislation passed in 1997.[88] By mid-year 2002, the federal prison system, with 161,681 inmates, for the first time in U.S. history, had more inmates than did any state system. By the end of 2017, the federal prison population stood at 183,058 inmates–down about 18% from the 219,298 inmates at the end of the peak year of 2013, as mentioned previously, but still the largest prison system in the United States (Texas was second at the end of 2017 with 162,523 inmates; see Table 10.1).[89] (As of February 27, 2020, the BOP reported 175,135 total federal inmates–about 4.3% fewer than at the end of 2017.)[90]

The BOP's inmate facilities are distinguished by their **security level**. An institution's security level is determined by two related factors: (1) the degree of external or perimeter security surrounding the prison and (2) the measures taken to preserve internal security within the institution. The BOP utilizes four security levels: minimum, low, medium, and high. Of the 167,213 federal prison inmates classified on February 15, 2020 (7,011 inmates were unclassified), 16.9% were in minimum-security facilities, 38.8% were in low-security facilities, 32.0% were in medium-security facilities, and 12.3% were in high-security facilities.[91]

security level A designation applied to a facility to describe the measures taken, both inside and outside, to preserve security and custody.

Minimum-security institutions also are known as Federal Prison Camps (FPCs) (see Figure 10.4). They have limited or no perimeter fencing, dormitory housing, and a relatively low staff-to-inmate ratio. Work and programs are the focus of these institutions. Several BOP institutions have a small, minimum-security satellite work camp next to the main facility (see Figure 10.4). These camps provide inmate labor to the main institution and to off-site work programs.[92]

Low-security Federal Correctional Institutions (FCIs) (see Figure 10.4) have double-fenced perimeters, mostly dormitory or cubicle housing, and higher staff-to-inmate ratios than minimum-security facilities. They also have strong work and treatment program elements.[93]

FIGURE 10.4 Locations of Institutions in Federal Bureau of Prisons

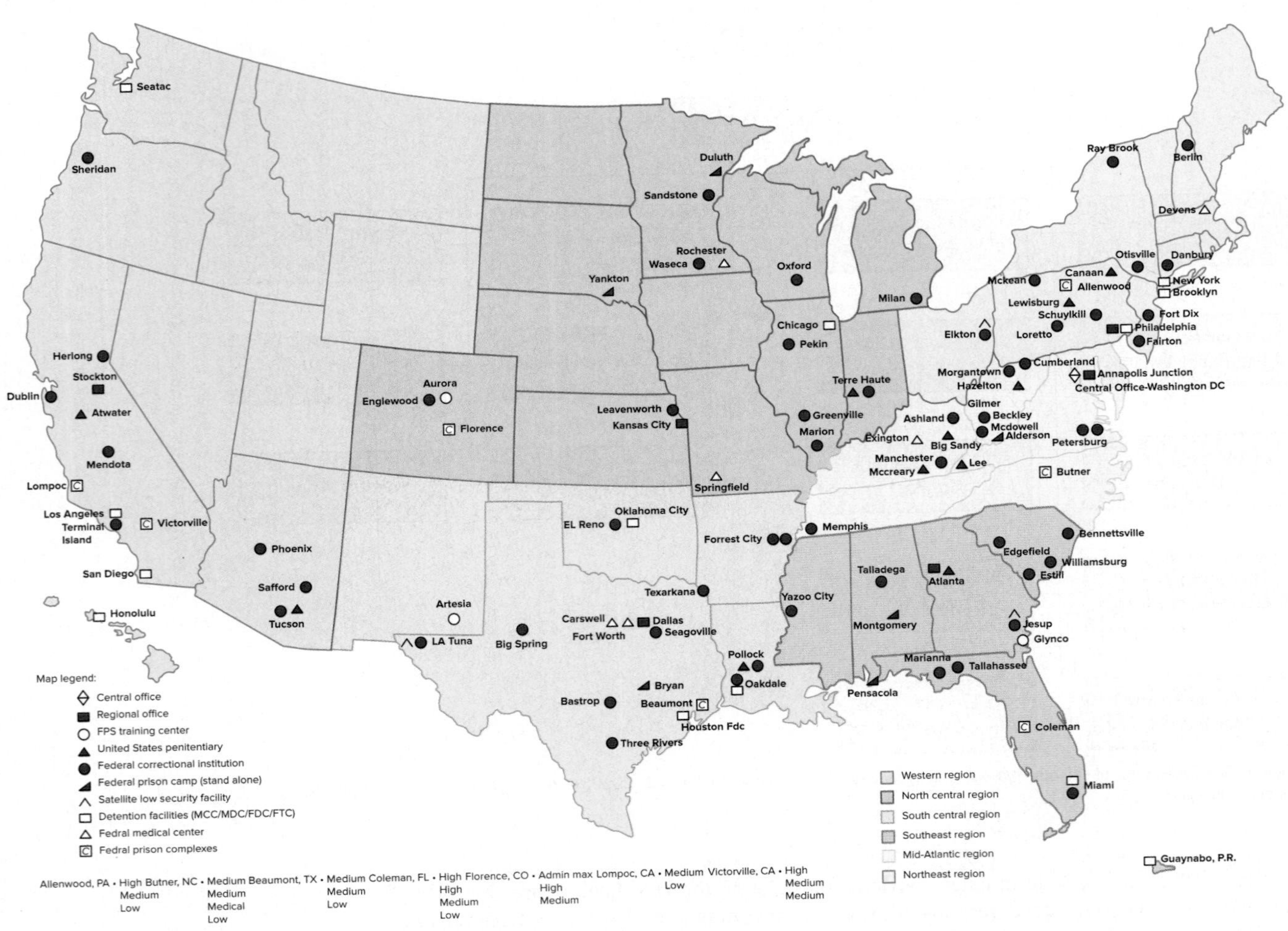

Prison work camps are located with most higher security facilities and complexes (satellite camps).

Source: The United States Department of Justice, "Organization, Mission and Functions Manual: Federal Bureau of Prisons," map updated October 24, 2018, accessed July 23, 2019, https://www.justice.gov/jmd/organization-mission-and-functions-manual-federal-bureau-prisons.

Medium-security FCIs (and United States Penitentiaries (USPs) designated to house medium-security inmates, see Figure 10.4) often have double fences with electronic detection systems, mostly cell-type housing, an even higher staff-to-inmate ratio than low-security FCIs, and stronger internal controls. They also have a variety of work and treatment programs.[94]

High-security institutions, also known as United States Penitentiaries (USPs) (see Figure 10.4), have walls or reinforced fences and multiple- and single-occupant cell housing. They also have the highest staff-to-inmate ratio and close control of inmate movement.[95]

The BOP also features Federal Correctional Complexes (FCCs) (see Figure 10.4), which are institutions with different missions and security levels that are clustered near each other. FCCs increase efficiency through the sharing of services, boost emergency preparedness by having additional resources within close proximity, and facilitate the ability of staff to gain experience at institutions with different security levels.[96]

Finally, the BOP has a variety of administrative facilities with special missions, such as the restraint of exceptionally dangerous, violent, or escape-prone inmates; the care of inmates with severe or chronic medical problems; or the confinement of pretrial offenders. Administrative facilities include Metropolitan Correctional Centers (MCCs), Metropolitan Detention Centers (MDCs), Federal Detention Centers (FDCs), Federal Medical Centers (FMCs), the Federal Transfer Center (FTC), the Medical Center for Federal

Prisoners (MCFP), and the Administrative-Maximum Security Penitentiary (ADX) (see Figure 10.4). Administrative facilities, except the ADX, are able to hold inmates in all security categories.[97]

A special type of high-security prison or administrative-maximum security penitentiary (ADX) in the BOP system is the "ultramaximum-" or "supermaximum-security" prison opened in November 1994, in Florence, Colorado (115 miles south of Denver). ADX Florence, nicknamed the "Alcatraz of the Rockies," is a $60 million state-of-the-art prison and part of the Florence Federal Correctional Complex (see Figure 10.4). This "supermax" or "control unit" prison was custom-built to confine the 400 most notorious male offenders and problem inmates in the federal system. It has never been filled to capacity. Among its 380 inmates, as of July 22, 2019, were Joaquin "El Chapo" Guzman Loera, former leader of the Sinaloa Drug Cartel; Ted Kaczynski, the "Unabomber"; Eric Robert Rudolph, Atlanta's Centennial Olympic Park bomber; Larry Hoover, former leader of Chicago's Gangster Disciples street gang; Terry Nichols, Timothy McVeigh's accomplice in the 1995 Oklahoma City bombing; Matthew Hale, former leader of the white supremacist group World Church of the Creator; Ramzi Yousef, organizer of the 1993 World Trade Center bombing; Robert Hanssen, former FBI agent and spy for the Soviet Union and Russia; Richard Reid, the "Shoe bomber" who tried to blow up a passenger jet in 2001; Dzhokhar Tsarnaev, 2013 Boston Marathon bomber; Zacarias Moussaoui, involved in the 9/11 hijackings; Umar Farouk AbdulMutallab, the "Underwear" bomber who tried to blow up a commercial airliner in 2009; Jose Padilla, a U.S. citizen and al-Qaeda enemy combatant who planned to set off radioactive "dirty bombs" in the United States; and Faisal Shahzad, who tried to detonate a car bomb in New York's Time Square in 2010.[98]

The security features of ADX Florence "supermax" prison are unprecedented. The prison is built into the side of a mountain and its perimeter is secured with 12 gun towers, razor wire, guard dogs, and laser beams. Within the perimeter are nine units. Motion sensors, floors equipped with pressure sensors that monitor foot traffic, 1,400 electronically controlled gates and doors, and 168 television monitors control movement inside the units. Each unit is self-contained and includes separate sick-call rooms, law libraries, and barber chairs. Inmates do not leave their cells for more than an hour a day. When they must leave their cells, they do so only with leg irons and handcuffs that are attached to a belt around the inmate's waist, and an escort of two or three guards per inmate. Sometimes an inmate's cuffed hands are further restricted inside a black box. Head counts are conducted at least six times a day. Inmates live in total isolation (solitary-cell confinement) with constant lockdowns, during which inmates are confined to their cells. Each 7-by-12 foot supposedly

Bureau of Prisons/The Gazette/AP Images

ADX Florence. *Is there a better way to deal with the most notorious and predatory inmates in the federal system? If yes, what is it?*

soundproof cell has a double-entry door with the classic barred cage door backed up by a windowed steel door that helps minimize voice contact among prisoners. A slot in the door allows for meal delivery with minimal or no interaction between inmates and guards. Each cell also has a narrow window about 4 inches wide, 42 inches high, and angled upward so only the sky is visible so that inmates cannot know their location in the prison. The cells and everything in them, including the sink-toilet, shower, desk and bed, are constructed of concrete. Inmates can watch television in their cells and have access to a commissary. Religious services, educational programs, psychiatric evaluations, and other services are provided through the door or by telecommunication. After 3 years of this type of confinement, a successful inmate gradually may regain social contact by being allowed to go to the cafeteria or to the recreation yard, which is sectioned off into a series of individual cages.[99]

Classification and Other Special Facilities

classification facility A facility to which newly sentenced offenders are taken so that their security risks and needs can be assessed and they can be assigned to a permanent institution.

When offenders are sentenced to the custody of the department of corrections in most states, they are transported initially to a **classification facility** (sometimes referred to as an assessment, reception, or diagnostic center). Stays at classification facilities are ordinarily short (e.g., 60 days). The process of classification entails assessing an offender's security risk and determining which program services (e.g., counseling and education) the offender needs. Assessment information is used in deciding to which institution in the jurisdiction an offender will go to begin his or her term and which problems (such as alcohol dependency) the offender must address while imprisoned. A variety of other factors influence those decisions, including the nature of the offense, the offender's prior record (if any), propensity toward violence and escape, and vulnerability to victimization by other inmates; the programs offered at the state's various institutions; as well as the levels of crowding at those institutions. The idea is to place inmates in facilities that can accommodate their risk-and-needs profiles.

Classification is not a one-time process; it occurs periodically throughout an inmate's sentence. Inmates routinely are monitored and reclassified to preserve institutional security and for purposes of transfer, programming, and release decisions.

Classification centers are not the only short-term-stay facilities. For example, the BOP administers a number of medical institutions that receive inmates with health problems and provide them with health care services. If an inmate's condition improves, he or she is returned to the institution of origin or sent to another institution to complete the sentence. Other short-term facilities in various jurisdictions provide services for offenders with mental disorders.

State Prisons

Although states vary in the way they organize institutional corrections, each state has a department of corrections or a similar administrative body to coordinate the various adult prisons in the state. Whether federal or state, correctional institutions are formal, bureaucratic organizations characterized by agency goals, rules and regulations, a staff chain of command, a staff division of labor, and similar features. Most adult prisons employ a quasi-military model of administration and management.

Several states in the United States now have as many or more prisons in operation as some entire Western nations. At the beginning of 2017, the states, combined, operated 1,002 adult prisons.[100] Figure 10.5 shows the number of prisons in each state in 2017. Note that Florida with 148 prisons had nearly 2.5 times more prisons than the state with the second most prisons, Texas, with 61 prisons. The 1,002 state prisons in 2017 were about 15% fewer than the 1,185 state prisons in 2005.[101]

Most state prisons are exclusively for male inmates; only a small number of prisons are solely for female inmates or for both male and female inmates (cocorrectional facilities). Women's prisons are described in their own section later in this chapter.

State prisons, like federal prisons, often are distinguished from one another by security level. As noted previously, an institution's security level is determined by two related factors: (1) the degree of external or perimeter security surrounding the prison and (2) the measures taken to preserve internal security within the prison. The simplest security level categorization

FIGURE 10.5 Number of State Prisons in the United States, 2017

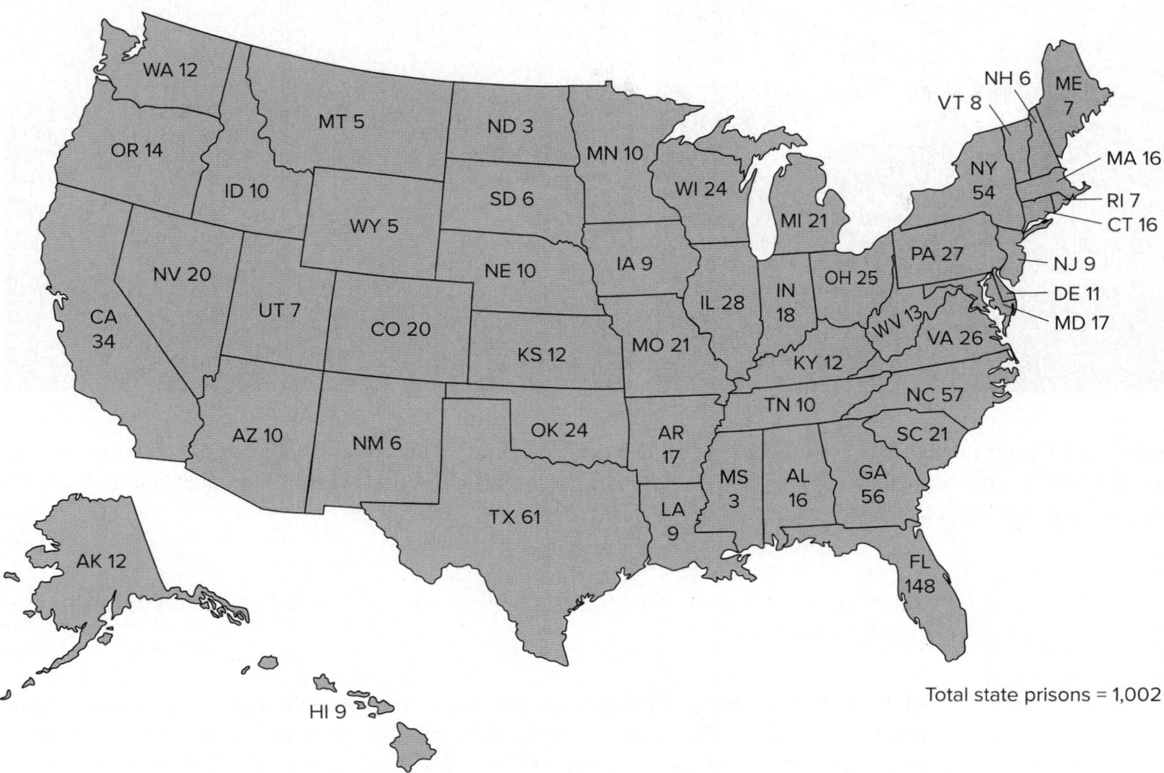

Source: Number of state prisons calculated from data in "State Prison Facilities," State Statistics Information, National Institute of Corrections, accessed July 16, 2019, https://nicic.gov/state-statistics-information?

typically used for state prisons is maximum, medium, and minimum. However, states vary in the security categorizations they use.

Maximum-Security Prisons Maximum-security prisons are characterized by very tight external and internal security. A high wall or razor-wire fencing, with armed-guard towers, electronic detectors, or both, usually surrounds the prison, although armed-guard towers are becoming obsolete. External armed patrol also is common. Some maximum-security institutions have a wide, open buffer zone between the outer wall or fence and the free community.

Most maximum-security prison designs follow the radial plan or the telephone pole plan. The Willow River/Moose Lake Correctional Facility in Minnesota is based on the radial plan, and the Oregon State Correctional Institution displays the telephone pole design (see photos below). Internal security consists of such features as cell-block living, restrictions on inmate movement, and the capability of closing off areas of the institution to contain riots and disruptions.

"Supermax" Prisons Currently, nearly every state is operating either one or more "supermax" units within existing maximum-security facilities or entire "supermax" facilities.[102] Pelican Bay State Prison in northern California and Red Onion State Prison in western Virginia are examples of ultramaximum- or supermaximum-security state prisons.

Medium-Security Prisons Compared with maximum-security prisons, medium-security prisons place fewer restrictions on inmate movement inside the facility. Cell blocks often coexist with dormitory- or barracks-type living quarters, and in some medium-security prisons, cells are relatively few. Typically, there is no wall for external security. Fences and

Supermax Prisons

According to a 2014 Amnesty International report, more than 40 states now operate supermax prisons. On any given day, about 80,000 U.S. prisoners are in solitary confinement.

Source: "Entombed: Isolation in the U.S. Federal Prison System," Amnesty International, 2014, accessed May 31, 2016, www.amnestyusa.org/sites/default/files/amr510402014en.pdf.

BanksPhotos/Getty Images

Source: Oregon Department of Corrections

The Willow River/Moose Lake Correctional Facility in Minnesota is an example of the radial prison design similar to the prison in the photo on the left, and the Oregon State Correctional Institution in the photo on the right is an example of the telephone pole prison design. *What are the advantages and disadvantages of each of these prison designs?*

towers exist but are less forbidding in appearance; razor wire may be replaced with less-expensive barbed wire.

Minimum-Security Prisons Compared with prisons at the other security levels, minimum-security prisons are smaller and more open. Inmates frequently are transferred to such prisons from more secure facilities after they have established records of good behavior or when they are nearing release. Dorm or barracks living quarters predominate, and often there are no fences. Some inmates may be permitted to leave the institution during the day to work (under *work release*) or study (under *study release*) in the community. Likewise, some inmates may be granted furloughs so that they can reestablish ties with family members, make living arrangements for their upcoming release, or establish employment contacts.

Custody Level versus Security Level *Custody level* should be distinguished from *security level*. Whereas institutions are classified by security level, individual inmates are classified by custody level. An inmate's **custody level** indicates the degree of precaution that needs to be taken when working with that inmate. Confusion arises because custody levels are sometimes designated by the same terms used to designate institutional security levels (maximum, medium, and minimum). However, the two levels are independent of each other. For example, some inmates with medium or minimum custody levels may be housed in maximum-security prisons.

custody level The classification assigned to an inmate to indicate the degree of precaution that needs to be taken when working with that inmate.

Women's Prisons

The BOP has 29 locations across the United States that have one or more female institutions.[103] Twenty of the institutions are designed for the long-term confinement of female inmates. Eight of the 20 female institutions are a part of larger facilities that contain at least one other female institution; for example, a low-security institution and a minimum-security institution in Danbury, Connecticut; a low-security institution and a minimum-security institution in Dublin, California; a low-security institution and a minimum-security institution in Aliceville, Alabama; and a minimum-security institution and a female inmate medical facility near Fort Worth, Texas. Because of these multi-institutional facilities, the BOP's 20 female institutions are in 16 locations around the nation (see Figure 10.6).

The BOP manages female inmates in five types of institutions: (1) minimum security, (2) low security, (3) medium security, (4) a federal medical center, and (5) the transfer center. At year-end 2016, the BOP had custody of a total of 10,567 female inmates. (As of February 15, 2020, the BOP had custody of 12,306 female inmates or 7.1% of all

FIGURE 10.6 Locations of Institutions for Women in Federal Bureau of Prisons

Key to the map

#	Name	Number of Female Institutions on Site	Security Level(s) of Female Institution(s)
1	Federal prison camp Alderson	1	Minimum
2	Federal correctional institution Aliceville	2	Low and minimum
3	Federal prison camp Bryan	1	Minimum
4	Federal medical center Carswell	2	Minimum and administrative
5	Federal correctional center Coleman	1	Minimum
6	Federal correctional institution Danbury	2	Low and minimum
7	Federal correctional institution Dublin	2	Low and minimum
8	Federal correctional institution Greenville	1	Minimum
9	Federal correctional institution Hazelton	1	Low
10	Federal medical center Lexington	1	Minimum
11	Federal correctional institution Marianna	1	Minimum
12	Federal correctional institution Pekin	1	Minimum
13	Federal correctional institution Phoenix	1	Minimum
14	Federal correctional institution Tallahassee	1	Low
15	Federal correctional institution Victorville II	1	Minimum
16	Federal correctional institution Waseca	1	Low

Source: "Review of the Federal Bureau of Prisons' Management of Its Female Inmate Population," Office of the Inspector General, U.S. Department of Justice, September 2018, 48, Appendix 2, accessed July 28, 2019, https://oig.justice.gov/reports/2018/e1805.pdf.

inmates in the BOP system.[104]) Thirteen minimum-security institutions, also known as Federal Prison Camps (FPCs), confined 4,584 of those female inmates or 44% of the total at the end of 2016. FPCs for females have limited or no perimeter fencing and dormitory-style housing.

The BOP managed 4,363 female inmates or 41% of the total at the end of 2016 in one of six low-security institutions, also known as Federal Correctional Institutions (FCIs) or Federal Satellite Lows. FCIs for females have perimeter fencing and either dormitory-style or cell-style housing.

Ten percent of all female inmates or 1,082 of them at the end of 2016 were confined in the one medical center devoted to female inmates at FMC Carswell near Fort Worth, Texas. FMC Carswell treats female inmates of all custody levels who have serious or chronic medical problems. FMC Carswell also serves as a high-security institution for a small number of the BOP's female inmates.

In addition to the aforementioned 20 institutions, at the end of 2016, the BOP confined 538 female inmates or 5% of the total in 12 detention centers or the transfer center. Detention centers are designed to confine pretrial detainees for short periods of time, and the transfer center confines inmates in transit to other BOP facilities. Unlike institutions designed for long-term confinement, detention centers and the transfer center house both male and female inmates.

Unfortunately, recent information about state prisons exclusively for women was not available at this writing. The last year data relevant to this discussion was available was 2010, about a decade ago. At that time, 86% of all state prisons in the United States were for men only, 9% were for women only, and 5% were for both men and women (cocorrectional facilities).[105] The following generalizations will have to suffice. Many states operate only one major prison for women. Prisons exclusively for women generally are smaller and house fewer inmates than institutions exclusively for men. Dorm and cottage plans are much more common than cell-block plans in institutions for women.

Jails and Lockups

lockup A very short-term holding facility that is frequently located in or very near an urban police agency so that suspects can be held pending further inquiry.

jail A facility, usually operated at the local level, that holds convicted offenders and unconvicted persons for relatively short periods.

A **lockup** is a very short-term (for instance, 24- to 48-hour) holding facility that frequently is located in or very near an urban police agency so that suspects can be held pending further inquiry into their cases by the police or the court. If there is cause, a suspect may be transferred from the lockup to the jail.

A **jail** is a facility that holds convicted offenders and unconvicted persons for relatively short periods. The modern term "jail" comes from the English term "gaol" (the pronunciation is identical), and English gaols have a history that dates to the 1100s. In contrast to prisons, most jails are administered at the county or city level of government. Excluding lockups, there are more jails (between 3,000 and 4,000) than any other kind of confinement facility in the United States. Although most jails in the United States are small (about half of them hold fewer than 50 people), some, such as those in Los Angeles and New York City, are very large (refer to Table 10.2).

Jail Functions Most people think of jails primarily as short-term holding facilities where suspects or defendants are detained pending further court processing, such as arraignment, trial, conviction, or sentencing. In practice, though, jails serve a catchall function in criminal justice and corrections. For example, jails:

- Receive individuals pending arraignment and hold them awaiting trial, conviction, and sentencing.
- Readmit probation, parole, and bail-bond violators and absconders.
- Temporarily detain juveniles pending transfer to juvenile authorities.
- Hold mentally ill persons pending their movement to appropriate health care facilities.
- Hold individuals for the military.
- Hold individuals for protective custody.
- Hold individuals for contempt (contempt refers to conduct that defies the authority or dignity of a court or legislature).
- Hold witnesses for the courts (witnesses sometimes are held in jails for their safety or if they are unlikely to appear in court when requested).
- Release convicted inmates to the community upon completion of sentence.
- Transfer inmates to federal, state, or other authorities.
- House inmates for federal, state, or other authorities because of crowding of their facilities.
- Sometimes operate community-based programs as alternatives to incarceration (those programs are described in Chapter 12).
- Hold inmates sentenced to short terms (generally 1 year or less).[106]

Because jails must be able to hold all types of people, including those who pose grave threats to the security and safety of others, the limited funds available through local taxes are directed primarily toward custody and security. Accommodations, services, and programs for inmates often suffer as a consequence.

Jail Populations Until 2009, the number of jail inmates had risen dramatically, increasing every year since at least 1982. The U.S. local jail population increased from 209,582 in 1982 to 785,533 at mid-year 2008, an increase of about 275%. However, at mid-year 2009, the number of jail inmates was 767,434, a decrease of 2.3% from the previous year. Since mid-year 2009, the number of jail inmates has zigzagged, increasing and decreasing. The local jail population continued to decline in 2010 (by 2.4%) and 2011 (by 1.8%) but increased 1.2% in 2012, decreased 1.8% in 2013, increased 1.8% in 2014, decreased 3.1% in 2015, increased 1.8% in 2016, and increased 0.6% in 2017. At year-end 2017, the local jail population was 721,300, a decrease of 2.9% from 2009. Thus, from the 2008 high of 785,533, the local jail population decreased 5.1% by year-end 2017 (see Figure 10.7).[107] The jail incarceration rate was 96 inmates per 100,000 U.S. residents in 1983; by June 30, 2007, that number had climbed to 259 per 100,000, an increase of 170%. However, beginning in 2008, the rate has declined nearly every year: 258 in 2008, 250 in 2009, 242 in 2010, and 236 in 2011. The rate increased slightly in 2012 to 237, but declined to 231 in 2013, before increasing to 233 in 2014 then decreasing to 226 in 2015, before increasing to 229 in 2016 and 2017. Thus, from a high of 259 inmates per 100,000 U.S. residents in 2007, the local jail incarceration rate decreased by 11.6% to 229 inmates per 100,000 U.S. residents by 2017. This decrease in the jail population is not particularly surprising because in recent years the growth rate has been slowing

FIGURE 10.7 Growth in Local Jail Population, 1982–2017

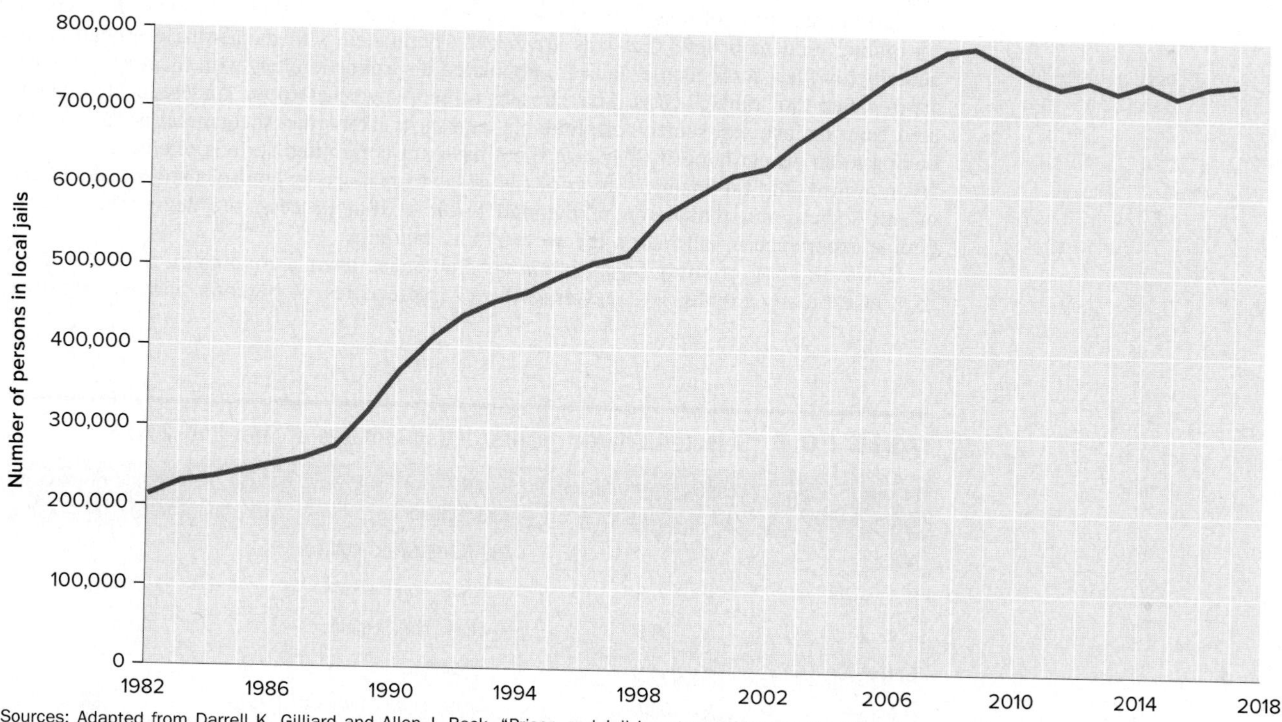

Sources: Adapted from Darrell K. Gilliard and Allen J. Beck, "Prison and Jail Inmates at Midyear 1997," in *Bureau of Justice Statistics Bulletin*, U.S. Department of Justice, January 1998; Craig A. Perkins, James J. Stephan, and Allen J. Beck, "Jails and Jail Inmates 1993–1994," in *Bureau of Justice Statistics Bulletin*, U.S. Department of Justice, April 1995; Allen J. Beck, "Prison and Jail Inmates at Midyear 1999," in *Bureau of Justice Statistics Bulletin*, U.S. Department of Justice, April 2000; Paige M. Harrison and Jennifer C. Karberg, "Prison and Jail Inmates at Midyear 2002," in *Bureau of Justice Statistics Bulletin*, U.S. Department of Justice, April 2003; Paige M. Harrison and Allen J. Beck, "Prison and Jail Inmates at Midyear 2004," in *Bureau of Justice Statistics Bulletin*, U.S. Department of Justice, April 2005; Paige M. Harrison and Allen J. Beck, "Prison and Jail Inmates at Midyear 2005," in *Bureau of Justice Statistics Bulletin*, U.S. Department of Justice, May 2006; William J. Sabol, Todd D. Minton, and Paige M. Harrison, "Prison and Jail Inmates at Midyear 2006," in *Bureau of Justice Statistics Bulletin*, U.S. Department of Justice, June 2007; William J. Sabol and Todd D. Minton, "Jail Inmates at Midyear 2007," in *Bureau of Justice Statistics Bulletin*, U.S. Department of Justice, June 2008; Todd D. Minton, "Jail Inmates at Midyear 2011—Statistical Tables," in *Bureau of Justice Statistics Bulletin*, U.S. Department of Justice, April 2012; Todd D. Minton and Zhen Zeng, "Jail Inmates at Midyear 2014," in *Bureau of Justice Statistics Bulletin*, U.S. Department of Justice, June 2015; Zhen Zeng, "Jail Inmates in 2017," in *Bureau of Justice Statistics Bulletin*, U.S. Department of Justice, April 2019.

Jail Statistics

To obtain the most recent jail statistics, go to the Bureau of Justice Statistics website at https://www.bjs.gov/content/pub/pdf/ji17.pdf.

How has the jail population changed between 2005 and 2017?

substantially. Between 2000 and 2013, for example, the jail population increased only an average of 1.3% a year and, except for 2005, 2010, 2012, 2014, and 2016, the growth rate of the jail population has decreased each year since 2002.[108]

Although less than 1% of the jail population in 2017 consisted of juveniles (under the age of 18), that still amounted to 3,500 juveniles. More than 90% of the juveniles were being held as adults.[109] The practice of holding juveniles in adult jails, where they are vulnerable to influence and victimization by adult criminals, is most common in rural areas, where there are no separate juvenile detention centers. That practice has been the target of much criticism and many policy initiatives for more than 2 decades. As a consequence, the number of juveniles held in adult jails generally has declined in recent years, except for 2006–2008 and 2009–2010: the number decreased by about 22% between mid-year 1995 and mid-year 2006 (from 7,800 to 6,102) but increased 26% between mid-year 2006 and mid-year 2008 (from 6,102 to 7,703). Between mid-year 2008 and mid-year 2009, the number of juveniles held decreased about 6%, but between mid-year 2009 and mid-year 2010, the number of juveniles held increased by about 5%. Since mid-year 2010, the number of juveniles held has decreased each year through mid-year 2015: about 22% between mid-year 2010 and mid-year 2011; 8.5% between mid-year 2011 and mid-year 2012; 14.8% between mid-year 2012 and mid-year 2013; 8.7% between mid-year 2013 and mid-year 2014; and 17% between mid-year 2014 and mid-year 2015. However, between mid-year 2015 and mid-year 2016, the number of juveniles held increased by 8.3%, but between mid-year 2016 and mid-year 2017, the number of juveniles held decreased by 10.3%.[110]

Table 10.7, which displays selected characteristics of the jail population, shows that, in 2017, most jail inmates were adult males and almost evenly split between non-Hispanic whites and all others. About 70% were charged with a felony, and nearly 65% were unconvicted and awaiting disposition of their case.

In a classic study of jails,[111] criminologist John Irwin found that, although members of the public tend to believe that jails are heavily populated with dangerous criminals, jails actually hold few such people. Jails are populated disproportionately with members of what Irwin called the "rabble" class. This consists of people who are poor, undereducated, alienated from mainstream society, disreputable, and more likely than the general population to belong to an ethnic minority. Most of them have not committed serious offenses. In short, Irwin argued that the main function of the jail is to manage or control marginal members of our society and that as an unintended result of that process, the degradation those people experience in jail makes them even more marginal.

However, another study of jail bookings challenges Irwin's conclusion that jails tend to house mostly rabble. By analyzing the characteristics of persons booked into two

TABLE 10.7 Selected Characteristics of Jail Inmates at Year-End 2017

Inmate Characteristic	Percentage of Inmates (Total = 745,200)	Inmate Characteristic	Percentage of Inmates (Total = 745,200)
AGE		**RACE/HISPANIC ORIGIN**	
Juvenile	0.5	White, non-Hispanic	49.7
Adult	99.5	Black, non-Hispanic	33.6
GENDER		Hispanic	14.5
Male	84.7	American Indian/Alaska Native	1.2
Female	15.3	Asian	0.6
MOST SERIOUS TYPE OF OFFENSE		Other (including two or more races)	0.4
Felony	69.4	**CONVICTION STATUS**	
Misdemeanor	26.1	Convicted	35.3
Other	4.5	Unconvicted	64.7

Source: Calculated from data in Zhen Zeng, "Jail Inmates in 2017," in *Bureau of Justice Statistics Bulletin*, U.S. Department of Justice (Washington, DC: U.S. Government Printing Office, April 2019), 6, Table 4, accessed July 27, 2019, https://www.bjs.gov/content/pub/pdf/ji17.pdf.

jails (one urban, one rural), researchers found that nearly 47% had been charged with felonies, and more than 90% had been booked for a felony or a class A misdemeanor. Moreover, in 1-day counts of those jails, the researchers discovered that 82.5% of those held had been either charged with or convicted of a felony offense. As the authors of the study observe, "These data do not appear to support Irwin's claims about detached and disreputable persons whose real problem is offensiveness, not serious criminality."[112] Note that the study critical of Irwin's theory only examined two jails. Data reported by the U.S. Bureau of Justice Statistics shows that in 2017, nearly 70% of jail inmates were either charged with or convicted of felonies; also not supporting Irwin's claims.[113]

Jail Architecture and Management Philosophies Jails traditionally have represented, and continue to represent, one of the most problematic aspects of criminal justice. Many jails in the nation are old buildings plagued by overcrowding, a lack of services and programs for inmates, inadequate staffing, and unsanitary and hazardous living conditions. Some of the interrelated reasons for those problems are (1) the limited and unstable nature of local taxes to fund and staff jails; (2) a general lack of public support for jail reform; (3) rapid rates of inmate turnover, which make it difficult to coordinate programs; and (4) the sheer diversity of the risks and needs of the inmates. Also, in many areas, the chief jail administrator is the local sheriff, an elected or appointed political official. Because jails are usually under local administration, they sometimes are affected by the erratic and corrupt elements that often characterize local politics.

With increasing pressure from the courts to reform jail conditions and management practices,[114] efforts at jail reform continue. Historically, jails have progressed through three general overlapping stages of architectural design. Each design reflects a different philosophy about how to operate a jail.[115] The earliest jails, which date back to the eighteenth century, were built in a linear design. These *first-generation jails* have inmates live together in cells, dormitories, or "tanks." The cells line corridors, which make the supervision of inmates difficult (see Figure 10.8). At regular intervals, guards (they did not become "correctional officers" until the late 1960s or early 1970s—see the FYI titled First-Generation Jails) walk up and down the corridors and observe inmates in their cells. To observe the inmates in the dormitories or tanks, guards periodically walk through the dormitories or along a perimeter catwalk, which separates them from the inmates by bars. (This type of supervision is referred to as "linear/intermittent surveillance.") A "48-man tank," for example, is a large cage that might have six inner cells, each with eight bunks, that open into a "dayroom" or "bullpen" equipped with two long metal picnic-like tables

First-Generation Jails

The first-generation jail in which Robert Bohm (an author of this textbook) worked in the early 1970s was located on the eleventh to fifteenth floors of the Jackson County Courthouse in Kansas City, Missouri. When Bohm was hired as a guard, the official title of the jail was the Jackson County Jail. During the time he worked there, the official title of the jail was changed to the Jackson County Department of Corrections, and Bohm became a "correctional officer." Physically, the jail remained the same, and Bohm's work duties did not change.

Source: Personal experience.

Karl Gehring/The Denver Post/Getty Images

In first-generation jails, correctional officers cannot observe all inmate housing areas from one location, so they can only provide intermittent surveillance. *What are the advantages and disadvantages of working in this type of jail?*

FIGURE 10.8 A Typical First-Generation Jail Design

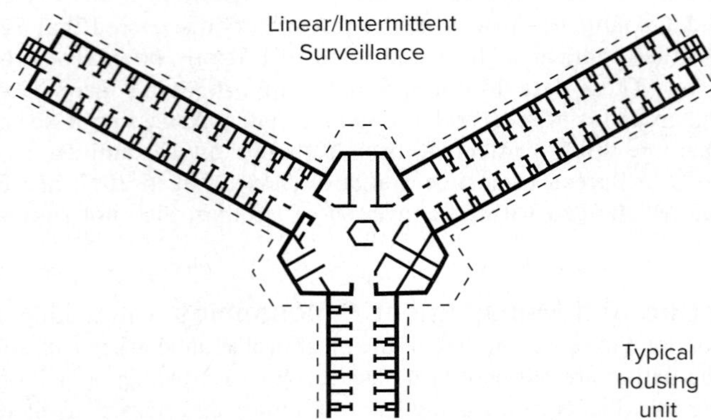

Source: U.S. Department of Justice, National Institute of Corrections, Jails Division, *Podular, Direct Supervision Jails: Information Packet*, January 1993, www.nicic.org/resources/topics/DirectSupervisionJails.aspx.

attached to the floor and, perhaps, one television for the entire tank. In the cells, the beds, sinks, and toilets are made of reinforced metal and are bolted to the floor or wall. There is little direct contact between the guards and inmates unless the guards have to respond to an incident, such as a fight or medical emergency. The first-generation jail is often a separate building or set of buildings surrounded by walls of reinforced concrete topped with razor wire (but see the FYI titled First-Generation Jails). Many first-generation jails resemble early prisons.

Second-generation jails began to emerge in the 1960s. In these jails, correctional officers constantly supervise inmates from secure control booths that overlook inmate living areas, called "pods" or "modules." (This type of supervision is referred to as *indirect* or *podular supervision* or *remote surveillance*.) Cells become rooms and are clustered around dayrooms or common living areas, where inmates can congregate for activities (see Figure 10.9). Inmate rooms have reinforced metal doors with unbreakable windows instead of bars and generally house only one or two inmates. Beds, sinks, toilets, desks, and tables (in the dayrooms) are made of concrete or reinforced metal bolted to the floor or wall. There is very little direct interaction between inmates and correctional officers. Officers generally communicate with inmates by using a public address or intercom system. When trouble occurs inside the living area, a response team is called to intervene. The facility's

FIGURE 10.9 A Typical Second-Generation Jail Design

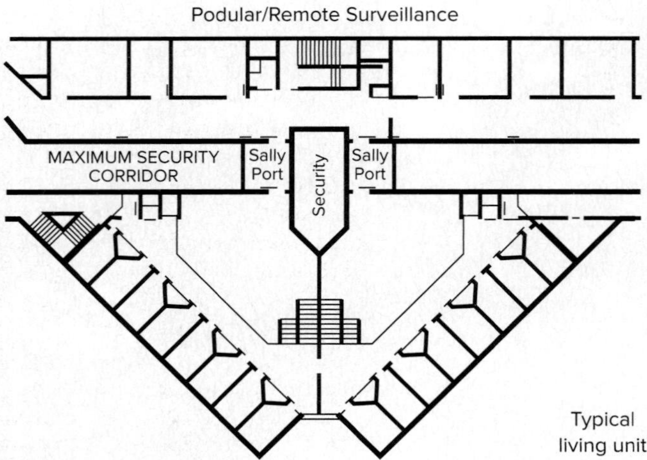

Source: U.S. Department of Justice, National Institute of Corrections, Jails Division, Podular, Direct Supervision Jails: Information Packet, January 1993, www.nicic.org/resources/topics/DirectSupervisionJails.aspx.

ZUMA Press Inc/Alamy Stock Photo

In second-generation jails, correctional officers continuously observe inmates from a secure control booth that overlooks inmate living areas, called "pods" or "modules." They provide indirect or remote supervision. *What are the advantages and disadvantages of working in this type of jail?*

perimeter walls continue to be made of reinforced concrete topped with razor wire or, in some cases, fences secured by razor wire.

Third-generation, *new-generation*, or *direct-supervision jails* emerged in the late 1970s. They use a "podular" or "modular" design similar to the one employed in second-generation jails in which inmates' rooms are arranged around a common area or dayroom. A third-generation jail may have one or more pods, with a pod typically containing 48 to 60 beds. However, in third-generation jails, there is no secure control booth for correctional officers, and there are no physical barriers between correctional officers and inmates (see Figure 10.10). Officers may have a desk or table for paperwork, but it is in the open dayroom area. Many new-generation jails have added amenities in inmate living areas, such as carpeting, upholstered furnishings, individual shower stalls, several television sets and viewing areas, collect-call telephones, game tables, and exercise equipment. Third-generation jails generally have strong perimeter security.

The pods in third-generation jails are self-contained to reduce costly and potentially dangerous inmate transport. Consider visitation, for example. In older jails, including many second-generation jails, inmates are moved from secure housing units to a visitation area. In many new-generation jails, visitors are taken to the housing unit by way of a nonsecure

FIGURE 10.10 **A Typical Third-Generation Jail Design**

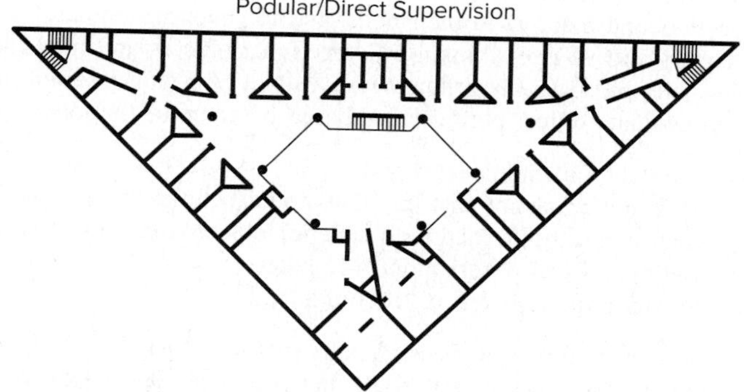

Source: U.S. Department of Justice, National Institute of Corrections, Jails Division, *Podular, Direct Supervision Jails: Information Packet*, January 1993, www.nicic.org/resources/topics/DirectSupervisionJails.aspx.

Third-generation jails employ a "podular" or "modular" design, but unlike in second-generation jails, correctional officers are stationed inside the housing unit where they directly supervise and interact with inmates. *What are the advantages and disadvantages of working in this type of jail?*

©Marilyn Humphries

Third-Generation Jails

To learn more about third-generation jails, access the Corrections Center of Northwest Ohio website at https://www.ccnoregionaljail.org/newgenerationjail.htm. Be sure and take the virtual tour.

What might be the features of the next-generation jails?

corridor and visit through a high-security glass barrier. The inmate stays securely in the housing unit and the visitor stays in the nonsecure corridor. Some new-generation jails are employing new technology for visitation. They are installing state-of-the-art video screens and phones in cubicles or carrels in inmate living areas and in remote video visitation centers for family members and friends. This arrangement also stops contraband from entering the institution by way of visitors. Attorneys, chaplains, or others visiting inmates for professional reasons, however, still are allowed to visit in person.[116] To reduce inmate transport, many new-generation jails also have their own courtroom, which is used primarily for arraignments and other pretrial proceedings.

The major difference between second- and third-generation jails is that third-generation jails are based on a different philosophy or management strategy called *podular* or *direct supervision*. In this strategy, correctional officers continue to constantly supervise inmates, but rather than isolate themselves from the inmates, they are stationed inside the pod to give the inmates the opportunity to directly and regularly interact with them. (Direct supervision also is employed in some prisons.) In the new-generation jail, each congregate living area is a staff post, usually with one correctional officer in direct control of 40 to 50 inmates.

Correctional officers in direct-supervision jails are more than just turnkeys; they are professionals who need skills in interpersonal communication, crisis intervention, and counseling. Perhaps that is why research has found that female officers do at least as well as male officers while working in male units. Officers are taught to view the pods as space that they control, to be leaders and mentors in their pods, and to encourage inmates to change their destructive behaviors and ways of thinking. Correctional officers are more independent in direct-supervision jails and have the primary responsibility for daily decision making in the pods. Supervisors evaluate officer performance based on some of the following objectives:

- Being in daily contact with inmates in pods.
- Refraining from doing or saying anything that would belittle or degrade an inmate.
- Reporting promptly critical information to superior officers (e.g., inmate escape plans).
- Engaging in continual visual observation of inmates.
- Investigating activity appearing out of the ordinary.[117]

Research suggests that direct-supervision jails provide a less-stressful, more positive, and safer environment for inmates and staff; may enhance supervision by reducing the number of unsupervised areas; and reduce inmate violations for contraband possession, destruction of property, escapes, disrespecting officers and staff, suicides, and violence, including sexual assaults.[118] Figure 10.11 shows the American Jail Association's Code of Ethics for Jail Officers. It reflects the professionalism expected of correctional officers who work in today's jails.

FIGURE 10.11 American Jail Association's Code of Ethics for Jail Officers

As an officer employed in a detention/correctional capacity, I swear (or affirm) to be a good citizen and a credit to my community, state, and nation at all times. I will abstain from questionable behavior which might bring disrepute to the agency for which I work, my family, my community, and my associates. My lifestyle will be above and beyond reproach and I will constantly strive to set an example of a professional who performs his/her duties according to the laws of our country, state, and community and the policies, procedures, written and verbal orders, and regulations of the agency for which I work.

On the Job I promise to:

KEEP	The institution secure so as to safeguard my community and the lives of the staff, inmates, and visitors on the premises.
WORK	With each individual firmly and fairly without regard to rank, status, or condition.
MAINTAIN	A positive demeanor when confronted with stressful situations of scorn, ridicule, danger, and/or chaos.
REPORT	Either in writing or by word of mouth to the proper authorities those things which should be reported, and keep silent about matters which are to remain confidential according to the laws and rules of the agency and government.
MANAGE	And supervise the inmates in an evenhanded and courteous manner.
REFRAIN	At all times from becoming personally involved in the lives of the inmates and their families.
TREAT	All visitors to the jail with politeness and respect and do my utmost to ensure that they observe the jail regulations.
TAKE	Advantage of all education and training opportunities designed to assist me to become a more competent officer.
COMMUNICATE	With people in or outside of the jail, whether by phone, written word, or word of mouth, in such a way so as not to reflect in a negative manner upon my agency.
CONTRIBUTE	To a jail environment which will keep the inmate involved in activities designed to improve his/her attitude and character.
SUPPORT	All activities of a professional nature through membership and participation that will continue to elevate the status of those who operate our nation's jails. Do my best through word and deed to present an image to the public at large of a jail professional, committed to progress for an improved and enlightened criminal justice system.

The American Jail Association's Board of Directors has approved the AJA Code of Ethics as part of an integral program to achieve a high standard of professional conduct among those officers employed in our nation's jails. Adopted by the American Jail Association Board of Directors on November 10, 1991. Revised May 19, 1993. Re-affirmed on May 3, 2008, by the AJA Board of Directors in Sacramento, California.

Source: American Jail Association, https://www.americanjail.org/files/Resolutions/AJA%20Resolution%20-%20Code%20of%20Ethics%20for%20Jail%20Officers.pdf. The American Jail Association, 1135 Professional Court, Hagerstown, MD 21740-5853.

THINKING CRITICALLY

1. What, if any, effect does a jail's architectural design have on inmates?
2. Do you agree that one of a jail's functions is to manage marginal members of society? Why or why not?

Institutional Security, Services, and Programs

In many ways, an incarceration facility is like a miniature society within the larger society. Institutions have many of the same features as the wider society, such as security procedures for maintaining order and preserving the safety of inhabitants, as well as a variety of services and programs meant to provide for inmate needs and encourage inmates to better themselves.

Security and Inmate Discipline

An orderly and safe environment is the foundation for all else that happens in an institution. When the environment is not stable, everything else tends to become secondary. For that reason, security procedures strongly affect the daily activities of both staff and inmates.

In any prison or jail, special security precautions are directed toward certain locations because of the importance of those locations to the institution's capacity for maintaining order. Examples of such locations include:

1. The front entry to the facility, through which all persons coming and going must pass.
2. The control room, which is usually located close to the front entry and is the heart of the institution's communication system.
3. The cell blocks or other quarters where the inmates live.
4. The dining area.
5. The area where the institution's confidential records and documents are maintained.
6. The indoor recreation areas and the outdoor recreation area or yard.
7. The sites where inmates work.

All institutions routinely employ a range of security procedures to maintain control over inmates. The classification of inmates by custody levels, mentioned earlier, is one such method. An inmate's custody level indicates the degree of precaution to be used when working with that inmate. In some facilities, certain inmates are given special custody designations that distinguish them from members of the general inmate population. Those inmates are not permitted to live among the general population of the facility. For example, inmates who are vulnerable to assault by other inmates may be designated for **protective custody**, meaning that they are to be kept segregated for their own safety. In contrast, inmates who represent a danger to other inmates or staff may be designated for **administrative segregation**, indicating that they must be kept in secure isolation so they cannot harm others. Inmates who display signs of serious mental disorders may be segregated in a similar fashion. The percentage of state prison inmates held in either protective custody or administrative segregation is small, probably about 5%, although no recent data are available to confirm this estimate.[119]

protective custody The segregation of inmates for their own safety.

administrative segregation The keeping of inmates in secure isolation so that they cannot harm others.

As a basis for security procedures, institutions publish written rules that regulate the daily activities of inmates and staff alike. During the course of daily activities, staff members routinely count inmates to detect escapes, and inmates' whereabouts are constantly monitored within the facility. There are standard procedures for staff to follow when transporting certain inmates within the facility or from the facility to an outside location, such as to another prison, to court, or to a hospital in the community. Much effort goes into controlling the property of the institution—for example, firearms, medicine, keys, tools, and basic commodities like clothes and food. In many prisons and jails, searches of inmates' clothing and bodies and shakedowns (usually random and warrantless searches) of their cells are commonplace in an effort to control the flow of contraband, especially handmade knives and other weapons. There are special procedures and even special staff units for responding to riots, escapes, and other disturbances.

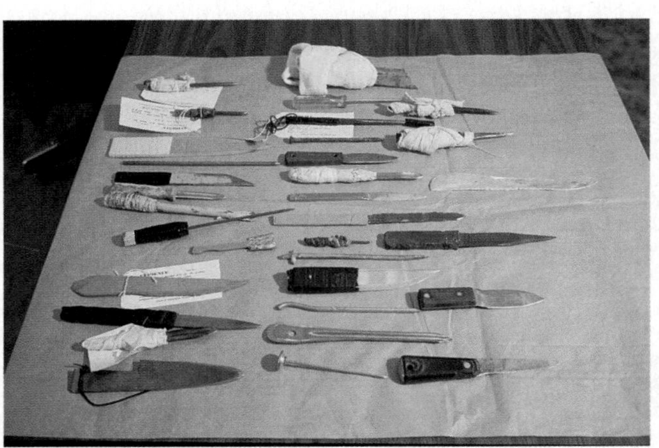

Matthew McVay/Stock Boston Inc.

Handmade weapons such as those shown here are contraband and make prisons and jails dangerous places. *Where do inmates get the materials for these weapons?*

The mail and phone conversations of inmates may be monitored if there is sufficient security justification. Inmates may visit with relatives and friends only at designated times and may visit only those people the institution has approved. Although a small number of prisons permit **conjugal visits**—in which an inmate and his or her spouse or significant other visit in private to maintain their personal relationship (see FYI titled Conjugal Visits)—routine visits usually occur in large, open rooms with other inmates, their visitors, and staff present. Visits are supervised closely because of the potential for contraband to enter the prison (but see the discussion of visitation in new-generation jails in the previous section).

Note that written rules and regulations are not the exclusive basis for institutional security. Written rules and regulations are part of an institution's formal bureaucratic structure, which was discussed earlier in this chapter. Within that formal structure, there develops an unwritten, informal structure of norms and relations that is vital to the operation of the facility. For example, most institutions have an elaborate **snitch system** in which staff members learn from inmate informants about the presence of contraband, the potential for disruptions, and other threats to security. Informants often receive special concessions and protection in exchange for snitching on other inmates. Such arrangements can be very elaborate, as illustrated by the research of criminologists James Marquart and Ben Crouch. In their study of a Texas prison, Marquart and Crouch found that certain elite inmates, known as building tenders, actually functioned as extensions of the uniformed guard force to achieve control of the institution.[120] The use of inmates (building tenders or trustees) to discipline other inmates was formally outlawed in 1984.

Inmates who violate formal institutional regulations may be subjected to disciplinary measures. Staff members typically have broad discretion when they detect rule violations. They simply may overlook the infraction or may issue an informal warning. Alternatively, they may file a disciplinary report so that formal sanctions can be considered. The report usually leads to a disciplinary hearing during which the charges and supporting evidence are presented to the institution's disciplinary committee or hearing officer. If the report is found to be valid, a number of sanctions are possible. For example, the inmate may have some of his or her privileges temporarily restricted (e.g., no commissary or store privileges for 30 days), may be placed in solitary confinement or "the hole" for a specified number of days, may forfeit some of the time that has been deducted from his or her sentence for good behavior, or may even be transferred to another facility. Because institutions have many rules governing the behavior of inmates, it should come as no surprise that rule infractions and disciplinary measures occur frequently.[121]

Eric Risberg/AP Images

Family visitation is extremely important to inmate morale. *What are other advantages and disadvantages of family visitation?*

Conjugal Visits

In 1918, James Parchmann, the warden at Mississippi State Penitentiary, introduced conjugal visits as an incentive for inmates to work harder. He even provided inmates with prostitutes. Twenty years ago, 17 states provided conjugal visits. Today, conjugal visitation survives only in four states: California, Connecticut, New York, and Washington. The BOP does not permit conjugal visits.

Source: Victoria Cavaliere, "How Conjugal Visits Work," vocativ, May 24, 2016, accessed August 4, 2019, https://www.vocativ.com/underworld/sex/conjugal-visits-work/index.html.

conjugal visits An arrangement whereby inmates are permitted to visit in private with their spouses or significant others to maintain their personal relationship.

snitch system A system in which staff learn from inmate informants about the presence of contraband, the potential for disruptions, and other threats to security.

Services and Programs

Many of the human services and programs found in the free society are duplicated within institutions. At a minimum, inmates must be fed, clothed, and provided with such basic shelter requirements as warmth, electricity, and plumbing, and their health care needs must be addressed.

Food services are an important part of an institution's operation. Waste and inefficiency must be avoided to control expense, but inmate demand and dietary standards must be met. Some institutions, particularly those under budgetary constraints, attempt to save money by serving only two meals a day or cheaper food. For example, in 2008, the sheriff of Polk County, Florida, announced that he would be serving cheaper food to his jail's 2,400 inmates to shave approximately $161,000 from his annual budget. Instead of corn bread, he would be serving crackers and would save $33,304; tea and juice would be replaced by water, saving $56,630; one slice of bread would be served instead of two slices, saving $25,116; two fresh eggs would be replaced with one frozen egg patty, saving $24,545; bologna sandwiches would replace peanut butter and jelly sandwiches, saving $11,076; and fresh milk would be replaced with powdered milk, saving $10,545. All these changes met the approval of a nutritionist, followed state jail nutrition guidelines, and still provided inmates with a 2,300- to 2,800-calorie daily diet.[122] All detention facilities must have a licensed dietician review their menus in order to be accredited by the American Correctional Association.[123] Some institutions are notorious for serving cheap food even under normal budgetary circumstances. For example, as of 2018, Alabama counties still operated under a 90-year-old law that requires sheriffs to spend only $1.75 a day on food for jail inmates. The law also allows the sheriffs to keep any money they do not spend. No one knows how much money sheriffs make on "food profits" because they are not required to report their private accounts to state auditors. However, in 2018, Etowah County Sheriff Todd Entrekin admitted that over 3 years he pocketed more than $750,000 that was budgeted to feed jail inmates and used $740,000 of it to buy a beach house. Entrekin earns about $93,000 annually, not counting food provision funds. Former inmates reported getting meat maybe once a month, and every other day getting just beans and vegetables. Sheriffs argue that they do not make much money on food service, and the state holds them personally liable for budget shortfalls and, possibly, lawsuits over jail food.[124]

Prisons and jails are buildings that, like all buildings, require maintenance and repair. Inmates as part of their job assignments perform a large portion of the maintenance and repair work.

Courts have held that inmates are entitled to medical and dental services. Prisons normally have an infirmary where less serious ailments can be treated. More serious problems necessitate transfer to either a prison with more extensive medical services or a hospital in the local community.

Institutions make an array of services available to inmates for their leisure time. We already have mentioned mail, phone, and visitation services. In addition, institutions operate commissaries where inmates can purchase such items as food, radios, reading materials, and arts and crafts supplies. A number of recreational facilities also are provided, such as weightlifting equipment, softball fields, basketball courts, game tables, and television viewing areas. Also, legal resources and religious services are available (discussed in more detail in the next chapter). Although most prisons still offer a substantial number of services to prisoners, over the past 2 decades, there has been a movement to reduce those services.

myth

The high cost of incarceration is a product of amenities provided to inmates, such as cable television, law libraries, and weightlifting equipment. Eliminating these "country club" add-ons will make prison costs more reasonable.

fact

Actually, four out of every five dollars of prison operating costs are for employee salaries and facility maintenance. In addition, debt service to finance prison construction triples the original cost of construction. A maximum-security prison in New York State, for example, costs about $100,000 per bed to build, but financing costs add another $200,000 per bed. Finally, costs have grown substantially as a result of the health care needs of an increasingly older inmate population.

Sources: *Seeking Justice: Crime and Punishment in America* (New York: Edna McConnell Clark Foundation, 1997), 9; also see Beth Pelz, Marilyn McShane, and Frank P. Williams III, "The Myth of Prisons as Country Clubs," in *Demystifying Crime & Criminal Justice,* 2nd ed., ed. R. M. Bohm and J. T. Walker (New York: Oxford, 2013), 267–77.

Inmates with Special Needs All institutions have special-needs populations. Although those populations generally consist of far fewer inmates than the general prison population, it still is necessary to provide services for them. For example, elderly inmates frequently need more medical attention than younger inmates. Correctional officials generally consider 50-year-old inmates elderly inmates or geriatrics because their lives often have been filled with violence, drug, alcohol and tobacco abuse, poor diets, and inadequate medical care.[125] The number of elderly prison inmates is increasing. At year-end 2007, 10.4% or about 148,000 state and federal inmates were age 50 or older, and 1.1% or about 15,500 state and federal inmates were 65 or older.[126] By year-end 2017, 20.3% or 292,279 state and federal inmates were age 50 or older, and 2.8% or 40,314 state and federal inmates were age 65 or older.[127] Thus, in a decade, the number of state and federal inmates age 50 or older increased approximately 97%, and the number of state and federal inmates age 65 or older increased nearly 160%. In 2018, the cost of housing elderly inmates was estimated to be $16 billion a year, which will continue to grow in the future if nothing is done.[128] In the federal prison system, each inmate age 55 or older costs the government about $69,000 a year in maintenance costs, two or three times as much as is spent on inmates younger than age 55. The average cost to house a state prison inmate age 50 and older is about $68,000, compared to about $34,000 to house an average state prison inmate.[129] Most of the added expense is related to health costs. One health-related problem that has not been addressed adequately is the decreased mobility of older inmates–a problem that current prisons rarely are designed to accommodate.[130]

Another special-needs population is the mentally ill. Since the deinstitutionalization of the mentally ill in the late 1950s in favor of treating them in the community and the disappointing performance of community alternatives, prisons and especially jails have become surrogate public mental hospitals. Approximately 20% of jail inmates, 15% of state prison inmates, and 4% of federal prison inmates are estimated to have a serious mental illness, such as schizophrenia, schizoaffective disorder, bipolar disorder, major depression, or brief psychotic disorder.[131] That means that in 2017, more than 350,000 inmates in the United States had a serious mental disorder: 149,040 jail inmates, 195,946 state prison inmates, and 7,831 federal prison inmates. However, the number of federal prison inmates with a serious mental condition likely is grossly underestimated (more about this later), as may be the number of jail inmates and state prison inmates. (A report by the Treatment Advocacy Center claims that, in 2016, "nearly 400,000 inmates in US jails and prisons were estimated to have a mental health condition.")[132] A 2015 source pegs the number of mentally ill inmates in prisons and jails at 450,000.[133] Regardless, if 350,000 individuals with serious mental illness were incarcerated in the United States in 2017, that would be nearly ten times the number of patients who were in the nation's state psychiatric hospitals.[134] Moreover, in 44 states, a jail or prison houses more individuals with a serious mental condition than the largest remaining state psychiatric hospital in that state, and in every U.S. county with both a county jail and a county psychiatric facility, more individuals with serious mental disorders are incarcerated than hospitalized, confirming the observation that prisons and especially jails have become the nation's de facto mental health care providers.[135]

In the nation's jails, where the problem is greatest, about twice the percentage of women than men have been diagnosed with a serious mental disorder. Nearly 15% of men or approximately 95,000 male jail inmates in 2017, and 30% of women or about 34,000 female jail inmates in 2017 were diagnosed with a serious mental disorder compared to about 3% of men and 5% percent of women in the general population.[136] Although estimates vary, the prevalence of serious mental conditions is likely at least two to four times higher for state prison inmates than it is for the general population.[137]

In addition to serious mental illness, nearly 68% of jail inmates and more than 50% of state prison inmates have a diagnosable substance abuse problem, compared to 9% of the general population. More than 70% of jail inmates with a substance abuse problem also have been found to have a serious mental illness condition.[138]

Research indicates that as many as 50% of Americans who suffer from serious mental illness will be arrested at some point in their lives.[139] Studies of the most frequently arrested people in New York, Los Angeles, and other major cities show that "they are far more likely than others to have mental illness, to require antipsychotic medications while incarcerated and to have a substance use problem."[140] Most mentally ill inmates who are arrested are charged with misdemeanors such as trespassing or disorderly conduct. They also are frequently charged with drug- or alcohol-related offenses. Inmates with more serious psychiatric problems, such as paranoid schizophrenia, sometimes are arrested for assault because they mistakenly believe that someone was following them or trying to hurt them. Police also arrest people with severe psychiatric problems to protect them from harm (so-called mercy bookings). This is especially true of women on the streets who are easily victimized, including being raped. Family members sometimes have mentally ill relatives arrested because it is the easiest and most effective way to get the mentally ill relative needed treatment. Thus, mentally ill people frequently are held in jails even though they have not been charged with a crime. Many of them are held under state laws that permit emergency detentions of people suspected of being mentally ill. They typically are incarcerated pending a psychiatric evaluation or until they can be transported to a psychiatric hospital. Sometimes, being incarcerated is the only way for mentally ill people to get needed treatment.[141]

Yet, despite the obvious need, only about 15% of prison or jail inmates receive appropriate treatment.[142] Prisons and jails simply are not equipped to do the job. The problem is nationwide and growing. For example, in 2017, 43% of jail inmates at New York's Rikers Island, the second largest jail in the United States, were diagnosed with a mental illness, compared to about 30% of jail inmates in 2010.[143] Table 10.8 shows the ten states that offer the most and the least access to mental health care for their incarcerated populations. Generally, states with the highest incarceration rates offer the least access to mental health care, while states with the lowest incarceration rates provide the most access to mental health care.

Previously, we mentioned that the 4% figure for federal inmates with a serious mental illness in 2017 probably was a gross underestimation. Here we explain why. In 2014, the Federal Bureau of Prisons changed its policy regarding the treatment of inmates with mental-health conditions because of intense criticism and legal pressure. The BOP promised better treatment.[144] However, according to information received by The Marshall Project through a Freedom of Information Act request, instead of expanding treatment, the BOP decreased the number of inmates designated for mental-health treatment by more than 35%. Prison staff was instructed to determine that inmates did not require mental-health treatment, including inmates with long histories of psychiatric illnesses. The major problem was that, although the BOP changed its policy, it was not provided the resources, especially increased mental-health staffing, to implement it, which created an incentive for employees to downgrade inmates to lower treatment levels or to not recommend treatment at all. One likely result was that the number of suicides, suicide attempts, and self-inflicted injuries

TABLE 10.8 The 10 States that Provide the Most and the Least Access to Mental Health Care for Their Incarcerated Populations, 2015

States Providing the Most Access (1 = Most)	States Providing the Least Access (1 = Least)
1. Vermont	1. Nevada
2. Massachusetts	2. Alabama
3. Maine	3. Mississippi
4. Connecticut	4. Tennessee
5. Minnesota	5. Georgia
6. New Hampshire	6. Texas
7. South Dakota	7. South Carolina
8. Rhode Island	8. Arkansas
9. Iowa	9. Florida
10. Alaska	10. Idaho

Source: Adapted from "Access to Mental Health Care and Incarceration," Mental Health America, accessed August 5, 2019, https://www.mentalhealthamerica.net/issues/access-mental-health-care-and-incarceration.

increased 18% from 2015, when the BOP began tracking such figures, through 2017. In short, because the new policy did not receive adequate resources and the staff response, the claim that only 4% of federal inmates have serious mental illnesses seems absurd.

The lack of resources for treating mentally ill inmates is not only a problem for the BOP; it also is a problem with prisons and jails, but less so for prisons than for jails. Incarcerating mentally ill inmates is expensive. The U.S. Department of Justice estimates that it costs billions of additional dollars a year to house inmates with psychiatric disorders. Although the BOP states that it does not track mental health care services costs systemwide and by institution, in 2016, it reported that, systemwide, it spent more than $80 million on just three items: approximately $72 million on psychology services, $5.6 million on psychotropic drugs, and $4.1 million on mental health care in residential reentry centers.[145] One source maintains that the cost of incarcerating people with mental illness is about three times the cost of incarcerating people without mental illness, and that incarcerating an inmate in jail costs about $31,000 a year, compared to about $10,000 a year in the community.[146] Another source relates that "on average, at the Cook County Jail [in Chicago], it cost $143 a day to incarcerate someone who is not mentally ill, but twice as much if the individual has a psychiatric condition."[147] Mentally ill inmates may require segregated housing for their own protection and the protection of other inmates, frequent consultation with psychologists and psychiatrists, efforts to ensure that they take prescribed medication, and special safety precautions (e.g., for a suicidal inmate, the use of paper sheets rather than cloth to prevent hanging).

Besides generally inadequate treatment services, the mentally ill do poorly in the criminal justice system in other ways. They are less likely to be released on bail, more likely to receive longer sentences, more likely to be placed in solitary confinement, less likely to be paroled, and more likely to commit suicide than are other prisoners without mental illness.[148] Because many of them are unable to describe their symptoms to correctional officers, their needs frequently are ignored. Even if they get treatment, jails and prisons often are brutal places for mentally ill inmates, where they are vulnerable to assault, rape, and exposure to infectious diseases. Still, for most prisoners, jail and prison remain their best opportunity for getting mental health care.

Without implying a cause-and-effect relationship, the nexus between people with serious mental illness and the criminal justice system manifests itself in at least three other disturbing ways. First, according to a comprehensive study of mass shootings in public spaces in 2017 by the Secret Service National Threat Assessment Center, 64% of the shooters experienced mental health symptoms prior to the attacks. The most common symptoms were psychosis-related, such as paranoia, hallucinations or delusions, and suicidal thoughts. In addition, prior to the attacks, 25% of the shooters had been hospitalized for their mental health problems or had been prescribed psychiatric medications.[149] Second, according to a study by the *Washington Post*, about 25% of the nearly 1,000 fatal police shootings in 2016 and 2017 involved a person with mental illness.[150] Third, about 50% of the executions in the United States between 2000 and 2015 involved people who had been diagnosed as adults with a mental illness and/or substance abuse problem. When a legal settlement required California to construct a psychiatric unit on its death row at San Quentin, the 40 beds were filled as soon as the facility opened.[151]

A third special-needs population consists of those inmates with HIV, AIDS, and other infectious diseases. Much controversy surrounds the presence of the human immunodeficiency virus (HIV) and acquired immune deficiency syndrome (AIDS) among institutional populations. As of year-end 2015, 1.3% of state and federal prison inmates (17,146) were reported by prison officials to be HIV-positive.[152] The total was down from 20,093 (1.5% of state and federal prison inmates) in 2010. The estimated rate of HIV/AIDS among state and federal prison inmates declined nearly 11%, from 1,453 cases per 100,000 inmates in 2010 to 1,297 cases per 100,000 inmates in 2015. The number of state prisoners who died from AIDS-related causes declined about 38% between 2010 and 2015, from 73 to 45 deaths. The rate of AIDS-related deaths among state prisoners with HIV/AIDS declined from 6 deaths per 100,000 inmates in 2010 to 4 deaths per 100,000 inmates in 2015. Since 1991, when HIV and AIDS data were first collected for state prison inmates, 1995 was the peak year for the number of AIDS-related deaths with 1,010, and 1994 was the peak year for the rate of AIDS-related deaths per 100,000 state prison inmates at 104. The AIDS-related death rate in state prisons fell below the rate in the general U.S. population in 2009. Only one federal prison inmate died from AIDS-related causes in 2015. The number of federal prison inmates who have died from AIDS-related causes since 1999, when these data were first collected, has fluctuated between a high of 27 in 2005 to a low of zero in 2014.

The declines between 2010 and 2015 were consistent with trends over the previous 15 years. The number of state and federal prison inmates with HIV peaked in 1998, when nearly 26,000 inmates were afflicted. From 1998 to 2015, inmates with HIV decreased each year from 0.1% to about 7%. The decrease from 1998 to 2015 is almost entirely the result of the more than 37% decline of HIV among state prison inmates, from 24,910 to 15,610. During the same period, federal prison inmates with HIV increased 44%, from 1,066 to 1,536.

Three states—Florida (2,571), Texas (2,082), and New York (1,820)—accounted for 41.5% of all HIV-infected state prison inmates at year-end 2015. New York had the largest percentage of its custody population (3.5%) that was HIV-positive or had confirmed AIDS in 2015. Louisiana was second with 3.4%, and Florida was third with 3.0%. Intravenous drug use is the key risk factor for AIDS among inmates.

HIV testing in prison—a sometimes controversial practice—may occur during the intake process, while in custody, or during the discharge process. In 2015, Texas and 14 other states, which incarcerated 34% of admitted inmates, tested all inmates for HIV as part of the prison intake process, a practice known as "mandatory testing." California and 16 other states, which incarcerated 31% of admitted inmates, offered an inmate the opportunity to "opt-out" of HIV testing. In this situation, all inmates were offered the test and the test was given unless the inmate refused to take it. In 2011, by comparison, 14 states had mandatory testing, and 13 states allowed inmates to opt-out. Florida and seven other states, which imprisoned 13% of admitted inmates in 2015, employed an "opt-in" policy, where newly admitted inmates were offered the opportunity to take the test but had to opt-in or ask to be given the test. In 2015, Connecticut was the only state that did not test for HIV at intake.

In 2015, if an inmate in custody asked for HIV testing, the BOP, Texas, and 40 other states provided it—a decrease from the 43 states and the BOP that provided testing upon request in 2011. The BOP, California, and 39 other states offered HIV testing if there was a clinical symptom in 2015—an increase of one state since 2011. The BOP provided testing in both years. The BOP, New York, and 36 other states provided HIV tests if inmates were party to an incident, such as an accident—up from 33 states and the BOP in 2011. The BOP, Ohio, and 27 states tested inmates for HIV if they were court ordered to do so—down from 34 states in 2011. Routine medical exams were another opportunity for inmates to receive an HIV test in Florida and 17 other states (but not the BOP) in 2015—an increase from 14 states and the BOP in 2011. The BOP, North Carolina, and 17 other states HIV tested all inmates in high-risk groups, such as injection drug users, and gay, bisexual, and other men who have sex with men. Eighteen states also HIV tested inmates in high-risk groups in 2011. Note that the BOP and many states HIV tested inmates in multiple situations.

During the prison discharge process, HIV testing was conducted most commonly when an inmate requested it. The BOP, California, and 19 other states provided the test upon request to 41% of inmates released in 2015—down from 24 states and the BOP in 2011. Florida and 11 other states, accounting for 21% of inmates released in 2015, offered HIV

Inmate Health Problems

To learn more about inmate health problems in the United States, including infectious diseases, go to the Centers for Disease Control and Prevention website at https://www.cdc.gov/correctionalhealth/health-data.html.

What does this website tell you about inmate health in the United States?

testing to all inmates as part of discharge planning–up from 11 states in 2011. Finally, Kentucky and five other states, which accounted for 8% of inmates discharged in 2015, did not provide HIV testing to any inmates upon release–the same as in 2011.

In addition to HIV/AIDS, other infectious diseases found in prisons and jails are hepatitis C (HCV), hepatitis B (HBV), tuberculosis, and various sexually transmitted diseases (STDs). Estimates indicate that 4% of prison and jail inmates, or approximately 90,000 inmates in 2017, had HIV; 15%, or about 335,000 inmates in 2017, had hepatitis C; and 3%, or about 67,000 inmates in 2017, had active tuberculosis.[153] The rates (not provided) for those infectious diseases are much higher in the inmate population than in the general population. As is the case with HIV, infectious diseases result primarily from needle sharing, drug use, and consensual and nonconsensual sex among inmates. Although effective prevention strategies for prisons and jails exist, such as needle exchange, condom distribution, and opioid substitution therapy, prison and jail officials generally are unwilling to implement them, believing they are incompatible with institutional security.[154] For example, in 2015, California became only the second state to provide condoms to all inmates that requested them. However, even in institutions that provide condoms to inmates, corrections officers sometimes limit distribution as a control mechanism.[155]

Inmate Rehabilitation Programs Inmates hoping to better themselves during their incarceration normally have the opportunity to participate in a number of rehabilitation programs. The particular programs offered vary across jurisdictions and institutions. Some examples:

1. Self-improvement programs offered by religious and civic groups (Alcoholics Anonymous, the Jaycees, a Bible club).
2. Work programs.
3. Education and vocational training.
4. Counseling and therapy.

FYI

Rehabilitating Criminals

A January 2016 public opinion poll conducted in Florida, North Carolina, Nevada, Kentucky, Missouri, and Wisconsin found that an average of 76% of voters in those states believed "the main goal of our criminal justice system should be rehabilitating criminals to become productive, law-abiding citizens." Voters agreeing with the statement ranged from a low of 71% in Nevada to a high of 79% in Florida.

Source: U.S. Justice Action Network, "New Poll: Voters in Key Election States Overwhelmingly Agree: Criminal Justice System Is Broken, Strongly Support Federal Reform." Accessed May 18, 2016, www.justiceactionnetwork.org/new-poll-voters-in-key-election-states-overwhelmingly-agree-criminal-justice-system-is-broken-strongly-support-federal-reforms/.

Work Programs Since the creation of the first houses of correction in Europe, there has been one constant: the belief that the imprisonment experience should improve inmates' work habits. Today, however, there is tremendous variation among institutions regarding work programs. In some institutions, all inmates who are physically able are required to work. In other institutions, the inmates who work are those who choose to do so. Likewise, there is great variation in the types of work inmates perform. Some inmates are employed to help the daily running of the institution and work in such areas as food services, maintenance and repair, laundry, health care, and clerical services. Other inmates work in factories at industrial tasks, such as wood or metal manufacturing. Still other inmates perform agricultural work.

About half of all state and federal prison inmates work full-time jobs, but they are not counted in standard labor surveys.[156] Most of those jobs do not teach marketable skills that can be used by inmates when they are released, but rather involve prison maintenance; for example, mopping cellblock floors; cleaning bathrooms; preparing and serving food in the dining hall; washing dishes, cutting grass; filing papers in the warden's office; and laundering tons of uniforms, underwear, and bed sheets. The average wage in state and federal prisons is 20 cents and 31 cents per hour, respectively. Texas, Georgia, and Arkansas do not pay inmates for their work and, in Texas, make them work under threat of punishment. Despite working full-time jobs, inmates are not considered employees and do not receive minimum wage, sick leave, worker's compensation for injuries, social security withholding, or overtime pay. They are under the jurisdiction and authority of the Occupational Safety and Health Administration (OSHA), but OSHA must notify a prison in advance before an inspection. What most people do not realize is that if prisons had to pay inmates minimum wage for their labor, prisons would become prohibitively expensive to operate. If prisons had to hire outside workers to do the jobs that inmates do, costs could increase 30 to 45 times. Prisons save hundreds of millions of dollars a year by exploiting inmate labor. Ironically, the savings are offset because, without a meaningful source of income, inmates' families frequently must rely on government social services while inmates are incarcerated (and in many cases when they are released as well).

Institutions also vary in the degree to which the private sector is involved in work programming. Although extensive ties between inmate labor and the free market characterized the early history of the prison in the United States, those ties diminished during the first half of the 1900s. That change was due to concern by organized labor about unemployment among free citizens, concern by businesses that did not employ convict labor about their ability to compete on the open market, and concern by prison reformers about exploitation of inmates. Because of a series of laws restricting the private use of inmate labor, the *state-use system* became—and still is—the predominant way of arranging inmate work programs. Under that system, inmates are allowed to produce goods and services for the state government.

However, in 1979, the federal Prison Industry Enhancement certification program (PIE) was established despite labor unions' vehement objections. PIE allowed for-profit companies to employ inmates in factories inside prisons for the first time since convict-lease programs were banned. PIE operates in 38 states. PIE products can be sold on the open market, and companies must pay the prevailing wage for their industry so as to not undercut private-sector wages. In practice, the prevailing wage usually is minimum wage. Unfortunately, the PIE program is tiny, less than 1% of inmates work in it. Proponents of the program argue that (1) paying inmates the prevailing wage addresses labor union complaints about prison labor undercutting their members' wages, (2) the program enables inmates to help support their families and also pay alimony and child support, (3) the program allows inmates to atone for their crimes by making it possible for them to pay restitution to their victims, and (4) the program helps inmates accumulate some savings so they can rebuild their lives when they are released.[157]

The BOP's main work program was established by Congress in 1934, and is known as UNICOR. UNICOR is the trade name for Federal Prison Industries (FPI) and is a wholly owned and self-sustaining government corporation that receives no funding from Congress or taxpayers. It serves primarily as a correctional program and not a business that prepares inmates for their eventual release from prison by teaching them marketable skills. In this program, inmates make more than 100 products, including office supplies, air filters, clothes, and lamps, and provide about 80 services, such as sewing and weaving, upholstery, welding, printing, and data entry, for agencies of the federal government. BOP officials view the inmate-produced products and services as secondary to the job skills training program. Nevertheless, in 2017, UNICOR had $483.8 million in net sales.[158]

Only about 8% or about 14,000 of the work-eligible federal inmates in 2019 participated in UNICOR; 7% of UNICOR participants were military veterans. The program has a wait list of about 25,000 inmates. A high school diploma or General Educational Development (GED) certificate is required for all work assignments above entry level. UNICOR inmates typically earn between 23 cents and $1.15 per hour and are required to contribute 50% of their earnings to their financial obligations. Thus, despite the low wage, in 2017, UNICOR inmates contributed nearly $1 million toward court-ordered fines, child and family support, and victim restitution.[159]

An evaluation of UNICOR and other vocational and apprenticeship work programs conducted by BOP's Office of Research and Evaluation followed more than 7,000 federal inmates for up to 12 years. The evaluation found that inmates who participated in the work programs were 24% less likely to recidivate (return to prison) for as long as 12 years following release from prison compared to similar inmates who did not participate. Inmates who participated in the work programs also were 14% more likely than non-participants to be employed 12 months following their release from prison. The evaluation discovered that young minorities, who have the greatest risk of recidivating, especially benefitted from the work programs. Other groups that participated in the work programs and significantly reduced their risk of recidivating were 18- to 24-year-old inmates of Hispanic or other minority ancestry and inmates with a ninth to eleventh grade education.[160]

In 2011, Congress permitted UNICOR to begin selling, within limits, products and services in the private sector. The rationale was that the change would better prepare inmates to work in a business environment upon their release from prison and, hopefully, it would keep some U.S. companies from outsourcing jobs overseas.[161] UNICOR now markets its "cost-effective labor pool" and a workforce with "Native English and Spanish language skills." It touts, for example, its call centers staffed by prisoners and its nationwide network of manufacturing locations.[162]

Education and Vocational Training It has long been assumed that rehabilitation can be facilitated by improving inmates' academic skills and providing them with job skills. Many offenders enter prison with deficits in their education. As shown in Table 10.5, in

myth

Prisons are "factories behind fences" that exist to provide companies with a huge slave labor force.

fact

Private companies using prison labor are not the source of most prison jobs. Only about 5,000 people in prison—less than 1%—are employed by private companies through the federal PIECP program, which requires them to pay at least minimum wage before deductions. A larger proportion—about 6% of people incarcerated in state prisons—work for state-owned "correctional industries," which pay much less.

Source: Wendy Sawyer and Peter Wagner, "Mass Incarceration: The Whole Pie 2019," Prison Policy Initiative, March 19, 2019, accessed July 21, 2019, https://www.prisonpolicy.org/reports/pie2019.html.

UNICOR

To learn more about UNICOR, visit the Federal Bureau of Prisons website at https://www.bop.gov/inmates/custody_and_care/unicor_about.jsp.

Is this a work program that should be adopted for use by all states? Why or why not?

2016, the highest education level attained by 62% of state prison inmates and nearly 57% of federal prison inmates was less than a high school diploma.[163] Perhaps not surprisingly, it is not at all uncommon to encounter adult inmates who are reading, writing, and performing math operations at an elementary school level.[164] Therefore, much prison education amounts to remedial schooling designed to prepare inmates to obtain their GEDs.

Seven general types of educational and vocational programs presently are available in prisons and jails throughout the United States. They are as follows:

- *Adult Basic Education (ABE)*: Basic skills training in math, reading, writing, and English as a Second Language (ESL).
- *Adult Secondary Education*: Instruction for the GED tests or another certificate of high school equivalency.
- *Vocational Education*: Training to prepare individuals for general positions of employment as well as skills related to specific jobs and/or industries.
- *College Coursework*: Advanced college coursework, where credits may be applied toward an associate, bachelor's, or master's degree.
- *Special Education*: Educational training designed for individuals with disabilities or other special needs.
- *Study Release*: Release of individuals from correctional supervision for participation in coursework or training offered outside of a prison or jail.
- *Life Skills/Competency-Based Education*: Wide variety of programs that focus on providing individuals with communication skills, job and financial skill development, education, interpersonal and family relationship development, as well as stress and anger management.[165]

Traditionally, some inmate vocational programs have operated as part of job assignments (on-the-job training), while others have been separate from job assignments. Either way, the goal has been to provide prisoners with job skills that will improve their marketability upon release. Most prison vocational training has been geared toward traditional blue-collar employment, such as welding and auto mechanics. Vocational programs offered in women's prisons often have been criticized for concentrating excessively on stereotypical women's jobs, such as cosmetology.[166]

Currently, the role of information technology in inmate education has been growing. In 2014, for example, 24 states reported offering Microsoft Office certification as part of their vocational education programs. However, the role of computers in inmate educational programs has been tricky. Although at least 39 states use desktop computers in inmate educational programs, and at least 17 states use laptops, Internet access and use of Internet-based

Research shows that inmates exposed to education programs have lower recidivism rates than nonparticipants. *Why do only 25% to 50% of inmates take part in some form of education?*

Spencer Grant/Alamy

education is sparse. In 30 states, only teachers and vocational instructors had access to live Internet technology; in 26 states, adult students did not have any access to Internet technology (simulated or live).[167]

At year-end 2015, only about 45% of all state and federal prisoners had earned a high school diploma or GED, but even inmates who had a high school diploma or GED often demonstrated lower basic skills than did members of the general public with a high school diploma or GED.[168] The 45% of prisoners with a high school diploma or GED in 2015 was a 25% increase from the approximately 20% of state and federal prisoners who had earned a high school diploma or GED in 2011-2012.[169] For prisoners with a high school diploma or GED, college courses are made available in some prisons through correspondence or study release arrangements or by bringing college instructors to the prison to offer courses. However, low educational levels render college courses inappropriate for many inmates.

At year-end 2015, 7.9% of state and federal prisoners had a high school diploma or GED and had taken some college courses or had earned a college degree–a decrease of about 15 percentage points from the nearly 23% of prisoners who had similar achievements in 2011-2012.[170] Another source reported that in 2012 and 2014, 6% of the incarcerated population aged 16-74 had either an associate degree (4%), a bachelor's degree (1%), or a graduate or professional degree (1%).[171]

The educational attainment of U.S. prisoners while incarcerated has not been particularly impressive. In 2014, for example, 58% of prisoners did not bother to pursue any educational advancement, although 70% of them expressed a desire to enroll in an academic program. Reasons for not enrolling in an academic program included the classes and programs offered are not useful (13%), do not have the qualifications necessary to enroll (10%), have a volunteer or work assignment they do not want to give up (9%), want to enroll in a higher level of classes than are available (8%), the quality of the program being offered is poor (7%), and the waiting list is too long (3%). The remaining 51% of prisoners who did not enroll provided other reasons, such as wanting to devote their time to working on their legal appeals, their imminent release from prison, being already enrolled in some type of class or training, or "not interested." Among the 42% of prisoners that gave education a try, 8% completed grades 7, 8, or 9; 21% earned a high school diploma or GED; 4% worked at pre-associate education; 7% received a certificate from a college or trade school; 2% achieved an associate's degree; but not one inmate completed a bachelor's degree or higher.[172]

One of the reasons that so few prisoners pursue postsecondary education in prison is that they cannot afford to do so, and neither the federal government nor most state governments provide funding for them. The federal government once did offer funding to prisoners for postsecondary education through the Pell Grant Program, but no more. The federal Pell Grant Program was created in 1972, to provide financial support to low-income undergraduate students, including prisoners.[173] In 1982, 27,000 Pell Grant-funded prisoners were enrolled in 350 postsecondary programs in prison. By 1993-1994, nearly 1,300 prisons had enrolled about 23,000 Pell Grant-funded prisoners in more than 770 postsecondary programs. However, in 1994, the Violent Crime Control and Law Enforcement Act was passed in a punitive atmosphere caused by a rising crime rate, and one of its many provisions was the revocation of Pell Grants for prisoners. Prior to that, in 1992, Congress already had excluded from receiving Pell Grants prisoners who were serving life sentences and those sentenced to death. An amendment added to the 1992 legislation also stipulated that Pell Grant funding now would augment rather than replace state funding. To meet this requirement, states had to maintain their fiscal year 1988 funding levels for tuition assistance to prisoners. Prisoners in those states that did not keep or increase that funding level were barred from receiving Pell Grants. Not surprisingly, by 2014, only 9% of prisoners completed a postsecondary education while in prison, another 7% received a certificate from a college or trade school in prison, and 2% completed an associate degree in prison.

A glimmer of hope for prisoners presented itself in 2015, when the Obama administration initiated a 3-year pilot program called the "Second Chance Pell Experimental Sites Initiative." The program permits 67 colleges and universities in 27 states to provide prisoners with postsecondary education funded by Pell Grants. By fall 2017, the 67 colleges and universities were offering 822 certificates, 69 associate degrees, and 24 bachelor's degrees and serving 5,053 prisoners. A preliminary evaluation finds that the pilot program is yielding positive results. By the second year of the pilot program, a total of 934 credentials had been awarded, including 701 certificates, 230 associate's degrees, and 23 bachelor's degrees.[174] Anecdotally, released prisoners are finding good-paying jobs in industries, such as advanced manufacturing, which enable them to support themselves and their families. Estimates indicate that if half of eligible prisoners complete relevant postsecondary coursework in prison, in the first year following their release, they could enjoy a more than $45 million increase in their combined earnings (in 2015 dollars) and pay more in taxes. Moreover, because of projected declines in recidivism rates, states could expect to save a combined nearly $366 million annually (in 2015 dollars) in incarceration costs. Perhaps best of all, the federal government would not have to significantly increase Pell Grant expenditures for states and families to realize these financial benefits. The reason is that prisoners represent only a tiny fraction of all Pell Grant recipients. Assuming that all eligible state prisoners received a Pell Grant in a single year–a highly improbable situation–cost would increase less than 10%. In 2018, a bill called the Restoring Education and Learning (REAL) Act had been introduced in both houses of Congress with bipartisan sponsorship that would repeal the ban on state and federal prisoners receiving Pell grants.[175] As of this writing (August 19, 2019), it has not been passed.

States and the federal government underfund inmate educational programs at their own peril. For example, a recent study found a 13-percentage-point reduction in the risk of recidivating for inmates who participated in educational programs compared to inmates who did not.[176] A reduction in recidivism was achieved for inmates participating in adult basic education, high school/GED programs, postsecondary education, and/or vocational/career technical education training programs. The educational programs also proved cost effective. The study discovered that for every dollar spent on inmate educational programs, five dollars, on average, was saved in reincarceration costs. An often-cited 2013 study that was updated in 2018 showed that prisoners who participated in a correctional education program had 48% lower odds of recidivating than prisoners who did not participate. Prisoners who participated in an educational or vocational program also had 12% higher odds than non-participants of obtaining a job following release.[177] A recent Indiana study discovered that while all inmate education programs reduced recidivism, inmates who took college courses recidivated at a rate of less than 5%, compared to a national recidivism rate of about 68% within three years of release.[178]

Counseling and Therapy A wide range of counseling techniques and therapy modalities are used in prisons across the nation. The description of specific techniques and modalities, however, is beyond the scope of this chapter.[179] Suffice it to say that the techniques and modalities used at a given institution ordinarily reflect the training and professional orientation of the treatment staff–caseworkers, religious counselors, social workers, psychologists, and psychiatrists.

That said, one correctional therapeutic intervention deserves a few words because it has proven particularly successful in reducing both prison misconduct and recidivism, while being cost effective. That intervention is **cognitive behavioral therapy** (CBT). CBT focuses on the connection between dysfunctional thought processes and harmful behavior. It employs timely reward and punishment, as well as role-playing and skill-building exercises. The goal of CBT is twofold. First is to improve decision-making and problem-solving skills and, second, to teach prisoners how to handle various types of outside stimuli. CBT attempts to reduce recidivism by targeting several risk factors, such as general antisocial thinking and chemical dependency. The most widely used and evaluated CBT programs for prisoners are Reasoning & Rehabilitation (R&R), Moral Reconation Therapy (MRT), and Thinking for a Change (TEAC).[180]

cognitive behavioral therapy A correctional therapeutic intervention that focuses on the connection between dysfunctional thought processes and harmful behavior.

Cognitive Behavioral Therapies

To learn more about cognitive behavioral therapies, access an annotated bibliography at https://info.nicic.gov/virt/sites/info.nicic.gov.virt/files/T4C%20Annotated%20Bibliography.pdf and read some of the references.

Why do you suppose cognitive behavioral therapies have been successful with prison inmates?

A distinction usually is drawn between individual counseling, which involves one-on-one interaction between the counselor and the inmate, and group counseling, which involves the interaction of the counselor with a small group of inmates. Those categories of treatment may overlap because an inmate receiving individual counseling also may be in group counseling, and many of the techniques and principles used in individual counseling also are applied to group settings. Still, the distinction has merit because individual counseling is more appropriate for some inmates (e.g., those with deep-seated problems who will require long-term help), and group counseling is more appropriate for other inmates (e.g., those who are defensive, manipulative, and prone to denying their problems).

Group counseling is more popular than individual counseling in institutional settings, primarily because it is more economical and because there are large numbers of inmates who share similar backgrounds and problems. In fact, the members of a group frequently are selected on the basis of their common backgrounds and problem behavior patterns. For example, a group may consist of persons who have substance abuse problems or those who are sex offenders. Some institutions employ a variant of group therapy known as **milieu therapy** (also called a *therapeutic community*). Rather than having inmates attend periodic group sessions (one or two hourly sessions per week), milieu therapy encompasses the total living environment of inmates so that the environment continually encourages positive behavioral change. In effect, the entire inmate population becomes the group, and inmates have active roles in helping other inmates change.

milieu therapy A variant of group therapy that encompasses the total living environment so that the environment continually encourages positive behavioral change.

Most counselors and therapists who work in prison spend a considerable portion of their time performing **crisis intervention**. Crisis intervention consists of a counselor's efforts to address some crisis that has erupted in an inmate's life (e.g., suicidal thoughts, rejection by the spouse, a mental breakdown, or a conflict between inmates). Institutions are stressful living environments, so inmate crises are common. The counselor's task is to assist inmates in restoring a state of emotional calm and greater stability so that problems can be addressed rationally before inmates harm themselves or others.

crisis intervention A counselor's efforts to address some crisis in an inmate's life and to calm the inmate.

Programs in Perspective We usually think of rehabilitation programs as serving one main objective: to help inmates better themselves. In practice, however, programs serve other, more subtle functions as well. Programs give inmates a way to occupy themselves; they help inmates manage time. They also help the institution achieve control over inmates. For example, inmates who are in counseling for displaying too much aggression toward other inmates may realize that unless they demonstrate "progress" in reducing their aggression, the counselor will not recommend them to the parole board for early release (in states where parole is available). Although such progress may be achieved quickly in prison, it may be short-lived upon release. Similarly, from the standpoint of inmates, the most desirable work assignments are those that give inmates access to valued resources (e.g., a kitchen job provides greater access to food). More desirable work assignments are routinely held out by the staff as privileges to be earned through good behavior.

myth

The services and programs available to inmates make prisons seem like country clubs. Therefore, services and programs should be severely cut back or eliminated so that inmates receive punishment while incarcerated.

fact

First, the services and programs available to inmates are sometimes of inferior quality when compared with those available in the free community. Second, programs and services perform a vital time management function. Institutions would be much more volatile and disorderly than they currently are without programs and services to fill time.

Source: See, for example, Beth Pelz, Marilyn McShane, and Frank P. Williams III, "The Myth of Prisons as Country Clubs," in *Demystifying Crime and Criminal Justice,* 2nd ed., ed. R. M. Bohm and J. T. Walker (New York: Oxford University Press, 2013), 267–77.

Institutional programs also are plagued by a variety of problems that hinder their ability to effect rehabilitation. For example, prison work assignments frequently do not parallel work in the free world. In prison, workdays often are short and interrupted. In many cases, there is little concern for the quantity and quality of work. Furthermore, some of the jobs lack any real counterpart in the free world. The classic example is license plate manufacturing.

In many prisons, educational programs suffer from a lack of funding. Vocational training focuses primarily on traditional blue-collar jobs and thus prepares inmates to enter jobs for which there is already abundant competition. A dilemma is created by the argument that inmates should not receive highly technical or professional training (such as training in computer programming) because many members of the free, law-abiding community cannot afford such training. That argument is based on the **less-eligibility principle**, the position that prisoners should receive no service or program superior to the services and programs available to free citizens without charge. Moreover, prison education and vocational programs can do little to create jobs. If an inmate returns to a community with a poor economy and high unemployment, the education and training received in prison may do little more than raise the inmate's expectations to unrealistic levels.

less-eligibility principle The position that prisoners should receive no service or program superior to the services and programs available to free citizens without charge.

For their part, counseling and therapy programs must operate against the harsh realities of the prison environment, where custody and security ordinarily take priority over rehabilitation. Despite those obstacles, some programs are able to bring about positive changes in offenders' attitudes and behavior. Psychologist Ted Palmer has observed that although no single type of treatment can be identified as the most effective, one feature that seems to characterize programs that consistently reduce offender recidivism is the quality of the program's implementation. Thus, when the integrity of treatment efforts is allowed to take priority over institutional concerns for security and custody, offending can be reduced.[181] Also, as criminologist Robert Johnson points out, many counseling programs place nearly exclusive emphasis on inmates' pasts and futures, with insufficient attention given to present coping patterns. Johnson recommends that programs begin teaching inmates to cope maturely and constructively with their present environment so that they will be better able to cope with life after release.[182]

THINKING CRITICALLY

1. Do you think inmate rehabilitation programs can truly make a difference? Why or why not?
2. Do you think inmates should earn money for the work they do while in prison? Why or why not?

SUMMARY

1. **Summarize the purposes of confinement in Europe before it became a major way of punishing criminals.**

   Confinement became a major way of punishing criminals in Europe in the 1600s and 1700s. Before that, it was used to (1) detain people before trial; (2) hold prisoners awaiting other sanctions, such as death and corporal punishment; (3) coerce payment of debts and fines; (4) hold and punish slaves; (5) achieve religious indoctrination and spiritual reformation (as during the Inquisition); and (6) quarantine disease (as during the bubonic plague).

2. **Describe how offenders were punished before the large-scale use of confinement.**

   Before the large-scale use of confinement, punishments were directed more at the offender's body and property. One basic goal was to inflict pain. Those punishments commonly were carried out in public to humiliate the offender and to deter onlookers from crime. Examples of such punishments are fines, confiscation of property, and diverse methods of corporal and capital punishment.

3. **Explain why confinement began to be used as a major way of punishing offenders in Europe.**

   Reforms of the Enlightenment era led to an emphasis on deterring and reforming criminals through confinement. Confinement also was advocated as a humane alternative to older punishments. In addition, during the 1500s and 1600s, workhouses were established as places where offenders could be sent to learn discipline and productive work habits.

4. **Describe the recent trends in the use of incarceration in the United States.**

   Until 2009, there had been a dramatic increase since the mid-1970s in the number of people incarcerated in the United States, partly because of the war on drugs. That increase was accompanied by much concern over rising costs and institutional crowding. In response to those problems, confinement alternatives to traditional incarceration were developed. One of the primary alternatives was contracting with the private sector for the construction or operation of some confinement facilities. Since 2009 and through 2017, the trend has reversed and the number of people incarcerated in the United States has been decreasing steadily.

5. **List some of the characteristics of the incarcerated population in the United States.**

   State and federal inmates are disproportionately male, non-Hispanic black or Hispanic, 25–44 years old, never married, have less than a high school education, a U.S. citizen, and no military service. State prison inmates more likely are incarcerated for violent offenses, while federal inmates more likely are incarcerated for drug and public order offenses. Jail inmates also are disproportionately male but evenly divided between non-Hispanic whites and non-Hispanic blacks and Hispanics. The most serious type of offense for which most jail inmates are incarcerated is felonies; still, more than half of jail inmates are unconvicted, usually awaiting trial or other case disposition.

6. **Describe how incarceration facilities are structured, organized, and administered by the government in the United States.**

   Incarceration facilities in the United States are administered primarily by the executive branch of government. They exist at all three levels of government. Prisons are administered at the federal level and state levels, while jails tend to be locally administered. Despite this diversification, all the institutions have somewhat similar administrative structures, ranging from the warden or superintendent at the top to the correctional (line) officers at the bottom.

7. **Name some of the common types of correctional facilities in the United States.**

   Common types of adult correctional facilities in the United States include (1) classification and special facilities; (2) "supermaximum-," maximum-, medium-, and minimum-security men's prisons; (3) women's prisons; and (4) jails and lockups.

8. **Identify some of the procedures that institutions employ to maintain security and order.**

   To maintain security and order, correctional institutions employ a number of procedures. Among them are the use of custody designations for inmates, inmate counts, property control, searches of inmates and their living quarters, and restrictions on inmate communication (mail, phone calls, and visits) with outsiders. Also, institutions commonly rely on inmate-informant, or snitch, systems to maintain security. Inmates found guilty of violating institutional rules may be subjected to a variety of disciplinary sanctions (such as loss of good time, restriction of privileges, and solitary confinement) to deter future rule infractions.

9. **List the services and programs that commonly are available to inmates.**

   Some of the services and programs available to inmates are subsistence services (food, clothing, and shelter), health care services, legal services, recreation, and religious services. There are also a number of programs designed to improve inmates' lives. They include low-paying work, education and vocational training, and group and individual counseling.

KEY TERMS

banishment 388
transportation 388
workhouses 388
penology 389
panopticon 390
Pennsylvania system 390
Auburn system 391
medical model 392
privatization 394
incarceration rate 402
security level 415
classification facility 418
custody level 420
lockup 422
jail 422
protective custody 430
administrative segregation 430
conjugal visits 431
snitch system 431
cognitive behavioral therapy 440
milieu therapy 441
crisis intervention 441
less-eligibility principle 441

REVIEW QUESTIONS

1. What did Cesare Beccaria, the Enlightenment thinker, mean when he said that a punishment should fit the crime?
2. What reforms in penal institutions did John Howard advocate in his book, *The State of the Prisons in England and Wales* (1777)?
3. What generally is considered the first state prison in the United States, and of what did the daily routine of inmates in this prison consist?
4. How did the Pennsylvania system of confinement differ from the Auburn system of confinement, and which system became the model followed by other states?
5. What were the main features of the reformatory?
6. According to John Irwin, what three types of penal institutions dominated different parts of the last century?
7. What is an incarceration rate, and why is it used?
8. How does the incarceration rate of the United States compare with the incarceration rates of other countries?
9. How do the authors of this textbook explain the prison overcrowding crisis in the United States?
10. What are some differences between the federal prison and state prison populations?
11. What is the official mission of the Federal Bureau of Prisons?
12. What are the purposes of inmate classification?
13. What are prison security and custody levels, and how do they differ?
14. What are the purposes of a jail?
15. Why do jails represent one of the most problematic aspects of criminal justice?
16. What are some objectives of inmate rehabilitation programs?
17. What is the less-eligibility principle, as applied to corrections?

IN THE FIELD

1. **Prison Tour** Many prisons and jails conduct tours for students in criminal justice courses. Arrange to visit one or more prisons and jails in your area. Compare and contrast what you see during your visit(s) with the material in this chapter. Pay special attention to inmate characteristics, the physical plant, and interaction between inmates and correctional officers. Also, if you tour both a prison and a jail, describe similarities and differences between the two institutions. Either present your findings orally in class, or put them in writing for your instructor.
2. **Create a Prison** Design a prison conceptually and, perhaps, three dimensionally. Decide what type of inmates the prison is to hold (men, women, or both; prisoners requiring maximum security, minimum security, and so on) and what the security level will be. Consider location, architectural design, bureaucratic structure, security and discipline procedures, and inmate services and programs.

ON THE WEB

1. **Federal Bureau of Prisons** Go to the Federal Bureau of Prisons website at www.bop.gov. Access "Statistics" under the "About Us" tab. Use the data provided to construct a current profile of inmates. Then access the "Population Statistics," on the same page as "Statistics," and search for data about any federal prisons in your state. Write a brief report on what you find.
2. **Incarceration Worldwide** Go to the World Prison Brief's data on the worldwide prison population rate from highest to lowest at www.prisonstudies.org/highest-to-lowest/prison_population_rate?field_region_taxonomy_tid=All. Review the data and write a brief essay explaining why rates of incarceration vary so greatly among countries of the world.

CRITICAL THINKING EXERCISES

PRISON RESEARCH

1. According to a recent report by the National Center on Addiction and Substance Abuse, 65% of all U.S. prison and jail inmates meet the Diagnostic and Statistical Manual of Mental Disorders (DSM IV) medical criteria for substance abuse addiction, and another 20%, while not meeting the strict DSM IV criteria, had histories of substance abuse, were under the influence of alcohol or other drugs at the time of their crime, committed their offense to get money to buy drugs, were incarcerated for an alcohol or drug law violation, or shared some combination of

these characteristics. Yet, only 11% of all inmates with substance abuse and addiction disorders receive any treatment during their incarceration. The report found that if all inmates who needed treatment and aftercare received such services, the nation would break even in a year if just over 10% remained substance- and crime-free and employed. Thereafter, for each inmate who remained sober, employed, and crime-free, the nation would reap an economic benefit of $90,953 per year. Drugs and alcohol are implicated in 78% of violent crimes; 83% of property crimes; and 77% of weapon, public order, and other crimes.

a. Why do the nation's prisons and jails not provide treatment for substance abuse and addiction to more inmates?
b. Should citizens pay more taxes to fund substance abuse and addiction treatment programs in prisons and jails? Why or why not?
c. If such programs were provided, should completion of substance abuse and addiction treatment programs be a requirement for release? Why or why not?

THE EX-CONVICT

2. To the average citizen, the ex-convict is an individual of questionable character. This image is reinforced by the fact that the only thing that usually is newsworthy about ex-convicts is bad news, such as when they are rearrested. Few citizens know a rehabilitated offender, for obvious reasons. Yet, the rehabilitated ex-convict faces a variety of civil disabilities. Not the least among those is the difficulty of getting a "good" job. On virtually every job application, there is the question: "Have you ever been convicted of a felony or misdemeanor or denied bond in any state?" In addition, prospective employers generally ask applicants to reveal their former substance abuse, which is a problem that most former offenders had. An affirmative answer to either question almost always eliminates a person's chance for employment.

a. How can ex-offenders shed the ex-convict status? Should they be allowed to shed it? Why or why not?
b. Do your answers to the questions in part (a) change depending upon the ex-convict's conviction offense?

NOTES

1. See Robert Johnson, *Hard Time: Understanding and Reforming the Prison* (Monterey, CA: Brooks/Cole, 1987).
2. Michel Foucault, *Discipline and Punish: The Birth of the Prison* (New York: Pantheon, 1978).
3. See Graeme Newman, *The Punishment Response* (Albany, NY: Harrow and Heston, 1985).
4. Todd R. Clear and George F. Cole, *American Corrections,* 3rd ed. (Belmont, CA: Wadsworth, 1994).
5. Graeme Newman and Pietro Marongiu, "Penological Reform and the Myth of Beccaria," *Criminology* 28 (1990): 325–46.
6. Ibid.
7. Clear and Cole, *American Corrections.*
8. David J. Rothman, *The Discovery of the Asylum: Social Order and Disorder in the New Republic* (Boston: Little, Brown, 1971).
9. See Paul Takagi, "The Walnut Street Jail: A Penal Reform to Centralize the Powers of the State," *Federal Probation* (December 1975): 18–26.
10. Rothman, *The Discovery of the Asylum.*
11. Clear and Cole, *American Corrections.*
12. Nicole Hahn Rafter, "Gender and Justice: The Equal Protection Issue," in *The American Prison: Issues in Research and Policy*, ed. L. Goodstein and D. L. MacKenzie (New York: Plenum, 1989), 89–109; see Nicole Hahn Rafter, *Partial Justice: Women in State Prisons, 1800–1935* (Boston: Northeastern University Press, 1985).
13. John Irwin, *Prisons in Turmoil* (Boston: Little, Brown, 1980).
14. Ibid., 3.
15. See Robert Martinson, "What Works? Questions and Answers About Prison Reform," *The Public Interest* 42 (1974): 22–54; see also The American Friends Service Committee, *Struggle for Justice: A Report on Crime and Justice in America* (New York: Hill & Wang, 1971).
16. Federal Bureau of Prisons, "Historical Information #1." Accessed July 27, 2019, https://www.bop.gov/about/history/timeline.jsp.
17. Ibid; "Chillicothe Reformatory Photograph," Ohio History Connection Selections. Accessed July 27, 2019, https://ohiomemory.org/digital/collection/p267401coll32/id/17036/; "Federal Reformatory for Women," Alderson West Virginia – A History." Accessed July 27, 2019, http://www.min7th.com/awv/ghsfrw.htm.
18. Federal Bureau of Prisons, "Historical Information #1."
19. Federal Bureau of Prisons, "Historical Information #2." Accessed July 22, 2019, https://www.bop.gov/about/history/.
20. Ibid.
21. Federal Bureau of Prisons, "Historical Information #1."
22. Ibid.
23. Ibid.
24. Federal Bureau of Prisons, "Historical Information #2."
25. Ibid.
26. Ibid.
27. Ibid.
28. Ibid.
29. Ibid.
30. Ibid; Federal Bureau of Prisons, "Past Inmate Population Totals." Accessed July 23, 2019, https://www.bop.gov/about/statistics/population_statistics.jsp#old_pops.
31. Ibid.
32. Stef W. Kight and Dan Primack, "Special Report: Profiting from Prison," *AXIOS*, June 18, 2019. Accessed July 9, 2019, https://www.axios.com/profiting-prison-c2bd43b2-4b2f-44ee-8f23-c6c9a14c1aaa.html.
33. Ibid; Jimmy Jenkins, "Whistleblower Says Corizon Health Administrators Directed Him to Cheat Arizona Prison Monitors," *KJZZ*, January 24, 2019. Accessed July 9, 2019, https://theshow.kjzz.org/content/752172/whistleblower-says-corizon-health-administrators-directed-him-cheat-arizona-prison.
34. Kight and Primack, "Special Report: Profiting from Prison"; Paul Egan, "Michigan to End Prison Food Deal with Aramark," *Detroit Free Press*, July 13, 2015. Accessed July 9, 2019, https://www.freep.com/story/news/local/michigan/2015/07/13/state-ends-prison-food-contract-aramark/30080211/; Paul Egan and Kathleen Gray, "Gov. Rick Snyder: State to End Problem-Plagued Privatization Experiment with Prison Food," *Detroit Free Press*, February 7, 2018. Accessed July 9, 2019, https://www.freep.com/story/news/local/michigan/2018/02/07/gov-rick-snyder-state-end-problem-plagued-privatization-experiment-prison-food/314693002/.
35. Kight and Primack, "Special Report: Profiting from Prison."

36. Calculated from data in Jennifer Bronson and E. Ann Carson, "Prisoners in 2017," in *Bureau of Justice Statistics Bulletin*, U.S. Department of Justice (Washington, DC: U.S. Government Printing Office, April 2019), 27–28, Table 17, accessed July 9, 2019, https://www.bjs.gov/content/pub/pdf/p17.pdf; E. Ann Carson and Elizabeth Anderson, "Prisoners in 2015," in *Bureau of Justice Statistics Bulletin*, U.S. Department of Justice (Washington, DC: U.S. Government Printing Office, December 2016), 3, Table 1, 28, Appendix Table 2, accessed January 9, 2017, www.bjs.gov/content/pub/pdf/p15.pdf; Danielle Kaeble, Lauren Glaze, Anastasios Tsoutis, and Todd Minton, "Correctional Populations in the United States, 2014," in *Bureau of Justice Statistics Bulletin*, U.S. Department of Justice (Washington, DC: U.S. Government Printing Office, Revised January 21, 2016), 22, Appendix Table 5, accessed May 17, 2016, www.bjs.gov/content/pub/pdf/cpus14.pdf.

37. *American Correctional Association 2013 Directory: Adult and Juvenile Correctional Departments, Institutions, Agencies, and Probation and Parole Authorities,* 74th edition. American Correctional Association, Alexandria VA. Calculated from data on pp. 28–31.

38. James J. Stephan, "Census of State and Federal Correctional Facilities, 2005," in *Bureau of Justice Statistics Bulletin,* U.S. Department of Justice (Washington, DC: U.S. Government Printing Office, October 2008), 4, accessed February 9, 2011, http://bjs.ojp.usdoj.gov/content/pub/pdf/csfcf05.pdf.

39. CoreCivic 2018 Annual Report. Accessed July 11, 2019, http://ir.corecivic.com/static-files/60371436-e930-40bc-8cfa-830d12b4edd0. Unless indicated otherwise, all information about CoreCivic is from this source.

40. The GEO Group, Inc. 2018 Annual Report. Accessed July 11, 2019, http://www.snl.com/interactive/newlookandfeel/4144107/GEOGroup2018AR.pdf. Unless indicated otherwise, all information about The GEO Group is from this source.

41. The GEO Group > Acquisitions, *Crunchbase*. Accessed July 14, 2019, https://www.crunchbase.com/organization/the-geo-group/acquisitions/acquisitions_list#section-overview; "The GEO Group Closes $415 Million Acquisition of B.I. Incorporated," *BusinessWire*, February 11, 2011. Accessed July 14, 2019, http://www.businesswire.com/news/home/20110211005372/en/GEO-Group-Closes-415-Million-Acquisition-B.I.

42. Geoff West, "Politicians Shun GEO Group Contributions," *OpenSecrets*, July 20, 2018. Accessed July 14, 2019, https://www.opensecrets.org/news/2018/07/politicians-shun-geo-contributions/.

43. GEO Group, OpenSecrets. Accessed July 14, 2019, https://www.opensecrets.org/orgs/totals.php?id=D000022003&cycle=2018.

44. The 2015 figures are from Corrections Corporation of America, 2015 Annual Report, Form 10-K and "Who We Are." Accessed January 9, 2017, www.cca .com; The GEO Group, Inc., "Management & Operations." Accessed January 9, 2017, www.geogroup.com/Management_and _Operations.

45. Kalena Thomhave, "Banning Private Prisons–and Prisoner Exploitation," *The American Prospect*, June 25, 2019. Accessed July 11, 2019, https://www.commondreams.org/views/2019/06/25/banning-private-prisons-and-prisoner-exploitation; Spencer Ackerman and Adam Rawnsley, "$800 Million in Taxpayer Money Went to Private Prisons Where Migrants Work for Pennies," *Daily Beast*, January 2, 2019. Accessed July 11, 2019, https://www.thedailybeast.com/dollar800-million-in-taxpayer-money-went-to-private-prisons-where-migrants-work-for-pennies?ref=scroll; Alex Baumgart, "Companies that Funded Trump's Inauguration Came up Big in 2017, *OpenSecrets*, January 19, 2018. Accessed July 13, 2019, https://www.opensecrets.org/news/2018/01/companies-that-funded-trumps-inauguration/.

46. Ibid.

47. Spencer Ackerman and Adam Rawnsley, "$800 Million in Taxpayer Money Went to Private Prisons Where Migrants Work for Pennies," *Daily Beast*, January 2, 2019. Accessed July 11, 2019, https://www.thedailybeast.com/dollar800-million-in-taxpayer-money-went-to-private-prisons-where-migrants-work-for-pennies?ref=scroll.

48. Ibid; Alene Tchekmedyian, "Immigrants Allege Forced Labor at Detention Facility," *Orlando Sentinel*, March 6, 2017, p. A6; "ICE Detention Center Paid Imprisoned Workers Less Than $1 per Day, Washington AG Says," *KTLA5 News*, September 20, 2017. Accessed July 13, 2019, https://ktla.com/2017/09/20/ice-detention-center-paid-imprisoned-workers-less-than-1-per-day-washington-ag-says/.

49. Unless indicated otherwise, information on the advantages of private prisons is from W. Thomas, "Private Adult Correctional Facility Census: A 'Real-Time' Statistical Profile," September 4, 2001, www.crim.ufl.edu/pcp; Crystal Ayres, "20 Privatization of Prisons Pros and Cons," *Vittana Personal Finance Blog*. Accessed July 14, 2019, https://vittana.org/20-privatization-of-prisons-pros-and-cons.

50. Unless indicated otherwise, information on the disadvantages of private prisons is from Donna Selman and Paul Leighton, *Punishment for Sale: Private Prisons, Big Business, and the Incarceration Binge* (Lanham, MD: Rowman & Littlefield, 2010); James Austin and Gary Coventry, "Emerging Issues on Privatized Prisons," in *Bureau of Justice Statistics Bulletin*, U.S. Department of Justice (Washington, DC: U.S. Government Printing Office, February 2001); Ayres, "20 Privatization of Prisons Pros and Cons."

51. Michael Rubinkam, "Judges: We Took $2M in Kickbacks," *Orlando Sentinel*, February 13, 2009, A4.

52. Spencer Ackerman and Adam Rawnsley, "$800 Million in Taxpayer Money Went to Private Prisons Where Migrants Work for Pennies," *Daily Beast*, January 2, 2019. Accessed July 11, 2019, https://www.thedailybeast.com/dollar800-million-in-taxpayer-money-went-to-private-prisons-where-migrants-work-for-pennies?ref=scroll.

53. Calculated from data in *Sourcebook of Criminal Justice Statistics Online*, Table 6.28.2012, accessed July 17, 2019, https://www.albany.edu/sourcebook/pdf/t6282012.pdf; Jennifer Bronson and E. Ann Carson, "Prisoners in 2017, in *Bureau of Justice Statistics Bulletin*, U.S. Department of Justice (Washington, DC: U.S. Government Printing Office, April 2019), 9, Table 5, accessed July 17, 2019, https://www.bjs.gov/content/pub/pdf/p17.pdf.

54. "Prisoners in 1992," in *Bureau of Justice Statistics Bulletin*, U.S. Department of Justice (Washington, DC: U.S. Government Printing Office, May 1993).

55. Bronson and Carson, "Prisoners in 2017."

56. Ibid.

57. Ibid.

58. Bronson and Carson, "Prisoners in 2017," 27, Table 17, 28, Table 18, 30, Table 20; Todd D. Minton and Mary Cowhig, "Jails in Indian Country, 2016," in *Bureau of Justice Statistics Bulletin*, U.S. Department of Justice (Washington, DC: U.S. Government Printing Office, December 2017), 2, Table 1. Accessed July 18, 2019, https://www.bjs.gov/content/pub/pdf/jic16.pdf; Zhen Zeng, "Jail Inmates in 2017," in *Bureau of Justice Statistics Bulletin,* U.S. Department of Justice (Washington, DC: U.S. Government Printing Office, April 2019), 2, Table 1. Accessed July 18, 2019, https://www.bjs.gov/content/pub/pdf/ji17.pdf; "Juveniles in Corrections," Statistical Briefing Book, Office of Juvenile Justice and Delinquency Prevention (U.S. Department of Justice, October 25, 2017), accessed July 19, 2019, https://www.ojjdp.gov/ojstatbb/corrections/qa08201.asp?qaDate=2017.

59. Zeng, "Jail Inmates in 2017," 2, Table 1.
60. Calculated from Zeng, "Jail Inmates in 2017," 9, Table 9.
61. Bronson and Carson, "Prisoners in 2017," 41, Appendix Table 1.
62. Todd D. Minton and Zhen Zeng, "Jail Inmates at Midyear 2014," in *Bureau of Justice Statistics Bulletin*, U.S. Department of Justice (Washington, DC: U.S. Government Printing Office, June 2015), 2, Table 1, accessed January 10, 2017, www.bjs.gov/content/pub/pdf/jim14.pdf.
63. Zeng, "Jail Inmates in 2017," 4, Table 2.
64. Tyjen Tsai and Paola Scommegna, "U.S. Has World's Highest Incarceration Rate," Population Reference Bureau, accessed May 21, 2016, www.prb.org/Publications/Articles/2012/us-incarceration.aspx.
65. Adam Liptak, "Inmate Count in U.S. Dwarfs Other Nations," *The New York Times*, April 23, 2008, A1.
66. James P. Lynch, "A Cross National Comparison of the Length of Custodial Sentences for Serious Crimes," *Justice Quarterly* 10 (1993): 639–60.
67. Liptak, "Inmate Count in U.S. Dwarfs Other Nations."
68. Calculated from data in "State Expenditure Report," National Association of State Budget Officers, 2018, 62, Table 32. Accessed July 19, 2019, https://higherlogicdownload.s3.amazonaws.com/NASBO/9d2d2db1-c943-4f1b-b750-0fca152d64c2/UploadedImages/SER%20Archive/2018_State_Expenditure_Report_S.pdf; Federal Prison System (BOP), U.S. Department of Justice. Accessed July 19, 2019, https://www.justice.gov/jmd/page/file/968276/download.
69. Ibid.
70. "Following the Money of Mass Incarceration," Prison Policy Initiative. Accessed July 27, 2019, https://www.prisonpolicy.org/profiles/US.html.
71. "Cost of Corrections," North Carolina Public Safety. Accessed August 4, 2019, https://www.ncdps.gov/adult-corrections/cost-of-corrections.
72. Data are from Office of the Assistant Secretary for Planning and Evaluation, "2017 Poverty Guidelines," U.S. Department of Health & Human Services. Accessed July 21, 2019, https://aspe.hhs.gov/2017-poverty-guidelines; Kayla Fontenot, Jessica Semega, and Melissa Kollar, "Income and Poverty in the United States: 2017," United States Census Bureau, Report Number P60-263, September 12, 2018, Table 3. Accessed July 21, 2019, https://www.census.gov/library/publications/2018/demo/p60-263.html.
73. Bronson and Carson, "Prisoners in 2017," 25–26, Table 16; Carson and Anderson, "Prisoners in 2015," 27, Appendix Table 1; Carson, "Prisoners in 2014," 31, Appendix Table 14.
74. Carson, "Prisoners in 2014," 4–5.
75. Sam Stanton, "Judges Appoint Prison Population Oversight Chief," *The Sacramento Bee*, April 9, 2014. Accessed May 23, 2016, www.sacbee.com/news/politics-government/article2595242.html.
76. Carson, "Prisoners in 2014," 3, Table 2, 20, accessed February 18, 2016; E. Ann Carson, "Prisoners in 2013," in *Bureau of Justice Statistics Bulletin,* U.S. Department of Justice (Washington, DC: U.S. Government Printing Office, September 2014), 3, Table 2, accessed May 23, 2016, www.bjs.gov/content/pub/pdf/p13.pdf.
77. "California Has Reduced Its Prison Population to the Court-Ordered Level, at Least for Now," Public Policy Institute of California. Accessed May 23, 2016, www.ppic.org/main/publication_show.asp?i=702.
78. Bronson and Carson, "Prisoners in 2017," 25, Table 16.
79. Nils Christie, *Crime Control as Industry: Toward Gulags Western Style?* (New York: Routledge, 1993).
80. Michael Tonry, "Crime and Human Rights—How Political Paranoia, Protestant Fundamentalism, and Constitutional Obsolescence Combined to Devastate Black America: The American Society of Criminology 2007 Presidential Address," *Criminology* 46 (2008): 10.
81. Franklin E. Zimring, "The Great American Lockup," *The Washington Post*, March 1991, 4–10, national weekly edition.
82. Calculated from data in E. Ann Carson, "Prisoners in 2016," in *Bureau of Justice Statistics Bulletin,* U.S. Department of Justice (Washington, DC: United States Government Printing Office, January 2018), 3, Table 1.
83. Ibid., 8, Table 6.
84. See, for example, John Kramer and Darrell Steffensmeier, "Race and Imprisonment Decisions," *Sociological Quarterly* 34 (1993): 357–76; Ronald L. Akers, *Criminological Theories: Introduction and Evaluation*, 2nd ed. (Los Angeles: Roxbury, 1997). For different views, see Edmund F. McGarrell, "Institutional Theory and the Stability of a Conflict Model of the Incarceration Rate," *Justice Quarterly* 10 (1993): 7–28; Cassia Spohn and Jerry Cederblom, "Race and Disparities in Sentencing: A Test of the Liberation Hypothesis," *Justice Quarterly* 8 (1991): 305–27.
85. Michelle Alexander, *The New Jim Crow: Mass Incarceration in the Age of Colorblindness* (New York: The New Press, 2011): 4, 6, and 11.
86. Federal Bureau of Prisons, "About Our Agency—Mission," accessed July 22, 2019, https://www.bop.gov/about/agency/agency_pillars.jsp.
87. Federal Bureau of Prisons, "Our Locations." Accessed July 23, 2019, https://www.bop.gov/locations/.
88. Federal Bureau of Prisons, "Historical Information," accessed July 22, 2019, https://www.bop.gov/about/history/timeline.jsp.
89. Bronson and Carson, "Prisoners in 2017," 3, Table 1; Federal Bureau of Prisons, "Past Inmate Population Totals," accessed July 23, 2019, https://www.bop.gov/about/statistics/population_statistics.jsp#old_pops.
90. Federal Bureau of Prisons, "Statistics." Accessed March 1, 2020, https://www.bop.gov/about/statistics/population_statistics.jsp.
91. Calculated from data in Federal Bureau of Prisons, "Prison Security Levels," accessed March 1, 2020, https://www.bop.gov/about/statistics/statistics_inmate_sec_levels.jsp.
92. Federal Bureau of Prisons, "About Our Facilities." Accessed July 23, 2019, https://www.bop.gov/about/facilities/federal_prisons.jsp.
93. Ibid.
94. Ibid.
95. Ibid.
96. Ibid.
97. Ibid.
98. Federal Bureau of Prisons, "Restricted Housing," July 22, 2019. Accessed July 23, 2019, https://www.bop.gov/about/statistics/statistics_inmate_shu.jsp; Jason Hanna, "'El Chapo' Is Likely Going to the Same Prison Where Ted Kaczynski and Dzhokhar Tsarnaev Are Held," *CNN*, February 13, 2019. Accessed July 23, 2019, https://www.cnn.com/2019/02/13/us/gallery/el-chapo-supermax-prisoners/index.html.
99. Jonathan Stempel, "After two escapes, 'El Chapo' may go to 'Supermax' prison to avoid a third," *Reuters*, July 17, 2019, https://www.reuters.com/article/us-usa-mexico-el-chapo-supermax/after-two-escapes-el-chapo-may-go-to-supermax-prison-to-avoid-a-third-idUSKCN1UC142; Eleanor Cummins, "Here's what makes ADX Florence the country's most secure prison," *Popular Science*, July 18, 2019, accessed July 24, 2019, https://www.popsci.com/adx-florence-prison-design/. Francis X. Clines, "A Futuristic Prison Awaits the Hard Core 400," *The New York Times,* October 17, 1994, 1A, national edition; C1 Staff, "5 things to know about the 'escape proof' supermax prison," CorrectionsOne.com, July 19, 2019, accessed July 22, 2019, https://www.correctionsone.com/escapes/articles/482288187-5-things-to-know-about-the-escape-proof-supermax-prison/.

100. Calculated from data in "State Prison Facilities," State Statistics Information, National Institute of Corrections. Accessed July 16, 2019, https://nicic.gov/state-statistics-information?.

101. James J. Stephan, "Census of State and Federal Correctional Facilities, 2005," Bureau of Justice Statistics, National Prisoner Statistics Program, U.S. Department of Justice (Washington, DC: U.S. Government Printing Office, October 2008), Appendix Table 2. Accessed July 16, 2019, https://www.bjs.gov/content/pub/pdf/csfcf05.pdf.

102. "Entombed: Isolation in the U.S. Federal Prison System," Amnesty International, 2014, accessed May 31, 2016, www.amnestyusa.org/sites/default/files/amr510402014en.pdf; Cathy Cockrell, "Probing the Haphazard Rise of Harsh Supermaximum Prisons," UC Berkeley News Center, December 7, 2010, accessed February 13, 2011, http://newscenter.berkeley.edu/2010/12/07/supermax/.

103. Unless indicated otherwise, information in this section is from "Review of the Federal Bureau of Prisons' Management of Its Female Inmate Population," Office of the Inspector General, U.S. Department of Justice, September 2018, 3–4. Accessed July 26, 2019, https://oig.justice.gov/reports/2018/e1805.pdf; Federal Bureau of Prisons; "Custody & Care." Accessed July 26, 2019, https://www.bop.gov/inmates/custody_and_care/female_offenders.jsp.

104. Federal Bureau of Prisons, "Inmate Gender." Accessed March 1, 2020, https://www.bop.gov/about/statistics/statistics_inmate_gender.jsp.

105. *American Correctional Association 2013* Directory, calculated from data on 28–29.

106. Paige M. Harrison and Allen J. Beck, "Prison and Jail Inmates at Midyear 2004," in *Bureau of Justice Statistics Bulletin*, U.S. Department of Justice (Washington, DC: U.S. Government Printing Office, April 2005), 7; Paige M. Harrison and Allen J. Beck, "Prison and Jail Inmates at Midyear 2005," in *Bureau of Justice Statistics Bulletin*, U.S. Department of Justice (Washington, DC: U.S. Government Printing Office, May 2006), 7.

107. Calculated from data in Zeng, "Jail Inmates in 2017," 2, Table 1, 4, Table 2; Minton and Zeng, "Jail Inmates in 2015," 3, Table 1; Todd D. Minton and Zhen Zeng, "Jail Inmates at Midyear 2014," in *Bureau of Justice Statistics Bulletin*, U.S. Department of Justice (Washington, DC: U.S. Government Printing Office, June 2015), 2, Table 1; Todd D. Minton and William J. Sabol, "Jail Inmates at Midyear 2008—Statistical Tables," in *Bureau of Justice Statistics Bulletin*, U.S. Department of Justice (Washington, DC: U.S. Government Printing Office, March 2009).

108. Ibid.

109. Zeng, "Jail Inmates in 2017," 5, Table 3, 6, Table 4.

110. Calculated from data in Zeng, "Jail Inmates in 2017," 5, Table 3; Minton and Zeng, "Jail Inmates in 2015," 4, Table 3; Minton and Zeng, "Jail Inmates at Midyear 2014," 3, Table 2.

111. John Irwin, *The Jail: Managing the Underclass in American Society* (Berkeley/Los Angeles: University of California Press, 1985).

112. John A. Backstrand, Don C. Gibbons, and Joseph F. Jones, "Who Is in Jail: An Examination of the Rabble Hypothesis," *Crime and Delinquency* 38 (1992): 219–29.

113. Zeng, "Jail Inmates in 2017," 6, Table 4.

114. See Dale K. Sechrest and William C. Collins, *Jail Management and Liability Issues* (Miami, FL: Coral Gables Publishing, 1989).

115. Unless indicated otherwise, information in this section on jail design is from Frank Schmalleger and John Ortiz Smykla, *Corrections in the 21st Century*, 3rd ed. (Columbus, OH: Glencoe/McGraw-Hill, 2007), 220–23; *Podular, Direct Supervision Jails: Information Packet*, U.S. Department of Justice, National Institute of Corrections, Jails Division, January 1993, www.nicic.org/resources/topics/DirectSupervisionJails.aspx; Steve Fishback, "The Design of the New Anchorage Jail," *Alaska Justice Forum* 18, no. 3 (Fall 2001), www.uaa.alaska.edu/just/forum/f183fa01/c_anchjail.html.

116. Doris Bloodsworth, "Jail Ready to Debut Visitation by Video," *Orlando Sentinel*, January 5, 2003, B1.

117. Mary Stohr-Gillman, Linda Zupan, Craig Curtis, Ben Menke, and Nicholas Lovrich, "The Development of a Behavioral-Based Performance Appraisal System," *American Jails* 5 (1992): 10–16.

118. Jeffrey D. Senese, "Evaluating Jail Reform: A Comparative Analysis of Popular/Direct and Linear Jail Inmate Infractions," *Journal of Criminal Justice* 25 (1997): 61–73; Linda L. Zupan, *Jails: Reform and the New Generation Philosophy* (Cincinnati, OH: Anderson, 1991); Schmalleger and Smykla, *Corrections in the 21st Century*, 98; Fishback, "The Design of the New Anchorage Jail."

119. *American Correctional Association 2011 Directory*, 36–37.

120. James W. Marquart and Ben M. Crouch, "Coopting the Kept: Using Inmates for Social Control in a Southern Prison," in *Prisons Around the World: Studies in International Penology*, ed. M. K. Carlie and K. I. Minor (Dubuque, IA: William C. Brown, 1992), 124–38.

121. "Prison Rule Violators," *Bureau of Justice Statistics Special Report*, U.S. Department of Justice (Washington, DC: U.S. Government Printing Office, December 1989).

122. Amy L. Edwards, "Crackers and Water: Sheriff Cooks Up $161,000 in Meal Cuts," *Orlando Sentinel*, July 10, 2008, A1.

123. Alysia Santo and Lisa Iaboni, "What's in a Prison Meal?" The Marshall Project, July 7, 2015. Accessed June 2, 2016, www.themarshallproject.org/2015/07/07/what-s-in-a-prison-meal#.omAsGtANR.

124. Camila Domonoske, "Alabama Sheriff Legally Took n$750,000 Meant To Feed Inmates, Bought Beach House," NPR, March 14, 2018, accessed August 4, 2019, https://www.npr.org/sections/thetwo-way/2018/03/14/593204274/alabama-sheriff-legally-took-750-000-meant-to-feed-inmates-bought-beach-house; Jay Reeves, "Enjoying Leftovers," The Advisor, Retirement Systems of Alabama, Vol. XLIII, No. 12, June 2018.

125. "Virginia Opens Special Prison for Aging Inmates," *Dallas Morning News,* July 3, 1999, accessed December 21, 2000, www.dallasnews.com/national/0703nat4prison.html.

126. Calculated from data in Sourcebook of criminal justice statistics Online, Table 6.33.2007. Accessed August 4, 2019, https://www.albany.edu/sourcebook/pdf/t6332007.pdf.

127. Calculated from data in Bronson and Carson, "Prisoners in 2017," 17, Table 8.

128. "The Rising Cost of Incarcerating the Elderly," The Crime Report, May 17, 2018. Accessed August 4, 2019, https://thecrimereport.org/2018/05/17/the-rising-cost-of-punishing-the-elderly/.

129. "At America's Expense: The Mass Incarceration of the Elderly," American Civil Liberties Union, June 2012. Accessed June 9, 2016, www.aclu.org/report/americas-expense-mass-incarceration-elderly?redirect=criminal-law-reform/report-americas-expense-mass-incarceration-elderly.

130. "The Rising Cost of Incarcerating the Elderly," *The Crime Report*, May 17, 2018, accessed August 4, 2019, https://thecrimereport.org/2018/05/17/the-rising-cost-of-punishing-the-elderly/.

131. "Serious Mental Illness Prevalence in Jails and Prisons," Treatment Advocacy Center, September 2016. Accessed August 4, 2019, https://www.treatmentadvocacycenter.org/evidence-and-research/learn-more-about/3695; "Federal Prisons: Information on Inmates with Serious Mental Illness and Strategies to Reduce Recidivism," United State Government Accountability Office, Report to Congressional Committees, February 2018, 2. Accessed August 7, 2019, https://www.gao.gov/assets/700/690090.pdf.

132. Doris A. Fuller, Elizabeth Sinclair, H. Richard Lamb, Judge James D. Cayce, and John Snook, "Emptying the 'New Asylums': A Beds Capacity Model to Reduce Mental Illness Behind Bars," Treatment Advocacy Center, January 2017. Accessed August 7, 2019, https://www.treatmentadvocacycenter.org/storage/documents/emptying-new-asylums.pdf.

133. Deborah L. Shelton, "How Sending the Mentally Ill to Jail Is a Cost to Us All," Takepart, May 15, 2015. Accessed August 7, 2019, http://www.takepart.com/article/2015/05/18/when-sickness-crime.

134. "Serious Mental Illness Prevalence in Jails and Prisons."

135. Ibid.

136. "Jailing People With Mental Illness," NAMI (National Alliance on Mental Illness). Accessed August 4, 2019, https://www.nami.org/learn-more/public-policy/jailing-people-with-mental-illness; "The Burden of Mental Illness Behind Bars," Vera Institute of Justice, June 21, 2016. Accessed August 4, 2019, https://www.vera.org/the-human-toll-of-jail/inside-the-massive-jail-that-doubles-as-chicagos-largest-mental-health-facility/the-burden-of-mental-illness-behind-bars.

137. "The Burden of Mental Illness Behind Bars."

138. Ibid.

139. Alisa Roth, "A 'Hellish World': The Mental Health Crisis Overwhelming America's Prisons."

140. Ibid.

141. Treatment Advocacy Center, "Fact Sheet: Criminalization of Americans with Severe Mental Illnesses," www.psychlaws.org/GeneralResources/Fact3.htm; Human Rights Watch, "Ill-Equipped: U.S. Prisons and Offenders with Mental Illness," www.hrw.org/_reports/2003/usa1003/.

142. "The Burden of Mental Illness Behind Bars"; "Jailing People with Mental Illness."

143. Alisa Roth, "A 'hellish World': The Mental Health Crisis Overwhelming America's Prisons," *The Guardian*, March 31, 2018. Accessed August 5, 2019, https://www.theguardian.com/society/2018/mar/31/mental-health-care-crisis-overwhelming-prison-jail.

144. Information on the BOP is from Christie Thompson and Taylor Elizabeth Eldridge, "Treatment Denied: The Mental Health Crisis in Federal Prisons," The Marshall Project, November 21, 2018. Accessed August 5, 2019, https://www.themarshallproject.org/2018/11/21/treatment-denied-the-mental-health-crisis-in-federal-prisons.

145. "Federal Prisons: Information on Inmates with Serious Mental Illness and Strategies to Reduce Recidivism."

146. Mary Gillberti, "Treatment, Not Jail: It's Time to Step Up," *NAMI*, May 5, 2015. Accessed August 7, 2019, https://www.nami.org/Blogs/From-the-Executive-Director/May-2015/Treatment,-Not-Jail-It's-Time-to-Step-Up.

147. Deborah L. Shelton, "How Sending the Mentally Ill to Jail Is a Cost to Us All."

148. Alisa Roth, "A 'Hellish World': The Mental Health Crisis Overwhelming America's Prisons."

149. "Mass Attacks in Public Spaces – 2017," Department of Homeland Security, United States Secret Service, National Threat Assessment Center, March 2018. Accessed August 4, 2019, https://mentalillnesspolicy.org/wp-content/uploads/secret_service_64_attacks_MI.pdf.

150. Alisa Roth, "A 'Hellish World': The Mental Health Crisis Overwhelming America's Prisons."

151. Ibid.

152. Unless indicated otherwise, information in this section is from Laura M. Maruschak and Jennifer Bronson, "HIV in Prisons, 2015–Statistical Tables," in Bureau of Justice Statistics, U.S. Department of Justice (Washington, DC: U.S. Government Printing Office, August 2017), accessed August 9, 2019, https://www.bjs.gov/content/pub/pdf/hivp15st.pdf; Laura M. Maruschak, "HIV in Prisons, 2001–2010," in *Bureau of Justice Statistics Bulletin*, U.S. Department of Justice (Washington, DC: U.S. Government Printing Office, September 2012), accessed November 21, 2012, http://bjs.ojp.usdoj.gov/content/pub/pdf/hivp10.pdf.

153. James Hamblin, "Mass Incarceration Is Making Infectious Diseases Worse," *The Atlantic*, July 18, 2016. Accessed August 10, 2019, https://www.theatlantic.com/health/archive/2016/07/incarceration-and-infection/491321/. Calculations made by authors.

154. Leonard S. Rubenstein, Joseph J. Amon, Megan McLemore, Patrick Eba, Kate Dolan, Rick Lines, Chris Beyrer, "HIV, Prisoners, and Human Rights," *The Lancet*, July 14, 2016. Accessed August 10, 2019, https://www.thelancet.com/journals/lancet/article/PIIS0140-6736(16)30663-8/fulltext.

155. James Hamblin, "Mass Incarceration Is Making Infectious Diseases Worse."

156. Information in this section is from Beth Schwartzapfel, "Taking Freedom: Modern-Day Slavery in America's Prison Workforce," Pacific Standard, May 7, 2018. Accessed August 10, 2019, https://psmag.com/social-justice/taking-freedom-modern-day-slavery.

157. Beth Schwartzapfel, "Taking Freedom: Modern-Day Slavery in America's Prison Workforce."

158. "UNICOR," Federal Bureau of Prisons. Accessed August 11, 2019, https://www.bop.gov/inmates/custody_and_care/unicor_about.jsp.

159. Ibid.

160. "FPI and Vocational Training Works: Post-Release Employment Project (PREP)," Federal Bureau of Prisons. Accessed August 11, 2019, https://www.bop.gov/resources/pdfs/prep_summary_05012012.pdf.

161. Alexia Fernandez Campbell, "The Federal Government Markets Prison Labor to Businesses as the 'Best-Kept Secret'," *Vox*, August 24, 2018. Accessed August 10, 2019, https://www.vox.com/2018/8/24/17768438/national-prison-strike-factory-labor.

162. Ibid.

163. Calculated from National Corrections Reporting Program, 1991–2015: Selected Variables, Year-End Population. Accessed August 12, 2019, http://www.icpsr.umich.edu/cgi-bin/SDA/NACJD/hsda3.

164. Calculated from data in Mariel Alper and Lauren Glaxe, "Source and Use of Firearms Involved in Crimes: Survey of Prison Inmates, 2016," U.S. Department of Justice, *Bureau of Justice Statistics Special Report* (Washington, DC: USGPO, January 2019), 6, Table 4, accessed February 26, 2020, https://www.bjs.gov/content/pub/pdf/suficspi16.pdf.

165. "Correctional Education," The National Reentry Resource Center. Accessed August 12, 2019, https://csgjusticecenter.org/corrections/correctional-education/.

166. Lois M. Davis, Jennifer L. Steele, Robert Bozick, Malcolm Williams, Susan Turner, Jeremy N. V. Miles, Jessica Saunders, and Paul S. Steinberg, Correctional Education in the United States: How Effective Is It, and How Can We Move the Field Forward? (Santa Monica, CA: RAND Corporation, 2014).

Accessed June 20, 2016, www.rand.org/pubs/research_briefs/RB9763.html.

167. Ibid.

168. Calculated from National Corrections Reporting Program, 1991–2015: Selected Variables, Year-End Population.

169. Patrick Oakford et al., "Investing in Futures: Economic and Fiscal Benefits of Postsecondary Education in Prison."

170. Calculated from National Corrections Reporting Program, 1991–2015: Selected Variables, Year-End Population; Beck, Berzofsky, and Krebs, "Sexual Victimization in Prisons and Jails Reported by Inmates, 2011–12."

171. Patrick Oakford et al., "Investing in Futures: Economic and Fiscal Benefits of Postsecondary Education in Prison."

172. Bobby D. Rampey, Shelley Keiper, Leyla Mohadjer, Tom Krenzke, Jianzhu Li, Nina Thornton, and Jacquie Hogan, "Highlights from the U.S. PIAAC Survey of Incarcerated Adults: Their Skills, Work Experience, Education, and Training," U.S. Department of Education, National Center for Education Statistics, November 2016, 24, Table 3.1. Accessed August 31, 2019, https://nces.ed.gov/pubs2016/2016040.pdf; Patrick Oakford et al., "Investing in Futures: Economic and Fiscal Benefits of Postsecondary Education in Prison."

173. Unless indicated otherwise, information in this section on Pell Grants is from Ibid.

174. Michael T. Nietzel, "Congress Should Pass The REAL Act and Make Prisoners Eligible for Pell Grants Again," *Forbes*, April 15, 2019. Accessed August 13, 2019, https://www.forbes.com/sites/michaeltnietzel/2019/04/15/congress-should-pass-the-real-act-and-make-prisoners-eligible-for-pell-grants-again/#636fbcaf2f0c.

175. Ibid.

176. Lois M. Davis et al., *Correctional Education in the United States: How Effective Is It, and How Can We Move the Field Forward?*

177. Lois M. Davis, Robert Bozick, Jennifer L. Steele, Jessica Saunders, and Jeremy N. V. Miles, "Evaluating the Effectiveness of Correctional Education: A Meta-Analysis of Programs That Provide Education to Incarcerated Adults," The RAND Corporation, 2013. Accessed August 12, 2019, https://www.rand.org/content/dam/rand/pubs/research_reports/RR200/RR266/RAND_RR266.pdf; Robert Bozick, Jennifer L. Steele, Lois M. Davis, and Susan Turner, "Does Providing Inmates with Education Improve Post-Release Outcomes? A Meta-Analysis of Correctional Education Programs in the United States," *Journal of Experimental Criminology* 14, no. 3 (2018), 389–428. Accessed August 12, 2019, https://perma.cc/NKE4- KDFK.

178. Donna Gordon Blankinship, "College Behind Bars: An Old Idea with Some New Energy," AP News Archive, March 1, 2016. Accessed June 20, 2016, www.apnewsarchive.com/2016/For-more-than-20-years-prisons-have-focused-on-punishment-over-rehabilitation-but-now-that-is-changing/id-2d2d0b63ef054c5298a27da86a8bde95.

179. For reference, see Anthony Walsh, *Correctional Assessment, Casework and Counseling*, 5th ed. (Alexandria, VA: American Correctional Association, 2010); also see Patricia van Voorhis, Michael Braswell, and David Lester, *Correctional Counseling and Rehabilitation*, 7th ed. (Cincinnati, OH: Anderson, 2009).

180. Grant Duwe, "The Use and Impact of Correctional Programming for Inmates on Pre- and Post-Release Outcomes," U.S. Department of Justice, National Institute of Justice, June 2017. Accessed August 14, 2019, https://www.ncjrs.gov/pdffiles1/nij/250476.pdf.

181. Ted Palmer, *A Profile of Correctional Effectiveness and New Directions for Research* (Albany, NY: State University of New York Press, 1994).

182. Johnson, *Hard Time: Understanding and Reforming the Prison.*

Design Elements: *Image of part of a court building with columns and statue*: Pixtal/AGE Fotostock RF; *Image of a door slightly open*: jadding/Shutterstock.com

11

Giles Clarke/Getty Images

Prison Life, Inmate Rights, Release, Reentry, and Recidivism

CHAPTER OUTLINE

LEARNING OBJECTIVES

After completing this chapter, you should be able to:

1. Distinguish between the deprivation and importation models of inmate society.
2. Explain how today's inmate society differs from those of the past.
3. Identify some of the special features of life in women's prisons.
4. Describe the profile of correctional officers, and explain some of the issues they face.
5. Identify prisoners' rights, and relate how they were achieved.
6. List the two most common ways that inmates are released from prison, and compare those two ways in frequency of use.
7. Summarize what recidivism research reveals about the success of the prison in achieving deterrence and rehabilitation.

Crime Story

On May 23, 2019, John Walker Lindh (pictured), the then-38-year-old so-called "American Taliban," was released from the federal prison in Terre Haute, Indiana, after serving 17 years of a 20-year sentence for providing support to the Taliban and carrying an explosive during the commission of a felony. Lindh was the first U.S.-born detainee in the U.S. "war on terrorism." He originally was charged with 10 counts, including conspiracy to murder U.S. citizens or U.S. nationals. Had he been convicted on all counts, he could have been sentenced to as much as three life terms plus 90 years in prison. However, when he pleaded not guilty to the charges, Justice Department attorneys offered him the 20-year plea deal because they feared that his confession would be inadmissible in court. Not only was Lindh not read his *Miranda* rights, but he also confessed under torture, according to his attorneys. At his 2002 sentencing, Lindh expressed remorse for his crimes. He tearfully told the judge: "I have never supported terrorism in any form, and I never will . . . I made a mistake by joining the Taliban . . . [and] [h]ad I realized then what I know now, I would never have joined them."

Lindh was raised north of San Francisco in Marin County, converted to Islam at age 16 after viewing the movie "Malcolm X," and went to Yemen and Pakistan at age 19 to study Arabic and the Koran. He met Osama bin Laden and for months had been fighting with the Taliban in its civil war in Afghanistan, when, on September 11, 2001, al-Qaeda terrorists flew planes into the World Trade Center and the Pentagon. Lindh was 20 years old.

A couple of months after 9/11, Lindh was captured with a few hundred other Taliban fighters retreating from the battlefield and imprisoned in northern Afghanistan. On November 25, 2001, the Taliban prisoners began an 8-day uprising, during which a CIA officer was killed, along with all but 86 prisoners and many of their jailers. Lindh survived and denied any role in the CIA officer's death. No evidence showed that he took an active part in the prison violence. For most of the uprising, he apparently was hiding in the basement. Nevertheless, the officer's parents still blame Lindh for their son's death.

He was returned to the United States to stand trial. He was 21 years old. A media blitz ensued, and Lindh was branded as a traitor and the "American Taliban." As his infamy grew, a frenzied public demanded the death penalty. Attempts to try Lindh for treason faltered for lack of evidence. In the end, all he really was guilty of was serving in a foreign army, but that is not what the public believed.

In prison, Lindh was treated as a terrorist. For the first 11 years of his sentence, prison officials forbade his speaking Arabic, the language of Muslim prayer, and any meaningful interaction with other Muslim inmates.

Although Lindh was released early for good behavior, he still must serve 3 years on probation and be subjected to several restrictions. He cannot hold a passport or leave the United States; he must receive mental health counseling; he needs permission to own Internet-capable devices; if he gets that permission, he must install monitoring software so the probation office is able to supervise his Internet use; he must refrain from viewing or possessing extremist or terrorist material; he only may conduct online communication in English; and he may not communicate with known extremists. The court imposed these restrictions because it believes that Lindh remains a potentially violent Islamist terrorist. Evidence for the belief included a 2015 handwritten letter that he sent to a producer at NBC Los Angeles in which he expressed support for the Islamic State militant group (ISIS), praising it for the "spectacular job" it was doing. Concern also was expressed about his prison correspondence with journalists, which was inspected routinely by prison officials. The National Counterterrorism Center (NCC) cited the correspondence in a 2016 U.S. intelligence document that concluded Lindh "continued to advocate for global jihad and to write and translate violent extremist texts."

Alexandria Sheriff's Office/AP Images

Critics of Lindh's release, and there are many, are concerned about where Lindh will live and what he will do. He faces a difficult time. President Trump, Secretary of State Pompeo, and other government officials publicly opposed his release. President Trump asked his lawyers whether Lindh's release could be blocked and, being told that it could not, declared that the United States will monitor him closely. Secretary Pompeo believes that Lindh's [early] release is "unexplainable and unconscionable." He would like a review of prison system policies.

Officials in the Bureau of Prisons (BOP) defend Lindh's early release, maintaining that they followed federal laws and guidelines. They explained that the Bureau collaborates with outside agencies "to reduce the risk terrorist offenders pose inside and outside of prisons." They boast, "No radicalized inmate has returned to federal prison on terrorism-related charges."

Supporters of Lindh argue that concern and criticism over his early release are misplaced. They believe that Lindh was imprisoned too long and point to many other Taliban fighters sent to GITMO as enemy combatants who were released after serving much less time. They contend that when Lindh's Islamic State support letter was written in 2015, he might not have been aware of the full extent of the Islamic State's atrocities since he had been in prison. Supporters ask, "Who really knows what Lindh believes now?" Of course, Lindh's critics ask the same question.

The subject of this chapter is "Living in Prison," but it also addresses the problems of prisoner release, reentry, and recidivism. Should Lindh have been released? Was his sentence too long? Where should he live? Should he be employed? If so, what should he do? Beyond Lindh, what should be done about the approximately 100 federal prisoners, including 62 U.S. citizens, sentenced for terrorism-related crimes that are scheduled for release in the next few years? How can their recidivism be prevented? Is long-term imprisonment, by itself, an adequate deterrent? These are questions that currently befuddle prison officials and members of Congress.

Living in Prison

When people think of prisons, they usually imagine the big-house, maximum-security prison for men. Indeed, the majority of scholarly research on prison life has focused on this type of institution. In this chapter, we survey some of that research. However, we caution the reader to be careful about generalizing findings from studies of the big house to all institutions. As we have already seen, institutions are quite diverse.

Inmate Society

total institution An institutional setting in which persons sharing some characteristics are cut off from the wider society and expected to live according to institutional rules and procedures.

In his classic book *Asylums,* sociologist Erving Goffman described prisons as total institutions. Goffman defined a **total institution** as "a place of residence and work where a large number of like-situated individuals, cut off from the wider society for an appreciable period of time, together lead an enclosed, formally administered round of life."[1] As already pointed out and as further implied by Goffman's definition, a prison represents a miniature, self-contained society.

Although prisons certainly are influenced by the outside world,[2] they also are separated and closed off from that world. A society of prisoners, like any society, possesses distinctive cultural features (values, norms, and roles), and because prisoners do not serve their time completely isolated from other inmates, the features of the inmate society influence the way prisoners adjust to prison life.

convict code A constellation of values, norms, and roles that regulate the way inmates interact with one another and with prison staff.

Central to the inmate society of traditional men's prisons in the United States is the convict code. The **convict code** is a constellation of values, norms, and roles that regulate the way inmates interact with one another and with prison staff. For example, a principle of the convict code is that individual inmates should mind their own affairs and do their own time. Other principles are that inmates should not inform the staff about the illicit activities of other prisoners and that inmates' overall attitude and behavior should be indifferent to the staff and loyal to other convicts. Conning and manipulation skills are highly valued under the code, as is the ability to show strength, courage, and toughness.

deprivation model A theory that the inmate society arises as a response to the prison environment and the painful conditions of confinement.

Two major theories of the origins of the inmate society have been advanced: the deprivation model and the importation model. Advocates of the **deprivation model**, which is sometimes called the "indigenous origins model," contend that the inmate society arises in response to the prison environment and the painful conditions of confinement.[3] Specifically, imprisonment deprives inmates of such things as material possessions, social acceptance, heterosexual relations, personal security, and liberty. This environment of shared deprivation gives inmates a basis for solidarity. The inmate society and its convict code represent a functional, collective adaptation of inmates to this environment. When an inmate enters prison for the first time, the inmate is socialized into the customs and principles of the inmate society, a process former correctional officer Donald Clemmer termed **prisonization.**[4] Clemmer believed that the longer inmates stayed in prison, the more "prisonized" they became and the more likely they were to return to crime after their release from prison.

prisonization The process by which an inmate becomes socialized into the customs and principles of the inmate society.

importation model A theory that the inmate society is shaped by the attributes inmates bring with them when they enter prison.

An alternative to the deprivation model is the **importation model**, which holds that the inmate society is shaped by factors external to the prison environment—specifically, the preprison experiences and socialization patterns that inmates bring with them when they enter prison.[5] For example, inmates who were thieves and persistently associated with other thieves before going to prison bring the norms and values of thieves into the prison. They

PF Bentley/pfpix.com

Conflict builds when inmates form competing gangs along ethnic, racial, and geographic lines. *Why do inmates form gangs in prison?*

will remain more loyal to those norms and values than to the staff while incarcerated. However, if an inmate primarily conformed to the law before entering prison, the inmate probably will exhibit greater loyalty to staff norms once in prison. In short, the deprivation model assumes that new inmates are socialized into the existing inmate society, which arises in response to the prison environment; the importation model suggests that the inmate society is the product of the socialization inmates experience before entering prison.

The deprivation and importation models were developed from studies of prisons of the pre-1960s era. Some scholars believe that the image of a fairly unified inmate society with an overarching convict code fails to characterize many of today's crowded and violent prisons. According to those scholars, several factors have rendered the inmate society fragmented, disorganized, and unstable:

1. Increasing racial heterogeneity.
2. The racial polarization of modern prisoners.
3. Court litigation.
4. The rise and fall of rehabilitation.
5. The increased politicization of inmates.

The order and stability provided by the old inmate subculture have been replaced by an atmosphere of conflict and tension in which inmates align themselves into competing gangs (such as the Aryan Brotherhood, the Mexican Mafia, and the Black Guerrilla Family) and other inmate organizations (such as the Lifer's Club and the Muslims).[6]

Violence and Victimization Experts generally agree that there is more physical violence by inmates in today's state prisons than there was in earlier periods. Although data are elusive, it is safe to say that, in recent years, inmates have assaulted thousands of other inmates and prison staff. In the case of lethal assaults, from 2001 through 2014, inmates killed approximately 1,000 other inmates and prison staff in state and federal prisons, or about 70 per year: 845 in state prisons or about 60 per year, and 130 in federal prisons, or about 9 per year.[7] Most of the killings involved inmates killing other inmates (inmates killing prison staff is addressed below). As the prison population was decreasing (see Figure 10.1), the number of inmates killing other inmates was increasing. From 2010 through 2014, inmates killed an average of nearly 80 other inmates in state and federal prisons (up from an average of 70 from 2001 through 2014): 70 in 2010, 70 in 2011, 85 in 2012, 90 in 2013, and 83 in 2014.[8] A preliminary count shows an additional 70 inmate-on-inmate killings in state prisons in 2015 (an increase from the average of 60 from 2001 through 2014).[9] From 2001 through 2014, California led the nation with 191 inmate-on-inmate killings, followed by Florida with 76, Texas with 56, Georgia with 53, and Oklahoma with 43.[10]

FYI

Attica Prison Riot

The deadliest prison riot in the nation's history occurred at the Attica Correctional Facility in upstate New York in September 1971. The uprising began when inmates took over one of the prison yards and held 49 guards hostage, demanding better living conditions within the facility. It ended 4 days later after state troopers, under orders from Governor Nelson Rockefeller, raided the prison, sparking violence that left 11 guards and 32 prisoners dead and more than 80 others wounded. On February 15, 2000, a U.S. District Court judge approved a settlement under which New York State, without admitting wrongdoing or liability, would pay $8 million to 1,280 inmates who were abused by law enforcement officials during the riot. The agreement concluded a long-stalled $2.8 billion class action lawsuit originally filed against prison and state officials in 1974.

Source: "$8 Million Award Ends Attica Prison Suit," in *Facts on File: World News Digest with Index*, no. 3097 (April 13, 2000), 248.

However, in 2017 and 2018, Alabama led the nation in inmate-on-inmate killings with 10 murders in 2017 and 9 in 2018, which were more than 600% greater than the national average from 2001 to 2014.[11] From 2014 through 2018, inmates killed 35 other inmates in Alabama prisons. The level of violence in Alabama's prisons has been increasing for more than a decade, while the prison population has been decreasing. Reasons for the increased violence include serious understaffing (a single officer typically was assigned to a dorm of 198 inmates, and, at times, eight officers were responsible for managing an entire prison with a population of more than 1,200 men), systemic classification failures, excessive violence by correctional staff against inmates, and other official misconduct and corruption.

As for inmates killing prison staff, in 2017 and 2018, inmates killed a dozen prison staff: eight in 2017 and four in 2018.[12] Four of the eight staff members killed in 2017 were killed in a single incident at the Pasqutoank Correctional Institution in North Carolina. Veronica Darden, Correction Enterprise Manager; Geoffrey Howe, Correction Maintenance Mechanic; and Wendy Shannon and Justin Smith, Corrections Officers were ambushed by four inmates in the prison sewing plant.[13]

Most experts agree that prison staff violence against inmates is less common today. However, the number of inmates assaulted and killed by prison staff in the last few years is unknown.

Commonly cited reasons for high rates of prison violence include improper management and classification practices by staff, high levels of crowding and competition over resources, the young age of most inmates in many prisons, and increases in racial tensions and prison gang activity. Although gang and racial conflicts surely play a role, it seems unlikely that the bulk of prison violence is extensively planned or gang orchestrated. One reason is that many prisons across the United States lack well-developed gang structures and serious gang problems, but they still are plagued by violence. Furthermore, what we know about violent crime in general suggests that much prison violence is probably spontaneous, motivated by particular circumstances. Some common perpetrator motives for physical violence in prison are (1) to demonstrate power and dominance over others; (2) to retaliate against a perceived wrong, such as the failure of another inmate to pay a gambling debt; and (3) to prevent the perpetrator from being victimized (e.g., raped) in the future.

A good deal of prison violence–but not all–has sexual overtones. Note that not all instances of sex in prison are violent and not all instances are homosexual in nature; also note that sexual encounters can involve both inmates and staff. Instances of prison sex can be further divided into three basic categories: (1) consensual sex for gratification, (2) prostitution, and (3) sexual assault. The third category obviously involves violence, and the first two sometimes have indirect links to violence. For example, a consensual sexual relationship

SCOTT MCCLOSKEY/AP Images

Prison violence and disorder generally result spontaneously from particular circumstances. *What are some of those circumstances?*

between two inmates may have started out as a forced one. Likewise, an inmate who is vulnerable to sexual assault may perform sexual favors for an aggressive, well-respected prisoner in exchange for protection from other inmates. In the language of the prison, such vulnerable inmates are known as "punks." Prison sexual encounters can be intraracial or interracial; however, sexual assaults are more often interracial than intraracial.

Before passage of the Prison Rape Elimination Act of 2003, no reliable measures of sexual victimization in prisons and jails nationwide existed. Congress attempted to rectify that problem by requiring the U.S. Department of Justice's Bureau of Justice Statistics (BJS) "to carry out, for each calendar year, a comprehensive statistical review and analysis of the incidence and effects of prison rape." BJS expanded its mandate to "all types of sexual activity–for example, oral, anal, or vaginal penetration, hand jobs, touching of the inmate's butt, thighs, penis, breasts, or vagina in a sexual way and other sexual acts." Also included are "nonconsensual sexual acts, abusive sexual contacts, and both willing and unwilling sexual activity with staff," as well as "staff sexual harassment," such as "demeaning references to an inmate's sex, derogatory comments about an inmate's body or clothing, or repeated profane or obscene language or gestures."[14]

Prior to 2007, data were obtained only from administrative records. That data showed that from 2005 through 2008, the number of sexual victimization allegations reported to prison and jail staff ranged from a low of 6,241 in 2005 to a high of 7,444 in 2008. The rate of allegations per 1,000 inmates ranged from 2.83 in 2005 to 3.18 in 2008. Of the 7,444 allegations in 2008, 31% were for inmate-on-inmate nonconsensual sexual acts, 19% were for inmate-on-inmate abusive sexual contacts, 34% were for staff sexual misconduct, and 16% were for staff sexual harassment. The number of substantiated allegations, however, was much lower. From 2005 through 2008, the number of substantiated allegations ranged from a low of 885 in 2005 to a high of 1,001 in 2007. Of the 7,444 allegations in 2008, only 12.5% were substantiated: 25% were for inmate-on-inmate nonconsensual sexual acts, 29% were for inmate-on-inmate abusive sexual contacts, 39% were for staff sexual misconduct, and 7% were for staff sexual harassment.[15]

Knowing that inmates might be reluctant to report sexual victimization to prison staff, particularly when staff committed incidents, in 2007, BJS researchers supplemented the data from administrative records with data from the first National Inmate Survey–a self-report sexual victimization survey. Researchers, who promised a sample of inmates in both state and federal prisons (jail inmates were not included) anonymity and confidentiality, conducted the survey; it revealed that about 60,500 inmates, or 4.5% of the nation's prisoners, experienced sexual victimization. The 60,500 inmates were more than 8 times the number of inmates in 2007 that reported to prison staff being victims of sexual victimization. The data showed that about 2.1% of inmates reported sexual victimization involving another inmate, 2.9% reported sexual victimization involving staff, and about 0.5% of inmates had been victimized by both other inmates and staff. Inmates also were asked the number of times they had experienced each type of sexual victimization, from which rates were calculated. Nationwide, 123 incidents of sexual victimization per 1,000 inmates were reported, compared to 2.95 incidents of sexual victimization per 1,000 inmates reported by correctional authorities in 2007. The 123 incidents per 1,000 inmates included 49 incidents of inmate-on-inmate sexual acts per 1,000 inmates and 75 incidents of unwilling sex with staff per 1,000 inmates. The rates excluded unwanted touching by other inmates and willing sexual contacts with staff.[16]

BJS completed the second National Inmate Survey between October 2008 and December 2009.[17] This second survey collected data from most inmates by employing a computer-assisted self-interview in which inmates followed instructions delivered via headphones and used a touch-screen to complete a computer-assisted questionnaire. Results of the second survey mirror those of the 2007 survey (but include data from jails): 4.4% of prison inmates and 3.1% of jail inmates reported being sexually victimized, which extrapolates into 88,500 prison and jail inmates nationwide.

The BJS conducted the third National Inmate Survey between February 2011 and May 2012.[18] In addition to inmates in prisons and jails, this survey interviewed inmates in confinement facilities operated by Immigration and Customs Enforcement (ICE), the U.S. Military, and correctional authorities in Native American country. The survey, for the first time, also provided estimates of sexual victimization of juveniles held in adult facilities. As in the previous two surveys, the third survey collected only allegations of sexual victimization, some of which undoubtedly may not be true. All surveys that promise anonymity and confidentiality, as these surveys did, suffer from the possibility that respondents may not report incidents that occurred or may report incidents that did not occur.

Results of the third survey mirror the results of the first two—that is, they are not statistically different from each other, and thus support the previous findings. In 2011–2012, 4.0% of all prison inmates and 3.2% of all jail inmates alleged that they had experienced sexual victimization in the past 12 months, or since admission to the facility, if less than 12 months. The percentages extrapolate to a total of 80,600 inmates nationwide.

Results of the third National Inmate Survey conducted in 2011 and 2012 were the last ones published by the BJS as of this writing (August 24, 2019). The BJS only conducts the surveys as funding permits. To fulfill Congress's PREA mandate "to carry out, for each calendar year, a comprehensive statistical review and analysis of the incidence and effects of prison rape," the BJS has developed two other instruments to measure inmate sexual victimization: the Survey of Sexual Victimization, which annually collects only administrative data on the incidence of sexual victimization (as did pre-2007 surveys) in adult and juvenile correctional facilities, and the National Survey of Youth in Custody (NSYC), which collects data, as funding permits, on the prevalence of sexual assault as reported by inmates in prisons and jails and by youth held in juvenile correctional facilities. The last NSYC survey was conducted in 2012, and its results were published in 2013.

The latest data available on sexual victimization in prisons and jails as of this writing are from the 2015 Survey of Sexual Victimization (data in the survey from other correctional facilities are not reported here).[19] In 2015, prison administrators reported 18,666 allegations of sexual victimization, which was a 180% increase from the 6,660 allegations reported in 2011. Jail administrators reported 5,809 allegations of sexual victimization in 2015, which was a 141% increase from the 2,047 allegations reported in 2011.[20] Rates of allegations increased from 4.5 allegations per 1,000 prison inmates in 2011 to 12.6 allegations per 1,000 prison inmates in 2015. Comparable rates for jail inmates were 2.73 in 2011 and 8.03 in 2015.[21] The large increases in prison and jail sexual victimization allegations are attributed to new Justice Department directives issued in 2012, which specified what data must be tracked and reported to BJS and required the education of correctional staff and inmates about sexual victimization, the investigation of each allegation, and the provision of medical and mental health care for victims.[22]

During the 4 years from 2012 through 2015, state and federal prison administrators reported a total of 47,066 allegations of sexual victimization.[23] Fifty-nine percent of the allegations involved staff sexually victimizing inmates, and 41% of the allegations involved inmates sexually victimizing other inmates. However, completed investigations only substantiated (had sufficient evidence to determine the incident occurred) 5.8% of staff-victimizing-inmate allegations and 8.6% of inmate-victimizing-inmate allegations.[24] The number of substantiated incidents of sexual victimization in state and federal prisons (including private prisons) increased about 44% from 605 in 2011 to 873 in 2015.[25]

From 2012 through 2015, jail administrators reported a total of 16,171 allegations of sexual victimization.[26] Fifty-nine percent of the allegations involved inmates sexually victimizing other inmates, and 41% of the allegations involved staff sexually victimizing inmates—exactly the opposite from the percentages and types of allegations reported by state and federal prison administrators. The number of allegations in jails substantiated by completed investigations was 14% for inmate-victimizing-inmate allegations and 9.6% for staff-victimizing-inmate allegations.[27] The number of substantiated incidents of sexual victimization in jails increased about 103% from 284 in 2011 to 576 in 2015.[28]

Physical victimization is not the only or even the most frequent kind of victimization that takes place in prison. Sociologist Lee Bowker identifies three other kinds: economic, psychological, and social.[29] As with physical victimization, these three kinds may be perpetrated by inmates against other inmates, by inmates against staff, or by staff against inmates. In addition, the four kinds of victimization are interrelated: combinations can occur, and one can lead to another.

sub-rosa economy The secret exchange of goods and services among inmates; the black market of the prison.

All societies have an economy, and most have a black-market component. The inmate society is no exception. The **sub-rosa**—that is, the secret or underground—**economy** of an institution consists of the exchange of goods and services that, although often illicit (contraband), are in high demand among inmates. Examples of such goods and services include food, clothes, alcohol and other drugs, pornography, weapons, loan services, protection, sex, and gambling. Engaging in illicit economic exchanges is commonly referred to as "playing the rackets" or "hustling."

Over the past decade or so, cell phones have become a popular type of contraband. They have been used in escapes, arranging drug deals, kidnappings, and murders. Thousands of illegal cell phones are confiscated every year. Some cell phones have been thrown over prison fences, including one incident where someone threw a football loaded with cell phones into the prison yard. Staff also smuggle cell phones into prisons. A California corrections officer reportedly made $150,000 in a single year by supplying inmates with cell phones. He was fired but not charged with any crime.

Recently, tobacco and tobacco products have become contraband, as many state correctional systems and the BOP have banned them and smoking. Figure 11.1 shows the states and the status of tobacco, tobacco products, and smoking in their jails and prisons. The ban

FIGURE 11.1 States and the Status of Smoking and Using Tobacco and Tobacco Products in Correctional Facilities, July 1, 2019

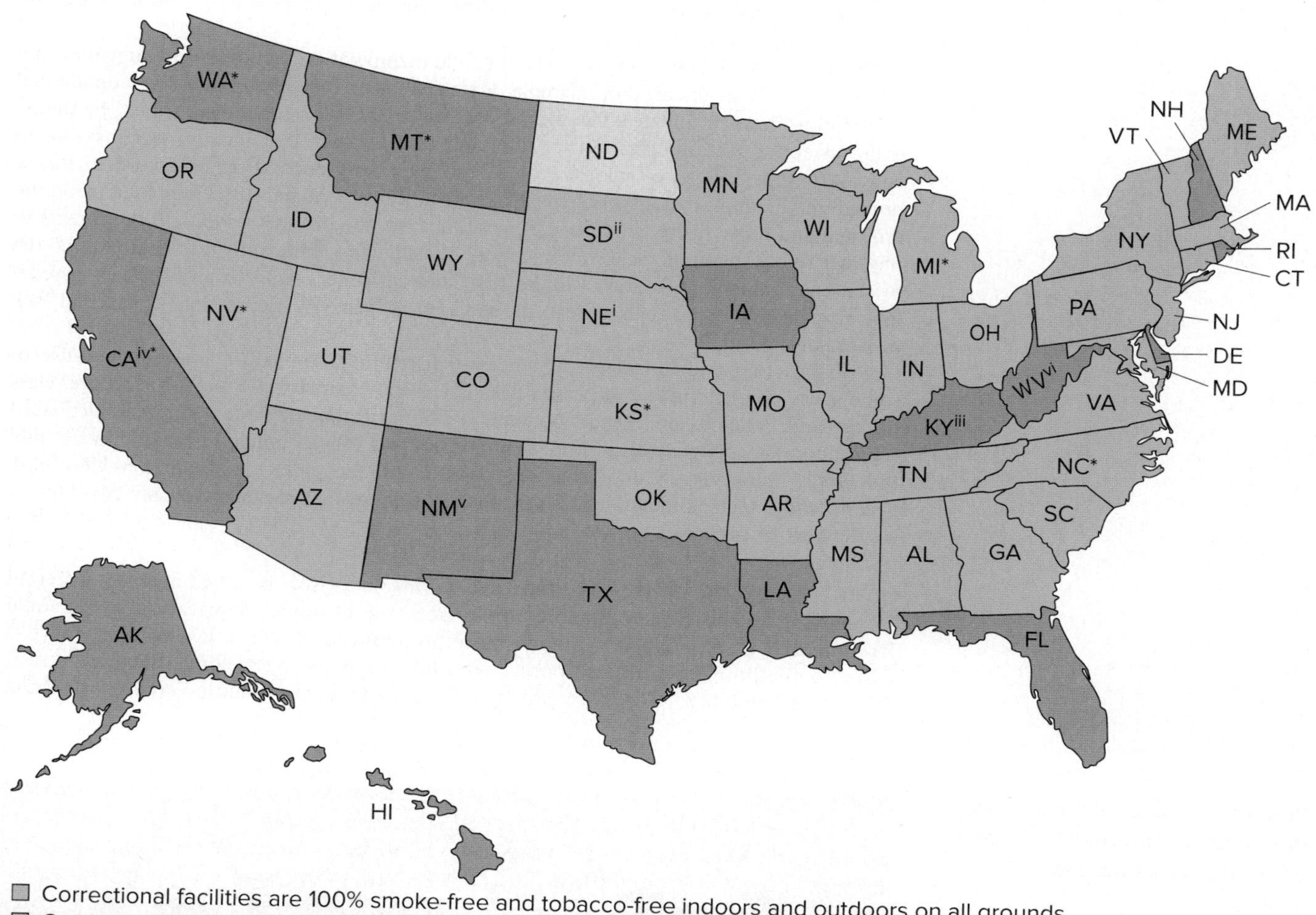

- Correctional facilities are 100% smoke-free and tobacco-free indoors and outdoors on all grounds
- Correctional facilities are 100% smoke-free indoors and outdoors on all grounds
- Correctional facilities are 100% smoke-free and tobacco-free indoors
- Correctional facilities are 100% smoke-free indoors

Notes:

[i] Nebraska policy does not apply to Community Corrections Centers in Omaha and Lincoln, which have their own policies.

[ii] In 2009, South Dakota's Department of Corrections eliminated an exemption for tobacco use in Native American religious ceremonies. In April 2014, the 8th Circuit Court of Appeals ruled that the removal of the religious exemption violates the constitutional rights of Native American inmates. The Department of Corrections may appeal the decision.

[iii] All Kentucky correctional facilities are 100% smoke-free and tobacco-free indoors except for Eddyville State Penitentiary.

[iv] All California correctional facilities are 100% smoke-free and tobacco-free indoors except for employee housing.

[v] At New Mexico correctional facilities, smoking and tobacco-use is permitted by employees in private residences and personal vehicles on facility property.

[vi] State law requires that inmates not use or possess tobacco products. Facilities run by the West Virginia Regional Jail and Correctional Facility Authority are tobacco-free. Facilities run by the West Virginia Division of Corrections are not smoke-free.

* Tobacco for ritual tribal or other religious use is permitted.

Source: "100% Smokefree and Tobacco-Free Correctional Facilities," American Nonsmokers' Rights Foundation, July 1, 2019, accessed August 25, 2019, http://no-smoke.org/wp-content/uploads/pdf/100smokefreeprisons.pdf.

on cigarettes has caused problems among inmate populations because cigarettes traditionally have served as their medium of exchange, since currency typically is contraband. The money inmates earn from prison work usually is maintained in accounts from which they may deduct only specified amounts for specified purposes. Hence, packs and cartons of cigarettes assumed the symbolic value of money in the sub-rosa economy. However, currency also may be used. Like any kind of contraband (e.g., drugs), currency can enter the institution via mail, visits, or staff. In place of cigarettes, two of the more popular new mediums of exchange used in prisons and jails are postage stamps and packets of ramen noodles.[30] A prison's sub-rosa economy is a major concern to prison officials because it sets the stage for various types of economic victimization, including theft, robbery, fraud, extortion, loan-sharking, and price-fixing by gangs and cliques.

Psychological victimization consists of subtle manipulation tactics and mind games that occur frequently in prison. For example, a staff member may obtain sensitive confidential information from an inmate's file and proceed to "mess" with the inmate's mind by threatening to convey the information to other inmates. Likewise, an inmate may threaten to convey to superiors some instance of a staff member's corruption or failure to follow procedure. Examples of psychological victimization are numerous. Combined with the crowded and noisy living conditions, separation from the outside world, and the threats to physical safety, psychological victimization contributes significantly to the stress of the prison environment. It is small wonder that many prisoners suffer from a host of psychological maladaptations, such as fear, paranoia, excess anxiety, depression, anger, and lack of trust in others.

Social victimization involves prejudice or discrimination against a person because of some social characteristic that person has. The social characteristic may be race, age, class background, religious preference, political position, or another factor. For example, racial segregation and discrimination against minorities was official policy in many prisons until the 1960s. Following the desegregation movement in the wider society, prisons, too, have become more racially integrated. Today, the segregation that remains typically is voluntary on the part of inmates.

Inmate Coping and Adjustment Living in prison is fundamentally different from living in the free community. In prison, for example, deprivation of personal freedom and material goods is much more pronounced. There is less privacy; there is more competition for scarce resources; and life is typified by greater insecurity, stress, and unpredictability. Moreover, prison life may encourage qualities counter to those

PATTI LONGMIRE/AP Images

Inmates develop a variety of coping mechanisms in prison. *If you were in prison, how would you cope?*

required for functioning effectively in the free community. Criminologist Ann Cordilia, expanding on the work of Erving Goffman, argues that prisons "desocialize" and alienate inmates by discouraging personal responsibility and independence, creating excessive dependency on authority, and diminishing personal control over life events.[31] One result is what Goffman called *self-mortification*, a subduing or deadening of self-identity and self-determination.[32]

Therefore, a person coming from the free community into prison must learn to cope with and adjust to the institutional environment. Building on the ideas of criminologist Hans Toch,[33] criminologist Robert Johnson identifies two broad ways that inmates cope with imprisonment.[34] Some inmates enter what Johnson describes as the public domain of prison culture. Predatory, violent convicts, many of whom have lengthy prior records of incarceration (i.e., are "state-raised youths"), dominate this domain. Those convicts seek power and status in the prison world by dominating and victimizing others. Johnson claims that most inmates do not enter the public domain and live by its norms. Instead, most enter the prison's private culture. To understand this method of coping and adjustment, we must realize that each inmate brings a unique combination of personal needs into prison. That is, inmates vary in their needs for such things as privacy, environmental stimulation, safety, and emotional feedback. Entering the private culture of the prison means finding in the diverse environment of the institution a niche that will accommodate the inmate's unique combination of needs. For example, an inmate who has strong needs for privacy, safety, and intellectual stimulation may arrange to have a job in the prison library. The library becomes the inmate's niche. The private domain of prison culture consists of many diverse niches.

From another perspective, criminologist John Irwin suggests that in coping and adjusting to prison, an inmate will develop a prison career or lifestyle.[35] Three such lifestyles are "doing time," "jailing," and "gleaning." Inmates who adopt the "doing time" lifestyle primarily are concerned with getting out of prison as soon as possible and avoiding hard time in the process. Avoiding hard time means maximizing comfort and minimizing discomfort in a way that does not threaten to extend the sentence or place the inmate in danger. Inmates with lengthy prior records of incarceration, who are more accustomed to life in an institution than life outside one, often embrace the "jailing" lifestyle. Those inmates are concerned with achieving positions of influence in the inmate society; prison is their world. "Gleaning" entails trying to take advantage of the resources available for personal betterment, such as obtaining marketable vocational skills. The idea is to prepare for life after release. Relatively few inmates adopt the gleaning lifestyle.

Life in Women's Prisons

Life in women's prisons is similar to life in men's prisons in some respects, but there also are important differences. For example, both female and male inmates must cope with the deprivations, stress, depersonalization, and authoritarian atmosphere of prison life. However, women's prisons usually are not characterized by the levels of violence, interpersonal conflict, and interracial tension found in men's institutions. Nor is the anti-staff mentality of male prisoners as common among female inmates. Consequently, the environments of women's institutions often are less oppressive.

Those observations should not be taken to mean that female inmates generally do easier time than male inmates. Indeed, it has been suggested that women experience imprisonment more negatively than men.[36] A major reason for this is separation from friends and family.[37] Female inmates are more likely than male inmates to have children and to have been living with those children immediately before incarceration. In recent years, for example, about 60% to 65% of female state prisoners had children under 18 years old, and about the same percentage of those women lived with their children before incarceration.[38]

Procedures for determining the custodial relationship between a female inmate and her child vary widely. In some cases, very young children of incarcerated women may live with their mothers in prison for a temporary period. In other cases, maternal custody might be terminated. The procedure depends on the jurisdiction and the particular circumstances. A common arrangement is for children to reside with their fathers, grandparents, or other relatives or friends during the period of incarceration. Visitation is often irregular or nonexistent, and the amount of visitation is dictated by institutional rules and the geographic distance between children and their mothers.[39]

Sean Cayton/The Image Works

Many women in prison are mothers or mothers-to-be, and the restrictions imprisonment imposes on them as they try to maintain relationships with their children affect their physical and emotional well-being. *How could the stress of a prison mother's parent/child relationship be reduced?*

FYI **Federal Female Prisoner Reforms**

The 2018 "First Step Act," with a few exceptions, prohibits restraints on federal prisoners during pregnancy, labor, and postpartum recovery. Also prisoners no longer must pay for feminine hygiene products. They now are provided for free. Both reforms are supported by a large percentage of Americans. A 2018 survey found that 86% of registered voters believe that pregnant prisoners should not be shackled or handcuffed while in labor or during delivery, and 90% of registered voters support providing free basic feminine hygiene products to female prisoners. The survey results presumably apply to both federal and state prisoners.

Source: Public Law No: 115-391; Robert Blizzard, "National Poll Results," *Public Opinion Strategies*, January 25, 2018. Accessed June 3, 2019, https://www.politico.com/f/?id=00000161-2ccc-da2c-a963-efff82be0001.

Another issue unique to women's prisons is pregnant inmates. In a first-of-its kind study of pregnancy in prisons, Johns Hopkins University School of Medicine researchers surveyed for 12 months during 2016–2017, a diverse sample of female inmates in 22 state systems and all federal prisons. The prisons in the survey housed 57% of imprisoned women in the United States. The researchers reported that 1,396 of the women surveyed were pregnant at intake. Of those pregnant women, 753 (54%) had live births while in prison (more than 90% of the women eventually had live births). Six percent of the live births in prison were preterm and 30% were delivered by cesarean section. The researchers also reported that there were 46 miscarriages, 11 abortions, 4 stillbirths, 3 newborn deaths, and 2 ectopic pregnancies. There were no maternal deaths. The researchers found that there were no mandatory standards for prenatal and pregnancy care for women in U.S. prisons.[40]

Pseudofamilies and Homosexuality A distinguishing feature of the inmate society in many women's prisons is the presence of make-believe families, known as pseudofamilies. Studies have discovered that some female inmates adopt family roles, such as mother, daughter, sister, husband, and father to form kinship networks.[41] The number of inmates who adopt masculine roles, known as "butches," appears to be small in comparison with the number who take on feminine roles.[42] Furthermore, kinship ties may cut across racial lines, reflecting that race is not the divisive factor in women's prisons that it is in men's institutions. Pseudofamily structures are central to inmate society in women's prisons, providing social interaction and emotional support. Sociologist Rose Giallombardo suggests that pseudofamilies provide imprisoned women with the family affiliation and bonding that they have been socialized to desire but have been deprived of by incarceration.[43]

Family activity and homosexual activity appear to be relatively independent of one another. A female prisoner who is involved in homosexuality may or may not be part of a family. Likewise, a female prisoner who is part of a make-believe family may or may not be involved in homosexuality. If she is, she may be having homosexual relations with someone who is part of the family or with someone who is not. The number of female prisoners who take part in family activities probably exceeds the number who participate in homosexuality. In addition, the majority of female inmates who participate in homosexuality are what sociologists David Ward and Gene Kassebaum call "turnouts," those who were heterosexual before incarceration and will return to heterosexuality upon release.[44] As in men's prisons, interracial homosexuality is quite common, but in contrast to men's prisons, the overwhelming majority of homosexual activity in women's institutions is consensual and is rooted in affection and attachment instead of dominance motives.

myth

Virtually all female prisoners are lesbians.

fact

Although there are no totally reliable data on the proportion of female inmates who engage in homosexuality while incarcerated, a reasonable estimate (based on the literature) is approximately half. This proportion is comparable to the proportion of male prisoners who do so.

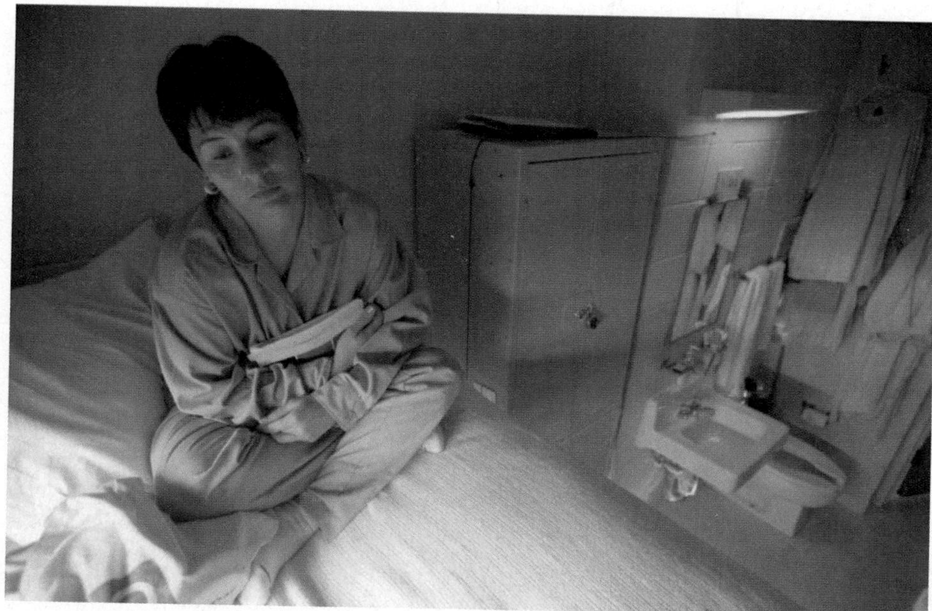

STORMI GREENER/Star Tribune/Getty Images

Female inmates are an increasing part of the nation's prison population. *Why?*

Inmate Roles Social scientist Esther Heffernan identified three roles that women commonly adopt when adjusting to prison. Heffernan's roles are very similar to roles identified by sociologist Clarence Schrag in his work on male inmates.[45] The "square" role is adopted by women who primarily were noncriminals before coming to prison; these women tend toward conventional behavior in prison. The square role discussed by Heffernan corresponds to the "square john" role identified by Schrag. Heffernan's "life" role is analogous to Schrag's "right guy" role. Inmates who assume the life role were habitual offenders before coming to prison, and they continue to adhere to anti-social and anti-authority norms in prison. Sophisticated professional criminals occupy the "cool" role, or what Schrag referred to as the "con politician." They try to do easy time by manipulating other inmates and the staff to their own advantage.

THINKING CRITICALLY

1. Is there an inmate sexual victimization problem in U.S. prisons and jails? If yes, what can be done about it?
2. In comparison to male prisons, why do you think that there is less violence, interpersonal conflict, and interracial tension in women's prisons?
3. Should very young children of incarcerated women be allowed to temporarily live with their mothers in prison? Why or why not?
4. Should pregnant women be imprisoned? Why or why not?

Correctional Officers

In 1976, criminologist Gordon Hawkins wrote, "It is in fact remarkable how little serious attention has been paid to prison officers in the quite extensive literature on prisons and imprisonment."[46] Although it has increased steadily since, research on prison staff remains sparse compared with research on inmates. Most studies of prison staff have concentrated on guards or correctional officers, and there is good reason for this. Correctional officers represent the majority of staff members in a prison, are responsible for the security of the institution, and have the most frequent and closest contact with inmates. Yet correctional officers should not be thought of as existing in isolation from the larger structure of the prison bureaucracy discussed in the previous chapter.

As of May 2018, there were approximately 440,000 correctional officers employed in the United States: about 57% were employed in state prisons, around 39% were employed in local jails, and roughly 4% were employed in federal prisons.[47] In 2012, by comparison, nearly 470,000 correctional officers were employed in the United States–about 6% more than in 2018.[48] Table 11.1 shows the approximate number of correctional officers employed by the BOP and the states in 2018, ranked from highest to lowest. Texas employed the most correctional officers (48,600 in 2017), while Vermont employed the fewest (550).[49] The BOP employed 18,674 correctional officers in 2019.[50]

TABLE 11.1 Approximate Number of Correctional Officers Employed by the BOP and the States in 2018, Ranked from Highest to Lowest*

State	Number of Correctional Officers
1. Texas*	48,600
2. New York	35,460
3. California	34,980
4. Florida	33,060
5. BOP	18,670
6. Pennsylvania	16,850
7. North Carolina	15,720
8. Georgia	15,500
9. Virginia	14,910
10. Illinois	14,300
11. Arizona	13,810
12. Ohio	13,310
13. New Jersey	11,240
14. Tennessee	9,670
15. Michigan	9,100
16. Missouri	8,420
17. Indiana	7,990
18. Louisiana	7,890
19. Washington	7,190
20. Colorado	6,990
21. Wisconsin	6,870
22. Maryland	6,780
23. South Carolina	6,200
24. Massachusetts	6,130
25. Kentucky	5,850
26. Arkansas	5,500
27. Mississippi	4,930
28. Minnesota	4,770
29. Alabama	4,680
30. Oklahoma	4,290
31. New Mexico	4,210
32. Oregon	4,200
33. Connecticut	3,610
34. Kansas	3,470
35. Iowa	3,030
36. Nevada	2,680
37. Nebraska	2,270

TABLE 11.1 *(Continued)*

State	Number of Correctional Officers
38. Utah	2,270
39. Idaho	1,900
40. Delaware*	1,602
41. Hawaii	1,520
42. Maine	1,440
43. South Dakota	1,330
44. Alaska	1,120
45. New Hampshire	1,110
46. Rhode Island	1,070
47. Montana	990
48. West Virginia	830
49. Wyoming	830
50. North Dakota	680
51. Vermont	550

*BOP data are for 2019, and Delaware and Texas are for 2017.

Sources: "Occupational Employment Statistics," U.S. Department of Labor, Bureau of Labor Statistics, accessed August 26, 2019, https://www.bls.gov/oes/current/oes333012.htm; "Federal Prison System (BOP)," U.S. Department of Justice, https://www.justice.gov/jmd/page/file/1033161/download.

In 2018, approximately 70% of correctional officers were male, and about 30% were female. Roughly 67% of correctional officers were white, around 29% were black or African American, about 1% were Asian, and nearly 14% were Hispanic or Latino.[51]

States vary in the requirements needed to become a correctional officer. Among the more common requirements are: (1) at least 18 years of age, (2) a valid driver's license, (3) passing physical and physiological examinations, and (4) good communication skills.[52] Another common requirement for correctional officers is at least a high school diploma or equivalent. Some states and localities require some college credits. Military or any other law enforcement experience often can be substituted for the education requirement. The BOP requires at least a bachelor's degree for an entry-level position and, typically, 3 years of full-time experience in a job involving counseling or supervision of individuals.

Female Correctional Officers

In 1978, Linda Allen became the first female correctional officer to work at a high-security institution (the U.S. penitentiary at McNeil Island, Washington, now a Washington state correctional facility). Today, it is commonplace for female correctional officers to work in federal and state high-security institutions.

Source: Harry E. Allen and Clifford E. Simonsen, *Corrections in America: An Introduction*, 8th ed. (Upper Saddle River, NJ: Prentice Hall, 1998), 539.

Beyond the aforementioned requirements, correctional officer candidates nearly always go through training, which must be in compliance with standards set by the American Correctional Association (ACA). Federal, state, and some local corrections departments, as well as private corrections companies, offer training. Some states have training academies that teach courses on basic safety and self-defense; operations; regulations and legal restrictions; institutional policies; custody and security processes, including interpersonal relations; and crisis management, including responses to disturbances, hostage situations, riots, and other dangerous situations. The length of academy training varies widely among states, from as little as 3 to 4 weeks to as long as 16 to 20 weeks.[53] Candidates also receive on-the-job training, usually by an experienced correctional officer. On-the-job training can last for several additional weeks.

The BOP requires new correctional officers to complete 200 hours of formal training during their first year on the job, including 120 hours of specialized training at the BOP residential trainings center. To stay up-to-date on the latest developments and procedures, the BOP also requires correctional officers to complete annual in-service trainings.

The mean annual wage for correctional officers in 2018 was $49,300 or $23.70 an hour (the median annual wage was $44,330 or $21.31 an hour). The median annual wage of the lowest paid 10% of correctional officers was $31,140 or $14.97 an hour, while the median annual wage of the highest 10% of correctional officers was $76,760 or $36.90 an hour. The mean annual wage for correctional officers employed in local jails was $48,670 or $23.40 an hour; for correctional officers employed in state prisons, $49,870 or $23.98 an hour; and for correctional officers employed in federal prisons, $57,540 or $27.66 an hour. Table 11.2 shows the approximate mean annual wage and the approximate entry-level wage paid to correctional

TABLE 11.2 Approximate Mean Annual Wage Paid to Correctional Officers Employed by the BOP and the States in 2018, Ranked from Highest to Lowest, and Approximate Entry-Level Wage in 2018*

State	Mean Annual Wage	Entry-Level Wage
1. California	$75,400	$52,620
2. New Jersey	70,280	40,850
3. Rhode Island	68,710	60,160
4. Massachusetts	67,920	53,900
5. Alaska	64,670	47,580
6. New York	64,490	43,660
7. Nevada	62,140	42,870
8. Hawaii	59,020	51,040
9. Oregon	58,450	43,220
10. Washington	57,740	43,130
11. BOP**	57,540	41,204
12. Illinois	56,070	43,890
13. Connecticut	54,420	47,900
14. Pennsylvania	54,140	38,480
15. Michigan	51,890	37,890
16. Minnesota	51,660	39,040
17. Iowa	50,750	35,780
18. Colorado	50,620	41,180
19. Maryland	49,220	43,060
20. New Hampshire	47,860	35,050
21. Utah	46,150	35,830
22. Vermont	45,810	36,060
23. Ohio	45,190	35,630
24. North Dakota	44,700	37,560
25. Wisconsin	44,230	35,650
26. Arizona	44,150	33,380
27. Florida	43,410	31,530
28. Texas***	42,380	38,830
29. West Virginia	42,290	25,680
30. Wyoming	42,290	33,510
31. Virginia	41,860	32,620
32. Montana	40,570	32,170
33. Maine	40,440	32,350
34. Nebraska	40,310	31,810
35. Idaho	39,710	30,760
36. South Dakota	39,310	29,020
37. North Carolina	38,500	31,510
38. South Carolina	37,560	28,740
39. Indiana	36,960	31,000
40. Alabama	36,760	24,880
41. Kansas	36,550	28,540

TABLE 11.2 *(Continued)*

State	Mean Annual Wage	Entry-Level Wage
42. New Mexico	36,250	29,380
43. Kentucky	35,940	22,070
44. Tennessee	35,470	26,470
45. Louisiana	34,370	23,780
46. Georgia	34,290	26,320
47. Arkansas	34,120	25,870
48. Oklahoma	33,060	25,080
49. Missouri	31,650	25,490
50. Mississippi	30,840	21,290

*No data were available for Delaware.

**BOP entry-level wage is from 2019 (Grade 7, Step 5). It does include locality supplement.

***Average entry-level wage for Texas from 2019.

Sources: Data for Mean Annual Wage from "Occupational Employment Statistics," U.S. Department of Labor, Bureau of Labor Statistics, accessed August 26, 2019. https://www.bls.gov/oes/current/oes333012.htm. Data of Entry-Level Wage from Chris Kolmar, "The 10 Best States for Correction Officers in 2019," ZIPPIA, accessed August 27, 2019. https://www.zippia.com/advice/best-states-for-correction-officers/. BOP entry-level wage is from "Salary Table 2019-GS," OPM.GOV, accessed August 27, 2019. https://www.opm.gov/policy-data-oversight/pay-leave/salaries-wages/2019/general-schedule/. Average entry-level wage for Texas from "Entry Level Correctional Officer Salary in Texas," ZipRecruiter, accessed August 27, 2019, https://www.ziprecruiter.com/Salaries/How-Much-Does-an-Entry-Level-Correctional-Officer-Make-a-Year--in-Texas.

officers employed by the BOP and the states (except Delaware) in 2018. Mean annual wages are ranked from highest to lowest. California paid the highest mean annual wage in 2018 ($75,400), while Mississippi paid the lowest ($30,840).[54] Rhode Island paid the highest entry-level wage in 2018 ($60,160), while Mississippi paid the lowest ($21,290).[55]

Correctional officers typically work 8 hours a day, 5 days a week on rotating shifts. Because prisons operate continuously, officers are required to work any 8-hour shift, day or night, weekends and holidays, and sometimes overtime.

Correctional officers face a number of conflicts in their work. Criminologists Richard Hawkins and Geoffrey Alpert observe that the job is characterized by both boredom and stimulus overload; officers assigned to the towers may experience the former, whereas officers assigned to work the cell blocks may experience the latter.[56] Much also has been written about the role ambiguity and role strain resulting from conflict between custody and treatment objectives. How does an officer supervise and discipline inmates and at the same time attempt to counsel and help them? Traditionally, the role of guards was clearer and less ambiguous; they were responsible for custody. With the advent of rehabilitation, guards became correctional officers, and the custodial role grew clouded with treatment considerations. In addition, a series of court decisions has given many officers the perception that they have lost power while inmates have gained it.[57] Officers generally have considerable discretion in discharging their duties within the constraints of rules, regulations, and policies. Yet, because they lack clear and specific guidelines on how to exercise their discretion, they feel vulnerable to second-guessing by their superiors and the courts.

A popular misconception has been that officers manage inmates with an iron fist and are therefore corrupted by the power they have over inmates. However, as sociologist Gresham Sykes realized, the limits on an officer's power may be the real source of corruption.[58] Although their formal role places them in a superior position over inmates, officers still must obtain a measure of voluntary compliance from inmates. In the modern prison, it is

A correctional officer's custodial role is of primary importance. *Why?*

A.Ramey/PhotoEdit

impossible to rely exclusively on force to gain compliance if bureaucratic inefficiency, mass inmate rebellion, and court litigation are to be avoided. Moreover, because of the nature of the prison environment, officers do not have many incentives they can use to solicit compliance, and they do not directly control many of the main rewards, such as release date.

Still, officers' performances often are judged by how smoothly the officers interact with inmates and get inmates to go along with their wishes. Consequently, many officers seek inmate compliance through informal negotiation and exchange, and this practice can promote corruption. For example, an officer may gain compliance by becoming friendly with particular inmates, by granting inmates concessions in exchange for their cooperation, and by overlooking certain rule infractions in exchange for compliance with other rules. A study of a federal prison by criminologist John Hewitt and his colleagues found that inmates reported committing, and officers reported observing, far more prison rule infractions than were documented in the institution's official disciplinary files.[59]

How do correctional officers respond to their roles and their work conditions? According to Hawkins and Alpert, some officers become alienated and cynical and withdraw from their work. Withdrawal can be figurative: An officer may minimize his or her commitment to the job and may establish a safe and comfortable niche in the prison, such as prolonged tower duty. Withdrawal also can be literal: Turnover and absenteeism are high in many prison systems. Other officers become overly authoritarian and confrontational in a quest to control inmates by intimidation. Those officers put up a tough facade similar to that displayed by some inmates. Still other officers respond by becoming corrupt (for example, selling drugs to inmates in response to a low salary).[60] Finally, a number of officers respond by adopting a human-services orientation toward their work.[61] Those officers seek to make prison a constructive place for themselves and for inmates. They try to deliver goods and services to inmates in a regular and responsive manner, to advocate and make referrals on behalf of inmates when appropriate, and to assist inmates in coping with prison by providing protection and counseling.

Efforts are underway to transform prison work from a mere job into a profession, but a number of problems and issues surround those efforts. First, factors such as low pay, the nature and prestige of the work, and the remote location of many prisons make recruitment of new officers difficult in some jurisdictions. Some people take prison work simply because they believe that nothing better currently is available and intend to leave as soon as a better job opens.

Second, the general lack of competition for prison jobs in many jurisdictions makes it difficult to impose restrictive criteria for choosing among applicants. Most states require applicants to be at least a certain age (18 or 21) and to have completed high school or the GED. Some states have additional requirements, such as no prior felony convictions, related work experience, and the passing of certain tests.

Third, in some jurisdictions, a backlash against affirmative action–which has increased female and minority representation among correctional officers–has resulted in tensions between the genders and races[62] and resentment by some white male officers. In jurisdictions where this backlash has occurred, efforts toward professionalization have been hampered.

Fourth, once officers have been recruited and hired, they need to be trained, and the move toward professionalization has been accompanied by increased attention to training. As noted previously, officers usually are trained at the outset of their jobs and then receive annual training thereafter. At present, however, training standards are not uniform across or even within jurisdictions. In addition, when turnover rates and training costs are high, training may start to be viewed as a waste of money and time.

Fifth, professionalization has been accompanied by unionism among officers. Although the number of states with correctional officer unions has decreased recently with the increase in the number of states with "right to work laws," in 2019, AFSCME Corrections United (ACU), an affiliate of the American Federation of State, County and Municipal Employees, AFL-CIO, represented 62,000 correctional officers and 23,000 other corrections employees working in state and federal prisons and county jails across the United States.[63] In addition, the American Federation of Government Employees (AFGE) represented 33,000 BOP correctional officers in 2018.[64] Correctional officer unions can help improve pay levels, job benefits, and work conditions, but they also can divide the superior and subordinate ranks. A major shortcoming of most correctional officer unions is that they have a very limited ability to engage in collective bargaining because strikes commonly are illegal.

Careers in Criminal Justice

Correctional Sergeant

My name is Pietro DeSantis II, and I am a correctional sergeant employed by the California Department of Corrections, assigned to the California Correctional Center (CCC) in Susanville, California. I have an associate of science degree from Lassen Community College in Susanville. I began my career as a correctional officer with the department in November 1986 and was promoted to sergeant in February 1996. During my tenure with corrections, I have also been chapter president and board member of the California Correctional Peace Officer Association and commissioner and curriculum review committee chair of the Correctional Peace Officer Standards and Training (CPOST) Commission.

A normal workday starts as soon as you arrive on grounds and start talking to staff that are either beginning or ending their shift. Normal shifts at CCC are first watch (2200–0600 hours), second watch (0600–1400 hours), and third watch (1400–2200 hours). A normal day on all three watches starts with a security check of our safety equipment and security locks. First watch staff normally read the incoming and outgoing mail, count the inmate population three times, and ensure workers for the culinary service are awake and report to work. Second watch staff normally feed breakfast and lunch to the inmates and supervise the inmate workers who clean the living quarters, inner perimeter yard areas, and the outside grounds. Third watch staff normally deliver the mail, count the inmates twice, feed them dinner, and house new arrivals from the transportation buses. As a sergeant, I could be assigned to supervise an office or unit with between 0 and 40 correctional officers.

Pietro DeSantis II

Before entering into a career within corrections, I strongly suggest you do everything possible to become an effective communicator.

Why do you think it is important for a correctional officer to be an effective communicator?

THINKING CRITICALLY

1. What do you think could be done to further professionalize corrections work?
2. What are some ways you can think of to add selectivity to correctional worker recruitment?

Inmate Rights and Prison Reform

Over the past 5 decades, the major means of reforming prisons has been court intervention. Until the middle of the twentieth century, the courts followed a **hands-off philosophy** toward prison matters. Under this hands-off philosophy, court officials were reluctant to hear prisoners' claims regarding their rights while they were incarcerated. Among other considerations, court officials questioned whether they had the necessary expertise to evaluate prison administration, and judges did not want to undermine the power of prison authorities. As a consequence of this inaction, prisoners, for all practical purposes, had no civil rights. (Remember that civil rights are those rights an individual possesses as a member of the state, especially those guaranteed against encroachment by the government.) This situation began to change as progress was made in civil rights in the wider society. Inmates were first granted rights of access to the courts, and the courts subsequently turned their attention to other rights.

hands-off philosophy A philosophy under which courts are reluctant to hear prisoners' claims regarding their rights while incarcerated.

Access to the Courts and Legal Services

Ordinarily, prisoners want to get their cases into the federal courts because they perceive those courts as more receptive to their claims than state courts. In its 1941 landmark ruling in *Ex parte Hull,* the U.S. Supreme Court granted inmates the right of unrestricted access to the federal courts. Just 3 years later (in *Coffin v. Reichard*, 6th Cir., 1944), a federal circuit court held that prisoners may challenge in federal court not only the fact of their confinement but also the conditions under which they are confined. But the most important U.S. Supreme Court ruling in this area (*Cooper v. Pate*) did not occur until 1964, when Thomas Cooper, a Stateville inmate, first successfully used Section 1983 of the Federal Civil Rights Act of 1871 to challenge the conditions of confinement. Prior to the ruling in *Cooper,* inmates had relied primarily on *habeas corpus* petitions to obtain access to the federal courts. ***Habeas corpus*** is a court order requiring that a confined person be brought to court so that his or her claims can be heard. For various technical reasons, it normally has been easier for prisoners to win federal cases under Section 1983 than under *habeas corpus*. In effect, then, the *Cooper* decision launched the prisoners' rights movement by opening the door to a flood of Section 1983 claims from prisoners. Calling *Cooper v. Pate* "the Supreme Court's first modern prisoners' rights case," law professor James Jacobs writes:

habeas corpus A court order requiring that a confined person be brought to court so that his or her claims can be heard.

> But for the prisoners' movement it was not the breadth of the decision that mattered but the Supreme Court's determination that prisoners have constitutional rights; prison officials were not free to do with prisoners as they pleased. And the federal courts were permitted, indeed obligated, to provide a forum where prisoners could challenge and confront prison officials. Whatever the outcome of such confrontations, they spelled the end of the authoritarian regime in American penology.[65]

However, by the 1990s, Congress had become exasperated with the flood of claims by prisoners challenging their conditions of confinement in the federal courts. In response, it passed the Prison Litigation Reform Act (PLRA) of 1995.[66] The intent of the legislation was to restrict and discourage litigation by prison inmates. The PLRA mandates that prisoners must exhaust all available prison administrative remedies–that is, prison grievance procedures–before they can bring lawsuits in federal court challenging prison conditions. If all administrative remedies are not exhausted–that is, pursued to their conclusion–then the federal courts are obligated to dismiss the lawsuits. The PLRA applies to civil actions arising under federal law that challenge conditions of confinement or the effects of actions by government officials on the lives of prison inmates. It does not apply to *habeas corpus* proceedings that challenge the fact or duration of prison confinement. The act also restricts the relief that courts may grant in successful lawsuits. Relief to correct the violation cannot extend any further than the particular plaintiff or plaintiffs that filed the lawsuit. Any relief that is granted must be narrowly drawn, extending no further than necessary to correct the violation of the federal right. It also must be the least intrusive means necessary to correct the violation. In granting relief, the court must give substantial weight to any adverse impact on public safety or the operation of a criminal justice system the relief may cause. In exercising its remedial powers, the court is not authorized to order the construction of prisons or the raising of taxes.

In 2018, state and federal prisoner petition filings in U.S. District Courts decreased about 30%, from 76,570 in 2017 to 53,965 in 2018.[67] The year-over-year decrease in filings is attributable primarily to the 77% drop in motions to vacate sentence, from 24,914 in 2017 to 5,734 in 2018. Such motions had increased 260% in 2017, after *Welch v. United States* (2016) established that *Johnson v. United States* (2015) applied retroactively and made prisoners serving sentences enhanced under an unconstitutional clause of the Armed Career Criminal Act of 1984 (ACCA) eligible to have their sentences vacated or remanded. The ACCA provided for a special mandatory prison term of 15 years for a felon who unlawfully possessed a firearm, and had three or more previous "violent felony" convictions.[68] Also, in 2018, state and federal prisoner general *habeas corpus* petitions increased only about 1%, from 17,223 in 2017 to 17,242 in 2018; *habeas corpus* petitions involving the death penalty increased nearly 12%, from 189 in 2017 to 211 in 2018; civil rights petitions decreased almost 12%, from 20,673 in 2017 to 18,216 in 2018; and petitions related to prison conditions dropped approximately 12%, from 10,947 in 2017 to 9,698 in 2018.[69]

To get their cases to court, prisoners need access to legal materials, and many of them need legal assistance from persons skilled in law. The U.S. Supreme Court has recognized those facts. In *Johnson v. Avery* (1969), the Court held that inmates skilled in legal matters–so-called **jailhouse lawyers**–must be permitted to assist other inmates in preparing cases unless the government provides a reasonable alternative. Furthermore, in *Bounds v. Smith* (1977), the U.S. Supreme Court ruled that inmates are entitled to either an adequate law library or adequate legal assistance. So if a correctional institution does not wish to allow a jailhouse-lawyer system and does not wish to provide adequate library facilities for inmates, the implication is that the correctional institution should allow inmates to consult with licensed attorneys to obtain legal assistance.

jailhouse lawyer An inmate skilled in legal matters.

Ty Wright/Bloomberg/Getty Images

Inmates are entitled to adequate law library facilities. *Why do inmates have this right and should they?*

Procedural Due Process in Prison

In Chapter 10, we noted that inmates can face disciplinary action for breaking prison rules. The U.S. Supreme Court has moved to ensure that inmates receive minimal elements of due process during the disciplinary process. Probably the most important case in this area is *Wolff v. McDonnell* (1974). In *Wolff*, the Court held that although inmates facing a loss of good time for a rule infraction are not entitled to the same due process protections as in a criminal trial, such inmates are entitled to (1) a disciplinary hearing by an impartial body, (2) written notice of the charges within 24 hours, (3) a written statement of the evidence relied on and the reasons for the disciplinary action, and (4) an opportunity to call witnesses and present documentary evidence provided that this does not jeopardize institutional security. The Court ruled that inmates are not entitled to confront and cross-examine people who testify against them or to have legal counsel.

First Amendment Rights

The First Amendment to the Constitution guarantees freedom of speech, press, assembly, petition, and religion. The U.S. Supreme Court has rendered numerous decisions affecting prisoners' rights to freedom of speech and expression and freedom of religion.

Free Speech The significant case of *Procunier v. Martinez* (1974) dealt with censorship of prisoners' outgoing mail but is generally regarded as applicable to other aspects of correspondence and expression. In *Procunier*, the Supreme Court ruled that censorship is legal only if it furthers one or more of the following substantial government interests: security, order, and rehabilitation. Moreover, the degree of censorship can be no greater than that required to preserve the government interest in question.

The Fortune Society

The Fortune Society is a not-for-profit community-based organization staffed mostly by ex-offenders. The organization is dedicated to educating the public about prisons and prison life, criminal justice issues, and the root causes of crime. You can visit the website of the Fortune Society at www.fortunesociety.org.

Do you think organizations such as this one are necessary? Why or why not?

Religious Freedom With the demise of the hands-off policy, the first substantive right won by inmates was freedom of religion, which was awarded to Black Muslims in *Cooper v. Pate*. As a rule, inmates are free to practice either conventional or unconventional religions in prison, and prison officials are obligated to provide accommodations. However, restrictions may be imposed, for example, where prison officials can demonstrate convincingly that religious practices compromise security or are unreasonably expensive. Granting special religious privileges also can be denied on the grounds that they will cause other groups to make similar demands. In general, the reasons given for

As a rule, inmates are free to practice their religion in prison. Religious symbols and practices that interfere with institutional security can be restricted. *Should religious practices be allowed in prisons? Why or why not?*

Lezlie Sterling/ZUMA Press/Newscom

restricting religious freedom must be compelling, and the restrictions imposed must be no more limiting than necessary.

Eighth Amendment Rights

As described in Chapter 4, the Eighth Amendment outlaws the imposition of cruel and unusual punishment. The courts have considered a number of issues under the umbrella of cruel and unusual punishment.

Medical Care In 1976, the Supreme Court decided *Estelle v. Gamble*. The Court ruled that inmates, under the Eighth Amendment, have a right to adequate medical care, but that inmates claiming Eighth Amendment violations on medical grounds must demonstrate that prison officials have shown deliberate indifference to serious medical problems. Under these conservative and subjective criteria, medical services and circumstances usually have to be really bad—even extreme—for inmates to win cases. Furthermore, the deliberate-indifference standard formulated in *Estelle* now applies to Eighth Amendment challenges to any condition of confinement, medical or otherwise.[70]

In the previously mentioned case of *Brown v. Plata* (2011) in Chapter 10, which dealt with overcrowding in California's prison system, the U.S. Supreme Court ordered the state of California to reduce its prison population to 137.5% of design capacity. For at least 11 years, California's prison system had operated at about 200% of design capacity. The order was intended to remedy persistent and systemic Eighth Amendment violations of prisoners' rights to the treatment of serious mental and medical conditions that have caused needless suffering and death. According to the Court, overcrowding is the primary cause of the violations [and] "has overtaken the limited resources of prison staff; imposed demands well beyond the capacity of medical and mental health facilities; and created unsanitary and unsafe conditions that make progress in the provision of care difficult or impossible to achieve." In rendering the order, the Court acknowledged the political and fiscal realities facing the state of California; specifically, the state's unwillingness and inability to allocate the resources necessary to reduce overcrowding. The Court considers the imposition of a prison population limit as a "last resort remedy" to prison problems.

FYI **Hepatitis C in Florida Prisons**

In April 2019, a federal judge ruled that Florida must expand treatment for hepatitis C to 20,000 to 40,000 more inmates who had been denied proper treatment partly because of a lack of funding. The court already had mandated the state to treat inmates with hepatitis C, but the new order applies to those inmates in the early stages of the disease. The federal judge stated that prison officials have been "deliberately indifferent" in caring for thousands of inmates infected with the virus. The ruling could cost the Florida Department of Corrections (FDC) as much as $20 million more than what it already is spending to treat inmates with the virus. Before the judge's order, the FDC had asked the legislature for $36.9 million to treat inmates infected with hepatitis C.

Source: Ana Ceballos, "Hepatitis C Ruling Could Add Millions to Florida Prison Health Costs," *Orlando Sentinel*, April 19, 2019. Accessed August 28, 2019, https://www.orlandosentinel.com/politics/os-ne-hep-c-prison-costs-20190419-story.html.

Staff Brutality Ironically, the Eighth Amendment has done little to protect inmates from staff brutality because brutality is normally construed as a tort rather than a constitutional issue.[71] (Recall that a tort is the breach of a duty to an individual that results in damage to him or her. It involves only duties owed to an individual as a matter of law.) However, whipping and related forms of corporal punishment have been prohibited under this amendment. Also, in *Hudson v. McMillian* (1992), the Supreme Court found that staff use of force against an inmate need not cause a significant physical injury to violate the Eighth Amendment. Despite these protections, staff brutality remains a problem.

Total Prison Conditions Totality-of-conditions cases involve claims that some combination of prison practices and conditions (e.g., crowding, lack of services and programs, widespread brutality, and labor exploitation) makes the prison, as a whole, unconstitutional under the Eighth Amendment. It is primarily in this area that the Eighth Amendment has been used. For example, in the famous case of *Holt v. Sarver* (1971), the entire Arkansas prison system was declared unconstitutional on grounds of totality of conditions and was ordered to implement a variety of changes. Totality-of-conditions rulings were later handed down against prisons in Alabama (*Pugh v. Locke,* 1976) and Texas (*Ruiz v. Estelle,* 1982).

Prisons have long had the right to provide only the minimal conditions necessary for human survival. Such conditions include the necessary food, shelter, clothing, and medical care to sustain life. Whether institutional crowding violates the Eighth Amendment typically is determined in relation to total prison conditions. Instead of being treated as a separate issue, crowding is viewed as part of the totality of conditions. The Supreme Court has been reluctant to side with inmates in their challenges to crowding. In both *Bell v. Wolfish* (1979) and *Rhodes v. Chapman* (1981), the Court refused to prohibit the placing of two inmates in cells designed for one person ("double-bunking") as a response to crowding.

FYI **Brubaker**

In *Holt v. Sarver* (1971), the Supreme Court ordered the reform of the Arkansas prison system. The movie *Brubaker*, starring Robert Redford, is a fictionalized account of the real story of Tom Murton, warden of Arkansas's Tucker & Cummins prison in 1967–1968. Murton's evidence was instrumental in the Supreme Court's decision. The U.S. District Court characterized the Arkansas prison as "a dark and evil world completely alien to the free world."

Other Prison Conditions Found to Violate the Eighth Amendment Federal courts have found all of the following prison conditions to be in violation of the Eighth Amendment's prohibition of cruel and unusual punishments:

- Inadequate ventilation.
- Excessive heat.
- Excessive cold.
- Lack of drinkable water (also lack of cold water where prison yard temperatures reached 100 degrees).
- Toxic or noxious fumes (pesticides sprayed into housing units; inadequate ventilation of toxic fumes in inmate workplaces).

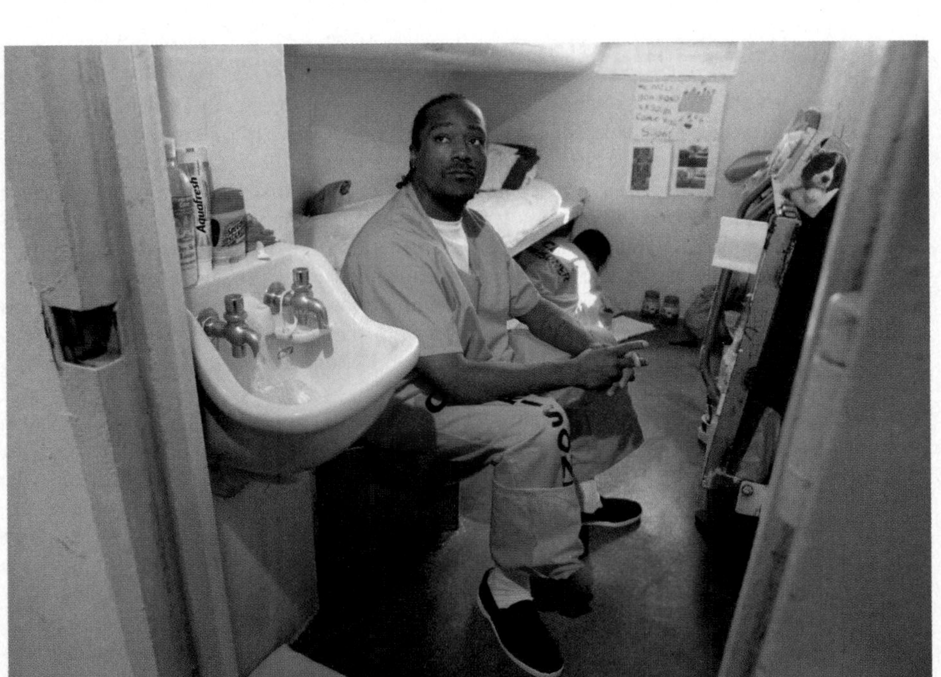

Lucy Nicholson/REUTERS/Newscom

In 1981, the U.S. Supreme Court refused to hold that double-bunking in prison cells was cruel and unusual punishment. *Should double-bunking be allowed in prisons? Why or why not?*

- Exposure to sewage (exposure to flooding and human waste).
- Exposure to second-hand tobacco smoke (cell mates smoking five packs of cigarettes a day).
- Excessive noise.
- Sleep deprivation.
- Sleeping on the floor (failure to provide inmates with a mattress and bed or bunk).
- Lack of fire safety.
- Risk of injury or death in the event of an earthquake.
- Inadequate food or unsanitary food service.
- Inadequate lighting or constant lighting.
- Exposure to insects, rodents, and other vermin.
- Defective plumbing.
- Deprivation of basic sanitation.
- Denial of adequate toilet facilities.
- Exposure to asbestos (exposure to "moderate levels of asbestos" did not violate the Eighth Amendment).
- Exposure to the extreme behavior of severely mentally ill prisoners (exposure to constant screaming and feces-smearing of mentally ill prisoners "contributes to the problems of uncleanliness and sleep deprivation, and by extension mental health problems, for other inmates").
- Miscellaneous unhealthy or dangerous conditions (unsafe conditions for prisoners performing electrical work; prisoner injured in vehicle accident after transport officers refused to fasten his seat belt).[72]

Fourteenth Amendment Rights

As described in Chapter 4, the Fourteenth Amendment guarantees U.S. citizens due process of law and equal protection under law. The due process clause inspired the Supreme Court's decision in the previously discussed case of *Wolff v. McDonnell,* for example. Likewise, the equal-protection clause has led the Supreme Court to forbid racial discrimination (*Lee v. Washington*, 1968) and has led state courts to target gender discrimination (*Glover v. Johnson*, ED Mich., 1979). Compared with the rights of male inmates, however, the rights of female prisoners remain underdeveloped. The inmates' rights movement has been primarily a male phenomenon, and the rights of female prisoners deserve more attention from the courts in the future.

FYI Sexually Violent Predator Laws

By a 5–4 decision in *Kansas v. Hendricks* (1997), the Supreme Court ruled that sexual predators judged to be dangerous may be confined indefinitely even after they finish serving their sentences. The decision allowed the state of Kansas to continue to hold an admitted pedophile, Leroy Hendricks, under the provisions of the state's Sexually Violent Predator Act. The Court's majority noted that such confinement, intended to protect society, does not violate the constitutional right to due process and is not double punishment for the same crime. At least 20 states and the federal government have laws similar to the Kansas Sexually Violent Predator Act.

Sources: *Kansas v. Hendricks,* 521 U.S. 346 (1997); AR Felthous and J Ko, "Sexually Violent Predator Law in the United States," *East Asian Archives of Psychiatry* 28 (2018) 159–173. Accessed August 28, 2019, https://www.easap.asia/index.php/find-issues/current-issue/item/826-1812-v28n4-p159.

The Limits of Litigation

The progress toward prison reform resulting from the inmates' rights movement should not be underestimated. Nevertheless, one can question on a number of grounds the almost exclusive reliance, during the past 5 decades, on court intervention to reform the nation's prisons. Perhaps the most basic and compelling argument is that the monies being spent by prison systems to defend against inmate lawsuits and to comply with court orders (when inmate suits are successful) could have been spent better to reform the unacceptable practices that sparked the suits in the first place. Meanwhile, those prison systems are unable to address problems that help generate further inmate lawsuits precisely because they are defending against new suits and trying to comply with past court orders. The result is that even more money must be spent for legal defense and compliance with court orders. This cycle is difficult to break.

Court litigation is not just an expensive way to reform prisons; it also is a very slow and piecemeal way. A high percentage of lawsuits filed by inmates are judged frivolous and are therefore dismissed. If an inmate's suit is not judged frivolous, the inmate must still win the case in court and many lose in the process. Even when inmates win a case, their success usually does not lead to wide-scale reforms. Successful, large-scale, class-action suits that have far-reaching impact on the prison system have become relatively uncommon as a result of the Prison Litigation Reform Act of 1995. Moreover, winning a case and then getting the desired changes implemented can take many years. Prison systems do not comply automatically with court decrees. Achieving compliance with court orders often requires sustained monitoring by court-appointed officials, as well as considerable negotiation and compromise by all parties involved to agree on a timetable and the specific nature of reforms. Finally, the transformation of traditional, ingrained practices in a prison system can render the prison environment chaotic and unstable, at least in the short run.

THINKING CRITICALLY

1. What do you think are legitimate complaints from prisoners about the conditions of their imprisonment?
2. What constitutional rights do you think should be extended to prisoners? Or do you think prisoners already have too many rights? Why?

Release, Reentry, and Recidivism

At least 95% of all inmates eventually will be released from prison. The 5% of inmates who will not be released are serving death or life sentences and likely will die in prison. In 2016 (the latest year for which data were available), 4,137 inmates died in state and federal prisons: 3,749 in state prisons and 388 in federal prisons (excludes deaths in private federal facilities but includes deaths in private state facilities; executions added).[73] Figure 11.2 shows the causes of death of the 3,749 state prison inmates who died in prison in 2016. As noted in Figure 11.2, about 86% of state prison inmates died of illnesses: 30.2% died of cancer; 27.5% of heart disease; 7.0% of liver disease; 6.0% of respiratory disease; 0.8% of AIDS-related illnesses; and 14.1% of other illnesses, such as cerebrovascular disease and influenza.[74] Besides illnesses, a much smaller percentage of state prison inmates died of accidents (1.1%), executions (0.5%), homicides (2.5%), suicides (6.8%), and drug or alcohol intoxication (2.8%) (see Figure 11.2).

Depending on the jurisdiction and the specific case, inmates may be released from prison in a number of ways. Examples include expiration of the maximum sentence allowed by law (known as "maxing out"); **commutation**, or reduction of the original sentence by executive authority; release at the discretion of a parole authority; and mandatory release. Of those ways, the two most common are release at the discretion of a parole authority and mandatory release.

The term **parole** has two basic meanings. It can refer to a way of being released from prison before the entire sentence has been served, or it can refer to a period of community supervision following early release. We are concerned here with the former meaning. In jurisdictions that permit parole release, inmates must establish eligibility for parole. Eligibility normally requires that inmates have served a given portion of their terms minus time served in jail prior to imprisonment and minus good time. **Good time** is time subtracted from the sentence for good behavior and other meritorious activity in prison. Once eligible, an inmate submits a parole plan stipulating such things as where he or she plans to live and work upon release. The parole authority, which may consist of a board, members of the board, or a representative of the board, then considers the inmate's plan. The parole authority also considers reports from institutional staff who have worked with the inmate. The decision to grant or deny parole is announced at a parole-grant hearing. Parole will be discussed more fully in the next chapter.

Most offenders who receive lengthy prison terms (e.g., 25 years) do not serve the entire term, and many do not even come close to doing so. The reason is the availability of parole, good time, and other mechanisms for reducing time served. Some opponents of those reductions in terms fail to understand that the mechanisms are essential to the operation of many current prison systems. Without them, many systems would be rendered dysfunctional by crowding. In addition, without early release incentives for inmates, it would be extremely difficult to maintain order in prisons.

Mandatory release is release under the provisions of law, not at the discretion of a parole board. The inmate is released after serving his or her sentence or a legally required portion of the sentence, minus good-time credits. Mandatory release is similar to parole in that persons let out under either arrangement ordinarily receive a period of community supervision by a parole officer or the equivalent.

commutation Reduction of the original sentence given by executive authority, usually a state's governor.

parole A method of prison release whereby inmates are conditionally released at the discretion of a board or other authority before having completed their entire sentences; can also refer to the community supervision received upon release.

good time Time deducted from an inmate's sentence by prison authorities for good behavior and other meritorious activities in prison.

mandatory release A method of prison release under which an inmate is released after serving a legally required portion of his or her sentence, minus good-time credits.

FIGURE 11.2 Causes of State Prison Inmate Deaths, 2016 (Total = 3,749)

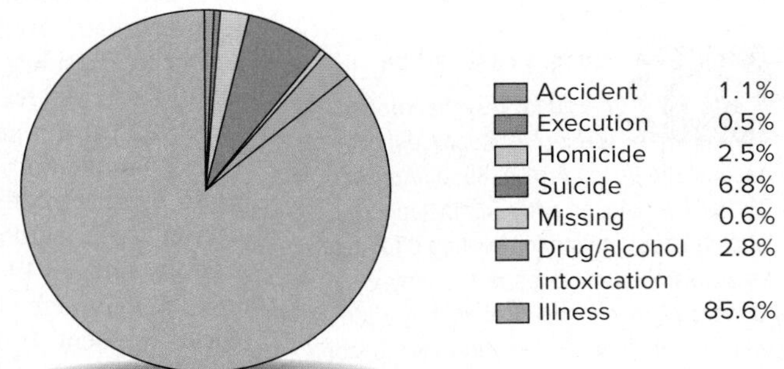

Sources: E. Ann Carson and Mary P. Cowhig, "Mortality in State Prisons, 2001-2016-Statistical Tables," U.S. Department of Justice, Bureau of Justice Statistics, February 2020, 6, Table 3, accessed March 5, 2020, https://www.bjs.gov/content/pub/pdf/msfp0116st.pdf; "Executions by Region," The Death Penalty Information Center, accessed March 5, 2020, https://deathpenaltyinfo.org/executions/executions-overview/number-of-executions-by-state-and-region-since-1976.

Most parole decisions are made by administrative boards whose members are appointed by state governors. An inmate's eligibility for release on parole depends on requirements set by law and on the sentence imposed by the court. *Should parole be abolished? Why or why not? If not, how could the decision-making process be improved?*

Nick Wolcott/Bozeman Daily Chronicle/AP Images

Their freedom from prison is conditioned on following the rules of their community supervision (e.g., obeying curfew, abstaining from drug use, and maintaining gainful employment) and on avoiding criminal activity.

In 1977, approximately 72% of inmates released from state prisons were released at the discretion of parole boards; only about 6% left prison under mandatory release that year. By contrast, in 2016, about 44% were released by parole boards, compared with approximately 27.5% who received mandatory release. This shifting pattern indicates the strong emphasis placed on determinate sentencing since the 1980s and the use of statistical guidelines to structure early-release decisions.

clemency A way by which inmates are released from prison that allows the governor of a state, or the president of the United States, when federal or military law is violated, to exercise leniency or mercy.

pardon A "forgiveness" for the crime committed that stops further criminal processing.

Clemency is another way by which inmates are released from prison. **Clemency** allows the governor of a state or the president of the United States, when federal or military law is violated, to exercise leniency or mercy. Many states have specialized administrative boards or panels authorized to assist the governor in making the clemency decision. Commutation is one of three types of clemency. As noted previously, commutation is a reduction of the original sentence given by executive authority, usually a state's governor or the president of the United States, in the case of federal crimes. The other two types of clemency are reprieve and pardon. Reprieve, or "stay of execution," temporarily postpones an execution and is used only in death penalty cases. **Pardon** is the most expansive type of clemency. With a pardon, the prisoner's crime is erased and his or her punishment is terminated. A pardoned individual is freed entirely from the criminal justice system and is treated as if he or she had never been charged or convicted of a crime.

Trump's First Pardon

President Trump pardoned former Maricopa County, Arizona Sheriff Joe Arpaio, 85, on August 25, 2017. Arpaio, an early Trump supporter, who had a reputation for treating offenders, especially Hispanic offenders, poorly, was found guilty of criminal contempt by a federal judge for ignoring a court order to stop racial profiling and detaining Hispanics simply because he thought they were in the country illegally.

Source: Kevin Liptak, Deniella Diaz, and Sophie Tatum "Trump Pardons Former Sheriff Joe Arpaio," *CNN Politics,* August 27, 2017. Accessed August 30, 2019, https://www.cnn.com/2017/08/25/politics/sheriff-joe-arpaio-donald-trump-pardon/index.html.

During President Trump's first 30.5 months in office, he granted 15 pardons and 6 commutations.[75] He received 1,032 pardon petitions and 5,797 commutation petitions. He denied 82 pardon petitions and 98 commutation petitions, while 562 pardon petitions and 3,623 commutation petitions were closed without presidential action.[76] By comparison, in his 96 months in office, President Obama granted 212 pardons and 1,715 commutations. He received 3,395 pardon petitions and 33,149 commutation petitions. He denied 1,708 pardon petitions and 18,749 commutation petitions, while 508 pardon petitions and 4,252 commutation petitions were closed without presidential action.[77] Note that in his first 30.5 months in office, President Trump granted 1.5% of the 1,032 pardon petitions and 0.1% of the 5,797 commutation petitions that he received. During President Obama's first 32.5 months in office, he granted 2% of the 825 pardon petitions and none of the 4,573 commutation petitions that he received. President Obama granted 70% of his pardons and 95% of his commutations during the last 15.5 months of his presidency.

President Obama received an usually large number of commutation petitions–by far more than any previous administration–during the end of his presidency because of a Justice Department initiative intended to correct past sentencing injustices and relieve federal prison overcrowding.[78] The program commenced on April 23, 2014, and targeted crack cocaine users and dealers sentenced under an especially harsh law that was eased by Congress in 2010. An estimated 7,000 to 8,000 overwhelmingly black inmates had been incarcerated but would not have been under the new law. The program was unusual because it actively solicited requests for clemency and provided an expedited process for getting applications to the president. Inmates were to be notified of the new policy and be able to complete an electronic survey form that would be used to evaluate their application. Inmates also were assigned free, volunteer lawyers. To be eligible for the new federal clemency program, inmates had to meet six criteria:

- Likely would have received a substantially lower sentence if convicted today.
- Nonviolent, low-level offenders without significant ties to large-scale criminal organizations, gangs, or cartels.
- Served at least 10 years of their prison sentence.
- Do not have a significant criminal history.
- Demonstrated good conduct in prison.
- No history of violence before or during time in prison.

As expected, the Obama administration received thousands of commutation petitions from inmates not eligible for the program, including inmates convicted of murder, terrorism, sex crimes, public corruption, and financial fraud. In the end, the administration reviewed 16,776 petitions from potentially eligible inmates–about half of all the petitions received–and, as noted previously, President Obama commuted about 1,700 sentences. The program was ended when President Obama left office on January 20, 2017, with 3,469 petitions from eligible inmates pending.[79]

When inmates are released from correctional institutions, the hope is that as a result of deterrence or rehabilitation (discussed in Chapter 9), they will not return to criminal activity. Unfortunately, the reintegration of inmates into society following their prison or jail terms–commonly referred to as **reentry**–has been an afterthought (here the focus is on prison inmates). Until recently, not much attention has been given to this critical process. However, since 95% of inmates eventually will be released from prison and recidivism figures (to be discussed next) remain stubbornly high, to not focus on reentry is short sided. Fortunately, today, reentry is considered an integral part of the "smart justice" movement (discussed further in Chapter 14).

reentry The reintegration of inmates into society following their prison or jail terms.

When inmates complete their prison sentences they receive, if needed, nothing more, and sometimes less, than an allotment of clothing, funds ("gate money"), and transportation. They also generally receive any money left in their prison accounts (from work or contributions from family or friends). Provisions of the BOP are typical, albeit more generous than the states. The BOP gives its departing inmates:

> (1) suitable clothing; (2) an amount of money, not more than $500, determined by the Director to be consistent with the needs of the offender and the public interest, unless the Director determines that the financial position of the offender is such that no sum should be furnished; and (3) transportation to the place of the prisoner's conviction, to the prisoner's bona fide residence within the United States, or to such other place within the United States as may be authorized by the Director.[80]

Table 11.3 shows what each state provides to departing inmates. No state can be considered generous; most states provide next to nothing. The provisions (or lack of) given to inmates leaving prison underscore the importance of reentry programs.

Inmates in 49 states (all states except Hawaii) leave prison with something else–a bill for the costs of their own incarceration–a bill that makes reentry even more difficult than it already is.[81] In 46 states, failure to repay incarceration-cost bills is an offense punishable by more incarceration. In most states, paying incarceration-cost bills is a higher priority than paying restitution to victims. Inmates able to pay these bills are fortunate; most inmates are not, and no matter how hard inmates work in prison, they never can earn enough to pay off the bill. Some states send former inmates bills; other states charge former inmates fines (sometimes called legal financial obligations, or LFOs); and still other states collect the cost of incarceration through windfall statutes, which allow the state to deduct money owed it from any inheritances, lottery winnings, or litigation proceeds the former inmate may receive. Incarceration-cost bills, which include processing fees and accrue interest if not paid in full, keep former inmates tethered to the criminal justice system–in many cases forever. In most states, former-inmate debtors cannot regain certain civil rights lost upon conviction, such as the right to vote, carry a weapon, serve on juries, or run for elected office.

TABLE 11.3 Departing Inmate Provisions, by State

State	Clothing	Funds	Transportation
Alabama	Clothing, appropriate to weather	$10	Bus ticket to county of conviction
Alaska	Personal property	$150, if eligible	–
Arizona	Set of clothes	$50	Bus or rail ticket to closest stop outside state line
Arkansas	Personal property in jacket	$50 check	Bus ticket, if needed
California	Clothing	$200, less cost of clothing	Public transportation
Colorado	Sweatpants and shirt	$100	Bus ticket to state line
Connecticut	–	$50, if eligible	Transportation, case-by-case basis
Delaware	–	$50, if eligible	Busfare, if released in urban area; cabfare, if released in rural area
Florida	–	$100	Bus ticket within state
Georgia	–	$25	Bus ticket
Hawaii	–	$100 for cash spending; $100 for clothing, at discretion of parole board	Plane ticket within state, if prison on different island
Idaho	Appropriate clothing	Brown bag lunch	Bus ticket, on occasion
Illinois	Own possessions	$20–$50 for meals, at administrator's discretion	Bus ticket, if needed
Indiana	Weather-appropriate clothing, if needed	Up to $75, at administrator's discretion	Bus ticket to planned residence
Iowa	Clothing	$100	Instate transportation
Kansas	Clothing	$100	–
Kentucky	Not specified	Not specified	Not specified
Louisiana	Clothing	$10	Bus ticket home, if needed
Maine	Clothing	Up to $50	Transportation to home or employment
Maryland	Personal property	–	–
Massachusetts*	–	–	–
Michigan	Clothing	"Reasonable maintenance" loan, must be paid back in 180 days	Bus ticket, within state
Minnesota	Clothing	$100	–
Mississippi	Civilian clothes	$15–$100, depending on time served	Bus ticket to county of conviction or state line
Missouri	Clothing	–	Bus ticket, if necessary
Montana	–	Unspecified	–
Nebraska	–	Up to $100	Bus ticket anywhere in 48 contiguous states
Nevada**	Clothing	Up to $100	Transportation anywhere in the United States
New Hampshire	–	$100	–
New Jersey***	–	–	–
New Mexico	–	$50	Bus ticket
New York	–	$40	Bus ticket to county of conviction
North Carolina	Personal property	–	Bus ticket home

TABLE 11.3 Departing Inmate Provisions, by State (*Continued*)

State	Clothing	Funds	Transportation
North Dakota	–	–	Bus ticket, if needed
Ohio	Weather-appropriate clothing, 3 sets of underwear and socks and other accumulated clothing	$25–$75	–
Oklahoma	–	$50, if indigent	Bus ticket to county of conviction
Oregon	Dress-outs, if no clothing	$25, if indigent	Bus voucher, if needed
Pennsylvania	–	–	–
Rhode Island	Personal belongings	–	–
South Carolina	Donated clothes	–	Bus ticket within state, if necessary
South Dakota	–	$50	Bus ticket or ride to county of commitment
Tennessee	Used clothing	$30	Least expensive transportation within state, if needed
Texas	–	$100	Bus ticket, if released to parole or mandatory supervision
Utah	–	More than $100, if needed	–
Vermont	–	–	Transportation to community service provider, if necessary
Virginia	Not specified	Not specified	Not specified
Washington	–	$40	Transportation, if required
West Virginia	–	Up to $300 from private trust fund	Bus ticket to county of conviction, money for such a ticket, or ride to bus station
Wisconsin	–	–	Bus ticket to their community, if necessary
Wyoming	Clothing, if needed	Some money for meals, if long bus trip	Bus ticket, if needed

–Not mentioned in source information.

*Massachusetts has an optional reentry program available to all inmates, which includes post-release supervision.

**In Nevada, costs may be taken legally from inmate's account.

***In New Jersey, inmates receive money from own accounts upon release. May make contact with Department of Labor and Social Services in their home counties for assistance.

Source: Adapted from "Hard Time: Life After Prison, Gate Money by State," accessed September 1, 2019, http://americanradioworks.publicradio.org/features/hardtime/gatemoney/.

As noted, most former inmates never are able to pay off their incarceration-cost bills. Most inmates were poor when they entered prison, and about half of former inmates are unemployed six months after their release from prison. Former inmates who find jobs generally earn very little–the median income for former inmates 1 year after their release from prison is only about $10,000–and only 55% of former inmates had any earnings at all. Thus, paying any bills is difficult. Incarceration-cost bills compete with essential living-cost bills, such as those for housing, food, utilities, and transportation. Former inmates in the United States owe about $50 billion for various criminal-justice costs, such as pretrial detention, court fees, and incarceration costs. As much as 60% of former inmates' income is estimated to go toward "criminal justice debt." All of this makes successful reentry extremely difficult for all but the best-connected former inmates.

The main goal of reentry programs is to improve offender outcomes following release from prison.[82] The reentry movement benefitted greatly from the Second Chance Act (SCA) of 2008, which provided millions of dollars in grants through the Justice Department's Bureau of Justice Assistance for the implementation and evaluation of reentry programs. Before the SCA, the federal government funded the Serious and Violent Offender Reentry Initiative with more than $100 million to implement reentry programs for high-risk offenders in all 50 states.

Most reentry programs attempt to modify offender behavior through cognitive behavioral interventions, often paired with educational or vocational training and assistance, substance

Break Down Barriers to Reentry

A 2018 national survey found that 90% of registered voters agree that we should break down barriers for people coming out of prison so they can get jobs, support their families, and stop being so dependent on government services. Support is strong among Republicans (91%), Independents (90%), and Democrats (89%).

Source: Robert Blizzard, "National Poll Results," Public Opinion Strategies, January 25, 2018. Accessed June 3, 2019, https://www.politico.com/f/?id=00000161-2ccc-da2c-a963-efff82be0001.

Reentry Programs

To learn more about reentry programs, visit the National Institute of Justice's website crimesolutions.gov at https://www.crimesolutions.gov/TopicDetails.aspx?ID=2. As of this writing, the website rates reentry programs as "very effective" (8), "very promising" (82), and "no effects" (46) (number of programs in parentheses).

How do the reentry programs rated "very effective" differ from the programs rated "no effects"?

recidivism The return to illegal activity after release from incarceration.

myth

People in prison for violent or sexual crimes are too dangerous to be released.

fact

People convicted of homicide are the least likely to be rearrested, and those convicted of rape or sexual assault have rearrest rates about 30%–50% lower than people convicted of larceny or motor vehicle theft. More broadly, people convicted of any violent offense are less likely to be rearrested in the years after release than those convicted of property, drug, or public order offenses.

Source: Wendy Sawyer and Peter Wagner, "Mass Incarceration: The Whole Pie 2019," Prison Policy Initiative, March 19, 2019. Accessed July 21, 2019, https://www.prisonpolicy.org/reports/pie2019.html.

abuse treatment, and/or life skills development. Some reentry programs begin after an inmate is released from prison, while other reentry programs begin in prison and transition to the community upon an inmate's release. Ideally, reentry programs address an inmate's multiple needs, such as housing, employment, education, addiction recovery, and mental and physical health treatment.

Evaluations of reentry programs' effectiveness indicate inconsistent results. Two of the biggest problems are that they rarely address a former inmate's multiple needs, and seldom are they individualized. The evaluation literature suggests that effective reentry programs must (1) start with accurate screening and assessment using validated actuarial screening instruments, (2) be based on individualized strategies that target specific criminogenic needs, (3) be implemented with reliability, dependability, and commitment, and (4) begin in prison and transition to the community upon an inmate's release to keep the inmate connected to the treatment milieu and to provide critical aftercare through effective case management.

Recidivism figures are a testament to the need for successful reentry programs. We conclude this chapter by considering **recidivism**–the return to illegal activity after release from incarceration. Measuring recidivism and attributing the lack of recidivism to the influence of correctional programs are complicated by scientific methodological problems. Nevertheless, until recently, numerous studies conducted during the past couple of decades in several different jurisdictions revealed that recidivism rates, when recidivism was defined in a similar way, had remained remarkably stable. The research found that a good predictor of whether an inmate would continue to commit crimes following release was the number of times the inmate had been arrested in the past. Prisoners with longer prior records were more likely to be rearrested than prisoners with shorter prior records. The research also showed that the first year of release is the most critical one. Nearly two-thirds of all the recidivism of the first 3 years occurred in the first year.

The U.S. Department of Justice's Bureau of Justice Statistics recently published a recidivism study that followed prison inmates released in 2005 in 30 states for an unprecedented 9 years (2005–2014).[83] The measure of recidivism used in the study was rearrest *both within and outside the state of release* (other studies have used reconviction or reincarceration as measures of recidivism). Rearrest as a recidivism measure is the least restrictive measure that could be used, and the measure that is likely to produce the highest rates of recidivism because it includes arrestees who later may have charges dropped or are found innocent of the crime for which they had been arrested.

The study found that 83.4% of the more than 400,000 state prison inmates released in 30 states in 2005 were arrested at least once during the 9 years after release either within or outside the state of release (15.4% were arrested outside the state of release); only 16.6% of released inmates were not arrested, or *desisted*, during the 9-year follow-up period. The more than 400,000 released inmates were arrested nearly 2 million times during the 9-year follow-up period, an average of about 5 arrests per released inmate. About 40% of the arrests occurred during years 1–3, approximately 31% occurred during years 4–6, and nearly 29% occurred during years 7–9. An estimated 23% of released inmates were responsible for about half of the nearly 2 million arrests during both 3-year and 9-year follow-up periods.

Released male inmates were more likely to be arrested than released female inmates, although the gap narrowed as time passed. Younger released inmates (24 or younger) were more likely to be arrested than older released inmates (especially those 40 or older). Released black or African American and American Indian or Alaska Native inmates were more likely to be arrested than released white, Hispanic/Latino, and Asian, Native Hawaiian, or other Pacific Islander inmates, especially in the first year, but differences narrowed in subsequent years. Released property offenders were more likely to be arrested than released violent, drug, or public order offenders (arrests for probation and parole violations, which accounted for only about 1% of all arrests, were included as public order offenses, and inmates who committed multiple offenses were categorized by the most serious offense). Note that the number of former inmates arrested declined dramatically in follow-up years 4–9 (see next paragraph), and that 83.4% of all released inmates, as noted previously, had been arrested at least once during the 9-year follow-up period, so within category differences were small.

As found in previous research, the first year following release is the most critical. In the study, 43.9% of inmates released from prison were arrested during their first year of freedom (only 3.3% outside the state of release). The percentage of released inmates arrested for the first time during subsequent years declined each year: during year 2,

16.2% were arrested; during year 3, 8.3%; during year 4, 5.1%; during year 5, 3.5%; during year 6, 2.4%; during year 7, 1.7%; during year 8, 1.3%; and during year 9, 1%.[84] Among the 43.9% of released inmates arrested during their first year of freedom, 38.9% were arrested again at least once during the next 8 years; only 4.9% of them were not arrested, or desisted, during the subsequent 8 years. (Some of them may not have been arrested because they already were incarcerated or had died.) More than two-thirds (68.4%) of released inmates were arrested during the first 3 years following release (only 7.7% outside the state of release), showing both the importance and the ineffectiveness of reentry programs during those first 3 years.

Obviously, those research findings are not good news for people who believe that imprisonment can achieve large-scale deterrence of crime and rehabilitation of criminals. More bad news for deterrence advocates comes from the research of sociologist Ben Crouch, who found that newly incarcerated offenders frequently express a preference for prison over probation sentences. Crouch explains:

> A fundamental irony emerges in our justice system. That is, the lawbreakers whom middle-class citizens are most likely to fear and want most to be locked away . . . tend to be the very offenders who view prison terms of even two or three years as easier than probation and as preferable. To the extent that these views among offenders are widespread, the contemporary demand for extensive incarceration (but often for limited terms) may foster two unwanted outcomes: less deterrence and more prisoners.[85]

One reason offenders may prefer imprisonment to probation is that if probation is revoked, the time spent on probation, even if it is several years, generally does not count as credit against the new prison sentence. Similar research by criminologist Kenneth Tunnell shows that most repeat property offenders are not threatened by the prospects of imprisonment.[86]

Unfavorable news for advocates of rehabilitation began trickling in even before the release in 1974 of the famous Martinson report, which cast grave doubts on the ability of prisons to reform offenders.[87] In a later study, criminologist Lynne Goodstein demonstrated that the inmates who adjusted most successfully to prison had the most difficulty adjusting to life in the free community upon release. About her research, Goodstein writes:

> These findings provide a picture of the correctional institution as a place which reinforces the wrong kinds of behaviors if its goal is the successful future adjustment of its inmates. In the process of rewarding acquiescent and compliant behavior, the prison may, in fact, be reinforcing institutional dependence.[88]

Of course, some people are deterred from crime by the threat of imprisonment, and the prison experience does benefit some inmates in the way rehabilitation advocates envision. Criminologists Frank Cullen and Paul Gendreau conducted a comprehensive review of research that evaluated the effects of correctional interventions on recidivism rates.[89] They found that across studies, the correctional interventions reduced recidivism rates, on average, 10%. The most successful interventions reduced recidivism 25%. They used cognitive-behavioral treatments, targeted known predictors of crime for change, and intervened mainly with high-risk offenders. Cognitive-behavioral treatments, as noted previously, emphasize the important role of thinking in how people feel and behave. Therapists help offenders identify the thinking that is causing their problematic feelings and behavior and then teach them how to replace those thoughts with thoughts that will lead to more desirable feelings and actions. Cullen and Gendreau also reported that punishment-oriented, correctional interventions generally had no effect on offender criminality.

Congress recently addressed the problems of reentry and recidivism, for at least federal prisoners, when it passed and President Trump signed on December 21, 2018, the "Formerly Incarcerated Reenter Society Transformed Safely Transitioning Every Person Act" (Public Law No: 115-391) or the "First Step Act," for short. The legislation, which was modeled after successful reform policies in Texas and Georgia, shifts the primary goal of federal prisons from retribution and punishment to rehabilitation, reentry, and recidivism reduction. It also aims to ameliorate the harsh sentencing policies that contributed to the imprisonment boom and to provide new protections to women and juveniles in the federal prison system. The following description is limited to only the issues of reentry and recidivism.

The lynchpin of the reform policy is a new and statistically validated risk of recidivism and needs assessment system, which is to be reviewed annually and revised and updated as needed. Equally important is an inventory of successful (or likely to be successful) evidence-based recidivism reduction programs and productive activities and an assessment of their most effective and efficient uses. Examples of such programs and activities are: (1) social learning and communication, interpersonal, anti-bullying, rejection response, and other life skills; (2) family relationship building, structured parent–child interaction, and parenting skills; (3) classes on morals and ethics; (4) academic classes; (5) cognitive behavioral treatment; (6) mentoring; (7) substance abuse treatment; (8) vocational training; (9) faith-based classes or services; (10) civic engagement and reintegration community services; (11) a prison job, including through a prison work program; (12) victim impact classes or other restorative justice programs; and (13) trauma counseling and trauma-informed support programs.

All federal prisoners are to be assigned to programs or productive activities based on their risks and needs and to be reassessed periodically to determine whether other programs or activities are more appropriate and when they are ready to be transferred to prerelease custody. Incentives and rewards are to be provided for successful participation in the programs and activities, including, but not limited to, phone and visitation privileges, transfer to an institution closer to release residence, increased commissary spending limits and product offerings, extended opportunities to access the email system, consideration of transfer to a preferred housing unit (including transfer to a different prison), and time credits, e.g., 10 days of credit for every 30 days of successful program participation. Prisoners convicted of numerous offenses are ineligible to receive time credits. Also, rewards and incentives can be reduced if prisoners violate prison, program, or activity rules.

The Attorney General is charged with developing and implementing training programs for Bureau of Prisons officers and employees responsible for administering the rehabilitation system. In order to expand programs and activities, prison wardens are encouraged to enter into partnerships with outside entities such as nonprofit and other private organizations and institutions of higher education. To carry out the "First Step Act," Congress authorized to be appropriated $50 million for each of fiscal years 2019 through 2023.

In sum, we must be careful not to hold unrealistic expectations about what imprisonment can accomplish in our society. Traditionally, we have placed undue confidence in the ability of imprisonment to control crime. But imprisonment is a reactive (versus proactive) response to the social problem of crime, and crime is interwoven with other social problems, such as poverty, inequality, and racism, in our wider society. We should not expect imprisonment to resolve or control those problems.

Another impediment to the successful reintegration of released inmates is their loss of civil rights. Depending on the jurisdiction, convicted felons lose a variety of civil rights. Among the most common rights forfeited are the rights to vote, hold public office, serve on a grand or petit jury, practice a profession that requires an occupational or professional license, be a parent (termination of parental rights), and carry or possess firearms. In many states, women convicted of drug offenses may be denied welfare benefits. If convicted of a federal felony, a person is ineligible for enlistment in any of the armed services, may have public housing benefits restricted under certain circumstances, and, depending on the crime, may lose veteran benefits, including pensions and disability. A majority of jurisdictions provide a means for restoring lost rights. Some rights are restored automatically. Other rights are restored by executive or judicial proceedings that often are based on the ex-offender's demonstration that he or she has been rehabilitated. Some rights are lost permanently.[90] It is difficult enough for released prison inmates to reenter society and succeed. The loss of civil rights and the hurdles that make them difficult to restore (when they can be restored) do not make it any easier.

FYI

Fair Chance Hiring

A 2018 national survey found that 65% of registered voters support fair chance hiring—allowing job applicants to explain their qualifications for a job before they are asked about their criminal histories; 33% do not support fair chance hiring.

Source: Robert Blizzard, "National Poll Results," Public Opinion Strategies, January 25, 2018. Accessed June 3, 2019, https://www.politico.com/f/?id=00000161-2ccc-da2c-a963-efff82be0001.

THINKING CRITICALLY

1. Do you think anything could be done to lower recidivism rates among offenders? If so, what?
2. In your opinion, should prisoners be allowed to reduce their time in prison through parole or good time? Why or why not?
3. What do you think is the biggest impediment to reentry, and what can be done about it?
4. Should convicted felons lose civil rights? If yes, which ones and why? If not, why not?

SUMMARY

1. Distinguish between the deprivation and importation models of inmate society.

The deprivation model of inmate society emphasizes the role of the prison environment in shaping the inmate society. The importation model of inmate society emphasizes attributes inmates bring with them when they enter prison.

2. Explain how today's inmate society differs from those of the past.

Compared with prisons of the past, the inmate society in today's prisons is much more fragmented and conflict ridden. Physical, psychological, economic, and social victimization are facts of life in many contemporary institutions. Inmates must learn to cope with this state of affairs. Some do so by becoming part of the prison's violent public culture. Others carve out niches and meet their needs in the prison's private culture.

3. Identify some of the special features of life in women's prisons.

Life in women's prisons is somewhat different from life in men's prisons. In particular, women's prisons often have less stringent security measures, are less violent, and are characterized by pseudo-family structures that help the inmates cope.

4. Describe the profile of correctional officers, and explain some of the issues they face.

Correctional officers who staff prisons are predominantly white and male. Correctional officers face a number of problems in their work, including low pay, work-related conflicts, and a potential for corruption. They also are subject to role conflict because of their dual objectives of custody and treatment. Some officers respond to those problems more constructively than others. Efforts are ongoing to transform prison work into a profession, and issues surrounding those efforts include recruitment and selection, the backlash against affirmative action and other hiring practices, training, and unionism.

5. Identify prisoners' rights, and relate how they were achieved.

The main way that prisoners have gained rights during the past 5 decades is through intervention by the courts. Until the 1960s, inmates had minimal rights. As the hands-off policy was lifted, prisoners gained a number of important rights: greater access to the courts, easier access to legal services in prison, and improved prison disciplinary procedures. They also gained certain rights under the First, Eighth, and Fourteenth Amendments.

6. List the two most common ways that inmates are released from prison, and compare those two ways in frequency of use.

Most inmates are released from prison at the discretion of a parole authority or under mandatory release laws. Mandatory release, a method of prison release under which an inmate is released after serving a legally required portion of his or her sentence minus good-time credits, has begun to rival parole release in frequency of use.

7. Summarize what recidivism research reveals about the success of the prison in achieving deterrence and rehabilitation.

When inmates are released from prison, it is hoped that they will not return to crime. Studies show, however, that the rate of return to crime, or recidivism, is high. Studies also show that recidivism rates have remained fairly constant for a long time. Those studies call the deterrence and rehabilitation rationales into question. As a society, our expectations of what incarceration can accomplish probably are unrealistic.

KEY TERMS

total institution 452
convict code 452
deprivation model 452
prisonization 452
importation model 452
sub-rosa economy 456
hands-off philosophy 467
habeas corpus 468
jailhouse lawyers 468
commutation 473
parole 473
good time 473
mandatory release 473
clemency 474
pardon 474
reentry 475
recidivism 478

REVIEW QUESTIONS

1. What did Erving Goffman mean when he wrote that prisons are *total institutions*?
2. What is the *convict code*, and why is it central to inmate society in traditional men's prisons?
3. What is *prisonization*?
4. What factors have rendered contemporary inmate society fragmented, disorganized, and unstable?
5. What are some of the reasons for high rates of prison violence?
6. What are four types of victimization that take place in prisons?
7. What are some of the general ways that inmates cope and adjust to the institutional environment, according to Johnson, Irwin, and Heffernan?
8. According to Hawkins and Alpert, what are four ways that correctional officers respond to their roles and their work conditions?
9. In *Wolff v. McDonnell* (1974), what rights were given to inmates facing disciplinary actions, and what rights were not extended?
10. What prison conditions are Eighth Amendment violations?
11. How has court intervention limited prison reform?
12. What is the goal of reentry programs, and how effective have they been?
13. What civil rights can convicted felons lose in some jurisdictions?

IN THE FIELD

1. **Rehabilitation Programs** Contact a local prison or jail, and arrange to investigate one or more rehabilitation programs. Obtain pertinent written material, and observe the program in operation as much as possible. Assess how successful the program is in achieving its goals.

   Consider these specific issues:

   a. What are the goals of the program?
   b. Are the goals of the program clearly defined in some document?
   c. Is the program administered in an effective way?
   d. Is instruction adequate?
   e. Is there adequate equipment (if applicable)?
   f. Will the skills learned in the program be important to inmates upon release?
   g. How likely are inmates to successfully complete the program?
   h. Does the program have an evaluation component that would allow officials to determine the program's success or failure?
   i. On what basis is the program's success or failure judged? After assessing the program, explain what changes, if any, you would recommend.

2. **Prison Films** Watch a film about prison life. Good choices are *Murder in the First* with Kevin Bacon and Christian Slater, *The Shawshank Redemption* with Tim Robbins and Morgan Freeman, *Birdman of Alcatraz* with Burt Lancaster, *Brubaker* with Robert Redford, *Cool Hand Luke* with Paul Newman, *The Green Mile* with Tom Hanks and Michael Clarke Duncan, and *Escape from Alcatraz* with Clint Eastwood. From what you have learned in this chapter, determine how realistic the portrayals are. Note common themes.

ON THE WEB

1. **Life in Prison** Visit the website of the Hawaii Department of Corrections at http://dps.hawaii.gov/hcf-virtual-tour/, and take a virtual prison tour of the Halawa Correctional Facility in Honolulu. Pretend you are an inmate and write an essay describing your experiences in the prison.

2. Find your state's felony-theft threshold. Do you think it is too low, too high, or about right? Write a short essay defending your position.

CRITICAL THINKING EXERCISES

RIGHT TO VOTE

1. More than 6 million Americans are denied the right to vote (disenfranchised) because they are felons or ex-felons. In 48 states and the District of Columbia, incarcerated felons are denied the right to vote (only Maine and Vermont permit prison inmates to vote). In 17 states and the District of Columbia, felons lose their voting rights only while incarcerated, and receive automatic restoration upon release. In 22 states, felons lose their voting rights during incarceration, and for a period of time after, typically while on parole and/or probation. Voting rights are restored automatically after this time period. Former felons also may have to pay outstanding fines, fees, or restitution before their voting rights are restored. Finally, in 9 states, felons lose their voting rights indefinitely for some crimes, or require a governor's pardon in order for voting rights to be restored, or face an additional waiting period after completion of sentence (including parole and probation) before voting rights can be restored.

   In the 2016 national election, an estimated 6.1 million people were disenfranchised as a result of a felony conviction. By comparison, only an estimated 1.2 million felons were disenfranchised in 1976. The 6.1 millions felons disenfranchised in 2016 represented about 2.5% of the total U.S. voting age population—1 of every 40 adults. In Alabama, Florida, Kentucky, Mississippi, Tennessee, and Virginia, more than 7% of the adult population was disenfranchised in 2016. Florida, alone, accounted for 27% of the disenfranchised population nationally in 2016, and its nearly 1.5 million individuals disenfranchised post-sentence in 2016 accounted for 48% of the national total for that category. More than 7.4% of the adult African American population—about 1 of every 13—was disenfranchised in 2016, compared to 1.8% of the non-African American population—about 1 of every 55. In four states, more than 20% of African Americans was disenfranchised in 2016: Florida (21%), Kentucky (26%), Tennessee (21%), and Virginia (22%).

   A recent national opinion poll found that 80% of the public generally favors restoring voting rights to ex-felons, while a low of 52% even favors restoring voting rights to sex offenders. Sixty percent support enfranchising felony parolees, and 66% favor enfranchising ex-felons convicted of a violent crime who have served their entire sentence.

   a. Should incarcerated felons have the right to vote? What about felony probationers or parolees? What about ex-felons who have completed their sentences? What about sex offenders who have completed their sentences? What about felons who were convicted of a violent crime and have completed their sentences? Justify your answers.
   b. If you believe that voting rights should be restored, should there be conditions? If so, what should they be?
   c. Currently, some states restore voting rights automatically; other states do so by executive or judicial action. If you believe that voting rights should be restored, what should the process entail?

INMATE ORGAN TRANSPLANTS

2. A California prison inmate received a donor heart. He had suffered congestive heart failure the month before. The inmate was

serving a 14-year sentence for robbery, and the transplant and treatment for his illness cost California taxpayers about $1 million. Medical professionals defended the transplant, saying the recipient met medical criteria. Prison officials cited a 1976 Supreme Court decision and a 1997 California ruling by the 9th U.S. Circuit Court of Appeals requiring them to meet inmate medical needs. Critics point out that, among other things, at the time of the surgery, more than 4,000 Americans were on a waiting list for heart transplants, and 700 would die that year waiting for one. The prison inmate died about a year after receiving the transplant. Prison authorities noted that he had failed to maintain rigorous medical routines following the transplant.

a. Should prison inmates be eligible for organ transplants?
b. Does it matter whether they can pay for them?
c. What does the law require prison officials to do about organ transplants?
d. Should the "less-eligibility" principle apply to organ transplants? Why or why not?
e. Should there be limits to the medical services provided to prison inmates and, if so, what should the limits be?

NOTES

1. Erving Goffman, *Asylums* (New York: Doubleday, 1961), xiii.
2. James B. Jacobs, *Stateville: The Penitentiary in Mass Society* (Chicago: University of Chicago Press, 1977).
3. See Donald Clemmer, *The Prison Community* (New York: Holt, Rinehart & Winston, 1940); Gresham M. Sykes, *The Society of Captives: A Study of a Maximum Security Prison* (Princeton, NJ: Princeton University Press, 1958).
4. Clemmer, *The Prison Community*.
5. John Irwin and Donald R. Cressey, "Thieves, Convicts, and the Inmate Culture," *Social Problems* 10 (1962): 142–55. Also see Clarence Schrag, "Some Foundations for a Theory of Corrections," in *The Prison: Studies in Institutional Organization and Change*, ed. D. R. Cressey (New York: Holt, Rinehart & Winston, 1961), 30–35.
6. See Richard Hawkins and Geoffrey P. Alpert, *American Prison Systems: Punishment and Justice* (Englewood Cliffs, NJ: Prentice Hall, 1989); John Irwin, *Prisons in Turmoil* (Boston: Little, Brown, 1980). Also see Leo Carroll, *Hacks, Blacks, and Cons: Race Relations in a Maximum Security Prison* (Lexington, MA: Heath, 1974); Geoffrey Hunt, Stephanie Riegel, Tomas Morales, and Dan Waldorf, "Changes in Prison Culture: Prison Gangs and the Case of the 'Pepsi Generation,'" *Social Problems* 40 (1993): 398–409; James B. Jacobs, *New Perspectives on Prisons and Imprisonment* (Ithaca, NY: Cornell University Press, 1983).
7. Calculated from data in Margaret E. Noonan, "Mortality in State Prisons, 2001-2014–Statistical Tables," U.S. Department of Justice, Bureau of Justice Statistics, December 2016, 4, Table 1, accessed August 23, 2019, https://www.bjs.gov/content/pub/pdf/msp0114st.pdf.
8. Ibid., Table 2.
9. Ibid., 14, Table 15.
10. Ibid., 12, Table 13.
11. "Alabama's Prisons are Deadliest in the Nation," Equal Justice Initiative, December 3, 2018, accessed August 24, 2019, https://eji.org/news/alabamas-prisons-are-deadliest-in-nation.
12. Calculated from data in "Fallen But Not Forgotten," Correctional Peace Officers Foundation, accessed August 24, 2019, http://cpof.org/fallen-officers/by-year/.
13. Ibid.
14. Allen J. Beck and Paige M. Harrison, "Sexual Victimization in State and Federal Prisons Reported by Inmates, 2007" in *Bureau of Justice Statistics Special Report*, U.S. Department of Justice (Washington, DC: U.S. Government Printing Office, December 2007); Paul Guerino and Allen J. Beck, "Sexual Victimization Reported by Adult Correctional Authorities, 2007-2008," in *Bureau of Justice Statistics Special Report*, U.S. Department of Justice (Washington, DC: U.S. Government Printing Office, January 2011), accessed February 17, 2011, http://bjs.ojp.usdoj.gov/content/pub/pdf/svraca0708.pdf (accessed February 17, 2011).
15. Ibid.
16. Ibid.
17. Allen J. Beck, Marcus Berzofsky, Rachel Caspar, and Christopher Krebs, "Sexual Victimization in Prisons and Jails Reported by Inmates, 2008-09," in *Bureau of Justice Statistics Bulletin*, U.S. Department of Justice (Washington, DC: U.S. Government Printing Office, August 2010), accessed February 19, 2011, http://bjs.ojp.usdoj.gov/content/pub/pdf/svpjri0809.pdf.
18. Information about the third National Inmate Survey is from Allen J. Beck, Marcus Berzofsky, Rachel Caspar, and Christopher Krebs, "Sexual Victimization in Prisons and Jails Reported by Inmates, 2011-12," in Bureau of Justice Statistics, U.S. Department of Justice (Washington, DC: U.S. Government Printing Office, May 2013). Accessed June 28, 2016, www.bjs.gov/content/pub/pdf/svpjri1112.pdf.
19. Ramona R. Rantala, "Sexual Victimization Reported by Adult Correctional Authorities, 2012-15," U.S. Department of Justice, Bureau of Justice Statistics Special Report (Washington, DC: U.S. Government Printing Office, July 2018), accessed August 24, 2019, https://www.bjs.gov/content/pub/pdf/svraca1215.pdf.
20. Calculated from data in ibid., 5, Table 1.
21. Ibid., 6, Table 2.
22. Ibid., 5.
23. Calculated from data in ibid., 7, Table 3.
24. Ibid., 8, Table 4.
25. Calculated from data in ibid., 9, Table 5.
26. Calculated from data in ibid., 7, Table 3.
27. Ibid., 8, Table 4.
28. Calculated from data in ibid., 9, Table 5.
29. Lee Bowker, *Prison Victimization* (New York: Elsevier, 1980).
30. Kenneth Foster, "In Prison, Stamps Are The Currency That Binds Everybody," *PRISON WRITERS*, accessed August 24, 2019, https://prisonwriters.com/stamps-as-currency-in-prison/; Aimee Picchi, "Forget cigarettes–there's a new prison currency," *CBS NEWS*, August 23, 2016, accessed August 24, 2019, https://www.cbsnews.com/news/forget-cigarettes-theres-a-new-prison-currency/.
31. Ann Cordilia, *The Making of an Inmate: Prison as a Way of Life* (Cambridge, MA: Schenkman, 1983).
32. Goffman, *Asylums*, 14.
33. Hans Toch, *Living in Prison: The Ecology of Survival* (New York: Free Press, 1977).
34. Robert Johnson, *Hard Time: Understanding and Reforming the Prison* (Monterey, CA: Brooks/Cole, 1987).
35. Irwin, *Prisons in Turmoil*.

36. David Ward and Gene Kassebaum, "Homosexuality: A Model of Adaptation in a Prison for Women," *Social Problems* 12 (1964): 159–77.
37. Joycelyn Pollock-Byrne, *Women, Prison, and Crime* (Pacific Grove, CA: Brooks/Cole, 1990).
38. Lauren E. Glaze and Laura M. Maruschak, "Parents in Prison and Their Minor Children," in *Bureau of Justice Statistics Special Report*, U.S. Department of Justice (Washington, DC: U.S. Government Printing Office, August 2008), accessed February 19, 2011, http://bjs.ojp.usdoj.gov/content/pub/pdf/pptmc.pdf.
39. Emily Halter, "Parental Prisoners: The Incarcerated Mother's Constitutional Right to Parent," *Journal of Criminal Law & Criminology* 108 (2018): 539–67. Accessed August 25, 2019, http://www.law.northwestern.edu/legalclinic/wrongfulconvictions/documents/parental-prisoners-by-emily-halter.pdf. Phyllis Jo Baunach, *Mothers in Prison* (New Brunswick, NJ: Transaction Books, 1985).
40. "First of its Kind Statistics on Pregnant Women in U.S. Prisons," *Johns Hopkins Medicine*, March 21, 2019, accessed March 4, 2020, https://www.hopkinsmedicine.org/news/newsroom/news-releases/first-of-its-kind-statistics-on-pregnant-women-in-us-prisons.
41. Rose Giallombardo, *Society of Women: A Study of a Women's Prison* (New York: Wiley, 1966); Esther Heffernan, *Making It in Prison: The Square, the Cool, and the Life* (New York: Wiley, 1972); Alice M. Propper, *Prison Homosexuality* (Lexington, MA: Heath, 1981).
42. Propper, *Prison Homosexuality*.
43. Giallombardo, *Society of Women*.
44. David Ward and Gene Kassebaum, *Women's Prison: Sex and Social Structure* (Chicago: Aldine, 1965).
45. Heffernan, *Making It in Prison*; Schrag, "Some Foundations for a Theory of Corrections."
46. Gordon Hawkins, *The Prison: Policy and Practice* (Chicago: University of Chicago Press, 1976), 85.
47. Calculated from data in "Occupational Employment Statistics," U.S. Department of Labor, Bureau of Labor Statistics, accessed August 26, 2019, https://www.bls.gov/oes/current/oes333012.htm; "Federal Prison System (BOP)," U.S. Department of Justice, https://www.justice.gov/jmd/page/file/1033161/download. The BOP data is from the FY 2019 Budget Request. Data for Delaware and Texas is from May 2017.
48. Calculated from data above and in "How to Become a Corrections Officer?" All Criminal Justice Degrees, accessed August 27, 2019, http://www.allcriminaljusticedegrees.com/correctional-officer/.
49. "Occupational Employment Statistics."
50. "Federal Prison System (BOP)."
51. "Labor Force Statistics from the Current Population Survey," U.S. Department of Labor, Bureau of Labor Statistics, accessed August 26, 2019, https://www.bls.gov/cps/cpsaat11.htm.
52. Unless indicated otherwise, information on becoming a correctional officer is from "How to Become a Corrections Officer?"
53. "General Expectations of CO Academy Training," Correctional OfficerEDU.org, https://www.correctionalofficeredu.org/correctional-officer-academy/.
54. "Occupational Employment Statistics."
55. Chris Kolmar, "The 10 Best States for Correction Officers in 2019," ZIPPIA, accessed August 27, 2019, https://www.zippia.com/advice/best-states-for-correction-officers/; "Salary Table 2019–GS," OPM.GOV, accessed August 27, 2019, https://www.opm.gov/policy-data-oversight/pay-leave/salaries-wages/2019/general-schedule/; "Entry Level Correctional Officer Salary in Texas," ZipRecruiter, accessed August 27, 2019, https://www.ziprecruiter.com/Salaries/How-Much-Does-an-Entry-Level-Correctional-Officer-Make-a-Year–in-Texas.
56. Hawkins and Alpert, *American Prison Systems*.
57. Ibid.
58. Sykes, *The Society of Captives*.
59. John D. Hewitt, Eric D. Poole, and Robert M. Regoli, "Self-Reported and Observed Rule-Breaking in Prison: A Look at Disciplinary Response," *Justice Quarterly* 3 (1984): 437–47.
60. Hawkins and Alpert, *American Prison Systems*.
61. Lucien X. Lombardo, *Guards Imprisoned: Correctional Officers at Work*, 2nd ed. (Cincinnati, OH: Anderson, 1989); also see Johnson, *Hard Time*.
62. Barbara A. Owen, "Race and Gender Relations Among Prison Workers," *Crime and Delinquency* 31 (1985): 147–59.
63. "Corrections," AFSCME, accessed August 27, 2019, https://www.afscme.org/union/jobs-we-do/corrections.
64. American Federation of Government Employees, "Union Representing 33,000 Federal Correctional Officers Opposes Prison Reform Bill," CISION PR Newswire, May 18, 2018, accessed August 27, 2019, https://www.prnewswire.com/news-releases/union-representing-33-000-federal-correctional-officers-opposes-prison-reform-bill-300651349.html.
65. Jacobs, *Stateville*, 36–37.
66. "Appendix B: Text of the Prison Litigation Reform Act of 1995," http://ojjdp.ncjrs.org/PUBS/walls/appen-b.html; The American Civil Liberties Union, "The Prison Litigation Reform Act (PLRA)," www.aclu.org/prison/gen/14769res20031113.html.
67. "Federal Judicial Caseload Statistics 2018," United States Courts, Statistics & Reports, Table C-2, accessed August 28, 2019, https://www.uscourts.gov/statistics-reports/federal-judicial-caseload-statistics-2018.
68. "Armed Career Criminal Act (ACCA) Law and Legal Definition," USLegal, accessed August 28, 2019, https://definitions.uslegal.com/a/armed-career-criminal-act-acca/.
69. Calculated from data in "Federal Judicial Caseload Statistics 2018."
70. *Wilson v. Seiter*, 501 U.S. 294 (1991).
71. Hawkins and Alpert, *American Prison Systems*.
72. ACLU National Prison Project, "Your Rights Regarding Environmental Hazards and Toxic Materials," www.aclu.org/prison/conditions/14661res20031114.html.
73. E. Ann Carson and Mary P. Cowhig, "Mortality in State Prisons, 2001–2016-Statistical Tables," U.S. Department of Justice, Bureau of Justice Statistics, February 2020, 5, Table 2, accessed March 5, 2020, https://www.bjs.gov/content/pub/pdf/msfp0116st.pdf; "Executions by Region," The Death Penalty Information Center, accessed March 5, 2020, https://deathpenaltyinfo.org/executions/executions-overview/number-of-executions-by-state-and-region-since-1976.
74. Calculated from data in ibid.
75. "Clemency Statistics," U.S. Department of Justice, accessed August 29, 2019, https://www.justice.gov/pardon/clemency-statistics.
76. Ibid.
77. Ibid.
78. "Clemency Initiative," U.S. Department of Justice, accessed August 30, 2019, https://www.justice.gov/pardon/clemency-initiative.
79. Ibid.
80. U.S. Code, Title 18. Crimes and Criminal Procedure, Part II. Criminal Procedure, Chapter 229. Postsentence Administration, Subchapter C. Imprisonment, Section 3624. Release of a prisoner, accessed September 1, 2019, https://uscode.house.gov/view.xhtml?req=(title:18%20section:3624%20edition:prelim).

81. Information on billing inmates for incarceration costs is from Chandra Bozelko and Ryan Lo, "You've Served Your Time. Now Here's Your Bill." HUFFPOST, September 16, 2018, accessed August 31, 2019, https://www.huffpost.com/entry/opinion-prison-strike-labor-criminal-justice_n_5b9bf1a1e4b013b0977a7d74; Juleyka Lantigua-Williams, "How Prison Debt Ensnares Offenders," The Atlantic, June 2, 2016, accessed September 2, 2019, https://www.theatlantic.com/politics/archive/2016/06/how-prison-debt-ensnares-offenders/484826/.

82. The following description of reentry is from Holly Ventura Miller, "Reentry in the United States," in the *Routledge Handbook of Corrections in the United States*, eds, O. H. Griffin III and V. H. Woodward (New York and London: Routledge, 2018), 481–93.

83. Mariel Alper, Matthew R. Durose, and Joshua Markman, "2018 Update on Prisoner Recidivism: A 9-year Follow-up Period (2005–2014)," in *Bureau of Justice Statistics Special Report*, U.S. Department of Justice (Washington, DC: U.S. Government Printing Office, May 2018) accessed September 3, 2019, https://www.bjs.gov/content/pub/pdf/18upr9yfup0514.pdf.

84. Calculated from data in ibid., 4, Table 2.

85. Ben M. Crouch, "Is Incarceration Really Worse? Analysis of Offenders' Preferences for Prison over Probation," *Justice Quarterly* 10 (1993): 67–88.

86. Kenneth D. Tunnell, "Choosing Crime: Close Your Eyes and Take Your Chances," *Justice Quarterly* 7 (1990): 673–90.

87. Robert Martinson, "What Works? Questions and Answers About Prison Reform," *The Public Interest* 42 (1974): 22–54.

88. Lynne Goodstein, "Inmate Adjustment to Prison and the Transition to Community Life," in *Correctional Institutions,* 3rd ed., eds. R. M. Carter, D. Glaser, and L. T. Wilkins (New York: Harper & Row, 1985), 285–302.

89. Francis T. Cullen and Paul Gendreau, "Assessing Correctional Rehabilitation: Policy, Practice, and Prospects," in *Criminal Justice 2000*, vol. 3, U.S. Department of Justice, Office of Justice Programs, www.ojp.usdoj.gov/nij/criminal_justice2000/vol3_2000.html.

90. "State Felon Voting Laws," ProCon.org, July 2, 2019. Accessed September 4, 2019, https://felonvoting.procon.org/view.resource.php?resourceID=000286; "Felony Voting Rights," National Conference of State Legislatures, December 21, 2018. Accessed September 4, 2019, http://www.ncsl.org/research/elections-and-campaigns/felon-voting-rights.aspx.

Design Elements: *Image of part of a court building with columns and statue*: Pixtal/AGE Fotostock RF; *Image of a door slightly open*: jadding/Shutterstock.com

Spencer Grant/PhotoEdit

12 Community Corrections

CHAPTER OUTLINE

Community Corrections: Definition and Scope
- Goals and Staff Roles
- The Importance of Community Corrections

Probation
- Historical Context
- Administration
- Process and Procedures
- Issues in Probation

Parole
- Historical Context
- Administration
- Process and Procedures
- Parole Issues

Intermediate Sanctions
- Intensive-Supervision Probation and Parole
- Day Reporting Centers
- Structured (or Day) Fines
- Home Confinement and Electronic Monitoring
- Halfway Houses

Temporary-Release Programs

LEARNING OBJECTIVES

After completing this chapter, you should be able to:

1. Define community corrections, and identify the goals and responsibilities of community corrections agencies and their staffs.
2. Define probation, and summarize the research findings on recidivism rates.
3. Distinguish parole from probation.
4. Explain the functions of a parole board.
5. Describe how intermediate sanctions differ from traditional community corrections programs.
6. Explain two major concerns about intensive-supervision probation and parole.
7. Explain what day reporting centers and structured fines are.
8. Explain what home confinement and electronic monitoring are.
9. Identify the goal of halfway houses, and compare halfway houses with other community corrections programs.
10. Summarize the purposes and outcomes of temporary-release programs.

Crime Story

On August 27, 2019, 32-year-old Philadelphia rapper and hip-hop superstar Meek Mill (pictured), whose real name is Robert Rihmeek Williams, finally was freed from criminal justice custody—a status that partially has defined him for most of his adult life. In exchange for the district attorney dropping all charges against him, Mill pleaded guilty to an old misdemeanor gun charge in a deal that resolved a January 2007 arrest in Philadelphia for various drug- and gun-related offenses, including felony firearm possession and aggravated assault of a police officer that kept him on parole, probation, or in prison for 12 years. At his trial in August 2008, Mill testified in his own defense and claimed that he was innocent of all but one of the charges. His arresting officer, who has since been discredited (see below), was the only witness in the case. Mill was convicted and, in January 2009, was sentenced to serve 11 to 23 months in prison, followed by 8 years of probation, which had been extended through 2023. He served 5 months of the original sentence and was released in early 2009 under a 5-year parole agreement.

Because of several technical violations, he would be sent to prison 3 times without committing a crime and serve 11 years on probation. In December 2012, Mill was found to have violated his probation, and the judge (the same trial judge) revoked his travel permit. In May 2013, the same trial judge found that Mill had violated his probation again for failing to report travel plans as required and for social media postings that resulted in death threats to the probation officer assigned to his case. The judge ordered Mill to take etiquette classes. In 2015, a federal jury rejected a lawsuit Mill filed against a Philadelphia police officer for what Mill claimed was a 10-hour, racially motivated traffic stop that caused him to miss the launch party for his 2012 debut album, "Dreams & Nightmares." His follow-up albums included the chart-topping "Dreams Worth More Than Money" and his 2018 "Championships," which included performances by Jay-Z, his mentor, and former girlfriend Nicki Minaj. In December 2015, the same trial judge found Mill guilty of violating probation again for "popping a wheelie" on a motorcycle in a music video in New York. In March 2017, Mill was arrested at an airport in St. Louis, Missouri, for assaulting two pedestrians. Shortly after his arrest and court appearance, he was given a court summons. In November 2017, his original trial judge sentenced him to 2 to 4 years in state prison for violating his parole conditions. After clashing repeatedly with his trial judge, who allegedly had a long-standing vendetta against Mill and wanted to get into the music business, Mill became active in the criminal justice reform movement, a frequent topic of his music, and founded an advocacy group, Reform Alliance, dedicated to helping the people that helped him. At the end of November 2017, the family of a man, who was shot and killed in the parking lot outside a Connecticut concert in December 2016, sued Mill and Roc Nation (an entertainment agency founded by Jay-Z). The man had just left the facility when he was killed by gunfire. The lawsuit claimed that Mill and Roc Nation allowed "thugs" to remain on the premises after "exhibiting disorderly, disruptive, argumentative, angry, and agitated behaviors toward patrons."

Mill had served five months of his 2- to 4-year sentence for his parole violation, when a three-judge panel of the Pennsylvania Superior Court granted him a new trial, bail, and removed the original trial judge from the case. The district attorney's office supported Mill's appeal and stated that it could not call the former arresting officer, the only witness to testify against Mill, to testify again after the department found newly discovered evidence indicating that the officer lied on the affidavit used to obtain the search warrant for Mill,

PETER FOLEY/EPA-EFE/Shutterstock

and at trial. Several years after the original trial, the FBI investigated the arresting officer for corrupt behavior in several cases, including falsifying evidence, physical assault, perjury, and theft while on duty. The officer failed an FBI polygraph test in 2014, and lied about the theft during an internal affairs investigation in 2017. The Philadelphia Police Department found the officer guilty of "conduct unbecoming" and unanimously called for his dismissal from the force. He resigned in disgrace.

Even though Mill repudiated the accusations that he pointed a gun at the police or sold drugs, the district attorney could not ignore that fact that Mill acknowledged that he had a gun and was guilty of a gun crime. Still, the district attorney believed that Mill's punishment was excessive, and that Mill had evolved in the last 10-plus years. On July 24, 2019, just hours before the Pennsylvania Superior Court threw out his conviction, Meek Mill and Jay-Z, who has since become an activist for criminal justice reform, announced they were launching a new label and starting a $50 million criminal justice reform group. Mill stated that his 12-year criminal justice odyssey was "mentally and emotionally challenging," but no different than what millions of other people like him face. About one-third of Pennsylvania's prison population is comprised of probation or parole violators. Approximately 40% of people sent to jail or prison in the United States each year are on probation or parole at the time of their incarceration.

This chapter is about community corrections, which includes probation and parole. At the same time that prison populations are decreasing, community corrections is experiencing rapid growth, and, as noted, a large percentage of prison and jail populations currently are comprised of probation and parole violators. Are probation and parole good substitutes for prison or jail? Justify your answer. Was Meek Mill treated fairly? Why or why not? Should offenders and inmates reject probation and parole because violations may extend their sentences, as was the case with Meek Mill? Justify your answer. What will happen if offenders and inmates reject probation and parole?

Community Corrections: Definition and Scope

community corrections The subfield of corrections in which offenders are supervised and provided services outside jail or prison.

Community corrections can be defined broadly as the subfield of corrections consisting of programs in which offenders are supervised and provided services outside jail or prison. For this reason, community corrections sometimes is referred to as "noninstitutional" corrections. Community corrections includes such programs as diversion, restitution, probation, parole, halfway houses, and various provisions for temporary release from prison or jail. Those programs are the subject of this chapter.

Note that *community corrections* is a generic term. Federal, state, and local jurisdictions differ widely in the way they organize and administer community corrections and in the specific procedures they use. Criminologist David Duffee's delineation of three varieties of community corrections illustrates this diversity.[1] Community-run correctional programs are controlled by local governments, with minimal connection to state and federal authorities. In effect, local officials determine how such programs will be run within the broad guidelines of state and federal laws. In community-placed programs, as in community-run ones, offenders are handled by agencies within the local district. But agencies in community-placed programs are connected to central state or federal authorities, or both. Consequently, central authority affects program operation, and community-placed programs tend to be more isolated from local affairs than community-run programs. Community-based correctional programs are a combination of the other two types. Connection to central authority for resources and other support services (a feature of community-placed programs) is combined with strong links between the program and the surrounding locality (a feature of community-run programs).

myth

Modern community correctional programs are invariably "soft on crime." They focus too much on rehabilitation, to the exclusion of punishment, deterrence, and incapacitation.

fact

Even though prison populations in the United States have decreased over the past decade or so, major prison crowding in several states has made it necessary to channel into the community, under supervision, many felons who formerly would have gone to (or stayed in) prison. The priorities and operations of community agencies have shifted to meet this challenge. Many community programs—especially the newer "intermediate sanctions," such as electronic monitoring—emphasize punishment, deterrence, and incapacitation as much as, if not more than, rehabilitation.

Goals and Staff Roles

As described in Chapter 9, the goals of sentencing, which also are the goals of corrections, include punishment, deterrence, incapacitation, and rehabilitation. Yet, as a subfield of corrections, community corrections traditionally has emphasized rehabilitation. Although community programs are concerned with supervising and controlling offenders and ensuring that they follow the rules of their sentences, there frequently is great emphasis on assisting offenders with personal problems and needs. Another emphasis is establishing stronger ties between the offender and the community, by helping the offender get and keep a job, for example. However, the traditional preoccupation of community corrections with rehabilitation had given way in the 1980s and 1990s to concern with the other goals.

The staff members of community correctional programs have two potentially competing roles that reflect different goals. The first amounts to a law enforcement role: seeing that offenders comply with the orders of community sentences. This means that staff members must supervise offenders, investigate possible rule infractions, and take action to address any serious or repeated rule violations.

The staff's other role is the social worker role: to help offenders identify and address their problems and needs. The social worker role has three aspects. The first is the direct provision of services, such as counseling, to offenders. The second is commonly described as the "resource broker" role. In this role, staff members identify particular

JANET HOSTETTER/AP Images

One aspect of probation and parole is client counseling. *What kinds of counseling problems are probation or parole officers likely to encounter?*

problems and needs and refer offenders to various community agencies for help. An important advantage of community corrections is that correctional agencies can draw on the services of other agencies in the locality. For example, a probation officer may refer a client to a local mental health center for counseling. The third is advocacy. A community may not offer services that a significant number of offenders require—for example, opportunities for vocational training. It is the staff's responsibility to advocate greater availability for services that are lacking and to work with community leaders to develop those services.

Many people believe that community correctional staff *should* occupy the dual roles of enforcing the law and helping clients. Those people maintain that supervision and control are necessary to facilitate rehabilitation and that rehabilitation enhances supervision and control. Opponents of that position point to the potential for conflict between roles. They argue that such conflict creates undue stress for staff and makes it difficult to accomplish anything of value.

The Importance of Community Corrections

A discussion of the definition and scope of community corrections would not be complete without stressing how other components of criminal justice depend on community programs. It is not feasible to send all convicted persons—or even all felons—to jail or prison; resources are simply too limited. The sheer number of cases would overwhelm the courts and institutional corrections were it not for the availability of community programs.

To illustrate, examine Figure 12.1, which presents data on the number of adults under the supervision of state and federal corrections agencies at the end of 1980, compared with the end of 2016. Note that at the end of 2016, the number of offenders in jail and prison combined is considerably smaller than the number serving community sentences on probation and parole. Approximately 4.5 million adults were on probation or parole at the end of 2016, compared with about 2.2 million in jail or prison. At year-end 2016, 1 in 38 adults in the United States was under correctional supervision; 1 in 55 was under community supervision.[2]

Figure 12.1 also shows that large increases in the prison and jail populations have been accompanied by large increases in the probation and parole populations. Between 1980 and 2017, the probation population increased by 229%, and the parole population by 297%. The comparable increases in the prison and jail populations were 371% and 306%, respectively. Nevertheless, 2016 marked the ninth consecutive year in which the number and the rate of

FYI

Shift Money to Community Corrections

A 2018 national survey found that 87% of registered voters agree that some money now being spent incarcerating nonviolent offenders should be shifted to alternatives such as electronic monitoring, community service, and probation. Agreement is strong among Republicans (80%), Independents (90%), Democrats (90%), and women (87%).

Source: Robert Blizzard, "National Poll Results," Public Opinion Strategies, January 25, 2018. Accessed June 3, 2019, https://www.politico.com/f/?id=00000161-2ccc-da2c-a963-efff82be0001.

FIGURE 12.1 Corrections Populations in the United States

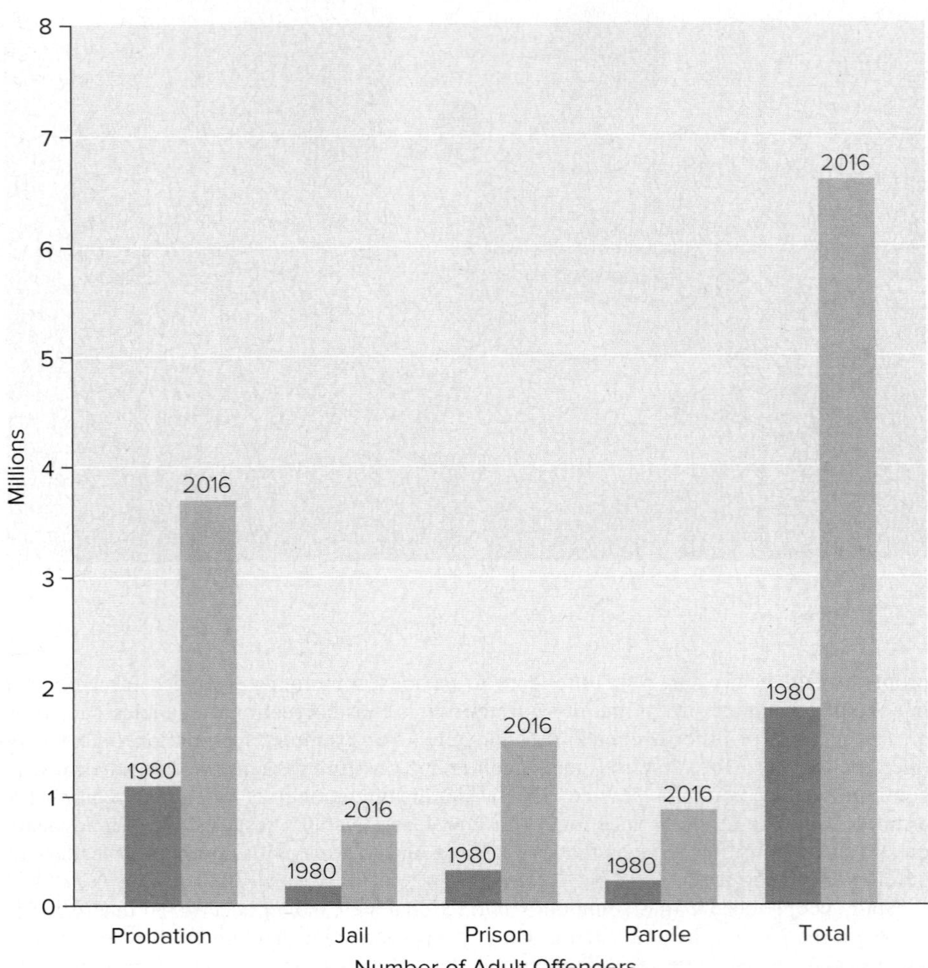

Sources: Danielle Kaeble and Mary Cowhig, "Correctional Populations in the United States, 2016," in *Bureau of Justice Statistics Bulletin*, U.S. Department of Justice (Washington, DC: U.S. Government Printing Office, April 2018), 2, Table 1, accessed September 9, 2019, https://www.bjs.gov/content/pub/pdf/cpus16.pdf. Figures for 1980 are from Ann L. Pastore and Kathleen Maguire, eds., Sourcebook of *Criminal Justice Statistics Online*, Table 6.1.2011, accessed September 9, 2019, https://www.albany.edu/sourcebook/pdf/t612011.pdf.

adults under community supervision declined. The decline in 2016 was driven by a decrease in the probation population, which fell 1.4% from the prior year; the parole population in 2016 increased 0.5% from the prior year. Probationers accounted for about 81% of adults under community supervision in 2016. The number of persons under community supervision at year-end 2016 was the lowest observed since 1999, and the community corrections rate was the same as it was in 1992.[3]

FYI

Funding Corrections

Of all adults under some type of correctional supervision in the United States, approximately 30% are confined in prisons or jails. Approximately 70% are serving community sentences. Approximately 85% of all funds allocated to corrections in the United States are spent to build and run institutions, and about 15% are spent on community corrections. Thus, 85% of the funds are spent on 30% of the correctional population, and 15% of the funds are left to accommodate 70% of that population.

Sources: Calculated from data in Danielle Kaeble and Mary Cowhig, "Correctional Populations in the United States, 2016," in *Bureau of Justice Statistics Bulletin*, U.S. Department of Justice (Washington, DC: U.S. Government Printing Office, April 2018), 2, Table 1, accessed September 10, 2019, https://www.bjs.gov/content/pub/pdf/cpus16.pdf; Jennifer Bronson, "Justice Expenditure and Employment Extracts 2015, Preliminary," in *Bureau of Justice Statistics*, U.S. Department of Justice (Washington, DC: U.S. Government Printing Office, June 2018), Table 10, jeee15t10.csv.

THINKING CRITICALLY

1. Which do you think is more beneficial to society: community corrections or prison? Why?
2. Which role of community corrections staff do you think is more important: the law enforcement role or the helping role? Why?

Probation

Probation can be defined as a sentence imposed by the courts on offenders who have either pleaded guilty or been found guilty. Instead of being incarcerated, an offender placed on probation is retained in the community under the supervision of a probation agency. The offender is provided with supervision and services. Continuation of probation (i.e., avoidance of incarceration) depends on the offender's compliance with the rules and conditions of the probation sentence. Probation is the most frequently imposed criminal sentence in the United States.

probation A sentence in which the offender, rather than being incarcerated, is retained in the community under the supervision of a probation agency and required to abide by certain rules and conditions to avoid incarceration.

Table 12.1 shows the states with the largest probation populations and the highest and lowest rates of persons under probation supervision per 100,000 adult U.S. residents at year-end 2016. The overall probation population declined for the ninth consecutive year in 2016–a decrease of about 14% from 2007 through 2016.[4] The decrease in the probation population from 2007 through 2016 accounted for about 85% of the decline in the correctional population since 2007.[5]

Probation should be distinguished from diversion, although, in broad terms, probation can be thought of as a type of posttrial diversion from incarceration. **Diversion** refers to organized and systematic efforts to remove people from the criminal justice process by placing them in programs that offer alternatives to the next, more restrictive stage of processing. Although diversion commonly is associated with juvenile justice, it also is used in adult criminal justice. Diversion can occur at any point from the initial police contact up to, and even in conjunction with, formal sentencing. For example, the police may divert a domestic-violence case by making a referral to a family-in-crisis center instead of making an arrest. Instead of filing charges against an arrestee who is an alcoholic, a prosecutor may divert the case to a detoxification and counseling agency. Those are examples of *pretrial* diversion. *Posttrial* diversion–for example, probation–occurs when an offender who has pleaded guilty or has been found guilty is placed in a program that is an alternative to a more restrictive sentence, such as incarceration.

diversion Organized, systematic efforts to remove individuals from further processing in criminal justice by placing them in alternative programs; diversion may be pretrial or posttrial.

There are seven types of probation:

1. *Informal or unsupervised probation* (also *court probation*) frees the probationer from the direct supervision of a probation officer, while still requiring the probationer to obey specific court-ordered conditions; often reserved for less serious, nonviolent crimes.
2. *Straight probation* (also *supervised* or *formal probation*) occurs when an offender is sentenced only to probation, with no incarceration or other form of residential placement; probationer must report to a probation officer on a regular basis.

TABLE 12.1 States with the Largest Probation Populations and the Highest and Lowest Rates of Persons Under Probation Supervision per 100,000 Adult U.S. Residents at Year-End 2016

States with the Largest Probation Populations		States with the Highest Rates of Supervision		States with the Lowest Rates of Supervision	
Georgia*	432,235	Georgia*	5,570	New Hampshire	366
Texas	374,285	Rhode Island	2,680	West Virginia	448
California	239,735	Ohio	2,624	Utah	568
Ohio	236,754	Idaho	2,578	Nevada	601
Florida	214,066	Minnesota	2,280	New York	628
Pennsylvania	180,492	Michigan*	2,276	Maine	632
Michigan*	175,965	Indiana	2,135	Kansas	758
New Jersey	140,589	Delaware	2,049	California	791
Illinois	113,989	New Jersey	2,015	New Mexico	798
Indiana	108,302	Colorado	1,870	South Carolina	839

*Based on 2015 data.

Sources: Danielle Kaeble, "Probation and Parole in the United States, 2016," in *Bureau of Justice Statistics Bulletin*, U.S. Department of Justice (Washington, DC: U.S. Government Printing Office, April 2018) 13–14, Appendix Table 2, accessed September 11, 2019, https://www.bjs.gov/content/pub/pdf/ppus16.pdf.

Alcoholics Anonymous and Al-Anon frequently are used as diversion programs for offenders who otherwise would be incarcerated. *Should some offenders be diverted from jail or prison? Why or why not?*

Pressmaster/Shutterstock

FYI **Characteristics of Adults on Probation**

At year-end 2016, 75% of all probationers in the United States were male and 25% were female. Fifty-five percent were white (non-Hispanic), 28% were black (non-Hispanic), 14% were Hispanic, and the remainder were of other races. Fifty-nine percent of probationers had committed felonies, 40% had committed misdemeanors, and 2% had committed other infractions.

Source: Danielle Kaeble, "Probation and Parole in the United States, 2016," in *Bureau of Justice Statistics Bulletin*, U.S. Department of Justice (Washington, DC: U.S. Government Printing Office, April 2018), 17, Appendix Table 4, accessed September 11, 2019, https://www.bjs.gov/content/pub/pdf/ppus16.pdf.

3. In *suspended-sentence probation*, the judge pronounces a jail or prison sentence but suspends the sentence on the condition that the offender performs well on probation.
4. With *split sentence probation*, the judge divides a single sentence into a relatively short jail term followed by probation supervision (e.g., a 5-year probation sentence with the first 6 months to be served in jail).
5. *Shock probation* usually involves two sentences. The offender is initially sentenced to prison but is soon (perhaps after 120 days) recalled to court and placed on probation.
6. *Residential probation* involves placement of the probationer in a structured but generally open living environment, such as a halfway house. When residential probation is used, it is common for the probationer to spend the early part of the sentence in the residential facility and then, upon successful discharge, to complete the probation sentence living in the free community.
7. *Community control probation* is the strictest form of probation. Effectively, it is a jail sentence without the jail. The probationer is required to remain at home and is monitored at all times, typically through use of an ankle monitor.

A probation agency has three fundamental objectives. The first is to assist the court in matters pertaining to sentencing. Conducting inquiries and furnishing the court with information about offenders' backgrounds and current situations accomplish this objective. The second is to promote community protection by supervising and monitoring the activities of persons sentenced to probation. The third is to promote the betterment of offenders by ensuring that they receive appropriate rehabilitation services. Thus, probation agencies work closely with the courts to assist offenders and, when possible, help them avoid incarceration without compromising the safety and security of the community.

Historical Context

Probation developed out of various practices used under English common law. One such practice, known as benefit of clergy, allowed certain accused individuals to appeal to the court for leniency in sentencing by reading from the Bible. Another practice was judicial reprieve, whereby a convicted offender could ask the judge to suspend the sentence on the condition that the offender display good future behavior. Those and other practices were important forerunners of modern probation.

The more immediate origins of modern probation lie in the efforts of John Augustus (1785–1859), a prosperous Boston shoemaker, who is considered the "father" of probation. Starting in the early 1840s, Augustus volunteered to stand bail and assume custody for select, less serious offenders in exchange for the judge deferring the sentence. Augustus was responsible for monitoring offenders' activities and later reporting to the judge on their performance in the community. If the judge was satisfied with an offender's community performance, charges were dropped; if not, sentencing proceeded. Augustus received no pay for his 18 years of court work. He used his own money and voluntary contributions from others to finance his efforts.[6]

Influenced by the efforts of Augustus, Massachusetts passed the first formal probation law in 1878. By 1920, a majority of the states allowed probation. However, it was not until 1957 that all states had probation statutes. The federal probation system was established in 1925.[7] The first federal probation officer was appointed in 1927 in the District of Massachusetts.[8] In more than 170 years, probation has grown from the efforts of a volunteer in Boston into the most frequently used sentence in criminal justice.

Administration

As mentioned earlier in the chapter, probation is administered in many different ways across the nation. The federal government administers its own probation system under the Administrative Office of the Courts, and each state has responsibility for determining how to administer probation. In some states, such as Kentucky, probation is administered at the state level as part of a department of corrections or other state agency. In other states, such as California and Indiana, probation administration is a local function. Still other states combine state and local administration. Under *probation subsidy*, which became popular in the 1960s, states agree to financially support locally administered probation services in exchange for the localities not sentencing all their offenders to the state prison system. The goal is to give localities a financial incentive to retain offenders in their communities when possible.

There are other variations in the way probation is administered. Depending on the jurisdiction, administrative responsibility may lie with the executive branch, the judicial branch, or both. Adult and juvenile probation services can be administered separately or jointly, as can misdemeanor and felony services. Finally, probation administration can be combined with parole administration, or the two may be separate.

FYI **Federal Probation Officers**

As of September 15, 2019, the Administrative Office of the U.S. Courts employed 5,463 probation officers (down about 31% from the 7,874 in 2016, but up approximately 9% from the 4,995 in 2008). Federal probation officers supervise offenders placed on probation and supervised release and conduct presentence investigations to assess the risk to the community of future criminal behavior, the harm caused by the offense, the need for restitution, and the defendant's ability to pay restitution.

Sources: Personal communication (September 19, 2019, July 8, 2016, December 18, 2008).

Process and Procedures

A probation sentence can be viewed as a process with an identifiable beginning and ending. The process consists of three basic stages:

1. Placement of an offender on probation by a judge.
2. Supervision and service delivery for the probationer by probation officers.
3. Termination of the probation.

Before and during the process, probation agency staff members employ a number of important procedures, which will be examined in the following subsections.

Placement on Probation In deciding whether an offender should be sentenced to probation, a judge usually considers a host of factors, such as (1) statutes outlining eligibility for probation, (2) structured sentencing guidelines (in jurisdictions where they are used), (3) recommendations from the prosecuting and defense attorneys, (4) the offender's freedom or detention in jail before and during trial, (5) the presentence investigation report prepared by the probation agency, and (6) characteristics of the offender and offense. Judges ordinarily give great consideration to the seriousness of the current offense and the prior legal record of the offender. Cases involving more serious offenses or offenders with more extensive prior records are less likely to receive probation.

The Presentence Investigation The **presentence investigation (PSI)** is conducted by the probation agency at the request of the judge, usually during the period between the finding or plea of guilt and sentencing. In performing a PSI, a probation officer conducts an inquiry into the offender's past and current social and psychological functioning, as well as the offender's prior criminal record. The main tasks of the inquiry are to estimate the risk the offender presents to the community and to determine the offender's treatment needs. The probation officer obtains the necessary information by interviewing the offender and others who know the person, such as an employer, family members, and victims. The officer also reviews relevant documents and reports. The information is assembled into a PSI report, which is submitted to the court.

presentence investigation (PSI) An investigation conducted by a probation agency or other designated authority at the request of a court into the past behavior, family circumstances, and personality of an adult who has been convicted of a crime, to assist the court in determining the most appropriate sentence.

myth

An essential function of the PSI report is to individualize justice by helping judges structure their sentencing decisions around the offender's unique circumstances.

fact

In an important study, criminologist John Rosecrance found that experienced probation officers try to recommend sentences that are consistent with judicial expectations, given the seriousness of the offense and the offender's prior criminal record. Rosecrance found that officers selectively structure information in the report to support the recommendations they already have decided on. Information that is inconsistent with those recommendations is minimized in the report. Rosecrance's research implies that the PSI report merely gives the sentencing process a guise of individualized justice.

Source: John Rosecrance, "Maintaining the Myth of Individualized Justice: Probation Presentence Reports," *Justice Quarterly* 5 (1988), 235–56.

A sample PSI checklist appears in Table 12.2. There are no universally accepted standards for PSI report format and content. However, reports typically contain these basic elements:

1. A face sheet of identifying demographic data.
2. A discussion of the instant or current offense as perceived by the police, the victim, and the offender.
3. A summary of the offender's prior legal record.
4. An overview of the offender's past and present social and psychological functioning.
5. The probation officer's evaluation of the offender and the officer's recommendation for an appropriate sentence.

The probation officer must convey the essential information in a concise and objective manner that supports the sentencing recommendation being offered.

TABLE 12.2 Checklist for Completing a Presentence Investigation Report

FACT SHEET OF IDENTIFYING DEMOGRAPHIC INFORMATION
Name, aliases, and address
Physical description
Social Security number
Date and place of birth
School history, occupation, marital status, family members
INSTANT OFFENSE
Offense and docket number
Name(s) of codefendants, if any
Prosecutor and defense attorney (names and addresses)
Means of conviction (trial, plea agreement, etc.)
Sentencing date
PRESENTENCE INVESTIGATION
Defendant's version of events relating to offense
Victim impact, restitution needs
Jail time served in connection with offense
Previous record
Social history (family, friends, self-image)
Educational history
Marital history
Employment history
Economic situation
Religious association
Outside interests
Health (including drug and alcohol use)
Present attitudes of defendant
Plea agreement (if any)
Fiscal impact of sentencing option(s) to victim and state
Psychiatric insights
Statement of probation officer
Sentencing recommendation
PROBATION PLAN
Length of probationary period
Probationer reporting requirements
Job placement
Postemployment education training needs
Community service requirements
Other probation requirements

Once prepared, the PSI report serves a variety of functions. It is useful in formulating supervision and treatment plans for persons who are given probation sentences. It also serves as the baseline for progress reports on probationers. When an offender is sent to prison, the report helps prison officials learn about and make decisions about the offender. However, the most well-known and immediate function of the PSI report is to assist judges in arriving at a proper sentence.

In most cases, there is a high degree of consistency between the sentencing recommendation in the PSI report and the actual sentence handed down by the judge. As shown in the myth/fact box on page 494, however, this does not mean that judges are always strongly influenced by probation officers' sentencing recommendations. In some cases, probation officers recommend sentences that they believe are consistent with judicial expectations.

In addition to questions about the actual effect of PSI reports on sentencing decisions, another issue is whether courts should use PSIs conducted by private individuals and agencies rather than by probation agencies. Private PSIs became popular in the 1960s. Private individuals and agencies may conduct PSIs under contract with the defense counsel or the court. Advocates of privately prepared PSIs argue that privatization reduces probation agency workloads and saves tax dollars (when such PSIs are commissioned by the defendant). Opponents argue that improper sentences may be recommended, that private PSIs can discriminate against the poor, and that the PSIs can reduce the credibility and funding of probation agencies.

Aaron Roeth Photography

As part of a presentence investigation, a probation officer interviews an offender to obtain background information for the court. *What kinds of background information, if any, should influence the length and type of sentence? Why?*

The PSI report usually is the property of the court for which it is prepared. Once the report is submitted to the court, a copy often is provided to the defendant, defense counsel, and the prosecuting attorney. Both prosecution and defense have the opportunity to review the report for accuracy and to bring any concerns to the court's attention. This process is known as *disclosure*, and there has been controversy about it. Those who oppose the practice of disclosure maintain that people with important information may not make it available if the offender will see it and know who provided it. Those who support disclosure, on the other hand, argue that it is a matter of fairness for the offender to see and ensure the accuracy of the information on which the sentence is based.

The Probation Order

When an offender is sentenced to probation, the court files the appropriate documents indicating the length and conditions of the probation sentence. Although the length of probation sentences varies among jurisdictions and among cases, it is common for adults placed on probation for misdemeanor offenses to receive 1- to 2-year terms. Those convicted of felonies and sentenced to probation generally receive longer terms (e.g., 5 years). Some states allow sentences of lifetime probation. In some cases, the court has the authority to grant early discharge from probation for commendable performance or to extend the term of probation.

Probation conditions are rules that specify what an offender is and is not to do during the course of a probation sentence. They are of crucial significance because the success or failure of the probationer is evaluated with respect to those rules. There are two types of conditions. *Standard* or *general conditions* apply to all persons placed on probation and

probation conditions Rules that specify what an offender is and is not to do during the course of a probation sentence.

pertain primarily to control and supervision of the offender. Figure 12.2 shows the standard or general conditions of probation for the state of New Hampshire.

Special conditions are imposed at the discretion of the judge and probation officials and are designed to address the offender's particular situation. Special conditions frequently deal with treatment matters. For example, an offender may be ordered to

FIGURE 12.2 Rules of the Superior Court of the State of New Hampshire, Terms and Conditions of Probation

The terms and conditions of probation, unless otherwise prescribed, shall be as follows:

The probationer shall:

a. Report to the probation or parole officer at such times and places as directed, comply with the probation or parole officer's instructions, and respond truthfully to all inquiries from the probation or parole officer;
b. Comply with all orders of the Court, board of parole or probation or parole officer, including any order for the payment of money;
c. Obtain the probation or parole officer's permission before changing residence or employment or traveling out of State;
d. Notify the probation or parole officer immediately of any arrest, summons, or questioning by a law enforcement officer;
e. Diligently seek and maintain lawful employment, notify probationer's employer of his or her legal status, and support dependents to the best of his or her ability;
f. Not receive, possess, control, or transport any weapon, explosive, or firearm, or simulated weapon, explosive, or firearm;
g. Be of good conduct, obey all laws, and be arrest-free;
h. Submit to reasonable searches of his or her person, property, and possessions as requested by the probation or parole officer and permit the probation or parole officer to visit his or her residence at reasonable times for the purpose of examination and inspection in the enforcement of the conditions of probation or parole;
i. Not associate with any person having a criminal record or with other individuals as directed by the probation or parole officer unless specifically authorized to do so by the probation or parole officer;
j. Not indulge in the illegal use, sale, possession, distribution, or transportation, or be in the presence, of controlled drugs, or use alcoholic beverages to excess;
k. Agree to waive extradition to the State of New Hampshire from any State in the United States or any other place and agree to return to New Hampshire if directed by the probation or parole officer; and
l. Comply with each of the following, or any other, special conditions as may be imposed by the Court, the parole board, or the probation or parole officer:
    1. Participate regularly in Alcoholics Anonymous to the satisfaction of the probation or parole officer;
    2. Secure written permission from the probation or parole officer prior to purchasing and/or operating a motor vehicle;
    3. Participate in and satisfactorily complete a specific designated program;
    4. Enroll and participate in mental health counseling on a regular basis to the satisfaction of the probation or parole officer;
    5. Not be in the unsupervised company of minors of one or the other sex at any time;
    6. Not leave the county without permission of the probation or parole officer;
    7. Refrain totally from the use of alcoholic beverages;
    8. Submit to breath, blood, or urine testing for abuse substances at the direction of the probation or parole officer; and
    9. Comply with designated house arrest provisions.

Source: Victim Services, https://www.nh.gov/nhdoc/divisions/victim/pandp_offender.html.

participate in drug abuse counseling. Special conditions can be creative. For example, in Massachusetts, Connecticut, Texas, and a few other states, a special condition of probation for some offenders is participation in a reading program called "Changing Lives Through Literature." Eligible offenders agree to attend a class on a college campus each week for 10 or 12 weeks. They read and discuss such books as *The Old Man and the Sea*, *Deliverance*, *The Sea Wolf*, *Of Mice and Men*, and *One Flew over the Cuckoo's Nest*. If they complete the program, their probation is reduced, usually by 3 months. An evaluation of the first four classes in Massachusetts found that only 18% of the men who completed the class committed crimes again, compared with 43% in a comparable group of offenders.[9]

In recent years, it has become increasingly common for jurisdictions to include restitution orders as part of probation, as either a standard condition or a special condition. **Restitution**, you may recall, usually means that the offender provides either the victim or the community with money or work service.

restitution Money paid or services provided by a convicted offender to victims, their survivors, or the community to make up for the injury inflicted.

In some areas, the court later may amend or modify conditions outlined in the initial probation order. The conditions can be made more or less restrictive, depending on the probationer's behavior. However, to be legal, it is necessary for all conditions, regardless of when they are imposed, to be clear, reasonable, permitted by the Constitution, and related to the rehabilitation of the offender or the protection of society, or both.[10]

myth

Expanding community supervision is the best way to reduce incarceration.

fact

The conditions of community supervision, especially probation, often are so restrictive that they set people up to fail. Probation, in particular, leads to unnecessary incarceration; until it is reformed to support and reward success rather than detect mistakes, it is not a reliable "alternative."

Source: Wendy Sawyer and Peter Wagner, "Mass Incarceration: The Whole Pie 2019," Prison Policy Initiative, March 19, 2019. Accessed July 21, 2019, https://www.prisonpolicy.org/reports/pie2019.html.

Supervision and Service Delivery Once offenders have been placed on probation, the probation agency must shift attention to supervision and service delivery. At this point, probationers must be assigned to probation officers. In making assignments, it is important to match officer and probationer characteristics. For example, a manipulative probationer who poses a risk to community security might be assigned to an officer with a law enforcement background. An offender who poses little risk but has pressing treatment needs might be assigned to an officer with more of a helping background.

Because not all persons placed on probation require the same amounts and types of supervision and services, an important task is to determine what is appropriate for each client. In recent years, this task has been facilitated by the development of risk-and-needs assessment instruments such as the one shown in Figure 12.3. The probation agency more objectively can determine the amount and type of supervision a probationer requires by examining the risk score. For instance, a probationer with a risk score in the maximum range might receive weekly contacts from the probation officer, whereas a probationer with a minimum-range risk score might receive monthly contacts. Likewise, the amount and kind of services are directed by the needs score. The point is to ensure that the highest levels of supervision are reserved for probationers who present the greatest risk to the community and that the highest levels of treatment services are reserved for those who present the greatest needs.

Once risk-and-needs assessment has been completed, the probation agency can formulate supervision and treatment plans for its clients. In effect, those plans further specify the conditions of probation and describe what is expected of probationers. Some of the requirements might include weekly contacts with the probation officer, random drug tests, weekly Alcoholics Anonymous meetings, or obtainment of a GED. Once supervision and treatment plans are set, the monitoring of probationers and the periodic filing of progress reports for the court can begin.

Adults on Probation

At year-end 2016, 1 in 68 U.S. adults were on probation.

Source: Danielle Kaeble, "Probation and Parole in the United States, 2016," in *Bureau of Justice Statistics Bulletin*, U.S. Department of Justice (Washington, DC: U.S. Government Printing Office, April 2018), 3, Table 2. Accessed September 16, 2019, https://www.bjs.gov/content/pub/pdf/ppus16.pdf.

Termination of Probation Ultimately, the probation agency must make recommendations to the court about how probation is to be terminated. Clients who generally have fulfilled the conditions of their sentences and have served their terms are recommended for successful discharges. If a client has violated the conditions of probation, the probation agency may recommend **revocation**, which entails repealing the probation sentence and substituting a more restrictive sentence, such as a jail or prison term. If probation is revoked, the judge can impose any sentence, including incarceration that was authorized for the offense for which the offender was originally placed on probation. Furthermore, the time already spent on probation, even though it may be several years, does not count as credit against the new sentence.[11]

revocation The repeal of a probation sentence or parole, and substitution of a more restrictive sentence, because of violation of probation or parole conditions.

Revocation can be recommended for two general categories of violations. One category involves commission of new offenses. The second category, known as **technical violations**, involves failure to abide by the technical rules of the sentence. For example, a probationer

technical violations Failure to abide by the technical rules or conditions of probation or parole (e.g., not reporting regularly to the probation officer), as distinct from commission of a new criminal act.

FIGURE 12.3 Sample Risk and Needs Assessment Instrument

DEPARTMENT OF CORRECTIONS
Division of Community Corrections
Doc-502(Rev. 1/03)

ADMISSION TO ADULT FIELD CASELOAD
ASSESSMENT OF OFFENDER RISK

WISCONSIN

OFFENDER NAME — Last	First	MI	DOC NUMBER
DATE PLACED ON PROBATION OR RELEASED ON PAROLE IN WISCONSIN (MM/DD/YY)	AGENT LAST NAME		AREA NUMBER
FACILITY OF RELEASE		CODE	DATE COMPLETED (MM/DD/YY)

(Select the appropriate answer and enter the associated weight in the score column.)

Item	Weight	Answer	SCORE
Number of Address Changes in last 12 Months: (Prior to incarceration for parolees)	0 2 3	None One Two or More	____
Percentage of Time Employed in Last 12 Months: (Prior to incarceration for parolees)	0 1 2 0	60% or more 40% – 59% Under 40% Not applicable	____
Alcohol Usage Problems: (Prior to incarceration for parolees)	0 2 4	No interference with functioning Occasional abuse; some disruption of functioning Frequent abuse; serious disruption; needs treatment	____
Other Drug Problems: (Prior to incarceration for parolees)	0 1 2	No interference with functioning Occasional abuse; some disruption of functioning Frequent abuse; serious disruption; needs treatment	____
Attitude:	0 3 5	Motivated to change; receptive to assistance Dependent or unwilling to accept responsibility Rationalizes behavior; negative; not motivated to change	____
Age at First Conviction: (or Juvenile Adjudications)	0 2 4	24 or older 20 – 23 19 or younger	____
Number of Prior Periods of Probation/Parole Supervision: (Adult or Juvenile)	0 4	None One or more	____
Number of Prior Probation/Parole Revocations: (Adult or Juvenile)	0 4	None One or more	____
Number of Prior Felony Convictions: (or Juvenile Adjudications)	0 2 4	None One Two or more	____
Convictions or Juvenile Adjudications for: (include current offense, Score must be either 0,2,3, or 5.)	0 2 3 5	None of the Offense(s) stated below Burglary, theft, auto theft, or robbery Worthless checks or forgery One or more from the above categories	____
Convictions or Juvenile Adjudication for Assaultive Offense within Last Five Years: (An offense which involves the use of a weapon, physical force or the threat of force)	15 0	Yes No	____
		TOTAL	____

Total all scores to arrive at the risk assessment score

Source: Mike Eisenberg, Jason Bryl, and Tony Fabelo, *Validation of the Wisconsin Department of Corrections Risk Assessment Instrument* (June 2009, revised), 4, accessed February 25, 2016, https://csgjusticecenter.org/wp-content/uploads/2012/12/WIRiskValidationFinalJuly2009.pdf. Used with permission.

might fail to report regularly to the probation officer or might leave the jurisdiction without the officer's consent. A recommendation of revocation is not automatic in the event of a violation, even if the violation is a new crime. Probation agents have considerable discretion on this matter. If the violation is a serious crime, a recommendation of revocation is very likely. However, a less serious offense or a technical violation may result in a warning, a tightening of probation conditions (perhaps the addition of a nightly curfew), or a brief jail term. Many probation agencies tend to let technical violations and petty offenses accumulate, with warnings and condition modifications along the way. However, repeated and excessive violations of this nature usually result in revocation recommendations.

If the probation agency asks the court to consider revocation of probation, the court must work within the guidelines of case law established by the U.S. Supreme Court. Two landmark cases are *Morrissey v. Brewer* and *Gagnon v. Scarpelli.* The Court's 1972 ruling in *Morrissey v. Brewer* dealt with parole revocation, but by virtue of the 1973 ruling in *Gagnon v. Scarpelli,* it also applies to probation revocation. *Morrissey* established that revocation is to be a two-stage process. In the first stage, there must be an informal, preliminary inquiry to establish probable cause that a violation has occurred. If probable cause is established, then in the second stage, there must be a formal court hearing to determine whether the violation warrants revocation. Offenders have certain rights at both stages. They include the right to notice of the hearing and charges, the right to be present at the hearing and to present evidence and witnesses, and the right to a detached and neutral hearing body.

The *Gagnon* decision extended the requirements of *Morrissey* to include probationers and also addressed the issue of right to counsel, which the *Morrissey* decision did not address. In *Gagnon,* the Court ruled that there is no absolute right to counsel at revocation proceedings. Whether the offender is provided with counsel is determined on a case-by-case basis; if there is no compelling reason for providing counsel, counsel is unnecessary. However, the Court added that in probation revocation hearings, counsel should be provided when probationers present a timely claim that (1) they did not commit the violation or (2) there are mitigating circumstances making revocation inappropriate. In other situations, attorneys are allowed at revocation hearings as long as defendants provide their own, at no cost to the state.

John Patriquin/Portland Press Herald/Getty Images

Committing a new crime or violating the conditions of probation may lead to revocation of probation. *How serious should a new crime or violation of probation conditions be before probation is revoked?*

Issues in Probation

Probation is an evolving, changing field with many controversial issues. Several such issues already have been considered in this chapter. We now turn our attention to some additional issues.

Probation Fees Increases in the probation population have been accompanied by increases in the financial costs of probation. Consequently, a trend has emerged toward having probationers pay fees (e.g., $30 per month) to help offset the cost of their supervision and treatment. Probation fees–or supervision fees, as they are sometimes called–are distinct from fines, court costs, and restitution.

Advocates of probation fees argue that fees can help contain the increasingly high costs of probation. Advocates further contend that because the majority of probationers can afford reasonable monthly fees, it is only fair that those probationers should be held responsible for supporting at least part of the services they receive. However, critics charge that fees are unfair for indigent offenders and that offenders should not have to pay for services they are mandated to receive. Critics also claim that the administrative costs associated with collecting fees can exceed the amount of money collected. Given the skyrocketing number of persons on probation and the high cost of supervision and treatment, it seems unlikely that the trend toward probation fees will decrease in the near future.

Probation Fees

In 1929, Michigan became the first jurisdiction in the United States to impose probation supervision fees.

Source: www.appa-net.org.

Legal Issues: Confidentiality and Privacy Counseling often is an aspect of probation. It can be argued that to protect the counseling relationship, the information a probationer shares with a counselor should not be divulged to outside parties. However, it also can be argued that the counselor, who could be the offender's probation officer or a counselor to whom the probation officer has referred the offender, has a duty to divulge information about certain activities, particularly new crimes and technical violations, to appropriate authorities, such as police officers and judges. To what degree, then, should confidentiality govern the relationship between probation officer and probationer? In *Minnesota v. Murphy* (1984), the U.S. Supreme Court held that this relationship is not governed by the same degree of confidentiality as that between attorney and client or between physician and patient. Counselors taking probationer referrals can reveal information about a probationer's illegal activities to the probation officer, and the probation officer can notify the police.

Another issue is the conditions under which a probation officer may search a probationer's home for evidence. In its decision in *Griffin v. Wisconsin* (1987), the Supreme Court held that a search warrant based on probable cause is unnecessary for a probation officer to search a probationer's home; reasonable grounds is a sufficient basis for the search. This decision and the *Murphy* decision show that probationers generally are entitled to fewer due process protections than are free citizens who are not on probation. Whether this is just and fair is open to debate.

recidivism The return to illegal activity after release from incarceration.

Caseload and Recidivism It is not unusual for probation officers in larger urban jurisdictions to have as many as 200 offenders in their caseloads. Large caseloads, in turn, have been criticized for contributing to **recidivism**—the return of probationers to crime during or after probation. Clearly, it is difficult for probation officers with large caseloads to give each case individual attention and to provide proper supervision and services for all their clients.

Interestingly, however, research generally has failed to confirm that smaller caseloads, such as 25 cases per officer, are associated with decreased recidivism. In the words of criminologist Jay S. Albanese and his colleagues, "It appears from many studies that the simple expedient of reducing caseloads will not of itself assure a reduction of recidivism."[12] In fact, it is reasonable to suppose that caseload reduction sometimes may be associated with an *increase* in recidivism detected by probation officers because officers with smaller caseloads are able to scrutinize each probationer's activities more closely. There is some support for this supposition.[13] However, to the extent that (1) large caseloads promote recidivism because of the inability of probation officers to provide sufficient supervision and services and (2) smaller caseloads inflate recidivism numbers because of greater scrutiny of probationers' activities, probation agencies are in a no-win situation with respect to the issue of optimal caseload.

How effective is probation in controlling recidivism? First, note that researchers attempting to address this question confront a variety of difficulties. One difficulty is deciding whether to define probationer recidivism in terms of technical violations, arrests for new crimes, new convictions, or revocations. The definition employed affects the amount of recidivism uncovered. Another important difficulty lies in accurately determining whether it is the probation experience or some additional factor that is responsible for the recidivism observed among a group of probationers. A simple finding that recidivism is low among a group of probationers does not necessarily mean that probation is containing recidivism. Other factors, such as improvements in the local economy or the selection of low-risk offenders for probation, may be responsible. To date, many studies of probation effectiveness have not been designed well enough to rule out the role of factors other than the probation experience. Difficulties like these must be kept in mind when reviewing studies on probation recidivism.

According to data provided by the U.S. Justice Department's Bureau of Justice Statistics, at year-end 2016, about 60% of the approximately 1.6 million adults discharged from probation whose outcome was known had successfully met the conditions of their supervision. The percentage of probationers successfully discharged has remained relatively constant since 2005. Also, at year-end 2016, about 14% of probationers were incarcerated because they had committed a new offense, a rule violation, or for some other reason. Another 3% of probationers at year-end 2016 had absconded (escaped). The remainder were discharged to a warrant or a detainer, transferred to another probation agency, died, or for some other reason. These figures also have remained relatively constant over the previous decade.[14]

In general, the recidivism figures associated with probation do not seem substantially higher or lower than those associated with incarceration. (See Chapter 11 for data on recidivism following incarceration.) Accordingly, it can be argued that when feasible, probation should be the preferred sentence because it costs less than imprisonment. Of course, this logic holds true only if probation does not culminate in revocation followed by incarceration. If revocation and subsequent incarceration do occur, the combined costs of probation followed by incarceration may well exceed the cost of incarceration alone.

FYI

Expunge Conviction

A 2018 national survey found that 79% of registered voters strongly support providing first-time, low-level nonviolent offenders under the age of 25 the ability to expunge their conviction after successful completion of court-imposed probation; 18% do not support the proposal. The proposal is supported by 71% of Republicans, 80% of Independents, and 84% of Democrats.

Source: Robert Blizzard, "National Poll Results," *Public Opinion Strategies*, January 25, 2018, accessed June 3, 2019, https://www.politico.com/f/?id=00000161-2ccc-da2c-a963-efff82be0001.

Careers in Criminal Justice

Probation Officer

My name is Art Silkowski. I am the Director of Probation and Drug Court Coordinator for the 34th District Court in Romulus, Michigan. The 34th District Court is located in Wayne County and serves numerous cities and townships, as well as the Detroit Metro Airport. The 34th District Court has jurisdiction of all city ordinance violations and state law misdemeanors, including misdemeanors reduced from felonies.

I have a Bachelor of Arts degree with a concentration in criminal justice from Oakland University located in Rochester, Michigan. While in my last semester of college, I completed an internship as a probation officer at the 52-4 District Court in Troy, Michigan. Upon completion of the internship and graduation, I was hired at that same court where I remained for approximately 2 years. In 2000, I was hired as the Deputy Director of Probation at the 34th District Court and eventually promoted to Director of Probation and Drug Court Coordinator.

Although I supervise numerous probation officers and staff members, my primary role is serving the community as a probation officer. As a probation officer, I am responsible for conducting presentence investigation (PSI) interviews, alcohol screening and assessments (ASA) that are required by statute for drinking and drug use violations, and supervising defendants after sentencing. If a PSI or an ASA is required prior to sentencing, I will consider the information obtained from the interview, police reports, the defendant's criminal history and driving record, the blood and/or breathalyzer results (if applicable), and so on, when making a recommendation. Recommendations may include probation, incarceration, alcohol monitoring, GPS tethering, work program, community service, transitional housing, inpatient treatment, outpatient treatment, education, drug and alcohol testing, and so on. Recommendations of incarceration without probation may be made if the defendant cannot be monitored safely in the community. If sentenced to probation, I will formalize a plan of action to assist the defendant and promote compliance. If a defendant is not compliant with their probation terms, a violation of probation is generated. If this happens, I may be required to testify in open court as to the violation and/or to provide a written recommendation.

Drug court has the same concepts as probation; however, drug court is designed for those individuals in need of immediate intervention and intense supervision. As the Drug Court Coordinator, my role is to complete applications for certification, grant funding, putting together a "team" of individuals including a judge, probation officers, prosecutors, a court-appointed attorney, and treatment providers.

Art Silkowski

The most rewarding part of my job are the people who come back months and even years later with their lives completely turned around. For any student who is interested in a career in this field, whether it be probation or not, the advice I have to give is reach out to local courts, police departments, etc. and inquire about an internship (even if it is unpaid) to get your foot in the door.

THINKING CRITICALLY

1. What do you think makes an offender an ideal candidate for probation?
2. What are the benefits and drawbacks of probation versus traditional jail or prison time?

Parole

Recall that probation refers to the court-imposed sentence in which the offender, rather than being imprisoned, stays in the local community under the supervision of a probation officer. Two basic differences between probation and **parole** are that (1) parole is not a court-imposed sentence, and (2) parole is used with persons leaving prison. Some states do not use the term "parole," but instead call it "postprison transfer" or "earned release."[15] For purposes of definition, parole can be divided into two components. *Parole release* is one mechanism for releasing persons from prison. It involves releasing the inmate from prison, at the discretion of a parole board or similar paroling authority, before his or her sentence expires. *Parole supervision*, the aspect of parole that often is confused with

parole A method of prison release whereby inmates are conditionally released at the discretion of a board or other authority before having completed their entire sentences; can also refer to the community supervision received upon release.

SPECNER TIREY/AP Images

Some offenders are released on parole before the end of their sentences, on the condition that they remain law-abiding and follow the rules designed to control their movement and to help them adjust to society. *Should parole be abolished? Why or why not?*

probation, occurs after parole release. Essentially, parole supervision is a community-based continuation of the prison sentence. It involves supervision of the released offender in the community, often for a period roughly equal to the time remaining in the prison sentence.

Probation and parole supervision have similar features, which is why the two are sometimes confused. For example, both involve specific rules and conditions that offenders must follow to avoid revocation, and both entail providing offenders with supervision and services. In some instances, one officer may supervise both probationers and parolees. However, *probationer* and *parolee* are two distinct legal statuses. It is not uncommon for parole rules and conditions to be somewhat stricter and for officers to be less tolerant of violations committed by parolees; revocation of parole may be sought quickly, even for a technical parole violation. In addition, parolees often face greater adjustment problems because of the stigma attached to their prison records and because of the time they have spent away from the free community.

Just as there are different types of probation (suspended-sentence probation, split-sentence probation, and so on), there are two general types of parole. In *straight parole*, offenders are released from prison directly into the community under the supervision of the parole agency. In *residential parole*, offenders serve part of the parole term in a community residential facility or halfway house. There are two variants of residential parole. In the first variant, offenders are released from prison into the residential facility, where they spend a temporary, transitional period before returning home. The idea is to make the release process gradual. In the second variant, a person who violates the conditions of parole is kept on parole and is placed in the residential facility for a period of structured living rather than having parole revoked and being returned to prison.

There are four fundamental objectives of parole. Two of these, discussed earlier, also are objectives of probation: (1) to provide community safety and (2) to promote offender betterment and reintegration into society. The other two objectives of parole are more subtle, often unstated, but no less important: (3) to relieve and contain prison crowding and (4) to control the behavior of prison inmates.

Since its inception, parole has functioned as a "safety valve" for institutional corrections; crowding levels can be better contained if more inmates are granted early release. One of the clearest manifestations of this objective is the growing popularity of emergency release laws that permit executive authorities (usually governors) to accelerate parole eligibility for selected inmates when prison crowding reaches a particular level. Parole also gives prison officials some control over the behavior of inmates. The prospect of early release gives inmates an incentive to cooperate with prison officials and to avoid infractions of prison rules.

Historical Context

As with probation, parole emerged from earlier practices. One of those is the *tickets-of-leave* concept pioneered around the mid-1800s by Captain Alexander Machonochie off the coast of Australia and by Sir Walter Crofton in Ireland. Under this concept, inmates, after serving a portion of their sentences and exhibiting good performance in prison, could be granted tickets of leave, whereby they were released into the community under supervision. Release was made conditional on continuation of good behavior, so the practice also was known as conditional pardon.

The idea of releasing some prisoners early to community supervision and making release contingent on good conduct found its way to the United States. The concept initially was implemented, along with indeterminate sentencing, in the 1870s by Zebulon Brockway at Elmira Reformatory in New York. Brockway's system allowed prison officials to grant parole release to inmates they perceived as ready for release. Parole spread rapidly after its inception at Elmira. By the turn of that century, 20 states had provisions for parole; by 1920, the majority of states had adopted such provisions.[16]

Administration

It is helpful to divide parole administration into two areas. The first area is the parole board, or a similar paroling authority, which makes parole release decisions. The second is the parole field service agency, which provides parole supervision in the community after release.

Like probation administration, parole administration varies a great deal across the nation. At the federal level, the paroling authority is the U.S. Parole Commission, and the Administrative Office of the Courts administers field services. Only those offenders sentenced or under supervision before November 1, 1987, are still eligible for federal parole supervision. Federal parole has been abolished for anyone who has committed a federal crime after November 1, 1987. Offenders imprisoned for federal crimes after November 1, 1987, must serve their entire sentences, less a maximum of 54 days a year good time, if granted.[17]

Federal Parole Population

At year-end 2016, the federal parole population was 114,385; the rate per 100,000 U.S. adult residents was 46, compared to the 349 average rate for federal and state parolees combined. In 2016, exits from federal parole exceeded entries by 2,639 (48,108 v. 45,469).

Source: Danielle Kaeble, "Probation and Parole in the United States, 2016," in *Bureau of Justice Statistics Bulletin*, U.S. Department of Justice (Washington, DC: U.S. Government Printing Office, April 2018), 18, Appendix Table 5. Accessed September 16, 2019, https://www.bjs.gov/content/pub/pdf/ppus16.pdf.

Each state is responsible for administering its own parole system. Parole at the state level generally is an executive branch function, and each state has its own paroling authority. In about a third of the states, parole boards and field service agencies are administratively separate; in about half of the states, the board administers field services.[18] Whether probation and parole field services are jointly administered by the same government agency varies among the states. It has become quite common, however, for states to combine probation and parole administration. In Kentucky, for example, the same unit of the Department of Corrections administers both.

Another area of difference between states is whether the paroling authority or board is (1) administratively autonomous and independent of prison officials or (2) administratively consolidated with the department of corrections. Most parole boards are autonomous.[19]

In many ways, the parole board is the centerpiece of parole administration; the board is very influential in establishing a jurisdiction's parole policies. A national survey of parole boards found that in most states, the governor, subject to legislative confirmation, appoints parole board members. Appointment terms in most states are 4 to 6 years, and the vast majority of states use renewable appointments. Some states have term limits and, in most states, parole board members serve staggered terms. The professional qualifications required for appointment differ across states. Interestingly, law in many jurisdictions requires no minimum professional qualifications. The number of parole board members also varies by state, with five and seven members being common. A minority of states employs part-time parole board members.[20]

Process and Procedures

Besides helping establish the jurisdiction's parole policies, the parole board generally is responsible for managing parole release processes and making decisions to terminate parole supervision (which is discussed in a later section).

Parole Release There are two types of parole release: "discretionary parole" and "mandatory parole." With discretionary parole, "a parole board has authority to release prisoners conditionally based on a statutory or administrative determination of eligibility." In other words, with discretionary parole, parole boards have complete discretion to grant or deny parole. Discretionary parole generally is employed in states with indeterminate sentencing statutes. With mandatory parole, "a parole board releases an inmate from prison conditionally after the inmate has served a specified portion of the original sentence minus any good time earned." Put differently, with mandatory parole, a state requires a parole board to grant parole at a certain time. Mandatory parole generally is used in jurisdictions with determinate sentencing statutes. In 2016, 18 states used discretionary parole, 17 states used mandatory parole, and 15 states used a combination of discretionary and mandatory parole. Figure 12.4 shows the type of parole used by each of the 50 states. In those states that employ a combination of parole schemes, the type of parole used depends on the type of crime for which the inmate was convicted. Offenders in every state except Alaska are subject to life imprisonment without opportunity of parole upon conviction for certain crimes.[21]

FIGURE 12.4 Type of Parole Release by State, 2016

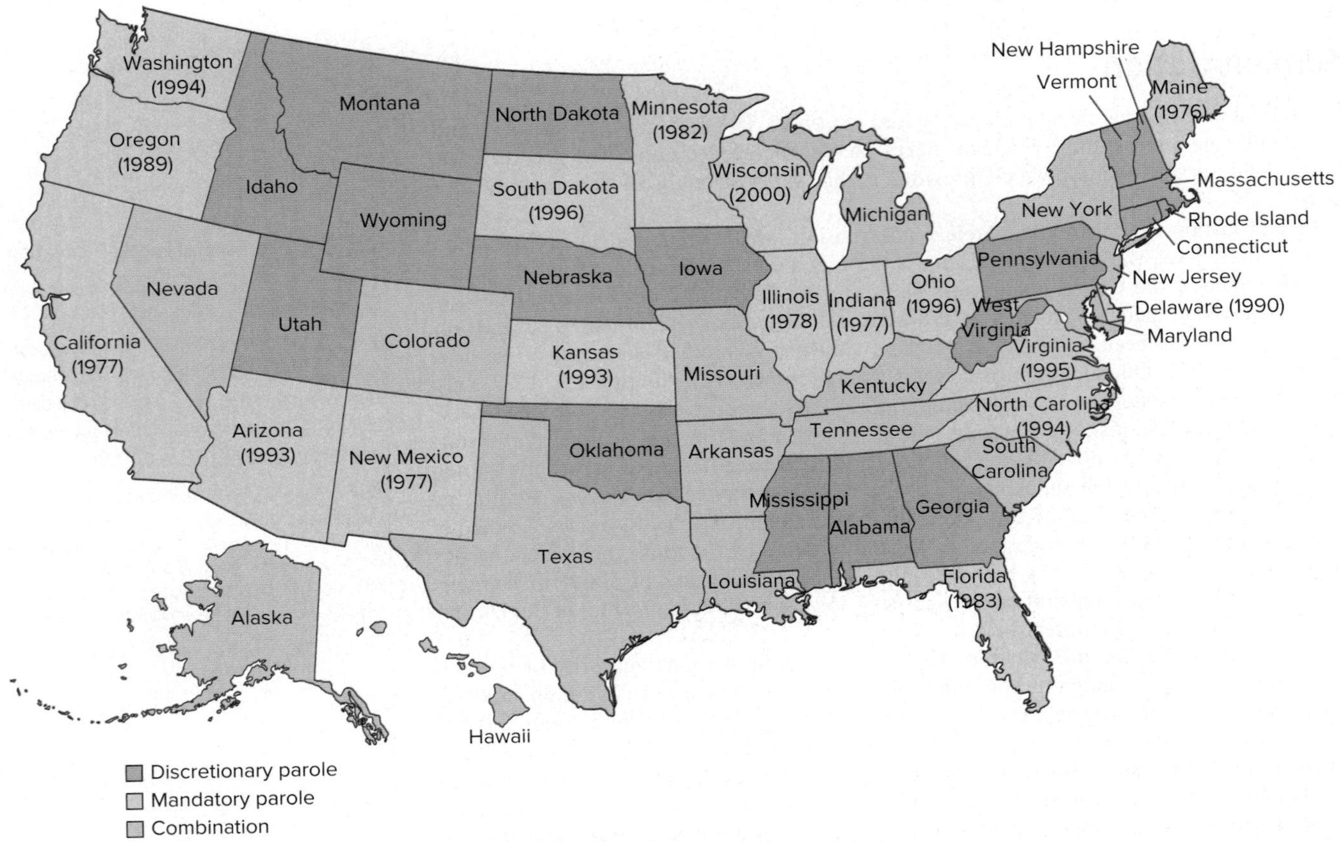

- Discretionary parole
- Mandatory parole
- Combination

Note: The years in parentheses represent the year that enabling legislation became effective.

Source: Edward E. Rhine

Parole Release

In 2016, nearly 62% of state inmates were released from prison by discretionary parole, while about 38% of the inmates were released by mandatory parole.

Source: Calculated from data in Danielle Kaeble, "Probation and Parole in the United States, 2016," in *Bureau of Justice Statistics Bulletin*, U.S. Department of Justice (Washington, DC: U.S. Government Printing Office, April 2018, 20), Appendix Table 6. Accessed September 16, 2019, https://www.bjs.gov/content/pub/pdf/ppus16.pdf.

Prior to appearing before the parole authority for a parole-grant hearing, a prisoner must first become eligible for parole and complete a parole plan describing such things as where he or she plans to live and work after release. In addition, prison staff members usually prepare a pre-parole report for the parole board. The report summarizes the characteristics of the inmate and his or her offense, reviews the inmate's adjustment to prison and his or her progress toward rehabilitation, and presents the inmate's parole plan. Reports sometimes contain a recommendation about whether the inmate should be paroled.

Depending on the jurisdiction and the particular case, the actual parole-grant hearing can be conducted by the full parole board, a partial board, or representatives of the board known as examiners or hearing officers. Roughly half of the states rely on hearing officers, who generally either make recommendations or prepare case summaries for parole board members.[22] Other concerned parties, such as prison staff, prosecuting attorneys, and victims, also may be present. Parole hearings typically are quite short and routine. Parole authorities review relevant documents, such as the inmate's parole plan, the pre-parole report, and victim statements, and often interview the inmate. Inmate interviews may be held in person, by video conference, or, far less likely, by telephone. Authorities then vote on whether to grant or to deny parole release. Parole decisions may be made by one board member, two board members, a panel that combines a board member with staff (e.g., a hearing officer), a panel of three board members, a majority of a full board, a super-majority of a full board, or a unanimous board. In addition, the number of persons required to take action may vary depending on the nature of a case or particular crime. In about half of the states, the inmate is notified of the outcome of voting at the hearing or immediately thereafter. In the remaining states, notification timeframes range from 1 week to more than 30 days. In most states, inmates are notified both verbally and through a written letter or document. Some states provide notification in writing only, while only one state provides notification solely by verbal communication.[23] If parole is denied, most states require the inmate to be provided with a written reason or reasons for the denial. In a majority of states, this information is made available to the public.[24] In most states, inmates are entitled to appeal or to request the parole board to reconsider its decision to deny release.[25] If parole is denied, the parole board in many states has the authority to order the inmate to serve the remainder of his or her sentence without additional hearings; in other states the inmate automatically is eligible for a future hearing. Regarding the latter, in April 2012, Charles Manson, then 77, was denied parole by a California parole board for the twelfth time. He was sentenced to death in 1970 for his role in the 1969 Tate-LaBianca murders. When the Supreme Court declared capital punishment unconstitutional in 1972, his death sentence was commuted automatically to life imprisonment with opportunity for parole. His next parole hearing was scheduled for 2027, but he died November 19, 2017, at age 83.[26]

The vast majority of parole authorities use structured instruments called **parole guidelines** to estimate the probability of recidivism and to direct their release decisions.[27] Those guidelines are similar to the sentencing guidelines used by judges and the risk-and-needs assessment instruments used in probation work. A sample parole guidelines risk worksheet and parole guidelines instrument are presented in Figure 12.5. The risk instrument accurately predicted that 39% of 33,000 offenders released, either discharged or paroled, from Georgia's prisons over a 3-year period were at high risk of being reconvicted of a new crime.[28]

parole guidelines Structured instruments used to estimate the probability of parole recidivism and to direct the release decisions of parole boards.

Parole authorities consider a variety of factors in determining whether to grant or deny an inmate parole. Those factors tend to be assigned different levels of importance by different parole board members. A recent analysis of national data found a near consensus among paroling authorities on 19 factors. They are listed below in the order of the number of states that consider the factor in their release decision making. In parentheses following the factor is the parole board chairs' ranking of the factor's importance.

1. Nature of the present offense (1); Severity of current offense (2)
2. Prior adult criminal record (3)
3. Institutional program participation (6)
4. Psychological report (10)
5. Inmate's disciplinary record (4)
6. Risk assessments or reports (5)
7. Previous parole adjustment (8)
8. Victim input (9)
9. History of illegal drug use (-)
10. Inmate's disposition or demeanor at hearing (13)
11. Previous probation adjustment (12)
12. Inmate family input (16)
13. Inmate testimony (14)
14. Prior juvenile criminal record (-)
15. Age at first conviction (-)
16. Treatment reports or discharge summaries (11)
17. Prosecutor input (17)
18. Offender's case plan as prepared by institutional staff (7)
19. Sentencing judge input (15)[29]

The most important factor is the nature or severity of the current or present offense, followed by prior adult criminal record. Note three factors considered by parole boards in their releasing decision making but not listed among parole board chairs' important factors are history of illegal drug use, prior juvenile criminal record, and age at first conviction.

The U.S. Supreme Court addressed due process in parole release decisions in *Greenholtz v. Inmates of the Nebraska Penal and Correctional Complex* (1979). Noting that parole release is an act of grace, the Court held that release on parole is distinct from parole revocation. Thus, the Court declined to apply the provisions outlined in *Morrissey v. Brewer*, discussed earlier in this chapter, to parole release. Consequently, what constitutes acceptable due process in release decision making is largely case specific. Most states, however, allow inmates to be represented by attorneys.

FIGURE 12.5 Georgia Parole Guidelines Risk Worksheet and Parole Decision Guidelines Instrument

Parole Guidelines Risk Worksheet

Risk Factor	Points
Current Prison Conviction Offense(s)*-score all types	
Other Offense (minimum score, if no conviction listed below)	0
Burglary	1
Drug Possession	1
Forgery	2
Obstruction	3
Theft	5
**If multiple convictions for the same offense, score only once per offense type.*	
Number of Felony Conviction Counts (including current prison episode)	
None	0
1–2	1
3	2
4–5	3
6–7	4
8	5
9 or more	6
Number of Prior Prison Incarceration Episodes	
None	0
1	1
2	3
3	4
4 or more	6
Age at Prison Admission (current prison episode)	
up to 20 years	0
over 20 to 40 years	–1
over 40 years	–2
History of Drug or Alcohol Problem	
No	0
Yes	2
Employed at Time of Arrest (full or part time)	
No	0
Yes	–3

Total Points	Risk Group
Up to 2 points	Low
3–6 points	Medium
7 or more points	High

Parole Decision Guidelines, Time to Serve GRID

Risk	Low (–5 to +12)			Medium (3–6)			High (7+)			Risk
CSL	Low	Mid	High	Low	Mid	High	Low	Mid	High	CSL
1	15	**17**	19	17	**20**	22	20	**22**	26	1
2	18	**20**	22	20	**22**	24	24	**26**	28	2
3	20	**22**	24	22	**24**	28	26	**28**	32	3
4	22	**24**	26	24	**28**	34	28	**32**	38	4
5	30	**34**	40	34	**42**	52	40	**50**	60	5
6	36	**40**	52	40	**50**	60	52	**65**	78	6
7	40	**44**	60	48	**60**	78	60	**76**	102	7
8	65% of sentence			75% of sentence			90% of sentence			8

GRID effective for cases considered on or after January 1, 2008

Notes: Risk = risk to re-offend; CSL = crime severity level (1 = least severe; 8 = most severe); numbers in figure represent months.

Source: "Parole Consideration, Eligibility & Guidelines," accessed July 6, 2016, http://pap.georgia.gov/parole-consideration-eligibility-guidelines.

In about half of all states, parole release remains an act of grace (at least for some crimes), and parole boards retain complete discretion over the parole decision. However, in the other half of states, as noted, parole boards are required to parole inmates after they have served a specific length of time, providing the inmates have met certain conditions or accomplished specific goals, such as successfully completing a program or meeting the requirements of the parole plan. In those states in which the parole board's enabling statutes or regulations create an expectation of release by indicating that the parole board "must" or "shall" release an inmate if certain conditions are met, for example, the board must follow various due process requirements before parole can be denied, such as finding that the inmate had not fulfilled the conditions for release or complied with stated expectations. Some states do not have an expectation of release but, instead, offer an increased probability of release if certain conditions are met. States that offer an increased probability of release hope that it will motivate inmates to comply with programming or intervention goals.[30]

Once inmates receive parole release, they begin the period of parole supervision, discussed shortly, which continues until it is terminated. In addition to its responsibility for release decisions, the parole board is responsible for parole discharge decisions (which are discussed later). In most states, the board can discharge individuals from parole supervision; this commonly is done upon the recommendation of the parole supervision agency. Alternatively, if no board action is taken, the individual is discharged from parole upon expiration of the legal sentence.

U.S. Adults on Parole

At year-end 2016, 1 in 287 U.S. adults was on parole.

Source: Danielle Kaeble, "Probation and Parole in the United States, 2016," in *Bureau of Justice Statistics Bulletin*, U.S. Department of Justice (Washington, DC: U.S. Government Printing Office, April 2018), 3, Table 2. Accessed September 16, 2019, https://www.bjs.gov/content/pub/pdf/ppus16.pdf.

Field Services—Supervision and Service Delivery In about half of all states, the parole board or paroling authority exercises full authority for parole supervision; in about a quarter of states, parole boards have partial authority; and, in the other quarter of states, parole boards have no authority or jurisdiction for parole supervision, which is the responsibility of the parole field service agency. In about half of all states, the parole board recommends a specific level of supervision for individual cases; in the other half of states it does not. In nearly all states, the parole board sets the standard and special conditions that govern parole supervision; in a few states, this is done by the field service agency. Parole boards can impose numerous conditions on a parolee. The five most common standard conditions are:

1. Obey all federal, state, and local laws.
2. Notify the parole officer of any change in residence.
3. Report to the parole officer as directed and answer all reasonable inquiries by the parole officer.
4. Refrain from possessing a firearm or other dangerous weapons, unless granted written permission.
5. Permit the parole officer to visit the parolee at home or elsewhere.

Other common standard conditions include:

6. Comply with requests for drug testing.
7. Obey all rules and regulations of the parole supervision agency.
8. Pay all court-ordered fines, restitution, or other financial penalties.
9. Maintain gainful employment.
10. Abstain from association with persons with criminal records.
11. Pay supervision fees.
12. Abstain from alcohol or frequenting bars.[31]

Parole Supervision Fees

Most states have parole supervision fees, but most of these states can waive them for the parolee. Monthly fees vary widely, ranging from a minimum of $10 to a maximum of $100; the overall average is $30–$35. A few states adjust fees on a sliding scale based on the parolee's economic situation.

Source: Ebony L. Ruhland, Edward E. Rhine, Jason P. Robey, and Kelly Lyn Mitchell, "The Continuing Leverage of Releasing Authorities: Findings from a National Survey," University of Minnesota, Robina Institute of Criminal Law and Criminal Justice, 2017, accessed September 13, 2019, https://robinainstitute.umn.edu/sites/robinainstitute.umn.edu/files/parole_executive_summary_web_11-15.pdf.

Nearly all states allow parole boards to modify parole conditions during the period of supervision; several states also allow parole officers or parole field service agencies to modify conditions, usually with board approval.[32] Nearly half of all states require parolees to serve a minimum period of time before they become eligible for final discharge. In a majority of states, the time parolees must serve usually is the period between their date of release and the expiration of the maximum prison sentence. For a majority of states, the length of supervision is not fixed.[33]

What has remained constant through the years is the basic task of field service agencies–that is, to provide control and assistance for persons reentering the community from prison. Virtually everything that was stated earlier in this chapter about probation supervision and service delivery applies to parole supervision and service delivery. The subtle differences between the two are overshadowed by the many similarities. However, one difference warrants mention. Specialization is more common in parole than in probation supervision. Under specialization, offenders who pose a similar threat to public safety or those who share similar treatment needs are grouped together and assigned to the same officers. Specialization by parole officers is used in slightly more than half the jurisdictions. The two most common areas for specialization are with sex offenders and substance-abusing offenders.[34]

Social scientist Richard McCleary conducted what is considered a classic sociological study of a parole supervision agency.[35] In his book *Dangerous Men*, McCleary observes that parole officers are very concerned with avoiding "trouble" from parolees. "Trouble" is defined broadly as anything that runs counter to the status quo of the parole agency. Trouble can threaten the agency's public image, thus bringing the agency pressure—and possibly increased structure—from political officials. The likelihood of trouble is reduced to the extent that officers can anticipate what clients might do in response to threats of various punishments and rewards from officers. "Dangerous men" are parolees who do not respond predictably to officers' threats and promises. Dangerous men are not necessarily prone to violent behavior; they are merely unpredictable. Because unpredictability implies possible trouble, officers try to identify dangerous men as soon as possible in the supervision process. The dangerous-man label is used sparingly because it means greater supervision and more documentation and paperwork for officers. However, the label can protect the supervising officer and the agency from subsequent criticism because, early on, it conveys that problems may be expected from a particular parolee. Concerned parties in the parole bureaucracy are alerted to what might happen, and the potential for "surprises" is reduced. Once parolees have acquired the dangerous-man label, the agency is quick to seek reasons to revoke their paroles. McCleary's research underscores the crucial significance of agency bureaucratic dynamics in shaping supervision practices.

Parole Discharge and Revocation In most states, parole boards grant final discharge from parole, and, in many states, the parole board has the power to grant an early final discharge from parole (before maximum expiration of sentence).[36] In most states, parolees can be discharged from parole even if they have not paid all fines and other fees, assuming they have completed all other conditions of supervision. However, in several states, they cannot be discharged from parole if they have outstanding financial obligations.[37]

Parole revocation also is the responsibility of the parole board and can occur in response to new crimes or technical violations. In nearly all states, a violation of any condition of parole constitutes grounds for parole revocation.[38] Parole revocation has long been recognized as a major contributor to prison-overcrowding and mass incarceration problems.

myth

If parole is revoked, return to prison is automatic.

Though reincarceration is common, other options usually are possible, such as placement in a halfway house or reinstatement of parole. Much discretion is involved.

As is true in the field of probation, parole officers enjoy considerable discretion when deciding whether to recommend revocation for violations. However, many parole officers and their field service agencies now use structured decision tools or progressive sanction grids or guidelines to aid them in their decision making. The two most common grid factors are the seriousness of the violation and the parolee risk level.[39]

Parole revocation is quite similar to probation revocation and is governed by the same case law (*Morrissey v. Brewer* and *Gagnon v. Scarpelli*). If revocation is sought, the two-stage hearing process required under *Morrissey* becomes applicable (see the discussion of probation revocation for details).

If parole is revoked, parole boards may impose one of several possible consequences, including those listed in a progressive sanctions grid or actuarial instrument, if the board uses one. If parole is revoked, a return to prison is not mandatory in a majority of states. Among possible consequences, ranked by the number of states for which it is an option, are:

1. Restore to parole status, modify conditions.
2. Restore to parole status, no change.
3. Reincarcerate for original term.
4. Send to an in-prison treatment program.
5. Restore to parole status and place in a community-based treatment facility.
6. Incarcerate for short jail term.
7. Discharge from parole.
8. Extradite to another state to serve new sentence.

9. Restore to parole status, extend term of supervision.
10. Incarcerate for new term.
11. Reincarcerate for balance of unfinished maximum sentence; parole board retains option to determine if and/or when another parole consideration will be held.
12. Add absconder time to extend term.
13. Cancel next review date for possible re-release.
14. Send to prison for remainder of mandatory parole period.[40]

Parole Issues

Since the 1970s, discretionary parole release has been among the most controversial issues in criminal justice. Proponents of parole release argue, for example, that early release provisions are essential for controlling prisoners' behavior and for containing institutional crowding. However, several criticisms have been directed at parole release. Some critics claim that parole undermines both retribution and deterrence because offenders are permitted to leave prison early, sometimes many years before finishing their maximum sentences. Similarly, critics argue that because prisons generally fail to reform offenders and it is impossible for parole boards to accurately predict which offenders will commit new crimes, parole does not sufficiently guarantee public safety. Other critics believe that parole is unfair to offenders. Those opponents charge that parole leads to significant disparities in time served in prison for offenders who should be serving equal amounts of time. For example, two offenders in the same jurisdiction with very similar prior legal backgrounds may be convicted of committing armed robbery under very similar circumstances. One offender may serve 5 years in prison and receive parole, whereas the other may serve twice as long before release. Another contention of critics is that linking the degree of participation in prison treatment programs to the possibility of early parole amounts to subtly coercing inmates into programs that often are of questionable effectiveness.

Those and other criticisms helped persuade a number of jurisdictions to curtail discretionary parole release. Instead, they moved from indeterminate sentencing to determinate sentencing, decreased reliance on parole release and increased reliance on mandatory release, and devised both sentencing and parole guidelines. There were even forceful calls and some initiatives undertaken to completely abolish parole release, although the movement to abolish parole seems to have peaked.[41]

Besides the issue of whether parole release should exist, there are a number of other pressing concerns surrounding parole. They include legal issues, strained parole resources, and parolee adjustment and recidivism.

Mikael Karlsson/Alamy Stock Photo

Parole officers frequently meet their clients at the clients' workplaces. *What problems, if any, might be created by this practice?*

National Parole Resource Center

To learn more about parole in the United States, visit the National Parole Resource Center website at https://nationalparoleresourcecenter.org.

What parole issues currently appear to be most important?

Legal Issues There are two general categories of legal issues that relate to parole: (1) parolees' civil rights and (2) the liabilities of parole officials.

As noted previously, individuals forfeit a variety of civil rights when they are convicted of a felony. Some of those include the right to vote, the right to hold public office and certain other jobs, the right to jury service, and the right to obtain some types of licenses and insurance. Although the specific rights forfeited vary by jurisdiction, in many instances, parolees must seek court action to restore their rights following release from incarceration. Only a few jurisdictions grant their parole boards authority to restore ex-offenders' rights.[42]

People who support denying released inmates certain rights argue that the practice is necessary to preserve public safety and maintain high moral standards in society. Opponents argue that it is unfair to continue penalizing individuals who have paid their debts. They point out that denial of civil liberties contributes to poor adjustment on parole. Despite continued resistance in some states to restoring ex-felons' civil rights, other states have embraced a "smart justice" model of criminal justice (see discussion in Chapter 14) that advocates several progressive policies, including the restoration of ex-felons' civil rights, especially their right to vote. Whether "smart justice" and civil rights restoration policies are adopted more widely remain to be seen.

Can parole board officials and field service agents be held legally liable if a parolee's actions cause harm to a victim or victims? The answer is a qualified yes. Even though the vast majority of paroling authorities enjoy some type of immunity from liability (through constitution, statute, or case law), such immunity has been eroding in recent years.

According to John Watkins Jr., attorney and criminal justice professor, three central elements must be proven to establish liability on the part of parole and probation officials.

1. The existence of a legal duty owed the public.
2. Evidence of a breach of the required standard of duty.
3. An injury or damage to a person or group proximately caused by the breach of duty.

Watkins further observes that as a general rule, tort liability applies to injuries, or damages, caused by the negligent or improper performance of ministerial functions but not to injuries caused by unsatisfactory performance of discretionary functions. A *discretionary* function is one "that may involve a series of possible choices from a wide array of alternatives, none of which may be absolutely called for in a particular situation." A *ministerial* function is "one regarding which nothing is left to discretion–a simple and definite duty imposed by law, and arising under conditions admitted or proved to exist."[43] An example of a breach of a ministerial function would be if a parole board was to completely disregard parole guidelines that the board legally is bound to consider before releasing inmates. Thus, if parole and probation officials wish to minimize the likelihood of being held liable, those officials should determine what legally is required of them in their jurisdiction and make good-faith efforts to abide by those requirements.

Strained Parole Resources Parole supervision agencies across the nation currently are attempting to manage large numbers of parolees. At year-end 2016, for example, approximately 874,800 adults were under active parole supervision. That represents a 0.5% increase from the year before and an increase of about 10% over the total number of parolees at year-end 2006.[44] The parole population was the only correctional population to increase from 2007 through 2016. It increased by about 48,700 parolees during the period, partially offsetting the overall decline in the correctional population during the 9-year period.[45] Table 12.3 shows the states with the largest parole populations and the highest and lowest rates of persons under parole supervision per 100,000 adult U.S. residents at year-end 2016. The U.S. government had a parole population of 114,385 at the end of 2016.[46] Moreover, an increasingly large proportion of parolees has been released from prison early because of pressures to relieve institutional crowding rather than because those persons have been judged good candidates for parole. Such parolees often require above-average levels of supervision and services. Because most jurisdictions are facing pressure to expand prison space, it is not surprising that parole resources, such as budgets and staffing, have failed to keep pace with those developments.

The long-term implications of this predicament often are unappreciated. Parolee recidivism normally is quite high, partly because of the strained resources of parole agencies. The typical public and political response to high parolee recidivism is to demand a "get-tough" stance that culminates in the return of large numbers of parolees to prison. This exacerbates

TABLE 12.3 States with the Largest Parole Populations and the Highest and Lowest Rates of Persons Under Parole Supervision per 100,000 Adult U.S. Residents at Year-End 2016

States with the Largest Parole Populations		States with the Highest Rates of Supervision		States with the Lowest Rates of Supervision	
Texas	111,287	Pennsylvania	1,097	Maine	2
Pennsylvania	111,087	Arkansas	1,038	Virginia	25
California	93,598	Louisiana	864	Florida	27
New York	44,426	Texas	537	Massachusetts	34
Louisiana	30,907	Wisconsin	453	Delaware	52
Illinois	29,428	Kentucky	448	Rhode Island	54
Oregon	24,711	South Dakota	410	Oklahoma	64
Arkansas	23,792	Idaho	402	Nebraska	76
Georgia	22,386	Mississippi	381	South Carolina	112
Wisconsin	20,401	Missouri	377	Connecticut	119

Source: Danielle Kaeble, "Probation and Parole in the United States, 2016," in *Bureau of Justice Statistics Bulletin*, U.S. Department of Justice (Washington, DC: U.S. Government Printing Office, April 2018), 18–19, Appendix Table 5, accessed September 16, 2019, https://www.bjs.gov/content/pub/pdf/ppus16.pdf.

both prison crowding and the tendency to rely on parole release to relieve crowding. The result is further strain on parole resources and, consequently, further recidivism.

Parolee Adjustment and Recidivism A key to successful community adjustment for inmates leaving prison is **reintegration** or **reentry**, which means rebuilding former prosocial ties to the community and establishing new ties. An important part of reintegration or reentry is finding a satisfactory job and obtaining adequate subsistence funds through legal means.

reintegration or **reentry** The process of rebuilding former ties to the community and establishing new ties after release from prison.

The stigma of a prison record can result in grim employment prospects, especially without the assistance of the parole agency. Yet, research suggests that obtaining employment improves adjustment to community life. In a study of the effects of postrelease employment on the emotional well-being of a sample of ex-felons in Texas and Georgia, it was found that employment enhanced emotional well-being by providing wages and improving perceptions of self-worth. The research discovered that sustained unemployment caused emotional distress. This distress reduced the motivation of releasees to search for jobs. In the words of the study's author, "Unemployment fed on itself by creating psychological stress which in turn reduced effectiveness in finding work."[47] Related evidence suggests that providing newly released ex-inmates with temporary unemployment benefits can decrease the recidivism rate, although it is important that such benefits do not create a work disincentive.[48]

States vary widely in their rates of parole success, which can be attributed to a number of factors such as variations in the parole populations (e.g., age at release, criminal history, and most serious offense), the level of supervision, and parole agency policies on the revocation of technical violators. Research shows that inmates released from prison on parole for the first time are more likely to succeed than re-releases. Re-releases are inmates who are being discharged from prison on parole who have served time for a previous violation of parole or conditional release or a new offense committed while under previous parole supervision.

However, the success rate for first-time releases declines the longer the inmate has been in prison. Success rates also vary by method of release. As noted previously, state inmates released by a parole board (*discretionary parole*) almost always have a higher rate of success than inmates released through *mandatory parole* (i.e., inmates who generally are sentenced under determinate sentencing statutes and are released from prison after serving a portion of their original sentence minus any good time earned). There also is a temporal element to parole success or failure. The chances of failure generally are highest during the early stages of parole. The first year appears to be the most critical if a successful discharge from parole is going to occur. Finally, parole success and failure have

FYI **Characteristics of Adults on Parole**

At year-end 2016, 87% of all parolees in the United States were male and 13% were female. Forty-five percent were white (non-Hispanic), 38% were black (non-Hispanic), 15% were Hispanic, and the remainder were of other races. Thirty percent of parolees had committed violent offenses, 21% had committed property offenses, 31% had committed drug offenses, and 17% had committed public order or other offenses.

Source: Danielle Kaeble, "Probation and Parole in the United States, 2016," in *Bureau of Justice Statistics Bulletin,* U.S. Department of Justice (Washington, DC: U.S. Government Printing Office, April 2018), 24, Appendix Table 8, accessed September 16, 2019, https://www.bjs.gov/content/pub/pdf/ppus16.pdf.

varied by parolee gender, race or ethnicity, age, and most serious offense for which he or she was incarcerated.

At year-end 2016 (the last year for which these data were available), 428,022 adults exited parole: 89% exited from state supervision and 11% exited from federal supervision.[49] Approximately 57% of all parolees discharged in 2016 successfully completed supervision—a substantial increase from the approximately 42% to 49% of parolees who successfully completed supervision and were discharged during the 1990s.[50] Twenty-seven percent of the 2016 parolees failed to complete supervision successfully and were returned to incarceration. Of those returned to incarceration, 60% simply had their parole revoked, 27.3% were returned with a new sentence, 2.4% were returned to receive treatment, and 10.3% were returned for another or unknown reason. Another nearly 2% of parolees absconded (escaped from supervision), about 1.5% of parolees died while under supervision, approximately 5% of parolees failed to successfully complete supervision but were not returned to incarceration, and for roughly 7.5% of parolees, their outcome was unknown or not reported.

THINKING CRITICALLY

1. How much weight do you think victim impact statements should get in the decision to grant parole?
2. Given what you have read, do you think there are ways to improve the parole process? If so, what can be done?

Intermediate Sanctions

The intermediate-sanction movement arose in the 1980s as a response to the prison overcrowding crisis. Offenders who would have been sent to—or remained in—prison had sufficient space been available now were being sentenced to these intermediate sanctions. At the same time, community corrections, of which intermediate sanctions are a part, was experiencing a general decline in philosophical support for rehabilitation, and a growing emphasis on punishing and controlling offenders.

intermediate sanctions Sanctions that, in restrictiveness and punitiveness, lie between traditional probation and traditional imprisonment or, alternatively, between imprisonment and traditional parole.

In restrictiveness and punitiveness, **intermediate sanctions** lie somewhere between traditional probation and traditional imprisonment or, alternatively, between imprisonment and traditional parole supervision (see Figure 12.6). Intermediate sanctions are designed to widen the range of incarceration alternatives and to calibrate that range according to the differential risks and needs of offenders.

As a rule, the newer intermediate sanctions are oriented less toward rehabilitation and more toward retribution, deterrence, and incapacitation than older community correctional programs. However, the distinction between intermediate sanctions and older, more traditional programs has become somewhat blurred because some traditional programs, such as restitution and halfway houses, have been incorporated into the intermediate-sanction category. The specific intermediate sanctions to be discussed in this section include intensive-supervision probation and parole, day reporting centers, structured fines (or day fines), home confinement and electronic monitoring, and halfway houses.

intensive-supervision probation and parole (ISP) An alternative to incarceration that provides stricter conditions, closer supervision, and more treatment services than do traditional probation and parole.

Intensive-Supervision Probation and Parole

Intensive-supervision probation and parole (ISP) provides stricter conditions, closer supervision, and more treatment services than traditional probation and parole. Offenders often

FIGURE 12.6 Relationship of Intermediate Sanctions to Traditional Sanctions

No Sanction	Traditional Probation	Intermediate Sanctions	Traditional Imprisonment	Intermediate Sanctions	Traditional Parole Supervision

are selected for ISP on the basis of their scores on risk-and-needs-assessment instruments. Alternatively, they may be placed on ISP after violating regular probation or parole. Most ISP programs are for nonviolent felons.

The ISP programs in existence across the nation are diverse, and some programs come much closer than others to providing supervision and services that are genuinely intensive. In theory, ISP programs have the following features:

1. Specially trained intensive-supervision officers with small caseloads (e.g., 25 cases per officer)
2. Inescapable supervision, such as multiple weekly contacts between officers and clients and frequent testing for drug use
3. Mandatory curfews
4. Mandatory employment or restitution requirements, or both
5. Mandatory or voluntary participation in treatment
6. Supervision fees to be paid by clients

ISP usually lasts 6 months to 2 years. Typically, offenders must pass through a series of phases that become progressively less restrictive during the period of intensive supervision.

The majority of states have implemented ISP programs for adult probationers and parolees.[51] However, of all persons placed on probation or parole, relatively few are placed under intensive supervision because of resource limitations and the small caseloads required by ISP. Although it is commonly argued that ISP costs less money than incarceration, provided it is not followed by incarceration due to revocation, ISP generally costs more than traditional probation or parole. For this reason, it is used rather sparingly.

Although a number of concerns have been raised about ISP, two of the most important are (1) the potential for net widening and (2) the lack of demonstrated reduction of recidivism.

Net widening takes various forms and can plague virtually any type of community correctional program. Net widening occurs when the offenders placed in a novel program like ISP are not the offenders for whom the program was intended. The consequence is that those in the program receive more severe sanctions than they would have received had the new program remained unavailable. Suppose, for example, that a jurisdiction facing a prison-crowding crisis establishes a new intensive probation program so that a substantial number of nonviolent offenders can be sentenced to ISP rather than prison. That is, the ISP target group consists of nonviolent felons who would have gone to prison had ISP not become available. If only persons who would formerly have gone to prison are placed on ISP, net widening has not taken place. However, if persons who would formerly have been

American Probation and Parole Association

The American Probation and Parole Association explores issues relevant to the field of community corrections. You can visit its website at www.appa-net.org/.

Would a career in probation or parole be of interest to you? Why or why not?

net widening A phenomenon that occurs when the offenders placed in a novel program are not the offenders for whom the program was designed. The consequence is that those in the program receive more severe sanctions than they would have received had the new program remained unavailable.

Earnie Grafton/ZUMAPRESS/Newscom

Intensive-supervision probation and parole (ISP) provides stricter conditions, closer supervision, and more treatment services than do traditional probation and parole. *Is ISP worth the extra expense? Why or why not?*

placed on regular probation are now placed on ISP, net widening has occurred. To the degree that such net widening is the result, ISP becomes a costly addition to ordinary probation instead of serving its intended function of providing a cost-efficient alternative to imprisonment.

Again, it is important to realize that almost all community correctional programs, especially those specifically designed to divert offenders from more severe sanctions, are vulnerable to net widening. Net widening can be avoided or minimized by establishing and following standard criteria for assigning offenders to programs. If the goal of an intensive probation program is to divert offenders from prison, then only prison-bound offenders should be placed in the program. Offenders who would have been placed on regular probation had the intensive program been unavailable should continue to be placed on regular probation.

When modern ISP programs were introduced in the early 1980s, there was much optimism that a cost-efficient means of diverting offenders from prison and controlling their recidivism had been found. Writing about the Georgia ISP program, social scientists Billie Erwin and Lawrence Bennett stated that "the recidivism rates are considerably better than for groups under regular probation and for those released from prison. [ISP] offenders commit fewer and less serious crimes."[52] Erwin and Bennett also estimated that Georgia saved $6,775 for each case diverted from prison into ISP.

Unfortunately, the initial optimism surrounding ISP has not been sustained according to subsequent, well-designed research by criminologist Joan Petersilia and her colleagues. In one study, Petersilia and Turner compared offenders who had been randomly assigned to either intensive or regular probation in three California counties. Efforts were made to ensure that only high-risk, serious offenders were included in the study, and offenders were followed for 6 months after program placement. Across counties, an average of 30% of the ISP cases had technical violations, and about 20% had new arrests. In two counties, ISP clients were significantly more likely than regular probationers to incur technical violations but neither more nor less likely to incur new arrests. In the other county, ISP cases and regular probationers did not differ significantly in technical violations or new arrests. Petersilia and Turner observe that in the first two counties, the increased supervision of ISP may have simply increased awareness of clients' technical violations. That is, regular probationers may have committed just as many technical violations, but more of the violations may have escaped detection because of less supervision. More important, Petersilia and Turner point out that if ISP simply increases awareness of technical violations and if the official response to such violations is swift revocation of ISP followed by incarceration, two things may occur. First, because offenders are no longer at risk of recidivism in the community once they are incarcerated, the artificial impression may be created that ISP is associated with low arrest rates. Second, ISP will fuel the prison crowding problem it is intended to relieve.[53]

myth

Incarceration for technical violations prevents crime.

fact

In the largest controlled experiment ever conducted nationwide on intensive-supervision probation and parole, researchers found that offenders who committed technical violations were no more likely to be arrested for new crimes than those who did not commit technical violations.

Source: Joan Petersilia and Susan Turner, *Evaluating Intensive Supervision Probation/Parole: Results of a Nationwide Experiment*, Research in Brief, National Institute of Justice, U.S. Department of Justice (Washington, DC: U.S. Government Printing Office, 1993).

In another study, Petersilia and Turner evaluated a national ISP demonstration project in 14 jurisdictions across nine states.[54] The 14 programs involved approximately 2,000 offenders. In each program, offenders were assigned randomly to either ISP or an alternative sanction, such as regular probation, prison, or regular parole. They then were followed for 1 year. The average results across jurisdictions are summarized in Table 12.4. ISP was associated with a substantially higher percentage of technical violations than the alternative sanctions. Again it appears that ISP may have increased detection of technical violations and thereby inflated incarceration rates. The data suggest that if ISP is to relieve prison crowding in any meaningful way, officials will have to become more reluctant to revoke ISP and impose incarceration for technical violations.

TABLE 12.4 Offender Recidivism Averages in Petersilia and Turner's Study of ISP in 14 Jurisdictions

Sanction	Percentage Arrested	Percentage with Technical Violations	Percentage Returned to Prison
ISP	37	65	24
Alternative	33	38	15

Renee C. Byer/ZUMA Press/Newscom

Day reporting centers, which may be publicly or privately operated, are corrections centers where offenders must report regularly to comply with a sentence. Most day center programs include employment, counseling, education, and community service components. *Should the use of day reporting centers be expanded? Why or why not?*

Day Reporting Centers

day reporting centers Facilities that are designed for offenders who would otherwise be in prison or jail and that require offenders to report regularly to confer with staff about supervision and treatment matters.

Day reporting centers, a relatively new facet of the intermediate-sanction movement, emerged in Great Britain in the 1970s and were pioneered in the United States in Massachusetts in 1986. The centers can be administered publicly or privately and are designed primarily for offenders who would otherwise be in jail or prison. This includes offenders such as those on ISP and those awaiting trial who did not receive release on recognizance or bail. Offenders are permitted to live at home but must report to the center regularly to confer with center staff about supervision and treatment matters. Program components commonly focus on work, education, counseling, and community service.[55]

The program objectives devised for offenders by the Metropolitan Day Reporting Center in Boston are illustrated in Figure 12.7. In June 2014 (the latest year for which these data were available), 4,413 offenders nationwide were being supervised by personnel of day reporting centers instead of being confined in jails (up from 3,683 in 2013 and 3,890 in 2012, but down from 5,552 in June 2010 and 6,492 in June 2009). The 2014 figure represents about 7% of offenders being supervised outside a jail facility (excluding persons on probation or parole).[56]

Structured (or Day) Fines

structured fines, or **day fines** Fines that are based on defendants' ability to pay.

Another relatively new intermediate sanction, at least in the United States, is structured fines, or day fines.[57] **Structured fines**, or **day fines**, differ fundamentally from the fines (called *tariff fines*) more typically imposed by American criminal courts. While tariff fines require a single fixed amount of money, or an amount of money within a narrow range, to be paid by all defendants convicted of a particular crime *without regard to their financial circumstances*, structured fines are based on defendants' ability to pay. The basic premise of structured fines is that "punishment by a fine should be proportionate to the seriousness of the offense and should have roughly similar impact (in terms of economic sting) on persons with differing financial resources who are convicted of the same offense."

The creation of a structured fine is a two-part process. In the first part, the number of fine units for a crime is determined from a scale that ranks crimes according to their seriousness. In the second part, the dollar amount of the fine is determined by multiplying the number of fine units by a proportion of the defendant's net daily income (hence, the term, *day fine*), adjusted to account for dependents and special circumstances.

No Time for Fine

The U.S. Supreme Court has ruled that an offender cannot be imprisoned for failure to pay a fine unless the default is willful, if imprisonment is not an authorized penalty for the offense of which the defendant was convicted.

Source: *Tate v. Short* (1971).

FIGURE 12.7 Metropolitan Day Reporting Center Participant Program Objectives

I, ____________________, agree to participate in the activities and to adhere to the schedule of this contract. I understand that this contract may be modified during my participation in the Day Reporting Center and I agree to those modifications as determined by my participation.

Objectives *Time frame*

Reporting: I will report in to the Center at the times designated by my case manager and as noted on my daily itinerary.

I will call in to the Center at the times designated by my ca[se] manager and as noted on my daily itinerary.

Supervision: I will be available for phone calls or house checks by Day Reporting Center staff at the locations listed on my daily itine[rary].

Employment: Employer: ________
Address: ________
Contact Person: ________
Telephone: ________
Hours: ________
Salary: ________

Financial: I agree to submit my pay stub to the Center for verification o[f] my employment. I further agree to work with my case manager on budgeting my money; including the payment of any court-ordered fines or other restitution.
Details: ________

Substance Abuse Treatment: I agree to participate in the following program[s to] address my substance abuse problems: ________

Urinalysis: I agree to submit to urinalysis testing upon the request of Day Reporting Center staff. ________

Curfew: I understand that I am to be at my residence from the hours of _____ P.M. to _____ A.M., and that I may be contacted by Day Reporting Center staff during that time. ________

Other Objectives: ________

________ ________
Client *Date*

I agree to assist ________________ to achieve the listed objectives of this contract in my capacity as case manager.

________ ________
Case Manager *Date*

FYI

Structured Fines

Regarding the use of structured fines, or day fines, in the United States, criminologists Norval Morris and Michael Tonry make this interesting observation: "It is paradoxical that a society that relies so heavily on the financial incentive in its social philosophy and economic practice should be so reluctant to use the financial disincentive as a punishment of crime."

Source: Norval Morris and Michael Tonry, eds., *Between Prisons and Probation: Intermediate Punishments in a Rational Sentencing System* (New York: Oxford University Press, 1990), 111.

Structured fines were first introduced in the 1920s in Sweden and soon thereafter were adopted by other Scandinavian countries. West Germany began employing them in the early 1970s. Western European nations have made day fines the sanction of choice in a large proportion of criminal cases, including many involving serious offenses. For example, in Germany, day fines are used as the only sanction for three-fourths of all offenders convicted of property crimes and for two-thirds of all offenders convicted of assaults.

The first structured-fine program in the United States began in 1988 in Richmond County (Staten Island), New York, as a demonstration project. Other structured-fine demonstration projects have been established in Maricopa County (Phoenix), Arizona; Bridgeport, Connecticut; Polk County (Des Moines), Iowa; and four counties in Oregon (Marion, Malheur, Josephine, and Coos). An evaluation of the Richmond County project, sponsored by the National Institute of Justice, showed very promising results.

Among the presumed advantages of structured-fine or day-fine programs are the following:

1. *Offender accountability*—"The offender is, quite literally, made to pay his or her debt to society."
2. *Deterrence*—"Structured fines provide an economic disincentive for criminal behavior."

3. *Fairness*–Structured fines are fairer than tariff fines because tariff fines frequently are too low to be meaningful to wealthier offenders and too high for poorer offenders to pay.
4. *Effective and efficient use of limited system resources*–"Structured fines are relatively inexpensive to administer compared with most other types of intermediate sanctions."
5. *Revenue*–"Structured fines can be more effective than tariff fines in generating revenue."
6. *Credibility for the court*–Because offenders pay in full in a very large proportion of cases, the sanction has credibility with the offender and the community.

Among the problems with structured fines are:

1. *Collection problems*–"If judges are not convinced that such fines will be paid in a high proportion of cases, or if offenders assume that the fines need not be paid, the usefulness of the structured fine as a criminal sanction is seriously eroded."
2. *The effect of other monetary penalties*–In most U.S. jurisdictions, defendants convicted of crimes have a variety of monetary penalties to pay–for example, penalty assessment fees, crime victim compensation fees, indigent defense fees, and probation fees. Those monetary sanctions create a high cost to a defendant even before a fine is imposed. A fine added to those costs may exceed the ability of poorer defendants to pay.

Home Confinement and Electronic Monitoring

In **home confinement** programs (also known as home incarceration, home detention, house arrest, and community control), offenders are required by the court to remain in their homes except for preapproved periods of absence. For example, offenders may be permitted to leave home to go to work, school, or church, or to run errands. Home confinement usually is considered more punitive than ISP but often is used in conjunction with ISP. For example, home confinement may be required during the initial phase of intensive probation or may be ordered if an offender violates probation or parole. Home confinement also can serve as an alternative to pretrial detention in jail.

home confinement A program that requires offenders to remain in their homes except for approved periods of absence; commonly used in combination with electronic monitoring.

electronic monitoring An arrangement that allows an offender's whereabouts to be gauged through the use of computer technology.

House arrest traditionally has been employed in the military and has a long history of limited use in criminal justice. However, home confinement programs did not gain wide popularity until **electronic monitoring** equipment became readily available during the 1980s. The first statewide home confinement program without electronic monitoring was implemented in Florida in the early 1980s. St. Louis was one of the cities to pioneer such programming in the early 1970s. Among the first jurisdictions to implement an electronic monitoring program was Palm Beach County, Florida, in 1984.[58]

Electronic monitoring allows an offender's whereabouts to be gauged through the use of computer technology. Before the use of electronic monitoring, it was time-consuming and expensive for officials to ensure that offenders were complying with home confinement orders. Electronic monitoring rendered home confinement more practical and cost effective because officials did not have to rely on personal phone calls and home visits to assess compliance with home detention orders.

Today, it is very common for home confinement programs to have electronic monitoring components. In June 2014 (the latest year for which these data were available), for example, 14,869 offenders nationwide who otherwise would have been confined in a jail facility were being supervised in a home detention program (down from 15,908 in June 2012, but up from 13,360 in June 2013 and 12,759 in June 2011). That represents about 23% of offenders being supervised outside a jail facility (excluding persons on probation or parole). Of those offenders participating in a home detention program, about 96% were being electronically monitored.[59]

The two dominant forms of electronic monitoring devices use GPS (global positioning system) and RF (radio frequency) technology. GPS is the newer and more expensive technology. GPS devices can track offenders continuously in real time, identifying their movements and whereabouts by transmitting location information

Home confinement requires offenders to remain in their homes except for preapproved periods of absence. Most programs use electronic monitoring equipment to verify that offenders are at their designated locations at particular times. *For what type of offenders are home confinement and electronic monitoring appropriate, and for what type of offenders are they not?*

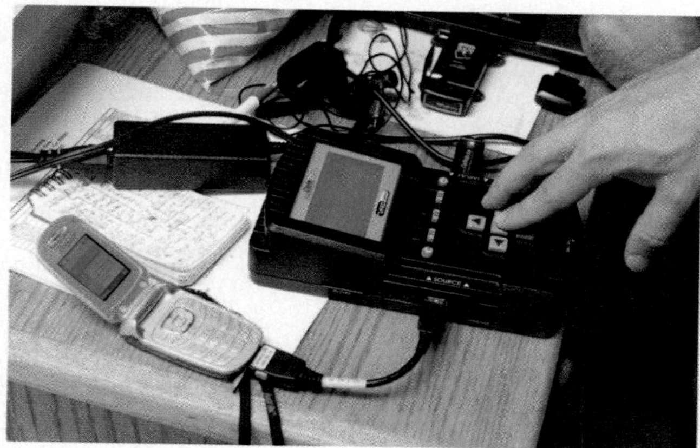

Rich Pedroncelli/AP Images

to monitoring centers and triangulating signals from satellites and cellular towers. The monitoring devices usually are ankle bracelets; however, some jurisdictions have replaced ankle bracelets with smartphones equipped with GPS-tracking capabilities.[60] RF is the older and less expensive technology. RF devices monitor offenders' presence or absence from a fixed location. They typically are used with offenders on house arrest or confinement and to enforce curfews by monitoring an offender's presence either continuously or during specified times. RF monitors consist of battery-powered transmitters, generally worn around ankles or wrists, and home-based receivers that can verify whether offenders are within a certain distance and alert monitoring centers if they are not.[61] Both types of monitoring have their problems. For example, GPS monitoring generates numerous false alarms, and the GPS signal will not penetrate most buildings.

Home confinement and electronic monitoring have not yet been evaluated extensively. However, a national survey found that electronic monitoring has expanded rapidly since its inception. For example, from 2005 through 2015, the number of accused and convicted criminal offenders monitored with electronic tracking devices in the United States increased about 147%, from approximately 53,000 to more than 131,000.[62] Almost all of the increase is the result of the nearly 3,000% increase in the use of GPS technology from about 2,900 units in 2005 to approximately 88,000 in 2015. During the same period, the number of RF units decreased 25% from more than 50,000 to fewer than 38,000.[63] Furthermore, there appears to be a trend toward using electronic monitoring with offenders who have committed more serious crimes. Still, the use of electronic monitoring remains relatively rare in the United States, as fewer than 2% of the correctional population is monitored electronically.[64]

As might be expected, home confinement and electronic monitoring are controversial practices. This is particularly true of monitoring. Supporters argue that home confinement and monitoring are cost-effective alternatives to incarceration in jail or prison that allow offenders to receive more supervision and punishment than they would receive on traditional probation or parole. Another argument is that offenders are spared the potentially harmful effects of institutionalization, one of the most important of which is the breaking of family ties.

Critics counter with a variety of claims. One important criticism is that home confinement and monitoring encroach on the constitutional rights to privacy, protection against self-incrimination, and protection against unreasonable search and seizure. Critics maintain that not only are the offender's rights threatened but also the rights of persons who share the offender's residence. However, it seems unlikely that the use of those sanctions will be restricted severely on constitutional grounds in the foreseeable future. As criminologists Rolando del Carmen and Joseph Vaughn observe, "A review of decided cases in probation and parole indicates that while the use of electronic devices raises constitutional issues, its constitutionality will most likely be upheld by the courts, primarily based on the concept of diminished rights [for offenders]."[65]

Some people worry that home confinement and electronic monitoring may discriminate against economically disadvantaged persons, who are less likely to have homes and telephone systems for the installation of monitoring devices. Others are concerned that offenders and manufacturers of monitoring equipment have become locked in a perpetual and costly battle to outwit one another. Manufacturers continually work to devise monitors that will resist offender tampering. Meanwhile, offenders seek ways to manipulate the latest monitoring innovations on the market. This situation, of course, is potentially lucrative for equipment vendors, but it is potentially costly for taxpayers and for offenders who must pay monitoring fees. A broader criticism is that a profit incentive is being created to subject more and more persons to government surveillance.

Some people object to monitoring on the grounds that it is impersonal, thus removing the human and helping elements from community corrections. Those critics suggest that by reducing the need for probation and parole officers' skill and professionalism, monitoring may undermine rehabilitation. As criminologist John Conrad wrote:

> Probation and parole were once intended to help men and women in the worst kind of trouble to find ways to lead law abiding lives. No more. Official services to offenders in the community have been transformed into a cat-and-mouse game in which surveillance, not service, has become the primary, if not the exclusive requisite. . . . Electronic surveillance is a recent and welcomed contribution of high technology to the community control of convicted offenders. We settle for temporary control in place of a serious program for changes in attitude, behavior and style of life.[66]

An investigation of Florida's house-arrest program provides empirical support for these criticisms. It also reveals some problems heretofore overlooked by house-arrest critics.[67] Florida's house-arrest program, which the Florida Department of Corrections manages and calls "community control," is one of the largest such programs in the United States. In 2016, the state supervised about 6,300 active program participants a day–about 16% of whom were electronically monitored. By contrast, in 2002, the state supervised about 10,000 active program participants a day–6% of whom were electronically monitored. Since its inception in 1983 and through July 2016, nearly 335,000 offenders have served time in the program (offenders who served two or more sentences under community control since 1983 were counted only once).

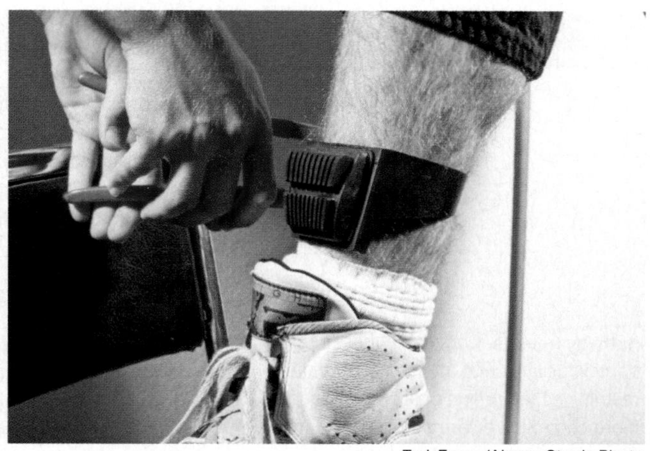

Ted Foxx /Alamy Stock Photo

Some electronic monitoring devices use global positioning system (GPS). *In what ways are GPS-equipped monitoring devices superior to conventional phone or radio signal devices?*

A principal reason for the program's creation was the prospect of saving money by reducing the prison population. Whether the program has achieved that goal is a difficult question to answer. On the one hand, participation in house arrest is significantly cheaper than the costs of incarceration. In 2016, house-arrest program participants who were electronically monitored cost the state $9.34 a day, while state prison inmates cost the state $51.65 a day. By that measure, the program has saved the state of Florida millions of dollars. However, if in the absence of the program, offenders would have been placed on regular probation, for which the average cost per probationer per day in 2016 was $4.44, then the program produced no cost savings.

The house-arrest program may have reduced the prison population, but that is not entirely clear, either. From June 1983 through June 2015, Florida's prison population increased each year except from June 1983 to June 1984, June 2011 to June 2012, and June 2014 to June 2015. From June 1983 through June 2015, the prison population increased about 260%. Still, the program may have reduced the prison population by approximately 10,000 inmates per year–the number participating in the house-arrest program. However, if the program participants would have been sentenced to regular probation instead of prison in the absence of the program, then the program did not reduce the prison population.

In any event, an early evaluation of Florida's house-arrest program found about a 66% success rate. That is, about two of every three program participants completed a year in the program without being removed for committing new crimes or violating other program rules. The other "unsuccessful" 34% of Florida's house-arrest participants were removed from the program and usually incarcerated during the first year for committing crimes or violating program rules such as testing positive for drugs, leaving home without authorization, failing to pay restitution, or possessing a firearm.

In addition, during the past three and a half decades, program participants committed hundreds of murders and sex crimes (many of the murderers and sex offenders were in absconder status at the time the murder or sex offense was committed). In fiscal year 2015–2016, 1,610 offenders were admitted to community control for violent offenses. Many of those offenders were placed in the program even though state law prohibited it. In Florida, offenders are not to be placed on house arrest if (1) they have been convicted of a crime that appears on a specific list of violent felonies, including robbery, kidnapping, and assault; and (2) they have a prior conviction from that same list. Judges who sentence such offenders to house arrest either don't know the law or ignore it. Some judges claim that the law makes no sense. They argue that the law allows some violent offenders barred from house arrest to be placed on probation, where they get even less supervision.

Another problem is that, as of June 30, 2016, 2,108 house-arrest participants fled the program. Although most of them eventually will be caught, some of them will not. The program also has been understaffed. The state legislature has set a limit at 25 cases per officer. Although community control officers had an average caseload of 21.5, as of June 30, 2016, the average caseload in 2008 was 31.6 and in 2002, it was 39. By law, officers are supposed to check on program participants three times a week. In practice, however, most participants were checked on only twice a week. For one of those checks, the participant reported to a department of corrections office.

House-arrest participants who wear electronic monitors are five times more likely to successfully complete their sentences and half as likely to flee the program. However, as noted previously, in 2016, only about 16% of house-arrest participants were electronically monitored. The problem is money. In 2016, it cost $9.34 a day to monitor an offender by satellite (down from $20.33 in 2002). By comparison, in 2016, it cost $4.44 a day to supervise a house-arrest participant without electronic monitoring. Although it would seem that electronic monitoring of house-arrest participants would be a good investment, the Florida Department of Corrections prefers to use the bulk of its resources on prisons.

Halfway Houses

halfway houses Community-based residential facilities that are less secure and restrictive than prison or jail but provide a more controlled environment than other community correctional programs.

Halfway houses (sometimes referred to as community correctional centers or residential community centers) are community-based residential facilities that are an alternative to confinement in jail or prison.[68] Although those facilities are an alternative to more secure and restrictive placements, they provide a much more controlled environment than is possible with traditional probation and parole, or even with ISP and home confinement. As the term *halfway* implies, the offender is midway between jail or prison and the free community.

The goal of halfway houses is to provide offenders with a temporary (e.g., 6-month) period of highly structured and supportive living so that they will be better prepared to function independently in the community upon discharge. To this end, most programs place a heavy emphasis on addressing offenders' educational and employment deficits. For example, persons who have not completed high school are encouraged to obtain the GED, individuals who have few marketable job skills are encouraged to complete vocational training, and those who lack employment are assisted in finding and keeping jobs.

Halfway houses have existed in the United States since the mid-1800s and, prior to the 1960s, were used primarily for persons coming out of prison and making the transition back into the community. During the 1960s, the number and functions of halfway house programs in the nation expanded. In addition to providing services for parolees and prereleasees (prisoners nearing their release dates), houses began to service probationers, pretrial detainees, and persons on furlough from prison. More recently, in the 1980s, halfway houses became an integral part of the intermediate-sanction movement. They now are sometimes used in conjunction with other intermediate sanctions. For example, as part of a probation sentence, an offender may complete a stay in a halfway house before being discharged to ISP, or a person who has violated a home confinement order may be placed in a halfway house program to give authorities added control.

Halfway houses are quite diverse. Not all houses are for offenders; some facilities serve other populations, such as those with mental disorders or drug addictions. Programs for offenders may serve a specific category of offenders, such as all probationers or all parolees, or some combination of offender groups. In addition, a halfway house's population may be all male, all female, or both male and female. Some houses rely heavily on referrals to other local agencies to ensure that offenders' treatment needs are met, while others provide extensive in-house treatment. Programs can be administered publicly or privately, but private administration is more common. Public programs may be administered at the federal, state, or local level of government.

Procedures Halfway house programming involves five basic procedures:

1. Referral
2. Administrative screening
3. Intake and orientation
4. Program participation
5. Termination of the stay

Referral of clients to a halfway house may be done by a correctional institution (as in the case of prereleasees), a court, or a probation or parole agency. The fundamental issue referral sources must confront is whether the offender will benefit from halfway house placement without compromising community safety. Thus, such factors as the offender's propensity for violence and degree of employment and educational deficits generally are given great weight in determining whether to make a referral.

When a referral is made, the next step is *administrative screening* in which halfway house administrators examine the case and decide whether to accept the offender into the program.

JOHN RUSSELL/AP Images

Halfway houses may be operated by private, for-profit or not-for-profit contractors or by public agencies. Some provide room, board, and help with employment. Others provide remedial education, individual or group counseling, and other types of life skills training. *Should halfway houses be restricted in the type of offender that is accepted? If yes, what types of offenders should be barred?*

Some house administrators have more authority to reject referrals than others. Administrators of private houses often have more discretion to decline referrals than do administrators of public programs. An important point to note is that referral sources and house administrators sometimes are motivated by differing objectives. For instance, correctional institution staff may be motivated by the desire to reduce crowding in their facilities, while house administrators may wish to maximize the rate of successful discharge from their programs.

If administrative screening results in a decision to accept the referral, the focus shifts to *intake and orientation*. This entails assessing the new resident's risks and needs as well as orienting him or her to the rules, expectations, and routines of the program. An important aspect of intake and orientation is determining exactly what the resident should accomplish during the stay to improve his or her life and reduce the likelihood of recidivism. That determination is usually written as a treatment plan with specific goals and objectives.

With respect to *program participation*, many halfway houses have a series of levels, or phases, through which residents must pass to receive a successful discharge. Typically, each level is associated with progressively more demanding goals and responsibilities for the resident to meet and also with more privileges and freedoms. For example, a requirement for an offender to move from the entry level to the second level may be the successful completion of a job skills class. On promotion to level two, the offender may receive one weekend furlough per month to spend with family and friends. Promotion to level three may carry additional furlough time and require the offender to obtain a job and complete a class on job retention skills.

As noted previously, halfway houses usually provide a variety of treatment services, such as employment counseling and training, life skills training, substance abuse intervention, and remedial education. Those services are provided either directly by house staff or indirectly through referrals to community agencies. To allow control and supervision, houses have a number of rules that govern both the in-house behavior of residents and residents' activities away from the facility. Those rules cover such matters as interaction between residents, care of facility property, curfews, and the use of alcohol. Progress through program levels requires completion of the goals and responsibilities associated with a particular level, participation in relevant treatment components, and compliance with house rules. Failure to achieve goals, to complete treatment components, or to abide by rules can result in the resident staying at a given level or being demoted to a lower level.

A resident's behavior must be monitored and periodically reviewed throughout the term of program participation because, ultimately, the house administration must decide the resident's *termination*, or discharge, status. Residents who have satisfactorily completed all of the required levels are discharged into the community, frequently under probation or parole supervision. If a resident fails to make satisfactory progress through levels in a reasonable time, compiles an excessive number of less-serious rule violations, or commits a serious rule infraction (such as another felony), the resident also can be discharged from the facility. An unsuccessful discharge can be granted at any point in a resident's stay and frequently is followed by incarceration.

Issues Several issues plague the use of halfway houses. Three of the most critical issues are (1) the relations between a halfway house and the local community of which it is a part, (2) the influence of privatization and politics, and (3) recidivism among persons who have been discharged from the house.

For a halfway house to be established in a community and to operate effectively, cooperative relations between the staff of the house and members of the community are essential. Halfway house staff members often depend on the surrounding community for such things as resident health care, counseling services, educational programming, and job placements. Furthermore, unfavorable reactions from community members toward residents simply will reinforce the sense of marginalization and alienation that many offenders already experience. To the extent that community members feel threatened by residents or have poor relations with staff, the halfway house program is in jeopardy.

Virtually all halfway houses must confront the issue of community resistance, and some must confront outright hostility from community members. There are at least two keys to doing this effectively. First, halfway house officials must actively cultivate support and assistance from community members by engaging in open, honest communication with them from the earliest possible stage of the halfway house's existence. Many fears that community members have of residents are founded on inaccurate, media-fueled stereotypes and can be reduced with realistic, factual information. Second, it is very helpful if community members can see that the halfway house is contributing something of value to the community rather than simply consuming resources. For example, house residents can be involved in well-publicized and highly visible community service projects that save tax dollars. Also, in houses where a high proportion of residents are employed, community residents should be reminded frequently that residents are contributing to the tax base instead of consuming public revenue as jail or prison inmates. If community members can be convinced that the halfway house has a positive, contributing side, they likely are to be more accepting of the house.

myth

When an offender halfway house is established in a neighborhood, crime rates increase and property values fall.

fact

Research on this subject has confirmed neither of those assumptions. Crime rates and property values tend to remain the same.

Just as there are ways to facilitate cooperative community relations, there also are ways to create unfavorable relations. One way to almost ensure poor relations is for the halfway house to be sprung on the community without any notice of officials' desire to establish the house and without community input in the early planning process. Another common mistake is for halfway house officials to be content with a low-profile image once the house is established. Believing that adequate community relations will exist as long as offenders are controlled and unfavorable media exposure is avoided, officials may try to minimize exposure of house operations to the public. The typical result is an atmosphere of secrecy, suspicion, and distrust. When the inevitable negative incident occurs, such as a crime by a resident against a community member, public outcry against the house is likely.

In June 2012, *The New York Times*, following a 10-month investigation, published a scathing critique of New Jersey's new halfway house system.[69] The expose revealed the political corruption that can occur with the privatization of community corrections. New Jersey, like many other states, had a crowded and expensive prison system. It responded to the problem by becoming one of the first states to divert more of its inmates to less costly privately run halfway houses. At the time, New Jersey had about two dozen halfway houses with approximately 3,500 beds. Some of New Jersey's halfway houses were as big as prisons and bore little resemblance to the small neighborhood halfway houses of the past where offenders were sent to serve community sentences or former prison inmates were sent to help ease their way back into society.

To some extent, New Jersey's experiment with privately run halfway houses has been successful. Approximately 10,000 prison inmates and parolees a year–about 40% of the New Jersey state prison population–now pass through the halfway house system, which does save the state money. The state of New Jersey spends approximately $125 to $150 a day to maintain a prison inmate, compared to about half that much for a halfway house resident. However, monetary savings come at a cost in safety. For example, the percentage of New Jersey violent offenders in halfway houses increased from 12% in 2006 to 21% in

2012, and some of them escaped and committed more violent crimes. In addition, until the Christie administration commissioned a 3-year study of its halfway house system in 2011, the state had never conducted a study to determine whether the halfway house system helped residents.

New Jersey's then new halfway houses encountered problems from the start, including problems with drugs, gang activity, and violence, especially sexual assaults. Perhaps the biggest problem with the halfway houses was escapes. Since 2005, more than 5,000 residents have escaped from the privately run halfway houses, and they escaped with some regularity. Some residents simply walked out the back, side, or emergency doors of the halfway house and never returned. Some residents fled before being returned to prison for violating halfway house rules. Other residents did not return to the halfway house from their work-release assignments–the most common type of escape. The halfway house's only recourse was to notify the police.

Many residents who escaped were recaptured within hours or days, or turned themselves in after having second thoughts. However, many were not caught for weeks or months or were caught only after they committed new crimes. Some are never caught–85 residents were still at large when *The Times* completed its investigation. When escapees were caught, they could be charged with felonies for escaping that carried 3- to 5-year sentences, but that rarely occurred. Instead, they were either returned to prison to complete their original sentences, or they were returned to the halfway house.

After *The Times* began its investigation, then New Jersey Governor Chris Christie started fining halfway houses for escapes. That reform presumably reduced the number of escapes to 181 during the first 5 months of 2012–a 35% reduction from a similar period in 2009, before Christie took office in January 2010. However, *The Times* report suggested that the Christie administration may have manipulated the statistics downward to make the situation look better. The report also noted that halfway house regulation remained lax.

The Christie administration had good reason to make sure the privately run halfway houses were seen in the best light. Governor Christie for years had championed the dominant company in New Jersey's privately run halfway house system–Community Education Centers. The company operated six of the largest halfway houses in New Jersey, with 1,900 of the 3,500 beds. Out of approximately $105 million spent by state and county agencies on halfway houses in New Jersey during the 2011 fiscal year, Community Education received about $71 million. A senior vice president of Community Education was William Palatucci, Governor Christie's good friend, political adviser, and former law partner. Governor Christie, himself, was a registered lobbyist for the company in 2000 and 2001, when he was a private lawyer. Community Education, its executives, and their family members had contributed over the last decade more than $200,000 in campaign contributions to Governor Christie and two other politicians connected to halfway house decisions. In early 2010, Christie hired the son-in-law of Community Education's chief executive as an assistant in the governor's office. Also, as United States attorney for New Jersey and then governor, Christie often visited the company's halfway houses and praised the company's work. The company highlighted the governor's visits in its publicity material, which probably helped it land government contracts in Colorado, Pennsylvania, and other states. The Christie administration denied that there had been any inappropriate political influence on halfway house decisions. As of June 2019, all of New Jersey's halfway houses remained privately run.[70]

Finally, how many persons placed in halfway houses successfully complete their stays, and how high is recidivism among former residents after discharge? Those questions were addressed in a study of 156 probationers admitted to a halfway house in Michigan.[71] The study found that nearly 60% of all probationers received unsuccessful discharges. However, other studies have discovered successful discharge figures as high as 60%.[72] The Michigan study tracked former halfway house residents for 7 years after their discharges. The researchers found that approximately 67% of the former residents were arrested at least once during the 7-year period for some type of criminal activity; about 60% were arrested at least once for felonies. However, persons who had successfully completed the halfway house program were significantly less likely to incur new arrests than those who had not done so. For instance, 44% of the persons who had received successful discharges were arrested for new felonies, compared with 68.8% of those who had received unsuccessful discharges.

In another study, the recidivism of offenders who had been placed in halfway house programs was compared with the recidivism of those placed on regular probation supervision. Using a 3-year follow-up period, the researchers found that 29.5% of the former halfway house residents and 30.7% of the regular probationers experienced new criminal

ICCA

To learn more about halfway houses, visit the website of the International Community Corrections Association (ICCA), formerly the International Halfway House Association, at www.iccaweb.org/.

Does working in a halfway house appeal to you? Why or why not?

convictions. Slightly more than 40% of the halfway house residents had failed to complete their programs successfully.[73]

Research clearly suggests that a large number of offenders placed in halfway houses do not successfully complete their programs. However, those who do receive successful discharges seem less likely to commit new crimes than those who do not. Yet, there is little reason to believe that halfway houses are associated with less recidivism than other types of community corrections programs.

THINKING CRITICALLY

1. Do you think people should have the right to reject the placement of a halfway house in their neighborhood? Why or why not?
2. What could be done to bring about good relations between a community and the halfway house located within the community?
3. How could the Christie administration have avoided the New Jersey halfway-houses scandal?

Temporary-Release Programs

temporary-release programs Programs that allow jail or prison inmates to leave the facility for short periods to participate in approved community activities.

Temporary-release programs allow inmates in jail or prison to leave the facility for short periods to participate in approved community activities. Those programs are designed to permit inmates to establish or maintain community ties, thereby gradually preparing them for reentry into society. The programs also give institutional authorities a means of testing the readiness of inmates for release, as well as a means of controlling institutional behavior. Not surprisingly, opportunities for temporary release create a major incentive for inmates to engage in the conduct officials desire.

Three common temporary-release programs are work release, study release, and furloughs. The persons involved in such programs may be prison or jail inmates, or halfway house residents. As should be apparent from the previous section of this chapter, temporary release is an integral aspect of halfway houses. In prisons and jails, temporary release is reserved for inmates who have demonstrated that they are appropriate candidates to participate in community-based activities.

Organized temporary-release programs are not new in American corrections. Although the programs date to the early 1900s, they did not gain widespread popularity until the 1960s, when the emphasis on community corrections began to grow. Today, almost all states offer temporary release.[74]

In work-release programs, inmates leave the facility for certain hours each day to work for employers in the free community. Work-release inmates are paid prevailing wages, but they usually are required to submit their paychecks to corrections officials to deduct for such things as dependent support and restitution orders before placing the balance in inmates' accounts. Inmates may withdraw some money while in the facility but usually are required to save a certain amount for their ultimate release from incarceration. A limitation of such programs is that most inmates must settle for low-skilled and low-paying jobs. This limitation exists because:

1. A large proportion of inmates lack the educational and vocational backgrounds needed to secure higher-paying employment.
2. There is more competition for higher-paying jobs.
3. Many employers are reluctant to hire work-release inmates, especially for jobs that pay well and require skill and responsibility.

Study-release inmates can leave prison for high school equivalency classes, vocational training, or college coursework. A limitation of study release is that, unlike work release, it does not generate financial resources. If inmates cannot pay for their own educational programs or obtain financial aid, study release must be funded by the jurisdiction. Even in the unlikely event that a jurisdiction has resources to fund study release, there still may be reluctance to do so because of the argument that inmates are getting free education at taxpayers' expense.

ZUMA Press Inc/Alamy Stock Photo

Temporary-release programs allow inmates in jail or prison to leave the facility for short periods to participate in approved community activities. *Should temporary release from jail or prison be allowed? Why or why not?*

In furlough programs, inmates are granted leaves of absence for brief periods (e.g., 48 hours) to accomplish specific things. Inmates may be granted furloughs to spend time with family members, to attend funerals, or to search for employment and housing before release.

One commonly voiced fear about temporary-release programs is that inmates will flee or commit serious crimes while away from the facility. In fact, research shows that most inmates neither flee nor commit serious offenses. The most frequent problems are late returns and the use of alcohol and illegal drugs.

Temporary-release programs became a significant issue in the U.S. presidential election campaign of 1988. Massachusetts Governor Michael Dukakis, the Democratic presidential nominee, was heavily criticized by Vice President George H. W. Bush, the Republican nominee, for allowing temporary release of inmates with records of violence. Willie Horton, a black man who had been serving a first-degree murder sentence at a Massachusetts institution, committed violent crimes against a white couple during his tenth furlough from the institution. Bush made much of the "liberal" prison release policies of Massachusetts in an attempt to discredit Dukakis.

Around the same time, an evaluation of temporary release in Massachusetts was being conducted. The evaluation was published in 1991.[75] The researchers studied persons released from Massachusetts facilities between 1973 and 1983. Offenders were tracked for 1 year after release, and recidivism was defined as return to prison. The recidivism of persons who participated in furloughs, prerelease-center (halfway house) programs, and both furloughs and prerelease centers was compared with the recidivism of persons who participated in neither. Using statistical controls for differences in offenders' backgrounds, researchers concluded that "prerelease programs following prison furloughs, and prison furloughs alone, appear to reduce dramatically the risks to public safety after release." This study implies that eliminating or significantly curtailing temporary-release programs is likely to result in a long-term decline in community safety and an increase in recidivism.

THINKING CRITICALLY

1. Do you think offenders should be allowed to take college courses, at taxpayers' expense, through study-release programs? Why or why not?
2. Do you think temporary-release programs help rehabilitate offenders? Why or why not?

SUMMARY

1. Define community corrections, and identify the goals and responsibilities of community corrections agencies and their staffs.

Although community corrections programs are very diverse across the nation, the common feature of those programs is that they provide supervision and treatment services for offenders outside jails and prisons. The dual emphasis on supervision and treatment creates a potential role conflict for community corrections staff. There is some indication, however, that the traditional focus on treatment has been declining relative to the growing emphasis on supervision.

2. Define probation, and summarize the research findings on recidivism rates.

The most commonly used type of community sentence is probation. Offenders placed on probation are supervised and provided with various services in the community instead of being incarcerated. In return, they are required to abide by the rules of the probation sentence. Overall, studies indicate that probation is about as effective as incarceration in controlling recidivism. Available data indicate that recidivism is quite high among felons sentenced to probation. Furthermore, research has not found a strong association between probation officers' caseloads and the likelihood of probationer recidivism. If anything, reduced caseloads seem to increase the probability that instances of recidivism will be detected.

3. Distinguish parole from probation.

Unlike probation, parole is not a court-imposed sentence. Parole is a mechanism of releasing persons from prison and a means of supervising them after release instead of an alternative to an incarceration sentence. Parole supervision often is confused with probation because the two share many similarities. Four major objectives of parole agencies are to (1) preserve community safety by supervising the behavior of parolees, (2) promote the betterment of parolees by responding to their treatment needs, (3) control prison crowding, and (4) control the behavior of prison inmates by providing early release opportunities in exchange for good prison conduct.

4. Explain the functions of a parole board.

In general, a parole board directs a jurisdiction's parole policies and manages the parole release and termination processes. Parole board members (or their representatives) conduct parole-grant hearings to determine which inmates should receive early release from prison. Boards also determine whether to revoke the parole of persons who have violated parole conditions. In most states, field service agencies are administratively independent of parole boards. A field service agency provides community supervision and treatment services for persons who have been granted parole release by the board and makes recommendations to the board concerning termination of parole.

5. Describe how intermediate sanctions differ from traditional community corrections programs.

Compared with traditional programs in community corrections, intermediate sanctions are oriented less toward rehabilitation and more toward retribution, deterrence, and incapacitation. They are more punitive and more restrictive. The recent popularity of intermediate sanctions is attributable largely to the record high levels of prison crowding that plague many jurisdictions and a corresponding need to devise acceptable alternatives to imprisonment.

6. Explain two major concerns about intensive-supervision probation and parole.

Two major concerns about ISP are (1) the potential for net widening and (2) the lack of demonstrated reduction of recidivism. Net widening occurs when offenders placed in a novel program such as ISP are not the offenders for whom the program was intended. The consequence is that those in the program receive more severe sanctions than they would have received had the new program been unavailable. Studies show that ISP is associated with a substantially higher percentage of technical violations than alternative sanctions such as regular probation, prison, or regular parole.

7. Explain what day reporting centers and structured fines are.

Day reporting centers allow offenders to live at home but require them to report to the center regularly to confer with center staff about supervision and treatment matters. Program components commonly focus on work, education, counseling, and community service. Structured fines, or day fines, differ fundamentally from the fines (called *tariff fines*) more typically imposed by American criminal courts. While tariff fines require a single fixed amount of money, or an amount within a narrow range, to be paid by all defendants convicted of a particular crime without regard to their financial circumstances, structured fines, or day fines, are based on defendants' ability to pay. The basic premise of structured fines is that punishment by a fine should be proportionate to the seriousness of the offense and should have a roughly similar economic impact on persons with differing financial resources who are convicted of the same offense.

8. Explain what home confinement and electronic monitoring are.

In home confinement programs (also known as home incarceration, home detention, and house arrest), offenders are required by the court to remain in their homes except for preapproved periods of absence. Electronic monitoring, which is generally coupled with home confinement, allows an offender's whereabouts to be gauged through the use of computer technology.

9. Identify the goal of halfway houses, and compare halfway houses with other community corrections programs.

The goal of halfway houses is to provide offenders with a temporary (e.g., 6-month) period of highly structured and supportive living so that they will be better prepared to function independently in the community upon discharge. To this end, most programs place a heavy emphasis on addressing offenders' educational and employment deficits. Research clearly suggests that a large number of offenders placed in halfway houses do not successfully complete their programs. However, those who do receive successful discharges seem less likely to commit new crimes than those who do not. Yet, there is little reason to believe that halfway houses are associated with less recidivism than other community corrections programs.

10. **Summarize the purposes and outcomes of temporary-release programs.**

By allowing incarcerated persons to temporarily leave their facilities to participate in approved activities in the community, temporary-release programs are intended to foster ties between inmates and their communities, thus gradually preparing inmates for return to society. These programs also give officials a way to judge the readiness of inmates for release, and because most inmates want to participate in temporary release, the programs give inmates an incentive to maintain good institutional behavior. Although fear and other forms of resistance from community members often plague the programs, there is evidence that participation in temporary release is associated with a decreased likelihood of recidivism upon release from incarceration. Some common types of temporary release include work release, study release, and furlough.

KEY TERMS

community corrections 488
probation 491
diversion 491
presentence investigation (PSI) 493
probation conditions 495
restitution 497
revocation 497
technical violations 497
recidivism 500
parole 501
parole guidelines 505
reintegration or reentry 511
intermediate sanctions 512
intensive-supervision probation and parole (ISP) 512
net widening 513
day reporting centers 515
structured fines, or day fines 515
home confinement 517
electronic monitoring 517
halfway houses 520
temporary-release programs 524

REVIEW QUESTIONS

1. What are David Duffee's three varieties of community corrections, and how do they differ?
2. What are three aspects of the helping role of community corrections staff?
3. Why is community corrections important?
4. What are seven types of probation?
5. What are three fundamental objectives of probation agencies?
6. What is a probation subsidy, and what is its goal?
7. What is the main task of the PSI?
8. What are two types of probation conditions, and how do they differ?
9. For what general types of violations can probation be revoked?
10. What two landmark Supreme Court cases define the procedural guidelines for revoking probation or parole?
11. What are two general types of parole, and what is their function?
12. What are the two most important factors parole authorities consider before granting release on parole?
13. What are some criticisms of parole release?
14. What are two ways that halfway house officials might effectively confront community resistance?

IN THE FIELD

1. **Tour a Probation or Parole Agency** Take a guided tour of the probation or parole agency in your community.
   a. Ask the guide to explain the various activities of the agency. Also ask to speak with various officers about their work and any controversial probation or parole issues that affect the local community.
   b. Request sample copies of agency documents, such as risk-and-needs-assessment instruments.
   c. Write a short essay describing what the agency documents reveal about the work of the agency.
2. **Evaluate Halfway House Rules** Obtain a copy of the rules of behavior for a halfway house.
   a. Evaluate the rules, and decide whether you think they are too strict, not strict enough, or on target.
   b. Rewrite any rules that you think need changing.

ON THE WEB

1. **State Community Corrections Programs** Visit the Colorado Department of Public Safety, Office of Community Corrections website, https://www.colorado.gov/pacific/dcj/community-corrections, and the North Carolina Department of Corrections, Division of Community Corrections website, www.ncdps.gov/index2.cfm?a=000003,002223. Read the descriptions of the two programs, and write a short paper analyzing the similarities and differences in the two state programs. Based on the differences, decide which program is likely to be most effective in reducing recidivism, and explain why.
2. **U.S. Parole Commission** Visit the website of the U.S. Parole Commission, www.justice.gov/uspc. Click on "Frequently Asked Questions" for the answers to many commonly asked questions about federal parole. Write a short essay explaining why you support or oppose the 1987 abolition of federal parole.

CRITICAL THINKING EXERCISES

PROBATION/PAROLE OFFICER

1. You are a probation or parole officer. Your caseload averages 100 clients.

   a. A client tells you that her boss is treating her unfairly at work because of her criminal record and probation or parole status. Your client is afraid of being fired and having her probation or parole revoked. What do you do?

   b. You discover that a client is using marijuana. You like the client, and other than the marijuana use, he has been doing well on probation or parole. What do you do?

   c. You have a problem client who is using drugs (marijuana and cocaine), hanging out with a "bad" group of people, and probably committing petty thefts to support her drug habit (although you have no hard evidence of this). You have warned the client to stop this behavior, but she has ignored your warnings. Furthermore, the client has threatened to harm you and your family should you revoke her probation or parole. What do you do?

   d. You have a client who, as a condition of his probation or parole, is required to earn his GED. Although the client has been attending classes regularly and seems to be trying very hard, his teacher informs you that the client just does not have the intellectual capacity to earn the GED. What do you do?

HALFWAY HOUSE

2. As an employee of your state department of corrections, you have been asked to establish a halfway house for parolees, all of whom are former substance abuse offenders, in the nice, middle-class community where you live. Assuming that you take the job, how would you address the following questions?

   a. Would you communicate to the community your intention of establishing the halfway house? Why or why not?

   b. How would you address community resistance to the halfway house should it arise?

   c. Would you ask community residents to aid you in your efforts? Why or why not?

   d. Besides community resistance, what other problems might arise in your effort to establish the halfway house? How would you handle them?

   e. Would you be willing to remain a resident of the community after the halfway house is established (especially if you have a family with small children)? Why or why not?

NOTES

1. David E. Duffee, "Community Corrections: Its Presumed Characteristics and an Argument for a New Approach," in *Community Corrections: A Community Field Approach*, ed. D. E. Duffee and E. F. McGarrell (Cincinnati, OH: Anderson, 1990), 1–41.
2. Danielle Kaeble and Mary Cowhig, "Correctional Populations in the United States, 2016," in *Bureau of Justice Statistics Bulletin*, U.S. Department of Justice (Washington, DC: U.S. Government Printing Office, April 2018), 4, Table 4. Accessed September 11, 2019, https://www.bjs.gov/content/pub/pdf/cpus16.pdf; Danielle Kaeble, "Probation and Parole in the United States, 2016," in *Bureau of Justice Statistics Bulletin*, U.S. Department of Justice (Washington, DC: U.S. Government Printing Office, April 2018), 3, Table 2. Accessed September 11, 2019, https://www.bjs.gov/content/pub/pdf/ppus16.pdf.
3. Danielle Kaeble and Mary Cowhig, "Correctional Populations in the United States, 2016."
4. Calculated from data in Danielle Kaeble, "Probation and Parole in the United States, 2016," 3, Table 1.
5. Danielle Kaeble and Mary Cowhig, "Correctional Populations in the United States, 2016," 3, Table 3.
6. John Augustus, *A Report of the Labors of John Augustus, for the Last Ten Years, in Aid of the Unfortunate* (Boston: Wright and Hasty, 1852); reprinted as John Augustus, *First Probation Officer* (New York: National Probation Association, 1939).
7. David J. Rothman, *Conscience and Convenience: The Asylum and Its Alternatives in Progressive America* (Boston: Little, Brown, 1980), 44; www.ohnd.uscourts.gov/U_S_Probation/U_S_Probation_Employment/u_s_probation_employment.html.
8. "Probation and Pretrial Services," United States Courts. Accessed September 11, 2019, https://www.uscourts.gov/services-forms/probation-and-pretrial-services.
9. Tom Condon, "Prison Experiment: Sentenced to Read," www.ctnow.com/templates/misc/printstory.jsp?slug5hc%2D.C.ondon1010%2Eartoct10.
10. Rolando V. del Carmen, "Legal Issues and Liabilities in Community Corrections," in *Contemporary Community Corrections*, ed. T. Ellsworth (Prospect Heights, IL: Waveland Press, 1992), 383–407.
11. www.defgen.state.vt.us/lawbook/ch38.html.
12. Jay S. Albanese, Bernadette A. Fiore, Jerie H. Powell, and Janet R. Storti, *Is Probation Working? A Guide for Managers and Methodologists* (New York: University Press of America, 1981), 65.
13. Ibid.
14. Calculated from data in Kaeble, "Probation and Parole in the United States, 2016," 15, Appendix Table 3.
15. National Institute of Corrections, "Parole Essentials: Practical Guides for Parole Leaders," accessed February 24, 2011, http://community.nicic.gov/blogs/parole/archive/2010/08/30/parole-s-function-purpose-and-role-in-the-criminal-justice-system.aspx.
16. Rothman, *Conscience and Convenience.*
17. U.S. Parole Commission, www.usdoj.gov/uspc/mission.html.
18. Ebony L. Ruhland, Edward E. Rhine, Jason P. Robey, and Kelly Lyn Mitchell, "The Continuing Leverage of Releasing Authorities: Findings from a National Survey," University of Minnesota, Robina Institute of Criminal Law and Criminal Justice, 2017, accessed September 11, 2019, https://robinainstitute.umn.edu/sites/robinainstitute.umn.edu/files/parole_executive_summary_web_11-15.pdf.
19. Ibid.
20. Ibid.
21. Data generously provided by Edward E. Rhine, July 18, 2016.
22. Ruhland et al., "The Continuing Leverage of Releasing Authorities: Findings from a National Survey."

23. Ibid.; National Institute of Corrections, "Parole Essentials: Practical Guides for Parole Leaders."

24. Ruhland et al., "The Continuing Leverage of Releasing Authorities: Findings from a National Survey."

25. Ibid.

26. CNNJustice, "Charles Manson Denied Parole, with Next Parole Hearing Set for 2027," accessed January 5, 2013, http://articles.cnn.com/2012-04-11/justice/justice_california-charles-manson_1_charles-manson-parole-hearing-slayings-of-pregnant-actress?_s5PM:JUSTICE; John Rogers, "'Evil' cult leader was convicted in '69 killings," The Orlando Sentinel, November 21, 2017, A6.

27. Ruhland et al., "The Continuing Leverage of Releasing Authorities: Findings from a National Survey."

28. "Georgia Parole Decision Guidelines," accessed December 18, 2008, www.pap.state.ga.us/opencms/export/sites/default/resources/Proposed_Parole _Decision_Guidelines.pdf.

29. Ruhland et al., "The Continuing Leverage of Releasing Authorities: Findings from a National Survey," Tables 5 and Chart 10.

30. Ibid.

31. Ibid.

32. Ibid.

33. Ibid. Ruhland et al., "The Continuing Leverage of Releasing Authorities: Findings from a National Survey."

34. National Institute of Corrections, "Parole Essentials: Practical Guides for Parole Leaders."

35. Richard McCleary, *Dangerous Men: The Sociology of Parole* (Beverly Hills, CA: Sage, 1978).

36. Ruhland et al., "The Continuing Leverage of Releasing Authorities: Findings from a National Survey."

37. Ibid.

38. Ibid.

39. Ibid.

40. Ibid.

41. Update generously provided by Edward E. Rhine, July 18, 2016; Peggy B. Burke, *Abolishing Parole: Why the Emperor Has No Clothes* (Lexington, KY: American Probation and Parole Association, 1995); Peggy B. Burke, "Issues in Parole Release Decision Making," in *Correctional Theory and Practice*, ed. C. A. Hartjen and E. E. Rhine (Chicago: Nelson-Hall, 1992), 213–32.

42. Burke, *Abolishing Parole.*

43. John C. Watkins Jr., "Probation and Parole Malpractice in a Noninstitutional Setting: A Contemporary Analysis," in *Federal Probation* 53 (1989): 29–34 (quotations from page 30); also see del Carmen, "Legal Issues and Liabilities in Community Corrections."

44. Danielle Kaeble and Mary Cowhig, "Correctional Populations in the United States, 2016," 2, Table 1.

45. Calculated from ibid.

46. Kaeble, "Probation and Parole in the United States, 2016," 18–19, Appendix Table 5.

47. Jeffrey K. Liker, "Wage and Status Effects of Employment on Affective Well-Being Among Ex-Felons," *American Sociological Review* 47 (1982): 264–83 (quotation from page 282).

48. Richard A. Berk and David Rauma, "Capitalizing on Nonrandom Assignment to Treatments: A Regression-Discontinuity Evaluation of a Crime-Control Program," *Journal of the American Statistical Association* 78 (1983): 21–27. And see Peter H. Rossi, Richard A. Berk, and K. J. Lenihan, *Money, Work, and Crime: Experimental Evidence* (New York: Academic Press, 1980).

49. Information about parole exits in 2016 is calculated from data in Kaeble, "Probation and Parole in the United States, 2016," 22, Appendix Table 7.

50. Ibid.; Timothy A. Hughes, Doris James Wilson, and Allen J. Beck, "Trends in State Parole, 1990–2000," *Bureau of Justice Statistics Special Report*, U.S. Department of Justice (Washington, DC: U.S. Government Printing Office, October 2001), 10–14.

51. John C. Runda, Edward E. Rhine, and Robert E. Wetter, *The Practice of Parole Boards* (Lexington, KY: Host Communications Printing, 1994). Also see U.S. General Accounting Office, *Intermediate Sanctions: Their Impacts on Prison Crowding, Costs, and Recidivism Are Still Unclear* (Washington, DC: General Accounting Office, 1990).

52. Billie S. Erwin and Lawrence A. Bennett, *New Dimensions in Probation: Georgia's Experience with Intensive Probation Supervision*, Research in Brief, National Institute of Justice, U.S. Department of Justice (Washington, DC: U.S. Government Printing Office, 1987), 4.

53. Joan Petersilia and Susan Turner, "Comparing Intensive and Regular Supervision for High-Risk Probationers: Early Results from an Experiment in California," *Crime and Delinquency* 36 (1990): 87–111.

54. Joan Petersilia and Susan Turner, *Evaluating Intensive Supervision Probation/Parole: Results of a Nationwide Experiment*, Research in Brief, National Institute of Justice, U.S. Department of Justice (Washington, DC: U.S. Government Printing Office, 1993). For further detail, see Joan Petersilia and Susan Turner, "Intensive Probation and Parole," in *Crime and Justice: A Review of Research*, vol. 17, ed. M. Tonry and A. J. Reiss (Chicago: University of Chicago Press, 1993), 281–335.

55. Dale G. Parent, *Day Reporting Centers for Criminal Offenders: A Descriptive Analysis of Existing Programs*, Issues and Practices, National Institute of Justice, U.S. Department of Justice (Washington, DC: U.S. Government Printing Office, 1990).

56. Todd D. Minton and Zhen Zeng, "Jail Inmates at Midyear 2014," *Bureau of Justice Statistics Bulletin,* U.S. Department of Justice (Washington, DC: U.S. Government Printing Office, June 2015), 9, Table 9, accessed February 18, 2016, www.bjs.gov/content/pub/pdf/jim14.pdf; Todd D. Minton, "Jail Inmates at Midyear 2011—Statistical Tables," Bureau of Justice Statistics, U.S. Department of Justice (Washington, DC: U.S. Government Printing Office, April 2012), 8, Table 9, accessed January 6, 2013, http://bjs.ojp.usdoj.gov/content/pub/pdf/jim11st.pdf.

57. Material in this section is from Bureau of Justice Assistance, U.S. Department of Justice, *How to Use Structured Fines (Day Fines) as an Intermediate Sanction* (November 1996).

58. J. R. Lilly and R. A. Ball, "A Brief History of House Arrest and Electronic Monitoring," *Northern Kentucky Law Review* 13, no. 3 (1987): 343–74.

59. Minton and Zeng, "Jail Inmates at Midyear 2014," 9, Table 9.

60. "Use of Electronic Offender-Tracking Devices Expands Sharply," *PEW*, September 7, 2016. Accessed September 17, 2019, https://www.pewtrusts.org/en/research-and-analysis/issue-briefs/2016/09/use-of-electronic-offender-tracking-devices-expands-sharply.

61. Ibid.

62. Ibid.

63. Ibid.

64. Ibid.

65. Rolando V. del Carmen and Joseph B. Vaughn, "Legal Issues in the Use of Electronic Surveillance in Probation," in *Contemporary*

Community Corrections, ed. T. Ellsworth (Prospect Heights, IL: Waveland Press, 1992), 426.

66. John P. Conrad, "Concluding Comments: VORP and the Correctional Future," in *Criminal Justice, Restitution, and Reconciliation*, ed. B. Galaway and J. Hudson (Monsey, NY: Criminal Justice Press, 1990), 229.

67. 2016 data generously provided by Thomas Seaman, Correctional Program Administrator, Office of Community Corrections, Electronic Monitoring Unit/Absconder Unit, Florida Department of Corrections; 2008 data generously provided by Shari Britton, Chief, Bureau of Probation & Parole Field Services, Florida Department of Corrections; Rene Stutzman, "State Takes Eyes Off Inmates," *Orlando Sentinel*, December 29, 2002, A1; Rene Stutzman, "Ankle Monitors Show a Higher Rate of Success," *Orlando Sentinel*, December 29, 2002, A15; Florida Department of Corrections, www.d.c.state.fl.us; Camille Graham Camp and George M. Camp, *The 2001 Corrections Yearbook: Adult Systems* (Middletown, CT: Criminal Justice Institute, 2002), 206.

68. Unless indicated otherwise, material about halfway houses is from Belinda Rodgers McCarthy, Bernard J. McCarthy Jr., and Matthew C. Leone, *Community-Based Corrections*, 4th ed. (Belmont, CA: Wadsworth, 2001).

69. Information on New Jersey's halfway house system is from a series of articles published by *The New York Times*, which can be found at http://topics.nytimes.com/top/features/timestopics/series/unlocked/index.html (accessed most recently on July 20, 2016).

70. S. P. Sullivan, "Top Dems want to force Murphy to close 2 prisons, send more inmates to halfway houses," *NJ.com*, June 26, 2019, accessed September 2019, https://www.nj.com/politics/2019/06/top-dems-want-to-force-murphy-to-close-2-prisons-send-more-inmates-to-halfway-houses-governor-blindsided.html.

71. David J. Hartmann, Paul C. Friday, and Kevin I. Minor, "Residential Probation: A Seven-Year Follow-Up Study of Halfway House Discharges," *Journal of Criminal Justice* 22 (1994): 503–15.

72. Patrick G. Donnelly and Brian E. Forschner, "Client Success or Failure in a Halfway House," *Federal Probation* 48 (1984): 38–44. Also see R. E. Seiter, H. Bowman Carlson, J. Grandfield, and N. Bernam, *Halfway Houses: National Evaluation Program: Phase I, Summary Report*, U.S. Department of Justice (Washington, DC: U.S. Government Printing Office, 1977).

73. Edward J. Latessa and Lawrence F. Travis, "Halfway House or Probation: A Comparison of Alternative Dispositions," *Journal of Crime and Justice* 14 (1991): 53–75.

74. McCarthy, McCarthy, and Leone, *Community-Based Corrections*. Unless otherwise noted, the material in the remainder of this section draws on pages 148–56 of this source.

75. Daniel P. LeClair and Susan Guarino-Ghezzi, "Does Incapacitation Guarantee Public Safety? Lessons from the Massachusetts Furlough and Prerelease Programs," *Justice Quarterly* 8 (1991): 9–36 (quotation from page 26).

Design Elements: *Image of part of a court building with columns and statue*: Pixtal/AGE Fotostock RF; *Image of a door slightly open*: jadding/Shutterstock.com

13

Mikael Karlsson/Alamy Stock Photo

Juvenile Justice

CHAPTER OUTLINE

LEARNING OBJECTIVES

After completing this chapter, you should be able to:

1. Describe some of the early institutions used to respond to wayward and criminal youths.
2. Explain the effects of important landmark U.S. Supreme Court cases on the juvenile justice system.
3. Identify and describe factors that influence the ways police process juvenile cases.
4. Summarize the rationale for the use of diversion in juvenile justice.
5. Describe the adjudication hearing in juvenile justice.
6. Describe the disposition hearing and the types of dispositions available to the juvenile court.
7. Identify the types and describe the effectiveness of community-based correctional programs for juveniles.
8. Summarize recent trends in juvenile incarceration.
9. Identify the types and describe the effectiveness of institutional programs for juveniles.

Crime Story

On May 7, 2019, 18-year-old senior Devon Erickson and 16-year-old sophomore Alec McKinney (pictured) opened fire in a British literature class at STEM School Highlands Ranch in Littleton, Colorado. Kendrick Castillo, an 18-year-old senior, who with two other students rushed Erikson in an attempt to stop the attack, was the only fatality. Eight students, all 15 or older, were injured. The charter school has about 1,800 students in grades kindergarten through high school and is just 15 minutes from Columbine. It was the fourth school shooting in Colorado and about the 230th school shooting in the United States since the Columbine massacre on April 20, 1999. This school shooting differs from others in several important ways. First, McKinney and Erickson were high on drugs during the shooting, which experts say almost never occurs. Nearly all school shooters, regardless of their drug use history, experts agree, "are clean and sober on the day [they decide] to kill people." Second, not since 20 years ago at Columbine has a second shooter taken part in a massacre. All other school shootings have involved a single assailant.

Another unique feature of this school shooting is that McKinney is transgender—the first known transgender suspected mass shooter in U.S. history. McKinney hated being in a female body and first started using male pronouns and the name of Alec in the eighth grade. He sometimes discussed hallucinating and his thoughts not being his own. He told police that he had been suicidal and homicidal since the age of 12. In 2018, McKinney gave Xanax to a girl who accidentally overdosed. She survived, but McKinney was placed into a juvenile diversion program.

McKinney told authorities that he sought revenge on students who "misgendered" him. He wanted them to "have to suffer from trauma like he has had to in his life." McKinney especially wanted to kill two students who repeatedly had bullied him about his gender. He told police that one student "hated him" and said he was disgusting for presenting himself as male; the other student referred to him as "she." McKinney had been self-harming himself by cutting his body from his ankles to his shoulders until he decided to harm others. His cuts were so deep that most of them needed stitches, but he did not receive them because he refused to tell his mother. Because of his drug use, his mother, who essentially raised him alone, would not allow him to start testosterone treatments.

McKinney and Erickson had discussed bullies on social media when they conceived of the attack on Snapchat. They only had known each for a few months, but they bonded quickly over music, drugs, and their mental health issues. McKinney, the leader behind the attack, despite being 2 years younger than his accomplice, had been planning the attack for a few weeks. He recruited Erickson and threatened him with an ax if he refused to go along. Erikson, the reluctant accomplice, was not faultless. McKinney's ex-boyfriend said of Erikson that his being involved in the shooting "fit into my concept of him." Erickson was known to regularly hold up his hands as finger guns and pretend to shoot people. A friend told reporters that Erickson had talked about causing a lot of harm and sadness.

On the morning of the shooting, Erickson left his photography class about 10:30 a.m. His ex-girlfriend texted him to see whether he was all right. He texted back that he had thrown up. He went home, told nobody about McKinney and his plan, and then came back to school. He left a second time supposedly to call the police, but before he could, McKinney messaged him and insisted that he come and pick him up from school, which he did. Then they went to Erickson's house. They snorted cocaine together, and McKinney used nail polish to write a final message on a closet wall inside Erickson's home where the gun safe was kept: "The voices win." They broke into Erickson's

Joe Amon/MediaNews Group/The Denver Post/Getty Images

parent's supposedly secure gun safe with an ax and a crowbar to get the weapons used in the attack: a Ruger .22-caliber rifle, a Glock 21 handgun, and a Beretta M9 semiautomatic pistol. Then they spray-painted Erickson's mother's car, writing "fuck society" and "666" and drew a pentagram, before setting the car on fire. Then they drove to school in Erickson's 2013 Honda Civic. The .22-caliber rifle and the Glock 21 handgun were in Erickson's guitar case. McKinney wore his favorite Nirvana hoodie and carried the Beretta M9 in his backpack. McKinney promised that he would not kill Erickson if Erickson instructed everyone to stay still and watched the classroom's other door. Erickson admitted not knowing the entire plan, but what he did know was that McKinney wanted to kill one specific student.

McKinney and Erickson returned to school at about 1:50 p.m. and entered through the middle school entrance, where McKinney "knew they would not be checked." They went to their British literature class, which was screening the movie *The Princess Bride*. Erickson told authorities that he saw McKinney reach for a gun, when he pulled the Glock from his guitar case, pointed it straight up, and shouted at people to get down. That is when Castillo and two other students ran at Erickson to stop him. Erickson claims that Costello and the other student hit him and that the impact caused the gun to discharge. An injured student held

Erickson down until police arrived. In the meantime, McKinney fired his gun until it was empty, injuring eight people. Students and a teacher tackled McKinney, but he got up, grabbed another gun, and ran out of the classroom intending to commit suicide. He could not fire the gun because he did not know how to unlock the safety. A school security guard ordered him to the ground, and he obeyed. The police took them both into custody. McKinney and Erickson are charged with 48 counts of murder, attempted murder, and other crimes, such as theft and arson. As of this writing, McKinney is being charged as an adult, although his attorneys are attempting to get his case moved to juvenile court, where the maximum sentence is 7 years.

The subject of Chapter 13 is juvenile justice, and the children over which it has jurisdiction. Are the charges against McKinney and Erickson appropriate? Why or why not? Are both offenders equally culpable? Why or why not? Should McKinney be tried as an adult? Why or why not? What would be an appropriate punishment for each offender? Defend your answer. What can be done to reduce school shootings?

Historical Development of Juvenile Justice

From a historical perspective, juvenile delinquency and a separate justice process for juveniles are recent concepts. So, too, are the ideas of childhood and adolescence. **Juvenile delinquency**, as you may recall from Chapter 2, is a special category of offense created for youths—that is, in most U.S. jurisdictions, persons between the ages of 7 and 18.

juvenile delinquency A special category of offense created for youths who, in most U.S. jurisdictions, are persons between the ages of 7 and 18.

Through most of recorded history, the young have not enjoyed the statuses of childhood and adolescence as special times during which they need nurturing and guidance for their healthy development. Before the sixteenth century, the young were viewed either as property or as miniature adults who, by the age of 5 or 6, were expected to assume the responsibilities of adults. They also were subject to the same criminal sanctions as adults. However, in the sixteenth and seventeenth centuries, a different view of the young emerged that recognized childhood as a distinct period of life and children as corruptible but worth correcting.[1] Youths began to be viewed not as miniature adults or as property, but rather as persons who required molding and guidance to become moral and productive members of the community. American colonists brought with them those new ideas about childhood, as well as European mechanisms for responding to violators of social and legal rules.

During the colonial period, the family was the basic unit of economic production and the primary mechanism through which social control was exerted. Survival depended on the family's ability to produce what it needed rather than relying on the production of others. Consequently, a primary responsibility of the family was overseeing the moral training and discipline of the young. During this period, two age-old mechanisms were employed to teach a trade to children who were difficult to handle or needed supervision in order to allow them an opportunity to earn a livelihood. One of those mechanisms was the **apprenticeship system**, which served as a primary means for teaching skilled trades to the children of the middle and upper classes. The other tradition was the binding-out system, which was reserved for poor children. Under the **binding-out system**, children were bound over to masters for care. However, under this system, masters were not required to teach the youths a trade. As a result, boys often were given farming tasks, and girls were assigned to domestic duties.[2]

apprenticeship system The method by which middle- and upper-class children were taught skilled trades by a master.

binding-out system Practice in which children were "bound over" to masters for care. However, under the binding-out system, masters were not required to teach youths a trade.

Religion, particularly in New England, was another powerful force that shaped social life in the colonies. Regular church attendance was expected, and religious beliefs dominated ideas about appropriate behavior. Present-day concerns about the separation of church and state were nonexistent. What was believed to be immoral was also unlawful and subject to punishment by the authorities. Punishments such as fines, whipping, branding, and the use of stocks and the pillory served as reminders to both young and old that violations of community norms would not go unpunished.[3]

By the early 1800s, the social organization of colonial life began to change as a result of economic and social developments. The family-based production unit that had characterized colonial social life was giving way to a factory-based system in the growing towns. As parents—particularly fathers—and children began to leave the home for work in a factory, fundamental changes occurred in the relationships between family members and in the role of the family in controlling the behavior of children. Further, as industry developed and as towns grew, communities became more diverse and experienced problems on a scale unheard of during earlier periods.

Until the early 1900s, children were subject to the same punishments as adults. *Should children be punished in the same way as adults? Why or why not?*

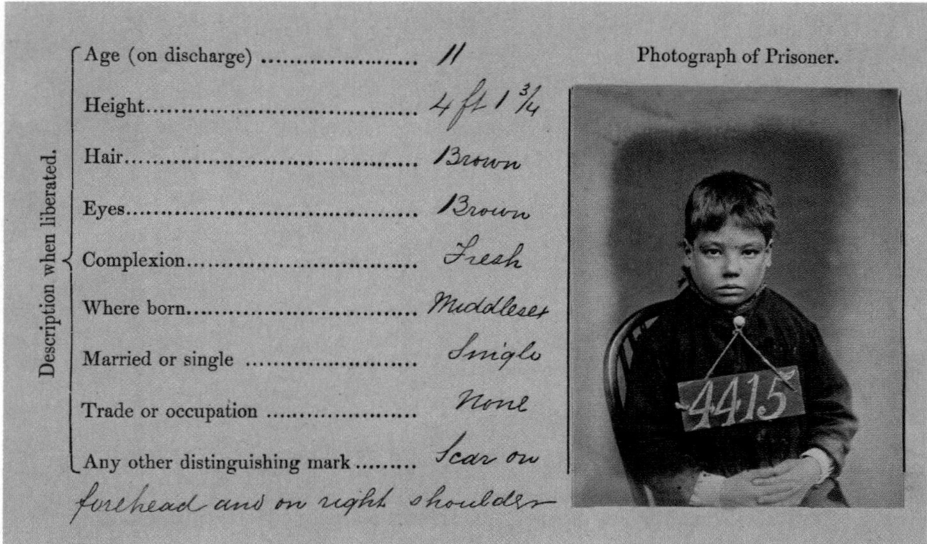

Description when liberated.

Age (on discharge)	11	Photograph of Prisoner.
Height	4 ft 1 3/4	
Hair	Brown	
Eyes	Brown	
Complexion	Fresh	
Where born	Middlesex	
Married or single	Single	
Trade or occupation	None	
Any other distinguishing mark	Scar on forehead and on right shoulder	

Public Record Office/HIP/The Image Works

The Development of Institutions for Youths

At the time of the American Revolution, Philadelphia had fewer than 20,000 residents, and other large towns, such as New York, Boston, Newport, and Charleston, had fewer than 15,000 inhabitants each. However, by 1820, the population of New York City was about 120,000 and was growing rapidly as a result of immigration. Immigration, in turn, was changing the composition of communities, which had been more homogeneous during colonial times.

houses of refuge The first specialized correctional institutions for youths in the United States.

Juvenile Crime Wave

Each generation of Americans seems to believe that the country is experiencing a "juvenile crime wave."

Source: Thomas J. Bernard and Megan C. Kurlychek, *The Cycle of Juvenile Justice*, 2nd ed. (New York: Oxford University Press, 2010).

The Houses of Refuge

Accompanying those changes in the social and economic life of the growing cities were a host of social problems, such as poverty, vagrancy, drunkenness, and crime, including crimes committed by children. In response to those conditions, the first correctional institutions specifically for youths were developed. Those institutions were called **houses of refuge**. The first was established in New York City in 1825, and houses of refuge soon spread to other cities such as Boston and Philadelphia.[4]

A primary goal of the houses of refuge was to prevent pauperism and to respond to youths who were ignored by the courts. Houses of refuge were meant to be institutions where children could be reformed and turned into hard-working members of the community. To accomplish this mission, youths were placed in houses of refuge for indeterminate periods or until their 18th or 21st birthday. Placement, moreover, did not require a court hearing. A child could be committed to a house of refuge by a constable, by a parent, or on the order of a city alderman.[5]

While in a house of refuge, children engaged in a daily regimen of hard work, military drills, and enforced silence, as well as religious and academic training. It also was common practice for outside contractors to operate shops within the houses of refuge. In those shops, children produced goods such as shoes or furniture; in return, the houses of refuge were paid 10 to 15 cents per youth each day. This arrangement allowed houses of refuge to pay a substantial percentage of their daily operating expenses.[6] When the youthful prisoners failed to meet production quotas, they often were punished. After "reformation," boys frequently were indentured to masters on farms or to tradesmen, and girls were placed in domestic service.[7]

placing out The practice of placing children on farms in the Midwest and West to remove them from the supposedly corrupting influences of their parents and the cities.

Placing Out

Soon after the establishment of houses of refuge, reformers began to recognize the inability of those institutions to accommodate the large numbers of children needing placement and to either reform or control youths. One early response to those problems was **placing out**, which involved placing children on farms in the West and Midwest. This was believed to have several advantages over the houses of refuge. First, placing out was seen as a way of removing children from the supposedly corrupting influences

Historical/Contributor/Getty Images

Houses of refuge were the first correctional institutions specifically for youths. *Should houses of refuge be used again for delinquent youths? Why or why not?*

of their parents, many of whom were immigrants, and especially from the corrupting influence of the cities, which reformers viewed as breeding grounds for idleness and crime. Second, many reformers recognized that the conditions in the houses of refuge were counterproductive to the goal of reform. Third, rural areas were assumed to be an ideal environment for instilling in children the values the reformers cherished, such as discipline, hard work, and piety.

Agents hired by charitable organizations would take children west by train and place them with farm families. Although some children were placed in caring homes and were treated as members of the family, others were not so lucky. Many of the children who were placed out were required to work hard for their keep, were abused, were not accepted as members of the family, and never saw their own families again.

Probation Another effort to deal with troubled children was initiated by a Boston shoemaker, John Augustus. As described in Chapter 12, Augustus spent considerable time observing the court and became convinced that many minor offenders could be salvaged. As a result of his concern and his willingness to work with offenders, Augustus was permitted to provide bail for his first probation client in 1841.

Augustus's first client was a drunkard who showed remarkable improvement during the period in which he was supervised. The court was impressed with Augustus's work and permitted him to stand bail for other minor offenders, including children.

After Augustus died, the Boston Children's Aid Society and other volunteers carried on his work. Then, in 1869, the state of Massachusetts formalized the existing volunteer probation system by authorizing visiting probation agents, who were to work with both adult and child offenders who showed promise. Under this arrangement, youths were allowed to return home to their parents, provided they obeyed the law.[8] In 1878, an additional law was passed in Boston that provided for paid probation officers.[9] Subsequently, several other states authorized the appointment of probation officers. However, it was not until after the turn of the twentieth century and the development of the first juvenile court that probation gained widespread acceptance.[10]

Reform Schools, Industrial Schools, and Training Schools By the late 1800s, the failure of houses of refuge was well known. Dislocations produced by the Civil War placed tremendous strain on houses of refuge and the placing-out system. Perhaps most disappointing, the number of problem youths was growing. In response, state and city governments developed their own institutions for problem children. Another response was the establishment of **reform, industrial,** or **training schools**, correctional facilities that focused on custody.[11] There were two types of these correctional facilities: cottage reformatories and institutional reformatories.

reform, industrial, or **training schools** Correctional facilities for youths, first developed in the late 1800s, that focused on custody. Today, those institutions are often called training schools, and although they may place more emphasis on treatment, they still rely on custody and control.

cottage reformatories Correctional facilities for youths, first developed in the late 1800s, that were intended to closely parallel family life and remove children from the negative influences of the urban environment. Children in those facilities lived with surrogate parents, who were responsible for the youths' training and education.

Cottage reformatories usually were located in rural areas to avoid the negative influences of the urban environment. They were intended to closely parallel family life. Each cottage contained 20 to 40 youths, who were supervised by cottage parents charged with the task of overseeing residents' training and education.[12]

In addition to cottage reformatories, larger, more institutional reformatories were developed in many states. Like the cottage reformatories, the institutional reformatories usually were located in rural areas in an effort to remove youths from the negative influences of city life. However, the institutions frequently were large and overcrowded.

Another development, in the late 1800s, was the establishment of separate institutions for females. Previously, girls had been committed to the same institutions as boys, although there was strict gender segregation in those institutions. Moreover, parents or relatives often committed girls for moral, as opposed to criminal, offenses. Those moral offenses consisted of such actions as "vagrancy, beggary, stubbornness, deceitfulness, idle and vicious behavior, wanton and lewd conduct, and running away."[13] The expressed goal of those institutions was to prepare girls to be good housewives and mothers. Yet, girls' institutions, like those for boys, were little different from the prisons of that era.

The reform, industrial, and training schools placed more emphasis on formal education than did the houses of refuge, but in many other respects there was little difference. Indeed, those institutions confronted many of the same problems as the houses of refuge. Moreover, the conditions in the reformatories were certainly no better than those in the houses of refuge, and, in many cases, they were worse.

Juveniles incarcerated at the turn of the twentieth century were put into job training programs that would help them when they were released. Typically, male offenders learned industrial trades; female offenders learned ironing, laundry work, and cooking. *Should these practices be revived? Why or why not?*

Source: George Grantham Bain Collection (Library of Congress). https://www.loc.gov/item/2014681565/.

Source: National Photo Company Collection (Library of Congress).

The Development of the Juvenile Court

By the end of the 1800s, a variety of institutions and mechanisms had been developed in response to problem children. Still, the problems presented by children who were believed to be in need of correctional treatment–such as homelessness, neglect, abuse, waywardness, and criminal behavior–proved difficult to solve. Consequently, during the late 1800s a new group of reformers, the *child savers*, began to advocate a new institution to deal with youth problems. This new institution was the juvenile court, and to understand how and why it developed, it is necessary to examine the social and legal context of the late 1800s and early 1900s.

FYI

Juvenile Courts

Each state has at least one court with juvenile jurisdiction, but it is called juvenile court in only a few states. Other names include district, superior, circuit, county, or family court. Regardless of the name, courts with juvenile jurisdiction generally are referred to as juvenile courts.

Source: Melissa Sickmund, "Juveniles in Court," in *Juvenile Offenders and Victims National Report Series Bulletin*, Office of Juvenile Justice and Delinquency Prevention, U.S. Department of Justice (Washington, DC: U.S. Government Printing Office, June 2003), 4.

The Social Context of the Juvenile Court The period from 1880 to 1920, which historians refer to as the Progressive Era, was a time of major change in the United States. Although industrialization and urbanization were well under way, and previous waves of immigrants had added to the country's population, the pace of industrialization, urbanization, and immigration quickened. The city of Chicago provides a good example of the changes experienced during the Progressive Era. Between 1890 and 1910, Chicago's population grew from one million to two million. Between 1880 and 1890, the number of factories nearly tripled. By 1889, nearly 70% of the inhabitants of the city were immigrants.[14]

Along with the dramatic changes in the size and diversity of Chicago came a number of social problems. Two of these problems were youth crime and waywardness. In an effort to respond to these problems, reformers again sought to save children from the crime-inducing conditions of the cities. Supported by important philanthropic and civic organizations, the child savers worked to improve jail and reformatory conditions. However, the primary outcome of the child-saving movement was the extension of governmental control over children's lives. The child savers argued for stricter supervision of children, and they improved legal mechanisms designed to regulate children's activities.[15] In short, the child savers believed that children needed to be protected and that the best institutions for protecting them were government agencies such as the police and the courts, as well as local charitable organizations.

The Legal Context of the Juvenile Court By the late 1800s, legal mechanisms for treating children differently and separately from adults had existed for some time. For example, jurisdictions had set the minimum age at which a child could be considered legally responsible for criminal behavior. Minimum ages for placement in adult penitentiaries also were enacted during the first half of the 1800s.[16] Moreover, special institutions for dealing with youths had been in existence since 1825, when the first house of refuge was established in New York City. Yet, cases involving juveniles still were heard in criminal courts, although many people, including the child savers, believed that criminal courts failed to respond adequately to problem youth.

The legal philosophy justifying state intervention in the lives of children, embodied in the doctrine of ***parens patriae*** ("the state as parent"), was given judicial endorsement in the case *Ex parte Crouse* (1838). Mary Ann Crouse had been committed to the Philadelphia House of Refuge by her mother against her father's wishes. Mary Ann's father contested his daughter's placement, arguing that she was being punished even though she had committed no criminal offense. However, the Pennsylvania Supreme Court ruled that Mary Ann's placement was legal because (1) the purpose of the Philadelphia House of Refuge was to reform youths and not to punish them; (2) formal due process protections provided to adults in criminal trials were unnecessary because Mary Ann was not being punished; and (3) when parents were unwilling or unable to protect their children, the state had a legal obligation to do so.[17]

parens patriae The legal philosophy justifying state intervention in the lives of children when their parents are unable or unwilling to protect them.

However, the right of the state to intervene in the lives of children did not go unchallenged. In *People v. Turner* (1870), for example, the Illinois Supreme Court ruled that Daniel O'Connell, who was committed to the Chicago House of Refuge against both his parents' wishes, was being punished and not helped by his placement. Like Mary Ann Crouse, who was committed by her mother against her father's wishes, Daniel O'Connell was institutionalized even though he had committed no criminal offense. He was placed in the house of refuge because he was perceived to be in danger of becoming a pauper or a criminal. However, Daniel's case differed from Mary Ann's because both his parents objected to his placement and, even more important, because the court ruled that Daniel's placement was harmful, not helpful. The court also decided that because placement in the house of refuge actually was punishment, due process protections were necessary.[18] The ruling, together with increasing concern over the willingness or ability of the criminal courts to protect or control youths, led reformers in Chicago to consider other mechanisms by which their aims might be achieved. The mechanism they created was the first juvenile court, which was established in Chicago in 1899 by passage of the Juvenile Court Act in Illinois.[19] Some scholars argue that the Juvenile Court Act was a means for the child savers, who were intent on salvaging poor children–especially poor immigrant children–to get around the *Turner* decision's requirement of due process protections for youths.

The Operation of Early Juvenile Courts The Juvenile Court Act of 1899 gave the Chicago juvenile court broad jurisdiction over persons under the age of 16 who were delinquent, dependent, or neglected. In addition–the act required:

1. The court to be overseen by a special judge.
2. Hearings to be held in a separate courtroom.
3. Separate records to be kept of juvenile hearings.[20]

It also made probation a major component of the juvenile court's response to offenders and emphasized the use of informal procedures at each stage of the juvenile court process. Indeed, this informality has been a hallmark of the juvenile court since its beginning, a key feature distinguishing it from criminal court proceedings. The major reason for this informality is that the juvenile court traditionally has focused not on the act but on the *whole child.*

In practice, the informality of the juvenile court allowed complaints against children to be made by almost anyone in the community. It also allowed juvenile court hearings to be held in offices instead of in traditional courtrooms and to be closed to the public, unlike criminal trials, which were open to the public. In the typical juvenile court hearing, the only persons present were the judge, the parents, the child, and the probation officer, who met and discussed the case. Also, few (if any) records were kept of hearings, proof of guilt was not necessary for the court to intervene in children's lives, and little or no concern for due process existed. Finally, judges exercised wide discretion in how they dealt with children, ranging from a warning to placement in an institution.[21]

The idea of a juvenile court spread rapidly after the passage of the Juvenile Court Act. Within a decade, 10 states had established special courts for children, and by 1925, all but 2 states had juvenile courts.[22] Moreover, those juvenile courts closely followed the model developed in Chicago of an informal court intended to "serve the best interests of children."

The use of informal procedures is characteristic of the juvenile court system. *Are the juvenile court's informal procedures desirable? Why or why not?*

Design Pics Inc./Alamy

Despite the growing popularity of the new courts, they did not go completely unchallenged. In the case *Commonwealth of Pennsylvania v. Fisher* (1905), for example, the Pennsylvania Supreme Court again examined the juvenile court's mission, the right to intervene, and the due process protections owed to children. In this case, Frank Fisher, a 14-year-old male, was indicted for larceny and committed to the house of refuge until his 21st birthday. Frank's father objected to his placement, claiming that Frank's 7-year sentence for a minor offense was more severe than he would have received in criminal court.[23]

In its ruling, the Pennsylvania Supreme Court upheld the idea of the juvenile court and in many respects repeated the arguments it had made in the *Crouse* decision. The court found that the state may intervene in families when parents are unable or unwilling to prevent their children from engaging in crime and that Frank was being helped by his placement in the house of refuge. It further ruled that due process protections were unnecessary when the state acted under its *parens patriae* powers.[24]

The *Fisher* case set the legal tone for the juvenile court from its beginnings until the mid-1960s, when new legal challenges began to be mounted. Those legal challenges primarily attempted to expand juveniles' due process protections. Critics of the juvenile courts recognized that despite their expressed goal of "serving the best interests of children," the established institutions of juvenile justice often did the opposite.

The Legal Reform Years: The Juvenile Court After *Gault* The 1960s and 1970s provided the social context for a more critical assessment of American institutions, including juvenile justice. Beginning in the mid-1960s and continuing through the mid-1970s, a number of cases decided by the U.S. Supreme Court altered the operation of the juvenile court. The most important of those cases was *In re Gault,* which expanded the number of due process protections afforded juveniles within the juvenile court. However, a number of other cases also helped define juveniles' rights within juvenile justice and contributed to the legal structure found in juvenile courts today. The first of those cases was *Kent v. United States* (1966).

In re

The Latin phrase *in re* in the description of court cases (e.g., *In re Gault*) means, literally, "in the matter of" or "concerning." It is used when a case does not involve adversarial parties.

Morris Kent was a 16-year-old juvenile on probation who was transferred to criminal court to stand trial on charges of robbery and rape. Although the juvenile court judge received several motions from Kent's attorney opposing the transfer, he made no ruling on them. Further, after indicating that a "full investigation" had been completed, the juvenile court judge transferred Kent to criminal court for trial.[25] Thus, an important decision had been made—the decision to try Kent as an adult—even though no hearing had been held. Kent's attorney had no opportunity to see or to question material that had been used to make the decision to transfer jurisdiction, and no reasons for the court's decision were given.

The *Kent* case is important for several reasons. It was the first major ruling by the U.S. Supreme Court that closely examined the operation of the juvenile courts. It also made clear the need for due process protections for juveniles who were being transferred to criminal court for trial. The Court noted that even though a hearing to consider transfer to criminal court is far less formal than a trial, juveniles still are entitled to some due process protections. Specifically, the Court ruled that before a juvenile court could transfer a case to criminal court, (1) there must be a hearing to consider the transfer, (2) the defendant must have the assistance of defense counsel if requested, (3) defense counsel must have access to social records kept by the juvenile court, and (4) the reasons for the juvenile court's decision to transfer must be stated.[26]

Having given notice that it would review the operations of the juvenile court, the Supreme Court heard another landmark case within a year of the *Kent* decision. This case, *In re Gault* (1967), went far beyond the ruling in *Kent* in its examination of juvenile court practices. The *Gault* case is important because it extended a variety of due process protections to juveniles. In addition, the facts of the case clearly demonstrate the potential for abuse in the informal practices of the traditional juvenile court.

Gerald Gault was 15 years old when he, along with a friend, was taken into custody by the Gila County, Arizona, Sheriff's Department for allegedly making an obscene phone call to a neighbor, Ms. Cook. At the time of his arrest, Gerald was on 6-month probation as a result of his presence when another friend had stolen a wallet from a woman's purse. Without notifying his mother, a deputy took Gerald into custody on Ms. Cook's oral complaint and transported him to the local detention unit.

When Ms. Gault heard Gerald was in custody, she went to the detention facility. The superintendent of the facility told her that a juvenile court hearing would be held the next day. On the following day, Gerald, his mother, and the deputy who had taken Gerald into custody appeared before the juvenile court judge in chambers. The deputy had filed a petition alleging that Gerald was delinquent. Ms. Cook, the complainant, was not present. Without being informed that he did not have to testify, Gerald was questioned by the judge about the telephone call and was sent back to detention. No record was made of this hearing, no one was sworn, and no specific charge was made other than an allegation that Gerald was delinquent. At the end of the hearing, the judge said he would "think about it." Gerald was released a few days later, although no reasons were given for his detention or release.

On the day of Gerald's release, Ms. Gault received a letter indicating that another hearing would be held a few days later about Gerald's delinquency. The hearing was held and, again, the complainant was not present. Again, there was no transcript or recording of the proceedings, and the parties later disputed what was said. Neither Gerald nor his parent was advised of a right to remain silent, a right to be represented by counsel, or any other constitutional rights. At the conclusion of the hearing, Gerald was found delinquent and

Bettmann/Contributor/Getty Images

In the 1967 landmark case of *In re Gault*, Supreme Court Justice Abe Fortas wrote, "Due process of law is the primary and indispensable foundation of individual freedom." *Do you agree with Justice Fortas? Why or why not?*

was committed to the state industrial school until he was 21 years old, unless released earlier by the court. This meant that Gerald received a sentence of up to 6 years for an offense that, if committed by an adult, would have been punished by a maximum sentence of 2 months in jail and a $50 fine.

When the case finally reached the Supreme Court, the Court held that a youth has procedural rights in delinquency hearings where there is the possibility of confinement in a state institution. Specifically, the Court ruled that juveniles have the following rights: (1) a right against self-incrimination, (2) a right to adequate notice of charges against them, (3) a right to confront and cross-examine their accusers, and (4) a right to assistance of counsel. In addition, the Court's ruling implied that juveniles also have the rights to sworn testimony and appeal.

The landmark *Gault* decision was not the last Supreme Court decision to influence juvenile court procedures. The Supreme Court further expanded protections for juveniles 3 years after *Gault*. In a 1970 case, *In re Winship*, the Court ruled that delinquency charges must be proven beyond a reasonable doubt where there was a possibility that a youth could be confined in a locked facility. Until the *Winship* ruling, the standard of proof typically employed at the adjudication stage of the juvenile justice process was a preponderance of the evidence, the level of proof employed in civil proceedings. **Adjudication** is the juvenile court equivalent of a trial in criminal court, or the process of rendering a judicial decision regarding the truth of the facts alleged in a petition. Under the preponderance-of-the-evidence standard of proof, juveniles could be adjudicated delinquent (found guilty) if the weight of the evidence was slightly against them, which is a much lower standard of proof than the standard required in criminal courts.

adjudication The juvenile court equivalent of a trial in criminal court, or the process of rendering a judicial decision regarding the truth of the facts alleged in a petition.

Note that both the *Gault* and *Winship* decisions not only increased procedural formality in juvenile court cases but also shifted the traditional focus from the "whole child" to the child's act. Once this shift occurred, it was only a short step to offense-based sentencing and the more punitive orientation that is characteristic of the juvenile justice system today.

The Supreme Court's extension of due process protections to juveniles slowed in the year following the *Winship* decision. In a 1971 ruling, *McKeiver v. Pennsylvania*, the Court held that juveniles were not entitled to a trial by jury. The Court cited several reasons for the decision:

1. The Court did not want to turn the juvenile court into a fully adversarial process.
2. The Court determined that bench trials could produce accurate determinations.
3. The Court felt that it was too early to completely abandon the philosophy of the juvenile court and its treatment mission.

To grant juveniles all of the protections accorded adults, the Court surmised, would make the juvenile court indistinguishable from the criminal court. Nevertheless, many states have required more due process protections for juveniles than have been mandated by the Supreme Court. For example, many state laws specify that juveniles have a right to trial (adjudication) by a jury. In fact, Texas requires that a jury hear all adjudications. Nevertheless, jury trials are rare in states that have this right, including Texas. This is because the right must be exercised or, as in Texas, the automatic right to a trial by jury can be waived. In practice, jury trials often are discouraged because they are time-consuming and costly.

The continued informality of the juvenile court may explain why very few youths contest charges against them and why a surprising number of youths who are not adjudicated delinquent (found guilty) are placed on probation. Indeed, most youths who appear before the juvenile court admit to the charges against them.[27] Moreover, data collected by the National Center for Juvenile Justice indicate that in 2017, about 35% of youths who were not adjudicated delinquent in juvenile courts still were placed on some form of probation, up from 28% in 2013 (also see Figure 13.2).[28]

Today, juveniles have been granted many, but not all, of the due process protections given adults in criminal trials. However, the daily operation of juvenile courts calls into question the extent to which court-mandated changes in juvenile justice procedures have influenced the traditional informality of the juvenile court. Juvenile court procedures continue to be characterized by an informality that most people would find unacceptable if it were applied to adults in criminal court.

THINKING CRITICALLY

1. Should youths in juvenile court proceedings be granted the same due process protections as adults in criminal trials? Why or why not?
2. Historically, what do you think was the most significant change in the treatment of juvenile offenders? How did the change occur?

The Processing of Juvenile Offenders

Juvenile delinquency in the United States is widespread, and people respond to it both formally and informally. Informal responses consist of actions taken by members of the public that do not rely on official agencies of juvenile justice. Formal responses, in contrast, rely on official agencies of juvenile justice such as the police and juvenile court. Figure 13.1 depicts the stages in the formal juvenile justice process. Before providing a detailed examination of the formal juvenile justice process, however, we take a brief look at the informal juvenile justice process.

Juvenile Justice by State

Access the Office of Juvenile Justice and Delinquency Prevention website at https://www.ojjdp.gov/ojstatbb/special_topics/stateprofile.asp to find descriptive information and analyses of the various juvenile justice systems in the United States. Compare your state's juvenile justice system to that of another state.

How are the two states similar? Different?

The Informal Juvenile Justice Process

An examination of the juvenile justice process usually begins with the police, who play a critical role in juvenile justice. However, any discussion of the juvenile justice process really should begin with the public because police involvement with juveniles typically is the result of citizen complaints.

The informal actions taken by citizens to respond to delinquency constitute an **informal juvenile justice** process that operates outside the official agencies of juvenile justice. Neighbors, business owners, teachers, and others who are not part of the formal juvenile justice apparatus handle many illegal behaviors by juveniles informally. The informal processing of juveniles is important because it is one of the mechanisms that operate to control youths' behavior. The more citizens rely on informal control and the more effective it is, the less necessary the formal processing of juveniles becomes.

informal juvenile justice The actions taken by citizens to respond to juvenile offenders without involving the official agencies of juvenile justice.

The informal processing of juveniles is important for another reason as well. A recent review of 29 experiments conducted over a 35-year period with 7,304 juveniles found that formal juvenile system processing did not have an effect on crime control. Rather, formal processing appeared to increase delinquency. In studies that compared formal processing with diversion programs, the negative effects of formal processing were even greater than when formal processing was compared to "doing nothing."[29]

Although many delinquent actions are handled informally, in other instances members of the public decide to call the police or the juvenile court and to request action from the formal agencies of juvenile justice. Thus, most youths become involved in the formal juvenile justice process when the people who make up the informal process decide, through exercising discretion, to involve the police or the juvenile court. This means that some combination of the public and the police plays a major role in determining who the clientele of juvenile justice will be.

Corbis/PictureQuest

Teachers are an important part of the informal juvenile justice process. *What can teachers do to keep students from becoming delinquent?*

The Formal Juvenile Justice Process

The police represent the primary gatekeepers to the formal juvenile justice process (see Figure 13.1). For example, in 2017, 81% of the approximately 466,000 petitioned delinquency cases referred to the juvenile courts came from police agencies.[30] This percentage has remained about the same for more than 2 decades. In the case of status offenses, however, the percent-

FIGURE 13.1 The Formal Juvenile Justice Process

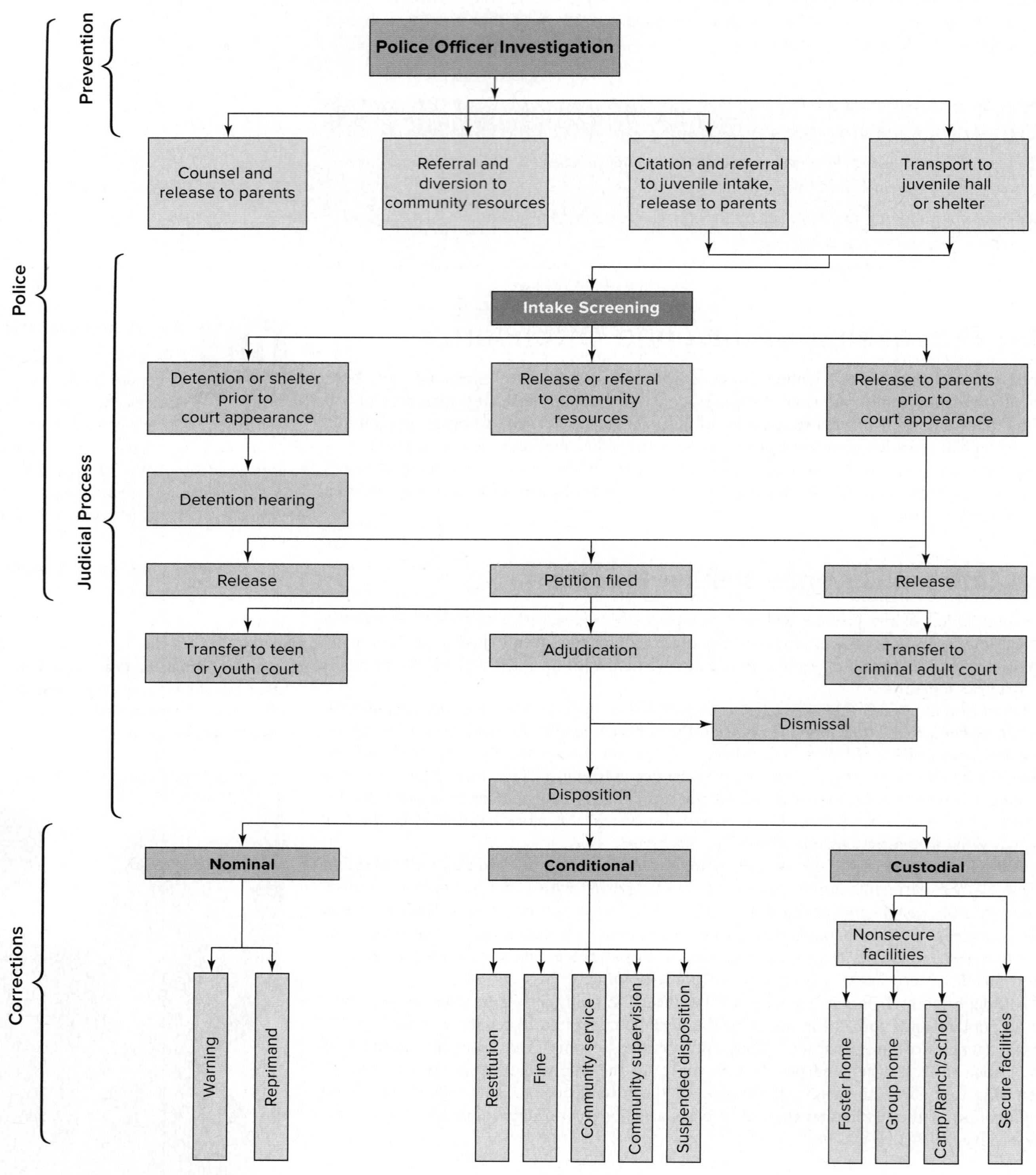

status offense An act that is not a crime when committed by adults but is illegal for minors (e.g., truancy or running away from home).

age of cases referred by police agencies is much smaller because other sources such as schools and parents refer many status offenses cases to juvenile courts. For example, in 2017, police departments referred only 20% of the approximately 88,000 petitioned status offenses to juvenile courts, down from 55% in 2013, and 34% in 2005.[31] **Status offenses**, as you may recall, are acts that are not crimes when committed by adults but are illegal for minors.

The Police Response to Juveniles Like citizens, the police exercise discretion in handling juvenile cases. Typical responses that police officers may employ are to:

1. Warn and release.
2. Refer to parents.
3. Refer to a diversionary program operated by the police or another community agency.
4. Refer to court.

In some communities, an officer may have a variety of options, while in other communities, available options are more limited.

Describing the typical police response to juveniles in trouble is difficult because there is considerable variation in the ways individual officers approach juvenile offenders. A number of factors influence the ways police officers handle juvenile suspects. Among them:

1. The seriousness of the offense.
2. The police organization.
3. The community.
4. The wishes of the complainant.
5. The demeanor of the youth.
6. The gender of the offender.
7. The race and social class of the offender.

Offense Seriousness The most important factor in the decision to arrest is the seriousness of the offense. Regardless of the subject's demeanor or other factors, as the seriousness of the offense increases, so does the likelihood of arrest. Most police-juvenile encounters involving felony offenses result in an arrest. However, most police-juvenile encounters involve minor offenses. When offenses are minor, a number of other factors influence police decision making.

The Police Organization As described in Chapter 6, police departments develop their own particular styles of operation. Those styles, in turn, help structure the way officers respond to juvenile offenders. For example, in one study that categorized police departments by the extent to which they employed a legalistic style of policing, characterized by a high degree of professionalism and bureaucratic structure, it was discovered that officers in most legalistic departments were more likely to arrest juvenile suspects than were officers in less legalistic departments.[32] Thus, the ways that police department personnel respond to juvenile offenders is, to some extent, a product of the organizational characteristics and policies developed by their individual departments. However, police departments do not operate in a political and social vacuum. The communities in which they operate also influence them.

myth

The commission of serious, violent crime by juveniles is widespread and growing.

fact

Regardless of age, the violent crime (murder, robbery, and aggravated assault) arrest rate grew substantially between 1980 and 1994. During this period, the relative increase was greater for juveniles than for adults. By 2017, arrest rates for violent crimes fell substantially from the 1994 peak for every age group younger than 45. Juveniles showed the largest decline—falling more than 65% in each age group from 10 through 17.

Source: Charles Puzzanchera, "Juvenile Arrests, 2017," *Juvenile Justice Statistics* (Laurel, MD: Office of Juvenile Justice and Delinquency Prevention, August 2019), 7. Accessed September 22, 2019, http://www.ncjj.org/pdf/Juvenile%20Arrests%20Bulletins/252713.pdf.

The Community Communities influence policing in a variety of ways. Through their interactions with residents of a community, police develop assumptions about the communities they protect, the people who live in those communities, and the ability and willingness of community residents to respond to crime and delinquency. Research suggests that police operate differently in lower-class communities than in wealthier communities. Police expect lower-class communities to have higher levels of crime and delinquency because more arrests are made in those communities and because they know that lower-class communities have fewer resources with which to respond informally to the array of problems experienced there. Consequently, when police have contact with juveniles in lower-class communities, formal responses become more likely because informal responses are believed to be ineffective in most cases. Indeed, research that has examined the effect of neighborhood socioeconomic status has found that as the socioeconomic status of the neighborhood increases, the likelihood that a police-juvenile encounter will end in arrest declines.[33]

The Wishes of the Complainant As noted earlier, many police–juvenile interactions occur because of a citizen complaint to the police. What police do in those situations often depends on whether the complainant is present, what the complainant would like the police to do, and, sometimes, who the complainant is.

The Demeanor of the Youth As one might expect, how youths behave toward the police can influence whether an arrest is made. Interestingly, youths who are either unusually antagonistic or unusually polite are more likely to be arrested. In contrast, youths who are moderately respectful to the police are less likely to be arrested–as long as the offense is not serious.[34]

The Gender of the Offender Gender also appears to influence the decision to arrest, particularly when status offenses are involved. A number of studies have found that female status offenders are more likely to be formally processed than male status offenders.[35]

When it comes to criminal offenses, however, research results are mixed. Some of the research shows that girls are less likely than boys to be arrested for criminal offenses, even when prior record and seriousness of offense are taken into account.[36] Other research suggests that any gender bias that existed in the past has diminished or disappeared.[37] In 2017, 29% of juvenile arrests involved females, the same percentage as in 2012.[38] Table 13.1 shows the percentage change in arrests of females and males under age 18 from 2008 through 2017. The table reveals that from 2008 through 2017, arrests of both juvenile females and males decreased in every offense category, and decreases in arrests of juvenile males were greater than for juvenile females in every category except burglary and larceny-theft. The decrease in driving under the influence was the same for both genders.

The Race and Social Class of the Offender There is strong evidence that a juvenile's race and social class influence police decision making. The evidence is clear that minority and poor youths are represented disproportionately in arrest statistics. For example, according to 2017 uniform crime reports (UCR) data, white youths accounted for about 62% of all arrests of persons under 18 years of age. They accounted for approximately 56% of arrests for index property offenses and about 46.5% of arrests for index violent offenses. In contrast, African American youths accounted for about 35% of all arrests of persons under 18 years of age, roughly 40% of index property-crime arrests, and approximately 51% of index violent-crime arrests. Other racial groups accounted for the remainder of arrests of persons under 18 years of age.[39] Thus, although white youths accounted for the majority

Daniel Wray/The Image Works

A juvenile's demeanor can influence arrest decisions. *How would you advise a juvenile to act if he or she is stopped by a police officer?*

TABLE 13.1 Percentage Change in Juvenile Arrests, 2008–2017, by Gender

Most Serious Offense	Female	Male
Violent Crime Index	−42%	−49%
Robbery	−38	−46
Aggravated assault	−43	−51
Simple assault	−42	−50
Property Crime Index	−67	−58
Burglary	−64	−63
Larceny-theft	−69	−59
Motor vehicle theft	−22	−37
Vandalism	−54	−67
Weapons	−50	−54
Drug abuse violations	−19	−52
Liquor law violations	−72	−75
DUI	−62	−62
Disorderly conduct	−63	−69

Source: Charles Puzzanchera, "Juvenile Arrests, 2017," *Juvenile Justice Statistics*, U.S. Department of Justice (Laurel, MD: Office of Juvenile Justice and Delinquency Prevention, August 2019), 8, accessed September 23, 2019, http://www.ncjj.org/pdf/Juvenile%20Arrests%20Bulletins/252713.pdf.

of juvenile arrests, African American youths were arrested in disproportionate numbers because they made up only 16% of the U.S. population aged 10–17.[40]

The UCR does not report arrests by social class. However, self-report data reveal that the prevalence of delinquency (i.e., the proportion of the youth population involved in delinquency) does not differ significantly between social classes when all types of offenses are considered. In other words, the proportions of middle-class and lower-class youths who engage in delinquency are similar. However, when different types of offenses are examined, significant social class differences appear. Middle-class youths have higher rates of involvement in such offenses as stealing from their families, cheating on tests, cutting classes, disorderly conduct, and drunkenness. Lower-class youths, in contrast, have higher rates of involvement in more serious offenses, such as felony assault and robbery.[41]

"Child Delinquents"

In 2017, the police arrested 176,559 youths age 15 and younger. Such very young offenders represented about 28% of the total number of juvenile arrestees. These numbers may underestimate the number of child delinquents because, in many jurisdictions, it is unusual for youth under age 12 to be arrested or referred to juvenile court. However, in 2014–15, 80 children younger than 9 years of age were arrested in Florida. Florida, like about two-thirds of states, has no minimum age for arrest. The youngest child arrested in Florida was a 4-year-old boy arrested in Orange County in 2010. He was the youngest of five children arrested on felony burglary and misdemeanor criminal mischief charges for breaking into and vandalizing a neighbor's shed. The other four children were 6, 8, 9, and 11.

Sources: Calculated from Federal Bureau of Investigation, *2017 Crime in the United States*, Table 36, accessed September 23, 2019, https://ucr.fbi.gov/crime-in-the-u.s/2017/crime-in-the-u.s.-2017/tables/table-36; Tonya Alanez, "Arresting Children Presents Challenges," *Orlando Sentinel* (December 25, 2016), p. B1.

Police Processing of Juvenile Offenders When a police officer encounters a juvenile who has committed an illegal act, the officer must decide what to do. One option is to make an arrest. (In some jurisdictions, this is referred to as "taking into custody.") For practical purposes, an arrest takes place whenever a youth is not free to walk away.

As a general rule, the basis for arresting a juvenile is the same as for arresting an adult. An officer needs to have probable cause. However, there are several differences between arrests of adults and juveniles. First, the police can arrest juveniles for a wider range of behaviors. For example, juveniles, but not adults, can be arrested for status offenses. (Technically, juveniles are not "arrested" for status offenses because such offenses are not crimes. Also, because status offenses are not crimes, the apprehension of juveniles for status offenses does not require probable cause.) Second, at least in some jurisdictions, juveniles are given the *Miranda* warnings in the presence of a parent, guardian, or attorney. This is not necessary, however, because the Supreme Court ruled in *Fare v. Michael C.* (1979) that parents or attorneys do not have to be present for juveniles to waive their rights. Third, in many jurisdictions, juveniles are more likely than adults who have committed similar offenses to be detained pending adjudication. In *Schall v. Martin* (1984), the Supreme Court ruled that preventive detention of juveniles is acceptable.

Concern about the detention of juveniles generally focuses on the use of preventive detention. Some critics argue that it amounts to punishment before a youth has been found guilty of an offense. The conditions juveniles sometimes are exposed to in detention units and adult jails raise additional concerns.

The exact procedures that the police must follow when taking a juvenile into custody vary from state to state and are specified in state juvenile codes. However, those codes typically require officers to notify a juvenile's parents that the juvenile is in custody. Police often ask parents to come to the police station. Sometimes, an officer transports a youth home prior to any questioning. When an officer feels that detention of a juvenile is appropriate, the juvenile is transported to a juvenile detention facility. If such facilities are not available, the youth in some jurisdictions may be taken to an adult jail. In cases where a juvenile is released to his or her parents, the juvenile and the parents are informed that the court will contact them at a later date about the case. After a juvenile is released, an officer completes the complaint, collecting any additional information needed, and then forwards it to the next stage of the juvenile justice process for further action. When a juvenile is detained in a juvenile detention facility or an adult jail, the processing of the complaint is expedited because juvenile codes require a detention hearing (and therefore, a specific complaint) to determine the appropriateness of detention.

FYI

Juvenile Diversion

Diversion is used in about 40% of juvenile cases.

Source: Stephen Bishop, "Juvenile Probation Officers Should Not Be Fixers, but Levers to Resources for Youth," Juvenile Justice Information Exchange, February 20, 2019, https://jjie.org/2019/02/20/juvenile-probation-officers-should-not-be-fixers-but-levers-to-resources-for-youth/.

Diversion The goal of juvenile diversion programs is to respond to youths in ways that avoid formal juvenile justice processing. Diversion can occur at any stage of the juvenile justice process, but, because its primary purpose is to avoid formal court processing, it most often is employed before adjudication.

Diversion programs are based on the understanding that formal responses to youths who violate the law, such as arrest and adjudication, do not always protect the best interests of youths or the community. Consequently, efforts to divert youths *from* the juvenile justice process by warning and releasing them, as well as efforts to divert youths *to* specific diversionary programs, such as counseling, have long been a part of juvenile justice practice. This is especially true for status offenders. Since the enactment of the Juvenile Justice and Delinquency Act of 1974, which stipulated that status offenders not be placed in secure detention facilities or secure correctional facilities, the number of status offenders diverted from formal juvenile justice processing has increased dramatically.

radical nonintervention A practice based on the idea that youths should be left alone if at all possible, instead of being formally processed.

Today, diversion strategies are of two basic types. Some are based on the idea of radical nonintervention. Others involve the referral of youths, and possibly parents, to a diversionary program. **Radical nonintervention** is based on the idea that youths should be left alone if at all possible instead of being formally processed. The police practice of warning and releasing some juvenile offenders is an example of radical nonintervention. The referral of juveniles to community agencies for services such as individual or family counseling is an example of efforts to involve youths in a diversion program.

Both juvenile justice and community agencies operate contemporary diversion programs. Interventions employed in those programs include providing basic casework services to youths; providing individual, family, and group counseling; requiring restitution; and imposing community service. Even though diversion programs frequently are touted as a way to reduce the number of youths involved in juvenile justice, some diversionary strategies contribute to net widening. Net widening occurs when a program handles youths who would have been left alone in the absence of the new program.

Recently, a relatively small number of juvenile offenders (estimates suggest about 140,000 cases a year or approximately 17% of the more than 800,000 delinquency cases in 2017) have been adjudicated in teen courts or youth courts instead of juvenile courts or adult courts.[42] Teen courts or youth courts are a new way of dealing with relatively young and usually first-time offenders charged with offenses such as theft; vandalism; misdemeanor assault; disorderly conduct; and possession of alcohol, marijuana, and tobacco. As of September 2019, there were more than 1,800 teen or youth court programs in the United States, up about 2,200% from the 78 in 1994. Teen or youth courts are operating in 48 states, the District of Columbia, and 34 Native American tribes. Only Vermont and Rhode Island do not have teen or youth courts.[43]

Teen Courts

To learn more about teen courts, visit the website of the National Association of Youth Courts, www.youthcourt.net.

Do you think teen courts are a good idea? Why or why not?

Teen or youth courts are based on one of four models:

1. *Adult judge*—An adult serves as judge and rules on legal matters and courtroom procedure. Youths serve as attorneys, jurors, clerks, bailiffs, and so on. The adult-judge model is the most common type of teen or youth court.
2. *Youth judge*—This is similar to the adult judge model, but a youth serves as judge. The youth-judge model is the third most common type of teen or youth court.

Rick Gebhard/AP Images

All states except Rhode Island and Vermont have at least one teen court. *Are teen courts a good idea? Why or why not?*

3. *Tribunal*–Youth attorneys present the case to a panel of three youth judges, who decide the appropriate disposition for the defendant. A jury is not used. The tribunal model is the least common type of teen or youth court.
4. *Peer jury*–This model does not use youth attorneys; the case is presented to a youth jury by a youth or adult. The youth jury then questions the defendant directly. The peer-jury model is the second most common type of teen or youth court.

More than 90% of teen courts do not determine guilt or innocence. Rather, they serve as diversion alternatives, and youths must admit to the charges against them to qualify for teen court. The most common disposition used in teen court cases is community service. Other frequently used dispositions include victim apology letters, apology essays, educational workshops, teen court jury duty, drug/alcohol classes, and monetary restitution. A recent study of nearly 500 Duval County, Florida, teen court participants, who completed their dispositions between 2009 and 2011, found that about 20% of them recidivated (were arrested for a new offense) within 1 year, and that males were four times more likely to recidivate than females.[44] The authors of the study also reviewed the results of 12 other studies of teen court participants conducted between 1987 and 2011, and reported an average recidivism rate of 24%, with a range from 10% to 49%. Recidivism typically was measured by rearrest, and follow-up periods ranged from 5 months to when the participant turned 18 years of age.[45]

Another new juvenile diversion program is restorative justice conferences or family group conferences.[46] Commonly used in Australia and New Zealand and increasingly being used throughout the world, restorative justice conferences bring together juvenile offenders, their victims, and other relevant parties (such as parents or guardians) with trained facilitators to discuss the offense and the harm done to victims and the other relevant parties. As is the case with teen courts, juvenile offenders who have been screened to participate in restorative justice conferences generally are nonviolent offenders who commit less serious offenses and have not had a previous referral.

At the conferences, victims and the other relevant parties have an opportunity to explain how they have been affected by the offense and to question the offending juveniles. At the end of the conference, the participants determine how the offending juveniles can make amends to the victims for the wrong done and then sign a reparation agreement. These agreements typically include an apology and some type of restitution to the victims. They also may require the offenders to perform community service or complete other requirements such as improving school attendance, finishing homework, or doing chores at home or school. If an agreement cannot be reached or a juvenile challenges the allegations, then the matter proceeds to juvenile court.

Advocates of restorative justice conferences and teen courts maintain that the programs provide greater benefits to all parties involved than does the traditional juvenile court system. Advocates argue that restorative justice conferences and teen courts hold juveniles more accountable for their offenses, enable offending juveniles to appreciate how their actions negatively affected others, involve and meet the needs of victims, and create a more supportive community for offending juveniles. Additional benefits include the timelier handling of cases, cost savings because of the heavy reliance on volunteers, and higher levels of victim satisfaction. Because there have been few rigorous evaluations of these programs, it is not possible to determine whether restorative justice conferences and teen courts are more effective at reducing recidivism than traditional juvenile court processing or other juvenile diversion programs. However, based on the research available, as well as anecdotal evidence, it appears that both restorative justice conferences and teen courts may be positive alternatives to the traditional juvenile court process, especially in jurisdictions that do not have a wide variety of intervention options for young, first-time juvenile offenders.

Another new diversion program for juvenile offenders is the issuance of a civil citation in lieu of arrest.[47] A civil citation, or an "appearance ticket," is a police-issued order to appear before a judge or other legal authority on a given date to defend against a stated charge. In civil-citation programs, police officers are given the discretion, usually with a victim's consent, to issue a civil citation to a first-time offender under 18 years of age. Receiving a civil citation allows a juvenile offender to avoid an arrest record and the problems associated with having such a record. Civil-citation programs typically allow police officers to issue citations for the following offenses: alcohol possession; disorderly conduct; fighting at school; loitering; misdemeanor drug possession; petty theft or shoplifting; resisting arrest; trespassing; vandalism; and violation of hunting, fishing, or boating laws. Young offenders who receive civil citations generally must reimburse victims for any financial losses, write a letter of apology, and perform community service. Participants also have been required to write a "letter of appreciation" to the police officer who directed them to the program. In some programs, participants are required to attend anger-management classes or substance-abuse programs, participate in family counseling, or work with a mentor.

A recent evaluation of Florida's civil-citation program shows that for fiscal year 2017–2018, 14,540 youth received civil citations, keeping them out of juvenile court.[48] Of the 14,540 youth given civil citations, 60% were male, 40% were female, 44% were white, 35% were black, 19% were Hispanic, 2% were of other races or ethnicities, 6% were 11 years of age and under, 20% were 12 or 13, 35% were 14 or 15, 38% were 16 or 17, and 2% were 18+. During fiscal year 2016–2017, 9,405 youth were released from the civil-citation program with 7,756 (82%) successfully completing the program. The overall recidivism rate for youth completing the program was 5%, with 3% of the youth committing a new offense. Recidivism was defined as all adjudications, adjudications withheld, and convictions for any violation of law within 12 months of program release. The recidivism rate for the civil-citation program is the lowest of any program type monitored by the Florida Department of Juvenile Justice. Of the youth who successfully completed the program, 58% were male, 31% were black, and 21% were Hispanic. The average length of stay in the program was 130.5 days, and the average age at admission was 15.6. Finally, a recent independent study claims that issuing civil citations could save Florida taxpayers at least $44 million a year statewide and help cash-strapped authorities use their limited resources on more dangerous juveniles.

Detention Sometimes a youth is held in a secure detention facility during processing. There are three primary reasons for this practice: (1) to protect the community from the juvenile, (2) to ensure that the juvenile appears at a subsequent stage of processing, and (3) to keep the juvenile safe. From 2005 through 2017, the number of delinquency cases involving detention decreased 48% to its lowest level during the 12-year period. Detention for drug offenses declined 57%; it also declined 51% for public order offenses, 47% for property offenses, and 43% for person offenses. However, the proportion of juveniles detained was slightly larger in 2017 (26%) than in 2005 (24%).[49] In 2017, 28% of male juveniles were detained compared to 21% of female juveniles. Also, in 2017, black and Hispanic youth were detained in larger proportions than white youth for all offense categories, and this was true for most years from 2005 through 2017.[50]

At mid-year 2017, 3,600 juveniles were confined in adult jails in the United States—about a 14% decrease from 2014, and an approximately 53% decrease from 2010. Of the 3,600 juveniles confined in adult jails at mid-year 2017, approximately 89% were tried or awaiting trial as adults.[51] In states that allow the jailing of juveniles, statutes usually prevent youths from being placed in cells with adults. This means that jail administrators must provide separate rooms and supervision for juveniles within their facilities. Federal rules require that a juvenile who is detained in an adult jail or lockup be held for no more than 6 hours and in a separate area out of sight or sound of adult inmates.[52] For juveniles, however, separation often means isolation, which is in turn related to an increased risk of self-destructive behavior. A recent study found that from 2000 to 2014, suicide rates were two to three times higher for youth in custody than youth in the general population, with more youth in confinement dying by suicide than any other reported cause of death.[53] The majority of suicides in and out of incarceration were white males, aged 20–24. However, suicides among black youth were much more likely to occur in custody. Incarcerated youth were more likely to die by hanging, strangulation, or suffocation and less likely to die by firearms than youth not incarcerated. One-third of youth suicides in custody occurred within 24 hours of incarceration, and 76% occurred within 7 days of incarceration.

Intake Screening When the decision to arrest a youth is made, or a social agency such as a school alleges that an offense has occurred, the next step in the juvenile justice process is **intake screening**. The purpose of intake screening is to make decisions about the continued processing of cases. Those decisions and others made during juvenile court processing of delinquency cases in 2017 are shown in Figure 13.2. The location of the intake screening and the educational background and training of the person who conducts it vary from jurisdiction to jurisdiction. Traditionally, probation officers have performed intake screening. However, in recent years, there has been a move toward involving the prosecuting attorney in the intake process. In fact, in some jurisdictions, such as Colorado and Washington, intake screening is now the responsibility of the prosecuting attorney's office.[54] In other states, such as Michigan, intake screening is done by an intake officer (probation officer) who works for the juvenile court, although the prosecuting attorney's office reviews most complaints for legal sufficiency.

intake screening The process by which decisions are made about the continued processing of juvenile cases. Decisions might include dismissing the case, referring the youth to a diversion program, or filing a petition.

Possible intake decisions might include dismissing the case or having youths and parents in for a conference or an informal hearing to collect additional information for making the intake decision. Other decisions include referral of the youth to a diversion program (e.g., informal probation or counseling at a community agency); filing a **petition**, a legal form of the police complaint that specifies the charges to be heard at the adjudication; and waiver or transfer of the case to criminal court.

petition A legal form of the police complaint that specifies the charges to be heard at the adjudication.

As in police decision making, a number of factors have been found to influence intake-screening decisions. For example, youths who have committed serious offenses, those with prior records, and those who are uncooperative are more likely to have petitions filed. In addition, lower-class minority males, at least in some jurisdictions, are more likely than other groups to receive more formal responses at intake. Finally, there is considerable evidence that female status offenders are likely to be treated more harshly at intake than are their male counterparts, at least in some jurisdictions.[55]

Transfer, Waiver, Remand, Bindover, or Certification to Criminal Court

Since the early days of the juvenile court, state legislatures have given juvenile court judges statutory authority to transfer certain juvenile offenders to criminal court. Currently, 45 states give their juvenile court judges statutory transfer authority. Only 5 states do not: Massachusetts, Montana, New Jersey, New Mexico, and New York.[56] In some jurisdictions, transfer is called waiver, remand, bindover, or certification. **Transfer, waiver, remand, bindover,** or **certification** may occur in cases where youths meet certain age, offense, and (in some jurisdictions) prior-record criteria. For example, in Illinois, Mississippi, North Carolina, and Wyoming, waiver may occur if a youth has reached 13 years of age and has committed any criminal offense. In Colorado, Missouri, and Vermont, a juvenile must be 12 years of age and must have committed one of a number of felonies specified by law. In Iowa, the minimum age for waiver is 10.[57] By contrast, in most other countries, juveniles as old as 13 or 14 cannot be criminally prosecuted no matter what their offense.[58] In many states, before a youth is transferred to an adult criminal court, a separate transfer or waiver hearing must be conducted to determine the waiver's appropriateness. A juvenile cannot be tried in juvenile court, and then transferred and retried again in an adult court. To do so would be a violation of double jeopardy.[59]

transfer, waiver, remand, bindover, or **certification** The act or process by which juveniles who meet specific age, offense, and (in some jurisdictions) prior-record criteria are transferred to criminal court for trial.

The judicial waiver process varies from state to state, but at the typical waiver hearing, probable cause must be shown. In addition, in many states, the prosecutor must show that:

1. The youth presents a threat to the community.
2. Existing juvenile treatment programs would not be appropriate.
3. Programs within the adult system would be more appropriate.

FIGURE 13.2 Juvenile Court Processing of Delinquency Cases, 2017, National Estimates

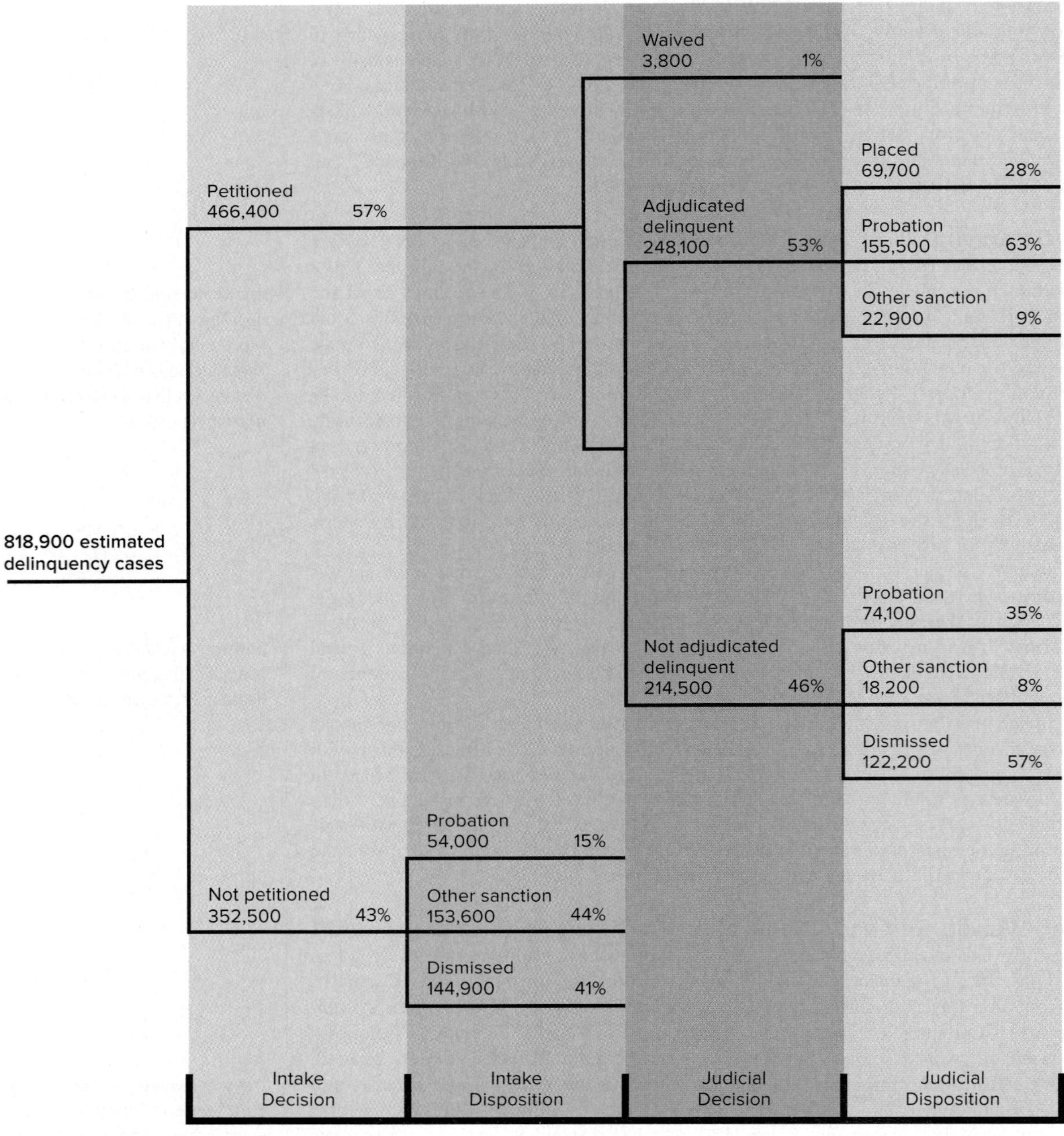

Note: Detail may not add to totals because of rounding.

Source: Sarah Hockenberry and Charles Puzzanchera, *Juvenile Court Statistics 2017* (Pittsburgh, PA: National Center for Juvenile Justice, June 2019), accessed September 24, 2019, http://www.ncjj.org/pdf/jcsreports/JCS2017report.pdf.

Historically, judicial transfer authority has been exercised infrequently. For example, between 1985 and 2017, the percentage of petition cases transferred to criminal court remained around 1%. However, because of changes in the total number of petition cases between 1985 and 2017, the number of delinquency cases waived by juvenile court judges rose about 70% between 1985 and 1994, the peak year (from 7,100 cases to 12,100 cases), then dropped about 69% between 1994 and 2017 (from 12,100 cases to 3,800 cases). The net result was that the number of cases judicially waived was about 46% lower in 2017 than in 1985.

The number of juvenile cases transferred to criminal court in recent years actually is much larger than the aforementioned data suggest. This is because there are multiple ways that juveniles can be transferred to adult courts in many states. As a result, it is estimated that as many as 250,000 juveniles a year are charged as adults nationwide.[60] However, because there are no comprehensive national data on all forms of juvenile transfer, the number of youths tried in criminal courts each year is unknown.

In addition to judicial transfer, some states use prosecutorial transfer. Prosecutorial transfers occur in states that have passed legislation giving prosecutors concurrent jurisdiction in some juvenile cases. In the 14 concurrent jurisdiction states (as of 2019), both the juvenile court and the criminal court have original jurisdiction in cases that meet certain age, offense, and (in some jurisdictions) prior-record criteria. In those states, prosecutors have discretion to file eligible cases in either court.[61] Today, because of the availability of prosecutorial transfer or *direct file*, judicial waiver almost never is used in Florida.[62] In fiscal year 2015–2016, for example, more than 99% of transferred cases in Florida were through direct file.[63] That represented 8.8% (1,178) of the 13,432 juveniles eligible for direct file in fiscal year 2015–2016.[64]

Besides an increase in the use of prosecutorial transfer, additional developments in this area include (1) an expansion of the number of transfer mechanisms; (2) simplification of the transfer process in some states, including the *exclusion* of offenders charged with certain offenses from juvenile court jurisdiction or their *mandatory* or *automatic waiver* to criminal court; and (3) the lowering by some states of the maximum age for juvenile court jurisdiction. Statutory exclusion ("legislature transfers") accounts for the largest number of transfers.[65] Twenty states have legislative transfers or mandatory or automatic waiver based on certain statutory criteria. All of these states set a minimum-age threshold for eligibility. For example, 19 states set 17 as the minimum-age threshold, while only one state, Wisconsin, set a minimum-age threshold of 16. In addition, in 35 states, juveniles who have been convicted as adults must be prosecuted in criminal court for any subsequent offenses (the "Once an adult, always an adult" provision). Those developments reflect a shift in many jurisdictions toward a more punitive orientation toward juvenile offenders. In addition, some jurisdictions have developed "reverse waivers" and "criminal court blended sentencing." In 28 states, reverse waivers allow juveniles whose cases are handled in criminal courts to petition to have the case heard in juvenile court. However, it is not clear how often reverse waivers are used. Regarding criminal court blended sentencing, 23 states give juveniles convicted in criminal court the opportunity to be sanctioned in the juvenile system.[66]

The Adjudication Hearing When a petition is filed at intake and the case is not transferred to criminal court, the next step is the adjudication hearing. As noted previously, the adjudication hearing (often referred to simply as the adjudication) is the juvenile court equivalent of a trial in criminal court. In some states, such as Michigan, adjudication is now called a trial. The adjudication hearing is an important event in the juvenile justice process because, if youths are adjudicated, they will have an official court record.

Before adjudication can take place, several preliminary actions are necessary. First, a determination must be made that the juvenile is old enough to be adjudicated delinquent. Twenty-eight states do not have a statutory minimum age for delinquency adjudication, so, theoretically, in those states, a 2- or 3-year-old could be adjudicated delinquent. Twenty-two states do have a minimum age. North Carolina has the youngest minimum age of 6, followed by Connecticut, Maryland, and New York with a minimum age of 7. Figure 13.3 shows the states with and without minimum ages for delinquency adjudication and the minimum ages for those states that require one. Once the age issue is determined, a petition must be filed, a hearing date must be set, and the necessary parties (such as the youth, the parents, and witnesses) must be given notice of the hearing. Notice typically is given through a summons or subpoena. A summons, which is an order to appear in court, is issued to the youth; copies also are given to the parents or guardians. The summons specifies the charges and the date, time, and location of the hearing, and it may list the youth's rights, such as a right to an attorney. Subpoenas are issued to witnesses, instructing them to appear on a certain day, time, and place to provide testimony or records.

When a juvenile and his or her attorney contest the charges specified in the petition, another critical event often takes place before adjudication–a plea bargain. Plea bargaining, including its problems, was discussed in detail in Chapter 8. Although the extent of plea bargaining in juvenile justice is unknown, it is a common practice in many juvenile courts. Nevertheless, a nationwide study of plea bargaining in juvenile courts found that "much of the country has either not addressed at all or has not fully developed standards regarding the guilty plea process in juvenile court." It also was discovered that there was considerable

Richard Sheinwald/AP Images

Nathaniel Abraham was 11 when he was charged with first-degree murder for shooting an 18-year-old stranger outside a convenience store in Pontiac, Michigan. *Should youths such as Abraham be transferred to criminal courts for trials? Why or why not?*

myth

Greater use of transfer or expansion of transfer criteria will reduce crime because of the adult sanctions that will be applied.

fact

Youths transferred to criminal court are more likely to commit further crimes than others retained in the juvenile justice system. There is no evidence that transfer has any deterrent effect.

Sources: Patrick Griffin, Sean Addie, Benjamin Adams, and Kathy Firestine, "Trying Juveniles as Adults: An Analysis of State Transfer Laws and Reporting," Office of Juvenile Justice and Delinquency Prevention, U.S. Department of Justice (Washington, DC: U.S. Government Printing Office, September 2011), 26; Patrick Griffin, "Different from Adults: An Updated Analysis of Juvenile Transfer and Blended Sentencing Laws, with Recommendations for Reform" (Pittsburgh, PA: National Center for Juvenile Justice, November 2008), 8, accessed February 28, 2011, www.ncjjservehttp.org/NCJJWebsite/pdf/MFC/MFC_Transfer_2008.pdf.

FIGURE 13.3 States With and Without Minimum Ages for Delinquency Adjudication, 2019

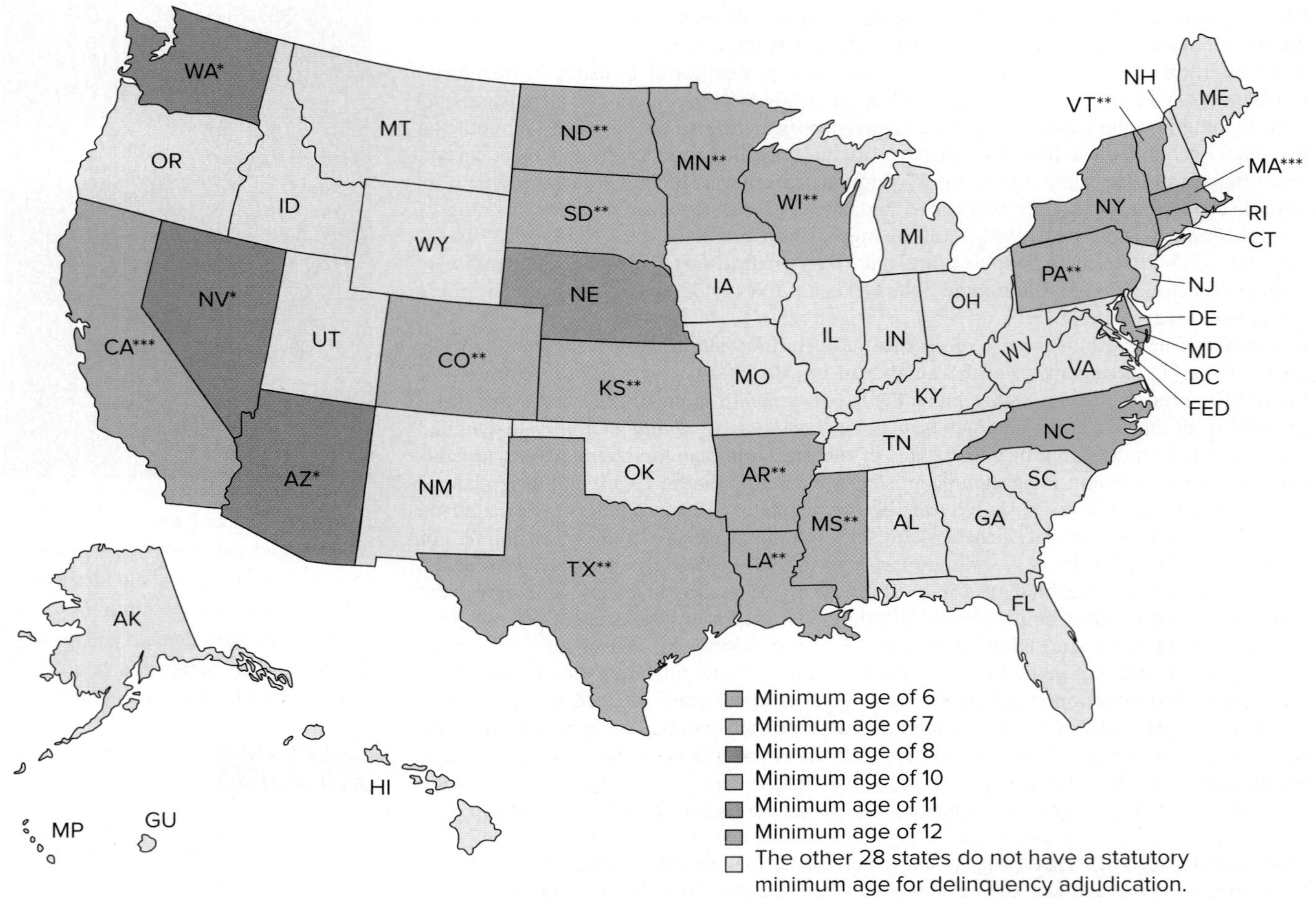

Notes:
* rebuttable presumption that children at least 8 but under 12 are incapable of committing a crime
** except for murder, for which there is no age limit
*** except for murder, rape by force, sodomy by force, oral copulation by force, and sexual penetration by force, for which there is no age limit

Source: "Minimum Age for Delinquency Adjudication-Multi-Jurisdiction Survey," *National Juvenile Defender Center*, last updated July 2019, accessed September 25, 2019, https://njdc.infor/practice-policy-resources/state-profiles/multi-jurisdiction-data/minimum-age-for-delinquency-adjudication-multi-jurisdiction-survey/.

variation in the ways that plea bargaining was carried out in various courts. For example, urban courts were more likely to institute formal procedures to regulate plea bargaining than were suburban or rural courts.[67]

There are two types of adjudications: contested ones (in which juveniles dispute the charges) and uncontested ones. Contested adjudications are similar to trials in criminal courts, which were described in Chapter 8. They typically employ the same rules of evidence and procedure. Most contested adjudications are bench adjudications in which the hearing officer–a judge, referee, court master, or commissioner–makes a finding of fact based on the evidence presented. A **hearing officer** is a lawyer empowered by the juvenile court to hear juvenile cases. In some jurisdictions, contested adjudications are jury trials. However, as noted earlier in this chapter, juveniles do not have a constitutional right to a jury trial, and even in states that give juveniles the right to a jury trial, jury trials are rare.

hearing officer A lawyer empowered by the juvenile court to hear juvenile cases.

Like their criminal court counterparts, the vast majority of juvenile court adjudications are uncontested. Uncontested adjudications generally are brief and consist of a reading of the charges, advice of rights, and possibly brief testimony by the youth or other parties, such as a probation officer. After this, the youth, or frequently the youth's attorney, admits the charges. In some states, an uncontested adjudication is called an arraignment.

The majority of cases that are not adjudicated are dismissed. Surprisingly, however, many juveniles whose cases are neither adjudicated nor dismissed still are placed on informal probation or treated in some other way. For example, of the estimated 214,500 delinquency cases filed but not adjudicated in 2017, 57% were dismissed. However, in 35% of the cases, the juvenile was placed on informal probation, and the remaining 8% were disposed of in other ways (see Figure 13.2). Still, the majority of cases in which petitions are filed are adjudicated. In 2017, for example, 53% of the delinquency cases in juvenile courts were adjudicated (see Figure 13.2).[68] Although 53% may not seem like a large percentage, keep in mind that many cases not adjudicated still receive some court supervision, such as informal probation.

FYI

Adjudications Decreasing

The annual number of delinquency cases in which youth were adjudicated delinquent steadily decreased 55% from 555,400 in 2005 to its lowest level in 2017 (248,100).

Source: Sarah Hockenberry and Charles Puzzanchera, "Juvenile Court Statistics 2017" (Pittsburgh, PA: National Center for Juvenile Justice, June 2019), 43. Accessed September 28, 2019, http://www.ncjj.org/pdf/jcsreports/JCS2017report.pdf.

Disposition

Disposition is the juvenile court equivalent of sentencing in criminal court. At the disposition hearing, the court makes its final determination of what to do with the juvenile officially labeled "delinquent." Some of the options available to juvenile courts are probation, placement in a diversion program, restitution, community service, detention, placement in foster care, placement in a long-term or short-term residential treatment program, placement with a relative, and placement in a public or private corrections facility. In addition, the court may order some combination of those dispositions, such as placement on probation, restitution, and a short stay in detention. However, disposition possibilities are limited by the available options in a particular jurisdiction. In a few jurisdictions, they also are limited by statutory sentencing guidelines. In practice, the disposition options available to most juvenile courts are quite narrow, consisting of probation or incarceration. When incarceration is used, an indeterminate period of commitment or incarceration is the norm.[69]

disposition The juvenile court equivalent of sentencing in criminal court. At the disposition hearing, the court makes its final determination of what to do with the juvenile officially labeled delinquent.

As part of the disposition, the court also enters various orders regarding the youth's behavior. Those orders consist of rules the youth must follow. In addition, the court may enter orders regarding parents, relatives, or other people who live in the home. For example, the court may order parents to attend counseling or a substance-abuse treatment program, a boyfriend to move out of the house, or parents to clean up their house and pay for court costs or services provided, such as counseling caseworker services. If parents fail to follow those orders, they may be held in contempt of court and placed in jail, at least in some jurisdictions.

In making the disposition, the hearing officer usually relies heavily on a presentence investigation report (sometimes called a predisposition report), which is completed by a probation officer or an investigator before the disposition. Presentence investigation reports for criminal courts were described in detail in Chapter 9. Although some research has failed to find evidence that extralegal factors, such as race or social class, influence disposition, the bulk of the research indicates that extralegal factors often play a role in disposition. For example, a number of studies have found that minority and lower-class youths are more likely to receive the most severe dispositions, even when seriousness of offense and prior record are taken into account.[70]

The most frequently used disposition in juvenile courts is probation, followed by placement. In 2017, for example, approximately 63% of the youths adjudicated delinquent were placed on probation; 28% received some type of commitment. Another 9% of youths adjudicated delinquent received some other sanction (see Figure 13.2).[71]

Because of heightened concerns about violent juvenile offenders, many states legislatively redefined the juvenile court's mission by deemphasizing the goal of rehabilitation and stressing the need for public safety, punishment, and accountability in the juvenile justice system. Along with this change in purpose has been a fundamental philosophical change in the focus of juvenile justice, from offender-based dispositions to offense-based dispositions, which emphasize punishment or incapacitation instead of rehabilitation. These changes are reflected in new disposition or sentencing practices, including (1) the use of *blended sentences*, which combine both juvenile and adult sanctions (see Figure 13.4); (2) the use of mandatory minimum sentences for specific types of offenders or offense categories; and (3) the extension of juvenile court dispositions beyond the offender's age of majority, that is, lengthening the time an offender is held accountable in juvenile court.[72]

Because of policies that allow juvenile offenders to be transferred to adult courts, children as young as 12 years old have been sentenced to life imprisonment without opportunity of parole (LWOP), meaning they would spend the rest of their lives in prison.[73] That is no longer the case. In 2010, in *Graham v. Florida*, the U.S. Supreme Court ruled that the Eighth Amendment's prohibition of cruel and unusual punishments does not permit a juvenile offender (someone under 18 years of age) to be sentenced to LWOP for a crime other than homicide.

FIGURE 13.4 Blended Sentencing Options Create a "Middle Ground" Between Traditional Juvenile Sanctions and Adult Sanctions

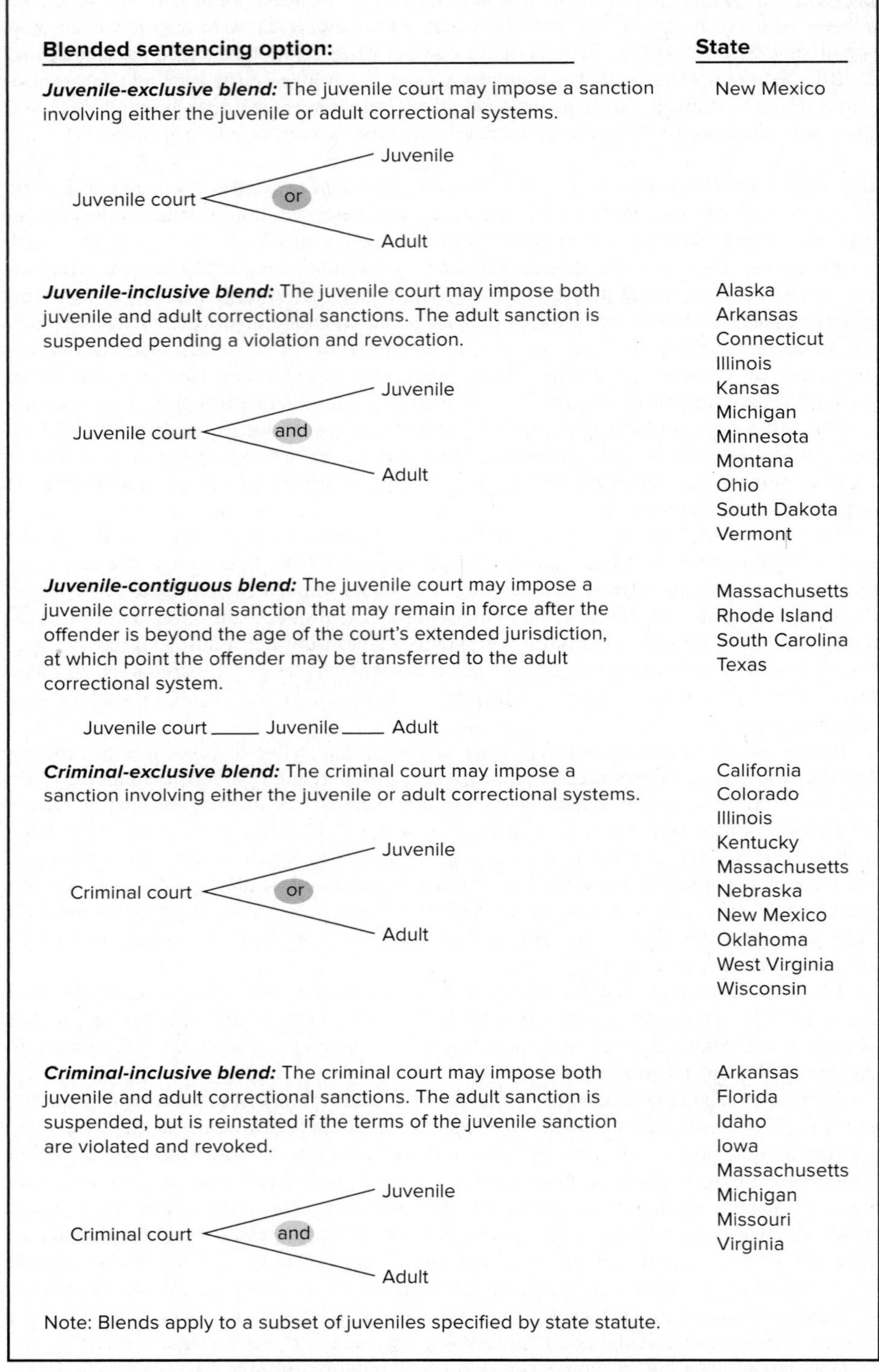

Source: Regan Comis, "Blended Sentencing," March 27, 2014, accessed August 1, 2016, www.leg.state.nv.us/Interim/77th2013/Exhibits/JuvJustTaskForce/E032714G.pdf.

At the time of the decision, about 100 people worldwide were serving LWOP sentences for such crimes, and all of them were in the United States; 77 of them were in Florida. In 2012, in *Miller v. Alabama*, the Court augmented its decision in *Graham* by holding that the Eighth Amendment forbids a sentencing scheme that mandates LWOP for juvenile homicide offenders, as the laws of 29 states did. Under the ruling, judges and juries still can impose an LWOP sentence as long as mitigating factors such as the juvenile's upbringing are taken into account. At the time of the decision, about 2,500 prisoners were serving LWOP sentences for murders committed before they turned 18. In 2016, in *Montgomery v. Louisiana*, the Court ruled that its decision in *Miller v. Alabama* should be applied retroactively. The Court previously had prohibited the death penalty for juveniles in *Roper v. Simmons* (2005).

THINKING CRITICALLY

1. Which do you think is a better way to handle juvenile crime—the informal or the formal process? Why?
2. Do you think juvenile offenders should be transferred to criminal courts for trial? Why or why not?

Correctional Programs for Juveniles

As noted, when youths are adjudicated, a number of disposition options are available to juvenile courts, although the options typically used in any one jurisdiction are fairly narrow. Three general types of dispositions are these:

1. Dismissal of the case, which is used in a small percentage of cases.
2. The use of a community-based program.
3. The use of an institutional program.

Because both community-based and institutional correctional programs for adults were the subjects of previous chapters of this book, only the features of those programs that are unique to juveniles will be discussed in the following sections.

OJJDP

Visit the Office of Juvenile Justice and Delinquency Prevention (OJJDP) website, www.ojjdp.gov/, to review the various programs offered by the organization.

Which programs do you think would help deter juvenile crime?

Community-Based Correctional Programs for Juveniles

Among the community-based correctional programs for juvenile offenders are diversion, pretrial release, probation, foster care, group home placement, and parole. Some of those programs are designed to provide services to youths in their own homes; others provide services to youths who have been removed from their homes, at least for short periods of time. Moreover, although all community-based programs are intended to control offenders and provide sanctions for their behavior, they also are designed to accomplish a variety of additional objectives. Those objectives include allowing youths to maintain existing ties with the community, helping them restore ties and develop new and positive ones with the community (reintegration), avoiding the negative consequences of institutional placement, providing a more cost-effective response to offenders, and reducing the likelihood of recidivism. Some of the community-based correctional programs for juveniles are examined in the following sections.

Probation Juvenile probation officers are referred to as "the workhorses of the juvenile justice system." Although the actual practice of juvenile probation varies from one jurisdiction to the next, probation officers usually perform four important roles in the juvenile justice process:

1. They perform intake screening.
2. They conduct presentence investigations.
3. They supervise offenders.
4. They provide assistance to youths placed on probation.

In some jurisdictions, the same individual performs all of those roles. In other jurisdictions, each probation officer specializes in only one of the roles.

Probation is the most frequently used correctional response for youths who are adjudicated delinquent in juvenile courts, just as it is for adults who are convicted of crimes in criminal courts. In 2017, it was the most severe disposition in 63% of all petitioned delinquency cases

Careers in Criminal Justice

Juvenile Probation Officer

My name is Todd C. Lunger. I am a juvenile probation officer for The Erie County Juvenile Probation Department. My educational path to this position did not include the traditional focus of criminal justice, but rather a focus in social work. I first completed my Bachelor's degree of Social Work at Edinboro University of Pennsylvania, and through the required internship while obtaining that degree, I was first introduced to The Erie County Juvenile Probation Department, where I was hired upon graduation. After a few years in the field, I decided it would be beneficial to further my education, so I completed my Master's degree of Social Work, and this degree propelled me to become a Certified Master Forensic Social Worker and a Licensed Social Worker in Pennsylvania.

Juvenile probation office duties are focused around caseload management and prioritization through the development of case plans that are based on client needs and risk level. These recommendations are prepared through client research and documented outcome studies that will aid in the recommendations presented to the court. After the decision is handed down from the court, the probation office's role changes to supervision and monitoring of clients' behavior, compliance with the Juvenile Act, supervision rules, and case plan goals in an effort to rehabilitate the juvenile offenders and reduce the risk of subsequent delinquent conduct. Direct supervision of juveniles in neighborhood settings (home, school, and community) is employed in order to facilitate the utilization of community resources and assure the accountability of juvenile offenders. Evidence-based practices and structured decision making are employed in all areas to ensure the best possible outcomes.

Todd Lunger

My passion for this position and the reasoning behind my nontraditional educational path are that juvenile probation has a treatment focused approach with the client's best outcome in the forefront. Juveniles are so resilient and workable, and I have found that the majority of them just need ongoing attention, services, and resources. However, this approach presents its own challenges and struggles in cases that require a child to be removed from a home, or when the courts adjudicate a juvenile to the adult system because he or she is not suitable for the rehabilitation approach of the juvenile system.

With the choice to enter this field it is best to understand that it requires a lot of research, compassion, and ongoing resource education with very few praises for the decisions you make. Take pride and satisfaction in the fact that you are making the best decisions for the client while making an impact on the future of an entire community through the rehabilitation of its youth. Remember that the client is the change agent, and sometimes a broken law can lead to a mended home.

that were adjudicated (248,100), compared to 64% (323,300) in 2013 (see Figure 13.2 for data from 2017).[74] In delinquency cases that were petitioned but not adjudicated, 35% (74,100) of juveniles were placed on probation, compared to 28% (70,900) in 2013.[75] In petitioned and adjudicated status offense cases, 56% (18,500) of juveniles were placed on probation in 2017, compared to 54% (26,100) in 2013, while 16% (8,700) of non-adjudicated status offenders received probation in 2017, compared to 12% (7,000) in 2013.[76] Remember that, in 2017, 15% (54,000) of the delinquency cases not petitioned received probation, compared to 23% (107,400) in 2013.[77]

Recent evidence suggests that "traditional community supervision—both as an alternative to residential supervision (probation) and as a means to continue supervision after release from a correctional institution (parole)—is ineffective."[78] Research also shows that programs that reduce delinquency "through counseling, skill building and restorative justice all reduce juvenile reoffending by an average of 10 percent or more, while supervision reduces reoffending by just 1 percent."[79]

A trend in juvenile probation is the development of intensive-supervision (probation) programs, which in some jurisdictions involves home confinement. Intensive-supervision programs are intended to ensure regular contact between probationers and probation officers. They also are intended to serve as an intermediate response that is more restrictive than standard probation but less restrictive than incarceration. However, like standard probation programs, the frequency of contact between probation officers and probationers varies considerably.[80]

Although there is wide variation in the meaning of intensive supervision, there is some indication that programs that provide frequent supervision of offenders, as well as services, are as effective as incarceration at reducing recidivism.[81] The same research also suggests that intensive-supervision programs are more cost effective than incarceration, provided they actually divert a sizable number of youths from institutions.

A trend in intensive supervision of juveniles is the use of home confinement, which was first used with juvenile offenders in the 1970s[82] and has grown in popularity since then. Today, home confinement programs use two mechanisms to monitor youths: frequent probation officer contacts and electronic monitoring. Electronic monitoring requires an offender to wear a tamper-resistant electronic device that automatically notifies the probation department if the juvenile leaves home or another designated location.

The use of home confinement employing electronic monitoring of juvenile offenders began in the 1980s and has grown substantially.[83] However, because electronic monitoring technology is diverse and rapidly developing, little evidence of its effectiveness exists. Still, there are good reasons that some jurisdictions find electronic monitoring attractive:

1. It eases the problem of detention overcrowding.
2. It allows youths to participate in counseling, education, and vocational programs without endangering public safety.
3. It allows youth to live with supervision in an environment more natural than an institution.
4. It allows court workers to better assess the ability of youth to live in the community under standard probation after they leave the program.[84]

Besides probation, juvenile courts in some jurisdictions employ several other types of community-based interventions with juvenile offenders, such as restitution or community service programs, wilderness probation programs, and day treatment programs. In practice, probation often is combined with one of those other community-based interventions.

Restitution Recall that restitution programs require offenders to compensate victims for damages to property or for physical injuries. The primary goal of restitution programs is to hold youth accountable for their actions. In practice, there are three types of restitution:

1. *Monetary restitution*, a cash payment to the victim for harm done.
2. *Victim-service restitution*, in which the youth provides some service to the victim.
3. *Community-service restitution*, in which the youth provides assistance to a community organization.

Despite the growing popularity of juvenile restitution programs, as well as some programs' effective reduction of recidivism, there are potential problems with some of those programs. Problems include poorly managed, informal programs with low compliance rates; high recidivism rates in some programs; and hearing officers making unrealistic restitution orders that juveniles cannot complete. There also is a potential for net widening, and restitution requirements can be subject to discretionary abuse, which has encouraged some jurisdictions to establish restitution guidelines. Thus, rather than achieving the goals of accountability, offender treatment, and victim compensation, restitution programs may fail to protect community safety. Moreover, they are likely to produce negative perceptions of juvenile justice by both offenders and victims.

Steve Skjold/Alamy Stock Photo

Some juvenile treatment programs strive to provide physically and emotionally challenging experiences to help youthful offenders gain confidence and learn responsibility for their actions. *What are the benefits and drawbacks of these kinds of programs?*

Wilderness Probation (Outdoor Adventure) Programs
Wilderness probation or outdoor adventure programs for juvenile offenders are based, in part, on ideas derived from programs such as Outward Bound. (Outward Bound is a company that sells outdoor experiences, such as mountain climbing, backpacking, and mountain biking.) A basic assumption of those programs is that learning is best accomplished by acting in an environment where there are consequences for one's actions. Consequently, the programs involve youths in a physically and sometimes emotionally challenging outdoor experience intended to help them develop confidence in themselves, learn to accept responsibility for themselves and others, and develop a relationship of trust with program staff. This is done by engaging youths in a variety of activities, such as

Wilderness Probation

You can learn more about wilderness probation programs by accessing the Rescue Youth website at https://www.rescueyouth.com/wilderness-schools/ and selecting Boarding School Directory from the menu. From the Boarding School Directory, select Wilderness Schools.

What do you think are some of the benefits and drawbacks of wilderness probation?

camping, backpacking, rock climbing, canoeing, sailing, negotiating rope courses, and a solo experience (spending one or more nights alone in the wilderness).[85]

Evaluations of several wilderness probation programs have shown that they can produce positive effects, such as increases in self-esteem and a decrease in criminal activity both during and after the program.[86] However, research also indicates that the positive effects of the programs may diminish over time or may be no greater than the effects produced by probation programs that provide regular and meaningful contacts between probation officers and probationers.[87] In 2016, there were about 30 wilderness probation programs operating in the United States, down about 81% from the 157 programs in 2002.[88]

Day Treatment Programs Day treatment programs for juvenile offenders operate in a number of jurisdictions around the United States. The programs often target serious offenders who would otherwise be candidates for institutionalization. They provide treatment or services to youths during the day and allow them to return home at night. Because they are viewed as alternatives to incarceration, they are believed to be cost effective. Because they provide highly structured programs for youths during the day, it is assumed that they protect community safety as well. The range of services or treatments can be quite varied and may include academic remediation, individual and group counseling, job skills training, job placement, and social skills training. Although some evidence suggests that day treatment programs are as effective as, or more effective than, institutional placement, some programs may be no more effective at reducing recidivism than standard probation.[89]

Foster Homes Foster homes are out-of-home placements intended to resemble a family setting as much as possible. Foster parents are licensed to provide care for one or more youths (usually one to three) and are paid a daily rate (*per diem*) for the costs of care. A court often uses foster placement when a youth's home life has been particularly chaotic or harmful. In such a case, foster care is used to temporarily separate the youth from the parents or guardian in an effort to resolve the problems that resulted in the youth's removal. Foster homes also are used to remove youths from particular neighborhoods and for some nonviolent offenders instead of more restrictive placements, such as institutionalization. When used in those ways, they are considered "halfway-in programs." In other instances, foster homes are used as transitions to home and are considered "halfway-out programs."

Although foster homes are used widely by juvenile courts in some jurisdictions, there are few sound evaluations of their effectiveness. The limited research that exists indicates that foster care generally is not effective and may even be counterproductive, when youths are subjected to abuse and neglect, for example. Problems such as frequent movement of youths from home to home and poor training and support for foster parents likely inhibit the effectiveness of many foster home placements. However, research on a more intensive form of foster care, multidimensional treatment foster care (MTFC), which uses close supervision of youths and provides intensive training and support for foster parents, has been found to be effective with many youths.[90]

Group Homes Similar to foster homes, group homes are open, nonsecure, community-based facilities used in both halfway-in and halfway-out programs. However, they are somewhat larger and frequently less family-like than foster homes. The purpose of many group homes is to avoid requiring youths to accept "substitute parents" because many youths are in the process of developing emotional independence from parental figures. Nevertheless, group homes generally are less impersonal than institutions and are less expensive than institutional placements. In addition, their location allows residents to take advantage of community services. Youths who live in group homes usually go to school in the home or community, or they work in the community. Treatment typically consists of group or individual counseling provided by group home staff or outside counselors. In 2016, there were about 344 group homes for juvenile offenders operating in the United States, down about 70% from the 1,135 group homes in 2002.[91]

Although group homes are used extensively in juvenile justice, relatively little sound, recent research has examined their effectiveness. One well-known but older study of a group home program, the Silverlake Experiment, found that youths placed in the program for approximately 6 months were no more likely to commit more crimes than similar youths who were assigned randomly to an institutional placement. In other words, placement in the program was neither more nor less effective than institutionalization at reducing recidivism.[92]

Richard Lord/The Image Works

Group homes for delinquent youths generally are less impersonal than institutions and are less expensive than institutional placements. *What are some other advantages of group homes? What are some disadvantages?*

Group homes are a key feature of New York City's new "Close to Home" (C2H) program, which is the most recent reform effort in the evolution of New York's juvenile justice system.[93] Modeled after similar programs in Wayne County (Detroit), Michigan and Missouri, the fundamental goal of C2H is the deinstitutionalization of New York's juvenile delinquents and the closure of most of the nearly three-dozen expensive state-run detention facilities operated by the New York State Office of Children and Family Services (OCFS) or by private providers contracted by OCFS. By 2011, the year before C2H was implemented, the cost of incarcerating a juvenile in a state-run detention facility had reached more than $250,000 per youth per year because many of the facilities still were operating while empty or near empty (between 2002 and 2011, the number of juveniles placed with OCFS decreased 62%, while the daily per-youth costs increased about 150%). Most of the juvenile detention facilities were in New York's upstate communities and were staffed predominantly by poor, white, local residents who annually supervised approximately 4,000 largely poor, minority youth primarily from New York City and the state's other urban areas. Outcomes were poor. One study found that by age 28, 71% of boys released from juvenile detention facilities spent some time in an adult jail or prison. Recidivism rates were in the 90 percentile. A major reason poorly performing and expensive facilities were kept open was to maintain the state union jobs in economically distressed upstate New York.

The structural change that made C2H possible was the transfer of the care and custody of all New York City youth adjudicated as juvenile delinquents from the state to the city, from the New York State Office of Children and Family Services to the New York City Administration for Children's Services. Under C2H, the vast majority of juveniles who otherwise would have been sent to a youth prison are supervised in the community, closer to home, school, and their families, either on probation or, when needed, in one of 29 foster or group homes with bed capacities ranging from 6 to 18. C2H residential programming includes three key elements: (1) education, (2) family engagement, and (3) release planning and aftercare. C2H probably would not have happened had it not been for the decrease in crime, especially youth crime, since the mid-1990s; the significant drop in the number of juveniles entering the juvenile justice system; the concomitant decline in people's fear of crime, particularly youth crime; and, of course, the astronomical cost of incarcerating juvenile delinquents and of operating empty or near empty detention facilities.

One of the first and most perplexing problems with the implementation of C2H was the large numbers of AWOLs (youth who were absent without leave). At first, C2H lacked residential options that provided a higher level of custody. The problem was addressed by creating "Children's Village," which was a separate facility that was used as a "time out" facility for juveniles who had behavioral problems. Youth sent to Children's Village stayed there for a short time and then were returned to the C2H facility or, in some cases, home.

Responses to the AWOL problem varied, but included improved tracking and monitoring of juveniles, closer collaboration with the police department, and termination of service provider contracts. Through these efforts, the problem with AWOLs was reduced significantly (the number of AWOLs fell from a high of more than 1,000 incidents in Year 1 of C2H to 136 in 2016, and is still declining). Another enduring problem with C2H is its failure to reduce racial disparities. In 2018, 90% of C2H admissions were either black (60%) or Latino (30%), compared to their overall representation in the New York City population of approximately 40%.

C2H has proven effective based on several measures. First, between 2012 and 2016, the number of New York City juveniles who were placed out-of-home decreased by 68%, compared to the 20% reduction for the rest of New York State. This was occurring while the number of low-level cases coming into the system was declining, resulting in a higher ratio of felony arrests. Second, by 2016, New York City no longer sent any juveniles from its family court (New York's name for its juvenile court) to state-sponsored youth prisons. Third, in 2019, the family court placed only about 100 New York City juveniles into any kind of residential facility, and only a dozen of those juveniles were in a locked facility. Fourth, during the 2016–2017 school year, 91% of C2H juveniles passed their academic classes. However, preliminary information suggests that the academic gains mostly disappear once juveniles return to their home schools. Fifth, in 2016, 82% of juveniles transitioned from C2H to a parent, other family member, or guardian. Sixth, in 2016–2017, 91% of juveniles who transitioned from C2H were enrolled with community-based programs, with 67% of those completing the program and the remainder still involved. Seventh, although there are no longitudinal data showing recidivism rates for juveniles in C2H, other evidence indicates that C2H has not jeopardized public safety. For example, in the 4 years preceding C2H, juvenile arrests in New York City decreased 24%; in the 4 years following the introduction of C2H in 2012, juvenile arrests in New York City dropped about 52%, compared to 41% in the rest of the state (where C2H was not implemented). Eighth, readmissions to C2H and violations of aftercare conditions have been few. For example, 36 or 14% of juveniles admitted to C2H residence in 2016 previously had been placed in a C2H facility. Of the 836 juveniles released from C2H placement between 2014 and 2016, 64 or 7.6% had their aftercare revoked for violations of release terms, such as a new arrest. Although C2H remains a work in progress, it is a program that other jurisdictions are seeking to emulate.

myth

Most of the youths housed in juvenile correctional institutions pose an immediate threat to public safety.

fact

As Jerome Miller notes in his book, *Last One Over the Wall: The Massachusetts Experiment in Closing Reform Schools* (1991), in Massachusetts only about 25% of youths committed to state correctional facilities had committed offenses against persons. Many of those offenses against persons actually did not involve physical violence or the threat of physical violence. Consequently, many youths placed in correctional facilities, including many of those placed for violent offenses, do not pose a grave threat to public safety. Nevertheless, incarceration remains a popular response to delinquency. For example, in 2017, only 42% of youths placed in juvenile correctional institutions had committed crimes against persons.

Sources: J. Miller, *Last One Over the Wall: The Massachusetts Experiment in Closing Reform Schools* (Columbus: Ohio State University Press, 1991); Charles Puzzanchera and Sarah Hockenberry, "Trends and characteristics of youth in residential placement, 2017," Office of Juvenile Justice and Delinquency Prevention, July 2019. Accessed September 28, 2019, https://www.ojjdp.gov/ojstatbb/snapshots/DataSnapshot_CJRP2017.pdf.

Institutional Programs for Juveniles

A variety of correctional institutions house juveniles within the United States, including detention centers, adult jails, shelter facilities (some of them are more community based), reception and diagnostic centers, ranches, forestry camps, farms, and training schools. These institutions hold a variety of youths, including those who are status offenders as well as those who have committed violent offenses against others. They are administered by either state or local governments or by private agencies.

Juvenile Correctional Institutions What distinguishes institutional programs from their community-based counterparts is that institutional programs typically restrict youths' access to the community more than community-based programs do. Indeed, institutional programs are the most restrictive placements available to juvenile courts. However, juvenile institutions vary in the extent to which they focus on custody and control.

For example, some juvenile institutions employ a variety of security hardware: perimeter fencing or walls; barbed or razor wire; and surveillance and detection devices, such as motion detectors, sound monitors, and security cameras. Those juvenile institutions, classified as secure facilities, closely monitor residents' movement within the facility and restrict residents' access to the community. Most public and private detention centers, reception and diagnostic centers, and state training schools are secure facilities. In contrast, other juvenile institutions rely much less on security devices. Those facilities have no perimeter fencing, and some do not lock entrances or exits at night. Classified as open institutions, they rely more heavily on staff than on physical security. Most private facilities, as well as most public shelters, ranches, forestry camps, and farms, are open institutions.[94]

In addition to differences in the use of security hardware, juvenile correctional institutions also differ in a number of other ways. Some of the institutions are privately operated institutions, although the majority are public institutions. The majority of both private and public institutions are small, housing 40 or fewer residents, although some large, state-operated institutions have a legal capacity of 800 or more. Some institutions are coeducational; others are single-gender institutions.

The Times/John Luke/AP Images

Juvenile boot camps are popular in some jurisdictions. *What are some of the advantages and disadvantages of juvenile boot camps?*

Institutions also differ in the average length of time that residents stay in the facility. Typically, youths stay longer in private facilities than in public facilities. However, some private and public facilities are for short-term placements, whereas others are for long-term placements. Institutions such as detention centers and diagnostic and reception centers typically are for short-term placements. Detention centers usually house youths awaiting adjudication or those who have been adjudicated and are awaiting disposition. In some cases, youths are placed in those institutions for a period of time as a disposition. Other institutions, such as ranches, forestry camps, farms, and training schools, generally are for long-term placements. Youths are placed in them as a result of disposition or possibly assessment at a reception and diagnostic center.

Juvenile institutions also differ in types of programming and quality of care. Almost all juvenile correctional institutions offer basic educational and counseling programs for their residents. Moreover, more than half of all institutions offer family counseling, employment counseling, peer group meetings, behavior modification using a point system, or behavioral contracts. However, there is considerable variation in the extent to which institutions offer more specialized educational or counseling programs for clients and the extent to which residents participate in the programs. For example, research shows that many institutions do not offer vocational training, GED courses, tutoring, suicide prevention, or programs for special offender types, such as violent offenders, sex offenders, or drug offenders.[95]

In response to the significant increase in juvenile arrests and repeat offenses from the mid-1980s to the mid-1990s (see Figure 13.5), a few states and many localities established juvenile boot camps, modeled after adult boot camps. Evaluations of juvenile boot camps show they do not reduce the recidivism of graduates any better than incarceration or probation. Some research has found that boot camp graduates are more likely to be rearrested or rearrested more quickly than other offenders. Most boot camp programs have high dropout rates.

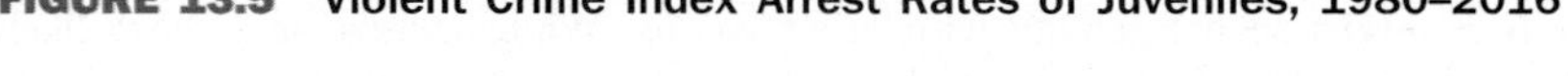

FIGURE 13.5 Violent Crime Index Arrest Rates of Juveniles, 1980–2016

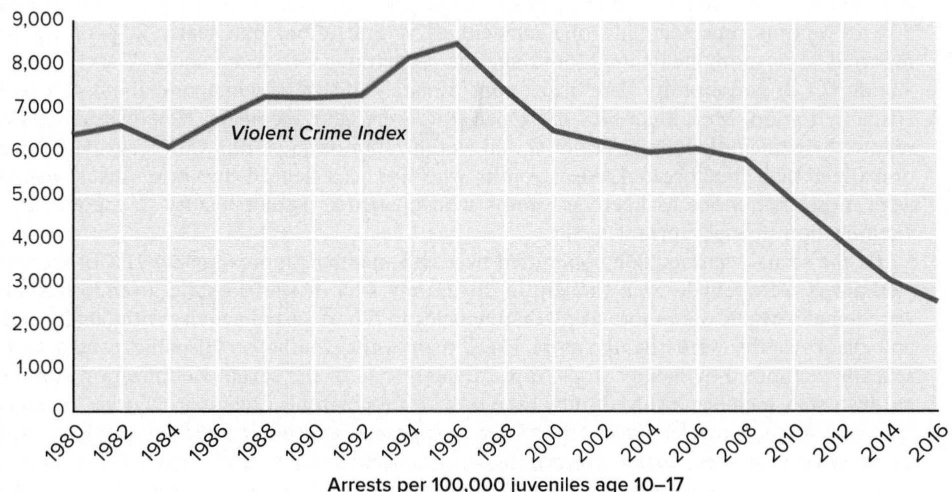

Note: In 2016, the violent crime index rate of juveniles reached its lowest level since at least 1980, and was 70% lower than its peak year of 1996. All the growth in the rate that began in the latter part of the 1980s was erased by 2000. Since 2006, the rate has decreased each subsequent year.

Source: *OJJDP Statistical Briefing Book*, online, accessed October 1, 2019, https://www.ojjdp.gov/ojstatbb/crime/ucr_trend.asp?table_in=1&selOffenses=1&rdoGroups=1&rdoData=r.

myth

A new type of violent juvenile offender, one for whom violence is a way of life—a "superpredator"—emerged in the late 1980s and early 1990s.

fact

Although evidence shows that in the early 1990s, juvenile arrest rates for violent crimes broke out of their historic range and increased to a level greater than in previous generations (see Figure 13.5), the evidence also shows that by 1995 arrest rates for juvenile violence had returned to a level comparable to that of the preceding generation. Ironically, because of the myth of the "superpredator," nearly every state legislature made it easier to handle juveniles as adult offenders.

Source: "Challenging the Myths," in *1999 National Report Series: Juvenile Justice Bulletin*, Office of Juvenile Justice and Delinquency Prevention, U.S. Department of Justice (Washington, DC: U.S. Government Printing Office, February 2000).

An examination of the history of juvenile institutions reveals that children often have been subjected to abuse and inhumane treatment in juvenile correctional institutions. Certainly, the overall quality of institutional life in correctional facilities today is vastly improved over that experienced by most youths placed in houses of refuge and early reform and training schools. Competent, caring, and professional administrators, who oversee skilled and caring staff in delivering a variety of high-quality services to their residents, administer many juvenile institutions.

In June 2013, the Justice Department's Bureau of Justice Statistics released a special report entitled "Sexual Victimization in Juvenile Facilities Reported by Youth, 2012."[96] The report was required by the Prison Rape Elimination Act of 2003, and it presents results from the second survey in what is supposed to be an annual effort. Approximately 9,700 adjudicated youths held in 273 state-owned or state-operated facilities and 53 locally or privately operated facilities were surveyed (approximately 8,700 youths completed the survey). The survey consisted of an audio computer-assisted self-interview in which youth, using a touch-screen, completed a computerized questionnaire. The survey was self-administered to ensure the confidentiality of the reporting youths and to encourage truthful reporting of victimization. The survey made use of audio technology to provide assistance to youths with literacy or language problems. Ninety-one percent of the surveyed youths were male, and 9% were female.

Results of the survey showed that about 10% of the surveyed youths experienced one or more incidents of sexual victimization by another youth (2.5%) or facility staff member (7.7%) in the previous 12 months or since admission, if less than 12 months. (This does not include the 0.7% of youths who reported sexual victimization by both another youth and facility staff. Also, the 10% of youths who experienced sexual victimization in the 2012 survey is less than the nearly 13% recorded in the 2008–2009 survey.) Of the sexual victimization committed by another youth, about 67% was nonconsensual and included "contact between the penis and the vagina or the penis and the anus; contact between the mouth and the penis, vagina, or anus; penetration of the anal or vaginal opening of another person by a hand, finger, or other object; and rubbing of another person's penis or vagina by a hand." In addition, approximately 24% of the sexual contact among youths included "kissing on the lips or other part of the body, looking at private body parts, showing something sexual like pictures or a movie, and engaging in some other sexual contact that did not involve touching." About two-thirds of the sexual victimizations among youths involved force or the threat of force; in about 18% of the incidents, the perpetrator gave the victim drugs or alcohol; in approximately 25% of incidents, the perpetrator offered the victim protection. About 20% of the victims reported being victimized by another youth, and about 6% of the youths victimized by staff reported being physically injured. Few physically injured youths reported seeking medical attention. Females were more likely than males to report forced sexual activity with other youths. White youths were more likely than black youths and Hispanic youths to be sexually victimized by another youth. About 30% of victims reported only one incident of victimization, but nearly 20% of victims reported 11 or more incidents. More than one perpetrator victimized about 37% of victims. About 57% of perpetrators were black, approximately 65% were white, and about 41% were Hispanic/Latino. More than 50% of perpetrators were gang members. Heterosexual youths reported significantly lower rates of sexual victimization than youths with a sexual orientation other than heterosexual. Also, youths who had experienced any prior sexual assault were more than twice as likely as youths with no sexual assault history to report sexual victimization in their current facility.

Of the sexual victimization committed by a staff member, approximately 92% of the staff victimizers were female even though, in 2012, only 44% of staff in state juvenile facilities were female. Males were more likely than females to report sexual activity with facility staff, and black youths were slightly more likely than white youths or Hispanic youths to be sexually victimized by facility staff. Approximately 45% of the sexual victimization committed by a staff member involved force (about 40% if touching is excluded). "Force" included "physical force, threat of force, other force or pressure, and other forms of coercion, such as being given money, favors, protections, or special treatment." Only about 6% of victims reported being injured, and less than 1% sought medical treatment. About 14% of youths victimized by staff experienced only one incident, but approximately 20% experienced 11 or more incidents. More than one staff member victimized nearly one-third of the victimized youths. It is important to remember that these survey results are based on the self-reports of incarcerated youths and, therefore, are subject to the criticisms of all self-report surveys (see the discussion of self-report crime surveys in Chapter 2).

Despite the long history of juvenile correctional institutions, there is surprisingly little information on the effectiveness of this response to juvenile offenders. Moreover, what is known is not encouraging. Although there is some indication that effective institutional programs for juveniles exist,[97] the bulk of the evidence indicates that many juvenile institutions have little effect on recidivism. For example, a review of the rearrest rates of youths released in states that rely heavily on institutions found that the percentage of youths rearrested ranged from 51% to more than 70%.[98] More recently, a review of rearrest rates that tracked youths for three or more years after release from juvenile institutions reported that 74% to 89% were arrested for a new crime.[99] The results may not be surprising, considering the quality of life in many juvenile institutions.

Recent Trends in Juvenile Incarceration Among recent trends in juvenile incarceration are (1) its decreased use, (2) the use of both public and private facilities, (3) the disproportionately large percentage of males and racial or ethnic minorities that are incarcerated, and (4) the continued use of adult jails and state prisons. The number of juvenile offenders incarcerated in publicly and privately operated facilities has declined steadily since 2000. For example, in 2017, 43,580 youths were held in residential placement facilities, about 20% fewer than in 2013. The 43,580 juvenile offenders incarcerated in 2017 were the fewest number of juvenile offenders incarcerated since at least 1993.[100] However, the use of juvenile incarceration varies considerably by state. For example, in 2015, the incarceration rate per 100,000 juveniles was 329 in West Virginia but only 38 in Connecticut. The juvenile incarceration rate for the United States in 2015 was 152 juveniles per every 100,000 juveniles in the population (down from 173 in 2013, and 225 in 2010).[101]

Nearly all juveniles placed in residential facilities in 2015 committed delinquency offenses.[102] The delinquency offenses included person offenses (38% of all offenses for which juveniles were incarcerated), property offenses (22%), drug offenses (5%), and public order offenses (13%). Juveniles also were placed in residential facilities for technical violations (18%) and status offenses (5%). Of the juveniles who were placed for delinquency offenses in 2016, 71% were placed in public facilities and 29% were placed in private facilities.[103] However, states varied greatly in the extent to which they used public or private residential facilities. In 2015, Hawaii, Maine, New Jersey, and Virginia held 100% of their juvenile offenders in public facilities and none in private facilities, while Pennsylvania held 76% of its juvenile offenders in private facilities (the most of any state), and only 24% in public facilities (the fewest of any state).[104]

A disproportionately large percentage of males and racial or ethnic minorities are incarcerated in juvenile facilities. Although juvenile males constituted about half of the U.S. juvenile population age 10 through 17 in 2017, 87% of all juveniles committed to institutions

Spencer Grant/Science Source

Historically, large juvenile institutions have proven ineffective at preventing youths from offending again. *What do you think could be done to make these institutions more effective?*

FIGURE 13.6 Racial and Ethnic Composition of Inmate Population of Public and Private Juvenile Incarceration Facilities, 2017

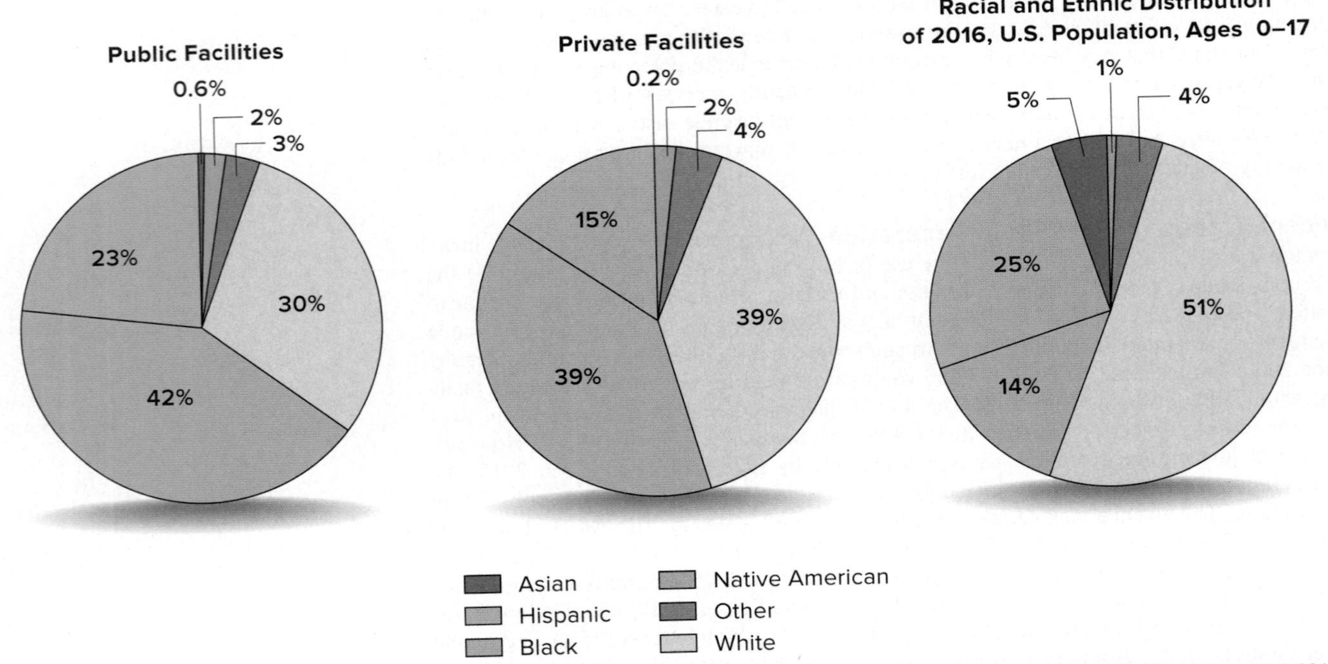

Sources: Based on data in M. Sickmund, T. J. Sladky, W. Kang, and C. Puzzanchera, "Easy Access to the Census of Juveniles in Residential Placement: 1997–2017," accessed October 1, 2019, https://www.ojjdp.gov/ojstatbb/ezacjrp/asp/display.asp; Child Trends, "Appendix 1-Percentage of Children Under 18, by Race and Hispanic Origin, Selected Years: 1980–2016 and Projections, 2020–2050," accessed October 1, 2019, https://www.childtrends.org/indicators/racial-and-ethnic-composition-of-the-child-population.

in 2017 were males. However, females accounted for about one-third of juveniles committed for status offenses. About 78% of all juveniles committed for status offenses were placed in private as opposed to public facilities.[105]

As for racial and ethnic minorities, in 2017, minorities (i.e., nonwhites and Hispanics) made up about 48% of the American population age 10 through 17.[106] Yet, minorities accounted for 62% of all juveniles in custody in 2017. About 75% of all juveniles in public facilities and about 25% of all juveniles in private facilities in 2017 were members of minority groups. Figure 13.6 illustrates the racial and ethnic composition of public and private juvenile incarceration facilities in 2017. Whites are underrepresented in both public and private facilities, blacks are overrepresented in both, Hispanics are slightly underrepresented in public facilities and underrepresented in private facilities, Asians are underrepresented in both, and Native Americans are overrepresented in both. The Juvenile Justice and Delinquency Prevention Act of 1974, as amended, requires states to determine whether minorities are disproportionately represented in confinement. If overrepresentation is found, the state must reduce it.

Many states also continue to hold juveniles in local adult jails and state prisons. Those juveniles are not included in the aforementioned total of 43,580 juveniles who were incarcerated in 2017. At mid-year 2017, local adult jails held about 3,600 juveniles, a decrease of about 14% from mid-year 2014, and a decrease of about 47% from mid-year 2005. Approximately 90% of the juveniles at mid-year 2017 were being held as adults, and 10% were being held as juveniles. Another 935 juveniles were being held in adult state prisons at year-end 2017, down nearly 10% from the 1,035 at year-end 2014. At year-end 2017, the BOP held 42 prisoners age 17 or younger (5%), while the states held 893 (95%).[107]

Aftercare Programs Aftercare involves the provision of services to assist youths in successfully making the transition from juvenile institutions to life back in the community. The services are the same as those provided by other types of community-based programs and may include foster care, shelter or group home placement, home placement, or efforts to help youths live on their own. Parole, too, is one form of aftercare.

Unfortunately, the quality of many aftercare programs is questionable, and in some cases, youths fail to receive any services after institutional release. Like probation supervision, parole supervision may involve very little contact between parole officers and parolees. Moreover, large caseloads carried by aftercare workers may prevent the provision of meaningful services.

In conclusion, twenty-first-century juvenile justice in the United States has several important problems to resolve. Many of the problems are the same ones that have plagued juvenile justice since the first specialized institutions for children were established in the early 1800s. The problems include:

1. Providing adequate due process protections to youths at all stages of the juvenile justice process.
2. Continuing to build on knowledge of effective correctional interventions and developing a range of effective and humane correctional responses, from diversion to institutional aftercare programs.
3. Eliminating the abusive treatment of youths placed in correctional programs.
4. Working out the appropriate balance between community-based and institutional correctional programs.
5. Conducting rigorous evaluations of juvenile justice agencies and programs.
6. Recognizing the limits of correctional responses in solving the juvenile crime problem.
7. Working out an appropriate balance between preventing and correcting delinquency.
8. Hiring and supporting a skilled, knowledgeable, and professional workforce that can adequately address existing problems and develop programs better able to address youth, family, and community problems.

THINKING CRITICALLY

1. Which of the community-based correctional programs described in the chapter (e.g., restitution, wilderness probation, and day treatment) do you think would be most effective at rehabilitating youths? Why?
2. What do you think are the pros and cons of institutional programs for juveniles?

SUMMARY

1. **Describe some of the early institutions used to respond to wayward and criminal youths.**
   Among the early institutions used to respond to wayward and criminal youths were houses of refuge, placing out, reform schools, industrial schools, and training schools.

2. **Explain the effects of important landmark U.S. Supreme Court cases on the juvenile justice system.**
   Landmark Supreme Court cases on juvenile justice include *Ex parte Crouse*, *People v. Turner*, *Commonwealth v. Fisher*, *Kent v. United States*, *In re Gault*, *In re Winship*, and *McKeiver v. Pennsylvania*. Some of the effects of those cases on juvenile justice are the right to adequate notice of charges, protection against compelled self-incrimination, the right to confront and to cross-examine accusers, and the right to the assistance of counsel.

3. **Identify and describe factors that influence the ways police process juvenile cases.**
   Among the factors that influence the ways that police process juvenile cases are (1) the seriousness of the offense, (2) the community, (3) the wishes of the complainant, (4) the demeanor of the youth, (5) the gender of the offender, (6) the race and social class of the offender, and (7) the police organization.

4. **Summarize the rationale for the use of diversion in juvenile justice.**
   Diversion programs in juvenile justice are based on the understanding that formal responses to youths who violate the law, such as arrest and adjudication, do not always protect the best interests of children or the community. Indeed, some formal responses may be harmful to many youths and may increase the likelihood of future delinquent behavior. This is because formal processing may cause youths to develop negative or delinquent self-images, may stigmatize youths in the eyes of significant others, or may subject youths to inhumane treatment.

5. **Describe the adjudication hearing in juvenile justice.**
   There are two types of adjudication: contested ones (in which juveniles dispute the charges) and uncontested ones. Contested adjudications are similar to trials in criminal courts. Most contested adjudications are bench adjudications in which the hearing officer makes a finding of fact based on the evidence presented. In some

jurisdictions, contested adjudications are jury trials. At an uncontested adjudication hearing, the youth, or frequently the youth's attorney, admits to the charges. The vast majority of juvenile court adjudications are uncontested.

6. Describe the disposition hearing and the types of dispositions available to the juvenile court.

The disposition is the juvenile court equivalent of sentencing in criminal court. At the disposition hearing, the court makes its final determination of what to do with the youth who is officially labeled delinquent. At the disposition, the court also enters various orders regarding the youth's behavior. Those orders consist of various rules the youth must follow. Also, the court may enter orders regarding parents, relatives, or other people who live in the home. Dispositions available to the juvenile court include probation, placement in a diversion program, restitution, community service, detention, placement in foster care, placement in a long-term or short-term residential treatment program, placement with a relative, and placement with the state for commitment to a state facility. Not all disposition alternatives are available in all jurisdictions.

7. Identify the types and describe the effectiveness of community-based correctional programs for juveniles.

Community-based correctional programs include diversion, pretrial release, probation, foster care, group home placements, and parole. Evaluations of some programs indicate that they are effective at reducing recidivism or that they are as effective as institutional placement. In contrast, evaluations of other community-based programs indicate that they have little effect on subsequent offenses.

8. Summarize recent trends in juvenile incarceration.

Among recent trends in juvenile incarceration are (1) its decreased use, (2) the use of both public and private facilities, (3) the disproportionately large percentage of males and racial or ethnic minorities that are incarcerated, and (4) the continuing incarceration of juveniles in adult jails and state prisons.

9. Identify the types and describe the effectiveness of institutional programs for juveniles.

In the United States, a variety of correctional institutions house juveniles, including detention centers, adult jails, shelter facilities, reception and diagnostic centers, ranches, forestry camps, farms, and training schools. There is some evidence that small, secure treatment facilities for violent or chronic offenders are effective at reducing recidivism. However, many institutions, particularly large state institutions, often have been found to have little positive effect on youths' subsequent delinquent behaviors. In fact, they may increase the likelihood that youths will commit further offenses.

KEY TERMS

juvenile delinquency 533
apprenticeship system 533
binding-out system 533
houses of refuge 534
placing out 534
reform, industrial, or training schools 536
cottage reformatories 536
parens patriae 537
adjudication 540
informal juvenile justice 541
status offenses 542
radical nonintervention 546
intake screening 549
petition 549
transfer, waiver, remand, bindover, or certification 549
hearing officer 552
disposition 553

REVIEW QUESTIONS

1. What changes occurred in the sixteenth and seventeenth centuries in the ways the young were viewed?
2. What were the purposes of houses of refuge?
3. What is *parens patriae*, and what was the legal context in which it arose?
4. What was the social and historical context in which the juvenile court was created?
5. Historically, what has been the fundamental difference between the procedures used in juvenile courts and those employed in criminal courts?
6. What is the informal juvenile justice process, and why is it important?
7. What are four typical responses that police officers employ when handling juvenile cases?
8. What are two relatively new juvenile diversion alternatives?
9. What are five possible intake decisions that might be made in the juvenile justice process?
10. What are five recent trends in the practice of transferring juvenile cases to criminal court, and what do they suggest about the current orientation of juvenile justice?
11. In practice, what two dispositions typically are available to juvenile court judges in most jurisdictions, and which one is used most frequently?
12. What are three new dispositional or sentencing practices employed by juvenile court judges?
13. What are six objectives of community-based correctional programs for juveniles?
14. What are four problems commonly found in juvenile correctional facilities?
15. What are some important problems that remain unresolved in juvenile justice in the United States in the twenty-first century?

IN THE FIELD

1. **Debate Juvenile Justice** With fellow students, family members, or friends, debate whether the juvenile justice system ought to be abolished and juvenile offenders treated as adults.
2. **Compare Correctional Facilities** Visit different juvenile correctional facilities in your community, both community-based facilities and institutional facilities. Compare what you observe at these facilities based on the following criteria:
   a. How do terms for similar offenses compare at each facility?
   b. How are juveniles treated?
   c. Are there educational or rehabilitation opportunities?
   d. How does the way of life compare?

ON THE WEB

1. **Juvenile Justice Information Exchange** Go to the Juvenile Justice Information Exchange website, http://jjie.org/. Click on Ideas and Opinions and select a subject that interests you. Read the entry, and then write a brief comment about it. Do you agree or disagree with the author's opinion? Defend your choice.
2. **Juvenile Delinquency Prevention** Go to the Office of Juvenile Justice and Delinquency Prevention Model Program Guides at www.ojjdp.gov/mpg. Select a program from "Recently Posted Programs" and explain how the program might be implemented in your community. Describe needed resources and possible impediments to the program's success.

CRITICAL THINKING EXERCISES

THE CASE OF PAUL

1. Paul, a 15-year-old, sexually assaulted, robbed, and killed Billy, an 11-year-old. Billy was missing for 2 days. Paul hid Billy's body for those 2 days in his family's garage before dumping it in a wooded area near the house. Neighbors describe Paul as quiet and introverted. His parents state that he grew increasingly violent after they kept him from a 43-year-old man named Smith who Paul had met in a chat room. Smith subsequently was charged with sexually assaulting Paul. The two apparently had sex in motels five times during the previous 4 months.
   a. Should Paul be handled by the justice system (either juvenile or adult) or be diverted, perhaps to a mental health facility? Why?
   b. Should he be charged with first-degree murder (as well as the other crimes)? Why or why not?
   c. Should he be transferred to criminal court and tried as an adult?
   d. If tried as an adult, should he enter a plea of not guilty by reason of insanity? A plea of guilty but insane? Why or why not?
   e. If Paul is convicted of any of the charges, what sentence should be imposed? Why?

THE CASE OF JAMES

2. Seventeen-year-old James lives in a nice lower-middle-class neighborhood in central Florida. A few years ago, James began running with a bad crowd. His parents tried a variety of punishments, but to no avail. James eventually was arrested for burglary, theft, and other crimes. A juvenile court judge placed him on probation and ordered him to receive counseling. James violated probation, stopped going to counseling sessions, and began committing crimes again. After a subsequent arrest, James was ordered into a residential rehabilitation program for several months. After finishing the program, he was placed in a special school from which he soon ran away. He was next placed in another residential program designed for youths "needing more structure." He ran away. After he was caught, a juvenile court judge returned him to the residential program for about 6 months.

   James's father states that the rehabilitation programs are a waste of time and money because they are disorganized and ineffectual. He fears James will kill or be killed, and he wants James to be locked up until the age of 19. However, space is limited in the most secure juvenile institutions. In Florida, youths must commit an average of 12 crimes before they are placed in such institutions.
   a. Should James be handled by the justice system (either juvenile or adult) or be diverted? Why?
   b. If he is diverted, what type of program would benefit him most? Why?
   c. If James is retained in the juvenile justice system, what disposition would be most appropriate and beneficial for him?

NOTES

1. Philippe Ariès, *Centuries of Childhood: A Social History of Family Life,* trans. Robert Baldick (New York: Random House, 1962).
2. Barry Krisberg and James F. Austin, *Reinventing Juvenile Justice* (Newbury Park, CA: Sage, 1993), 9.
3. Harry Elmer Barnes, *The Story of Punishment,* 2nd ed. (Montclair, NJ: Patterson Smith, 1972).
4. David J. Rothman, *The Discovery of the Asylum* (Boston: Little Brown, 1971).
5. Thomas J. Bernard and Megan C. Kurlychek, *The Cycle of Juvenile Justice*, 2nd ed. (New York: Oxford University Press, 2010), 53; Rothman, *The Discovery of the Asylum*, 207.
6. Robert M. Mennel, *Thorns and Thistles* (Hanover, NH: University Press of New England, 1973); Steven L. Schlossman, *Love and the American Delinquent* (Chicago: University of Chicago Press, 1977).
7. Alexander Pisciotta, "Treatment on Trial: The Rhetoric and Reality of the New York House of Refuge, 1857–1935," *American Journal of*

Legal History 29 (1985): 151–81; Rothman, *The Discovery of the Asylum*, 231.

8. Clemens Bartollas and Stuart J. Miller, *Juvenile Justice in America* (Englewood Cliffs, NJ: Regents/Prentice Hall, 1994), 136.

9. LaMar T. Empey and Mark C. Stafford, *American Delinquency: Its Meaning and Construction,* 3rd ed. (Belmont, CA: Wadsworth, 1991), 368.

10. Belinda R. McCarthy and Bernard J. McCarthy, *Community-Based Corrections,* 2nd ed. (Pacific Grove, CA: Brooks/Cole, 1991), 98.

11. Krisberg and Austin, *Reinventing Juvenile Justice*, 23–24; Hastings Hart, *Preventive Treatment of Neglected Children* (New York: Russell Sage, 1910), 70.

12. Bartollas and Miller, *Juvenile Justice in America*, 209; John T. Whitehead and Steven P. Lab, *Juvenile Justice: An Introduction* (Cincinnati, OH: Anderson, 1990), 47.

13. Cited in Meda Chesney-Lind and Randall G. Shelden, *Girls, Delinquency, and Juvenile Justice* (Pacific Grove, CA: Brooks/Cole, 1992), 111.

14. Harold Finestone, *Victims of Change* (Westport, CT: Greenwood Press, 1976).

15. Anthony M. Platt, *The Child Savers: The Invention of Delinquency* (Chicago: University of Chicago Press, 1969), 99.

16. Ibid., 101–2.

17. Bernard and Kurlychek, *The Cycle of Juvenile Justice*, 58–59.

18. Ibid., 59–61.

19. Platt, *The Child Savers*, 134–36; Bernard and Kurlychek, *The Cycle of Juvenile Justice*, 76–78.

20. Empey and Stafford, *American Delinquency*, 58–59.

21. Bartollas and Miller, *Juvenile Justice in America*, 92; Empey and Stafford, *American Delinquency*, 59.

22. Krisberg and Austin, *Reinventing Juvenile Justice*, 30.

23. Bernard and Kurlychek, *The Cycle of Juvenile Justice*, 83–85.

24. Ibid., 84.

25. Walter Wadlington, Charles H. Whitebread, and Samuel M. Davis, *Cases and Materials on Children in the Legal System* (Mineola, NY: Foundation Press, 1983), 202; Bernard and Kurlychek, *The Cycle of Juvenile Justice*, 97.

26. M. A. Bortner, *Delinquency and Justice: An Age of Crisis* (New York: McGraw-Hill, 1988), 60.

27. Bernard and Kurlychek, *The Cycle of Juvenile Justice*, 129–132.

28. Sarah Hockenberry and Charles Puzzanchera, "Juvenile Court Statistics 2017" (Pittsburg, PA: National Center for Juvenile Justice, June 2019), 52. Accessed September 22, 2019, http://www.ncjj.org/pdf/jcsreports/JCS2017report.pdf; Office of Juvenile Justice and Delinquency Prevention, *Easy Access to Juvenile Court Statistics: 1985–2013*, accessed February 26, 2016, www.ojjdp.gov/ojstatbb/ezajcs/.

29. Anthony Petrosino, Carolyn Turpin-Petrosino, and Sarah Guckenburg, "Formal System Processing of Juveniles: Effects on Delinquency," *Campbell Systematic Reviews* (2010): 1–89, accessed January 16, 2013, www.pjdc.org/wp/wp-content/uploads/Review_System_Process_Effect_Juvenile_Delinquency_100129.pdf.

30. Sarah Hockenberry and Charles Puzzanchera, *Juvenile Court Statistics* 2017, 31.

31. Ibid., 76.

32. Douglas A. Smith, "The Organizational Context of Legal Control," *Criminology* 22 (1984): 19–38.

33. Robert J. Sampson, "Effects of Socioeconomic Context on Official Reaction to Juvenile Delinquency," *American Sociological Review* 51 (1986): 876–85; also see A. Cicourel, *The Social Organization of Juvenile Justice* (New York: Wiley, 1968).

34. Donald Black and Albert J. Reiss, "Police Control of Juveniles," *American Sociological Review* 35 (1970): 63–77; Richard J. Lundman, Richard E. Sykes, and John P. Clark, "Police Control of Juveniles: A Replication," in *Juvenile Delinquency: A Justice Perspective*, ed. Ralph Weisheit and Robert G. Culbertson (Prospect Heights, IL: Waveland Press, 1978), 107–15; Irving Piliavan and Scott Briar, "Police Encounters with Juveniles," *American Journal of Sociology* 70 (1964): 206–14.

35. Marvin D. Krohn, James P. Curry, and Shirley Nelson-Kilger, "Is Chivalry Dead? An Analysis of Changes in Police Dispositions of Males and Females," *Criminology* 21 (1983): 417–37; Katherine Teilmann and Pierre H. Landry, "Gender Bias in Juvenile Justice," *Journal of Research in Crime and Delinquency* 18 (1981): 47–80; see also William G. Staples, "Law and Social Control in Juvenile Justice Dispositions," *Journal of Research in Crime and Delinquency* 24 (1987): 7–22; Meda Chesney-Lind, "Judicial Paternalism and the Female Status Offender: Training Women to Know Their Place," *Crime and Delinquency* 23 (1977): 121–30.

36. Robert J. Sampson, "Sex Differences in Self-Reported Delinquency and Official Records: A Multiple-Group Structural Modeling Approach," *Journal of Quantitative Criminology* 1 (1985): 345–67; Dale Dannefer and Russell K. Schutt, "Race and Juvenile Justice Processing in Court and Police Agencies," *American Journal of Sociology* 87 (1982): 1113–132.

37. Merry Morash, "Establishment of a Juvenile Police Record: The Influence of Individual and Peer Group Characteristics," *Criminology* 22 (1984): 97–111; Krohn, Curry, and Nelson-Kilger, "Is Chivalry Dead?"; D. Elliott and H. L. Voss, *Delinquency and Dropout* (Lexington, MA: Lexington Books, 1974).

38. Charles Puzzanchera, "Juvenile Arrests, 2017," Juvenile Justice Statistics, U.S. Department of Justice (Laurel, MD: Office of Juvenile Justice and Delinquency Prevention, August 2019), 8. Accessed September 23, 2019, http://www.ncjj.org/pdf/Juvenile%20Arrests%20Bulletins/252713.pdf; Charles Puzzanchera, "Juvenile Arrests 2012," in Juvenile Offenders and Victims: National Report Series, Office of Juvenile Justice and Delinquency Prevention, U.S. Department of Justice (Washington, DC: U.S. Government Printing Office, December 2014), 5, accessed February 26, 2016, www.ojjdp.gov/pubs/248513.pdf.

39. Federal Bureau of Investigation, *2017 Crime in the United States*, Table 43B, accessed September 23, 2019, https://ucr.fbi.gov/crime-in-the-u.s/2017/crime-in-the-u.s.-2017/topic-pages/tables/table-43.

40. Charles Puzzanchera, "Juvenile Arrests, 2017," 8.

41. Delbert S. Elliott and David Huizinga, "Social Class and Delinquent Behavior in a National Youth Panel," *Criminology* 21 (1983): 149–77.

42. Sarah Hockenberry and Charles Puzzanchera, "Juvenile Court Statistics 2017," 7; Information about teen courts or youth courts is from Jeffrey Butts, Dean Hoffman, and Janeen Buck, "Teen Courts in the United States: A Profile of Current Programs," Office of Juvenile Justice and Delinquency Prevention Fact Sheet #118, Office of Juvenile Justice and Delinquency Prevention, U.S. Department of Justice (Washington, DC: U.S. Government Printing Office, October 1999); Jeffrey A. Butts and Janeen Buck, "Teen Courts: A Focus on Research," in *Juvenile Justice Bulletin*, Office of Juvenile Justice and Delinquency Prevention, U.S. Department of Justice (Washington, DC: U.S. Government Printing Office, October 2000); Jeffrey A. Butts, Janeen Buck, and Mark B. Coggeshall, *The Impact of Teen Court on Young Offenders* (Washington, DC: The Urban Institute, Office of Juvenile Justice and Delinquency Prevention, U.S. Department of Justice, April 2002); National Youth

Court Center, www.youthcourt.net; National Association of Youth Courts, accessed December 29, 2008, www.youthcourt.net/content/view/7/14/; The National Association of Youth Courts, "Facts and Stats," accessed February 27, 2011, www.youthcourt.net/?page_id524.

43. Personal communication with Scott Peterson, CEO and Board President of Global Youth Services, September 24, 2019.

44. Brenda Vose and Kelly Vannan, "A Jury of Your Peers: Recidivism Among Teen Court Participants," *Journal of Juvenile Justice*, 3 (Fall 2013): 97–109, accessed July 31, 2016, www.journalofjuvjustice.org/JOJJ0301/JOJJ0301.pdf.

45. Ibid., p. 99, Table 1.

46. Information on restorative justice conferences or family group conferences is from Edmund F. McGarrell, "Restorative Justice Conferences as an Early Response to Young Offenders," in *Juvenile Justice Bulletin*, Office of Juvenile Justice and Delinquency Prevention, U.S. Department of Justice (Washington, DC: U.S. Government Printing Office, August 2001); *OJJDP Research 2000*, Office of Juvenile Justice and Delinquency Prevention, U.S. Department of Justice (Washington, DC: U.S. Government Printing Office, May 2001).

47. Stephen Hudak, "Citations for Minor Crimes Help Juveniles Avoid Criminal Record," *Orlando Sentinel*, May 1, 2012, A1.

48. Jamaal Harrison, "2018 Comprehensive Accountability Report: Civil Citation," Florida Department of Juvenile Justice, December 2018. Accessed September 24, 2019, http://www.djj.state.fl.us/docs/car-reports/(2017-18-car)-civil-citation-(mg).pdf?sfvrsn=2.

49. Sarah Hockenberry and Charles Puzzanchera, "Juvenile Court Statistics, 2017," 32.

50. Ibid.

51. Calculated from data in Zhen Zeng, "Jail Inmates in 2017," *Bureau of Justice Statistics Bulletin,* U.S. Department of Justice (Washington, DC: U.S. Government Printing Office, April 2019), 5, Table 3, accessed September 24, 2019, https://www.bjs.gov/content/pub/pdf/ji17.pdf.

52. Howard N. Snyder and Melissa Sickmund, *Juvenile Offenders and Victims: 1999 National Report*, National Center for Juvenile Justice, Office of Juvenile Justice and Delinquency Prevention, U.S. Department of Justice (Washington, DC: U.S. Government Printing Office, September 1999), 97.

53. Carolyn Crist, "Suicide-risk screening might cut deaths among incarcerated youth," *Reuters*, January 31, 2019. Accessed September 24, 2019, https://www.reuters.com/article/us-health-youth-prison-suicide/suicide-risk-screening-might-cut-deaths-among-incarcerated-youth-idUSKCN1PP2LH.

54. Patricia McFall Torbet, *Juvenile Probation: The Workhorse of the Juvenile Justice System,* OJJDP Juvenile Justice Bulletin, March 1996, Table 3, accessed February 29, 2016, www.ncjrs.gov/pdffiles/workhors.pdf; Ted Rubin, "The Emerging Prosecutor Dominance of the Juvenile Court Intake Process," *Crime and Delinquency* 26 (1980): 299–318.

55. For a review of this literature, see Chesney-Lind and Shelden, *Girls, Delinquency, and Juvenile Justice*, 137–39.

56. Julie Harrington, Martijn Niekus, and Jian Cao, "An Economic and Fiscal Analysis of Direct File Reform Proposals—Final Report," Southern Poverty Law Center, February 2019, 13, Table 3. Accessed September 25, 2019, https://www.splcenter.org/sites/default/files/djj_3-4-2019_final_report.pdf.

57. "Juveniles Tried as Adults" [2016], Statistical Briefing Book, Office of Juvenile Justice and Delinquency Prevention, U.S. Department of Justice. Accessed September 24, 2019, https://www.ojjdp.gov/ojstatbb/structure_process/qa04110.asp?qaDate=2016&text=yes&maplink=link1.

58. Michael Tonry, "Crime and Human Rights—How Political Paranoia, Protestant Fundamentalism, and Constitutional Obsolescence Combined to Devastate Black America: The American Society of Criminology 2007 Presidential Address," *Criminology* 46 (2008): 8.

59. "Juvenile Waiver," *LegalMatch*. Accessed September 25, 2019, https://www.act4jj.org/sites/default/files/ckfinder/files/ACT4JJ%20Youth%20In%20Adult%20System%20Fact%20Sheet%20Aug%202014%20FINAL.pdf.

60. "Youth in the Adult System Fact Sheet," Act 4 Juvenile Justice, A Campaign of the Juvenile Justice & Delinquency Prevention Coalition, August 2014. Accessed September 24, 2019, https://www.act4jj.org/sites/default/files/ckfinder/files/ACT4JJ%20Youth%20In%20Adult%20System%20Fact%20Sheet%20Aug%202014%20FINAL.pdf.

61. Anne Teigen, *Juvenile Age of Jurisdiction and Transfer to Adult Court Laws,* National Conference of State Legislatures, January 11, 2019, accessed September 25, 2019, http://www.ncsl.org/research/civil-and-criminal-justice/juvenile-age-of-jurisdiction-and-transfer-to-adult-court-laws.aspx.

62. Harrington, Niekus, and Cao, "An Economic and Fiscal Analysis of Direct File Reform Proposals—Final Report," 25; *Two-Year Trends & Conditions*, Florida Department of Juvenile Justice. Accessed February 29, 2016, www.djj.state.fl.us/research/fast-facts/trends-conditions.

63. Julie Harrington, Martijn Niekus, and Jian Cao, "An Economic and Fiscal Analysis of Direct File Reform Proposals—Final Report," 26.

64. Ibid., 34, Table 17.

65. Melissa Sickmund and Charles Puzzanchera (eds.), *Juvenile Offenders and Victims: 2014 National Report* (Pittsburgh, PA: National Center for Juvenile Justice, 2014), 103, accessed August 1, 2016, www.ojjdp.gov/ojstatbb/nr2014/downloads/chapter4.pdf.

66. "Jurisdictional boundaries," Juvenile Justice Geography, Policy, Practice & Statistics, 2016. Accessed September 25, 2019, http://www.jjgps.org/jurisdictional-boundaries.

67. Joseph B. Sanborn Jr., "Pleading Guilty in Juvenile Court," *Justice Quarterly* 9 (1992): 127–50.

68. Hockenberry and Puzzanchera, *Juvenile Court Statistics 2017,* 52.

69. Martin L. Forst, Bruce A. Fisher, and Robert B. Coates, "Indeterminate and Determinate Sentencing of Juvenile Delinquents: A National Survey of Approaches to Commitment and Release Decision-Making," *Juvenile and Family Court Journal* (Summer 1985): 1–12.

70. Jeffrey Fagan, Ellen Slaughter, and Richard Hartstone, "Blind Justice: The Impact of Race on the Juvenile Justice Process," *Crime and Delinquency* 33 (1987): 244–58; Belinda R. McCarthy and Brent L. Smith, "The Conceptualization of Discrimination in the Juvenile Justice Process," *Criminology* 24 (1986): 41–64.

71. Hockenberry and Puzzanchera, *Juvenile Court Statistics 2017*, 52.

72. Snyder and Sickmund, *Juvenile Offenders and Victims: 1999 National Report*; Patricia Torbert, Richard Gable, Hunter Hurst IV, Imogene Montgomery, Linda Szymanski, and Douglas Thomas, *State Responses to Serious and Violent Juvenile Crime*, Office of Juvenile Justice and Delinquency Prevention, U.S. Department of Justice (Washington, DC: U.S. Government Printing Office, July 1996), chap. 3.

73. Tonry, "Crime and Human Rights—How Political Paranoia, Protestant Fundamentalism, and Constitutional Obsolescence Combined to Devastate Black America."

74. Hockenberry and Puzzanchera, "Juvenile Court Statistics 2017," 52; Sarah Hockenberry and Charles Puzzanchera, "Juvenile Court Statistics 2013" (Pittsburgh, PA: National Center for Juvenile Justice, July 2015), 52. Accessed September 26, 2019, http://www.ncjj.org/pdf/jcsreports/jcs2013.pdf.

75. Ibid.

76. Ibid., 84.

77. Ibid.

78. "Transforming Juvenile Probation: A Vision for Getting It Rights," The Annie E. Casey Foundation, 2018. Accessed September 28, 2019, https://www.aecf.org/m/resourcedoc/aecf-transforming juvenileprobationsummary-2018.pdf#page=3.

79. Ibid.

80. Troy L. Armstrong, "National Survey of Juvenile Intensive Supervision," *Criminal Justice Abstracts* 20 (1988): 342–48 (Part I) and 497–523 (Part II).

81. Richard G. Wiebush, "Juvenile Intensive Supervision: The Impact on Felony Offenders Diverted from Institutional Placement," *Crime and Delinquency* 39 (1993): 68–89; William H. Barton and Jeffrey A. Butts, "Viable Options: Intensive Supervision Programs for Juvenile Delinquents," *Crime and Delinquency* 36 (1990): 238–56.

82. Richard A. Ball, Ronald Huff, and Robert Lilly, *House Arrest and Correctional Policy: Doing Time at Home* (Newbury Park, CA: Sage, 1988).

83. See Marc Renzema and David T. Skelton, "Use of Electronic Monitoring in the United States: 1989 Update," *Research in Brief* (Washington, DC: National Institute of Justice, 1990); Daniel Ford and Annesley K. Schmidt, "Electronically Monitored Home Confinement," *Research in Action* (Washington, DC: National Institute of Justice, 1985).

84. Joseph B. Vaughn, "A Survey of Juvenile Electronic Monitoring and Home Confinement Programs," *Juvenile and Family Court Journal* 40 (1989): 1–36.

85. Kevin I. Minor and Preston Elrod, "The Effects of a Probation Intervention on Juvenile Offenders' Self-Concepts, Loci of Control, and Perceptions of Juvenile Justice," *Youth and Society* 25 (1994): 490–511; Gerald L. Golins, *Utilizing Adventure Education to Rehabilitate Juvenile Delinquents* (Las Cruces, NM: Educational Resources Information Center, Clearinghouse on Rural Education and Small Schools, 1980).

86. See, for example, R. Callahan, "Wilderness Probation: A Decade Later," *Juvenile and Family Court Journal* 36 (1985): 31–51; John Winterdyk and Ronald Roesch, "A Wilderness Experiential Program as an Alternative for Probationers: An Evaluation," *Canadian Journal of Criminology* 24 (1982): 39–49.

87. H. Preston Elrod and Kevin I. Minor, "Second Wave Evaluation of a Multi-Faceted Intervention for Juvenile Court Probationers," *International Journal of Offender Therapy and Comparative Criminology* 36 (1992): 247–62; John Winterdyk and Curt Griffiths, "Wilderness Experience Programs: Reforming Delinquents or Beating Around the Bush?" *Juvenile and Family Court Journal* 35 (1984): 35–44; Winterdyk and Roesch, "A Wilderness Experiential Program."

88. Calculated from data in Charles Puzzanchera, Sarah Hockenberry, Anthony Sladky, and Wei Kang (2018), "Juvenile Residential Facility Census Databook," Available at: https://www.ojjdp.gov/ojstatbb/jrfcdb/.

89. Ted Palmer, *The Re-Emergence of Correctional Intervention* (Newbury Park, CA: Sage, 1992); Office of Juvenile Justice and Delinquency Prevention, *Project New Pride: Replication* (Washington, DC: Office of Juvenile Justice and Delinquency Prevention, 1979); LaMar T. Empey and Maynard L. Erickson, *The Provo Experiment: Evaluating Community Control of Delinquency* (Lexington, MA: Lexington Books, 1972).

90. Patricia Chamberlain and John B. Reid, "Comparison of Two Community Alternatives to Incarceration for Chronic Juvenile Offenders," *Journal of Consulting and Clinical Psychology* 66 (1998): 624–633; Patricia Chamberlain, Leslie D. Leve, and David S. DeGarmo, "Multidimensional Treatment Foster Care for Girls in the Juvenile Justice System: 2-year Follow-up of a Randomized Clinical Trial," *Journal of Consulting and Clinical Psychology* 75 (2007): 187–93; J. Mark Eddy, Rachel B. Whaley, and Patricia Chamberlain, "The Prevention of Violent Behavior by Chronic and Serious Male Juvenile Offenders: A 2-year Follow-up of a Randomized Clinical Trial," *Journal of Emotional and Behavioral Disorders* 12 (2004): 2–8; Leslie D. Leve, Patricia Chamberlain, and John B. Reid, "Intervention Outcomes for Girls Referred from Juvenile Justice: Effects on Delinquency," *Journal of Consulting and Clinical Psychology* 73 (2005): 1181–85.

91. Calculated from data in Charles Puzzanchera, Sarah Hockenberry, T.J. Sladky, and Wei Kang (2018), "Juvenile Residential Facility Census Databook," Available at: https://www.ojjdp.gov/ojstatbb/jrfcdb/.

92. LaMar T. Empey and Steven G. Lubeck, *The Silverlake Experiment: Testing Delinquency Theory and Community Intervention* (Chicago: Aldine Publishing Company, 1971).

93. Information about New York City's C2H program is from Marsha Weissman, Vidhya Ananthakrishnan, and Vincent Schiraldi, "Moving Beyond Youth Prisons: Lessons from New York City's Implementation of Close to Home," Columbia University Justice Lab, February 2019. Accessed September 29, 2019, https://thecrimereport.org/wp-content/uploads/2019/02/close-to-home-report-.pdf.

94. Terrence P. Thornberry, Stewart E. Tolnay, Timothy J. Flanagan, and Patty Glynn, *Office of Juvenile Justice and Delinquency Prevention Report on Children in Custody 1987: A Comparison of Public and Private Juvenile Custody Facilities* (Washington, DC: Office of Juvenile Justice and Delinquency Prevention, 1991).

95. Ibid.; Dale G. Parent, "Conditions of Confinement," *Juvenile Justice* 1, no. 1 (Spring/Summer 1993): 2–7, www.ncjrs.org/_pdffiles/jjjs93.pdf.

96. Allen J. Beck, David Cantor, John Hartge, and Tim Smith, *Sexual Victimization in Juvenile Facilities Reported by Youth, 2012: National Survey of Youth in Custody, 2012,* Bureau of Justice Statistics, U.S. Department of Justice (Washington, DC: U.S. Government Printing Office, June 2013), accessed August 3, 2016, www.bjs.gov/content/pub/pdf/svjfry12.pdf.

97. See, for example, Barry Krisberg, *Juvenile Justice: Improving the Quality of Care* (San Francisco: National Council on Crime and Delinquency, 1992); Carol J. Garrett, "Effects of Residential Treatment on Adjudicated Delinquents," *Journal of Research in Crime and Delinquency* 22 (1985): 287–308.

98. Barry Krisberg, Robert DeComo, and Norma C. Herrera, *National Juvenile Custody Trends 1978–1989* (Washington, DC: Office of Juvenile Justice and Delinquency Prevention, 1992), 2; also see Steven Lab and John T. Whitehead, "A Meta-Analysis of Juvenile Correctional Treatment," *Journal of Research in Crime and Delinquency* 26 (1989): 276–95; Norman G. Hoffmann, Ana M. Abrantes, and Ronald Anton, "Criminals, Troubled Youth, or a Bit of Both," *Addiction Professional* 1, no. 4 (July 2003); Snyder and Sickmund, *Juvenile Offenders and Victims: 1999 National Report*, 234.

99. Richard A. Mendel, *No Place for Kids: The Case for Reducing Juvenile Incarceration* (Baltimore, MD: The Annie E. Casey Foundation, 2011).

100. *OJJDP Statistical Briefing Book* online, April 23, 2019. Accessed October 1, 2019, https://www.ojjdp.gov/ojstatbb/corrections /qa08201.asp?qaDate=2017; *OJJDP Statistical Briefing Book* online, October 2, 2015, accessed August 3, 2016, www.ojjdp.gov/ojstatbb /corrections/qa08601.asp?qaDate=2013; Sickmund et al., "Easy Access to the Census of Juveniles in Residential Placement"; Melissa Sickmund, "Juveniles in Residential Placement, 1997–2008," OJJDP Fact Sheet, U.S. Department of Justice, February 2010, accessed March 7, 2011, www.ncjrs.gov/pdffiles1/ojjdp/229379.pdf.

101. *OJJDP Statistical Briefing Book* online, June 1, 2017. Accessed October 1, 2019, https://www.ojjdp.gov/ojstatbb/corrections /qa08601.asp?qaDate=2015&text=yes; *OJJDP Statistical Briefing Book* online, October 02, 2015; *OJJDP Statistical Briefing Book* online, December 9, 2011, www.ojjdp.gov/ojstatbb/corrections /qa08601.asp?qa Date52010.

102. *OJJDP Statistical Briefing Book* online, June 1, 2017. Accessed October 1, 2019, https://www.ojjdp.gov/ojstatbb/corrections /qa08301.asp?qaDate=2015&text=yes.

103. *OJJDP Statistical Briefing Book* online, March 27, 2018. Accessed October 1, 2019, https://www.ojjdp.gov/ojstatbb/corrections /qa08502.asp?qaDate=2016.

104. *OJJDP Statistical Briefing Book* online, June 1, 2017. Accessed October 1, 2019, https://www.ojjdp.gov/ojstatbb/corrections /qa08603.asp?qaDate=2015&text=yes.

105. Melissa Sickmund, T.J. Sladky, Wei Kang, and Charles Puzzanchera, "Easy Access to the Census of Juveniles in Residential Placement," 2019. Accessed October 1, 2019, https://www.ojjdp.gov /ojstatbb/ezacjrp/asp/display.asp.

106. Calculated from data in ibid.

107. Zhen Zeng, "Jail Inmates in 2017," *Bureau of Justice Statistics Bulletin*, U.S. Department of Justice (Washington, DC: Government Printing Office, April 2019), 5, Table 3. Accessed October 1, 2019, https://www.bjs.gov/content/pub/pdf/ji17.pdf; Jennifer Bronson and E. Ann Carson, "Prisoners in 2017," *Bureau of Justice Statistics Bulletin*, U.S. Department of Justice (Washington, DC: Government Printing Office, April 2019), 20, Table 11. Accessed October 1, 2019, https://www.bjs.gov/content/pub/pdf/p17.pdf; Minton and Zeng, "Jail Inmates at Midyear 2014," p. 3, Table 2; E. Ann Carson, "Prisoners in 2014," *Bureau of Justice Statistics Bulletin,* U.S. Department of Justice (Washington, DC: Government Printing Office, September 2015), 31, Appendix Table 6, accessed August 4, 2016, www.bjs.gov /content/pub/pdf/p14.pdf.

Snap Stills/Shutterstock

The Future of Criminal Justice in the United States

CHAPTER OUTLINE

LEARNING OBJECTIVES

After completing this chapter, you should be able to:

1. Describe the possible future of law enforcement if the crime control model dominates and the possible future if the due process model dominates.
2. Describe the possible future of the administration of justice if the crime control model dominates and the possible future if the due process model dominates.
3. Identify perhaps the most divisive issue that will confront correctional policy makers in the future.
4. Describe the possible future of corrections.
5. List some of the cost-reduction strategies likely to be advocated in corrections in the future.
6. List specific reforms of the "smart justice" movement and identify its overarching goal.
7. Describe some of the challenges faced by criminal justice in the Age of Terrorism.

Crime Story

How about a career as a "peacemaker"? Peacemakers are employed by the Community Relations Service (CRS)—a little-known agency of the U.S. Department of Justice. The CRS was created by Title X of the Civil Rights Act of 1964 to address community conflicts and tensions arising from differences of race, color, and national origin. It is the only federal agency dedicated to assisting state and local units of government, private and public organizations, and community groups with preventing and resolving racial and ethnic tensions, incidents, and civil disorders and with restoring racial stability and harmony. With the passage of the Matthew Shepard and James Byrd Jr. Hate Crimes Prevention Act in 2009, the CRS also has been charged with working with communities to prevent and respond to alleged violent hate crimes committed on the basis of actual or perceived race, color, national origin, gender, gender identity, sexual orientation, religion, or disability. The CRS may respond to requests for help from appropriate state or local officials, such as police chiefs, mayors, school superintendents, or other interested parties, or it may proactively offer its services.

The CRS facilitates the development of viable, mutual understandings and agreements as alternatives to coercion, violence, or litigation. It also assists communities in developing local mechanisms, conducting training, and engaging in other proactive measures to prevent racial or ethnic tension and violent hate crimes. The CRS does not take sides among disputing parties and, in promoting the principles and ideals of nondiscrimination, applies skills that allow parties to come to their own agreement. In performing this mission, the CRS deploys highly skilled professional conciliators, who are able to assist people of diverse backgrounds. Each year, the CRS's conciliators bring hundreds of communitywide conflicts to peaceful closure across the United States and its territories.

In fiscal year 2018, for example, the CRS completed 292 cases. Thirty percent of those cases dealt with administration of justice cases (number of cases in parentheses): police–community relations (38); conflict over excessive use of force/police misconduct (25); bias-based policing/racial profiling conflict (12); tensions over hate crimes or bias incidents (7); corrections/prison conflict (4); and conflict involving tribes or reservations (2). Twenty-six percent of the cases addressed education cases: student racial conflicts and tensions, K–12 (29); college or university conflict/tensions/disturbances (14); hate or bias-motivated incidents, K–12 (12); hate or bias-motivated incidents at colleges/universities (10); school disturbance (5); conflict over disparities in treatment/opportunities, K–12 (4); and conflict involving tribes or reservations (2). Finally, 44% of the cases had to do with general community relations: hate crimes or bias incidents (59); special events and gatherings tension/conflict (22); conflict over disparities in access to services (18); hate group activity and gatherings/events/demonstrations (10); conflict from changes in demographics/other transitions (5); post 9/11-related tension and conflict (4); community development conflict (3); youth-related conflict (2); housing conflict (2); private or public sector employment conflict (1); cross-cultural and language-based conflict (1); conflict involving tribes or reservations (1).

The CRS is headquartered in Washington, DC, and its current leader is Gerri Ratliff (pictured), who was appointed deputy director on January 9, 2017. The CRS's 19 conflict resolution specialists are stationed in 10 regional and 4 field offices across the country, where they are available on a 24-hour basis. Regional offices are located in Boston, New York, Philadelphia, Atlanta, Chicago, Dallas, Kansas City, Denver,

Source: U.S. Department of Justice.

Los Angeles, and Seattle. Field offices are located in Miami, Detroit, Houston, and San Francisco. For each situation, the CRS mediators will first assess the situation, which includes hearing everyone's perspective. After gaining a good understanding of the situation, the CRS mediators will fashion an agreement among local officials and leaders on the services the CRS will provide to help resolve the conflict or prevent further violence. The recipients of CRS's services, for which it does not charge a fee, may decline those services at any time.

The CRS has no law enforcement authority (CRS mediators do not carry guns or badges) and does not impose sanctions, investigate, prosecute, or assign blame and fault. All CRS mediators are required by law to conduct their activities in confidence, without publicity, and are prohibited from disclosing confidential information.

The CRS offers several unpaid internship programs. For information about the internship programs, go to https://www.justice.gov/crs/join-our-team#internwithus.

Chapter 14 examines the future of criminal justice, and it is likely that in the future DOJ's peacemakers will play an increasing role in addressing social conflicts of a criminal justice nature. Do you think the peacemakers are a viable alternative to traditional criminal justice intervention in social conflicts? Would you be interested in a career as a DOJ peacemaker?

The Future of Criminal Justice

In the preceding chapters of this book, you have learned about the nature of crime and its consequences in the United States, theories of crime and delinquency causation, criminal law and its application, and the historical development of criminal and juvenile justice. You also have examined in detail the current operation of criminal and juvenile justice and have analyzed problems associated with each. In this last chapter, you will explore the directions that American criminal and juvenile justice might take in the future and consider some predictions. Excluded from this chapter are predictions about future trends in the most common types of crime and delinquency because that subject is too far afield. We conclude the chapter with a summary of the criminal justice response to terrorism, the defining crime of the early twenty-first century.

The principal guide that informs predictions of the future directions of criminal and juvenile justice is Herbert Packer's model of the U.S. criminal justice process, which was introduced in Chapter 1.[1] For us, Packer's model has had considerable explanatory power for more than 50 years, and it seems that its explanatory power will remain strong in the future. However, in making predictions, this text's authors will not limit themselves to Packer's vision of possible futures for criminal and juvenile justice but will, instead, peer into their own crystal balls–and those of others–to venture some predictions in areas where Packer's model is silent.

The Future of Law Enforcement

The manner in which law is enforced in the future will vary substantially depending on whether it is driven by crime control model principles or due process model principles (see the discussion of these models in Chapter 1). Either way, advances in technology will play a huge role in law enforcement's future. This section addresses those issues.

Law Enforcement Under Crime Control If the future of law enforcement increasingly reflects the principles and policies of the crime control model, then you might expect fewer limitations on how the police attempt to combat crime. For example, the practice of detaining and arresting suspects for investigation without probable cause is likely to increase, as is the length of time a suspect may be held before being charged. The Supreme Court may augment those practices by granting good-faith exceptions to the Fifth Amendment protection against compelled self-incrimination. If the crime control model is fully embraced, it is possible that the Court's decision in *Miranda v. Arizona* could be overturned and that coerced confessions would be admissible at trial.

The investigative abilities of the police should be improved and made easier by the expansion of community policing. Although the current meaning of community policing remains vague (see Chapter 6), it ultimately will be defined by the way it actually is employed. If it is employed according to crime control principles, American citizens can expect greater intrusion into their lives. Privacy will be sacrificed for efficiency in crime control. The same probably is true if CompStat is adopted widely (see Chapter 5).

Advances in electronic surveillance promise greater efficiency in crime control. Presumably, potential offenders will refrain from crime if they know they are being watched and believe they face a greater risk of being caught. Police, it is assumed, will learn more quickly of dangerous situations and, equipped with crucial information, be able to respond in the safest way possible. Video documentation of criminal activity and the identification of perpetrators and witnesses also may help in investigations and prosecutions, benefiting crime victims whose cases can be closed with the help of video evidence and incapacitating more offenders from committing future crimes. Electronic surveillance also may increase perceptions of safety among law-abiding citizens, encouraging them to make greater use of public spaces where they can serve as informal guardians and potential witnesses.[2]

bionics The replacing of human body parts with mechanical parts.

Advances in electronic surveillance also will facilitate greater intrusion into people's lives. For example, advances in **bionics** (the replacing of human body parts with mechanical parts) may someday produce bionic eyes powerful enough for law enforcement officers to see for miles, or through walls, and bionic ears sensitive enough to enable law enforcement officers, from a considerable distance, to clearly hear conversations held behind closed doors.[3] Already, computer-controlled supersensitive listening and video devices are able to accomplish the same things.[4] Miniaturization and remote-control technology will allow the creation of tiny eavesdropping aerial drones equipped with a camera and microphone that can fly into open spaces to record what is happening.[5] As noted in Chapter 4, that, too, is a reality.

The FBI's Next Generation Identification (NGI) system that integrates facial recognition software and other biometric markers into its identification data programs became fully operational on September 15, 2014 (see the discussion of NGI in Chapter 6).[6] NGI is a vast storehouse of "iris scans, photos searchable with face recognition technology, palm prints, and measures of gait and voice recordings alongside records of fingerprints, scars, and tattoos." A component of NGI, the Universal Face Workstation, already contains more than 30 million facial images (including nearly 16 million unique individuals), obtained from civil and criminal mug shot photos. NGI is part of the FBI's Facial Analysis, Comparison, and Evaluation Services, which not only has access to the NGI database, but also has access to the State Department's Visa and Passport databases, the Defense Department's biometric database, and the driver's license databases of at least 16 states. Together, the FBI now has access to more than 640 million photographs, most of which are of Americans and foreigners who have not committed crimes.[7] One wonders how long it will take for the FBI to tap into the millions of images stored on social media and photo sharing websites such as Facebook. NGI not only is shared by federal agencies, such as the Departments of Homeland Security, State, Justice, and Defense, but also is used by state governments.

Trusting Law Enforcement to Use Facial Recognition Technology Responsibly

A 2019 Pew Research Center poll found that 56% of U.S. adults say they trust law enforcement agencies "a great deal" (17%) or "somewhat" (39%) to use facial recognition technology responsibly, while 29% trust law enforcement agencies "not too much" (17%) or "not at all" (12%). Thirteen percent of U.S. adults had not heard of facial recognition technology.

Source: Aaron Smith, "More Than Half of U.S. Adults Trust Law Enforcement to Use Facial Recognition Technology Responsibly," *Pew Research Center*, September 5, 2019, assessed October 5, 2019, https://www.pewinternet.org/2019/09/05/more-than-half-of-u-s-adults-trust-law-enforcement-to-use-facial-recognition-responsibly/.

Police are in the vanguard of amassing biometric data. For example, the New York City Police Department began photographing suspects and arrestees' irises in 2010, routinely telling suspects that the scans were mandatory even though no law requires iris scans. Police departments nationwide now are equipping themselves with Mobile Offender Recognition and Information Systems (MORISs), which are an iPhone add-on that allows patrol officers to scan the irises and faces of individuals and match them against government databases. To build these databases, some judges are requiring biometric data from defendants in exchange for more lenient sentences. Some Occupy Wall Street protesters, who were arrested for trespassing and disorderly conduct, were given bail amounts based on whether or not they consented to an iris scan during their booking. To avoid inclusion in the government's massive identification database, the hacktivist group Anonymous advises that when going out in public, a person should wear a transparent plastic mask, tilt one's head at a 15-degree angle, wear obscuring makeup, and wear a hat outfitted with infrared LED lights.[8] Though the tactics seem extreme, sadly, they may be necessary in the near future to maintain one's privacy in public.

To the extent that the crime control model is followed, there may be fewer or, perhaps, no limitations on the future uses of electronic surveillance. According to this scenario, author George Orwell's vision of a society monitored totally by Big Brother (the state and its agents) may become a reality in the United States, as it did to varying degrees in the totalitarian nations of eastern Europe, communist China, and the former Soviet Union. In other words, to facilitate efficient crime control, according to the logic of the crime control model, it is conceivable that every move you make, every word you say, and, possibly, every thought you think will be recorded for future incrimination, if the need arises.

New software is being developed that allows law enforcement agencies nationwide to share data in a timely manner.[9] The new software combines several public, private, and criminal databases and allows them to be searched simultaneously. An advanced feature of the software is the ability to take vague bits of information, such as a partial license tag number, and get lists of possible suspects or vehicles in seconds. The software is intended primarily for terrorism investigations, but it also may be used to help solve child abductions or other time-sensitive crimes. Another feature of the software is the ability to use digital driver's license photos to quickly generate a photo lineup that can be shown to witnesses. Critics of the new software contend that it infringes on civil liberties and privacy rights and could lead to racial profiling. It may even be possible, as depicted in Steven Spielberg's movie *The Minority Report*, that certain people or devices will be able to see into the future to identify criminals before they could do any harm. In any event, advanced electronic surveillance devices and other new technologies in law enforcement (to be discussed in more detail later in this chapter) should be available sooner than normally might be expected because of the war on terrorism.

Law Enforcement Under Due Process If Americans see a shift to the principles and policies of the due process model, they should expect existing limitations on how the police combat crime to remain intact or even be expanded. Certainly, the practice of detaining and arresting suspects for investigation without probable cause will end or at least will be decreased substantially. The present 48-hour limitation—with certain

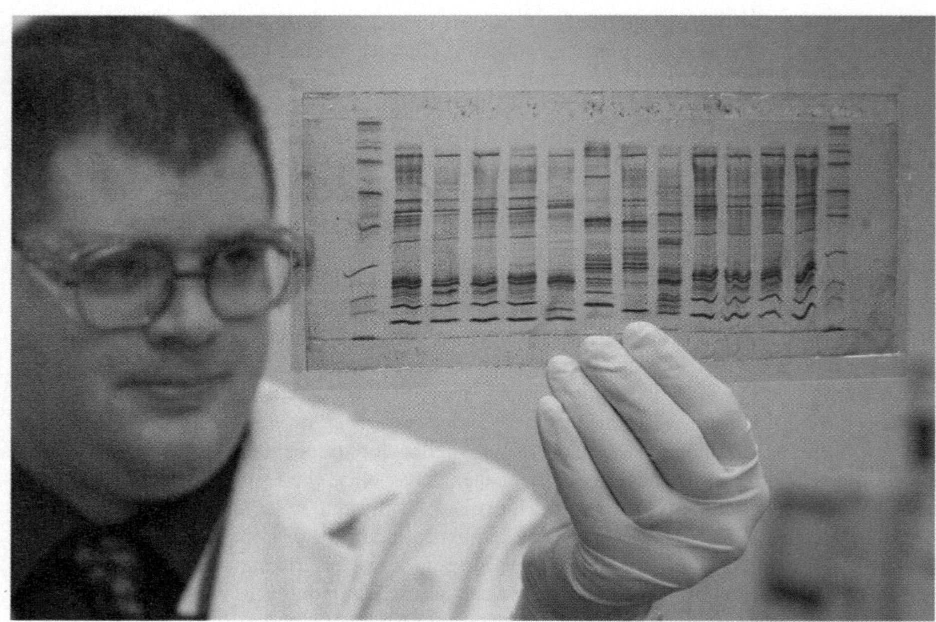

Source: DHS/Frontline Photography/CBP Laboratories Photography.

DNA profiling will be used increasingly in the future. *How should the DNA database be collected and used?*

DNA Profiling

To learn more about how DNA profiling actually works, visit the Science Learning Hub website at http://biology-animations.blogspot.co.nz/2011/07/dna-fingerprinting-animation.html and watch the video.

Do you think DNA profiling will play an important role in the future of law enforcement? Why?

exceptions—on the length of time a suspect may be held before being charged will be standard operating procedure. Good-faith exceptions to protections against compelled self-incrimination will not be allowed, and current good-faith exceptions to the exclusionary rule may be eliminated. Adherence to the *Miranda* requirements will become even more routine among police officers than it is currently. Perhaps, suspects will no longer be allowed to waive their rights.

To the extent that the due process model shapes future law enforcement practice, electronic surveillance will not be allowed under any circumstances or at least will be strictly limited to a few types of crimes (e.g., crimes that threaten national security, such as terrorism, treason, and espionage).

As described more fully in Chapter 6, a relatively recent development in criminal investigation is DNA profiling. Although this technology still is controversial, it probably will become a routine law enforcement tool in the near future. Perhaps the most thorny issue with this new technology is how the DNA database will be collected and used. If the principles of the due process model prevail in the future, then the DNA database probably will comprise DNA samples taken only from booked suspects (and criminal justice personnel upon hiring), as in the current practice with fingerprints. Such a policy is unlikely to be very controversial or to attract much public resistance. However, if policy reflects the principles of the crime control model, then the DNA database may comprise DNA samples taken shortly after birth from all infants born in the United States. This latter strategy clearly would make the DNA database more complete and, therefore, a more efficient tool in the effort to solve crimes. It also would be much more controversial and probably would be resisted by a large segment of the population as a violation of privacy rights. Similar concerns will accompany the collection of biometric data.

Advances in law enforcement technology are inevitable, but technology employed according to due process principles will be used to protect citizens and law enforcement personnel. For example, robots will be employed with greater frequency to perform a variety of law enforcement duties. Law enforcement personnel have used robots since the mid-1970s for handling bombs, and they currently are being used in other law enforcement situations, usually in SWAT or other dangerous circumstances.[10] In the future, they may be used routinely in hostage situations and in the arrest of dangerous suspects.

Jeff Gritchen/Digital First Media/Orange County Register/Getty Images

Some law enforcement agencies use robots to handle bombs and other dangerous assignments. *Should all law enforcement agencies have access to robots? If so, how should they be financed and distributed?*

Some police departments are beginning to install digital video systems in their patrol cars to record and store pictures of every encounter their officers have.[11] Although many police departments already have camera systems in their cars to provide evidence in arrests and to protect themselves from lawsuits, the new systems record and store the data in computers. Also, unlike analog video systems, which usually are not activated until after a violation has occurred, the new digital systems allow the police to record the events preceding the infraction and the infraction itself while it is taking place. The new systems also come with small microphones that officers wear on their belts so they can record what is said outside their cars.

When officers begin their shifts, they plug a portable hard drive into a computer mounted between the front seats of their patrol cars. They then activate a microphone and bi-directional camera that is mounted on the car's visor. The microphone and camera record continuously but only save 3 to 4 minutes at a time. When officers turn on their pursuit lights, the system automatically saves and stores the last few minutes and whatever comes next until the system is turned off. Once their shifts are completed, officers take the portable hard drive to police headquarters and upload all the stored data to a central server.[12] Some police departments have plans or already have put small wireless video cameras on police officers' lapels and have installed or are planning to install equipment in patrol cars that will allow officers to remotely monitor video feeds from banks.[13]

According to a 2015 national public opinion poll, 88% of Americans want police to be required to wear body cameras while on duty. The strongest support was from blacks, with 73% responding that they "strongly supported" requiring the cameras, compared to 62% of people overall. The poll found that 88% of the public believes cameras will help hold police accountable, and 67% thinks that cameras will reduce police brutality. Also, 60% of Americans oppose "communities of color" being disproportionately targeted by police surveillance.[14]

As noted previously, digital video surveillance cameras, many equipped with microphones, also increasingly are being used to provide around-the-clock surveillance in the public areas of cities and towns throughout the United States.[15] Many of these systems have been financed with Department of Homeland Security grants for the purpose of combating terrorism. They are being installed at possible attack targets, high-crime areas, and heavily trafficked areas. Most cameras have a 360-degree range of motion, can be remotely rotated, and zoom in and out. Few police departments, however, have written policies governing their use, formal training for users, or methods of evaluating whether they achieve results. Nevertheless, despite the privacy concerns of civil libertarians, studies showing that cameras are not effective in deterring terrorism or crime, and questions about whether live monitoring is the best use of limited police resources, most Americans support the use of surveillance cameras. As for the concern with limited police resources and using "live" personnel to monitor the cameras, intelligent "video analytics" software has been developed to monitor live feeds. The software commands the cameras to be focused on specific things such as a crowd collecting in a particular neighborhood or the sound of a gunshot. It also allows one operator to monitor numerous cameras. London, England, began installing cameras in the early 1990s and has the most extensive network, with about 500,000 cameras in its "Ring of Steel." Recently, the British Home Office approved funding to add thousands of "head-cams" mounted on the hats of police officers. With a countless number of private-surveillance cameras, it is estimated that the average Londoner is observed 300 times a day.

A related development gaining popularity in the United States and elsewhere is automatic license number plate recognition systems (ALPR or, more commonly, ANPR, systems), which are replacing ordinary traffic-surveillance cameras in many locations. ANPR systems use infrared (IR) cameras and sophisticated software to recognize the letters and numbers on a plate and compare them with a database to detect vehicles of interest. Most systems now consist of multiple cameras, including both the IR cameras and color cameras

Digital video surveillance cameras increasingly are used to provide around-the-clock surveillance in the public areas of cities and towns throughout the United States. *Is this a good idea? Why or why not?*

Simon Marcus Taplin/Getty Images

FYI **Persistent Surveillance System**

Another possible solution to the "live" monitoring of cameras problem is the so-called "persistent surveillance system." The device has the capacity to fly high over a 35-square mile target area taking one photo a second. If an event of interest occurs, such as a shooting or an armed robbery, the device can backtrack a frame at a time, tracing the suspect to his or her pre-crime locale.

Source: "Fight for privacy only beginning," *Orlando Sentinel*, July 6, 2017, A14.

to snap photos of vehicles. Systems can be mounted almost anywhere, and there are aerial units capable of reading a plate from an elevation of more than 1,600 feet. Most of the systems can detect, capture, and compare 1,500 to 3,000 plates per minute, even under seemingly difficult conditions, such as at night, in the rain, through a chain-link fence, and a plate with mild mud splatters. Patrol cars typically are equipped with at least three cameras to cover oncoming traffic, multilane traffic, and parked cars. An officer can record oncoming plates at a combined speed of 120 mph or scan parked cars in a parking lot or on a neighborhood street without slowing down or taking his or her eyes off the road. The system can record the identification, photo, GPS location, and timestamp for any vehicle. The units have been used for tracking stolen cars, felony warrants, Amber Alerts, and parking violations.[16]

New Law Enforcement Technology Some other ideas currently on drawing boards include the following:

- A pocket-sized, voice-activated voice-stress analyzer that police officers could use to determine whether suspects or witnesses experienced stress during questioning, indicating possible dishonest answers (to be used like current polygraph machines).
- An ultra-small, two-way cellular phone (like a Dick Tracy wrist communicator), possibly implanted in officers' larynxes, that would allow police officers to be in constant contact with headquarters, fellow officers, or anyone around the world.
- A wireless interoperability system, which connects the radio frequencies of various emergency federal, state, and local first responders and provides smooth, fast, and accurate real-time communications for emergency personnel who may need to coordinate activities at a large incident.[17]
- A universal translator (i.e., an ultra-small computer) that could instantly translate speech from one language to another, allowing police officers to question suspects, witnesses, or crime victims without the language barrier they frequently confront today.[18]
- An ultra-wideband device that will allow police officers to detect motions through surfaces such as walls (e.g., to determine whether there are people inside a room before they knock down a door).[19]
- Video-equipped, nearly unnoticeable pilotless drones for aerial surveillance that are guided by a remote operator or by GPS that can hover above a target or just outside a window while transmitting real-time video.[20]
- A video stabilization system that electronically converts useless, unstable surveillance video into clear, court-presentable evidence.[21]
- A "smart" gun that would electronically disable itself if taken away from a police officer during a struggle.
- A microwave device to shut off a car's ignition, stopping fleeing suspects without the risk of a high-speed chase.
- A superSticky foam that could be sprayed on armed suspects, neutralizing them by temporarily gluing their arms to their bodies.
- Spikes embedded in retractable panels beneath roads that could be raised by remote control to blow out a getaway car's tires.[22]
- An exoskeleton suit using nanotechnology and artificial muscles to allow officers to run with minimal effort over prolonged periods at a speed of up to 20 mph with a top speed of 35 mph for shorter distances and to lift items up to four times their own weight.[23]
- A comprehensive, integrated modular tactical uniform that offers ballistic, chemical, and biological protection for special operations police officers such as SWAT officers and hazardous materials specialists.[24]
- Augmented reality technology that overlays computer-generated images onto a person's real-world vision that could be used by officers to have patrol car operator data and regional traffic management information on heads-up display to make driving safer and more efficient, especially during pursuit and rapid response situations; to identify friend-or-foe to reduce or eliminate friendly fire casualties by visually highlighting fellow officers both on and off duty; to project a display of officer location, activity, and status information on a three-dimensional map of the community; to manage the coordinated use of robots, aerial drones, and police officers to enhance surveillance activities; and to employ realistic training scenarios to simulate dangerous police environments while blending real-world equipment and fellow trainees into the scenario.[25]
- An Oculus Rift VR headset that police commanders can use to pinpoint officers' locations, find out their status during a call, or view a live feed of body cameras while officers are equipped with virtual reality gear.[26]

- Jet packs that officers could wear on their backs so that they could patrol by air.[27]
- A handheld scanner that will allow remote body-cavity searches.[28]
- A mini-buster secret compartment detector, which is a handheld device that senses density in solid objects and will allow police officers to scan over the body of a motor vehicle to locate hidden compartments used to smuggle contraband, terrorist devices, or other illegal items.[29]
- A handheld nonintrusive cargo inspection device that could be used at seaports, truck inspection facilities, airports, and ports of entry to reveal the presence of contraband in a sealed container and identify the contents (drugs, weapons, biological agents, or explosives) without expending costly time and resources searching by hand.[30]
- Nanosized computer chips that can be placed in the neural networks (such as the human brain) of police officers to give them access to billions of gigabytes of instantly accessible data (such as criminal records of suspects).[31]

New technology also promises to improve the police response to specific crimes, such as domestic violence and drunk driving. For example, in the case of domestic violence, one of the best ways of preventing recurrences is by having a court issue a protective order to keep the abuser separated from potential victims. Unfortunately, a police officer called to the scene of a domestic disturbance can verify the existence of a protective order only by calling the county clerk's office, assuming the disturbance occurred during business hours. Even then, the officer may not be able to verify the existence of a protective order if the county clerk has not received the order from the court. If the disturbance occurs after hours and the officer is unaware of the protective order, the officer's response to the situation–for example, not arresting the abuser–could result in tragedy. In the future, through comprehensive databases, all police officers will have access to every protective order at all times and virtually from the moment it has been issued.[32]

Similarly, in the case of drunk driving, courts routinely suspend a person's driver's license pending trial. The court then sends notification of the suspension by mail to the Bureau of Motor Vehicles (BMV); it often takes weeks for the suspension to show up in the BMV's computers. It is conceivable that the drunk driver could leave the courtroom, get in his or her car, and drive to the nearest bar. An officer who stopped the drunk driver for a taillight violation, for example, would check the computer, not see that the driver's license was suspended, and send the driver on his or her way. In the future, courts will routinely send suspension orders by computer directly to BMVs and law enforcement agencies and save lives in the process.[33]

The future of law enforcement in the United States likely is to be very different from law enforcement today. New styles of policing, such as community policing, will be embraced and new technologies employed. The form that the changes ultimately take, however, will depend substantially on whether there is a dramatic shift toward the principles and policies of the crime control model or toward the principles and policies of its due process counterpart. In the next section, possible future developments in the administration of justice are considered.

The Future of the Administration of Justice

Depending on whether the crime control model or the due process model dominates the future of criminal justice, there likely are to be many changes in the ways justice is administered. Some of those changes may involve the right to counsel at various stages of the criminal justice process, the preliminary hearing, the grand jury, pretrial detention, plea bargaining, sentencing, appeals, and juvenile justice. The criminal courts also will have to adapt to the changing demographics of the population. Most of the conflicts currently processed by the criminal courts likely are to be handled by alternative dispute resolution programs that utilize mediation and arbitration. Restorative justice may play a greater role in the future. As recent events have illustrated so vividly, the criminal courts will have to be better prepared to deal with all sorts of disasters in the future. Finally, advances in technology undoubtedly will change the administration of justice.

Right to Counsel If, in the future, the administration of justice in the United States is more in line with the crime control model, then the right to court-appointed and privately retained legal counsel at critical pretrial and posttrial stages may be scaled back significantly. Advocates of the crime control model consider legal representation at any stage, other than perhaps at trial, a luxury and an unnecessary impediment to the efficient operation of the process. Advocates of the crime control model argue that prior to trial (e.g., during interrogation), providing counsel to a suspect only hampers the ability of police to investigate a case. After trial, during the appellate process, the availability of legal counsel further reduces the speed with which a case can be brought to closure. Crime control model supporters maintain that in both situations, a crafty lawyer may be able to win the freedom of a factually guilty client by means of a legal technicality.

If the principles and policies of the due process model dominate the future of the administration of justice, however, then it is likely that the current right to counsel at a variety of critical stages in the process will be retained and perhaps even extended somewhat (e.g., to appeals beyond the first one). Advocates of the due process model believe that legal counsel is crucial throughout the process and must be made available immediately after arrest. Otherwise, the likelihood that innocent people will be subject to harassment, or worse, by agents of the state (police officers and prosecutors) is increased to an intolerable level.

Preliminary Hearing Along with a greatly reduced role for legal counsel in the administration of justice, crime control model enthusiasts advocate the abolition of the preliminary hearing. They argue that it is a waste of time and money to conduct a preliminary testing of the evidence and the charges imposed because prosecutors have no reason to pursue cases that unlikely are to lead to conviction. Recall from the discussion in Chapter 8 that prosecutors' reputations and chances of achieving higher political office depend at least partially on the proportion of convictions they are able to obtain. Besides, for several reasons that include the sheer volume of cases that must be handled in the lower courts, the evidence against a suspect is rarely tested at the preliminary hearing anyway. Today, the preliminary hearing is used mostly for making decisions about bail.

Due process model advocates, however, believe that the preliminary hearing is a critical stage in the administration of justice. Although they concede that this hearing currently is not being used as it is supposed to be, they argue that it needs reform instead of abandonment. As described previously, due process model advocates are skeptical about the motivations of prosecutors in criminal cases. History certainly has recorded numerous incidences of prosecutorial misbehavior. Because most criminal cases are resolved through plea bargaining rather than criminal trial, the preliminary hearing may be the only opportunity for a judicial officer to scrutinize the prosecutor's work. Elimination of the preliminary hearing (as it is supposed to operate in theory) would substantially reduce the effectiveness of the entire adversarial process of justice.

Grand Jury Instead of elimination of the preliminary hearing, due process model enthusiasts are more inclined to argue for elimination of the grand jury. As described in Chapter 8, the grand jury is another vehicle by which an impartial body examines the evidence and charges brought by a prosecutor against a suspect. However, in practice, the grand jury is even more of a sham than the preliminary hearing because it has become nothing more than a tool or a rubber stamp for the prosecutor. Invariably, a grand jury will indict or not indict a suspect if that is what the prosecutor wants. Moreover, abuses of the process are more likely to occur in grand jury proceedings, which are conducted in private—away from the watchful eyes of defense attorneys and the public—than in preliminary hearings, which allow the presence of defense attorneys and are open to the public.

Los Angeles Lakers star Kobe Bryant arrives for his preliminary hearing in 2003. *Should the preliminary hearing be abolished, as crime control model proponents advocate? Why or why not?*

Ed Andrieski-Pool/Getty Images

Pretrial Detention If the crime control model dominates the administration of justice in the future, it is likely that the use of pretrial detention will be expanded. Advocates of the crime control model maintain that expanded use of pretrial detention will encourage more factually guilty offenders to plead guilty—they would have nothing to gain by not doing so—and will better protect society from the crimes those people are likely to commit if not confined. As for the protection of society, perhaps in the future the pretrial detention of potentially dangerous suspects will be unnecessary because of technical advances in electronic monitoring. For example, the use of electronically monitored house arrest, in which suspects are connected to monitors by electrodes (perhaps surgically implanted in their bodies) that shock them intermittently when they are outside designated areas, may be sufficient to ensure that suspects do not pose a threat to society while awaiting trial.[34] If this type of technology is employed, it is likely that suspects would be given a choice between electronic monitoring or jail. Most suspects probably would choose monitoring.

Due process model supporters argue that pretrial detention should be used sparingly if at all. They believe that people accused of crimes should be entitled to remain free until they are found guilty unless they are unlikely to appear when required or they pose a significant threat to society. Regarding those two exceptions to pretrial release, due process model advocates believe that nonappearance problems can be handled adequately through the existing bail system. They do admit, however, that the present bail system discriminates against the poor. That discrimination is another problem that must be rectified.

As for the pretrial detention of defendants who presumably pose a significant threat to society, due process model advocates doubt that prosecutors and judges can predict accurately which defendants actually pose a significant threat and which do not. The result is that many defendants who do not need to be jailed prior to trial will be jailed–or electronically monitored–because of inevitable false predictions. Due process model enthusiasts also suggest that both exceptions to pretrial release–nonappearance and potential threat to society–could be handled by methods other than pretrial detention–for example, by a prompt trial as required by the Sixth Amendment.

Plea Bargaining As mentioned in Chapter 1, a key to the crime control model's efficiency in the administration of justice is heavy reliance on plea bargaining. Thus, if the crime control model dominates the administration of justice in the future, Americans should expect even fewer criminal trials than today. If there is, instead, a dramatic shift to the principles and policies of the due process model, plea bargaining probably will be discouraged, and the number of criminal cases that go to trial will increase substantially. For due process model supporters, a principal problem with plea bargaining is that after a guilty plea has been accepted, the possibility of any further judicial examination of earlier stages of the process is eliminated. In other words, with the acceptance of a guilty plea, there is no longer any chance that police or prosecutorial errors before trial will be detected. Plea bargaining probably will not be eliminated entirely, however, even in a legal atmosphere dominated by due process model principles and policies, because eliminating it would be impractical. The time and expense involved in subjecting every criminal case to a trial would be prohibitive.

Sentencing As discussed in Chapter 9, many jurisdictions have eliminated or significantly reduced judicial discretion in sentencing by enacting mandatory sentencing statutes and using sentencing guidelines. This policy seems to reflect crime control model principles, although a due process model argument could be made that the restriction of judicial discretion in sentencing reduces unwarranted judicial disparity in sentencing. In any event, if the trend continues, it is likely that in the area of sentencing, judges in the future will function simply as automatons, empowered only to apply the specific sentence dictated by law.

Appeals If the crime control model dominates the future of the administration of justice, it is likely that appeals–following the few trials that occur–will be strongly discouraged and limited. Crime control model advocates consider appeals a remote and marginal part of the process. They do recognize that the appellate process can correct errors in procedure or in the determination of factual guilt, and they endorse that role. They believe, however, that such errors occur only occasionally. For crime control model supporters, the appellate process can impede the efficient operation of the administration of justice.

If the due process model dominates the administration of justice in the future, there probably will be no limitations on the right to appeal. Due process model advocates, in contrast to their crime control model counterparts, believe that the right to appeal is a significant and indispensable part of the process in which earlier errors and infringements of rights can be corrected. Perhaps even more important, due process model enthusiasts maintain that the prospect, if not the actual act, of appellate review can deter the commission of similar errors and other forms of misbehavior in subsequent cases. Due process model proponents argue that without the real prospect of appellate review, there would be no effective deterrent of police, prosecutorial, or judicial misbehavior or error.

Changing Demographics of the Population Because the population of the United States in the future will have greater cultural and racial diversity, criminal courts may be forced to adapt their routines and personnel to those changing demographics. Among the changes that may be required in the future are the following:

1. A greater sensitivity of court personnel to cultural diversity issues.
2. An increased ability of court personnel to communicate effectively with non-English-speaking people (this will require the employment of interpreters and multilingual staff, as well as the provision of other language services, to ensure that due process rights are protected).
3. A more culturally and racially diverse workforce that better reflects the demographic characteristics of the population it serves.[35]

FYI **Defendant with No Language Skills**

Juan Jose Gonzalez Luna, 42, who is deaf, mute, illiterate, and does not know sign language, was arrested October 8, 2010, on drug trafficking charges. Police seized more than 2 pounds of cocaine from his car. How will the courts deal with him? Many judges declare such defendants incompetent to stand trial and order confinement to institutional language programs, hoping the defendant can be taught American Sign Language (ASL) and eventually be tried. A time-consuming and costly alternative with limited success is "relay interpreting," which uses two interpreters in the courtroom—one who translates from spoken English to ASL, and another who uses makeshift gestures to communicate with the defendant. The second interpreter must be a person who was born deaf because that person is more attuned to thinking in strictly visual terms.

Source: Jeremy Roebuck, "When Right to Stay Silent Is Moot, Justice Flies Blind," *Orlando Sentinel*, January 16, 2011, A10.

State and federal courts have a continuing need for qualified interpreters, especially in languages other than Spanish, and this need will only grow in the future. For example, on January 19, 2000, the New Mexico Supreme Court ruled that people cannot be disqualified from serving as jurors simply because they do not speak English. The decision requires the use of translators in New Mexico courtrooms.[36] Some states are so desperate for court interpreters that they have created training programs to increase the skills of borderline candidates. To identify people who have the minimally required knowledge, skills, and abilities to be a court interpreter, three major programs offer oral performance examinations. The first is the Federal Court Interpreter Certification Examination program, which was established in 1980 and tests and certifies Spanish interpreters for the federal courts. The second is the Council of Language Access Coordinators (CLAC) (formerly the Consortium for Language Access in the Courts and the Consortium for State Court Interpreter Certification), which was founded in 1995 and currently develops and shares test instruments in 16 languages to certify state court interpreters. The third is the National Association of Judiciary Interpreters and Translators (NAJIT), which was founded in 1978 and created a Spanish performance examination in 2001. To meet the growing demand for interpreters, judges and court administrators are likely to hire staff with bilingual skills for noncourtroom services and then arrange their calendars so they can use their bilingual staff inside the courtroom. Another strategy is for different agencies within a jurisdiction to share interpreters.[37] Table 14.1 shows starting and maximum compensation for salaried court interpreters in select states in 2019.

TABLE 14.1 Compensation for Salaried Court Interpreters, Select States, 2019

	Starting Salary	Maximum Salary
Arizona	$35,500	$76,000
California	71,000	84,261
Connecticut	50,874	79,398
Iowa	41,059	62,379
Kansas	34,270	80,400
Kentucky	33,000	56,200
Massachusetts	65,000	93,000
Minnesota	50,000	75,000
New Jersey	63,000	101,000
New Mexico	50,000	64,000
New York	55,000	76,000
North Carolina	50,000	75,000
Ohio	36,000	60,000
Oregon	47,748	85,632
Pennsylvania	30,000	65,000
Washington	50,000	75,000
Wisconsin	25,000	50,000

Note: Salaries are approximate for some states and are for spoken-language interpreters only.

Source: Based on data from National Center for State Courts, "Language Access Programs by State," accessed October 6, 2019, https://www.ncsc.org/Services-and-Experts/Areas-of-expertise/Language-access/Resources-for-Program-Managers/LAP-Map/Map.aspx.

Careers in Criminal Justice

Court Interpreter

My name is Patricia Greco. I am a certified Spanish interpreter for the California Courts working in Orange County, California. I work with a variety of cases, from traffic trials to felony cases. Most of the work involves interpreting for the defendants in open court, but I also interpret for witnesses or victims delivering an impact statement prior to sentencing. As of 2016, court interpreters work the civil cases in California as well.

I grew up speaking Spanish and English at the same time, so I can speak both without any element of foreign accent. Growing up I lived in Honduras for 6 years, Ecuador for 4 years, and Argentina for 4 years, and I traveled to the neighboring countries. This gave me the opportunity to learn different nuances and idioms as well as the various forms of the Spanish language from different parts of the world. This helped me tremendously with the accuracy of my interpretation.

I went to high school at the American School in Tegucigalpa, Honduras. After that I went to Switzerland to learn French and German at a boarding school in Interlaken, followed by my studies in linguistics at the Catholic University in Quito, Ecuador. Finally, I studied to be an interpreter for the United Nations through a 3-year program with a UN interpreter in Buenos Aires.

Returning to the United States, I was contracted to work on many technical translation projects, and I worked conference interpreting. Eventually, I decided to try out with the courts, so I took some classes at the local college and eventually passed the test to get my certification in Spanish.

I have been working at the courts for 4 years now as a staff interpreter. I enjoy being able to assist the people who do not feel comfortable enough to communicate in English. I love this job because it allows me to interact and contribute with the most difficult cases, without being involved in the case personally. Sometimes it is difficult to see and hear some of the bad choices people have made in committing some of the crimes that I interpret for. At times, it feels like I am in the middle of some TV show except that this is real life. All in all, it is a pleasure to work with the attorneys, judges, and court staff as a Certified Spanish interpreter.

Patricia Greco

As the number of people age 65 and over increases, the operation of criminal courts probably will have to change in other ways as well. (It is predicted that by 2030, 71 million people in the United States will be age 65 and over–double the number in the year 2000.[38]) In the first place, the types of cases routinely heard in criminal courts are likely to change somewhat and to reflect more and more the problems confronted by the elderly. For example,

St Petersburg Times/Tampa Bay Times/ZUMAPRESS.com/Newscom

Courts already are adapting to the future by employing signers for hearing-impaired persons. *In what other ways will courts of the future have to change?*

crimes committed by and against the elderly, such as elder abuse, domestic violence, and family violence, probably will increase, as will consideration of such issues as the right to die and the appropriate use of new medical technologies.[39] An older population also will put new demands on the way criminal courts conduct business. At the very least, courts will have to better accommodate people with physical disabilities such as hearing and visual impairments. For example, to serve hearing-impaired clients, a text telephone (telecommunications device for the deaf, or TDD) or an interpreter probably will become a standard fixture in courtrooms of the future. Likewise, court personnel might be trained to establish eye contact when addressing people with hearing impairments, to provide verbal prompts to people with visual impairments, and to sit down when talking to people who are confined to wheelchairs.[40] Interstate court collaboration will be needed to ensure that court orders are recognized and oversight is provided when older persons move from one state to another. For those unable to come to the courthouse, remote-access technology, such as videoconferencing, will need to be enhanced.[41]

Juvenile Justice If the crime control model dominates the future, criminal courts probably will have to handle an increasing number of juvenile offenders, too, as the practice of transferring juvenile offenders to criminal court for trial increases (see the discussion in Chapter 13). Moreover, to the extent that crime control model principles guide the practice, Americans should expect younger and younger juvenile offenders to be transferred. If this trend becomes the norm, it is conceivable that sometime in the future the juvenile court, and possibly the entire juvenile justice system, could be eliminated entirely and juvenile offenders would be treated similarly to adult criminal offenders.

Ironically, even if due process model principles guide the future, juvenile justice might be eliminated anyway. Juvenile courts currently are providing juvenile offenders with a greater number of procedural rights and are adopting more adversarial procedures (reflecting due process model principles). They also are increasingly replacing paternalistic, treatment-oriented correctional strategies with more punitive ones. As a result of those changes, there may be no practical need for a separate juvenile justice process in the future. In other words, if juvenile offenders are going to be treated exactly the same as adult criminal offenders, why have a separate juvenile justice process?

Alternative Dispute Resolution Programs and Restorative Justice
Regardless of which model of criminal justice dominates the future, it is likely that most conflicts—with the possible exception of the more serious ones—dealt with today through the adversarial process of justice will be handled differently through alternative dispute resolution programs. In other words, criminal courts will become the arenas of last resort for most conflicts, used only when all other methods of resolving disputes have been exhausted. The two alternative methods of dispute resolution that probably will be used most widely in the future are mediation and arbitration.

Gene Stephens, a criminologist who specializes in the future of crime and criminal justice, proposes a model, based on current neighborhood justice centers, that integrates mediation and arbitration into a process he calls "participatory justice."[42] **Mediation** brings disputants together with a third party (a mediator) who is trained in the art of helping people resolve disputes. The job of the mediator is to help the disputants talk out their problems, to offer suggestions about possible resolutions, and, if possible, to achieve consensus about how the dispute can be resolved to everyone's satisfaction. The agreed-on resolution is then formalized into a binding consent agreement. **Arbitration**, in contrast, brings disputants together with a third party (an arbitrator) who has the skills to listen dispassionately to evidence presented by both sides of a conflict. The arbitrator asks probing and relevant questions of each side and arrives at an equitable solution to the dispute. The process is concluded when the arbitrator imposes a resolution of the dispute, which is binding on all parties.

mediation A dispute resolution process that brings disputants together with a third party (a mediator) who is trained in the art of helping people resolve disputes to everyone's satisfaction. The agreed-upon resolution is then formalized into a binding consent agreement.

arbitration A dispute resolution process that brings disputants together with a third party (an arbitrator) who has the skills to listen objectively to evidence presented by both sides of a conflict, to ask probing and relevant questions of each side, and to arrive at an equitable solution to the dispute.

For Stephens, the goal of both methods is to resolve any conflicts through a consent agreement completed by all parties involved in the dispute. Mediation is the primary method of resolution. Arbitration is a backup procedure used only when mediation fails or is inappropriate (e.g., in contract murder cases). In Stephens's model, appeals from mediation would be unnecessary because of the consent agreement, while appeals from arbitration would be limited to a single appeal to a three-member board of arbitrators. A variety of resolutions, whether by consent agreement or imposed arbitration, would be available. They

Careers in Criminal Justice

BARJ Coordinator

My name is Donald J. Haldemann. I am the balanced and restorative justice (BARJ) coordinator for the Delaware County Juvenile Court in southeastern Pennsylvania. I also supervise our Victim Services Unit with a staff of six. I have a bachelor's degree in history, a master's degree in administration of justice, and a juris doctor degree from Widener Law School in Delaware.

I worked with 1,300 juveniles as a juvenile probation officer for 21 years before embarking on the quest for restorative justice. In 1995, Pennsylvania's Juvenile Act was amended to include the concept of restorative justice.

Slowly, with grant money from the Pennsylvania Commission on Crime and Delinquency, individual counties hired restorative justice specialists, and I jumped at the opportunity to join this elite fraternity. So, what is restorative justice and how can you join the movement?

A crime is considered to be an offense against the "State" under the traditional model, but restorative justice views crime as a harm against individual victims and the community. Offenders need to accept responsibility for the harm they have caused and participate in victim and community restoration.

On a daily basis, my job consists largely of educating my staff, criminal justice professionals, social agencies, schools, and communities on restorative justice and how it can be a win/win proposition for all stakeholders in the process.

I can't possibly list all the positives about my job. I get to promote a philosophy of justice that is based on practices used in cultures all over the world for hundreds of years. What I find most difficult about my job is the reluctance of many criminal justice "players" to welcome a new way of looking at crime, opting instead to hang on to a process that has served no one's best interest. Offenders have not been helped to accept responsibility, victims have been ignored, and communities have been trained to believe that crime is a government problem and not their concern.

If you are thinking about becoming a restorative justice coordinator, I recommend you take a restorative justice training course like the one sponsored by Florida Atlantic University in partnership with the National Institute of Corrections. The more exposure you have to restorative justice philosophy and practitioners, the better.

Courtesy of Donald J. Haldermann

Do you share this BARJ coordinator's enthusiasm for this philosophy of justice? Why or why not?

include monetary restitution, community service, therapy, and even incarceration. Violations of consent agreements would be handled through arbitration, and violations of imposed arbitration resolutions would be dealt with through increasingly coercive measures–probably ending with incarceration. Stephens maintains that participatory justice would be cheaper, faster, and more equitable than the current adversary process of justice.[43]

Participatory justice is a form of restorative justice, an alternative to the punitive justice currently used in the United States and many other countries. The primary goals of **restorative justice** are to restore the health of the community, meet victims' needs, repair the harm done, and require the offender to contribute to those repairs. As described in Chapter 3, another form of restorative justice that may be adopted in the United States in the future is Australian criminologist John Braithwaite's "reintegrative shaming."[44] **Reintegrative shaming**, you may recall, is a strategy in which disappointment is expressed for the offender's actions and the offender is shamed and punished; more important, following the expression of disappointment and shame is a concerted effort on the part of the community to forgive the offender and reintegrate him or her back into society. Braithwaite contends that the practice of reintegrative shaming is one of the primary reasons for Japan's relatively low crime rate.

restorative justice A process whereby an offender is required to contribute to restoring the health of the community, repairing the harm done, and meeting victims' needs.

reintegrative shaming A strategy in which disappointment is expressed for the offender's actions, the offender is shamed and punished, and, more important, following the expression of disappointment and shame is a concerted effort on the part of the community to forgive the offender and reintegrate him or her back into society.

Clearly, the future of the administration of justice in the United States should be quite different from what it is today. The exact differences will depend on which model's principles and policies dominate. Areas in which differences and changes are likely to be most pronounced include the availability of legal counsel at various stages in the process; retention or abolition of the preliminary hearing, the grand jury, and juvenile justice; and the extent of pretrial detention, plea bargaining, alternative dispute resolution procedures, judicial sentencing flexibility, and the appellate process. Courts of the future also will have to adapt to the needs of an increasingly older and more racially and culturally diverse population.

Better Preparation for Disasters One of the lessons learned from the tragedies of the 1995 bombing of the Alfred P. Murrah Federal Building in Oklahoma City; September 11, 2001; and the 2005 hurricanes is that criminal courts must be better prepared to protect court assets–people, facilities, and records–from future disasters. Currently, few courts have continuity of operations plans (COOPs).[45] However, experiences from the aforementioned disasters, especially the hurricanes, have provided some ideas of how courts can respond or how others can help the courts respond when the next disaster strikes. For example, after Hurricane Katrina, a special assessment team was sent to the Gulf States by the Administrative Office of the U.S. Courts to determine local court needs. To institutionalize the response, the College of William & Mary Law School and the National Center for State Courts' Center for Legal and Court Technology (formerly the Courtroom 21 Project) developed a conceptual plan for a national state court corps of first responders. Lawyers who needed court information, such as where to contact court officials, what special orders were in place, and how to seek extensions or continuances, were helped by representatives from the local, federal, and Louisiana bar associations. In addition, the Communications Center for Displaced Attorneys was established to help direct e-communications for relocated lawyers with Internet access.[46]

The aftermath of the hurricanes also revealed the importance of electronic filing of court documents. By filing court documents electronically, backup files can more easily and efficiently be sent to alternate sites for protection, thereby minimizing the disruption of court business. Among the problems with electronic filing, however, are the possibility of deliberate cyberattacks and power and data outages.[47] The need for intranet data communication networks also became apparent in the aftermath of the hurricanes. During emergencies, the networks allow the courts to maintain essential functions by providing remote access via private broadband Internet and dial-up services. The critical need for laptop computers during emergencies was realized as well.

Technology and the Administration of Justice Before considering the future of corrections in the United States, we emphasize that advances in technology are likely to influence the administration of justice in the future, just as they will affect the future of law enforcement. The possible substitution of electronically monitored house arrest for pretrial detention and the use of technology during disasters already have been examined. In addition, in many jurisdictions in the United States today, it is no longer necessary for defendants to physically appear in court for the initial appearance, preliminary hearing, arraignment, or even trial. Interactive television can be used to administer justice to defendants without incurring the risks involved in physically transporting them to a courtroom. This development has allowed jails to be built in more remote areas of municipalities,

Patrick Farrell/AP Images

The use of interactive television in the administration of justice is likely to be expanded in the future. *Besides those mentioned, what are some of the other benefits of having interactive television in the courtroom?*

affording greater protection to the citizens of a community and reducing the costs of constructing jails because they need not be built in expensive urban locations. Courts also are allowing witnesses to provide videotaped depositions when they are physically unable to be present in court and allowing children to testify via interactive television in some child sex abuse cases.[48] Those practices are likely to continue and to be expanded.

Developments in neuroscience also promise to transform the administration of justice.[49] For example, during the penalty phase of capital trials, lawyers frequently use the results of brain scans to argue mitigation—that a neurological impairment prevented a defendant from controlling his or her behavior. In fact, a Florida court recently held that the failure to admit neuroscience evidence during capital sentencing is grounds for a reversal. As a result of neuroscientific evidence, jurors have been persuaded to sentence defendants to life imprisonment rather than to death. Despite studies that question the reliability of neuroimaging evidence, courts also have admitted brain-imaging evidence during criminal trials to support claims that defendants are insane.

There are many other applications of neuroscience. For example, preliminary research suggests that brain activity in the prefrontal cortex is critical for selecting among punishments. Defense attorneys and prosecutors could use this information in the selection of jurors, or legislatures or sentencing commissions could use the information for devising sentencing guidelines acceptable to the public. Eyewitness identifications could be aided by using neuroscience. Brain scans can show activity in the parahippocampus, the area of the brain that responds strongly to places and the recognition of scenes, and in the fusiform area of the brain, which is responsible for facial recognition. Findings from neuroscience were used in the 2005 decision in *Roper v. Simmons*, the case in which the Supreme Court struck down the death penalty for juveniles (i.e., those under age 18 at the time of their crimes). The leading brief in the case, filed by the American Medical Association and other groups, argued that "adolescent brains are not fully developed" in the prefrontal regions; therefore, adolescents cannot control their impulses as adults do and should not be held fully culpable "for the immaturity of their neural anatomy." The brain region associated with deliberate problem solving and self-control is the dorsolateral prefrontal cortex. Currently, two lie-detection technologies rely on neuroimaging. The first, developed in the 1980s, is known as "brain fingerprinting" and measures changes in the frequency of brain waves. The second technology uses functional magnetic resonance imaging (fMRI) to distinguish between the brain activity of liars and truth tellers. So far, neither technology is more reliable than polygraphs, which are not considered reliable enough to use in most legal cases. Figure 14.1 shows the parts of the brain associated with various criminal justice issues.

FIGURE 14.1 Brain Areas with Implications for the Administration of Justice

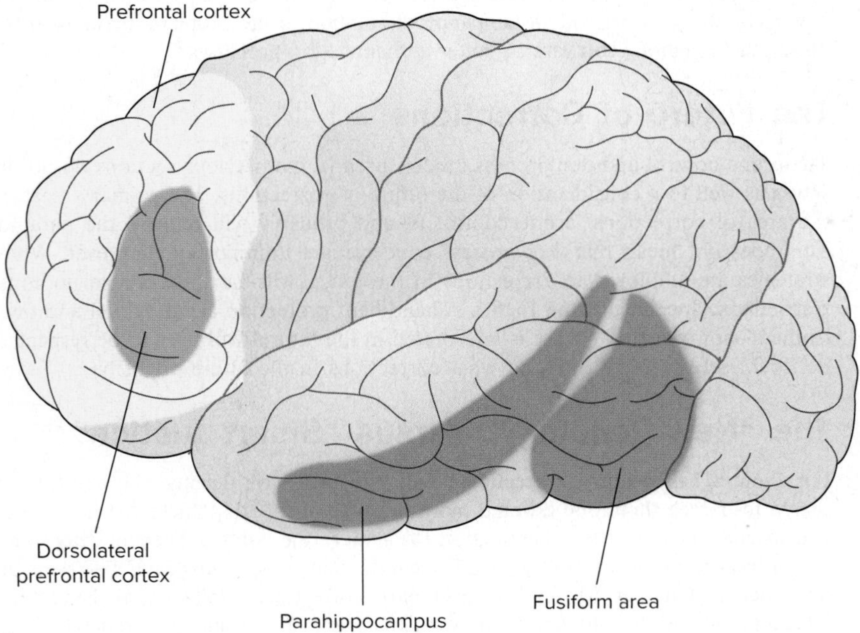

If the brain causes all behavior, including criminal behavior, then the concepts of retribution and criminal responsibility, which assume that criminal behavior is freely chosen, seriously must be reconsidered. Neuroscience also may change the focus of rehabilitation to repairing defective brains. For example, a new technique called transcranial magnetic stimulation (TMS) has been employed to stimulate or inhibit specific areas of the brain. TMS temporarily can alter how people feel and what they think.

Perhaps, someday, determination of guilt or innocence, as well as the truthfulness of testimony, will become a routine matter of simply employing mind-reading technology. For example, the transfer to the public domain, via computer or some other device, of ribonucleic acid (RNA) structures that form memories will make deception and privacy of thoughts nearly impossible.[50] Sometime in the future, it even may be possible to retrieve memories of the events of a crime from the stored RNA memory chains (or "memory banks") of deceased victims and witnesses.[51] Imagine the constitutional questions raised by such technological abilities.

Perhaps as important as the new technologies is the development of national court technology standards. Draft versions of standards already are circulating for integrated, computerized case management systems (CMSs); electronic filing of court documents; and the electronic exchange of common forms and documents among courts, law enforcement, and other justice organizations. The nationwide exchange of information is the goal of the Global Justice XML Data Model (GJXDM). By way of uniform data semantics and structure, the GJXDM will allow organizations using different computer systems and databases to share information (XML is a software- and hardware-independent tool for transmitting information). The National Information Exchange Model (NIEM), a by-product of the GJXDM, is being developed through the sponsorship of the Departments of Justice and Homeland Security. The NIEM will include nonjustice agencies and will aid courts with exchanging information with all relevant partners.[52]

These developments, of course, will raise significant issues about privacy versus public access, as well as data security concerns. For example, as more and more court records are accessed over the Internet, they become more susceptible to identity thieves. (Identity theft is discussed in Chapter 6.) Courts therefore must give greater consideration to what is included in a court record, and new policies and procedures must be created regarding access to court records. Of most concern are personal identifiers (such as Social Security number, city and date of birth, mother's maiden name, children's names, street address), third-party identifications (victims, witnesses, informants, jurors), and unique identifying numbers (operators' licenses, financial accounts, state identifications). States are responding to this problem in different ways. Some states are obscuring information by using only the last four digits of a Social Security number; asking only for the year of birth; identifying children by initials; and asking only for city, state, and ZIP code. Other states are creating two records: a public record and a private record for sensitive information. Still other states are editing out sensitive information. Critics contend these methods have not been entirely effective. The hope is that new technology will provide a tamper-proof solution to the problem, perhaps in the form of more reliable editing software or better authentication processes.[53]

The Future of Corrections

The crime control and due process models used in the previous sections do not lend themselves as well to a consideration of the future of corrections. The primary reason is that in the area of corrections, crime control is, and probably will remain, the paramount goal. This does not mean that due process concerns are unimportant, however. Whatever new strategies are employed in corrections in the future will have to conform to constitutional restrictions. Specifically, the Eighth Amendment protection against cruel and unusual punishment—however that phrase is interpreted in the future—will have to be respected and will no doubt set the outer limits of what corrections in the future might be.

The "New Penology" Versus "Smart Justice"

For about a half century, correctional policy makers have debated the question of whether scarce resources should be devoted more to punishment (to achieve the goals of retribution and incapacitation) or to rehabilitation (to achieve the goals of specific deterrence and successful reintegration or reentry). Until recently, punishment advocates had won the day. As described in Chapter 10, for the most part, correctional dollars have been spent on the incarceration and punishment of an ever-increasing prison population; rehabilitation has been little more than an afterthought. The near-total dominance of punishment over rehabilitation in the United States inspired law professors Malcolm Feeley and Jonathan Simon in the early 1990s to coin the term "new penology" to describe this development.[54] This new penology has abandoned rehabilitation in favor of efficiently managing large numbers of prisoners. Success for this new penology is not measured by reductions in recidivism (a standard measure of correctional success used in the past), but rather by how efficiently correctional systems manage prisoners within budgetary constraints.[55]

A recent challenge to the new penology is the "smart justice" movement. Although smart justice means different things to different people, most of its advocates view it as a comprehensive and evidence-based approach to (1) helping former offenders live as law-abiding citizens once they are released from prison, (2) reducing recidivism, (3) making communities safer, and (4) saving taxpayers money (or at least to allocate tax monies to other priorities). The critical focus of the smart justice movement generally is tough-on-crime laws that, according to them, have caused unjust sentences, mass incarceration, and wasteful spending. Targeted are laws that require offenders to serve 85% of their sentences ("truth-in sentencing"), sentencing guidelines, mandatory-minimum sentences, and other laws that eliminate judicial discretion, such as "three strikes" and "10-20-Life" laws.

Specific reforms advocated by at least some advocates of the "smart justice" movement include (1) ending the War on Marijuana, (2) eliminating so-called "tough-on-crime" laws, (3) incentivizing smart practices and eradicating the waste and abuse caused by private corporate profiteers, (4) abolishing "debtors' prisons," and (5) promoting alternatives to incarceration for special populations, such as the mentally ill and drug-law violators. All of these reforms share in common the overarching goal of ending mass incarceration. Few, if any, of the reforms are new; most have been promoted for some time, albeit with little success. All of the reforms have been described in previous chapters of this book. What is new about the "smart justice" movement is the economic circumstances in which it emerged.

Those economic circumstances were all related to the 2007–2009 economic recession. The recession caused states and the federal government to rethink their priorities and budget allocations. Cash-strapped states and, to a lesser extent, the federal government began to seriously consider whether they could afford the increasing costs of incarceration. Had it not been for the economic recession of 2007–2009, in short, the decades-long increase in the incarceration rate and the concomitant increase in incarceration costs likely would have continued unabated, although perhaps at a slower rate. The recession caused legislatures to reconsider their penal priorities, and "smart justice" was created to help them do just that.

Lizzie Himmel/Contributor/Getty Images

The federal supermaximum-security prison in Florence, Colorado, was one of the first of the "new penology" institutions that abandoned rehabilitation in favor of efficiently managing a large population of violent offenders. *What are some of the problems of abandoning rehabilitation as a goal of punishment?*

Available data suggest that the recent decline in the U.S. prison incarceration rate during the economic recession was not a coincidence. As noted in Chapter 10, between 1980 and 2008, the prison incarceration rate rose 264% from 139 prisoners per 100,000 residents to an all-time high of 506 prisoners per 100,000 residents. However, in 2008, at the height of the economic recession, the incarceration rate did not increase but remained at 506 prisoners per 100,000 residents. Then, in 2009, the incarceration rate declined for only the second time since 1968, from the high of 506 prisoners per 100,000 residents in 2007 and 2008, to 504 prisoners per 100,000 residents in 2009. The incarceration rate has decreased each year since 2009 and, in 2017 (the last year for which data were available), stood at 440 prisoners per 100,000 residents, down about 13% from the 2007 and 2008 high.[56] The downward pressure on the incarceration rate has begun, and the "smart justice" movement is ready to offer viable, evidence-based alternatives.

Smart Justice

Learn more about the "smart justice" movement by accessing the ACLU's Campaign for Smart Justice website at www.aclu.org/feature/campaign-smart-justice or the National Association of Counties, "Smart Justice: Creating Safer Communities" at www.naco.org/sites/default/files/documents/Smart%20Justice.pdf, among many Internet sources.

Do you think "smart justice" will replace the "new penology" as the guiding philosophy of corrections? Defend your answer.

Cost-Reduction Strategies Future expenditures of tax dollars on corrections by governments at all levels likely will be made grudgingly, after much wrangling and debate. Every attempt will be made to carry out corrections functions as inexpensively as possible. Because the vast majority of correctional clientele will be members of the underclass, as is the case today, there will be little public resistance to low-cost management strategies. Indeed, corrections in the future is likely to take on "a kind of waste management function."[57]

Justice Department Reverses Plan to End Use of Private Prisons

On August 18, 2016, the U.S. Department of Justice announced plans to end its use of private prisons. According to then-Deputy Attorney General Sally Yates, "They simply do not provide the same level of correctional services, programs, and resources; they do not save substantially on costs; and as noted in a recent report by the Department's Office of Inspector General, they do not maintain the same level of safety and security." The directive applied only to federal prisons. Then, on February 23, 2017, new Attorney General Jeff Sessions reversed the plan, arguing that it went against long-standing Justice Department policy and practice. Sessions directed the BOP to "return to its previous approach."

Sources: Matt Zapotosky and Chico Harlan, "Justice Department Says It Will End Use of Private Prisons," *The Washington Post* (August 18, 2016), accessed September 1, 2016, www.washingtonpost.com/news/post-nation/wp/2016/08/18/justice-department-says-it-will-end-use-of-private-prisons/?utm_term=.83f6d2f36e06; "U.S. Attorney General Expresses Support for Private Prisons," *Orlando Sentinel* (February 24, 2017), p. A6.

If "smart justice" reforms are widely adopted, the cost-reduction strategies likely to be advocated in the future are various alternatives to incarceration. For example, most persons convicted of minor offenses may be required to perform community service such as litter control and maintenance and construction of government buildings and grounds.[58] However, as described in Chapter 12, the cost savings of alternatives to incarceration tend to be illusory.

Another cost-reduction strategy likely to receive increasing support in the future is the privatization of corrections. As described in Chapter 10, the private sector has been involved in corrections in various ways for a long time. Keenly aware of the ongoing decrease in the incarceration rate, private companies that have been operating entire jails and prisons have been strategically shifting resources to community-based alternatives to incarceration, such as probation, electronic monitoring, and halfway houses. However, a problem with the operation of community-based alternatives by private companies—and a reason that cost savings from this strategy may prove minimal at best—is that private companies are in business to make a profit. If the companies are publicly owned, there is tremendous pressure on management to maximize shareholder value. As a result, it can be expected that private or public correctional companies will do all in their power to protect and enhance their interests by lobbying government officials for more favorable terms of operation and by using marketing to produce a greater demand for their products (i.e., more offenders sentenced to community-based alternatives to incarceration).

A third cost-reduction strategy—potentially the most effective one—is the use of new technology. For example, in the case of prisons and jails, the use of cameras to watch prisoners and robots to service them, as well as the use of sensing devices on an institution's perimeter, would reduce the need for most correctional officers. The use of ultrasound may be the solution to costly prison and jail riots. "Piping high-pitched sound over improved intercom systems would momentarily render everyone in the affected area unconscious and allow staff to enter, disarm, and regain custody."[59] If the United States continues to enforce the death penalty, ultrasound may provide a more humane and cost-effective method of execution. In the near future, ultrasound could be provided at levels that would literally dematerialize the offender (eliminating the costs of disposing of the body).[60] Another possibility is instant death by laser ray.[61] The costs of building or enlarging more jails and prisons because of overcrowding would be reduced significantly in the future through the use of **cryonics** (freezing) and other forms of human hibernation.[62] Many prisoners could be "stored" in a small amount of space with the use of such technologies. In fact, they could literally be stacked on top of each other in coffin-like containers. Prisoners in the future also may be incarcerated in self-supporting undersea or space prisons, as the old practice of transporting prisoners is revived.[63]

cryonics A process of human hibernation that involves freezing the body.

Prison overcrowding could someday lead to the use of cryonics and other forms of human hibernation. *What problems might arise if cryonics were used to deal with prison overcrowding?*

Jeff Topping/Stringer/Getty Images

In the case of community-based alternatives to incarceration, you already have learned about the use of electronically monitored house arrest for pretrial detainees (using electrodes connected to a monitor, perhaps surgically implanted, that would intermittently shock clients while they were outside a designated area). The same technology could be used for probationers and parolees, reducing the number of probation and parole officers needed. In addition, a subliminal-message player might be implanted in probationers, parolees, and prisoners that would provide 24-hour anticrime messages, such as "Obey the law" or "Do what is required of you," or synthesized body chemicals could be implanted that would keep offenders under constant control.[64] The costs related to future violations could be reduced dramatically through such methods.

However, one must keep in mind that several possible factors might significantly reduce or negate the tremendous cost savings anticipated from new correctional technologies. For example, the success of the strategy may create incentives for net widening, which will reduce or eliminate any anticipated cost savings.[65] In addition, technology has its own costs: Experts must be employed, staff must be trained, equipment must be serviced, and systems must be upgraded as new technologies are introduced.[66]

You may have noticed that this discussion did not cover an important subject: crime and delinquency prevention—the effort to prevent criminal and delinquent acts from being committed in the first place. Although crime and delinquency prevention probably will be a priority in the future, it will be mostly the responsibility of government agencies and philanthropic organizations outside the criminal justice process (e.g., social, health, and welfare agencies). Because the focus of this book is criminal justice in the United States, we do not speculate here about future crime and delinquency prevention strategies.

THINKING CRITICALLY

1. What impact do you think DNA profiling will have on the future of law enforcement?
2. Do you think that people who don't speak English should be allowed to serve as jurors? Why or why not?
3. When it comes to corrections, what do you think more money should be devoted to: punishment or rehabilitation? Why?

Criminal Justice in the Age of Terrorism

The tragic events of September 11, 2001, have transformed the United States in many ways. A new kind of fear has gripped the American people, who no longer feel as safe as they did before that infamous day. The most notable change to occur is the war on terrorism–a financially costly war that conceivably has no end. The war on terrorism, in turn, has altered criminal justice in the United States, as the fear of terrorism has joined the fear of crime in the nation's collective nightmares.

However, terrorism and crime are not separate phenomena. Although they may think of themselves as "freedom fighters," according to the FBI's counterterrorism policy, terrorists are criminals. Furthermore, as the U.S. government and its allies have increasingly succeeded in freezing their financial assets, terrorist groups have turned to different types of crime to support their operations. Many of those crimes are the same ones committed by transnational organized crime groups: bootlegging cigarettes; counterfeiting compact discs, movies, and other products; drug trafficking; and smuggling human beings for profit to name a few.[67]

Legislative Responses to Terrorism

As for the response to terrorism, one of the more controversial developments in this new Age of Terror was the passage of the USA PATRIOT Act less than 2 months after the planes slammed into the World Trade Center and the Pentagon. As described in Chapter 4, the act gave the government broad new powers to address the menacing threat. For example, in terrorism-related matters, it expanded the government's search and surveillance powers and eliminated safeguards such as judicial oversight, the need for a warrant and probable cause, the ability to challenge certain searches in court, and public accountability. The act also created the new crime of "domestic terrorism," which is defined so broadly that it could include members of controversial activist groups such as Operation Rescue, the Earth Liberation Front, Greenpeace, or the World Trade Organization protesters, as long as they committed acts "dangerous to human life" to "influence the policy of a government by intimidation or coercion." Another provision of the act allows the attorney general to detain noncitizens in the United States indefinitely without trial if there is reason to believe they endanger national security and a foreign country will not accept them for deportation. Critics contend that in removing traditional checks on law enforcement, the USA PATRIOT Act and other similar legislation threaten the very rights and freedoms that the war on terrorism is trying to preserve.

Another legislative response to 9/11 was passage of the Homeland Security Act of 2002. As described in Chapter 5, the act established the Department of Homeland Security (DHS), whose fundamental mission is to prevent terrorist attacks in the United States and to minimize the damage and assist in the recovery from terrorist acts that do occur. The development of the DHS, which is ongoing, involves a massive reorganization of the federal law enforcement bureaucracy. Federal law enforcement agencies have been moved from their traditional departments, such as Justice or Treasury, to the new DHS. Other agencies have been combined or

APA Images/Shutterstock

The United States is hated by people all over the world. *Why?*

reconfigured. For example, the border and agency functions of the U.S. Customs Service, the Federal Protective Service, and the former Immigration and Naturalization Service have been reorganized into the Bureau of Immigration and Customs Enforcement under the DHS. New agencies also have been created, such as the Bureau of Citizenship and Immigration Services, the Office of Private Sector Liaison, and the State, Local, Tribal, and Territorial Coordinating Council. Each of these new agencies is part of the DHS.

Terrorism and Law Enforcement

As also discussed in Chapter 5, the FBI was *not* moved to the DHS; it remains in the Justice Department, but has undergone some fundamental changes in response to 9/11. For example, its top priority has shifted from being a federal police agency to being an intelligence and counterterrorism agency. The management hierarchy at the FBI's headquarters in Washington has been restructured to support counterterrorism efforts. One-quarter of the FBI's then 11,000 agents were reassigned to work on counterterrorism. A National Joint Terrorism Task Force was established at FBI headquarters to coordinate the flow of information with task forces in each of the Bureau's 56 field offices, and new FBI offices were opened in foreign countries where terrorist groups have a presence.

Local law enforcement officers also have become soldiers in the war on terrorism, as explained in Chapter 6. As "first responders" and the "front line of defense," local police officers and sheriffs' deputies not only must be able to respond to terrorist acts, but they also must be proactive in preventing them. Their job is especially difficult because they must be ever vigilant without engaging in racial or ethnic profiling. Local law enforcement agencies are increasingly trying to train their personnel for their new roles but, according to recent reports, most of them are drastically underfunded and dangerously unprepared.

Terrorism and the Administration of Justice

The war on terrorism has revealed the existence of a secretive federal court, previously unknown to most Americans. The Foreign Intelligence Surveillance Court, or FISA court, which is described in Chapter 4, has been in operation since the late 1970s. The court has approved thousands of Justice Department requests to conduct secret searches and surveillance of people in the United States who are suspected of having links to foreign agents or powers that often involve terrorism and espionage.

Until recently, American citizens who were suspected of being terrorists or of having ties to terrorist groups in the wake of 9/11 were arrested and detained without being charged with any crimes and were denied legal representation and access to the criminal courts to challenge their detention. Military tribunals instead of federal courts tried non-U.S. citizens accused of terrorism by the U.S. government, if they were charged and tried at all. In 2008, in *Boumediene v. Bush*, the U.S. Supreme Court ruled that the then approximately 260 detainees at the supermaximum-security prison at Guantánamo Bay Naval Base in Cuba had the constitutional privilege of *habeas corpus*. This meant that federal courts have jurisdiction over habeas petitions filed by foreign citizens detained at the U.S. Naval Base at Guantánamo Bay, Cuba, and the detainees at Guantánamo Bay have the protection of the Fifth Amendment right not to be deprived of liberty without due process of law and of the Geneva Conventions.

Terrorism and Corrections

The inmates at Guantánamo Bay prison, originally from 42 different countries, are being held until they no longer have useful intelligence to provide, are cleared of criminal wrongdoing, or no longer pose a threat to the United States. Some of them may be held until the war on terrorism is over, whenever that might be. The U.S. government has not filed criminal charges against most of them or assigned them prisoner-of-war status, leaving them in legal limbo. If and when Camp Delta is closed, the remaining prisoners will be moved to other prisons either in the United States or perhaps abroad.

Final Remarks

Few people question the threat to U.S. citizens' lives and well-being posed by twenty-first-century terrorism. The tragic events of September 11, 2001, provided a frightening wake-up call for most Americans. Unfortunately, there is no established roadmap to follow in fighting the war on terrorism. There are few precedents to guide decision makers in the difficult task of preserving domestic security. Critics contend that some government officials have been overzealous in their efforts. Some critics question the motives of particular government officials. Perhaps the most frequent criticism of the war on terrorism is that some of the legislation enacted and tactics allowed threaten the very rights and freedoms that the war on terrorism is trying to preserve. Some people worry that provisions of the USA PATRIOT Act and other similar legislation designed for the war on terrorism–provisions that override or set aside traditional procedural safeguards–will become a permanent part of the way criminal justice is administered in the United States. Some people are concerned that law enforcement strategies implemented to fight the war on terrorism–strategies that remove traditional constraints on law enforcement activity–will become commonplace in the ways law enforcement officers carry out their duties. There is no doubt that the collapse of the World Trade Center's twin towers changed the United States in unanticipated ways. The challenge faced by the United States in this Age of Terrorism is to defend domestic security without sacrificing the rights and freedoms of American citizens.

As we write these last words, the United States and most of the world are experiencing the initial stages of the coronavirus pandemic. This terrorism threat, and it seems fair to call it that, already is causing panic and major social dislocations and promises more of the same as it spreads through society. The dislocations caused by the pandemic will manifest themselves in many known and unknown ways, but two ways appear certain: The pandemic will affect crime and the criminal justice system, and crime and the criminal justice system will affect the pandemic. Already scams related to the pandemic are being reported, people are ignoring social distancing directives, courts are closing or are closed, and criminal justice authorities are trying to reduce jail and prison populations in various ways to contain the pandemic's spread. By the time the next edition of this book is published, we should have a much better idea of the pandemic's scope and consequences. We hope that they are minimal.

THINKING CRITICALLY

1. Do you think there should be legal limits on the ways the United States fights the war on terrorism? If so, what should they be? If not, why not?
2. Do you think it is acceptable to suspend traditional legal rights and freedoms in the war on terrorism? Why or why not? If your answer is yes, which legal rights and freedoms should be suspended, and which ones should not? Defend your choices.

SUMMARY

1. **Describe the possible future of law enforcement if the crime control model dominates and the possible future if the due process model dominates.**

   If law enforcement in the future increasingly reflects the principles and policies of the crime control model, then Americans might expect fewer limitations on how the police attempt to combat crime. However, if there is a shift to the principles and policies of the due process model, Americans should expect existing limitations on how the police combat crime to remain intact or even to be expanded.

2. **Describe the possible future of the administration of justice if the crime control model dominates and the possible future if the due process model dominates.**

   If the crime control model dominates the administration of justice in the future, then the right to legal counsel at critical pretrial and posttrial stages may be scaled back significantly, the preliminary hearing may be abolished, and the use of pretrial detention may be expanded; there may be fewer criminal trials and more plea bargaining, and appeals may be strongly discouraged and limited. If the due process model dominates, then the current right to counsel at a variety of critical stages in the process is likely to be maintained or, perhaps, extended somewhat (e.g., to appeals beyond the first one); the grand jury may be eliminated; plea bargaining probably will be discouraged, and the number of criminal cases that go to trial is therefore likely to increase substantially; and there probably will be no limitations on the right to appeal.

3. **Identify perhaps the most divisive issue that will confront correctional policy makers in the future.**

   Perhaps the most divisive issue that will confront correctional policy makers in the future is whether increasingly scarce resources will be devoted more to punishment (to achieve the goals of retribution and incapacitation) or to rehabilitation (to achieve the goals of specific deterrence and successful reintegration or reentry).

4. **Describe the possible future of corrections.**

The future of corrections will depend on whether the "new penology" or "smart justice" movement prevails. Either way, future expenditures of tax dollars on corrections by governments at all levels likely will be made grudgingly after much wrangling and debate. Every attempt will be made to carry out corrections functions as inexpensively as possible. Because the vast majority of correctional clientele will be members of the underclass, as is the case today, there will be little public resistance to low-cost management strategies. Indeed, corrections in the future is likely to take on "a kind of waste management function," especially if the "new penology" dominates corrections.

5. **List some of the cost-reduction strategies likely to be advocated in corrections in the future.**

Among the cost-reduction strategies likely to be advocated in corrections in the future are various alternatives to incarceration. However, cost savings from those alternatives may be illusory. Another cost-reduction strategy likely to receive increasing support in the future is the privatization of corrections. A third cost-reduction strategy, and potentially the most effective one, is the use of new technology.

6. **List specific reforms of the "smart justice" movement and identify its overarching goal.**

Specific reforms of the "smart justice" movement include (1) ending the War on Marijuana, (2) eliminating so-called "tough-on-crime" laws, (3) incentivizing smart practices and eradicating the waste and abuse caused by private corporate profiteers, (4) abolishing "debtors' prisons," and (5) promoting alternatives to incarceration for special populations, such as the mentally ill and drug-law violators. The overarching goal of the "smart justice" movement is ending mass incarceration.

7. **Describe some of the challenges faced by criminal justice in the Age of Terrorism.**

A major challenge for criminal justice is to fight the war on terrorism without threatening the very rights and freedoms that the war on terrorism is trying to preserve. Another challenge is to create the institutions, such as the DHS, to fight the war on terrorism. The FBI has the challenge of shifting its priorities from being a federal police agency to being an intelligence and counterterrorism agency. Local law enforcement agencies have the challenge of preventing terrorist acts and responding to them when they occur, without engaging in racial or ethnic profiling.

KEY TERMS

bionics 574
mediation 584
arbitration 584
restorative justice 585
reintegrative shaming 585
cryonics 590

REVIEW QUESTIONS

1. How might community policing based on the crime control model of criminal justice differ from community policing based on the due process model?
2. How might bionics affect law enforcement in the future?
3. What is perhaps the thorniest issue regarding the use of DNA technology as a law enforcement tool?
4. What is a principal problem with plea bargaining for due process model supporters?
5. How might the changing demographics of the U.S. population affect the operation of criminal courts?
6. What is the principal difference between mediation and arbitration?
7. What are restorative justice and reintegrative shaming, and how do they differ from punitive justice?
8. In what ways do the "new penology" and the "smart justice" movement differ?
9. What is cryonics, and how might it be used in corrections in the future?

IN THE FIELD

1. **Criminal Justice in the Movies** There are numerous science fiction films that depict criminal justice in the future, including *Blue Thunder, Escape from New York, Demolition Man, Blade Runner, Minority Report,* and the *RoboCop* series. Watch one or more of these films, and consider how plausible they seem in light of what you have read in this chapter. Have any of the innovations of these futuristic justice systems become a part of the current American justice system? Are any of these innovations discussed in this chapter?
2. **Ideas About Improving Job Performance** Interview several criminal justice practitioners, such as police officers, prosecutors, defense attorneys, judges, probation or parole officers, or corrections officers. Ask them what "futuristic" advancements in procedure and technology would improve their job performances. You probably will have to prompt them with ideas presented in this chapter. Share your results with your classmates through an oral or written presentation.

ON THE WEB

1. **Explore New Technologies** Go to the Justice Technology Information Center (Justnet) at https://www.justnet.org. View the "Topical List" and look for new technologies. Survey the new technologies available, and write a three-page scenario of how criminal justice will change when the new technology is implemented.
2. **Evaluate Future Policing Trends** Go to the Police Executive Research Forum website and under Publications in the menu bar select Free Online Documents at https://www.policeforum.org/free-online-documents. From the list of documents, select a recent one and write a short essay describing the problems that may be encountered if the programs and policies described in the document are implemented.

CRITICAL THINKING EXERCISES

CORRECTIONS IN THE FUTURE

1. It is the year 2028, and you have been asked to serve on a citizens' committee charged with providing input about the creation of a new supermaximum-security federal prison. The prison will be built either several hundred feet beneath the ocean's surface off the east coast of the United States or on the moon. The rationale for these locations is to better isolate prisoners from the law-abiding population and to take advantage of prison labor for the high-risk jobs of either marine farming or the mining of new and useful materials in space. The technology and the resources to build the prison, although expensive, are available.
   a. Which location would be preferable? Why?
   b. Should public tax dollars be used to fund the project, or should a private corporation be allowed to build and operate the facility at its own expense and keep any profits from the farming or mining business?
   c. What special problems might arise because of the location of the prison?

LAW ENFORCEMENT IN THE FUTURE

2. It is the year 2028 in Washington, DC, and public service officers, formerly called police officers, are flying routine patrol with the aid of their new jet packs. While flying over a condo near the city's center, two of the officers, using their bionic eyes and ears, detect what appear to be a half-dozen men plotting to bomb the White House. Surveying the condo from their sky perch, the officers see, stored in a bedroom closet, enough of a new, illegal, and largely undetectable hydrogen-based explosive to do the job. The officers, using the ultra-small two-way communication devices implanted in their larynxes, communicate to headquarters what they have seen. They await further orders.
   a. Do the officers have probable cause to obtain a search warrant from a magistrate or to make an arrest? If you were a proponent of Packer's crime control model, what would your answer be? If you were a proponent of Packer's due process model, what would your answer be?
   b. How might the legal issues of the right to privacy, the admissibility of evidence, the exclusionary rule, and the plain-view doctrine affect law enforcement officers' use of this new technology?
   c. What restraints, if any, should be imposed on law enforcement officers' use of this new technology? If you were a proponent of Packer's crime control model, what would your answer be? If you were a proponent of Packer's due process model, what would your answer be?

NOTES

1. Herbert Packer, *The Limits of the Criminal Sanction* (Stanford, CA: Stanford University Press, 1968).
2. Nancy G. LaVigne, Samantha S. Lowry, Joshua A. Markman, and Allison M. Dwyer, "Evaluating the Use of Public Surveillance Cameras for Crime Control and Prevention," Justice Policy Center (Washington, DC: The Urban Institute, September 2011), accessed January 17, 2013, www.cops.usdoj.gov/Publications/e071112381_EvalPublicSurveillance.pdf.
3. Gene Stephens, "High-Tech Crime Fighting: The Threat to Civil Liberties," *The Futurist* (July–August 1990): 20–25.
4. Gene Stephens, "Law Enforcement," in *Encyclopedia of the Future*, ed. G. T. Kurian and G. T. T. Molitor (New York: Simon & Schuster, Macmillan, 1996), 538; Jeff Kunerth, "You Have No Right to Privacy You Seek," *Orlando Sentinel*, August 8, 1999, A1.
5. Scott Parks, "High-Tech's Time Is on Horizon in Security Business," *Orlando Sentinel*, October 7, 2001, A10; Jeremiah Marquez, "A New Weapon Against Crime?" *Orlando Sentinel*, June 21, 2006, A17.
6. "FBI Announces Full Operational Capability of the Next Generation Identification System," September 15, 2014, accessed August 10, 2016, www.fbi.gov/news/pressrel/press-releases/fbi-announces-full-operational-capability-of-the-next-generation-identification-system; John W. Whitehead, "Smile, the Government Is Watching: Next Generation Identification," *Right Side News*, September 17, 2012, accessed January 16, 2013, www.rightsidenews.com/2012091731028/editorial/us-opinion-and-editorial/smile-the-government-is-watching-next-generation-identification.html.
7. Drew Harwell, "FBI, ICE find state driver's license photos are a gold mine for facial-recognition searches," *The Washington Post*, July 7, 2019, accessed October 5, 2019, https://www.washingtonpost.com/technology/2019/07/07/fbi-ice-find-state-drivers-license-photos-are-gold-mine-facial-recognition-searches/.

8. "FBI Announces Full Operational Capability of the Next Generation Identification System," September 15, 2014; John W. Whitehead, "Smile, the Government Is Watching: Next Generation Identification," *Right Side News,* September 17, 2012.

9. Ethan Horowitz, "Speedy Database Helps Cops Put Pieces Together," *Orlando Sentinel,* August 24, 2003, B1.

10. Gene Stephens, "Drugs and Crime in the Twenty-First Century," *The Futurist* (May–June 1992): 19–22.

11. "Police to Use IBM's Digital Video System," *Orlando Sentinel,* May 18, 2003, H3.

12. Ibid.

13. Ibid.

14. Lydia Wheeler, "Poll: Overwhelming Majority Want Police to Wear Body Cameras," *The Hill,* October 26, 2015, accessed August 11, 2016, http://thehill.com/regulation/technology/258085-majority-of-americans-want-set-rules-for-police-body-cameras-poll-shows.

15. Unless indicated otherwise, information on surveillance cameras is from Charlie Savage, "U.S. Doles Out Millions for Street Cameras," August 12, 2007, *Boston Globe,* www.boston.com/news/nation/washington/articles/2007/08/12/us_doles_out_millions_for_street_cameras/; Demian Bulwa, "Future Fuzzy for Use of Public Surveillance Cameras," *SFGate,* July 23, 2006, www.sfgate.com/cgi-bin/article.cgi?f5/c/a/2006/07/23/CAMERAS.TMP; Allison Klein, "Police Go Live Monitoring D.C. Crime Cameras," *The Washington Post,* www.washingtonpost.com/wp-dyn/content/article/2008/02/10/AR2008021002726_pf.html; Alex Johnson, "Smile: More and More, You're on Camera," *MSNBC,* www.msnbc.msn.com/id/25355673/; J. Douglas Walker, "Information Technology Advances Push the Privacy Boundaries Again," in *Future Trends in State Courts 2008,* ed. C. Flango, A. McDowell, C. Campbell, and N. Kauder (Williamsburg, VA: National Center for State Courts, 2008), 40–43; LaVigne et al., "Evaluating the Use of Public Surveillance Cameras for Crime Control and Prevention."

16. Walker, "Information Technology Advances Push the Privacy Boundaries Again," 41.

17. Ben Reed, "Future Technology in Law Enforcement," *Law Enforcement Bulletin* 77, no. 5 (May 2008), accessed January 26, 2009, www.fbi.gov/publications/leb/2008/may2008/may2008leb.htm.

18. Stephens, "Law Enforcement," 539; "Police Test Voice Translator Box," *Orlando Sentinel,* December 5, 1999, A17.

19. "New Ultra-Wideband Could Make Walls See-Through," *Orlando Sentinel,* June 15, 2000, A3.

20. Walker, "Information Technology Advances Push the Privacy Boundaries Again," 40.

21. Reed, "Future Technology in Law Enforcement."

22. Jack Cheevers, "Beyond 'RoboCop': Concern for Officer Safety Fuels Innovation," *Charlotte Observer,* August 26, 1994, 14A.

23. Reed, "Future Technology in Law Enforcement."

24. Ibid.

25. Ibid.

26. Marco Santana, "Law Enforcement Going Virtual to Fight Crime," *Orlando Sentinel,* August 17, 2016, A8.

27. Stephens, "Law Enforcement," 539.

28. Terry D. Anderson, Kenneth D. Gisborne, Marilyn Hamilton, Pat Holiday, John C. LeDoux, Gene Stephens, and John Welter, *Every Officer Is a Leader: Transforming Leadership in Police, Justice, and Public Safety* (Boca Raton, FL: St. Lucie Press, 2000), 368.

29. Reed, "Future Technology in Law Enforcement."

30. Ibid.

31. Ibid.

32. Randall T. Shepard, "Indiana Court Technology Is About Service, Not Bytes and Bandwidth," in *Future Trends in State Courts 2008,* ed. Carol R. Flango, Chuck Campbell, and Neal Kauder (Williamsburg, VA: National Center for State Courts, 2006), 26–28.

33. Ibid., 27.

34. Stephens, "High-Tech Crime Fighting: The Threat to Civil Liberties."

35. Anita Neuberger Blowers, "The Future of American Courts," in *The Past, Present, and Future of American Criminal Justice,* ed. B. Maguire and P. Radosh (New York: General Hall, 1995); Carol R. Flango, Chuck Campbell, and Neal Kauder, eds., *Future Trends in State Courts 2006* (Williamsburg, VA: National Center for State Courts, 2006).

36. "Court: Jurors Don't Have to Speak English to Serve," *Orlando Sentinel,* January 20, 2000, A10.

37. Flango et al., *Future Trends in State Courts 2008*; National Center for State Courts, "Consortium for Language Access in the Courts," accessed March 8, 2011, www.ncsconline.org/d_research/CourtInterp/CICourtConsort.html; National Association of Judiciary Interpreters & Translators, accessed March 8, 2011, www.najit.org/about/about.php.

38. Ibid.

39. Blowers, "The Future of American Courts."

40. Ibid.

41. Richard Van Duizend, "The Implications of an Aging Population for the State Courts," in *Future Trends in State Courts 2008,* ed. Carol R. Flango, Chuck Campbell, and Neal Kauder (Williamsburg, VA: National Center for State Courts, 2006), 76–80.

42. Gene Stephens, "Crime and Punishment: Forces Shaping the Future," *The Futurist* (January–February 1987): 18–26.

43. Ibid.

44. John Braithwaite, *Crime, Shame and Reintegration* (Cambridge, UK: Cambridge University Press, 1989).

45. Flango et al., *Future Trends in State Courts 2008.*

46. Ibid.

47. William E. Raftery, "Weathering the storm," *Judicature* 101 (Winter 2017), accessed October 6, 2019, https://judicialstudies.duke.edu/wp-content/uploads/2018/01/JUDICATURE101.4-disaster.pdf.

48. Blowers, "The Future of American Courts"; Flango et al., *Future Trends in State Courts 2008.*

49. Material on neuroscience is from Jeffrey Rosen, "The Brain on the Stand," *The New York Times Magazine,* March 11, 2007, 49–53, 70, 77, 82, 84.

50. Stephens, "High-Tech Crime Fighting: The Threat to Civil Liberties."

51. Ibid.

52. Flango et al., *Future Trends in State Courts 2008*; Brian A. Jackson, Duren Banks, John S. Hollywood, Dulani Woods, Amanda Royal, Patrick W. Woodson, and Nicole J. Johnson, *Fostering Innovation in the U.S. Court System,* Rand Corporation, 2016, accessed October 8, 2019, https://www.jstor.org/stable/10.7249/j.ctt1d41ddx.5?refreqid=excelsior%3A6776004ec6c3f0dc74edb37698c12f24&seq=1#metadata_info_tab_contents.

53. Ibid.

54. Malcolm M. Feeley and Jonathan Simon, "The New Penology: Notes on the Emerging Strategy of Corrections and Its Implications," *Criminology* 30 (1992): 470.

55. Ibid.

56. Jennifer Bronson and E. Ann Carson, "Prisoners in 2017," in *Bureau of Justice Statistics Bulletin,* U.S. Department of Justice (Washington, DC: U.S. Government Printing Office, April 2019), 9, Table 5, accessed October 8, 2019, https://www.bjs.gov/content/pub/pdf/p17.pdf.

57. Feeley and Simon, "The New Penology," 470.

58. Gene Stephens, "Prisons," in *Encyclopedia of the Future*, ed. G. T. Kurian and G. T. T. Molitor (New York: Simon & Schuster, Macmillan, 1996), 751.

59. Stephens, "High-Tech Crime Fighting: The Threat to Civil Liberties."

60. Ibid.

61. Stephens, "Prisons," 751.

62. Stephens, "High-Tech Crime Fighting: The Threat to Civil Liberties."

63. Stephens, "Prisons."

64. Stephens, "High-Tech Crime Fighting: The Threat to Civil Liberties."

65. Francis T. Cullen and John P. Wright, "The Future of Corrections," in *The Past, Present, and Future of American Criminal Justice,* ed. B. Maguire and P. Radosh (New York: General Hall, 1995).

66. Ibid.

67. Josh Meyer, "Diminished al-Qaeda Remains a Threat to U.S.," *Orlando Sentinel*, September 11, 2003, A8.

Design Elements: *Image of part of a court building with columns and statue*: Pixtal/AGE Fotostock RF; *Image of a door slightly open*: jadding/Shutterstock.com

Glossary

actus reus Criminal conduct–specifically, intentional or criminally negligent (reckless) action or inaction that causes harm. (2)

adjudication The juvenile court equivalent of a trial in criminal court, or the process of rendering a judicial decision regarding the truth of the facts alleged in a petition. (13)

administrative segregation The keeping of inmates in secure isolation so that they cannot harm others. (10)

aggravating factors, aggravating circumstances, or **special circumstances** In death sentencing, facts or situations that increase the blameworthiness for a criminal act. (9)

aggressive patrol The practice of having an entire patrol section make numerous traffic stops and field interrogations. (6)

allocution The procedure at a sentencing hearing in which the convicted defendant has the right to address the court before the sentence is imposed. During allocution, a defendant is identified as the person found guilty and has a right to deny or explain information contained in the PSI if his or her sentence is based on it. (9)

anomie For Durkheim, the dissociation of the individual from the collective conscience. For Merton, the contradiction between the cultural goal of achieving wealth and the social structure's inability to provide legitimate institutional means for achieving the goal. For Cohen, the inability of juveniles to achieve status among peers by socially acceptable means. (3)

appellate jurisdiction The power of a court to review a case for errors of law. (8)

apprenticeship system The method by which middle- and upper-class children were taught skilled trades by a master. (13)

arbitration A dispute resolution process that brings disputants together with a third party (an arbitrator) who has the skills to listen objectively to evidence presented by both sides of a conflict, to ask probing and relevant questions of each side, and to arrive at an equitable solution to the dispute. (14)

arraignment A pretrial stage; its primary purpose is to hear the formal information or indictment and to allow the defendant to enter a plea. (1)

arrest The seizing or the taking of a person into custody by lawful authority, either actual physical custody, as when a suspect is handcuffed by a police officer, or constructive custody, as when a person peacefully submits to a police officer's control. (4)

arrest warrant A written order directing law enforcement officers to arrest a person. The charge or charges against a suspect are specified on the warrant. (8)

atavist A person who reverts to a savage type. (3)

Auburn system An early system of penology, originating at Auburn Penitentiary in New York, in which inmates worked and ate together in silence during the day and were placed in solitary cells for the evening. (10)

bail bond or **bail** Usually a monetary guarantee deposited with the court that is supposed to ensure that the suspect or defendant will appear at a later stage in the criminal justice process. (1)

banishment A punishment, originating in ancient times, that required offenders to leave the community and live elsewhere, commonly in the wilderness. (10)

bench or **summary trial** A trial before a judge without a jury. (1)

bench warrant or ***capias*** A document that authorizes a suspect's or defendant's arrest for not appearing in court as required. (8)

beyond a reasonable doubt The standard of proof necessary to find a defendant guilty in a criminal trial. (4)

bifurcated trial A two-stage trial (unlike the one-stage trial in other felony cases) consisting of a guilt phase and a separate penalty phase. (9)

binding-out system Practice in which children were "bound over" to masters for care. However, under the binding-out system, masters were not required to teach youths a trade. (13)

biological inferiority According to biological theories, a criminal's innate physiological makeup produces certain physical or genetic characteristics that distinguish criminals from noncriminals. (3)

bionics The replacing of human body parts with mechanical parts. (14)

booking The administrative recording of an arrest. Typically, the suspect's name, the charge(s) for which the person was arrested, and perhaps the suspect's fingerprints or photograph are entered in the police blotter. (1)

civil law One of two general types of law practiced in the United States (the other is criminal law); a means of resolving conflicts between individuals. It includes personal injury claims (torts), the law of contracts and property, and subjects such as administrative law and the regulation of public utilities. (4)

class struggle For radical criminologists, the competition among wealthy people and among poor people and between rich people and poor people, which causes crime. (3)

classical theory A product of the Enlightenment, based on the assumption that people exercise free will and are thus completely responsible for their actions. In classical theory, human behavior, including criminal behavior, is motivated by a hedonistic rationality in which actors weigh the potential pleasure of an action against the possible pain associated with it. (3)

classification facility A facility to which newly sentenced offenders are taken so that their security risks and needs can be assessed and they can be assigned to a permanent institution. (10)

clear and convincing evidence The standard of proof required in some civil cases and, in federal courts, the standard of proof necessary for a defendant to make a successful claim of insanity. (4)

clemency A way by which inmates are released from prison that allows the governor of a state, or the president of the United States, when federal or military law is violated, to exercise leniency or mercy. (11)

cognitive behavioral therapy A correctional therapeutic intervention that focuses on the connection between dysfunctional thought processes and harmful behavior. (10)

collective conscience The general sense of morality of the times. (3)

college academies Schools where students pursue a program that integrates an associate's degree curriculum in law enforcement or criminal justice with the state's required peace officer training. (7)

community corrections The subfield of corrections in which offenders are supervised and provided services outside jail or prison. (12)

community policing A contemporary approach to policing that actively involves the community in a working partnership to control and reduce crime. (5)

commutation Reduction of the original sentence given by executive authority, usually a state's governor. (9)

complaint A charging document specifying that an offense has been committed by a person or persons named or described; usually used for misdemeanors and ordinance violations. (8)

CompStat A technological and management system that aims to make the police better organized and more effective crime fighters. It combines innovative crime analysis and geographic information systems; that is, crime mapping, with the latest management principles. (5)

conditional release A form of release that requires that a suspect/defendant maintain contact with a pretrial release program or undergo regular drug monitoring or treatment. (8)

confession An admission by a person accused of a crime that he or she committed the offense charged. (4)

conflict theory A theory that assumes that society is based primarily on conflict between competing interest groups and that criminal law and the criminal justice system are used to control subordinate groups. Crime is caused by relative powerlessness. (3)

conjugal visits An arrangement whereby inmates are permitted to visit in private with their spouses or significant others to maintain their personal relationship. (10)

constable The peacekeeper in charge of protection in early English towns. (5)

constable-watch system A system of protection in early England in which citizens, under the direction of a constable, or chief peacekeeper, were required to guard the city and to pursue criminals. (5)

contraband An illegal substance or object. (4)

contract security Protective services that a private security firm provides to people, agencies, and companies that do not employ their own security personnel or that need extra protection. (5)

convict code A constellation of values, norms, and roles that regulate the way inmates interact with one another and with prison staff. (11)

copicide A form of suicide in which a person gets fatally shot after intentionally provoking police officers. (7)

cottage reformatories Correctional facilities for youths, first developed in the late 1800s, that were intended to closely parallel family life and remove children from the negative influences of the urban environment. Children in those facilities lived with surrogate parents, who were responsible for the youths' training and education. (13)

crime control model One of Packer's two models of the criminal justice process. Politically, it reflects traditional conservative values. In this model, the control of criminal behavior is the most important function of criminal justice. (1)

crime index offenses cleared The number of offenses for which at least one person has been arrested, charged with the commission of the offense, and turned over to the court for prosecution. (2)

crime index An estimate of crimes committed. (2)

crime rate A measure of the incidence of crime expressed as the number of crimes per unit of population or some other base. (2)

criminal anthropology The study of "criminal" human beings. (3)

criminal law One of two general types of law practiced in the United States (the other is civil law); "a formal means of social control [that uses] rules . . . interpreted [and enforced] by the courts . . . to set limits to the conduct of the citizens, to guide the officials, and to define . . . unacceptable behavior." (4)

criminal sanctions or **criminal punishment** Penalties that are imposed for violating the criminal law. (9)

criminalization process The way people and actions are defined as criminal. (3)

criminological theory The explanation of criminal behavior, as well as the behavior of police, attorneys, prosecutors, judges, correctional personnel, victims, and other actors in the criminal justice process. (3)

crisis intervention A counselor's efforts to address some crisis in an inmate's life and to calm the inmate. (10)

cryonics A process of human hibernation that involves freezing the body. (14)

custody level The classification assigned to an inmate to indicate the degree of precaution that needs to be taken when working with that inmate. (10)

cybercrime The use of computer technology to commit crime. (6)

dark figure of crime The number of crimes not officially recorded by the police. (2)

day reporting centers Facilities that are designed for offenders who would otherwise be in prison or jail and that require offenders to report regularly to confer with staff about supervision and treatment matters. (12)

defendant A person against whom a legal action is brought, a warrant is issued, or an indictment is found. (1)

deprivation model A theory that the inmate society arises as a response to the prison environment and the painful conditions of confinement. (11)

determinate sentence A sentence with a fixed period of incarceration, which eliminates the decision-making responsibility of parole boards. (9)

differential association Sutherland's theory that persons who become criminal do so because of contacts with criminal definitions and isolation from anticriminal patterns. (3)

directed patrol Patrolling under guidance or orders on how to use patrol time. (6)

discretion The exercise of individual judgment, instead of formal rules, in making decisions. (7)

disposition The juvenile court equivalent of sentencing in criminal court. At the disposition hearing, the court makes its final determination of what to do with the juvenile officially labeled delinquent. (13)

diversion Organized, systematic efforts to remove individuals from further processing in criminal justice by placing them in alternative programs; diversion may be pretrial or posttrial. (12)

doctrine of fundamental fairness The rule that makes confessions inadmissible in criminal trials if they were obtained by means of either psychological manipulation or "third-degree" methods. (4)

doctrine of legal guilt The principle that people are not to be held guilty of crimes merely on a showing, based on reliable evidence, that in all probability they did in fact do what they are accused of doing. Legal guilt results only when factual guilt is determined in a procedurally regular fashion, as in a criminal trial, and when the procedural rules designed to protect suspects and defendants and to safeguard the integrity of the process are employed. (1)

domestic terrorism Perpetrated by individuals and/or groups inspired by or associated with primarily U.S.-based movements that espouse extremist ideologies of a political, religious, social, racial, or environmental nature. (6)

double jeopardy The trying of a defendant a second time for the same offense when jeopardy attaches in the first trial and a mistrial was not declared. (4)

dual court system The court system in the United States, consisting of one system of state and local courts and another system of federal courts. (8)

due process model One of Packer's two models of the criminal justice process. Politically, it embodies traditional liberal values. In this model, the principal goal of criminal justice is at least as much to protect the innocent as it is to convict the guilty. (1)

due process of law The rights of people suspected of or charged with crimes. Also, the procedures followed by courts to ensure that a defendant's constitutional rights are not violated. (4)

duress Force or coercion as an excuse for committing a crime. (2)

eight index crimes The Part I offenses in the FBI's uniform crime reports. They were (1) murder and nonnegligent manslaughter, (2) forcible rape, (3) robbery, (4) aggravated assault, (5) burglary, (6) larceny-theft, (7) motor vehicle theft, and (8) arson, which was added in 1979. (2)

electronic monitoring An arrangement that allows an offender's whereabouts to be gauged through the use of computer technology. (12)

entrapment A legal defense against criminal responsibility when a person, who was not already predisposed to it, is induced into committing a crime by a law enforcement officer or by his or her agent. (2)

***ex post facto* law** A law that (1) declares criminal an act that was not illegal when it was committed, (2) increases the punishment for a crime after it is committed, or (3) alters the rules of evidence in a particular case after the crime is committed. (2)

excessive force A measure of coercion beyond that necessary to control participants in a conflict. (7)

exclusionary rule The rule that illegally seized evidence must be excluded from trials in federal courts. (4)

exonerations Cases in which a person was wrongly convicted of a crime and later cleared of all the charges based on new evidence of innocence. (4)

extinction A process in which behavior that previously was positively reinforced is no longer reinforced. (3)

felony A serious offense punishable by confinement in prison for more than 1 year or by death. (1)

feminist theory A perspective on criminality that focuses on women's experiences and seeks to abolish men's control over women's labor and sexuality. (3)

field interrogation A temporary detention in which officers stop and question pedestrians and motorists they find in suspicious circumstances. (6)

flat-time sentencing Sentencing in which judges may choose between probation and imprisonment but have little discretion in setting the length of a prison sentence. Once an offender is imprisoned, there is no possibility of reduction in the length of the sentence. (9)

frisking Conducting a search for weapons by patting the outside of a suspect's clothing, feeling for hard objects that might be weapons. (4)

full enforcement A practice in which the police make an arrest for every violation of law that comes to their attention. (7)

general deterrence The attempt to prevent people in general or society at large from engaging in crime by punishing specific individuals and making examples of them. (9)

general jurisdiction The power of a court to hear any type of case. (8)

GIS crime mapping A technique that involves the charting of crime patterns within a geographic area. (6)

good time Time deducted from an inmate's sentence by prison authorities for good behavior and other meritorious activities in prison. (9)

grand jury Generally a group of 12 to 23 citizens who meet in closed sessions to investigate charges coming from preliminary hearings or to engage in other responsibilities. A primary purpose of the grand jury is to determine whether there is probable cause to believe that the accused committed the crime or crimes. (1)

grand jury indictment A written accusation by a grand jury charging that one or more persons have committed a crime. (8)

grass eaters Officers who occasionally engage in illegal and unethical activities, such as accepting small favors, gifts, or money for ignoring violations of the law during the course of their duties. (7)

habeas corpus A court order requiring that a confined person be brought to court so that his or her claims can be heard. (11)

halfway houses Community-based residential facilities that are less secure and restrictive than prison or jail but provide a more controlled environment than other community correctional programs. (12)

hands-off philosophy A philosophy under which courts are reluctant to hear prisoners' claims regarding their rights while incarcerated. (11)

harm The external consequence required to make an action a crime. (2)

hearing officer A lawyer empowered by the juvenile court to hear juvenile cases. (13)

highway patrol model A model of state law enforcement services in which officers focus on highway traffic safety, enforcement of the state's traffic laws, and the investigation of accidents on the state's roads, highways, and property. (5)

home confinement A program that requires offenders to remain in their homes except for approved periods of absence; commonly used in combination with electronic monitoring. (12)

houses of refuge The first specialized correctional institutions for youths in the United States. (13)

hung jury The result when jurors cannot agree on a verdict. The judge declares a mistrial. The prosecutor must decide whether to retry the case. (8)

imitation or **modeling** A means by which a person can learn new responses by observing others without performing any overt act or receiving direct reinforcement or reward. (3)

importation model A theory that the inmate society is shaped by the attributes inmates bring with them when they enter prison. (11)

incapacitation The removal or restriction of the freedom of those found to have violated criminal laws. (8)

incarceration rate A figure derived by dividing the number of people incarcerated by the population of the area and multiplying the result by 100,000; used to compare incarceration levels of units with different population sizes. (10)

indeterminate sentence A sentence with a fixed minimum and maximum term of incarceration, rather than a set period. (9)

indictment A document that outlines the charge or charges against a defendant. (8)

informal juvenile justice The actions taken by citizens to respond to juvenile offenders without involving the official agencies of juvenile justice. (13)

information A document that outlines the formal charge(s) against a suspect, the law(s) that have been violated, and the evidence to support the charge(s). (1)

initial appearance A pretrial stage in which a defendant is brought before a lower court to be given notice of the charge(s) and advised of her or his constitutional rights. (1)

inquisitorial trial system A legal system where the court or a part of the court is actively involved in investigating the facts of the case, as opposed to an adversarial system where the role of the court is primarily that of an impartial referee between the prosecution and the defense. (8)

insanity Mental or psychological impairment or retardation as a defense against a criminal charge. (2)

institution of social control An organization that persuades people, through subtle and not-so-subtle means, to abide by the dominant values of society. (1)

intake screening The process by which decisions are made about the continued processing of juvenile cases. Decisions might include dismissing the case, referring the youth to a diversion program, or filing a petition. (13)

intensive-supervision probation and parole (ISP) An alternative to incarceration that provides stricter conditions, closer supervision, and more treatment services than do traditional probation and parole. (12)

intermediate sanctions Sanctions that, in restrictiveness and punitiveness, lie between traditional probation and traditional imprisonment or, alternatively, between imprisonment and traditional parole. (12)

internal affairs investigations unit The police unit that ferrets out illegal and unethical activity engaged in by the police. (7)

international terrorism Perpetrated by individuals and/or groups inspired by or associated with designated foreign terrorist organizations or nations (state-sponsored). (6)

jail A facility, usually operated at the local level, that holds convicted offenders and unconvicted persons for relatively short periods. (10)

jailhouse lawyer An inmate skilled in legal matters. (11)

job stress The harmful physical and emotional outcomes that occur when the requirements of a job do not match the capabilities, resources, or needs of the worker. (7)

jurisdiction A politically defined geographical area. The right or authority of a justice agency to act with regard to a particular subject matter, territory, or person. The authority of a court to hear and decide cases. (8)

just deserts The punishment rationale based on the idea that offenders should be punished automatically, simply because they have committed a crime–they "deserve" it–and the idea that the punishment should fit the crime. (9)

juvenile delinquency A special category of offense created for youths who, in most U.S. jurisdictions, are persons between the ages of 7 and 18. (13)

labeling theory A theory that emphasizes the criminalization process as the cause of some crime. (3)

learning theory A theory that explains criminal behavior and its prevention with the concepts of positive reinforcement, negative reinforcement, extinction, punishment, and modeling, or imitation. (3)

left realists A group of social scientists who argue that critical criminologists need to redirect their attention to the fear and the very real victimization experienced by working-class people. (3)

legal definition of crime An intentional violation of the criminal law or penal code, committed without defense or excuse and penalized by the state. (2)

legality The requirement (1) that a harm must be legally forbidden for the behavior to be a crime and (2) that the law must not be retroactive. (2)

less-eligibility principle The position that prisoners should receive no service or program superior to the services and programs available to free citizens without charge. (10)

limbic system A structure surrounding the brain stem that, in part, controls the life functions of heartbeat, breathing, and sleep. It also is believed to moderate expressions of violence; such emotions as anger, rage, and fear; and sexual response. (3)

lockup A very short-term holding facility that is frequently located in or very near an urban police agency so that suspects can be held pending further inquiry. (10)

mala in se Wrong in themselves. A description applied to crimes that are characterized by universality and timelessness. (2)

mala prohibita Offenses that are illegal because laws define them as such. They lack universality and timelessness. (2)

mandatory release A method of prison release under which an inmate is released after serving a legally required portion of his or her sentence, minus good-time credits. (11)

mandatory sentencing Sentencing in which a specified number of years of imprisonment (usually within a range) is provided for particular crimes. (9)

meat eaters Officers who actively seek ways to make money illegally while on duty. (7)

mediation A dispute resolution process that brings disputants together with a third party (a mediator) who is trained in the art of helping people resolve disputes to everyone's satisfaction. The agreed-upon resolution is then formalized into a binding consent agreement. (14)

medical model A theory of institutional corrections, popular during the 1940s and 1950s, in which crime was seen as symptomatic of personal illness in need of treatment. (10)

mens rea Criminal intent; a guilty state of mind. (2)

mere suspicion The standard of proof with the least certainty; a "gut feeling." With mere suspicion, a law enforcement officer cannot legally even stop a suspect. (4)

merit system A system of employment whereby an independent civil service commission, in cooperation with the city personnel section and the police department, sets employment qualifications, performance standards, and discipline procedures. (7)

milieu therapy A variant of group therapy that encompasses the total living environment so that the environment continually encourages positive behavioral change. (10)

misdemeanor A less serious crime generally punishable by a fine or by incarceration in jail for not more than 1 year. (1)

mitigating factors, mitigating circumstances, or **extenuating circumstances** In death sentencing, facts or situations that do not justify or excuse a criminal act but reduce the degree of blameworthiness and thus may reduce the punishment. (9)

myths Beliefs based on emotion rather than analysis. (1)

national crime victimization surveys (NCVS) A source of crime statistics based on interviews in which respondents are asked whether they have been victims of any of the FBI's index offenses (except murder, nonnegligent manslaughter, and arson) or other crimes during the past 6 months. If they have, they are asked to provide information about the experience. (2)

necessity defense A legal defense against criminal responsibility used when a crime has been committed to prevent a more serious crime. (2)

negative reinforcement The removal or reduction of a stimulus whose removal or reduction increases or maintains a response. (3)

negligence The failure to take reasonable precautions to prevent harm. (2)

neoclassical theory A modification of classical theory in which it was conceded that certain factors, such as insanity, might inhibit the exercise of free will. (3)

net widening A phenomenon that occurs when the offenders placed in a novel program are not the offenders for whom the program was designed. The consequence is that those in the program receive more severe sanctions than they would have received had the new program remained unavailable. (12)

nolle prosequi (nol. pros.) The notation placed on the official record of a case when prosecutors elect not to prosecute. (8)

nolo contendere Latin for "no contest." When defendants plead *nolo*, they do not admit guilt but are willing to accept punishment. (8)

nonenforcement The failure to routinely enforce prohibitions against certain behaviors. (2)

norm or **social more** Any standard or rule regarding what human beings should or should not think, say, or do under given circumstances. (2)

offenses known to the police A crime index, reported in the FBI's uniform crime reports, composed of crimes that are both reported to and recorded by the police. (2)

operational styles The different overall approaches to the police job. (6)

original jurisdiction The authority of a court to hear a case when it is first brought to court. (8)

overcriminalization The prohibition by the criminal law of some behaviors that arguably should not be prohibited. (2)

panopticon A prison design consisting of a round building with tiers of cells lining the inner circumference and facing a central inspection tower. (10)

pardon A "forgiveness" for the crime committed that stops further criminal processing.

parens patriae The legal philosophy justifying state intervention in the lives of children when their parents are unable or unwilling to protect them. (13)

parole A method of prison release whereby inmates are conditionally released at the discretion of a board or other authority before having completed their entire sentences; can also refer to the community supervision received upon release. (1)

parole guidelines Structured instruments used to estimate the probability of parole recidivism and to direct the release decisions of parole boards. (12)

patriarchy Men's control over women's labor and sexuality. (3)

peacemaking criminology An approach that suggests that the solutions to all social problems, including crime, are the transformation of human beings, mutual dependence, reduction of class structures, creation of communities of caring people, and universal social justice. (3)

Peel's Principles of Policing A dozen standards proposed by Robert Peel, the author of the legislation resulting in the formation of the London Metropolitan Police Department. The standards are still applicable to today's law enforcement. (5)

penal code The criminal law of a political jurisdiction. (4)

penal sanction An ideal characteristic of criminal law: the principle that violators will be punished or at least threatened with punishment by the state. (4)

Pennsylvania system An early system of U.S. penology in which inmates were kept in solitary cells so that they could study religious writings, reflect on their misdeeds, and perform handicraft work. (10)

penology The study of prison management and the treatment of offenders. (10)

personal jurisdiction A court's authority over the parties to a lawsuit. (8)

petition A legal form of the police complaint that specifies the charges to be heard at the adjudication. (13)

placing out The practice of placing children on farms in the Midwest and West to remove them from the supposedly corrupting influences of their parents and the cities. (13)

plea bargaining or **plea negotiating** The practice whereby the prosecutor, the defense attorney, the defendant, and–in many jurisdictions–the judge agree on a specific sentence to be imposed if the accused pleads guilty to an agreed-upon charge or charges instead of going to trial. (1)

police cadet program A program that provides persons aged 18 to 21 a chance to experience the challenges and rewards of a police career. Often cadets are paid and work part or full time. Cadet programs are designed to assist cadets in transitioning into the position of full-time police officer. (7)

politicality An ideal characteristic of criminal law, referring to its legitimate source. Only violations of rules made by the state, the political jurisdiction that enacted the laws, are crimes. (4)

positive reinforcement The presentation of a stimulus that increases or maintains a response. (3)

posses Groups of able-bodied citizens of a community, called into service by a sheriff or constable to chase and apprehend offenders. (5)

postmodernism An area of critical thought that, among other things, attempts to understand the creation of knowledge and how knowledge and language create hierarchy and domination. (3)

power differentials The ability of some groups to dominate other groups in a society. (3)

precedent A decision that forms a potential basis for deciding the outcomes of similar cases in the future; a by-product of decisions made by trial and appellate court judges, who produce case law whenever they render a decision in a particular case. (4)

preliminary hearing A pretrial stage used in about one-half of all states and only in felony cases. Its purpose is for a judge to determine whether there is probable cause to support the charge or charges imposed by the prosecutor. (8)

preponderance of evidence Evidence that more likely than not outweighs the opposing evidence, or sufficient evidence to overcome doubt or speculation. (4)

presentence investigation (PSI) An investigation conducted by a probation agency or other designated authority at the request of a court into the past behavior, family circumstances, and personality of an adult who has been convicted of a crime, to assist the court in determining the most appropriate sentence. (12)

presentence investigation reports Reports, often called PSIs or PSIRs, that are used in the federal system and the majority of states to help judges determine the appropriate sentence. They are also used in classifying probationers, parolees, and prisoners according to their treatment needs and security risk. (9)

presumptive sentencing Sentencing that allows a judge to retain some sentencing discretion, subject to appellate review. The legislature determines a sentence range for each crime. (9)

preventive detention Holding suspects or defendants in jail without giving them an opportunity to post bail because of the threat they pose to society. (8)

preventive patrol Patrolling the streets with little direction; between responses to radio calls, officers are "systematically unsystematic" and observant in an attempt to both prevent and ferret out crime. Also known as *random patrol*. (6)

prisonization The process by which an inmate becomes socialized into the customs and principles of the inmate society. (11)

privatization The involvement of the private sector in the construction and the operation of confinement facilities. (10)

probable cause The amount of proof necessary for a reasonably intelligent person to believe that a crime has been committed or that items connected with criminal activity can be found in a particular place. It is the standard of proof needed to conduct a search or to make an arrest. (1)

probation A sentence in which the offender, rather than being incarcerated, is retained in the community under the supervision of a probation agency and required to abide by certain rules and conditions to avoid incarceration. (1)

probation conditions Rules that specify what an offender is and is not to do during the course of a probation sentence. (12)

procedural law The body of law that governs the ways substantive laws are administered; sometimes called *adjective* or *remedial* law. (4)

proof evident, presumption great The standard of proof required for a judicial officer to deny bail in cases involving capital felonies. (4)

proportionality review A review in which the appellate court compares the sentence in the case it is reviewing with penalties imposed in similar cases in the state. The object is to reduce, as much as possible, disparity in death penalty sentencing. (9)

proprietary security In-house protective services that a security staff provides for the entity that employs it. (5)

protective custody The segregation of inmates for their own safety. (10)

psychopaths, sociopaths, or **antisocial personalities** Persons characterized by no sense of guilt, no subjective conscience, and no sense of right and wrong. They have difficulty in forming relationships with other people; they cannot empathize with other people. (3)

public safety officers Police department employees who perform many police services but do not have arrest powers. (7)

punishment The presentation of an aversive stimulus to reduce a response. (3)

racial profiling The stopping and/or detaining of individuals by law enforcement officers based solely on race. (7)

radical nonintervention A practice based on the idea that youths should be left alone if at all possible, instead of being formally processed. (13)

radical theories Theories of crime causation that are generally based on a Marxist theory of class struggle. (3)

reasonable suspicion A standard of proof that is more than a gut feeling. It includes the ability to articulate reasons for the suspicion. With reasonable suspicion, a law enforcement officer is legally permitted to stop and frisk a suspect. (4)

recidivism The return to illegal activity after release from incarceration.

reentry The reintegration of inmates into society following their prison or jail terms. (11)

reform, industrial, or **training schools** Correctional facilities for youths, first developed in the late 1800s, that focused on custody. Today, those institutions are often called training schools, and although they may place more emphasis on treatment, they still rely on custody and control. (13)

regularity An ideal characteristic of criminal law: the applicability of the law to all persons, regardless of social status. (4)

rehabilitation The attempt to "correct" the personality and behavior of convicted offenders through educational, vocational, or therapeutic treatment and to return them to society as law-abiding citizens. (8)

reintegration The process of rebuilding former ties to the community and establishing new ties after release from prison. (12)

reintegrative shaming A strategy in which disappointment is expressed for the offender's actions, the offender is shamed and punished, and, more important, following the expression of disappointment and shame is a concerted effort on the part of the community to forgive the offender and reintegrate him or her back into society. (14)

relative deprivation Refers to inequalities (in resources, opportunities, material goods, etc.) that are defined by a person as unfair or unjust. (3)

relative powerlessness In conflict theory, the inability to dominate other groups in society. (3)

release on own recognizance (ROR) A release secured by a suspect's written promise to appear in court. (8)

restitution Money paid or services provided by a convicted offender to victims, their survivors, or the community to make up for the injury inflicted. (9)

restorative justice A process whereby an offender is required to contribute to restoring the health of the community, repairing the harm done, and meeting victims' needs. (14)

retribution A justification for punishment that implies repayment for an offense committed. (9)

revenge The punishment rationale expressed by the biblical phrase, "An eye for an eye, and a tooth for a tooth." People who seek revenge want to pay back offenders by making them suffer for what they have done. (9)

revocation The repeal of a probation sentence or parole, and substitution of a more restrictive sentence, because of violation of probation or parole conditions. (12)

role The rights and responsibilities associated with a particular position in society. (6)

role conflict The psychological stress and frustration that results from trying to perform two or more incompatible responsibilities. (6)

role expectation The behavior and actions that people expect from a person in a particular role. (6)

rules of discovery Rules that mandate that a prosecutor provide defense counsel with any exculpatory evidence (evidence favorable to the accused that has an effect on guilt or punishment) in the prosecutor's possession. (8)

searches Explorations or inspections, by law enforcement officers, of homes, premises, vehicles, or persons, for the purpose of discovering evidence of crimes or persons who are accused of crimes. (4)

security level A designation applied to a facility to describe the measures taken, both inside and outside, to preserve security and custody. (10)

seizures The taking of persons or property into custody in response to violations of the criminal law. (4)

selective enforcement The practice of relying on the judgment of the police leadership and rank-and-file officers to decide which laws to enforce. (7)

self-incrimination Being a witness against oneself. If forced, it is a violation of the Fifth Amendment. (4)

self-report crime surveys Surveys in which subjects are asked whether they have committed crimes. (2)

shire reeve In medieval England, the chief law enforcement officer in a territorial area called a *shire*; later called the *sheriff.* (5)

slave patrols The earliest form of policing in the South. They were a product of the slave codes. The plantation slave patrols have been called "the first distinctively American police system." (5)

snitch system A system in which staff learn from inmate informants about the presence of contraband, the potential for disruptions, and other threats to security. (10)

social contract An imaginary agreement to sacrifice the minimum amount of liberty necessary to prevent anarchy and chaos. (3)

social control theory A view in which people are expected to commit crime and delinquency unless they are prevented from doing so. (3)

social disorganization The condition in which the usual controls over delinquents are largely absent; delinquent behavior is often approved of by parents and neighbors; there are many opportunities for delinquent behavior; and there is little encouragement, training, or opportunity for legitimate employment. (3)

special jurisdiction The power of a court to hear only certain kinds of cases. (8)

special or **specific deterrence** The prevention of individuals from committing crimes again by punishing them. (9)

specificity An ideal characteristic of criminal law, referring to its scope. Although civil law may be general in scope, criminal law should provide strict definitions of specific acts. (4)

stare decisis The principle of using precedents to guide future decisions in court cases; Latin for "to stand by decided cases." (4)

state police model A model of state law enforcement services in which the agency and its officers have the same law enforcement powers as local police but can exercise them anywhere within the state. (5)

status offense An act that is not a crime when committed by adults but is illegal for minors (e.g., truancy or running away from home). (13)

statute of limitations A law establishing a time limit for prosecuting a crime, based on the date when the offense occurred. (1)

structured fines, or **day fines** Fines that are based on defendants' ability to pay. (12)

subject matter jurisdiction The power of a court to hear a particular type of case. (8)

subpoena A written order issued by a court that requires a person to appear at a certain time and place to give testimony. It can also require that documents and objects be made available for examination by the court. (8)

sub-rosa economy The secret exchange of goods and services among inmates; the black market of the prison. (11)

substantive law The body of law that defines criminal offenses and their penalties. (4)

summary or **bench trial** A trial before a judge without a jury. (8)

system A smoothly operating set of arrangements and institutions directed toward the achievement of common goals. (1)

tech prep (technical preparation) A program in which area community colleges and high schools team up to offer 6 to 9 hours of college law enforcement courses in the 11th and 12th grades, as well as one or two training certifications, such as police dispatcher or local corrections officer. Students who graduate are eligible for police employment at age 18. (7)

technical violations Failure to abide by the technical rules or conditions of probation or parole (e.g., not reporting regularly to the probation officer), as distinct from commission of a new criminal act. (12)

temporary-release programs Programs that allow jail or prison inmates to leave the facility for short periods to participate in approved community activities. (12)

terrorism The systematic use of terror or unpredictable violence against governments, publics, or individuals to attain a political objective; the unlawful use of force and violence against persons or property to intimidate or coerce a government, the civilian population, or any segment thereof, in furtherance of political or social objectives; or premeditated, politically motivated violence perpetrated against noncombatant targets by subnational groups or clandestine agents, usually intended to influence an audience. (6)

theory An assumption (or set of assumptions) that attempts to explain why or how things are related to each other. (3)

three I's of police selection Three qualities of the American police officer that seem to be of paramount importance: intelligence, integrity, and interaction skills. (7)

tithing system A private self-help protection system in early medieval England, in which a group of 10 families, or a *tithing*, agreed to follow the law, keep the peace in their areas, and bring law violators to justice. (5)

tort A violation of the civil law. (4)

total institution An institutional setting in which persons sharing some characteristics are cut off from the wider society and expected to live according to institutional rules and procedures. (11)

traffic accident investigation crews In some agencies, the special units assigned to all traffic accident investigations. (6)

transfer, waiver, remand, bindover, or **certification** The act or process by which juveniles who meet specific age, offense, and (in some jurisdictions) prior-record criteria are transferred to criminal court for trial. (13)

transportation A punishment in which offenders were transported from their home nation to one of that nation's colonies to work. (10)

trial *de novo* A trial in which an entire case is reheard by a trial court of general jurisdiction because there is an appeal and there is no written transcript of the earlier proceeding. (8)

undercriminalization The failure to prohibit some behaviors that arguably should be prohibited. (2)

uniform crime reports (UCR) A collection of crime statistics and other law enforcement information gathered under a voluntary national program administered by the FBI. (2)

uniformity An ideal characteristic of criminal law: the enforcement of the laws against anyone who violates them, regardless of social status. (4)

unsecured bond An arrangement in which bail is set but no money is paid to the court. (8)

utility The principle that a policy should provide "the greatest happiness shared by the greatest number." (3)

venire The pool from which jurors are selected. (8)

venue The place of the trial. It must be geographically appropriate. (4)

victim-impact statements Descriptions of the harm and suffering that a crime has caused victims and their survivors. (9)

voir dire The process in which potential jurors who might be biased or unable to render a fair verdict are screened out. (8)

warrant A written order from a court directing law enforcement officers to conduct a search or to arrest a person. (4)

workhouses European forerunners of the modern U.S. prison, where offenders were sent to learn discipline and regular work habits. (10)

writ of *certiorari* A written order, from an appellate court to a lower court whose decision is being appealed, to send the records of the case forward for review. (8)

writ of *habeas corpus* An order from a court to an officer of the law to produce a prisoner in court to determine if the prisoner is being legally detained or imprisoned. (8)

Case Index

Subject Index

Italic locators (*10*) signify illustrations; bold locators (**10**) signify definitions. *f* after a locator signifies a figure; *t* signifies a table.